Liquid Sloshing Dynamics

Theory and Applications

The problem of liquid sloshing in moving or stationary containers remains of great concern to aerospace, civil, and nuclear engineers; physicists; designers of road tankers and ship tankers; and mathematicians. Beginning with the fundamentals of liquid sloshing theory, this book takes the reader systematically from basic theory to advanced analytical and experimental results in a self-contained and coherent format. It presents liquid sloshing effects on space vehicles, storage tanks, road vehicle tanks and ships, and elevated water towers under ground motion.

The book is divided into four sections. Part I deals with the theory of linear liquid sloshing dynamics; Part II addresses the nonlinear theory of liquid sloshing dynamics, Faraday waves, and sloshing impacts; Part III presents the problem of linear and nonlinear interaction of liquid sloshing dynamics with elastic containers and supported structures; and Part IV considers the fluid dynamics in spinning containers and microgravity sloshing.

This book will be invaluable to researchers and graduate students in mechanical and aeronautical engineering, designers of liquid containers, and applied mathematicians.

Raouf A. Ibrahim is a professor of mechanical engineering at Wayne State University. From 1963 until 1971, he worked as a research engineer at the Aerospace Research Center of the rockets industry in Egypt, then gained his Ph.D. in Mechanical Engineering from the University of Edinburgh in 1974. From 1976 to 1979, he worked as a senior research specialist at Sakr Factory in Cairo and as an adjunct assistant professor at Cairo University. In 1979, he moved to the United States and worked at Shaker Research Corporation, before joining Texas Tech University in 1980 as assistant, associate, and then full professor. In 1987, he joined Wayne State University and continued his research activities in nonlinear random vibration, liquid sloshing dynamics, friction-induced vibration, and flutter of aeroelastic structures. In 1994 he was named the Arthur Carr Professor of Engineering and in 1995 he was awarded the Board of Governors Outstanding Professor Award. He has published more than 90 papers in refereed journals, and a research monograph. He is a Fellow of the ASME and Associate Fellow of the AIAA.

Liquid Sloshing Dynamics

Theory and Applications

Raouf A. Ibrahim
Wayne State University
Department of Mechanical Engineering

CAMBRIDGE
UNIVERSITY PRESS

CAMBRIDGE
UNIVERSITY PRESS

University Printing House, Cambridge CB2 8BS, United Kingdom

Cambridge University Press is part of the University of Cambridge.

It furthers the University's mission by disseminating knowledge in the pursuit of education, learning and research at the highest international levels of excellence.

www.cambridge.org
Information on this title: www.cambridge.org/9780521838856

First published 2005

A catalogue record for this publication is available from the British Library

ISBN 978-0-521-83885-6 Hardback

Contents

Foreword

When I was a very young boy I was enchanted by airplanes. The very idea that such a machine, with no apparent motions of its own – except, of course, for that tiny rotating thing at the front – could fly through the air was amazing. I could see that birds and insects could all fly with great dexterity, but that was because they could flap their wings, and thus support their weight as well as maneuver. And fish could even "fly" through water by motions of their body. How exciting it was then, to begin to learn something about how objects interact with the fluids surrounding them, and the useful consequences of those flows. That ultimately led, of course, to the broad study of fluid dynamics, with all of its wonderful manifestations.

There is hardly a single aspect of our daily lives, and indeed even of the entire universe in which we live, that is not in some way governed or described by fluid dynamics – from the locomotion of marine animals to the birth and death of distant galaxies. As a major field of technical and scientific knowledge, there are vast bodies of literature devoted to almost every facet of fluid behavior: laminar and turbulent flows, discontinuous (separated) flows, vortex flows, internal waves, free surface waves, compressible fluids and shock waves, multi-phase flows, and many, many others. With such a countless array of fluid phenomenon before us, what then leads to the focus of the present work?

Almost since the earliest philosophers began the development of what we now call "rational mechanics," curiosity about how liquids behave when contained in some form of vessel has been strong. Beginning with the oscillations of water in lakes and harbors occurring as the result of earthquakes, similar phenomena surround us at almost every turn, arising from all aspects of our modern technology and civilization as well as from nature herself. Thus, geophysicists and seismologists, engineers, mathematicians, and many other scientific workers have devoted themselves to the study of these fascinating subjects for many years. In terms of modern technology, the behavior of liquids in contained systems is of importance in the tankage of ocean-going ships and offshore platforms, in airplanes, automotive vehicles, railway cars, nuclear power plants, and, of course, rockets and spacecraft, and countless others as well. One would think, however, that the general question of the behavior of liquids in such systems is a rather simple one to answer – not so. Our everyday experience in carrying a cup of coffee or a bowl of soup may be frustrating unless we are very careful as to how we move, but may still deceive us into believing that the "sloshing" of the liquid is simple. Indeed, depending on the amplitudes and frequencies of the motions of the container, the responses of the liquid free surface can be almost bewildering in their complexity!

Interest in this subject came almost to a crescendo, of course, with the advent of large liquid propelled rockets nearly half a century ago. At that time almost no one imagined that so many forms of liquid behavior could manifest themselves or that such a tremendous body of technical literature would evolve as a consequence of trying to describe, understand, and predict these fascinating phenomena. Over the years, a fairly large number of monographs and literature summaries appeared in an effort to catalog what is known, and while these have each been useful in their own way, none has been exhaustive. In this work, the author has tried to do just that.

This is a monumental work in that it attempts to cover almost every aspect of liquid sloshing dynamics and cites approximately three thousand references. I would hope that its readers will find in it whatever they are looking for – if not, I do not know where else they should look and so they might have to discover it for themselves, perhaps using the methods and analytical tools described herein.

H. Norman Abramson
San Antonio, March 2004

Acknowledgment

The story of this book goes back to my first job as a Research Engineer at the Rocket Research Center, Cairo, Egypt, in 1963. I was involved in estimating the trajectories of liquid propellant rockets. This task required the knowledge of the actual values of mass moment of inertia and center of mass versus the time of rocket trajectory. The difficulty was mainly in the liquid part which constitutes almost 85% of the total weight of the rocket. I used to ignore the contribution of the liquid propellant in roll oscillations. This inability on my part motivated me to enroll for the Masters degree (by research) at Cairo University. Professor M. I. Rashid, my M.S. thesis advisor, encouraged me to conduct an extensive literature review on the entire subject of liquid propellant sloshing dynamics in addition to the main task of solving the Navier–Stokes equations to estimate the effective boundary layer thickness in roll due to tank walls and immersed pipes. Upon completion of my Masters degree, I was awarded a scholarship to study for my Ph.D. at the University of Edinburgh under the supervision of Professor A. D. S. Barr to work on the autoparametric interaction of liquid sloshing dynamics with supported structure dynamics. At that time I compiled over 900 references on liquid sloshing dynamics and related topics.

When I joined academia at Texas Tech University in 1981, I continued my research on liquid sloshing under random excitation with the help of my graduate students Drs Soundararajan, Hun Heo, Wenlung Li, J. Gau, and Messrs G. Lattore and R. Henriech. The work was supported by the National Research Foundation and dealt with the random response of liquid free surface under different types of random excitation. When I moved to Wayne State University, the work was extended jointly with Dr Valery Pilipchuk to consider the liquid sloshing impact supported by the National Science Foundation (NSF). Dr Mohamed El-Sayad joined me to conduct his Ph.D. on the same problem. During the same time, Professor Takashi Ikeda spent several weeks at Wayne State University to conduct Monte Carlo simulation on parametric random excitation of liquid containers supported on elastic structures in the presence of internal resonance. Both Professor Ikeda and Dr Pilipchuk joined me to write an extensive literature review on liquid sloshing dynamics citing over 1200 references. Upon completing the review article I was strongly motivated to write the present book to cover every aspect of the subject. However, as soon as I began, I found the task was formidable and beyond my own capacity. It was my wife, Sohair, who encouraged me to accept the challenge and to complete the task at my own pace. I thank her so deeply for her patience and longsuffering during the last four years. Indeed without her support I could not have completed the book.

Upon completion of the first draft of the manuscript, I asked Dr Norman Abramson, who I consider to be the father of liquid sloshing, to write a Foreword for the book. May God grant him health and many years. I extend my sincere thanks to my colleagues and friends who have helped me. In particular I thank Professors Victor Berdichevsky at Wayne State University, and Professor W. Eidel of the Universitat der Bundeswehr, Germany, who helped me prepare the proof of some results.

Last but not least, I would like to thank the editorial staff of Cambridge University Press. In particular, I thank Dr Phil Meyler who handled the editorial process, and Ms Emily Yossarian, Ms Carol Miller, and Mr Keith Westmoreland who have been working with me during the production process.

Introduction

Sloshing means any motion of the free liquid surface inside its container. It is caused by any disturbance to partially filled liquid containers. Depending on the type of disturbance and container shape, the free liquid surface can experience different types of motion including simple planar, nonplanar, rotational, irregular beating, symmetric, asymmetric, quasi-periodic and chaotic. When interacting with its elastic container, or its support structure, the free liquid surface can exhibit fascinating types of motion in the form of energy exchange between interacting modes. Modulated free surface occurs when the free-liquid-surface motion interacts with the elastic support structural dynamics in the neighborhood of internal resonance conditions. Under low gravity field, the surface tension is dominant and the liquid may be oriented randomly within the tank depending essentially upon the wetting characteristics of the tank wall.

The basic problem of liquid sloshing involves the estimation of hydrodynamic pressure distribution, forces, moments and natural frequencies of the free-liquid surface. These parameters have a direct effect on the dynamic stability and performance of moving containers.

Generally, the hydrodynamic pressure of liquids in moving rigid containers has two distinct components. One component is directly proportional to the acceleration of the tank. This component is caused by the part of the fluid moving with the same tank velocity. The second is known as "convective" pressure and represents the free-surface-liquid motion. Mechanical models such as mass-spring-dashpot or pendulum systems are usually used to model the sloshing part.

A liquid's motion inside its container has an infinite number of natural frequencies, but it is the lowest few modes that are most likely to be excited by the motion of a vehicle. Most studies have therefore concentrated on investigating forced harmonic oscillations near the lowest natural frequencies, predicted by the fluid field linear equations. However, nonlinear effects result in the frequency of maximum response being slightly different from the linear natural frequency and dependent on amplitude. Nonlinear effects include amplitude jump, parametric resonance, chaotic liquid surface motion, and nonlinear sloshing mode interaction due to the occurrence of internal resonance among the liquid sloshing modes. The nonlinearities associated with free-surface motion inside moving containers are different from those nonlinear water waves in ocean and canals. The theory of nonlinear dispersive waves originated by Stokes (1947) and solitons[1] observed by Russell (1844) is well documented in Debnath (1994) and will not be addressed in this book.

[1] The solitary wave represents not a periodic wave, but the propagation of a single isolated symmetrical hump of unchanged form.

Analytical solutions are limited to regular geometric tank shapes such as cylindrical, and rectangular. The nature of sloshing dynamics in cylindrical tanks is better understood than for prismatic tanks. However, analytical techniques for predicting large-amplitude sloshing are still not fully developed. Such loads are extremely important in the design stage of supporting structures and internal components of vehicle tanks. In addition, much of the sloshing technology developed for space applications is not applicable to road tankers. The reason is that more emphasis has been placed on frequencies and total forces as they relate to control system requirements. Accordingly, the effects of local peak impact pressure on structural requirements have not been studied to any extent. Further, the excitation amplitudes considered in space applications are too small for road vehicle simulation. To avoid catastrophic sloshing in space vehicles, the control system frequencies, the vehicle elastic structure frequencies, and the fluid-slosh frequencies must be fairly widely separated.

Sloshing phenomena in moving rectangular tanks can usually be described by considering only two-dimensional fluid flow if the tank width is much smaller than its breadth. Sloshing in spherical or cylindrical tanks, however, is usually described by three-dimensional flow. Tanks with two-dimensional flow are divided into two classes: low and high liquid fill depths. The low fill depth case is characterized by the formation of hydraulic jumps and traveling waves for excitation periods around resonance. At higher fill depths, large standing waves are usually formed in the resonance frequency range. When hydraulic jumps or traveling waves are present, extremely high impact pressures can occur on the tank walls. Impact pressures are only measured experimentally and cannot be estimated theoretically or numerically.

Early attempts to analyze liquid waves in oscillating containers include those of Hough (1895), Honda and Matsushita (1913), Jeffries (1924), Sen (1927), Goldsborough (1930), Westergaard (1933), Binnie (1941, 1955), K. W. Smith (1947, 1956), C. Smith (1948), Taylor (1950, 1954), Luskin and Lapin (1952), Schy (1952), Moiseev (1952a,b,c,d, 1953, 1954, 1956), Senda and Nakagawa (1954), Nakagawa (1955, 1956), Birkhoff (1956), Narimanov (1956, 1957c), Okhotsimski (1956), Heinrich and Kaufman (1956), Sretanskii (1956, 1957), Housner (1957), Krein and Moiseev (1957), and Shved (1959). The fundamental theory of liquid surface waves is documented in several references (see, e.g., Lamb, 1945, Stoker, 1957, Kochin, *et al.*, 1964, Thomson, 1965, Brodkey, 1967 and Barber and Ghey, 1969). Different aspects of the subject were reviewed by Sloane (1960), Cooper (1960), Abramson (1961a, 1963b, 1966b, 1968), Ring (1964), Abramson and Kana (1967), Brown (1982b), Rammerstofer, *et al.* (1990), and Ibrahim, *et al.* (2001). Some research monographs entirely devoted to different space problems of liquid sloshing include Abramson (1966a), Moiseev (1968), Moiseev and Rumyantsev (1968), Babskii, *et al.* (1976a), Narimanov, *et al.* (1977), Myshkis, *et al.* (1987, 1992), Walter (1987), and Monti (2001).

The problem of liquid sloshing in moving or stationary containers remains of great concern to aerospace, civil, and nuclear engineers, physicists, designers of road tankers and ship tankers, and mathematicians. Civil engineers and seismologists have been studying liquid sloshing effects on large dams, oil tanks and elevated water towers under ground motion. They also mounted liquid tanks on the roofs of multistory buildings as a means to control building oscillations due to earthquakes. Since the early 1960s, the problem of liquid sloshing dynamics has been of major concern to aerospace engineers studying the influence of liquid-propellant sloshing on the flight performance of jet vehicles, and new areas of research activities have emerged. The modern theory of nonlinear dynamics has indeed promoted further studies and uncovered complex nonlinear phenomena. These include rotary sloshing, Faraday waves, nonlinear liquid sloshing interaction with elastic structures, internal resonance effects, stochastic sloshing dynamics, hydrodynamic sloshing impact dynamics, g-jitter under microgravity field, dynamics of liquid bridges, cross-waves, and spatial resonance. The dynamic stability of liquefied natural gas tankers and ship cargo tankers, and liquid hydrodynamic impact loading are problems of current interest to the designers of such systems. In populated cities, gasoline and other flammable liquid tankers are prone to rollover accidents while entering and exiting highways. This is a difficult mathematical problem to solve analytically, since the dynamic boundary condition at the free surface is nonlinear and the position of the free surface varies with time in a manner not known a priori. A liquid free surface in partially filled containers can experience a wide spectrum of motions such as planar, nonplanar, rotational, quasi-periodic, chaotic, and disintegration. Other important contributions include the development of digital computer codes to solve complex problems that were difficult to handle in the past.

The purpose of this book is to present the basic theories and results developed in different applications in a coherent format. It highlights the major achievements and results reported in the literature. It begins with the fundamentals of liquid sloshing theory to take the reader systematically from the basic theory to advanced analytical and experimental results. Each chapter includes an extensive literature review for its major sections, followed by analytical description and main results. The book is supported by an extensive bibliography of technical journal papers, NASA reports, research monographs, M.S. and Ph.D. theses, and conference proceedings.

The book can be divided into four major parts. Part I deals with the theory of linear liquid sloshing dynamics (Chapters 1 to 3). Part II addresses the nonlinear theory of liquid sloshing dynamics (Chapters 4, 6, and 7). Note that Chapter 5 is in Part II but links linear and nonlinear equivalent mechanical models, and thus it also belongs to Part I. Part III presents the problem of linear and nonlinear interaction of liquid sloshing dynamics with elastic containers and supported structures (Chapters 8 to 10). Part IV considers the fluid dynamics in spinning containers and microgravity sloshing (Chapters 11 and 12). The book is closed by an extensive list of references that exceeds 2600.

The fluid field equations are developed in the first chapter with the purpose of determining the natural frequencies and the corresponding mode shapes in different container geometries. The analytical results are compared with those measured experimentally. Within the framework of the linear theory, the hydrodynamic loads and moments generated under harmonic excitations are evaluated in Chapter 2. A unique feature in this chapter is the study of forced excitation of magnetic fluids. Analytical treatments of some nontraditional tank geometries that were regarded as difficult in the past are presented.

Chapter 3 presents the influence of viscous damping and sloshing suppression devices in closed containers. The damping and modal analysis in a circular cylindrical container is analytically treated. The influence of sloshing suppression devices is addressed based on reported experimental results. The basic concept of Stokes boundary layers over an oscillating flat plate is introduced as a foundation to the estimation of inertia and damping parameters. It introduces the contribution of periodic boundary layers of fluids, in upright circular containers experiencing roll oscillations, to the system inertia and damping parameters. The analysis is extended to determine the contribution of immersed rods and pipes.

Chapter 4 deals with weakly nonlinear sloshing dynamics under lateral excitation. The associated phenomena include nonplanar unstable motion of the free surface accompanied by rotation of the nodal diameter (rotary sloshing) and chaotic sloshing. These phenomena can be uncovered using the theory of weakly nonlinear oscillations for quantitative analysis and the modern theory of nonlinear dynamics for stability analysis. In circular containers, the problem is strongly analogous to the dynamics of a spherical pendulum whose support point experiences lateral sinusoidal excitation. Some nonlinear theorems have been developed to predict not only rotary sloshing, but also other complex motions such as period doubling and chaos. The main sources of nonlinearity in the fluid field equations are the free-surface boundary conditions. The study is devoted to circular cylinders and rectangular containers. There is a critical fluid depth, which separates two regimes of liquid nonlinearity, hard and soft. A unique feature in this chapter is the random excitation of rigid containers, which has a direct application to liquid containers subjected to earthquakes. The chapter includes the problem of self-induced sloshing usually encountered in nuclear reactors. Self-induced sloshing implies the interaction between the liquid free-surface dynamics and submerged fluid jet dynamics.

Chapter 5 introduces the concept of equivalent mechanical models. A realistic representation of the liquid dynamics inside closed containers can be approximated by an equivalent mechanical system. The equivalence is taken in the sense of equal resulting forces and moments acting on the tank wall. By properly accounting for the equivalent mechanical system, the problem of overall dynamic system behavior can be formulated more simply. For small oscillations in which the fluid free surface remains planar without rotation of its nodal diameter the equivalent mechanical model can be represented by a set of pendulums or mass-spring-dashpot systems. For relatively large-amplitude oscillations, in which the free liquid surface experiences nonplanar motion, the equivalent mechanical model is a simple pendulum describing relatively large motion. To model nonplanar and rotational sloshing one has to replace the simple pendulum by a compound or spherical pendulum. The theory of the spherical pendulum will be presented for promoting our understanding of rotary sloshing. For strongly nonlinear motion associated with hydrodynamic pressure impacts, the modeling can be achieved by an equivalent pendulum of high restoring force. This modeling will be considered in details in Chapter 7.

Chapter 6 deals with the problem of parametric sloshing or Faraday's waves. Parametric sloshing refers to the motion of the liquid free surface due to an excitation applied perpendicular to the plane of the undisturbed free surface. Generally, parametric oscillation occurs in dynamical systems as a result of time-dependent variation of such parameters as inertia, damping, or stiffness. The early research efforts focused on the estimation of instability boundaries of liquid in the presence of viscous and surface tension forces or in their absence. While the damping forces raise the threshold excitation amplitude, they do not bring the free-surface amplitude into a bounded value. It is the inherent nonlinearity of the surface-wave amplitude that can limit the growth of the amplitude. Both nonlinear theory and experimental tests revealed "soft" or "hard" spring characteristics depending on the liquid depth ratio. With the advent of the modern theory of nonlinear dynamics, many complex surface dynamic phenomena have been uncovered. These include modal competition, quasi-periodic motion, chaotic motion, and mixed mode motion in the presence of internal resonance. These phenomena are presented for circular and rectangular tanks and the corresponding motion was identified in bifurcation diagrams under slow variations of certain control parameters such as excitation amplitude and frequency. Most of these phenomena may take place under random parametric excitation in terms of excitation bandwidth and spectral density function.

Chapter 7 presents the basic concepts of the theory of hydrodynamic slamming and impacting. The material of this chapter has not been addressed in other research monographs on liquid sloshing dynamics. It begins with the analogy of shock waves that take place in a gas column. Analytical expressions of hydrodynamic impact forces and moments acting on the tank walls, due to sinusoidal pitching excitation, are given in closed approximate forms and compared with experimental measurements. Under lateral excitation, the modal interaction is presented. Hydrodynamic pressure impacts due to sudden acceleration of the tank are displayed according to experimental measurements. A phenomenological modeling of impact forces and damping forces is proposed and used in the analytical modeling of elastic structures carrying liquid containers. The impact forces are replaced by a high-power force, which covers both elastic and rigid impact cases. Three nontraditional analytical techniques, namely saw-tooth time transformation, nonsmooth coordinate transformation, and Lie group transformations are briefly outlined. An application of a mechanical system simulating an elastic structure carrying a liquid container that allows high-power impact forces will be adopted. The response will be obtained using the method of multiple scales and the results will be compared with those obtained by nontraditional techniques. The analysis involves the simultaneous occurrence of parametric resonance and internal resonance conditions. Sloshing impact in ship tankers and road tankers are addressed together with a brief account of numerical algorithms used for large surface motion.

Chapter 8 presents the linear problem of liquid interaction with its elastic container. The first step in studying this interaction is to estimate the linear eigenvalue problem and the linear response to external excitations. The coupling may take place between the liquid free-surface dynamics and either the tank bending oscillations or the breathing modes (or shell modes). Breathing vibrations of the tank are essentially radial, such that both flexure and stretching deformations of the wall occur while the longitudinal axis of the tank remains straight. The mathematical formulation of interacting forces is formulated for a general motion of the liquid containers. The linear analysis deals mainly with the estimation of the coupled natural frequencies of liquid–structure systems. Specific cases are addressed. These include: (i) rigid tank walls with an elastic bottom modeled as either a stretched membrane or elastic plate, (ii) a rigid bottom with elastic walls experiencing either bending deformation or breathing vibrations, and (iii) interaction with a completely elastic container. The nonlinear interaction under external and parametric excitations is treated in Chapter 9. The nonlinear interaction covers three different groups: (1) nonlinear free-surface interaction with linear shell deformation, (2) linear free-surface interaction with nonlinear shell deformation, and (3) nonlinear surface interaction with nonlinear shell deformation. Storage tanks interacting with their liquid and the numerical algorithms developed to solve for the liquid–tank interaction are reviewed. Chapter 10 deals with the interaction of elastic support structure and tuned liquid sloshing absorbers. This problem was not addressed in other books. This type of interaction takes place between the free-liquid-surface motion and the supported elastic structure dynamics based on the assumption that the liquid container is rigid. Under the base motion of the supporting structure, the fluid container experiences motion in a certain trajectory governed by the excitation and the liquid response. The free-liquid-surface motion will result in hydrodynamic forces that are fed back to the supporting structure. Civil engineers took advantage of this by using liquid tanks placed on the top of large buildings to act as vibration absorbers. A detailed treatment of sloshing absorbers such as tuned liquid-sloshing absorbers, and liquid-column vibration absorbers is presented. Nonlinear interaction of liquid free surface with support structure is then treated in the presence of parametric excitation, internal resonance, and combination resonance, under deterministic and random excitations.

Chapter 11 presents the problem of liquid dynamics in spinning tanks. This problem is usually encountered in the study of stability and control of rockets, space vehicles, liquid-cooled gas turbines, centrifuges, earth with its oceans and fluid core. The spinning and unavoidable wobbling and precession of spacecraft vehicles cause the contained liquid to oscillate, and thus generate dynamic forces, which can destabilize the spacecraft. Note that the kinetic energy of the spinning spacecraft is dissipated by the liquid motion and thus destabilizes the vehicle and causes it to tumble. Spacecraft vehicles are designed to spin in order to gain gyroscopic stiffness during the transfer from low earth orbit. This spin helps to control the location of liquid propellant in its container. Another class of problems deals with the dynamics and stability of rotating rigid bodies as they are applied to the evolution of celestial bodies and astronavigation control. Stabilization is achieved when the spacecraft spins about its axis of minimum moment of inertia. A satellite that spins about its axis of minimum moment of inertia may experience instability if energy is dissipated. This is similar to a spinning top on a rough surface and as a result of friction the top's nutation angle increases as it seeks to conserve angular momentum. Other topics addressed in this chapter include the dynamics of fluid-filled spinning containers, inviscid fluid in a partially filled upright cylinder, inertia waves in rotating fluid, and rotating liquids in microgravity.

The last chapter deals with the physics and dynamics of liquid sloshing in microgravity. In view of the diversity of microgravity fluid physics, this chapter will address selected topics supported by literature reviews of the main problems. It begins with the mechanics of free liquid surface under microgravity, stability of static and dynamic free liquid surface, and contact line and contact angle. The modal analysis includes the estimation of free-surface natural frequencies and associated surface shapes under micro- and zero-gravity environments. The problem of forcing sloshing for slipping and anchored contact lines will be addressed for simple cases. The influence of g-jitter and liquid handling are discussed briefly. Capillary systems including Marangoni flow and liquid bridges will be treated for static and dynamic cases. The chapter also includes the problem of the thermocapillary effects of fluid flow and liquid bridges, and the sloshing of cryogenics, and briefly discusses hydroelastic oscillations under microgravity.

Raouf A. Ibrahim
Detroit, Michigan
ibrahim@eng.wayne.edu

Part I

Linear sloshing dynamics

1

Fluid field equations and modal analysis in rigid containers

1.1 Introduction

The theory of liquid sloshing dynamics in partially filled containers is based on developing the fluid field equations, estimating the fluid free-surface motion, and the resulting hydrodynamic forces and moments. Explicit solutions are possible only for a few special cases such as upright cylindrical and rectangular containers. The boundary value problem is usually solved for modal analysis and for the dynamic response characteristics to external excitations. The modal analysis of a liquid free-surface motion in a partially filled container estimates the natural frequencies and the corresponding mode shapes. The knowledge of the natural frequencies is essential in the design process of liquid tanks and in implementing active control systems in space vehicles. The natural frequencies of the free liquid surface appear in the combined boundary condition (kinematic and dynamic) rather than in the fluid continuity (Laplace's) equation.

For an open surface, which does not completely enclose the field, the boundary conditions usually specify the value of the field at every point on the boundary surface or the normal gradient to the container surface, or both. The boundary conditions may be classified into three classes (Morse and Fesbach, 1953):

1) the *Dirichlet boundary conditions*, which fix the value of the field on the surface;
2) the *Neumann boundary conditions*, which fix the value of the normal gradient on the surface; and
3) the *Cauchy conditions*, which fix both value of the field and normal gradient on the surface.

Each class is appropriate for different types of equations and different boundary surfaces. For example, Dirichlet conditions on a closed surface uniquely specify a solution of Laplace's equation inside the closed surface.

The variational formulation based on Hamilton's principle is regarded as the most powerful tool for developing the fluid field equations. This approach has been proposed and used by Lawrence, *et al.* (1958), Troesch (1960), Bogoryad (1962), Borisova (1962), Petrov (1962a,b,c), Moiseev (1964), Moiseev and Petrov (1966), Luke (1967), Whitham (1967), Lukovskii (1967, 1976), Moiseev and Rumyantsev (1968), Limarchenko (1978a, 1980, 1983b), Lukovskii and Timokha (1992, 1995), and Rocca, *et al.* (1997). The method of integral equations was adopted for containers whose wetted walls are not straight vertical but curved such as spherical containers and horizontal cylindrical containers (see, e.g., Budiansky, 1960 and McIver, 1989). Some analytical and approximate approaches to estimating the sloshing frequencies were developed by Housner (1963a), Evans (1990), and Evans and Linton (1993).

Henrici, *et al.* (1970) presented an extensive treatment of liquid sloshing in a half-space bounded above by a rigid plane that contains either a circular or infinite-strip aperture. They obtained both upper and lower bounds of the natural frequencies. Troesch and Troesch (1972) and Miles (1972) discussed some features of the spectrum of the eigenvalues of liquid sloshing in a half-space with an emphasis on their upper bounds. Banning, *et al.* (1966) built an apparatus for demonstrating the dynamics of liquid sloshing.

The dynamic behavior of liquid propellant free surface was addressed by Ehrlich (1959), Abramson (1961b, 1965), Eulitz and Glaser (1961), Eulitz (1963), Bonneau (1964), Buchanan and Bugg (1966), Fontenot (1968), Martin (1971), and Dodge and Garza (1971). The modal analysis in a circular cylindrical container was originally treated by Poisson (1828) but the results were not interpreted because the theory of Bessel's function was not sufficiently developed at that time. The equations of motion of a liquid in rigid rectangular and rigid circular tanks of uniform depth and with linearized boundary conditions were also given by Rayleigh (1887), Steklov (1902), and Lamb (1945). The solution of the Laplace equation using the method of separation of variables is somewhat less powerful for cases where the liquid depth is variable and other methods, such as the Ritz method, should be used. Bratu (1971) studied the oscillations of liquid masses in reservoirs.

The free-surface mode shapes for containers with axial symmetry were determined by Borisova (1962), Bonneau (1964), Moiseev and Petrov (1965, 1966), Pfeiffer (1967a,b), Einfeldt, *et al.* (1969), McNeil and Lamb (1970), Henrici, *et al.* (1970), Pshenichnov (1972), and Boyarshina and Koval'chuk (1986) determined the normal modes and natural frequencies of the free surface in an inclined cylinder. Trotsenko (1967) studied the liquid oscillations in a cylindrical tank with annular baffle. For a spherical tank, the problem is analytically more complex and approximate solutions for the natural frequencies were obtained by Bauer (1958a), Budiansky (1960), Leonard and Walton (1961), Riley and Trembath (1961), Lukovskii (1961a,b), Chu (1964a), Boudet (1968), McIver (1989), El-Rahib and Wagner (1981), and Bauer and Eidel (1989b). The natural frequency of horizontal circular canals and spherical containers is determined from an integral equation, which is usually discretized into a matrix form for numerical calculations (Barnyak, 1997).

Bauer (1964b) and Mooney, *et al.* (1964a,b) analyzed the free-surface oscillations in a quarter tank and in a tank with annular sector cross-section. In both cases, the natural frequency of the free surface was found to have the same expression as the circular cylindrical tank but with different roots of the Bessel function. The liquid sloshing frequencies for different container geometries were evaluated by Miles (1964, 1972), Kuttler and Sigillito (1969), Fox and Kuttler (1981, 1983), Meserole and Fortini (1987), and McIver and McIver (1993). The influence of movable devices and internal pipes on the natural frequencies of the free surface was determined by Siekmann and Chang (1971b) and Drake (1999). Bauer and Eidel (1999c) considered different configurations of cylindrical containers.

Based on the two-dimensional analysis of liquid motion in rectangular tanks, the natural frequency depends essentially on the liquid depth to width ratio. The effect of liquid depth is diminished as the mode order increases. Graham and Rodriguez (1952) solved the three-dimensional velocity potential for which the natural frequency depends on the three major dimensions of the fluid. Ghali (1965) determined the nonlinear dependence of the natural frequencies on the wave motion amplitude. The influence of damping on the natural frequency was studied experimentally by Ghali (1965), Scarsi and Brizzolara (1970), Scarsi (1971), and Schilling and Siekmann (1980). It was found that for higher viscosities (of kinematic viscosity $\nu = 2.5$ poise) the resonance frequency is slightly higher than the predicted value for an ideal liquid.

The purpose of this chapter is to develop the fluid field equations with reference to an inertial frame and with respect to a moving coordinate system. The variational approach is demonstrated as a tool to derive the boundary-value problem in one treatment. The modal analysis of liquid free-surface motion is formulated for different tank shapes. The analysis includes the estimation of the velocity potential function, fluid free-surface natural frequencies and mode shapes. The influence of surface tension is included in some cases. However, its effect is dominant in a microgravity field as will be demonstrated in Chapter 12. Note that analytical solutions in a closed form are only obtained for regular tank shapes whose walls are upright straight. For other tank geometries with variable depth, one can determine the natural frequencies and mode shapes either experimentally or numerically.

1.2 Fluid field equations

The analytical description of the fluid field equations is documented for different cases of tank geometries by Ewart (1956), Bauer (1962a, 1966c, 1969a), Lomen (1965a), Abramson (1966a), Ibrahim (1969), Khandelwal (1980), Kornecki (1983), and Bauer (1999). The general equations of motion for a fluid in closed containers can be simplified by assuming the container rigid and impermeable. Furthermore, the fluid is assumed inviscid, incompressible, and initially irrotational. Capillary or surface tension effects will be ignored in a gravitational field. However, the effect of surface tension will be introduced for some simple cases. The free-surface oscillations can be generated by giving an initial impulse, or an initial disturbance to the free surface.

This section considers the general case of a tank moving along some trajectory in space. The formulation is applicable to free and forcing liquid free-surface oscillations. It is convenient to refer the fluid motion to a moving coordinate system as the variables are measured relative to the moving frame. It is also useful to write the fluid equations of motion with reference to stationary and moving coordinates as shown in Figure 1.1. In the present analysis, the tank is allowed to move in planar curvilinear motion without rotation. Let $O'X'Y'Z'$ be the stationary Cartesian coordinate frame. The Euler equations of motion of the fluid are written in the vector form

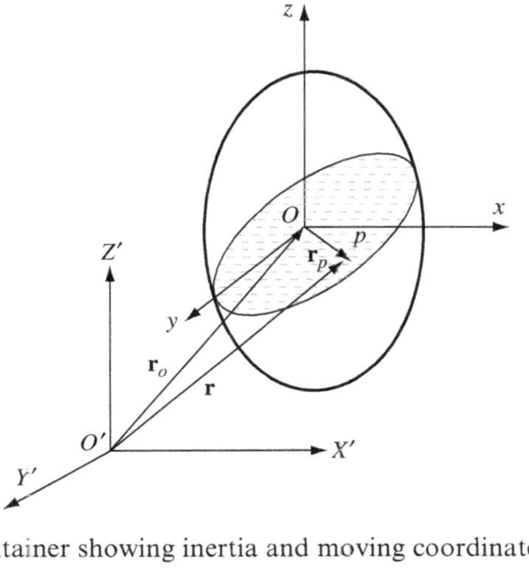

Figure 1.1 Moving liquid container showing inertia and moving coordinates.

$$\frac{\partial}{\partial t}\boldsymbol{q}+(\boldsymbol{q}\cdot\nabla)\boldsymbol{q}=-\frac{1}{\rho}\nabla P-\nabla(gZ') \tag{1.1}$$

where $\boldsymbol{q}$ is the fluid velocity, $\partial\boldsymbol{q}/\partial t$ is the local acceleration of the flow at the point whose coordinates are not allowed to vary (this acceleration is measured by a fixed observer), $(\boldsymbol{q}\cdot\nabla)\boldsymbol{q}$ is the convective acceleration for a fluid particle drifting with the stream at a velocity $\boldsymbol{q}$ in the flow direction (this acceleration is measured by an observer moving with the particle p), P is the fluid pressure, ρ is the fluid density, gZ' is the gravitational potential, and ∇ is an operator given for different coordinate frames in the appendix to this chapter. Note that the convective acceleration $(\boldsymbol{q}\cdot\nabla)\boldsymbol{q}$ may also be written in the form (Thomson, 1965, p. 44)

$$(\boldsymbol{q}\cdot\nabla)\boldsymbol{q}=\frac{1}{2}\nabla q^2-\boldsymbol{q}\times(\nabla\times\boldsymbol{q})=\frac{1}{2}\nabla q^2 \tag{1.2}$$

For irrotational flow the curl of the velocity vanishes, that is, $\nabla\times\boldsymbol{q}=0$.

For irrotational fluid motion, there exists a velocity potential function, Φ, whose (negative) gradient gives the fluid velocity,

$$\mathbf{q}=-\nabla\Phi \tag{1.3}$$

The negative sign is optional and in some cases it may be removed provided the analysis preserves the sign convention. Introducing relations (1.2) and (1.3) into equation (1.1) gives

$$\nabla\left(\frac{P}{\rho}+\frac{1}{2}q^2+gZ'-\frac{\partial\Phi}{\partial t}\right)=0 \tag{1.4}$$

Upon integrating equation (1.4) one obtains

$$\frac{P}{\rho}+\frac{1}{2}q^2+gZ'-\frac{\partial\Phi}{\partial t}=C(t) \tag{1.5}$$

where $C(t)$ is an arbitrary function of time.

Equation (1.5) is the general form of Kelvin's equation for an unsteady fluid flow. In this equation the potential function Φ is a function of space and time, and its derivative with respect to time measures the unsteadiness of the flow. However, $\partial\Phi/\partial t$ is interpreted as the work done on a unit mass of the fluid whose coordinates are (X, Y, Z). Furthermore, equation (1.5) is only valid for incompressible flow for which the continuity condition $\nabla\cdot\boldsymbol{q}=0$ yields Laplace's equation, which after introducing equation (1.2) takes the form

$$\nabla^2\Phi=0 \tag{1.6}$$

Let $Oxyz$ be another coordinate frame fixed to the tank such that the Oxy plane coincides with the undisturbed free surface. Let $\boldsymbol{V}_0$ be the velocity of the origin O relative to the fixed origin O'. In this case, the time rate of change of the velocity potential Φ at a point fixed in the stationary frame $O'X'Y'Z'$ as measured by an observer in the moving frame $Oxyz$ is $(\partial/\partial t-\ \boldsymbol{V}_0\cdot\nabla)\Phi$, since this point will appear to have a velocity $-\boldsymbol{V}_0$ with respect to the observer. Accordingly, the pressure equation (1.5) takes the form

$$\frac{P}{\rho}+\frac{1}{2}q^2+gZ'-\frac{\partial\Phi}{\partial t}+\boldsymbol{V}_0\cdot\nabla\Phi=C(t) \tag{1.7a}$$

The fluid particle velocity $\boldsymbol{q}_{\text{rel}}$ relative to the moving coordinate is

$$\boldsymbol{q}_{\text{rel}} = \boldsymbol{q} - \boldsymbol{V}_0 = -\nabla\Phi - \boldsymbol{V}_0 \tag{1.8}$$

Expressing $\boldsymbol{q}$ in terms of $\boldsymbol{q}_{\text{rel}}$ and $\boldsymbol{V}_0$, using relation (1.8), gives

$$\frac{P}{\rho} + \frac{1}{2}q_{\text{rel}}^2 + gZ' - \frac{\partial\Phi}{\partial t} - \frac{1}{2}V_0^2 = C(t) \tag{1.7b}$$

Equation (1.7a) is written in terms of the total fluid velocity as measured by the fixed coordinate and equation (1.7b) is given in terms of the fluid relative velocity to the tank moving coordinates. At the free surface, the pressure is equivalent to the ambient pressure or can be set to zero in equation (1.7a). This gives the dynamic boundary condition

$$\frac{1}{2}(\nabla\Phi \cdot \nabla\Phi) + g\eta - \frac{\partial\Phi}{\partial t} + \mathbf{V}_0 \cdot \nabla\Phi = 0 \tag{1.9}$$

where the function $C(t)$ has been absorbed in the potential function, Φ.

The vertical velocity of a fluid particle located on the free surface $z = \eta(r, \theta, t) = \eta(x, y, t)$ should be equated to the vertical velocity of the free surface itself. This condition is known as the kinematic free-surface condition and is given by the following expression

$$-\frac{\partial\Phi}{\partial z} = \frac{\partial\eta}{\partial t} + \mathbf{q}_{\text{rel}} \cdot \nabla\eta \tag{1.10}$$

At the wetted rigid wall and bottom, the velocity component normal to the boundary must have the same value of the corresponding velocity component of the solid boundary at the point in question. For example, if the tank is allowed to move in the vertical plane then the velocity vector in terms of Cartesian and cylindrical coordinates may be written in the form, respectively

$$\mathbf{V}_0 = \dot{X}_0\boldsymbol{i} + \dot{Z}_0\boldsymbol{k} \tag{1.11a}$$

$$\mathbf{V}_0 = \left(\dot{X}_0 \cos\theta\right)\mathbf{i}_r - \left(\dot{X}_0 \sin\theta\right)\mathbf{i}_\theta + \dot{Z}_0\mathbf{i}_z \tag{1.11b}$$

The boundary conditions at the wall and bottom for Cartesian and cylindrical coordinates are, respectively,

$$-\left.\frac{\partial\Phi}{\partial z}\right|_{z=-h} = \dot{Z}_0, \quad -\left.\frac{\partial\Phi}{\partial x}\right|_{x=a} = \dot{X}_0 \tag{1.12a}$$

$$-\left.\frac{\partial\Phi}{\partial z}\right|_{z=-h} = \dot{Z}_0, \quad -\left.\frac{\partial\Phi}{\partial r}\right|_{r=R} = \dot{X}_0 \cos\theta \tag{1.12b}$$

It is possible to split the total velocity potential function, Φ, into a disturbance potential function, $\tilde{\Phi}$, and a potential function, Φ_o, which defines the motion of the tank, that is,

$$\Phi = \tilde{\Phi} + \Phi_o \tag{1.13}$$

The function Φ_o can be determined by integrating equation (1.11) as

$$\Phi_o = -\dot{X}_0 r\cos\theta - \dot{Z}_0 z - \frac{1}{2}\int (\dot{X}_0^2 + \dot{Z}_0^2)\, dt \tag{1.14}$$

Introducing (1.13) and (1.14) into the free-surface boundary conditions gives

$$\frac{1}{2}(\nabla\tilde{\Phi}\cdot\nabla\tilde{\Phi}) + (g+\ddot{Z}_0)\eta - \frac{\partial\tilde{\Phi}}{\partial t} + \ddot{X}_0 r\cos\theta = 0 \tag{1.15}$$

$$-\frac{\partial\tilde{\Phi}}{\partial z} = \frac{\partial\eta}{\partial t} - \frac{\partial\eta}{\partial r}\frac{\partial\tilde{\Phi}}{\partial r} - \frac{1}{r^2}\frac{\partial\eta}{\partial\theta}\frac{\partial\tilde{\Phi}}{\partial\theta} \tag{1.16}$$

The corresponding conditions for a rectangular tank are

$$\frac{1}{2}(\nabla\tilde{\Phi}\cdot\nabla\tilde{\Phi}) + (g+\ddot{Z}_0)\eta - \frac{\partial\tilde{\Phi}}{\partial t} + \ddot{X}_0 x = 0 \tag{1.17}$$

$$-\frac{\partial\tilde{\Phi}}{\partial z} = \frac{\partial\eta}{\partial t} - \frac{\partial\eta}{\partial x}\frac{\partial\tilde{\Phi}}{\partial x} - \frac{\partial\eta}{\partial y}\frac{\partial\tilde{\Phi}}{\partial y} \tag{1.18}$$

One may introduce the effect of surface tension, σ, by including the pressure change across the displaced free liquid surface as described by the Laplace–Young equation

$$p_s = \sigma\left(\frac{1}{R_1} + \frac{1}{R_2}\right) \tag{1.19}$$

where R_1 and R_2 are the principal radii of curvature. The complete formulation of the boundary value problem in terms of the disturbance potential function is summarized as follows:

(1) *for a cylindrical container*:

$$\nabla^2\tilde{\Phi} = 0 \tag{1.20a}$$

$$\left.\frac{\partial\tilde{\Phi}}{\partial r}\right|_{r=R} = 0, \quad \left.\frac{\partial\tilde{\Phi}}{\partial z}\right|_{z=-h} = 0 \tag{1.20b, c}$$

$$\frac{1}{2}(\nabla\tilde{\Phi}\cdot\nabla\tilde{\Phi}) + (g+\ddot{Z}_0)\eta - \frac{\partial\tilde{\Phi}}{\partial t} + \frac{\sigma}{\rho}\left(\frac{1}{R_1} + \frac{1}{R_2}\right) + \ddot{X}_0 r\cos\theta = 0, \quad \text{at } z = \eta(r,\theta,t) \tag{1.20d}$$

$$-\frac{\partial\tilde{\Phi}}{\partial z} = \frac{\partial\eta}{\partial t} - \frac{\partial\eta}{\partial r}\frac{\partial\tilde{\Phi}}{\partial r} - \frac{1}{r^2}\frac{\partial\eta}{\partial\theta}\frac{\partial\tilde{\Phi}}{\partial\theta}, \quad \text{at } z = \eta(r,\theta,t) \tag{1.20e}$$

The curvature κ for cylindrical coordinates is given by the expression

$$\begin{aligned}\kappa &= -\left(\frac{1}{R_1} + \frac{1}{R_2}\right)\\ &= -\frac{\eta_{rr}(1+(\eta_\theta^2/r^2)) + (1+\eta_r^2)((\eta_r/r) + (\eta_{\theta\theta}/r^2)) - 2\,\eta_r\,(\eta_\theta/r^2)(\eta_{rr} + (\eta_\theta/r))}{\left[1+\eta_r^2 + (\eta_\theta^2/r^2)\right]^{3/2}}\end{aligned} \tag{1.21}$$

This expression can be linearized in the form

$$\kappa = -\left[\eta_{rr} + \frac{\eta_r}{r} + \frac{\eta_{\theta\theta}}{r^2}\right] \tag{1.22}$$

(2) *for a rectangular container*:

$$\nabla^2\tilde{\Phi} = 0 \tag{1.23a}$$

$$\left.\frac{\partial\tilde{\Phi}}{\partial x}\right|_{x=\pm a/2} = 0, \quad \left.\frac{\partial\tilde{\Phi}}{\partial y}\right|_{x=\pm b/2} = 0, \quad \left.\frac{\partial\tilde{\Phi}}{\partial z}\right|_{z=-h} = 0 \tag{1.23b, c, d}$$

$$\frac{1}{2}(\nabla\tilde{\Phi}\cdot\nabla\tilde{\Phi}) + (g+\ddot{Z}_0)\eta - \frac{\partial\tilde{\Phi}}{\partial t} + \frac{\sigma}{\rho}\left(\frac{1}{R_1}+\frac{1}{R_2}\right) + \ddot{X}_0 x = 0, \quad \text{at } z=\eta(x,y,t) \tag{1.23e}$$

$$-\frac{\partial\tilde{\Phi}}{\partial z} = \frac{\partial\eta}{\partial t} - \frac{\partial\eta}{\partial x}\frac{\partial\tilde{\Phi}}{\partial x} - \frac{\partial\eta}{\partial y}\frac{\partial\tilde{\Phi}}{\partial y}, \quad \text{at } z=\eta(x,y,t) \tag{1.23f}$$

The curvature, κ, for cylindrical coordinates is given by the expression

$$\kappa = -\left(\frac{1}{R_1}+\frac{1}{R_2}\right) = -\frac{\eta_{xx}\left(1+\eta_y^2\right) + \eta_{yy}(1+\eta_x^2) - 2\eta_x\eta_y\eta_{xy}}{\left[1+\eta_x^2+\eta_y^2\right]^{3/2}}. \tag{1.24}$$

This expression can be linearized in the form

$$\kappa = -\left[\eta_{xx} + \eta_{yy}\right] \tag{1.25}$$

For other container geometries, such as spherical, prolate and oblate spherical, and elliptic containers, the continuity equation and other related operators are listed in the appendix to this chapters. Note that the velocity potential function, $\tilde{\Phi}$, must satisfy Laplace's equation, $\nabla^2\tilde{\Phi} = 0$, which is a linear partial differential equation. The nonlinearity in the boundary value problem only exists in the free-surface boundary conditions on $z=\eta$. If one is interested in the modal analysis then one should drop the nonlinear and nonconservative terms from the free-surface boundary conditions. If the potential function is obtained analytically in a closed form, then the natural frequencies of the fluid free surface are obtained by using the dynamic free-surface condition based on the fact that $\tilde{\Phi}$ is harmonic in time. Another powerful approach is to use the variational formulation together with the Rayleigh–Ritz method. The next section describes an alternative approach based on the variational principle.

1.3 Variational formulation

The variational approach is based on establishing the superlative of a certain function that describes the system behavior. The Lagrangian, $L=T-V$, has to be minimized (or maximized), where T and V are the kinetic and potential energies of the system, respectively. The variational principle, or Hamilton's principle, is

$$\delta I = \delta\int_{t_1}^{t_2}(T-V)\,dt = 0 \tag{1.26}$$

Basically Hamilton's principle states that the actual path in the configuration space yields the value of the definite integral stationary with respect to all arbitrary variations of the path between two instants of time t_1 and t_2 provided the path variations vanish at these two end points. For any actual motion of the system, the system will move so that the time average of the difference between the kinetic and potential energies will be a minimum. This formulation is very powerful since it brings in one statement the fluid field equations and the associated boundary conditions. Substituting for the kinetic energy, $T = \int_v (\rho/2)|\nabla\Phi|^2\, dv$, and potential energy, $V = \int_S (\eta/2)(\rho g \eta\, dS)$, where v is the fluid volume, and S is the fluid free surface, in equation (1.26), and taking the variation gives

$$\begin{aligned}\delta I &= \delta \int_{t_1}^{t_2} dt \left\{ \int_v \frac{\rho}{2} |\nabla\Phi|^2\, dv - \int_S \frac{\eta}{2}(\rho g \eta\, dS) \right\} \\ &= \rho \int_{t_1}^{t_2} dt \left\{ \int_v \nabla\Phi\nabla\delta\Phi\, dv - g \int_S \eta\delta\eta\, dS \right\} = 0 \end{aligned} \tag{1.27}$$

The volume integral can be transformed into a surface integral using Green's formula, $\int_v \nabla\Phi\nabla\delta\Phi\, dv = \int_S \Phi \frac{\partial\delta\Phi}{\partial n}\, dS$. Furthermore, one may use the relationship

$$\nabla\Phi = \boldsymbol{n}\frac{\partial\Phi}{\partial n} = -\boldsymbol{n}\frac{\partial\eta}{\partial t}$$

(see, e.g., Thomson, 1965), where $\boldsymbol{n}$ is the unit vector along the normal at the point in question to the equi-potential surface of Φ. In this case, the variational takes the form

$$\rho \int_{t_1}^{t_2} dt \left\{ \int_S \left\{ \Phi \frac{\partial\delta\Phi}{\partial n} - g\eta\delta\eta \right\} dS = -\rho \int_{t_1}^{t_2} dt \left\{ \int_S \left\{ \Phi\delta\frac{\partial\eta}{\partial t} + g\eta\delta\eta \right\} dS = 0 \right.\right.$$

Integrating by parts gives

$$\rho \int_{t_1}^{t_2} \int_S \left\{ -\frac{\partial\Phi}{\partial t} + g\eta \right\} \delta\eta\, dS\, dt = 0$$

This statement yields the linearized dynamic free surface condition

$$-\frac{\partial\Phi}{\partial t} + g\eta = 0 \tag{1.28}$$

Moiseev and Rumyantsev (1968) introduced the Neumann operator H, which makes the velocity potential, Φ, harmonic inside the fluid volume domain, v. The harmonic property is based on the fact that the integral of free-surface velocity, $\dot{\eta}(s)$, vanishes over the free surface, that is, $\int_S \dot{\eta}(s)\, ds = 0$. The function Φ satisfies the following conditions on the fluid boundary $\frac{\partial\Phi}{\partial n} = 0$ at the container walls, and $\frac{\partial\Phi}{\partial n} = -\dot{\eta}$ at the free surface.

In this case, one can write the correspondence as $\Phi = H\dot{\eta}$, where H is the integral operator

$$H\dot{\eta} = \int_S H(s, v)\dot{\eta}(s)\, ds = \Phi(v) \tag{1.29}$$

Note that the kernel is the Green function of Neumann's problem for the region v. This kernel is known to be symmetric (see, e.g., Mikhlin, 1964, and Gunter, 1965). The kernel is logarithmic in the case of the plane problem, that is, when the volume is reduced to a plane surface. The kernel is a polar function in the case of the three-dimensional problem. Thus, H is a fully continuous selfadjoint operator, and one can write the following representation

$$\Phi(v) = H\frac{\partial \Phi}{\partial z} = -H\frac{\partial \eta}{\partial t} \tag{1.30}$$

Taking the time derivative for both sides and using equation (1.28), gives

$$g\eta + H\frac{\partial^2 \eta}{\partial t^2} = 0 \tag{1.31}$$

The average energy function can be written in terms of the Neumann operator in the form

$$I_1 = \frac{\rho}{2}\int_{t_1}^{t_2}\int_S \left\{ H\left(\frac{\partial \eta}{\partial t}\right)^2 - g\eta^2 \right\} ds\, dt \tag{1.32}$$

Alternatively, one can write the average energy function in terms of the scalar potential function in the form, after using equation (1.28)

$$I = \int_{t_1}^{t_2} dt \left\{ \int_v \frac{\rho}{2}|\nabla \Phi|^2\, dv - \frac{1}{g}\int_S \Phi^2\, ds \right\} \tag{1.33}$$

Equations (1.32) and (1.33) can be used for estimating the natural frequencies for the liquid free surface. At the free surface, S, both the velocity potential function and the free-surface wave height can be expressed in terms of time and space as follows

$$\Phi(s, t) = F(s)\cos \omega t, \quad \text{and} \quad \eta(s, t) = G(s)\sin \omega t \tag{1.34}$$

where ω is the free-surface natural frequency. Substituting equations (1.34) in the average energy functions given by equations (1.32) and (1.33) and integrating over time t from $t_1 = 0$ to $t_2 = 2\pi/\omega$, gives

$$I_1 = \lambda \int_S HG \cdot G\, ds - \int_S G^2\, ds \tag{1.35}$$

$$I_2 = \int_v \frac{\rho}{2}|\nabla F|^2\, dv - \lambda \int_S F^2\, ds \tag{1.36}$$

where $\lambda = \omega^2/g$. The natural frequencies of the fluid free surface are determined by using the Rayleigh–Ritz method. The method is based on introducing a linear combination of a complete set of functions whose coefficients form a set of linear variational parameters. For example, the function F may be expressed by the following summation of the trial functions f_n

$$F = \sum_{n=1}^{N} a_n f_n \tag{1.37}$$

Substituting equation (1.37) into equation (1.36), it follows from equation (1.27) that the coefficients a_n should satisfy a system of algebraic homogeneous equations for the coefficients a_n by setting $\partial I/\partial a_n = 0$, for $n = 1, 2, \ldots, N$. This yields the following set of equations

$$\sum_{m=1}^{N} a_{mn}(A_{mn} - \lambda B_{mn}) = 0 \tag{1.38}$$

where $A_{mn} = \int_v \nabla f_n \cdot \nabla f_m \, dv$, $B_{mn} = \int_S f_n f_m \, ds$, $A_{nm} = A_{mn}$, and $B_{nm} = B_{mn}$.
A nontrivial solution of equations (1.38) exists only if the determinant of the coefficients of a_n vanishes

$$|A_{mn} - \lambda B_{mn}| = 0 \tag{1.39}$$

Equation (1.39) is the frequency equation. The first term can be determined by the variational method. The procedure involves the inclusion of the first few terms of the series (1.37) and solving the secular equation exactly, increasing the number of terms and solving again. The process continues until the values of the natural frequencies do not change. In view of the absence of any rigorous method of convergence and the difficulty of selecting the coordinate functions, the process may yield some errors. Morse and Feshbach (1953) introduced a number of perturbational techniques to determine the natural frequencies of dynamical systems using the Rayleigh–Ritz method. Moiseev (1970) and Moiseev and Rumyantsev (1968) indicated that the value of λ_1 is "only slightly sensitive" to the selection of the functions f_n. This means that if the function F_1, which brings about a minimum of the functional λ_1, is replaced by another function F_1^* (such that $\int \nabla F_1 \nabla F_1^* \, dv \neq 0$) then the value of λ_1 will not change significantly.

The system of coordinate functions $\{f_n\}$ can be selected quite roughly. It is only necessary to make sure that this system is complete. Hence, it is expedient to take these functions as the eigenfunctions of the problem of liquid sloshing in some volume that contains the specified volume but has a simple shape. This will be demonstrated for the case of liquid sloshing in conical containers later in this chapter.

1.4 Normal modes in an upright cylindrical container

Circular cylindrical tanks appear as bare containers or with partitioning walls. One can raise the natural frequencies of the fluid by partitioning the tank with longitudinal walls, or by introducing annular rings or a concentric tank. This section summarizes the main results of different tank geometries shown in Figures 1.2(a)–(d).

1.4.1 Bare wall cylindrical tank

With reference to Figure 1.2(a), a possible solution of the Laplace equation, which satisfies the wall and bottom boundary conditions listed in relations (1.20b, c), is

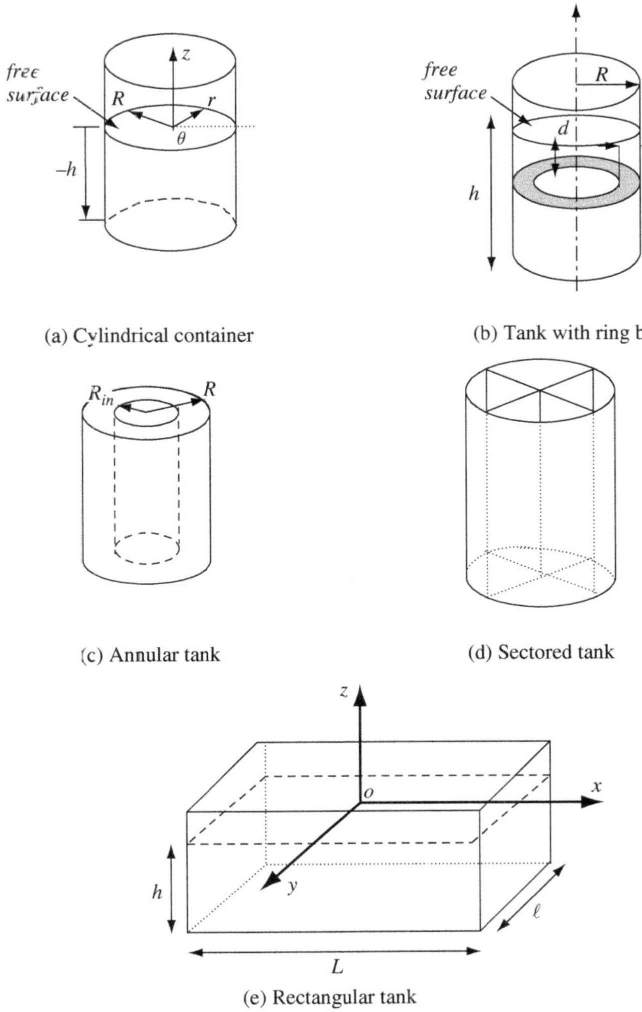

(a) Cylindrical container (b) Tank with ring baffle

(c) Annular tank (d) Sectored tank

(e) Rectangular tank

Figure 1.2 Different tank geometries.

$$\tilde{\Phi}(r,\theta,z,t)=\sum_{m=0}^{\infty}\sum_{n=1}^{\infty}[\alpha_{mn}(t)\cos m\theta+\beta_{mn}(t)\sin m\theta]J_m(\lambda_{mn}r)\frac{\cosh[\lambda_{mn}(z+h)]}{\cosh\lambda_{mn}h} \tag{1.40}$$

where α_{mn} and β_{mn} are time dependent to be determined from the free-surface initial conditions, $J_m(.)$ is the Bessel function of the first kind of order m, $\lambda_{mn}=\xi_{mn}/R$ are the roots of $\partial J_m(\lambda_{mn}r)/\partial r|_{r=R}=0$. Bauer (1963a) developed a set of tables and graphs for zeros of cross product Bessel functions. These roots are also well documented in Abramowitz and Segun (1968).

The analysis can be significantly simplified if the fluid field equations are linearized for small displacements. The normal mode frequencies are determined from the linearized free-surface boundary condition

$$\frac{\partial\tilde{\Phi}}{\partial t} - g\eta = 0 \tag{1.41}$$

Differentiating once with respect to time and using equation (1.3) gives

$$\frac{\partial^2\tilde{\Phi}}{\partial t^2} + g\frac{\partial\tilde{\Phi}}{\partial z} = 0 \tag{1.42}$$

If the functions α_{mn} and β_{mn} are expressed as harmonics, $\sin\omega_{mn}t$, one can obtain the natural frequencies of the liquid free-surface by substituting equation (1.40) into (1.42)

$$\omega_{mn}^2 = \frac{g\xi_{mn}}{R}\tanh(\xi_{mn}h/R) \tag{1.43a}$$

The above expression approaches a constant value for $h/R > 2$, given by the following relation

$$\omega_{mn}^2 = \frac{g\xi_{mn}}{R}, \qquad \text{for } \xi_{mn}h/R \geq 2.65 \tag{1.43b}$$

If one considers surface tension, the linearized dynamic free-surface condition combined with the kinematic condition (1.42) gives, after using Laplace's equation,

$$\frac{\partial^2\hat{\Phi}}{\partial t^2} + g\frac{\partial\tilde{\Phi}}{\partial z} + \frac{\sigma}{\rho}\frac{\partial^3\hat{\Phi}}{\partial z^3} = 0 \tag{1.44}$$

In this case the natural frequency is given by the expression

$$\omega_{mn}^2 = \left[\frac{g\xi_{mn}}{R} + \frac{\sigma\xi_{mn}^3}{\rho R^3}\right]\tanh(\xi_{mn}h/R) \tag{1.45}$$

This result is valid for the slip contact line and it reveals that the surface tension causes an increase in the normal mode frequencies.

The fluid surface elevation η measured from the undisturbed free surface is obtained from equation (1.41) using (1.40)

$$\eta = \frac{1}{g}\sum_{m=0}^{\infty}\sum_{n=1}^{\infty}\left[\bar{\alpha}_{mn}\cos m\theta + \bar{\beta}_{mn}\sin m\theta\right]J_m(\lambda_{mn}r)\cosh(\lambda_{mn}h)(\omega_{mn}\cos\omega_{mn}t) \tag{1.46}$$

where $\bar{\alpha}_{mn}$ and $\bar{\beta}_{mn}$ are constant coefficients to be determined from initial conditions. The symmetric mode shape takes the form

$$\eta(r,\theta,t) = \frac{1}{g}\sum_{m=0}^{\infty}\sum_{n=1}^{\infty}\bar{\alpha}_{mn}\cos m\theta J_m(\lambda_{mn}r)\cosh(\lambda_{mn}h)(\omega_{mn}\cos\omega_{mn}t) \tag{1.47}$$

For the first mode, $m=0$, the motion is symmetric about the origin in the form of angular ridges and furrows. In this case, the roots of $dJ_0(\xi_{0n}r/R)/dr|_{r=R}=0$ are $\xi_{0n}=3.832$, 7.0156, 10.173, ..., π(n + 0.25). The corresponding nodal circles are determined by setting $\eta=0$, or $J_0(\xi_{0n}r/R)=0$, which gives $\xi_{0n}r/R=2.404$, 5.517, 8.648, ... The corresponding radii of nodal circles are $r=0.628R$, $0.786R$, $0.85R$.

For asymmetric sloshing modes the wave height takes the form

$$\eta(r,\theta,t) = \frac{1}{g}\sum_{m=0}^{\infty}\sum_{n=1}^{\infty}\left[\bar{\beta}_{mn}\sin m\theta\right]J_m(\lambda_{mn}r)\cosh(\lambda_{mn}h)(\omega_{mn}\cos\omega_{mn}t) \tag{1.48}$$

For the first mode, $m=1$, the roots of $dJ_1(\xi_{1n}r/R)/dr|_{r=R}=0$ are $\xi_{1n}=1.841$, 5.335, 8.535, 11.205, 14.850, …, and $\xi_{1n}=\xi_{1(n-1)}+\pi$, for $n>5$. In this case, we have a nodal diameter perpendicular to the direction of rocking the container.

McCarty and Stephens (1960) found excellent agreement between the analytical and measured values of the natural frequency for the first three modes. They also included other results obtained by Eulitz (1958). Werner and Coldwell (1961) experimentally validated the analytical models for the inertias and natural frequencies of fuel sloshing in circular cylindrical tanks. Further analysis of liquid sloshing in cylindrical tanks was given by Kuttler and Sigillito (1984).

Bauer (1992d) considered the case of strong surface tension for which the contact line remains anchored at its location, that is, $\eta(R,\theta,t)=0$. This condition results in the following free-surface displacement

$$\eta(r,\theta,t) = e^{i\omega t}\bar{\eta}(r,\theta) = \sum_{m=0}^{\infty}\sum_{n=1}^{\infty}\frac{\alpha_{mn}}{i\omega}\lambda_{mn}\cos m\theta\sinh(\lambda_{mn}h)J_m(\lambda_{mn}r)\,e^{i\omega t} \tag{1.49}$$

The dynamic condition takes the form

$$\frac{\partial^2\bar{\eta}}{\partial r^2}+\frac{1}{r}\frac{\partial\bar{\eta}}{\partial r}+\frac{1}{r^2}\frac{\partial^2\bar{\eta}}{\partial\theta^2}-\frac{\rho g}{\sigma}\bar{\eta} = i\frac{\omega\rho}{\sigma}\sum_{m=0}^{\infty}\sum_{n=1}^{\infty}\alpha_{mn}\cos m\theta\cosh(\lambda_{mn}h)J_m(\lambda_{mn}r) \tag{1.50}$$

The solution of this partial differential equation must preserve the volume of the fluid

$$\int_0^{2\pi}\int_0^{R}\bar{\eta}(r,\theta)r\,dr\,d\theta = 0 \tag{1.51}$$

Bauer (1992d) obtained the solution of equation (1.50) in the form

$$\begin{aligned}\bar{\eta}(r,\theta) = {} & \sum_{m=0}^{\infty}C_m I_m(\beta r/R)\cos m\theta - i\frac{\omega}{g}\alpha_{00}\\ & - i\frac{\omega}{\sigma}R^2\rho\sum_{m=0}^{\infty}\sum_{n=1}^{\infty}\alpha_{mn}\frac{\cosh(\lambda_{mn}h)}{(\xi_{mn}^2+\beta^2)}J_m(\lambda_{mn}r)\cos m\theta\end{aligned} \tag{1.52}$$

where I_m is the modified Bessel function, $\beta^2=\rho g R^2/\sigma$, and C_m are constants of integration.

For axisymmetric oscillations, $\partial/\partial\theta=0$, $m=0$, equation (1.52) takes the form

$$\bar{\eta}(r,\theta) = C_0 I_0(\beta r/R) - i\frac{\omega}{g}\alpha_{00} - i\frac{\omega}{\sigma}R^2\rho\sum_{n=1}^{\infty}\alpha_{0n}\frac{\cosh(\lambda_{0n}h)}{(\xi_{0n}^2+\beta^2)}J_0(\lambda_{0n}r) \tag{1.53}$$

The stick condition $\eta(R,\theta,t)=0$ gives

$$C_0 I_0(\beta) - \frac{\varpi}{\beta^2}\bar{\alpha}_{00} - \varpi\sum_{n=1}^{\infty}\bar{\alpha}_{0n}\frac{\cosh(\lambda_{0n}h)}{(\xi_{0n}^2+\beta^2)}J_0(\xi_{0n}) = 0 \tag{1.54}$$

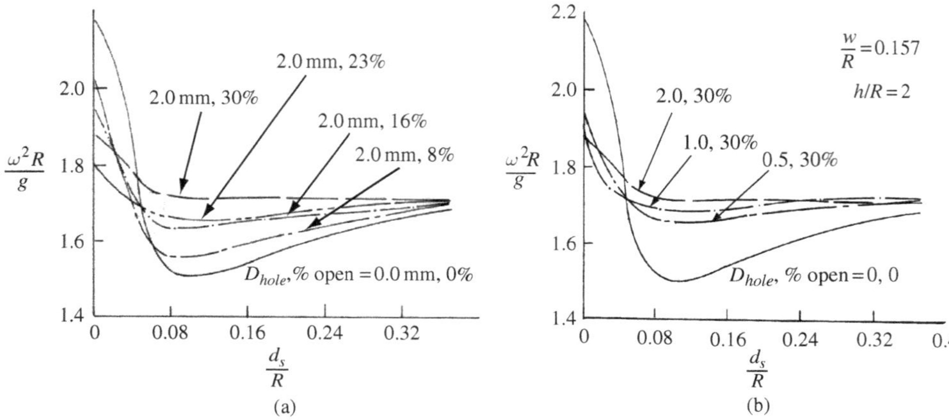

Figure 1.3 (a) Effect of % perforation on the liquid natural frequencies as function of baffle depth, (b) effect of perforation hole size on the liquid natural frequencies as function of baffle depth. (Garza and Abramson, 1963)

where $\varpi = \omega/\sqrt{\sigma/(\rho R^3)}$, and $\bar{\alpha}_{0n} = i\alpha_{0n}/\left(R\sqrt{\sigma/(\rho R^3)}\right)$, $n = 0, 1, 2, \ldots$

Expanding the modified Bessel function I_0 into a Dini-series

$$I_0(\beta r/R) = \frac{2}{\beta}\, I_1(\beta) + \sum_{n=1}^{\infty} \frac{2\beta I_1(\beta) J_0(\lambda_{0n} r)}{(\xi_{0n}^2 + \beta^2) J_0(\lambda_{0n})} \tag{1.55}$$

Introducing expansion (1.54) into (1.52) and comparing coefficients of the Bessel–Fourier series with those of equation (1.49) gives

$$2I_1(\beta)\beta C_0 - \varpi\bar{\alpha}_{00} = 0 \tag{1.56a}$$

$$2\beta C_0 \frac{I_1(\beta)}{J_0(\xi_{0n})}\varpi - \bar{\alpha}_{0n}\left[\varpi^2 \cosh(\lambda_{0n}h) - \xi_{0n}\left(\xi_{0n}^2 + \beta^2\right)\sinh(\lambda_{0n}h)\right] = 0 \tag{1.56b}$$

Equations (1.54) and (1.56) constitute $(n+2)$ homogeneous algebraic equations for the constants C_0, $\bar{\alpha}_{00}$ and $\bar{\alpha}_{0n}$, $n = 1, 2, \ldots$ Setting the truncated determinant of the coefficients to zero gives the frequency equation of the axisymmetric natural oscillations. For asymmetric oscillations one should follow the same procedure.

1.4.2 Cylindrical tank with ring baffles

Usually ring baffles, see Figure 1.2(b), are introduced to control the free surface motion and to raise its natural frequency. Garza and Abramson (1963) conducted a series of experimental tests on a circular cylindrical tank with ring baffles and found the free-surface natural frequencies depend on the ring baffle area and its location, d_s, below the free-surface level. Figures 1.3(a) and (b) shows the dependence of the liquid first mode natural frequency on the location of the baffle d_s/R for baffle width, w, to tank radius ratio, $w/R = 0.157$. Figure 1.3(a) shows the influence of percentage-perforated area for a hole diameter of 2.0 mm, while Figure 1.3(b) shows the effect of the hole diameter on the natural frequency of the free surface for a fixed opening ratio. It is seen that the liquid frequency exhibits a maximum value when the baffle is located at the fluid free surface $d_s/R=0$. It then decreases to a minimum value near a baffle depth close to $d_s/R=0.1$. This observed drop might be attributed to the fact that the baffle restricts the motion of the liquid to that of an equivalent shallow tank of the same diameter. The effect diminishes as the baffle moves to a larger depth. Above depth ratio 0.1 the frequency increases gradually until it reaches the first resonant frequency for a bare-wall cylindrical tank. Another feature may be observed for baffle depths greater than 0.06, and with holes of 2 mm diameter, is that the free-surface frequency increases as the percentage of perforated area is increased.

In some cases, it may be desirable to maintain the highest possible resonant frequency with a given baffle system. This can be achieved by using ring baffles with an axial spacing that must be less than 0.08.

1.4.3 Annular tank

Introducing a concentric tank of radius R_{in}, as shown in Figure 1.2(c), divides the fluid flow into two regions, the concentric tank and the ring tank. Bauer (1960b, 1961c) studied the effect of a concentric tank and derived the following velocity potential function

$$\tilde{\Phi}(r,\theta,z,t)=\sum_{m=0}^{\infty}\sum_{n=1}^{\infty}[A_{mn}\cos m\theta+B_{mn}\sin m\theta]\sin\omega_{mn}t\frac{C_m(\xi_{mn}r/R)}{Y'_m(\xi_{mn})}\times\frac{\cosh[\xi_{mn}(z+h)/R]}{\cosh\xi_{mn}h/R} \tag{1.57}$$

where a prime denotes differentiation with respect to r, Y_m (.) is the Bessel function of the second kind of order m, ξ_{mn} are zeros of the determinant

$$\Delta_m(\xi_{mn})=\begin{vmatrix} J'_m(\xi_{mn}) & Y'_m(\xi_{mn}) \\ J'_m(k\xi_{mn}) & Y'_m(k\xi_{mn}) \end{vmatrix}=0 \tag{1.58}$$

$k=R_{in}/R$, and $C_m(\xi_{mn}r/R)$ is given by the expression

$$C_m(\xi_{mn}r/R)=J_m(\xi_{mn}r/R)Y'_m(\xi_{mn})-J'_m(\xi_{mn}r)Y_m(\xi_{mn}r/R) \tag{1.59}$$

The roots of $\Delta_m(\xi_{mn})=0$ are required to determine the liquid natural frequencies. McMahon (1894) gave asymptotic expansions for most of the roots. Curves showing the roots for $m=0$, 1, 2, 3, and 4 are given by Kirkham (1958). The dominant mode in most applications is the first mode, $m=1$, because it exerts most fluid forces onto the tank wall. The corresponding natural frequencies are given by the expression

$$\omega_{1n}^2=\frac{g}{R}\xi_{1n}\tanh(\xi_{1n}h/R) \tag{1.60}$$

The fluid free-surface displacement measured from the undisturbed surface is:

$$\eta=\frac{1}{g}\sum_{m=0}^{\infty}\sum_{n=1}^{\infty}[A_{mn}\cos m\theta+B_{mn}\sin m\theta]\omega_{mn}\cos\omega_{mn}t\frac{C_m(\xi_{mn}r/R)}{Y'_m(\xi_{mn})} \tag{1.61}$$

Table 1.1 *Roots of the determinant $\Delta_m(\xi_{mn})=0$ for different values of radii ratio k*

$k\backslash n$	0.01	0.1	0.2	0.3	0.4	0.5	0.6	0.7	0.8	0.9
0	1.8408	1.8035	1.7051	1.5821	1.4618	1.3537	1.2621	1.1824	1.1134	1.0531
1	5.3291	5.1371	4.9608	5.1374	5.6592	6.5649	8.0411	10.592	15.778	11.4469
2	8.5305	8.1992	8.4331	9.8083	10.6833	12.7066	15.8013	21.0041	31.4513	62.8481
3	11.6951	11.3588	12.1650	13.6837	15.8481	18.9427	23.6239	31.4557	47.1504	94.2645
4	14.8461	14.6344	15.9932	18.1159	21.0488	25.2025	31.4633	41.9189	62.8510	125.675
5	17.9898	17.9864	19.8616	22.5707	26.2641	31.4721	39.3076	52.3845	78.5549	157.088
6	21.1289	21.3837	23.7502	27.0369	31.4859	37.7459	47.1552	62.8523	94.2601	188.502
7	24.2647	24.8081	27.6498	31.5092	36.7118	44.0223	55.0047	73.3212	109.966	219.917
8	27.3979	28.2497	31.5563	35.9855	41.9403	50.3005	62.8553	83.791	125.673	251.332
9	30.5288	31.7027	35.4675	40.4643	47.1704	56.5797	70.7066	94.2612	141.379	282.747

Source: Bauer (1963a).

The first ten zeros of ξ_{1n} are listed for different values of radii ratio k in Table 1.1. Figure 1.4(a) and (b) shows the natural frequencies in the inner and outer tanks for radii ratios $k=0.5$ and 0.62, respectively. For these ratios, the liquid masses are approximately equal and the natural frequencies are spread favorably in such a way that the relative phase shift between the inner and outer tanks makes it possible to reduce the total hydrodynamic sloshing forces and moments.

1.4.4 Cylindrical quarter tank

When an upright cylindrical tank is divided by two orthogonal walls in the form of a crucifix as shown in Figure 1.2(d) each quarter can be treated independently. In this case the boundary conditions at the tank solid boundaries are

$$\left.\frac{\partial\tilde{\Phi}}{\partial r}\right|_{r=R}=0,\quad \left.\frac{\partial\tilde{\Phi}}{\partial z}\right|_{z=-h}=0,\quad \left.\frac{1}{r}\frac{\partial\tilde{\Phi}}{\partial\theta}\right|_{\theta=0,\,\pi/2}=0 \tag{1.62}$$

in addition to the free-surface condition (1.42). Bauer (1963e, 1964d) solved the boundary value problem and obtained the following velocity potential function

$$\tilde{\Phi}(r,\theta,z,t)=\sum_{m=0}^{\infty}\sum_{n=1}^{\infty}[A_{mn}\cos 2m\theta]\sin(\omega_{mn}t)\frac{\cosh[\xi_{mn}(z+h)/R]}{\cosh\xi_{mn}h/R}J_{2m}(\xi_{mn}r/R) \tag{1.63}$$

The natural frequency of the fluid free surface is

$$\omega_{mn}^2=\frac{g}{R}\xi_{mn}\tanh(\xi_{mn}h/R),\qquad m,\ n=0,\ 1,\ 2,\ \dots \tag{1.64}$$

The corresponding wave height is

$$\eta(r,\theta,z,t)=\frac{1}{g}\sum_{m=0}^{\infty}\sum_{n=1}^{\infty}[A_{mn}\cos 2m\theta]\omega_{mn}\cos(\omega_{mn}t)J_{2m}(\xi_{mn}r/R) \tag{1.65}$$

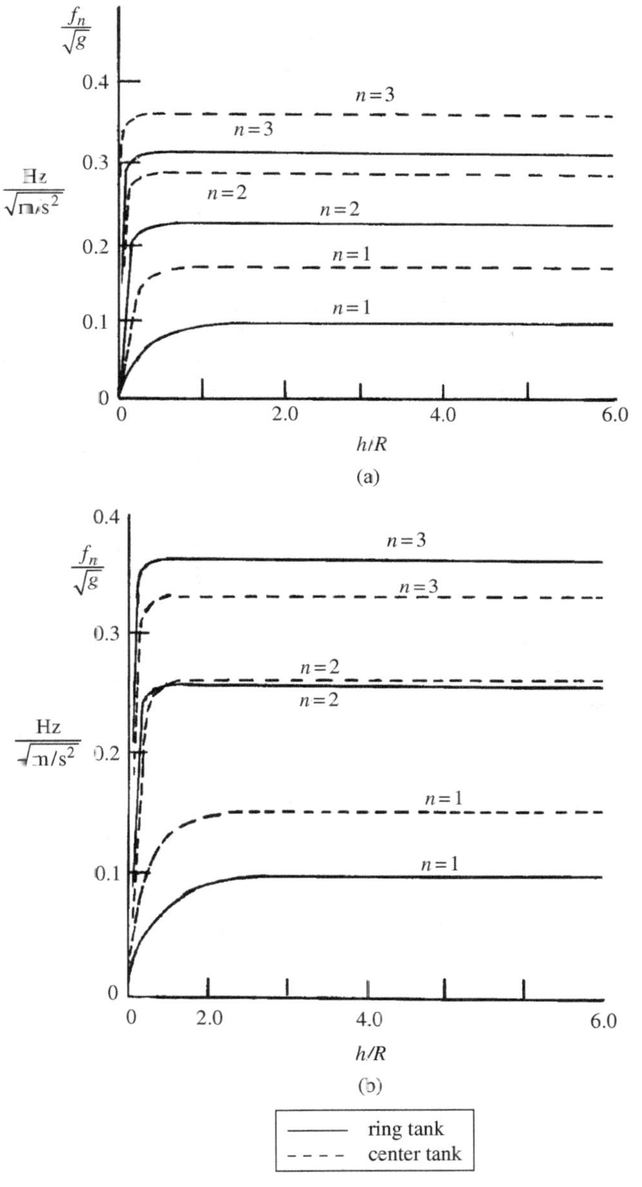

Figure 1.4 (a) Dependence of natural frequencies on liquid depth ratio in annular tank $k = R/R_{in} = 0.5$, $R = 128$, (b) dependence of natural frequencies on liquid depth ratio in annular tank $k = R/R_{in} = 0.62$, $R = 128$. (Bauer, 1960b)

where ξ_{mn} are the roots of $dJ_{2m}(\xi_{mn}\, r/R)/dr|_{r=R} = 0$, and are given in Table 1.2

Sector walls perforation effect

Note that if the tank with perforated sector walls is under lateral excitation, the resonance frequency is significantly affected by the excitation amplitude due to liquid intermixing from one sector to another. Because of the complexity of liquid sloshing in sector tanks with

Table 1.2 *Roots of* $dJ_{2m}(\xi_{mn}\, r/R)/dr|_{r=R} = 0$

$n\backslash m$	0	1	2	3	4	5	6	7	8	9
0	3.832	3.054	5.318	8.105	9.648	11.716	13.821	15.917	18.104	20.189
1	7.016	6.706	9.282	11.735	14.115	16.448	18.745	21.015	23.264	25.495
2	10.173	9.969	12.682	15.268	17.774	20.223	22.629	25.002	27.347	29.670
3	13.324	13.370	15.964	18.637	21.587	23.761	26.246	28.694	31.112	33.504
4	16.471	16.348	19.196	21.932	24.587	27.182	29.729	32.237	34.712	37.160
5	19.616	19.513	22.401	25.184	27.889	30.535	33.131	35.689	38.212	40.707
6	22.760	22.672	25.590	28.410	31.155	33.842	36.481	39.079	41.643	44.178
7	25.904	25.826	28.768	31.618	34.397	37.118	39.792	42.426	45.052	47.595
8	29.047	28.978	31.939	34.813	37.620	40.371	43.075	45.740	48.371	50.971
9	32.189	32.127	35.104	38.000	40.830	43.607	46.338	49.030	51.687	54.315

Source: Bauer (1963e).

perforated walls, the natural frequencies were measured experimentally by Abramson, *et al.* (1962a), Garza (1964), and Abramson and Garza (1965). The measured results revealed a decrease of resonance frequencies as the size of perforated holes increased. Figure 1.5(a), (b) and (c) shows the dependence of the first mode natural frequency parameter $\omega_1^2 R/g$ on an equivalent Reynolds number $d2R\sqrt{gd}/(\nu X_0)$, where ν is the fluid kinematic viscosity, d is the diameter of perforations, and X_0 is the excitation amplitude, for 23% and 30% open. Figure 1.5(a), (b) and (c) are is obtained for 45°, 60°, and 90° sector tanks, respectively. It is seen that the values of the resonance frequency could be maintained equal to or greater than that of solid wall sector up to a certain Reynolds number, above which the resonance frequencies drop to experimental values for an uncompartmented cylindrical tank.

1.5 Normal modes in a rectangular container

The fluid behavior in rectangular tanks may be treated as two- or three-dimensional flow depending on the tank length. If the length is very long, $\ell \to \infty$, a two-dimensional flow treatment is adequate. Consider first the case where the motion is two-dimensional along the x-axis, see Figure 1.2(e). In this case, the potential function $\tilde{\Phi}$ must satisfy the Laplace equation

$$\frac{\partial^2 \tilde{\Phi}}{\partial x^2} + \frac{\partial^2 \tilde{\Phi}}{\partial z^2} = 0 \tag{1.66}$$

The solution of equation (1.66) subject to the boundary conditions $\frac{\partial \tilde{\Phi}}{\partial x}|_{x=\pm L/2} = 0$ and $\frac{\partial \tilde{\Phi}}{\partial z}|_{z=-h} = 0$, is

$$\tilde{\Phi}(x, y, z, t) = \sum_{m=1}^{\infty} [\alpha_m(t) \cos k_m x + \beta_m(t) \sin k_m x] \cosh[k_m(z + h)] \tag{1.67}$$

where $k_m = (2m - 1)\pi/L$ for asymmetric modes, and $k_m = 2m\pi/L$ for symmetric modes. Following the steps of the previous case, the natural frequencies of the free surface are given by the expression

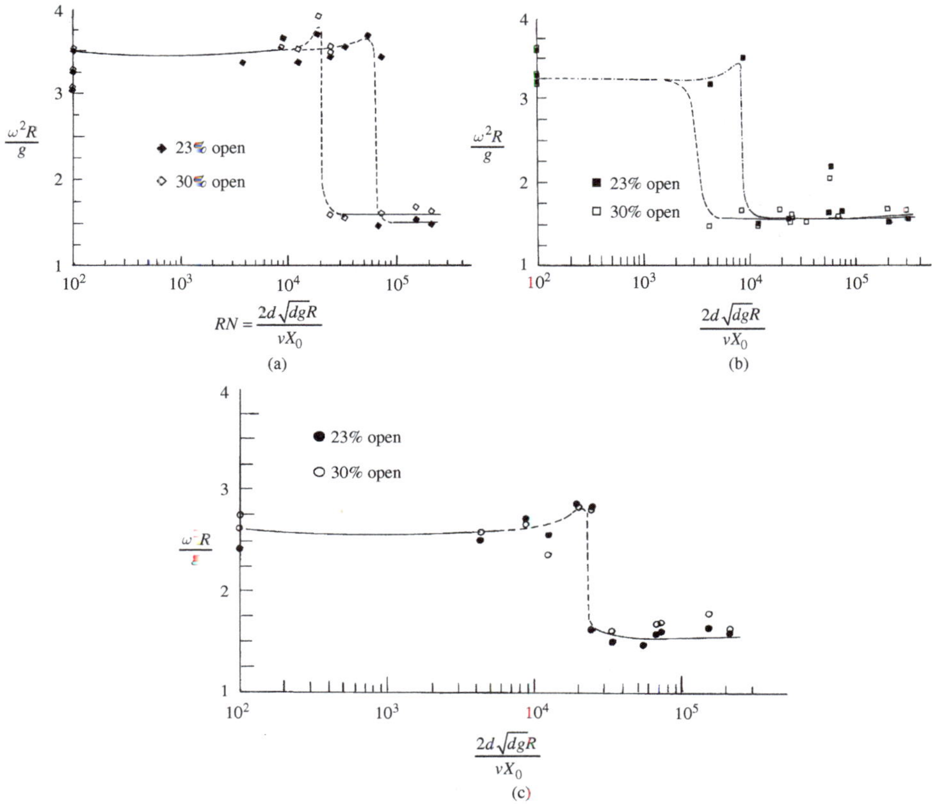

Figure 1.5 (a) Dependence of the lowest liquid resonant frequency on Reynolds number for 45°-sector perforated tank, (b) dependence of the lowest liquid resonant frequency on Reynolds number for 60°-sector perforated tank, (c) dependence of the lowest liquid resonant frequency on Reynolds number for 90°-sector perforated tank. (Abramson and Garza, 1965)

$$\omega_m^2 = gk_m \tanh(k_m h) \tag{1.68}$$

For deep liquids, $h/L > 2$, $\tanh(k_{mn}h) \approx 1$.

If the surface tension is considered, then, for a slipping contact line, the natural frequency takes the form

$$\omega_m^2 = \left[gk_m + \frac{\sigma}{\rho} k_m^3 \right] \tanh(k_m h) \tag{1.69}$$

For the case of an anchored contact line, one may follow the same procedure as that of the circular cylindrical tank.

The surface elevation is

$$\eta(x,t) = \frac{1}{g} \sum_{m=1}^{\infty} \left[\bar{\alpha}_m \cos k_m x + \bar{\beta}_m \sin k_m x \right] \omega_m \cos \omega_n t \cosh(k_m h) \tag{1.70}$$

This equation combines symmetric and asymmetric standing waves. The symmetric mode shape is

$$\eta(x,t) = \frac{1}{g}\sum_{m=1}^{\infty}[\bar{\alpha}_m \cos k_m x]\omega_m \cos\omega_n t \cosh(k_m h) \tag{1.71}$$

At a given value of x the surface of a liquid moves up and down, and for a given value of time t, the form of the surface is a cosine curve. A wave of this type is not propagated.

The asymmetric mode shape is given by the expression

$$\eta(x,t) = \frac{1}{g}\sum_{m=1}^{\infty}[\bar{\beta}_m \sin k_m x]\omega_m \cos\omega_n t \cosh(k_m h) \tag{1.72}$$

With reference to Figure 1.2(e), Graham and Rodriguez (1952) solved the three-dimensional velocity potential

$$\tilde{\Phi}(x,y,z,t) = \sum_{m=0}^{\infty}\sum_{n=1}^{\infty}\overline{\alpha}_{mn}(t)\cos\left(\frac{2m\pi x}{L}\right)\cos\left(\frac{2n\pi y}{\ell}\right)\cosh[k_{mn}(z+h)] \tag{1.73}$$

where L and ℓ are the tank width and breadth, respectively, $k_{mn} = \pi\sqrt{((2m)^2/L^2) + ((2n)^2/\ell^2)}$, m and n are positive integers. The corresponding free-surface natural frequency is

$$\omega_{mn}^2 = gk_{mn}\tanh(k_{mn}h) \tag{1.74}$$

If the surface tension is considered then expression (1.74) takes the following form for the case of a slipping contact line

$$\omega_{mn}^2 = \left[gk_{mn} + \frac{\sigma}{\rho}k_{mn}^3\right]\tanh(k_{mn}h) \tag{1.75}$$

The wave height η can be written in the form

$$\eta(x,y,t) = \frac{1}{g}\sum_{m=0}^{\infty}\sum_{n=1}^{\infty}\bar{\alpha}_{mn}\omega_{mn}\cos\left(\frac{2m\pi x}{L}\right)\cos\left(\frac{2n\pi y}{\ell}\right)\cosh(k_m h) \tag{1.76}$$

Note that the time dependence of the wave height results in a variation of the position of the container's center of mass (see, e.g., Lorell, 1952). Bauer and Eidel (2000) considered the normal modes of liquid-free-surface oscillation in a long rectangular tank with structural obstructions.

Several techniques have been developed to measure the wave height of free liquid surface during sloshing (Whittenburry, *et al.*, 1959, Wilner, *et al.*, 1960, Crews and Earth, 1963, Kana, 1964b, Muenz and Marcello, 1964, Sturtevant, 1966 and Lindstrom, *et al.*, 1969). The fundamental natural frequency in a rectangular tank was measured experimentally by Addington (1960) and the measured results were within 5% accuracy compared to the analytical values. Addington also examined the effect of baffles on the natural frequencies and suggested that the baffles should be positioned in a region of the highest velocity, such that the surface waves are trapped but allow dissipation of energy at the shallowest depth possible.

1.6 Normal modes in containers with variable depth

The analysis of liquid sloshing dynamics in containers with curved walls or walls that are not parallel to the container longitudinal axis is not a simple task. Examples of such containers include conical, spherical, horizontal cylinder (circular canal), toroidal (ring with circular cross-section) tanks. Generally, these cases do not have a closed form solution and the natural frequencies are estimated numerically or by using approximate techniques. For example, the actual geometry of the container can be approximated by a geometry that permits an exact solution provided the free-surface area and liquid volume are preserved. Alternatively, one may select a proper coordinate system in which most of the boundaries of the liquid form coordinate surfaces such that the boundary conditions are satisfied exactly. Budiansky (1960) used a sophisticated integral equation technique and obtained the liquid natural frequency for cases of nearly empty, half full, and nearly full tanks. Petrov (1961, 1962a,b,c, 1963) considered different problems of liquid oscillations in horizontal cylindrical tanks and arbitrary tanks. Further analytical and experimental investigations were carried out by McCarty and Stephens (1960), McCarty, *et al.* (1960), Mikishev and Dorozhkin (1961), Stofan and Armstead (1962), Sumner (1963, 1966), and Barron and Chang (1989).

This section presents liquid modal analyses in different tank geometries. These include canals with 45°-straight walls, horizontal cylinders, spherical tanks, prolate and oblate spheroidal tanks, conical containers, toroidal containers, upright elliptic tanks, and paraboloid containers.

1.6.1 Canal with 45°-straight walls

This classical case was considered by Lamb (1945, Section 258) who analyzed two-dimensional oscillations of a liquid bounded by two walls each inclined at 45° to the vertical z-axis. With reference to Figure 1.5(a), the walls are defined by the two straight-line equations, $z = \pm y$. Lamb assumed the following velocity potential function for symmetric modes

$$\Phi(y, z, t) = A(\cosh ky \cos kz + \cos ky \cosh kz) \sin \omega t \tag{1.77}$$

Introducing this function in the free-surface boundary condition (1.42) yields the frequency equation

$$\omega^2(\cosh ky \cos kh + \cos ky \cosh kh) = gk(-\cosh ky \sin kh + \cos ky \sinh kh) \tag{1.78}$$

This condition requires

$$\omega^2 \cos kh = -gk \sin kh \quad \text{and} \quad \omega^2 \cosh kh = gk \sinh kh \tag{1.79}$$

This means that the wave number k is obtained by solving the transcendental equation

$$\tanh kh = -\tan kh \tag{1.80}$$

Having obtained the values of k, one can determine the corresponding natural frequencies from one of the expressions of relations (1.79), that is,

$$\omega^2 = gk \tanh kh \tag{1.81}$$

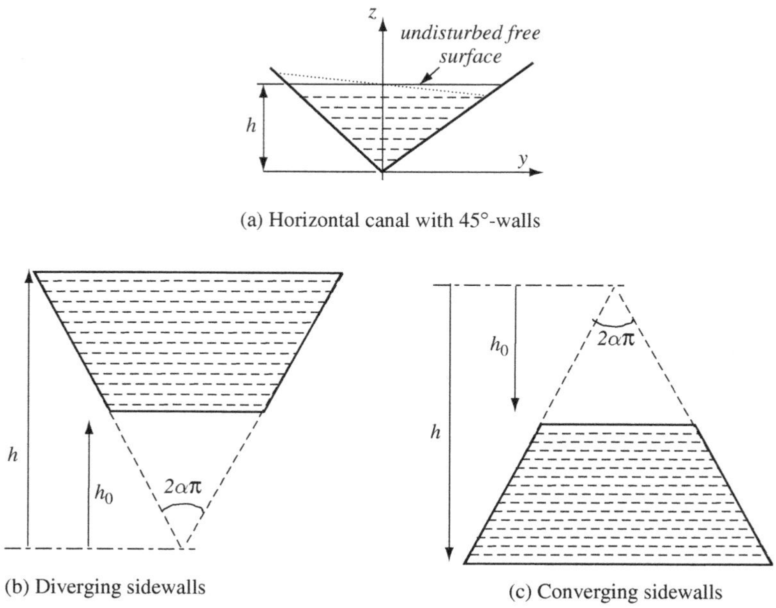

Figure 1.6 Horizontal canals with diverging and converging sidewalls.

The first root of equation (1.80) is $kh = 2.365$ and the corresponding natural frequency is

$$\omega = 1.5244\sqrt{g/h} \tag{1.82}$$

For asymmetric sloshing modes the following velocity potential function may be considered

$$\Phi(y, z, t) = -A(\sinh ky \sin kz + \sin ky \sinh kz) \sin \omega t \tag{1.83}$$

The dynamic free-surface condition requires the following relations

$$\omega^2 \sin kh = gk \cos kh \quad \text{and} \quad \omega^2 \sinh kh = gk \cosh kh \tag{1.84}$$

The corresponding wave number, k, is given by solving the transcendental equation

$$\tanh kh = \tan kh \tag{1.85}$$

For the first mode, the free surface is always plane and one can set $\tan kh \approx kh$, and from relations (1.84) the first natural frequency of the asymmetric sloshing mode is

$$\omega^2 = g/h \tag{1.86}$$

For wedge-shaped containers, Kelland (1839) considered the symmetric natural frequencies in a 90°-wedge-shaped container, while Greenhill (1887) estimated the asymmetric frequencies. Macdonald (1894) analyzed the free-surface oscillations in a container of apex angle 120°. Bauer (1981c, 1999) treated different geometries of wedge-shaped containers of finite length, ℓ, and arbitrary apex angle $2\pi\alpha < 2\pi$. Without going into detail, the following are the basic formulas for the natural frequencies of diverging and converging sidewalls shown in Figure 1.6(a).

For a wedged-shaped container with diverging sidewalls, Figure 1.6(b), the natural frequencies of the free surface are

$$\omega_{mn}^2 = \frac{n\pi}{\ell}\left[g + \frac{\sigma}{\rho R^2}\left(\frac{n^2\pi^2h^2}{\ell^2} + \frac{m^2}{4\alpha^2}\right)\right]$$
$$\times \frac{\left[I'_{m/2\alpha}(n\pi h/\ell)K'_{m/2\alpha}(n\pi h_0/\ell) - I'_{m/2\alpha}(n\pi h_0/\ell)K'_{m/2\alpha}(n\pi h/\ell)\right]}{\left[I_{m/2\alpha}(n\pi h/\ell)K'_{m/2\alpha}(n\pi h_0/\ell) - I'_{m/2\alpha}(n\pi h_0/\ell)K_{m/2\alpha}(n\pi h/\ell)\right]} \tag{1.87}$$

where h and h_0 are the heights of the upper and lower levels of the liquid volume, I and K are the modified Bessel functions of the first and second kinds, respectively, and R is the radius of curvature of the free surface.

For the case of a container with converging sidewalls, Figure 1.6(c), the gravitational term g is replaced by $-g$, and h by h_0. If the free-surface oscillations do not depend on the angular coordinate of the wedge, the natural frequencies are obtained by setting $m=0$, and one may use the well-known equalities $I_0' = I_1$ and $K_0' = -K_1$. For free-surface oscillations independent of the axial coordinate, $r=0$, and in terms of the ratio $k = h_0/h$ the natural frequencies for the diverging walls are

$$\omega_{m0}^2 = \frac{m}{2\alpha h}\left[g + \frac{\sigma}{\rho R^2}\left(\frac{m^2}{4\alpha^2} - 1\right)\right]\left[\frac{1 - k^{m/\alpha}}{1 + k^{m/\alpha}}\right] \tag{1.88}$$

For a converging wall, the natural frequencies are

$$\omega_{m0}^2 = \frac{m}{2\alpha hk}\left[g + \frac{\sigma}{\rho R^2 k^2}\left(\frac{m^2}{4\alpha^2} - 1\right)\right]\left[\frac{1 - k^{m/\alpha}}{1 + k^{m/\alpha}}\right] \tag{1.89}$$

In all cases, the surface tension results in an increase of the fluid natural frequencies.

1.6.2 Horizontal cylindrical and spherical containers

Horizontal cylinders and spherical tanks partially filled with liquid are difficult to analyze to determine the free-liquid natural frequencies and mode shapes. The difficulty arises due to the fact the walls are not straight and parallel to the axis of symmetry. Budiansky (1960) and Chu (1964b) used conformal mapping to formulate the problem as an integral equation, which was discretized for numerical solution. The method was restricted to particular liquid filling conditions. The restriction in the analysis is due to the lack of Green's function of the second kind (Neumann function) for the spherical bowl. Green's function of the first kind for spherical containers is documented in Bateman (1944). It is not an easy task to obtain a simple expression for Green's function of the second kind because the normal derivative on the spherical surface is a combination of two derivatives in toroidal coordinates (see the appendix). Chu (1964a) devised a numerical scheme to determine the kernel function, which forms one component of the Neumann function. The method allows the estimation of the natural frequencies of a free-liquid surface in a spherical tank to any arbitrary liquid filling.

Moiseev and Petrov (1965) developed a variational technique and estimated the liquid modal analysis for the lowest frequency as a function of liquid depth. Kuttler and Sigillito

(1984) estimated the liquid natural frequencies in a horizontal cylindrical tank following an approach similar to the Rayleigh–Ritz method and the bounds of the eigenvalues of elliptic operators (Kutler and Sigillito, 1969, 1978). McIver (1989) selected coordinate systems in which the container walls and the free surface coincide with coordinate lines or a coordinate surface. The eigenvalue problem was formulated in terms of integral equations, which were solved numerically. For a horizontal cylindrical container, McIver used two-dimensional bipolar coordinates and obtained a homogeneous Fredholm integral equation of the second kind. For a spherical container, he used toroidal coordinates and formulated the eigenvalue problem in terms of a pair of coupled integral equations. This complicated formulation is mainly due to the fact that Laplace's equation is not fully separable when expressed in toroidal coordinates.

Barnyak (1997) and Barnyak and Barnyak (1996) determined the normal mode frequencies of viscous liquid partially filling a circular horizontal cylinder.

Horizontal cylinder

Lamb (1945, Sections 72 and 259) presented an energy approach to determine the natural frequency of the first transverse mode of liquid free surface in a half-filled horizontal cylinder, that is, $h = -R$. He obtained the following expression for the velocity potential function

$$\Phi(r,\theta) = \frac{1}{2}\omega r^2 \sin 2\theta - \sum A_{2n+1} \sin(2n+1)\theta \left(\frac{r}{R}\right)^{(2n+1)} \tag{1.90}$$

where $A_{2n+1} = (-1)^{n+1}(\omega R^2/\pi)\{(1/(2n-1)) - (2/(2n+1)) + (1/(2n+3))\}$.

The kinetic energy of the fluid free surface experiencing small oscillations is

$$T = -\frac{\rho}{2}\int_s \frac{\partial\Phi}{\partial n}\, ds = -\rho\omega \int_0^R \Phi(r,\pi/2) r\, dr$$
$$= \frac{1}{2}\pi\rho R^4 \left(\frac{4}{\pi^2} - \frac{1}{4}\right)\omega^2$$

From this expression one can write the effective mass moment of inertia of the liquid about the center of the cylinder

$$I_{o(\text{eff})} = \pi\rho R^4 \left(\frac{4}{\pi^2} - \frac{1}{4}\right)$$

which is much less than the moment of inertia of the frozen liquid.

The potential energy is

$$U = \frac{1}{3}\rho g R^3 \theta^2$$

Equating the maximum kinetic energy to the maximum potential energy, and assuming sinusoidal oscillation, gives

$$\omega_1^2 = \frac{8\pi}{48 - 3\pi^2}\frac{g}{R} = 1.366\,56\frac{g}{R} \tag{1.91}$$

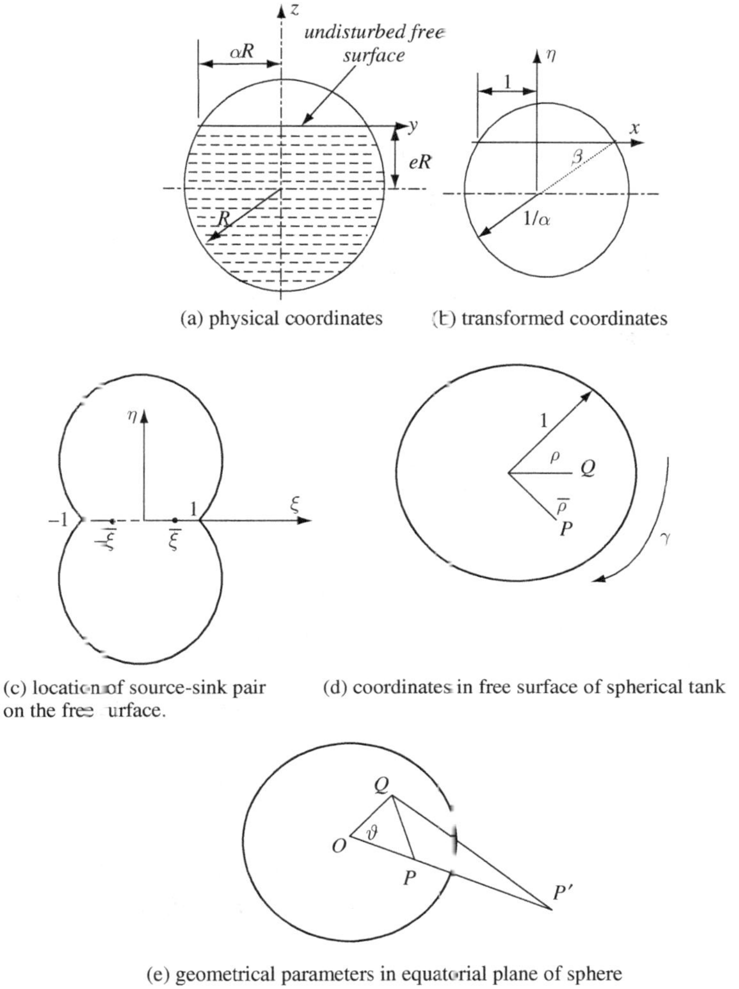

(a) physical coordinates (b) transformed coordinates

(c) location of source-sink pair on the free surface. (d) coordinates in free surface of spherical tank

(e) geometrical parameters in equatorial plane of sphere

Figure 1.7 Geometries and coordinate system of horizontal canal and spherical tank.

Conformal mapping With reference to Figure 1.7(a), Budiansky (1960) formulated the boundary value problem of free-liquid-surface asymmetric motion in the two-dimensional cross-section of a horizontal canal. In terms of the x, y coordinates and the velocity potential function $\phi(x, y)\sin\omega t$, the flow field equations are

$$\nabla^2\phi(x, y) = 0 \tag{1.92}$$

subject to the boundary conditions:
At the walls (given by (93) and the free surface condition (94))

$$\partial\phi/\partial n = 0 \tag{1.93}$$

where n is the normal to the wetted surface. The kinematic and dynamic free-surface conditions are combined to give the free-surface condition

$$\frac{\partial \phi}{\partial y} = \frac{\omega^2}{g}\phi. \tag{1.94}$$

Introducing the following coordinate transformation

$$\xi = x/\alpha R, \ \eta = y/\alpha R \tag{1.95}$$

where the factor $\alpha = a/R$, $0 \le \alpha \le 1$, and a is half the width of the free surface on the x-axis. Figure 1.7(b) shows the new coordinates and the angle β, defined by the relationship $\sin\beta = eR/R = e$, $-1 \le e \le 1$. The angle β also defines the fluid depth and it varies from $\pi/2$ for the completely filled condition to $-\pi/2$ for the empty case. In view of the geometrical symmetry, the natural modes of the free surface involve free-surface displacements that are either symmetric or anti-symmetric. The analysis is restricted for the anti-symmetric modes since the symmetric modes are not generated under lateral excitation. The free-surface condition (1.94) becomes

$$\frac{\partial \phi}{\partial \eta} = \alpha\lambda\phi, \quad \text{at } \eta = 0 \tag{1.96}$$

where $\lambda = \omega^2 R/g$ are the associated natural frequencies.

Budiansky considered the fluid-filled container bounded by the original wetted surface plus its reflection about the ξ as shown in Figure 1.7(c). A two-dimensional sink of strength $2w(\bar{\xi})\,d\bar{\xi}$ at $\bar{\xi}$ and a source of equal strength at $-\bar{\xi}$ over a strip of length $d\bar{\xi}$ are introduced to describe the motion of the free surface, where $w(\bar{\xi})$ represents the mode shape of the free surface over the strip $d\bar{\xi}$ at $\bar{\xi}$. The resulting velocity potential function for the internal flow is

$$\phi(\xi,\eta) = \int_0^1 w(\bar{\xi})\Phi(\xi,\eta;\bar{\xi})\,d\bar{\xi} \tag{1.97}$$

where $\Phi(\xi,\eta;\bar{\xi})$ is the normalized velocity potential function whose value at $\xi = 0$ is zero, that is, $\Phi(0,\eta;\bar{\xi}) = 0$, and it is odd for asymmetric motion, that is, $\Phi(\xi,\eta;\bar{\xi}) = -\Phi(-\xi,\eta;\bar{\xi})$. At the free surface, $\eta = 0$, condition (1.96) must be satisfied and yields the characteristic free-surface shapes, $w(\xi)$,

$$w(\xi) = \lambda\alpha \int_0^1 w(\bar{\xi})\Phi(\xi,0;\bar{\xi})\,d\bar{\xi}, \quad 0 \le \xi \le 1 \tag{1.98}$$

The evaluation of the above integral requires knowledge of the kernel function $\Phi(\xi,0;\bar{\xi}) \equiv G(\xi,\bar{\xi})$, which is determined by using conformal mapping through the successive transformations, $z \to \zeta \to W \to \varpi \to f$, shown in Figure 1.8,

$$f = \left[\frac{1-z}{2(1+z)}\right]^{\gamma}, \qquad z = \xi + i\eta \tag{1.99}$$

where $\gamma = 2\pi/(\pi + 2\beta)$. Let f_1 and f_2 correspond to $\bar{\xi}$ and $-\bar{\xi}$, respectively. Introducing the complex potential,

$$F = G + i\psi = -\frac{1}{\pi}\log\left[\frac{f-f_1}{f-f_2}\right] + g(\bar{\xi}) \tag{1.100}$$

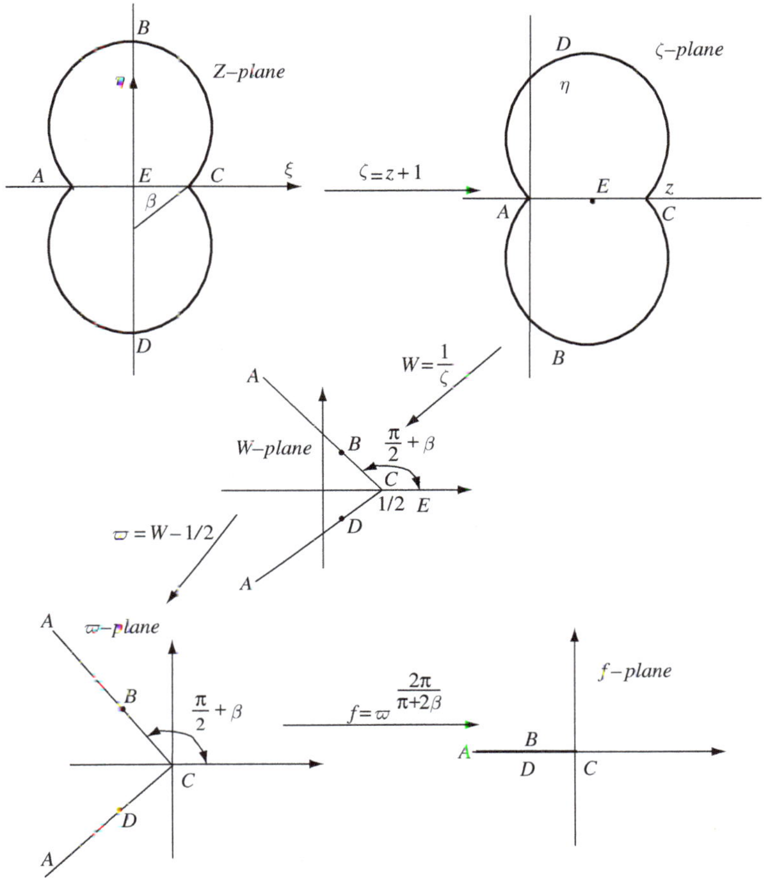

Figure 1.8 Successive transformation of conformal mapping sequence $z - \zeta - W - \varpi - f$ planes.

where ψ is the stream function, of a sink at f_1 and a source at f_2 each of strength 2, g is an arbitrary function of $\bar{\xi}$. Substituting equation (1.99) into (1.100) gives the corresponding potential function

$$F(z) = -\frac{1}{\pi}\log\left\{\frac{[(1-z)/(1+z)]^{\gamma} - [(1-\xi)/(1+\xi)]^{\gamma}}{[(1-z)/(1+z)]^{\gamma} - [(1-\bar{\xi})/(1+\bar{\xi})]^{\gamma}}\right\} + g(\xi) \tag{1.101a}$$

The function g is selected such that $F(0) = 0$, and thus one can write

$$\begin{aligned} F(z) &= -\frac{1}{\pi}\log\left\{\frac{[(1-z)/(1+z)]^{\gamma} - [(1-\bar{\xi})/(1+\bar{\xi})]^{\gamma}}{[(1-z)/(1+z)]^{\gamma} - [(1+\bar{\xi})/(1-\bar{\xi})]^{\gamma}}\right\} \\ &\quad\times\left\{\frac{1-[(1+\bar{\xi})/(1-\bar{\xi})]^{\gamma}}{1-[(1-\bar{\xi})/(1+\bar{\xi})]^{\gamma}}\right\} \\ &= -\frac{1}{\pi}\log\left\{\frac{(1-z)^{\gamma}(1+\bar{\xi})^{\gamma} - (1+z)^{\gamma}(1-\bar{\xi})^{\gamma}}{(1+z)^{\gamma}(1+\bar{\xi})^{\gamma} - (1-z)^{\gamma}(1-\bar{\xi})^{\gamma}}\right\} \end{aligned} \tag{1.101b}$$

The real part $G(\xi, \bar{\xi}) = \text{Re}\, F(z)$ is

$$G(\xi, \bar{\xi}) = -\frac{1}{\pi}\log\left\{\frac{(1+\xi)^{\gamma}(1-\bar{\xi})^{\gamma} - (1-\xi)^{\gamma}(1+\bar{\xi})^{\gamma}}{(1+\xi)^{\gamma}(1+\bar{\xi})^{\gamma} - (1-\xi)^{\gamma}(1-\bar{\xi})^{\gamma}}\right\} \tag{1.102}$$

Note that the function G has singularity at $\xi = \bar{\xi}$ and $G(\xi, \bar{\xi}) = G(\bar{\xi}, \xi) = -G(-\xi, \bar{\xi})$. The kernel function given by equation (1.102) is then substituted in the integral equation (1.99) that gives the characteristic of free-surface shapes $w(\xi)$ and associated eigenvalues, $\lambda\alpha$. For ξ near $\bar{\xi}$, with $\xi \neq 1$, the kernel takes the approximate value

$$G(\xi, \bar{\xi}) \approx -\frac{1}{\pi}\log|\xi - \bar{\xi}| - \frac{1}{\pi}\log\left|\frac{2\gamma(1-\xi^2)^{\gamma-1}}{(1+\xi)^{2\gamma} - (1-\xi)^{2\gamma}}\right|. \tag{1.103}$$

At the wall, $\xi = 1$, expression (1.103) takes the form

$$G(1, \overline{\chi}) \approx -\frac{\gamma}{\pi}\log|1-\xi| + \gamma \log 2 \tag{1.104}$$

where $G(0, \bar{\xi}) = G(\xi, 0) = 0$, and $w(0) = 0$. In order to generate a numerical algorithm to determine the normal modes, the integral on the right-hand side of equation (1.98) may be discretized in the form

$$\sum_{j=1}^{N-1} w(j\Delta) \int_{j\Delta-(\Delta/2)}^{j\Delta+(\Delta/2)} G(\xi, \bar{\xi})\, d\bar{\xi} + w(1) \int_{1-(\Delta/2)}^{1} G(\xi, \bar{\xi})\, d\bar{\xi} \tag{1.105}$$

where $\Delta = 1/N$. Suppose $\xi = n\Delta$, then

$$\int_{j\Delta-(\Delta/2)}^{j\Delta+(\Delta/2)} G(\xi, \bar{\xi})\, d\bar{\xi} \approx \Delta G(n\Delta, j\Delta) \qquad n \neq j \neq N \tag{1.106a}$$

$$\int_{1-(\Delta/2)}^{1} G(\xi, \bar{\xi})\, d\bar{\xi} \approx \frac{\Delta}{2} G(n\Delta, 1) \qquad n \neq N \tag{1.106b}$$

$$\begin{aligned}\int_{n\Delta-(\Delta/2)}^{n\Delta+(\Delta/2)} G(\chi, \bar{\xi})\, d\bar{\xi} &\approx -\frac{1}{\pi}\log\left|\frac{2\gamma(1-(n\Delta)^2)^{\gamma-1}}{(1+n\Delta)^{2\gamma} - (1-n\Delta)^{2\gamma}}\right| \\ &\quad + \int_{-(\Delta/2)}^{(\Delta/2)} -(1/\pi)\log|t|\, dt \\ &= \Delta\left\{-\frac{1}{\pi}\log\left|\frac{2\gamma(1-(n\Delta)^2)^{\gamma-1}}{(1+n\Delta)^{2\gamma} - (1-n\Delta)^{2}\gamma}\right| \right. \\ &\quad \left. + (1/\pi)[1 - \log(\Delta/2)]\right\}, \quad n \neq N\end{aligned}$$

$$\int_{1-(\Delta/2)}^{1} G(1,\bar{\xi})\, d\bar{\xi} \approx \frac{\Delta}{2}\{[(\gamma/\pi)\log 2] + (\gamma/\pi)[1 - \log(\Delta/2)]\} \tag{1.106d}$$

The integral equation (1.98) is approximated by the matrix equation

$$[A][T]\{w\} = \frac{N}{\lambda\alpha}\{w\} \tag{1.107}$$

where $\{w\}$ is the column vector with elements $w(n\Delta)$, $(n = 1, 2, \ldots, N)$, T is the integrating diagonal matrix

$$[T] = \begin{bmatrix} 1 & & & & \\ & 1 & & & \\ & & \cdot & & \\ & & & \cdot & \\ & & & & \cdot \\ & & & & & 1/2 \end{bmatrix}$$

and $[A]$ is a symmetric square matrix with the general elements

$$\begin{aligned} A_{jn} &= G(j\Delta, n\Delta) & n \neq j \neq N \\ &= \chi(n\Delta) + (1/\pi)[1 - \log(\Delta/2)] & j = n \neq N \\ &= \frac{\gamma}{\pi}\log 2 + (1/\pi)[1 - \log(\Delta/2)] & j = n \neq N \end{aligned}$$

In the absence of a logarithmic singularity, the integration scheme is reduced to the trapezoidal rule. The orthogonality condition for distinct eigenfunctions w_k and the integral

$$w_l \int_0^1 w_k(\xi) w_l(\xi)\, d\xi = 0 \tag{1.108a}$$

is approximated by the expression

$$\{w_k\}^T[T]\{w_l\} = 0 \tag{1.108b}$$

where the superscript T denotes transpose.

Consider the special case of a half-full container, $e = 0$ and $\alpha = 1$, the kernel takes the form

$$G(\xi, \bar{\xi}) = -\frac{1}{\pi}\log\left|\frac{(\xi - \bar{\xi})(1 - \xi\bar{\xi})}{(\xi + \bar{\xi})(1 + \xi\bar{\xi})}\right| \tag{1.109}$$

A one-term Galerkin's solution for the lowest frequency is applied to equation (1.98)

$$\int_0^1 [w(\xi)]^2\, d\xi = \lambda\alpha \int_0^1 \int_0^1 G(\xi, \bar{\xi}) w(\bar{\xi}) w(\xi)\, d\xi\, d\bar{\xi} \tag{1.110}$$

Table 1.3 *Liquid free-surface eigenvalues in circular canal for different fluid depth*

e	α	$\alpha\lambda_1$	$\alpha\lambda_2$	$\alpha\lambda_3$	λ_1	λ_2	λ_3
−1.0	0.0	0.0	0.0	0.0	1.0	6.0	15.0
−0.8	0.6	0.627	3.23	6.51	1.045	5.38	10.85
−0.6	0.8	0.879	3.98	7.3	1.099	4.97	9.13
−0.4	0.917	1.068	4.34	7.63	1.165	4.74	8.33
−0.2	0.980	1.224	4.56	7.82	1.249	4.65	7.99
0.0	1.0	1.360	4.70	7.96	1.360	4.70	7.96
0.2	0.98	1.482	4.81	8.06	1.513	4.91	8.23
0.4	0.917	1.596	4.89	8.15	1.742	5.34	8.89
0.6	0.8	1.706	4.97	8.22	2.13	6.22	10.23
0.8	0.6	1.822	5.05	8.30	3.04	8.42	13.84
1.0	0.0	2.018	5.20	8.44	∞	∞	∞

Source: Budiansky (1960).

For small liquid-free-surface oscillations, the mode shape may be expressed by the planar shape, $w = \xi$. Carrying out the integration (1.110), gives

$$\lambda_1 \approx \frac{1}{3}\left[\frac{2}{\pi} - \frac{\pi}{8}\right] = 1.367 \rightarrow \omega_1 = 1.169\sqrt{g/R} \tag{1.111}$$

This result is identical to the one reported by Lamb and given by relation (1.91).

The first three modes and frequencies were evaluated numerically for different values of fluid depth ratio, $e = h/R$, using the 20×20 matrix formulation as described above. The numerical results are listed in Table 1.3 and plotted in Figure 1.9 by solid curves as a function of the fluid depth ratio e. It is seen that the fundamental mode has its lowest frequency in the nearly empty state, whereas the minimum frequency of a given higher mode appears to occur slightly below the half-full condition. As the fluid depth increases, the natural frequencies tend to shift to more closely packed spectra and the ratio λ_{n+1}/λ_n decreases monotonically with increasing depth ratio. McCarty and Stephen (1960) obtained some numerical experimental results for the dependence of the liquid natural frequency parameter $\omega_n\sqrt{R/g}$ on the depth ratio $h/2R$. Figures 1.10 and 1.11 show this dependence for transverse and longitudinal sloshing modes, respectively.

Bipolar coordinate transformation McIver (1989) formulated the fluid field equations of a horizontal circular cylinder in terms of bipolar coordinates. Figure 1.7(a) shows a horizontal circular cylindrical tank partially filled with liquid to the depth h and the undisturbed free liquid surface coincides with the x-axis, instead of y, such that the positive y-axis, instead of z, is directed downward. With reference to the transformation to the bipolar coordinates η, ξ, outlined in the appendix of this chapter, the coordinates of the container center are $(x, y) = (0, a\cot\xi)$ where a is half the width of the free surface on the x-axis and

$$x + iy = a\tanh[(\eta + i\xi)/2] \tag{1.112}$$

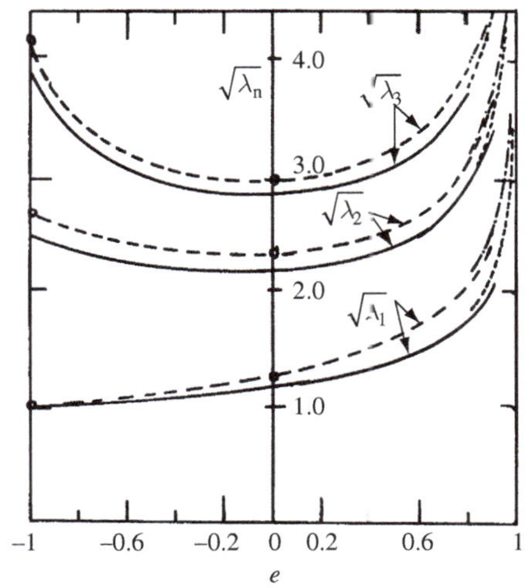

Figure 1.9 Dependence of the liquid natural frequencies on the liquid depth parameter e for circular canal (solid) and spherical tank (dashed). (Budiansky, 1960)

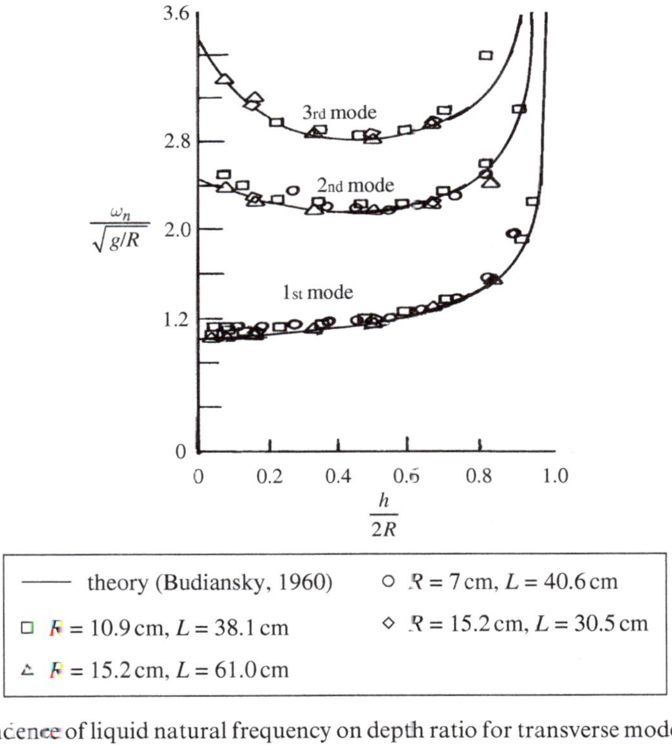

Figure 1.10 Dependence of liquid natural frequency on depth ratio for transverse modes in circular canal. (McCarty and Stephens, 1960)

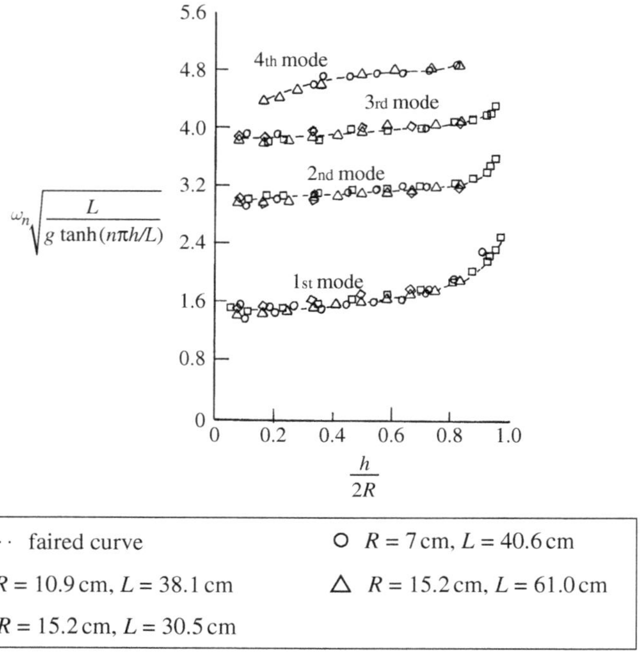

Figure 1.11 Dependence of liquid natural frequency on depth ratio for longitudinal modes in circular canal. (McCarty and Stephens, 1960)

Note that the cylinder surface coincides with the coordinate, $\xi = \xi_0$, where $0 < \xi_0 \leq \pi$, and ξ_0 is given by the relationship

$$\cos\xi_0 = 1 - \frac{h}{R} \tag{1.113}$$

The free surface coincides with $\xi = 0$. Laplace's equation for the velocity potential function for the two-dimensional case in terms of bipolar coordinates, η and ξ, is

$$\frac{\partial^2\Phi}{\partial\eta^2} + \frac{\partial^2\Phi}{\partial\xi^2} = 0, \quad -\infty < \eta < \infty, \quad 0 < \xi < \xi_0 \tag{1.114}$$

It is seen that equation (1.114) is separable and can be solved by setting $\Phi(\eta, \xi) = H(\eta)Z(\xi)$, which gives $(H''/Z = -Z''/H = \tau)$, where τ is a constant. The solution must satisfy the following boundary conditions:
The velocity normal to the wetted wall must vanish, that is,

$$\left.\frac{\partial\Phi}{\partial\xi}\right|_{\xi=\xi_0} = 0 \tag{1.115}$$

The combined free-surface condition is

$$\frac{\partial\Phi}{\partial\xi} + \frac{\lambda}{1+\cosh\eta}\Phi = |_{\xi=0}0. \tag{1.116}$$

where $\lambda = \omega^2 a/g$. Two types of solutions are considered: anti-symmetric and symmetric.

(1) *Anti-symmetric sloshing modes*
For anti-symmetric sloshing modes, the general solution of equation (1.114) satisfying the wall boundary condition (1.115) may be written in the integral form

$$\Phi(\eta,\xi)=\int_0^\infty A(\tau)\cosh[\tau(\eta-\eta_0)]\sin(\tau\xi)\,d\tau \tag{1.117}$$

Substituting equation (1.117) into the free-surface boundary condition (1.116) and carrying out Fourier sine transform with respect to ξ, gives

$$A(\tau)(\tau')\sinh(\tau'\eta_0)=\lambda\int_0^\infty A(\tau)\cosh(\tau'\eta_0)I_A(\tau,\tau')\,d\tau, 0<\tau'<\infty \tag{1.118}$$

where

$$I_A(\tau,\tau')=\frac{2}{\pi}\int_0^\infty\frac{\sin(\tau\xi)\sin(\tau'\xi)}{1+\cosh\xi}\,d\xi=\frac{\tau-\tau'}{\sinh[\pi(\tau-\tau')]}-\frac{\tau+\tau'}{\sinh[\pi(\tau+\tau')]}. \tag{1.119}$$

The integral formula $\int_0^\infty\frac{\cos(\tau\xi)}{\cosh\xi}\,d\xi=\frac{\tau\pi}{2\sinh(\tau\pi/2)}$ is taken from Gradshteyn and Ryzhik (1980, formula 3.982(1)). Introducing the new variable

$$B(\tau)=A(\tau)\sqrt{\tau\sinh\tau\xi_0\cosh\tau\xi_0} \tag{1.120}$$

equation (1.120) takes the form

$$B(\tau')=\lambda\int_0^\infty B(\tau)K_A(\tau,\tau')\,d\tau,\quad 0<\tau'<\infty \tag{1.121}$$

where

$$K_A(\tau,\tau')=\frac{I_A(\tau,\tau')}{\sqrt{\tau\tau'\tanh\tau\beta_0\tanh\tau'\beta_0}}. \tag{1.122}$$

The eigenvalue problem has been reduced to the integral equation (1.121) as in the previous analysis by Budiansky (1960). The symmetry of the kernel given by expression (1.122) guarantees the real values of the natural frequencies.

(2) *Symmetric sloshing modes*
For symmetric sloshing modes, the solution of equation (1.114) satisfying condition (1.115) takes the form

$$\Phi(\eta,\xi)=M+\int_0^\infty A(\tau)\cosh[\tau(\eta-\eta_0)]\cos(\tau\xi)\,d\tau \tag{1.123}$$

where M is a constant, which may be obtained by setting the mean level of the free surface to zero. The combined free-surface condition (1.116), after carrying out the Fourier cosine transform and introducing $B(\tau)$, gives,

Table 1.4 *Natural frequencies for different values of liquid depth ratio h/R*

h/R	Antisymmetric	Symmetric	h/R	Antisymmetric	Symmetric
0.2	1.04385	2.92908	1.2	1.50751	3.21640
	5.35498	8.03025		4.85091	6.46747
	10.76724	13.48837		8.07834	9.68639
	16.1798	18.8477		11.2932	12.8989
0.4	1.09698	2.89054	1.4	1.73463	3.53751
	4.93704	6.99058		5.27678	6.99993
	9.00749	11.00134		8.72206	10.43884
	12.9835	14.9595		12.1571	13.8722
0.6	1.16268	2.88924	1.6	2.12372	4.14328
	4.69867	6.46064		6.13932	8.10314
	8.19875	9.92610		10.08074	12.04189
	11.6490	13.3691		14.0138	15.9749
0.8	1.24606	2.93246	1.8	3.02140	5.62694
	4.60670	6.23613		8.31388	10.90612
	7.85373	9.46499		13.55955	16.15857
	11.07407	12.6813		18.7997	21.4033
1.0	1.35573	3.03310	2.0	2.00612	3.4533
	4.65106	6.23920		5.12530	6.62861
	7.81986	9.39668		8.25995	9.7839
	19.9718	12.5457		11.3982	12.9330

Source: McIver (1989).

$$B(\tau') = \lambda \int_0^\infty B(\tau) K_s(\tau, \tau')\, d\tau, \quad 0 < \tau' < \infty \tag{1.124}$$

where

$$K_s(\tau, \tau') = \frac{I_s(\tau, \tau')}{\sqrt{\tau\tau' \tanh \tau\beta_0 \tanh \tau'\beta_0}} \tag{1.125}$$

and

$$I_s(\tau, \tau') = \frac{2}{\pi} \int_0^\infty \frac{\cos(\tau\xi)\cos(\tau'\xi)}{1 + \cosh \xi}\, d\xi = \frac{\tau - \tau'}{\sinh[\pi(\tau - \tau')]} + \frac{\tau + \tau'}{\sinh[\pi(\tau + \tau')]} \tag{1.126}$$

McIver (1989) estimated the natural frequencies $\omega_{mn}^2 R/g$ corresponding to the lowest four symmetric and four anti-symmetric modes and the results are listed in Table 1.4 The natural frequencies are given for discrete values of liquid depth ratio h/R. McIver employed the Gaussian–Laguerre quadrature formula (see "Toroidal coordinate transformation" later) for discretizing the integral equations.

Spherical container

Conformal mapping With reference to Figure 1.7(a), the cylindrical coordinate frame (r, θ, z) is introduced, where $x = r\cos\theta$, $y = r\sin\theta$, and the plane x–y coincides with the undisturbed surface whose radius is αR. If the linear coordinates r and z are normalized with respect to the radius, one can establish the new nondimensional coordinate frame (ρ, η, θ), where $\rho = r/\alpha R$, and $\eta = z/\alpha R$. The Laplace equation takes the form

$$\frac{\partial^2\Phi}{\partial\rho^2} + \frac{1}{\rho}\frac{\partial\Phi}{\partial\rho} + \frac{\partial^2\Phi}{\partial\eta^2} + \frac{1}{\rho^2}\frac{\partial^2\Phi}{\partial\theta^2} = 0 \tag{1.127}$$

The solution of this equation subject to the free-surface boundary condition (1.96) may be written in the form

$$\Phi(\rho, \eta, \theta) = \sum_{k=1}^{\infty} \Psi(\rho, \eta) \cos k\theta \tag{1.128}$$

The analysis is restricted to the modes associated with $k = 1$, which are induced by lateral accelerations in the x-direction. Budiansky developed an integral equation similar to equation (1.98) by considering the volume of the fluid enclosed by the wetted surface and its reflection about the undisturbed free surface, $\eta = 0$. With reference to Figure 1.7(d), which represents the plane of the undisturbed free surface, a distribution of three-dimensional sinks along the annulus $(\bar{\rho}, \bar{\rho} + d\bar{\rho})$ of strength $2f(\bar{\rho})\cos\gamma$ per unit area is introduced. The resulting velocity potential is $f(\bar{\rho})\Omega(\rho, \eta, \theta; \bar{\rho})\bar{\rho}\, d\bar{\rho}$. Just below the plane $\eta = 0$ the velocity potential is $f(\bar{\rho})\cos\theta$, and the vertical velocity just below that plane is zero everywhere except in the annulus. At $\theta = 0$ the free-surface condition (1.96) gives

$$f(\rho) = \lambda\alpha \int_0^1 \Omega(\rho, 0, 0; \bar{\rho})\bar{\rho} f(\bar{\rho})\, d\bar{\rho} \tag{1.129}$$

where the vertical velocity just below the plane $\eta = 0$ is zero everywhere except in the annulus in which its value is $f(\bar{\rho})\cos\theta$. Setting $g(\rho) = \sqrt{\rho f(\rho)}$, one can write

$$g(\rho) = \lambda\alpha \int_0^1 H(\rho, \bar{\rho}) g(\bar{\rho})\, d\bar{\rho} \tag{1.130}$$

where $\alpha = a/R$ is bounded between the values 0 and 1, a is half the width of the free surface on the x-axis, and

$$H(\rho, \bar{\rho}) = \sqrt{\bar{\rho}\rho}\,\Omega(\rho, 0, 0; \bar{\rho}) \tag{1.131}$$

For a nearly full tank, the symmetrized region is the entire space and the potential of a sink in the plane $\eta = 0$ is unmodified by the presence of a rigid boundary at $\eta = 0$. Since the potential of a simple sink (source) of strength m is $\phi = m/(4\pi r)$, the potential of a sink of strength 2 at $(\bar{\rho}, 0, \gamma)$ is

$$\phi = \frac{1}{2\pi} \frac{1}{\sqrt{(\rho\cos\theta - \bar{\rho}\cos\gamma)^2 + (\rho\sin\theta - \bar{\rho}\sin\gamma)^2 + \eta^2}} \tag{1.132}$$

Accordingly one may write

$$\Omega(\rho, 0, 0; \bar{\rho}) = \frac{1}{2\pi} \int_0^{2\pi} \frac{\cos\gamma \, d\gamma}{\sqrt{\rho^2 + \bar{\rho}^2 - 2\rho\bar{\rho}\cos\gamma}} \tag{1.133}$$

The kernel $H(\rho, \bar{\rho})$ is given by the expression

$$H(\rho, \bar{\rho}) = \frac{\sqrt{\rho\bar{\rho}}}{\pi} \int_0^{\pi} \frac{\cos\gamma \, d\gamma}{\sqrt{\rho^2 + \bar{\rho}^2 - 2\rho\bar{\rho}\cos\gamma}} \tag{1.134a}$$

Let $\rho < \bar{\rho}$, $\mu = \rho/\bar{\rho}$, $q = 2\sqrt{m}/(m+1)$, the above kernel takes the form

$$H(\mu) = \frac{q}{\pi} \int_0^{\pi/2} \frac{[2\sin^2(\gamma/2) - 1] \, d(\gamma/2)}{\sqrt{1 - q^2 \sin^2(\gamma/2)}} \tag{1.134b}$$

Evaluating this integral in terms of complete elliptic integrals, gives

$$H = \frac{1}{\pi}\left\{ \left[\frac{2}{q} - q\right] K(q) - \frac{2}{q} E(q) \right\} \tag{1.135}$$

where

$$K(q) = \int_0^{\pi/2} \frac{d\theta}{\sqrt{1 - q^2 \sin^2\theta}}, \quad E(q) = \int_0^{\pi/2} \sqrt{1 - q^2 \sin^2\theta} \, d\theta \tag{1.136}$$

are the complete elliptic integrals of the first and second kind, respectively. The kernel (1.135) has a logarithmic singularity at $\rho = \bar{\rho}$.

For a half-full tank, the kernel function is evaluated on the basis of the known Neumann function or Dirichlet function of the second kind for the sphere. With reference to Figure 1.7(e) the function of the position Q is given by the expression

$$N(Q) = \frac{1}{2\pi}\left\{ \frac{1}{\overline{PQ}} + \frac{1}{\overline{PQ}\,\overline{QP'}} + \ln \frac{2}{1 - \overline{OQ}\,\overline{OP}\cos\vartheta + \overline{PQ}\,\overline{QP'}} \right\} \tag{1.137a}$$

This function is harmonic and has a sink of strength 2 at P in the unit sphere. Furthermore, the normal derivative of $N(Q)$ has the constant value of $-(1/2\pi)$ at the surface of the sphere. With reference to Figure 1.7(e), let Q be the point $(\rho, 0, 0)$ in the equatorial plane of the sphere, and let P be the point $(\bar{\rho}, 0, \gamma)$. In terms of these points the position function (1.137a) takes the form

$$N(\rho,0,0)=\frac{1}{2\pi}\left\{\frac{1}{\sqrt{\rho^2+\bar{\rho}^2-2\rho\bar{\rho}\cos\gamma}}+\frac{1}{\sqrt{(\rho\bar{\rho})^2+1-2\rho\bar{\rho}\cos\gamma}}\right.$$
$$\left.+\ln\frac{2}{1-\rho\bar{\rho}\cos\gamma+\sqrt{(\rho\bar{\rho})^2+1-2\rho\bar{\rho}\cos\gamma}}\right\} \tag{1.137b}$$

and the kernel $H(\rho,\bar{\rho})$ takes the form

$$H(\rho,\bar{\rho})=\frac{\sqrt{\rho\bar{\rho}}}{\pi}\left\{\int_0^\pi\frac{\cos\gamma\,d\gamma}{\sqrt{\rho^2+\bar{\rho}^2-2\rho\bar{\rho}\cos\gamma}}+\int_0^\pi\frac{\cos\gamma\,d\gamma}{\sqrt{(\rho\bar{\rho})^2+1-2\rho\bar{\rho}\cos\gamma}}\right.$$
$$\left.+\int_0^\pi\cos\gamma\left[\ln\frac{2}{1-\rho\bar{\rho}\cos\gamma+\sqrt{(\rho\bar{\rho})^2+1-2\rho\bar{\rho}\cos\gamma}}\right]d\gamma\right\} \tag{1.137c}$$

In view of the cosine variation of local sink strength at the radius $\bar{\rho}$, the normal flow at the spherical surface vanishes as expected, although the Neumann function (1.137a) for a single sink produces a constant flux at the surface. The evaluation of the first two integrals in equation (1.137c) follows immediately from the results of a nearly full container. The last integral may be evaluated by setting $s=\rho\bar{\rho}$, and letting

$$\begin{aligned}I(s)&=\int_0^\pi\cos\gamma\left[\ln\frac{2}{1-s\cos\gamma+\sqrt{s^2+1-2s\cos\gamma}}\right]d\gamma\\&=-\int_0^\pi\cos\gamma\ln\left(1-s\cos\gamma+\sqrt{s^2+1-2s\cos\gamma}\right)d\gamma\end{aligned} \tag{1.138}$$

Subtracting the zero expression $\int_0^\pi\cos\gamma\,\ln s\,d\gamma=0$ from (1.138) and differentiating both sides with respect to s, gives

$$I'(s)=\frac{1}{s}\int_0^\pi\frac{\cos\gamma\,d\gamma}{\sqrt{s^2+1-2s\cos\gamma}} \tag{1.139a}$$

Equation (1.139a) can be written in terms of complete elliptic integrals (1.136) in the form

$$I'(s)=\frac{2}{s^2}[K(s)-E(s)] \tag{1.139b}$$

Integrating equation (1.139b) with respect to s gives

$$I(s)=\int_0^s\frac{2}{s^2}[K(s)-E(s)]\,ds=\frac{2}{s}\left[E(s)-(1-s^2)K(s)\right] \tag{1.140}$$

Table 1.5 *Liquid free-surface eigenvalues in spherical tank for different fluid depth*

e	α	$\alpha\lambda_1$	$\alpha\lambda_2$	$\alpha\lambda_3$	λ_1	λ_2	λ_3
−1	0.0	0.0	0.0	0.0	1.0	7.0	17.0
0	1.0	1.565	5.34	8.66	1.565	5.34	8.56
1	0.0	2.78	5.99	9.25	∞	∞	∞

Source: Budiansky (1960).

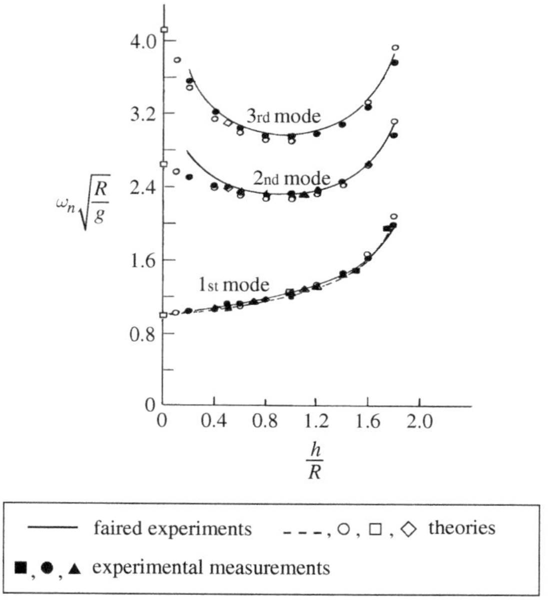

Figure 1.12 Dependence of liquid natural frequencies on depth ratio in a spherical tank as obtained experimentally and analytically. (Abramson, 1966a)

Substituting this result in terms of ρ and ρ, equation (1.137c) takes the form

$$H(\rho,\bar{\rho}) = \frac{1}{\pi}\left\{\left(2\sqrt{\bar{\rho}/\rho}\right)[K(\rho/\bar{\rho}) - E(\rho/\bar{\rho})] + 2(\rho\bar{\rho})^{3/2}K(\rho\bar{\rho})\right\} \tag{1.141}$$

with $\rho < \bar{\rho}$, and $H(\rho,\bar{\rho}) = H(\bar{\rho},\rho)$.

The solution of the integral equation (1.110) with kernel equations (1.135) and (1.141) was obtained numerically as outlined for the case of the circular canal. The values of λ_n for nearly full ($e=1$) and half-full ($e=0$) for the first three modes are listed in Table 1.5. Figure 1.9 shows the asymptotic behavior of $\sqrt{\lambda_n}$ near $e=1$ by the dot-dash curves. And the values of $\sqrt{\lambda_n}$ for $e=0$ and 1 are shown as circles.

Abramson (1966a) documented a set of results reported by Mikishev and Dorozhkin (1961), Leonard and Walton (1961), Stofan and Armstead (1962), Abramson, *et al.* (1963), and Chu (1964a). A comparison of theoretical and measured values of the first three natural frequencies is shown in Figure 1.12. Based on the measured natural frequency of the first mode the

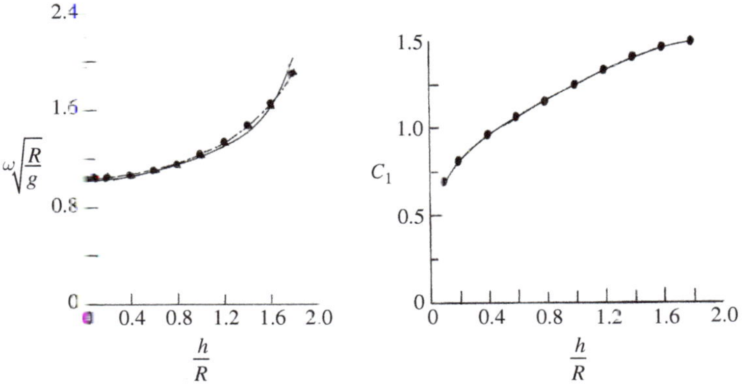

(a) First mode natural frequency (b) Constant C_1

Figure 1.13 Dependence of the liquid first mode natural frequency and constant C_1 on depth ratio. (Abramson, 1966a)

following empirical equations were obtained by Mikishev and Dorozhkin (1961) for certain ranges of the liquid depth parameter h/R

$$\sqrt{\lambda_1} = \frac{\sqrt{(1.84)^2 - (1.46 - (h/R))^2} - 0.56}{\sqrt{2(h/R) - (h/R)^2}}, \quad 0.1 < h/R < 1.0 \tag{1.142a}$$

$$\sqrt{\lambda_1} = \frac{\sqrt{(1.84)^2 - (1.97 - (h/R))^2} - 0.34}{\sqrt{2(h/R) - (h/R)^2}}, \quad 1 < h/R < 2.0 \tag{1.142b}$$

$$\sqrt{\lambda_1} = \frac{C_1}{\sqrt{2(h/R) - (h/R)^2}}, \quad 0.1 < h/R < 2.0 \tag{1.142c}$$

Figure 1.13 shows the dependence of the first mode natural frequency, $\sqrt{\lambda_1} = \omega\sqrt{R/g}$, and the constant C_1 in relation (1.142c), on the depth ratio h/R, as measured by McCarty and Stephens (1960), Mikishev and Dorozhkin (1961), and Abramson and Ransleben (1961e). The following empirical formula was obtained by Rattayya (1965) for the first mode

$$\lambda_1 = \frac{1}{3}\frac{240 - 220(h/R) + 72(h/R)^2 - 9(h/R)^3}{80 - 100(h/R) + 44(h/R)^2 - 9(h/R)^3 + (h/R)^4}\sin\left[\frac{\pi}{4}\left(2 - \frac{h}{R}\right)\left(1 + \frac{h}{3R}\right)\right] \tag{1.143}$$

Table 1.6 shows a comparison of the first natural frequency of a liquid free surface in a spherical tank as reported by Abramson (1966a). Riley and Trembath (1961) developed a computer program to compute the sloshing in axially symmetric tanks of a general shape. Their program applies a Rayleigh–Ritz procedure using three deep-tank eigenfunctions and five odd powers of the radius vector to minimize the quantity (Hieatt and Riley, 1959 and Troesch, 1960),

Table 1.6 *Comparison of the first natural frequency $\sqrt{\lambda_1}$ of a liquid in a spherical tank as reported by different sources*

H/R	Mikishev and Dorozhkin (1961)	Budiansky (1960)	Abramson, *et al.* (1963)	Chu (1964a)	Rattayya (1965)
0.05	1.02	1.0	1.01		1.01
0.5	1.09	1.06	1.08	1.1	1.1
1.0	1.22	1.25	1.22		1.22
1.8	1.92	2.04	2.03		1.83

Source: Abramson, *et al.* (1963).

$$\iiint_{\text{Volume}} (\nabla\Phi)^2\, dv \Big/ \iint_S \Phi^2\, ds$$

Their algorithm computes the lowest three eigenvalues and the corresponding mode shapes of the free surface.

Toroidal coordinate transformation With reference to Figure 1.7(a), McIver (1989) introduced cylindrical coordinates, (r, θ, z), where the z-axis is pointed vertically downwards and the origin is at the center of the free-surface circle of radius a. The toroidal coordinates, η, ξ, ϕ, outlined in the appendix to this chapter, are related to the Cartesian coordinates through the following transformations

$$x = a\frac{\sinh\eta\cos\varphi}{\cosh\eta - \cos\xi}, \quad y = a\frac{\sinh\eta\sin\varphi}{\cosh\eta - \cos\xi}, \quad z = a\frac{\sin\xi}{\cosh\eta - \cos\xi} \tag{1.144}$$

$$r + iz = a\,\tanh[(\eta + i\xi)/2], \quad 0 \le \eta \le \infty, \quad -\pi < \xi \le \pi \tag{1.145}$$

If x is replaced by r, then the geometry represents a vertical cross-section containing the axis of the spherical tank. The circle of intersection of the container with the free surface is at $\eta = \infty$ and the z-axis coincides with $\eta = 0$. It is obvious that the free surface coincides with the coordinate surface $\xi = 0$ and the container wall with the coordinate surface $\xi = \xi_0$, $0 < \xi \le \pi$. The fluid depth ratio h/R is related to ξ_0 by equation (1.113) but in toroidal coordinates. The boundary conditions are the same as those given by equations (1.115) and (1.116) for the horizontal cylinder. Note that Laplace's equation in toroidal coordinates (see the appendix to this chapter) is not fully separable. The general solution of Laplace's equation with azimuthal wave number $m = 0, 1, 2, \ldots$ is

$$\Phi(\eta, \xi, \varphi) = \cos m\varphi\sqrt{\cosh\eta + \cos\xi} \times \int_0^\infty [\bar{A}(\tau)\cosh\tau\xi + \bar{B}(\tau)\sinh\tau\xi]\,P^m_{-\frac{1}{2}+ir}(\cosh\eta)\,d\tau \tag{1.146}$$

where $P^\mu_\nu(\cosh\eta)$ is the associated Legendre function of the first kind with degree μ, order ν, and argument $\cosh\eta$. For the limiting case of a fluid-filled half-space bounded by a solid plane containing a circular aperture, $\xi_0 = \pi$, further conditions are necessary to ensure solutions having no flow to infinity. Henrici, *et al.* (1970) indicated that there is no such flow provided

$$\int\int_S \Phi\, ds = \int_0^{2\pi}\int_0^{\infty} \frac{\Phi a^2 \sinh\eta}{(1+\cosh\eta)^2}\, d\eta\, d\varphi = 0, \quad \beta = 0 \tag{1.147}$$

where S is the free-surface area. For $m \neq 0$, condition (1.147) is satisfied trivially because of the form of variation in φ. For $m = 0$, the behavior of the gradient of the velocity potential for large $R = \sqrt{r^2 + z^2}$, that is, $(\eta, \xi) = (0, \pi)$, may be obtained from equation (1.146) using the approximate form of relation (1.145), that is,

$$\eta + i(\xi - \pi) = \frac{2a}{r + iz} \tag{1.148}$$

McIver (1989) found that the condition (Henrici, *et al.*, 1970)

$$|\nabla\Phi| = O(R^{-2}), \quad R \to \infty \tag{1.149}$$

is violated unless the function $\bar{A}(\tau)$ satisfies a condition determined by substituting equation (1.146) into equation (1.147). This process gives

$$\int_0^{\infty} \frac{\bar{A}(\tau)\, d\tau}{\sinh\tau\pi} = 0 \tag{1.150}$$

where the integral

$$\int_0^{\infty} \frac{P_{-\frac{1}{2}+ir}(\cosh\eta)\sinh\eta}{(1+\cosh\eta)^{3/2}}\, d\eta = \frac{2^{3/2}\pi}{\sinh\tau\pi} \tag{1.151}$$

is obtained from equation 7.135(2) of Gradshteyn and Ryzhik, 1980. Condition (1.150) is satisfied by taking

$$\bar{A}(\tau) = A(\tau) + \sqrt{2}M\,\text{sech}\,\tau\pi \tag{1.152}$$

where $M = -2^{5/2}\int_0^{\infty} \frac{A(\tau)\, d\tau}{\sinh\tau\pi}$ is a constant. Substituting equation (1.152) into equation (1.146) gives

$$\Phi(\eta, \xi, \varphi) = M\delta_{m0} + \cos m\varphi\sqrt{\cosh\eta + \cos\xi} \times \int_0^{\infty} [A(\tau)\cosh\tau\xi + B(\tau)\sinh\tau\xi] P^m_{-\frac{1}{2}+ir}(\cosh\eta)\, d\tau \tag{1.153}$$

where δ_{mn} is the Kronecker delta. The introduction of M might be interpreted as shifting the mean free surface by a prescribed amount to ensure that it coincides with $\xi = 0$ and thus condition (1.147) is satisfied. Substituting equation (1.153) into the wall boundary condition (1.115), gives

$$\int_0^\infty \tau[A(\tau)\sinh\tau\xi_0 + B(\tau)\cosh\tau\xi_0]P^m_{-\frac{1}{2}+ir}(\cosh\eta)\,d\tau$$
$$= \frac{1}{2}\frac{\sin\xi_0}{\cosh\eta + \cos\xi_0}\int_0^\infty [A(\tau)\cosh\tau\xi_0 + B(\tau)\sinh\tau\xi_0]P^m_{-\frac{1}{2}+ir}(\cosh\eta)\,d\tau \tag{1.154}$$

Substituting equation (1.153) into the free-surface boundary condition (1.116) gives

$$\int_0^\infty \tau B(\tau)P^m_{-\frac{1}{2}+ir}(\cosh\eta)\,d\tau + \frac{\lambda}{(1+\cosh\eta)^{3/2}}$$
$$\times\left[-2^{5/2}\delta_{mn}\int_0^\infty \frac{\tau A(\tau)}{\sinh\tau\pi}\,d\tau + \sqrt{1+\cosh\eta}\int_0^\infty A(\tau)P^m_{-\frac{1}{2}+ir}(\cosh\eta)\,d\tau\right] = 0 \tag{1.155}$$

McIver (1989) applied the Mehr–Fock transform of order m (see Sneddon, 1972) to equations (1.154) and (1.155). The transform of equation (1.154) is

$$A(\tau')\sinh\tau'\xi_0 + B(\tau')\cosh\tau'\xi_0 = \frac{1}{2}\tanh\tau'\pi\sin\xi_0$$
$$\times\int_0^\infty [A(\tau)\cosh\tau\xi_0 + B(\tau)\sinh\tau\xi_0]g_m(\tau)I_m(\xi_0;\tau,\tau')\,d\tau \tag{1.156}$$

The transform of equation (1.155) is, after using relation (1.151),

$$B(\tau') = -\lambda\tanh\tau'\pi\int_0^\infty A(\tau)g_m(\tau)\left[I_m(0;\tau,\tau') - \delta_{mn}\frac{16\tau\tau'}{\sinh\tau\pi\sinh\tau'\pi}\right]d\tau \tag{1.157}$$

where (see Gradshteyn and Ryzhik, 1980)

$$g_m(\tau) = \left(\tau^2 + \frac{1}{4}\right)\left(\tau^2 + \frac{9}{4}\right)\ldots\left(\tau^2 + \frac{1}{4}(2m-1)^2\right). \tag{1.158a}$$

$$I_m(\xi_0;\tau,\tau') = \int_0^\infty \frac{P^{-m}_{-\frac{1}{2}+ir}(\cosh\eta)P^{-m}_{-\frac{1}{2}+ir}(\cosh\eta)\sinh\eta}{\cosh\eta + \cos\xi_0}\,d\tau \tag{1.158b}$$

$$P^m_{-\frac{1}{2}+ir}(\cosh\eta) = (-1)^m g_m(\tau)P^{-m}_{-\frac{1}{2}+ir}(\cosh\eta) \tag{1.158c}$$

Introducing the new functions

$$C(\tau) = [A(\tau)\cosh\tau\xi_0 + B(\tau)\sinh\tau\xi_0]\sqrt{\frac{g_m(\tau)\tanh\tau\xi_0}{\tanh\tau\pi}} \tag{1.159a}$$

$$D(\tau) = A(\tau)\sqrt{\frac{g_m(\tau)\tanh\tau\xi_0}{\tanh\tau\pi}} \tag{1.159b}$$

Equations (1.156) and (1.157) take the form

$$C(\tau')\coth^2\tau'\xi_0 - D(\tau')\coth\tau'\xi_0\,\text{cosech}\,\tau'\xi_0 = \frac{1}{2}\sin\xi_0\int_0^\infty C(\tau)H_m(\xi_0;\tau,\tau')\,d\tau \tag{1.160}$$

$$C(\tau')\coth\tau'\xi_0\,\text{cosech}\,\tau'\xi_0 - D(\tau')\coth^2\tau'\xi_0 = -\lambda\int_0^\infty D(\tau)L_m(\xi_0;\tau,\tau')\,d\tau \tag{1.161}$$

where

$$H_m(\xi_0;\tau,\tau') = G_m(\tau)G_m(\tau)I_m(\xi_0;\tau,\tau') \tag{1.162a}$$

$$L_m(\xi_0;\tau,\tau') = G_m(\tau)G_m(\tau)\left[I_m(0;\tau,\tau') - \delta_{m0}\frac{16\tau\tau'}{\sinh\tau\pi\sinh\tau'\pi}\right] \tag{1.162b}$$

$$G_m(\tau) = \sqrt{\frac{g_m(\tau)\tanh\tau\pi}{\tanh\tau\xi_0}} \tag{1.162c}$$

The natural frequencies are determined by estimating the values of $\lambda=\omega^2a/g$ for which the coupled integral equations (1.160) and (1.161) possess nontrivial solution for the functions $C(\tau)$ and $D(\tau)$. McIver (1989) adopted an algorithm for solving the integral equations by using an appropriate quadrature rule and obtained a matrix eigenvalues problem. The integrals in equations (1.160) and (1.161) are approximated using the Gauss–Laguerre quadrature formula in the form

$$\int_0^\infty f(\tau)\,d\tau \approx \sum_{i=1}^{N} w_i f(\tau_i) \tag{1.163}$$

where w_i, $i=1, 2, \ldots, N$, are adjusted weights containing the factor $e^{2\pi\tau}$ to account for the exponential decay of the integrands with large τ. This approximation yields the matrix equations

$$\mathbf{SC} - \mathbf{TD} = \delta\mathbf{HC} \tag{1.164}$$

$$\mathbf{TC} - \mathbf{SD} = -\lambda\mathbf{LD} \tag{1.165}$$

where $\delta = \frac{1}{2}\sin\xi_0$. The elements of the diagonal matrices **S** and **T** are

$$S_{ii} = \coth^2\tau_i\xi_0,\quad T_{ii} = \coth\tau_i\xi_0\text{cosech}\,\tau_i\xi_0 \tag{1.166a, b}$$

The elements of the matrices **H** and **L** are

$$H_{ij} = w_jH_m(\xi_0;\tau_i,\tau_j),\quad L_{ij} = w_jL_m(\xi_0;\tau_i,\tau_j) \tag{1.167a, b}$$

The elements of the column vectors **C** and **D** are

$$C_i = C(\tau_i),\quad D_i = D(\tau_i) \tag{1.168a, b}$$

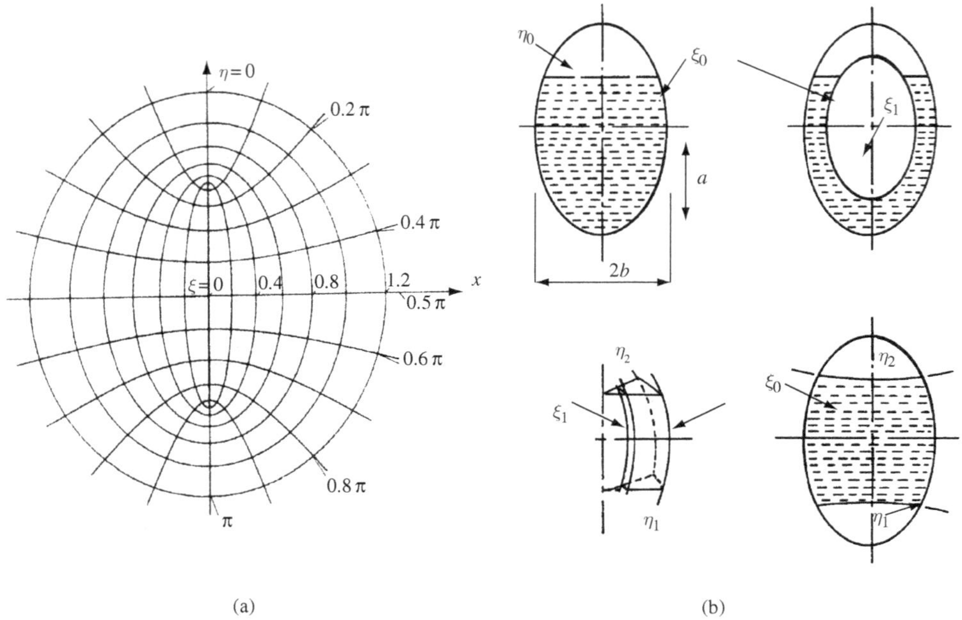

Figure 1.14 (a) Geometry of prolate container showing the coordinate system (b) four different geometries of prolate spheroidal containers.

Eliminating the vector **C** from equations (1.164) and (1.165) gives the matrix eigenvalues equation

$$\left[\mathbf{T}[\delta\mathbf{H} - \mathbf{S}]^{-1}\mathbf{T} + \mathbf{S}\right]\mathbf{D} = \lambda\mathbf{L}\mathbf{D} \tag{1.169}$$

The elements of matrices **H** and **L** depend on the integral (1.158b), which can be estimated numerically. Table 1.7 lists the numerical values of the natural frequency parameter $\omega_{ij}^2 R/g$ for the first four azimuthal wave numbers $m = 0$, 1, 2, 3 and the first lowest modes. The natural frequencies are estimated for discrete values of fluid depth ratio h/R.

1.6.3 Prolate spheroidal tank

Prolate containers are usually subdivided by annular and radial walls and different configurations as shown in Figure 1.14. They also have the advantage of a remarkable reduction in the dynamic contribution of sloshing loads, higher values of liquid natural frequencies, and smaller liquid mass participating in the liquid motion. Bauer and Eidel (1989b) examined the free oscillations of the free surface in an annular prolate spheroidal sector. A limiting case is the slim ellipse, which resembles a special case of a cylinder.

With reference to the appendix to this chapter and Figure 1.14(a), which shows the prolate spheroidal coordinates ξ, η, and ϕ, the surfaces $\xi = \xi_0$ and $\xi = \xi_1$ represent the annular walls such that any particular value of ξ gives a measure of the semi-axes ratio of the ellipse, b/a, while $\phi = 0$ and $\phi = 2\pi\alpha$ are sector walls. Note that as $\xi \to 1$ the prolate spheroid approaches a regular spherical container. The bottom may be represented by the surface $\eta = \eta_1$ while the

Table 1.7 *Natural frequency parameter $\omega_{ij}^2 R/g$ for spherical container for the first four azimuthal wave numbers $m = 0, 1, 2, 3$ and the first four lowest modes*

h/R	$m=0$	$m=1$	$m=2$	$m=3$
0.2	3.82612	1.07232	2.10792	3.12949
	9.25613	6.20081	8.39523	10.48832
	14.75561	11.88212	14.2944	15.58021
	20.118	17.3589	19.8090	22.1585
0.4	3.70804	1.15826	2.23491	3.28209
	7.91895	5.67422	7.42178	9.05998
	11.94118	9.85513	11.6525	13.36738
	15.9077	13.8685	15.6994	17.4637
0.6	3.65014	1.26251	2.38767	3.46642
	7.26596	5.36832	6.88669	8.31214
	10.74498	8.94181	10.5082	12.0064
	14.1964	12.4233	14.0217	15.5636
0.8	3.65836	1.39239	2.57671	3.6961
	6.98858	5.24058	6.65230	7.98544
	10.23113	8.55088	10.1555	11.4186
	13.4553	11.7995	13.2950	14.7386
1.0	3.74617	1.56016	2.81969	3.99416
	6.97636	5.27555	6.65941	7.97281
	10.14748	8.50444	9.94129	11.31996
	13.3042	11.6835	13.1499	14.5669
1.2	3.9381	1.7881	3.1492	4.4031
	7.2188	5.4930	6.9145	8.2674
	10.452	8.7793	10.251	11.665
	13.673	12.021	13.521	14.971
1.4	4.3010	2.1232	3.6336	5.0123
	7.8006	5.973	7.509	8.971
	11.2559	9.476	11.058	12.579
	14.698	12.938	14.547	16.104
1.6	5.0075	2.6864	4.4512	6.0547
	9.01565	6.9571	8.7390	10.433
	12.9748	10.956	12.782	14.536
	16.9191	14.916	16.769	18.562
1.8	6.76418	3.9593	6.3155	8.4631
	12.1139	9.4535	11.858	14.137
	17.396	14.755	17.210	19.565
	22.657	20.022	22.509	24.912
2.0	4.1213	2.7547	4.1213	5.4000
	7.3421	5.8922	7.3421	8.7183
	10.5171	9.0328	10.517	11.941
	13.6773	12.174	13.677	15.129

Source: McIver (1989).

undisturbed free surface is at $\eta = \eta_0$. Note that the hyperbolas are selected to provide a description for possible bottom curved shapes.

For incompressible and irrotational flow, the continuity equation is (see the appendix to this chapter),

$$\nabla^2 \Phi = \frac{1}{c^2(\sinh^2 \xi + \sin^2 \eta)\sinh \xi}\frac{\partial}{\partial \xi}\left(\sinh \xi \frac{\partial \Phi}{\partial \xi}\right) + \frac{1}{c^2(\sinh^2 \xi + \sin^2 \eta)\sin \eta}\frac{\partial}{\partial \eta}\left(\sin \eta \frac{\partial \Phi}{\partial \eta}\right) + \frac{1}{c^2 \sinh^2 \xi \sin^2 \eta}\frac{\partial^2 \Phi}{\partial \phi^2} = 0 \tag{1.170a}$$

Introducing the new variables $\bar{\xi} = \cosh \xi$, $\bar{\eta} = \cos \eta$, $a/2 = c$, $1 \le \bar{\xi}$, $-1 \le \bar{\eta} \le 1$ and their relations to the x, y, z coordinates:

$$x = \frac{a}{2}\sqrt{(\bar{\xi}^2 - 1)(1 - \bar{\eta}^2)}\cos \phi, \; y = \frac{a}{2}\sqrt{(\bar{\xi}^2 - 1)(1 - \bar{\eta}^2)}\sin \phi, \quad \text{and} \quad z = \frac{a}{2}\bar{\xi}\bar{\eta}$$

Laplace's equation can be written in terms of the new variables

$$\nabla^2 \Phi = \frac{4}{a^2(\bar{\xi}^2 - \bar{\eta}^2)}\left[\frac{\partial}{\partial \bar{\xi}}(\bar{\xi}^2 - 1)\frac{\partial \Phi}{\partial \bar{\xi}} + \frac{\partial}{\partial \bar{\eta}}(1 - \bar{\eta}^2)\frac{\partial \Phi}{\partial \bar{\eta}} + \frac{\bar{\xi}^2 - \bar{\eta}^2}{(\bar{\xi}^2 - 1)(1 - \bar{\eta}^2)}\frac{\partial \Phi}{\partial \phi^2}\right] \tag{1.170b}$$

The surface, $\bar{\xi} = \text{const}$, is a prolate spheroid with interfocal distance a. The major axis is $a\bar{\xi}/2$, and the minor axis is $(a\sqrt{\bar{\xi}^2} - a)/2$. The surfaces, $\bar{\eta} = \text{const}$, are two sheets of a hyperboloid of revolution with foci at $z = \pm a/2$ asymptotic to the cone which has its generating line at an angle $\theta = \cos^{-1} \bar{\eta}$. The surface $\phi = \text{const}$ is a plane through the vertical axis at an angle ϕ to the x, z plane.

Equation (1.170a) must be solved subject to the following boundary conditions

$$\left.\frac{\partial \Phi}{\partial \xi}\right|_{\xi = \xi_0, \xi_1} = 0 \quad \text{at the annular container walls} \tag{1.171a}$$

$$\left.\frac{\partial \Phi}{\partial \phi}\right|_{\phi = 0, 2\alpha\pi} = 0 \quad \text{at the container sector wall} \tag{1.171b}$$

$$\left.\frac{\partial \Phi}{\partial \eta}\right|_{\eta = \eta_1} = 0 \quad \text{at the tank bottom} \tag{1.171c}$$

The combined free-surface condition, which is obtained from the unsteady Bernoulli equation and the kinematic condition, is

$$\frac{\partial^2 \Phi}{\partial t^2} - \frac{g \cosh \xi \sin \eta}{c(\sinh^2 \xi + \sin^2 \eta)}\frac{\partial \Phi}{\partial \eta} = 0 \quad \text{at } \eta = \eta_0 \tag{1.171d}$$

The presence of the coordinate ξ in the free-surface condition (1.171d) creates some difficulty. However, for a small value of ξ, one can set $\cosh \xi \approx 1$ and $\sinh \xi \approx 0$. In this case, condition (1.171d) takes the approximate form

$$\frac{\partial^2 \Phi}{\partial t^2} - \frac{g}{c \sin \eta} \frac{\partial \Phi}{\partial \eta} = 0 \tag{1.171e}$$

Now, the solution of Laplace's equation (1.170b) may be obtained using the separation of variables in the form

$$\Phi(\bar{\xi}, \bar{\eta}, \phi) = F(\phi) Z(\bar{\xi}) H(\bar{\eta}) \tag{1.172}$$

where

$$\frac{d^2 \Phi}{d\phi^2} = -m^2 \Phi \tag{1.173a}$$

$$\frac{d}{d\bar{\eta}} \left[(1 - \bar{\eta}^2) \frac{dH}{d\bar{\eta}} \right] + n(n+1)H - \frac{m^2}{(1 - \bar{\eta}^2)} H = 0, \quad -1 \leq \bar{\eta} \leq 1 \tag{1.173b}$$

$$\frac{d}{d\bar{\xi}} \left[(1 - \bar{\xi}^2) \frac{dZ}{d\bar{\xi}} \right] + n(n+1)Z - \frac{m^2}{(1 - \bar{\xi}^2)} Z = 0, \quad \bar{\xi} \geq 1 \tag{1.173c}$$

The solutions of equation (1.173a), for periodic boundary conditions on ϕ, are in the form of $\cos m\phi$ and $\sin m\phi$ (if the tank is not sectored), where m is zero or a positive integer. Equations (1.173b) and (1.173c) are solved by using the spherical harmonics P_n^m and their second solutions Q_n^m, defined as follows

$$P_n^m(x) = \frac{(1 - x^2)^{m/2}}{2^2 n!} \frac{d^{m+n}}{dx^{m+n}} (x^2 - 1)^n, \quad m, n = 0, 1, 2, \ldots, \quad m \leq n \tag{1.174a}$$

$$Q_n^m(x) = (-1)^m (x^2 - 1)^{m/2} \frac{d^m}{dx^m} Q_n(x) \tag{1.174b}$$

$$Q_n(x) = \frac{1}{2^n n!} \frac{d^n}{dx^n} \left[(x^2 - 1)^n \ln\left(\frac{x+1}{x-1}\right) \right] - \frac{1}{2} P_n(x) \ln\left(\frac{x+1}{x-1}\right) \tag{1.174c}$$

For the function H to be finite in equation (1.173b), n must be zero or a positive integer and H must be proportional to the <anchor role="index" id="i00-117-050-286/> Legendre function of the first kind, $P_n^m(\bar{\eta})$. The variable $\bar{\xi}$ goes from $+1$, along the line between the two foci, to infinity. There is no solution which is finite over the entire range for most values of n and m. For the motion of fluid inside this type of container the solution for the potential function must satisfy the boundary conditions (1.171).

For a simple prolate spheroidal tank with no truncation or walls, the velocity potential function in terms of the coordinates ξ, η, ϕ takes the form, after dropping condition (1.171b),

$$\Phi(\xi, \eta, \phi, t) = \sum_{m=0}^{\infty} \sum_{n=1}^{\infty} A_{mn} \exp(i\omega_{mn} t) P_{\lambda_{mn}}^m(\cosh \xi) P_{\lambda_{mn}}^m(-\cos \eta) \cos(m\phi) \tag{1.175}$$

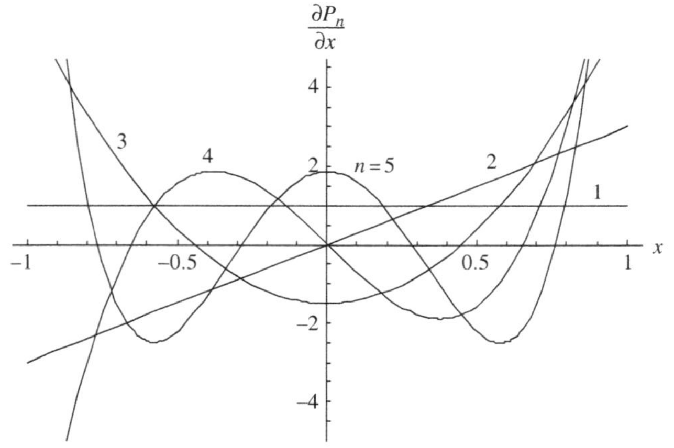

Figure 1.15 Plot of $\partial P_n(x)/\partial x$ showing the zero roots.

In order to determine the natural frequency we introduce equation (1.175) into equation (1.171d). The corresponding approximate natural frequency is

$$\omega_{mn}^2 = -\frac{g}{c}\frac{P'^m_{\lambda_{mn}}(-\cos\eta_0)}{P^m_{\lambda_{mn}}(-\cos\eta_0)} \tag{1.176}$$

where λ_{mn} are the roots of $P'^m_{\lambda}(\cosh\xi_0)=0$. The zeros of this function are obtained graphically using MATHEMATICA by plotting the following expression

$$\frac{\partial P_n(x)}{\partial x} = \frac{-nP_{n-1}(x)+nxP_n(x)}{-1+x^2} = 0 \tag{1.177}$$

Figure 1.15 shows a plot for the derivative of this function for $n=1, 2, \ldots, 5$.

For a sectored tank, one may assume the following solution (see Bauer and Eidel, 1989b)

$$\Phi(\xi,\eta,\phi,t) = \sum_{m=0}^{\infty}\sum_{n=1}^{\infty} A_{mn}\exp(i\omega_{mn}t)L_{mn}(\xi)\bar{L}_{mn}(\eta)\cos(m\phi/2\alpha) \tag{1.178}$$

where ω_{mn} is the natural frequency of the sloshing mode mn,

$$L_{mn}(\xi) = P^{m/2\alpha}_{\lambda_{mn}}(\cosh\xi)Q'^{m/2\alpha}_{\lambda_{mn}}(\cosh\xi_0) - P'^{m/2\alpha}_{\lambda_{mn}}(\cosh\xi_0)Q^{m/2\alpha}_{\lambda_{mn}}(\cosh\xi) \tag{1.179a}$$

$$\bar{L}_{mn}(\eta) = P^{m/2\alpha}_{\lambda_{mn}}(\cosh\eta)Q'^{m/2\alpha}_{\lambda_{mn}}(\cosh\eta_1) - P'^{m/2\alpha}_{\lambda_{mn}}(\cosh\eta_1)Q^{m/2\alpha}_{\lambda_{mn}}(\cosh\eta) \tag{1.179b}$$

Functions $P^{m/2\alpha}_{\lambda_{mn}}$ and $Q^{m/2\alpha}_{\lambda_{mn}}$ are the associated Legendre functions of the first and second kind, and λ_{mn} are the roots of $L_{mn}(\xi_1)=0$, which are obtained numerically. The constants A_{jk} are obtained from the initial conditions. In order to determine the natural frequencies of the free surface, equation (1.178) is introduced into the free-surface condition (1.171d)

$$-\sum_{m=0}^{\infty}\sum_{n=1}^{\infty}\omega_{mn}^2 A_{mn}\exp(i\omega_{mn}t)L_{mn}(\xi)\bar{L}_{mn}(\eta)\cos(m\phi/2\alpha)$$

$$=\sum_{m=0}^{\infty}\sum_{n=1}^{\infty}\frac{g\cosh\xi\sin\eta_0}{c(\sinh^2\xi+\sin^2\eta_0)}A_{mn}\exp(i\omega_{mn}t)L_{mn}(\xi)$$

$$\times\left(\frac{\partial\bar{L}_{mn}(\eta)}{\partial\eta}|_{\eta=\eta_0}\right)\cos(m\phi/2\alpha) \tag{1.180}$$

The right-hand side is expanded into a series of functions L_{mn} using the orthogonality property

$$\int_{\xi_0}^{\xi_1} L_{mn}(\xi)L_{mk}(\xi)\sinh\xi\, d\xi = \begin{cases} 0 & \text{if } n\neq k \\ S_{mn} & \text{if } n=k \end{cases} \tag{1.181}$$

where

$$S_{mn}=\frac{\sinh\xi_1 L_{mn}(\xi_1)(\partial^2 L_{mn}(\xi_1)/\partial\lambda_{mn}\partial\xi)-\sinh\xi_0 Lmn(\xi_0)(\partial^2 L_{mn}(\xi_0)/\partial\lambda_{mn}\partial\xi)}{2\lambda_{mn}+1}.$$

The result of the expansion yields an infinite hierarchy set of linear homogenous equations, and the natural frequencies are evaluated by solving the determinant:

$$\det\left|\gamma_{n\ell}\left(m,\frac{g}{c},\xi_0,\xi_1,\eta_0,\eta_1\right)+\omega_{mn}^2\delta_{n\ell}\right|=0 \tag{1.182}$$

where $m=1, 2, \ldots,\quad \ell=1,2,\ldots,\quad j=0, 1, 2, \ldots,\quad \delta_{n\ell}$ is the Kronecker symbol, and

$$\gamma_{n\ell}(m,\frac{g}{c},\xi_0,\xi_1,\eta_0,\eta_1)=\frac{g\sin\eta_0\bar{L}'_{m\ell}(\eta_0)}{c\bar{L}_{jk}(\eta_0)}\frac{\int_{\xi_0}^{\xi_1}\frac{\cosh\xi\sinh\xi}{(\sinh^2\xi+\sin^2\eta_0)}L_{mn}(\xi)L_{m\ell}(\xi)\,d\xi}{\int_{\xi_0}^{\xi_1}L_{mn}^2(\xi)\sinh\xi\,d\xi} \tag{1.183}$$

The dependency on the coordinate ξ may be neglected for $0<\xi<\xi_0\ll 1$. In this case one can obtain the approximate expression for the natural frequency

$$\omega_{mn}^2=-\frac{g}{c}\frac{\bar{L}'_{mn}(\eta_0)}{\sin\eta_0 L_{mn}(\eta_0)} \tag{1.184}$$

The dependence of the natural frequency ratio, $\omega_{jk}/\sqrt{g/c}$, on the coordinate η_0/π, which represents a fluid depth parameter, and for the sloshing modes $jk=(0, 1), (0, 2), (1, 1)$ and $(1, 2)$ is shown in Figures 1.16(a,b) for $b/a=0.2$ and 0.5, respectively. With reference to Figure 1.16(a) the dashed curves represent the natural frequencies as estimated using the simplified expression given by equation (1.184). The solid curves represent the enhanced results for the natural frequencies by expanding all functions that depend on ξ into series of associated Legendre functions in the free-surface condition (1.171d). This method is lengthy and yields an infinite hierarchy determinant, which is truncated to order three. Another set of results

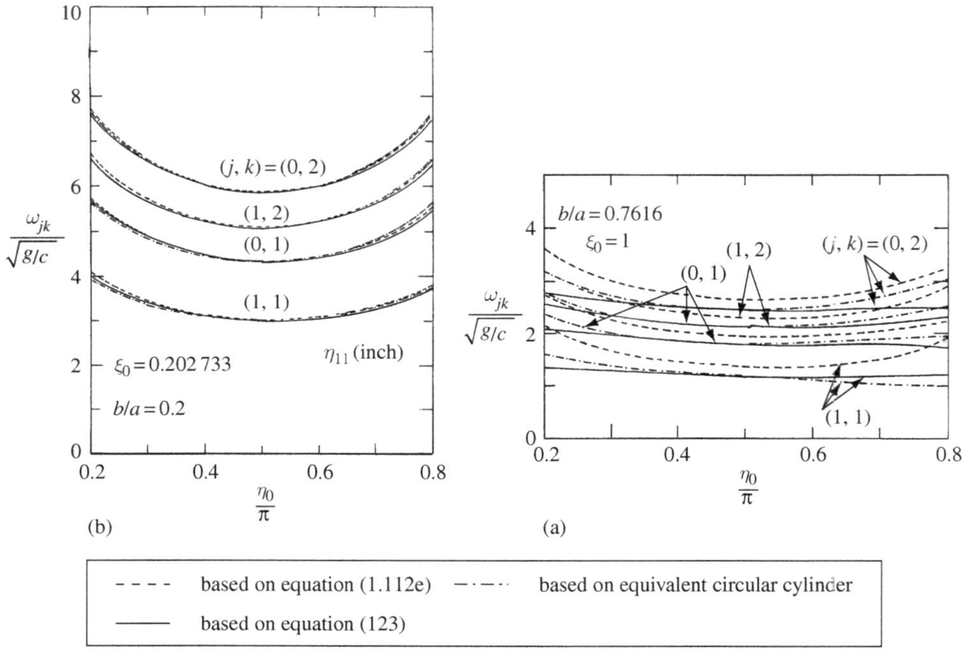

Figure 1.16 Dependence of liquid natural frequencies on height coordinate in prolate spheroidal tank. (Bauer and Eidel, 1989b)

shown by dash-dot curves in Figures 1.16 was obtained by employing the natural frequency expression of an equivalent circular cylindrical container whose radius is

$$R_e = c \cosh \xi_0 \sin \eta_0$$

and liquid depth

$$h_e = h[1 + (4a - h)/(2a - h)]/6$$

The equivalent natural frequency is given in the form

$$\omega_{jk}^2 = \frac{g\varepsilon_{jk}}{R_e} \tanh\left(\varepsilon_{jk} h_e / R_e\right) \tag{1.185}$$

The equivalence is taken on the basis of equal free-surface area and equal fluid volume. The values of ε_{jk} are the zeros of the derivative of Bessel functions of the first kind and jth-order, that is, $J'_j(\varepsilon_{jk}) = 0$. It is seen that all methods agree very well in the vicinity of $\eta_0 = \pi/2$ where the actual free surface is close to the coordinate plane η_0 at which the free-surface condition has been satisfied. For large and small values of η_0, the deviation is noticed but not significant. Another feature is that as the liquid depth decreases, that is, by increasing η_0, the frequency decreases to depth level $\eta_0 = \pi/2$, above which it increases again. Note that the natural frequency for a relatively larger fluid depth, for example, $\eta_0 = 0.2\pi$, is larger than for a smaller fluid depth, as may be compared with $\eta_0 = 0.8\pi$. The mode $j = 1$ is associated with a nodal line at $\pi/2$ and $3\pi/2$. A similar trend is also observed in Figure 1.16(b) for $b/a = 0.7616$, but there is a

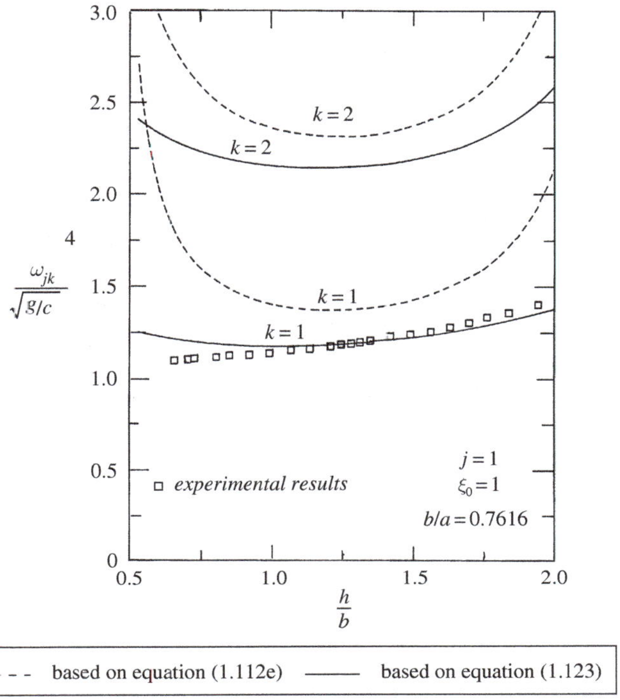

Figure 1.17 Dependence of liquid natural frequencies on depth ratio in prolate spheroidal tank. (Bauer and Eidel, 1989b)

drop in the natural frequencies. Figure 1.17 shows another set of results including those measured experimentally as a function of the fluid depth ratio h/b for $b/a = 0.7616$, that is, $\xi_0 = 1.0$. The measured results agree very well with the equivalent cylinder expression given by relation (1.185).

1.6.4 Oblate spheroidal tank

Geometrically, oblate spheroidal tanks are flattened spheres and, depending on the direction of excitation, they can assume different orientations, referred to as longitudinal, transverse, and horizontal, as shown in Figure 1.18. Leonard and Walton (1961) experimentally measured the natural frequencies and mode shapes in spheroidal tanks of different orientations. The dependence of the measured natural frequencies of the first three sloshing modes on the fluid depth ratio, $h/2b$ or $h/2a$, is shown in Figure 1.19 for the three tank orientations.

For the horizontal orientation case, the measured results were compared with the natural frequencies of a liquid in an upright circular cylinder of radius, r, equal to the radius of the liquid surface and a liquid depth $h_c = h(3b - h)/[3(2b - h)]$, which yields the same liquid volume. For this case, the natural frequency is given by an expression similar to relation (1.43a), that is,

$$\omega_n = \sqrt{\frac{g\xi_n}{r}\tanh(\xi_n h_c/r)} \tag{1.186}$$

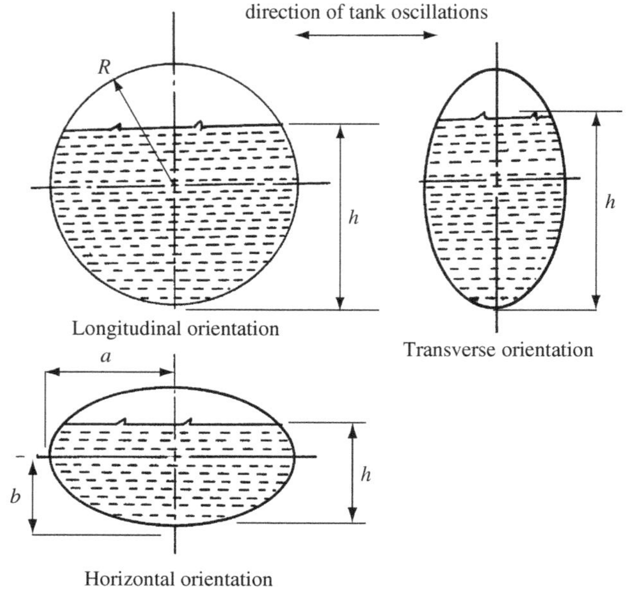

Figure 1.18 Orientation and dimensions of spheroidal tanks.

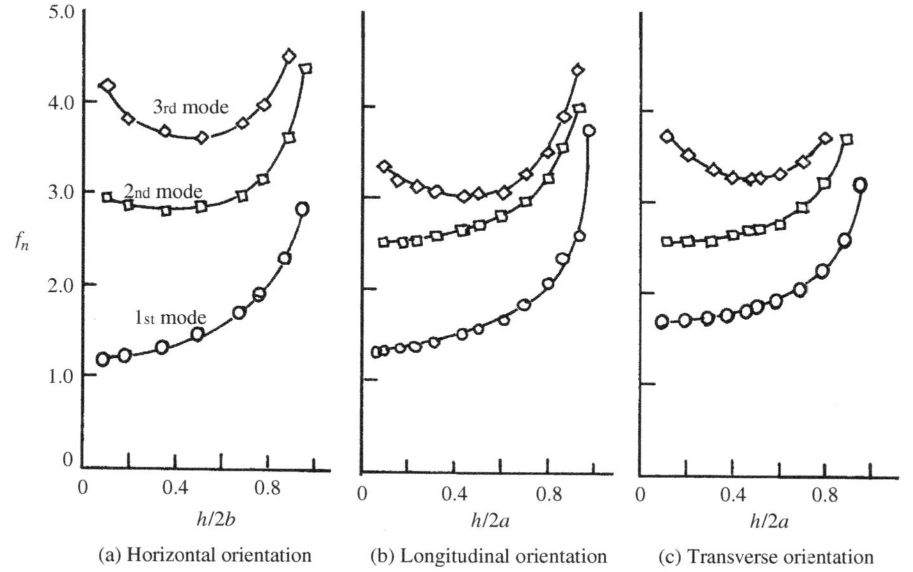

Figure 1.19 Dependence of liquid natural frequencies on depth ratio for three orientations of a spheroidal tank. (Leonard and Walton, 1961)

where ξ_n is the nth zero of $dJ_1(\xi_n\tilde{r}/r)/d\tilde{r}|_{\tilde{r}=r} = 0$. Leonard and Walton (1961) plotted the ratio $\sigma_n = \omega_{n(\text{measured})}/\omega_n$ versus the fluid depth ratio $h/2b$ and the results are shown in Figure 1.20. Any deviation from unity reflects the accuracy of the equivalent analytical expression given by relation (1.186). It is seen that the deviation is significant for the first mode particularly as the fluid depth

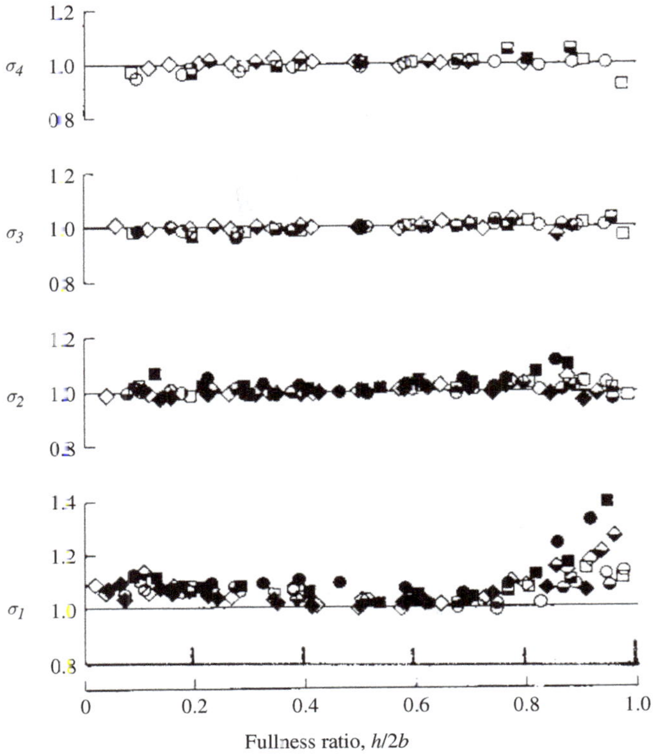

Figure 1.20 Dependence of the frequency parameter $\sigma_n = \omega_n / \sqrt{\frac{r}{g}\xi_n \tanh(\xi_n h_c/r)}$ on the filling ratio in spheroids. (Leonard and Walton, 1961)

ratio exceeds 0.8. Rattayya (1965) calculated the natural frequencies and mode shapes in horizontally oriented oblate spheroidal tanks using a variational approach. Figure 1.21 shows the dependence of the frequency parameter $\omega\sqrt{a/g}$ on the fluid depth ratio for the first three modes and for different values of the ratio of the ellipse axes b/a as estimated analytically by Rattayya.

Mikishev and Dorozhkin (1961) developed the following two empirical formulas for the liquid-free-surface natural frequencies in longitudinal and transverse orientations, respectively

$$\omega_{n,l} = \sqrt{\frac{g}{r} k_{l,n} \tanh(h_e k_{l,n}/r)} \tag{1.187}$$

$$\omega_{n,t} = \sqrt{\frac{g}{r} k_{t,n} \tanh(h_e k_{t,n}/r)} \tag{1.188}$$

where r is the radius of the free surface, $h_e = (h/3)(3a - h)/(2a - h)$ and the parameters $k_{l,n}$ and $k_{t,n}$ are proportional to the positive parametric zeros of the first derivative of the Mathieu function (see, for example, Chu, 1960a, and Abramowitz and Segun, 1968). Approximate values for the first mode parameters $k_{l,1}$ and $k_{t,1}$ are plotted in Figure 1.22. Figures 1.23 and 1.24 show the dependence of the natural frequency parameter on the fluid depth ratio for different values of the ellipse axes ratio b/a, for longitudinal and transverse orientations, respectively.

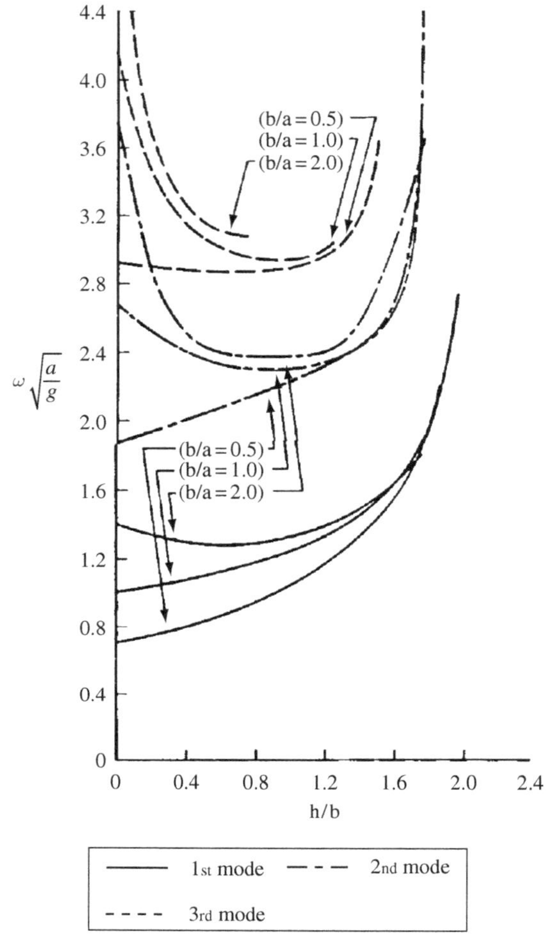

Figure 1.21 Dependence of sloshing frequency on the liquid depth in oblate spheroidal tanks (Rattayya, 1965)

1.6.5 Conical container

Closed form solutions for the case of a conical container have been obtained by Green (1957 1959), Green and Landau (1957), Levin (1957, 1963), and Bauer (1982i). It was noted that when the tank is very deep, the bottom shape has no influence on the liquid-free-surface motion and the bottom can be treated as being flat, based on equal liquid volumes (Abramson and Ransleben, 1961a). On the other hand, for extremely shallow liquid depths, the bottom shape governs the fluid motion and the problem is reduced to being two-dimensional. The two limiting cases were analyzed by Lawrence, *et al.* (1958) using a variational approach to obtain an approximate solution for tanks of intermediate depth.

First mode analysis

Simple formulations to estimate the first mode natural frequency in a conical container were developed by Troesch (1960) and Levin (1963). For the case of a conical container whose vertex

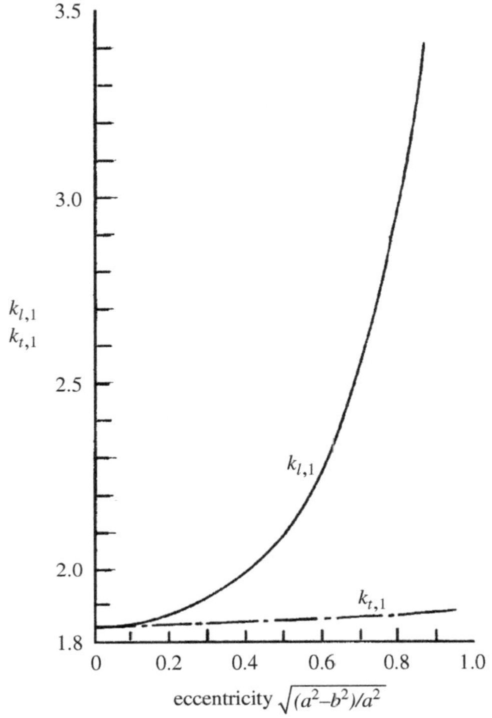

Figure 1.22 Dependence of the parameters $k_{l,1}$ and $k_{t,1}$ on the eccentricity. (Leonard and Walton 1961)

is below the liquid free surface, see Figure 1.25, a possible solution of the velocity potential function in terms of cylindrical coordinates may be written in the form

$$\Phi = \exp(i\omega t)F(r, z)\cos m\theta \tag{1.189}$$

In this case, Laplace's equation takes the form

$$r^2\frac{\partial^2 F}{\partial r^2} + r\frac{\partial F}{\partial r} + r^2\frac{\partial^2 F}{\partial z^2} - m^2F = 0, \quad z < 0 \text{ and } r > 0 \tag{1.190}$$

This equation should be solved subject to the boundary conditions

$$\frac{\partial F}{\partial n} = 0 \quad \text{on the container wall} \tag{1.191}$$

$$\frac{\partial F}{\partial z} - \frac{\omega^2}{g}F = 0 \quad \text{on the free surface} \tag{1.192}$$

For the case of free-surface motion with one nodal diameter, that is, $m = 1$, Troesch (1960) used a general series expansion and chose the following form for F that satisfies equation (1.190)

$$F(r, z) = r + Arz \tag{1.193}$$

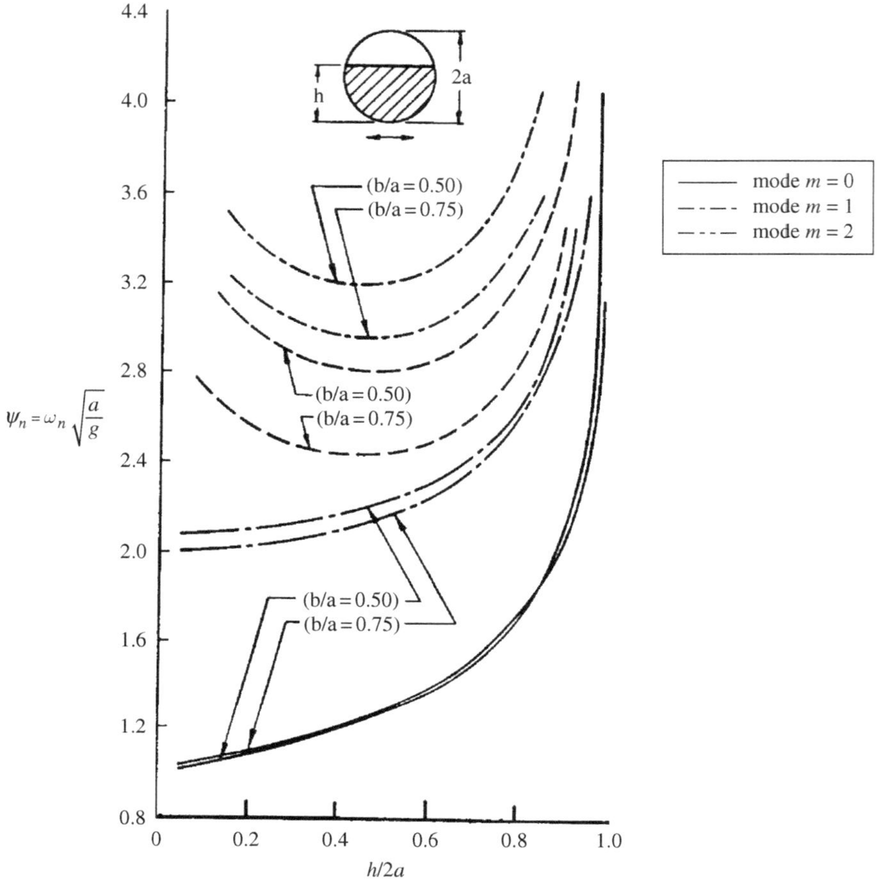

Figure 1.23 Dependence of frequency parameter with depth ratio in spheroids with longitudinal orientation.

Substituting equation (1.193) into (1.192) gives $A = \omega^2/g$. The shape of the container wall belonging to the function F should be compatible with the ordinary differential equation

$$\frac{\partial F}{\partial z} dr - \frac{\partial F}{\partial r} dz = 0 \tag{1.194}$$

Substituting equation (1.193) into equation (1.194), and integrating yields the shape of the container

$$r = \pm\left(z + \frac{g}{\omega^2}\right) \tag{1.195}$$

where the constant of integration is set to zero. Equation (1.195) represents the shape of a 45° cone with a depth of $h = g/\omega^2$, and radius $R = g/\omega^2$ at the free surface, ($z = 0$). In this case, the frequency of the free surface is

$$\omega = \sqrt{g/h} \tag{1.196}$$

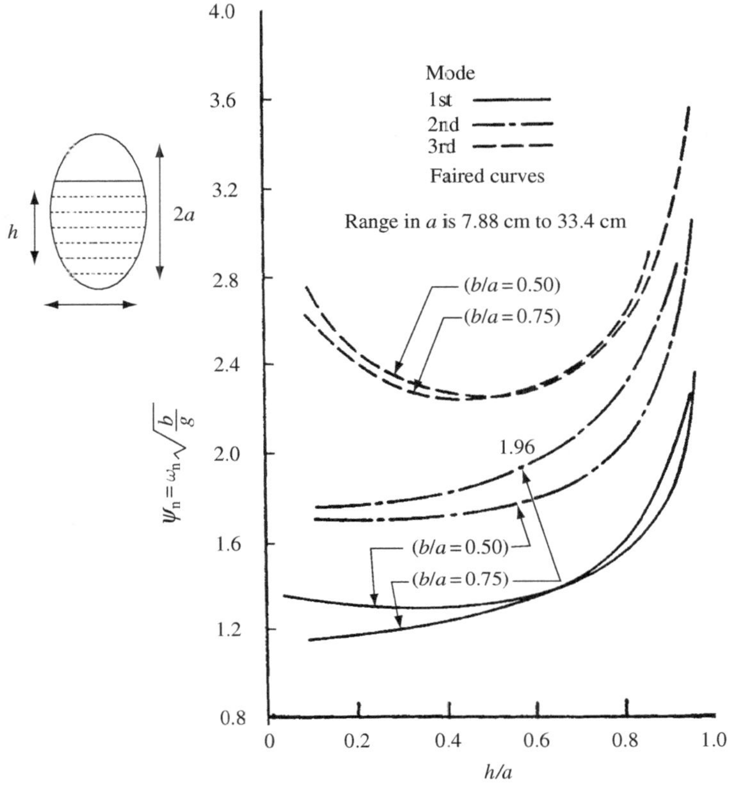

Figure 1.24 Dependence of liquid frequency parameter on the depth ratio in spheroids with transverse orientation. (Leonard and Walton, 1961)

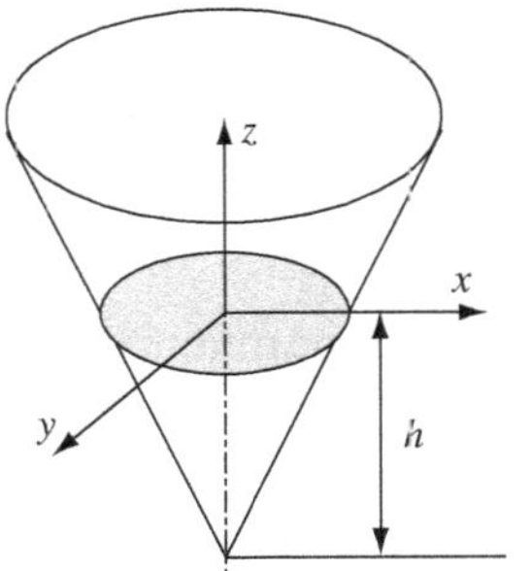

Figure 1.25 Coordinate system of a conical tank.

Levin (1963) obtained the same result by using a direct approach based on the geometry of the container and the appropriate coordinate system. Lawrence, *et al.* (1958) developed a variational approach based on Hamilton's variational formulation to obtain approximate estimation of sloshing frequencies and mode shapes of axially symmetrical containers This approach is described in the next section.

Variational approach

It is possible to follow the variational method outlined in Section 1.3. The fluid wave height η can be expressed in terms of the velocity potential function using the linearized Bernoulli equation $\eta = \frac{1}{g}\frac{\partial \Phi}{\partial t}$. Upon substituting this expression in equation (1.27), and performing the variation gives

$$0 = \delta \int_{t_1}^{t_2} dt \left\{ - \int_{v} dv\, \nabla^2 \Phi\, \delta\Phi + \int_{S_w} dS_w \frac{\partial \Phi}{\partial n}\, \delta\Phi + \int_{S_f} dS_f \left(\frac{\partial \Phi}{\partial n} + \frac{1}{g}\frac{\partial^2 \Phi}{\partial t^2} \right) \delta\Phi \right\}$$
$$- \frac{1}{g} \int_{S_f} dS_f \frac{\partial \Phi}{\partial t}\, \delta\Phi |_{t_1}^{t_2} \tag{1.197}$$

Note that the fluid density ρ has been absorbed in the definition of the Lagrangian $L = T - V$. Furthermore, the last term in the above equation vanishes at the end points of the time interval. For the variation δI to vanish for all admissible variations of the potential function Φ the following equations can be extracted from equation (1.197)

$$\nabla^2 \Phi = 0 \tag{1.198a}$$

$$\frac{\partial \Phi}{\partial n} = 0 \quad \text{on the wetted wall} \tag{1.198b}$$

$$\frac{\partial \Phi}{\partial n} + \frac{1}{g}\frac{\partial^2 \Phi}{\partial t^2} = 0 \quad \text{on the free surface} \tag{1.198c}$$

These equations constitute the fluid field equations. The sloshing modes and natural frequencies may be determined as solutions of the extremum problem for the integral I

$$I = \frac{1}{2} \int_{t_1}^{t_2} dt \left\{ \int_{v} |\nabla \Phi|^2\, dv - \frac{1}{g} \int_{S} \left(\frac{\partial \Phi}{\partial t} \right)^2 ds \right\} \tag{1.199}$$

Now, let the velocity potential function be harmonic in time and given by the expression

$$\Phi(r, z, \theta, t) = F(r, z, \theta) \sin \omega t \tag{1.200}$$

Substituting equation (1.200) into equation (1.199) gives

$$I = \frac{1}{2} \int_{v} |\nabla F|^2\, dv - \frac{\omega^2}{2g} \int_{S} F^2\, dS \tag{1.201}$$

A constant factor $\pi\rho/\omega$ in equation (1.201) has been set to unity as it is absorbed in the Lagrangian. For axially symmetric containers, one may use the cylindrical coordinates and express the potential function in the form

$$F = G(r, z) \cos m\theta \tag{1.202}$$

Substituting (1.202) into equation (1.201) gives

$$I = \int_0^{R_0} r\,dr \int_{-h}^{0} \left(\left(\frac{\partial G}{\partial r}\right)^2 + \left(\frac{\partial G}{\partial z}\right)^2 + \frac{m^2}{r^2} G^2 \right) dz - \frac{\omega^2}{g} \int_0^{R_0} r G^2\,dr \tag{1.203}$$

where the constant factor $\pi/2$ has been omitted.

In order to estimate the mode shapes and natural frequencies one may substitute equation (1.200) into equations (1.198) or take the variation of equation (1.201). In both cases the following equations are obtained

$$\nabla^2 F = 0 \tag{1.204a}$$

$$\frac{\partial F}{\partial n} = 0 \quad \text{on the wetted wall} \tag{1.204b}$$

$$\frac{\partial F}{\partial n} - \frac{\omega^2}{g} F = 0 \quad \text{on the free surface} \tag{1.204c}$$

Multiplying each of the above equations by the function F and integrating over the corresponding domains by parts, gives

$$\int_{S_w} F \frac{\partial F}{\partial n}\,dS_w + \int_{S_f} F \frac{\partial F}{\partial n}\,dS_f - \int_{v} |\nabla F|^2\,dv = 0 \tag{1.205a}$$

$$\int_{S_w} F \frac{\partial F}{\partial n}\,dS_w = 0 \tag{1.205b}$$

$$\int_{S_f} F \frac{\partial F}{\partial n}\,dS_f - \frac{\omega^2}{g} \int_{S_f} F^2\,dS_f = 0 \tag{1.205c}$$

Introducing equation (1.205b) into (1.205a) and solving the resulting equation together with equation (1.205c) gives the following expression for the natural frequency

$$\omega^2 = g \int_{v} |\nabla F|^2\,dv \Big/ \int_{S_f} F^2\,dS_f \tag{1.206}$$

It is not difficult to show that this value is a minimum when the integral expression (1.201c) is a superlative. The natural frequency given by expression (1.206) can be estimated once the potential function F is defined. Lawrence, *et al.* (1958) determined the natural frequencies for two cases of tank filling, the shallow and deep tank fillings.

For the shallow tank, the depth of the fluid is sufficiently smaller than the distance between mode crests, and equation (1.206) can be written in the form

$$\omega^2 = g \frac{\int\limits_{S_f} dS_f \int\limits_{-h}^{0} dz \left((\partial F/\partial r)^2 + (\partial F/r\partial\theta)^2 + (\partial F/\partial z)^2 \right)}{\int\limits_{S_f} F^2 \, dS_f} \tag{1.207}$$

One may assume that the potential function F is independent of z and, thus, it is convenient to denote this special function by $E(r, \theta)$ and the corresponding sloshing frequency by ω_e. In this case, expression (1.207) takes the form

$$\omega_e^2 = gh \frac{\int\limits_{S_f} \left((\partial E/\partial r)^2 + (\partial E/r\partial\theta)^2 \right) dS_f}{\int\limits_{S_f} E^2 \, dS_f} \tag{1.208}$$

For axially symmetric containers, the potential is a function of r only, $G(r, \theta) = \bar{G}(r)$, and one may express equation (1.203) in the form

$$I = \int\limits_{0}^{R_0} \left(\left[\left(\frac{\partial \bar{G}}{\partial r} \right)^2 + \frac{m^2}{r^2} \bar{G}^2 \right] - \frac{\omega^2}{g} \bar{G}^2 \right) r \, dr \tag{1.209}$$

To solve this equation, the function $\bar{G}(r)$ may be represented by the polynomial, (Lamb, 1945)

$$\bar{G}(r) = \sum_{n=0}^{N} A_n (r/R_0)^{m+2n} \tag{1.210}$$

It is important to note that the presence of m in the exponent is consistent with the fact that when $m \neq 0$ the singularity of equation (1.203) at $r = 0$ requires that $\bar{G}(r)$ contain no constant terms. The depth, h, will be represented by a polynomial in the form

$$h = h_0 \sum_{i=0}^{K} B_i (r/R_0)^i \tag{1.211}$$

The polynomial coefficients A_n are determined such that the integral (1.209) has a stationary value. This condition requires

$$\frac{\partial I}{\partial A_n} = 0, \quad n = 0, 1, 2, \ldots, N \tag{1.212}$$

This condition yields a set of $N + 1$ simultaneous homogeneous algebraic equations of the general form

$$\sum_{\ell=0}^{N} \left\{ \left[\sum_{i=0}^{K} \frac{m^2 + (n+\ell)m + 2n\ell}{m + n + \ell + i/2} k_i \right] - \frac{\beta^2/2}{m + n + \ell + 1} \right\} A_\ell = 0 \tag{1.213}$$

where $\beta^2 = \dfrac{\omega^2}{g} \dfrac{R_0^2}{h_0}$. These equations are consistent if the following determinant vanishes

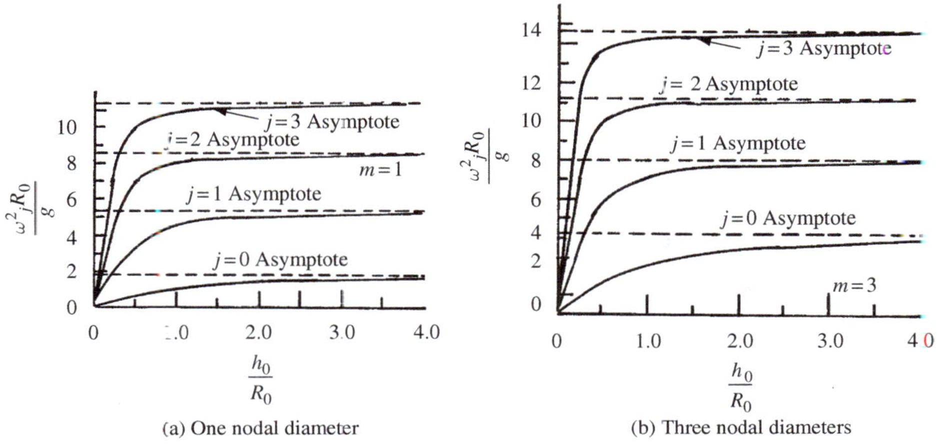

Figure 1.26 Dependence of liquid natural frequencies in a conical tank. (Lawrence, *et al.*, 1958)

$$|D_{n\ell}| = 0 \tag{1.214}$$

where

$$D_{n\ell} = \left[\sum_{i=0}^{K} \frac{m^2 + (n + \ell)m + 2n\ell}{m + n + \ell + i/2} k_i \right] - \frac{\beta^2/2}{m + n + \ell + 1}$$

The solution of equation (1.214) provides $N+1$ frequencies. The corresponding mode shapes are given by equations (1.210) and (1.213).

For the case of a conical bottom, the height is described by the expression

$$h = h_0 \left(1 - \frac{r}{R_0} \right) \tag{1.215}$$

For $m = 1$ and 3 (i.e., sloshing modes with one and three nodal diameters), Lawrence, *et al.* (1958) numerically determined the dependence of the first four ($j = 0, 1, 2, 3$) natural frequencies $\omega_j^2 R_0/g$ on the depth ratio h_0/R_0 and the results are plotted in Figure 1.26(a, b). Mikishev and Doroshkin (1961) and Dokuchaev (1964) extended the work of Lawrence, *et al.* and determined analytically and experimentally the natural frequencies in upright and inverted cones. For cones with small semi-vertex angles Mikishev and Doroshkin (1961) expressed the natural frequencies by the empirical relationship

$$\frac{\omega_j^2 R_0}{g} = 1.84 C_3^2 \tag{1.216}$$

where the constant C_3 depends on the semi-vertex angle and is plotted in Figure 1.27 for the upward and downward cones. This result is valid for $h_0/R_0 > 2.75$ and liquid oscillation amplitudes less than $0.01 R_0$.

Transformation to conical spherical coordinates

The analysis of free oscillations of a liquid free surface in conical tanks was considered by Bauer (1999) in terms of spherical coordinates, (r, ϑ, φ), where r emanates from the vertex of

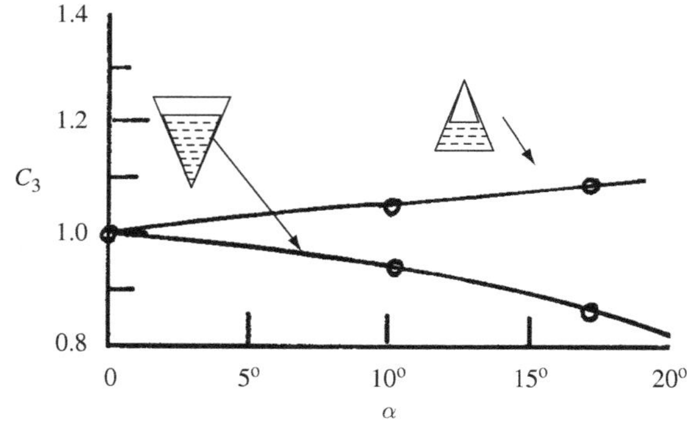

Figure 1.27 Dependence of the constant C_3 on the cone semi-angle. (Mikishev and Dorozhkin, 1961)

the cone (center of the sphere), ϑ is the angular coordinate with the vertical, and φ the angular coordinate around the vertical. For an incompressible and inviscid flow, the liquid flow is governed by the Laplace equation

$$\frac{1}{r^2}\frac{\partial}{\partial r}\left(r^2\frac{\partial\Phi}{\partial r}\right)+\frac{1}{r^2\sin\vartheta}\frac{\partial}{\partial\vartheta}\left(\sin\vartheta\frac{\partial\Phi}{\partial\vartheta}\right)+\frac{1}{r^2\sin^2\vartheta}\frac{\partial^2\Phi}{\partial\varphi^2}=0 \tag{1.217}$$

The solution of equation (1.217) for conical tanks, whose wall converges downward (**∇**), may be written in the form

$$\Phi(r,\vartheta,\varphi,t)=\sum_{m=0}^{\infty}\sum_{n=0}^{\infty}A_{mn}\left(\frac{r}{a}\right)^{\lambda_{mn}}P^m_{\lambda_{mn}}(\cos\vartheta)\cos m\varphi\, e^{i\omega t} \tag{1.218}$$

where $P^m_{\lambda_{mn}}(\cos\vartheta)$ are associated Legendre functions of the first kind of degree λ and of angular mode m, and a is the radius to the fluid surface. The velocity potential function Φ must satisfy the combined free-surface condition involving the effect of surface tension, at $r=a$

$$\frac{\partial^2\Phi}{\partial t^2}+g\frac{\partial\Phi}{\partial r}-\frac{\sigma}{\rho a^2}\left[2\frac{\partial\Phi}{\partial r}+\frac{1}{\sin\vartheta}\frac{\partial}{\partial\vartheta}\left(\sin\vartheta\frac{\partial^2\Phi}{\partial r\partial\vartheta}\right)+\frac{1}{\sin^2\vartheta}\frac{\partial^3\Phi}{\partial\varphi^2\partial r}\right]=0 \tag{1.219}$$

For free-liquid oscillations, the velocity normal to wall, $\vartheta=\alpha$, vanishes, that is,

$$\frac{1}{r}\frac{\partial\Phi}{\partial\vartheta}=0 \tag{1.220}$$

This condition yields

$$\frac{\partial}{\partial\vartheta}\left[P^m_{\lambda_{mn}}(\cos\vartheta)\right]\bigg|_{\vartheta=\alpha}=0 \tag{1.221}$$

The roots λ_{mn} of equation (1.221) depend on the cone vertex angle β and are listed by Bauer (1999) for several cone angles. Samples of these roots are given for $m=(0, 1)$ and $n=(1, 2)$ in Table 1.8. Applying the surface condition (1.219), gives the natural frequencies of the free surface

$$\omega^2_{mn}=\frac{g\lambda_{mn}}{a}+\frac{\sigma}{\rho a^3}\lambda_{mn}(\lambda_{mn}-1)(\lambda_{mn}+2) \tag{1.222}$$

Table 1.8 *Zeros of the derivative of the associated Legendre function,* $dP^m_{\lambda_{mn}}(\cos\vartheta)/d\vartheta = 0$

α	m	$n = 1$	$n = 2$
$\alpha = 10°$	0	21.4598	39.6995
	1	10.0835	30.0567
$\alpha = 15°$	0	14.1446	26.3023
	2	6.5842	19.8793
$\alpha = 20°$	0	10.4885	19.6044
	1	4.8432	14.7931
$\alpha = 25°$	0	8.2960	15.5864
	1	3.8056	11.7434
$\alpha = 30°$	0	6.8354	12.9083
	2	3.1196	9.7121
$\alpha = 35°$	0	5.7930	10.9958
	1	2.6347	8.2626

Source: Bauer, 1999

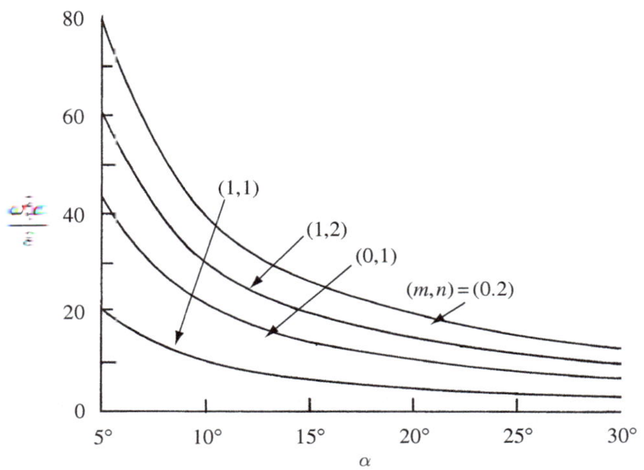

Figure 1.28 Dependence of liquid natural frequencies on vertex angle in a conical tank (∇). (Bauer, 1982i, 1999)

The dependence of the natural frequency ratio $\omega_{mn}{}^2/(g/a)$ on the cone semi-vertex angle α is shown in Figure 1.28 for $(m, n) = (0, 1), (0, 2), (1, 1)$, and $(1, 2)$. It is seen that as the vertex angle increases the natural frequency decreases.

For inverted conical containers (Δ), the following boundary conditions must be satisfied

$$\frac{\partial\Phi}{\partial r} = 0 \quad \text{at the tank bottom } (r = a) \tag{1.223}$$

At $\vartheta = \alpha$

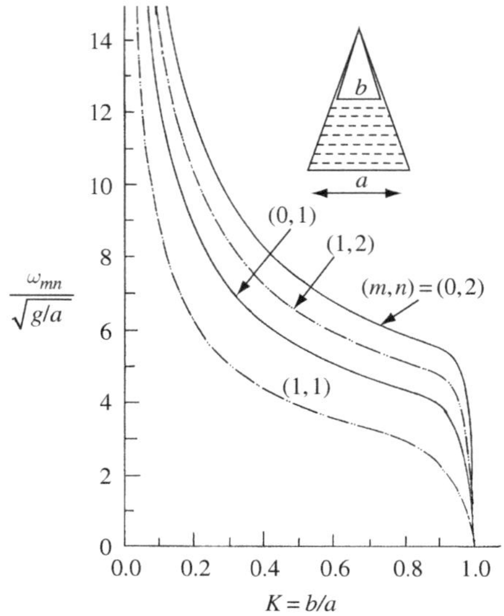

Figure 1.29 Dependence of liquid natural frequencies on the depth ratio. (Bauer, 1982i, 1999)

$$\frac{1}{r}\frac{\partial \Phi}{\partial \vartheta} = 0 \tag{1.224}$$

The combined free-surface condition, ignoring surface tension effect is

$$\frac{\partial^2 \Phi}{\partial t^2} - g\frac{\partial \Phi}{\partial r} = 0 \quad \text{at } r = b \tag{1.225}$$

The solution of Laplace's equation is

$$\Phi(r, \vartheta, \varphi, t) = \sum_{m=0}^{\infty}\sum_{n=0}^{\infty}\left[A_{mn}\left(\frac{r}{a}\right)^{\lambda_{mn}} + \frac{\lambda_{mn}}{(\lambda_{mn}+1)}\left(\frac{r}{a}\right)^{-(\lambda_{mn}+1)}\right] \times P_{\lambda_{mn}}^{m}(\cos\vartheta)\cos m\varphi\, e^{i\omega t} \tag{1.226}$$

where λ_{mn} are obtained from $(\partial/\partial\vartheta)\left[P_{\lambda_{mn}}^{m}(\cos\vartheta)\right]\Big|_{\vartheta=\alpha} = 0$. The natural frequencies are given by the expression

$$\omega_{mn}^2 = \frac{g\lambda_{mn}}{ak}\frac{\left(1 - k^{2\lambda_{mn}+1}\right)}{\left(\lambda_{mn}/1(+\lambda_{mn})\right) + k^{2\lambda_{mn}+1})}, \qquad k = b/a \tag{1.227}$$

Figure 1.29 shows the dependence of the natural frequency ratio, $\omega_{mn}/\sqrt{g/a}$, on the radii ratio, $k = b/a$, which measures the fluid depth, for the modes $(m, n) =$ (0, 1), (0, 2), (1, 1), and (1, 2) , and for semi-cone angle $\alpha = 15°$. As k increases the fluid natural frequencies drop from infinity to zero as $k \rightarrow 1$.

1.6.6 Toroidal containers

Toroidal tanks exist in the form of a ring whose cross-section is circular hollow. Figure 1.30 shows a schematic diagram of a toroidal tank in three possible orientations with respect to a

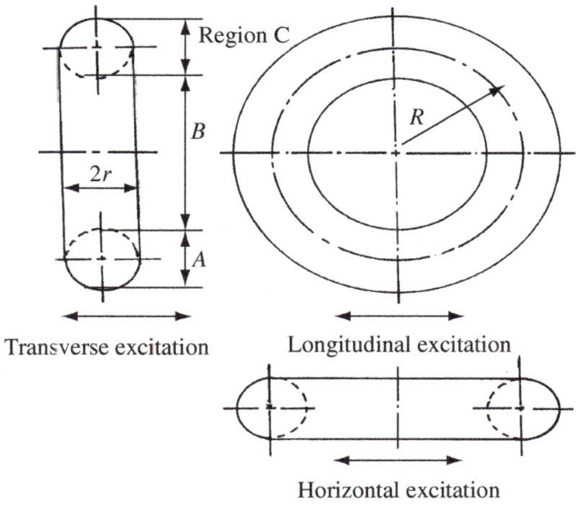

Figure 1.30 Toroidal tank.

given excitation, longitudinal, transverse, and horizontal. This type of container is found to be very convenient for carrying liquid propellant in large space vehicles such as the Space Transportation System (STS). In particular, a reusable liquid hydrogen/liquid oxygen-fueled orbit transfer vehicle (OTV) for boosting large satellites to geosynchronous orbit must have a large volume for the low-density liquid hydrogen fuel. In view of the lack of analytical closed form solutions, McCarty, *et al.* (1960) and Sumner (1963) measured the natural frequencies and mode shapes of liquid for different tank orientations. Meserole and Fortini (1987) and Shi (1987) conducted experimental and analytical studies to derive equivalent models for the slosh modes.

In adopting a nondimensional frequency parameter, certain regular tank shapes that approximate every boundary region of the toroidal tank may be considered. The following are different cases considered by McCarty, *et al.* (1960).

Horizontal modes

A frequency parameter may be taken as the ratio of the measured natural frequency to the natural frequency of a liquid contained in an upright annular circular cylinder having the same inner and outer radii of the toroid, (r_i and r_o), and a liquid depth h_c necessary to produce the same liquid volume in the toroid. This frequency is given by equation (1.60) and should read in the present case as

$$\omega_{ne} = \sqrt{\frac{g}{r_o}\xi_{1n}\tanh(\xi_{1n}h_c/r_o)} \tag{1.228}$$

and the frequency parameter is

$$\lambda_n = \omega_n/\omega_{ne} \tag{1.229}$$

where ω_n is the measured frequency. Figure 1.31 shows the dependence of the frequency parameter λ_n on the fluid depth ratio $h/2r$ for the first four sloshing modes and different values of minor and major radii. It is seen that at a given fluid fullness ratio, the frequency parameter for a given mode is the same for all toroids examined. This implies that the frequency

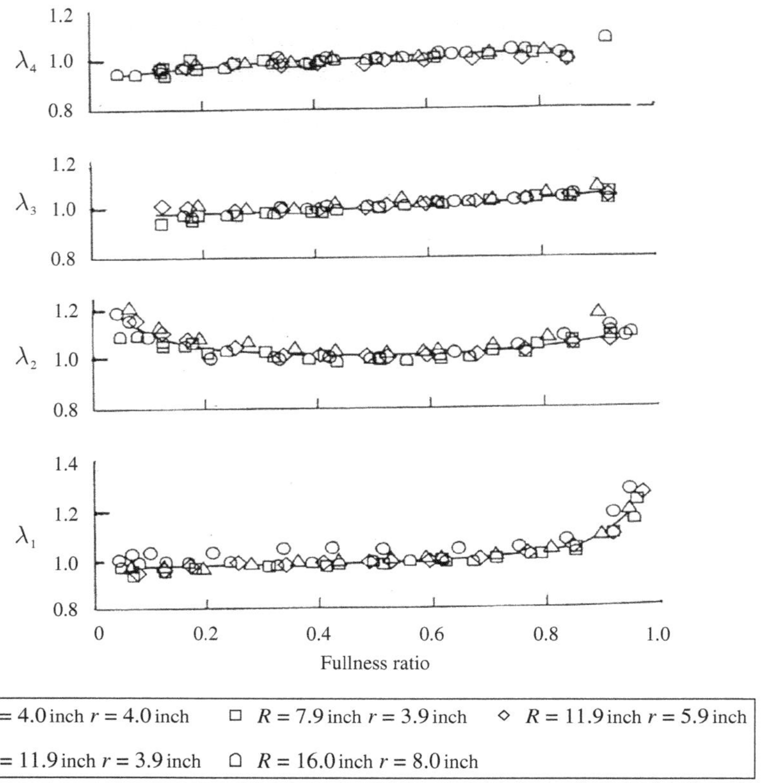

Figure 1.31 Dependence of liquid natural frequency parameter $\lambda_n = \omega_n/\omega_{ne}$ on fullness ratio for horizontal toroidal tanks. (McCarty, *et al.*, 1960)

parameter λ_n is independent of the tank dimensions. However, at the near-full and near-empty conditions, the value of λ_n deviates from 1.

Vertical modes

The liquid boundaries will depend on the fluid depth, and McCarty, *et al.* divided the tank into three regions labeled "A" bounded by the region ($0 < h < 2r$), "B" bounded by ($2r < h < 2R$), and "C" given by ($2R < h < 2(R + r)$), as show in Figure 1.30.

Under transverse disturbance, the liquid in region A or C has boundaries that are somewhat similar to spheres or horizontal cylinders undergoing transverse oscillations. In this case, the following frequency parameter can be used

$$\sigma_n = \omega_n \sqrt{r/g} \tag{1.230}$$

In region B, the liquid boundary is similar to that of an upright circular cylinder, and the corresponding frequency is similar to relation (1.43a) and should read as

$$\omega_{ne} = \sqrt{\frac{g\xi_{1n}}{r}\tanh(\xi_{1n}h/r)}, \quad \text{or } \sqrt{\xi_{1n}g/r} \quad \text{as } \tanh(\xi_{1n}h/r) \to 1 \tag{1.231}$$

where the symbols follow equation (1.43a) but with radius r.

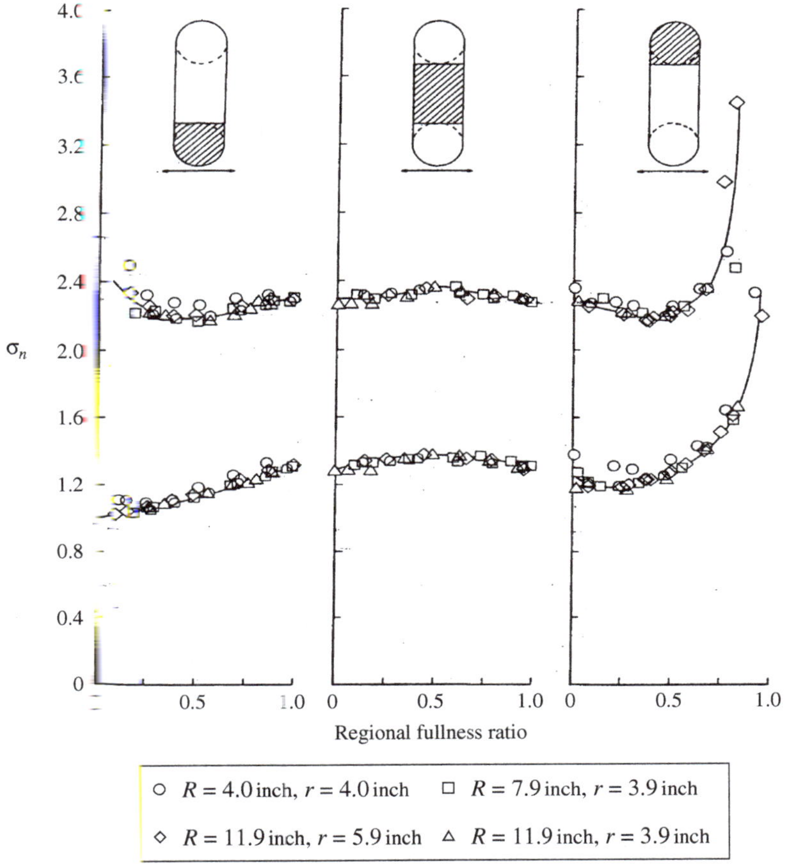

Figure 1.32 Dependence of liquid natural frequency parameter $\sigma_n = \omega_n\sqrt{r/g}$ on fullness ratio for transverse modes of vertical toroidal tanks. (McCarty, *et al.*, 1960)

Figure 1.32 shows the dependence of the frequency parameter, σ_n, on the regional fluid fullness ratio [region "A" $h/2r$, $(h-2r)/2(R-r)$, region "C" $(h-2R)/2r$] for the first two modes.

Under longitudinal disturbance, the frequency parameter in region A or C is taken as the ratio of the measured natural frequency to the frequency of a simple pendulum whose length equals the distance from the center of the toroid to the center of mass of a mass distributed uniformly along the arc of the peripheral circle subtended by the liquid surface as shown in Figure 1.33. The corresponding frequency of the pendulum is

$$\omega_p = \sqrt{\frac{g\phi}{(R+r)\sin\phi}} \tag{1.232}$$

where ϕ is the angle between the vertical and the radius to the point of intersection of the fluid surface with the peripheral circle. The resulting frequency parameter of the first mode is

$$\psi_1 = \omega_1/\omega_p \tag{1.233}$$

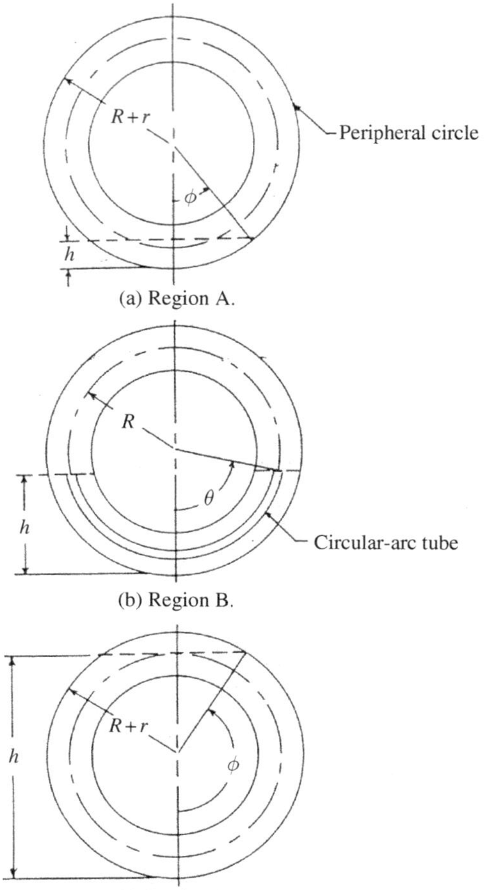

Figure 1.33 Regions of liquid in vertical toroidal tanks for longitudinal sloshing modes.

For higher modes in these two toroidal regions, the frequency parameter can be established through similarity of liquid boundaries in a sphere of radius $R+r$. The resulting parameter takes the form

$$\psi_n = \frac{\omega_n}{\sqrt{(g/(R+r))}} \tag{1.234}$$

where ω_1 and ω_n are measured natural frequencies of the first and nth modes, respectively. Figure 1.34(a) shows the dependence of the frequency parameter ψ_n on the fluid regional fullness ratio for the first four modes for region A. The values of this parameter for each of the three regions are plotted separately as a function of the region fullness ratio. It is seen that for a given fullness ratio, only the first mode frequency parameter is independent of the tank size. However, for higher modes, the frequency parameter is dependent on the radii ratio R/r. For a given fullness ratio, as R/r increases the frequency parameter increases as well. Figure 1.31(b) shows the frequency depth relationship for region B for the first four modes. This region is also characterized by the dependence of the frequency parameter on the radii ratio R/r for a given

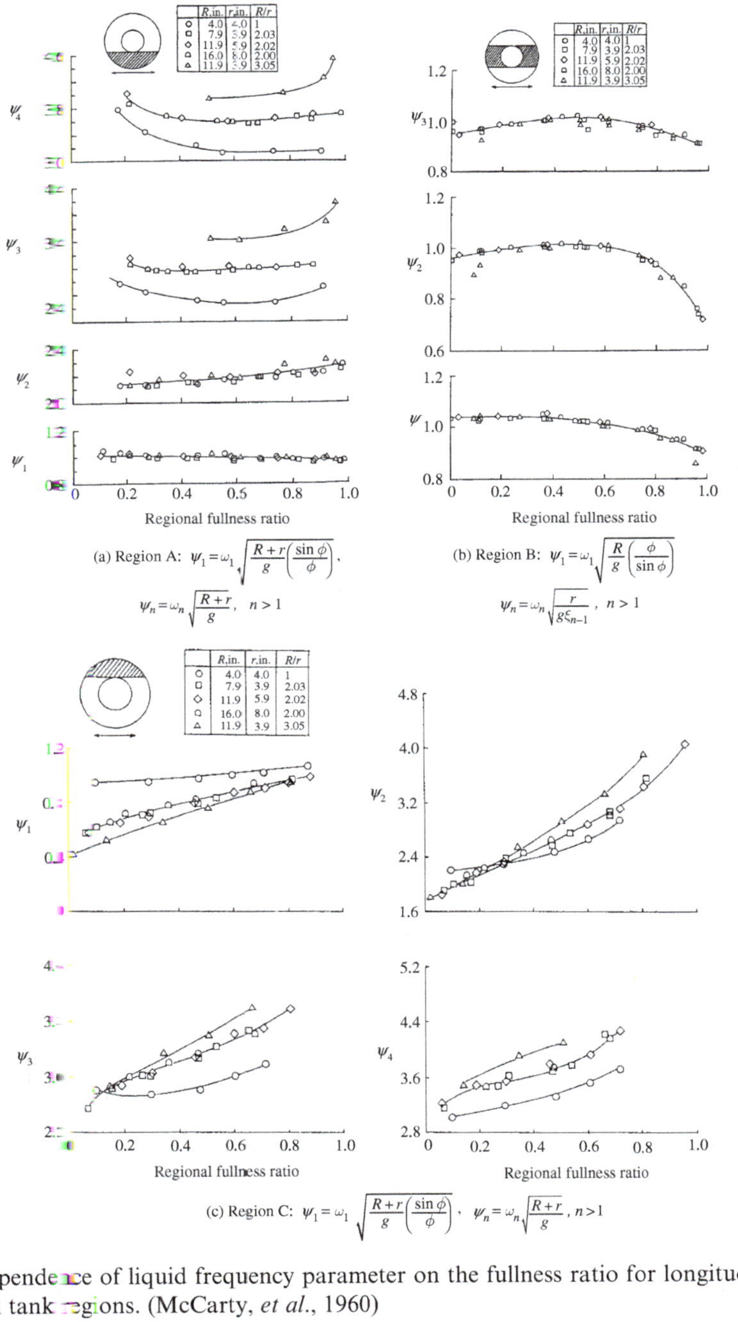

(a) Region A: $\psi_1 = \omega_1 \sqrt{\frac{R+r}{g}\left(\frac{\sin\phi}{\phi}\right)}$, $\psi_n = \omega_n \sqrt{\frac{R+r}{g}}$, $n > 1$

(b) Region B: $\psi_1 = \omega_1 \sqrt{\frac{R}{g}\left(\frac{\phi}{\sin\phi}\right)}$, $\psi_n = \omega_n \sqrt{\frac{r}{g\xi_{n-1}}}$, $n > 1$

(c) Region C: $\psi_1 = \omega_1 \sqrt{\frac{R+r}{g}\left(\frac{\sin\phi}{\phi}\right)}$, $\psi_n = \omega_n \sqrt{\frac{R+r}{g}}$, $n>1$

Figure 1.34 Dependence of liquid frequency parameter on the fullness ratio for longitudinal modes of vertical toroidal tank regions. (McCarty, *et al.*, 1960)

fullness ratio. For regions A and C, the frequency parameter is not effective in combining the effects of various tank size and geometry.

For region B, the frequency can be nondimensionalized by using the frequency of the circular arc tube of small cross-section shown in Figure 1.33, that is,

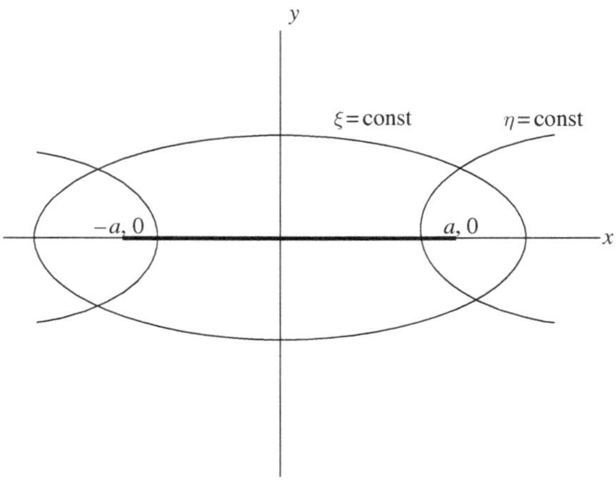

Figure 1.35 Elliptic cylindrical coordinates.

$$\omega_{1e} = \sqrt{\frac{g}{r}}\sqrt{\frac{\sin\theta}{\theta}} \tag{1.235}$$

where θ is the angle between the vertical and the radius R that terminates in the liquid surface. The corresponding frequency parameter is

$$\psi_1 = \omega_1/\omega_{1e} \tag{1.236}$$

For higher modes, the reference frequency of the liquid in an upright circular cylinder is taken. The order of the analogous cylindrical frequency is decreased by 1 because the manometer-type mode not present in the upright cylinder is considered to be the fundamental mode for this region. It is assumed that the equivalent fluid depth is sufficiently large to ensure that $\tanh(\xi_{1n}h/r) \to 1$. The corresponding frequency parameter is

$$\psi_n = \omega_n\sqrt{\frac{r}{g\xi_{n-1}}}, \qquad n > 1$$

Figure 1.34(b) presents the dependence of the frequency ratio, ψ_n, on the fluid regional fullness ratio for the first three modes for the mid-region of vertical toroids. For a given fullness ratio, the frequency parameter is independent of the tank dimensions.

1.6.7 Upright elliptic containers

Elliptic containers are different to circular ones, because the ellipticity of the container cross-sections affects the natural frequencies of the free liquid surface (Chu, 1960a and Bauer and Knölker, 1981). Bauer (1999) presented a brief analysis of free oscillations of liquid free in elliptic containers. With reference to Figure 1.35, the relationships between elliptic and Cartesian coordinates are (Fogiel, 1994)

$$x = a\cosh\xi\cos\eta, \;\; y = a\sinh\xi\sin\eta, \;\; z = z \tag{1.237}$$

where $\xi \geq 0$, $0 \leq \eta \leq 2\pi$, $-\infty < z < \infty$, and at the side wall, $\xi = \xi_0$, the equation of the ellipse takes the form

$$\frac{x^2}{a^2 \cosh^2 \xi_0} + \frac{y^2}{a^2 \sinh^2 \xi_0} = 1 \tag{1.238}$$

The Laplace equation is

$$\nabla^2 \Phi(\xi, \eta, z, t) = \frac{\partial^2 \Phi}{\partial \xi^2} + \frac{\partial^2 \Phi}{\partial \eta^2} + a^2(\cosh \xi - \cos \eta) \frac{\partial^2 \Phi}{\partial z^2} = 0 \tag{1.239}$$

The velocity potential function may be written in the form

$$\Phi(\xi, \eta, z, t) = \phi(\xi, \eta) \cosh(k(z+h)) e^{i\omega_n t} \tag{1.240}$$

Equation (1.240) satisfies the boundary condition at the tank bottom, $\partial\Phi/\partial z|_{z=-h} = 0$, where k is a factor to be determined from other boundary conditions. Equation (1.239) takes the form

$$\frac{\partial^2 \phi}{\partial \xi^2} + \frac{\partial^2 \phi}{\partial \eta^2} + a^2 k^2 (\cosh \xi - \cos \eta) \phi = 0 \tag{1.241}$$

The solution of (1.241) may be written in the form

$$\phi(\xi, \eta) = Z(\xi) H(\eta) \tag{1.242}$$

Introducing (1.242) into equation (1.241) gives the following ordinary differential equations

$$-\frac{d^2 G(\xi)}{d\xi^2} + (\nu - k^2 a^2 \cosh \xi) G(\xi) = 0 \tag{1.243a}$$

$$\frac{d^2 H(\eta)}{d\eta^2} + (\nu - k^2 a^2 \cos \eta) H(\eta) = 0 \tag{1.243b}$$

Equation (1.243b) is Mathieu's equation, while equation (1.243a) is a modified Mathieu's equation since it has imaginary argument. The solutions of equation (1.243b) are periodic in η since η varies from 0 to 2π. The solutions of equation (1.243a) are not periodic in ξ. Therefore we need the periodic eigenfunctions for η together with two solutions for ξ, one is proportional to the sine and cosine of Mathieu's functions $\text{Se}_m(\eta, ka)$, and $\text{Ce}_m(\eta, ka)$, respectively, and the other solution is independent of $\text{Se}_m(\eta, ka)$, and $\text{Ce}_m(\eta, ka)$. The second solution is given by the radial function of the first kind (Bessel functions) $\text{Je}_m(\xi, ka)$ and the second kind (Neumann functions) $\text{Ne}_{2m}(\xi, ka)$, (Morse and Feshbach, 1953). Two general solutions of equation (1.239) may be written in the form

$$\Phi(\eta, \xi, z, t) = \sum_{m=1}^{\infty} [A_m \text{Se}_m(\eta, ka) + B_m \text{Ce}_m(\eta, ka)] \begin{Bmatrix} \text{Je}_m(\xi, ka) \\ \text{Ne}_m(\xi, ka) \end{Bmatrix} \cosh(k(z+h))\, e^{i\omega_n t} \tag{1.244}$$

These solutions must satisfy the boundary condition at the tank wall

$$\left. \frac{\partial \Phi}{\partial \xi} \right|_{\xi = \xi_0} = 0 \tag{1.245}$$

The natural frequency is obtained from the combined free-surface condition

$$\frac{\partial^2 \Phi}{\partial t^2} + g\frac{\partial \Phi}{\partial z} = |_{z=0} 0 \tag{1.246}$$

Bauer (1999) obtained the following expression for the free-surface natural frequency for oscillations in the x-direction

$$\omega_{mn}^2 = \frac{2g}{eR_1}\sqrt{ak_{mn}}\tanh\left(\frac{2\sqrt{ak_{mn}}}{e}\frac{h}{R_1}\right) \tag{1.247a}$$

For oscillations in the y-direction, the natural frequency is given by the expression

$$\omega_{mny}^2 = \frac{2g}{eR_1}\sqrt{a\bar{k}_{mn}}\tanh\left(\frac{2\sqrt{a\bar{k}_{mn}}}{e}\frac{h}{R_1}\right) \tag{1.247b}$$

where $e = \sqrt{R_1^2 - R_2^2}/R_1$ is the ellipse eccentricity, R_1 and R_2 are the major and minor axes of the ellipse, q_{mn} and $\bar{q}_{mn}$ are the roots of the modified Mathieu functions, $\mathrm{Ne}'_m(\xi_0, q_{mn}) = 0$ and $\mathrm{Ne}'_m(\xi_0, a\bar{k}_{mn}) = 0$, respectively. These functions are defined in Gradshteyn and Ryzhik (1980, Section 8.61). The natural frequencies in the x-direction are smaller than those in the y-direction depending on the eccentricity e.

1.6.8 Paraboloid container

Paraboloid containers usually form the lower portion of cylindrical containers. Bauer (1984d, 1999) used coordinate surfaces as boundaries to approximate the liquid-free-surface motion. The approximation is accurate if the shape of the free surface is close to the actual free surface in the container. With reference to the paraboloid coordinates, ξ, η, φ, shown in Figure 1.36, the free surface at $\xi = \xi_0$ exhibits the value $\xi_0 \gg 1$, when the container wall at $\eta = \eta_0 \ll 1$. The coordinates, ξ, η, φ, are related to the Cartesian coordinates by the following relations

$$x = \xi\eta\cos\varphi, \quad y = \xi\eta\sin\varphi, \quad z = \frac{1}{2}(\xi^2 - \eta^2) \tag{1.248a}$$

The radius in the xy plane is

$$r = \sqrt{x^2 + y^2} = \xi\eta \tag{1.248b}$$

The spatial radius is

$$R = \sqrt{x^2 + y^2 + z^2} = \frac{1}{2}(\xi^2 + \eta^2) \tag{1.248c}$$

The container wall, $\eta = \eta_0$, satisfies the equation

$$z = -\frac{\eta_0^2}{2} + \frac{1}{2\eta_0^2}(x^2 + y^2) \tag{1.248d}$$

The container bottom, $\xi = \xi_0$, and the free surface, $\xi = \xi_1$, are described by the following relation

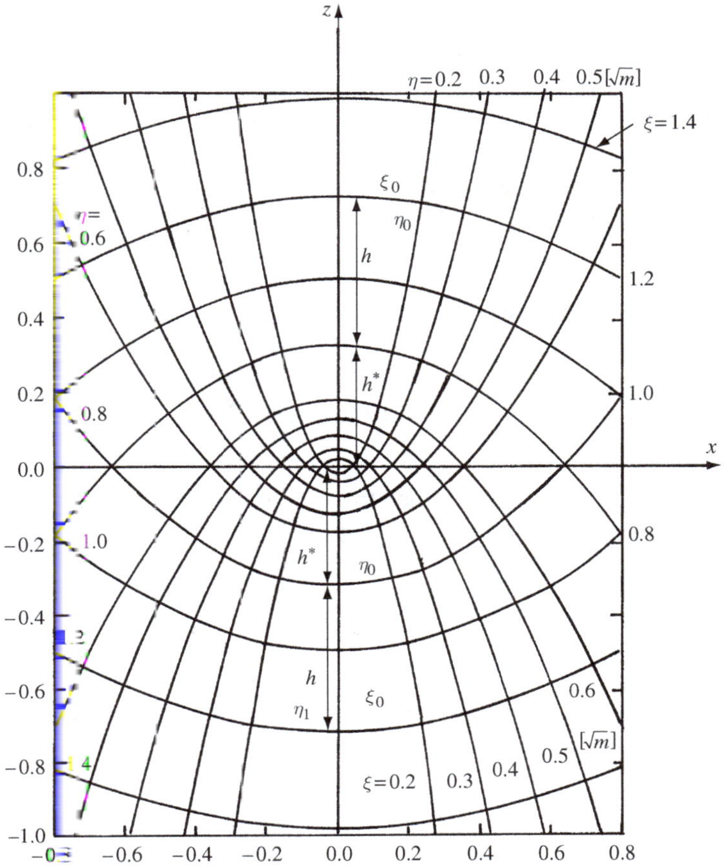

Figure 1.36 Parabolic coordinate system in the plane of $\varphi = 0$.

$$z = \frac{\xi_{0(1)}^2}{2} - \frac{1}{2\xi_{0(1)}^2}\left(x^2 + y^2\right) \tag{1.248e}$$

For $\xi_0 \gg 1$, and $\eta_0 \ll 1$, the liquid free surface may be approximated by the following relationship

$$z = \frac{\xi_0^2}{2} - \frac{\eta^2}{2}, \quad 0 \le \eta \le \eta_0 \tag{1.248f}$$

The relationships between the unit vectors $(\mathbf{e}_\xi, \mathbf{e}_\eta, \mathbf{e}_\varphi)$ of the rotational parabolic coordinates and the Cartesian unit vectors $(\mathbf{i}, \mathbf{j}, \mathbf{k})$ are

$$\mathbf{i} = \frac{\eta \cos\varphi}{\sqrt{\xi^2 + \eta^2}}\mathbf{e}_\xi + \frac{\xi \cos\varphi}{\sqrt{\xi^2 + \eta^2}}\mathbf{e}_\eta - \sin\varphi \mathbf{e}_\varphi \tag{1.249a}$$

$$\mathbf{j} = \frac{\eta \sin\varphi}{\sqrt{\xi^2 + \eta^2}}\mathbf{e}_\xi + \frac{\xi \sin\varphi}{\sqrt{\xi^2 + \eta^2}}\mathbf{e}_\eta + \cos\varphi \mathbf{e}_\varphi \tag{1.249b}$$

$$\mathbf{k} = \frac{\xi}{\sqrt{\xi^2+\eta^2}}\mathbf{e}_\xi - \frac{\eta}{\sqrt{\xi^2+\eta^2}}\mathbf{e}_\eta \tag{1.249c}$$

The fluid field equations are governed by Laplace's equation and boundary conditions:

$$\frac{\partial^2\Phi}{\partial\xi^2} + \frac{1}{\xi}\frac{\partial\Phi}{\partial\xi} + \frac{\partial^2\Phi}{\partial\eta^2} + \frac{1}{\eta}\frac{\partial\Phi}{\partial\eta} + \frac{(\xi^2+\eta^2)}{\xi^2\eta^2}\frac{\partial^2\Phi}{\partial\varphi^2} = 0 \tag{1.250a}$$

$$\frac{\partial\Phi}{\partial\eta} = 0 \quad \text{at the side wall } \eta = \eta_0 \tag{1.250b}$$

$$\frac{\partial^2\Phi}{\partial t^2} + g\frac{1}{\xi}\frac{\partial\Phi}{\partial\xi} = 0 \quad \text{at the free surface } \xi = \xi_0 \tag{1.250c}$$

Condition (1.250c) represents the combined free-surface condition of kinematic and dynamic conditions. The solution of equation (1.250a) that satisfies the wall boundary condition is

$$\Phi(\xi,\eta,\varphi,t) = \sum_{m=0}^{\infty}\sum_{n=1}^{\infty} A_{mn} I_m(\varepsilon_{mn}\xi/\eta_0) J_m(\varepsilon_{mn}\eta/\eta_0)\cos m\varphi\, e^{i\omega t} \tag{1.251}$$

Substituting equation (1.251) into equation (1.250c) gives the natural frequency of the free surface

$$\omega_{mn}^2 = \frac{g\varepsilon_{mn}}{\xi_0\eta_0}\frac{I'_m(\varepsilon_{mn}\xi_0/\eta_0)}{I_m(\varepsilon_{mn}\xi_0/\eta_0)} \tag{1.252}$$

where $I_m(\ )$ is the modified Bessel function of first kind and mth-order. The liquid depth h measured from the tank bottom at the axis of symmetry is expressed in terms ξ_0 and η_0 as $h = (\xi_0^2+\eta_0^2)/2$, and equation (1.252) may be written in the form

$$\frac{\omega_{mn}^2}{g/\eta_0^2} = \frac{mI_m\left(\varepsilon_{mn}\sqrt{2h/\eta_0^2-1}\right) + \varepsilon_{mn}I_{m+1}\left(\varepsilon_{mn}\sqrt{2h/\eta_0^2-1}\right)\sqrt{2h/\eta_0^2-1}}{(2h/\eta_0^2-1)I_m\left(\varepsilon_{mn}\sqrt{2h/\eta_0^2-1}\right)} \tag{1.253}$$

Figure 1.37 shows the dependence of the natural frequency ratio, $\omega_{mn}\eta_0/\sqrt{g}$, on the fluid depth ratio, h/η_0^2, for sloshing modes $m=0$, 1, 2 and $n=1$, 2, 3, with the mode $m,n=1,1$ having the lowest natural frequency. The measured natural frequency of the lowest mode m, $n=1$, 1 is shown by solid circles.

The results of special cases of paraboloid tanks are briefly summarized based on the work of Bauer (1999).

Sectional truncated paraboloid tank

The sectional truncated tank has a bottom at $\xi=\xi_1$, and the presence of the bottom imposes the boundary condition

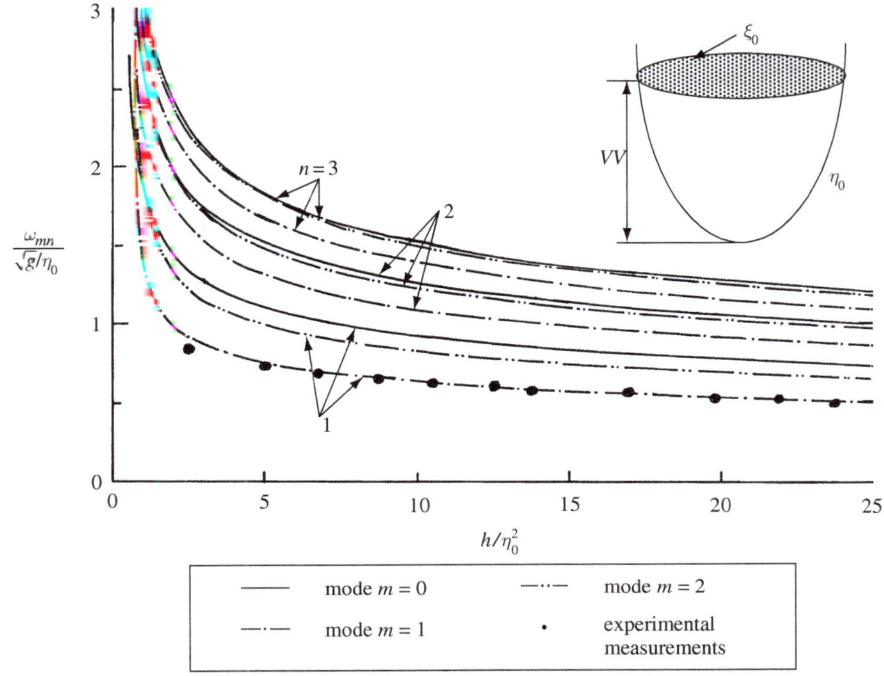

Figure 1.37 Dependence of liquid natural frequency ratio on the fluid depth ratio in paraboloid container for various modes. (Bauer, 1984d, 1999)

$$\frac{\partial \Phi}{\partial \xi} = 0 \quad \text{at } \xi = \xi_1 \tag{1.254}$$

The solution of Laplace's equation (1.25a) that satisfies the wall and bottom boundary conditions is

$$\Phi(\xi, \eta, \varphi, t) = \sum_{m=0}^{\infty} \sum_{n=1}^{\infty} A_{mn} \left[I_m\left(\frac{\varepsilon_{mn}\xi}{\eta_0}\right) K'_m\left(\frac{\varepsilon_{mn}\xi_1}{\eta_0}\right) - I'_m\left(\frac{\varepsilon_{mn}\xi_1}{\eta_0}\right) K_m\left(\frac{\varepsilon_{mn}\xi}{\eta_0}\right)\right] \times J_m(\varepsilon_{mn}\eta/\eta_0) \cos m\varphi \, e^{i\omega t} \tag{1.255}$$

Substituting equation (1.255) into the combined free-surface boundary condition (1.250c) gives the natural frequency of the free surface

$$\omega_{mn}^2 = \frac{g\varepsilon_{mn}}{\xi_0 \eta_0} \frac{\{I'_m(\varepsilon_{mn}\xi_0/\eta_0)K'_m(\varepsilon_{mn}\xi_1/\eta_0) - I'_m(\varepsilon_{mn}\xi_1/\eta_0)K'_m(\varepsilon_{mn}\xi_0/\eta_0)\}}{[I_m(\varepsilon_{mn}\xi_0/\eta_0)K'_m(\varepsilon_{mn}\xi_1/\eta_0) - I'_m(\varepsilon_{mn}\xi_1/\eta_0)K_m(\varepsilon_{mn}\xi_0/\eta_0)]} \tag{1.256}$$

The dependence of the natural frequency ratio, $\omega_{mn}\eta_0/\sqrt{g}$, on the fluid depth ratio, h/η_0^2, is shown in Figure 1.38 for $m = 0, 1, 2$ and $n = 1, 2, 3$. Note that h represents the fluid depth

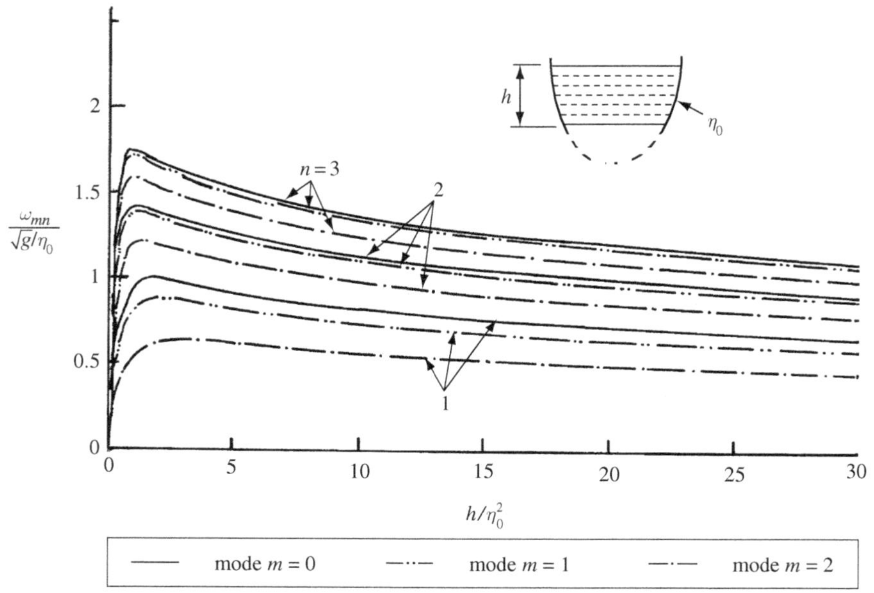

Figure 1.38 Dependence of liquid natural frequency ratio on the fluid depth ratio in sectional truncated paraboloid container for various modes.

and h^* is the distance between the tank bottom and the apex of the coordinate η_0 on the axis of symmetry. Both h and h^* are related to the paraboloid coordinates through the relationships

$$\frac{\xi_0}{\eta_0} = \sqrt{\frac{2(h+h^*)}{\eta_0^2} - 1}, \quad \text{and} \quad \frac{\xi_1}{\eta_0} = \sqrt{\frac{2h^*}{\eta_0^2} - 1} \tag{1.257a, b}$$

Sectional paraboloid annular sector tank

The sectional paraboloid annular sector tank is bounded by walls at $\eta = \eta_0$, $\eta = \eta_1 < \eta_0$, a bottom at $\xi = \xi_1 < \xi_0$, and additional side walls at $\varphi = 0$, and $\varphi = 2\pi\alpha$. The following boundary conditions must be satisfied

$$\frac{\partial \Phi}{\partial \eta} = 0 \quad \text{at the walls} \quad \eta = \eta_0 \quad \text{and} \quad \eta = \eta_1 \tag{1.258a}$$

$$\frac{\partial \Phi}{\partial \xi} = 0 \quad \text{at the bottom } \xi = \xi_1 \tag{1.258b}$$

$$\frac{\partial \Phi}{\partial \varphi} = 0 \quad \text{at the sector walls} \quad \varphi = 0 \quad \text{and} \quad \varphi = 2\pi\alpha \tag{1.258c}$$

The solution of Laplace's equation that satisfies the boundary conditions (1.258) is

$$\Phi(\xi,\eta,\varphi,t)=\sum_{m=0}^{\infty}\sum_{n=1}^{\infty}A_{mn}\left[I_{m/2\alpha}\left(\frac{\bar{\varepsilon}_{mn}\xi}{\eta_0}\right)K'_{m/2\alpha}\left(\frac{\bar{\varepsilon}_{mn}\xi_1}{\eta_0}\right)-I'_{m/2\alpha}\left(\frac{\bar{\varepsilon}_{mn}\xi_1}{\eta_0}\right)K_{m/2\alpha}\left(\frac{\bar{\varepsilon}_{mn}\xi}{\eta_0}\right)\right]$$
$$\times\left\{J_{m/2\alpha}\left(\frac{\bar{\varepsilon}_{mn}\eta}{\eta_0}\right)Y'_{m/2\alpha}(\bar{\varepsilon}_{mn})-J'_{m/2\alpha}(\bar{\varepsilon}_{mn})Y_{m/2\alpha}\left(\frac{\bar{\varepsilon}_{mn}\eta}{\eta_0}\right)\right\}\cos\left(\frac{m}{2\alpha}\varphi\right)e^{i\omega t} \tag{1.259}$$

where $\bar{\varepsilon}_{mn}$ are the roots of the equation

$$J'_{m/2\alpha}(\bar{\varepsilon}_{mn})Y'_{m/2\alpha}\left(\bar{\varepsilon}_{mn}\frac{\eta_1}{\eta_0}\right)-J'_{m/2\alpha}\left(\bar{\varepsilon}_{mn}\frac{\eta_1}{\eta_0}\right)Y'_{m/2\alpha}(\bar{\varepsilon}_{mn})=0 \tag{1.260}$$

Substituting equation (1.259) into the combined free-surface condition (1.250c) gives the natural frequencies of the free surface

$$\omega_{mn}^2=\frac{g\bar{\varepsilon}_{mn}}{\xi_0\eta_0}\frac{\{I'_{m/2\alpha}(\bar{\varepsilon}_{mn}\xi_0/\eta_0)K'_{m/2\alpha}(\bar{\varepsilon}_{mn}\xi_1/\eta_0)-I'_{m/2\alpha}(\bar{\varepsilon}_{mn}\xi_1/\eta_0)K'_{m/2\alpha}(\bar{\varepsilon}_{mn}\xi_0/\eta_0)\}}{[I_{m/2\alpha}(\bar{\varepsilon}_{mn}\xi_0/\eta_0)K'_{m/2\alpha}(\bar{\varepsilon}_{mn}\xi_1/\eta_0)-I'_{m/2\alpha}(\bar{\varepsilon}_{mn}\xi_1/\eta_0)K_{m/2\alpha}(\bar{\varepsilon}_{mn}\xi_0/\eta_0)]} \tag{1.261}$$

The coordinate ξ_1 may be expressed in terms of the bottom distance from the apex as given by relation (1.257b).

Annular paraboloid sector tank

This configuration does not possess a bottom at $\xi=\xi_1$ and the corresponding natural frequency is given by the following equation

$$\omega_{mn}^2=\frac{g\bar{\varepsilon}_{mn}}{\xi_0\eta_0}\frac{I'_{m/2\alpha}(\bar{\varepsilon}_{mn}\xi_0/\eta_0)}{I_{m/2\alpha}(\bar{\varepsilon}_{mn}\xi_0/\eta_0)} \tag{1.262}$$

Paraboloid sector tank

In the absence of the annular wall at $\eta=\eta_1$ and bottom at $\xi=\xi_1$ the velocity potential function takes the form

$$\Phi(\xi,\eta,\varphi,t)=\sum_{m=0}^{\infty}\sum_{n=1}^{\infty}A_{mn}I_{m/2\alpha}(\varepsilon_{mn}\xi/\eta_0)J_{m/2\alpha}(\varepsilon_{mn}\eta/\eta_0)\cos\left(\frac{m}{2\alpha}\varphi\right)e^{i\omega t} \tag{1.263}$$

Substituting equation (1.263) into the combined free-surface condition (1.250c) gives the natural frequencies of the free surface in the form

$$\omega_{mn}^2=\frac{g\varepsilon_{mn}}{\xi_0\eta_0}\frac{I'_{m/2\alpha}(\varepsilon_{mn}\xi_0/\eta_0)}{I_{m/2\alpha}(\varepsilon_{mn}\xi_0/\eta_0)} \tag{1.264}$$

where ε_{mn} are the roots of $J'_{m/2\alpha}(\varepsilon)=0$. Figures 1.39 and 1.40 show the dependence of the natural frequency ratio, $\omega_{mn}\eta_0/\sqrt{g}$, on the fluid depth ratio, h/η_0^2, for $m=0,1,2$, $n=1,2,3$, and two sectored tanks $\alpha=1/4$ and $1/8$, respectively. It is seen that in all cases, the natural frequencies decrease with the fluid depth except for very small fluid depth ratio $h/\eta_0^2<1$ where they experience a steep rise then gradually decrease.

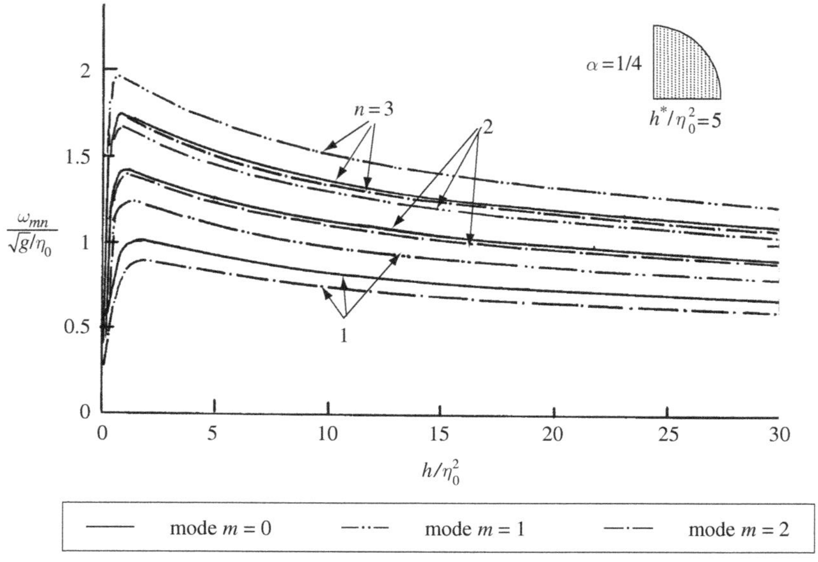

Figure 1.39 Dependence of liquid natural frequency ratio on the fluid depth ratio in 90°-sector paraboloid container for various modes. (Bauer, 1984d, 1999)

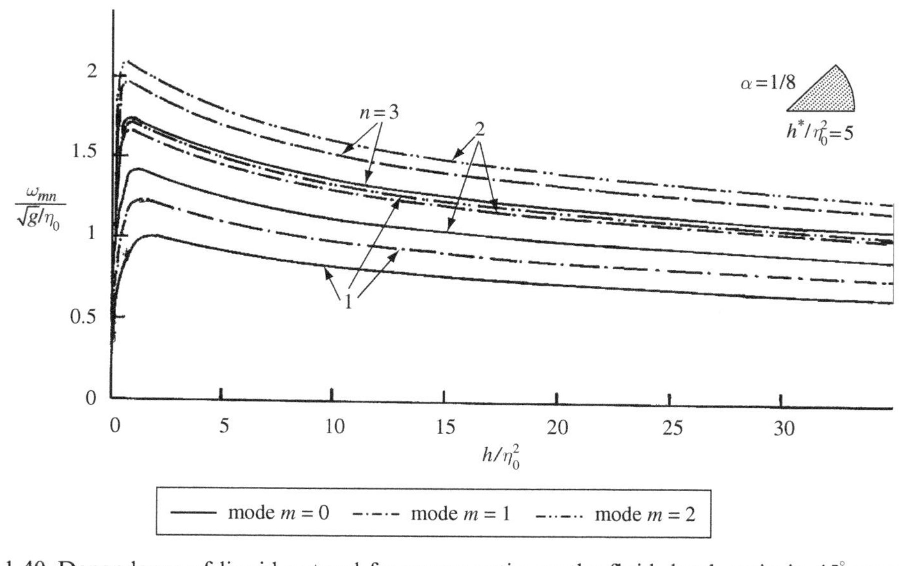

Figure 1.40 Dependence of liquid natural frequency ratio on the fluid depth ratio in 45°-sector paraboloid container for various modes. (Bauer, 1984d, 1999)

1.7 Closing remarks

The fluid field equations of liquid free surface in a partially filled container have been developed using two approaches: the Euler equations of motions together with associated boundary conditions, and variational formulation. The boundary value problem has been developed for various container geometries and the corresponding normal modes and natural frequencies were obtained in closed forms, numerically and experimentally. The natural frequencies appear as a parameter in the free-surface boundary condition rather than Laplace's equation of continuity. Closed form solutions were obtained for upright circular cylindrical and rectangular containers. The surface tension was included and its effect raised the natural frequencies of the free-surface. For containers whose walls are not parallel to the longitudinal vertical axis, such as horizontal cylinders and spherical containers, the natural frequencies were determined using conformal mapping or transformation to bipolar or toroidal coordinates. In both cases the natural frequencies were obtained in terms of integrals equations, which were discretized for numerical solutions. Other container geometries, such as convergent or divergent conical containers, were treated through simple one-mode analysis, a variational approach or spherical coordinate transformation.

Modal analysis of special cases such as moonpools and liquid containers with corrugated bottoms were considered by Molin (2001) and Kukner and Baykal (1999). The case of skewed tanks was studied by Abzug (1996). For axisymmetric containers the eigenvalue problem of the free surface for arbitrary shape containers was determined by Siekmann and Schilling (1974a,b). The influence of liquid sloshing on the mass matrix was examined by Steven (1976).

The nonconformal transformation technique together with a variational modal approach for treating circular conical tanks has been developed by Gavrilyuk, Lukovskii, and Timokha (2001). Two approximate schemes were developed. The first one employs a projective scheme within harmonic polynomials as the basis. This method was implemented by Lukovskii *et al.* (1984). Lukovskii and Timokha (2000b) used harmonic polynomials and combined them with a nonconformal mapping technique. However, the solutions did not satisfy zero Neumann condition on the walls over mean free surface. Alternatively, the solid harmonics satisfy zero Neumann condition everywhere on the conical surface. The second scheme was adopted for spectral problems in the conical domain and was employed by Dokuchaev (1964) and Bauer (1982i). The influence of a corrugated bottom and inserted pipes on liquid-free-surface behavior is analyzed based on two-dimensional flow by Kurner and Baykal (1999) and Drake (1999), respectively.

Numerical algorithms such as finite difference and finite element methods have been developed by Ehrlich, *et al.* (1961), Lomen (1965b), and Lewin, *et al.* (1997). The analysis and numerical algorithms did not consider the influence of liquid viscosity on the liquid natural frequencies. This will be considered in Chapter 3. However, the results presented in this chapter serve as the basis for the design of liquid containers subjected to external excitations whose frequencies may be close to the lowest natural frequency of the free surface.

Appendix

A1.1 Curvilinear coordinates

For orthogonal curvilinear coordinates ξ_1, ξ_2, ξ_3, with unit vectors e_1, e_2, e_3, line element $ds^2 = (dx)^2 + (dy)^2 + (dz)^2 = \sum_i h_i^2 (d\xi_i)^2$, the scale factors h_i for the coordinates ξ_i are defined by the following expressions

$$h_i^2 = \left[\left(\frac{\partial x}{\partial \xi_i}\right)^2 + \left(\frac{\partial y}{\partial \xi_i}\right)^2 + \left(\frac{\partial z}{\partial \xi_i}\right)^2\right] = \left[\left(\frac{\partial \xi_i}{\partial x}\right)^2 + \left(\frac{\partial \xi_i}{\partial y}\right)^2 + \left(\frac{\partial \xi_i}{\partial z}\right)^2\right]^{-1}$$

In general h_i varies from point to point in space. Note that $h_i\, d\xi_i$ represents the length of displacement corresponding to $d\xi_i$ and the rate of displacement along ξ_i due to a displacement along the x-axis will be $h_i(\partial\xi_i/\partial x)$ which is the direction cosine.

The differential operators of common use in liquid sloshing dynamics are

$$\operatorname{grad}\Phi = \nabla\Phi = \sum_i \mathbf{e}_i \frac{1}{h_i}\frac{\partial\Phi}{\partial\xi_i}$$

$$\operatorname{div}\operatorname{grad}\Phi = \nabla^2\Phi = \frac{1}{h_1h_2h_3}\sum_i \frac{\partial}{\partial\xi_i}\left[\frac{h_1h_2h_3}{h_i^2}\frac{\partial\Phi}{\partial\xi_i}\right]$$

$$\operatorname{curl}\boldsymbol{A} = \nabla\times\boldsymbol{A} = \frac{1}{h_1h_2h_3}\sum_{\ell,m,n} h_\ell e_\ell\left[\frac{\partial}{\partial\xi_m}(h_nA_n) - \frac{\partial}{\partial\xi_n}(h_mA_m)\right],$$

$$\ell, m, n = 1,2,3, \quad \text{or} \quad 2,3,1, \quad \text{or} \quad 3,1,2$$

$$\operatorname{div}\boldsymbol{A} = \nabla\cdot\boldsymbol{A} = \frac{1}{h_1h_2h_3}\sum_{\ell,m,n} \frac{\partial}{\partial\xi_n}\left(h_1h_2h_3\frac{A_n}{h_n}\right)$$

and the volume element is $dV = h_1h_2h_3\, d\xi_1\, d\xi_2\, d\xi_3$

A1.2 Cylindrical coordinates

The coordinates $\xi_1 = r,\ \xi_2 = \theta,\ \xi_3 = z$, and we have $h_1 = 1$, $h_2 = r$, $h_3 = 1$ $dV, = r\, dr\, d\theta\, dz$

$$\operatorname{grad}\Phi = \nabla\Phi = \mathbf{e}_r\frac{\partial\Phi}{\partial r} + \mathbf{e}_\theta\frac{1}{r}\frac{\partial\Phi}{\partial\theta} + \mathbf{e}_z\frac{\partial\Phi}{\partial z}$$

$$\operatorname{div}\ \operatorname{grad}\Phi = \nabla^2\Phi = \frac{1}{r}\frac{\partial}{\partial r}\left(r\frac{\partial\Phi}{\partial r}\right) + \frac{1}{r^2}\frac{\partial^2\Phi}{\partial\theta^2} + \frac{\partial^2\Phi}{\partial z^2}$$

$$\operatorname{curl}\boldsymbol{A} = \nabla\times\boldsymbol{A}$$

$$= \mathbf{e}_r\left(\frac{1}{r}\frac{\partial A_z}{\partial\theta} - \frac{\partial A_\theta}{\partial z}\right) + \mathbf{e}_\theta\left(\frac{\partial A_r}{\partial z} - \frac{\partial A_z}{\partial r}\right) + \mathbf{e}_z\left(\frac{1}{r}\frac{\partial(rA_\theta)}{\partial r} - \frac{1}{r}\frac{\partial A_r}{\partial\theta}\right)$$

$$\operatorname{div}\boldsymbol{A} = \nabla\cdot\boldsymbol{A} = \frac{1}{r}\frac{\partial(rA_r)}{\partial r} + \frac{1}{r}\frac{\partial A_\theta}{\partial\theta} + \frac{\partial A_z}{\partial z}$$

A1.3 Spherical coordinates

The coordinates $\xi_1 = r, \xi_2 = \theta, \xi_3 = \phi$, and we have $h_1 = 1$, $h_2 = r$, $h_3 = r\sin\theta$, $dV = r^2\sin\theta\, dr\, d\theta\, d\phi$

$$\operatorname{grad}\Phi = \nabla\Phi = \mathbf{e}_r\frac{\partial\Phi}{\partial r} + e_\theta\frac{1}{r}\frac{\partial\Phi}{\partial\theta} + \mathbf{e}_z\frac{1}{r\sin\theta}\frac{\partial\Phi}{\partial\phi}$$

$$\text{div grad}\,\Phi = \nabla^2\Phi = \frac{1}{r^2}\frac{\partial}{\partial r}\left(r^2\frac{\partial\Phi}{\partial r}\right) + \frac{1}{r^2\sin\theta}\frac{\partial}{\partial\theta}\left(\sin\theta\frac{\partial\Phi}{\partial\theta}\right) + \frac{1}{r^2\sin^2\theta}\frac{\partial^2\Phi}{\partial\phi^2}$$

$$\begin{aligned}\text{curl}\,\boldsymbol{A} &= \nabla\times\boldsymbol{A}\\ &= \frac{\boldsymbol{e}_r}{r\sin\theta}\left(\frac{\partial}{\partial\theta}(A_\phi\sin\theta) - \frac{\partial A_\theta}{\partial\phi}\right) + \frac{\boldsymbol{e}_\theta}{r}\left(\frac{1}{\sin\theta}\frac{\partial A_r}{\partial\phi} - \frac{\partial(rA_\phi)}{\partial r}\right)\\ &\quad + \frac{\boldsymbol{e}_\phi}{r}\left(\frac{\partial(rA_\theta)}{\partial r} - \frac{\partial A_r}{\partial\theta}\right)\end{aligned}$$

$$\text{div}\,\boldsymbol{A} = \nabla\cdot\boldsymbol{A} = \frac{1}{r^2}\frac{\partial(r^2A_r)}{\partial r} + \frac{1}{r\sin\theta}\frac{\partial(\sin\theta A_\theta)}{\partial\theta} + \frac{1}{r\sin\theta}\frac{\partial A_\phi}{\partial\phi}$$

A1.4 Prolate spheroidal coordinates

With reference to Figure 1.14, two sets of coordinate surfaces are obtained by revolving the confocal ellipses and hyperbolas about the z-axis and the coordinate ϕ specifies the rotation. The figure includes the special case of a spherical tank when the two semi-axes of an ellipse are equal. The relationships between these coordinates and Cartesian coordinates are

$$x = c\sinh\xi\sin\eta\cos\phi,\quad y = c\sinh\xi\sin\eta\sin\phi,\ \text{and}\ z = c\cosh\xi\cos\eta$$

where $\xi \geq 0$, $0 \leq \eta \leq \pi$, $0 \leq \phi < 2\pi$, and c is a scale factor between Cartesian and spheroidal coordinates. The transformation parameters h_i are given by the following expressions:

$$h_1^2 = h_2^2 = c^2(\sinh^2\xi + \sin^2\eta),\ \text{and}\ h_3^2 = c^2\sinh^2\xi\sin^2\eta$$

It is not difficult to show that the Laplace's equation of continuity takes the form

$$\begin{aligned}\nabla^2\Phi &= \frac{1}{c^2(\sinh^2\xi+\sin^2\eta)\sinh\xi}\frac{\partial}{\partial\xi}\left(\sinh\xi\frac{\partial\Phi}{\partial\xi}\right)\\ &\quad + \frac{1}{c^2(\sinh^2\xi+\sin^2\eta)\sin\eta}\frac{\partial}{\partial\eta}\left(\sin\eta\frac{\partial\Phi}{\partial\eta}\right)\\ &\quad + \frac{1}{c^2\sinh^2\xi\sin^2\eta}\frac{\partial^2\Phi}{\partial\phi^2} = 0\end{aligned}$$

Introducing the new variables $\bar{\xi} = \cosh\xi$, $\bar{\eta} = \cos\eta$, $a/2 = c$, $1 \leq \bar{\xi}$, $-1 \leq \bar{\eta} \leq 1$, which are related to the x, y, z coordinates as

$$\begin{aligned}x &= \frac{a}{2}\sqrt{(\bar{\xi}^2-1)(1-\bar{\eta}^2)}\cos\phi,\\ y &= \frac{a}{2}\sqrt{(\bar{\xi}^2-1)(1-\bar{\eta}^2)}\sin\phi,\ \text{and}\\ z &= \frac{a}{2}\bar{\xi}\bar{\eta}\end{aligned}$$

Laplace's equation can be written in terms of the new variables

$$\nabla^2\Phi = \frac{4}{a^2\left(\bar{\xi}^2-\bar{\eta}^2\right)}\left[\frac{\partial}{\partial\bar{\xi}}\left(\bar{\xi}^2-1\right)\frac{\partial\Phi}{\partial\bar{\xi}}+\frac{\partial}{\partial\bar{\eta}}\left(1-\bar{\eta}^2\right)\frac{\partial\Phi}{\partial\bar{\eta}}+\frac{\bar{\xi}^2-\bar{\eta}^2}{(\bar{\xi}^2-1)(1-\bar{\eta}^2)}\frac{\partial\Phi}{\partial\phi^2}\right]$$

The surface $\bar{\xi}=$ const is a prolate spheroid with interfocal distance a, major axis is $a\bar{\xi}/2$, and minor axis is $\left(a\sqrt{\bar{\xi}^2}-a\right)/2$. The surfaces, $\bar{\eta}=$ const, are two sheets of a hyperboloid of revolution with foci at $z=\pm a/2$, asymptotic to the cone which has its generating line at an angle $\theta=\cos^{-1}\bar{\eta}$. The surface $\phi=$ const is a plane through the a-axis at an angle ϕ to the x, z plane.

A1.5 Oblate spheroidal coordinates

With reference to Figure 1.14, two sets of coordinate surfaces are obtained by revolving the confocal ellipses and hyperbolas about the x-axis and the coordinate ϕ specifies the rotation. The relationships between these coordinates and Cartesian coordinates are

$$\begin{aligned} x &= c\cosh\xi\cos\eta\cos\phi, \\ y &= c\cosh\xi\cos\eta\sin\phi, \text{ and} \\ z &= c\sinh\xi\sin\eta \end{aligned}$$

such that $\xi\geq 0$, $-\pi/2\leq\eta\leq\pi/2$, $0\leq\phi<2\pi$.

The transformation parameters h_i are given by the following expressions

$$\begin{aligned} h_1^2 &= h_2^2 = c^2\left(\sinh^2\xi+\sin^2\eta\right), \text{ and} \\ h_3^2 &= c^2\cosh^2\xi\cos^2\eta \end{aligned}$$

It is not difficult to show that the Laplace's equation of continuity takes the form

$$\begin{aligned} \nabla^2\Phi = {} & \frac{1}{c^2\left(\sinh^2\xi+\sin^2\eta\right)\cosh\xi}\frac{\partial}{\partial\xi}\left(\cosh\xi\frac{\partial\Phi}{\partial\xi}\right) \\ & + \frac{1}{c^2\left(\sinh^2\xi+\sin^2\eta\right)\cos\eta}\frac{\partial}{\partial\eta}\left(\cos\eta\frac{\partial\Phi}{\partial\eta}\right) \\ & + \frac{1}{c^2\cosh^2\xi\cos^2\eta}\frac{\partial^2\Phi}{\partial\phi^2} \\ = {} & 0 \end{aligned}$$

A1.6 Bipolar coordinates (ξ,η,z)

Bipolar coordinates are useful for the analysis of liquid sloshing in a horizontal circular cylinder (or two parallel cylinders) of finite radius whose axis is parallel to the horizontal plane. By describing the fluid field equations in terms of bipolar coordinates the Laplace equation can be separated. With reference to Figure 1.41, points A and B are located on the x-axis at $+a$ and $-a$, respectively. The coordinates of point P with respect to A and B are

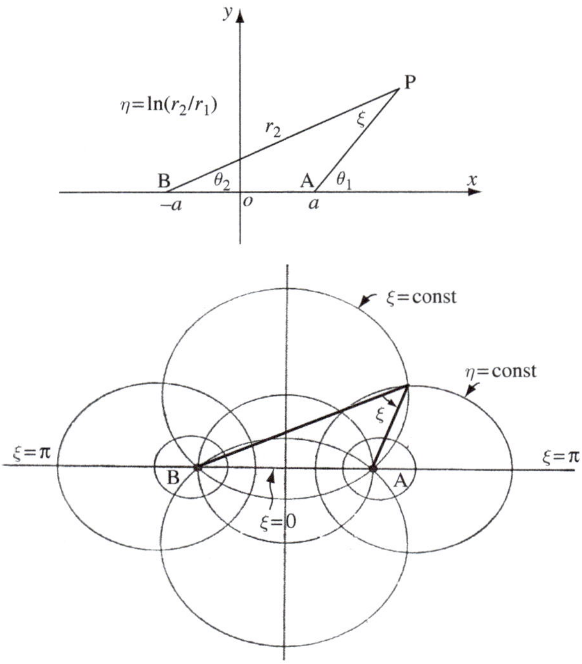

Figure 1.41 Family of circles passing through poles A and B and their orthogonal circles. A bipolar coordinate system ξ and η is established if the figure is translated at right angles to the plane of the paper (horizontal circular cylinder), a toroidal coordinate system is generated if the figure is rotated about the vertical axis (spherical cylinders).

(r_1, θ_1) and (r_2, θ_2) respectively, where r_1 and r_2 are the bipolar coordinates of P. If P describes a circle passing through A and B, the angle $\xi = \theta_1 - \theta_2$ is constant. The system of circles described by point P is orthogonal for all circles possessing constant radii ratio, that is, $r_2/r_1 = \text{const}$. Setting $\xi = \theta_1 - \theta_2$ and $\eta = \ln(r_2/r_1)$, one can write

$$\eta - i\xi = \ln(r_2\, e^{i\theta_2}) - \ln(r_1\, e^{i\theta_1}) \tag{A1.1}$$

and

$$e^{\eta - i\xi} = \frac{z+a}{z-a} \tag{A1.2}$$

Let

$$\zeta = \xi + i\eta, \text{ and } e^{-i\zeta} = \frac{z+a}{z-a} \tag{A1.3}$$

The coordinates ξ, η, are called coaxial. From (A3) one writes

$$\frac{z}{a} = \frac{e^{-i\zeta}+1}{e^{-i\zeta}-1} = i\ \cot(\zeta/2) \quad \text{or} \quad z = ia\ \cot(\zeta/2) \tag{A1.4}$$

The curve $\xi = \text{const}$ is a circle of center $(0,\ a\cot\xi)$ and its radius is $a\,\text{cosec}\,\xi$. The curve $\eta = \text{const}$ is a circle whose center is the point $(a\coth\eta,\ 0)$ and whose radius is $a\,\text{cosech}\,\eta$. Since $\bar{z} = -ia\cot(\zeta/2)$, one writes

$$\frac{x}{a} = \frac{i}{2}\left(\cot(\zeta/2) - \cot(\bar{\zeta}/2)\right),\ \frac{y}{a} = \frac{i}{2}\left(\cot(\zeta/2) + \cot(\bar{\zeta}/2)\right) \tag{A1.5}$$

Since $2\sin(\zeta/2)\sin(\bar{\zeta}/2) = \cos i\eta - \cos\xi = \cosh\eta - \cos\xi$, one writes

$$\frac{x}{a} = \frac{\sinh\eta}{\cosh\eta \pm \cos\xi},\ \frac{y}{a} = \frac{\sin\xi}{\cosh\eta \pm \cos\xi} \tag{A1.6}$$

On the real axis, $\xi = 0$, except for points between A and B for which $\xi = \pi$. Note that $\xi < \pi$ when $y > 0$. If $\xi = \xi_0$ on the arc of a circle through AB for which $y > 0$, then $\xi = \xi_0 + \pi$ on the arc of the same circle for which $y < 0$. At infinity, points PA and PB are parallel and nearly equal. Thus $\eta \to 0$, when $P \to \infty$, and $\xi = 0$ or $\xi \to 2\pi$ according to whether $y > 0$ or $y < 0$. At A and B, η is infinity.

It is not difficult to show that the two-dimensional bipolar coordinates ξ and η are related to the Cartesian coordinates through the complex representation,

$$x + iy = a\tanh[(\eta + i\xi)/2] \tag{A1.7}$$

where the two relations of hyperbolic functions are used

$$\sinh\xi \pm \sinh\eta = 2\sinh\frac{\xi \pm \eta}{2}\cosh\frac{\xi \mp \eta}{2}$$

$$\cosh\xi + \cosh\eta = 2\cosh\frac{\xi + \eta}{2}\cosh\frac{\xi - \eta}{2}$$

Also one can write

$$h_\xi^2 = h_\eta^2 = \frac{a^2}{(\cosh\eta + \cos\xi)^2},\ h_z^2 = 1 \tag{A1.8}$$

$$\eta = \tanh^{-1}\left[\frac{2ax}{a^2 + x^2 + y^2}\right],\ \xi = \tan^{-1}\left[\frac{2ay}{a^2 - x^2 - y^2}\right] \tag{A1.9}$$

$$r = a\sqrt{\frac{\sinh^2\eta + \sin^2\xi}{(\cosh\eta + \cos\xi)^2}} \tag{A1.10}$$

Where ξ represents the angular coordinate, $0 \le \xi < 2\pi$, and its value is governed by the depth of the fluid, and η represents the radial coordinate, where its range 0 to ∞ covers the positive half of the x, y plane, and its negative range corresponds to the negative range of x.

The portion of the x-axis from $x = -a$ to $x = a$ corresponds to the line $\xi = 0$, $-\infty < \eta < \infty$. The rest of the x-axis corresponds to $\xi = \pi$. Note that $a < x < \infty$ corresponds to $\infty > \eta > 0$, and $-\infty < x < -a$ to $0 > \eta > -\infty$

$$x^2 + (y - a\cot\xi)^2 = a^2\mathrm{cosec}^2\xi,$$

$$y^2 + (x - a\coth\eta)^2 = a^2\mathrm{cosech}^2\eta, \tag{A1.11}$$

$$z = z$$

Every ξ circle passes through the points $(\pm a, 0)$ as shown in Figure 1.32.

$$\nabla^2\Phi = \frac{(\cosh\xi + \cos\eta)^2}{a^2}\left[\frac{\partial^2\Phi}{\partial\xi^2} + \frac{\partial^2\Phi}{\partial\eta^2}\right] + \frac{\partial^2\Phi}{\partial z^2} \tag{A1.12}$$

A1.7 Toroidal coordinates (η, ξ, φ)

This coordinate system is obtained by rotating the bipolar axes about the perpendicular bisector of the line between the two poles and these result in the following transformation

$$x = a\frac{\sinh\eta\cos\varphi}{\cosh\eta - \cos\xi},$$

$$y = a\frac{\sinh\eta\sin\varphi}{\cosh\eta - \cos\xi}, \tag{A1.13}$$

$$z = a\frac{\sin\xi}{\cosh\eta - \cos\xi}$$

$$h_\xi^2 = h_\eta^2 = \frac{a^2}{(\cosh\eta - \cos\xi)^2}, \tag{A1.14}$$

$$h_3^2 = \frac{a^2\sinh^2\eta}{(\cosh\eta - \cos\xi)^2}$$

$$\nabla^2\Phi = \frac{(\cosh\eta - \cos\xi)^3}{a^2}\frac{\partial}{\partial\xi}\left[\frac{1}{\cosh\eta - \cos\xi}\frac{\partial\Phi}{\partial\xi}\right]$$

$$+ \frac{(\cosh\eta - \cos\xi)^3}{a^2\sinh\eta}\frac{\partial}{\partial\eta}\left[\left(\frac{\sinh\eta}{\cosh\eta - \cos\xi}\right)\frac{\partial\Phi}{\partial\eta}\right] + \frac{(\cosh\eta - \cos\xi)^2}{a^2\sinh^2\eta}\frac{\partial^2\Phi}{\partial\varphi^2} \tag{A1.15}$$

2
Linear forced sloshing

2.1 Introduction

The previous chapter treated the free oscillations of liquid free surface in different container geometries. The fluid field equations were developed and the natural frequencies were determined from the free-surface boundary conditions. The knowledge of liquid-free-surface natural frequencies is important in the design of liquid containers subjected to different types of excitations. In the design process, it is important to keep the liquid natural frequencies away from all normal and nonlinear resonance conditions. The excitation can be impulsive, sinusoidal, periodic, or random. Its orientation with respect to the tank can be lateral, parametric, pitching/ yaw, roll, or a combination. Under forced excitation, it is important to determine the hydrodynamic loads acting on the container, and their phase with respect to the excitation. The hydrodynamic forces are estimated by integrating the pressure distribution over the wetted area. One should also determine the free-surface wave height, which affects the location of the center of mass. In the neighborhood of resonance, the free surface experiences different types of nonlinear behavior and this will be considered in Chapters 4, 6, and 7.

Under impulsive excitation, liquid sloshing dynamics was studied by Morris (1938), Werner and Sundquist (1949), Jacobsen (1949), Jacobsen and Ayre (1951), and Hoskins and Jacobsen (1957). Bauer (1965a) considered the response of liquid propellant to a single pulse excitation. Transient and steady state response to sinusoidal excitation of a liquid free surface was studied by Sogabe and Shibata (1974a,b).

The dynamics of cylindrical tanks with arbitrary cross-sections undergoing translational and pitching motions was considered by Case and Parkinson (1956a) and Abramson and Ransleben (1961e). Bauer and Eidel (1999c) studied the liquid sloshing in different configurations of a circular cylindrical container. Liquid sloshing dynamics in cylindrical containers describing Lissajous oscillations was treated by Bauer (1958d).

The water pressure distribution on dams subjected to earthquake motion was determined by Canard (1956), Bustamante and Flores (1966), and Chopra (1967). Rundgren (1958) determined analytically and experimentally the water wave forces. Momoi (1965) and Troesch (1967, 1970) studied the fluid motion in trapezoidal and isoperimetric containers. Case and Parkinson (1956b, 1957) considered the liquid viscosity or "boundary layers" effects. Chalhoub and Kelly (1990) conducted a shake table test of a cylindrical water container for analysis of base-isolated structures.

A formal solution of liquid motion in tanks of arbitrary shape undergoing small horizontal oscillations and pitching was obtained by Trembath (1957) and W. H. Chu (1971). The solution was given in terms of certain integrals, which can be evaluated either algebraically or numerically depending on the problem. A general treatment of an arbitrary tank geometry undergoing forced motion was given in terms of the velocity potential function by Brooks (1959). Chu (1960a,b), Rattayya (1965), and Figarov (1971) obtained some results for an elliptic cylindrical tank. A Lagrangian formulation was developed by Budiansky (1960) to examine the sloshing dynamics under horizontal excitation of spherical and circular horizontal tanks. The same method was also adopted by Debonginie (1986) for liquid sloshing and related problems. An important task of forced sloshing is to determine the hydrodynamic forces in different engineering applications (see, e.g., Lorell, 1951, Epperson and Brown, 1957, Epperson, *et al.*, 1961, Housner, 1963b, Hutton and Yeh, 1964, Hutton and Bhuta, 1965, Roberts, *et al.*, 1966, Bhuta, *et al.*, 1968, Feng 1973, Bass, 1975, Mathiessen, 1976, and Park, *et al.*, 1995).

This chapter deals with the liquid response to lateral and pitching excitations for typical tank configurations such as upright cylindrical and rectangular containers. The analysis is extended to cover annular and sectored cylindrical tanks, spherical containers, and prolate and oblate spheroidal tanks. The analytical results are supported by reported experimental measurements. The lateral excitation of magnetic fluid is considered and the analysis is similar to the analysis of regular fluid except that the unsteady Bernoulli's equation includes the magnetic field effect.

2.2 Upright cylindrical containers

The problem of liquid forced oscillations in an upright circular cylindrical tank of uniform depth experiencing horizontal or pitching motion was analytically and experimentally studied by Wedemeyer and Reese (1953), Brown (1954), Reese (1955), Reese and Sewal (1955), Schmitt (1956,1957), Chobotov and Fowler (1957), Sewal (1957), Bauer (1958b,c,g,i, 1959a, 1966d), Bauer and Reinfurth (1959), Eide (1964), Huleux (1964), Takahara, *et al.* (1995), and Takahara and Kimura (1997). These studies essentially estimated the liquid effective mass moment of inertia, the forces and moments acting on the tank, and the free surface wave height.

Different approaches were adopted. For example, Kachigan (1955), Schmitt (1956), and Armstrong and Kachigan (1961) employed the Laplace transform to determine the liquid-free-surface response to arbitrary excitation in translation and pitching. Miles (1956) considered a more general formulation taking into account the influence of varying liquid depth in a cylindrical container of arbitrary cross-section. Miles also included the liquid viscosity in his formulation. Experimental investigations were conducted by Abramson and Ransleben (1961a,b,d,e), Mikishev and Dorozhkin (1961), and Barron and Chang (1989). Chu (1960a) studied the effect of ellipticity of the tank cross-section. Abramson (1961c,d,1963a,b), Abramson and Ransleben (1961a,b,d), and Abramson and Garza (1965), compared the analytical and measured pressure distribution on the tank wetted area, net horizontal force, and moment for steady state horizontal and pitching excitation of an upright circular cylindrical tank. Kana (1964a) and Guyett (1967), measured the hydrodynamic forces acting on a circular cylindrical tank experiencing harmonic motion.

In this section, the liquid hydrodynamic forces and moments acting on an upright circular tank are estimated under translational and pitching excitations. Under roll excitation, the fluid does not participate in the tank motion since it is assumed to be inviscid. If the fluid viscosity is considered, then one has to solve the Navier–Stokes equations to estimate the thickness of the fluid participating in the tank motion. This case will be considered in Chapter 3.

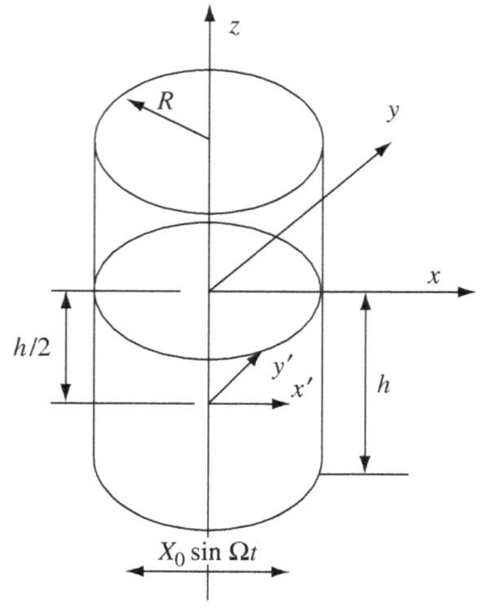

Figure 2.1 Cylindrical tank under sinusoidal lateral excitation.

2.2.1 Lateral excitation

Consider an upright circular container under sinusoidal excitation along the x-axis,

$$X(t) = X_0 \sin \Omega t \tag{2.1}$$

where X_0 and Ω are the excitation amplitude and frequency, respectively, see Figure 2.1. Under the assumption that the amplitudes of excitation and fluid response are small, the linearized fluid field equations take the form

$$\nabla^2 \tilde{\Phi} = 0 \qquad \text{inside the fluid domain} \tag{2.2a}$$

$$\left.\frac{\partial \tilde{\Phi}}{\partial r}\right|_{r=R} = 0, \quad \left.\frac{\partial \tilde{\Phi}}{\partial z}\right|_{z=-h} = 0 \tag{2.2b, c}$$

$$g\eta - \frac{\partial \tilde{\Phi}}{\partial t} + \ddot{X} r \cos\theta = 0, \quad \text{at } z = \eta(r, \theta, t) \tag{2.2d}$$

$$-\frac{\partial \tilde{\Phi}}{\partial z} = \frac{\partial \eta}{\partial t} \quad \text{at } z = \eta(r, \theta, t) \tag{2.2e}$$

The dynamic and kinematic free surface conditions (2.2d, e) can be combined in one equation

$$g\frac{\partial \tilde{\Phi}}{\partial z} + \frac{\partial^2 \tilde{\Phi}}{\partial t^2} = \dddot{X} r \cos\theta \tag{2.3}$$

A typical solution of equation (2.2a) subject to conditions (2.2b, c) can be written in the form

$$\tilde{\Phi} = \sum_{n=1}^{\infty} [A_{1n}(t)\cos\theta + B_{1n}(t)\sin\theta] J_1(k_n r) \frac{\cosh[k_{1n}(z+h)]}{\cosh k_{1n} h} \tag{2.4}$$

The generalized coordinates, A_{1n} and B_{1n}, depend on time. These functions are determined by satisfying the free-surface condition (2.3) and expressing the coordinate r in a Fourier–Bessel series

$$r = \sum_{n=1}^{\infty} F_n J_1(k_{1n} r) \tag{2.5}$$

where $F_n = \dfrac{2R}{(k_{1n}^2 R^2 - 1) J_1(k_{1n} R)}$.

Introducing equations (2.4) and (2.5) into the free-surface condition (2.2d), gives

$$\sum_{n=1}^{\infty} \left[\ddot{A}_{1n}(t) + \omega_{1n}^2 A_{1n}(t) - \frac{\ddot{x} F_n}{\cosh k_{1n} h} \right] J_1(k_n r) \cos\theta + \left[\ddot{B}_{1n}(t) + \omega_{1n}^2 B_{1n}(t) \right] J_1(k_n r) \sin\theta = 0 \tag{2.6}$$

where $\omega_{1n}^2 = \dfrac{g \xi_{1mn}}{R} \tanh(\xi_{1mn} h/R)$.

Equation (2.6) is satisfied if the generalized coordinates, A_{1n} and B_{1n}, satisfy the linear set of differential equations

$$\ddot{A}_{1n}(t) + \omega_{1n}^2 A_{1n}(t) = \frac{\ddot{x} F_n}{\cosh k_{1n} h} \tag{2.7a}$$

$$\ddot{B}_{1n}(t) + \omega_{1n}^2 B_{1n}(t) = 0 \tag{2.7b}$$

The steady-state solutions of these equations are

$$A_{1n}(t) = -\frac{X_0 \Omega^3 F_n}{(\omega_{1n}^2 - \Omega^2) \cosh k_{1n} h} \cos \Omega t, \; B_{1n}(t) = 0 \tag{2.8a, b}$$

Substituting (2.8a,b) in equation (2.4), gives

$$\tilde{\Phi} = -X_0 \Omega \cos\theta \cos \Omega t \sum_{n=1}^{\infty} \left[\frac{2R}{(\xi_{1n}^2 - 1)} \frac{\Omega^2}{(\omega_{1n}^2 - \Omega^2)} \frac{J_1(\xi_{1n} r/R)}{J_1(\xi_{1n})} \frac{\cosh[\xi_{1n}(z+h)/R]}{\cosh(\xi_{1n} h/R)} \right] \tag{2.9}$$

The total potential function is the sum of the fluid perturbed function, $\tilde{\Phi}$, and the tank potential function $\Phi_o = -X_0 r \Omega \cos\theta \cos \Omega t$, that is,

$$\Phi = -X_0 \Omega \cos\theta \cos\Omega t \times \left\{ r - \sum_{n=1}^{\infty} \left[\frac{2R}{(\xi_{1n}^2 - 1)} \frac{\Omega^2}{(\omega_{1n}^2 - \Omega^2)} \frac{J_1(\xi_{1n} r/R)}{J_1(\xi_{1n})} \frac{\cosh[\xi_{1n}(z+h)/R]}{\cosh(\xi_{1n} h/R)} \right] \right\} \tag{2.10}$$

Substituting (2.10) into (2.2d) gives the surface wave height

$$\eta = \frac{X_0 \Omega^2}{g} \cos\theta \cos \Omega t \left[r + \sum_{n=1}^{\infty} \frac{2R}{(\xi_{1n}^2 - 1)} \frac{\Omega^3}{(\omega_{1n}^2 - \Omega^2)} \frac{J_1(\xi_{1n} r/R)}{J_1(\xi_{1n})} \right] \tag{2.11}$$

The maximum wave height occurs at $r = R$, $\theta = 0$, and $\Omega t = \pi/2$, and is given by the following expression

$$\eta_{\max} = \frac{X_0\Omega^2}{g}\left[R + \sum_{n=1}^{\infty}\frac{2R}{(\xi_{1n}^2 - 1)}\frac{\Omega^3}{(\omega_{1n}^2 - \Omega^2)}\right] \tag{2.12}$$

Note that zero wave height (nodes) occurs at $r = 0$, $\theta = \pi/2, 3\pi/2, \ldots$ Furthermore, as the excitation frequency, Ω, approaches any sloshing frequency, ω_{1n}, both $\tilde{\Phi}$ and η become arbitrarily large (resonance). In such a case, these solutions are not valid and one should consider the nonlinear free-surface boundary conditions.

The hydrodynamic pressure at any point due to liquid sloshing (by neglecting the hydrostatic pressure, $\rho g z$) may be determined from the pressure equation

$$\begin{aligned} p &= \rho\frac{\partial\tilde{\Phi}}{\partial t} \\ &= \rho X_0\Omega^2\cos\theta\cos\Omega t \\ &\times\left\{r + \sum_{n=1}^{\infty}\frac{2R}{(\xi_{1n}^2 - 1)}\frac{\Omega^2}{(\omega_{1n}^2 - \Omega^2)}\frac{J_1(\xi_{1n}r/R)}{J_1(\xi_{1n})}\frac{\cosh[\xi_{1n}(z+h)/R]}{\cosh\xi_{1n}h/R}\right\} \end{aligned} \tag{2.13}$$

The maximum pressure occurs on the wall at $r = R$, $\theta = 0$, and $\Omega t = \pi/2$. It is given by the expression

$$\begin{aligned} \frac{p_w}{\rho g R(X_0/R)} &= \frac{\Omega^2 R}{g} \\ &\times\left\{1 + \sum_{n=1}^{\infty}\frac{2}{(\xi_{1n}^2 - 1)}\frac{[\Omega^2 R/\xi_{1n}g\tanh(\xi_{1n}h/R)]}{(1 - \Omega^2 R/\xi_{1n}g\tanh(\xi_{1n}h/R))}\left[\frac{\cosh[\xi_{1n}(z+h)/R]}{\cosh\xi_{1n}h/R}\right]\right\} \end{aligned} \tag{2.13a}$$

Similarly, the pressure distribution on the bottom at $z = -h$, $\theta = 0$, $\Omega t = \pi/2$, is

$$\begin{aligned} \frac{p_b}{\rho g R(X_0/R)} &= \frac{\Omega^2 R}{g} \\ &\times\left\{\frac{r}{R} + \sum_{n=1}^{\infty}\frac{2}{(\xi_{1n}^2 - 1)\cosh(\xi_{1n}h/R)}\frac{[\Omega^2 R/\xi_{1n}g\tanh(\xi_{1n}h/R)]}{(1 - \Omega^2 R/\xi_{1n}g\tanh(\xi_{1n}h/R))}\frac{J_1(\xi_{1n}r/R)}{J_1(\xi_{1n})}\right\} \end{aligned} \tag{2.13b}$$

Figure 2.2 shows the wall pressure distribution for fluid depth ratio $h/R = 2$ and for different values of excitation frequency parameter (Froude number) $\Omega^2 R/g$. It appears approximately that the lower third of the fluid behaves essentially as a rigid mass, while most of the sloshing effects occur near the surface. This observation is the basis for developing mechanical equivalent models that will be considered in Chapter 5. Abramson and Ransleben (1961b) conducted a series of experimental tests to compare the analytical results given by equations (2.13a) and (2.13b) with those measured experimentally, see Figure 2.3. Note that the analytical results do not include any damping effects, which are reflected in the measured results. One may observe some interesting phenomena that take place over the frequency range between the first mode natural frequency and the frequency, at which the net force vanishes, that is, $1.85 \leq \Omega^2 R/g \leq 2.35$. The pressure distributions over the first and second modes indicate that the pressures near the surface are 180° out of phase with those near the bottom

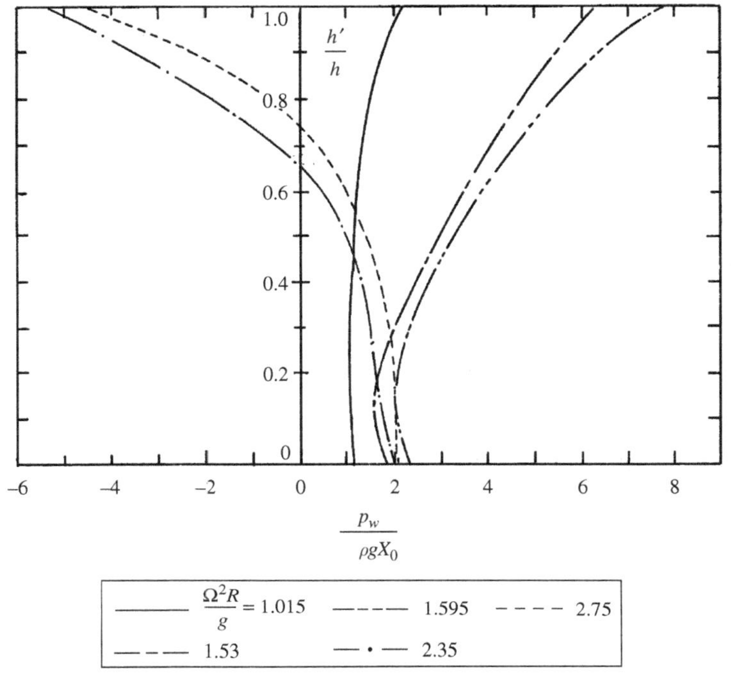

Figure 2.2 Wall pressure distribution in a circular tank under lateral excitation for $h/R = 2.0$. (Abramson and Ransleben, 1961b)

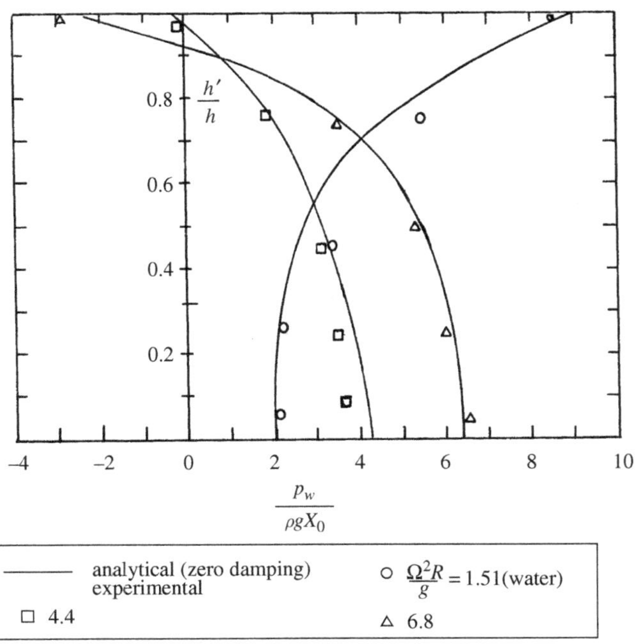

Figure 2.3 Comparison between analytical and measured wall distribution for $h/R = 2.0$. (Abramson and Ransleben, 1961b)

(which are in phase with the displacement). Indeed, the distribution at $\Omega^2 R/g = 2.35$ experiences a balance of the in- and out-of-phase portions and the resulting total force is close to zero, although the actual pressure magnitudes near the surface are relatively high. A similar scenario also exists for higher modes. This may be attributed to the fact that the shearing action between the fluid layers, which are moving in opposite directions, introduces the distortion observed in the range where the out-of-phase portion is larger in magnitude than the in-phase portion.

The net force components acting on the tank wall and bottom are obtained by integrating the pressure over the corresponding area of the boundary. Resolving along $\theta = 0$, one can obtain the force acting along the x-axis

$$F_x = \int_{\theta=0}^{2\pi} \int_{z=-h}^{0} p \cos\theta \, R \, d\theta \, dz$$

$$= m_f X_0 \Omega^2 \sin \Omega t \left[1 + \sum_{n=1}^{\infty} \frac{1}{\xi_{1n} h} \frac{2R}{(\xi_{1n}^2 - 1)} \frac{\Omega^2}{(\omega_{1n}^2 - \Omega^2)} \tanh(\xi_{1n} h/R) \right] \tag{2.14}$$

where $m_f = \rho \pi h R^2$ is the total mass of the fluid. The force acting along the y-axis is

$$F_y = \int_{\theta=0}^{2\pi} \int_{z=-h}^{0} p \sin\theta R \, d\theta \, dz = 0 \tag{2.15}$$

The force acting on the bottom is

$$F_b = \int_{\theta=0}^{2\pi} \int_{r=0}^{R} pr \, dr \, dz = 0 \tag{2.16}$$

These results indicate that there is a net force exerted along the direction of excitation as given by relation (2.14). In the perpendicular direction, the pressure distribution is symmetrical about the tank circumference such that the integration over it vanishes. The pressure at the tank bottom produces no forces in the direction of excitation. However, it will contribute a moment about an axis perpendicular to the xz plane. Figure 2.4 shows the dependence of the horizontal hydrodynamic force ratio, $F_x/\rho g R^2 X_0$, on the frequency parameter, $\Omega^2 R/g$, for fluid depth ratio $h/R = 2$. Figure 2.4 also includes the inertia force of the frozen fluid, which is linearly proportional to the frequency ratio, $\Omega^2 R/g$. The points indicated by circles and squares were obtained experimentally by Abramson and Ransleben (1961b) for excitation amplitude ratios, $X_0/R = 0.1666$ and 0.0048 respectively. The agreement between the predicted and measured results is good to within 10% only over excitation frequencies away from resonance where the linear theory fails due the occurrence of nonlinear phenomena.

The hydrodynamic moment due to liquid forces acting on the wall about the y-axis with the origin at the undisturbed free surface is

$$M_{yw} = \int_{\theta=0}^{2\pi} \int_{-h}^{0} p_w R \cos\theta \, dz \, d\theta$$

$$= m_f X_0 \Omega^2 \sin \Omega t \left[\frac{h}{2} + \sum_{n=1}^{\infty} \frac{1}{h} \left(\frac{R}{\xi_{1n}} \right)^2 \frac{2}{(\xi_{1n}^2 - 1)} \frac{\Omega^2}{(\omega_{1n}^2 - \Omega^2)} \frac{1 - \cosh(\xi_{1n} h/R)}{\cosh(\xi_{1n} h/R)} \right] \tag{2.17}$$

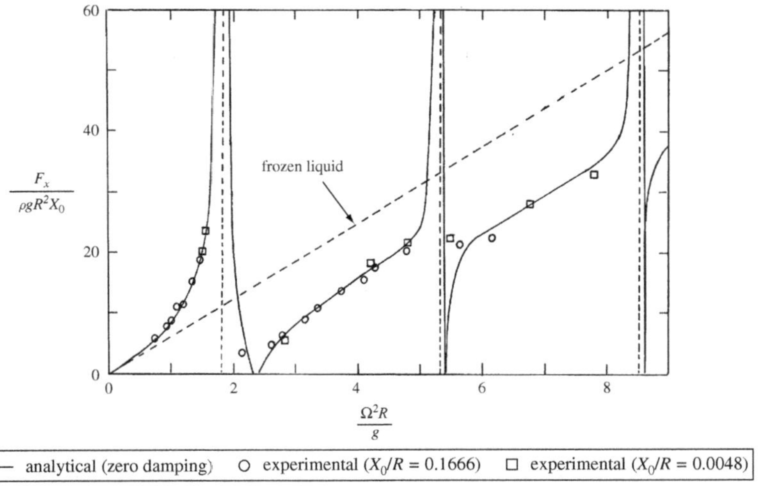

Figure 2.4 Dependence of liquid sloshing on lateral excitation frequency in a circular tank for $h/R = 2.0$. (Abramson, 1961d)

On the other hand, the moment of the forces acting on the tank bottom about the y-axis is

$$M_{yb} = \int_{\theta=0}^{2\pi}\int_0^R p_b r\cos\theta\, r\, dr\, d\theta$$
$$= m_f X_0 \Omega^2 \sin\Omega t\left[\frac{R^2}{4h} + \sum_{n=1}^{\infty}\frac{1}{h\xi_{1n}^2}\frac{2R^2}{(\xi_{1n}^2-1)}\frac{\Omega^2}{(\omega_{1n}^2-\Omega^2)}\frac{1}{\cosh(\xi_{1n}h/R)}\right] \tag{2.18}$$

The total moment about the y-axis is the sum of relations (2.17) and (2.18)

$$M_y = m_f X_0 \Omega^2 \sin\Omega t\left[\frac{h}{2} + \frac{R^2}{4h} + \sum_{n=1}^{\infty}\left(\frac{R}{\xi_{1n}}\right)^2 \times \frac{2\Omega^2}{(\xi_{1n}^2-1)(\omega_{1n}^2-\Omega^2)}\frac{[2-\cosh(\xi_{1n}h/R)]}{h\cosh(\xi_{1n}h/R)}\right] \tag{2.19}$$

It is not difficult to show that the moment of the hydrodynamic forces about an axis through the center of mass of the solidified fluid is

$$M_{ycg} = m_f X_0 \Omega^2 h \sin\Omega t \times\left[\frac{R^2}{4h^2} + \Omega^2\sum_{n=1}^{\infty}\frac{\left\{\xi_{1n}\tanh(\xi_{1n}h/R) - (2R/h) + \dfrac{4R/h}{\cosh(\xi_{1n}h/R)}\right\}}{\xi_{1n}^2(\xi_{1n}^2-1)(2R/h)(\omega_{1n}^2-\Omega^2)}\right] \tag{2.20}$$

Figure 2.5 shows a comparison between predicted and measured hydrodynamic moments about the y-axis through the undisturbed fluid center.

The location of the pressure center z_{cp} of the oscillating modes can be determined by dividing the moment due to the wall pressure distribution by the fluid force along the x-axis,

$$z_{cp} = \frac{Myw}{F_x} = \frac{R}{\xi_{1n}}\left(\frac{1-\cosh(\xi_{1n}h/R)}{\sinh(\xi_{1n}h/R)}\right) = -\frac{R}{\xi_{1n}}\tanh\left(\frac{\xi_{1n}h}{2R}\right) \tag{2.21}$$

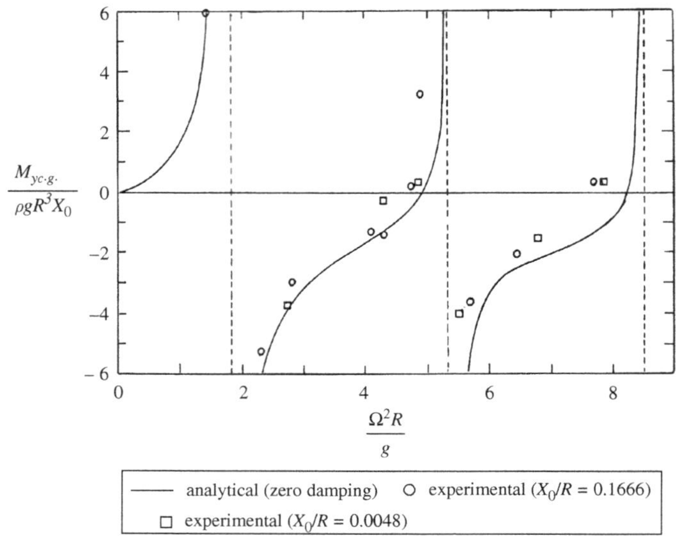

Figure 2.5 Hydrodynamic moment about the centre of gravity of undisturbed liquid due to lateral excitation of a cylindrical tank $h/R = 2.0$. (Abramson, 1961d)

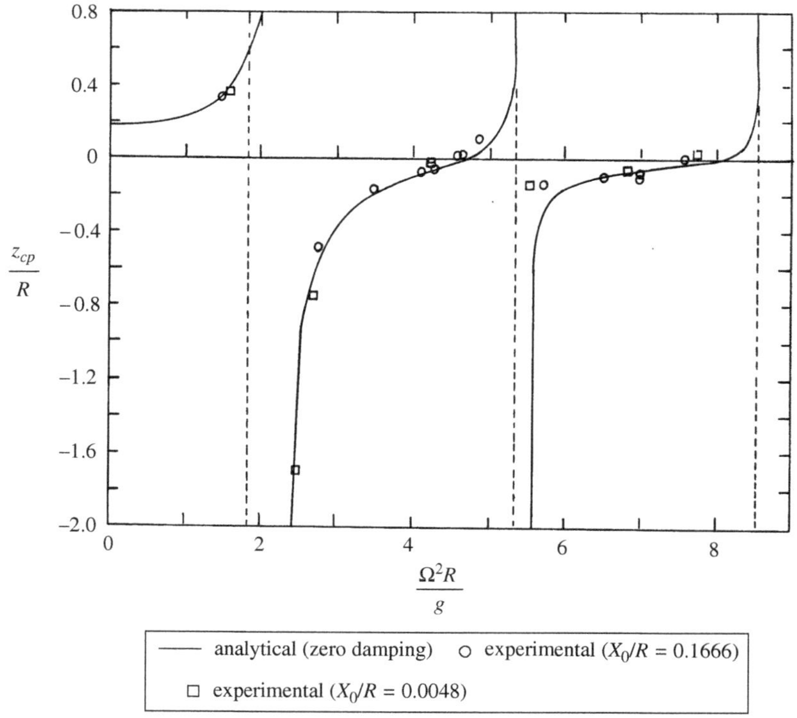

Figure 2.6. Dependence of the center of pressure on excitation frequency in a circular tank under lateral excitation, $h/R = 2.0$. (Abramson, 1961d)

For a deep fluid tank, $\xi_{1n}h/R > 2$, the center of pressure for the first sloshing mode is close to the value $-0.545R$. Figure 2.6 shows the dependence of the location of the pressure center z_{cp}/R on the frequency parameter $\Omega^2 R/g$.

2.2.2 Pitching excitation

Consider the pitching excitation of the tank about the y-axis passing through the center of mass of the solidified fluid as shown in Figure 2.1,

$$\psi(t) = \psi_0 \sin \Omega t \tag{2.22}$$

where ψ_0 and Ω are the amplitude and frequency of the tank motion, respectively. The boundary conditions in terms of the total velocity potential function Φ are:

$$\frac{\partial \Phi}{\partial r} = -\psi_0 z \Omega \cos \Omega t \cos \theta \quad \text{at } r = R \tag{2.23a}$$

$$-\frac{\partial \Phi}{\partial z} = -\psi_0 r \Omega \cos \Omega t \cos \theta \quad \text{at } z = -h/2 \tag{2.23b}$$

$$\frac{\partial^2 \Phi}{\partial t^2} + g \frac{\partial \Phi}{\partial z} = 0 \quad \text{at } z = h/2 \tag{2.23c}$$

The solution of Laplace's equation can be decomposed into three components, that is,

$$\Phi = \Phi_1 + \tilde{\Phi}_2 + \tilde{\Phi}_3 \tag{2.24}$$

where Φ_1 satisfies condition (2.23a) and this gives

$$\frac{\partial}{\partial r}\left(\tilde{\Phi}_2 + \tilde{\Phi}_3\right) = 0, \quad \text{at } r = R \tag{2.23d}$$

and $(\Phi_1 + \tilde{\Phi}_2)$ satisfies condition (2.23b), and thus at the free surface it gives zero velocity. On the other hand, $\tilde{\Phi}_3$ gives zero velocity at $z = -h/2$, that is,

$$\frac{\partial \tilde{\Phi}_3}{\partial z} = 0, \ z = -h/2 \quad \text{and} \quad \frac{\partial(\Phi_1 + \tilde{\Phi}_2)}{\partial z} = 0, \ z = h/2 \tag{2.23e, f}$$

The following functions satisfy the above conditions

$$\Phi_1 = -\Omega \psi_0 r z \cos \Omega t \cos \theta \tag{2.25a}$$

$$\tilde{\Phi}_2 = \Omega \psi_0 \cos \Omega t \cos \theta \sum_{n=1}^{\infty} \frac{J_1(k_n r)}{J_1(k_n R)} \{A_n \cosh k_n z + B_n \sinh k_n z\} \tag{2.25b}$$

$$\tilde{\Phi}_3 = \Omega \psi_0 \cos \Omega t \cos \theta \sum_{n=1}^{\infty} C_n \frac{J_1(k_n r)}{J_1(k_n R)} \cosh k_n \left(z + \frac{h}{2}\right) \tag{2.25c}$$

where k_n are the roots of $\left.\frac{dj_1(k_n r)}{dr}\right|_{r=R} = 0$. Introducing equations (2.25) into condition (2.23b) and using relation (2.5), gives

$$A_n \sinh \xi_{n1} \frac{h}{2R} - B_n \cosh \xi_{n1} \frac{h}{2R} = -\frac{4R^2}{\xi_{n1}(\xi_{n1}^2 - 1)} \tag{2.26}$$

Introducing equations (2.25a) and (2.25b) into (2.23f), gives

$$A_n \sinh \xi_{n1} \frac{h}{2R} + B_n \cosh \xi_{n1} \frac{h}{2R} = \frac{2R^2}{\xi_{n1}(\xi_{n1}^2 - 1)} \tag{2.27}$$

Solving equations (2.26) and (2.27), gives

$$A_n = -\frac{R^2}{\xi_{n1}(\xi_{n1}^2 - 1)} \frac{1}{\sinh \xi_{n1} \frac{h}{2R}}, \quad B_n = \frac{3R^2}{\xi_{n1}(\xi_{n1}^2 - 1)} \frac{1}{\cosh \xi_{n1} \frac{h}{2R}} \tag{2.28a, b}$$

Substituting equations (2.28a, b) into equation (2.25b) gives

$$\tilde{\Phi}_2 = -\Omega \psi_0 \cos \Omega t \cos \theta \sum_{n=1}^{\infty} \frac{R^2}{\xi_{n1}(\xi_{n1}^2 - 1)} \frac{J_1(k_n r)}{J_1(k_n R)} \left\{ \frac{\cosh \xi_n \frac{z}{R}}{\sinh \xi_{n1} \frac{h}{2R}} - \frac{3 \sinh \xi_n \frac{z}{R}}{\cosh \xi_{n1} \frac{h}{2R}} \right\} \tag{2.29}$$

The constant C_n in equation (2.25c) is obtained by using the free-surface condition (2.23c) together with equation (2.23f), and the following result is obtained

$$C_n = -\frac{\Omega^2}{(\omega_n^2 - \Omega^2)} \frac{R^2}{(\xi_n^2 - 1)} \left\{ \frac{(\xi_n h/R) + \coth(\xi_n h/R) - 3 \tanh(\xi_n h/2R)}{\xi_n \cosh(\xi_n h/2R)} \right\} \tag{2.30}$$

Summing up the above results the total velocity potential function can be written in the form

$$\begin{aligned}\Phi = -\Omega \psi_0 \cos \Omega t \cos \theta \Bigg\{ & rz + \sum_{n=1}^{\infty} \frac{R^2}{(\xi_n^2 - 1)} \frac{J_1\left(\xi_n \frac{r}{R}\right)}{J_1(\xi_n)} \left\{ \frac{\cosh \xi_n \frac{z}{R}}{\sinh \xi_n \frac{h}{2R}} - 3 \frac{\sinh \xi_n \frac{z}{R}}{\cosh \xi_n \frac{h}{2R}} \right\} \\ & + \sum_{n=1}^{\infty} \frac{\Omega^2}{\omega_n^2 - \Omega^2} \frac{R^2}{(\xi_n^2 - 1)} \left\{ \frac{(\xi_n h/R) + \coth(\xi_n h/2R) - 3 \tanh(\xi_n h/2R)}{\xi_n \cosh(\xi_n h/2R)} \right\} \\ & \times \frac{J_1(\xi_n r/R)}{J_1(\xi_n)} \cosh \frac{\xi_n}{R}\left(z + \frac{h}{2}\right) \Bigg\}\end{aligned} \tag{2.31}$$

The elevation, $\eta = \frac{1}{g}\frac{\partial \Phi}{\partial t}$, of the free surface above the mean level $z = +h/2$ is

$$\eta = \frac{\Omega^2 \psi_0}{g} R^2 \sin \Omega t \cos\theta \left\{ \frac{h}{2R}\frac{r}{R} + \sum_{n=1}^{\infty} \frac{1}{(\xi_n^2 - 1)} \frac{J_1(\xi_n(r/R))}{J_1(\xi_n)} \right.$$
$$\times \left\{ \coth\xi_n \frac{h}{2R} - 3\tanh\xi_n \frac{h}{2R} \right\} + \sum_{n=1}^{\infty} \frac{\Omega^2}{\omega_n^2 - \Omega^2} \frac{1}{(\xi_n^2 - 1)}$$
$$\times \left\{ \frac{(\xi_n h/R) + \coth(\xi_n h/2R) - 3\tanh(\xi_n h/2R)}{\xi_n \cosh(\xi_n h/2R)} \right\}$$
$$\left. \times \frac{J_1(\xi_n r/R)}{J_1(\xi_n)} \cosh \frac{\xi_n h}{R} \right\} \tag{2.32}$$

The oscillating part of the hydrodynamic pressure, $p = \rho \partial \Phi / \partial t$, is

$$p = \rho \Omega^2 \psi_0 R^2 \sin \Omega t \cos\theta \left\{ \frac{rz}{R^2} + \sum_{n=1}^{\infty} \frac{1}{(\xi_n^2 - 1)} \frac{J_1(\xi_n(r/R))}{J_1(\xi_n)} \right.$$
$$\times \left\{ \frac{\cosh \xi_n (z/R)}{\sinh \xi_n (h/2R)} - \frac{3\sinh \xi_n (z/R)}{\cosh \xi_n (h/2R)} \right\}$$
$$+ \sum_{n=1}^{\infty} \frac{\Omega^2}{\omega_n^2 - \Omega^2} \frac{1}{(\xi_n^2 - 1)} \left\{ \frac{(\xi_n h/R) + \coth(\xi_n h/R) - 3\tanh(\xi_n h/2R)}{\xi_n \cosh(\xi_n h/2R)} \right\}$$
$$\left. \times \frac{J_1(\xi_n r/R)}{J_1(\xi_n)} \cosh \frac{\xi_n}{R}\left(z + \frac{h}{2}\right) \right\} \tag{2.33}$$

The pressure at the wall $r = R$ is

$$p_w = \rho \Omega^2 \psi_0 R^2 \sin \Omega t \cos\theta \left\{ \frac{z}{R} + \sum_{n=1}^{\infty} \frac{1}{(\xi_n^2 - 1)} \left(\frac{\frac{\cosh(\xi_n z/R)}{\sinh(\xi_n h/2R)} - 3\sinh(\xi_n z/R)}{\cosh(\xi_n h/2R)} \right) \right.$$
$$+ \sum_{n=1}^{\infty} \left(\frac{(\xi_n h/R) + (\xi_n h/2R) - 3\tanh(\xi_n h/2R)}{\xi_n \cosh(\xi_n h/R))} \right.$$
$$\left. \times \frac{1}{(\xi_n^2 - 1)} \frac{\Omega^2}{(\omega_n^2 - \Omega^2)} \cosh \frac{\xi_n}{R}\left(z + \frac{h}{2}\right) \right\} \tag{2.34a}$$

The pressure at the bottom, $z = -h/2$, is

$$p_b = \rho \Omega^2 \psi_0 R^2 \sin \Omega t \cos\theta \left\{ -\frac{rh}{2R^2} + \sum_{n=1}^{\infty} \frac{1}{(\xi_n^2 - 1)} [\coth(\xi_n h/2R) + 3\tanh(\xi_n h/2R)] \right.$$
$$+ \sum_{n=1}^{\infty} \left(\frac{(\xi_n h/R) + \coth(\xi_n h/R) - 3\tanh(\xi_n h/R)}{\xi_n \cosh(\xi_n h/R)} \right)$$
$$\left. \frac{1}{(\xi_n^2 - 1)} \frac{\Omega^2}{(\omega_n^2 - \Omega^2)} \frac{J_1(\xi_n r/R)}{J_1(\xi_n)} \right\} \tag{2.34b}$$

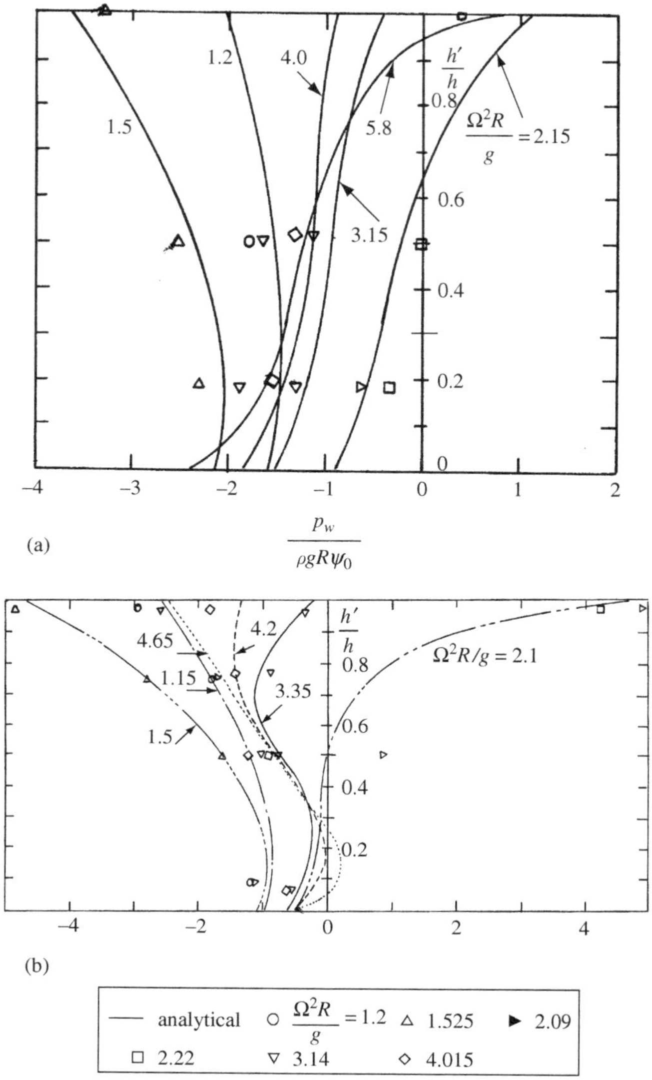

Figure 2.7 (a) Pressure distribution in a circular tank under pitching oscillation for $h/R = 1.0$, (b) pressure distribution in a cylindrical tank under pitching oscillations for $h/R = 2.0$. (Abramson and Ransleben, 1961e)

Figure 2.7(a) and (b) shows the pressure distribution for fluid depth ratios $h/R = 1$ and 2, respectively, and for different excitation frequencies.

The hydrodynamic sloshing force along $\theta = 0$ is

$$F_x = \int_{\theta=0}^{2\pi} \int_{-h/2}^{h/2} p_w R \cos\theta \, d\theta \, dz$$

$$= -\pi \psi_0 \Omega^2 \rho R^2 \sin \Omega t \sum_{n=1}^{\infty} \frac{1}{\xi_n} (2A_n + C_n) \sinh\left(\frac{\xi_n h}{2R}\right) \tag{2.35}$$

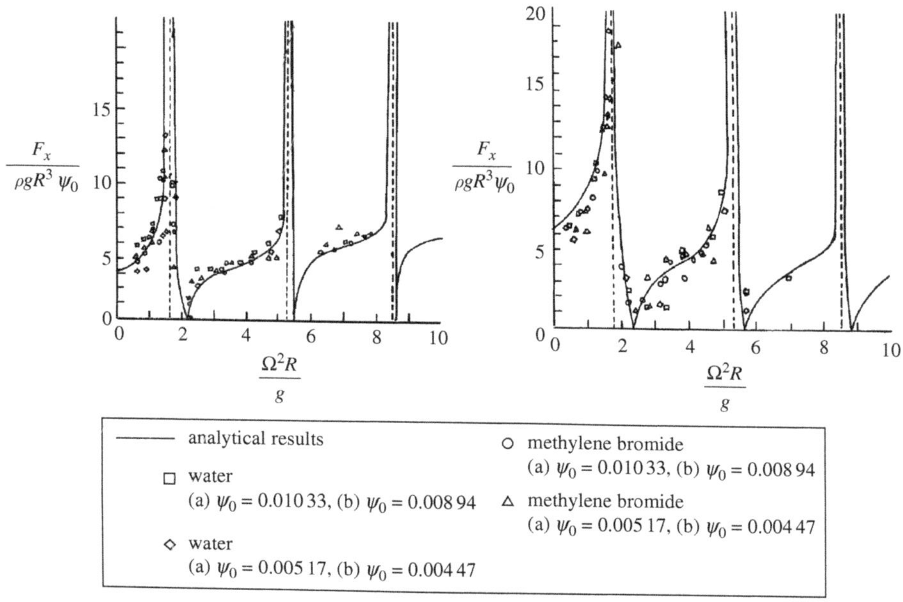

Figure 2.8 Total force response exerted on a circular tank under pitch excitation. (Abramson and Ransleben, 1961e)

where A_n and C_n are given by relations (2.28a) and (2.30). The dependence of the force on the excitation frequency is shown in Figure 2.8(a) and (b) for depth ratios, $h/R = 1$ and 2, respectively. The excitation frequency range shown covers the first three sloshing modes as indicated by the vertical dashed lines.

It is not difficult to show that the force components along the y-axis ($\theta = \pi/2$) and on the bottom are zero. The moment of the hydrodynamic forces acting on the tank walls about y-axis is

$$M_{yw} = \int_0^{2\pi} \int_{-h/2}^{h/2} p_w R z \, dz \cos\theta \, d\theta$$

$$= \pi\rho\Omega^2 \psi_0 R^3 h^2 \sin\Omega t \left\{ \frac{h}{12R} - \sum_{n=1}^{\infty} \frac{2B_n}{\xi_n h^2} \left[\frac{h}{2R} \cosh\left(\xi_n \frac{h}{2R}\right) - \frac{1}{\xi_n} \sinh\left(\xi_n \frac{h}{2R}\right) \right] \right.$$
$$\left. - \frac{C_n}{\xi_n h^2} \left[\frac{h}{2R} \sinh\left(\xi_n \frac{h}{R}\right) - \frac{1}{\xi_n} \left\{ \cosh\left(\xi_n \frac{h}{R}\right) - 1 \right\} \right] \right\} \tag{2.36}$$

The moment of the hydrodynamic forces acting on the tank bottom about the y-axis is

$$M_{yb} = \int_0^{2\pi} \int_{-h/2}^{h/2} p_b r^2 \, dr \cos\theta \, d\theta$$

$$= -\pi\rho\Omega^2 \psi_0 R^3 h^2 \sin\Omega t \left\{ \frac{R}{8h} + \sum_{n=1}^{\infty} \frac{1}{\xi_n h^2} \left[A_n \cosh\left(\xi_n \frac{h}{2R}\right) \right.\right.$$
$$\left.\left. - B_n \sinh\left(\xi_n \frac{h}{2R}\right) + C_n \right] \right\} \tag{2.37}$$

Summing up the two components (2.36) and (2.37), the total moment about the y-axis may be written in the nondimensional form

$$\frac{M_{y\,\text{total}}}{m_f\Omega^2\psi_0 Rh\sinh\Omega t} = \frac{h}{12R} - \frac{R}{8h} + \sum_{n=1}^{\infty}\left(\frac{R}{h}\right)^2 (1/\xi_n^3(\xi_n^2-1))$$

$$\times\left[\coth\left(\xi_n\frac{h}{2R}\right) + 9\tanh\left(\xi_n\frac{h}{2R}\right)\right]$$

$$+\frac{R}{h}\sum_{n=1}^{\infty}\frac{\Omega^2}{(\omega_n^2-\Omega^2)(1/\xi_n^2(\xi_n^2-1)}\left[\xi_n\frac{h}{R}+\coth\left(\xi_n\frac{h}{2R}\right) - 3\tanh\left(\xi_n\frac{h}{2R}\right)\right]$$

$$\times\left[\tanh\left(\xi_n\frac{h}{2R}\right) - \frac{2R}{h\xi_n} + \frac{4R/h}{\xi_n\cosh(\xi_n h/R)}\right] \tag{2.38}$$

Abramson and Ransleben (1961c) obtained another expression using equivalent models. The excitation of an upright circular cylindrical tank studied in these two sections may result in a significant motion of the fluid center of mass, and thus affect the stability of the vehicle. In order to reduce such an effect or to shift the liquid natural frequencies, one may divide the tank into compartments by either radial or concentric vertical walls. These cases will be considered in the next three sections.

2.3 Annular sectored tank

This case has been analytically treated by Bauer (1963c,e, 1964b). The geometry of an annular-general sectored tank is shown in Figure 2.9 together with the coordinates systems at the free surface and at the center of mass of the solidified liquid. The sector angle is $2\pi\beta$, and the tank inner and outer radii are R_{in} and R, respectively.

2.3.1 Lateral excitation

Under translational excitation, $X(t) = X_0\sin\Omega t$, the boundary conditions are, see Bauer (1963c,e, 1964b),

$$\left.\frac{\partial\Phi}{\partial z}\right|_{z=-h} = 0 \tag{2.39a}$$

$$\left.\frac{\partial\Phi}{\partial r}\right|_{r=R_{\text{in}}=R} = \Omega X_0\cos\theta\cos\Omega t \tag{2.39b}$$

$$\left.\frac{1}{r}\frac{\partial\Phi}{\partial\theta}\right|_{\theta=0} = 0 \tag{2.39c}$$

$$\left.\frac{1}{r}\frac{\partial\Phi}{\partial\theta}\right|_{\theta=2\beta\pi} = -\Omega X_0\sin(2\beta\pi)\cos\Omega t \tag{2.39d}$$

$$\frac{\partial^2\Phi}{\partial t^2} + g\frac{\partial\Phi}{\partial z} = 0 \quad \text{at the free surface} \tag{2.39e}$$

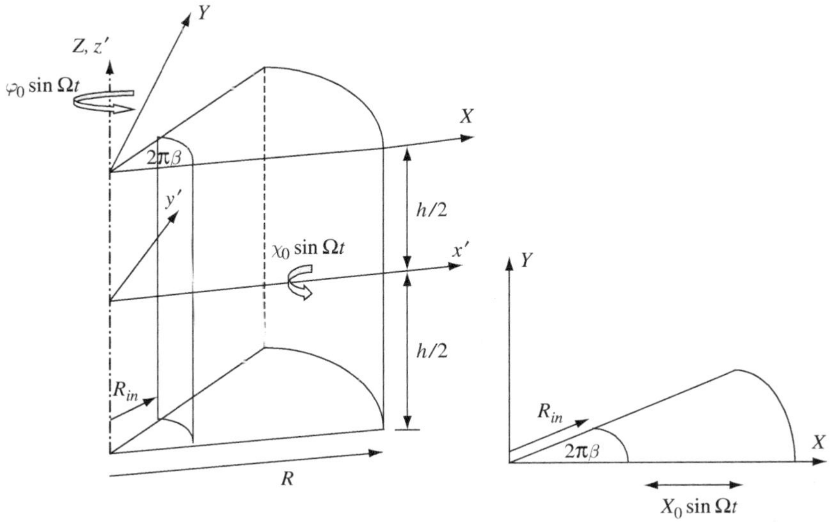

Figure 2.9. Annular sectored tank showing the coordinate systems.

These conditions belong to a single sector and the total contribution of all sectors should be superimposed to obtain the overall hydrodynamic forces and moments. Decomposing the velocity potential function into perturbed and rigid body components,

$$\Phi = \left[\tilde{\Phi} + \Omega X_0 r \cos\theta\right] \cos \Omega t \tag{2.40}$$

the corresponding boundary conditions take the new form

$$\left.\frac{\partial \tilde{\Phi}}{\partial z}\right|_{z=-h} = 0 \tag{2.41a}$$

$$\left.\frac{\partial \tilde{\Phi}}{\partial r}\right|_{r=R_{\text{in}}=R} = \Omega X_0 \cos\theta \cos \Omega t \tag{2.41b}$$

$$\left.\frac{1}{r}\frac{\partial \tilde{\Phi}}{\partial \theta}\right|_{\theta=0,\ 2\beta\pi} = 0 \tag{2.41c}$$

$$g\frac{\partial \tilde{\Phi}}{\partial z} - \Omega^2 \tilde{\Phi} = \Omega^3 X_0 r \cos\theta \qquad \text{at the free surface} \tag{2.41d}$$

A possible solution of Laplace's equation that satisfies the wetted boundaries may be written in the form

$$\tilde{\Phi} = \sum_{m=0}^{\infty} \sum_{n=1}^{\infty} A_{mn} C\left(\xi_{mn} \frac{r}{R}\right) \cos\left(\frac{m\theta}{2\beta}\right) \frac{\cosh\left(\xi_{mn} \dfrac{(h+z)}{R}\right)}{\cosh\left(\xi_{mn} \dfrac{h}{R}\right)} \tag{2.42}$$

where $C(\xi_{mn}(r/R))$ is defined by equation (1.44). To determine the coefficients A_{mn} from the free-surface condition (2.41e), one needs to expand the following functions into the Fourier series

$$\cos\theta = \sum_{m=0}^{\infty} a_m \cos\left(\frac{m}{2\beta}\theta\right), \tag{2.43a}$$

$$r = \sum_{m=0}^{\infty}\sum_{n=0}^{\infty} b_{mn} C\left(\xi_{mn}\frac{r}{R}\right) \tag{2.43b}$$

where

$$a_0 = \frac{\sin 2\pi\beta}{2\pi\beta},$$
$$a_m = \frac{2(2\pi\beta)(-1)^{m+1}\sin 2\pi\beta}{(m\pi)^2 - (2\pi\beta)^2} \tag{2.44a}$$

$$b_{mn} = \frac{R\int\limits_{k\xi_{mn}}^{\xi_{mn}} \left(\xi_{mn}\frac{r}{R}\right) C\left(\xi_{mn}\frac{r}{R}\right) d\left(\xi_{mn}\frac{r}{R}\right)}{\int\limits_{k\xi_{mn}}^{\xi_{mn}} \left(\xi_{mn}\frac{r}{R}\right) C^2\left(\xi_{mn}\frac{r}{R}\right) d\left(\xi_{mn}\frac{r}{R}\right)}$$
$$= \frac{2RN_2(\xi_{mn})}{\left[\frac{4}{\pi^2\xi_{mn}^2} - k^2C^2(k\xi_{mn})\right] - \frac{m^2}{4\beta^2\xi_{mn}^2}\left[\frac{4}{\pi^2\xi_{mn}^2} - C^2(k\xi_{mn})\right]}, \quad m = 0,\ 1,\ 2,\ \ldots \tag{2.44b}$$

the radii ratio is $k = R_{in}/R$, and $N_2(\xi_{mn}) = \frac{1}{\xi_{mn}^3}\int\limits_{k\xi_{mn}}^{\xi_{mn}} \left(\xi_{mn}\frac{r}{R}\right)^2 C\left(\xi_{mn}\frac{r}{R}\right) d\left(\xi_{mn}\frac{r}{R}\right)$.

The total velocity potential function takes the form

$$\Phi = \sum_{m=0}^{\infty}\sum_{n=0}^{\infty} \Omega X_0 \cos\Omega t$$
$$\times \left\{ r\cos\theta + \sum_{m=0}^{\infty}\sum_{n=0}^{\infty} a_m b_{mn} \frac{\Omega^2}{\omega_n^2 - \Omega^2} \frac{\cosh\left(\xi_{mn}((h+z)/R)\right)}{\cosh\left(\xi_{mn}(h/R)\right)} C\left(\xi_{mn}\frac{r}{R}\right)\cos\left(\frac{m\theta}{2\beta}\right)\right\} \tag{2.45}$$

where the first term accounts for the potential of the rigid body motion. The hydrodynamic pressure distribution, $p = -\rho(\partial\Phi/\partial t) - gz$, at depth $-z$ is

$$p = \Omega^2 \rho X_0 \sin\Omega t$$
$$\times \left\{ r\cos\theta + \sum_{m=0}^{\infty}\sum_{n=0}^{\infty} a_m b_{mn} \frac{\Omega^2}{\omega_n^2 - \Omega^2} C\left(\xi_{mn}\frac{r}{R}\right)\right.$$
$$\left.\times \frac{\cosh\left(\xi_{mn}((h+z)/R)\right)}{\cosh\left(\xi_{mn}(h/R)\right)} \cos\left(\frac{m\theta}{2\beta}\right)\right\} - \rho g z \tag{2.46}$$

At the outer wall, $r = R$, the function $C(\xi_{mn}) = 2/\pi\ \xi_{mn}$. The pressure distribution at the tank bottom is obtained by setting $z/R = -h/R$. The fluid hydrodynamic forces along the x- and y-axes are obtained by integrating the pressure distribution over the wetted boundary

$$F_x = \int_0^{2\beta\pi}\int_{-h}^{0} (Rp_{r=R} - R_{\text{in}}p_{r=R_{\text{in}}})\cos\theta\, d\theta\, dz - \int_{R_{\text{in}}}^{R}\int_{-h}^{0} p_{\theta=2\pi\beta}\sin(2\pi\beta)\, dr\, dz$$

$$= m_f X_0 \Omega^2 \sin\Omega t\left\{1 + \sum_{m=0}^{\infty}\sum_{n=0}^{\infty}\frac{2(-1)^{m+1}\alpha_m b_{mn}\sin(2\beta\pi)}{(2\beta\pi)R(1-k^2)(\xi_{mn}h/R)}\left(\frac{\Omega^2}{\omega_n^2 - \Omega^2}\right)\tanh\left(\xi_{mn}\frac{h}{R}\right)\right.$$
$$\left.\times\left[\frac{(2\beta\pi)^2}{\pi^2 m^2 - (2m\beta)^2}\left(\frac{2}{\pi\xi_{mn}} - kC(k\xi_{mn})\right) + N_o(\xi_{mn})\right]\right\} \tag{2.47}$$

$$F_y = \int_0^{2\beta\pi}\int_{-h}^{0} (Rp_{r=R} - R_{\text{in}}p_{r=R_{\text{in}}})\sin\theta\, d\theta\, dz + \int_{R_{\text{in}}}^{R}\int_{-h}^{0} p_{\theta=2\beta\pi}\cos(2\beta\pi)\, dr\, dz$$
$$- \int_{R_{\text{in}}}^{R}\int_{-h}^{0} p_{\theta=0}\, dr\, dz$$

$$= 2m_f X_0 \Omega^2 \sin\Omega t \sum_{m=0}^{\infty}\sum_{n=0}^{\infty}\frac{[1-(-1)^{m+1}\cos(2\beta\pi)]a_m b_{mn}}{(2\beta\pi)R(1-k^2)(\xi_{mn}h/R)}\left(\frac{\Omega^2}{\omega_n^2 - \Omega^2}\right)\tanh\left(\xi_{mn}\frac{h}{R}\right)$$
$$\times\left[\frac{4\beta^2}{[m^2 - 4\beta^2]}\left(\frac{2}{\pi\xi_{mn}} - kC(k\xi_{mn})\right) + N_o(\xi_{mn})\right] \tag{2.48}$$

where $N_o(\xi_{mn}) = \frac{1}{\xi_{mn}}\int_{k\xi_{mn}}^{\xi_{mn}} C(\xi_{mn}r/R)\, d(\xi_{mn}r/R)$, and m_f is the total mass of the fluid.

The first term in equation (2.47) represents the inertia force of the solidified liquid. The liquid moment about the y-axis that passes through the center of mass of the undisturbed fluid, $(0, 0, -h/2)$, is

$$M_{y'} = \int_0^{2\beta\pi}\int_{-h}^{0} (Rp_{r=R} - R_{\text{in}}p_{r=R_{\text{in}}})\left(\frac{h}{2}+z\right)\cos\theta\, d\theta\, dz$$
$$- \int_{R_{\text{in}}}^{R}\int_{-h}^{0} p_{\theta=2\beta\pi}\sin(2\beta\pi)\left(\frac{h}{2}+z\right) dr\, dz + \int_0^{2\beta\pi}\int_{R_{\text{in}}}^{R} p_{\text{bot}} r^2\cos\theta\, d\theta\, dz$$
$$= m_f X_0 \Omega^2 R\sin\Omega t\left\{\frac{1+k^2}{4h/R}\left[1+\frac{\sin(2\beta\pi)\cos(2\beta\pi)}{(2\beta\pi)}\right]\right.$$
$$+\frac{(-1)^{m+1}\sin(2\beta\pi)a_m b_{mn}}{(2\beta\pi)\xi_{mn}(1-k^2)R}\left(\frac{\Omega^2}{\omega_n^2-\Omega^2}\right)$$
$$\times\left[\left(\tanh\left(\xi_{mn}\frac{h}{R}\right)+\frac{2}{\xi_{mn}h/R}\left(\frac{1}{\cosh(\xi_{mn}h/R)}-1\right)\right)\right.$$
$$\left.\times\left[N_0(\xi_{mn})+\frac{4\beta^2}{m^2-4\beta^2}\left(\frac{2}{\pi\xi_{mn}}-kC(k\xi_{mn})\right)\right]\right.$$

$$+\frac{8\beta^2\xi_{mn}^2 N_2(\xi_{mn})}{[m^2-4\beta^2](\xi_{mn}(h/R))\cosh(\xi_{mn}(h/rR))}\Bigg]\Bigg\}$$

$$+\frac{2}{3}m_f gR\left(\frac{1-k^3}{1-k^2}\right)\frac{(1-\sin 2\beta\pi)}{2\beta\pi} \tag{2.49}$$

The moment about the x-axis is

$$M_{x'} = -\int_0^{2\beta\pi}\int_{-h}^{0}(Rp_{r=R}-R_{\text{in}}p_{r=R_{\text{in}}})\left(\frac{h}{2}+z\right)\sin\theta\, d\theta\, dz$$

$$+\int_{R_{\text{in}}}^{R}\int_{-h}^{0}p_{\theta=0}\left(\frac{h}{2}+z\right)dr\,dz - \int_{R_{\text{in}}}^{R}\int_{-h}^{0}p_{\theta=2\beta\pi}\cos(2\beta\pi)\left(\frac{h}{2}+z\right)dr\,dz$$

$$-\int_0^{2\beta\pi}\int_{R_{\text{in}}}^{R}p_{\text{bot}}r^2\sin\theta\,d\theta\,dz$$

$$= -m_f X_0\Omega^2 R\sin\Omega t\left\{\frac{(1+k^2)\sin^2(2\beta\pi)}{(4h/R)(2\beta\pi)}\right.$$

$$-\frac{[1-(-1)^m\cos(2\beta\pi)]a_m b_{mn}}{(2\beta\pi)\xi_{mn}(1-k^2)}\left(\frac{\Omega^2}{\omega_n^2-\Omega^2}\right)$$

$$\times\left[\left(\tanh\left(\xi_{mn}\frac{h}{R}\right)+\frac{2}{(\xi_{mn}h/R)}\left(\frac{1}{\cosh(\xi_{mn}h/R)}-1\right)\right)\right.$$

$$\times\left[N_0(\xi_{mn})+\frac{4\beta^2}{m^2-4\beta^2}\left(\frac{2}{\pi\xi_{mn}}-kC(k\xi_{mn})\right)\right]$$

$$\left.\left.+\frac{8\beta^2\xi_{mn}^2N_2(\xi_{mn})}{[m^2-4\beta^2]\xi_{mn}(h/R)\cosh(\xi_{mn}(h/R))}\right]\right\}$$

$$-\frac{2}{3}m_f gR\left(\frac{1-k^3}{1-k^2}\right)\frac{(1-\cos 2\beta\pi)}{2\beta\pi} \tag{2.50}$$

where $m_f = \pi(\pi\beta)h(R^2-R_{in}^2)$.

Note that the moment about the x'-axis does not vanish as in the case of a bare wall cylinder. Since the origin of the coordinate frame does not pass through the center of mass of the solidified liquid, the last expression in equations (2.49) and (2.50) represents the static moment of the liquid. Furthermore, in view of singularities that occur in the velocity potential (2.42), the results are not valid for $\beta = 1/4$ and $3/4$. For these two cases, special modifications must be developed.

2.3.2 Pitching excitation

Let the excitation about the y'-axis be $\psi(t) = \psi_0\sin\Omega t$ and about the x'-axis be $\chi(t) = \chi_0\sin\Omega t$, where ψ_0 and χ_0 are amplitudes of the angles of pitch and yaw. The axes x' and y' are located mid way between the tank flat bottom and the undisturbed free surface of the liquid. In this case the boundary conditions are

$$\frac{\partial \Phi}{\partial r} = \begin{Bmatrix} -\Omega \psi_0 z \cos\theta \cos\Omega t \\ -\Omega \chi_0 z \sin\theta \cos\Omega t \end{Bmatrix} \quad \text{at } r = R_{in} \text{ and } R \tag{2.51a}$$

$$\frac{\partial \Phi}{\partial z} = \begin{Bmatrix} \Omega \psi_0 r \cos\theta \cos\Omega t \\ \Omega \chi_0 r \sin\theta \cos\Omega t \end{Bmatrix} \quad \text{at } z = -h/2 \tag{2.51b}$$

$$\frac{1}{r}\frac{\partial \Phi}{\partial \theta} = \begin{Bmatrix} 0 \\ -\Omega \chi_0 \cos\Omega t \end{Bmatrix} \quad \text{at the tank sector wall } \theta = 0 \tag{2.51c}$$

$$\frac{1}{r}\frac{\partial \Phi}{\partial \theta} = \begin{Bmatrix} \Omega \psi_0 z \sin(2\beta\pi)\cos\Omega t \\ -\Omega \chi_0 z \cos(2\beta\pi)\cos\Omega t \end{Bmatrix} \quad \text{at the tank sector wall } \theta = 2\beta\pi \tag{2.51d}$$

$$\frac{\partial^2 \Phi}{\partial t^2} - g\frac{\partial \Phi}{\partial z} = 0 \quad \text{at the free surface} \tag{2.51e}$$

The velocity potential function may be split into a rigid body potential and a disturbance potential, that is,

$$\Phi = \left\{ -\Omega r z \begin{Bmatrix} \psi_0 \cos\theta \\ \chi_0 \sin\theta \end{Bmatrix} + \tilde{\Phi} \right\} \cos\Omega t \tag{2.52}$$

The velocity potential function $\tilde{\Phi}$ that satisfies the tank wall conditions may be written in the form (Bauer, 1963c,e)

$$\tilde{\Phi}(r,\theta,z) = \sum_{m=0}^{\infty}\sum_{n=1}^{\infty} \left\{ A_{mn} \cosh\left(\xi_{mn}\frac{z}{R}\right) + B_{mn} \sinh\left(\xi_{mn}\frac{z}{R}\right) \right\} C\left(\xi_{mn}\frac{r}{R}\right) \cos\left(\frac{m\theta}{2\beta}\right) \tag{2.53}$$

where the coefficients A_{mn} and B_{mn} are obtained by using the tank bottom condition (2.51b) and the free-surface condition (2.51e). Their values are given by the following expressions

$$A_{mn} = R\frac{b_{mn}}{\xi_{mn} \cosh\left(\xi_{mn}\frac{h}{R}\right)} \frac{\Omega^2}{(\omega_n^2 - \Omega^2)} \left[2\sinh\left(\xi_{mn}\frac{h}{2R}\right) - \xi_{mn}\left(\frac{h}{2R} + \frac{g}{R\Omega^2}\right)\cosh\left(\xi_{mn}\frac{h}{2R}\right) \right] \begin{Bmatrix} a_m \\ c_m \end{Bmatrix} \tag{2.54a}$$

$$B_{mn} = R\frac{b_{mn}}{\xi_{mn} \cosh(\xi_{mn}(h/R))} \frac{\Omega^2}{(\omega_n^2 - \Omega^2)} \left[\xi_{mn}\left(3\frac{g}{R\Omega^2} - \frac{h}{2R} + \right)\sinh\left(\xi_{mn}\frac{h}{2R}\right) - 2\cosh\left(\xi_{mn}\frac{h}{2R}\right) \right] \begin{Bmatrix} a_m \\ c_m \end{Bmatrix} \tag{2.54b}$$

where b_{mn} is given by equation (2.44b). The total velocity potential function takes the form

$$\Phi(r,\theta,z) = \Omega\cos\Omega t\begin{Bmatrix}\psi_0\\ \chi_0\end{Bmatrix}\left[\begin{Bmatrix}-rz\cos\theta\\ -rz\cos\theta\end{Bmatrix} + \sum_{m=0}^{\infty}\sum_{n=1}^{\infty}\left\{A_{mn}\cosh\left(\xi_{mn}\frac{z}{R}\right)\right.\right.$$

$$\left.\left. + B_{mn}\sinh\left(\xi_{mn}\frac{z}{R}\right)\right\}C\left(\xi_{mn}\frac{r}{R}\right)\cos\left(\frac{m\theta}{2\beta}\right)\right] \tag{2.55}$$

The liquid force along the x-axis is

$$F_x = -m_f g\begin{Bmatrix}0\\ \chi_0\end{Bmatrix}\sin\Omega t$$

$$-4m_f\Omega^2\begin{Bmatrix}\psi_0\\ \chi_0\end{Bmatrix}\sin\Omega t\frac{A_{mn}\sinh\left(\xi_{mn}\frac{h}{2R}\right)[1-(-1)^m\cos 2\beta\pi]}{2\beta\pi R\left(\xi_{mn}\frac{h}{R}\right)[1-k^2]}$$

$$\times\left[N_0(\xi_{mn}) + \frac{(2\pi\beta)^2}{(m\pi)^2-(2\pi\beta)^2}\left(\frac{2}{\pi\xi_{mn}} - \left(\xi_{mn}\frac{h}{R}\right)C_{(m/2\beta)}(k\xi_{mn})\right)\right] \tag{2.56}$$

The first expression represents the inertia force. The moment about the x'-axis is

$$M_y = m_f gR\frac{(1+k^2)}{4h/R}\begin{Bmatrix}\psi_0\left[1+\dfrac{\sin(2\beta\pi)\cos(2\beta\pi)}{(2\beta\pi)}\right]\\ \chi_0\dfrac{\sin^2(2\beta\pi)}{(2\beta\pi)}\end{Bmatrix}\sin\Omega t$$

$$+m_f R^2\Omega^2\begin{Bmatrix}\psi_0\\ \chi_0\end{Bmatrix}\sin\Omega t\left\{\begin{Bmatrix}((1+k^2)/8)\left[1+\dfrac{\sin(2\beta\pi)\cos(2\beta\pi)}{(2\beta\pi)}\right] - \frac{1}{12}(h/R)^2\\ \dfrac{(1+k^2)\sin^2(2\beta\pi)}{8(2\beta\pi)}\end{Bmatrix}\right.$$

$$+\frac{2(-1)^{m+1}\sin(2\beta\pi)}{(2\beta\pi)\xi_{mn}(1-k^2)R^2}\left[B_{mn}\left\{\cosh\left(\xi_{mn}\frac{z}{2R}\right) - \frac{2R}{\xi_{mn}z}\sinh\left(\xi_{mn}\frac{z}{2R}\right)\right\}\right.$$

$$\times\left[N_0(\xi_{mn}) + \frac{(2\beta\pi)^2}{\pi^2m^2-(2m\beta)^2}\left(\frac{2}{\pi\xi_{mn}} - kC(k\xi_{mn})\right)\right]$$

$$\left.\left. + \frac{(2\beta\pi) +^2 \xi_{mn}^2 N_2(\xi_{mn})}{\left[\pi^2m^2-(2m\beta)^2\right](\xi_{mn}(h/R))}\left\{A_{mn}\cosh\left(\xi_{mn}\frac{z}{2R}\right) - B_{mn}\sinh\left(\xi_{mn}\frac{z}{2R}\right)\right\}\right]\right\}$$

$$+\frac{2}{3}m_f gR\left(\frac{1-k^3}{1-k^2}\right)\frac{\cos(2\beta\pi)}{2\beta\pi} \tag{2.57}$$

The moment about the x'-axis is

$$M_x = m_f g\, R \sin\Omega t \frac{(1+k^2)}{4h/R}\left\{\begin{array}{c}\psi_0(\sin^2(2\beta\pi)/(2\beta\pi))\\ \chi_0[1-(\sin(2\beta\pi)\cos(2\beta\pi)/(2\beta\pi))]\end{array}\right\}$$
$$- m_f R^2\Omega^2\left\{\begin{array}{c}\psi_0\\ \chi_0\end{array}\right\}\sin\Omega t\left\{\left\{\begin{array}{c}((1+k^2)\sin^2(2\beta\pi)/8(2\beta\pi))\\ ((1+k^2)/8)[1+(\sin(2\beta\pi)\cos(2\beta\pi)/(2\beta\pi))]-\frac{1}{12}(h/R)^2\end{array}\right\}\right.$$
$$-\frac{2[1-(-1)^m\cos(2\beta\pi)]}{(2\beta\pi)\xi_{mn}(1+k^2)R^2}\left[B_{mn}\left\{\cosh\left(\xi_{mn}\frac{z}{2R}\right)-\frac{2R}{\xi_{mn}z}\sinh\left(\xi_{mn}\frac{z}{2R}\right)\right\}\right.$$
$$\times\left[N_0(\xi_{mn})+\frac{(2\beta\pi)^2}{\pi^2m^2-(2m\beta)^2}\left(\frac{2}{\pi\xi_{mn}}-kC(k\xi_{mn})\right)\right]$$
$$\left.\left.+\frac{(2\beta\pi)^2\xi_{mn}^2N_2(\xi_{mn})}{\left[\pi^2m^2-(2m\beta)^2\right](\xi_{mn}(h/R))}\left\{A_{mn}\cosh\left(\xi_{mn}\frac{z}{2R}\right)-B_{mn}\sinh\left(\xi_{mn}\frac{z}{2R}\right)\right\}\right]\right\}$$
$$-\frac{2}{3}m_f gR\left(\frac{1-k^3}{1-k^2}\right)\frac{(1-\cos 2\beta\pi)}{2\beta\pi} \tag{2.58}$$

Since the pitch and yaw axes do not pass through the center of mass of the solidified liquid, the last expression in equations (2.57) and (2.58) represents the static moment.

2.3.3 Roll excitation

Unlike the bare wall cylindrical tanks, the roll excitation of compartmented cylindrical tanks is very important since a large amount of liquid participates in the tank motion. Furthermore, there will be a significant roll moment due to hydrodynamic forces acting on the sector walls. Let $\varphi(t) = \varphi_0 \sin\Omega t$ be the roll excitation about the z-axis. Taking the origin of the coordinate frame at the undisturbed free surface, the boundary conditions are

$$\frac{\partial\Phi}{\partial r} = 0 \quad \text{at } r = R_{\text{in}} \text{ and } R \tag{2.59a}$$

$$\frac{\partial\Phi}{\partial z} = 0 \quad \text{at } z = -h \tag{2.59b}$$

$$\frac{1}{r}\frac{\partial\Phi}{\partial\theta} = \Omega\varphi_0 r\cos\Omega t \quad \text{at the tank sector walls } \theta = 0,\ 2\beta\pi \tag{2.59c}$$

$$\frac{\partial^2\Phi}{\partial t^2} + g\frac{\partial\Phi}{\partial z} = 0 \quad \text{at the free surface} \tag{2.59d}$$

These boundary conditions can be satisfied by two potential functions

$$\Phi(r,\theta,z,t) = \left[\Phi_0(r,\varphi) + \tilde{\Phi}(r,\theta,z,t)\right]\sin\Omega t \tag{2.60}$$

where $\Phi_0(r,\varphi)$ is determined such that it satisfies the boundary conditions at the tank sidewalls, $r = R_{in}$, R and $\theta = 0, 2\beta\pi$. The function $\tilde{\Phi}$ must satisfy Laplace's equation with homogeneous boundary conditions.

The solution represents the flow in an infinitely long tank. The other boundary conditions will be satisfied by the disturbance potential function, $\tilde{\Phi}(r,\theta,z,t)$. In this case the boundary conditions for these two potential functions are

$$\frac{\partial \Phi_0}{\partial r} = 0, \quad \text{at } r = R_{in} \text{ and } R, \quad \frac{1}{r}\frac{\partial \Phi_0}{\partial \theta} = 0 \quad \text{at } \theta = 0,\ 2\beta\pi \tag{2.61a, b}$$

$$\frac{\partial \tilde{\Phi}}{\partial r} = 0, \quad \text{at } r = R_{in} \text{ and } R, \quad \frac{1}{r}\frac{\partial \tilde{\Phi}}{\partial \theta} = 0 \quad \text{at } \theta = 0,\ 2\beta\pi \tag{2.61c, d}$$

$$\frac{\partial \tilde{\Phi}}{\partial z} = 0 \text{ at } z = -h, \quad g\frac{\partial \tilde{\Phi}}{\partial z} - \Omega^2\frac{\partial^2 \tilde{\Phi}}{\partial t^2} = \Omega^2 \Phi_0 \quad \text{at } z = 0 \tag{2.61e, f}$$

The velocity potential $\Phi_0(r,\varphi)$ can be written in the form

$$\Phi_0(r,\varphi) = \Omega\varphi_0 r^2(\theta - \beta\pi) + \Phi_1 \tag{2.62}$$

Since $\nabla^2\Phi_0 = 0$, one can obtain the Poisson equation

$$\nabla^2\Phi_1 = -4\Omega\varphi_0(\theta - \beta\pi) \tag{2.63}$$

subject to the boundary conditions

$$\frac{\partial \Phi_1}{\partial r} = -2\Omega\varphi_0 r(\theta - \beta\pi) \quad \text{at } r = R_{in} \text{ and } R \tag{2.64a}$$

$$\frac{1}{r}\frac{\partial \Phi_1}{\partial \theta} = 0 \quad \text{at } \theta = 0,\ 2\beta\pi \tag{2.64b}$$

The following solution satisfies the boundary conditions in θ

$$\Phi_1(r,\theta) = \sum_{m=0}^{\infty} \Re_m(r)\cos\left(\frac{m\theta}{2\beta}\right) \tag{2.65}$$

Substituting equation (2.65) into equation (2.63) gives an infinite number of ordinary differential equations for the functions $\Re_m(r)$ provided the coordinate θ on the right-hand side is represented by a cosine series

$$\theta - \beta\pi = \sum_{m=0}^{\infty} C_m \cos\left(\frac{m\theta}{2\beta}\right) \tag{2.66}$$

where $C_0 = C_{2m} = 0$ and $C_{2m-1} = -\frac{8\beta}{\pi}\frac{1}{(2m-1)^2}$. The resulting differential equations, for $\beta \neq 1/4$ and $3/4$, are

$$\frac{d^2\Re_0}{dr^2} + \frac{1}{r}\frac{d\Re_0}{dr} = 0 \tag{2.67a}$$

$$\frac{d^2\Re_{2m}}{dr^2} + \frac{1}{r}\frac{d\Re_{2m}}{dr} - \frac{m^2}{\beta^2 r^2}\Re_{2m} = 0, \text{ for } m = 1, 2, \ldots \tag{2.67b}$$

$$\frac{d^2\Re_{2m-1}}{dr^2} + \frac{1}{r}\frac{d\Re_{2m-1}}{dr} - \frac{(2m-1)^2}{4\beta^2 r^2}\Re_{2m-1} = \frac{32\Omega\varphi_0\beta}{(2m-1)^2}, \quad \text{for } m = 1, 2, \ldots, \ \beta \neq 1/4 \text{ or } 3/3 \tag{2.67c}$$

Subject to the boundary conditions in r, the solutions of equations (2.67) are

$$\Re_0(r) = 0, \qquad \Re_{2m}(r) = 0, \tag{2.68a, b}$$

$$\Re_{2m-1}(r) = \frac{32\Omega\varphi_0 R^2\beta^2}{\pi(2m-1)\left((2m-1)^2 - 4\beta^2\right)} \times \left[\left(\frac{r}{R}\right)^{(2m-1)/2\beta}\frac{\left(1 - k^{(2m-1)/2\beta+2}\right)}{\left(1 - k^{(2m-1)/2\beta}\right)} - \left(\frac{R}{r}\right)^{(2m-1)/2\beta}\frac{\left(k^2 - k^{(2m-1)/2\beta}\right)}{\left(1 - k^{(2m-1)/2\beta}\right)}k^{(2m-1)/2\beta} - \frac{4\beta}{(2m-1)}\left(\frac{r}{R}\right)^2\right] \tag{2.69}$$

For $\beta \neq 1/4$ and $3/4$, the velocity potential function $\Phi_0(r,\varphi)$ takes the form

$$\Phi_0(r,\theta) = \Omega\varphi_0 r^2(\theta - \beta\pi) + \frac{8\Omega\varphi_0 R^2(2\beta\pi)^2\cos((2m-1)\theta/2\beta)}{\pi(2m-1)\left((2m-1)^2\pi^2 - 4(2\beta\pi)^2\right)} \times \left[\left(\frac{r}{R}\right)^{(2m-1)/2\beta}\frac{\left(1 - k^{(2m-1)/2\beta+2}\right)}{\left(1 - k^{(2m-1)/2\beta}\right)} - \left(\frac{R}{r}\right)^{(2m-1)/2\beta}\frac{\left(k^2 - k^{(2m-1)/2\beta}\right)}{\left(1 - k^{(2m-1)/2\beta}\right)}k^{(2m-1)/2\beta} - \frac{2(2\beta\pi)}{\pi(2m-1)}\left(\frac{r}{R}\right)^2\right] \tag{2.70}$$

The solution for the disturbed potential function $\nabla^2\tilde{\Phi} = 0$ that satisfies the homogeneous boundary conditions of the container walls may be written in the form

$$\tilde{\Phi}(r,\theta,z) = D_{mn}\frac{\cosh\left(\xi_{mr}(h+z)/R\right)}{\cosh\left(\xi_{mn}h/R\right)}C(\xi_{mn}r/R)\cos((2m-1)\theta/2\beta) \tag{2.71a}$$

The coefficient D_{mn} is obtained by satisfying the free-surface boundary condition. The result of this process yields the following form

$$\tilde{\Phi}(r,\theta,z) = \frac{8\Omega\varphi_0 R^2(2\beta\pi)^2}{\pi}\frac{\cosh(\xi_{mn}(h+z)/R)}{\cosh(\xi_{mn}h/R)}\, C\left(\frac{\xi_{mn}r}{R}\right)\cos\left(\frac{(2m-1)\theta}{2\beta}\right)$$
$$\times \frac{\Omega^2}{(\omega^2_{2m-1,n}-\Omega^2)\left((2m-1)^2\pi^2-4(2\beta\pi)^2\right)(2m-1)}$$
$$\times \left\{\frac{l_{2m-1,n}(1-k^{(2m-1)/2\beta+2})-q_{2m-1,n}(k^2-k^{(2m-1)/2\beta})k^{(2m-1)/2\beta}}{(1-k^{(2m-1)/2\beta})}\right.$$
$$\left.-\frac{\pi(2m-1)}{2(2\beta\pi)}g_{2m-1,n}\right\} \tag{2.71b}$$

where ξ_{mn} are the roots of $dC_{(2m-1)/2\beta}(\xi_{mn}r/R)/dr|_{r=R}=0$, and the values of the coefficients with subscript $2m-1$, n are the coefficients of the Bessel series

$$\left(\frac{r}{R}\right)^2 = \sum_{n=0}^{\infty} g_{2m-1,n}C_{(2m-1)/2\beta}(\xi_{2m-1,n}r/R),$$

$$\left(\frac{R}{r}\right)^2 = \sum_{n=0}^{\infty} h_{2m-1,n}C_{(2m-1)/2\beta}(\xi_{2m-1,n}r/R)$$

$$\left(\frac{r}{R}\right)^{(2m-1)/2\beta} = \sum_{n=0}^{\infty} l_{2m-1,n}C_{(2m-1)/2\beta}(\xi_{2m-1,n}r/R),$$

$$\left(\frac{R}{r}\right)^{(2m-1)/2\beta} = \sum_{n=0}^{\infty} q_{2m-1,n}C_{(2m-1)/2\beta}(\xi_{2m-1,n}r/R) \tag{2.72}$$

where

$$g_{2m-1,n} = \frac{\int_{k\xi_{2m-1,n}}^{\xi_{2m-1,n}} (\xi_{2m-1,n}r/R)^3 C_{(2m-1)/2\beta}(\xi_{2m-1,n}r/R)\, d(\xi_{2m-1,n}r/R)}{\xi^2_{2m-1,n}I^*} \tag{2.73a}$$

$$h_{2m-1,n} = \xi^2_{2m-1,n}\frac{\int_{k\xi_{2m-1,n}}^{\xi_{2m-1,n}} C_{(2m-1)/2\beta}(\xi_{2m-1,n}r/R)\,\frac{d(\xi_{2m-1,n}r/R)}{(\xi_{2m-1,n}r/R)}}{I^*} \tag{2.73b}$$

$$l_{2m-1,n} = \frac{\int_{k\xi_{2m-1,n}}^{\xi_{2m-1,n}} (\xi_{2m-1,n}r/R)^{(2m-1)/2\beta+1} C_{(2m-1)/2\beta}(\xi_{2m-1,n}r/R)\, d(\xi_{2m-1,n}r/R)}{\xi^{(2m-1)/2\beta}_{2m-1,n}I^*} \tag{2.73c}$$

$$q_{2m-1,n} = \xi^{(2m-1)/2\beta}_{2m-1,n}\frac{\int_{k\xi_{2m-1,n}}^{\xi_{2m-1,n}} (\xi_{2m-1,n}r/R)^{1-(2m-1)/2\beta} C_{(2m-1)/2\beta}(\xi_{2m-1,n}r/R)\, d(\xi_{2m-1,n}r/R)}{I^*} \tag{2.73d}$$

with $I^* = \int_{k\xi_{2m-1,n}}^{\xi_{2m-1,n}} (\xi_{2m-1,n}r/R) C^2_{(2m-1)/2\beta}(\xi_{2m-1,n}r/R)\, d(\xi_{2m-1,n}r/R)$.

The total velocity potential function is

$$\begin{aligned}
&\Phi(r,\theta,z,t)\\
&= \Omega\varphi_0 R^2 \sin\Omega t \left\{ \left(\frac{r}{R}\right)^2 (\theta - \beta\pi) + \frac{8(2\beta\pi)\cos\left(\frac{(2m-1)\theta}{2\beta}\right)}{\pi(2m-1)\left((2m-1)^2\pi^2 - 4(2\beta\pi)^2\right)} \right.\\
&\quad \times \left[\left(\frac{r}{R}\right)^{((2m-1)/2\beta)} \frac{\left(1 - k^{((2m-1)/2\beta)+2}\right)}{\left(1 - k^{((2m-1)/2\beta)}\right)} \right.\\
&\quad \left. - \left(\frac{R}{r}\right)^{((2m-1)/2\beta)} \frac{\left(k^2 - k^{((2m-1)/2\beta)}\right)}{\left(1 - k^{((2m-1)/2\beta)}\right)} k^{((2m-1)/2\beta)} - \frac{2(2\beta\pi)}{\pi(2m-1)} \left(\frac{r}{R}\right)^2 \right]\\
&\quad + \frac{8(2\beta\pi)^2}{\pi} \frac{\Omega^2 C(\xi_{mn}r/R)}{(\omega^2_{2m-1,n} - \Omega^2)} \frac{\cosh(\xi_{mn}(h+z)/R)}{\cosh(\xi_{mn}h/R)}\\
&\quad \times \cos\left(\frac{(2m-1)\theta}{2\beta}\right) \frac{1}{\left((2m-1)^2\pi^2 - 4(2\beta\pi)^2\right)(2m-1)}\\
&\quad \times \left\{ \frac{l_{2m-1,n}\left(1 - k^{((2m-1)/2\beta)+2}\right) - q_{2m-1,n}\left(k^2 - k^{((2m-1)/2\beta)}\right) k^{((2m-1)/2\beta)}}{\left(1 - k^{((2m-1)/2\beta)}\right)} \right.\\
&\quad \left.\left. - \frac{\pi(2m-1)}{2(2\beta\pi)} g_{2m-1,n} \right\} \right\}
\end{aligned} \tag{2.74}$$

Note that the first term satisfies the boundary conditions at the sector walls while the infinite series vanishes. For the boundary conditions at $r = R_{in}$ and R, the double series vanishes as well and the simple summation together with the first term vanishes after differentiation with respect to r. The resulting hydrodynamic forces along the x- and y-axes are

$$\begin{aligned}
\begin{Bmatrix} F_x \\ F_y \end{Bmatrix} &= \Omega^2 \varphi_0 R m_f \sin\Omega t \left\{ \left(\frac{1-k^3}{1-k^2}\right) \begin{Bmatrix} (2/3)\sin(2\beta\pi) + 2((\cos(2\beta\pi) - 1)/(2\beta\pi)) \\ 2(\sin(2\beta\pi))/2\beta\pi - (2/3)[1 + \cos(2\beta\pi)] \end{Bmatrix} \right.\\
&\quad + 32\beta \begin{Bmatrix} \sin(2\beta\pi) \\ -[1 + \cos(2\beta\pi)] \end{Bmatrix} \frac{1}{\left((2m-1)^2\pi^2 - 4(2\beta\pi)^2\right)(2m-1)(1-k^2)}\\
&\quad \times \left[\frac{(2\beta\pi)^2}{\left((2m-1)^2\pi^2 - (2\beta\pi)^2\right)} \right.\\
&\quad \times \left[1 - 2k^{((2m-1)/2\beta)-2} + k^{((2m-1)/2\beta)} - 2k^{((2m-1)/2\beta)+1} + k^{((2m-1)/2\beta)+3} \right.\\
&\quad \left. + \frac{k^3}{1 - k^{((2m-1)/2\beta)}} - \frac{4\beta(1-k^3)}{(2m-1)} \right]
\end{aligned}$$

$$
+\frac{(2\beta\pi)^2/\left((2m-1)^2\pi^2-(2\beta\pi)^2\right)}{1-k^{((2m-1)/2\beta)}}\times\left\{\left(\frac{2m-1}{2\beta}-1\right)\right.
$$

$$
\times\left(1-k^{((2m-1)/2\beta)+1}\right)\left(1-k^{((2m-1)/2\beta)+2}\right)
$$

$$
\left.+\left(\frac{2m-1}{2\beta}+1\right)\left(k^{((2m-1)/2\beta)}-k\right)\left(k-k^{((2m-1)/2\beta)}\right)\right\}-\left.\frac{4\beta(1-k^3)}{3(2m-1)}\right]
$$

$$
+32\beta\begin{Bmatrix}\sin(2\beta\pi)\\-[1+\cos(2\beta\pi)]\end{Bmatrix}\frac{\Omega^2\tanh(\xi_{2m-1,n}z/R)}{(\omega^2_{2m-1,n}-\Omega^2)(\xi_{2m-1,n}z/R)}
$$

$$
\times\frac{\left[((2\beta\pi)^2/(\pi^2(2m-1)^2-(2\beta\pi)^2)\left((2/(\pi\xi_{2m-1,n}))-kC(k\xi_{2m-1,n})\right)+N_0\right]}{\left((2m-1)^2\pi^2-4(2\beta\pi)^2\right)(2m-1)(1-k^2)}
$$

$$
\times\left\{\frac{l_{2m-1,n}(1-k^{((2m-1)/2\beta)+2})-q_{2m-1,n}(k^2-k^{(2m-1)/2\beta)})k^{((2m-1)/2\beta)}}{(1-k^{(2m-1)/2\beta})}\right.
$$

$$
\left.\left.-\frac{(2m-1)}{4\beta}g_{2m-1,n}\right\}\right\} \tag{2.75}
$$

The hydrodynamic moments about the x- and y-axes are

$$
\begin{Bmatrix}M_x\\M_y\end{Bmatrix}=\Omega^2\varphi_0R^2m_f\sin\Omega t\left\{\frac{1}{5h/R}\left(\frac{1-k^5}{1-k^2}\right)\begin{Bmatrix}-[\sin(2\beta\pi)+2(\cos(2\beta\pi)-1)/(2\beta\pi)]\\2\sin(2\beta\pi)/(2\beta\pi)-(2/3)[1+\cos(2\beta\pi)]\end{Bmatrix}\right.
$$

$$
+\frac{32\beta}{(h/R)(1-k^2)}\begin{Bmatrix}\sin(2\beta\pi)\\1+\cos(2\beta\pi)\end{Bmatrix}\frac{1}{\left((2m-1)^2\pi^2-4(2\beta\pi)^2\right)(2m-1)}
$$

$$
\times\left[\frac{(2\beta\pi)^2}{\left((2m-1)^2\pi^2-(2\beta\pi)^2\right)}\right.
$$

$$
\times\left\{2\beta\frac{\dfrac{(1-2k^{((2m-1)/2\beta)+2})(1-k^{((2m-1)/2\beta)+3})}{(2m-1+6\beta)}+2\beta\dfrac{(k^{((2m-1)/2\beta)}-k^3)(k^2-2k^{((2m-1)/2\beta)})}{(2m-1-6\beta)}}{1-k^{((2m-1)/2\beta)}}\right.
$$

$$
\left.\left.-\frac{4\beta(1-k^5)}{5(2m-1)}\right\}\right]+8\beta\begin{Bmatrix}\sin(2\beta\pi)\\1+\cos(2\beta\pi)\end{Bmatrix}
$$

$$
\times\Omega^2\frac{\left\{\dfrac{l_{2m-1,n}(1-k^{((2m-1)/2\beta)+2})-q_{2m-1,n}(k^2-k^{((2m-1)/2\beta)})k^{((2m-1)/2\beta)}}{(1-k^{((2m-1)/2\beta)})}-\dfrac{(2m-1)}{4\beta}g_{2m-1,n}\right\}}{\left(\omega^2_{2m-1,n}-\Omega^2\right)\xi_{2m-1,n}\left((2m-1)^2\pi^2-4(2\beta\pi)^2\right)(2m-1)(1-k^2)}
$$

$$
\times\left\{\left[\tanh(\xi_{2m-1,n}z/R)+\frac{2R}{\xi_{2m-1,n}z}\left(\frac{1}{\cosh(\xi_{2m-1,n}z/R)-1}\right)\right]\right.
$$

$$
\times\left[\frac{(2\beta\pi)^2}{\left((2m-1)^2\pi^2-(2\beta\pi)^2\right)}\times\left(\frac{2}{\pi\xi_{2m-1,n}}-kC_{((2m-1),n/2\beta)}(k\xi_{2m-1,n})\right)+N_0(\xi_{2m-1,n})\right]
$$

$$+\frac{(2\beta\pi)^2}{\left((2m-1)^2\pi^2-(2\beta\pi)^2\right)}\frac{2\xi_{2m-1,n}RN_2(\xi_{2m-1,n})}{z\cosh(\xi_{2m-1,n}z/R)}\bigg\}\bigg\}$$

$$+\frac{2m_f gR}{3}\left(\frac{1-k^3}{1-k^2}\right)\left\{\begin{matrix}-\sin(2\beta\pi)/2\beta\pi\\ [1-\cos(2\beta\pi)/2\beta\pi]\end{matrix}\right\} \tag{2.76}$$

The hydrodynamic moments about the z-axes are

$$M_z=\Omega^2\varphi_C R^2 m_f\sin\Omega t\left\{\frac{1+k^2}{2}+\frac{64\beta}{\left((2m-1)^2\pi^2-4(2\beta\pi)^2\right)(2m-1)(1-k^2)}\times\left[\frac{\beta(1-k4)}{(2m-1)}\right.\right.$$

$$\left.-2\beta\frac{\frac{(1-2k^{((2m-1)/2\beta)+2})(1-k^{((2m-1)/2\beta)+3})}{(2m-1+4\beta)}+\frac{(k^{((2m-1)/2\beta)}-k^3)(k^2-2k^{((2m-1)/2\beta)})}{(2m-1-4\beta)}}{1-k^{((2m-1)/2\beta)}}\right]$$

$$+64\beta\frac{N_1(\xi_{2m-1,n})}{(\xi_{2m-1,n}z/R)}\frac{\Omega^2}{(\omega^2_{2m-1,n}-\Omega^2)}\frac{\tanh(\xi_{2m-1,n}z/R)}{\left((2m-1)^2\pi^2-4(2\beta\pi)^2\right)(2m-1)}$$

$$\times\left\{\frac{l_{2m-1,n}(1-k^{((2m-1)/2\beta)+2})-q_{2m-1,n}(k^2-k^{((2m-1)/2\beta)})k^{((2m-1)/2\beta)}}{(1-k^{((2m-1)/2\beta)})}\right.$$

$$\left.\left.-\frac{(2m-1)}{4\beta}g_{2m-1,n}\right\}\right\} \tag{2.77}$$

The first term in equation (2.77) represents the moment of the solidified liquid. The above results are valid for $\beta\neq 1/4$ and $3/4$. Bauer (1964e) analyzed the liquid oscillations in different container geometries describing roll oscillations. He also estimated the effective mass moment of inertia.

2.4 90°-sector cylindrical tank

For a sector tank of any angle, the ratio of the inner to outer radii is zero. For a 90°-sector tank, Bauer (1963e, 1964a) determined velocity potential functions under translational, pitching and yawing, and roll oscillations.

2.4.1 Lateral excitation

Under translational excitation, $X_0\sin\Omega t$, along the x-axis that passes through half the liquid depth the velocity potential function is given by the following expression (Bauer, 1963e),

$$\Phi(r,\theta,z,t)=\Omega X_0\sin\Omega t\left[r\cos\theta+a_m b_{mn}\frac{\Omega^2}{\omega^2_{mn}-\Omega^2_{2m}}J_{2m}\left(\varepsilon_{mn}\frac{r}{R}\right)\right.$$

$$\left.\cos(2m\theta)\frac{\cosh\left(\varepsilon_{mn}((h+z)/R)\right)}{\cosh\left(\varepsilon_{mn}(h/R)\right)}\right] \tag{2.78}$$

where ε_{mn} are the roots of $dJ_{2m}(\varepsilon_{mn}r/R)/dr|_{r=R}=0$, $a_0=2/\pi$, $a_m=(-1)^{m+1}/(\pi(m^2-(1/4)))$,

$$b_{mn}=\frac{16R\varepsilon_{mn}(m^2-(1/4))}{(\varepsilon^2_{mn}-4m^2)J^2_{2m}(\varepsilon_{mn})}\sum_{k=0}^{\infty}\frac{J_{2m+2k+1}(\varepsilon_{mn})}{(2m+2k-1)(2m+2k+3)}$$

The first term (rigid body potential) satisfies the boundary conditions at the container walls. For the second part (disturbance potential) the boundary conditions vanish at the walls. The free-surface displacement measured from the undisturbed position is

$$\eta = \frac{\Omega^2 R}{g} X_0 \sin(\Omega t)\left[\frac{r}{R}\cos\theta + \sum_{m=0}^{\infty}\sum_{n=1}^{\infty} a_m b_{mn}\frac{\Omega^2}{\omega_{mn}^2-\Omega^2} J_{2m}\left(\varepsilon_{mn}\frac{r}{R}\right)\cos(2m\theta)\right] \tag{2.79}$$

The hydrodynamic force acting along the x-direction is obtained by integrating the pressure distribution over the tank walls and bottom

$$\begin{aligned} F_x &= R\int_0^{\pi/2}\int_{-h}^{0} p_{\text{wall}}\cos\theta\, d\theta\, dz - \int_0^{R}\int_{-h}^{0} p_{\theta=\pi/2}\, dr\, dz \\ &= m_f\Omega^2 X_0 \sin\Omega t\left[1+\frac{a_m b_{mn}}{\pi R}\frac{4\Omega^2(-1)^{m+1}}{(\omega_{mn}^2-\Omega^2)}\tanh(\varepsilon_{mn}h/R)\right. \\ &\qquad\left.\times\left\{\frac{J_{2m}(\varepsilon_{mn})}{(4m^2-1)}+L_0(\varepsilon_{mn})\right\}\right] \end{aligned} \tag{2.80}$$

where $L_0(\varepsilon_{mn}) = \frac{2}{\varepsilon_{mn}}\sum_{k=0}^{\infty} J_{2m+2k+1}(\varepsilon_{mn})$.

The hydrodynamic force along the y-direction is

$$F_y = m_f\Omega^2 X_0 \sin\Omega t\frac{4a_m b_{mn}}{\pi R}\frac{\Omega^2\tanh(\varepsilon_{mn}h/R)}{(\omega_{mn}^2-\Omega^2)}\left\{\frac{J_{2m}(\varepsilon_{mn})}{(4m^2-1)}+L_0(\varepsilon_{mn})\right\} \tag{2.81}$$

Figure 2.10(a) and (b) show the dependence of the force components, $F_x/(X_0\rho R^3\sin\Omega t)$ and $F_y/(X_0\rho R^3\sin\Omega t)$, on the excitation frequency Ω.

The hydrodynamic moment about the x'-axis that passes through the point (0, 0, $-h/2$) is

$$\begin{aligned} M_{y'} &= R\int_0^{\pi/2}\int_{-h}^{0} p_{\text{wall}}\left(\frac{h}{2}+z\right)\cos\theta\, d\theta\, dz - \int_0^{R}\int_{-h}^{0} p_{\theta=\pi/2}\left(\frac{h}{2}+z\right) dr\, dz + \int_0^{\pi/2}\int_0^{R} p_{\text{bottom}} r^2\cos\theta\, d\theta\, dr \\ &= m_f\Omega^2 R X_0\sin\Omega t\left[\frac{R}{4h}+\frac{a_m b_{mn}}{\pi R\varepsilon_{mn}}\frac{\Omega^2(-1)^{m+1}}{(\omega_{mn}^2-\Omega^2)}\left\{\left[\frac{J_{2m}(\varepsilon_{mn})}{(4m^2-1)}+L_0(\varepsilon_{mn})\right]\right.\right. \\ &\qquad\times\left[\tanh(\varepsilon_{mn}h/R)+\frac{2R}{\varepsilon_{mn}h}\left(\frac{1}{\cosh(\varepsilon_{mn}h/R)}-1\right)\right] \\ &\qquad\left.\left.+\frac{2\varepsilon_{mn}^2 R L_2(\varepsilon_{mn})}{(4m^2-1)\varepsilon_{mn}h\cosh(\varepsilon_{mn}h/R)}\right\}\right]+\frac{4m_f gR}{3\pi} \end{aligned}$$

The last expression is the static moment because the reference axis does not pass through the center of mass. The moment about the x'-axis that passes through the point (0, 0, $-h/2$) is

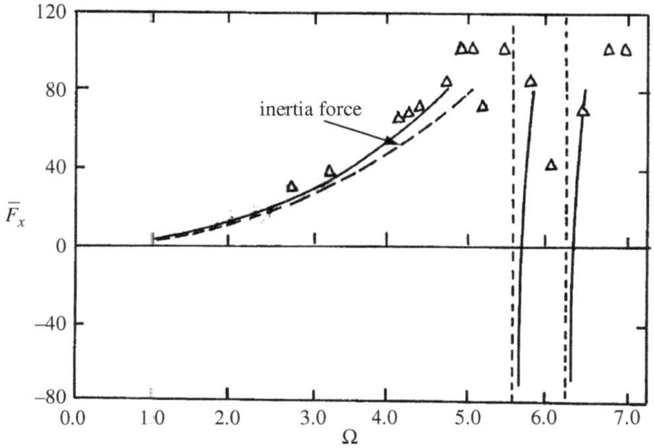

(a) $\overline{F}_x = F_x / [\rho X_0 R^3 \sin \Omega t]$

Δ experimental results

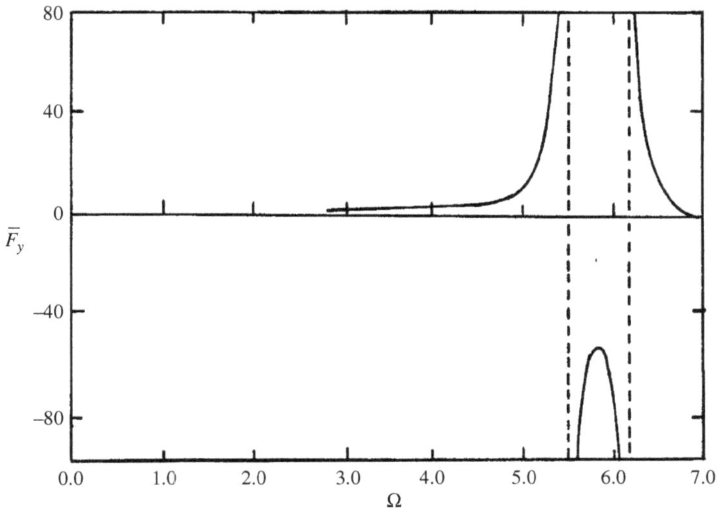

(b) $F_y = \overline{F}_y / [\rho X_0 R^3 \sin \Omega t]$

Figure 2.10 Dependence of hydrodynamic forces along (a) x-, and (b) y-axes on excitation frequency under lateral excitation along x-direction in 90°-sector cylindrical tank for $g/R = 10\,\mathrm{s}^{-2}$,and $h/R = 4.0$. (Bauer, 1964a)

$$\begin{aligned} M_x = m_f \Omega^2 R X_0 \sin \Omega t \Bigg[\frac{R}{\pi h} - \frac{2a_m b_{mn}}{\pi R \varepsilon_{mn}} \frac{\Omega^2}{(\omega_{mn}^2 - \Omega^2)} \\ \times \Bigg\{ \left[\frac{J_{2m}(\varepsilon_{mn})}{(4m^2-1)} + L_0(\varepsilon_{mn}) \right] \\ \times \left[\tanh(\varepsilon_{mn} h/R) + \frac{2R}{\varepsilon_{mn} h} \left(\frac{1}{\cosh(\varepsilon_{mn} h/R)} - 1 \right) \right] \\ + \frac{2\varepsilon_{mn}^2 R L_2(\varepsilon_{mn})}{(4m^2-1)\varepsilon_{mn} h \cosh(\varepsilon_{mn} h/R))} \Bigg\} \Bigg] - \frac{4 m_f g R}{3\pi} \end{aligned} \tag{2.83}$$

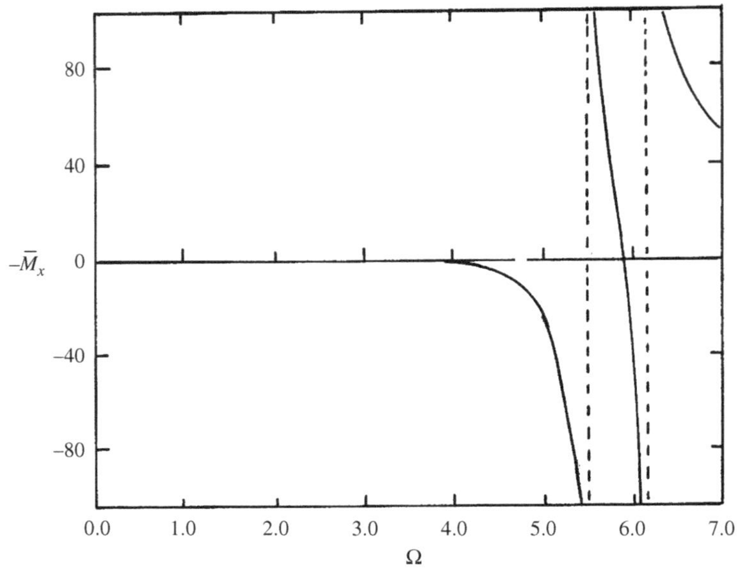

(a) $\bar{M}_x = M_x/[\rho X_0 R^4 \sin \Omega t]$

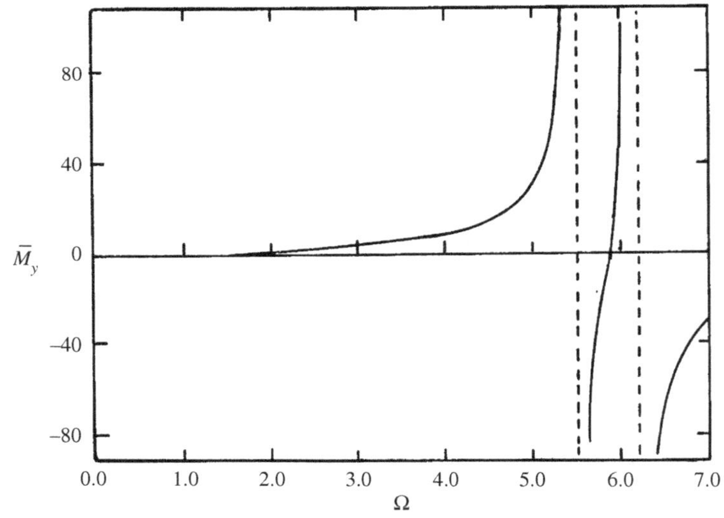

(b) $\bar{M}_y = M_y/[\rho X_0 R^4 \sin \Omega t]$

Figure 2.11 Dependence of hydrodynamic moments about (a) x-, and (b) y-axes on excitation frequency under lateral excitation along x-direction in 90°-sector cylindrical tank for $g/R = 10\,\text{sec}^{-2}$, and $h/R = 4.0$. (Bauer, 1964a)

where $L_2(\varepsilon_{mn}) = \dfrac{4(m^2-1)}{\varepsilon_{mn}} \sum_{k=0}^{\infty} \dfrac{J_{2m+2k+1}(\varepsilon_{mn})}{(2m+2k-1)(2m+2k+3)}$.

Figure 2.11(a) and (b) shows the dependence of the moment components $-M_{x'}/(X_0 \rho R^4 \sin \Omega t)$ and $M_{y'}/(X_0 \rho R^4 \sin \Omega t)$ on the excitation frequency.

2.4.2 Pitching excitation

Under pitching excitation, $\psi(t) = \psi_0 \sin \Omega t$, the velocity potential is

$$\Phi(r,\theta,z,t) = -\Omega\psi_0 \cos\Omega t \left\{ rz\cos\theta - \sum_{m=0}^{\infty}\sum_{n=0}^{\infty}\left[A_{mn}\cosh\left(\varepsilon_{mn}\frac{z}{R}\right) + B_{mn}\sinh\left(\varepsilon_{mn}\frac{z}{R}\right)\right] \times J_{2m}(\varepsilon_{2m-1,n}r/R)\cos(2m\theta)\right\} \tag{2.84}$$

where the coefficients A_{mn} and B_{mn} are given by relations (2.54a,b), but after replacing ξ_{mn} by ε_{mn}. The surface wave height is given by the expression, $\eta_{z=-h/2} = (1/g)\partial\Phi/\partial t$,

$$\eta = \frac{\Omega^2 R}{g} R\psi_0 \sin\Omega t\left\{\frac{r}{R}\frac{h}{2R}\cos\theta - \sum_{m=0}^{\infty}\sum_{n=0}^{\infty}\left[A_{mn}\cosh\left(\varepsilon_{mn}\frac{h}{2R}\right) + B_{mn}\sinh\left(\varepsilon_{mn}\frac{h}{2R}\right)\right] \times J_{2m}(\varepsilon_{mn}r/R)\cos(2m\theta)\right\} \tag{2.85}$$

The pressure distribution is

$$\begin{aligned} p &= -\rho\frac{\partial\Phi}{\partial t} + g\rho\left[\frac{h}{2} - z + \psi_0(R - r\cos\theta)\sin\Omega t\right] \\ &= \rho\Omega^2\psi_0\sin\Omega t\left[-rz\cos\theta + \sum_{m=0}^{\infty}\sum_{n=0}^{\infty}\frac{2Ra_m b_{mn}\sinh(\varepsilon_{mn}z/R)}{\varepsilon_{mn}\cosh(\varepsilon_{mn}h/2R)}\cos 2m\theta\right] \\ &\quad + g\rho\left[\frac{h}{2} - z + \psi_0(R - r\cos\theta)\sin\Omega t\right] \end{aligned} \tag{2.86}$$

The forces along the x- and y-axes are, respectively,

$$\begin{aligned} F_x = 8m_f\frac{\Omega^2 R}{\pi}R\psi_0\sin\Omega t &\times \sum_{m=0}^{\infty}\sum_{n=0}^{\infty}(-1)^{m+1}\frac{A_{mn}}{\varepsilon_{mn}h}\sinh\left(\varepsilon_{mn}\frac{h}{2R}\right)\left[\frac{J_{2m}(\varepsilon_{mn})}{(4m^2-1)} + L_0(\varepsilon_{mn})\right] \\ &- m_f g\psi_0\sin(\Omega t) \end{aligned} \tag{2.87}$$

$$\begin{aligned} F_y = -8m_f\frac{\Omega^2 R}{\pi}R\psi_0\sin\Omega t\sum_{m=0}^{\infty}\sum_{n=0}^{\infty}\frac{A_{mn}}{\varepsilon_{mn}h}\sinh\left(\varepsilon_{mn}\frac{h}{2R}\right)\left[\frac{J_{2m}(\varepsilon_{mn})}{(4m^2-1)} + L_0(\varepsilon_{mn})\right] \\ - m_f g\,\psi_0\sin(\Omega t) \end{aligned} \tag{2.88}$$

Figure 2.12(a) and (b) shows the dependence of force components on the excitation frequency for liquid depth ratio $h/2R = 2.0$. The hydrodynamic moments about the x'- and y'- axes, that pass through the point $(0,0,-h/2)$, are

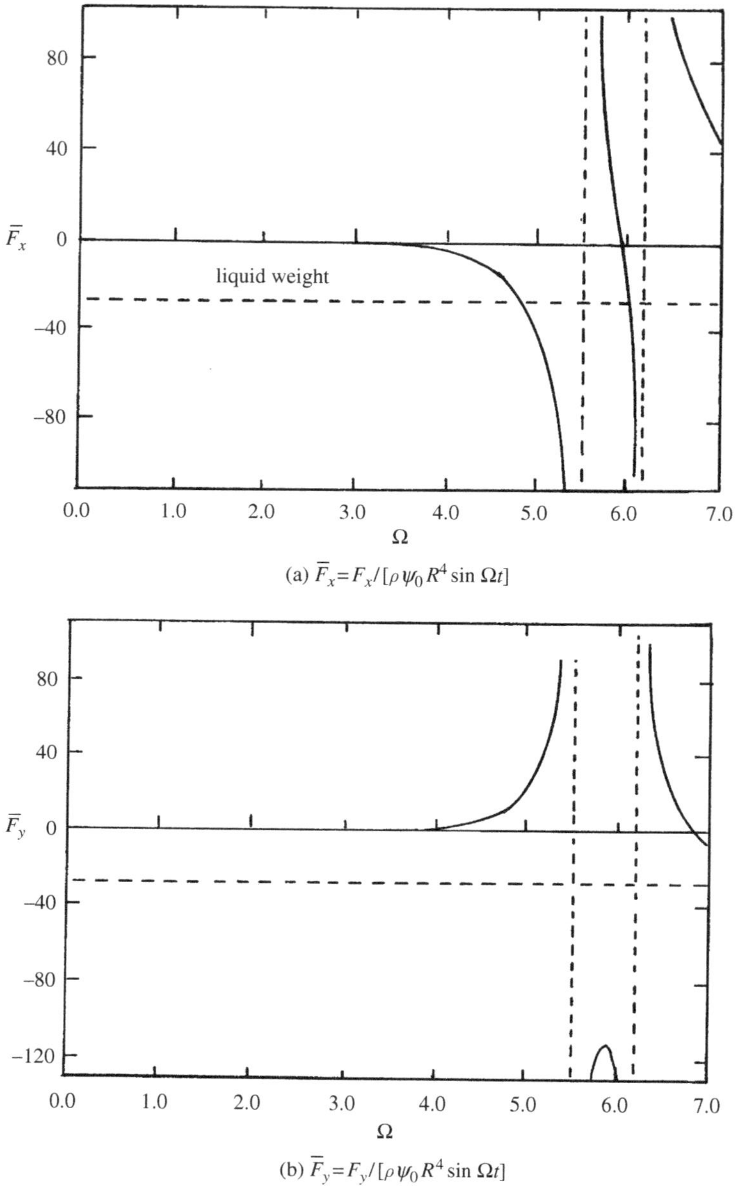

Figure 2.12 Dependence of hydrodynamic forces along (a) x-, and (b) y-axes on excitation frequency under pitching excitation about y-axis in 90° -sector cylindrical tank for $g/R = 10\text{s}^{-2}$, and $h/R = 4\ 0$. (Bauer, 1964a)

$$M_x = -\, m_f R^2 \Omega^2 \psi_0 \sin \Omega t$$
$$\times \left\{ \frac{1}{4\pi} - \sum_{m=0}^{\infty} \sum_{n=0}^{\infty} \frac{4}{\pi \varepsilon_{mn}} \left[B_{mn} \left\{ \cosh\left(\varepsilon_{mn} \frac{h}{2R}\right) - \frac{2R}{\varepsilon_{mn} h} \sinh\left(\varepsilon_{mn} \frac{h}{2R}\right) \right\} \right. \right.$$

$$\times\left\{\frac{J_{2m}(\varepsilon_{mn})}{(4m^2-1)}+L_0(\varepsilon_{mn})\right\}+\frac{\varepsilon_{mn}^2 L_2(\varepsilon_{mn})}{(4m^2-1)(\varepsilon_{mn}h/R)}$$
$$\times\left\{A_{mn}\cosh\left(\varepsilon_{mn}\frac{h}{2R}\right)-B_{mn}\sinh\left(\varepsilon_{mn}\frac{h}{2R}\right)\right\}\Bigg]\Bigg\}-\frac{m_f Rg}{2\pi(h/R)}\psi_0\sin\Omega t$$
$$-\frac{4m_f Rg}{3\pi} \tag{2.89}$$

$$M_y=m_f R^2\Omega^2\psi_0\sin\Omega t\left\{\frac{1}{8}-\frac{h^2}{12R^2}+\sum_{m=0}^{\infty}\sum_{n=0}^{\infty}\frac{4(-1)^{m+1}}{\pi\varepsilon_{mn}}\right.$$
$$\times\left[B_{mn}\left\{\cosh\left(\varepsilon_{mn}\frac{h}{2R}\right)-\frac{2R}{\varepsilon_{mn}h}\sinh\left(\varepsilon_{mn}\frac{h}{2R}\right)\right\}\left\{\frac{J_{2m}(\varepsilon_{mn})}{(4m^2-1)}+L_0(\varepsilon_{mn})\right\}\right.$$
$$\left.\left.+\frac{\varepsilon_{mn}^2 L_2(\varepsilon_{mn})}{(4m^2-1)(\varepsilon_{mn}h)/R}\times\left\{A_{mn}\cosh\left(\varepsilon_{mn}\frac{h}{2R}\right)-B_{mn}\sinh\left(\varepsilon_{mn}\frac{h}{2R}\right)\right\}\right]\right\}$$
$$-\frac{m_f Rg}{4h/R}\psi_0\sin\Omega t+\frac{4m_f Rg}{3\pi} \tag{2.90}$$

The effective mass moment of inertia of the liquid is

$$I_y=m_f R^2\left\{\frac{h^2}{12R^2}-\frac{1}{4}+\frac{2(-1)^m}{\pi R\varepsilon_{mn}^2}\left\{1-\frac{2R}{\xi_{mn}h}\tanh\left(\xi_{mn}\frac{h}{2R}\right)\right\}\right.$$
$$\times\left\{\frac{J_{2m}(\varepsilon_{mn})}{(4m^2-1)}+L_0(\varepsilon_{mn})\right\}$$
$$\left.-\frac{2\varepsilon_{mn}^2 L_2(\varepsilon_{mn})\tanh((\xi_{mn}h)/2R)}{(4m^2-1)((\xi_{mn}h)/R)}\right\} \tag{2.91}$$

Figure 2.13(a) and (b) shows the dependence of moment components of excitation frequency for liquid depth ratio $h/2R = 2.0$.

2.4.3 Roll excitation

Under roll excitation, $\varphi(t) = o_0 \sin \Omega t$, the velocity potential is

$$\Phi(r,\theta,z,t)=\Omega\varphi_0\cos\Omega t\left\{r^2\left(\theta-\frac{\pi}{4}\right)+\frac{2R^2\cos[(4m-2)\theta]}{\pi(2m-1)\left[(2m-1)^2-1\right]}\right.$$
$$\times\left[\left(\frac{r}{R}\right)^{4m-2}-\frac{(r/R)^2}{(2m-1)}\right]+\frac{2R^2}{\pi}\left(\frac{r}{R}\right)^2\left[\ln\left(\frac{r}{R}\right)+\frac{1}{2}\right]\cos 2\theta$$
$$+\frac{2R^2}{\pi}\frac{\left[f_{2m-1,n}-e_{2m-1,n}(2m-1)\right]}{(2m-1)\left[(2m-1)^2-1\right]}\frac{\Omega^2 J_{4m-2}\left(\varepsilon_{2m-1,n}\frac{r}{R}\right)}{(\omega_{2m-1,n}^2-\Omega^2)}$$

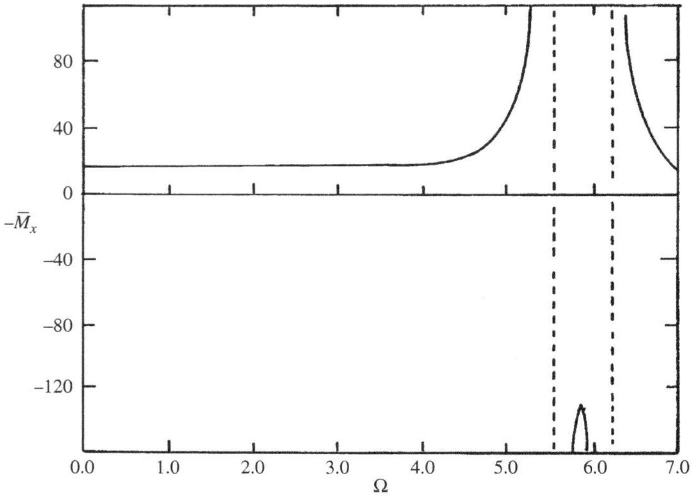

(a) $\bar{M}_x = M_x / [\rho \psi_0 R^4 \sin \Omega t]$

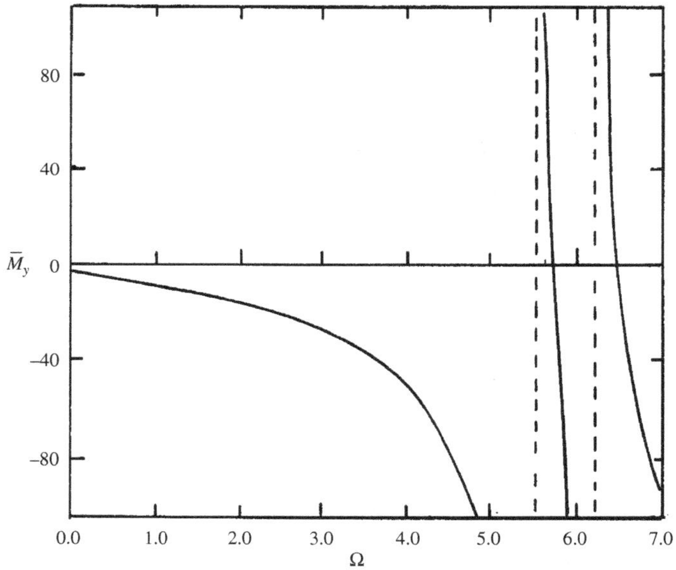

(b) $\bar{M}_y = M_y / [\rho \psi_0 R^4 \sin \Omega t]$

Figure 2.13 Dependence of hydrodynamic moments about (a) x-, and (b) y-axes on excitation frequency under pitching excitation about y-axis in 90°-sector cylindrical tank for $g/R = 10\,\mathrm{s}^{-2}$, and $h/R = 4.0$. (Bauer, 1964a)

$$\times \frac{\cosh\left(\varepsilon_{2m-1,n}\dfrac{(h+z)}{R}\right)}{\cosh\left(\varepsilon_{2m-1,n}\dfrac{h}{R}\right)} \cos((4m-2)\theta)$$
$$\left. + \frac{R^2}{\pi} \frac{(2f_n - e_n)\Omega^2 J_2(\varepsilon_{2m-1,n}\frac{r}{R})}{(\omega_{2m-1,n}^2 - \Omega^2)} \frac{\cosh\left(\varepsilon_{2m-1,n}\dfrac{(h+z)}{R}\right)}{\cosh\left(\varepsilon_{2m-1,n}\dfrac{h}{R}\right)} \cos 2\theta \right\} \tag{2.92}$$

where

$$f_n = \frac{-8}{(\bar{\varepsilon}_n^2 - 4)J_2^2(\bar{\varepsilon}_n)} \sum_{k=0}^{\infty} \frac{J_{2k+4}(\bar{\varepsilon}_n)}{(k+1)(k+3)}$$

$$f_{2m-1,n} = \frac{-4(2m-1)}{\left[\bar{\varepsilon}_{2m-1,n}^2 - 4(2m-1)^1\right] J_{4m-2}^2(\bar{\varepsilon}_{2m-1})}$$

$$e_n = \frac{4}{(\bar{\varepsilon}_n^2 - 4)J_2(\bar{\varepsilon}_n)}$$

$$e_{2m-1,n} = \frac{2m(2m-1)\bar{\varepsilon}_{2m-1,n}}{(m-1)\left[\bar{\varepsilon}_{2m-1,n}^2 - 4(2m-1)^2\right] J_{4m-2}^2(\bar{\varepsilon}_{2m-1,n})}$$

$$\times \sum_{k=0}^{\infty} \frac{(4m+2k-1)J_{4m+2k-1}(\bar{\varepsilon}_{2m-1,n})}{(2m+2k-2)(2m+2k-1)(2m+k+1)}$$

The values of $\bar{\varepsilon}_n$ are the roots of $dJ_2\,(\bar{\varepsilon}_n r/R)/dr|_{r=R} = 0$, and the values of $\bar{\varepsilon}_{2m-1,n}$ are the roots of $dJ_{4m-2}(\bar{\varepsilon}_{2m-1,n} r/R)/dr\,_{r=R} = 0$. The fluid-free-surface displacement is

$$\begin{aligned}
\eta(r,\theta,t) = \frac{\Omega^2 R}{g} R\varphi_0 \sin\Omega t \Bigg\{ & \frac{r^2}{R^2}\left(\theta - \frac{\pi}{4}\right) + \frac{2}{\pi}\left(\frac{r}{R}\right)^2 \left[\ln\left(\frac{r}{R}\right) + \frac{1}{2}\right] \cos 2\theta \\
& + \frac{2}{\pi} \frac{\cos((4m-2)\theta)}{(2m-1)\left[(2m-1)^2 - 1\right]} \left[\left(\frac{r}{R}\right)^{4m-2} - \frac{(r/R)^2}{(2m-1)}\right] \\
& + \frac{\Omega^2}{\pi(\omega_{2m-1,n}^2 - \Omega^2)} [2f_n - e_n(2m-1)] J_2\left(\bar{\varepsilon}_{2m-1,n}\frac{r}{R}\right) \cos 2\theta \\
& + \frac{2}{\pi} \frac{\Omega^2}{(\omega_{2m-1,n}^2 - \Omega^2)} \frac{(f_{2m-1,n} - (2m-1)e_{2m-1,n}) J_{4m-2}(\bar{\varepsilon}_{2m-1,n}(r/R))}{(2m-1)\left[(2m-1)^2 - 1\right]} \\
& \times \cos((4m-2)\theta) \Bigg\}
\end{aligned} \tag{2.93}$$

The hydrodynamic force components along the x- and y-directions are

$$F_x = -F_y = m_f \Omega^2 R\varphi_0 \sin(\Omega t) \left\{ \frac{2}{3} - \frac{4}{\pi} + \frac{16}{9\pi^2} + \frac{8}{\pi^2} \frac{\left(\frac{1}{4m-1} - \frac{1}{3(2m-1)}\right)}{(2m-1)[(2m-1)^2 - 1]} \right.$$

$$+ \frac{1}{8m(2m-1)^2\left[(2m-1)^2 - (1/4)\right]}$$

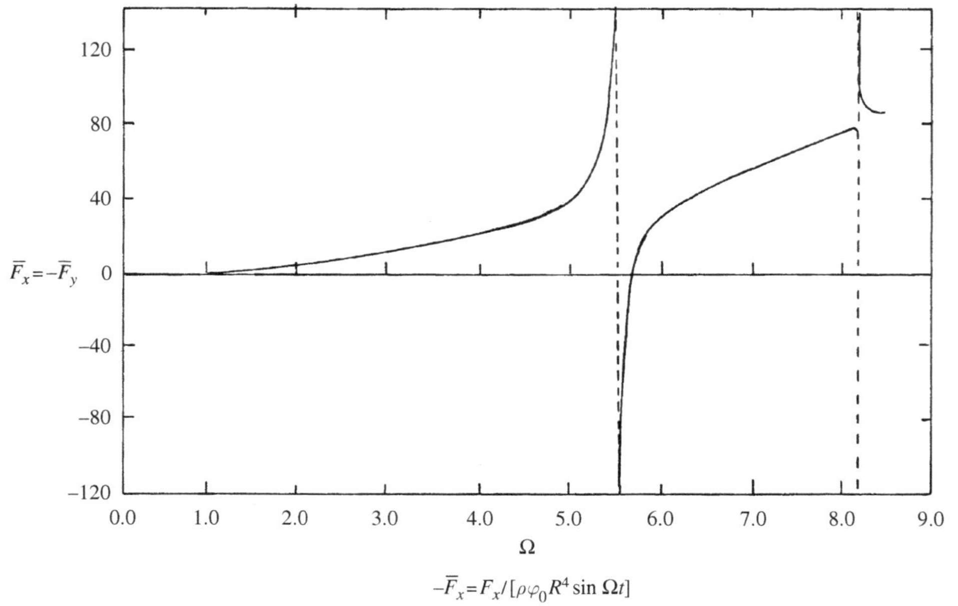

Figure 2.14 Dependence of hydrodynamic forces along x- and y-axes on excitation frequency under roll excitation about z-axis in 90°-sector cylindrical tank for $g/R = 10\,\mathrm{s}^{-2}$, and $h/R = 4.0$. (Bauer, 1964a)

$$+\frac{4}{\pi^2}\frac{(2f_n - e_n)\Omega^2}{(\omega^2_{2m-1,n} - \Omega^2)}\frac{\tanh(\bar{\varepsilon}_{2m-1,n}(h/R))}{(\bar{\varepsilon}_{2m-1,n}h/R)}\left[\frac{1}{3}J_2(\bar{\varepsilon}_n) + L_0(\bar{\varepsilon}_n)\right]$$

$$+\frac{8R}{\pi^2 h}\frac{\Omega^2(f_{2m-1,n}/(2m-1) - e_{2m-1,n})}{(\omega^2_{2m-1,n} - \Omega^2)\bar{\varepsilon}_{2m-1,n}}\frac{\tanh(\bar{\varepsilon}_{2m-1,n}(h/R))}{\left[(2m-1)^2 - 1\right]}$$

$$\left.\times\left[\frac{J_{4m-2}(\bar{\varepsilon}_{2m-1,n})}{4\left[(2m-1)^2 - (1/4)\right]} + L_0(\bar{\varepsilon}_{2m-1,n})\right]\right\} \tag{2.94}$$

Figure 2.14 shows the force components $F_x/(\varphi R^4(\rho_0 \sin\Omega t) = -F_y/(\rho R^4\varphi_0 \sin\,\Omega t)$ on excitation frequency. The hydrodynamic moments about the x'- and y'-axes, through the point $(0,0,-h/2)$ are

$$M_{x'} = M_{y'} = m_f\Omega^2 R^2\varphi_0\sin(\Omega t)\left\{\frac{R}{5h}\left(1 - \frac{4}{\pi}\right) + \frac{4R}{25\pi^2 h}\right.$$

$$+\frac{12R}{5\pi^2 h}\left[\frac{2(m-1)}{4[(2m-1)^2 - (1/4)][(2m-1)^2 - 1](2m + (1/2))(2m-1)^2}\right]$$

$$+\frac{2}{\pi^2}\frac{(2f_n - e_n)\Omega^2}{(\omega^2_{2m-1,n} - \Omega^2)\bar{\varepsilon}^2_n}\left[\tanh(\bar{\varepsilon}_{2m-1,n}\frac{h}{R}) + \frac{2}{\bar{\varepsilon}_{2m-1,n}(h/R)}\left(\frac{1}{\cosh(\bar{\varepsilon}_{2m-1,n}(h/R))} - 1\right)\right]$$

$$\times\left[\frac{1}{3}J_2(\bar{\varepsilon}_n)+L_0(\bar{\varepsilon}_n)\right]+\frac{2}{3}\frac{\bar{\varepsilon}_n^2 L_2(\bar{\varepsilon}_n)}{(\bar{\varepsilon}_{2m-1,n}(h/R))\cosh(\bar{\varepsilon}_{2m-1,n}(h/R))}$$

$$+\frac{4}{\pi^2}\frac{\Omega^2(f_{2m-1,n}/(2m-1)-e_{2m-1,n})}{(\omega_{2m-1,n}^2-\Omega^2)\left[(2m-1)^2-1\right]\bar{\varepsilon}_{2m-1,n}}\left[\left\{\tanh(\bar{\varepsilon}_{2m-1,n}(h/R))+\frac{2}{\bar{\varepsilon}_{2m-1,n}(h/R)}\right.\right.$$

$$\left.\times\left(\frac{1}{\cosh(\bar{\varepsilon}_{2m-1,n}h/R)}-1\right)\right\}\left[\frac{J_{4m-2}(\bar{\varepsilon}_{2m-1,n})}{4\left[(2m-1)^2-(1/4)\right]}+L_0(\bar{\varepsilon}_{2m-1,n})\right]$$

$$\left.\left.+\frac{\bar{\varepsilon}_{2m-1,n}^2 L_0(\bar{\varepsilon}_{2m-1,n})}{2\left[(2m-1)^2-(1/4)\right](\bar{\varepsilon}_{2m-1,n}h/R)\cosh(\bar{\varepsilon}_{2m-1,n}(h/R))}\right]\right\}+\frac{4m_f Rg}{3\pi} \tag{2.95}$$

The moment about the z-axis is

$$M_z=m_f\Omega^2R^2\varphi_0\sin(\Omega t)\left\{\frac{1}{2}-\frac{1}{\pi^2}-\frac{1}{\pi^2m^2(2m-1)^2}\right.$$
$$-\frac{8}{\pi^2}\frac{(2f_n-e_n)\,\Omega^2\tanh(\bar{\varepsilon}_{2m-1,n}(h/R))L_1(\bar{\varepsilon}_n)}{(\omega_{2m-1,n}^2-\Omega^2)\bar{\varepsilon}_n^2}$$
$$\left.-\frac{16}{\pi^2}\frac{\Omega^2(f_{2m-1,n}/2m-1-e_{2m-1,n})L_1(\bar{\varepsilon}_{2m-1,n})\tanh(\bar{\varepsilon}_{2m-1,n}(h/R))}{(\omega_{2m-1,n}^2-\Omega^2)\left[(2m-1)^2-1\right]\bar{\varepsilon}_{2m-1,n}(h/R)}\right\} \tag{2.96}$$

where

$$L_0=\frac{2}{\bar{\varepsilon}_{2m-1,n}}\sum_{k=0}^{\infty}J_{4m+2k-1}(\bar{\varepsilon}_{2m-1,n})$$

$$L_1=\frac{2m-1}{\bar{\varepsilon}_{2m-1,n}}\sum_{k=0}^{\infty}\frac{(4m+2k-1)J_{4m+2k-1}(\bar{\varepsilon}_{2m-1,n})}{(2m+k-1)(2m+k)}$$

$$L_2=\frac{2(4m-1)(4m-3)}{\bar{\varepsilon}_{2m-1,n}}\sum_{k=0}^{\infty}\frac{J_{4m+2k-1}(\bar{\varepsilon}_{2m-1,n})}{(4m+2k+1)(4m+2k-3)}$$

The effective mass moment of inertia of the liquid is

$$I_{z\mathrm{eff}}=\frac{1}{2}m_fR^2\left[1-\left(\frac{4}{3}-\frac{16}{\pi^2}\ln 2\right)\right]=\frac{1}{2}m_fR^2(0.7903) \tag{2.97}$$

The first term represents the moment of inertia of the solidified liquid. Figure 2.15 and 2.16 show the moment components, $-M_{x'}/(\rho R^5\varphi_0\sin\Omega t)=-M_{y'}/(\rho R^5\varphi_0\sin\Omega t)$, and $M_{z'}/(\rho R^4\varphi_0\sin\Omega t)$, on excitation frequency. The frequency range covers the first two sloshing modes. The dash-dot curve in Figure 2.16 represents the liquid moment of a nonsloshing solid liquid in an infinitely long quarter tank. The dash curve belongs to the nonsolid liquid in an infinite quarter tank. The solid curves exhibit the response moment of the sloshing liquid.

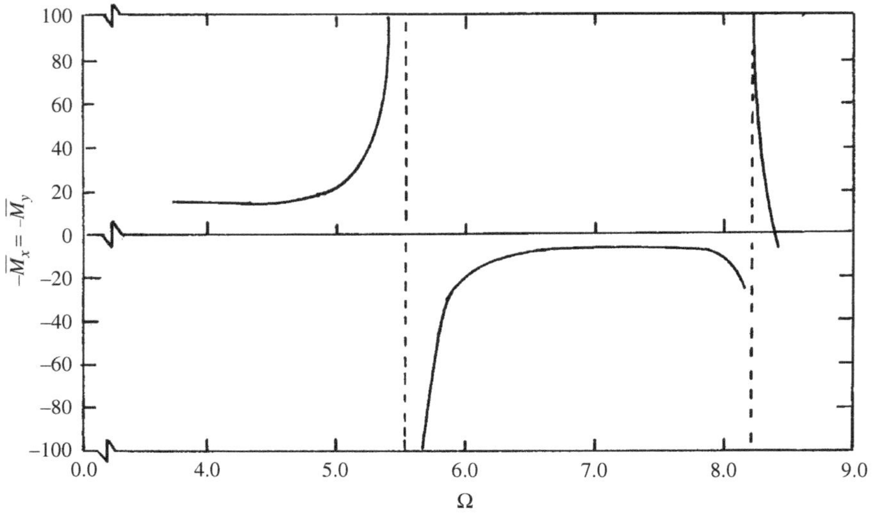

Figure 2.15 Dependence of hydrodynamic moments about x- and y-axes on excitation frequency under roll excitation about z-axis in 90°-sector cylindrical tank for $g/R = 10\,\mathrm{s}^{-2}$, and $h/R = 4.0$. (Bauer, 1964a)

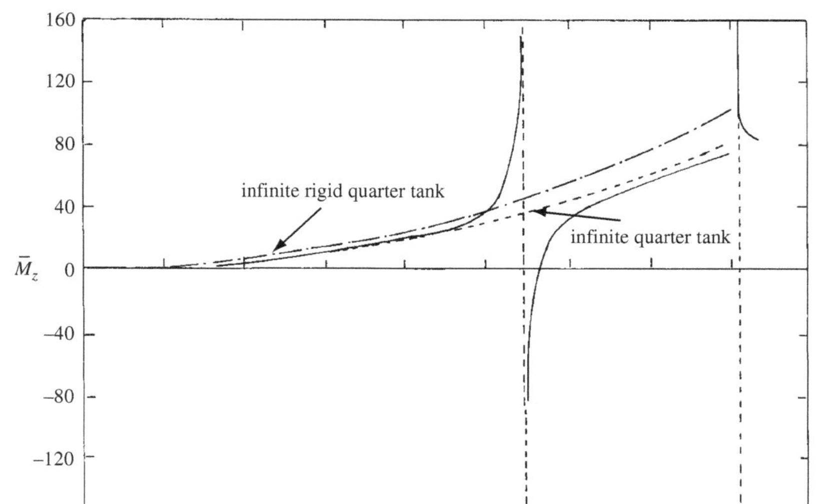

Figure 2.16 Dependence of hydrodynamic moment about z-axis on excitation frequency under roll excitation about z-axis in 90°-sector cylindrical tank for $g/R = 10\,\mathrm{s}^{-2}$, and $h/R = 4.0$. (Bauer, 1964a)

2.5 Annular containers

The liquid-free-surface natural frequencies in annular tanks are modified in such a manner that the frequency of the lowest mode is decreased and additional modes with relatively widely spaced frequencies are introduced. Campbell (1953), Bauer (1960b, 1961c), Petrov (1961), and Aslam, *et al.* (1979) treated the dynamics of fluid free surface under different types of excitations. In this section, the response of the fluid free surface under lateral and pitching excitations will be considered.

2.5.1 Lateral excitation

Consider an annular tank, shown in Figure 2.17 with radii R and R_{in}, under lateral excitation along the x-axis, $X_0 \sin \Omega t$. The velocity potential function is, (Bauer, 1960b)

$$\Phi(r, \theta, z, t) = \Omega X_0 R \cos(\Omega t) \cos \theta \times \left\{ \frac{r}{R} + \sum_{n=0}^{\infty} \frac{\bar{A}_n \cosh(\xi_{n1}(z+h)/R)}{\cosh(\xi_{n1} h/R)} C_1\left(\xi_{n1}\frac{r}{R}\right) \frac{\Omega^2}{(\omega_{n1}^2 - \Omega^2)} \right\} \tag{2.98}$$

where $\bar{A}_n = \dfrac{2[(2/\pi\xi_{n1}) - kC_1(k\xi_{n1})]}{(2/\pi\xi_{n1})^2(\xi_{n1}^2 - 1) + C_1^2(k\xi_{n1})(1 - k^2\xi_{n1}^2)}, \quad k = R_{in}/R,$

The waveform of the fluid free surface, measured from the undisturbed level is

$$\eta = \frac{\Omega^2 R}{g} X_0 \sin \Omega t \cos \theta \left\{ \frac{r}{R} + \sum_{n=0}^{\infty} \bar{A}_n C_1\left(\xi_{n1}\frac{r}{R}\right) \frac{\Omega^2}{(\omega_{n1}^2 - \Omega^2)} \right\} \tag{2.99}$$

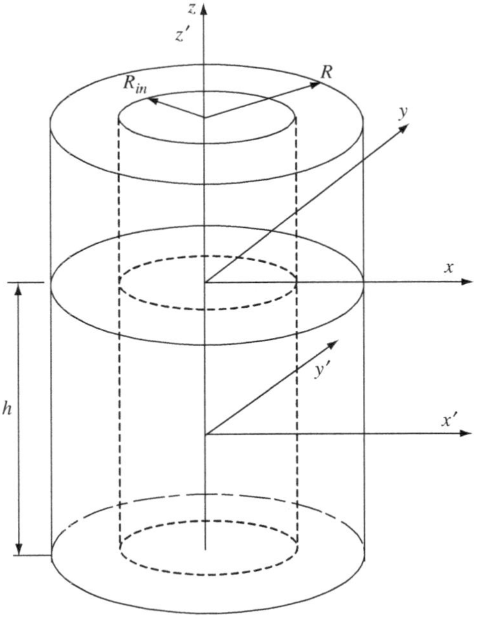

Figure 2.17 Annular tank showing the coordinate system.

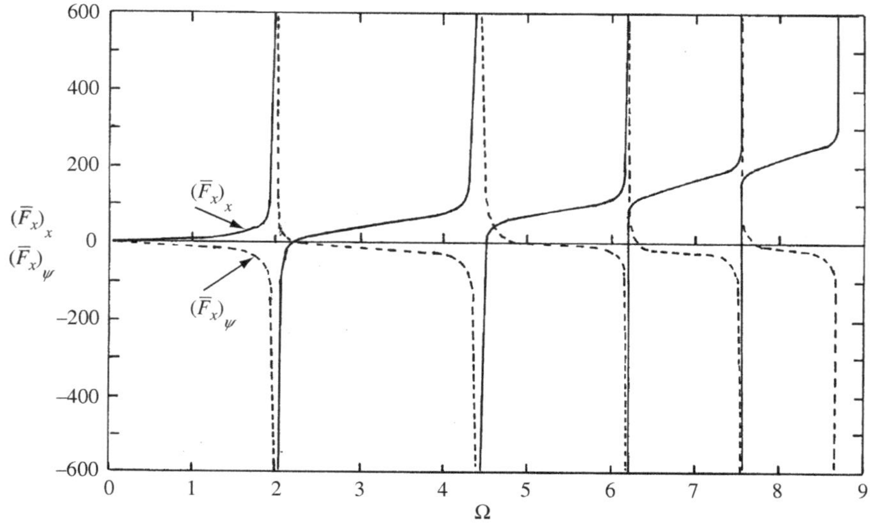

Figure 2.18 Dependence of hydrodynamic forces along x-axis on excitation frequency under translational and pitching excitations in an annular cylindrical tank for $k = 0.5$, $g/R = 3\,\mathrm{s}^{-2}$, and $h/R = 2.0$. (Bauer, 1960b)

The pressure distribution in the tank at depth $-z$ is

$$p = \rho\Omega^2 R X_0 \sin\Omega t\left\{\frac{r}{R} + \sum_{n=0}^{\infty}\frac{\bar{A}_n\cosh\left(\xi_{n1}(z+h/R)\right)}{\cosh\left(\xi_{n1}h/R\right)}C_1\left(\xi_{n1}\frac{r}{R}\right)\frac{\Omega^2}{(\omega_{n1}^2-\Omega^2)}\right\} - \rho gz \quad (2.100)$$

The hydrodynamic force along the x-direction is obtained by integrating equation (2.100) over the projected wetted area

$$F_x = m_f\Omega^2 X_0\sin\Omega t\left\{1 + \sum_{n=0}^{\infty}\frac{\bar{A}_n\tanh(\xi_{n1}h/R)}{(1-k^2)(\xi_{n1}h/R)}\left[\frac{2}{\pi\xi_{n1}} - kC_1(k\xi_{n1})\right]\frac{\Omega^2}{(\omega_{n1}^2-\Omega^2)}\right\} \quad (2.101)$$

It is not difficult to show that the force along the y-direction vanishes. Figure 2.18 shows the dependence of the hydrodynamic force $(\bar{F}_x)_x = F_x/[\rho X_0 R^3\sin\Omega t]$ along the x-direction, shown by solid curves on the excitation frequency Ω.

The hydrodynamic moment with respect to the center of mass of the undisturbed liquid about the y'-axis is

$$M_{y\,\mathrm{c.g.}} = m_f\Omega^2 R X_0\sin\Omega t\left\{\frac{1+k^2}{4h/R} + \sum_{n=0}^{\infty}\frac{\bar{A}_n((2/\pi\xi_{n1}) - kC_1(k\xi_{n1}))}{2\xi_{n1}(1-k^2)}\right.$$
$$\left.\times\frac{\Omega^2}{(\omega_{n1}^2-\Omega^2)}\left[\tanh\left(\xi_{n1}\frac{h}{R}\right) + \frac{2}{(\xi_{n1}h/R)}\left(\frac{2}{\cosh(\xi_{n1}h/R)} - 1\right)\right]\right\} \quad (2.102)$$

The first term represents the moment due to the lateral displacement of the center of mass for a plane free surface. The moment about the z-axis is zero because the fluid is assumed to be inviscid.

2.5.2 Pitching excitation

Under pitching excitation, $\psi_0 \sin \Omega t$, about the y'-axis passing through the liquid center of gravity, the velocity potential function is

$$\begin{aligned}\Phi(r,\theta,z,t) = \Omega\psi_0 R^2 \cos(\Omega t)\cos\theta\Bigg\{\frac{r}{R}\frac{z}{R} + \sum_{n=0}^{\infty}\frac{\bar{A}_n C_1(\xi_{n1}(r/R))}{\cosh(\xi_{n1}h/R)}\frac{\Omega^2}{(\omega_{n1}^2-\Omega^2)} \\ \times\Bigg[\left(\frac{g}{\Omega^2 R}+\frac{1}{2}\frac{h}{R}\right)\cosh\left[\frac{\xi_{n1}}{R}\left(\frac{h}{2}+z\right)\right] \\ -\frac{4g}{\Omega^2 R}\sinh\left(\xi_{n1}\frac{z}{R}\right)\sinh\left(\xi_{n1}\frac{h}{2R}\right) \\ -\frac{2}{\xi_{n1}}\sinh\left[\frac{\xi_{n1}}{R}\left(\frac{h}{2}-z\right)\right]\Bigg]\Bigg\}\end{aligned} \tag{2.103}$$

The wave height of the fluid free surface measured in the inertial system is

$$\begin{aligned}\eta(r,\theta,t) = -\Omega^2\psi_0 R\sin(\Omega t)\cos\theta\Bigg\{\frac{((h/2R)(r/R))}{(g/R)} \\ +\sum_{n=0}^{\infty}\frac{\bar{A}_n\Omega^2 C_1(\xi_{n1}(r/R))}{(\omega_{n1}^2-\Omega^2)}\left(\frac{(h/2R)}{(g/R)}-\frac{1}{\Omega^2}+\frac{2}{\Omega^2\cosh(\xi_{n1}h/R)}\right)\Bigg\}\end{aligned} \tag{2.104}$$

The free-surface displacement in the tank is $\tilde{\eta} = \eta - \psi_0 r \sin\Omega t\cos\theta$. The pressure distribution in the tank at depth z is

$$\begin{aligned}p &= \rho\frac{\partial\Phi}{\partial t} + \rho g\left[\frac{h}{2} - z - r\psi_0\sin\Omega t\cos\theta\right] \\ &= -\rho\Omega^2\psi_0 R^2\sin(\Omega t)\cos\theta\Bigg\{((h/R)(z/R)) \\ &\quad + \sum_{n=0}^{\infty}\frac{\bar{A}_n C_1(\xi_{n1}(r/R))}{\cosh(\xi_{n1}h/R)}\frac{\Omega^2}{(\omega_{n1}^2-\Omega^2)}\Bigg[\left(\frac{g}{\Omega^2 R}+\frac{1}{2}\frac{h}{R}\right)\cosh\left[\frac{\xi_{n1}}{R}\left(\frac{h}{2}+z\right)\right] \\ &\quad -\frac{4g}{\Omega^2 R}\sinh\left(\xi_{n1}\frac{z}{R}\right)\sinh\left(\xi_{n1}\frac{h}{2R}\right)-\frac{2}{\xi_{n1}}\sinh\left[\frac{\xi_{n1}}{R}\left(\frac{h}{2}-z\right)\right]\Bigg]\Bigg\} \\ &\quad + \rho g\left[\frac{h}{2} - z - r\psi_0\sin\Omega t\cos\theta\right]\end{aligned} \tag{2.105}$$

The fluid force component along the x-axis is

$$\begin{aligned}F_x = -m_f\psi_0\sin(\Omega t)\Bigg\{g + \Omega^2 R\sum_{n=0}^{\infty}\frac{\bar{A}_n\Omega^2\{(2/\xi_{n1}) - kC_1(k\xi_{n1})\}}{\xi_{n1}(\omega_{n1}^2-\Omega^2)(1-k^2)} \\ \times\left[\left\{\frac{g/(R\Omega^2)}{(h/R)}+\frac{1}{2}\right\}\tanh\left(\xi_{n1}\frac{h}{R}\right)+\frac{2}{(\xi_{n1}h/R)}\left\{\frac{2}{\cosh(\xi_{n1}h/R)}-1\right\}\right]\Bigg\}\end{aligned} \tag{2.106a}$$

$$F_y = 0 \tag{2.106b}$$

The first term represents the gravitational force component. Figure 2.18 shows the dependence of the hydrodynamic force $(\bar{F}_x)_\psi = F_x/[\rho \psi_0 R^4 \sin \Omega t]$ along the x-direction, by dash curves on the excitation frequency, Ω.

The moment of the fluid hydrodynamic force about the y'-axis through the center of mass of the solidified liquid is

$$\begin{aligned} M_{y'} = & -m_f \psi_0 R \sin(\Omega t) \left\{ g \frac{(1+k^2)}{(4h/R)} + \Omega^2 R \left\{ \frac{1}{12}(h/R)^2 - \frac{1}{8}(1+k^2) \right. \right. \\ & + \sum_{n=0}^{\infty} \frac{\bar{A}_n \Omega^2((2/\pi\xi_{n1}) - kC_1(k\xi_{n1}))}{\xi_{n1}(\omega_{n1}^2 - \Omega^2)(1-k^2)} \left[\left(\frac{h}{4R} - \frac{3g}{2R\Omega^2} - \frac{2}{\xi_{n1} h/R} \right) \times \tanh\left(\xi_{n1} \frac{h}{R} \right) \right. \\ & + \frac{1}{\xi_{n1} \cosh(\xi_{n1} h/R)} \left(2 - \frac{4g}{R\Omega^2} \frac{1}{(H/R)} \right) \\ & \left. \left. \left. + \left(\frac{5g}{R\Omega^2} \frac{1}{\xi_{n1}(H/R)} + \frac{1}{2\xi_{n1}} \right) \right] \right\} \right\} \end{aligned} \tag{2.107a}$$

$$M_{x'} = 0 \tag{2.107b}$$

The first term in equation (2.107a) represents the moment due to the lateral displacement of the center of mass for the liquid plane free surface.

2.6 Rectangular containers

The forced excitation of a rectangular tank was studied by Graham (1951), Graham and Rodriguez (1952), Eide (1964), Ghali (1965), Momoi (1965), Bauer (1966c,f, 1968a), Bauer and Villenueva (1966, 1967a,b), Scarsi and Brizzolara (1970), Kimura, *et al.* (1996b) for different types of excitations including translation, pitching, yawing, and roll. They used different techniques to estimate hydrodynamic pressure, forces and moments. This section presents the response of fluid motion in a rectangular container under sinusoidal translational, pitching, and roll excitations.

2.6.1 Lateral excitation

Consider a rectangular rigid tank subjected to lateral sinusoidal excitation $x(t)$. For a two-dimensional fluid motion and under the assumption of small excitation amplitude and small fluid response, the linearized fluid field equations take the form

$$\frac{\partial^2 \tilde{\Phi}}{\partial x^2} + \frac{\partial^2 \tilde{\Phi}}{\partial z^2} = 0 \tag{2.108a}$$

$$\left. \frac{\partial \tilde{\Phi}}{\partial x} \right|_{x=\pm L/2} = 0, \tag{2.108b}$$

$$g\eta - \frac{\partial \tilde{\Phi}}{\partial t} + \ddot{X}(t)x = 0, \quad \text{at } z = \eta(x,t) \tag{2.108c}$$

$$-\frac{\partial\tilde{\Phi}}{\partial z}=\frac{\partial\eta}{\partial t},\quad \text{at } z=\eta(x,t) \tag{2.108d}$$

The dynamic and kinematic free-surface conditions (2.108c, 2.108d) can be combined to give one condition

$$g\frac{\partial\tilde{\Phi}}{\partial z}+\frac{\partial^2\tilde{\Phi}}{\partial t^2}=\dddot{X}x \tag{2.109}$$

The total potential function is the sum of fluid perturbed function $\tilde{\Phi}$, (which is the solution of Laplace's equation (2.108a) that satisfies the boundary conditions (2.108b–d), plus the tank potential function $\Phi_0=-X_0x\Omega\cos\Omega t$. However, one must express the term x in the Fourier series

$$x=\frac{4L}{\pi^2}\sum_{n=0}^{\infty}\frac{(-1)^n}{(2n+1)^2}\sin\left((2n+1)\frac{\pi x}{L}\right) \tag{2.110}$$

The process of evaluating the velocity potential function follows the same steps as in the case of a cylindrical tank with the result

$$\Phi=-X_0\Omega\cos\Omega t$$

$$\times\left\{x+\sum_{n=0}^{\infty}\left[\frac{(-1)^n\sin\left((2n+1)\frac{\pi x}{L}\right)\cosh\left((2n+1)\frac{\pi(z+h)}{L}\right)}{\pi^2(2n+1)^2\cosh\left((2n+1)\frac{\pi h}{L}\right)}\frac{4L\Omega^2}{(\omega_n^2-\Omega^2)}\right]\right\} \tag{2.111}$$

where $\omega_n^2=(2n+1)\pi\,(g/L)\tanh\,((2n+1)\pi h/L)$ is the square of the fluid-free-surface natural frequency. The free-surface wave height is determined by substituting equation (2.111) into (2.108c)

$$\eta=\Omega^2\frac{X_0}{g}\sin\Omega t\left\{x-\sum_{n=0}^{\infty}\frac{(-1)^n4L}{\pi^2(2n+1)^2}\frac{\Omega^2}{(\omega_n^2-\Omega^2)}\sin\left((2n+1)\frac{\pi x}{L}\right)\right\} \tag{2.112}$$

The surface waves consist of a simple wave expressed by the first term in equation (2.112) and a group of symmetric sine waves. The pressure at any point inside the fluid domain (neglecting the hydrostatic pressure, ρgz) is

$$p=\rho\frac{\partial\Phi}{\partial t}$$

$$=\rho X_0\Omega^2\sin\Omega t\left\{x+\sum_{n=0}^{\infty}\left[\frac{(-1)^n\sin\left((2n+1)\frac{\pi x}{L}\right)\cosh\left((2n+1)\frac{\pi(z+h)}{L}\right)}{\pi^2(2n+1)^2\cosh\left((2n+1)\frac{\pi h}{L}\right)}\right.\right.$$

$$\left.\left.\times\frac{4L\Omega^2}{(\omega_n^2-\Omega^2)}\right]\right\} \tag{2.113}$$

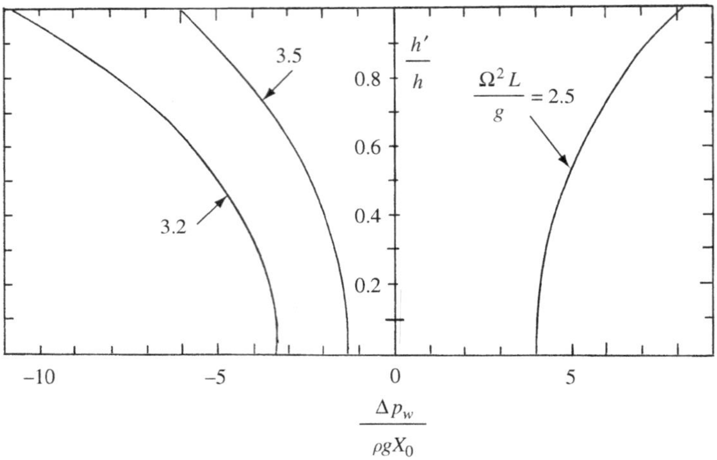

Figure 2.19 Wall pressure distribution corresponding to the first anti-symmetric mode in a rectangular tank under lateral excitation. (Bauer, 1966f)

The hydrodynamic pressure at any point on the wall side, $x = L/2$, is

$$p_w = \rho X_0 \Omega^2 \sin \Omega t \left\{ \frac{L}{2} + \sum_{n=0}^{\infty} \left[\frac{\cosh\left((2n+1)\dfrac{\pi(z+h)}{L}\right)}{\pi^2(2n+1)^2 \cosh\left((2n+1)\dfrac{\pi h}{L}\right)} \frac{4L\Omega^2}{(\omega_n^2 - \Omega^2)} \right] \right\} \tag{2.114}$$

Figure 2.19 shows the net wall pressure distribution parameter, $\Delta p_w/\rho g X_0$, for different values of excitation frequency parameter, $\Omega^2 L/g$. One may also observe a similar scenario of the phase shift discussed in the previous section.

The hydrodynamic pressure at any point on the bottom, $z = -h$, is

$$p_b = \rho X_0 \Omega^2 \sin \Omega t \left\{ x + \sum_{n=0}^{\infty} \left[\frac{\sin\left((2n+1)\dfrac{\pi x}{L}\right)}{\pi^2(2n+1)^2 \cosh\left((2n+1)\dfrac{\pi h}{L}\right)} \frac{4L\Omega^2}{(\omega_n^2 - \Omega^2)} \right] \right\} \tag{2.115}$$

The total hydrodynamic force exerted by the fluid on the container walls is determined by integrating the pressure on the wall over the entire area of the wall sides, $y \in \pm\ell/2$ and $z \in [-h, 0]$,

$$\begin{aligned} F_{xw} &= 2 \int_{-h}^{0} dz \int_{-\ell/2}^{\ell/2} p_w \, dy \\ &= m_f X_0 \Omega^2 \sin \Omega t \left\{ 1 + \sum_{n=0}^{\infty} \left[\frac{8 \tanh\left((2n+1)\dfrac{\pi h}{L}\right)}{(h/L)\pi^3(2n+1)^3} \frac{\Omega^2}{(\omega_n^2 - \Omega^2)} \right] \right\} \end{aligned} \tag{2.116}$$

where $m_f = \rho L\ell h$ = the mass of the total fluid volume.

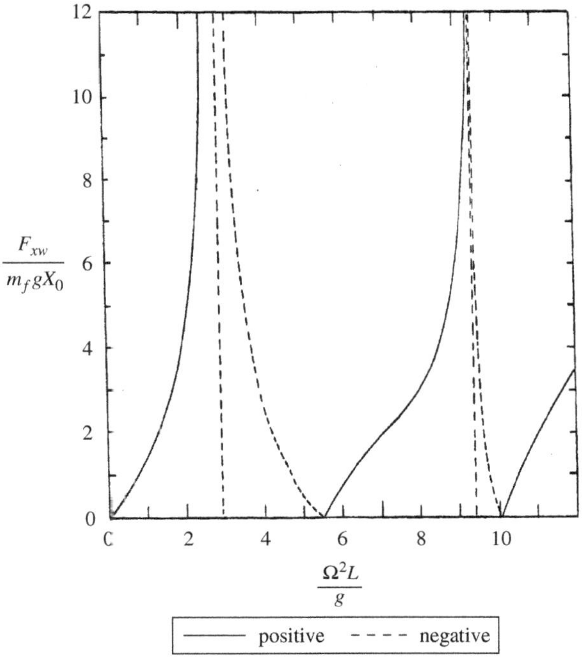

Figure 2.20 Total liquid force response in a rectangular tank under translational excitation for $h/L = 0.5$. (Bauer, 1966f)

It is not difficult to show that the total force acting on the tank bottom is zero. Figure 2.20 shows the dependence of the hydrodynamic force, $F_{xw}/\rho \ell h g X_0$, for the first two modes.

The moment due to the forces on the wall about y-axis is determined as follows

$$\begin{aligned} M_{yw} &= -2\int_{-h}^{0} p_w \ell z\, dz \\ &= m_f X_0 h \Omega^2 \sin \Omega t \left\{ \frac{1}{2} + \sum_{n=0}^{\infty} \left[\frac{8\Omega^2}{(\omega_n^2 - \Omega^2)(h/L)^2 \pi^4 (2n+1)^4} \right.\right. \\ &\qquad \left.\left. \times \left(1 - \frac{1}{\cosh((2n+1)\pi h/L)} \right) \right] \right\} \end{aligned} \tag{2.117a}$$

The moment due to the hydrodynamic pressure acting on the bottom about the y-axis is

$$\begin{aligned} M_{yb} &= \int_{-L/2}^{L/2} p_w \ell x\, dx \\ &= m_f \frac{X_0 h \Omega^2}{(h/L)^2} \sin \Omega t \left\{ \frac{1}{12} - \sum_{n=0}^{\infty} \left[\frac{8\Omega^2}{(\omega_n^2 - \Omega^2)\pi^4 (2n+1)^4 \cosh((2n+1)\pi h/L)} \right] \right\} \end{aligned} \tag{2.117b}$$

The total moment about the y-axis is the sum of (2.117a) and (2.117b), that is,

$$M_{yt} = m_f X_0 h \Omega^2 \sin \Omega t \left\{ \frac{1}{2} + \frac{1}{12(h/L)^2} + \sum_{n=0}^{\infty} \left[\frac{8\Omega^2}{(\omega_n^2 - \Omega^2)(h/L)^2 \pi^4 (2n+1)^4} \right] \right\} \tag{2.118}$$

One may also be interested in determining the moment about an axis parallel to the y-axis through the center of mass of the solidified fluid, that is,

$$M_{\bar{y}cg} = m_f X_0 h \Omega^2 \sin \Omega t \left\{ \frac{1}{2(h/L)^2} + \sum_{n=0}^{\infty} \frac{8 \tanh((2n+1)\pi h/L)}{(h/L)\pi^3 (2n+1)^3} \right. \times \left[\frac{1}{2} - \frac{\tanh((2n+1)\pi h/L)}{(h/L)\pi(2n+1)} + \frac{1}{(2n+1)\pi(h/L)\sinh((2n+1)\pi h/L)} \right] \left. \times \frac{\Omega^2}{(\omega_n^2 - \Omega^2)} \right\} \tag{2.119}$$

The location of the center of pressure, z_{cp}, on the tank wall, measured from the free surface, is

$$z_{\mathrm{cp}} = -\frac{h}{(h/L)} \frac{\tanh[(2n+1)(\pi h/2L)]}{\pi(2n+1)} \tag{2.120}$$

If the liquid oscillations produce large forces on the tank walls, baffles may be inserted to reduce them. The baffles should be placed in regions that exhibit high velocities. For example, for the fundamental mode the highest vertical velocities of the liquid free surface occur at the ends of the tank, and the highest horizontal velocities occur at the center of the tank. Baffles should be most effective in these locations.

2.6.2 Pitching excitation

Under pitching excitation, $\psi(t) = \psi_0 \sin \Omega t$, about the y-axis through the center of mass of the solidified liquid, the boundary conditions are

$$\frac{\partial \Phi}{\partial x} = -\psi_0 \Omega z \cos \Omega t \quad \text{at } x = \pm L/2 \tag{2.121a}$$

$$\frac{\partial \Phi}{\partial z} = -\psi_0 \Omega x \cos \Omega t \quad \text{at } z = -h/2 \tag{2.121b}$$

$$\frac{\partial^2 \Phi}{\partial t^2} + g \frac{\partial \Phi}{\partial z} = 0 \quad \text{at } z = h/2 \tag{2.121c}$$

The solution of Laplace's equation that satisfies the above boundary conditions is

$$\Phi = \Omega \psi_0 \cos \Omega t \left\{ \sum_{n=0}^{\infty} \frac{(-1)^n 4}{\pi^3 (2n+1)^3} \left[\frac{h^2 \sin((2n+1)\pi z/h) \sinh((2n+1)\pi x/h)}{\cosh((2n+1)\pi L/2h)} \right. \right. + \left. \frac{L^2 \sin((2n+1)\pi x/L) \cosh((2n+1)(z-(h/2))\pi/L)}{\sinh((2n+1)\pi h/L)} \right] + \sum_{n=0}^{\infty} \frac{(-1)^n 4L}{\pi^2 (2n+1)^2} \left(\frac{\Omega^2}{\omega_n^2 - \Omega^2} \right) \left[\frac{h}{2} - \frac{2L \tanh((2n+1)\pi h/2L)}{\pi(2n+1)} + \frac{g}{\omega_n^2} \right] \left. \times \frac{\sin((2n+1)\pi x/L) \cosh((2n+1)(z+(h/2)\pi/L)}{\cosh((2n+1)\pi h/L)} \right\} \tag{2.122}$$

The first term in this equation describes the velocity potential of the rigid body system. The horizontal force along the x-axis is

$$F_x = (\rho g h L \ell) \frac{\Omega^2 h}{g} \psi_0 \sin \Omega t \left\{ \frac{L^2}{12h^2} + \sum_{n=0}^{\infty} \frac{8 \tanh((2n+1)\pi h/L)}{\pi^3 (2n+1)^3 h/L} \left(\frac{\Omega^2}{\omega_n^2 - \Omega^2} \right) \times \left[\frac{1}{2} - \frac{2L \tanh((2n+1)\pi h/2L)}{\pi(2n+1)h} + \frac{g}{h\omega_n^2} \right] \right\} \tag{2.123}$$

where

$$\omega_n^2 = (2n+1) \frac{\pi g}{L} \tanh\left((2n+1) \frac{\pi h}{L} \right) \tag{2.124}$$

The moment acting on the container about the y-axis is

$$M_y = (\rho g h^2 L \ell /) \frac{\Omega^2 h}{g} \psi_0 \sin \Omega t \left\{ \frac{I_{fy}}{m_f h^2} + \frac{gL^2}{12h^3 \Omega^2} + \sum_{n=0}^{\infty} \frac{8L \tanh((2n+1)\pi h/L)}{\pi^3 (2n+1)^3 h} \times \left[\frac{1}{2} - \frac{2L \tanh((2n+1)\pi h/2L)}{\pi(2n+1)h} + \frac{g}{2h\omega_n^2} \right] \frac{g}{h\omega_n^2} + \sum_{n=0}^{\infty} \frac{8L \tanh((2n+1)\pi h/L)}{\pi^3 (2n+1)^3 h} \left[\frac{1}{2} - \frac{2L \tanh((2n+1)\pi h/2L)}{\pi(2n+1)h} + \frac{g}{h\omega_n^2} \right]^2 \left(\frac{\Omega^2}{\omega_n^2 - \Omega^2} \right) \right\} \tag{2.125}$$

where I_{fy} is the effective mass moment of inertia of the fluid about the y-axis given by the following expression

$$I_{fy} = I_{sy} \left\{ 1 - \frac{4}{1 + (h/L)^2} + \frac{768}{\pi^3 h (L^2 + h^2)/L^3} \sum_{n=0}^{\infty} \frac{\tanh((2n+1)\pi h/2L)}{(2n+1)^5} \right\} \tag{2.126}$$

where I_{sy} is the mass moment of inertia of the solidified liquid about the y-axis.

2.6.3 Roll excitation

Under roll excitation $\varphi(t) = \varphi_0 \sin \Omega t$ about the z-axis, the boundary conditions are

$$\frac{\partial \Phi}{\partial x} = \varphi_0 \Omega y \cos \Omega t \quad \text{at } x = L/2 \tag{2.127a}$$

$$\frac{\partial \Phi}{\partial y} = -\varphi_0 \Omega x \cos \Omega t \quad \text{at } y = \pm \ell/2 \tag{2.127b}$$

$$\frac{\partial \Phi}{\partial z} = 0 \quad \text{at } z = -h/2 \tag{2.127c}$$

The solution of Laplace's equation subject to these boundary conditions is

$$\Phi = -\Omega\varphi_0 \cos\Omega t\left\{\sum_{n=0}^{\infty}\frac{(-1)^n 4}{\pi^3(2n+1)^3}\left[\frac{\ell^2\sin((2n+1)\pi y/\ell)\sinh((2n+1)\pi x/\ell)}{\cosh((2n+1)\pi L/2\ell)}\right.\right.$$
$$\left.-\frac{L^2\sin((2n+1)\pi x/L)\sinh((2n+1)\pi y/L)}{\cosh((2n+1)\pi\ell/2L)}\right]+\sum_{m=0}^{\infty}\sum_{n=0}^{\infty}(-1)^{n+m}\left(\frac{\Omega^2}{\omega_{mn}^2-\Omega^2}\right)$$
$$\times\frac{16L\ell\left[\ell^2(2n+1)^2-\ell^2(2m+1)^2\right]\operatorname{sech}\left(\frac{\pi h}{L\ell}\sqrt{\ell^2(2n+1)^2+L^2(2m+1)^2}\right)}{\pi^4(2m+1)^2(2n+1)^2[\ell^2(2n+1)^2+L^2(2m+1)^2]}$$
$$\times\sin\left((2n+1)\frac{\pi x}{L}\right)\sin\left((2m+1)\frac{\pi y}{\ell}\right)$$
$$\left.\times\cosh\left[\frac{\pi}{L\ell}\sqrt{\ell^2(2n+1)^2+L^2(2m+1)^2}\left(z+\frac{h}{2}\right)\right]\right\} \tag{2.128}$$

The two-dimensional part of the potential function satisfies the boundary conditions on the tank side walls while the three-dimensional part is needed to satisfy the free-surface condition. The roll moment about the z-axis is

$$M_z = \Omega^2\varphi_0\sin\Omega t\left\{\frac{32\rho h}{\pi^5}\sum_{n=0}^{\infty}\frac{1}{(2n+1)^5}[L^4\tanh((2n+1)\pi\ell/2L)\right.$$
$$+\ \ell^4\tanh((2n+1)\pi L/2\ell)]-\frac{1}{12}\rho L\ell h(L^2+\ell^2)+\sum_{m=0}^{\infty}\sum_{n=0}^{\infty}\frac{\omega_{mn}^2}{g}\left(\frac{\Omega^2}{\omega_{mn}^2-\Omega^2}\right)$$
$$\left.\times\frac{64\rho L^3\ell^3[\ell^2(2n+1)^2-L^2(2m+1)^2]^2}{\pi^8[\ell^2(2n+1)^2+L^2(2m+1)^2]^2(2m+1)^4(2m+1)^4}\right\} \tag{2.129a}$$

This expression can be written in terms of the moment of inertia of the liquid in the form

$$M_z = I_{sz}\left(\frac{h}{L}\right)^2\Omega^2\varphi_0\sin\Omega\left\{\frac{I_{fz}}{I_{sz}}+\sum_{m=0}^{\infty}\sum_{n=0}^{\infty}\left(\frac{\Omega^2}{\omega_{mn}^2-\Omega^2}\right)\right.$$
$$\times\frac{768[\ell^2(2n+1)^2-L^2(2m+1)^2]^2}{\pi^7\ell^2(h/L)(L^2+\ell^2)(2m+1)^4(2n+1)^4}$$
$$\left.\times\frac{\tanh\left((\pi h/L\ell)\sqrt{\ell^2(2n+1)^2+L^2(2m+1)^2}\right)}{[(2n+1)^2+(L/\ell)^2(2m+1)^2]^{3/2}}\right\} \tag{2.129b}$$

where the resonant frequencies ω_{mn} are given by the expression

$$\omega_{mn}^2 = g\frac{\pi}{L\ell}\sqrt{\ell^2(2n+1)^2+L^2(2m+1)^2}\tanh\left(\frac{\pi h}{L\ell}\sqrt{\ell^2(2n+1)^2+L^2(2m+1)^2}\right) \tag{2.130}$$

and I_{fz} is the effective liquid mass moment of inertia about the z-axis given by the expression

$$I_{fz} = I_{sz}\left\{1 - \frac{4}{1+(L/\ell)^2} + \frac{768}{\pi^5 L(L^2+h^2)/\ell^3}\sum_{n=0}^{\infty}\frac{\tanh((2n+1)\pi h/2L)}{(2n+1)^5}\right\} \tag{2.131}$$

2.7 Spherical containers

Budiansky (1960) developed a Lagrangian formulation to determine the slosh forces in spherical containers under lateral excitation. He introduced a displacement potential function Ψ whose gradient with respect to the coordinate system gives the fluid displacement relative to the tank. The potential energy during motion is

$$\text{PE} = \frac{1}{2}\int_{S_f} \rho g\left(\frac{\partial \Psi}{\partial z}\right)^2 ds \tag{2.132}$$

where S_f denotes the liquid-free-surface area. The displacement potential Ψ may be expressed in terms of the velocity potentials ϕ_n associated with the natural modes of liquid oscillations in the series form

$$\Psi = \sum a_n \phi_n(x, y, z) \tag{2.133}$$

where a_n are the generalized coordinates of the sloshing modes. The linearized free-surface condition may be written in the form

$$\frac{\partial \phi_n}{\partial z} = \frac{\omega_n^2}{g}\phi_n \tag{2.134}$$

where ω_n is the natural frequency of the nth sloshing mode. Introducing equations (2.131) and (2.132) into (130), gives

$$\text{PE} = \frac{1}{2}\int_{S_f} \rho g\left[\sum (a_n\omega_n^2/g)\phi_n\right]^2 ds = \frac{\rho}{2g}\int_{S_f}\sum a_n^2\omega_n^4\phi_n^2\, ds \tag{2.135a}$$

For any two sloshing mode frequencies, ω_m and ω_n, with corresponding potentials ϕ_m and ϕ_n we write

$$\int_{v} \nabla\phi_m \nabla\phi_n\, dv = -\int_{v} \phi_m\nabla^2\phi_n\, dv + \int_{S} \phi_m\nabla\phi_n\cdot\mathbf{n}\, ds + \int_{S_f}\phi_m\frac{\partial\phi_n}{\partial z}\, ds \tag{2.136}$$

where v denotes the liquid volume. By definition of $\nabla^2\phi_n = 0$, and $\nabla\phi_n\cdot\mathbf{n} = 0$, the first two integrals on the right-hand side vanish and equation (2.136) takes the form

$$\int_{v} \nabla\phi_m\cdot\nabla\phi_n\, dv = \int_{S_f}\phi_m\frac{\partial\phi_n}{\partial z}\, ds \tag{2.137}$$

Introducing equation (2.130) into equation (2.137) gives

$$\int_v \nabla\phi_m \cdot \nabla\phi_n \, dv = \int_{S_f} \frac{\omega_n^2}{g} \phi_m\phi_n \, ds = \int_{S_f} \frac{\omega_m^2}{g} \phi_m\phi_n \, ds = 0, \quad \text{for } m \neq n \tag{2.138}$$

For $m = n$ we write

$$\int_v |\nabla\phi_n|^2 \, dv = \frac{\omega_n^2}{g} \int_{S_f} \phi_n^2 \, ds \tag{2.139}$$

This orthogonality property enables one to write the potential energy (2.135) in the form

$$\text{PE} = \frac{\rho}{2g} \int_{S_f} \sum a_n^2 \omega_n^4 \phi_n^2 \, ds \tag{2.135b}$$

Introducing the parameter $\alpha_n = \int_{S_f} \phi_n^2 \, ds$, the potential energy may be written in the form

$$\text{PE} = \frac{\rho}{2g} \sum \alpha_n a_n^2 \omega_n^4 \tag{2.135c}$$

The kinetic energy of the container-liquid system is

$$\text{KE} = \frac{1}{2} m_c \dot{X}^2 + \frac{\rho}{2} \int_v \left\{ \left[\dot{X} + \sum \dot{a}_n \frac{\partial\phi_n}{\partial x} \right]^2 + \left[\sum \dot{a}_n \frac{\partial\phi_n}{\partial y} \right]^2 + \left[\sum \dot{a}_n \frac{\partial\phi_n}{\partial z} \right]^2 \right\} dv \tag{2.140}$$

where m_c is the container mass. Introducing equation (2.139) into (2.140) gives

$$\text{KE} = \frac{1}{2}(m_c + m_l)\dot{X}^2 + \frac{\rho_l}{2g} \int_{S_f} \sum \omega_n^2 \dot{a}_n^2 \phi_n^2 \, ds + \rho_l \dot{X} \int_v \sum \dot{a}_n \frac{\partial\phi_n}{\partial x} \, dv \tag{2.141}$$

where m_ℓ is the total mass of the liquid. Introducing the parameter $\beta_n = \int_v (\partial\phi_n/\partial x) \, dv$, the kinetic energy may be written in the form

$$\text{KE} = \frac{1}{2}(m_c + m_l)\dot{X}^2 + \frac{\rho_l}{2g} \sum \alpha_n \omega_n^2 \dot{a}_n^2 + \rho_l \dot{X} \sum \beta_n \dot{a}_n \tag{2.142}$$

The potential functions ϕ_n may be written in terms of the sloshing natural modes of spherical coordinates (r, θ, z)

$$\phi_n = F_n(r, z) \cos k\theta \tag{2.143}$$

In terms of the nondimensional coordinates (ρ, η, θ), where $\rho = r/\alpha R$ and $\eta = z/\alpha R$, see Figure 1.7, the parameters α_n and β_n are evaluated as follows

$$\beta_n = \frac{\omega_n^2}{g} \int_0^{\alpha R} \int_0^{2\pi} r(r\cos\theta) F_n(r, 0) \cos k\theta \, dr \, d\theta \tag{2.144a}$$

Obviously, β_n will vanish for all $k \neq 1$. The values of F_n at the free surface are proportional to $f_n(\rho) = g_n(\rho)/\sqrt{\rho}$, where $g_n(\rho)$ is the eigenfunctions (1.131) of the integral equation (1.130) of

Chapter 1, and $f_n(\rho)$ is normalized such that $f_n(1) = 1$. The wave height at the wall is $\eta = \omega_n^2 a_n/g$ and the parameter β_n becomes

$$\beta_n = (\alpha R)^3 \frac{\omega_n^2}{g} \int_0^1 \int_0^{2\pi} \rho^2 f(\rho) \cos^2\theta \, d\rho \, d\theta = \pi(\alpha R)^3 \frac{\omega_n^2}{g} \int_0^1 \rho^2 f(\rho) \, d\rho \tag{2.144b}$$

Similarly

$$\alpha_n = (\alpha R)^2 \int_0^1 \int_0^{2\pi} \rho (f(\rho))^2 \cos^2\theta \, d\theta \, d\rho = \pi(\alpha R)^2 \int_0^1 \rho (f(\rho))^2 \, d\rho \tag{2.145}$$

Using Lagrange's equation with respect to the generalized coordinates $\eta_n = \omega_n^2 a_n/g$ and the container displacement $x(t) = X_0 \sin \Omega t$, gives

$$\frac{d^2\eta_n}{dt^2} + \omega_n^2 \eta_n = -\lambda_n^2 \alpha \frac{D_n}{E_n} \frac{d^2 x}{dt^2} \tag{2.146}$$

$$F_s = -m_l \frac{d^2 X_0}{dt^2} - \pi \rho_l (\alpha R)^3 \sum_{n=1}^{\infty} D_n \frac{d^2 \eta_n}{dt^2} \tag{2.147}$$

where D_n and E_n are coefficients that depend on the liquid depth and are given by the following expressions

$$D_n = \int_0^1 \rho^2 f_n(\rho) \, d\rho, \quad E_n = \int_0^1 \rho (f_n(\rho))^2 \, d\rho = \int_0^1 (g_n(\rho))^2 \, d\rho \tag{2.148a, b}$$

These coefficients are shown in Figure 2.21(a) and (b).

Several experimental studies were conducted by Stofan and Armstead (1962), Abramson, Chu and Garza (1963), Sumner and Stofan (1963), Chu (1964b), Sumner (1964, 1965, 1966) to measure the slosh force under different excitation amplitudes and frequencies, and different values of liquid depth. The dependence of the analytical and experimental slosh force parameter, $F_s/\rho_\ell g d^3$ (X_0/d), on the excitation frequency parameter, $\Omega\sqrt{R/g}$, where d and R are container diameter and radius, respectively, was obtained by Stofan and Armstead (1962). Their results are shown in Figure 2.22(a) and (b) for two values of liquid-depth ratio, $h/2R = 0.5$ and 0.4, respectively. Their experimental results were obtained for water and mercury over an excitation frequency range that covers the first two natural frequencies. Abramson, Chu and Garza (1963) measured the slosh force under lateral excitation for baffled and unbaffled spherical tanks. Sumner and Stofan (1963) conducted an experimental investigation to measure the viscous slosh-damping characteristics of several liquids having a wide range of kinematic viscosity. Sumner (1965) measured the horizontal slosh force for an excitation frequency equal to the fundamental frequency of the liquid free surface at each liquid-depth ratio considered in the tests. These forces are thus considered the maximum forces that could be obtained for a given excitation amplitude. Figure 2.23(a) shows the dependence of the first-mode slosh-force parameter, $F_s/\rho g D^3$, where ρ is the liquid density and D is the diameter of the container, on the fluid depth for different values of excitation amplitude X_0/D. It is seen that

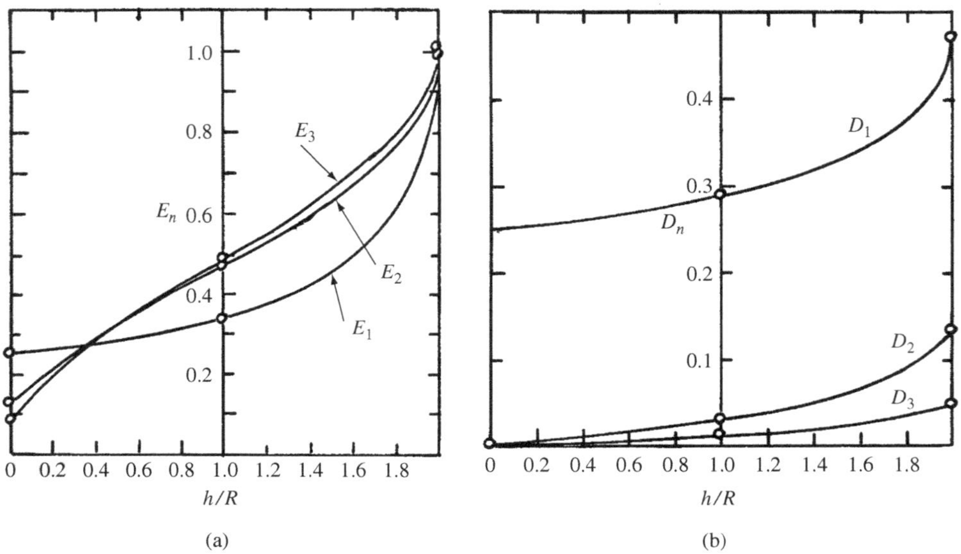

Figure 2.21 Dependence of the parameters E_n and D_n on the fluid depth ratio for the force response estimation in a spherical tank. (Budiansky, 1960)

the maximum slosh forces for a given excitation amplitude occur when each tank is nearly half-full. The slight observed dip at liquid-depth ratios of 0.6 to 0.75 and for excitation amplitude parameter X_0/D less than 0.001 56 was mainly due to the fact that the liquid near the surface tended to separate from the tank wall as the wave height of the liquid surface approached the maximum height that could be obtained. Figure 2.23(b) shows the dependence of the sloshing force on the fluid depth in a spherical tank for different values of excitation amplitude.

2.8 Prolate and oblate spheroidal containers

The forced excitation of an oblate spheroidal container was experimentally studied by Sumner (1965). The case of prolate spheroidal tanks was studied analytically and experimentally by Bauer and Eidel (1989b). With reference to the Laplace equation of the prolate spheroidal tank described in Chapter 1, the boundary conditions under translational excitation, $x(t) = X_0 \sin \Omega t$, are

$$\left.\frac{\partial \Phi}{\partial \xi}\right|_{\xi=\xi_0,\xi_1} = cX_0 \cosh \xi \sin \eta \cos \phi \cos \Omega t \quad \text{at the annular container walls} \tag{2.149a}$$

$$\left.\frac{\partial \Phi}{\partial \phi}\right|_{\phi=0,2\alpha\pi} = -c\Omega X_0 \cos \Omega t \quad \text{at the container sector wall} \tag{2.149b}$$

$$\left.\frac{\partial \Phi}{\partial \eta}\right|_{\eta=\eta_1} = c\Omega X_0 \sinh \xi \cos \eta \cos \phi \cos \Omega t \quad \text{at the tank bottom} \tag{2.149c}$$

The combined free-surface condition, which is obtained from the unsteady Bernoulli equation and the kinematic condition,

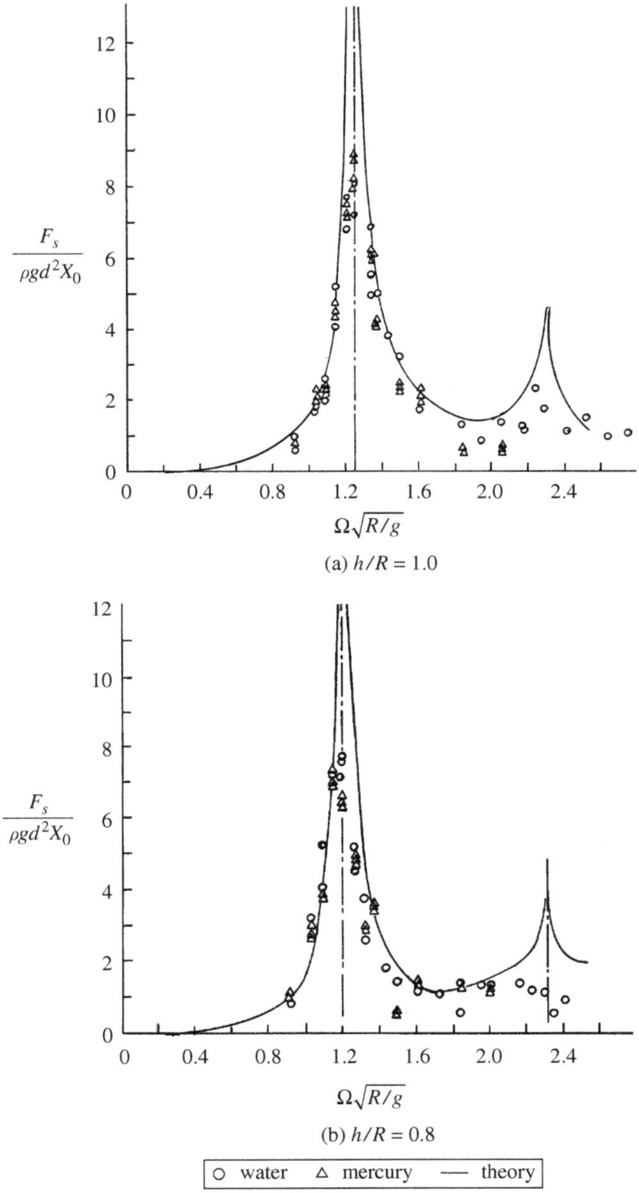

Figure 2.22 Dependence of liquid force response on excitation frequency parameter in a spherical tank for two liquid depth ratios. (Stofan and Armstead, 1962)

$$\frac{\partial^2\Phi}{\partial t^2} - \frac{g\cosh\xi\sin\eta}{c(\sinh^2\xi + \sin^2\eta)}\frac{\partial\Phi}{\partial\eta} = 0 \quad \text{at } \eta = \eta_0 \tag{2.149d}$$

The velocity potential function can be written in terms of container motion and the fluid motion as

$$\Phi(\xi, \eta, \phi, t) = \tilde{\Phi}(\xi, \eta, \phi)\sin\Omega t + \Omega X_0 c\sinh\xi\sin\eta\cos\phi\cos\Omega t \tag{2.150}$$

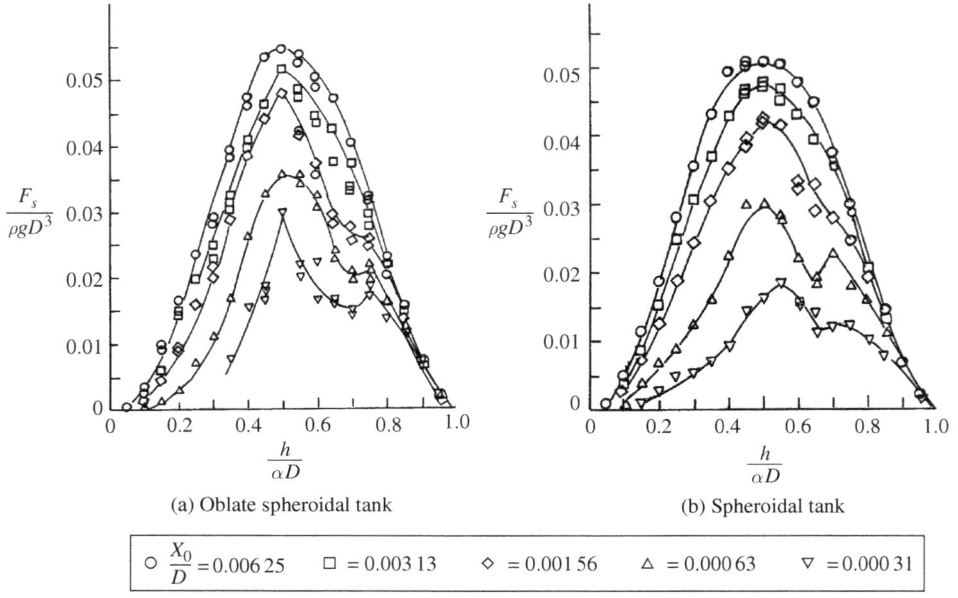

Figure 2.23 Dependence of the first mode slosh force parameter under different lateral excitation amplitudes. (Sumner, 1965)

The boundary conditions in terms of the perturbed velocity potential function are

$$\left.\frac{\partial \tilde{\Phi}}{\partial \xi}\right|_{\xi=\xi_0,\xi_1} = 0 \quad \text{at the annular container walls} \tag{2.151a}$$

$$\left.\frac{\partial \tilde{\Phi}}{\partial \phi}\right|_{\phi=0,2\alpha\pi} = 0 \quad \text{at the container sector wall} \tag{2.151b}$$

$$\left.\frac{\partial \tilde{\Phi}}{\partial \eta}\right|_{\eta=\eta_1} = 0 \quad \text{at the tank bottom} \tag{2.151c}$$

The free-surface boundary condition, at $\eta = \eta_0$, is

$$\Omega^2 \tilde{\Phi} + \frac{g \cosh \xi \sin \eta_0}{c\left(\sinh^2 \xi + \sin^2 \eta_0\right)} \frac{\partial \tilde{\Phi}}{\partial \eta_0} = \frac{-i\Omega X_0 c \sinh \xi \cos \phi \sin \eta_0}{\sinh^2 \xi + \sin^2 \eta_0} \left[\frac{g}{c} \cosh \xi \cos \eta_0 + \Omega^2 \left(\sinh^2 \xi + \sin^2 \eta_0\right)\right] \tag{2.151d}$$

The Laplace equation of the total potential yields the Laplace equation for the perturbed potential $\tilde{\Phi}$, that is, $\nabla^2 \tilde{\Phi} = 0$, whose solution satisfying conditions (2.151a–c) is given in the form

$$\tilde{\Phi}(\xi, \eta, \phi) = \sum_{m=0}^{\infty} \sum_{n=1}^{\infty} A_{mn} L_{mn}(\cosh \xi) \bar{L}_{mn}(\cos \eta) \cos \frac{m\phi}{2\alpha} \tag{2.152}$$

where the functions $L_{mn}(\cosh \xi)$ and $\bar{L}_{mn}(\cos \eta)$ (given by equations (1.199a,b)), are Legendre functions, and the unknown constants A_{mn} are determined by applying the free-surface condition (2.142d). The ξ-dependent expressions in equation (2.142d) are expanded into a series of associated Legendre functions, while cos φ is expanded into the Fourier series

$$\cos \phi = \sum_{m=0}^{\infty} a_m \cos \frac{m\phi}{2\alpha} \tag{2.153}$$

where

$$a_0 = \frac{\sin(2\pi\alpha)}{2\pi\alpha} \quad \text{and} \quad a_m = 4\alpha(-1)^{m+1} \frac{\sin(2\pi\alpha)}{\pi(m^2 - 4\alpha^2)}, \quad \text{for } m = 1, 2, \ldots \tag{2.154}$$

This yields the infinitely nonhomogeneous set of linear equations in A_{mn} as unknowns

$$\Omega^2 A_{mn} + \sum_{k=1}^{\infty} \gamma_{nk}(m, g/c, \xi_0, \xi_1, \eta_0, \eta_1) A_{mk}$$
$$= -i\Omega X_0 a_m \sin \eta_0 \sum_{k=1}^{\infty} \left[g \cos \eta_0 \mu_{nk}^{(2)}(m, \xi_0, \xi_1, \eta_0) + \Omega^2 c \mu_{nk}^{(3)}(m, \xi_0, \xi_1, \eta_0) \right] \tag{2.155}$$

with $m = 0, 1, \ldots$ and $n = 1, 2, \ldots$, where γ_{nk} is give in Chapter 1 (1.183) and

$$\mu_{nk}^{(2)}(m, \xi_0, \xi_1, \eta_0) = \frac{\int_{\xi_0}^{\xi} (\cosh \xi \sinh^2 \xi / \sinh^2 \xi + \sin^2 \eta_0) L_{mn}(\xi) L_{mk}(\xi)\, d\xi}{\bar{L}_{mn}(\eta_0) \int_{\xi_0}^{\xi} L_{mn}^2(\xi) \sinh \xi \, d\xi} \tag{2.156a}$$

$$\mu_{nk}^{(3)}(m, \xi_0, \xi_1, \eta_0) = \frac{\int_{\xi_0}^{\xi} \sinh^2 \xi L_{mn}(\xi) L_{mk}(\xi)\, d\xi}{\bar{L}_{mn}(\eta_0) \int_{\xi_0}^{\xi} L_{mn}^2(\xi) \sinh \xi \, d\xi} \tag{2.156b}$$

The solution of the algebraic equations (2.155) gives the integration constants A_{mn} as a function of the forcing frequency Ω. This completes the determination of the velocity potential function and may give an estimate of the pressure distribution from the Bernoulli equation

$$p = -\rho \frac{\partial \Phi}{\partial t} - \rho g c \cosh \xi (\cos \eta - \cos \eta_0) \tag{2.157}$$

The first term on the right-hand side of equation (2.157) represents the hydrodynamic pressure while the second term is the hydrostatic pressure. Bauer and Eidel (1989b) numerically determined the pressure distribution as a function of the forcing frequency parameter $\Omega/\sqrt{g/c}$. Their results are shown in Figure 2.24 for the fundamental mode $m = n = 1$ at the location $\phi = 0$, $\xi = \xi_0$ and $\eta_0 = \pi/2$. Note the pressure axis gives the absolute value of the pressure function

$$\frac{\{p - c\rho g \ \cosh \xi_0 (\cos \eta_0 - \cos \eta)\}}{\rho g \ X_0 \cos \phi \ \sin \Omega t}$$

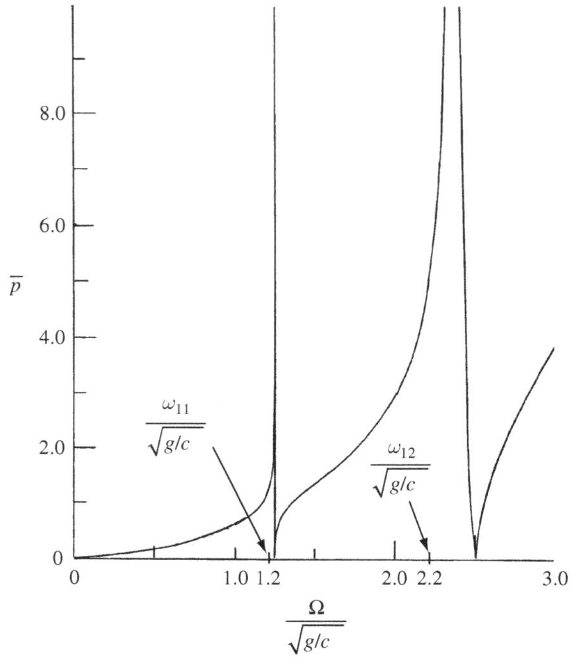

$$\bar{p} = \frac{[p - c\rho g \cosh \xi_0 (\cos \eta_0 - \cos \eta)]}{\rho g X_0 e^{i\Omega t} \cos \varphi}$$

Figure 2.24 The pressure–frequency response in a prolate spherical container with $b/a = 0.7616$, at $\eta_0 = \pi/2$, $\phi = 0$, $m = 1$, $\xi_0 = 1$ showing the first two modes. (Bauer and Eidel, 1989b)

Figure 2.24 reveals two singularities corresponding to the resonance frequencies $\omega_{11}/\sqrt{g/c} \approx 1.2$ and $\omega_{12}/\sqrt{g/c} \approx 2.3$. Integrating the pressure component over the wetted undisturbed area of the container sidewalls gives the liquid force exerted on the prolate spheroidal container. The liquid force along the x-direction is

$$F_x = \int_0^{2\pi} \int_{\eta_0}^{\pi} p|_{\xi=\xi_0} c^2 \sinh \xi_0 \sin \eta \sqrt{\sinh^2\xi + \sin^2\eta} \cos \phi \, d\phi \tag{2.158}$$

The liquid force may also be numerically estimated. For oblate spheroidal tanks, Sumner (1965) measured the horizontal slosh forces for an excitation frequency close to the fundamental liquid frequency. Figure 2.23(b) shows the maximum slosh force parameter $F_s/\rho g\alpha D^3$, where α is the ratio of tank height to diameter, b/D, for a given excitation amplitude. As in the case of spherical tanks, the maximum slosh forces for a given excitation amplitude were found when the tank was at or very near a half-full condition. The observed slight dip that appear in the faired curves at liquid-depth ratios of 0.6 to 0.75 and for excitation-amplitude parameters equal to or less than 0.001 56 is due to the fact that the liquid near the surface tends to separate from the tank wall as the wave height of the liquid surface approaches the maximum height.

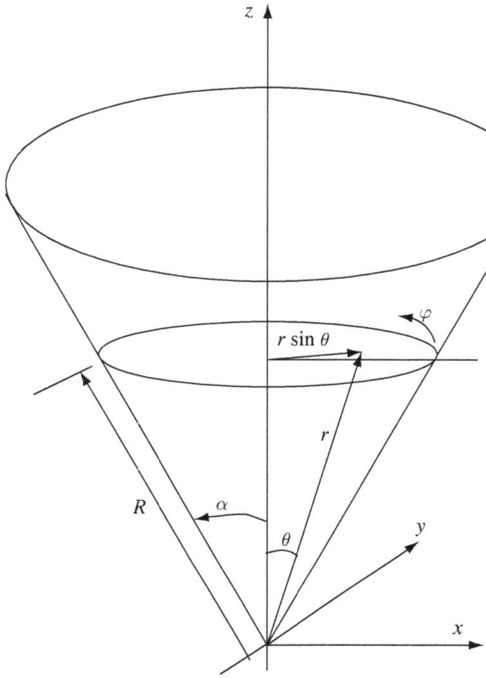

Figure 2.25 Inverted conical cone showing the coordinate systems.

2.9 Conical containers

The bottom of some circular containers may take a conical shape and in order to estimate the hydrodynamic forces one may replace the container by an equivalent flat bottom. For a conical container on its own the treatment may require a choice of a coordinate system in which most of boundary conditions may exactly be satisfied and only one or two boundary conditions may approximately be described by coordinate surfaces. Harper (1958) considered arbitrary forced excitation of conical containers. For the inverted conical container shown in Figure 2.25, the following spherical coordinates are introduced

$$\begin{aligned} x &= r\sin\theta\cos\varphi, \\ y &= r\sin\theta\sin\varphi, \\ z &= r\cos\theta \end{aligned} \tag{2.159}$$

The Laplace equation in terms of spherical coordinates is

$$\frac{1}{r^2}\frac{\partial\Phi}{\partial r}\left(r^2\frac{\partial\Phi}{\partial r}\right)+\frac{1}{r\sin\theta}\frac{\partial}{\partial\theta}\left(\sin\theta\frac{\partial\Phi}{\partial\theta}\right)+\frac{1}{r^2\sin^2\theta}\frac{\partial^2\Phi}{\partial\varphi^2}=0 \tag{2.160}$$

The solution of equation (2.160) must satisfy the boundary conditions. For example, at the free surface the following combined condition must be satisfied (Bauer, 1999)

$$\frac{\partial^2\Phi}{\partial t^2}+g\frac{\partial\Phi}{\partial r}=0 \quad \text{at } r=R \tag{2.161}$$

Under forced excitation along the x-axis, the following wall boundary condition must be satisfied

$$\frac{1}{r}\frac{\partial \Phi}{\partial \theta} = i\Omega X_0\, e^{i\Omega t} \cos\theta \cos\varphi, \quad \theta = \alpha \tag{2.162}$$

Introducing the transformation

$$\Phi(r,\theta,\varphi,t) = e^{i\Omega t}[\phi(r,\theta) + i\Omega X_0 r \sin\theta] \cos\varphi \tag{2.163}$$

where $\phi(r, \theta)$ is obtained by solving Laplace's equation, $\nabla^2\phi(r,\theta) = 0$, $\phi\,(r,\theta)$ has to satisfy the boundary conditions

$$\left.\frac{\partial \phi}{\partial \theta}\right|_{\theta=\alpha} = 0 \tag{2.164a}$$

and the combined free-surface condition

$$g\frac{\partial \phi}{\partial r} - \Omega^2\,\psi = i\Omega X_0\left(\Omega^2 R - g\right) \sin\theta \tag{2.164b}$$

A possible solution of $\phi\,(r, \theta)$ that satisfies the wall boundary condition (2.164a) is

$$\phi(r,\theta) = \sum_{n=1}^{\infty} A_n \left(\frac{r}{R}\right)^{\lambda_n} P^1_{\lambda_n}(\cos\theta) \tag{2.165}$$

where $P^1_{\lambda_n}(\cos\theta)$ is the associated Legendre function. The eigenvalues λ_n were obtained in Table 1.6. In order to satisfy the free-surface condition (2.164b), $\sin\theta$ must be expanded into a series of associated Legendre functions, that is,

$$\sin\theta = \sum_{n=1}^{\infty} \beta_n P^1_{\lambda_n}(\cos\theta) \tag{2.166}$$

Introducing the orthogonality relation

$$\int_1^{\cos\alpha} P^1_{\lambda_n}(x) P^1_{\lambda_m}(x)\, dx$$

$$= \left\{ \begin{array}{ll} 0 & \text{for } \lambda_m \neq \lambda_n \\ -\dfrac{\sin^2\alpha}{(2\lambda_n+1)}\left[P^1_{\lambda_n}(\cos\alpha)\dfrac{\partial^2}{\partial\lambda_n \partial x}\right]\left(P^1_{\lambda_n}(\cos\alpha)\right) \equiv F_n(\alpha) & \text{for } \lambda_m = \lambda_n \end{array} \right\} \tag{2.167}$$

gives the coefficient

$$\beta_n = -\frac{(2\lambda_n+1)}{F_n(\alpha)} \int_1^{\cos\alpha} \sqrt{1-x^2}\, P^1_{\lambda_n}(x)\, dx \tag{2.168}$$

Introducing the following relationship

$$P^1_{\lambda_n}(x) = -\sqrt{1-x^2}\,\frac{dP_{\lambda_n}(x)}{dx} \tag{2.169}$$

the integral in relation (2.168) is

$$\int_{\cos\alpha}^{1} (1-x^2)\frac{dP_{\lambda_n}(x)}{dx}\,dx = -\sin^2\alpha P_{\lambda_n}(\cos\alpha) + 2\int_{\cos\alpha}^{1} xP_{\lambda_n}(x)\,dx$$

$$= -\sin^2\alpha P_{\lambda_n}(\cos\alpha) + 2\frac{\sin\alpha}{(\lambda_n-1)(\lambda_n+2)} \times \left[\sin\alpha P_{\lambda_n}(\cos\alpha) + \cos\alpha P^1_{\lambda_n}(\cos\alpha)\right] \tag{2.170}$$

Thus the value of β_n takes the form

$$\beta_n = -\frac{(2\lambda_n+1)}{F_n(\alpha)}\sin^2\alpha P_{\lambda_n}(\cos\alpha)\left[\frac{2}{(\lambda_n-1)(\lambda_n+2)} - 1\right] + \frac{2\cos\alpha\cos\alpha P^1_{\lambda_n}(\cos\alpha)}{(\lambda_n-1)(\lambda_n+2)} \tag{2.171}$$

The velocity potential function can be written in the form

$$\Phi(r,\theta,\varphi,t) = i\Omega X_0 R e^{i\Omega t}\left[\frac{r}{R}\sin\theta + \sum_{n=1}^{\infty}\frac{(\Omega^2-(g/R))\beta_n}{(\omega_n^2-\Omega^2)}\left(\frac{r}{R}\right)^{\lambda_n} P^1_{\lambda_n}(\cos\theta)\right]\cos\varphi \tag{2.172}$$

The free-surface wave height is, $\eta(\theta,\varphi,t) = -\frac{1}{g}\frac{\partial\Phi}{\partial t}\Big|_{r=R}$

$$\eta(\theta,\varphi,t) = \frac{\Omega^2}{g}X_0\,e^{i\Omega t}R\cos\varphi\left[\sin\theta + \sum_{n=1}^{\infty}\frac{(\Omega^2-(g/R))}{(\omega_n^2-\Omega^2)}\left(\frac{r}{R}\right)^{\lambda_n} P^1_{\lambda_n}(\cos\theta)\right] \tag{2.173}$$

The pressure distribution is

$$p(r,\theta,\varphi,t) = -\rho\frac{\partial\Phi}{\partial t} + \rho g(R\cos\alpha - r\cos\theta)$$

$$= \rho\Omega^2 X_0 R e^{i\Omega t}\cos\varphi\left[\frac{r}{R}\sin\theta + \sum_{n=1}^{\infty}\frac{(\Omega^2-(g/R))\beta_n}{(\omega_n^2-\Omega^2)}\left(\frac{r}{R}\right)^{\lambda_n} P^1_{\lambda_n}(\cos\theta)\right] + \rho g(R\cos\alpha - r\cos\theta) \tag{2.174}$$

The hydrodynamic force acting on the container is obtained by integrating the pressure distribution over the wetted wall in the x-direction

$$F_x = \int_0^{2\pi}\int_0^R p|_{\theta=\alpha} r\sin\alpha\cos\varphi\,d\varphi\,dr$$

$$= m_f\Omega^2 X_0\,e^{i\Omega t}\cos\alpha\left[1 + 3\sum_{n=1}^{\infty}\frac{(\Omega^2-(g/R))\beta_n P^1_{\lambda_n}(\cos\theta)}{(\omega_n^2-\Omega^2)(\lambda_n+2)\sin\alpha}\right] \tag{2.175}$$

where $m_f = (1/3)\pi\rho R^3\sin^2\alpha\cos\alpha$ is the fluid mass. The first term in equation (2.175) represents the contribution of the inertia force due to the solidified liquid.

2.10 Paraboloid containers

The liquid hydrodynamic forces and moments under forced excitation are evaluated for a paraboloid container based on the work of Bauer (1984d, 1999). The fluid field equations are governed by Laplace's equation and boundary conditions. Laplace's equation, in terms of paraboloid coordinates defined in Chapter 1, is

$$\frac{\partial^2 \Phi}{\partial \xi^2}+\frac{1}{\xi}\frac{\partial \Phi}{\partial \xi}+\frac{\partial^2 \Phi}{\partial \eta^2}+\frac{1}{\eta}\frac{\partial \Phi}{\partial \eta}+\frac{\left(\xi^2+\eta^2\right)}{\xi^2\eta^2}\frac{\partial^2 \Phi}{\partial \varphi^2}=0 \tag{2.176}$$

Under lateral excitation in the x-direction, $x(t)=X_0\, e^{i\Omega t}$, the boundary conditions pertaining to a truncated annular paraboloid sector container are

$$\frac{\partial \Phi}{\partial \eta}=i\Omega\xi X_0\, e^{i\Omega t}\cos\varphi \quad \text{at the walls } \eta=\eta_0, \eta_1 \tag{2.177a}$$

$$\frac{\partial \Phi}{\partial \xi}=i\Omega\eta X_0\, e^{i\Omega t}\cos\varphi \quad \text{at the bottom } \xi=\xi_1 \tag{2.177b}$$

$$\frac{\partial \Phi}{\partial \varphi}=-i\Omega\xi\eta X_0\, e^{i\Omega t}\sin\varphi, \quad \text{at the side sectored walls } \varphi=0,\ 2\alpha\pi \tag{2.177c}$$

$$\frac{\partial^2 \Phi}{\partial t^2}+g\frac{1}{\xi}\frac{\partial \Phi}{\partial \xi}=0 \quad \text{at the free surface } \xi=\xi_0 \tag{2.177d}$$

Condition (2.177dc) represents the combined free-surface condition of kinematic and dynamic conditions. In order to solve Laplace's equation (2.176), the velocity potential function is split into perturbed and the rigid body components

$$\Phi(\xi,\eta,\varphi,t)=\left\{\tilde{\Phi}(\xi,\eta,\varphi)+i\Omega X_0\xi\eta\cos\ \varphi\right\}e^{i\Omega t} \tag{2.178}$$

The perturbed velocity potential function satisfies the boundary conditions

$$\frac{\partial \tilde{\Phi}}{\partial \eta}=0, \quad \text{at } \eta=\eta_0, \eta_1, \tag{2.179a}$$

$$\frac{\partial \tilde{\Phi}}{\partial \xi}=0, \quad \text{at } \xi=\xi_1, \tag{2.179b}$$

$$\frac{\partial \tilde{\Phi}}{\partial \varphi}=0, \quad \text{at } \varphi=0,\ 2\alpha\pi \tag{2.179c}$$

The combined free-surface boundary condition takes the form

$$g\frac{1}{\xi}\frac{\partial \tilde{\Phi}}{\partial \xi}-\Omega^2\tilde{\Phi}=i\Omega X_0\frac{\eta}{\xi}\left(\Omega^2\xi^2-g\right)\cos\varphi \quad \text{at } \xi=\xi_0 \tag{2.179d}$$

The solution of Laplace's equation, $\nabla^2\tilde{\Phi}=0$, that satisfies the boundary conditions (2.179a–c) may be written in the form

$$\tilde{\Phi}(\xi,\eta,\varphi)=\sum_{m=0}^{\infty}\sum_{n=1}^{\infty}A_{mn}\bar{L}_{m/2\alpha}(\xi)L_{m/2\alpha}(\eta)\cos\left(\frac{m}{2\alpha}\varphi\right) \tag{2.180}$$

where

$$\bar{L}_{m/2\alpha}(\xi)=I_{m/2\alpha}\left(\varepsilon_{mn}\frac{\xi}{\eta_0}\right)K'_{m/2\alpha}\left(\varepsilon_{mn}\frac{\xi_1}{\eta_0}\right)-I'_{m/2\alpha}\left(\varepsilon_{mn}\frac{\xi_1}{\eta_0}\right)K_{m/2\alpha}\left(\varepsilon_{mn}\frac{\xi}{\eta_0}\right) \tag{2.181a}$$

$$L_{m/2\alpha}(\xi)=J_{m/2\alpha}\left(\varepsilon_{mn}\frac{\eta}{\eta_0}\right)Y'_{m/2\alpha}(\varepsilon_{mn})-J'_{m/2\alpha}(\varepsilon_{mn})Y_{m/2\alpha}\left(\varepsilon_{mn}\frac{\eta}{\eta_0}\right) \tag{2.181b}$$

The coefficients A_{mn} are estimated from the free-surface boundary condition (2.179d). This requires the right-hand side of equation (2.179d) to be expressed in terms of the following series

$$\cos\varphi=\sum_{m=0}^{\infty}a_m\cos\left(\frac{m}{2\alpha}\varphi\right),\quad \mu=\sum_{n=1}^{\infty}b_{mn}L_{m/2\alpha}(\varepsilon_{mn}\eta/\eta_0) \tag{2.182a, b}$$

where

$$a_0=\frac{\sin 2\alpha\pi}{2\alpha\pi},\quad a_m=\frac{4\alpha(-1)^{m+1}\sin 2\alpha\pi}{\pi(m^2-4\alpha^2)} \tag{2.183a, b}$$

$$b_{mn}=\frac{\int_{\eta_1}^{\eta_0}\eta^2L_{m/2\alpha}(\varepsilon_{mn}\eta/\eta_0)\,d\eta}{\int_{\eta_1}^{\eta_0}\eta L^2_{m/2\alpha}(\varepsilon_{mn}\eta/\eta_0)\,d\eta} \tag{2.183c}$$

Substituting the series (2.182a, b) into equation (2.179d) gives A_{mn}

$$A_{mn}=i\frac{\Omega X_0(\Omega^2\xi_0^2-g)a_{mn}b_{mn}}{\xi_0(\omega_{mn}^2-\Omega^2)\bar{I}_{m/2\alpha}\left(\varepsilon_{mn}\dfrac{\xi}{\eta_0}\right)} \tag{2.184}$$

where ω_{mn}^2 are the squares of the fluid-free-surface natural frequencies. The pressure distribution is

$$p(\xi,\eta,\varphi,t)=-\rho\frac{\partial\Phi}{\partial t}+\frac{1}{2}\rho g(\xi_0^2-\xi^2) \tag{2.185}$$

The hydrodynamic force along the x-axis is obtained by integrating the pressure distribution over the wetted area, that is, from ξ_1 to ξ_0, and from η_1 to η_0:

$$F_x=\int_0^{2\alpha\pi}\int_{\xi_1}^{\xi_0}\xi^2\left(\eta_0 p|_{\eta=\eta_0}-\eta_1 p|_{\eta=\eta_1}\right)\cos\varphi\,d\xi\,d\varphi-\int_0^{2\alpha\pi}\int_{\eta_1}^{\eta_0}\eta^2\xi_1 p|_{\xi=\xi_1}\cos\varphi\,d\eta\,d\varphi$$
$$-\int_{\xi_1}^{\xi_0}\int_{\eta_1}^{\eta_0}(\xi^2+\eta^2)p|_{\varphi=2\alpha\pi}\sin(2\alpha\pi)\,d\xi\,d\eta \tag{2.186}$$

Consider the case of a simple paraboloid container (nonannular, nonsectored, and non-truncated), the corresponding velocity potential function is

$$\Phi(\xi, \eta, \varphi, t) = \left\{ i\Omega X_0 \xi\eta + \sum_{n=1}^{\infty} A_{1n} I_1(\varepsilon_{1n}\xi/\eta_0) J_1(\varepsilon_{1n}\eta/\eta_0) \right\} \cos\varphi \, e^{i\Omega t} \tag{2.187}$$

where the coefficient A_{1n} is given by the expression

$$A_{1n} = i \frac{2\Omega X_0 \left(\Omega^2 \xi_0^2 - g\right)\eta_0}{\xi_0 \left(\omega_{1n}^2 - \Omega^2\right)\left(\varepsilon_{1n}^2 - 1\right) J_1(\varepsilon_{1n}) I_1(\varepsilon_{1n}\xi_0/\eta_0)} \tag{2.188}$$

where η has been expanded in the Dini series

$$ta = 2\eta_0 \sum_{n=1}^{\infty} \frac{J_1(\varepsilon_{1n}\eta/\eta_0)}{\left(\varepsilon_{1n}^2 - 1\right) J_1(\varepsilon_{1n})} \tag{2.189}$$

and the square of the natural frequency of the surface is

$$\omega_{1n}^2 = \frac{g\varepsilon_{1n}}{\xi_0\eta_0} \frac{I_1'(\varepsilon_{1n}\xi_0/\eta_0)}{I_1(\varepsilon_{1n}\xi_0/\eta_0)} \tag{2.190}$$

The free-surface wave height is

$$z = -\frac{1}{g}\frac{\partial\Phi}{\partial t}\bigg|_{\xi=\xi_0} = \Omega^2 \frac{X_0}{g} \left\{ \xi_0\eta + 2\sum_{n=1}^{\infty} \frac{\left(\Omega^2\xi_0 - (g/\xi_0)\right)\eta_0 J_1(\varepsilon_{1n}\eta/\eta_0)}{\left(\omega_{1n}^2 - \Omega^2\right)\left(\varepsilon_{1n}^2 - 1\right) J_1(\varepsilon_{1n})} \right\} \cos\varphi \, e^{i\Omega t} \tag{2.191}$$

The pressure distribution is

$$\begin{aligned} p =& \rho\Omega^2 X_0 \left\{ \xi\eta + 2\eta_0 \sum_{n=1}^{\infty} \frac{\left(\Omega^2\xi_0 - (g/\xi_0)\right) I_1(\varepsilon_{1n}\xi/\eta_0) J_1(\varepsilon_{1n}\eta/\eta_0)}{\left(\omega_{1n}^2 - \Omega^2\right)\left(\varepsilon_{1n}^2 - 1\right) J_1(\varepsilon_{1n}) I_1(\varepsilon_{1n}\xi_0/\eta_0)} \right\} \cos\varphi \, e^{i\Omega t} \\ &+ \frac{1}{2}\rho g\left(\xi_0^2 - \xi^2\right) \end{aligned} \tag{2.192}$$

Figure 2.26 shows the dependence of the pressure on the excitation frequency at the location $\psi = 0$, $\xi = 1\,m^{1/2}$, and $\eta = \eta_0 = 0.2\,m^{1/2}$, for liquid depth $\xi_0 = 1.077m^{1/2}$, as obtained from $h = (\xi_0^2 + \eta_0^2)/2$. The predicted result is shown by solid curves and covers the first two resonance frequencies. The experimental results are shown by small circles.

The hydrodynamic force is obtained by integrating the pressure over the wetted area

$$\begin{aligned} F_x &= \eta_0 \int_0^{2\pi} \int_0^{\xi_0} \xi^2 p|_{\eta=\eta_0} \cos\varphi \, d\xi \, d\varphi \\ &= m\Omega^2 X_0 \frac{\xi_0^2}{\xi_0^2 + \eta_0^2} \left\{ 1 + 8\frac{\eta_0}{\xi_0} \sum_{n=1}^{\infty} \frac{\left(\Omega^2 - (g/\xi_0^2)\right) I_2(\varepsilon_{1n}\xi/\eta_0)}{\varepsilon_{1n}\left(\omega_{1n}^2 - \Omega^2\right)\left(\varepsilon_{1n}^2 - 1\right) I_1(\varepsilon_{1n}\xi_0/\eta_0)} \right\} e^{i\Omega t} \end{aligned} \tag{2.193}$$

where $m = \frac{\pi\rho}{4}\left(\xi_0^2 + \eta_0^2\right)\xi_0^2\eta_0^2$ is the liquid mass.

Figure 2.27 shows the dependence of the hydrodynamic force on excitation frequency over the first four natural frequencies. The dash curve represents the inertia force of the solidified liquid.

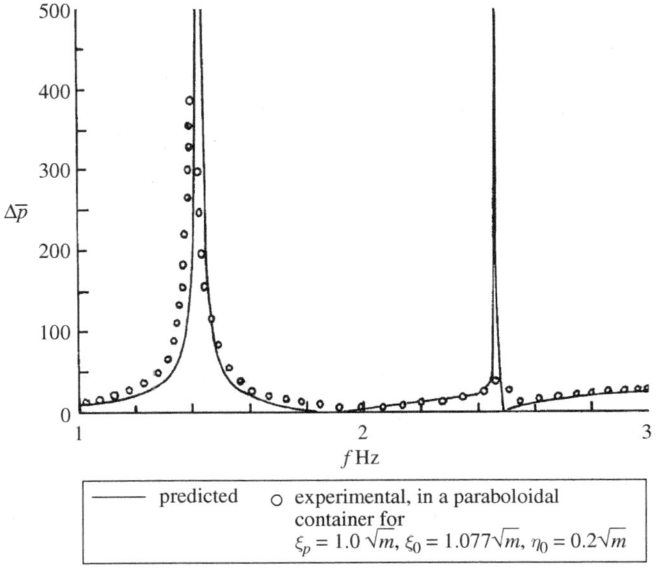

Figure 2.26 Dependence of pressure differential parameter, $\Delta\bar{p} = \dfrac{|p - p_{\text{stat}}|}{\rho X_0 \, e^{i\Omega t} \cos\varphi}[1/s^2]$, on excitation frequency. (Bauer, 1999)

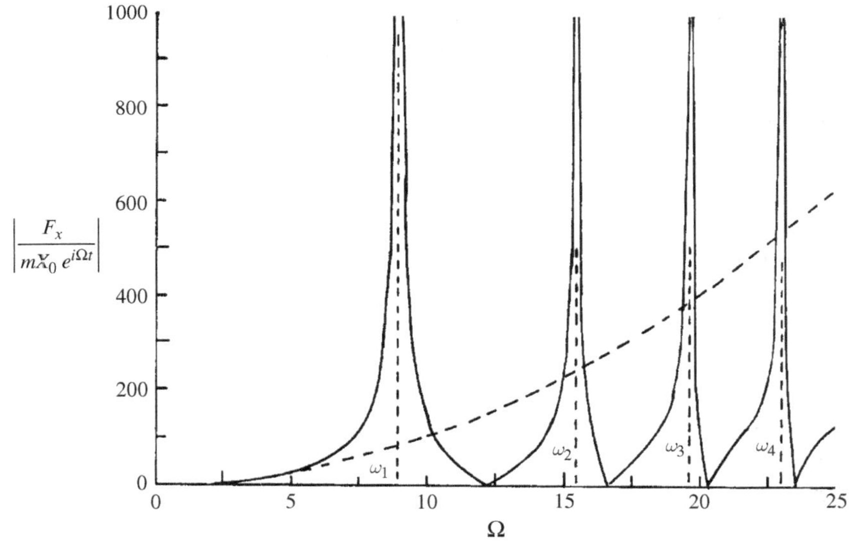

Figure 2.27 Dependence of hydrodynamic force on excitation frequency for $\xi_p = 1.0\sqrt{m}$, $\xi_0 = 1.077\sqrt{m}$, $\eta_0 = 0.2\sqrt{m}$. (Bauer, 1999)

2.11 Sloshing of magnetic fluids

A magnetic fluid is a stable colloidal dispersion of surfactant-coated magnetic particles in a liquid carrier. The surfactant coating inhibits coalescence of the particles. A colloid is a term coined by the Scottish chemist Thomas Graham, (1805–1969), and refers to a substance made up of very small, insoluble, nondiffusible particles that remain in suspension in a surrounding medium. The diameter of the solid magnetic particles is in the range of 5 ~ 15 nm. Magnetic fluids were developed to control the position of liquid propellant in a gravity-free state (Papell, 1965). Zelazo and Melcher (1969) and Omura, *et al.* (1999) studied the dynamic behavior of a magnetic fluid in an oscillating container in order to examine the influence of a tangential magnetic field on the resonance surface waves. Dodge and Garza (1970b, 72) demonstrated a simulation of liquid sloshing in low gravity by using a magnetic fluid. Sudo, *et al.* (1987a,b,c) studied the interfacial instability of magnetic fluids in a rectangular container in the presence of a magnetic field. Okubo, *et al.* (1990), Ohaba, Tomonori, *et al.* (1998), Ohaba, Sawada, *et al.* (1998, 1999) investigated the stability of the surface of a magnetic fluid layer under the influence of a vertical alternating magnetic field. Sawada, *et al.* (1987, 1993, 1998), Kikura, *et al.* (1991), and Matsuura, *et al.* (1995) examined the effect of the magnetic field on the resonance frequency of two-layer liquid sloshing of a magnetic fluid and silicon oil in a rectangular container.

Experimental measurements of magnetic liquid dynamic behavior using optical methods such as laser Doppler anemometry, or flow visualization techniques such as particle image velocimetry cannot be used because magnetic fluids are opaque. For this reason, most experiments were restricted to measuring the pressure distribution and free surface behavior (Sudo, *et al.* (1997), Sawada *et al.* (1999), and Sawada, *et al.* (1999, 2002a,b) studied the lateral sloshing behaviors of magnetic fluids in rectangular and cylindrical containers, respectively.

Consider an upright cylindrical container partially filled with a magnetic fluid subjected to sinusoidal lateral excitation $x(t) = X_0 \sin \Omega t$. The fluid field equations of magnetic liquids in moving containers are the same as nonmagnetic liquids except that the unsteady Bernoulli equation has an additional term due to the magnetic field. For irrotational, inviscid, and incompressible flow, a velocity potential function Φ exists which satisfies the continuity equation

$$\frac{\partial^2 \Phi}{\partial r^2} + \frac{1}{r}\frac{\partial \Phi}{\partial r} + \frac{1}{r^2}\frac{\partial^2 \Phi}{\partial \theta^2} + \frac{\partial^2 \Phi}{\partial z^2} = 0 \tag{2.194}$$

The unsteady Bernoulli equation takes the form (Rosenweig, 1987)

$$\frac{p}{\rho} + \frac{\partial \Phi}{\partial t} + gz - \frac{\mu_0}{\rho H}\frac{dH}{dz}\int_0^H M dH = \Omega^2 X_0 r \sin \Omega t \tag{2.195}$$

where μ_0 is the magnetic permeability of vacuum, M is the magnetization, H (amp-turns/m) is the magnetic field. If the magnetization and magnetic field are assumed to be parallel, the magnetic term takes the form

$$\frac{\mu_0}{\rho}\int_0^H M\, dH = \frac{\mu_0}{2\rho}\chi_m H^2 \tag{2.196}$$

where χ_m is the susceptibility of the magnetic fluid. The magnetic field induction is $B = H\mu$ (weber/m^2, or newton/ampere-meter, or tesla). The magnetic field $H(z)$ may be approximated by the following expression

$$H(z) = H_0\, e^{-\alpha(z+h)} \tag{2.197}$$

where H_0 is the magnetic field intensity at the bottom of the tank, $z = -h$, and α is a constant determined from experimental measurements. The kinematic and dynamic free surface boundary conditions are, respectively,

$$\frac{\partial \eta}{\partial t} = \left.\frac{\partial \Phi}{\partial z}\right|_{z=0} \tag{2.198}$$

$$\frac{\partial^2 \Phi}{\partial t^2} + \left[g + \frac{\mu_0}{\rho}\alpha\chi H_0^2 e^{-2\alpha h}\right]\left.\frac{\partial \Phi}{\partial z}\right|_{z=0} = \Omega^3 X_0 r \cos\Omega t \tag{2.199}$$

Condition (2.199) is obtained from equation (2.195) by taking the time derivative of both sides at the free surface where the pressure is considered constant. The expression in the bracket represents the effective gravity including the magnetic field. The other boundary conditions are at the sidewalls and bottom of the tank, where the normal component of the fluid vanishes, that is,

$$\left.\frac{\partial \Phi}{\partial r}\right|_{r=R} = 0, \qquad \text{and} \qquad \left.\frac{\partial \Phi}{\partial z}\right|_{z=-h} = 0 \tag{2.200a, b}$$

A possible solution of equation (2.194), satisfying the boundary conditions (2.198)–(2.200), may be written in the form

$$\begin{aligned}\Phi(r,\theta,z,t) = {} & \Omega X_0 R \cos\theta \cos\Omega t \\ & \times \left\{\frac{r}{R} - 2\sum_{n=1}^{\infty}\frac{\omega_{1n}^2\Omega^2}{\left(\omega_{1n}^2-\Omega^2\right)^2}\frac{J_1(\lambda_{1n}r)}{J_1(\lambda_{1n}R)}\frac{\cosh\left[\lambda_{1n}(z+h)\right]}{\cosh(\lambda_{1n}h)}\right\}\end{aligned} \tag{2.201}$$

where

$$\omega_{1n}^2 = \left[g + \frac{\mu_0}{\rho}\alpha\chi H_0^2 e^{-2\alpha h}\right]\lambda_{1n}\tanh(\lambda_{1n}h) \tag{2.202}$$

and λ_{1n} are the roots of $J'_1(\lambda_{1n}r)|_{r=R} = 0$. In order to determine the fluid-free surface wave height, $\eta(r, \theta, t)$, the solution (2.201) is used in equation (2.198) and integrating the result with respect to time gives

$$\eta(r,\theta,t) = \frac{X_0\Omega^2 R\cos\theta\sin\Omega t}{\left[g + \left(\mu_0\alpha\chi H_0^2 e^{-2\alpha h}/\rho\right)\right]}\left\{\frac{r}{R} - 2\sum_{n=1}^{\infty}\frac{\omega_{1n}^2\Omega^2}{\left(\omega_{1n}^2-\Omega^2\right)^2}\frac{J_1(\lambda_{1n}r)}{J_1(\lambda_{1n}R)}\right\} \tag{2.203}$$

Figure 2.28 shows the amplitude–frequency response of the free surface for four different values of the magnetic field induction. The excitation frequency ratio, Ω/ω_0, is normalized with respect to the first resonance frequency, ω_0, for magnetic field induction at the center, $B_0 = 0$ mT (mT = millitesla), calculated from equation (2.202). For low excitation frequency, $\Omega < \omega_0$, the free surface oscillates with a nodal diameter perpendicular to the excitation direction. As the excitation frequency increases the surface displacement increases until it reaches resonance at which the free surface forms a collapse wave. Any slight increase in the excitation frequency results in rotation of the nodal diameter. The rotation direction changes irregularly. Any further increase of the excitation frequency causes a further decrease in the free-surface displacement. This is associated with unstable swirling over a narrow range of excitation frequency above which the swirling motion has fixed direction. The free surface reaches a local maximum again and this regime is maintained over a limited range of excitation

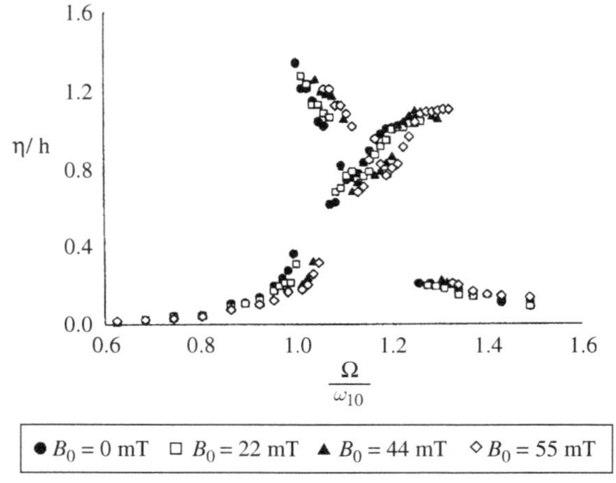

Figure 2.28 Amplitude frequency response of the free surface of magnetic fluid under lateral excitation for different values of the magnetic field induction B_0. (Sawada, *et al.*, 2002a)

frequency. Above that range, the free surface oscillates in a planar form with a nodal diameter perpendicular to the excitation direction. These observations are due to the nonlinear coupling of the sloshing modes along to orthogonal diameters and will be examined in Chapter 4.

Sudo, *et al.* (1997) conducted a series of lateral excitation tests on magnetic fluids (kerosene-based ferricolloid HC-50, and water-based ferricolloid W-35) and measured the fluid-free-surface amplitude for three different values of the magnetic gradient $dH/dz = -3.871 \times 10^6$ A/m^2, 0 A/m^2, and 3.871×10^6 A/m^2. The measured and predicted results are shown in Figure 2.29(a)–(c). The experimental maximum surface displacement η_{max} was taken as the distance between the undisturbed surface and the wave crest. It is seen that the experimental measurements are higher than those predicted by the linear analysis. The difference is manifested near the resonance frequency. It was reported that the measurements were taken for the actual upward displacement, which is usually greater than the downward displacement near resonance. It is seen that the natural frequency of the first anti-symmetric sloshing mode is higher for negative magnetic gradient than the one with zero or positive gradient as shown in Figure 2.29(a)–(c). The magnetic field shifts the resonance frequency to values higher than those in the absence of the magnetic field. The reason for that increase is due to the associated increase in the effective gravity given by the expression in brackets in equation (2.202).

2.12 Closing remarks

The forced excitation of liquid tanks of different geometries has been treated based on the linear theory of small oscillations. The excitation has been considered either lateral or rotational about the pitch or yaw axes. For upright circular cylindrical containers, the free surface was assumed to oscillate in a planar form without rotation about the nodal diameter. Hydrodynamic forces, moments, and wave heights were estimated for upright cylindrical containers with and without compartments. Two- and three-dimensional rectangular containers, and spherical, prolate and oblate spheroidal containers, and paraboloid containers were also considered.

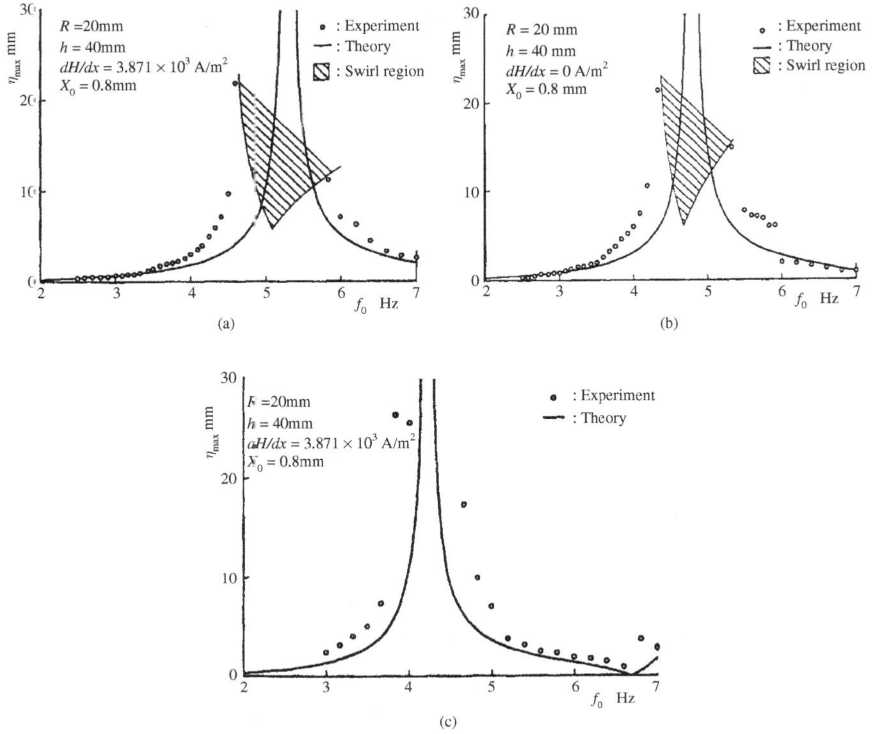

Figure 2.29 Amplitude–frequency response of magnetic fluid free surface in a cylindrical container for three different values of magnetic gradient. (Sudo, *et al.*, 1997)

The free surface response to off-axis accelerations was treated by Bowman (1966a). Numerical modeling of steep forced water waves was developed by Bredmose, *et al.* (2003). Numerical simulations of strong liquid waves will be addressed in Chapter 8.

The dynamic sloshing loads are of great importance to the stability and trajectories of liquid propellant rockets. The influence of liquid sloshing loads on the stability of aerospace vehicles was studied by Podesta (1956), Epperson and Brown (1957), Miles and Young (1958), Bauer (1961a,b,d, 1963b,d, 1964a, 1966b, 1968b), Lugovski (1961), Epperson, *et al.* (1961), Rumyantsev (1964c), Blacklock (1965), Abramson (1966a, Chapter 7 by Bauer), Eggleston (1968), Moiseev (1968), Chernousko (1969), and Gerrits (2001). They also determined the dynamic loads resulting from liquid sloshing in missile tanks. The stability of rigid bodies with an ellipsoid cavity completely filled with ideal fluid was determined by Ishlinskiy and Temochenko (1966).

This chapter did not consider the case of random excitation of liquid containers. This case was considered by Mathiesesen (1976), and will be treated in Chapter 4. The arbitrary excitation of conical containers was analyzed by Harper (1958) while the transient response was studied by Liu and Lou (1990). The parametric excitation will be considered in Chapter 6. The roll excitation of a bare wall cylindrical tank about its vertical axis, together with viscous effects, will be examined in Chapter 3. Other topics that were not addressed include the forced excitation of a two-liquid layer (Honda and Tajima, 1979), and liquid sloshing in a moving tank with a time-dependent discontinuous boundary and with draining (Yeh, 1965 and Yeh and Graham, 1969).

3
Viscous damping and sloshing suppression devices

3.1 Introduction

The estimation of the free-surface natural frequencies of inviscid liquids has been presented in Chapter 1. The inclusion of viscous damping due to boundary layers and contamination of the free surface is very important to bringing the sloshing wave height to a bounded level at resonance. The unsteady flow of a liquid contained in or surrounding bodies could be classified into two main categories:

(1) Boundary layers that are created when a body starts from rest either impulsively or with acceleration.
(2) Boundary layers that are periodic due to fluctuations of the stream or oscillations of the body.

The kinematic viscosity, ν, is a fundamental parameter for measuring the rate of vorticity and momentum by molecular transport. In fact, it is known that the vorticity is present at a distance, δ, outside the boundary of the body, whereas the time required for diffusing the vorticity or momentum through this distance is of order δ^2/ν, which is referred to as the diffusion time. The diffusion time, δ^2/ν, is of order $(1/\Omega)$ and the boundary layers thickness is proportional to $\sqrt{\nu/\Omega}$, where Ω is the frequency of roll oscillation. In the particular case, when the boundary layers thickness, δ, is very small compared with a reference body length, l, and the Reynolds number, $\Omega\, l^2/\nu \gg 1$, the boundary layers approximation is applicable.

Boussinesq (1878) introduced the influence of viscous damping in studying progressive and standing waves in closed containers. Keulegan (1959) extended Boussinesq's work in order to calculate the attenuation of solitary waves. The damping of progressive waves in a channel of finite depth but with infinite width was considered by Biesel (1949). Ursell (1952) treated the case of damping due to vertical walls with infinite depth, while Hunt (1952, 1964) calculated the combined effects of finite width and finite depth. In identifying the viscous damping rate in surface waves, Ursell (1952) found an example in which two different approaches gave different answers. The first approach was based on energy dissipation and the second relied on the pressure working on the edge of the sidewall boundary layers. The discrepancy between theoretical predictions and experimental measurements of damping led Benjamin and Ursell (1954) to consider the influence of capillary hysteresis and surface contamination. Mei and Liu (1973) resolved this discrepancy by examining the energy transfer in the neighborhood of the free-surface meniscus. With due care near the meniscus, a boundary-layer-Poincaré method was employed to give an alternative derivation for the rate attenuation.

Case and Parkinson (1957) determined the damping of surface waves of small amplitude in a partially filled cylindrical tank. They considered viscous dissipation in an assumed laminar boundary layer as the primary cause of damping. Van Dorn (1966) conducted experimental investigations and attributed the differences between the predicted and measured results to a surface film produced by spontaneous contamination. He found that "while the observed attenuation agreed with that computed for solid boundaries when the water was fresh, the former tended to increase with time to some higher limiting value, usually within an hour." Surfactants, such as oil on water, can lead to appreciable damping surface waves (see, e.g., Miles, 1967 for a complete historical account). Miles (1967) concluded that both surface contamination and capillary hysteresis might contribute significantly to the damping of surface waves in closed basins. Mikishev and Rabinovich (1968) determined the damping factor using the boundary layer theory due to low viscous fluid in rigid containers. Bogoryad (1990) determined the shear stresses on a wetted surface of a vessel due to a viscous capillary layer of fluid.

The influence of liquid viscosity on the sloshing modes natural frequencies was studied analytically and experimentally for rectangular tanks by Scarsi (1971). Scarsi found that the fluid viscosity affects the values of the natural frequencies only for low values of fluid depth ratio. Su (1981) considered the effect of fluid viscosity on the vibration of elastic shells by studying the axisymmetric free-surface oscillations of a fluid-filled spherical shell. Henderson and Miles (1994) and Martel, *et al.* (1998) calculated the natural frequencies and damping ratios for surface waves in a circular cylinder based on the assumption of a fixed contact line, Stokes boundary layers, and either a clean or a fully contaminated surface. They indicated that the damping of surface waves in a closed tank, which almost exceeds theoretical predictions, is due to four sources: (i) viscous dissipation at a rigid boundary of the tank, (ii) viscous dissipation at the free surface, which may be covered by a viscoelastic film, (iii) viscous damping in the interior fluid, and (iv) capillary hysteresis at the contact line. The interior damping is negligible for water in a tank whose lateral dimension is comparable to the wavelength.

The inherent liquid viscosity in tanks without baffles has a very limited effect in reducing the sloshing amplitude. Ring and vertical cruciform baffles, floating lids and mats, and flexible baffles are very effective in controlling the liquid wave motion and preventing vehicle instability during maneuvering. The effectiveness of damping devices should not be characterized by the amplitude decrement only. Miles (1958b), Schwind, *et al.* (1964, 1967), and Cole (1966b) obtained semi-empirical relationships for the damping contributed by a flat annular ring baffle. Buchanan (1968) and Shih and Buchanan (1971) determined the drag force acting on an oscillating plate in liquids at low Reynolds numbers. The determination of structural loads acting either on the vehicle structure or on the suppression devices as a result of liquid sloshing is discussed in Bhuta, *et al.* (1968). Abramson (1969) presented the state-of-the-art of slosh suppression in moving liquid containers.

Experimental, and a few analytical, studies were carried out to determine the damping in tanks of various shapes fitted with various types of baffles by Howell and Ebler (1956), Bauer (1959b, 1960b, 1962c, 1963c,e, 1964b), Cole and Gambucci (1961a,b), Silveira, *et al.* (1961), Stephens, *et al.* (1963), and Stephens and Scholl (1967). These studies presented the total damping of the fundamental anti-symmetric mode caused by two main sources. The first is due to the relative motion between the liquid and the tank wall and the second is due to the relative motion between the liquid and baffles. Langner (1963) and Schwind, *et al.* (1967) considered the design optimization of sloshing suppression devices. The total fluid sloshing force should not exceed a certain prescribed maximum value under all possible combinations of liquid level, tank orientation, acceleration, and external tank excitation. Another important requirement is to suppress sloshing effects throughout certain designated frequency ranges over which the liquid oscillations might influence the fundamental vibration mode of the vehicle.

Garza (1963, 1964, 1966), Garza and Abramson (1963), Abramson (1964), and Garza and Dodge (1967) carried out a series of experiments to find out the effect of different types of slosh suppression devices on the liquid natural frequencies, damping, and liquid forces. Liu (1964) and Scholl, *et al.* (1972) measured the hydrodynamic pressure distribution acting on a ring-baffle. Abramson, *et al.* (1962b, 1963) and Abramson and Garza (1964, 1965) conducted a series of tests on baffles with perforated holes of different sizes and various open area percentages. Their results indicated a large nonlinear effect on the liquid damping ratio. The damping provided by a cruciform was measured by Silveira, *et al.* (1961) for two orientations of the baffle with respect to the motion of the tank.

Liquid sloshing loads are usually reduced by partitioning the tank. The optimal compartment length is half the length of the free-surface wave. This concept was proposed by Berlamont and Vanderstappen (1979) for water tower tanks and by Popov, *et al.* (1993a) for road containers.

Other types of suppression devices were considered in the literature such as annular ring tanks, floating cans, floating lids and sound suppressors (Eulitz, 1958, 1961, Abramson and Ransleben, 1961d, Bauer, 1966a). Sayar (1971) and Sayar and Baumgarten (1982) studied different types of linear and nonlinear liquid slosh dampers. Agrawal (1987) discussed the possible interaction of liquid slosh modes with attitude control systems in a dual-spin spacecraft. Other methods included an inverse U-tube (Hayama and Iwabuchi, 1985, 1986, Hayama, *et al.*, 1989, Fujita, *et al.*, 1991), wire rope (Kobayash, 1982) and moving baffles using active feedback control (Su and Wang, 1991). Kobayashi, *et al.* (1995) studied the effectiveness of bulkheads in suppressing the liquid sloshing in cylindrical and co-axial double cylindrical vessels. Kobayashi, *et al.* showed that the sloshing mode is dissociated into two uncoupled modes of the U-tube mode and the bulkhead mode, both contributing to suppressing sloshing.

Huerta, *et al.* (1989) and Hiramatsu, *et al.* (1989a,b) analytically and experimentally studied the influence of immersed tubes and blocks in cylindrical water tanks. Huerta, *et al.* (1989) used the arbitrary Lagrangian–Eulerian finite element formulation while Hiramatsu, *et al.* (1989a,b) used the three-dimensional boundary element method to determine the liquid mode shapes. Numerical and experimental results revealed that immersed tubes reduced the liquid sloshing loads and flow velocity.

The interaction of liquid sloshing with membrane rubber, which acts as anti-sloshing system, was studied numerically and experimentally by Nishino and Mochio (1995). Their analysis was based on the assumption of membrane large deformation and its material nonlinearity. The influence of liquid viscosity and convective terms were considered. The dependence of the liquid sloshing amplitude response on the excitation base acceleration was found to be less than the linear values, and was associated with an increase of the sloshing frequency.

For an upright circular cylindrical tank describing roll oscillations, the fluid adjacent to the solid walls participates with the motion of the container. Its effective inertia and damping depends on its viscosity and tank roll frequency. The system roll dynamics is affected by the contribution of the fluid to the inertia and damping of the system. The effect of liquid forces on the stability of roll oscillations of missiles was examined by Clark (1959), Rosenblat (1959), Chautard (1963), Chernous'ko (1965, 1966, 1967a,b,c) and Chermykh (1966). These forces are usually determined by solving the Navier–Stokes equations subject to the inherent boundary conditions. The resulting solution gives an account of the thickness of the boundary layers that participates in container roll oscillations.

One may be interested in determining the thickness of the boundary layers over an infinite flat plate, which oscillates horizontally. This problem is well documented by Schlichting (1960) and Rosenhead (1963). The unsteady flow behavior over an oscillating flat plate was first examined by Stokes (1851), Lord Rayleigh (1883a), and Kestin and Persen (1954). This problem has been referred to as "Stokes' second problem."

This chapter considers the influence of viscosity on the liquid sloshing damping factor in upright cylindrical containers. The influence of viscosity and surface tension on the free-surface natural frequency and the hydrodynamic forces under forced lateral excitation will be analyzed. When the inherent damping due to liquid viscosity is not enough to reduce hydrodynamic forces near resonance, one must introduce baffles, or divide the tank into compartments. The effect of different types of baffles is usually determined experimentally except for a very few special cases where semi-empirical formulas have been obtained. The contribution of periodic boundary layers of fluids, in upright circular containers experiencing roll oscillations, to the system inertia and damping will be presented analytically and experimentally. The analysis will be extended in order to determine the contribution of immersed rods parallel to the tank axis.

3.2 Viscous damping in liquid containers

3.2.1 Damping in a circular cylindrical container

The fluid field equation of viscous fluid is given by the linearized Navier–Stokes equation for an incompressible fluid

$$\frac{d\mathbf{u}}{dt} \approx \frac{\partial \mathbf{u}}{\partial t} = -\vec{\nabla}\left(gz + \frac{p}{\rho}\right) + \nu\nabla^2\mathbf{u} \tag{3.1}$$

The fluid motion is subject to the continuity equation

$$\nabla\mathbf{u} = \mathbf{0} \tag{3.2}$$

and the boundary condition

$$\mathbf{u} = \mathbf{0} \quad \text{on the tank wetted boundaries} \tag{3.3}$$

where p is the pressure.

The fluid velocity is described in terms of the gradient of a scalar potential function, Φ, plus the curl of a vector potential function, $\mathbf{A}$, in the Helmholtz form

$$\mathbf{u} = -\nabla\Phi + \nabla\times\mathbf{A} \tag{3.4}$$

Substituting expression (3.4) into the continuity equation (3.2) and the boundary condition (3.3) gives, respectively,

$$\nabla^2\Phi = 0 \tag{3.5a}$$

$$-\nabla\Phi + \nabla\times\mathbf{A} = 0 \quad \text{at the tank wetted boundaries} \tag{3.5b}$$

The dynamic free-surface boundary condition is obtained by satisfying equation (3.1) at the free surface $z = \eta(r, \theta)$.

Substituting equation (3.4) into equation (3.1) and equating the gradients and curls on both sides gives the dynamic free-surface boundary condition

$$\frac{\partial \Phi}{\partial t} = gz - \frac{p}{\rho} + \text{const} \tag{3.6}$$

and the differential equation for the vector potential is

$$\nu \nabla^2 \mathbf{A} = \frac{\partial \mathbf{A}}{\partial \mathbf{t}} \tag{3.7}$$

Condition (3.7) is analogous to the diffusion equation. Because the first time derivative of $\boldsymbol{A}$ occurs instead of the second derivative, solutions of this equation are not of the waves type but are critically damped, dying out in time and attenuated in spatial motion (Morse and Feshbach, 1953). Note that the pressure is not affected by these waves. Here viscosity causes vorticity to diffuse away as expected.

The free-surface boundary condition requires that the vertical velocity of a fluid particle should equal the vertical velocity of the free surface itself

$$\frac{\partial \Phi}{\partial z} = \frac{\partial \eta}{\partial t} \tag{3.8}$$

It is useful to establish the influence of liquid viscosity on the liquid-free-surface motion. The boundary condition at the rigid walls (3.5b) together with the continuity equation (3.5a), the diffusion equation (3.7), and the kinematic free surface condition (3.8) make it possible, in principle, to obtain an explicit solution for the scalar potential function Φ and the vector function $\mathbf{A}$. However, further simplifications can be introduced based on the nature of the boundary layers. It is expected that Φ is essentially the velocity potential in the inviscid flow, Φ_c, that is, the potential for which $\partial\Phi/\partial n = 0$ at the rigid walls. Case and Parkinson (1957) expressed Φ in terms of two components, that is, $\Phi = \Phi_c + \Phi_b$, where Φ_b is due to an additional contribution which is very small. In this case, the following boundary conditions at the rigid tank walls are obtained

$$[\mathbf{n} \times (\nabla \times \mathbf{A})] = (\mathbf{n} \times \ \nabla\Phi_c) \quad \text{at rigid walls} \tag{3.9a}$$

$$[\mathbf{n} \ \ \nabla \times \mathbf{A}] = \left(\frac{\partial \Phi_b}{\partial n}\right) \quad \text{at rigid walls} \tag{3.9b}$$

The velocity potential function, Φ_c, is assumed to be known from the inviscid problem and by using equations (3.7) and (3.9a) the vector potential function $\mathbf{A}$ can be determined. Having obtained $\mathbf{A}$ one may use the continuity equation (3.5a) and equation (3.9b) to determine the additional potential function Φ_b using successive approximation. Note that the region of nonvanishing $\mathbf{A}$ is confined in the immediate vicinity of the boundaries. If one assumes the case of harmonic time dependence of $\mathbf{A} \propto e^{i\omega t}$, equation (3.7) takes the form

$$\left(\nabla^2 + \frac{i}{l^2}\right)\mathbf{A} = 0 \tag{3.10}$$

where $l = \sqrt{\nu/\omega}$ is the boundary layer thickness. Outside the boundary layer, the vector potential function **A** vanishes exponentially. For free oscillation in which the damping is small, the frequency ω is large in comparison with the damping coefficient α in the exponential decay factor $e^{-\alpha t}$. In the body of the liquid, including the free surface, away from the boundary layer, **A** can be taken to be zero. Accordingly, the boundary conditions are satisfied at the free surface in terms of Φ_c, which may be written in the form

$$\Phi_c = \sum_m \sum_n N_{mn}\dot{q}_{mn} \cos n\theta J_n(\lambda_{mn} r) \frac{\cosh \lambda_{mn}(z+h)}{\lambda_{mn} \sin \lambda_{mn} h} \tag{3.11}$$

where

$$N_{mn} = \frac{1}{\sqrt{\frac{1}{2}\pi R^2 J_n^2(\lambda_{mn}R)\left(1 - \frac{n^2}{(\lambda_{mn}R)^2}\right)}} \tag{3.12}$$

The energy dissipation may be estimated by using the well-known result of Lamb (1945)

$$\frac{d\bar{E}}{dt} = -2\bar{F} \tag{3.13}$$

where $\bar{E}$ is the average of the sum of kinetic T and potential Π energies per cycle, and $2\bar{F}$ is the dissipation function given by the expression (Case and Parkinson, 1957)

$$2\bar{F} \approx \mu \int_v \left(\vec{\nabla} \times (\vec{\nabla} \times \mathbf{A})\right)^2 dv - \mu \int_{S_t} \frac{\partial}{\partial n}\left(\vec{\nabla}\Phi_c\right)^2 ds \tag{3.14}$$

where μ is the liquid dynamic viscosity.

The first integral in equation (3.14) is a volume integral while the second integral is a surface integral over the equilibrium free surface. The vector potential function is expected to be insignificant away from the rigid walls. The differential equation for the vector potential function, **A**, in the vicinity of the tank walls can be integrated and the result is introduced in the first integral of equation (3.14). The second integral gives the dissipation due to the free surface S_t.

The kinetic and potential energies are given by the two expressions

$$T = \frac{1}{2}\rho \int_v (\mathbf{u}.\mathbf{u})\, dv = \frac{1}{2}\rho \dot{q}_{mn}^2 \frac{\coth \lambda_{mn} h}{\lambda_{mn}} \tag{3.15}$$

$$\Pi = \frac{1}{2}\rho g q_{mn}^2 \tag{3.16}$$

where q_{mn} is the surface coordinate corresponding to the symmetric surface waves ($\cos n\theta$). For damped oscillations the wave amplitude may be written in the form

$$q_{mn} = q^*_{mn}\, e^{-(i\omega_{mn}+\alpha)t}, \quad \omega_{mn} >> \alpha \tag{3.17}$$

Introducing equation (3.17) into equation (3.15) and (3.16), and taking the time derivative per one cycle, gives

$$\frac{d\bar{E}}{dt} = -\alpha\rho g q^{*2}_{mn}\, e^{-2\alpha t} \tag{3.18}$$

Introducing Φ_c given by equation (3.11) into the second integral of (3.14), gives

$$2F_t = -\mu \int\limits_{S_t} \frac{\partial}{\partial n}\left(\vec{\nabla}\Phi_c\right)^2 ds = 2\mu\lambda^2_{mn} g q^{*2}_{mn}\, e^{-2\alpha t} \tag{3.19}$$

Equating the two expressions (3.18) and (3.19) gives

$$\alpha_t = 2\nu\lambda^2_{mn} \tag{3.20}$$

where ν is the kinematic viscosity. Relation (3.20) is a measure of the damping coefficient of the free surface or the fluid body.

The differential equation of the vector potential function described by equation (3.10) can be solved in terms of known functions based on the assumption that the boundary layer thickness l is very small compared to the radius of the cylinder. The following are possible components of the vector potential near the sidewalls

$$A_r = \frac{-n}{\lambda_{mn}R}\frac{J_n(\lambda_{mn}R)}{e^{\sqrt{i(R-r)/l}}}\left(\frac{\dot{q}_{mn}N_{mn}}{\lambda_{mn}\sinh\lambda_{mn}h}\right)\cos n\theta\,\sinh(\lambda_{mn}(z+h)) \tag{3.21a}$$

$$A_\theta = \frac{-l\lambda_{mn}}{\sqrt{i}}\frac{J_n(\lambda_{mn}R)}{e^{\sqrt{i(R-r)/l}}}\left(\frac{\dot{q}_{mn}N_{mn}}{\lambda_{mn}\sinh\lambda_{mn}h}\right)\left[1-\frac{n^2}{(\lambda_{mn}R)^2}\right]\sin n\theta\,\sinh(\lambda_{mn}(z+h)) \tag{3.21b}$$

$$A_z = 0 \tag{3.21c}$$

Introducing expressions (3.21) into the first integral in equation (3.14), gives

$$2\bar{F}_s = \mu\int\limits_V\left(\vec{\nabla}\times(\vec{\nabla}\times\mathbf{A})\right)^2 dV = \frac{\mu\omega^2_{mn}}{\lambda_{mn}R}\frac{q^{*2}_{mn}\,e^{-2\alpha t}}{\left[1-\left(\frac{n}{\lambda_{mn}R}\right)^2\right]}$$
$$\times\left\{\coth\lambda_{mn}h\left[1+\frac{n^2}{(\lambda_{mn}R)^2}\right]-\frac{\lambda_{mn}h}{\sinh^2\lambda_{mn}h}\left[1-\frac{n^2}{(\lambda_{mn}R)^2}\right]\right\} \tag{3.22}$$

Equating the right-hand side of equation (3.22) to the right-hand side of equation (3.18) gives the damping parameter due to sidewalls

$$\alpha_s = \frac{1}{2R}\sqrt{\frac{\nu\omega_{mn}}{2}}\left\{\frac{1+(n/\lambda_{mn}R)^2}{1-(n/\lambda_{mn}R)^2}-\frac{2\lambda_{mn}h}{\sinh 2\lambda_{mn}h}\right\} \tag{3.23}$$

Similarly, the contribution to the dissipation from the tank bottom is obtained at $z=-h$ in the form

$$A_r = \frac{nl}{\sqrt{i}\lambda_{mn}R}\frac{J_n(\lambda_{mn}r)}{e^{\sqrt{i(z+h)/l}}}\left(\frac{\dot{q}_{mn}N_{mn}}{\sinh\lambda_{mn}h}\right)\cos n\theta \tag{3.24a}$$

$$A_\theta = \frac{-l}{\sqrt{i}\lambda_{mn}}\frac{1}{e^{\sqrt{i(z+h)/l}}}\left(\frac{\dot{q}_{mn}N_{mn}}{\sinh\lambda_{mn}h}\right)\sin n\theta\,\frac{d}{dr}J_n(\lambda_{mn}r) \tag{3.24b}$$

$$A_z = 0 \tag{3.24c}$$

Introducing expressions (3.24) into the first integral in equation (3.14), gives the dissipation $2\bar{F}_b$ due to the tank bottom

$$2\bar{F}_b = \frac{\mu\lambda_{mn}q_{mn}^{*2}g}{\sqrt{2l}\sinh 2\lambda_{mn}h}\,e^{-2\alpha t} \tag{3.25}$$

Equating the right-hand side of equation (3.25) with the right-hand side of equation (3.18) gives the damping parameter due to the tank bottom

$$\alpha_b = \frac{1}{2R}\sqrt{\frac{\nu\omega_{mn}}{2}}\frac{2\lambda_{mn}R}{\sinh 2\lambda_{mn}h} \tag{3.26}$$

The logarithmic decrements, δ_i, due to the free surface (3.20), sidewalls (3.23), and tank bottom (3.26) are obtained by using the definition $\delta_i = 2\pi\alpha_i/\omega_{mn}$. Thus, the following three expressions are obtained

$$\delta_t = 4\pi\nu\frac{\lambda_{mn}^2}{\omega_{mn}} \to O(l^2/R^2) \tag{3.27a}$$

$$\delta_s = \frac{\pi}{R}\sqrt{\frac{\nu}{2\omega_{mn}}}\left\{\frac{1+(n/\lambda_{mn}R)^2}{1-(n/\lambda_{mn}R)^2} - \frac{2\lambda_{mn}h}{\sinh 2\lambda_{mn}h}\right\} \to O(l/R) \tag{3.27b}$$

$$\delta_b = \frac{\pi}{R}\sqrt{\frac{\nu}{2\omega_{mn}}}\frac{2\lambda_{mn}R}{\sinh 2\lambda_{mn}h} \to O(l/R) \tag{3.27c}$$

where

$$\omega_{mn}^2 = g\lambda_{mn}\tanh(\lambda_{mn}h)\left\{1+\frac{\sigma\lambda_{mn}^2}{\rho g}\right\} \tag{3.28}$$

is the mn-sloshing mode natural frequency, where σ is the surface tension.

Figure 3.1 shows the dependence of the total logarithmic decrement, $\delta = \delta_t + \delta_s + \delta_b$, on the fluid depth ratio h/R, for $n=1$, $\lambda_{mn} = 1.841/R$, and for three values of the tank radius $R = 1.5$ inch, 3.0 inch, and 10 inch. For all three cases, the damping decreases significantly as the fluid depth increases up to $h/R = 1$ and then reaches asymptotic values. One may observe that the contribution of the fluid body given by expression (3.27a) is small compared to the wall

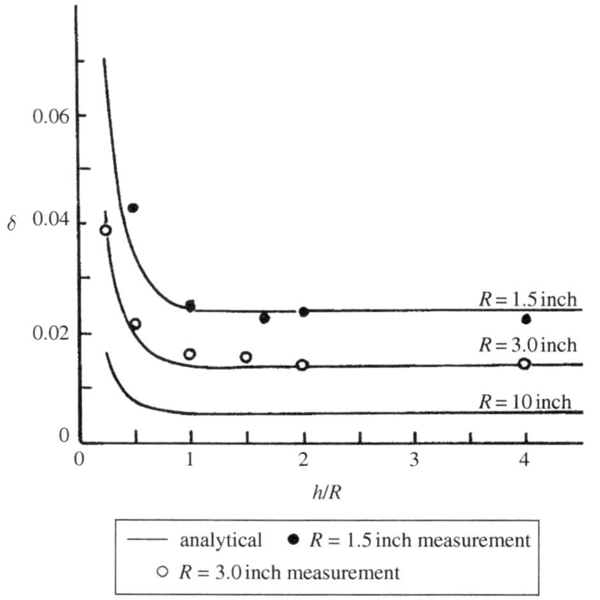

Figure 3.1 Dependence of the logarithmic decrement on fluid depth ratio for different values of tank radius. (Case and Parkinson, 1957)

damping particularly for a large cylinder radius. Figure 3.1 also includes experimental measurements for $R = 1.5$ inch, and 3.0 inch.

The damping due to the free surface (3.27a) is small compared to the wall damping particularly for large containers. Thus, an approximate expression for the damping of the fundamental anti-symmetric mode may be written in the form

$$\delta = 3.52 \frac{\sqrt{\nu}\{1 + 2(1 - h/R)\sinh^{-1}(3.68h/R)\}}{R^{3/4}g^{1/4}\tanh^{1/4}(1.84h/R)} \tag{3.29}$$

For deep-tank filling, $h/R \gg l$, the damping factor is invariant with the depth ratio, and is given by the simple expression

$$\delta_0 = 3.52 \frac{\sqrt{\nu}}{R^{3/4}g^{1/4}}\bigg|_{h/R>1} \tag{3.30}$$

Berlot, *et al.* (1957) verified equation (3.30) using dimensional analysis. Berlot, *et al.* indicated that if two cylindrical tanks are filled to the same depth ratio and are subjected to the same excitation, they exhibit similar sloshing behavior provided that the Reynold's number $C^{-1} = \sqrt{gR^3/\nu}$ is preserved. Figure 3.2 shows the analytical and experimental ratio, δ/δ_0, as a function of the fluid depth ratio according to Miles (1958b) and Stephens, *et al.* (1962), respectively.

The effects of liquid draining efflux on damping were analytically studied by Nelson (1960), Lindholm, *et al.* (1962b), Miles (1962b), and Hara (1990). The draining of the liquid from a vertical tank was found to produce a small but always stabilizing damping of the free-surface

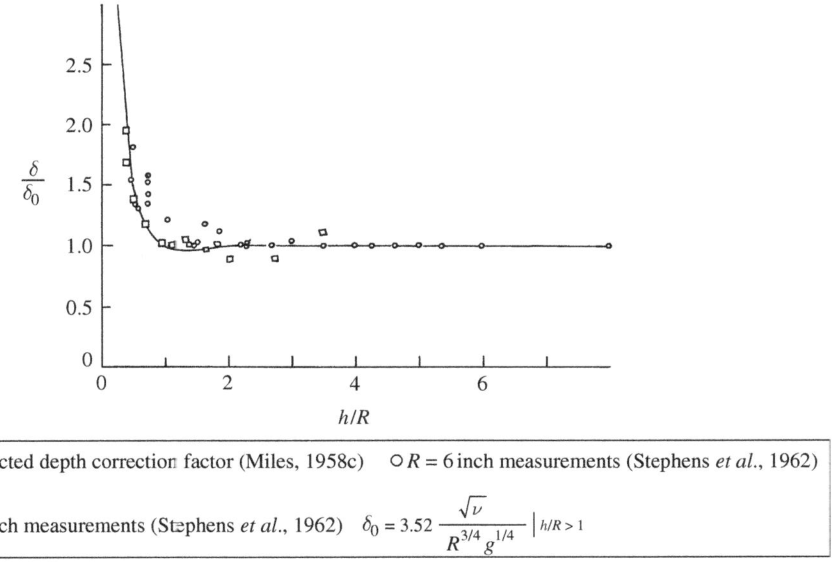

Figure 3.2 Predicted and measured damping factor ratio in a circular container. (Stephens, *et al.*, 1962)

oscillations. In a right-circular cylinder the damping attributed solely to draining may be approximated from Miles' (1962b) analysis for $h/R \gg 1$ by the following formula

$$\delta_{\text{drain}} = 2\pi\dot{h}\sqrt{\frac{1.84h/R}{gh}}\, e^{-3.68h/R} \tag{3.31}$$

where $\dot{h} = c\sqrt{2gh}$ is the rate of fluid drain, c is the orifice contraction, and has the value $c = 0.6\, A/A_1$ for a square-edged circular orifice of area A, and A_1 is the quiescent surface area.

Miles (1956), Bauer (1957, 1958f), Stephens, *et al.* (1962), Cole (1966a), and Bauer and Eidel (1998b) examined the dependence of the damping on the liquid depth in a cylindrical tank. Miles (1967) and Mei and Liu (1973) considered all possible sources of energy dissipation of surface waves in closed containers. The problem of viscous fluid motion in rigid containers was analyzed by Krein (1964), Krushinskaya (1965), Victorov (1965), Krein and Laptev (1968), and Krein and Kan (1969a,b). The oscillations of immiscible liquids in a rectangular tank were analyzed to determine the damping effectiveness by Bauer (1984a).

3.2.2 Damping in other containers

It is not difficult to measure the logarithmic decrement of the first asymmetric sloshing mode. Basically, the damping factor depends on the liquid height, liquid kinematic viscosity, and tank diameter (or tank width for rectangular cross-section). We will focus on viscous damping in containers without baffles. Dimensional analysis and empirical correlations showed that the damping ratio of the first asymmetric mode in a circular cylindrical tank is given by the empirical relationship, Abramson (1966a),

Table 3.1 *Values of C_1 in equation (3.32c)*

Tank shape		C_1	n_1
Circular cylinder	$h/R \geqslant 1.0$	0.79	0.5
	0.5	1.11	0.5
	0.1	3.36	0.5
Rectangular	$h/w \geqslant 1$	≈ 1.0	0.5
Spherical	¾ full	0.66	0.359
	½ full	0.39	0.359
	¼ full	0.32	0.359
Upright conical		0.81	0.5

$$\zeta = \frac{2.89}{\pi}\sqrt{\frac{\nu}{R^{3/2}g^{1/2}}}\left[1 + \frac{0.318}{\sinh(1.84h/R)}\left(\frac{1-(h/R)}{\cosh(1.84h/R)} + 1\right)\right] \tag{3.32a}$$

where ν is the kinematic viscosity, R is the tank radius, g is the gravitational acceleration. For $h/R > 1$ relation (3.32a) takes the form

$$\zeta = \frac{2.89}{\pi}\sqrt{\frac{\nu}{R^{3/2}g^{1/2}}} = \frac{C}{\sqrt{G_A}} \tag{3.32b}$$

where $G_A = R^{3/2}\, g^{1/2}/\nu$ is called the Galileo number, and C is a numerical coefficient whose value depends upon the tank geometry and liquid depth. This relation can be generalized for other tank shapes

$$\zeta = C_1\left(\frac{\nu}{d^{3/2}\sqrt{g}}\right)^{n_1} \tag{3.32c}$$

where d is the characteristic dimension of the tank, the width of a rectangular container, or the radius for cylindrical or spherical containers. The constant C_1 and the exponent n_1 take the values listed in Table 3.1 for the indicated tank geometry.

Stofan and Pauli (1962), Stofan and Irvine (1963), and Sumner and Stofan (1963) measured the viscous damping coefficient in spherical tanks. Viscous damping in a half-full sphere may be estimated from the following semi-empirical equation due to Stofan and Armstead (1962)

$$\zeta = 0.021\left(\frac{10^4}{2\sqrt{2}G_A}\right) \tag{3.32d}$$

For other tank shapes, it is recommended that $C_1 \approx 1$ and $n_1 \approx 0.5$ are used.

In order to minimize the sloshing hydrodynamic forces acting on the tank, it is desirable to suppress the liquid sloshing amplitude. Popov, *et al.* (1993b) studied numerically the sloshing dynamic characteristics of a viscous liquid in an arbitrary tank geometry. Shemer and Kit (1988) determined the energy dissipation of nonlinear sloshing waves in a rectangular channel.

The influence of liquid viscosity on the sloshing response of tanks containing heavy liquid was numerically and analytically studied by Mieda, *et al.* (1993) and Tang, *et al.* (1993). Uras (1995) used a finite element (FE) computer code and found that the dynamic response reaches the steady state faster as the viscosity becomes larger. The fundamental sloshing frequency for different cases stayed virtually unaffected by the liquid viscosity. For a small tank, a 5% difference was reported between the values of the fundamental frequency of a liquid with (1 cP) and (1000 cP) viscosity. The experimental measurements by Chiba, *et al.* (1995) confirmed these numerical results. They measured the liquid natural frequency, damping ratio, harmonic and seismic liquid response amplitudes for liquids with viscosities of 1 cP, 10 cP, 200 cP and 1000 cP. The damping factor and liquid sloshing response amplitude were significantly affected by the viscosity especially for small tank sizes. As the tank size increases the effect of viscosity decreases.

Jitu, *et al.* (1994) presented experimental and analytical results of modal parameters of coaxial cylinders with a viscous liquid in the horizontal and vertical orientations. They found that as liquid viscosity increases, the damping ratio of a sloshing mode increases. Generally, the seismic response of the coaxial cylinder was influenced by the liquid's viscosity. However, in practice, the liquid's viscosity did not have a particular effect on the seismic design of the coaxial cylinder in the horizontal direction. The seismic response of the pressure in the vertical excitation increased as the liquid's viscosity increased.

3.3 Free and forced oscillations of viscous fluids

3.3.1 Modal analysis of viscous fluids

The previous section treated the boundary layer effect so as to determine the logarithmic decrement. Moiseev (1961) studied the boundary value problem for the linearized Navier–Stokes equations for liquid with low viscosity. Schulkes and Cuvelier (1991) numerically solved the Navier–Stokes equations using a finite element algorithm to estimate the free oscillations and the rotation-induced internal modes for a rotating viscous fluid with surface tension. It was found that when the angular velocity of the container is large, the solution results in mixed internal–free-surface oscillation modes. Bauer and Eidel (1997a–c) considered the axisymmetric oscillations of viscous liquid in slowly rotating circular cylindrical containers. Their analysis did not take into account the influence of adhesive conditions at the sidewall of the tank, and thus the contact line of the free surface with the sidewall was treated either as a slipping or anchored edge. A large amount of damping is created by the sidewall, since most of the sloshing motion takes place in the upper part of the container close to the free surface. For small liquid depth, $h/R<1$, the contribution of the adhesion at the bottom is expected to be much larger than that at the sidewall. For larger fluid depth, $h/R>1$, the contribution of the bottom diminishes and the damping is mainly contributed by the sidewall and the internal damping of the viscous flow. Bauer and Eidel (1999b, 2002a) extended their work to include the influence of sidewall and internal damping in cylindrical containers.

Consider a viscous incompressible flow in an upright circular cylindrical container. The fluid flow is governed by the Navier–Stokes equations in cylindrical coordinates, r, θ, z, located at the undisturbed free surface

$$\frac{\partial u}{\partial t}=-\frac{1}{\rho}\frac{\partial p}{\partial r}+\nu\left[\frac{\partial^2 u}{\partial r^2}+\frac{1}{r}\frac{\partial u}{\partial r}-\frac{u}{r^2}+\frac{1}{r^2}\frac{\partial^2 u}{\partial\theta^2}-\frac{2}{r^2}\frac{\partial v}{\partial\theta}+\frac{\partial^2 u}{\partial z^2}\right] \tag{3.33a}$$

$$\frac{\partial v}{\partial t}=-\frac{1}{\rho}\frac{\partial p}{r\partial\theta}+\nu\left[\frac{\partial^2 v}{\partial r^2}+\frac{1}{r}\frac{\partial v}{\partial r}-\frac{v}{r^2}+\frac{1}{r^2}\frac{\partial^2 v}{\partial\theta^2}+\frac{2}{r^2}\frac{\partial u}{\partial\theta}+\frac{\partial^2 v}{\partial z^2}\right] \tag{3.33b}$$

$$\frac{\partial w}{\partial t} = -\frac{1}{\rho}\frac{\partial p}{\partial z} + \nu\left[\frac{\partial^2 w}{\partial r^2} + \frac{1}{r}\frac{\partial w}{\partial r} + \frac{1}{r^2}\frac{\partial^2 w}{\partial \theta^2} + \frac{\partial^2 w}{\partial z^2}\right] - g \tag{3.33c}$$

where u, v, and w are the fluid velocity components along r, θ, and z, respectively. For incompressible fluid, the flow must satisfy the continuity equation

$$\frac{\partial u}{\partial r} + \frac{u}{r} + \frac{1}{r}\frac{\partial v}{\partial \theta} + \frac{\partial w}{\partial z} = 0 \tag{3.34}$$

The solution of equation (3.34) may be written in the form

$$u(r,\theta,z,t) = \sum_{m=0}^{\infty} U_m(r,z)\, e^{\lambda t + im\theta}$$

$$v(r,\theta,z,t) = \sum_{m=0}^{\infty} V_m(r,z)\, e^{\lambda t + im\theta}$$

$$w(r,\theta,z,t) = \sum_{m=0}^{\infty} W_m(r,z)\, e^{\lambda t + im\theta}$$

$$p(r,\theta,z,t) = p_0 + \sum_{m=0}^{\infty} P_m(r,z)\, e^{\lambda t + im\theta} - \rho g z \tag{3.35}$$

where $\lambda = \zeta + i\omega$ is the complex natural frequency, where the real part measures the damping and the imaginary part represents the damped natural frequency. Introducing equation (3.35) into equations (3.33) and (3.34), gives

$$\frac{\partial^2 U_m}{\partial r^2} + \frac{1}{r}\frac{\partial U_m}{\partial r} - \frac{(m^2+1)}{r^2}U_m - \frac{\lambda}{\nu}U_m + \frac{\partial^2 U_m}{\partial z^2} - im\frac{2}{r^2}V_m = -\frac{1}{\mu}\frac{\partial P_m}{\partial r} \tag{3.36a}$$

$$\frac{\partial^2 V_m}{\partial r^2} + \frac{1}{r}\frac{\partial V_m}{\partial r} - \frac{(m^2+1)}{r^2}V_m - \frac{\lambda}{\nu}V_m + \frac{\partial^2 V_m}{\partial z^2} - im\frac{2}{r^2}U_m = -\frac{im}{\mu r}P_m \tag{3.36b}$$

$$\frac{\partial^2 W_m}{\partial r^2} + \frac{1}{r}\frac{\partial W_m}{\partial r} - \frac{m^2}{r^2}W_m - \frac{\lambda}{\nu}W_m + \frac{\partial^2 W_m}{\partial z^2} = -\frac{1}{\mu}\frac{\partial P_m}{\partial z} \tag{3.36c}$$

$$\frac{\partial U_m}{\partial r} + \frac{U_m}{r} + \frac{im}{r}V_m + \frac{\partial W_m}{\partial z} = 0 \tag{3.37}$$

The solution of these equations must satisfy the following boundary conditions:

- At the tank walls and bottom the fluid velocity must vanish, that is,

$$u(r,\theta,z,t) = v(r,\theta,z,t) = w(r,\theta,z,t) = 0 \quad \text{at } r = R \text{ and } z = -h \tag{3.38}$$

- At the free surface, we have the kinematic condition

$$\frac{\partial \eta}{\partial t} = w|_{z=0} \tag{3.39}$$

- At the free surface we have the dynamic condition

$$\bar{p}(r,\theta,t) - 2\mu\frac{\partial w}{\partial z} + \sigma\left[\frac{\partial^2 \eta}{\partial r^2} + \frac{1}{r}\frac{\partial \eta}{\partial r} + \frac{1}{r^2}\frac{\partial^2 \eta}{\partial \theta^2} - \frac{\rho g}{\sigma}\eta\right] = \text{const}|_{z=0} \tag{3.40}$$

where $\nu = \mu / \rho$, ρ is the fluid density, $\eta = \eta\ (r, \theta, t)$ is the free-surface elevation above $z=0$, σ is the surface tension, and $\bar{p}\ (r, \theta, t)$ is the net pressure difference between the gas pressure outside the liquid free surface and the pressure just inside the surface.

- At the free surface, the shear stresses τ_{rz} and $\tau_{\theta z}$ vanish, that is,

$$\left[\frac{\partial w}{\partial r} + \frac{\partial u}{\partial z}\right]_{z=0} = 0, \qquad \text{and} \qquad \left[\frac{\partial v}{\partial z} + \frac{1}{r}\frac{\partial w}{\partial \theta}\right]_{z=0} = 0 \tag{3.41a, b}$$

Applying divergence on both sides of the Stokes equations (3.36), gives the Laplace equation for the pressure distribution

$$\frac{\partial^2 P_m}{\partial r^2} + \frac{1}{r}\frac{\partial P_m}{\partial r} - \frac{m^2}{r^2}P_m + \frac{\partial^2 P_m}{\partial z^2} = 0 \tag{3.42}$$

The functions $U_m\ (r, z)$, $V_m(r, z)$, $W_m(r, z)$, and $P_m(r, z)$ satisfy the bottom boundary condition if they are selected in the form

$$\{U_m(r,z)\ V_m(r,z)\ W_m(r,z)\ P_m(r,z)\}^T = \{U(r)\ V(r)\ W(r)\ P(r)\}^T \cosh[k(z+h)] \tag{3.43}$$

where T denotes transpose. Substituting (3.43) in equations (3.36) and (3.37), gives

$$\frac{\partial^2 U}{\partial r^2} + \frac{1}{r}\frac{\partial U}{\partial r} + \left[k^2 - \frac{\lambda}{\nu} - \frac{(m^2+1)}{r^2}\right]U - im\frac{2}{r^2}V = \frac{1}{\mu}\frac{\partial P}{\partial r} \tag{3.44a}$$

$$\frac{\partial^2 V}{\partial r^2} + \frac{1}{r}\frac{\partial V}{\partial r} + \left[k^2 - \frac{\lambda}{\nu} - \frac{(m^2+1)}{r^2}\right]V + im\frac{2}{r^2}U = \frac{im}{\mu r}P \tag{3.44b}$$

$$\frac{\partial^2 W}{\partial r^2} + \frac{1}{r}\frac{\partial W}{\partial r} + \left[k^2 - \frac{\lambda}{\nu} - \frac{m^2}{r^2}\right]W = \frac{k}{\mu}P \tag{3.44c}$$

$$\frac{\partial U}{\partial r} + \frac{U}{r} + \frac{im}{r}V + kW = 0 \tag{3.45}$$

A possible solution for $P(r)$ can be obtained from equation (3.42) in the form

$$P(r) = \frac{\mu}{R}DJ_m(kr) \tag{3.46}$$

where D is a constant, $k = \xi_{mn}/R$, ξ_{mn} will be determined later, and μ has been introduced in order to simplify the subsequent steps. Introducing the following complex transformation for the radial and circumferential velocity components

$$U = \frac{1}{2}(F+G), \quad \text{and} \quad V = \frac{1}{2i}(F-G) \tag{3.47a}$$

such that

$$F = U + iV, \quad \text{and} \quad G = U - iV \tag{3.47b}$$

Adding and subtracting equations (3.44a) and (3.44b) and introducing (3.47b), gives

$$\frac{\partial^2 F}{\partial r^2}+\frac{1}{r}\frac{\partial F}{\partial r}+\left[k^2-\frac{\lambda}{\nu}-\frac{(m+1)^2}{r^2}\right]F=\frac{1}{\mu}\left[\frac{\partial P}{\partial r}-\frac{m}{r}P\right]=-\frac{k}{R}DJ_{m+1}(kr) \tag{3.48a}$$

$$\frac{\partial^2 G}{\partial r^2}+\frac{1}{r}\frac{\partial G}{\partial r}+\left[k^2-\frac{\lambda}{\nu}-\frac{(m-1)^2}{r^2}\right]G=\frac{1}{\mu}\left[\frac{\partial P}{\partial r}+\frac{m}{r}P\right]=\frac{k}{R}DJ_{m-1}(kr) \tag{3.48b}$$

The continuity equation takes the form

$$\frac{dF}{dr}+\frac{m+1}{r}F+\frac{dG}{dr}-\frac{m-1}{r}G=-2kW \tag{3.49}$$

The solution of equations (3.48) may be written in the form

$$F(r)=AJ_{m+1}\left(\sqrt{\xi_{mn}^2-\bar{\lambda}}(r/R)\right)+D(\xi_{mn}/\bar{\lambda})J_{m+1}(\xi_{mn}r/R) \tag{3.50a}$$

$$G(r)=BJ_{m-1}\left(\sqrt{\xi_{mn}^2-\bar{\lambda}}(r/R)\right)-D(\xi_{mn}/\bar{\lambda})J_{m-1}(\xi_{mn}r/R) \tag{3.50b}$$

where $\bar{\lambda}=\lambda R^2/\nu$ is a dimensionless parameter equivalent to the Reynolds number. The solution of equation (3.44c) may also be written in the form

$$W(r)=CJ_m\left(\sqrt{\xi_{mn}^2-\bar{\lambda}}(r/R)\right)-D(\xi_{mn}/\bar{\lambda})J_m(\xi_{mn}r/R) \tag{3.50c}$$

Substituting equations (3.50) into the continuity equation (3.45), after making use of (3.47a), gives

$$C=\frac{\sqrt{\xi_{mn}^2-\bar{\lambda}}}{2\xi_{mn}}(A-B) \tag{3.51}$$

The original velocity components, u, v, and w, and the pressure, p, can be written as functions of r, θ, z, and time, in the form

$$\begin{aligned}u(r,\theta,z,t)=\frac{1}{2}\sum_{m=0}^{\infty}\sum_{n=1}^{\infty}\Bigg\{&A_{mn}J_{m+1}\left(\sqrt{\xi_{mn}^2-\bar{\lambda}}(r/R)\right)\\&+B_{mn}J_{m-1}\left(\sqrt{\xi_{mn}^2-\bar{\lambda}}(r/R)\right)\\&-(2\xi_{mn}/\bar{\lambda})D_{mn}J'_m(\xi_{mn}r/R)\Bigg\}\cosh[\xi_{mn}(z+h)/R]\,e^{im\theta}\,e^{\lambda t}\end{aligned} \tag{3.52a}$$

$$\begin{aligned}v(r,\theta,z,t)=-\frac{i}{2}\sum_{m=0}^{\infty}\sum_{n=1}^{\infty}\Bigg\{&A_{mn}J_{m+1}\left(\sqrt{\xi_{mn}^2-\bar{\lambda}}(r/R)\right)\\&-B_{mn}J_{m-1}\left(\sqrt{\xi_{mn}^2-\bar{\lambda}}(r/R)\right)\\&+(2mR/r\bar{\lambda})D_{mn}J_m(\xi_{mn}r/R)\Bigg\}\cosh[\xi_{mn}(z+h)/R]\,e^{im\theta}\,e^{\lambda t}\end{aligned} \tag{3.52b}$$

$$w(r,\theta,z,t) = -\frac{1}{2}\sum_{m=0}^{\infty}\sum_{n=1}^{\infty}\left\{\frac{\sqrt{\xi_{mn}^2-\bar{\lambda}}}{\xi_{mn}}(A_{mn}-B_{mn})J_m\left(\sqrt{\xi_{mn}^2-\bar{\lambda}}(r/R)\right)\right.$$
$$\left.+(2\xi_{mn}/\bar{\lambda})D_{mn}J_m(\xi_{mn}r/R)\right\}\sinh[\xi_{mn}(z+h)/R]\,e^{im\theta}\,e^{\lambda t} \quad (3.52c)$$

$$p(r,\theta,z,t) = p_0 - \rho g z$$
$$+(\mu/R)\sum_{m=0}^{\infty}\sum_{n=1}^{\infty} D_{mn}J_m(\xi_{mn}r/R)\cosh[\xi_{mn}(z+h)/R]\,e^{im\theta}\,e^{\lambda t} \quad (3.53)$$

The values of ξ_{mn} are determined by satisfying the boundary condition at $r=R$, see condition (3.38). Applying (3.38) to equations (3.52) gives the following transcendental equation whose roots give the values of ξ_{mn},

$$\begin{vmatrix} J_{m+1}\left(\sqrt{\xi^2-\bar{\lambda}}\right) & J_{m-1}\left(\sqrt{\xi^2-\bar{\lambda}}\right) & -\xi J_m'(\xi) \\ J_{m+1}\left(\sqrt{\xi^2-\bar{\lambda}}\right) & -J_{m-1}\left(\sqrt{\xi^2-\bar{\lambda}}\right) & mJ_m(\xi) \\ \sqrt{\xi^2-\bar{\lambda}}J_m\left(\sqrt{\xi^2-\bar{\lambda}}\right) & -\sqrt{\xi^2-\bar{\lambda}}J_m\left(\sqrt{\xi^2-\bar{\lambda}}\right) & \xi^2 J_{m+1}(\xi) \end{vmatrix} = 0 \quad (3.54)$$

This transcendental equation is solved numerically for a given mode m to give the corresponding roots of ξ. The three conditions that generated equation (3.54) involve three coefficients, A_{mn}, B_{mn}, and D_{mn}, so one can express two of them in terms of the third, that is,

$$B_{mn} = -\beta_{mn}A_{mn}, \quad D_{mn} = -\bar{\lambda}\gamma_{mn}A_{mn} \quad (3.55\text{a, b})$$

where the parameters β_{mn} and γ_{mn} are obtained, after using the same properties of Bessel functions, in the form

$$\beta_{mn} = \frac{J_{m+1}\left(\sqrt{\xi_{mn}^2-\bar{\lambda}}\right)J_{m-1}(\xi_{mn})}{J_{m-1}\left(\sqrt{\xi_{mn}^2-\bar{\lambda}}\right)J_{m+1}(\xi_{mn})}, \quad \gamma_{mn} = \frac{J_{m+1}\left(\sqrt{\xi_{mn}^2-\bar{\lambda}}\right)}{\xi_{mn}J_{m+1}(\xi_{mn})} \quad (3.56\text{a, b})$$

In order to determine the natural frequencies of the free surface, the free-surface conditions must be satisfied. First the shear stresses (3.41a, b) give the following two conditions

$$\sum_{n=1}^{\infty}\xi_{mn}A_{mn}\sinh(\xi_{mn}h/R)\left\{\frac{\xi_{mn}^2-\bar{\lambda}}{\xi_{mn}^2}(1+\beta_{mn})J_m'\left(\sqrt{\xi_{mn}^2-\bar{\lambda}}(r/R)\right)\right.$$
$$-J_{m+1}\left(\sqrt{\xi_{mn}^2-\bar{\lambda}}(r/R)\right)+\beta_{mn}J_{m-1}\left(\sqrt{\xi_{mn}^2-\bar{\lambda}}(r/R)\right)$$
$$\left.-4\xi_{mn}\gamma_{mn}J_m'(\xi_{mn}r/R)\right\} = 0 \quad (3.57\text{a})$$

$$\sum_{n=1}^{\infty} \xi_{mn} A_{mn} \sinh(\xi_{mn} h/R) \left\{ J_{m+1}\left(\sqrt{\xi_{mn}^2 - \bar{\lambda}}(r/R)\right) + \beta_{mn} J_{m-1}\left(\sqrt{\xi_{mn}^2 - \bar{\lambda}}(r/R)\right) + \frac{m\sqrt{\xi_{mn}^2 - \bar{\lambda}}}{\xi_{mn}^2 (r/R)} (1+\beta_{mn}) J_m\left(\sqrt{\xi_{mn}^2 - \bar{\lambda}}(r/R)\right) - \frac{4m\gamma_{mn}}{(r/R)} J_m(\xi_{mn} r/R) \right\} = 0 \tag{3.57b}$$

Taking the time derivative of the dynamic free-surface condition (3.40) and using the kinematic condition (3.39), gives

$$\lambda \bar{p} - 2\mu\lambda \frac{\partial w}{\partial z} + \sigma \left[\frac{\partial^2 w}{\partial r^2} + \frac{1}{r}\frac{\partial w}{\partial r} + \frac{1}{r^2}\frac{\partial^2 w}{\partial \theta^2} - \frac{\rho g}{\sigma} w \right] = 0 \tag{3.58a}$$

Expressing the first three terms in the bracket in terms of the pressure gradient by using equation (3.33a), gives

$$\lambda \bar{p} - 2\mu\lambda \frac{\partial w}{\partial z} + \sigma \left[\frac{1}{\mu}\frac{\partial p}{\partial z} + \frac{\lambda}{\nu} w - \frac{\partial^2 w}{\partial z^2} - \frac{\rho g}{\sigma} w \right] = 0 \tag{3.58b}$$

Substituting the last expression of equation (3.53) for $\bar{p}$, and equation (3.52c) for w, gives

$$\sum_{n=1}^{\infty} A_{mn} \cosh(\xi_{mn} h/R) \left\{ \frac{\sqrt{\xi_{mn}^2 - \bar{\lambda}}}{\xi_{mn}} (1+\beta_{mn}) \times \left[\bar{\lambda}\xi_{mn} + \frac{1}{2(\mathrm{Oh})}(\xi_{mn}^2 - \bar{\lambda} + \mathrm{Bo}) \tanh(\xi_{mn} h/R) \right] J_m\left(\sqrt{\xi_{mn}^2 - \bar{\lambda}}(r/R)\right) - [\bar{\lambda}(\bar{\lambda} + 2\xi_{mn}^2) + \frac{\xi_{mn}}{(\mathrm{Oh})}(\xi_{mn}^2 + \mathrm{Bo}) \tanh(\xi_{mn} h/R)] \gamma_{mn} J_m(\xi_{mn} r/R) \right\} = 0 \tag{3.58c}$$

where $\mathrm{Oh} = \dfrac{\mu\nu}{\sigma R}$ is the capillary number (also known as Ohnesorge number) and it provides a measure of the surface deflection in response to capillary induced stresses, $\mathrm{Bo} = \dfrac{\rho g R^2}{\sigma}$ is the Bond number, which measures the ratio of gravitational to surface tension forces.

Equations (3.57a,b) and (3.58c) constitute infinite series for each mode m and they depend on the radial coordinate, r/R. These equations have an infinite number of unknown coefficients, A_{mn}, for each mode. Truncating these equations to a finite number of $n = 3N-1$, and satisfying each equation at N points, $r/R = j/N$, where $j = 0, 1, 2, \ldots, (N-1)$, gives $3N-1$ algebraic equations for the unknowns $A_{m1}, A_{m2}, \ldots, A_{m,3N-1}$. The truncated system of homogeneous algebraic equations represents the eigenvalue problem of the free oscillation of the liquid free surface. To estimate the damped natural frequencies $\bar{\lambda}_{mn} = (\zeta_{mn} + i\omega_{mn})R^2/\nu$, the determinant of the coefficients of the algebraic equations gives the frequency equation, which was solved numerically by Bauer and Eidel (1999a,b) for specific values of ξ.

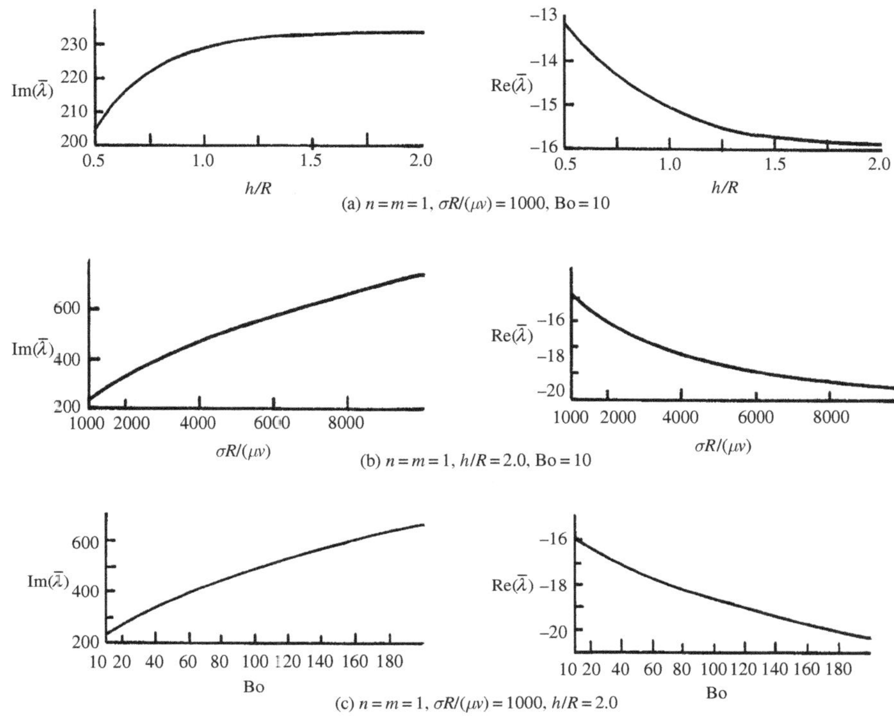

Figure 3.3 Dependence of the natural frequency and damping on (a) the fluid depth, (b) surface tension viscosity parameter, and (c) Bond number. (Bauer and Eidel, 1999)

Figure 3.3(a) shows the dependence of the real and imaginary parts of the lowest sloshing mode $m = n = 1$, on the fluid depth ratio, h/R, for Ohnesorge number $\mathsf{Oh} = \mu\nu/(\sigma R) = 10^{-3}$, and Bond number $\mathsf{Bo} = \rho g R^2/\sigma = 10$, and $N = 37$. It is seen that with the increase of the fluid depth ratio, the natural damped frequency increases. On the other hand the damping parameter is always negative and increases in magnitude (with negative sign) as the fluid depth increases. The dependence of the natural frequency components on the inverse of Ohnesorge number $1/\mathsf{Oh} = \sigma R/(\mu\nu)$ is shown in Figure 3.3(b) for $\mathsf{Bo} = 10$, $h/R = 2$, and $N = 31$. As the surface tension increases the damped natural frequency increases and the damping factor increases to higher negative values. The dependence of the natural frequency components on the Bond number is shown in Figure 3.3(c) for $\mathsf{Oh} = 10^{-3}$, $h/R = 2$, and $N = 31$.

3.3.2 Lateral excitation of viscous fluids

Under a lateral excitation along the x-axis, $x(t) = X_0\, e^{i\Omega t}$, the Stokes equations (3.33) take the form

$$\frac{\partial u}{\partial t} = -\frac{1}{\rho}\frac{\partial p}{\partial r} + \nu\left[\frac{\partial^2 u}{\partial r^2} + \frac{1}{r}\frac{\partial u}{\partial r} - \frac{u}{r^2} + \frac{1}{r^2}\frac{\partial^2 u}{\partial \theta^2} - \frac{2}{r^2}\frac{\partial v}{\partial \theta} + \frac{\partial^2 u}{\partial z^2}\right] - X_0\Omega^2\, e^{i\Omega t}\cos\theta \tag{3.59a}$$

$$\frac{\partial v}{\partial t} = -\frac{1}{\rho r}\frac{\partial p}{\partial \theta} + \nu\left[\frac{\partial^2 v}{\partial r^2} + \frac{1}{r}\frac{\partial v}{\partial r} - \frac{v}{r^2} + \frac{1}{r^2}\frac{\partial^2 v}{\partial \theta^2} + \frac{2}{r^2}\frac{\partial u}{\partial \theta} + \frac{\partial^2 v}{\partial z^2}\right] + X_0\Omega^2\, e^{i\Omega t}\sin\theta \tag{3.59b}$$

$$\frac{\partial w}{\partial t} = -\frac{1}{\rho}\frac{\partial p}{\partial z} + \nu\left[\frac{\partial^2 w}{\partial r^2} + \frac{1}{r}\frac{\partial w}{\partial r} + \frac{1}{r^2}\frac{\partial^2 w}{\partial \theta^2} + \frac{\partial^2 w}{\partial z^2}\right] - g \tag{3.59c}$$

In addition to the continuity equation (3.34), the pressure distribution may be written in the form

$$p(r, \theta, z, t) = p_0 + \bar{p}(r, \theta, z, t) - \rho g z - \rho\Omega^2 X_0 r\, e^{i\Omega t}\cos\theta \tag{3.60}$$

The solution of these equations may be written in the form

$$u(r, \theta, z, t) = \sum_{m=0}^{\infty} U_m(r, z)\, e^{i\Omega t + im\theta}, \quad v(r, \theta, z, t) = \sum_{m=0}^{\infty} V_m(r, z)\, e^{i\Omega t + im\theta}$$

$$w(r, \theta, z, t) = \sum_{m=0}^{\infty} W_m(r, z)\, e^{i\Omega t + im\theta}, \quad \bar{p}(r, \theta, z, t) = \sum_{m=0}^{\infty} P_m(r, z)\, e^{i\Omega t + im\theta} \tag{3.61}$$

Following the same procedure as in the previous case by introducing equations (3.61) into equations (3.59) and (3.60), and introducing the solution (3.43) and the following transformation

$$U = \frac{1}{2}(F + G), \quad \text{and} \quad V = \frac{1}{2}(F - G) \tag{3.62a}$$

such that

$$F = U + V, \quad \text{and} \quad G = U - V \tag{3.62b}$$

The Stokes equations (3.59) can be written in terms of the real functions F and G in the form, for $m = 1$,

$$\frac{\partial^2 F}{\partial r^2} + \frac{1}{r}\frac{\partial F}{\partial r} + \left[k^2 - \frac{i\Omega}{\nu} - \frac{4}{r^2}\right]F = \frac{1}{\mu}\left[\frac{\partial P}{\partial r} - \frac{1}{r}P\right] = -\frac{k}{R}DJ_2(kr) \tag{3.63a}$$

$$\frac{\partial^2 G}{\partial r^2} + \frac{1}{r}\frac{\partial G}{\partial r} + \left[k^2 - \frac{i\Omega}{\nu}\right]G = \frac{1}{\mu}\left[\frac{\partial P}{\partial r} + \frac{1}{r}P\right] = \frac{k}{R}DJ_0(kr) \tag{3.63b}$$

$$\frac{\partial^2 W}{\partial r^2} + \frac{1}{r}\frac{\partial W}{\partial r} + \left[k^2 - \frac{i\Omega}{\nu} - \frac{1}{r^2}\right]W = \frac{k}{\mu}P = \frac{k}{R}DJ_1(kr) \tag{3.63c}$$

where expression (3.46) for the pressure has been used for $m = 1$. The continuity equation (3.34) takes the form

$$\frac{dF}{dr} + \frac{2}{r}F + \frac{dG}{dr} = -2kW \tag{3.64}$$

The dynamic free-surface boundary condition (3.58a) takes the form

$$i\Omega\bar{p} - 2i\mu\Omega\frac{\partial w}{\partial z} + \sigma\left[\frac{\partial^2 w}{\partial r^2} + \frac{1}{r}\frac{\partial w}{\partial r} + \frac{1}{r^2}\frac{\partial^2 w}{\partial \theta^2} - \frac{\rho g}{\sigma}w\right] = i\Omega^3 \rho r X_0 \cos\theta \tag{3.65}$$

The solution of equations (3.63) in terms of the Reynolds number $\Omega^* = \Omega R^2/\nu$ may be written in the form

$$F(r) = AJ_2\left(\sqrt{\xi^2 - i\Omega^*}(r/R)\right) + D(\xi/i\Omega^*)J_2(\xi r/R) \tag{3.66a}$$

$$G(r) = BJ_0\left(\sqrt{\xi^2 - i\Omega^*}(r/R)\right) - D(\xi/i\Omega^*)J_0(\xi r/R) \tag{3.66b}$$

$$W(r) = CJ_1\left(\sqrt{\xi^2 - i\Omega^*}(r/R)\right) - D(\xi/i\Omega^*)J_1(\xi r/R) \tag{3.66c}$$

where $C = -\dfrac{\sqrt{\xi^2 - i\Omega^*}}{2\xi}(A - B)$. Following the same steps as in the previous section, the values of ξ (Ω^*) are obtained from the roots of the following equation

$$\begin{vmatrix} J_2\left(\sqrt{\xi^2 - i\Omega^*}\right) & J_0\left(\sqrt{\xi^2 - i\Omega^*}\right) & \xi J_1'(\xi) \\ J_2\left(\sqrt{\xi^2 - i\Omega^*}\right) & -J_0\left(\sqrt{\xi^2 - i\Omega^*}\right) & J_1(\xi) \\ \sqrt{\xi^2 - i\Omega^*}J_1\left(\sqrt{\xi^2 - i\Omega^*}\right) & -\sqrt{\xi^2 - i\Omega^*}J_1\left(\sqrt{\xi^2 - i\Omega^*}\right) & \xi^2 J_1(\xi) \end{vmatrix} = 0 \tag{3.67}$$

The original velocity components, u, v, and w, and the pressure p can be written as functions of r, θ, z, and time, in the form

$$u(r,\theta,z,t) = \frac{e^{i\Omega t}}{2}\cos\theta \sum_{n=1}^{\infty}\left\{A_n J_2\left(\sqrt{\xi_n^2 - \bar{\lambda}}(r/R)\right) + B_n J_0\left(\sqrt{\xi_n^2 - \bar{\lambda}}(r/R)\right.\right.$$
$$\left.\left. -(2\xi_n/i\Omega^*)D_n J_1'(\xi_n r/R)\right\}\cosh(\xi_n(z+h)/R)\right) \tag{3.68a}$$

$$v(r,\theta,z,t) = \frac{e^{i\Omega t}}{2}\sin\theta \sum_{n=1}^{\infty}\left\{A_n J_2\left(\sqrt{\xi_n^2 - i\Omega^*}(r/R)\right)\right.$$
$$\left. - B_n J_0\left(\sqrt{\xi_n^2 - i\Omega^*}(r/R)\right) + (2R/ir\Omega^*)D_n J_1(\xi_n r/R)\right\}$$
$$\times \cosh[\xi_n \times (z+h)/R] \tag{3.68b}$$

$$w(r,\theta,z,t) = -\frac{e^{i\Omega t}}{2}\cos\theta \sum_{n=1}^{\infty}\left\{\frac{\sqrt{\xi_n^2 - i\Omega^*}}{\xi_n}(A_n - B_n)J_1\left(\sqrt{\xi_n^2 - i\Omega^*}(r/R)\right)\right.$$
$$\left. + (2\xi_n/i\Omega^*)D_n J_1(\xi_n r/R)\right\}\sinh[\xi_n(z+h)/R] \tag{3.68c}$$

$$p(r,\theta,z,t) = p_0 - \rho g z + (\mu/R)\cos\theta \sum_{n=1}^{\infty} D_n J_1(\xi_n r/R)\cosh[\xi_{mn}(z+h)/R]\, e^{i\Omega t}$$
$$- \rho\Omega^2 X_0 r\, e^{i\Omega t}\cos\theta \tag{3.69}$$

where

$$B_n = -\beta_n A_n, \quad D_n = -i\Omega^* \gamma_n A_n \tag{3.70a, b}$$

The parameters β_{mn} and γ_{mn} are

$$\beta_n = \frac{J_2\left(\sqrt{\xi_n^2 - i\Omega^*}\right)J_0(\xi_n)}{J_0\left(\sqrt{\xi_n^2 - i\Omega^*}\right)J_2(\xi_n)}, \quad \gamma_n = \frac{J_2\left(\sqrt{\xi_n^2 - i\Omega^*}\right)}{\xi_n J_2(\xi_n)} \tag{3.71a, b}$$

Substituting the velocity components (3.68) into the shear stress conditions (3.41) of the free surface, gives

$$\sum_{n=1}^{\infty} \xi_n A_n \sinh(\xi_n h/R) \left\{ \frac{\xi_n^2 - i\Omega^*}{\xi_n^2} (1+\beta_n) J_1' \left(\sqrt{\xi_n^2 - i\Omega^*} (r/R) \right) - J_2 \left(\sqrt{\xi_n^2 - i\Omega^*} (r/R) \right) + \beta_n J_0 \left(\sqrt{\xi_n^2 - i\Omega^*} (r/R) \right) - 4\xi_n \gamma_n J_1'(\xi_n r/R) \right\} = 0 \tag{3.72a}$$

$$\sum_{n=1}^{\infty} \xi_n A_n \sinh(\xi_n h/R) \left\{ J_2 \left(\sqrt{\xi_n^2 - i\Omega^*} (r/R) \right) + \beta_n J_0 \left(\sqrt{\xi_n^2 - i\Omega^*} (r/R) \right) + \frac{\sqrt{\xi_n^2 - i\Omega^*}}{\xi_n^2 (r/R)} (1+\beta_n) J_1 \left(\sqrt{\xi_n^2 - i\Omega^*} (r/R) \right) - \frac{4\gamma_n}{(r/R)} J_1(\xi_n r/R) \right\} = 0 \tag{3.72b}$$

Substituting the velocity components (3.68) and the third expression in equation (3.69) for the pressure $\bar{p}$ into the dynamic free surface condition (3.65), gives the following nonhomogeneous algebraic equation

$$\sum_{n=1}^{\infty} A_{mn} \frac{R}{\nu} \cosh(\xi_n h/R) \left\{ \frac{\sqrt{\xi_n^2 - i\Omega^*}}{\xi_n} (1+\beta_n) [i\Omega^* \xi_n + \frac{1}{2(\mathrm{Oh})} (\xi_n^2 - i\Omega^* + \mathrm{Bo}) \tanh(\xi_n h/R)] J_1 \left(\sqrt{\xi_n^2 - i\Omega^*} (r/R) \right) - [i\Omega^* (i\Omega^* + 2\xi_n^2) + \frac{\xi_n}{(\mathrm{Oh})} (\xi_n^2 + \mathrm{Bo}) \tanh(\xi_n h/R)] \gamma_n J_1(\xi_n r/R) \right\} = i\Omega^{*3} \frac{X_0}{R} \frac{r}{R} \tag{3.72c}$$

Equations (3.72) constitute three nonhomogeneous algebraic equations, which can be solved numerically after truncation as carried out in the previous section. The numerical solution provides values for the coefficients A_n, B_n, and D_n. The velocity and pressure distribution given by equations (3.68) and (3.69) can also be determined. Upon integrating the kinematic free-surface condition (3.39), one can determine the fluid free-surface displacement

$$\eta(r,\theta,t) = -\frac{e^{i\Omega t}}{2\Omega} \cos\theta \sum_{n=1}^{\infty} \left\{ \frac{\sqrt{\xi_n^2 - i\Omega^*}}{\xi_n} (A_n - B_n) J_1 \left(\sqrt{\xi_n^2 - i\Omega^*} (r/R) \right) + (2\xi_n / i\Omega^*) D_n J_1(\xi_n r/R) \right\} \sinh[\xi_n (z+h)/R] \tag{3.73}$$

The hydrodynamic force component along the x-direction is obtained by integrating the pressure distribution and viscous stresses over the projected area

$$F_x = \int_{-h}^{0} \int_{0}^{2\pi} \left[p\cos\theta - 2\mu \frac{\partial u}{\partial r} \cos\theta + \mu \left(r\frac{\partial}{\partial r}(v/r) + \frac{1}{r}\frac{\partial u}{\partial \theta} \right) \sin\theta \right]_{r=R} R\, d\theta\, dz + \int_{0}^{R} \int_{0}^{2\pi} \mu \left[\left(\frac{\partial v}{\partial z} + \frac{1}{r}\frac{\partial w}{\partial \theta} \right) \sin\theta - \left(\frac{\partial u}{\partial r} + \frac{\partial w}{\partial r} \right) \cos\theta \right]_{z=-h} r\, dr\, d\theta \tag{3.74a}$$

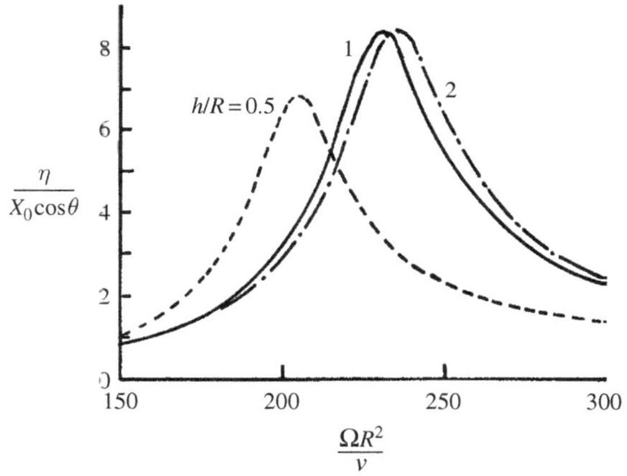

Figure 3.4 Dependence of fluid surface displacement on the excitation frequency estimated at $r/R = 0.8$ for $\sigma R/(\mu\nu) = 1000$, Bo = 10. (Bauer and Eidel, 1999a)

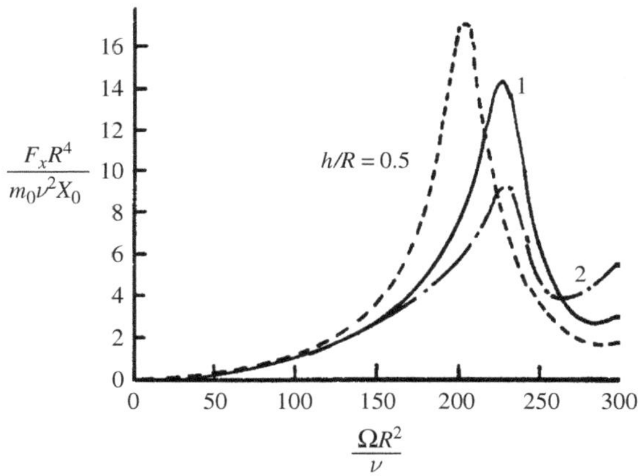

Figure 3.5 Dependence of fluid force in x-direction on the excitation frequency for $\sigma R/(\mu\nu) = 1000$, Bo = 10. (Bauer and Eidel, 1999a)

$$= -e^{i\Omega t}\left[m_0\Omega^2 X_0 - \pi\mu R \sum_{n=1}^{\infty} A_n \sinh(\xi_n h/R)\left\{(1+3\beta_n)\frac{\sqrt{\xi_n^2 - i\Omega^*}}{2\xi_n}\right.\right.$$
$$\left.\left.\times J_1\left(\sqrt{\xi_n^2 - i\Omega^*}\right) - (2\xi_n^2 - i\Omega^*)\gamma_n \frac{J_1(\xi_n)}{\xi_n}\right\}\right] \tag{3.74b}$$

where m_0 is the total fluid mass. In evaluating (3.74b), the second integral in (3.74a) was dropped since the adhesion stresses vanish, $\tau_{rz} = \tau_{r\theta} = 0$, in the bottom.

Bauer and Eidel (1999b) evaluated the fluid surface height ratio, $\eta/(X_0 \cos\theta)$, at $r = 0.8R$, and horizontal force ratio, $F_x/(m_0 \nu^2 X_0/R^4)$, for three different values of depth ratio $h/R = 0.5$, 1, and 2, Oh $= 10^{-3}$, Bo $= 10$, and $N = 30$. The dependence of η and F_x on the excitation frequency parameter, $\Omega^* = \Omega R^2/\nu$, are shown in Figures 3.4 and 3.5, respectively. It is seen that the resonance frequency increases as the fluid depth increases. The free surface displacement increases also with the fluid depth ratio h/R at the resonance frequency. However, the lateral force decreases with the fluid depth at resonance. In both cases, the fluid viscosity and surface tension bring the fluid dynamic response into bounded values at resonance as expected.

3.4 Suppression devices

As the size of liquid containers increases the hydrodynamic forces and moments become very large particularly in the neighborhood of resonance. In an attempt to avoid structural failure or undesirable dynamic behavior, one must introduce some means to suppress or to reduce the sloshing dynamic loads. This may be achieved by introducing baffles of different configurations. It has been shown in Chapter 1 that such baffles change the fluid natural frequencies depending on the shape, size, and position of baffles. Anti-slosh baffles are expected to prevent the total liquid force from exceeding a certain prescribed maximum value under all possible filling conditions, tank orientation, and external excitation. There have been extensive activities devoted to achieving optimum design of suppression devices (Langner, 1963, Abramson, 1966a, Schwind, *et al.*, 1967, and Sharma, *et al.*, 1992). Muto, *et al.* (1988) conducted experimental investigation to study the suppression effects of water restraint plates on sloshing in a water pool. Abramson (1969) summarized the primary parameters influencing the design of slosh-suppression devices as follows:

(1) The vehicle's mission profile and trajectory.
(2) The damping requirements for a given container, or liquid-slosh-motion amplitudes at various liquid levels.
(3) The physical characteristics of the tank such as its geometry, elastic deformation, and insulation.
(4) The liquid filling and draining requirements.
(5) The physical properties of the liquid.
(6) The handling, slosh, and impact loads that must be sustained by the devices.

Figure 3.6 shows some common types of baffles and anti-slosh devices. These are classified as follows

(1) Horizontal baffle rings, which can be movable, or fixed, or rings with radial clearance as shown in Figure 3.6(a)–(c).
(2) Conical baffles, which are placed upright or inverted as shown in Figure 3.6(d) and (e).
(3) Radial or sectored baffles in the form of complete sectored baffles as shown in Figure 3.6(f)–(i), or cruciform as shown in Figure 3.6(j).
(4) Annular tank shown in Figure 3.6(k).
(5) Z-ring conic section baffles shown in Figure 3.6(l).
(6) Floating cans shown in Figure 3.6(m).
(7) Floating lid devices shown in Figure 3.6(n).

Some detailed experimental and a few analytical studies were conducted to determine the damping present in tanks of various shapes fitted with various types of baffles (Howell and Ebler, 1956, Miles, 1958b, Bauer, 1958c,i, Eulitz, 1958, Silveira, *et al.*, 1961, Stephens, *et al.*, 1962, 1963, and Stephens and Scholl, 1967). These studies focused on the total damping of the fundamental anti-symmetric sloshing mode, which is due to two main sources. The first is due to the relative motion between the liquid and the tank wall and the second is due to the relative motion between the fluid and baffles. Experimental studies can be carried out at full-scale and on smaller scale models. Various similitude parameters have been shown to be important for different kinds of tests.

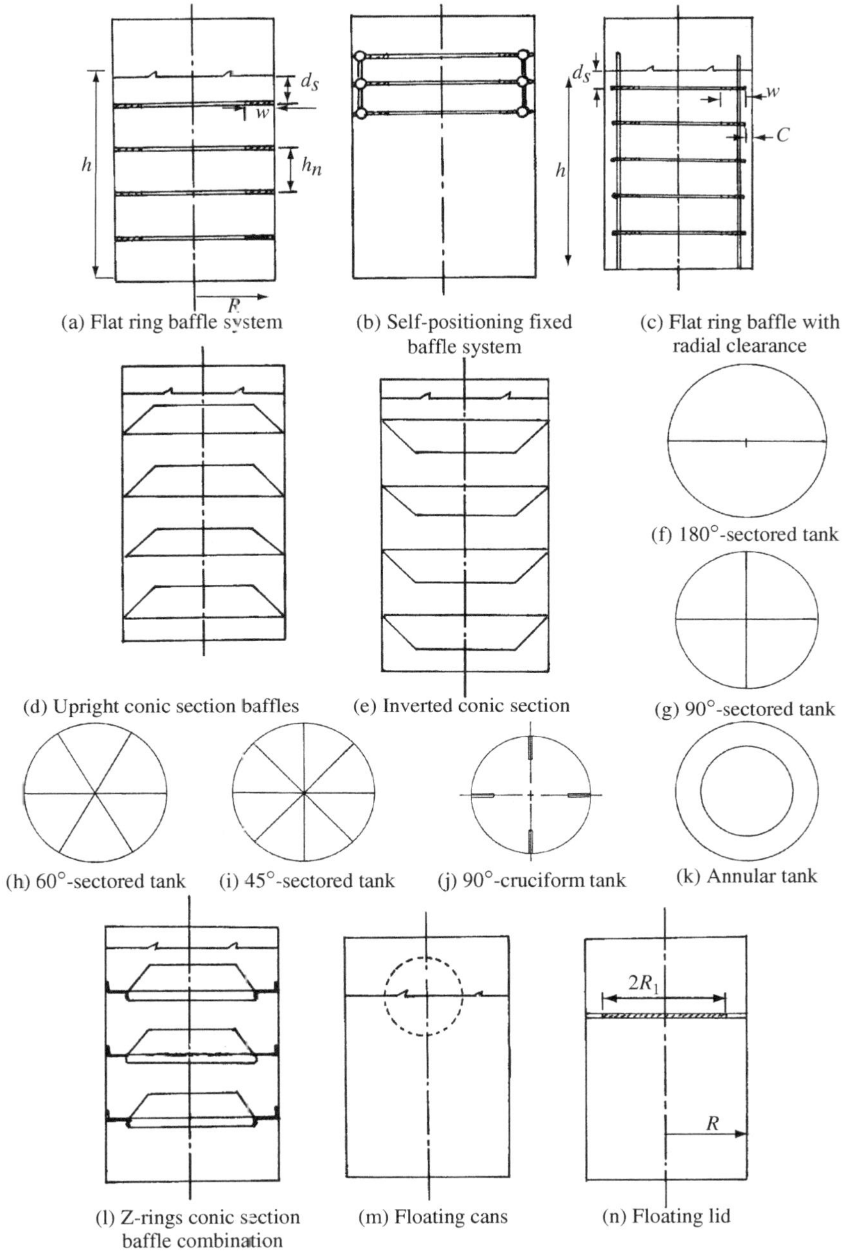

Figure 3.6 Sloshing suppression devices.

3.4.1 Damping in a circular cylinder with baffles

Flat ring baffles

The analytical determination of damping due to baffles is not an easy task and the most effective way is to measure it experimentally. However, Miles (1958b) developed a semi-empirical relationship for the damping ratio contributed by a flat annular ring baffle based on the work of Keulegan and Carpenter (1958). He assumed a small radial width of the baffle compared to the tank radius. The local flow in the neighborhood of the ring was assumed to be unaffected by the presence of the free surface or tank bottom. Furthermore, Miles (1958b) restricted his analysis to the first anti-symmetric mode. His analysis resulted in the following expression for the damping factor due to a ring baffle

$$\zeta = C_D \phi(\lambda R, \lambda d, \lambda h) \frac{A_b}{A} \frac{\eta_1}{R} \tag{3.75}$$

where C_D is the local drag coefficient, A_b is the baffle surface area, A is the tank cross-section area, η_1 is the maximum amplitude of liquid wave height during one cycle, $\lambda R = 1.84$, d is the fluid height above the baffle, h is the fluid total depth. The function ϕ (...) is given by the expression

$$\phi = \frac{(4/3\pi)^3 \lambda R \tanh(\lambda h)}{[1 - (1/\lambda R)^2]} \left[\frac{\sinh \lambda (z+h)}{\sinh \lambda h} \right]^3 \tag{3.76}$$

For $h/R > 1$, relation (3.76) may be approximated as

$$\phi \simeq 0.5\, e^{-5.52 d/R} \tag{3.77}$$

The drag coefficient on a flat plate in an oscillating fluid was measured by Keulegan and Carpenter (1958) and their results were found to be independent of the Reynolds number over the range $5\times 10^3 < \mathrm{Re} < 14\times 10^3$. A less reliable criterion is the critical Reynolds number ($\mathrm{Re}_{cr} \sim 10^3$) above which the flow is fully turbulent and C_D is constant in steady flow. Keulegan and Carpenter (1958) found that C_D depends strongly on the period parameter, $\tau = U_m T/w$, where U_m is the time-wise maximum velocity, T is the period, and w is the plate width. Miles fitted the experimental data of Keulegan and Carpenter (1958), and obtained the empirical relations

$$C_D = \frac{15}{\sqrt{U_m T/w}}, \quad \text{for } 2 \le U_m T/w \le 20 \tag{3.78a}$$

$$C_D = 2, \quad \text{for } U_m T/w \ge 100 \tag{3.78b}$$

These expressions are substituted in equation (3.75) to give the damping factor corresponding to each time scale. The first drag coefficient (3.78a) appears to be more accurate and the corresponding damping factor is

$$\zeta = 2.83\, e^{-4.5 d/R} \left(2 - \frac{w}{R}\right)^{3/2} \left(\frac{w}{R}\right)^{3/2} \left(\frac{\eta_1}{R}\right)^{1/2} \tag{3.79}$$

This formula reveals that the damping factor has a maximum value at zero baffle depth d/R as reflected in Figure 3.7 for three different sizes of fixed ring baffle. Figure 3.7 shows a comparison between the estimated (according to Miles) and measured (by Stephens, *et al.*, 1962) damping factors for different values of baffle width and wave height amplitude. The exact location of the maximum damping was not well defined. It is seen that the damping was reduced when the baffle is flush with the liquid surface since the baffle is in contact with the

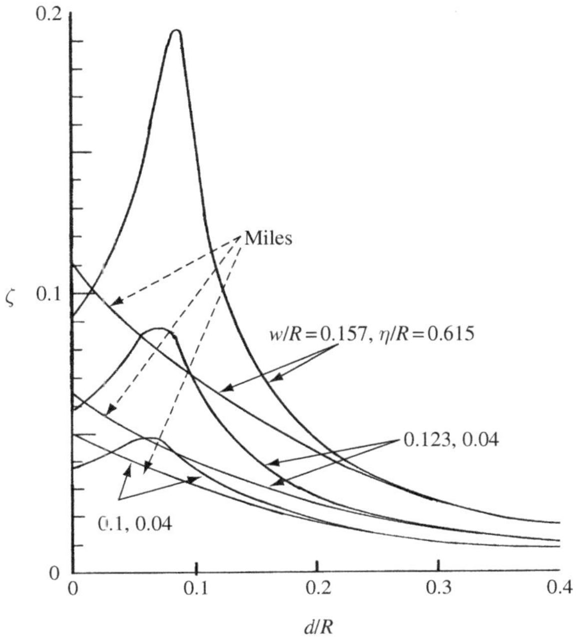

Figure 3.7 Dependence of damping factor on baffle location for different values of baffle width and wave height as predicted by Miles (1958b) and measured by Stephens, *et al.* (1962)

fluid for only a portion of the period of one sloshing cycle. The damping increases as the distance between the equilibrium surface and the baffle is increased, reaching a maximum at the minimum depth below the surface at which the ring does not break the fluid free surface during its motion. As the distance between the baffle and the equilibrium surface is further increased, the damping decreases rapidly approaching the value of the damping of the tank without baffles. This observation was verified experimentally by O'Neil (1960) and Abramson and Garza (1965).

The observed discrepancy between Miles' formula and the measured results at $d/R=0$ is attributed to the fact that the drag coefficient used by Miles is applicable only for baffles immersed from both sides in the liquid, which differs from the case where the liquid is bounded by one side only. The case of the liquid bounded from one side of the baffle results in a cavity between the baffle and the free streamlines, and the drag coefficient in this case follows that of the Helmholtz motion, that is, (see, e.g., Thomson, 1965), $C_D=2\pi/(4+\pi)=0.88$. This value is less than that used by Keulegan and Carpenter (1958). Note that fluid flow over a flat plate has the freedom to move with a sufficient area above the obstacle such that the damping provided will be caused by the total area of the baffle. In the case of a circular ring baffle, on the other hand, the flow area is finite and contracted at the baffle position. Flow movement is constrained and hence the liquid will acquire a higher kinetic energy, which is greater than that of the flat plate. Thus, for zero depth, $d/R=0$, only one half of the baffle is effective and the drag must be reduced to 50% of that used by Miles. However, due to contraction of the area, the drag is greater than that of Miles by 50%. In the case of $d/R>0$, the baffle area becomes completely effective, which, together with the increase of kinetic energy due to the contraction of the area,

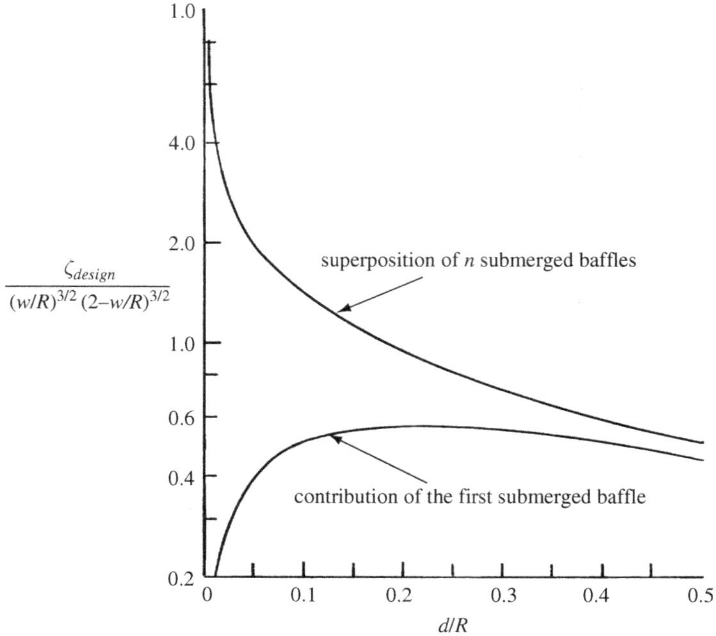

Figure 3.8 Dependence of damping ratio on depth ratio for single and system ring baffles. (Langer, 1963)

gives higher damping than that of Miles. At greater values of d/R, Miles' equation approaches the measured drag due to similar conditions between those of Keulegan and Carpenter and those conducted on a circular ring baffle.

A series of annular ring baffles can be used in the tank to provide a high degree of damping for all liquid levels. The minimum damping provided by such a system of multiple ring baffles is controlled by proper spacing of the baffles. Langner (1963) extended Miles' formula to a set of submerged baffles and obtained the following expression for the damping factor of the jth baffle

$$\zeta = 2.83\, e^{-2.3(h/NR)(2j-1)}\left(2-\frac{w}{R}\right)^{3/2}\left(\frac{w}{R}\right)^{3/2}\left(\frac{h}{2NR}\right)^{1/2} \tag{3.80}$$

where N is the number of submerged ring baffles. Assuming that the total damping from a system of N submerged ring baffles can be obtained simply by superimposing the damping contributions of the individual baffles, the design damping for such a system can be written in the form

$$\frac{\zeta}{\left(2-\frac{w}{R}\right)^{3/2}\left(\frac{w}{R}\right)^{3/2}} = 2\left(\frac{h}{NR}\right)^{1/2}\sum_{j=1}^{N} e^{-2.3(h/NR)(2j-1)} \tag{3.81}$$

Figure 3.8 presents the dependence of the damping ratio on the spacing parameter h/NR. The figure includes the damping contribution due to the first submerged ring baffle. Garza and Abramson (1963) conducted a series of tests with a set of two baffles where one is located above, and the other below, the liquid free surface. Figure 3.9 shows that the baffle above the

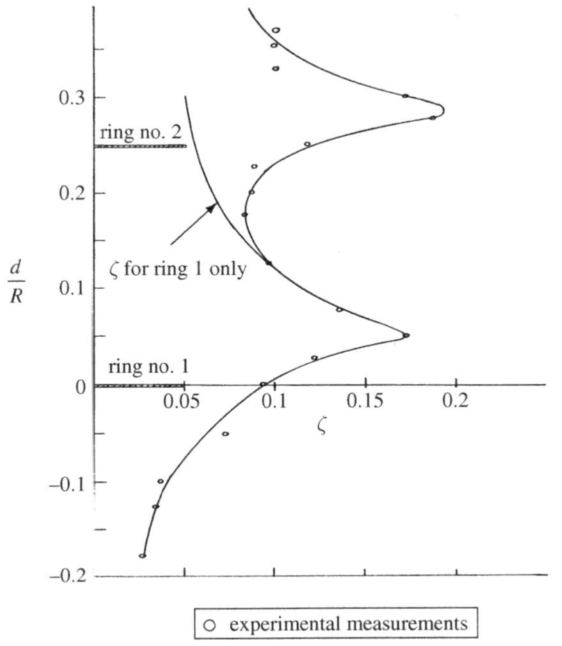

Figure 3.9 Effect of double ring baffles on the damping factor as function of baffle depth, $X_0/R = 0.008\,34$, $w/R = 0.157$. (Garza and Abramson, 1963)

liquid surface is effective only when its distance from the liquid surface is less than $d/R = 0.125$ and that the submerged baffle is effective over the range $d/R = 0$ to 0.33. From these results the baffle spacing corresponding to some minimum acceptable damping for a series of ring baffles can be determined. The spacing necessary to maintain the minimum acceptable damping is the depth of the submerged baffle corresponding to $\zeta_{\min}$ plus the additional damping from the upper baffle is $d/R = 0.125$.

O'Neill (1960) examined the range of validity of Miles' theory. Bauer (1962c) modified Miles' results to account for the effect of the baffle portion nonimmersed in the fluid during one cycle of sloshing. Cole (1966a,b) modified Miles' equation to incorporate the effects of tank and baffle geometry. The results of various model tests that reflect such effects to one degree or another were reported by Abramson and Ransleben (1960) and Stephens and Scholl (1967). Buchanan (1966) and Buchanan (1968) summarized the data on the influence of the Reynolds number on baffle damping. One interesting feature of these experimental studies was that the baffle-damping effectiveness decreases with an increase in the acceleration of the vehicle. Buzhinskii (1990, 1998a,b) determined the vortex damping of sloshing in tanks with plate baffles.

Flexible ring baffles

Several experimental studies demonstrated that flexible ring baffles offer a substantial advantage in increasing damping effectiveness and reducing a baffle's weight (see, e.g., Silveira, *et al.* 1961, Sumner, 1964, Schwind, *et al.*, 1964, 1967, Stephens, 1966, Stephens and Scholl, 1967, and Dodge, 1971). Schwind, *et al.* (1967) extended Miles' analysis to determine the damping

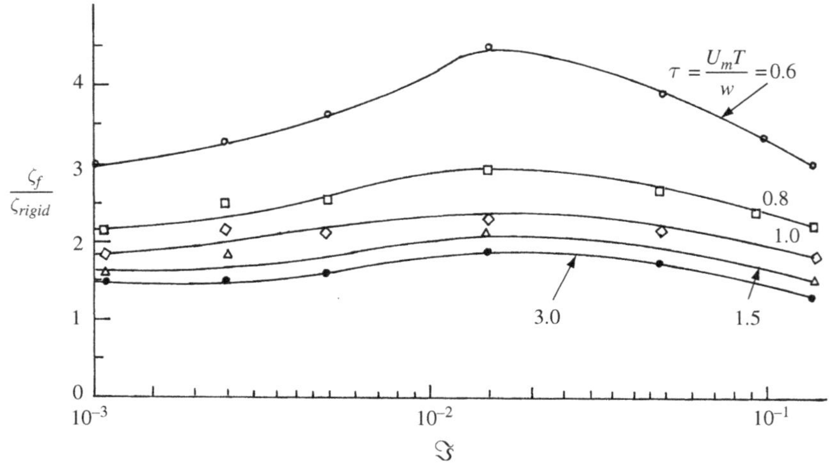

Figure 3.10 Effect of baffle flexibility, $\Im = \left(\frac{w}{t'}\right)^3 (1-\nu^2) \frac{\rho w^2}{ET^2}$, on the relative damping ratio for different values of time parameter τ. (Schwind, *et al.*, 1967)

coefficient provided by a flexible baffle fitted in a circular tank. The local energy dissipation per unit baffle length per cycle was given by the expression

$$E_b = C_L \rho w T U_0^3 \cos^3 \theta \tag{3.82}$$

where C_L is a coefficient depending on the baffle flexibility parameter $\Im$,

$$\Im = \left(\frac{w}{t'}\right)^3 \left(\frac{1-\gamma^2}{E}\right) \frac{\rho w^2}{T^2} f(w/R) \tag{3.83}$$

where t' is the baffle thickness, γ[1] is Poisson's ratio, E is Young's modulus of elasticity, T is the period of one sloshing cycle, ρ is the liquid density. Schwind, *et al.* (1967) integrated equation (3.83) around the tank at the radius of the mid point of the baffle and obtained the energy dissipation over one cycle using the dependence of C_L on the flexibility parameter $\Im$. They divided the result by the total kinetic energy of the liquid free surface to give the damping factor

$$\zeta_f = \frac{1.841}{0.553}\left(2 - \frac{w}{R}\right)\left(\frac{w}{R}\right)\left(\frac{\eta}{R}\right)\left[\frac{\sinh \lambda(z+h)}{\sinh \lambda h}\right]^3 f(\tau, \Im) \tag{3.84}$$

where $f\left(\frac{UT}{w}, \Im\right)$ is a function of the time parameter $\tau = U_m T/w$ and baffle flexibility parameter $\Im$.

The effectiveness of the flexible baffle is measured by dividing the damping factor (3.84) by the damping factor offered by a rigid ring baffle. Figure 3.10 shows the dependence of the relative damping, $\zeta_f/\zeta_{\text{rigid}}$, on the flexibility parameter for different values of the time parameter, $\tau = U_m T/w$. It is seen that the relative damping is always greater than unity over the entire range of the flexibility parameter. The value increases as the time parameter decreases. There is an optimum value of the flexibility parameter near the value $\Im \approx 0.02$.

The damping associated with an elastic cantilever baffle (w-mil Mylar) was examined by Stephens (1966) in a stationary rectangular tank 10 inch wide and 12 inch long partially filled

[1] In Chapters 8 and 9, the symbol for Poisson's ratio is ν. In this chapter, the symbol ν is allocated to kinematic viscosity.

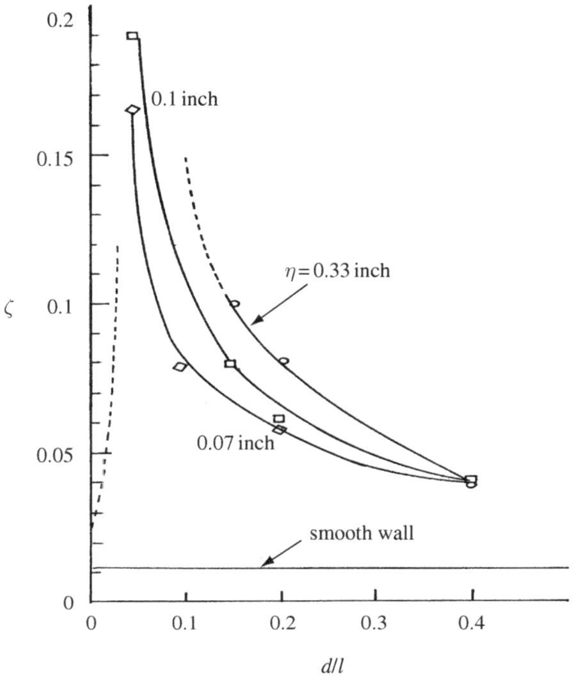

Figure 3.11 Variation of damping factor due to a flexible cantilever baffle in a rectangular tank, l is the tank length. (Stephens and Scholl, 1967)

with water to a depth of 12 inch. The dependence of the damping factor afforded by the elastic baffle on the baffle depth ratio is shown in Figure 3.11 for different values of surface wave height η. The value of damping in the absence of baffles in a smooth wall tank is shown by the constant value of 0.012.

Experimental measurements of flat annular ring baffles

Stephens, *et al.* (1962) conducted a series of experimental tests to measure the damping characteristics for the fundamental anti-symmetric mode of water in tanks of 12 inch and 30 inch diameter. The prime variables considered in their experiments were the baffle width, w, baffle depth from the free surface, d, amplitude of liquid wave height, η, kinematic viscosity of the liquid, ν, perforated baffle area, and the effective number of baffles.

The damping provided by three different sizes of fixed ring baffles as function of the baffle depth ratio was already presented in Figure 3.7. The viscosity did not have any apparent effect for the range considered in the experimental tests ($\nu = 0.5$ to $1.6\ \mathrm{ft}^2/\mathrm{s}$). Figure 3.12 shows the dependence of the measured damping on the wave amplitude for two different values of fluid depth ratio $h/R = 2.0$ and 2.5. The figure includes the curves estimated using Miles' equation. The damping was measured under sinusoidal lateral excitation of three different values of excitation amplitude $X_0/R = 0.003\,68$, 0.008 34, and 0.016 66. Figure 3.13 shows the measured damping factor and the corresponding predicted curves using Miles' theory and both reveal that the damping increases with excitation amplitude.

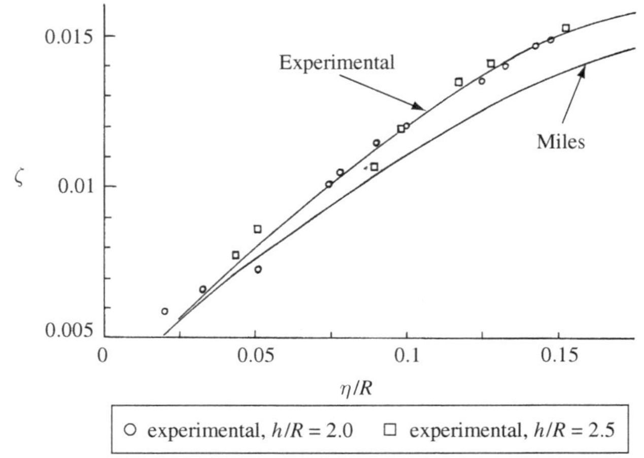

Figure 3.12 Dependence of damping factor on liquid wave amplitude for ring baffles of $w/R = 0.076$ and $d/R = 0.333$. (Stephens, *et al.*, 1962)

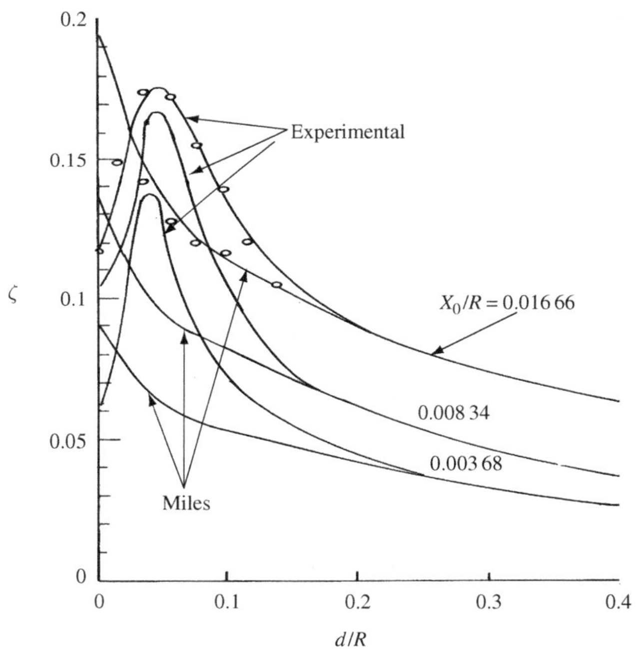

Figure 3.13 Effect of excitation amplitude of the damping factor due to solid ring baffle of $w/R = 0.157$. (Stephens, *et al.*, 1962)

The damping effectiveness of perforated baffles is largely dependent upon the perforation hole size and the percentage area removed by perforation. Figure 3.14 shows the dependence of the damping ratio on the baffle depth for various values of hole-sizes with a fixed area of 30% removed, and fixed baffle width of $w/R = 0.157$. For a fixed area removed, the damping effectiveness is improved as the perforation hole size is decreased. Figure 3.15 shows the same dependence for various values of percentage perforation area and fixed hole-size. It is

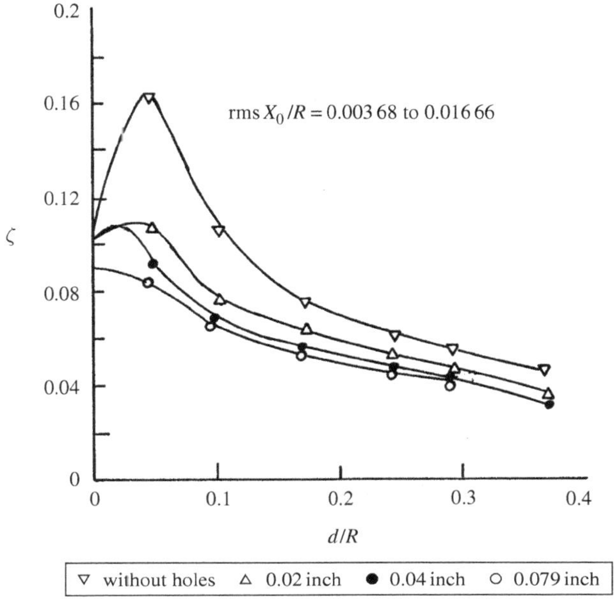

Figure 3.14 Effect of perforation hole size on the damping factor due to ring baffles with 30% opening, $w/R = 0.157$, and various hole diameters. (Stephens, *et al.*, 1962)

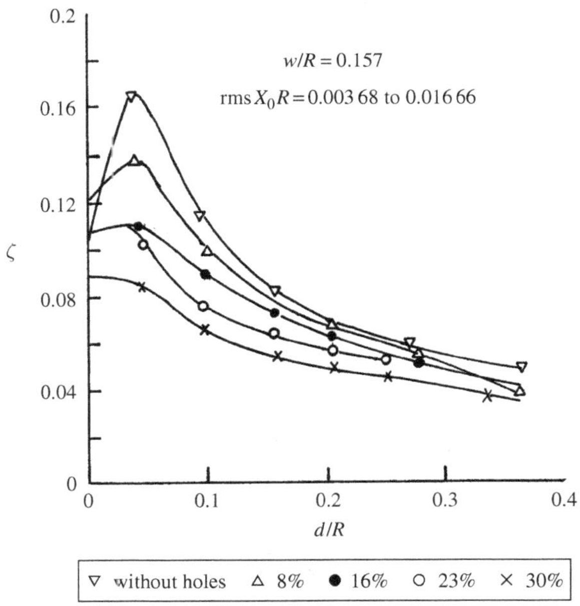

Figure 3.15 Effect of percentage perforation on damping factor due to baffles of hole diameter of 0.079 inch. (Stephens, *et al.*, 1962)

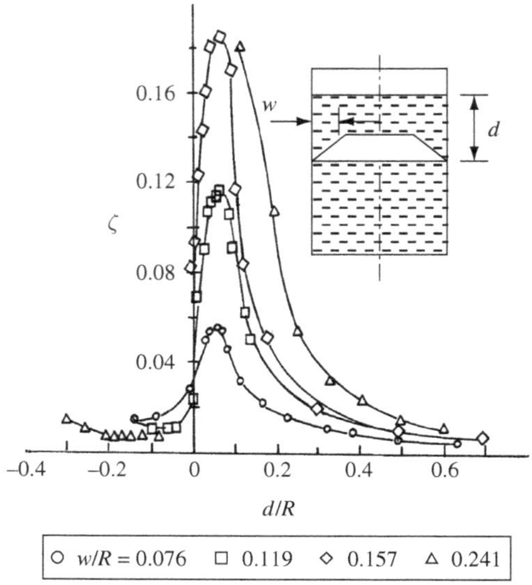

Figure 3.16 Effect of upright conic baffle size on damping factor for $h/R = 2$, $R = 6$ inch. (Silveira, *et al.*, 1961)

seen that the damping increases abruptly with decreasing perforation area for perforation areas less than 16%.

Abramson, Chu, and Garza (1963) and Sumner (1964) measured the damping ratio for various baffle configurations in spherical tanks. These baffles included perforated flat rings oriented as lines of longitude, and thus, for shallow liquid filling, had the same form as cruciform baffles. Experimental data on damping effectiveness were given for baffles with the north–south axis of the tank being directed both vertically and horizontally. It was found that the latter leads to a much greater reduction in slosh force. The damping provided by flat-ring baffles oriented as lines of latitude in oblate and prolate spheroids was experimentally measured by Stephens, *et al.* (1963). White (1964) and Sumner *et al.* (1966) measured slosh-damping effectiveness provided by baffles in specific vehicle-tank geometries such as S-IVB LO_2 and Centaur LH_2.

Conical baffles

Conical baffles are used either in upright or inverted positions. Silveira, *et al.* (1961) studied the damping of liquid sloshing due to different types of baffles including solid and perforated conical sections. The trends and maximum values of damping factor are similar to those of fixed rings and typical experimental results are shown in Figure 3.16 for the upright conic section. Figure 3.17 shows the variation of damping with the baffle location for an inverted conic section baffle. The maximum value of damping is seen to occur when the baffle is approximately half submerged in the liquid. This shift in the baffle location for maximum damping is apparently due to the almost complete restriction of the fluid motion in the region between the baffle and the tank wall. The results show that the effectiveness of the baffle is less than that of either the ring or the upright conic baffle. The effect of a 50% reduction in the blockage area for two perforated upright conical baffles is shown in Figure 3.18. The sharp peaks for the solid conic-section baffles are reduced for the perforated baffles. The damping of a combination of Z-conical baffles shown

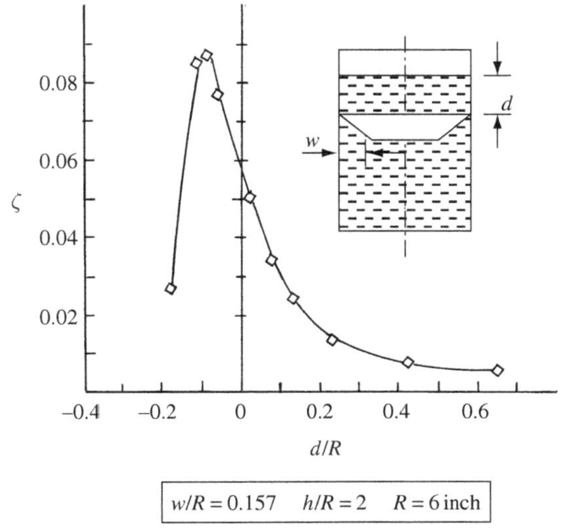

Figure 3.17 Dependence of damping factor on baffle location for inverted conic section baffle. (Silveira, *et al.*, 1961)

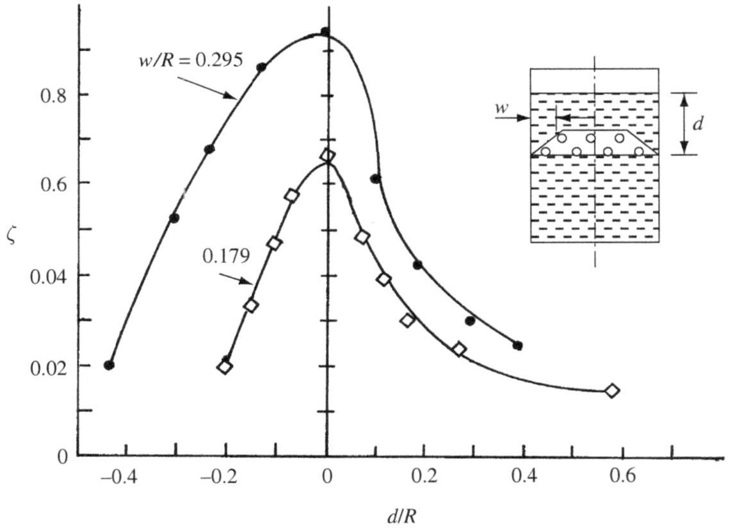

Figure 3.18 Dependence of damping factor on baffle location due to upright conic baffle with 50% perforated area. (Silveira, *et al.*, 1961)

in Figure 3.6(l), was measured experimentally by Abramson, *et al.* (1960, 1961) and Figure 3.18 shows the dependence of the damping factor on the equivalent Reynolds number for different values of liquid depth ratio.

Radial baffles

Radial baffles take the form of either sector walls or cruciform. Solid sector walls with 45°, 60°, and 90° were considered by Garza (1964). According to Garza's experimental results, the liquid

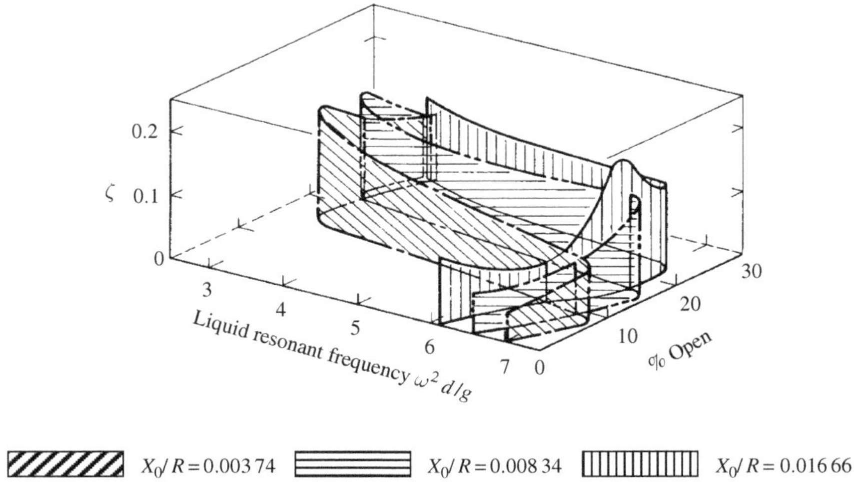

Figure 3.19 Dependence of damping factor on liquid resonant frequency and percentage opening in a 45° sectored tank. (Garza, 1964)

damping at resonance frequency contributed by these sectored tanks is approximately 0.04. However, at frequencies below resonance the liquid sloshing was effectively damped.

Abramson, *et al.* (1962b, d) and Abramson and Garza (1965) conducted experimental tests on perforated sectored walls with hole sizes varying from 0.02 inch to 0.079 inch and various values of open area varying from 8% to 30%. Their results indicated a nonlinear effect on the liquid damping ratio. The liquid viscosity and excitation amplitude have effects equal to those of the perforated hole-size and percentage of open area. It was found that perforated partitions with less than 10% open area will increase the damping ratio to approximately 0.1, while maintaining the liquid resonant frequency corresponding to a solid wall compartmented tank. For partitions with open areas greater than 10% the damping is greater than 0.1, but the corresponding liquid resonant frequencies approach that of an uncompartmented tank. Figure 3.19 shows three-dimensional plots of the damping ratio as a function of frequency parameter, $\omega^2 R/g$, and percentage open area for a 45° sectored tank.

The damping provided by cruciform baffles was measured by Silveira, *et al.* (1961) for two orientations of the baffle with respect to the motion of the liquid in the fundamental mode. For a 90° position, where two sections are located on the nodal diameter and two sections are 90° from the nodal diameter, the measured damping ratios were 0.011 48 and 0.025 for width ratios $w/R = 0.169$ and 0.337, respectively. In the 45° position, the damping ratios were 0.0115 and 0.0227 for $w/R = 0.169$ and 0.337, respectively. Evans and McIver (1987) determined the natural frequencies in a container with vertical baffles.

Expulsion bags and diaphragms

Under a low gravitational environment, positive expulsion bags and diaphragms are used as liquid-management devices in control and life-support systems. These devices have not been used in large propellant containers. Stofan and Pauli (1962) and Stofan and Sumner (1963) found that the damping ratio increases with increasing diaphragm thickness and excitation amplitude and decreasing tank diameter. They also found that peak slosh forces occurred at successively higher values of excitation frequency as diaphragm thickness increases. Levine, *et al.* (1966) discussed some design problems pertaining to the use of expulsion bags and diaphragms. Allingham (1968) proposed the use of metallic bellows to resolve some problems encountered in the expulsion bags. Dodge and Kana (1987) considered upright and inverted bladdered tanks.

3.5 Periodic boundary layers

3.5.1 Stokes boundary layer over an oscillating flat plate

For the case of a laminar motion of a viscous fluid on an oscillating plate, the Stokes boundary layer is governed by the diffusion equation (simplified Navier–Stokes equation)

$$\frac{\partial V}{\partial t} = \nu \frac{\partial^2 V}{\partial Z^2} \tag{3.85}$$

where Z is the vertical distance measured from the plate surface, $\nu = \mu/\rho$ is the fluid kinematic viscosity, and ρ is the fluid density.

If the plate experiences a harmonic motion, $e^{i\Omega t}$, the fluid velocity adjacent to the plate also oscillates with the same frequency and equation (3.85) takes the form

$$\frac{d^2 V(z)}{dZ^2} = \frac{i\Omega}{\nu} V(z) \tag{3.86}$$

The solution of this equation may be written in the form

$$V(Z) = A\, e^{(1+i)\sqrt{(\Omega/2\nu)}Z} + B\, e^{-(1+i)\sqrt{(\Omega/2\nu)}Z} \tag{3.87}$$

As $Z \to +\infty$, the velocity of the fluid vanishes and this requires $A = 0$. At $Z = 0$, the velocity must be the same as the velocity of the plate $V = V_0 e^{i\Omega t}$. Accordingly, one may write the velocity profile in the form

$$V(Z, t) = V_0\, e^{-Z\sqrt{\Omega/(2\nu)}} \cos\left(\Omega t - Z\sqrt{\frac{\Omega}{2\nu}}\right) \tag{3.88}$$

It is obvious that two fluid layers, a distance $Z = 2\pi\sqrt{2\nu/\Omega}$ apart, will move in phase. This distance is known as the *depth of penetration* of the viscous wave. The layer that is moving with the plate surface has a thickness of order $\sqrt{2\nu/\Omega}$, which decreases as the frequency increases. Relation (3.87) represents a transversal oscillatory wave that propagates inwards from the boundary with velocity $\sqrt{2\nu\Omega}$ with rapidly decaying amplitude. Figure 3.20 shows a sequence of velocity profiles at different time increments. This is known as the *Stokes' second problem*. It differs from the Stokes' first problem for which the plate is given a step function velocity. The resulting shear stress is

$$\begin{aligned} \tau_p &= -\mu \left.\frac{\partial V}{\partial Z}\right|_{Z=0} = \mu V_0 \sqrt{\frac{\Omega}{2\nu}} \{\cos \Omega t - \sin \Omega t\} = \rho V_0 \sqrt{\Omega \nu} \cos(\Omega t + \pi/4) \\ &= -\bar{m}_{\text{eff}} \dot{V} - \bar{C}_{\text{eff}} V \end{aligned} \tag{3.89}$$

where $\bar{m}_{\text{eff}} = \rho\sqrt{\nu/(2\Omega)}$ is the effective fluid mass per unit area, and $\bar{C}_{\text{eff}} = \mu\sqrt{\Omega/(2\nu)}$ is the effective damping coefficient per unit area.

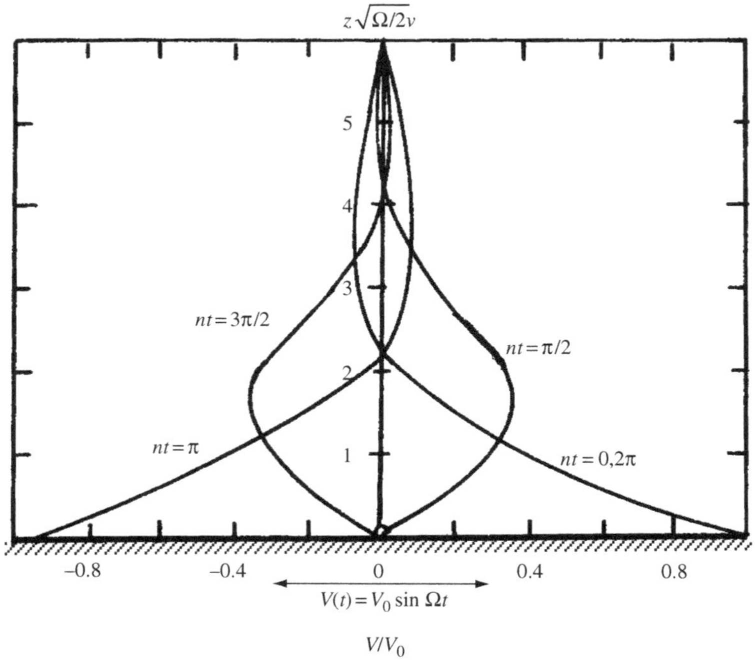

Figure 3.20 Velocity distribution in the neighborhood of an oscillating plate

Note that the force has its maximum at intervals of one-eighth of a period before the oscillating plate passes through its mean position.

If the fluid above the plate has a finite depth, say h, then the boundary conditions impose the following relations for the constants of equation (3.87)

$$A + B = V_0, \quad \text{and} \quad 0 = A\, e^{(1+i)\sqrt{(\Omega/2\nu)}h} + B\, e^{-(1+i)\sqrt{(\Omega/2\nu)}h} \tag{3.90}$$

Accordingly, the velocity profile is given by the expression

$$V(Z,t) = V_0 \frac{\sinh\left((1+i)(h-Z)\sqrt{\Omega/2\nu}\right)}{\sinh\left((1+i)h\sqrt{\Omega/2\nu}\right)} e^{i\Omega t} \tag{3.91}$$

The corresponding shear stress is

$$\tau_p = -\mu(1+i)V_0\sqrt{\frac{\Omega}{2\nu}}\coth\left((1+i)\sqrt{\frac{\Omega}{2\nu}}h\right)e^{i\Omega t} \tag{3.92a}$$

Taking the real part of this expression gives

$$\tau_p = -\mu V_0\sqrt{\frac{\Omega}{\nu}}\frac{\sinh\left(2\sqrt{\frac{\Omega}{2\nu}}h\right)\cos(\Omega t + \pi/4) + \sin\left(2\sqrt{\frac{\Omega}{2\nu}}h\right)\sin(\Omega t + \pi/4)}{\cosh\left(2\sqrt{\frac{\Omega}{2\nu}}h\right) - \cos\left(2\sqrt{\frac{\Omega}{2\nu}}h\right)} \tag{3.92b}$$

For larger values of $h\sqrt{\Omega/2\nu}$, the above expression becomes identical to equation (3.89). For a small fluid depth ratio $h\sqrt{\Omega/2\nu}$, equation (3.92b) takes the form

$$\tau_p = \frac{\mu V_0}{h} \cos(\Omega t) \tag{3.92c}$$

The case of a randomly oscillating plate was examined by Drolet and Viñals (1998). They assumed that the plate velocity is a Gaussian white noise with autocorrelation function, $R_V(\tau) = 2D\delta(\tau)$, and obtained the following expression for the fluid velocity mean square

$$E[V^2(Z,t)] = \frac{2D\nu}{\pi Z^2}\left(1 + \frac{Z^2}{2\nu t}\right) e^{-Z^2/2\nu t} \tag{3.93}$$

The velocity disturbance propagates into the fluid diffusively with a diffusion coefficient ν. For long times the velocity does not decay exponentially away from the wall but rather as a power law,

$$E[V^2(Z, t \to \infty)] = \frac{2D\nu}{\pi Z^2} \tag{3.94}$$

It is seen that the mean square velocity is linearly proportional to the excitation intensity, D.

Martel, *et al.* (1998) estimated the natural frequencies and damping rate of surface waves in a circular cylinder with pinned-end boundary conditions in terms of the gravitational Reynolds number, ($C^{-1} = \sqrt{gR^3/\nu}$), Bond number, ($\mathrm{Bo} = \rho g R^2/\sigma$, where σ is the surface tension (dynes/cm, or 10^{-5} N/cm)), and the slenderness ratio of the tank in the limited case of $C \to 0$. Miles and Henderson (1998) introduced a simplification to the work of Martel, *et al.* (1998) that omits the second-order boundary layer effects.

3.5.2 Periodic boundary layers in cylindrical tanks

Estimating the thickness of periodic boundary layers in cylindrical tanks is important to determining the effective mass moment of inertia and damping coefficient of liquid propellant rockets. Reismann (1958), Bauer (1959c), Rashed (1965), and Ibrahim and Latorre (1988) analytically and experimentally investigated the periodic boundary layers of liquid propellant inside a cylindrical tank subjected to roll harmonic excitation. The results of Reismann revealed that both the liquid inertia and damping coefficients are functions of the effective boundary layers thickness. For an upright circular cylindrical tank, filled with a viscous fluid and subjected to roll oscillation, the fluid inertia and damping are mainly due to the flat bottom and tank walls.

Cylinder bottom contribution

The flow generated in a semi-infinite mass of fluid by viscous action due to harmonic oscillations, $\varphi(t) = \varphi_0 \sin \Omega t$, of a horizontal circular plate about its center is given by the solution of the Navier–Stokes equation (3.85) subject to the appropriate boundary conditions. The solution of this equation is given in many references (see, e.g., Lamb, 1945, Schlichting, 1960, Rosenhead, 1963, and Rashed, 1965) in the form

$$V = r\Omega e^{-\sqrt{\Omega/2\nu} Z}\left\{\cos\left(\sqrt{\frac{\Omega}{2\nu}} Z\right)\cos\Omega t + \sin\left(\sqrt{\frac{\Omega}{2\nu}} Z\right)\sin\Omega t\right\}\varphi_0 \tag{3.95}$$

This solution indicates that the flow is damped away from the bottom surface. The velocity at any point is composed of two parts. The first is in phase with the motion of the disc, while the second is 90° out of phase. The shear stress, $\tau_b = \mu \partial V/\partial Z$, takes the form

$$\tau_b = \mu\left(\sqrt{\frac{\Omega}{2\nu}}Z\right)V + \frac{\rho}{2\left(\sqrt{\frac{\Omega}{2\nu}}Z\right)}\frac{\partial V}{\partial t} \tag{3.96}$$

The resulting shear torque T_b about an axis through the bottom center is

$$T_b = \int_0^{2\pi}\int_0^R (\tau_b)_{z=0} r^2\, dr\, d\theta = I_b\ddot{\varphi} + C_b\dot{\varphi} \tag{3.97}$$

where I_b and C_b are the corresponding effective mass moment of inertia and damping of the fluid participating in the tank bottom motion. They are given by the expressions

$$I_b = \frac{\pi}{2}\rho R^4\sqrt{\frac{\nu}{2\Omega}},\quad C_b = \frac{\pi}{2}\mu R^4\sqrt{\frac{\Omega}{2\nu}} \tag{3.98a, b}$$

It is seen that both inertia and damping are frequency dependent but of opposite effect.

Special case: annular ring baffle contribution

For an annular ring baffle of inner radius R_b with both sides immersed in a viscous fluid, the effective moment of inertia and damping coefficient can be obtained from the previous case by adjusting the area and multiplying by a factor of 2. The following results are obtained

$$I_{\text{baffle}} = \pi\rho\sqrt{\frac{\nu}{2\Omega}}(R^4 - R_b^4),\quad C_{\text{baffle}} = \pi\mu\sqrt{\frac{\Omega}{2\nu}}(R^4 - R_b^4) \tag{3.99a, b}$$

These results are based on deep fluid depth. As the roll frequency approaches zero, the moment of inertia approaches infinity and the analysis must be modified as indicated in the next section.

Fluid depth effect

Bauer (1959c) eliminated the problem arising from zero roll frequency by subtracting the torque from that part of the liquid above the fluid height h. For a liquid height h above the tank bottom one may write the velocity distribution in the form

$$V = ir\varphi_0\Omega\, e^{-\sqrt{i\Omega/\nu}h}\, e^{i\Omega t} \tag{3.100}$$

At this height we place the fluid free surface. The fluid free surface has the shape of a pulsating rotational paraboloid that depends on the roll frequency. The torque in the upper part of the fluid above the fluid height h due to the oscillating tank bottom at $z = 0$ is

$$\begin{aligned} T_{bu} &= \frac{\pi}{2}R^4\mu\varphi_0\Omega\sqrt{\frac{\Omega}{2\nu}}(-1+i)\, e^{-\sqrt{i\Omega/\nu}h}\, e^{i\Omega t} \\ &= \frac{\pi}{2}R^4\mu\varphi_0\Omega\sqrt{\frac{\Omega}{2\nu}}\, e^{-\sqrt{\Omega/2\nu}h} \\ &\quad \times\left\{\left[\sin\left(\sqrt{\frac{\Omega}{2\nu}}h\right) - \cos\left(\sqrt{\frac{\Omega}{2\nu}}h\right)\right] + i\left[\cos\left(\sqrt{\frac{\Omega}{2\nu}}h\right) + \sin\left(\sqrt{\frac{\Omega}{2\nu}}h\right)\right]\right\} e^{i\Omega t} \end{aligned} \tag{3.101}$$

The torque transmitted into the fluid of height h above the tank bottom is

$$\begin{aligned} T_{b\text{mod}} &= T_b - T_{bu} \\ &= \frac{\pi}{2} R^4 \mu \varphi_0 \Omega \sqrt{\frac{\Omega}{2\nu}} \left\{ -\left\{ 1 - e^{-\sqrt{\Omega/2\nu}h} \left[\sin\left(\sqrt{\frac{\Omega}{2\nu}}h\right) - \cos\left(\sqrt{\frac{\Omega}{2\nu}}h\right) \right] \right\} \right. \\ &\qquad \left. + i \left\{ 1 - e^{-\sqrt{\Omega/2\nu}h} \left[\cos\left(\sqrt{\frac{\Omega}{2\nu}}h\right) + \sin\left(\sqrt{\frac{\Omega}{2\nu}}h\right) \right] \right\} \right\} e^{i\Omega t} \\ &= I_{b\text{mod}} \ddot{\varphi} + C_{b\text{mod}} \dot{\varphi} \end{aligned} \tag{3.102}$$

where the effective moment of inertia and damping coefficients are

$$\begin{aligned} I_{b\text{mod}} &= \frac{\pi}{2} R^4 \frac{\mu}{\Omega} \sqrt{\frac{\Omega}{2\nu}} \left\{ 1 - e^{-\sqrt{\Omega/2\nu}h} \left[\cos\left(\sqrt{\frac{\Omega}{2\nu}}h\right) - \cos\left(\sqrt{\frac{\Omega}{2\nu}}h\right) \right] \right\} \\ &= \frac{\pi}{2} \rho R^4 h_{eff} \end{aligned} \tag{3.103a}$$

$$C_{b\text{mod}} = \frac{\pi}{2} R^4 \mu \sqrt{\frac{\Omega}{2\nu}} \left\{ 1 - e^{-\sqrt{\Omega/2\nu}h} \left[\cos\left(\sqrt{\frac{\Omega}{2\nu}}h\right) + \sin\left(\sqrt{\frac{\Omega}{2\nu}}h\right) \right] \right\} \tag{3.103b}$$

where the effective fluid height is

$$h_{\text{eff}} = \sqrt{\frac{\Omega}{2\nu}} \left\{ 1 - e^{-\sqrt{\Omega/2\nu}h} \left[\cos\left(\sqrt{\frac{\Omega}{2\nu}}h\right) - \cos\left(\sqrt{\frac{\Omega}{2\nu}}h\right) \right] \right\} \tag{3.104}$$

As the roll frequency decreases $\Omega \to 0$, the effective moment of inertia approaches the moment of inertia of the solidified fluid since the effective fluid depth approaches the actual fluid height h. Thus one may write $\lim_{\Omega \to 0} I_b = m_f R^2/2$, and $\lim_{\Omega \to 0} C_b = 0$.

Cylinder wall contribution

It is convenient to analyze this case in terms of complex algebra by assuming that the roll oscillation is expressed in the form $\varphi(t) = \varphi_0 \text{Exp}(i\Omega t)$. For axisymmetric flow, where $\partial \mathbf{u}/\partial \theta = 0$, $\partial \mathbf{u}/\partial z = 0$, and the radial and vertical velocity components vanish equally, the Navier–Stokes equation takes the form

$$\frac{\partial u_\theta}{\partial t} = \nu \left(\frac{\partial^2 u_\theta}{\partial r^2} + \frac{1}{r}\frac{\partial u_\theta}{\partial r} - \frac{u_\theta}{r^2} \right) \tag{3.105}$$

Subject to the boundary conditions

$$\text{at } r = R, u_\theta = iR\Omega\varphi_0 \exp(i\Omega t) = iU \cos \Omega t, \quad \text{and} \quad \text{at } r = 0, u_\theta = 0 \tag{3.106}$$

where $U = R\Omega\varphi_0$. Substituting $U(r)\cos\Omega t$ for $u_\theta(r,t)$, where $U(r) = r\Omega\varphi_0$, in equation (3.105), gives

$$\frac{d^2U(r)}{dr^2} + \frac{1}{r}\frac{dU(r)}{dr} - \left(\frac{1}{r^2} + i\frac{\Omega}{\nu} \right) U(r) = 0 \tag{3.107}$$

Equation (3.107) has the following solution

$$U(r) = C_1 I_1\left(\sqrt{i\frac{\Omega}{\nu}}\, r\right) + C_2 K_1\left(\sqrt{i\frac{\Omega}{\nu}}\, r\right) \tag{3.108}$$

where $i = \sqrt{-1}$, C_1 and C_2 are arbitrary constants of integration, $I_1(\,)$ is the modified Bessel function of the first kind of order one given by the following expression

$$I_1(x) = \frac{x}{2\Gamma(2)}\left[1 + \frac{x^2}{2(4)} + \frac{x^4}{2(4)(4)(6)} + \cdots\right] = \sum_{k=0}^{\infty} \frac{(x/2)^{2k-1}}{k!\,\Gamma(2+k)} \tag{3.109a}$$

and $K_1(\,)$ is the modified Bessel function of the second kind of order one given by the expression

$$K_1(x) = \lim_{n\to 1} \frac{\pi}{2\sin n\pi}\left[I_{-n}(x) - I_n(x)\right] \tag{3.109b}$$

where $I_{-n}(x) = I_n(x)$, for $n = 0, 1, 2, \ldots$ It is not convenient to manipulate Bessel functions with complex arguments and thus one may express $I_1(\,)$ and $K_1(\,)$ in terms of the Kelvin's functions. From the following general relations one can write similar relations for order one:

$$e^{\frac{1}{2}in\pi} I_n\left(x\, e^{\frac{1}{4}i\pi}\right) = \mathrm{Ber}_n(x) + i\,\mathrm{Bei}_n(x)$$

Thus

$$I_1\left(\sqrt{i\frac{\Omega}{\nu}}\, r\right) = \mathrm{Bei}_1\left(\sqrt{\frac{\Omega}{\nu}}\, r\right) - i\,\mathrm{Ber}_1\left(\sqrt{\frac{\Omega}{\nu}}\, r\right) \tag{3.110a}$$

and

$$e^{-\frac{1}{2}in\pi} K_n\left(x\, e^{\frac{1}{4}i\pi}\right) = \mathrm{Ker}_n(x) + i\,\mathrm{Kei}_n(x)$$

Thus

$$K_1\left(\sqrt{i\frac{\Omega}{\nu}}\, r\right) = i\mathrm{Ker}_1\left(\sqrt{\frac{\Omega}{\nu}}\, r\right) - \mathrm{Kei}_1\left(\sqrt{\frac{\Omega}{\nu}}\, r\right) \tag{3.110b}$$

where

$$\mathrm{Ber}_1(x) = \sum_{k=0}^{\infty} \frac{(x/2)^{2k+1}}{k!\Gamma(k+2)} \cos\left((3+2k)\frac{\pi}{4}\right) \tag{3.111a}$$

$$\mathrm{Bei}_1(x) = \sum_{k=0}^{\infty} \frac{(x/2)^{2k+1}}{k!\Gamma(k+2)} \sin\left((3+2k)\frac{\pi}{4}\right) \tag{3.111b}$$

$$\begin{aligned}\mathrm{Ker}_1(x) = &-\{\ln(x/2) + \gamma\}\mathrm{Ber}_1(x) + \frac{\pi}{4}\mathrm{Bei}_1(x)\\ &+ \frac{1}{2}\sum_{k=0}^{\infty} \frac{(x/2)^{2k+1}}{k!(k+1)!}\left\{\left(1 + \frac{1}{2} + \frac{1}{3} + \cdots + \frac{1}{k}\right)\right.\\ &\left. + \left(1 + \frac{1}{2} + \frac{1}{3} + \cdots + \frac{1}{(k+1)}\right)\right\} \cos\left((3+2k)\frac{\pi}{4}\right)\end{aligned} \tag{3.111c}$$

$$\mathrm{Kei}_1(x) = -\{\ln(x/2) + \gamma\}\mathrm{Bei}_1(x) - \frac{\pi}{4}\mathrm{Ber}_1(x) + \frac{1}{2}\sum_{k=0}^{\infty}\frac{(x/2)^{2k+1}}{k!(k+1)!}\left\{\left(1 + \frac{1}{2} + \frac{1}{3} + \cdots + \frac{1}{k}\right) + \left(1 + \frac{1}{2} + \frac{1}{3} + \cdots + \frac{1}{(k+1)}\right)\right\}\sin\left((3+2k)\frac{\pi}{4}\right) \tag{3.111d}$$

and $\gamma = 0.577\,215\,6$ is the Euler constant.

Since $\mathrm{Ker}_1(x) = \mathrm{Kei}_1(x) \to \infty$ as $x \to 0$, the constant $C_2 = 0$. In this case, the solution (3.108) takes the form

$$U(r) = C_1\left\{\mathrm{Bei}_1\left(\sqrt{\frac{\Omega}{\nu}}\,r\right) - i\mathrm{Ber}_1\left(\sqrt{\frac{\Omega}{\nu}}\,r\right)\right\} \tag{3.112}$$

Using the boundary conditions (3.106), the constant C_1 is

$$C_1 = \frac{R\Omega\varphi_0}{\mathrm{Bei}_1\left(\sqrt{\frac{\Omega}{\nu}}\,R\right) - i\mathrm{Ber}_1\left(\sqrt{\frac{\Omega}{\nu}}\,R\right)}$$

Accordingly, the velocity (3.112) distribution along r takes the form

$$U_1(r) = R\Omega\varphi\frac{\left\{\mathrm{Bei}_1\left(\sqrt{\frac{\Omega}{\nu}}\,r\right) - i\mathrm{Ber}_1\left(\sqrt{\frac{\Omega}{\nu}}\,r\right)\right\}}{\left\{\mathrm{Bei}_1\left(\sqrt{\frac{\Omega}{\nu}}\,R\right) - i\mathrm{Ber}_1\left(\sqrt{\frac{\Omega}{\nu}}\,R\right)\right\}} \tag{3.113a}$$

Let $\Re = \sqrt{\frac{\Omega}{\nu}}\,r$ and $\lambda = \sqrt{\frac{\Omega}{\nu}}\,R$. Equation (3.113a) can be written in the form

$$U_1(r) = R\Omega\varphi\frac{[\mathrm{Ber}_1\Re\mathrm{Ber}_1\lambda + \mathrm{Bei}_1\Re\mathrm{Bei}_1\lambda] + i[\mathrm{Bei}_1\Re\mathrm{Ber}_1\lambda + \mathrm{Ber}_1\Re\mathrm{Bei}_1\lambda]}{(\mathrm{Ber}_1\lambda)^2 + (\mathrm{Bei}_1\lambda)^2} \tag{3.113b}$$

Evaluating the shear stress, τ_w, at $r = R$, and for sufficiently large values of the parameter $r\sqrt{\Omega/\nu}$, the Kelvin's functions can be replaced by their asymptotic values, that is,

$$\tau_w = \rho\sqrt{\frac{\nu}{2\Omega}}\frac{du(R)}{dt} + \mu\left[\sqrt{\frac{\Omega}{2\nu}} - \frac{1}{R}\right]u(R) \tag{3.114}$$

The torque about the z-axis is

$$T_w = I_w\ddot{\varphi} + C_w\dot{\varphi} \tag{3.115}$$

where inertia and damping coefficients per unit fluid depth are given by the following expressions

$$I_w = 2\pi\rho R^3\sqrt{\frac{\nu}{2\Omega}}, \quad C_w = 2\pi\mu R^3\left[\sqrt{\frac{\Omega}{2\nu}} - \frac{1}{R}\right] \tag{3.116a, b}$$

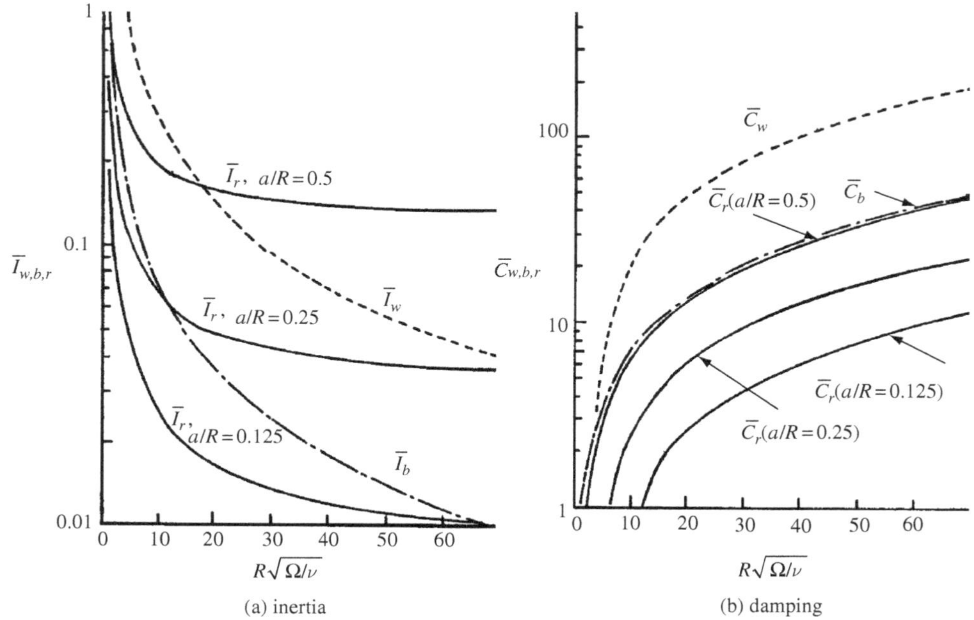

Figure 3.21 Effective fluid (a) mass moment of inertia and (b) damping due to tank wall, bottom, and immersed rod. Inertia is normalized with respect to the inertia of solidified liquid, while damping is normalized to reference damping coefficient. (Ibrahim and Latorre, 1988)

For small values of the frequency parameter $R\sqrt{\Omega/\nu}$, the following exact expression are used

$$I_w = 2\pi\rho R^4 \frac{(\mathrm{Ber}_1\lambda \mathrm{Bei}'_1\lambda - \mathrm{Bei}_1\lambda \mathrm{Ber}'_1\lambda)}{\lambda\{\mathrm{Ber}_1^2\lambda + \mathrm{Bei}_1^2\lambda\}} \tag{3.117a}$$

$$C_w = 2\pi\mu R^2 \frac{\left(\mathrm{Ber}_1\lambda \mathrm{Ber}'_1\lambda + \mathrm{Bei}_1\lambda \mathrm{Bei}'_1\lambda\right)\lambda}{\{\mathrm{Ber}_1^2\lambda + \mathrm{Bei}_1^2\lambda\}} - 1 \tag{3.117b}$$

where $\begin{Bmatrix} \mathrm{Ber}'_1\lambda \\ \mathrm{Bei}'_1\lambda \end{Bmatrix} = \frac{d}{d\Re}\begin{Bmatrix} \mathrm{Ber}_1\Re \\ \mathrm{Bei}_1\Re \end{Bmatrix}\Bigg|_{\Re=\lambda}$.

Note that both inertia and damping coefficients are dependent on the excitation frequency such that the inertia decreases with the frequency while the damping increases with the frequency. Figure 3.21(a) and (b) shows the dependence of these parameters, by dash curves, on the frequency parameter $R\sqrt{\Omega/\nu}$, which in effect represents another form of Reynolds number.

3.5.3 *Effect of immersed rods*

Bluff bodies immersed in liquid and experiencing harmonic oscillation usually give rise to another class of unsteady flow. If the spatial amplitude of the oscillating body is small compared with its nominal dimension (say the diameter of the body) then the linearization is permissible. However, the boundary layers approximation should not be employed, even for large values of Reynolds number. This is because one needs to determine the pressure distribution, which leads eventually to the forces from the fluid. This problem was originally considered by Stokes (1851, 1905), who derived an expression for the force on the cylinder. His result was shown (Rosenhead, 1963, p. 169) to be valid if either (i) the amplitude of oscillation is much smaller than the radius of the cylinder, and the viscous penetration depth Z is much smaller than the radius R, or (ii) the ratio of radius to the penetration depth is of order 1 and the Reynolds number is much less than 1. Williams and Hussey (1972) indicated that if the end effects of the cylinder are considered, both observations and Stokes' predictions agree very well. They also reported the measurements of the fluid forces in the range $0.286 < Z/R < 4.13$. Within this range their results were found to be in good agreement with Stokes' theory when the Reynolds number is greater than 1.

Morison, *et al.* (1950) developed a formula for the hydrodynamic forces acting on a pile due to liquid surface waves. The forces acting on a cylinder oscillating in water were estimated by Jen (1968) and Hamann and Datton (1971). The surface waves introduce two types of hydrodynamic forces, inertia and drag. The inertia force is the sum of two components. The first is buoyancy force acting on the structure in the fluid due to a pressure gradient generated from the flow acceleration. The buoyancy force is equal to the mass of the fluid displaced by the structure multiplied by the acceleration of the flow. The second inertia component is due to the added mass, which is proportional to the relative acceleration between the structure and the fluid, and accounts for the flow entrained by the structure. The drag force is the sum of the viscous and pressure drags produced by the relative velocity between the structure and the flow. This type of hydrodynamic drag is proportional to the square of the relative velocity. Carrier and Di Pima (1956) studied the torsional oscillations of a sphere in a viscous fluid. Keulegan and Carpenter (1958) estimated the forces on cylinders and plates in an oscillating fluid flow. Chen (1981) reviewed the results of a fluid damping force for circular cylinders vibrating in stationary fluid, cross flow, and parallel flow.

A torsion pendulum immersed in a viscous liquid is an excellent example of the limitations of the classical quasi-stationary approximation. The irreversible viscous dissipation of the liquid represents a damping albeit not of the simple quasi-stationary type. If the pendulum is very slender, the symmetry and simplicity of the geometry implies that the Navier–Stokes equation is linear and the flow in the damper is a nonstationary circular Couette flow due to the mass transport by the surface velocity. If the oscillations are very rapid, the vorticity diffusion distance, that is, the Stokesian boundary layer thickness, is small compared with the distance between the solid and the vessel. In such a case, a major part of the liquid is particularly at rest, which contradicts the quasi-stationary approximation. This general coupled problem is usually described by three linear equations. The motion of the liquid is described by a second-order partial differential equation, the motion of the pendulum is described by a second-order ordinary differential equation, and the relation between shear stress and strain rate is described by a first-order partial differential equation. These equations are linearly coupled and have been solved by Rehbinder and Apazidis (1995). It was found that the classical quasi-stationary approximation is valid if the vorticity propagation distance is much larger than the width of the gap between the pendulum and the container. If this is not the case, the quasi-stationary approximation deviates considerably from the exact solution. If the width of the gap between the pendulum and the container is larger than about four times the thickness of the boundary layer, the exponential part of the damping factor of the amplitude increases with increasing damping coefficient.

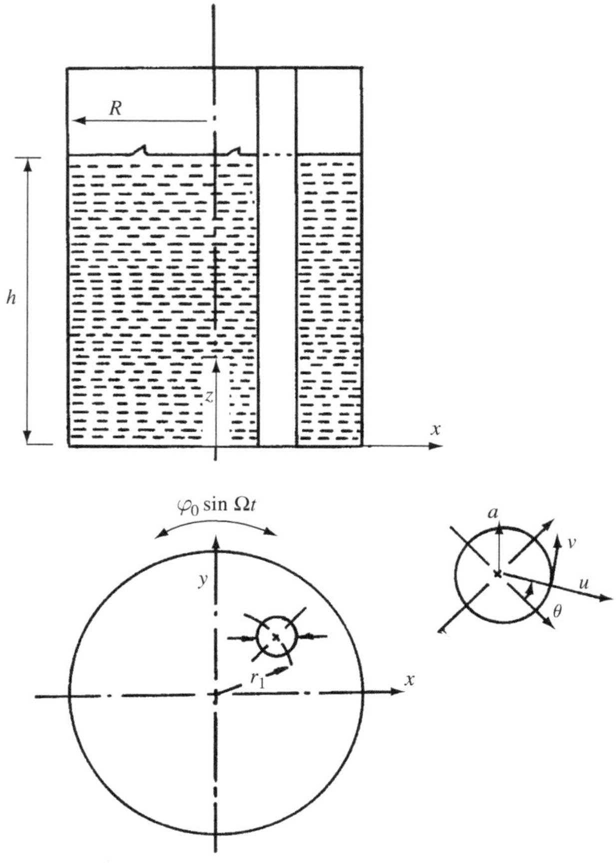

Figure 3.22 Schematic diagram of a tank with fitted circular rod showing the coordinate frame.

Ibrahim and Latorre (1988) carried out analytical and experimental investigations to determine the fluid inertia and damping introduced by an immersed circular rod of radius a at a distance r_1 from the center of the tank filled with a viscous fluid. They assumed that the spatial amplitude, $r_1\varphi_0$, is very small, that is, $r_1\varphi_0 \ll R$, and thus one can use the linearized Navier–Stokes equations along r and θ axes. However, the boundary layers approximation will not be performed, even for large values of Reynolds numbers, since the objective is to determine the pressure distribution and the torsional torque. The Navier–Stokes equations in polar coordinates, r and θ, are

$$\frac{\partial u}{\partial t} = -\frac{1}{\rho}\frac{\partial p}{\partial r} + \nu\left(\nabla^2 u - \frac{u}{r^2} - \frac{2}{r^2}\frac{\partial v}{\partial \theta}\right) \tag{3.118a}$$

$$\frac{\partial v}{\partial t} = -\frac{1}{\rho r}\frac{\partial p}{\partial \theta} + \nu\left(\nabla^2 v - \frac{v}{r^2} + \frac{2}{r^2}\frac{\partial v}{\partial \theta}\right) \tag{3.118b}$$

where $\nabla^2 = \frac{\partial^2}{\partial r^2} + \frac{1}{r}\frac{\partial}{\partial r} + \frac{1}{r^2}\frac{\partial^2}{\partial \theta^2}$, $r \geq a$, r and θ are measured from the rod center as shown in Figure 3.22. The velocity components, u and v, can be written in terms of the stream function Ψ as

$$u = \frac{1}{r}\frac{\partial \Psi}{\partial \theta}, \quad v = -\frac{\partial \Psi}{\partial r} \tag{3.119a, b}$$

Let

$$\chi(r\theta, t) = \nu \nabla^2 \Psi - \frac{\partial \Psi}{\partial t} \tag{3.120}$$

Introducing equations (3.119) and (3.120) into equations (3.118), gives

$$\frac{\partial u}{\partial t} = -\frac{1}{\rho}\frac{\partial p}{\partial r} + \frac{1}{r}\frac{\partial \chi}{\partial \theta} + \frac{\partial u}{\partial t}, \quad \frac{\partial v}{\partial t} = -\frac{1}{\rho r}\frac{\partial p}{\partial \theta} - \frac{\partial \chi}{\partial r} + \frac{\partial v}{\partial t} \tag{3.121a, b}$$

Eliminating the pressure term from equations (3.121), gives

$$\frac{\partial}{\partial r}\left(r\frac{\partial \chi}{\partial r}\right) + \frac{1}{r}\frac{\partial^2 \chi}{\partial \theta^2} = 0, \quad \text{or} \quad \nabla^2 \chi = 0 \tag{3.122}$$

$$\nabla^2\left(\nabla^2 - \frac{1}{\nu}\frac{\partial}{\partial t}\right)\Psi = 0 \tag{3.123}$$

This equation can be solved in terms of two stream functions $\Psi = \Psi_1 + \Psi_2$, where

$$\nabla^2 \Psi_1 = 0 \quad \text{and} \quad \left(\nabla^2 - \frac{1}{\nu}\frac{\partial}{\partial t}\right)\Psi_2 = 0 \tag{3.124a, b}$$

The solution of these equations may be written in the form

$$\Psi_1 = f_1(r)\sin\theta\, e^{i\Omega t}, \quad \Psi_2 = f_2(r)\sin\theta\, e^{i\Omega t} \tag{3.125a, b}$$

Substituting these solutions into equations (3.124), gives

$$r^2\frac{d^2 f_1(r)}{dr^2} + r\frac{df_1(r)}{dr} - f_1(r) = 0 \tag{3.126a}$$

$$\frac{d^2 f_2(r)}{dr^2} + \frac{1}{r}\frac{df_2(r)}{dr} - \left(\frac{1}{r^2} + \frac{i\Omega}{\nu}\right)f_2(r) = 0 \tag{3.126b}$$

The solutions of these equations are, respectively,

$$f_1(r) = A_1 r + \frac{A_2}{r} \tag{3.127a}$$

$$f_2(r) = B_1 I_1\left(\sqrt{\frac{i\Omega}{\nu}}r\right) + B_2 K_1\left(\sqrt{\frac{i\Omega}{\nu}}r\right) \tag{3.127b}$$

where A_i and B_i, $i = 1$ and 2 are constants of integration. Since Ψ must vanish at $r \to \infty$, the constants $A_1 = B_1 = 0$. The total stream function Ψ is

$$\Psi = \left[\frac{A_2}{r} + B_2 K_1\left(\sqrt{\frac{i\Omega}{\nu}}r\right)\right]\sin\theta\, e^{i\Omega t} \tag{3.128}$$

The constants A_2 and B_2 are determined from the boundary conditions. The velocity of the fluid at the rod must be $r_1\,\Omega\,e^{i\Omega t}$ parallel to the direction of oscillation, such that

$$u = r_1\Omega\cos\theta\, e^{i\Omega t}, \quad v = -r_1\Omega\sin\theta\, e^{i\Omega t} \tag{3.129a, b}$$

Using equations (3.119), (3.128) and (3.129), the constants B_1 and B_2 are obtained in the form

$$A_2 = r_1 a^2\Omega \frac{\left\{\left(\sqrt{\frac{i\Omega}{\nu}}a\right)K_1'\left(\sqrt{\frac{i\Omega}{\nu}}a\right) - K_1\left(\sqrt{\frac{i\Omega}{\nu}}a\right)\right\}}{\left\{K_1\left(\sqrt{\frac{i\Omega}{\nu}}a\right) + \left(\sqrt{\frac{i\Omega}{\nu}}a\right)K_1'\left(\sqrt{\frac{i\Omega}{\nu}}a\right)\right\}},$$

$$B_2 = \frac{2r_1 a\Omega}{\left\{K_1\left(\sqrt{\frac{i\Omega}{\nu}}a\right) + \left(\sqrt{\frac{i\Omega}{\nu}}a\right)K_1'\left(\sqrt{\frac{i\Omega}{\nu}}a\right)\right\}}$$

where $\frac{d}{dr}K_1\left(\sqrt{\frac{i\Omega}{\nu}}r\right)\Big|_{r=a} = \sqrt{\frac{i\Omega}{\nu}}K_1'\left(\sqrt{\frac{i\Omega}{\nu}}a\right)$

Substituting in equation (3.128) gives

$$\Psi = \frac{r_1\Omega a\sin\theta\, e^{i\Omega t}}{\left\{K_1\left(\sqrt{\frac{i\Omega}{\nu}}a\right) + \left(\sqrt{\frac{\Omega}{\nu}}a\right)K_1'\left(\sqrt{\frac{i\Omega}{\nu}}a\right)\right\}} \times \left[\frac{a}{r}\left\{\left(\sqrt{\frac{i\Omega}{\nu}}a\right)K_1'\left(\sqrt{\frac{i\Omega}{\nu}}a\right) - K_1\left(\sqrt{\frac{i\Omega}{\nu}}a\right)\right\} + 2K_1\left(\sqrt{\frac{i\Omega}{\nu}}r\right)\right] \tag{3.130}$$

The resistance force to the motion of the rod, arising from the fluid viscosity, is

$$F = ha\int_0^{2\pi}(p_{rr}\cos\theta - p_{r\theta}\,\sin\theta)\,d\theta \tag{3.131}$$

where

$$p_{rr} = -p - \frac{2}{3}\mu\left(\frac{\partial u}{\partial r} + \frac{1}{r}\frac{\partial v}{\partial\theta} + \frac{u}{r}\right) + 2\mu\frac{\partial u}{\partial r}, \quad p_{r\theta} = \mu\left(\frac{1}{r}\frac{\partial u}{\partial\theta} + \frac{\partial v}{\partial r} - \frac{v}{r}\right) \tag{3.132a, b}$$

At $r = a$, the following parameters are evaluated

$$\frac{\partial u}{\partial r}\Big|_{r=a} = \left(\frac{\partial u}{\partial r} + \frac{1}{r}\frac{\partial v}{\partial\theta} + \frac{u}{r}\right)\Big|_{r=a} = 0$$

$$\frac{\partial u}{\partial\theta}\Big|_{r=a} = -r\sin\theta\, e^{i\Omega t}$$

$$\left(\frac{1}{r}\frac{\partial u}{\partial\theta} - \frac{v}{r}\right)\Big|_{r=a} = 0$$

$$\left(\frac{1}{r^2}\frac{\partial^2 u}{\partial\theta^2} + \frac{1}{r}\frac{\partial\Psi}{\partial r}\right)\Big|_{r=a} = 0 \tag{3.133}$$

Adding equations (3.124a) and (3.124b) gives

$$\nabla^2\Psi - \frac{1}{\nu}\frac{\partial}{\partial t}\Psi_2 = 0 \tag{3.134}$$

Combining (3.133) and (3.134) gives

$$\frac{\partial v}{\partial r} = -\frac{1}{\nu}\frac{\partial \Psi_2}{\partial t} \tag{3.135}$$

Introducing equations (3.133–3.135) into equations (3.132) gives

$$p_{rr} = -p \quad \text{and} \quad p_{r\theta} = -\rho\frac{\partial \Psi_2}{\partial t} \tag{3.136a, b}$$

Equation (3.131) becomes, after using (3.136),

$$F = ha\int_0^{2\pi}\left(-p\cos\theta + \rho\frac{\partial \Psi_2}{\partial t}\sin\theta\right)d\theta \tag{3.137}$$

Integrating the first term by parts and using relations (3.120), (3.125), and (3.127) yields the following result from the above integration

$$F = -i(\pi\rho h a^2)r_1\Omega^2\varphi_0\left\{1 - \frac{4K_1\left(\sqrt{\frac{i\Omega}{\nu}}\,a\right)}{\left\{K_1\left(\sqrt{\frac{i\Omega}{\nu}}\,a\right) + \left(\sqrt{\frac{i\Omega}{\nu}}\,a\right)K_1'\left(\sqrt{\frac{i\Omega}{\nu}}\,a\right)\right\}}\right\} \tag{3.138}$$

where $\pi\rho\, ha^2$ represents the mass of the fluid displaced by length h of the rod. One may also express the Bessel function with imaginary argument in terms of Kelvin's functions as shown in the previous section. Using asymptotic expansion of the Bessel function with large argument, the following result is obtained

$$F = \left(\pi\rho h a^2\right)r_1\Omega^2\varphi_0\left\{\left(1 + \frac{2}{a}\sqrt{\frac{2\nu}{\Omega}}\right)\sin\Omega t - \left(\frac{2}{a}\sqrt{\frac{2\nu}{\Omega}} + \frac{2\nu}{\Omega a^2}\right)\cos\Omega t\right\} \tag{3.139}$$

The torque about the Z-axis is

$$T_{\text{rod}} = r_1 F = I_r\ddot{\varphi} + C_r\dot{\varphi} \tag{3.140}$$

The following expressions for the inertia and damping due to the rod are obtained

$$I_r = \pi\rho a^2 h r_1^2\left[\frac{4}{a}\sqrt{\frac{\nu}{2\Omega}} + 1\right], \quad C_r = \pi\rho a^2 h\Omega\left[\frac{4}{a}\sqrt{\frac{\nu}{2\Omega}} - \frac{2\nu}{\Omega a^2}\right] \tag{3.141a, b}$$

These results reflect the dependence of inertia and damping coefficients on the excitation frequency. Note that the amount of liquid participating with the tank motion is proportional to the thickness $\sqrt{\nu/2\omega}$. Figure 3.21 (a) and (b) shows the dependence of the effective liquid moment of inertia and damping coefficient on the frequency parameter $\lambda = R\sqrt{\Omega/\nu}$ due to the tank bottom, tank wall, and immersed rod.

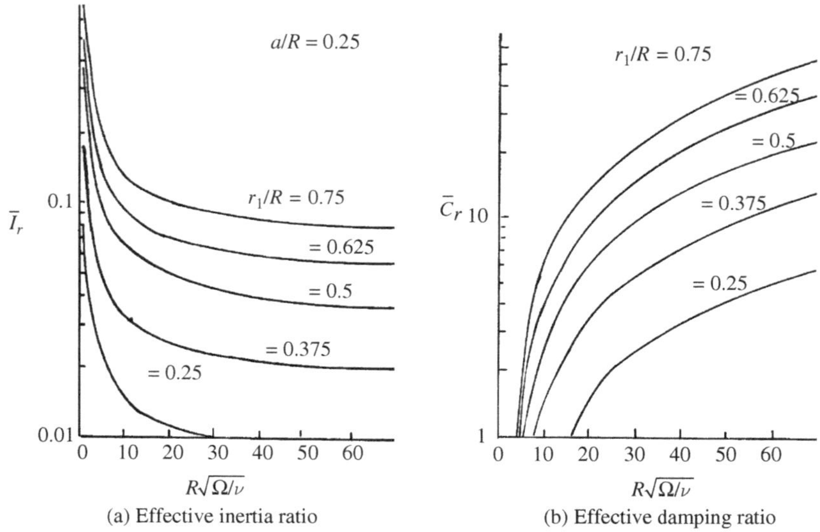

Figure 3.23 Effective inertia and damping parameters due to immersed rod. (Ibrahim and Latorre, 1983)

The above results were in good agreement with those measured experimentally. To have reliable experimental results, the experimental tank should not sit on ball bearings because the resulting friction of the bearings is higher than the effect of the fluid. The tank was suspended by a thin shaft in the form of a torsional pendulum, which avoided additional friction after filling the tank.

The general trend is that the mass moment of inertia decreases with the tank frequency while damping increases. At very small frequency range, $\lambda < 10$, the analytical results give unrealistic values for the mass moment of inertia. This invalidity is also reported by Reismann (1958) who obtained the values according to exact formulation (3.117). As the rod radius increases both the effective moment of inertia and damping coefficient increase. Figure 3.23 (a) and (b) shows the effect of the rod location r_1/R upon $\bar{I}_r$ and $\bar{C}_r$ for $a/R = 0.25$. It is evident that the farthest rod appreciably contributes to $\bar{I}_r$ and $\bar{C}_r$. However, these results are only valid as long as the proximity effects of the rod and wall are negligible, that is, if the ratio $(r_1 + a)/R$ does not approach unity.

It is noticed that the inertia torque, $I_r\ddot{\varphi}$, results from a force that is in phase with the acceleration. This force arises because, in moving the rod, part of the fluid is moved as well. The mass of the moving part depends on the frequency parameter λ (the square root of the Reynolds number). The damping torque, $C_r\dot{\varphi}$, results from a force that is in phase with the velocity of the rod. It is interesting to note that the mechanisms of these forces are, to a certain extent, similar to those analyzed by Morison, *et al.* (1950) for the forces exerted by the surface waves on an immersed pile.

3.6 Closing remarks

Control of liquid sloshing in moving containers is considered to be one of the major design issues of moving liquid systems. The inherent damping due to liquid viscosity is very useful in small size containers. The damping ratio in bare wall cylindrical containers owes its origin to three sources. These are the free surface, sidewalls, and bottom. Generally the damping depends on the fluid physical properties and the tank geometry. Most of the analytical approaches are based on ideal assumptions, which are rarely met in practice. It is important to measure the damping factor, and semi-empirical expressions for different tank geometries have been reported in the literature, see equation (3.32c). The modal analysis and forced response of viscous fluids in upright circular containers has been presented in Section 3.3. As expected, the fluid viscosity and surface tension bring the fluid dynamic forces into bounded values near the resonance frequency.

In relatively large liquid containers, the liquid sloshing is best controlled by introducing baffles and partitions. Section 3.4 has been devoted to summarizing the main analytical and experimental results of typical baffles reported in the literature. Most of the work relied heavily on experimental results in lateral and pitching tank oscillations. Under roll excitations, Section 3.5 introduced the concept of periodic boundary layers. This has been demonstrated by the Stokes' second problem. The thickness of periodic boundary layers has been estimated for an upright circular cylindrical tank experiencing roll oscillations. The influence of fitted pipes on the effective mass moment of inertia and damping factor was estimated in terms of roll frequency and liquid viscosity.

This chapter did not discuss the numerical simulations of viscous flow in moving containers. Numerical algorithms have been applied to free flows with accurate resolution of viscous effects below significant surface deformations (see, e.g., Lught and Ohring, 1992, Dommermuth, 1993, and Ananthakrishnan and Yeung, 1994). The viscous free surface numerical simulation algorithms have been limited by some simplifications, which have been discussed and reviewed by Floryan and Rasmussen (1989) and Tsai and Yue (1996). Hodges and Street (1999) developed a numerical simulation of the unsteady, three-dimensional, viscous Navier–Stokes equations for turbulent nonlinear free-surface flows. They applied the algorithm to simulations of a laminar standing wave and turbulent open channel flow with a finite amplitude surface wave. The finite volume numerical approach is based on a free-surface/moving-grid, curvilinear coordinate system developed by Zang, *et al.* (1994). It employs second-order-accurate discretization in time and space of the primitive variable in curvilinear coordinates.

This chapter did not address the use of active control of liquid sloshing. Hara and Shibata (1986, 1987), Hara and Saito (1988), Hara (1992), and Qiu and Scherer (1994) developed active control of liquid sloshing using intermittent gas-bubble injection. Koide and Goto (1988) presented a feedback control system to control liquid sloshing in storage tanks. The literature includes other applications of sloshing suppression control such as pouring machines (see, e.g., Terashima and Yano, 2001 and Warnitchai and Pinkaew, 1998).

Part II

Nonlinear and parametric sloshing dynamics

4
Weakly nonlinear lateral sloshing

4.1 Introduction

The linear theory of liquid sloshing is adequate for determining the natural frequencies and wave height of the free surface. Under translational excitation, the linear theory is also useful for predicting the liquid hydrodynamic pressure, forces, and moments as long as the free surface maintains a planar shape with a nodal diameter that remains perpendicular to the line of excitation. However, it does not take into account the important vertical displacement of the center of gravity of the liquid for large amplitudes of free-surface motion. It also fails to predict complex surface phenomena observed experimentally near resonance. These phenomena include nonplanar unstable motion of the free surface associated with rotation of the nodal diameter (rotary sloshing) and chaotic sloshing. These phenomena can be uncovered using the theory of weakly nonlinear oscillations for quantitative analysis and the modern theory of nonlinear dynamics for stability analysis. The main sources of nonlinearity in the fluid field equations appear in the free-surface boundary conditions.

The early work of nonlinear lateral sloshing is based on asymptotic expansion techniques. For example, Penney and Price (1952) carried out a successive approximation approach where the potential function is expressed as a Fourier series in space with coefficients that are functions of time. These coefficients are approximated by Fourier time series using the method of perturbation. The resulting solution is expressed in terms of a double Fourier series in space and time. This method was applied by Skalak and Yarymovich (1962) and Dodge, *et al.* (1965). Later, Moiseev (1958) constructed normal mode functions and characteristic numbers by integral equations in terms of Green's function of the second kind (Neumann function). Chu (1968) generalized Moiseev's method by employing a perturbation technique using the characteristic functions to determine the subharmonic response to an oscillatory axial excitation. Faltinsen (1974, 1978), Solaas (1994), and Solaas and Faltinsen (1997) used Moiseev's theory and obtained steady-state solutions of the nonlinear sloshing in a two-dimensional rectangular tank. Hutton (1962) expanded the dynamic and kinematic free-surface equations in Taylor series about a stationary surface position. This method was modified and used in many sloshing problems by Rogge and Weiss (1965), Weiss and Rogge (1965), Abramson, Chu and Kana (1966), Woodward (1966), Woodward and Bauer (1970), and Ibrahim and Barr (1975a).

An overview of the early work of nonlinear effects in lateral sloshing was presented in Abramson (1966a, Chapter 3). In a series of papers (Abramson, Garza and Kana, 1962d, Abramson, Chu and Garza, 1962a, and Abramson and Garza, 1965) dealing with liquid sloshing in compartmented tanks, it was shown that the measured natural frequencies of the liquid free surface are dependent on the excitation amplitude. The liquid hydrodynamic force amplitudes also exhibit the well-known jump phenomena at each resonant peak. Under this condition, the free-surface shape is no longer planar and its upward displacement is greater than the downward displacement. Thus, for excitations near resonance, one should carry out nonlinear analysis to calculate the hydrodynamic forces.

Kimura and Ohashi (1978), Komatsu (1987), Kimura, *et al.* (1992, 1993, 1996a,b), and Liu and Huang (1994) studied the nonlinear liquid motion in an arbitrary axi-symmetric container, circular cylindrical containers, and rectangular tanks subjected to lateral and pitching excitations. Yin, *et al.* (1999) used a variational principle proposed by Luke (1967) to determine the potential function and free-surface elevation in a circular cylinder subjected to pitching excitation. They confirmed the earlier findings that the free-surface planar motion surface follows the characteristics of a soft nonlinear oscillator, while the rotary motion follows a hard nonlinear oscillator. Shankar and Kidambi (2002) employed the modal method to determine nonlinear sloshing amplitude. Machida, *et al.* (1998) analyzed nonlinear sloshing with inlet and outlet flow in a rectangular tank subjected to horizontal harmonic excitation. Bhuta, *et al.* (1964) examined the coupled oscillations of liquid-free-surface normal modes.

Further studies on lateral nonlinear sloshing including rotational and vortexing were conducted by Anosov (1966), Chester (1968), Chester and Bones (1968), Moody and Reynolds (1972), Chen (1977), Miles (1985b), Ockendon, *et al.* (1986, 1996), and Ockendon and Ockendon (2001). For a rectangular tank, the nonlinear liquid surface elevation was examined by Hayama, *et al.* (1983), Su (1987), Lepelletier and Raichlen (1988), and Armenio and Rocca (1996). Wang, *et al.* (1996) and Pawell (1997) experimentally studied surface waves in a water-filled circular basin excited by applying two shakers at opposite sides of the basin. The forcing frequency was adjusted near one of the natural frequencies of the free surface. Standing waves first appeared with a number of nodes around the tank wall. As the excitation amplitude increases, waves break, and the water in the tank rises high in the air. A weakly nonlinear analysis of this phenomenon based on a two-time scale expansion was developed by Sun, *et al.* (1995) where the forcing frequency was in the neighborhood of the natural sloshing frequency. Chang and Shen (2000) examined the chaotic motions and internal resonance of nonlinear surface waves generated by a harmonic excitation applied to the tank side. The excitation consisted of a component with a frequency near to twice the fundamental resonance frequency and another component close to the fundamental frequency. At certain critical tank radii, twice the fundamental frequency was found to be close to another mode frequency indicating the presence of internal resonance.

The nonlinear response was analyzed using the shallow-water wave theory by Shimizu and Hayama (1987), Ishibashi and Hayama (1989), and Gardarsson (1997). Chaotic surface waves due to resonance horizontal oscillation were reported by Funakoshi and Inoue (1987, 1988). The hysteresis phenomenon associated with the nonlinear free-surface motion near resonance was studied by Shemer and Chamesses (1990). The nonlinear liquid response spectra under horizontal and vertical excitations were determined by Yamada, *et al.* (1987). Large-amplitude motions of a liquid vapor interface in an accelerating container were studied by Moore and Perko (1965), and Perko (1969). Sudo and Hashimoto (1986), Sudo, *et al.* (1988), and Sudo and Ohaba (1998) experimentally measured the nonlinear response of the free liquid surface in a cylindrical tank under lateral excitation. They observed planar surface waves, swirl motion, and chaotic surface responses for different excitation amplitudes. Kobayashi, *et al.* (1989) determined the free liquid surface response in horizontal cylindrical tanks.

The nonlinear analysis of liquid sloshing in inverted conical (V-shape) containers has been considered by Lukovskii and Bilyk (1985), Bauer and Eidel (1988b), Lukovskii and Timokha (2000b), and Gavrilyuk, *et al.* (2000, 2001). Bauer and Eidel (1988b) expanded the nonlinear kinematic and dynamic free-surface conditions into Taylor series about the undisturbed free-surface level. Instead of determining the response of the fluid free surface under lateral excitation, Bauer and Eidel estimated the dependence of the free-surface natural frequency on the wave height amplitude, which establishes the backbone of the amplitude frequency response. They found that the liquid exhibits softening behavior. Gavrilyuk, *et al.* (2001) employed a nonconformal transformation technique originally developed by Lukovskii and Timokha (2000b). Eidel (1989) extended the analysis of Bauer and Eidel (1988b,c) to the case of prolate spheroidal containers. He found that the liquid exhibits both softening and hardening nonlinear characteristics, depending on the liquid height and sloshing mode.

Nonlinear free-surface motion causes nonlinear hydrodynamic forces acting on the tank walls. Wang and Hung (1986) and Chen (1994, 1997) estimated the nonlinear hydrodynamic pressures due to earthquake excitation. Matsuzaki and Fung (1977) and Chen, Haroun, and Liu (1996) estimated the unsteady fluid dynamic forces on storage liquid containers.

This chapter presents an overview of rotary sloshing and its analogy to the spherical (conical) pendulum and stretched strings. In view of its versatility, we will present the modified Hutton's theory together with some experimental results. The stability analysis of steady-state solutions, and transitions from regular to chaotic motion in the context of strange attractors are outlined based on the work of Miles (1984c,d) and Koval'chuk and Podchasov (1996). Experimental and analytical results of response statistics of a liquid free surface under random lateral excitation are also presented. Nonlinear phenomena of standing longitudinal waves in rectangular tanks are addressed near the critical liquid depth where the free surface switches from hard to soft spring characteristics. Cross-waves in rectangular tanks will not be addressed but the reader can consult the work reviewed in Section 4.5. Nonlinear sloshing of magnetic fluids is presented for rectangular tanks in Section 4.6. Sections 4.7 and 4.8 deal with the nonlinear sloshing dynamics in conical and prolate spheroidal containers, respectively. Section 4.9 addresses the main experimental results of self-induced sloshing in pool-type tanks of nuclear reactors.

4.2 Rotary sloshing

4.2.1 Experimental observations

Rotary sloshing is usually observed in both deep and shallow liquid tanks under lateral sinusoidal excitation. While deep liquid exhibits soft nonlinear spring characteristics, shallow liquid experiences hard behavior. The main experimental observations are summarized in the next two sections.

Rotary sloshing in deep liquid tanks

Under lateral harmonic excitation, the liquid free surface may exhibit two types of nonlinearities. The first is large-amplitude response, and the second involves different forms of liquid behavior produced by coupling or instabilities of various sloshing modes. The most important of these is the rotary sloshing or swirl motion. This type of motion usually occurs very near the lowest liquid natural frequency (see, e.g., Eulitz, 1957, Berlot, 1959, Graham, 1960, Ransleben and Abramson, 1960, Eulitz and Glaser, 1961, Abramson, *et al.*, 1962b, 1966). Rotary sloshing may occur in containers of circular and noncircular cross-sections, such as rectangular tanks, indicating that the container geometry does not have a significant effect on the occurrence of such motion.

Moiseev (1958) presented an analytical investigation and discussed qualitatively some sloshing tests. He used the nonlinear free-surface conditions and applied the method presented by Stoker (1957). In this method, the velocity potential and wave height were expanded in a power series in which each term of the series represents a potential function. This series is substituted into the nonlinear free-surface condition to give a set of differential equations that must be solved successively. Abramson (1961a) observed that this motion invariably occurs whether or not the liquid has any initial rotational motion. It can, however, be initiated at an earlier excitation frequency by introducing some additional rotational motion of the liquid, as could happen during the drainage of the container in the form of a vortex. Abramson (1961a) described the rotary sloshing by the following scenario: "at the onset of this motion, the first mode of sloshing began to transform itself from translational into rotational motion, increasing in angular velocity in one direction until it reached a maximum. Then the angular velocity decreased to essentially zero, reversed its direction, and increased in angular velocity in the new direction."

The rotary sloshing is analogous to the behavior of the forced spherical (or conical) pendulum (Berlot and Freed, 1956, Howell, 1957, Troesch, 1957, Freed, 1957, Berlot, 1959, Miles, 1962a, 1984c, Tritton, 1986a,b) and also to the resonant nonplanar motion of a stretched string (Miles, 1984f). The early study by Miles (1962a) of a spherical pendulum, whose point of suspension describes a horizontal harmonic displacement, indicated that the steady-state response of the pendulum might depart from the plane oscillations in the neighborhood of resonance. In particular, if the driving frequency is in a certain range, it is impossible to predict just how the pendulum will move. Consecutive experiments under identical conditions will produce different patterns of motion. The linearized differential equations for the transverse motions of a spherical pendulum are uncoupled, however the nonlinear equations are coupled. In view of this nonlinear coupling, energy may be transferred between the two degrees of freedom (say x and y). Retaining terms up to third-order and solving the resulting nonlinear differential equations for steady-state harmonic solutions in the neighborhood of resonance, the following regimes were reported:

(1) Stable simple harmonic planar motion over external detuning parameter $\nu < -0.945$ or $\nu > 0.757$, where $\nu = \varepsilon^{-2/3}((\Omega/\omega)^2 - 1)$, ω is the pendulum natural frequency, and Ω is the excitation frequency.
(2) Stable nonplanar harmonic motion for $\nu > 0.164$.
(3) Unstable harmonic motion over a finite frequency range $-0.945 < \nu < 0.154$. This range of instability decreases with increasing damping.

Hutton (1962, 1964), Gillard (1963), Weiss and Rogge (1965), and Rogge and Weiss (1965) reported three types of fluid motion in circular cylindrical tanks: stable planar, stable nonplanar, and unstable motion near resonance. Stable planar motion is a steady-state liquid motion with a constant peak wave height and a stationary single nodal diameter perpendicular to the direction of excitation. Stable nonplanar motion is a steady-state liquid motion with a constant peak wave height associated with a single nodal diameter that rotates at a constant rate around the tank vertical axis, this motion occurred primarily above the natural frequency of the free surface, and unstable motion that never attains a steady-state harmonic response. The liquid motion exhibits a softening restoring characteristic in the stable planar motion and a hardening effect in the nonplanar regime.

The problem was further investigated by Verhagen and WijnGaarden (1965) and Schilling and Siekmann (1982), who found that the peak wave height, nodal diameter rotation rate and its direction change continuously with time. Verhagen and Wijngarden (1965) observed that a hydraulic jump is formed in a rectangular tank subjected to lateral excitation. The jump travels periodically back and forth between the walls of the container. This hydraulic jump is a nonlinear phenomenon, analogous to the shock wave appearing in one-dimensional gas flow under similar resonance conditions. Later, Miles (1984d) introduced three control parameters, representing damping, external detuning, and fluid depth ratio. He examined the stability of the fixed solutions as the control parameters vary and found four possible regimes. These regimes are (i) planar harmonic motion, (ii) nonplanar harmonic motion, (iii) periodically modulated sinusoid (limit cycle), and (iv) a chaotically modulated sinusoid. Each regime can take place over certain ranges of the control parameters.

Semenov (1975) examined the flow of a liquid from an oscillating cylindrical tank with a curvilinear bottom. Zhou and Graebel (1990) employed computational fluid mechanics techniques to examine the axisymmetric draining of a cylindrical tank. Su and Lian (1992) observed various types of instability associated with the development of vortex formation during draining from a tank bottom. Vortex formation during draining may become unstable to the extent that it forms a reverse flow in its core, yet the vorticity appeared to be uniform as opposed to chaotically diffuse during the "break-down."

Waterhouse (1994) discussed a critical liquid depth below which third-order analysis fails. He provided a detailed asymptotic analysis of the free liquid surface response near the critical depth using a fifth-order theory based on Moiseev's theory. At the critical depth the response switches from being a "hard-spring" to a "soft-spring." It was shown that the branches in the amplitude-frequency plane coincide with a third-order analysis only for small amplitude. New turning points on the branches occur at a critical value of the amplitude–frequency response.

Rotary sloshing in shallow liquid tanks

For shallow tanks, with low liquid depth, all the liquid takes part in the dynamic response to lateral excitation. There is a critical liquid depth where the free-surface exhibits the "hard–soft" spring characteristics of a rotating wave. Such critical depth in a circular container was estimated by DiMaggio and Rehm (1965) to be $h/R = 0.51$, and by Narimanov, *et al.* (1977) as 0.492.

Kana (1987) reported some experimental observations of the liquid free-surface behavior in a Centaur G-tank with conical walls of base diameter 24.74 inch and liquid depth 2.99 inch. In the vicinity of the first anti-symmetric slosh mode (about 0.29 Hz), the tank was excited below this frequency with a constant amplitude. It was found that pure normal slosh never occurred, but the response appeared to be normal slosh with a distinct clockwise rotary wave motion superimposed on it. Furthermore, the normal component of the slosh wave displacement was nearly in-phase with excitation when the excitation frequency was well below resonance, but the rotary wave motion lagged the excitation by approximately 180°. In other words, the rotary wave up-displacement appeared to occur spatially out-of-phase with the designated positive direction for excitation displacement. At the same time, the rotary component did not exhibit a rotating planar surface but appeared to be formed by a concentrated mound of fluid in the form of a hydraulic jump. As the excitation frequency was slowly increased, both normal and rotary slosh combined to form a complex resultant whose phase shifted with frequency. Eventually, the rotary slosh component became dominant, and between about 0.29 Hz and 0.31 Hz (i.e., slightly above the normal slosh resonance), the up wave lagged the excitation displacement by about 270°. A further slow increase of frequency resulted in pure normal slosh with the disappearance of the rotary wave.

If the excitation frequency 0.305 Hz was approached from below, then large-amplitude rotary response was observed. However, if the frequency was set at 0.305 Hz and the excitation amplitude was increased up to 0.2 inch slowly, the motion of the free surface was observed to be nearly normal slosh with a very low amplitude. During this state if a manual perturbation of a clockwise rotation was applied then the free surface was observed to experience stable large-amplitude rotation. If the perturbation was given in the counterclockwise sense, it eventually subsided. Kana (1987) concluded that two different stable states of clockwise rotary response were possible at this frequency, and the counterclockwise rotation was found to be unstable everywhere. This observation motivated Kana to develop a compound pendulum consisting of a spherical pendulum and a simple pendulum to encompass different sloshing regimes. His model will be described in Chapter 5.

Kana (1989) conducted another set of experimental tests on the same tank but with a liquid depth of 6.8 inch. Measurements were taken for the liquid-weight transfer functions in-line and cross-axis to the excitation. The reason for changing the liquid depth was to provide optimum response measurements. Similar observations to those described earlier were reported. From a frequency below resonance to somewhat above the resonance, a very large counterclockwise rotary response occurred at the same excitation frequency. The clockwise phase lag of this motion relative to the excitation axis varied from in-line to nearly 90° at the maximum response amplitudes, and large cross-axis weights were measured. Although the rotary motion always developed in a counterclockwise direction when the frequencies were approached from below, a sustained clockwise motion of the same response amplitude could be developed by manual perturbations. The phase of this motion relative to the excitation axis varied to nearly −90° for the large amplitudes.

Reaction force and excitation–acceleration measurements were obtained for steady-state excitation of 0.2 inch peak amplitude at various frequencies. Figure 4.1 shows the dependence of the cross-axis weight whose coincident (CO) and quadrature (QUAD) parts provide phase relative to the excitation displacement on the excitation frequency ratio, Ω/ω_{11}. Figure 4.2 shows corresponding results for the in-line weight. The measurements reveal a quite complex type of response, although the combined system displays a maximum in the imaginary part and nearly zero in the real part at the same frequency (the linear natural frequency is 0.447 Hz).

4.2.2 *Analysis of rotary sloshing*

Asymptotic expansion equations

As mentioned in the introduction of this chapter, several analytical studies have treated the problem of rotating sloshing. This section is adopted from Hutton (1962). For an upright circular cylindrical container partially filled with incompressible fluid, the fluid field equations are the same as those developed in Chapter 1. However, the main sources of nonlinearity in the

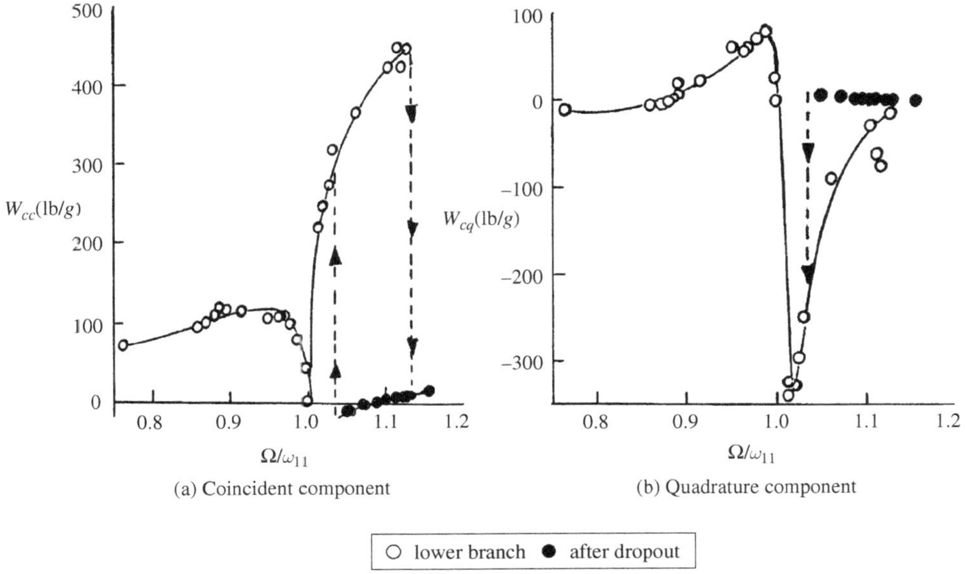

Figure 4.1 Dependence of measured cross-axis load components on excitation frequency ratio. (Kana, 1989)

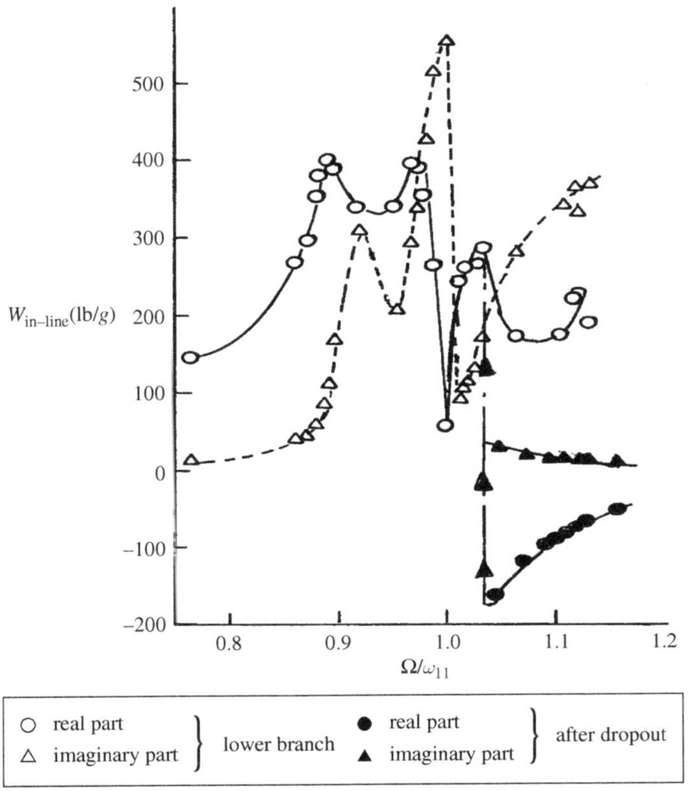

Figure 4.2 Dependence of measured effective liquid in-line load on excitation frequency ratio. (Kana, 1989)

fluid field equations are the free-surface boundary conditions. Under lateral excitation, $x(t)$, the dynamic and kinematic free surface boundary conditions, at $z=\eta$, may be written in the form, respectively,

$$-\frac{1}{2}\left(\tilde{\Phi}_r^2+\frac{1}{r^2}\tilde{\Phi}_\theta^2+\tilde{\Phi}_z^2\right)-g\eta+\tilde{\Phi}_t-\ddot{x}(t)r\cos\theta=0|_{z=\eta} \tag{4.1}$$

$$\eta_t-\eta_r\tilde{\Phi}_r-\frac{1}{r^2}\eta_\theta\tilde{\Phi}_\theta+\tilde{\Phi}_z=0|_{z=\eta} \tag{4.2}$$

where a subscript denotes differentiation with respect to the subscripted variable and a dot denotes differentiation with respect to time.

Expanding equations (4.1) and (4.2), about $z=\eta=0$, in Taylor series, gives

$$\begin{aligned}&\left\{-\frac{1}{2}\left(\tilde{\Phi}_r^2+\frac{1}{r^2}\tilde{\Phi}_\theta^2+\tilde{\Phi}_z^2\right)+\tilde{\Phi}_t-\ddot{x}(t)r\cos\theta\right\}_{\eta=0}\\&+\left\{\tilde{\Phi}_{tz}-\tilde{\Phi}_r\tilde{\Phi}_{rz}-\frac{1}{r^2}\tilde{\Phi}_\theta\tilde{\Phi}r_{\theta z}-\tilde{\Phi}_z\tilde{\Phi}_{zz}-g\right\}_{\eta=0}\eta\\&+\left\{\tilde{\Phi}_{tzz}-\tilde{\Phi}_{rz}^2-\tilde{\Phi}_r\tilde{\Phi}_{rzz}-\frac{1}{r^2}\tilde{\Phi}_{\theta z}^2-\frac{1}{r^2}\tilde{\Phi}_\theta\tilde{\Phi}r_{\theta zz}-\tilde{\Phi}_{zz}^2-\tilde{\Phi}_z\tilde{\Phi}_{zzz}\right\}_{\eta=0}\\&\times\frac{\eta^2}{2}+\cdots=0\end{aligned} \tag{4.3}$$

$$\begin{aligned}&\left\{\eta_t-\eta_r\tilde{\Phi}_r-\frac{1}{r^2}\eta_\theta\tilde{\Phi}_\theta+\tilde{\Phi}_z\right\}_{\eta=0}+\left\{-\eta_r\tilde{\Phi}_{rz}-\frac{1}{r^2}\eta_\theta\tilde{\Phi}_{\theta z}+\tilde{\Phi}_{zz}\right\}_{\eta=0}\eta\\&+\left\{-\eta_r\tilde{\Phi}_{rzz}-\frac{1}{r^2}\eta_\theta\tilde{\Phi}_{\theta zz}+\tilde{\Phi}_{zzz}\right\}_{\eta=0}\frac{\eta^2}{2}+\cdots=0\end{aligned} \tag{4.4}$$

Let the velocity potential function and wave height be expressed in terms of Fourier series in space with coefficients that are functions of time

$$\hat{\Phi}(r,\theta,z,t)=\sum_{m=0}^{\infty}\sum_{n=1}^{\infty}\alpha_{mn}(t)J_m(\lambda_{mn}r)\cos m\theta\frac{\cosh[\lambda_{mn}(z+h)]}{\cosh\lambda_{mn}h} \tag{4.5}$$

$$\eta(r,\theta,t)=\sum_{m=0}^{\infty}\sum_{n=1}^{\infty}a_{mn}(t)J_m(\lambda_{mn}r)\cos m\theta \tag{4.6}$$

where $\alpha_{mn}(t)$ and $a_{mn}(t)$ are the generalized coordinates and depend on time.

Substituting equations (4.5) and (4.6) into equations (4.3) and (4.4) and keeping terms up to third order, gives

$$
\begin{aligned}
&- r\ddot{x}(t)\cos\theta + (\dot{\alpha}_{mn} - g a_{mn})\cos(m\theta)J_m(\lambda_{mn}r) \\
&- \frac{1}{2}\{\alpha_{mn}\alpha_{pq}\cos(m\theta)\cos(p\theta)\lambda_{mn}\lambda_{pq}J_m(\lambda_{mn}r)J_p(\lambda_{pq}r) \\
&\quad + \frac{mp}{r^2}\alpha_{mn}\alpha_{pq}\sin(m\theta)\sin(p\theta)J_m(\lambda_{mn}r)J_p(\lambda_{pq}r) \\
&\quad + \alpha_{mn}\alpha_{pq}\cos(m\theta)\cos(p\theta)\lambda_{mn}\lambda_{pq}J_m(\lambda_{mn}r)J_p(\lambda_{pq}r)\tanh(\lambda_{mn}h)\tanh(\lambda_{pq}h) \\
&\quad - 2\dot{\alpha}_{mn}a_{pq}\cos(m\theta)\cos(p\theta)\lambda_{mn}J_m(\lambda_{mn}r)J_p(\lambda_{pq}r)\tanh(\lambda_{mn}h)\} \\
&- [\alpha_{mn}\alpha_{pq}a_{kl}\cos(m\theta)\cos(p\theta)\cos(k\theta)\lambda_{mn}\lambda_{pq}^2J_m(\lambda_{mn}r)J_p(\lambda_{pq}r)J_k(\lambda_{kl}r)\tanh(\lambda_{pq}h) \\
&+ \frac{mp}{r^2}\alpha_{mn}\alpha_{pq}a_{kl}\sin(m\theta)\sin(p\theta)\cos(k\theta)\lambda_{pq}J_m(\lambda_{mn}r)J_p(\lambda_{pq}r)J_k(\lambda_{kl}r)\tanh(\lambda_{pq}h) \\
&+ \alpha_{mn}\alpha_{pq}a_{kl}\cos(m\theta)\cos(p\theta)\cos(k\theta)\lambda_{mn}\lambda_{pq}^2J_m(\lambda_{mn}r)J_p(\lambda_{pq}r)J_k(\lambda_{kl}r)\tanh(\lambda_{mn}h) \\
&- \frac{1}{2}\dot{\alpha}_{mn}a_{pq}a_{kl}\cos(m\theta)\cos(p\theta)\cos(k\theta)\lambda_{mn}^2J_m(\lambda_{mn}r)J_p(\lambda_{pq}r)J_k(\lambda_{kl}r)] + \text{H.O.T.}
\end{aligned} \tag{4.7}
$$

with ($m, n, p, q, k, l = 0, 1, 2, \ldots$), where H.O.T. means higher order terms

$$
\begin{aligned}
&(\dot{a}_{mn} + \lambda_{mn}\tanh(\lambda_{mn})\alpha_{mn})\cos(m\theta)J_m(\lambda_{mn}r) \\
&- \Bigg\{\alpha_{mn}a_{pq}\cos(m\theta)\cos(p\theta)\lambda_{mn}\lambda_{pq}J'_m(\lambda_{mn}r)J'_p(\lambda_{pq}r) \\
&\quad - \alpha_{mn}a_{pq}J_m(\lambda_{mn}r)J_p(\lambda_{pq}r)\left(\frac{1}{r^2}mp\sin(m\theta)\sin(p\theta) + \lambda_{mn}^2\cos(m\theta)\cos(p\theta)\right)\Bigg\} \\
&- [\alpha_{mn}a_{pq}a_{kl}\cos(m\theta)\lambda_{mn}^2\lambda_{pq}\cos(m\theta)\cos(p\theta)\cos(k\theta) \\
&\times J'_m(\lambda_{mn}r)J'_p(\lambda_{pq}r)J_k(\lambda_{kl}r)\tanh(\lambda_{mn}h) \\
&+ \frac{mp}{r^2}\alpha_{mn}a_{pq}a_{kl}\lambda_{mn}\sin(m\theta)\sin(p\theta)\cos(k\theta) \\
&\times J_m(\lambda_{mn}r)J_p(\lambda_{pq}r)J_k(\lambda_{kl}r)\tanh(\lambda_{mn}h) \\
&- \frac{1}{2}\alpha_{mn}a_{pq}a_{kl}\lambda_{mn}^3\cos(m\theta)\cos(p\theta)\cos(k\theta) \\
&\times J_m(\lambda_{mn}r)J_p(\lambda_{pq}r)J_k(\lambda_{kl}r)\tanh(\lambda_{mn}h)] + \text{H.O.T.}
\end{aligned} \tag{4.8}
$$

where repeated indices indicate summation.

The radial component r in the first expression in equation (4.7) could be expanded in Dini-series as

$$r = F_n J_1(\lambda_{1n}r) \tag{4.9}$$

Pre-multiplying both sides of series (4.9) by $rJ_1\lambda_{1_n}r$ and integrating from $r=0$ to $r=R$, gives the value of F_n in the form

$$F_n = \frac{2R}{(\lambda_{1n}^2R^2 - 1)J_1(\lambda_{1n}R)} \tag{4.10}$$

For $n=1$, $F_1 = 1.4386R$.

Equations (4.7) and (4.8) can be rewritten in terms of cos $(m\theta)\, J_m(\lambda_{mn}r)$ as

$$-\ddot{x}(t)\left(\frac{2R}{(\lambda_{1n}^2R^2-1)J_1(\lambda_{1n}R)}\right)\cos\,\theta J_1(\lambda_{1n}r)+(\dot{\alpha}_{mn}-ga_{mn})\cos(m\theta)J_m(\lambda_{mn}r)$$
$$+F(r,\theta,\alpha_{mn},\dot{\alpha}_{mn},a_{mn})=0 \tag{4.11}$$

where F represents all other nonlinear terms in equation (4.7). Consider α_{mn}, a_{mn} and their time derivatives in F as parameters and expand F in a Fourier–Bessel series

$$F=f_{mn}(r,\theta,\alpha_{mn},\dot{\alpha}_{mn},a_{mn})\cos(m\theta)J_m(\lambda_{mn}r) \tag{4.12}$$

where

$$f_{mn}=\frac{\int_0^R\int_0^{2\pi}Fr\cos(m\theta)J_m(\lambda_{mn}r)\,d\theta\,dr}{\int_0^R\int_0^{2\pi}\cos^2(m\theta)J_m^2\lambda_{mn}r)\,d\theta\,dr} \tag{4.13}$$

Substituting equation (4.12) into equation (4.11) gives

$$-\ddot{x}(t)\left(\frac{2R}{(\lambda_{1n}^2R^2-1)J_1(\lambda_{1n}R)}\right)\cos\theta\,J_1(\lambda_{1n}r)$$
$$+(\dot{\alpha}_{mn}-ga_{mn}+f_{mn})\cos(m\theta)J_m(\lambda_{mn}r)=0 \tag{4.14}$$

Similarly, equation (4.8) can be written in the form

$$(\dot{a}_{mn}+\lambda_{mn}\tanh(\lambda_{mn})\alpha_{mn})\cos(m\theta)J_m(\lambda_{mn}r)+H(r,\theta,\alpha_{mn},a_{mn},\dot{a}_{mn})=0 \tag{4.15}$$

Expanding the function H in the Fourier–Bessel series

$$H=h_{mn}(r,\theta,\alpha_{mn},a_{mn},\dot{a}_{mn})\cos(m\theta)J_m(\lambda_{mn}r) \tag{4.16}$$

where

$$h_{mn}=\frac{\int_0^R\int_0^{2\pi}Hr\cos(m\theta)J_m(\lambda_{mn}r)\,d\theta\,dr}{\int_0^R\int_0^{2\pi}\cos^2(m\theta)J_m^2(\lambda_{mn}r)\,d\theta\,dr} \tag{4.17}$$

and introducing expansion (4.16) into equation (4.14), gives

$$(\dot{a}_{mn}+\lambda_{mn}\tanh(\lambda_{mn})\alpha_{mn}+h_{mn})\cos(m\theta)J_m(\lambda_{mn}r)=0 \tag{4.18}$$

Note that the functions cos $(m\theta)\, J_m\,(\lambda_{mn}\, r)$ in equations (4.14) and (4.18) form an orthogonal set such that their coefficients for every mode (m, n) constitute an equation in the generalized coordinates α_{mn} and a_{mn}:

$$-\ddot{x}(t)F_n + \dot{\alpha}_{1n} - ga_{1n} + f_{1n}(r, \theta, \alpha_{1n}, \dot{\alpha}_{1n}, a_{1n}) = 0, n = 1, 2, \ldots \quad (4.19a)$$

$$\dot{\alpha}_{mn} - ga_{mn} + f_{mn}(r, \theta, \alpha_{mn}, \dot{\alpha}_{mn}, a_{mn}) = 0 \quad (4.19b)$$

$$\dot{a}_{mn} + \lambda_{mn} \tanh(\lambda_{mn})\alpha_{mn} + h_{mn} = 0 \quad (4.19c)$$

The coupling of modes appears if nonlinear terms in α_{mn} and a_{mn} are considered. Considerable simplifications are possible in evaluating the integrals (4.13) and (4.17) by using orthogonality and recurrence relations listed in the appendix to this chapter. Equations (4.14) and (4.18) are still unbounded due to the presence of an infinite number of modes, and consequently constitute an infinite number of equations. The order of magnitude of α_{mn} and a_{mn} has to be known in order to decide the appropriate number of significant modes. This will limit the problem to a finite number of equations corresponding to the number of significant modes.

In their studies of free-standing waves, Penney and Price (1952) assumed that the forced waves are very nearly free waves. Skalak and Yarimovich (1962) showed the validity of that assumption and found that large-amplitude surface waves can be maintained by small values of excitation amplitude. They also developed an analysis of the orders of magnitude of α_{mn} and a_{mn} for forced waves. Dodge, Kana and Abramson (1965) found that the (0, 1), and (2, 1) modes appeared as the secondary modes when the primary was the first anti-symmetricmode (1, 1). The mode (0, 2) appeared as the secondary mode when the primary was the first symmetric mode (0, 1), and so on for other modes. These relations can be summarized as follows:

(1) Taking the first anti-symmetric mode as the predominant, and letting $\bar{\eta} = \eta/R$ be the nondimensional amplitude of the free surface, we write:

$$\begin{aligned} &a_{11} = O(\bar{\eta}), \quad \alpha_{11} = O(\bar{\eta}), \quad a_{01} = O(\bar{\eta}^2), \quad \alpha_{01} = O(\bar{\eta}^2) \\ &a_{21} = O(\bar{\eta}^2), \quad \alpha_{21} = O(\bar{\eta}^2), \quad a_{mn} = O(\bar{\eta}^m), \quad \alpha_{mn} = O(\bar{\eta}^m) \end{aligned} \quad (4.20a)$$

(2) Taking the first symmetric mode as the primary:

$$\begin{aligned} a_{01} &= O(\bar{\eta}), \qquad \alpha_{01} = O(\bar{\eta}) \\ a_{02} &= O(\bar{\eta}^2), \qquad \alpha_{02} = O(\bar{\eta}^2) \\ a_{0n} &= O(\bar{\eta}^n), \qquad \alpha_{0mn} = O(\bar{\eta}^n) \end{aligned} \quad (4.20b)$$

By induction, the order of magnitude of other higher modes can be generated. Based on this scheme, one can generate any set of coupled equations for a finite number of modes.

Response and stability analyses

Since the only unknown in equations (4.1) and (4.2) is the free-surface wave height, Hutton (1962) combined them into one equation. First, equation (4.1) is written in the form

$$g\eta = \Gamma(r, \theta, \eta, t) = -\frac{1}{2}\left(\tilde{\Phi}_r^2 + \frac{1}{r^2}\tilde{\Phi}_\theta^2 + \tilde{\Phi}_z^2\right) + \tilde{\Phi}_t^2 - \ddot{x}(t) r \cos\theta \quad (4.21)$$

Taking the partial derivatives of the wave height with respect to time and the polar coordinates, (r, θ), gives

$$g\eta_t = \Gamma_t + \Gamma_\eta \eta_t \rightarrow \eta_t(g - \Gamma_\eta) = \Gamma_t$$

$$\Gamma_t = -\tilde{\Phi}_r\tilde{\Phi}_{rt} - \frac{1}{r^2}\tilde{\Phi}_\theta\tilde{\Phi}_{\theta t} - \tilde{\Phi}_z\tilde{\Phi}_{zt} + \tilde{\Phi}_{tt} - \ddot{x}r\cos\theta \tag{4.22a}$$

$$g\eta_r = \Gamma_r + \Gamma_\eta \eta_r \rightarrow \frac{\partial\eta}{\partial r}(g - \Gamma_\eta) = \Gamma_r$$

$$\Gamma_r = -\tilde{\Phi}_r\tilde{\Phi}_{rr} + \frac{1}{r^3}\tilde{\Phi}_\theta^2 - \frac{1}{r^2}\tilde{\Phi}_\theta\tilde{\Phi}_{\theta r} - \tilde{\Phi}_z\tilde{\Phi}_{zr} + \tilde{\Phi}_{zt} - \ddot{x}\cos\theta \tag{4.22b}$$

$$g\eta_\theta = \Gamma_\theta + \Gamma_\eta \eta_\theta \rightarrow \eta_\theta(g - \Gamma_\eta) = \Gamma_\theta$$

$$\Gamma_\theta = -\tilde{\Phi}_r\tilde{\Phi}_{r\theta} - \frac{1}{r^2}\tilde{\Phi}_\theta\tilde{\Phi}_{\theta\theta} - \tilde{\Phi}_z\tilde{\Phi}_{\theta z} + \tilde{\Phi}_{\theta t} + \ddot{x}r\sin\theta \tag{4.22c}$$

where $\Gamma_\eta = \tilde{\phi}_{zt} - \tilde{\phi}_r\tilde{\phi}_{rz} - (1/r^2)\tilde{\phi}_\theta\tilde{\phi}_{\theta z} - \tilde{\phi}_z\tilde{\phi}_{zz}$.

Multiplying both sides of equation (4.2) by $(g - \Gamma_\eta)$ and substituting the resulting terms from relations (4.22), equation (4.2) takes the form

$$\begin{aligned}
&\tilde{\Phi}_{tt} + g\tilde{\Phi}_z - 2\tilde{\Phi}_r\tilde{\Phi}_{rt} - \frac{2}{r^2}\tilde{\Phi}_\theta\tilde{\Phi}_{\theta t} - 2\tilde{\Phi}_z\tilde{\Phi}_{zt} + \tilde{\Phi}_r^2\tilde{\Phi}_{rr} + \tilde{\Phi}_z^2\tilde{\Phi}_{zz} + \frac{1}{r^4}\tilde{\Phi}_\theta^2\tilde{\Phi}_{\theta\theta} \\
&- \frac{1}{r^3}\tilde{\Phi}_r\tilde{\Phi}_\theta^2 + \frac{2}{r^2}\tilde{\Phi}_r\tilde{\Phi}_\theta\tilde{\Phi}_{r\theta} + \frac{2}{r^2}\tilde{\Phi}_z\tilde{\Phi}_\theta\tilde{\Phi}_{z\theta} + 2\tilde{\Phi}_r\tilde{\Phi}_z\tilde{\Phi}_{rz} \\
&- \ddot{x}r\cos\theta + \ddot{x}\left(\tilde{\Phi}_r\cos\theta - \frac{1}{r}\tilde{\Phi}_\theta\sin\theta\right) = F(r,\theta,z=0,t) = 0
\end{aligned} \tag{4.23}$$

The complete statement of the boundary-value problem consists of equations (4.3), (4.4), $\nabla^2\tilde{\Phi} = 0$, and the boundary conditions at the tank walls and bottom, $\partial\tilde{\Phi}/\partial n = 0$, where n denotes normal to the solid boundary. A possible solution for the velocity potential function must satisfy the boundary conditions. In view of the nonlinear nature of the free-surface boundary conditions (4.3) and (4.4), Hutton (1962) assumed the following solution that exhibits nonlinear coupling among different sloshing modes

$$\begin{aligned}
\tilde{\Phi} &= \varepsilon^{1/3}[\Psi_1(r,\theta,z,\tau)\cos\Omega t + X_1(r,\theta,z,\tau)\sin\Omega t] \\
&+ \varepsilon^{2/3}[\Psi_0(r,\theta,z) + \Psi_2(r,\theta,z)\cos 2\Omega t] \\
&+ \varepsilon[\Psi_3(r,\theta,z)\cos 3\Omega t + X_3(r,\theta,z)\sin 3\Omega t]
\end{aligned} \tag{4.24}$$

where the excitation is taken to be sinusoidal, that is, $x(t) = \varepsilon_0\sin\Omega t$, and $\dot{x}(t) = \varepsilon_0\Omega\cos\Omega t = \varepsilon\cos\Omega t$, with $\varepsilon = \varepsilon_0\,\Omega$, and $\tau = \varepsilon^{2/3}(\Omega t)/2$. Note that the functions Ψ_n and X_n satisfy Laplace's equation. These functions must satisfy the boundary conditions at the tank walls and bottom, that is, the normal liquid velocity at these solid boundaries vanishes. The time varying generalized coordinates associated with these functions are selected to approximately satisfy the free-surface condition (4.3). This is done by introducing equation (4.24) into (4.3) and equating coefficients of $\varepsilon^{1/3}$, $\varepsilon^{2/3}$, and ε to zero. The coefficients of $\varepsilon^{1/3}$ will contain $\sin\Omega t$

and $\cos\Omega t$, the coefficients of $\varepsilon^{2/3}$ are associated with $\sin 2\Omega t$ and $\cos 2\Omega t$ in addition to terms independent of Ωt, while those of ε contain $\sin\Omega t$ and $\cos\Omega t$ plus higher harmonic terms. Introducing the following transformation

$$\tau = \varepsilon^{2/3}(\Omega t)/2 \text{ and } p_{11}^2 = \Omega^2(1 - \nu\varepsilon^{2/3}) \tag{4.25}$$

where ν is a detuning parameter. Substituting equation (4.24) into equation (4.3) and setting the coefficient $\varepsilon^{1/3}$ to zero, gives

$$\left(g\Psi_{1z} - p_{11}^2\Psi_1\right)\cos\Omega t + \left(gX_{1z} - p_{11}^2 X_1\right)\sin\Omega t = 0 \quad \text{on} \quad z = 0 \tag{4.26}$$

This equation is satisfied if

$$\Psi_{1z} = \frac{p_{11}^2}{g}\Psi_1, \quad \text{and} \quad X_{1z} = \frac{p_{11}^2}{g}X_1 \tag{4.27a,b}$$

where

$$\Psi_1 = \{f_1(\tau)\cos\theta + f_2(\tau)\sin\theta\}J_1(\lambda_{11}r)\frac{\cosh(\lambda_{11}(z+h))}{\cosh(\lambda_{11}h)} \tag{4.28a}$$

$$X_1 = \{f_2(\tau)\cos\theta + f_4(\tau)\sin\theta\}J_1(\lambda_{11}r)\frac{\cosh(\lambda_{11}(z+h))}{\cosh(\lambda_{11}h)} \tag{4.28b}$$

and that $p_{11}^2 = g\lambda_{11}\tanh(\lambda_{11}h)$.

Setting the coefficients of $\varepsilon^{2/3}$ to zero, results in second harmonic terms and constant terms. Setting these terms to zero, and manipulating relations (4.27a,b) gives

$$X_1\Psi_{1z} = X_{1z}\Psi_1 \quad \text{and} \quad X_1\Psi_{1zz} = X_{1zz}\Psi_1 \tag{4.29a,b}$$

one obtains the following three equations

$$\Psi_{0z} = 0 \tag{4.30a}$$

$$4p_{11}^2\Psi_2 - g\Psi_{2z} = 2p_{11}\left\{X_{1r}\Psi_{1r} + \frac{1}{r^2}X_{1\theta}\Psi_{1\theta} + \frac{1}{2}\left[3\tanh^2(\lambda_{11}h) - 1\right]\lambda_{11}^2 X_1\Psi_1\right\} \tag{4.30b}$$

$$4p_{11}^2 X_2 - gX_{2z} = p_{11}\left\{X_{1r}^2 - \Psi_{1r}^2 + \frac{1}{r^2}\left(X_{1\theta}^2 - \Psi_{1\theta}^2\right) + \frac{1}{2}\left[3\tanh^2(\lambda_{11}h) - 1\right]\lambda_{11}^2\left(X_1^2 - \Psi_1^2\right)\right\} \tag{4.30c}$$

Equation (4.30a) yields a constant value for Ψ_0, and the other two functions Ψ_2 and x_2 can be selected in the form

$$\Psi_2 = \sum_{n=1}^{\infty} \hat{A}_{0n} J_0(\lambda_{0n} r) \frac{\cosh[\lambda_{0n}(z+h)]}{\cosh(\lambda_{0n} h)} + \sum_{n=1}^{\infty} \left[\hat{A}_{2n} \cos 2\theta + \hat{B}_{2n} \sin 2\theta\right] J_2(\lambda_{2n} r) \frac{\cosh[\lambda_{0n}(z+h)]}{\cosh(\lambda_{0n} h)} \tag{4.31a}$$

$$X_2 = \sum_{n=1}^{\infty} \hat{C}_{0n} J_0(\lambda_{0n} r) \frac{\cosh[\lambda_{0n}(z+h)]}{\cosh(\lambda_{0n} h)} + \sum_{n=1}^{\infty} \left[\hat{C}_{2n} \cos 2\theta + \hat{D}_{2n} \sin 2\theta\right] J_2(\lambda_{2n} r) \frac{\cosh[\lambda_{0n}(z+h)]}{\cosh(\lambda_{0n} h)} \tag{4.31b}$$

where $\left.\dfrac{dJ_0(\lambda_{0n} r)}{dr}\right|_{r=R} = 0$, $\left.\dfrac{dJ_2(\lambda_{0n} r)}{dr}\right|_{r=R} = 0$. The generalized coordinates in equations (4.31) can be determined by substituting (4.28a, b) and (4.31a, b) into equations (4.30b, c). This process requires the application of Fourier–Bessel techniques in which the following relations are used:

$$\int_0^R r J_0(\lambda_{0m} r) J_0(\lambda_{0n} r)\, dr = \begin{Bmatrix} 0 & \text{for} \quad m \neq n \\ \frac{1}{2} R^2 J_0^2(\lambda_{0n} R) & \text{for} \quad m = n \end{Bmatrix} \tag{4.32a}$$

$$\int_0^R r J_2(\lambda_{2m} r) J_2(\lambda_{2n} r)\, dr = \begin{Bmatrix} 0 & \text{for} \quad m \neq n \\ \dfrac{\lambda_{2n}^2 R^2 - 4}{2\lambda_{2n}^2} J_{2n}^2(\lambda_{2n} R) & \text{for} \quad m = n \end{Bmatrix} \tag{4.32b}$$

Introducing the potential function equation (4.24) into the free-surface condition (4.3), yields two ordinary differential equations. Carrying out the Rayleigh–Ritz procedure by first multiplying by $J_1(\lambda_{11} r)\cos\theta\, r\, dr\, d\theta$, and integrating over $0 \leq r < R$, and $0 \leq \theta \leq 2\pi$, then multiplying by $J_1(\lambda_{11} r)\sin\theta\, r\, dr\, d\theta$, and integrating over the same region, and setting the first harmonic terms in Ω to zero in both equations, yields the following four first-order nonlinear differential equations

$$\frac{df_1}{d\tau} = -\nu f_2 - K_1 f_2 \left(f_1^2 + f_2^2 + f_3^2 + f_4^2\right) + K_2 f_3 (f_2 f_3 - f_1 f_4) \tag{4.33a}$$

$$\frac{df_2}{d\tau} = F_1 + \nu f_1 + K_1 f_1 \left(f_1^2 + f_2^2 + f_3^2 + f_4^2\right) + K_2 f_4 (f_2 f_3 - f_1 f_4) \tag{4.33b}$$

$$\frac{df_3}{d\tau} = -\nu f_4 - K_1 f_4 \left(f_1^2 + f_2^2 + f_3^2 + f_4^2\right) - K_2 f_1 (f_2 f_3 - f_1 f_4) \tag{4.33c}$$

$$\frac{df_4}{d\tau} = \nu f_3 + K_1 f_3 \left(f_1^2 + f_2^2 + f_3^2 + f_4^2\right) - K_2 f_2 (f_2 f_3 - f_1 f_4) \tag{4.33d}$$

where K_1, K_2, and F_1 are constants given in Hutton (1962). The steady-state solutions of equations (4.33) are obtained by setting the left-hand sides to zero. One solution yields the planar motion of the free surface with a constant peak wave height and a single, stationary nodal diameter perpendicular to the direction of excitation. This motion is given by the solution

$$f_1 = \gamma, \quad \text{and} \quad f_2 = f_3 = f_4 = 0 \tag{4.34}$$

where the amplitude γ is given by the solution of the cubic algebraic equation

$$K_1\gamma^3 + \nu\gamma + F_1 = 0 \tag{4.35}$$

The stability of the steady-state harmonic solution f_i^0 is determined by imposing a small perturbation to the solution and examining the subsequent motion, that is,

$$f_i(\tau) = f_i^0 + C_i\, e^{\lambda\tau} \tag{4.36}$$

where C_i are constants and λ is the eigenvalue that determines the stability of the fixed solution. Introducing equation (4.36) into equations (4.33) and neglecting higher-order terms of C_i yields a set of homogeneous equations in C_i whose nontrivial solutions exist only if the determinant of the coefficients is zero. Setting the determinant to zero and solving for the eigenvalues of λ one can examine the stability from the roots of λ. If all the real parts of λ are negative, the solution is stable, otherwise it is unstable. The boundary between stable and unstable planar motion corresponds to $\lambda_i = 0$. The roots will vanish for the following parameter values

$$\gamma = \left(\frac{F_1}{2K_1}\right)^{1/3} \rightarrow \nu_1 = \nu = -3K_1\left(\frac{F_1}{2K_1}\right)^{2/3} \tag{4.37a}$$

$$\gamma = -\left(\frac{F_1}{K_2}\right)^{1/3} \rightarrow \nu_2 = \nu = (K_2 - K_1)\left(\frac{F_1}{K_2}\right)^{2/3} \tag{4.37b}$$

The steady-state solutions given over the frequency range $\nu_2 - \nu_1$ are unstable.

The second solution is referred to as nonplanar motion characterized by a steady-state liquid free-surface motion with a constant peak wave height and a single nodal diameter that rotates at a constant rate about the axis of the cylindrical container. This solution is

$$f_1 = \gamma, \quad \text{and} \quad f_2 = f_3 = 0, \; f_4^2 = \gamma^2 + \frac{F_1}{K_2\gamma} \tag{4.38a}$$

where the amplitude γ is given by the solution of the cubic algebraic equation

$$(K_2 - K_1)\gamma^3 - \nu\gamma - \frac{K_1}{K_2}F_1 = 0 \tag{4.38b}$$

For the solution of f_4 to be real and nonzero, the condition $\gamma^2 + \dfrac{F_1}{K_2\gamma} > 0$ must be satisfied.

The velocity potential function corresponding to the planar solution takes the form

$$\begin{aligned}\tilde{\Phi} = {} & \gamma\varepsilon^{1/3}\cos\theta J_1(\lambda_{11}r)\cos\Omega t\frac{\cosh(\lambda_{11}(z+h))}{\cosh(\lambda_{11}h)} \\ & -\frac{1}{2}\varepsilon^{2/3}\gamma^2\sin 2\Omega t\left\{\sum_{n=1}^{\infty}\Omega_{0n}J_0(\lambda_{0n}r)\frac{\cosh(\lambda_{0n}(z+h))}{\cosh(\lambda_{0n}h)}\right. \\ & \left. + \sum_{n=1}^{\infty}\Omega_{2n}J_2(\lambda_{2n}r)\cos 2\theta\frac{\cosh(\lambda_{2n}(z+h))}{\cosh(\lambda_{2n}h)}\right\}\end{aligned} \tag{4.39}$$

where

$$\Omega_{0n} = \frac{2K_0\left(I^n_{01} + I^n_{02} + I^n_{03}\right)\sqrt{g\lambda_{11}\tanh(\lambda_{11}h)}}{R^2\{4g\lambda_{11}\tanh(\lambda_{11}h) - g\lambda_{0n}\tanh(\lambda_{0n}h)\}J_0^2(\lambda_{0n}R)}$$

$$\Omega_{2n} = \frac{2\lambda_{2n}^2 K_0\left(I^n_{21} + I^n_{22} + I^n_{23}\right)\sqrt{g\lambda_{11}\tanh(\lambda_{11}h)}}{\left(R^2\lambda_{2n}^2 - 4\right)\{4g\lambda_{11}\tanh(\lambda_{11}h) - g\lambda_{2n}\tanh(\lambda_{2n}h)\}J_2^2(\lambda_{2n}R)}$$

$$K_0 = \frac{1}{2\lambda_{11}^2}\left(3\tanh^3(\lambda_{11}) - 1\right),$$

$$I^n_{i1} = \frac{1}{\lambda_{11}}\int_0^{\lambda_{11}R} uJ_i(\lambda_{in}u/\lambda_{11})\left[\frac{d}{du}J_1(u)\right]^2 du, \quad i = 0,\ 2$$

$$I^n_{i2} = \frac{1}{\lambda_{11}}\int_0^{\lambda_{11}R} \frac{1}{u}J_i(\lambda_{in}u/\lambda_{11})J_1^2(u)\,du,$$

$$I^n_{i3} = \frac{1}{\lambda_{11}}\int_0^{\lambda_{11}R} uJ_i(\lambda_{in}u/\lambda_{11})J_1^2(u)\,du \tag{4.40}$$

The free-surface wave height corresponding to the first harmonic is

$$\eta_{\substack{r=R\\ \theta=0}} = \left\{\frac{\Omega\varepsilon}{g}\left[\varepsilon^{-2/3}\gamma J_1(\lambda_{11}R) + R + \frac{\lambda_{11}^3}{8g}\tanh(\lambda_{11}h)\right.\right.$$
$$\left.\times\left(1 + 5\tanh^2(\lambda_{11}h)\right)\left(\gamma J_1(\lambda_{11}R)\right)^3\right] - \frac{\lambda_{11}}{4g}\gamma^3\tanh(\lambda_{11}h)J_1(\lambda_{11}R)$$
$$\left[\sum_{n=1}^{\infty}(\lambda_{0n}\tanh(\lambda_{0n}h) + 2\lambda_{11}\tanh(\lambda_{11}h))\Omega_{0n}J_0(\lambda_{0n}R)\right.$$
$$\left.\left. + \sum_{n=1}^{\infty}(\lambda_{2n}\tanh(\lambda_{2n}h) + 2\lambda_{11}\tanh(\lambda_{11}h))\Omega_{2n}J_2(\lambda_{2n}R)\right]\right\}\sin\Omega t \tag{4.41}$$

The hydrodynamic pressure distribution is obtained by using the nonlinear Bernoulli's equation

$$p = -\rho\left[\frac{\partial\tilde{\Phi}}{\partial t} + gz + \frac{1}{2}\left(\nabla\tilde{\Phi}.\nabla\tilde{\Phi}\right) + \ddot{x}r\cos\theta\right] + p_0 \tag{4.42}$$

The above expression is used to determine the hydrodynamic force acting on the tank walls along the x-axis as (Abramson, *et al.*, 1966)

$$F_x = \int_{-h}^{\eta}\int_0^{2\pi} pR\cos\theta\,d\theta\,dx\big|_{r=R} \cong F_1 + F_2 \tag{4.43}$$

where

$$F_1 = \int_{-h}^{0}\int_{0}^{2\pi} pR\cos\theta\, d\theta\, dx\big|_{r=R}$$

$$= -\rho R\Big\{-\varepsilon^{1/3}\frac{\pi\Omega A}{\lambda_{11}}\sin(\Omega t)\sinh(\lambda_{11}h) + \frac{\varepsilon}{2R^2}[\sin(\Omega t)+\sin(3\Omega t)]$$

$$\times\sum_{n=1}^{\infty}\frac{\pi A_\theta N_\theta}{2(\lambda_{11}^2-\lambda_{2n}^2)}[\lambda_{11}\sinh(\lambda_{11}h)\cosh(\lambda_{2n}h) - \lambda_{2n}\cosh(\lambda_{11}h)\sinh(\lambda_{2n}h)]$$

$$+\frac{\varepsilon}{2}[\sin(\Omega t)+\sin(3\Omega t)]\sum_{n=1}^{\infty}\frac{\pi A_z}{2}\Bigg[\frac{2M_z}{2(\lambda_{11}^2-\lambda_{0n}^2)}[\lambda_{11}\cosh(\lambda_{11}h)\sinh h(\lambda_{0n}h)$$

$$-\lambda_{0n}\sinh(\lambda_{11}h)\cosh(\lambda_{2n}h)]\frac{N_z}{2(\lambda_{11}^2-\lambda_{2n}^2)}[\lambda_{11}\cosh h(\lambda_{11}h)\sinh h(\lambda_{2n}h)$$

$$-\lambda_{2n}\sinh(\lambda_{11}h)\cosh(\lambda_{2n}h)]\Bigg] - \varepsilon\Omega R\pi h\sin(\Omega t)\Big\} \tag{4.43a}$$

$$F_2 = \int_{0}^{\eta}\int_{0}^{2\Pi} pR\cos\theta\, d\theta\, dx\big|_{r=R}$$

$$= -\int_{0}^{2\Pi} R\left[\eta p(R,\theta,0) + \frac{\eta^2}{2}p_z(R,\theta,0) + O(\eta^4)\right]\cos\theta\, d\theta$$

$$\cong -\int_{0}^{2\Pi}\left[-\frac{p^2(R,\theta,0)}{p_z(R,\theta,0)} + \frac{1}{2}\frac{p^2(R,\theta,0)}{p_z(R,\theta,0)}\right]R\cos\theta\, d\theta = \int_{0}^{2\pi}\frac{1}{2}\frac{p^2(R,\theta,0)}{p_z(R,\theta,0)}R\cos\theta\, d\theta$$

where the relationship $\eta \approx -\frac{1}{2}\frac{p(R,\theta,0)}{p_z(R,\theta,0)}$ has been used.

The result of the second force component is

$$F_2 = \frac{1}{2}\frac{\rho R}{g}\pi\varepsilon\Big\{\frac{3}{4}\frac{\Omega^2}{g}\sin^3(\Omega t)[\gamma J_1(\lambda_{11}R)]^3\lambda_{11}\tanh(\lambda_{11}h)$$

$$-2\Omega^2\sin(\Omega t)\cos(2\Omega t)\gamma J_1(\lambda_{11}R)$$

$$\times\left[-\sum_{n=1}^{\infty}\gamma^2\Omega_{0n}J_0(\lambda_{0n}R) - \frac{1}{2}\sum_{n=1}^{\infty}\gamma^2\Omega_{2n}J_2(\lambda_{2n}R)\right]$$

$$-\frac{\Omega}{8R^2}\sin(\Omega t)(1+\cos(2\Omega t))[\gamma J_1(\lambda_{11}R)]^3$$

$$-\frac{3\Omega}{8}\sin(\Omega t)(1+\cos(2\Omega t))[\gamma J_1(\lambda_{11}R)]^3[\lambda_{11}\tanh(\lambda_{11}h)]^2\Big\} \tag{4.43b}$$

where

$$M_z = -\frac{1}{2}\gamma^2\Omega_{0n}\frac{\lambda_{0n}J_0(\lambda_{0n}R)}{\cosh(\lambda_{0n}h)}$$
$$N_\theta = \gamma^2\Omega_{2n}\frac{J_2(\lambda_{2n}R)}{\cosh(\lambda_{2n}h)}$$
$$N_z = -\frac{1}{2}\gamma^2\Omega_{2n}\frac{\lambda_{2n}J_2(\lambda_{2n}R)}{\cosh(\lambda_{2n}h)}$$
$$A = \gamma\frac{J_1(\lambda_{11}R)}{\cosh(\lambda_{11}h)}$$
$$A_\theta = -A$$
$$A_z = \lambda_{11}A$$

Numerical and experimental results

Hutton (1962) obtained quantitative results based on the following parameters $R = 5.938$ inch, $h = 8.907$ inch, $\lambda_{11}R = 1.841\,19$, $J_1\ (\lambda_{11}\ R) = 0.581\,865$. The first coefficient of the Dini-series (4.9) is

$$F_1 = \frac{2R}{(\lambda_{11}^2R^2 - 1)J_1(\lambda_{11}R)} = 8.539\,92$$
$$\tanh(\lambda_{11}h) = 0.992\,05$$
$$\omega_{11} = \sqrt{g\lambda_{11}\tanh(\lambda_{11}h)} = 10.897 \text{ rad/s} \tag{4.44}$$

The corresponding amplitude–frequency equation (4.35) for the planar motion is

$$\nu = -\frac{8.5399}{\gamma} - 0.485\,28 \times 10^{-5}\gamma^2 \tag{4.45}$$

Figure 4.3 shows the dependence of the amplitude γ on the detuning frequency parameter ν. The stable harmonic planar motion is shown by a solid curve and the unstable response by a dot-dash curve. The limits of the detuning parameter ν separating stable and unstable regions are $\nu = -0.1337$ and $\nu = 0.064\,59$. From relation (4.37a), the lower limit boundary between stable and unstable motion is

$$\gamma = \left(\frac{F_1}{2K_1}\right)^{1/3} = 95.82, \quad \text{and} \quad \nu = -3K_1\left(\frac{F_1}{2K_1}\right)^{2/3} = -0.1337 \tag{4.46a, b}$$

The upper limit boundary is computed from relation (4.37b) as follows

$$\gamma = -\left(\frac{F_1}{K_2}\right)^{1/3} = -85.41, \quad \text{and} \quad \nu = (K_2 - K_1)\left(\frac{F_1}{K_2}\right)^{2/3} = 0.064\,59 \tag{4.47a, b}$$

The amplitude–frequency relation for the nonplanar motion is obtained from relation (4.38a) and (4.38b) as

$$\nu = -\frac{3.0235}{\gamma} + 4.001 \times 10^{-6}\gamma^2, \quad \text{and} \quad f_4^2 = \gamma^2 + \frac{6.2305}{\gamma} \tag{4.48a, b}$$

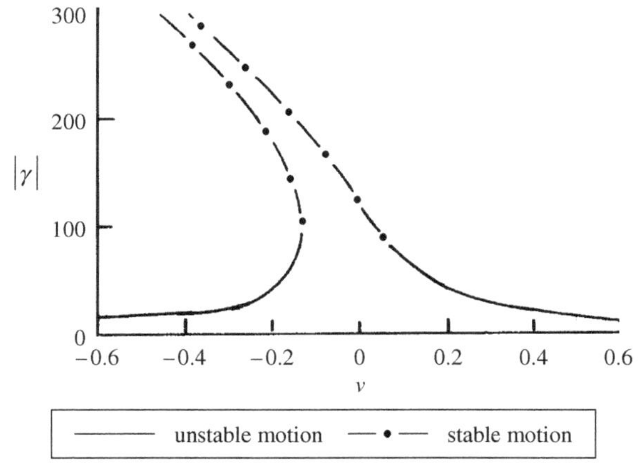

Figure 4.3 Planar motion amplitude-frequency response. (Hutton, 1962)

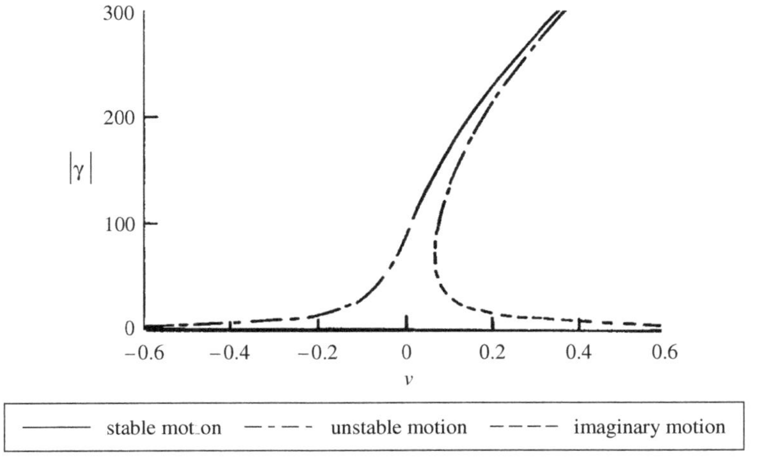

Figure 4.4 Nonplanar motion amplitude-frequency response. (Hutton, 1962)

Figure 4.4 represents the harmonic nonplanar response according to relation (4.5a). The ranges of stable and unstable motion are indicated. Figure 4.5 shows the dependence of $|f_4|$ on the frequency parameter ν according to relation (4.48b) after using (4.48a). Figure 4.6 shows the unstable regions for both planar and nonplanar motions as function of the peak tank velocity ΩX_0.

Hutton (1962) compared the predicted stability results with those measured experimentally as shown in Figure 4.7. The predicted results agree with experimental points for the planar region and small values of ΩX_0. However, poor agreement is seen for the nonplanar region. The observed discrepancy was due to the difficulty in precisely determining the point at which the fluid motion changes character. As the frequency is decreased, the transition is gradual from the steady-state harmonic nonplanar motion to the nonharmonic nonplanar motion existing in the instability region. Abramson, Chu and Kana (1966) conducted a series of experimental tests on a circular tank of inside diameter 7.7 inch and liquid depth ratio,

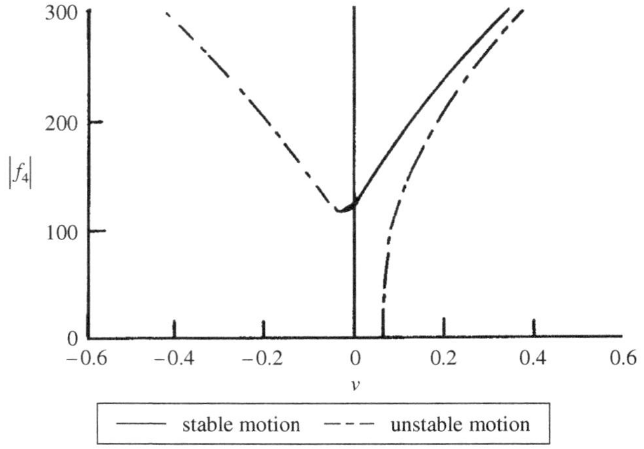

Figure 4.5 Nonplanar motion amplitude-frequency response. (Hutton, 1962)

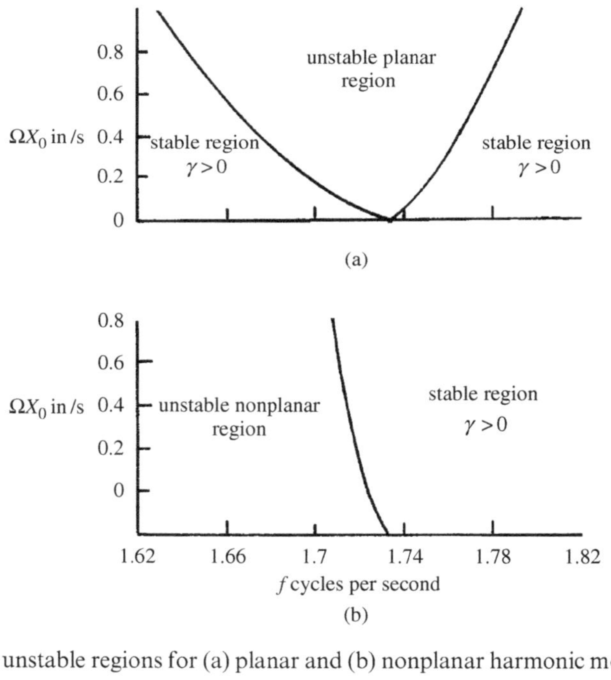

Figure 4.6 Stable and unstable regions for (a) planar and (b) nonplanar harmonic motions. (Hutton, 1962)

$h/R = 2.0$. Figure 4.8 shows the average liquid amplitude, $\eta_0/2R$, responses in the first antisymmetric slosh mode as a function of the excitation frequency parameter, $\Omega^2(2R)/g$. The average liquid free surface displacement at the wall, $\eta_0/2R$, was used since the actual upward displacement from the mean liquid level is usually greater than the downward displacement, in this mode. It is taken as half of the peak-to-peak amplitude measured near the wall, in line with the axis of excitation. In obtaining the results of Figure 4.8, the excitation amplitude was kept constant while the excitation frequency was allowed to vary. At first, data were taken

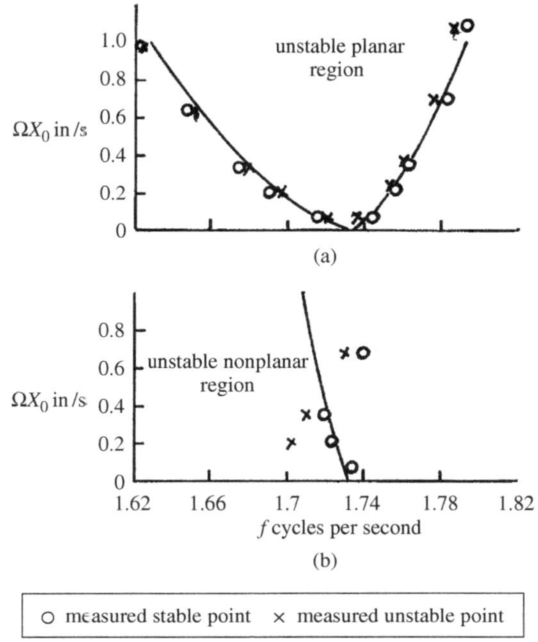

Figure 4.7 Stable and unstable regions for (a) planar and (b) nonplanar harmonic motions. Comparison with experimental measurements. (Hutton, 1962)

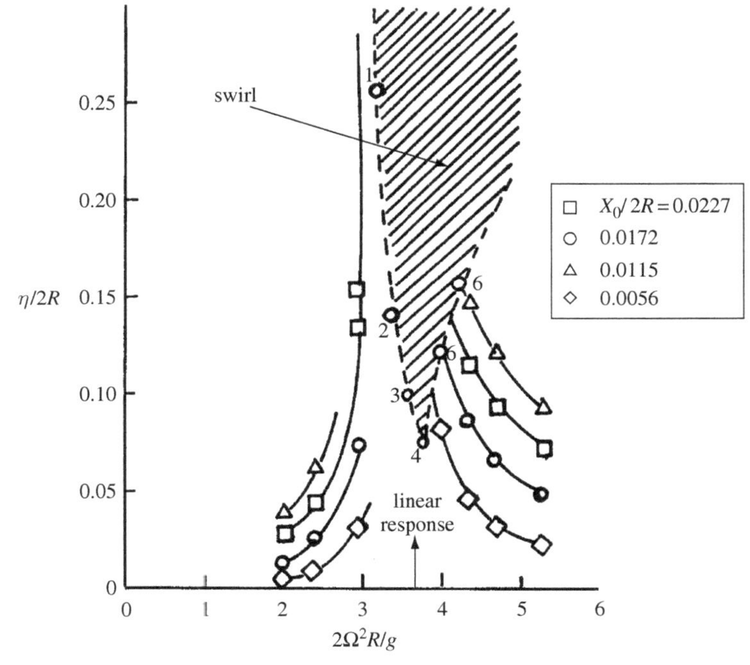

Figure 4.8 Dependence of liquid free-surface response on excitation frequency parameter near the first anti-symmetric mode showing swirl region for $h/2\,R = 1$. Points 1–6 are measured for different values of excitation amplitude. (Abramson, Chu and Kana, 1966)

only for liquid motions arising basically from the first anti-symmetric slosh mode, with no swirl occurring. Subsequently, the boundary of the swirl region was defined experimentally by maintaining constant excitation frequency and slowly increasing excitation amplitude until the swirl motion was observed. This swirl boundary shown in Figure 4.8 is dependent upon both excitation frequency and amplitude since in certain instances it was found that large-amplitude breaking waves could be produced without swirl. Within the swirl region, the motion of the liquid has a phase angle close to 90° with respect to the input displacement.

Miles (1984c) was motivated by the work of Lorenz (1963) and revisited the spherical pendulum problem. The main control parameters are the damping $\alpha = 2\zeta\,\varepsilon^{-2/3}(\omega/\Omega)$, and detuning ν parameter, where ζ is the damping ratio. Miles also employed the basic concepts of the modern theory of nonlinear dynamics for the liquid forcing sloshing in upright circular containers and his main results are summarized in the next section.

Chaotic sloshing

Chaotic sloshing is a form of unsteady free-surface motion, and owes its origin to nonlinearity and sensitivity to initial conditions. Miles (1976a, b) constructed the Lagrangian and Hamiltonian for nonlinear gravity waves in a cylindrical container in terms of the generalized coordinates of the free-surface displacement, $\mathbf{q} \equiv \{q_n(t)\}$. He solved the kinematic boundary-value problem through a variational formulation and the truncation and inversion of an infinite matrix. The results were applied to weakly coupled oscillations, using the time-averaged Lagrangian, and to resonantly coupled oscillations, using Poincaré's action-angle formulation. Later, Miles (1984d) employed the Lagrangian formulation to study the liquid free-surface motion in a cylindrical tank subjected to a simple harmonic horizontal excitation. Longuet-Higgins (1985) and Dias and Kharif (1999) studied bifurcation problems in gravity waves.

The free surface wave height may be expressed in terms of the modal expansion

$$\eta = \eta_n(t)\boldsymbol{\psi}_n(r,\theta) \tag{4.49}$$

where r and θ are the polar coordinates, the repeated indices indicate summation over participating modes, η_n are the generalized coordinates, and the mode shape functions $\boldsymbol{\psi}_n(r, \theta)$ are members of the complete set of orthogonal eigenfunctions determined from the equation

$$(\nabla^2 + k_n^2)\boldsymbol{\psi}_n = 0, \qquad \left.\frac{\partial \boldsymbol{\psi}_n}{\partial r}\right|_{r=R} = 0, \quad \text{and} \quad \int_0^R\int_0^{2\pi} \boldsymbol{\psi}_m\boldsymbol{\psi}_n r\,dr\,d\theta = \delta_{mn}\pi R^2 \tag{4.50a, b, c}$$

where δ_{mn} is the Kronecker delta, k_n is the eigenvalue. Square brackets indicate a spatial average, that is,

$$[f(r,\theta)] \equiv \frac{1}{\pi R^2}\int_0^R\int_0^{2\pi} fr\,dr\,d\theta \tag{4.51}$$

$$\boldsymbol{\psi}_j(r,\theta) \equiv \boldsymbol{\psi}_{ij}^{c,s} = N_{ij}^{-1}J_j(k_{ij}r)(\cos i\theta, \sin j\theta), \quad i = 0, 1, \ldots, \quad j = 1, 2, \tag{4.52}$$

$$N_{ij}^2 = \frac{1}{2}(1+\delta_{0i})\left\{1-\left(\frac{i}{k_{ij}R}\right)^2\right\}J_i^2(k_{ij}R) \tag{4.53}$$

Note that each of the eigenfunctions requires three indices; the azimuthal wave number i, the radial wave number j, and, except for the axisymmetric modes, the superscript c or s to distinguish between cosine and sine azimuthal variations. The use of a single index in expansion (4.49) is merely an abbreviation, and k_{ij} is one of the infinite, discrete set of eigenvalues determined from the roots of the condition $\frac{d}{dr}J_i(k_{ij}r)|_{r=R} = 0$.

The single subscripts 1 and 2 are allocated for the dominant modes according to the definition

$$\boldsymbol{\psi}_{1,2}(r,\theta) \equiv \boldsymbol{\psi}_{11}^{c,s} = N^{-1}J_1(kr)(\cos\theta, \sin\theta) \tag{4.54}$$

where $N \equiv N_{11} = 0.3455$, $k \equiv k_{11} = 1.8412/R$.

The kinetic energy corresponding to equation (4.49) is given by

$$T = \frac{1}{2}\rho\int\int\int(\nabla\Phi)^2\, ds\, dz = \frac{1}{2}\rho S a_{mn}\dot{\eta}_m\dot{\eta}_n \tag{4.55}$$

where S is the cross-sectional area of the circular cylinder, the inertial coefficients a_{mn} have the dimension of length and may be approximated by the quadratic truncation

$$a_{mn} = \delta_{mn}a_m + a_{\ell mn}\eta_\ell + \frac{1}{2}a_{j\ell mn}\eta_j\eta_\ell \tag{4.56}$$

where $a_n = \frac{1}{k_n}\coth k_n h \equiv \frac{g}{\omega_n^2}$, ω_n represents the natural frequency of the nth mode and a_n represents the length of an equivalent pendulum,

$$a_{\ell mn} = C_{\ell mn} - D_{\ell mn}a_m a_n \tag{4.57a}$$

$$a_{j\ell mn} = -D_{j\ell mn}(a_m + a_n) + 2D_{jmi}D_{\ell ni}a_i a_m a_n \tag{4.57b}$$

$$\begin{aligned} C_{\ell mn} &= [\psi_\ell \psi_m \psi_n] \\ D_{\ell mn} &= [\psi_\ell \nabla\psi_m \cdot \nabla\psi_m] \\ D_{j\ell mn} &= \left[\psi_j\psi_\ell\nabla\psi_m\cdot\nabla\psi_m\right] \end{aligned} \tag{4.57c–e}$$

The correlation integrals listed in (4.57c–e) are evaluated in Miles (1984b).

The potential energy is

$$V = \rho\int\int dS\int_0^\eta [\dot{\boldsymbol{u}}\cdot\boldsymbol{x} + gz]\, dz = \rho S\left(-Q_n\eta_n + \frac{1}{2}g\eta_n\eta_n\right) \tag{4.58}$$

where $\dot{u} = -\Omega^2 X_0\cos(\Omega t)$ is the horizontal acceleration of the container, and

$$Q_n = -\dot{\boldsymbol{u}}\cdot\mathbf{x}_n, \quad \boldsymbol{x}_n = \frac{1}{S}\int\int x\boldsymbol{\psi}_n\, ds, \quad \text{with } x_1 = \frac{1}{\pi R^2}\int_0^R\int_0^{2\pi} x\boldsymbol{\psi}_1\, \mathrm{r}\, dr\, d\theta = 0.4968R \tag{4.59}$$

The Lagrangian takes the form, after dividing by ρS

$$L \equiv T - V = Q_1\eta_1 + \frac{1}{2}a_n(\dot{\eta}_n^2 - \omega_n^2\eta_n^2) + \frac{1}{2}a_{\ell mn}\eta_\ell\dot{\eta}_m\dot{\eta}_n + \frac{1}{4}a_{j\ell mn}\eta_j\eta_\ell\dot{\eta}_m\dot{\eta}_n \tag{4.60}$$

One can generate a discrete set of differential equations from equation (4.60) for any number of modes. In view of the fourth-order nonlinearity in the Lagrangian, the equations of motion will contain terms of first- and third-order in the dominant modes, $\eta_{1,2}$. The linear inertial and gravitational terms for the dominant modes approximately cancel in the neighborhood of resonance ($\Omega \approx \omega_1$), and a balance between the nonlinear terms and external force. Secondary modes are induced by quadratic terms in η_1 and η_2. Thus, the dominant modes may be posed as slowly varying sinusoids with the carrier frequency Ω. Quadratic terms are associated with frequencies 0 and 2Ω. Accordingly, one may introduce the following form of solution for the dominant modes

$$\eta_n = \varepsilon\lambda\{p_n(\tau)\cos\Omega t + q_n(\tau)\sin\Omega t\}, \quad (n = 1, 2) \tag{4.61}$$

and the following form for the secondary modes

$$\eta_n = \varepsilon^2\lambda\{A_n(\tau)\cos 2\Omega t + B_n(\tau)\sin 2\Omega t + C_n(\tau)\}, \quad (n \neq 1, 2) \tag{4.62}$$

where $\lambda = (1/k)\tanh(kh)$ is a reference length, $\tau = \varepsilon^2\Omega t/2$ is a nondimensional time parameter, and p_n, q_n, A_n, B_n, and C_n are slowly varying nondimensional amplitudes.

Equations (4.61) and (4.62) are substituted into the Lagrangian expression (4.60). This is followed by taking the average $\langle L\rangle$ over a 2π interval of Ωt, while holding τ fixed, that is

$$\begin{aligned}\langle L\rangle =& \frac{1}{2}\varepsilon^4 g\lambda^2\left[\frac{1}{2}(\dot{p}_n q_n - p_n\dot{q}_n) + \frac{1}{4}\frac{\lambda^2}{a_1}\left\{a_{1111}(E_1^2 + E_2^2)\right.\right.\\ &+ 2a_{1122}\left(E_1E_2 + \frac{1}{2}M^2\right) + 2(a_{1212} + a_{1221})\left.\left(E_1E_2 - \frac{1}{2}M^2\right)\right\}\\ &+ \frac{1}{2}\frac{\lambda}{a_1}\left\{\left(2a_{nm\ell} - \frac{1}{2}a_{\ell nm}\right)(A_\ell(p_mp_n - q_mq_n) + B_\ell(p_mq_n + p_nq_m))\right.\\ &+ \left.\left. a_{\ell mn}C_\ell(p_mp_n + q_mq_n)\right\} + \frac{1}{2}b_1(A_\ell^2 + B_\ell^2) - C_\ell^2\right]\end{aligned} \tag{4.63a}$$

where

$$\begin{aligned} b_\ell &= \frac{4\Omega^2 - \omega_\ell^2}{\omega_\ell^2}, \quad E_n = \frac{1}{2}(p_n^2 + q_n^2), \quad E = E_1 + E_2,\\ M &= p_1q_2 - p_2q_1, \quad \varepsilon^3 = \frac{X_0x_1}{a_1\lambda} = k^2X_0x_1\end{aligned} \tag{4.64}$$

The values E and M are measures of the energy and the angular momentum in the dominant modes, respectively. Requiring $\langle L\rangle$ to be stationary with respect to each of A_ℓ, B_ℓ and C_ℓ, yields

$$\{A_\ell, B_\ell\} = -\frac{1}{2}\frac{\lambda}{a_1b_\ell}\left(2a_{nm\ell} - \frac{1}{2}a_{\ell mn}\right)\{p_mp_n - q_mq_n, p_mq_n + p_nq_m\} \tag{4.65a}$$

$$C_\ell = -\frac{1}{4}\frac{\lambda}{a_1} a_{\ell mn}(p_m p_n + q_n q_m) \tag{4.65b}$$

The average $\langle L \rangle$ can be written in the compact form

$$\langle L \rangle = \frac{1}{2}\varepsilon^4 g \lambda^2 \left\{ \frac{1}{2}(\dot{p}_n q_n - p_n \dot{q}_n) + H(p_1, q_1, p_2, q_2) \right\} \tag{4.63b}$$

where n is summed over 1 and 2, and

$$H = p_1 + \beta E + \frac{1}{2}AE^2 + \frac{1}{2}BM^2 \tag{4.64a}$$

$$\beta = \varepsilon^{-2}((\Omega/\omega_1)^2 - 1) \tag{4.64b}$$

$$A = -P_0 - P_2 + Q_0 + Q_2 + R_0 + R_2 - 0.567 \tanh^2(kh) \tag{4.64c}$$

$$B = P_0 - P_2 - Q_2 - R_0 + R_2 - 1.468 \tanh^2(kh) \tag{4.64d}$$

$$P_i = \frac{(3\tanh^2(kh) + 1 - 0.5\kappa_{ij}^2 - 2\kappa_{ij}^2\nu_{ij}^2)^2 I_{ij}^2}{4(4\nu_{ij}^2 - 1)}, \quad R_i = \frac{1}{4}\kappa_{ij}^2\nu_{ij}^2 I_{ij}^2 \tag{4.64e, f}$$

$$Q_i = \frac{1}{2}\left(\tanh^2(kh) - 1 + 0.5\kappa_{ij}^2\right)^2 I_{ij}^2, \quad V_{ij} = \frac{\Omega}{\sqrt{k_{ij}\tanh(k_{ij}h)}} \tag{4.64g, h}$$

$$I_{ij} = \frac{(1 + \delta_{0i})}{2k^2R^2N^2N_{ij}} \int_0^{kR} J_1^2(x) J_i(k_{ij}x/k)x\,dx, \quad k = k_1 = k_2, \quad N = N_{11} \tag{4.64i, j, k}$$

Both A and B have simple poles at $h/R = 0.1523$ due to the possible occurrence of internal resonance between the primary modes ($k_{11}R = 1.8412$) and the dominant axi-symmetric mode ($k_{01}R = 3.8317$). Accordingly, the present analysis does not apply in the neighborhood of this resonance condition. Other internal resonance conditions exist among the 02 and 22 modes but they are extremely narrow.

The evolution equations are obtained using the canonical equations

$$\dot{p}_n = -\frac{\partial H}{\partial q_n}, \quad \dot{q}_n = \frac{\partial H}{\partial p_n} \tag{4.65a, b}$$

Here H is regarded as the Hamiltonian. At this stage, one can introduce damping to the right hand of equations (4.65) in the form

$$\alpha = 2\zeta\varepsilon^{-2} \tag{4.66}$$

where ζ is the damping ratio.

Substituting equation (4.64a) into equations (4.65a,b) and incorporating damping, gives the following set of evolution equations in terms of three control parameters, namely the damping, α, detuning parameter, β, and the fluid depth ratio, h/R, which is impeded in A and B

$$\dot{p}_1 = -\alpha p_1 - (\beta + AE)q_1 + BMp_2 \tag{4.67a}$$

$$\dot{q}_1 = -\alpha q_1 + (\beta + AE)p_1 + BMq_2 + 1 \tag{4.67b}$$

$$\dot{p}_2 = -\alpha p_2 - (\beta + AE)q_2 - BMp_1 \tag{4.67c}$$

$$\dot{q}_2 = -\alpha q_2 + (\beta + AE)p_2 - BMq_1 \tag{4.67d}$$

These equations are invariant under the reflection, $(p_2, q_2) \to -(p_2, q_2)$, due to the symmetry with respect to the plane of excitation, $(\theta = 0)$.

The evolution equations (4.67) are similar to those of spherical pendulum (Miles, 1984c) if $A = 1/4$ and $B = -3/4$. They also reduce to the corresponding equations for a resonantly forced string (Miles, 1984f) if $A = -3$ and $B = 1$. It was found that the response of the pendulum is chaotic in various intervals of β if α is sufficiently small, whereas the response of the string appears to be regular for all values of α and β. The divergence of equations (4.67) is

$$\Delta = \frac{\partial \dot{p}_1}{\partial p_1} + \frac{\partial \dot{q}_1}{\partial q_1} + \frac{\partial \dot{p}_2}{\partial p_2} + \frac{\partial \dot{q}_2}{\partial q_2} = -4\alpha$$

It follows that an elementary volume in the space (p_1, p_2, q_1, q_2) contracts in the same way as does $e^{-4\alpha\tau}$ and every trajectory ultimately must be confined to a limiting subspace of dimension less than four, which may be a fixed point (dimension zero), a closed curve (dimension one), a torus (dimension two or three), or a strange attractor (fractional dimension). Every trajectory ultimately must lie within the hypersphere, $E \leq 1/2\alpha^2$, as demonstrated by Miles (1984c).

The steady-state solution, or fixed points, of equations (4.67) is obtained by setting the left-hand sides of equations (4.67) to zero. The resulting solution is

$$p_1 = -2\left(\beta E + AE^2 + BM^2\right), \quad q_1 = 2\alpha E \tag{4.68a, b}$$

$$p_2 = 2\alpha M, \qquad q_2 = -2(\beta + (A + B)E)M \tag{4.68c, d}$$

The planar motion, $(p_2 = q_2 = 0)$ is realized if

$$BM = 0, \quad \text{and} \quad A^2E^3 + 2A\beta E^2 + (\alpha^2 + \beta^2)E - \frac{1}{2} = 0 \tag{4.69a, b}$$

The nonplanar motion, $(p_2 \neq 0, q_2 \neq 0)$, takes place if

$$B^2M^2 = -(AE + \beta)\{(A + 2B)E + \beta\} - \alpha^2 > 0 \tag{4.70a}$$

$$\begin{aligned} &A(A + B)^2E^3 + (A + B)(3A + B)\beta E^2 \\ &\quad + \left\{A\alpha^2 + (3A + 2B)\beta^2\right\}E + \beta(\alpha^2 + \beta^2) + \frac{1}{4}B = 0 \end{aligned} \tag{4.70b}$$

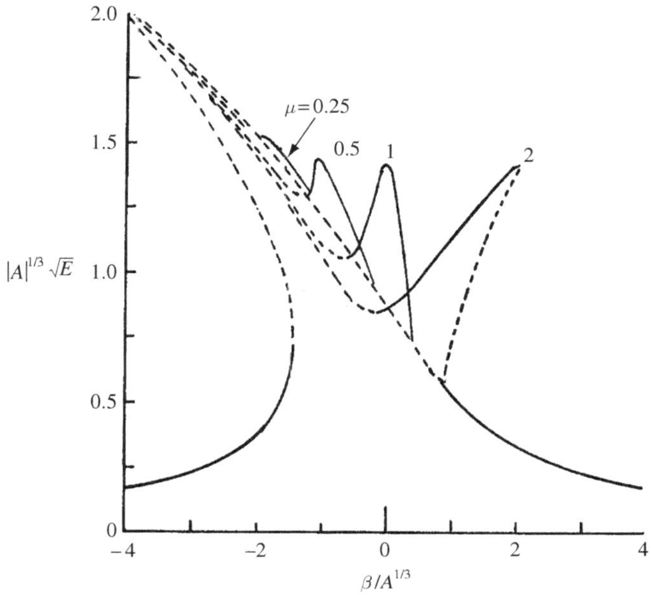

Figure 4.9 Resonance curves for damping parameter $\alpha/A^{1/3}=0.125$ and different values of the parameter $\mu=-B/A$. (Miles, 1984d)

The dependence of the response amplitude parameter, E, on the external detuning parameter, β, for fixed values of damping and liquid depth parameters is shown in Figure 4.9. The response curves are plotted for different values of the parameter $\mu=-B/A$. The Poincaré-bifurcation points at which there is a change in the number of fixed points, correspond to the turning points, at which $d\beta/dE=0$, of the resonance curves and to the termination of the nonplanar resonance curves. The planar resonance curve, which is determined from equation (4.69b), has a maximum at

$$E=-\frac{\beta}{A}=\frac{1}{2\alpha^2} \tag{4.71}$$

and corresponds to a soft/hard spring characteristics for $A>/<0$, although the planar resonance curve shown in Figure 4.9 leans to the left for either sign of A. The turning points occuring at β_1 and β_3 are given by the two expressions

$$\beta_1 \approx -\frac{1}{2}\frac{A}{\alpha^2}-\frac{1}{2}\frac{\alpha^4}{A}, \quad \beta_3 \approx -\frac{3}{2}A^{1/3}+\frac{1}{2}\frac{\alpha^2}{A^{1/3}} \tag{4.72a, b}$$

and the planar resonance curve is single-valued for $\alpha > 0.687|A|^{1/3}$.

The nonplanar resonance curve, determined using equations (4.70), has a maximum at

$$E=\frac{1}{4\alpha^2}, \quad \beta=-\frac{1}{4\alpha^2}(A+B) \tag{4.73a, b}$$

The nonplanar curve terminates on the planar resonance curve at the bifurcation points β_2 and β_4. These points are determined by equating the roots of M^2 obtained from equation (4.70a) and that obtained from equation (4.69b). Eliminating β from these two equations, gives

$$E^3(1-\alpha^2 E)=\frac{1}{8B^2} \tag{4.74}$$

This equation has two positive real roots for sufficiently small α^2. The corresponding bifurcation points are

$$\beta_n=-AE_n-\frac{1}{4BE_n^2}, \quad n=2,\ 4 \tag{4.75}$$

where $E_2 > E_4$. Solving equations (4.74) and (4.75) for $\alpha^2 \ll 1$, gives

$$\beta_2 \approx -\frac{1}{2}\frac{A}{\alpha^2}+\frac{1}{B^2}\left(\frac{1}{2}A-B\right)\alpha^4 \tag{4.76a}$$

$$\beta_4 \approx -\frac{1}{B^{2/3}}\left(\frac{1}{2}A+B\right)-\frac{2}{3B^{4/3}}\left(\frac{1}{4}A-B\right)\alpha^2 \tag{4.76b}$$

These two bifurcation points coincide at

$$\beta_2=\beta_4=-\frac{1}{(4B)^{2/3}}(2A+B), \quad \text{for } \alpha=\frac{\sqrt{3}}{2}\left|\frac{1}{2}B\right|^{1/3}\equiv\alpha^* \tag{4.77a, b}$$

For any damping parameter greater than α^*, equation (4.74) has no positive real roots and the nonplanar resonance curve disappears. It follows from equations (4.72a) and (4.76a) that $\beta_2-\beta_1>0$ for $0<\alpha^2\ll 1$.

Equation (4.70b) has two turning points, β_5 and β_6, if

$$\alpha>\frac{\sqrt{3}}{2}\left|\frac{(A+B)B}{4A}\right|^{1/3} \tag{4.78}$$

However, either or both β_5 and β_6 may be excluded as turning points of the nonplanar resonance curve since $E>0$ and $M^2>0$. Solving for $d\beta/dE=0$, for $\alpha^2\ll 1$, gives

$$\beta_5 \approx -\frac{1}{4\alpha^2}(A+B)-\left(\frac{A}{B}\right)^2\frac{\alpha 4}{(A+B)}, \tag{4.79a}$$

$$\beta_6 \approx -3\left(\frac{A}{4B}\right)^{2/3}(A+B)^{1/3}+\frac{\alpha^2}{2^{2/3}(A+B)^{1/3}}\left(\frac{A}{B}\right)^{4/3} \tag{4.79b}$$

Miles (1984d) studied the stability of the fixed points and indicated that Hopf-bifurcation points are possible at two particular values of the detuning parameter β. The dependence of these values on the damping parameter is shown in Figure 4.10 for different values of the parameter $\mu=-B/A$. The turning points of the planar resonance curve are shown by the dash curve. Limit cycles and/or chaotic motions were found impossible for all values of fluid depth ratio in the range $0.3<h/R<0.5$.

The asymptotic trajectories that do not lead to any stable equilibrium points are nonplanar or either periodic or chaotic. Numerical integrations of equations (4.67) were performed to determine the power spectra and the corresponding phase plane (p_1, p_2) projections for $\mu=-B/A=1.3772$ ($h/\mathrm{R}\to\infty$), $\alpha^2/A^{2/3}=1/8$ and for four values of the detuning parameter

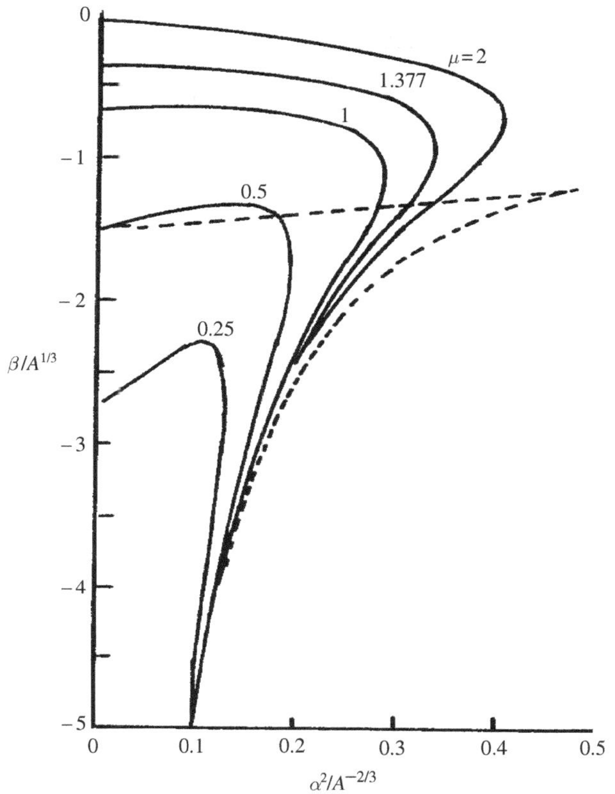

Figure 4.10 Hopf bifurcation points $\bar{\beta}_{1,2*}$ for different values of μ (solid curves) and turning points $\bar{\beta}_{1,3}$ of the planar resonance curve (dash curves). (Miles, 1984d)

$\bar{\beta} = \beta / A^{1/3} = -0.52, -0.54, -0.7, -0.85$. The main features of the response behavior for other values are summarized in Table 4.1.

McGoldrick (1970), Bryant (1989), Krasnopolskaya and Shvets (1990, 1992), Hammack and Henderson (1993), Bryant and Stiassnie (1994), Hsieh, *et al.* (1995), Sun, *et al.* (1995), and Puchka and Kholopova (1996) extended the work of Miles (1984d) to study the interaction of the dynamics of the driving mechanism and the liquid free-surface dynamics. Bryant reported the development of nonlinear progressive waves in a circular container. Krasnopolskaya and Shvets (1990, 1992) estimated the regions of the static characteristics of the driving motor over which the largest Lyapunov exponent is positive. Over these regions the liquid free-surface motion is chaotic.

Sine sweep excitation

The early studies of liquid rotational motion near resonance motivated Narimanov, *et al.* (1977), Lukovskii (1990), and Kubenko, *et al.* (1992) to study different aspects of nonlinear sloshing dynamics. They obtained a set of differential equations describing the free-liquid-surface amplitude and phase. Koval'chuk, *et al.* (1989), Koval'chuk and Kubenko (1991), Limarchenko, *et al.* (1992), and Koval'chuk and Podchasov (1996) considered the regular traveling wave modes of a liquid under quasi-stationary vibrations of its container. The

Table 4.1 *Type of trajectory for* $\mu = 1.3772$, $\alpha^2/A^{2/3} = 1/8$

$\bar{\beta} = \beta/A^{1/3}$	$\log_2 T$	Trajectory
0.8 (–0.2) –0.2	5	Fixed point
–0.4	6	Weak limit cycle
–0.5	6	Weak limit cycle
–0.52	10	Limit cycle
–0.53	8	Limit cycle – weak period doubling
–0.54	10	Limit cycle-period doubling
–0.56, –0.58, –0.60	7	Limit cycle-period doubling
–0.7	10	Chaotic + periodic
–0.8	7	Chaotic
–0.85	10	Chaotic
–0.9	7	Chaotic
–1.43	7	Chaotic
–1.44	6	Fixed point

excitation frequency was allowed to slowly increase with time, passing through resonance. They expressed the free-surface wave height, in terms of the polar coordinates r and θ, in the form

$$\eta(r, \theta, t) = a(t)\sin[\theta + \alpha(t)]R_1(r) \quad \text{with } \dot{\alpha}(t) \neq 0 \tag{4.80}$$

where $a(t)$ and $\alpha(t)$ are the amplitude and phase, both are unknown functions of time, and $R_1(r)$ is the eigenfuction of the homogeneous boundary-value problem

$$\frac{d^2R_1(r)}{dr^2} + \frac{1}{r}\frac{dR_1(r)}{dr} + \left(\lambda_1^2 - \frac{1}{r^2}\right)R_1(r) = 0 \tag{4.81a}$$

$$\left.\frac{dR_1(r)}{dr}\right|_{r=R} = 0, \quad R_1(r)|_{r=0} = \infty \tag{4.81b}$$

where R is the tank radius and λ_1 is the first eigenvalue of the anti-symmetric sloshing mode.

Under translational harmonic excitation, $x(t) = X_0 \cos \Omega t$, with constant frequency close to resonance $\Omega = \omega_1$, Koval'chuk and Kubenko (1991) studied the free-surface motion using asymptotic expansion (see, e.g., Kubenko, *et al.*, 1992). From the free-surface conditions, they obtained the following coupled differential equations of the amplitude and phase, up to cubic terms,

$$\ddot{a} + \left(\omega_1^2 - \dot{\alpha}^2\right)a = G_1(a, \dot{a}, \alpha) + f(t)\cos\alpha \tag{4.82a}$$

$$a\ddot{\alpha} + 2\dot{a}\dot{\alpha} = G_2(a, \dot{a}, \alpha) - f(t)\sin\alpha \tag{4.82b}$$

where

$$G_1 = b_1 a\left(\dot{a}^2 - \omega_1^2 a^2\right) + (b_1 - 2b_2)a^3\dot{\alpha}^2 \tag{4.83a}$$

$$G_2 = 2b_2 a^2 \dot{a}\dot{\alpha}, \omega_1^2 = g\frac{\lambda_1}{R}\tan h\frac{\lambda_1 h}{R}, \quad f(t) = \gamma\ddot{x}(t), \tag{4.83b}$$

b_1, b_2, and γ are constant parameters.

Koval'chuk, *et al.* (1989) introduced the following coordinate transformation

$$a = \sqrt{u(t) + v(t)\sin\psi(t)}, \quad \psi(t) = 2[\Omega t + \vartheta(t)] \tag{4.84a, b}$$

$$\alpha = \beta(t) + \text{arctg}\left\{\left[u(t)\tan\frac{\psi(t)}{2} + v(t)\right]\frac{1}{\sqrt{u^2(t) - v^2(t)}}\right\} \tag{4.84c}$$

$$\dot{a} = \frac{\Omega v(t)}{a}\cos\psi(t), \quad \dot{\alpha} = \Omega\frac{\sqrt{u^2(t) - v^2(t)}}{\sqrt{u(t) + v(t)\sin\psi(t)}} \tag{4.84d, e}$$

Using the averaging method the following first-order differential equations are obtained

$$\frac{du}{dt} = \frac{\Omega\gamma X_0}{2\sqrt{u}}\left[\sqrt{u^2 - v^2}\cos\beta\sin\vartheta + (u\cos\vartheta - v\sin\vartheta)\sin\beta\right] \tag{4.85a}$$

$$\frac{dv}{dt} = \frac{\Omega\gamma X_0}{2\sqrt{u}}\left[\sqrt{u^2 - v^2}\cos\beta\cos\vartheta + (u\sin\vartheta - v\cos\vartheta)\sin\beta\right] \tag{4.85b}$$

$$\frac{d\vartheta}{dt} = \frac{1}{2\Omega}\left[(\omega_1^2 - \Omega^2) + \frac{3}{2}ub_1\left(\omega_1^2 - \frac{\Omega^2}{3}\right)\right]$$
$$+ \frac{\Omega\gamma X_0}{2\sqrt{u}}\left[\sqrt{u^2 - v^2}\cos\beta\sin\vartheta - (u\cos\vartheta - v\sin\vartheta)\sin\beta\right] \tag{4.85c}$$

$$\frac{d\beta}{dt} = -\frac{\Omega\gamma X_0}{4uv\sqrt{u}}\left[\{(u^2 - v^2)\sin\vartheta - 2uv\cos\vartheta\}\cos\beta.\right.$$
$$\left. -(u\cos\vartheta + v\sin\vartheta)\sqrt{u^2 - v^2}\sin\beta\right] \tag{4.85d}$$

The steady-state solutions are obtained by setting the left-hand sides to zero. The following solutions provide possible traveling waves that propagate in the circular direction

$$u = \mp v + \frac{\gamma^2 X_0^2 \Omega^4}{b_1^2 v^2 (\omega_1^2 - \Omega^2)^2} \tag{4.86a}$$

$$(\omega_1^2 - \Omega^2) = -\frac{3}{2}ub_1\left(\omega_1^2 - \frac{\Omega^2}{3}\right) + \frac{\gamma^2 X_0^2 \Omega^4}{b_1^2 v^2 (\omega_1^2 - \Omega^2)^2} \tag{4.86b}$$

$$\tan\beta = \frac{\sqrt{u^2 - v^2}}{v \mp u}, \quad \cos 2\vartheta = 0 \tag{4.86c, d}$$

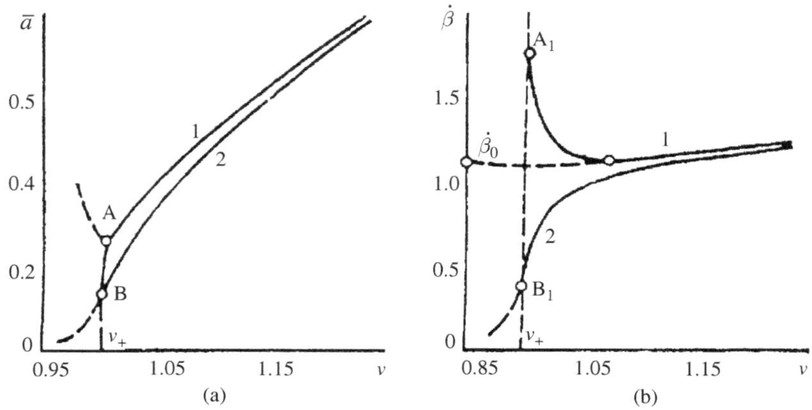

Figure 4.11 Amplitude (a) and phase (b) frequency response curves near resonance. (Koval'chuk and Podchasov 1996)

where the upper sign corresponds to $\tan\vartheta = 1$ and the lower sign to $\tan\vartheta = -1$.

The above solution is stable over an excitation frequency given by the following inequality

$$(\omega_1^2 - \Omega^2) < \frac{b_1}{2(4)^{1/3}}\left(\frac{-\Omega^2\gamma X_0}{b_1(\omega_1^2 - \Omega^2)}\right)^{2/3}\left(\left(\omega_1^2 - \Omega^2\right) - 9\left(\omega_1^2 - \frac{\Omega^2}{3}\right)\right) \tag{4.87a}$$

If $(\omega_1^2 - \Omega^2)$ exceeds the above boundary, the liquid free surface would experience chaotic motion within the frequency range

$$\frac{b_1}{2(4)^{1/3}}\left(\frac{-\Omega^2\gamma X_0}{b_1(\omega_1^2 - \Omega^2)}\right)^{2/3}\left(\left(\omega_1^2 - \Omega^2\right) - 9\left(\omega_1^2 - \frac{\Omega^2}{3}\right)\right) < \frac{9}{2}\left(\Omega^4\gamma^2 X_0^2 \frac{b_1}{18}\left(\omega_1^2 - \frac{\Omega^2}{3}\right)\right)^{1/3} \tag{4.87b}$$

The free surface may possess a standing wave if

$$\frac{b_1}{2(4)^{1/3}}\left(\frac{-\Omega^2\gamma X_0}{b_1(\omega_1^2 - \Omega^2)}\right)^{2/3}\left((\omega_1^2 - \Omega^2) - 9\left(\omega_1^2 - \frac{\Omega^2}{3}\right)\right) > \frac{9}{2}\left(\Omega^4\gamma^2 X_0^2 \frac{b_1}{18}\left(\omega_1^2 - \frac{\Omega^2}{3}\right)\right)^{1/3} \tag{4.87c}$$

Figure 4.11(a) shows the dependence of the maximum amplitude wave height parameter, $\bar{a} = a_{max}/R$, and phase rate parameter, $\dot{\beta} = \dot{\alpha}_{max}/\omega_1$, on the excitation frequency ratio, $\nu = \Omega/\omega_1$, for liquid depth ratio, $h/R = 3$. Upper curves 1 refer to the maximum steady-state amplitudes and phase velocities, while curves 2 correspond to their minimum values. Stable solutions are indicated by solid curves while dot curves belong to unstable solutions. It is seen that the maximum amplitudes of the free surface monotonically increase with the excitation frequency, with their minimum values occur at $\nu = 1$. Figure 4.11(b) shows the phase velocity $\dot{\beta}$ where the

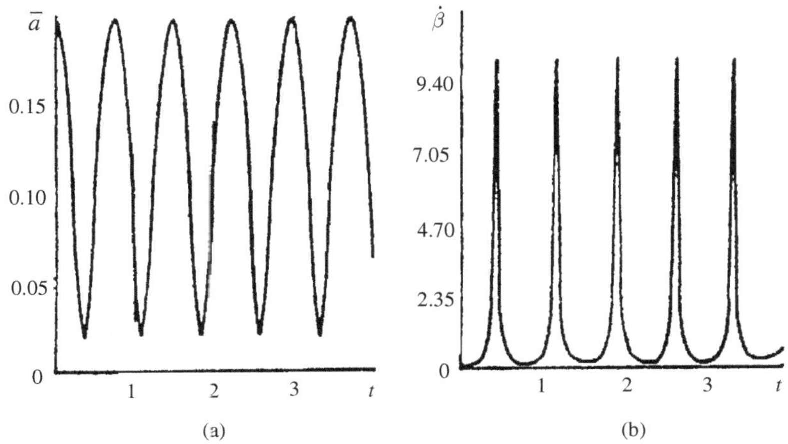

Figure 4.12 Amplitude (a) and phase (b) time history records near resonance during rotational sloshing. (Koval'chuk and Podchasov 1996)

maximum branch decreases with the excitation frequency and reaches its minimum value at $\nu = 1.07$ above which it experiences a slight increase. Careful inspection of the amplitude and phase velocity plots indicates that the maximum and minimum branches approach each other as the excitation frequency increases. This means that the free-surface parameters achieve constant nonoscillatory values. The solution branches over the region, $\nu < 1$, shown by dash curves, correspond to a plane or chaotic motion as indicated by Kubenko, *et al.* (1992).

Under sine sweep excitation, $x(t) = X_0 \cos(\Omega_0 + \varepsilon t)t$, where ε is a small positive parameter and with $\Omega_0 = 0.9\omega_1$, Koval'chuk and Podchasov (1996) studied the behavior of the free surface with initial conditions selected such that $\dot{\alpha}(0) \neq 0$. They solved equations (4.82) numerically for $\dot{\alpha}(0) = 0.1\omega_1$, $a(0) = 0.2R$, $\dot{a}(0) = 0$, and $\alpha(0) = 0$. Figures 4.12(a) and (b) shows the time history records of the amplitude and phase velocity. Both amplitude and phase velocity exhibit periodic motions and both experience energy exchange. The phase velocity assumes only positive values implying rotation in one direction compatible with the one given initially. The displacement of the wave crest in the circular direction corresponds to a "pulsed" mode over short intervals of time when the wave amplitudes are small. The corresponding position of the crest along the circular coordinate θ changes abruptly and the wave motion is "retarded." These features also occur in the case of quasi-periodic motion of the liquid container.

4.3 Random excitation

4.3.1 Background

In practice, the excitation is not periodic but random in nature, for example, liquid storage tanks subjected to earthquake motion, random sea waves acting on liquid cargo ships, and g-jitter in a microgravity field. In this case, the excitation must be modeled as a random process. The analysis of random processes includes probability theory and the theory of stochastic differential equations. A random process is specified in terms of statistical functions such as mean squares, probability density functions, autocorrelation functions, and power spectral density functions.

The prediction of these statistics for nonlinear liquid sloshing is not a simple task. However, experimental measurements can be processed to estimate these functions. Errors are usually encountered in the data processing of experimental measurements. Analytical methods include Markov methods (for which the Fokker–Planck–Kolmogorov equation and the Itô stochastic calculus are applicable), Gaussian and non-Gaussian closure schemes, stochastic averaging methods, equivalent linearization methods, perturbation techniques, and Volterra–Weiner functional analysis. These methods are well documented (see, e.g., Lin, 1967, Ibrahim, 1985, and Roberts and Spanos, 1990). They have been applied to systems with various types of nonlinearities. The liquid sloshing nonlinearity is essentially of the inertia type, and, thus, some of the mentioned methods cannot be applicable. Numerical simulations are usually applied when analytical methods fail or are used to verify the validity of analytical results if experimental results are not available.

The early attempts to study the stochastic excitation of free liquid surface motion included experimental investigations conducted by Dalzell (1967b) with the purpose of studying the swirl instability. Aslam, *et al.* (1979), Sakata, *et al.* (1983, 1984a, b, 1991), Utsumi, *et al.* (1983, 1984, 1986, 1987), and Kimura, *et al.* (1995) examined analytically the nonstationary response of nonlinear liquid motion in a cylindrical tank subjected to lateral random excitation. The excitation was modeled by an amplitude modulated nonwhite random process with a dominant frequency. The response statistical moment equations of three liquid sloshing modes formed an infinite hierarchy of coupled equations and were truncated through a heuristic approach. The nonlinear solution of these equations revealed a nonzero mean response of the liquid wave height although the excitation has a zero mean random process. Brocchini, *et al.* (1997) reported violent free-surface motion in their experimental investigation. Mathiessen (1976) determined the sloshing loads under random pitching excitation.

4.3.2 Experimental observations and results

Experimental investigations of random excitation of nonlinear systems are few although they are valuable in revealing complex dynamic behavior (Ibrahim, 1992). In this section we will discuss some experimental results of random excitation of liquid containers reported by Dalzell (1967b). Typical excitation spectra used in Dalzell (1967b) are shown in Figure 4.13. Under very narrow band-1 random excitation, with a half-power bandwidth of 0.11 rad/s, the free surface experienced similar behavior to that of sinusoidal excitation only when the excitation level was high. Violent rotation was observed to build up, decay, and change direction in a manner analogous to sinusoidal excitation. When the bandwidth was increased to 0.45 rad/s, "narrow band-2" the same qualitative behavior of the free surface was observed, however, the rotation was perceptibly less violent. For band-pass excitation (with half-power bandwidth = 1.77 rad/s), very little rotation was observed, and this disappears when the excitation becomes wide band.

The force spectra resulting from narrow band-2 excitation of two different levels are shown in Figure 4.14. For the higher excitation level, rotation was observed and the spectrum of force component normal to the direction of excitation is almost identical with the spectrum of force parallel to the excitation. When the excitation energy level was reduced from 5.83×10^{-3} to 7.52×10^{-4}, the spectrum of the force normal to the direction of excitation is about 1/100th the spectrum of force parallel to the excitation. The dependence of the ratio of the peak spectral

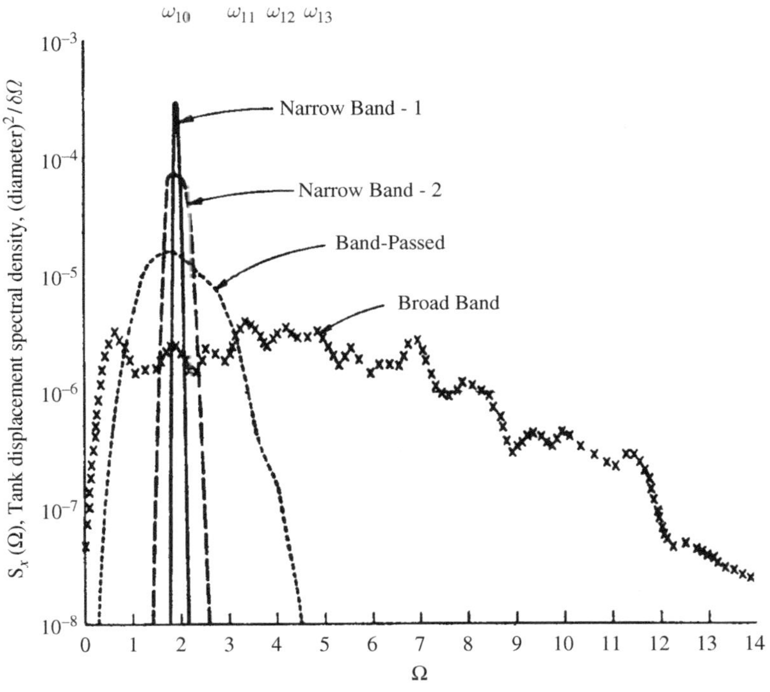

Figure 4.13 Excitation spectra for different bandwidths. (Dalzell, 1967b)

density of the force normal to the direction of excitation, $S_{F_{90}}$, to the peak spectral density of the force parallel to excitation, S_{F_0}, that is, $S_{F_{90}}/S_{F_0}$, on the tank displacement spectral density, (diameter)$^2/(\delta\Omega)$, is shown in Figure 4.15. When the ratio of force spectra $S_{F_{90}}/S_{F_0}$ is 1.0, violent rotation takes place, and diminishes as the ratio is significantly reduced.

Under Gaussian random excitation, the liquid sloshing force parallel to the excitation direction was found to be close to a normal distribution. However, the distribution of the liquid force component perpendicular to the direction of excitation (where rotational motion of the free surface took place under high narrow band excitation) deviated significantly from normality.

4.3.3 Stochastic analysis of earthquake excitation of liquid rigid tanks

Consider an upright cylindrical tank subjected to an amplitude modulated nonwhite random excitation, $f(t)$, along the x-axis, with a dominant frequency. Sakata, *et al.* (1984a) used the variational principle together with Galerkin's method to derive discrete dynamical equations describing the generalized coordinates of the velocity potential function and wave height. The velocity potential function and wave height are expressed by the following series

$$\Phi = \sum_{m=0}^{\infty}\sum_{n=1}^{\infty} \alpha_{mn}(t)\cos m\theta J_m(\lambda_{mn}r)\frac{\cosh\{\lambda_{mn}(z+h)\}}{\cosh(\lambda_{mn}h)} \tag{4.88}$$

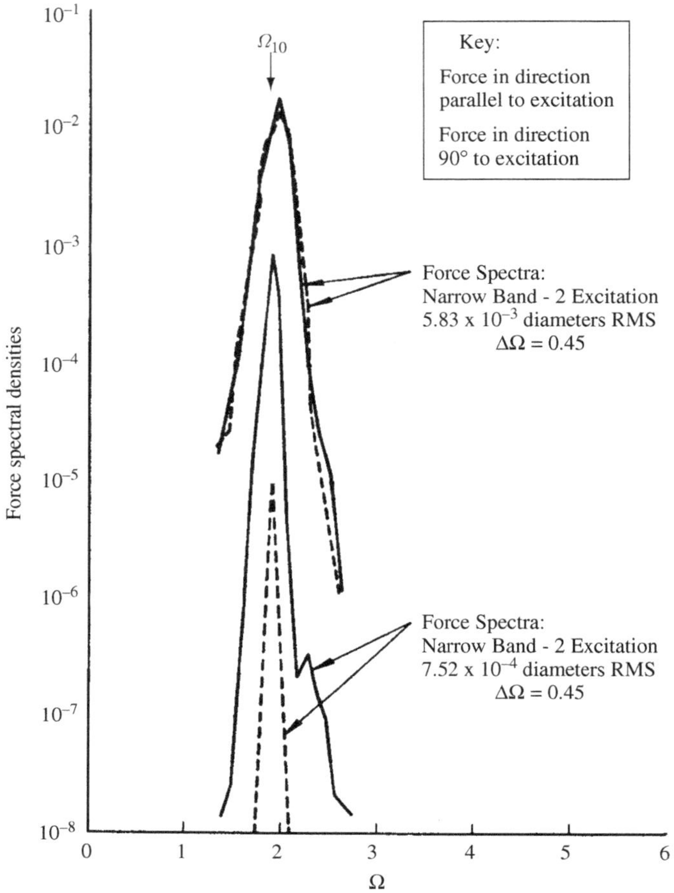

Figure 4.14 Liquid force spectra under two different excitation levels. (Dalzell, 1967b)

$$\eta=\sum_{m=0}^{\infty}\sum_{n=1}^{\infty}a_{mn}(t)\cos m\theta J_m(\lambda_{mn}r) \tag{4.89}$$

where m and n are the circumferential and radial mode numbers, respectively. Introducing the following nondimensional parameters

$$\begin{aligned}
&\{\bar r,\ \bar x,\ \bar z,\ \bar h\}=\{r,\ x,\ z,\ h\}/R,\quad \tau=\omega_{11}t,\quad \bar\Phi=\Phi/(\omega_{11}R^2),\quad \bar\eta=\eta/R,\\
&\bar\lambda_m=\lambda_{m1}R,\quad \bar f(\tau)=f(t)/R,\quad f''(\tau)=\ddot f(t)/(\omega_{11}^2R),\quad \varpi_m=\omega_{m1}/\omega_{11}=\sqrt{\frac{\Psi_m}{\Psi_1}},\\
&\Psi_m=\bar\lambda_m\tanh(\lambda_{m1}h),\quad \bar a_m(\tau)=a_{m1}(t)/(\omega_{11}R^2),\quad \bar a_m(\tau)=\alpha_{m1}(t)/R
\end{aligned} \tag{4.90}$$

and applying Galerkin's method to the dynamic and kinematic boundary conditions after using relations (4.88) and (4.89), one obtains the following set of ordinary differential equations for the generalized coordinates $\bar\alpha_m$ and $\bar a_m$ (for $m=0$, 1, 2):

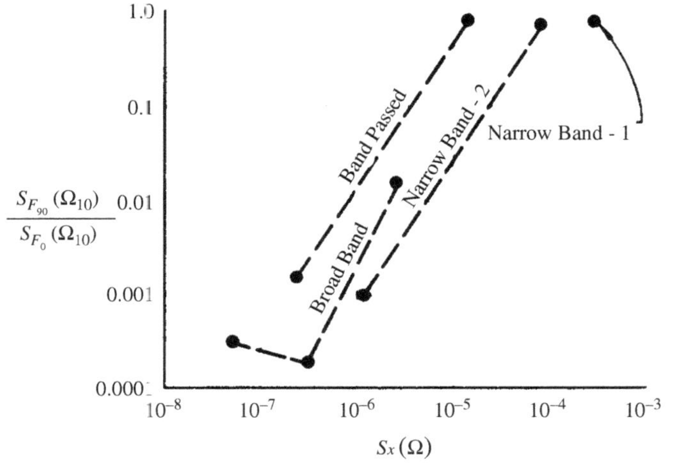

Figure 4.15 Dependence of the ratio of force spectra peaks (normal and parallel to excitation) on tank displacement spectral density at the first anti-symmetric mode frequency. (Dalzell, 1967b)

$$\alpha_1' = -\frac{1}{\tau_1}a_1 - S_1 f''(\tau) + H^{(1)}\alpha_1^2 a_1 + H^{(2)}a_1^3 + H^{(11)}\alpha_1\alpha_0 + H^{(12)}a_1 a_0 + H^{(13)}\alpha_1\alpha_2 + H^{(14)}a_1 a_2 \tag{4.91a}$$

$$a_1' = \psi_1\alpha_1 + H^{(3)}\alpha_1 a_1^2 + H^{(15)}\alpha_1 a_0 + H^{(16)}a_1\alpha_0 + H^{(17)}\alpha_1 a_2 + H^{(18)}a_1\alpha_2 \tag{4.91b}$$

$$\alpha_0' = -\frac{1}{\psi_1}a_0 + H^{(21)}\alpha_1^2 + H^{(22)}a_1^2 \tag{4.91c}$$

$$a_0' = \Psi_0\alpha_0 + H^{(23)}\alpha_1 a_1 \tag{4.91d}$$

$$\alpha_2' = -\frac{1}{\psi_1}a_2 + H^{(24)}\alpha_{\_}^2 + H^{(25)}a_1^2 \tag{4.91e}$$

$$a_2' = \psi_2\alpha_2 + H^{(26)}\alpha_1 a_{\_} \tag{4.91f}$$

where the over bar has been dropped, S_1 is a constant given in the form of a definite integral, and $H^{(i)}$ are the nonlinear coefficients that depend on the liquid depth ratio h/R and are listed in Sakata, *et al.* (1984a). Equations (4.91) were derived after expanding the velocity function in Taylor series around $z=0$. The following orders are made

$$f'(\tau) \approx O(\varepsilon), \qquad \alpha_1, a_1 \approx O(\varepsilon^{1/3}), \qquad \alpha_0, a_0, \alpha_2, a_2 \approx O\left(\varepsilon^{2/3}\right) \tag{4.92}$$

where ε denotes the order of excitation, and terms up to order ε are retained in equations (4.91). Eliminating the variables α_i from equations (4.91), gives a system of second-order differential equations for the generalized coordinates, a_i,

$$a_1'' + 2\zeta_1 a_1' + a_1 + P^{(1)} a_1^3 + P^{(2)} a_1 a_1''^2 + Q_0^{(1)} a_1 a_0 + Q_0^{(2)} a_1' a_0' + Q_2^{(1)} a_1 a_2 + Q_2^{(1)} a_1' a_2' = -\psi_1 S_1 f''(\tau) \tag{4.93a}$$

$$a_0'' + 2\zeta_0 \varpi_0 a_0' + \varpi_0^2 a_0 = R_0^{(1)} a_1^2 + R_0^{(1)} {a_1'}^2 \tag{4.93b}$$

$$a_2'' + 2\zeta_2 \varpi_2 a_2' + \varpi_2^2 a_2 = R_2^{(1)} a_1^2 + R_2^{(1)} {a_1'}^2 \tag{4.93c}$$

where linear damping terms, with damping factors ζ_i, have been introduced to account for energy dissipation, and the new coefficients are related to the coefficients of equations (4.91). Note that the first mode $m = 1$ is directly excited by the external excitation, while modes $m = 0, 2$ are indirectly excited through nonlinear coupling with mode $m = 1$.

The base acceleration can be modeled by an amplitude modulated nonwhite random process, that is,

$$f''(\tau) = e(\tau) n(\tau), \; E[n(\tau)] = 0, \; R_n(\Delta\tau) = E[n(\tau) n(\tau + \Delta\tau)] \tag{4.94}$$

where E denotes expectation, $e(t)$ is a deterministic envelope function, and $n(t)$ is a Gaussian stationary nonwhite noise with zero mean and autocorrelation function $R_n(\Delta\tau)$. Expressing the coordinates $a_m(\tau)$ and their time derivative $a_m'(\tau)$, for $m = 0, 2$ in terms of the convolution integrals after using equations (4.93b) and (4.93c)

$$a_m'(\tau) = \int_0^\tau h_m(\tau - \hat{\tau}) \left(R_0^{(1)} a_1^2(\hat{\tau}) + R_0^{(1)} {a_1'}^2(\hat{\tau}) \right) d\hat{\tau}, \tag{4.95a}$$

$$a_m'(\tau) = \int_0^\tau h_m'(\tau - \hat{\tau}) \left(R_0^{(1)} a_1^2(\hat{\tau}) + R_0^{(1)} {a_1'}^2(\hat{\tau}) \right) d\hat{\tau} \tag{4.95b}$$

where $h(\tau)$ is the unit impulse function given by the expression

$$h(\tau) = \frac{1}{\varpi_m \sqrt{1 - \zeta_m^2}} e^{-\zeta_m \varpi_m \tau} \sin\left(\varpi_m \sqrt{1 - \zeta_m^2}\, \tau \right) \tag{4.95c}$$

the following sets of statistical moments are developed based on relations (4.92)

(i) Moments of first mode $m = 1$

$$E[a_1], E[a_1'] \approx O(\varepsilon^{1/3}) \tag{4.96a}$$

$$E[a_1^2], E[a_1 a_1'], E[{a_1'}^2] \approx O(\varepsilon^{2/3}) \tag{4.96b}$$

$$E[a_1^3], E[a_1^2 a_1'], E[a_1 {a_1'}^2], E[{a_1'}^3] \approx O(\varepsilon^{3/3}) \tag{4.96c}$$

$$E[a_1^4], E[a_1'^3 a_1], E[a_1^2 a_1'^2], E[a_1 a_1'^3], E[a_1'^4] \approx O(\varepsilon^{4/3}) \tag{4.96d}$$

(ii) Moments of first mode $m = 0, 2$

$$E[a_m], E[a'_m] \approx O(\varepsilon^{2/3}) \tag{4.97a}$$

$$E[a_m^2], E[a_m a'_m], E[a'^2_m] \quad \text{and}$$

$$E[a_0 a_2], E[a_0 a'_2], E[a'_0 a_2], E[a'_0 a_2] \approx O(\varepsilon^{4/3}) \tag{4.97b}$$

(iii) Mixed moments of mode $m = 1$ and $m = 0, 2$

$$E[a_1 a_m], E[a_1 a'_m], E[a'_1 a_m], E[a'_1 a'_m] \approx O(\varepsilon^{3/3}) \tag{4.98a}$$

$$E[a_1^2 a_m], E[a_1, a'_1 a_m], E[a'^2_1 a_m], E[a_1^2 a'_m], E[a_1, a'_1 a'_m], E[a'^2_1 a'_m] \approx O(\varepsilon^{4/3}) \tag{4.98b}$$

Under Gaussian excitation, one may assume that the response coordinates a_1 and a'_1 do not deviate significantly from normality since the nonlinear terms in equation (4.93a) are of order ε and much smaller than the linear terms of $O(\varepsilon^{1/3})$. On the other hand, all other coordinates of modes $m = 0, 2$ will be non-Gaussian since these modes are indirectly excited through nonlinear terms of mode $m = 1$. In view of these properties, some of the moments presented in the order lists (4.96) to (4.98) can be expressed in terms of products of lower-order moments. The first-order moments in (4.96a) are zero, since the excitation is assumed to have zero mean, and the nonlinear terms have symmetry with respect to a_1 and a'_1. The third-order moments in (4.96c) are zero, and the fourth-order moments (4.96d) are expressed as the product of second-order moments represented by (4.96b) due to the Gaussian property of the responses a_1 and a'_1. Furthermore, the coupled second-order moments (4.98a) may be shown to be zero by multiplying equations (4.95a) and (4.95b) by a_1 and a'_1, taking expectations and then using the Gaussian property of a_1, a'_1, and $f'(\tau)$.

The response moments of the first mode can be determined from equation (4.93a) using the equivalent linearization technique. Equation (4.93a) may be written in the equivalent linearized form

$$a''_1 + 2\zeta_e a'_1 + k_e a_1 = -\psi_1 S_1 f''(\tau) \tag{4.99}$$

where ζ_e and k_e are the equivalent linear coefficients and are functions of time. These are expressed in terms of the moments listed in (4.96b) and (4.98b) as

$$\zeta_e = \zeta_1 + P^{(2)} E[a_1 a'_1] + \frac{1}{2D} \sum_{m=0,2} \Big[Q_m^{(1)} \{ E[a_1^2] E[a_1 a'_1 a_m] - E[a_1 a'_1] E[a_1^2 a_m] \} \\ + Q_m^{(2)} \{ E[a_1^2] E[a'^2_1 a'_m] - E[a_1 a'_1] E[a_1 a'_1 a'_m] \} \Big] \tag{4.100a}$$

$$k_e = 1 + 3P^{(1)} E[a_1^2] + P^{(2)} E[a'^2_1] + \frac{1}{2D} \sum_{m=0,2} \Big[Q_m^{(1)} \{ E[a'^2_1] E[a_1^2 a_m] \\ - E[a_1 a'_1] E[a_1 a'_1 a_m] \} + Q_m^{(2)} \{ E[a'^2_1] E[a_1 a'_1 a'_m] - E[a_1 a'_1] E[a'^2_1 a'_m] \} \Big] \tag{4.100b}$$

where $D = E[a_1^2]E[a_1'^2] - \left(E[a_1 a_1']\right)^2$.

The moment equations are derived from equation (4.99) and are expressed as

$$E'[a_1^2] = 2E[a_1 a_1'] \tag{4.101a}$$

$$E'[a_1 a_1'] = E[a_1'^2] - 2\zeta_e E[a_1 a_1'] - K_e E[a_1^2] - \psi_1 S_1 E[a_1 f''] \tag{4.101b}$$

$$E'[a_1'^2] = -4\zeta_e E[a_1'^2] - 2K_e E[a_1' a_1] - 2\psi_1 S_1 E[a_1' f''] \tag{4.101c}$$

The time dependence of the equivalent linear coefficients should be considered in the nonstationary response. One takes the time sequence $\tau_0, \tau_1, \tau_2, \ldots$ and assumes that the equivalent linear coefficients take on constant values determined from the moments at $\tau = \tau_{i-1}$, during each time interval $[\tau_{i-1}, \tau_i)$. Introducing the following notations

$$\zeta_e^{(i)} = \zeta_e(\tau) = \zeta_e(\tau_{i-1}), k_e^{(i)} = k_e(\tau) = k_e(\tau_{i-1}), \tau \in [\tau_{i-1}, \tau_i) \tag{4.102}$$

In equations (4.101), the excitation terms $E[a_1(\tau) f'(\tau)]$ and $E[a_1'(\tau) f'(\tau)]$, $\tau \in [\tau_{n-1}, \tau_{ni})$ are determined from the following recurrence relations (Sakata *et al.*, 1983 and Kimura *et al.*, 1995)

$$E[a_1(\tau_i) f''(\tau)] = p_e^{(i)}(\Delta\tau) E[a_1(\tau_{i-1}) f''(\tau)] + q_e^{(i)}(\Delta\tau) E[a_1'(\tau_{i-1}) f''(\tau)] + \int_{\tau_{i-1}}^{\tau_i} h_e^{(i)}(\tau_i - \hat{\tau}) E[f''(\tau) f''(\hat{\tau})]\, d\hat{\tau} \tag{4.103a}$$

$$E[a_1'(\tau_i) f''(\tau)] = p_e'^{(i)}(\Delta\tau) E[a_1(\tau_{i-1}) f''(\tau)] + q_e'^{(i)} \Delta\tau E[a_1'(\tau_{i-1}) f''(\tau)] + \int_{\tau_{i-1}}^{\tau_i} h_e'^{(i)}(\tau_i - \hat{\tau}) E[f''(\tau) f''(\hat{\tau})]\, d\hat{\tau} \tag{4.103b}$$

$$E[a_1(\tau) f''(\tau)] = p_e^{(n)}(\tau - \tau_{n-1}) E[a_1(\tau_{ni-1}) f''(\tau)] + q_e^{(i)}(\tau - \tau_{n-1}) E[a_1'(\tau_{n-1}) f''(\tau)] + \int_{\tau_{n-1}}^{\tau} h_e^{(n)}(\tau - \hat{\tau}) E[f''(\tau) f''(\hat{\tau})]\, d\hat{\tau} \tag{4.104a}$$

$$E[a_1'(\tau) f''(\tau)] = p_e'^{(n)}(\tau - \tau_{n-1}) E[a_1(\tau_{ni-1}) f''(\tau)] + q_e'^{(i)}(\tau - \tau_{n-1}) \times E[a_1'(\tau_{n-1}) f''(\tau)] + \int_{\tau_{n-1}}^{\tau} h_e'^{(n)}(\tau - \hat{\tau}) E[f''(\tau) f''(\hat{\tau})]\, d\hat{\tau} \tag{4.104b}$$

where $(i = 1, 2, n-1)$, $\Delta\tau = \tau_i - \tau_{i-1}$, $h_e^{(i)}$ is the unit impulse response function during the time interval $[\tau_{i-1}, \tau_i)$, $p_e^{(i)}$ and $q_e^{(i)}$ are functions corresponding to the free vibration during the same interval. These parameters are expressed as follows

$$
\begin{aligned}
h_e^{(i)}(\tau) &= -\frac{\Psi_1 S_1}{\varpi_e^{(i)}} e^{-\zeta_e^{(i)}\tau} \sin \varpi_e^{(i)}\tau \\
p_e^{(i)}(\tau) &= e^{-\zeta_e^{(i)}\tau}\left(\cos(\varpi_e^{(i)}\tau) + \frac{\zeta_e^{(i)}}{\varpi_e^{(i)}}\sin(\varpi_e^{(i)}\tau)\right) \\
q_e^{(i)}(\tau) &= \frac{1}{\varpi_e^{(i)}} e^{-\zeta_e^{(i)}\tau}\sin(\varpi_e^{(i)}\tau) \\
\varpi e^{(i)} &= \sqrt{k_e^{(i)} - \zeta_e^{(i)2}}
\end{aligned}
\tag{4.105}
$$

Giving the initial conditions, $E[a_1(\tau_0)f'(\tau_0)]$ and $E[a_1'(\tau_0)f'(\tau_0)]$, one can determine the moments $E[a_1(\tau)f'(\tau)]$ and $E[a_1'(\tau)f'(\tau)]$ at the reference time $\tau \in [\tau_{n-1}, \tau_{ni})$ by substituting into the equations (4.104) the values of $E[a_1(\tau_{n-1})f'(\tau)]$ and $E[a_1'(\tau_{n-1})f'(\tau)]$ as determined from equations (4.103).

The moment equations of the first- and second-orders listed in (4.97a,b) are constructed from equations (4.93b) and (4.93c), and given in the form

(1) First-order moments for modes $m = 0, 2$ are

$$E'[a_m] = E[a_m'] \tag{4.106a}$$

$$E'[a_m'] = -2\zeta_m\varpi_m E[a_m'] - \varpi_m^2 E[a_m] + R_m^{(1)} E[a_1^2] + R_m^{(2)} E[a_1'^2] \tag{4.106b}$$

(2) Second-order moments for modes $m = 0, 2$ are

$$E'[a_m^2] = 2E[a_m a_m'] \tag{4.107a}$$

$$E'[a_m a_m'] = E[a_m'^2] - 2\zeta_m\varpi_m E[a_m a_m'] - \varpi_m^2 E[a_m^2] + R_m^{(1)} E[a_1^2 a_m] + R_m^{(2)} E[a_1'^2 a_m] \tag{4.107b}$$

$$E'[a_m'^2] = -4\zeta_m\varpi_m E[a_m'^2] - 2\varpi_m^2 E[a_m a_m'] + 2R_m^{(1)} E[a_1^2 a_m'] + 2R_m^{(2)} E[a_1'^2 a_m'] \tag{4.107c}$$

$$E'[a_0 a_2] = E[a_0' a_2] + [a_0 a_2'] \tag{4.107d}$$

$$E'[a_0 a_1'^2] = -2\zeta_2\varpi_2 E[a_0 a_2'] - \varpi_2^2 E[a_0 a_2] + E[a_0' a_2'] + R_2^{(1)} E[a_1^2 a_0] + R_2^{(2)} E[a_1'^2 a_0] \tag{4.107e}$$

$$E'[a_0' a_2] = -2\zeta_0\varpi_0 E[a_0' a_2] - \varpi_0^2 E[a_0 a_2] + E[a_0' a_2'] + R_0^{(1)} E[a_1^2 a_2] + R_0^{(2)} E[a_1'^2 a_2] \tag{4.107f}$$

$$
\begin{aligned}
E'[a_0' a_2'] = &-2(\zeta_0\varpi_0 + \zeta_2\varpi_2)E[a_0' a_2'] - \varpi_0^2 E[a_0 a_2'] - \varpi_2^2 E[a_0' a_2] + R_0^{(1)} E[a_1^2 a_2'] \\
&+ R_0^{(2)} E[a_1'^2 a_2'] - R_2^{(1)} E[a_1^2 a_0'] + R_2^{(2)} E[a_1'^2 a_0']
\end{aligned}
\tag{4.107g}
$$

Third-order moment equations of moments listed in (4.98a,b) are determined from equations (4.93b,c) and (4.99) and are expressed in the form

$$E'[a_1^2 a_m] = 2E[a_1 a_1' a_m] + E[a_1^2 a_m'] \tag{4.108a}$$

$$E'[a_1 a_1' a_m] = 2E[a_1'^2 a_m] - 2\zeta_e E[a_1 a_1' a_m] - k_e E[a_1^2 a_m] + E[a_1 a_1' a_m'] \tag{4.108b}$$

$$E'[a_1'^2 a_m] = -4\zeta_e E[a_1'^2 a_m] - 2k_e E[a_1 a_1' a_m] + E[a_1'^2 a_m'] \tag{4.108c}$$

$$\begin{aligned} E'[a_1^2 a_m'] = {} & 2E[a_1 a_1' a_m'] - 4\zeta_e \varpi_m E[a_1^2 a_m'] - \varpi_m^2 E[a_1^2 a_m] \\ & + 3R_m^{(1)} \left(E[a_1^2]\right)^2 + R_m^{(2)} \left\{ E[a_1^2] E[a_1'^2] + 2\left(E[a_1 a_1']\right)^2 \right\} \end{aligned} \tag{4.108d}$$

$$\begin{aligned} E'[a_1 a_1' a_m'] = {} & E[a_1'^2 a_m'] - 2(\zeta_e + \varpi_m \zeta_m) E[a_1 a_1' a_m'] - k_e E[a_1^2 a_m'] \\ & - \varpi_m^2 E[a_1 a_1' a_m] + 3R_m^{(1)} E[a_1^2] E[a_1 a_1'] + 3R_m^{(2)} E[a_1 a_1'] E[a_1'^2] \end{aligned} \tag{4.108e}$$

$$\begin{aligned} E'[a_1'^2 a_m'] = {} & -2(2\zeta_e + \zeta_m \varpi_m) E[a_1'^2] - 2k_e E[a_1 a_1' a_m'] - \varpi_m^2 E[a_1'^2 a_m] \\ & + 3R_m^{(2)} \left(E[a_1'^2]\right)^2 + R_m^{(1)} \left\{ E[a_1'^2] E[a_1'^2] + 2\left(E[a_1 a_1']\right)^2 \right\} \end{aligned} \tag{4.108f}$$

where higher-order terms have been neglected.

Equations (4.101), and (4.106) through (4.108) form a closed set of first-order differential equations, which can be solved numerically. The mean and variance responses of the liquid free-surface displacement are determined using equation (4.89) as follows

$$\mu = E[\eta] = E[a_0] J_0(\lambda_0 r) + E[a_2] J_2(\lambda_2 r) \cos 2\theta \tag{4.109a}$$

$$\begin{aligned} \sigma^2 = \left(E[\eta^2] - E[\eta]\right)^2 = {} & E[a_1^2] J_1^2(\lambda_1 r) \cos^2\theta \\ & + E[a_0^2] J_0^2(\lambda_0 r) + E[a_2^2] J_2^2(\lambda_2 r) \cos^2 2\theta \\ & + 2E[a_0 a_2] J_0(\lambda_0 r) J_2(\lambda_2 r) \cos 2\theta - \mu^2 \end{aligned} \tag{4.109b}$$

Note that this nonlinear analysis predicts nonzero mean for the free-surface wave height when the excitation has a zero mean. Recall that the linear theory yields zero mean and variance $\sigma^2 = E[a_1^2] J_1^2(\lambda_1 r) \cos^2\theta$.

Sakata, *et al.* (1984a) considered an exponential decaying harmonic correlation function of the random process, $n(\tau)$, given in the form

$$R_n(\Delta\tau) = R_0\, e^{-\gamma|\Delta\tau|} \cos(\nu\Delta\tau) \tag{4.110a}$$

The power spectral density of this process is

$$S_n(\omega) = R_0 \frac{\gamma(\omega^2 + \nu^2 + \gamma^2)}{\pi\left\{(\omega^2 - \nu^2 - \gamma^2)^2 + 4\gamma^2\omega^2\right\}} \tag{4.110b}$$

where γ is the exponential decay constant and ν is the dominant frequency. These parameters are expressed by the following ratios

$$A = \gamma/\zeta_1, \qquad B = \nu\Big/\sqrt{1 - \zeta_1^2} \tag{4.111}$$

The relationship between the nondimensional root-mean-square (rms) of R_0 and the dimensional value, R_0^*, is

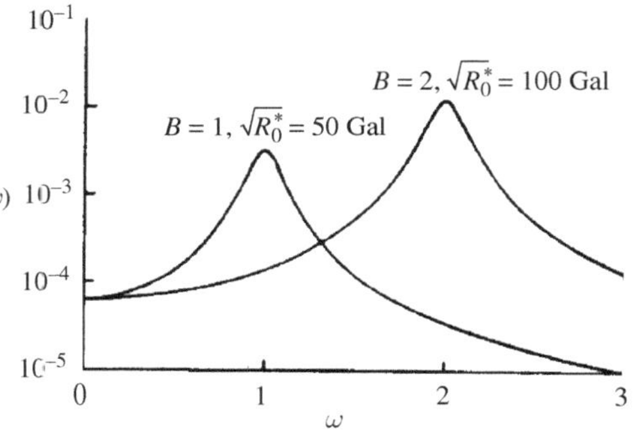

Figure 4.16 Power spectra of excitation for damping factor $\zeta_1 = 0.01$, $A = 10$, and two different values of center frequency parameter B. (Sakata, *et al.*, 1984a)

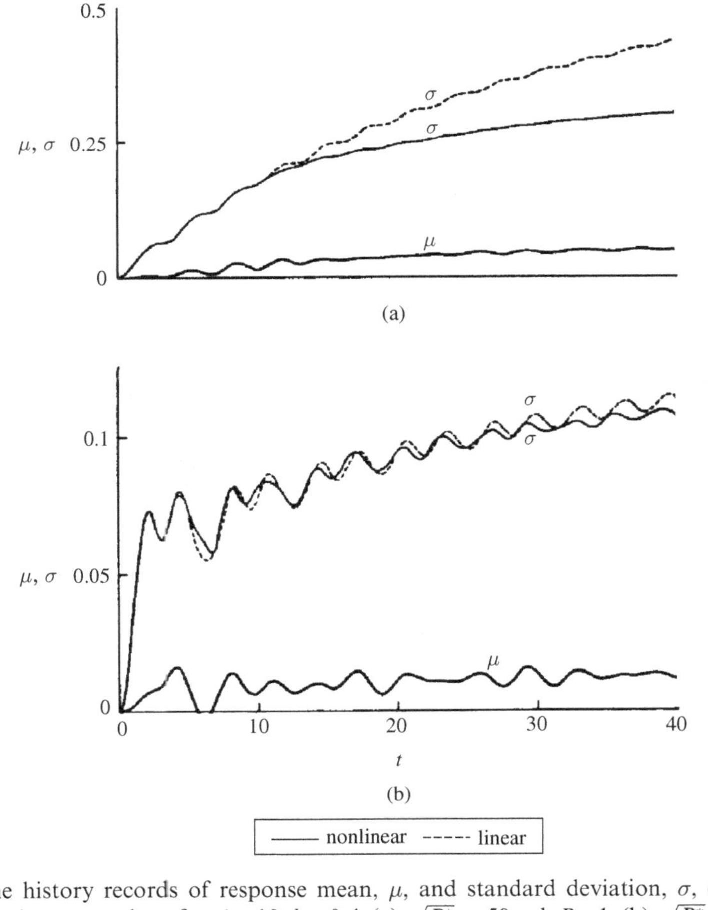

Figure 4.17 Time history records of response mean, μ, and standard deviation, σ, of liquid surface displacement to unit step envelope for $A = 10$, $h = 0.4$. (a) $\sqrt{R_0^*} = 50$ gal, $B = 1$, (b) $\sqrt{R_0^*} = 100$ gal, $B = 2$. (Sakata, *et al.*, 1984a)

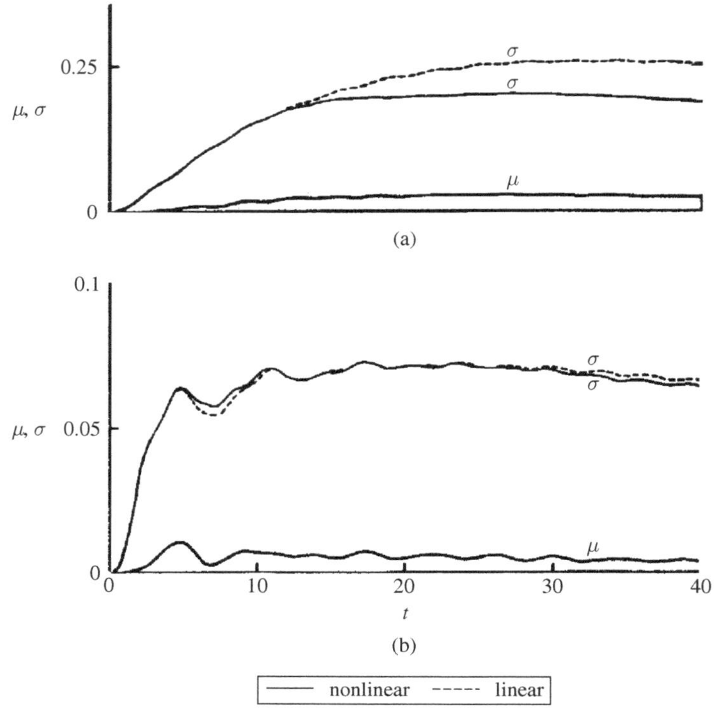

Figure 4.18 Time history records of response mean, μ, and standard deviation, σ, of liquid surface displacement to exponential envelope for $A=10$, $h=0.4$. (a) $\sqrt{R_0^*}=50$ Gal, $B=1$, (b) $\sqrt{R_0^*}=100$ Gal, $B=2$. (Sakata, *et al.*, 1984a)

$$\sqrt{R_0} = \sqrt{R_0^*}/(R\omega_1^2) = \sqrt{R_0^*}/(g\psi_1) \tag{4.112}$$

Recall that R is the tank radius and g is the gravitational acceleration.

Figure 4.16 shows the power spectra of excitation for two different values of dominant frequency parameter $B=1$ and 2. The value of $B=1$ corresponds to the resonance case and $B=2$ corresponds to the case of an actual earthquake-liquid storage tank system where the dominant frequency is considerably higher than the damped natural frequency of the linear liquid motion. A unit step function, $U(\tau)$, or an exponential function is used as an envelope function, that is,

$$e(\tau) = U(\tau), \quad e(\tau) = \frac{\left(e^{-a\tau} - e^{-b\tau}\right)}{\max\left(e^{-a\tau} - e^{-b\tau}\right)} \tag{4.113a, b}$$

Figures 4.17 and 4.18 show the transient mean and standard deviation of liquid surface displacement response to the unit step envelope function and the exponential envelope function, respectively. The dash curves refer to response statistics obtained by using linear analysis, for the case of small liquid surface amplitudes. Note that the mean response vanishes according to the linear theory. The rms values of the input noise are $\sqrt{R_0^*}=50$ Gal for $B=1$, and $\sqrt{R_0^*}=100$ Gal for $B=2$, where gal refers to a unit of acceleration, 1 Gal $=1$ cm/s^2. Figures 4.17(a) and 4.18(a) reveal that a small earthquake excitation may result in a significant response variance if resonance occurs. The variance predicted by the linear theory is relatively higher than the one estimated by the nonlinear analysis.

4.4 Nonlinear phenomena in rectangular tanks

4.4.1 Background

Nonlinear theories of forced sloshing in a rectangular tank were developed by Bauer (1965b, Verhagen and WijnGaarden (1965), Faltinsen (1974), Khosropour, *et al.* (1995), Young-Sun and Chung-Bang (1996), Lukovskii and Timokha (1999, 2000a), and Faltinsen and Timokha (2001). These studies pertain to lateral excitation of the whole tank and the nonlinear effects were manifested as a soft spring characteristic. Another group of studies were carried out on rectangular tanks with a wave maker fitted at one or two ends of the tank. The generated surface waves in these studies belonged to standing waves. Two-dimensional standing waves of finite amplitude may be shown to occur in rectangular tanks with flap wave generators. Depending on the length and width of the tank and the forcing frequency of the wave generator, two classes of resonant standing waves may occur. The first class is the well-known synchronous, resonant forced longitudinal standing wave whose crest is parallel to the wave maker. The second type is the subharmonic, parametrically excited transverse cross-wave whose crest is at right angles to the wave maker. Cross-waves generally possess half the frequency of the wave maker and reach a steady state at finite amplitude.

Penney and Price (1952) developed a Fourier series expansion carried up to fifth-order to determine the wave shape. They indicated that the maximum wave was a sharp-crested form with a 90° angle in contrast to Stokes' (1880) result of a 120° angle for the maximum progressive wave. The finding of Stokes was a correction of Rankine's (1865) erroneous deduction of a 90° angle. Penney and Price indicated that the finite amplitude affects the frequency of the wave in such a manner as to decrease the frequency. This is in contrast to the frequency increase found by Stokes for finite-amplitude progressive waves. Later, Sir G. Taylor (1954) conducted an experiment on a rectangular tank with wave generator flaps at the two ends of a rectangular tank. He observed that extremely small synchronous motions of the generator flaps were able to generate high amplitude waves when they were sufficiently close to resonance. For high amplitudes of water motion, he found that the frequency decreases with the amplitude as predicted by Penney and Price with fair precision of the 90° angle. Beyond the 90°-angle maximum form, he observed the occurrence of spontaneous instabilities and transverse wave motions of the free surface.

Cross-waves may be generated by applying parametric excitation using a plane wave maker at subharmonic frequency (Miles, 1985b). Lin and Howard (1960), Garrett (1970), Bernard and Pritchard (1972), Bernard, *et al.* (1977), and Lichter and Shemer (1986) conducted experimental and analytical investigations on cross-waves generated in a tank with a rigid wall opposite to a wave maker. Standing surface waves of finite amplitude were studied by Mack (1960, 1962), Verma and Keller (1962), Mack, *et al.* (1967), and Rajappa (1970a), and Vanden Broeck (1984). Other standing waves were reported by Amick and Tolland (1987) and Agon and Golzman (1996). For certain discrete fluid depths, Mack (1962) reported the occurrence of a coupled motion between higher modes at a frequency equal to an integral multiple of the primary frequency. Parametric instability in this case is interpreted in terms of the work done by the wave maker against transverse stresses associated with cross-waves.

Linear theory of sloshing waves fails to describe correctly the wave response in the vicinity of the cutoff frequencies (Shemer, *et al.* 1987). In order to account for the finite wave amplitude (observed experimentally) at the cutoff frequency, nonlinear effects have to be considered although dissipation can be crucial under certain circumstances (Kit, *et al.* (1987). Bernard, *et al.* (1977) showed that resonance waves with crests parallel to the channel walls and with wavelengths $\lambda = 2b/n$ (n is the mode number) might appear in a semi-infinite channel of depth h and width b. In some experiments of Bernard, *et al.* (1977) the maximum of the wave height distribution along the channel was found to be shifted downwards from the wave maker. Miles (1985a) suggested that this type of distribution might be attributed as a trapped soliton. A steady soliton-like solution was experimentally observed by Wu, *et al.* (1984). Some analyses of this nonpropagating soliton were made by Larraza and Putterman (1984) and Miles (1984c) by employing the nonlinear Schrödinger equation.

An example of parametrically excited hydrodynamic solitary waves in a Faraday water trough was given by Wu, *et al.* (1984) and later studied by Wang, *et al.* (1996). This kind of surface cross-wave in the transverse direction occurs with the frequency of the dominant mode close to one-half of the vertical excitation. The solitary wave possesses some very special characteristics, for example, a single solitary wave situated some distance from both ends of the trough will keep standing definitely in the absence of any external disturbance. Some features of the collision of solitary waves include the endless and periodic collision of a like polarity pair of solitary waves. Wang, *et al.* (1996) conducted experimental and numerical investigations to study the dynamical behavior of two interacting solitary waves.

Garrett (1970) showed that the mechanism for the excitation of transverse cross-waves is indeed a parametric resonance. By taking the average over the longitudinal waves, he obtained a Mathieu equation governing the amplitude of cross-waves. Later, Miles (1988a) used a Lagrangian formulation and obtained a Hamiltonian system governing the slow modulation of the cross-wave amplitude. Miles included the nonlinear interaction between the motion of the wave-maker and the cross-wave to second-order and the self-interaction of cross-waves to third order. The equation was found equivalent to that governing the parametric excitation of surface waves in vertical excitation (Miles, 1984a).

Kit, *et al.* (1987) studied resonant waves of the second mode for a wide range of wave maker amplitudes and frequencies. A transition from steady state to unsteady-state (modulated) regimes was reported. The frequency of the wave maker at which the transition takes place depends on whether the frequency is increasing or decreasing. This reveals the occurrence of a certain type of hysteresis. The experiments conducted by Kit, *et al.* (1987) were performed with a wave maker equipped with roughness elements. For a smooth wave maker, practically no hysteresis in transition between steady state and modulated regimes took place. However, the hysteresis was shifted down to frequencies where only a steady wave-field existed in the tank. Detailed measurements of two different steady-state wave distributions in the tank were reported in Shemer, *et al.* (1989).

Shemer (1990) and Tsai, *et al.* (1990) obtained averaged equations for the resonant standing waves in the form of Duffing oscillators. The numerical solution of the nonlinear Schrödinger equation obtained by Shemer (1990) revealed that bifurcations from a single steady state to multiple stationary solutions were realized for several values of damping coefficients along the tank and at the wave maker. Limit-cycle or fixed-point solutions were also obtained. They observed the occurrence of chaotic motions near resonance in some cases. Tsai, *et al.* (1990) and

Zhu *et al.* (2003) extended the analysis to include simultaneous resonance of the synchronous longitudinal and subharmonic cross-waves and their internal interactions. The interaction resulted in a complicated and varied dynamical system. Three different cases of wave maker frequency were considered. The first was when the wave maker frequency is close to the natural frequency of the longitudinal standing waves in which the cross-waves are of higher order of amplitude. The second case was when the cross-waves are parametrically in resonance and the longitudinal waves are not close to resonance. The third case was when the driving frequency approximates both a natural frequency of the directly forced longitudinal standing wave and twice that of the standing cross-wave. The two standing waves in this case are of the same order. For a broad range of physical parameters such as water depth, wave maker amplitude, width-to-length ratio and frequency detuning, the nonlinear interactions of the two waves exhibited chaotic wave motions.

Two-dimensional irrotational flow in rectangular containers of typical depth-to-breadth ratio, $h/L \leq 0.6$, may reveal different resonant fluid characteristics. Dean and Dalrymple (1992) classified three domains of fluid depth-to-breadth ratio: finite domain, $h/L \geq 0.24$, intermediate, $0.1 \leq h/L \leq 0.24$, and shallow, $h/L \leq 0.1$. Each domain is characterized by its own resonant behavior. The modifications of the sloshing behavior associated with decreasing depth ratio, h/L, and increasing excitation amplitude were examined using the modal methods (Faltinsen and Timokha, 2002a,b). These methods involve Fourier representation of the free-surface wave height and the velocity potential with time-dependent coefficients.

This section will examine the nonlinear phenomena of longitudinal standing waves that take place near a critical depth of the liquid in a rectangular tank.

4.4.2 Longitudinal standing waves

Longitudinal standing waves can be generated in a rectangular tank using a wave maker. The nonlinear analysis of such waves has received extensive theoretical and experimental studies. Tadjbakhsh and Keller (1960), motivated by the work of Penney and Price (1952) and Sir G. Taylor (1954), conducted a perturbation analysis up to third order to determine the surface profile, potential function, pressure, and frequency of the surface motion. They found that the dependence of the frequency on wave amplitude has a switching feature from hard to soft spring characteristics. The switching is such that the frequency increases with amplitude for depths less than a certain multiple of the wavelength and decreases with increasing amplitude for greater depths. They obtained expansions for the nondimensional frequency of the wave, $\omega = \omega^*/\sqrt{2\pi g/\lambda}$, and the wave height ratio, $\eta = \eta^*/a^*$, in the form

$$\omega = \omega_0 + \frac{1}{2}\varepsilon^2\omega_2 + O(\varepsilon^3) \tag{4.114}$$

$$\eta = \eta_0(x,\tau) + \varepsilon\eta_1(x,t) + \frac{1}{2}\varepsilon^2\eta_2(x,\tau) + O(\varepsilon^4) \tag{4.115}$$

where ω^* is the wave frequency along the x-axis, η^* is the free-surface displacement from the mean level, a^* is a measure of wave amplitude which is taken as the amplitude of the Fourier component of the wave motion identical in type with infinitesimal mode, $x = 2\pi x^*/\lambda$, $\tau = t\omega^*$ is a nondimensional time parameter, $\omega_0 = \sqrt{\tanh(2\pi/\lambda h)}$, $h = 2\pi h^*/\lambda$ is a nondimensional liquid depth, and $\varepsilon = 2\pi\, a^*/\lambda$. All quantities with a star, $*$, denote dimensional parameters. The higher order terms ω_2, η_1, and η^2 are correction functions to the infinitesimal solution.

The additional frequency component ω_2 was given in terms of ω_0 in the form

$$\omega_2 = \frac{1}{32}\left(\frac{9}{\omega_0^7} - \frac{12}{\omega_0^3} - 3\omega_0 - 2\omega_0^5\right) \tag{4.116}$$

The fundamental and first-order terms in the wave height expansion (4.115) were given by the following expressions

$$\eta_0(x,t) = \sin t \cos x \tag{4.117a}$$

$$\eta_1(x,t) = \frac{1}{8}\left[\left(\omega_0^2 - \frac{1}{\omega_0^2}\right) + \left(\frac{1}{\omega_0^2} - \frac{3}{\omega_0^6}\right)\cos 2t\right]\cos 2x \tag{4.117b}$$

This new feature stimulated the interest of Lin and Howard (1960), Fultz (1962), Hayama, *et al.* (1983), and Yeh (1986) who conducted experimental investigations to verify Tadjbakhsh and Keller's (1960) results. Fultz (1962) observed that the frequency effect reversal appears to occur at a depth ratio of 0.14, somewhat less than the predicted ratio of 0.17 by Tadjbakhsh and Keller (1960). Figures 4.19(a) and (b) shows the dependence of the wave height, $\Delta\eta'_m = \Delta\eta_m/\lambda$, on the excitation frequency ratio, $f' = \Omega/\omega_0$, for fluid depth ratios, $h' = h/\ell = 0.15$ and 0.13, respectively, where $\Delta\eta_m$ denotes the difference between the maximum and minimum wave height η_m at an antinode of the standing waves, Ω is the wave-generator frequency. Each figure includes a set of amplitude–frequency curves for different excitation angular amplitudes, in radians.

Ockendon and Ockendon (1973) and Waterhouse (1994) determined amplitude-frequency response diagrams at regions close to the critical depth where the response switches from "hard" to "soft" characteristics. In terms of the following nondimensional parameters

$$\bar{h} = \pi h/\ell, \quad \tau = \Omega t, \quad \varepsilon = \pi X_0/\ell \tag{4.118}$$

where the tank length is taken as $\ell = \pi l$ and the fluid depth as $\bar{h}l$, the fluid field equations in a rectangular tank take the form

$$\nabla^2\bar{\Phi} = 0 \tag{4.119a}$$

$$\frac{\partial\bar{\Phi}}{\partial x} = \sin\tau \quad \text{on} \quad x = -\varepsilon\cos\tau,\; x = \pi - \varepsilon\cos\tau \tag{4.119b}$$

$$\frac{\partial\bar{\Phi}}{\partial z} = 0 \quad \text{on} \quad z = -\bar{h} \tag{4.119c}$$

$$\nabla\bar{\Phi}(x,z,\tau) = \nabla\bar{\Phi}(x,z,\tau + 2\pi) \tag{4.119d}$$

$$\int_0^{\pi} \bar{\eta}\, dx = 0 \tag{4.119e}$$

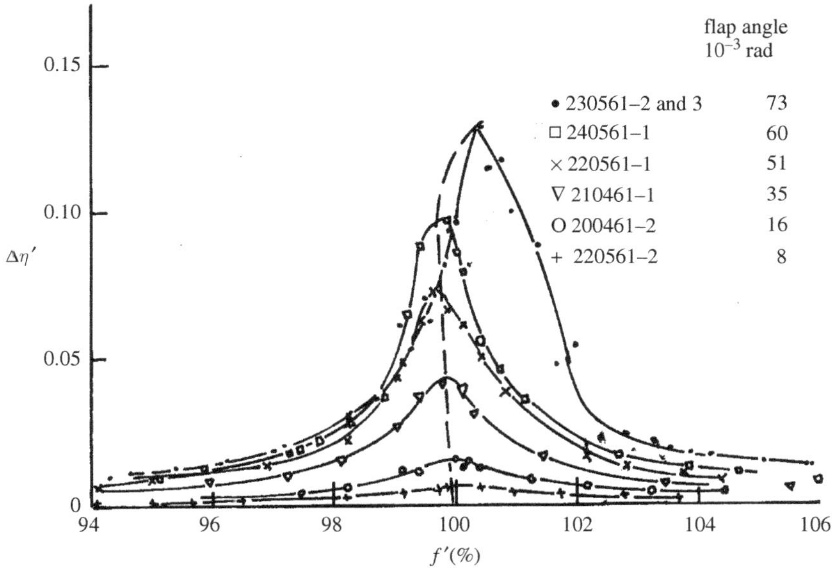

(a) $h' = 0.15$

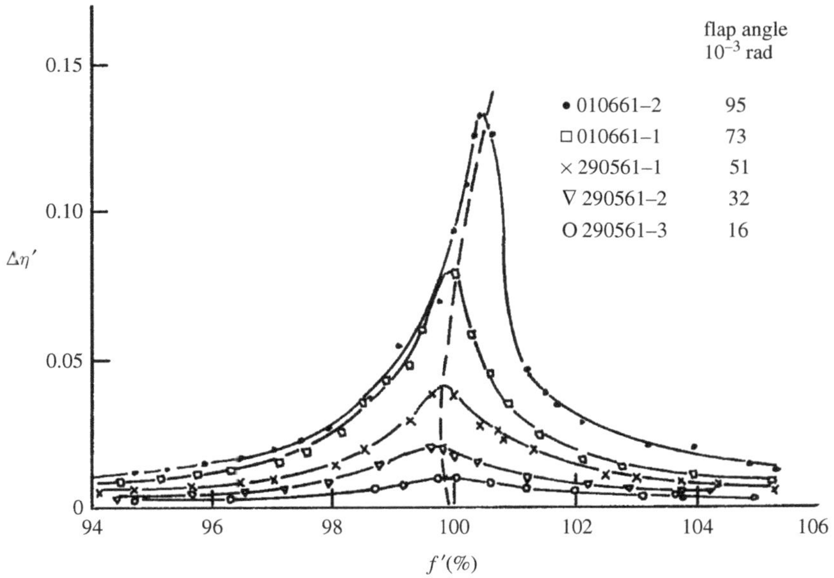

(b) $h' = 0.13$

Figure 4.19 Dependence of total wave amplitude parameter $\Delta\eta' = \Delta\eta/\lambda$ on excitation frequency ratio (%) for two different fluid depth ratios (a) $h' = 0.15$ and (b) $h' = 0.13$. (Fultz, 1962)

On $z = \varepsilon\bar{\eta}$, the kinematic free-surface condition is

$$\frac{\partial \bar{\Phi}}{\partial z} = \frac{\partial \bar{\eta}}{\partial \tau} + \varepsilon \frac{\partial \bar{\Phi}}{\partial x}\frac{\partial \bar{\eta}}{\partial x} \tag{4.119f}$$

and the dynamic free-surface condition is

$$\bar{\eta} + \frac{l\Omega^2}{g}\left[\frac{\partial \bar{\Phi}}{\partial \tau} + \frac{\varepsilon}{2}\left\{\left(\frac{\partial \bar{\Phi}}{\partial x}\right)^2 + \left(\frac{\partial \bar{\Phi}}{\partial z}\right)^2\right\}\right] = 0 \tag{4.119g}$$

where the dimensional velocity potential function is $\Phi = \varepsilon l^2 \Omega \bar{\Phi}$, and wave height is $\eta = \varepsilon l \bar{\eta}$. The fundamental natural frequency is

$$\omega^2 = (g/l)\tanh \bar{h} = (\pi g/\ell)\tanh(\pi h/\ell) \tag{4.120}$$

Define the frequency detuning parameter, ν , as

$$l\Omega^2 = (1+\nu) g \tanh \bar{h} \tag{4.121}$$

The dynamic free-surface condition can be rewritten in the form

$$\bar{\eta} + (1+\nu)\tanh \bar{h}\left[\frac{\partial \bar{\Phi}}{\partial \tau} + \frac{\varepsilon}{2}\left\{\left(\frac{\partial \bar{\Phi}}{\partial x}\right)^2 + \left(\frac{\partial \bar{\Phi}}{\partial z}\right)^2\right\}\right] = 0 \tag{4.122}$$

Away from resonance, the liquid free surface is governed by the linear solution of Laplace's equation (4.119a)

$$\bar{\Phi} \approx \frac{4}{\pi \nu \cosh \bar{h}} \cos x \cosh(z+\bar{h}) \sin \tau \tag{4.123}$$

As $\nu \to 0$, the response is $O(\varepsilon/\nu)$. At exact resonance, the nonlinearities become predominant. In this case, one may introduce the following asymptotic expansions

$$\bar{\Phi} = \varepsilon^{-2/3}\bar{\Phi}_0 + \varepsilon^{-1/3}\bar{\Phi}_1 + \bar{\Phi}_2 + \cdots \tag{4.124}$$

$$\bar{\eta} = \varepsilon^{-2/3}\bar{\eta}_0 + \varepsilon^{-1/3}\bar{\eta}_1 + \bar{\eta}_2 + \cdots \tag{4.125}$$

Solving for $\bar{\Phi}_0$ and $\bar{\Phi}_1$, and using the solvability condition imposed on $\bar{\Phi}_2$ gives

$$\bar{\Phi} \approx \varepsilon^{-2/3}\bar{A}\cos x \cosh(z+\bar{h}) \sin \tau \tag{4.126}$$

where the generalized coordinate $\bar{A}$ satisfies the algebraic equation

$$H(\bar{h})\bar{A}^3 + \varepsilon^{-2/3}\nu\bar{A}\sinh \bar{h} - \frac{4}{\pi}\tanh \bar{h} = 0 \tag{4.127}$$

and

$$H(\bar{h}) = -\frac{1}{32}\,\mathrm{sech}^2\bar{h}\,\mathrm{cosech}\,\bar{h}(9 + 15\,\mathrm{sech}^2\bar{h} - 8\sinh \bar{h}) \tag{4.128}$$

According to equation (4.127), the response behaves like an undamped Duffing oscillator with "soft" or "hard characteristics depending on whether $H(\bar{h}) > 0$ or < 0, respectively. When $H(\bar{h}) = 0$, that is, at $\bar{h} = \bar{h}_0 = 1.06$, the expansion breaks down as $\bar{A}$ grows large over a narrow detuning band near the origin. Ockendon and Ockendon (1973) found that as $\bar{h} \to \bar{h}_0$ it is necessary to rescale the variables and pose new expansions for $\bar{\Phi}$ and $\bar{\eta}$. First the fluid depth parameter is expressed as the critical value plus a small value, that is,

$$\bar{h} = \bar{h}_0 + \varepsilon^c \bar{h}_1 \tag{4.129}$$

where c is a positive constant to be determined. Substituting (4.129) into equation (4.127) reveals that $\varepsilon^c \bar{A}^3 \approx O(1)$ together with $\varepsilon^{-2/3} \nu \bar{A} \approx O(1)$ for all the terms. Thus, by setting $\bar{A} = \varepsilon^{-c/3} A$, and $\nu \approx O(\varepsilon^{(c+2)/3)})$, we write

$$\bar{\Phi} \approx \varepsilon^{-(c+2)/3} A \cos x \cosh(z + \bar{h}) \sin \tau \tag{4.130}$$

For any expansion of $\bar{\Phi}$ and $\bar{\eta}$ it is necessary that the nonlinear terms generate $\cos x \sin \tau$ terms of $O(1)$ so that the boundary conditions (4.119b) can be satisfied. This means that the solution must involve odd powers of $\bar{\Phi}_0$, that is, the first term in an expansion for $\bar{\Phi}$ must be of the form $\varepsilon^{2n/(2n+1)}$, $n = 0, 1, 2, \ldots$ Comparing this with equation (4.130) reveals that the simplest system is obtained when $n = 2$ and $(c + 2)/3 = 4/5$, or when $c = 2/5$. Thus, equation (4.129) takes the form

$$\bar{h} = \bar{h}_0 + \varepsilon^{2/5} \bar{h}_1, \quad \text{and} \quad \varepsilon^{-4/5} \nu \approx O(1), \bar{h}_1 \approx O(1) \tag{4.129a}$$

The following expansions are also considered

$$\bar{\Phi} \approx \varepsilon^{-4/5} \bar{\Phi}_0 + \varepsilon^{-3/5} \bar{\Phi}_1 + \varepsilon^{-2/5} \bar{\Phi}_2 + \varepsilon^{-1/5} \bar{\Phi}_3 + \bar{\Phi}_4 + \cdots \tag{4.131}$$

$$\bar{\eta} \approx \varepsilon^{-4/5} \bar{\eta}_0 + \varepsilon^{-3/5} \bar{\eta}_1 + \varepsilon^{-2/5} \bar{\eta}_2 + \varepsilon^{-1/5} \bar{\eta}_3 + \bar{\eta}_4 + \cdots \tag{4.132}$$

Note that near the critical depth $\bar{h}_0$ the response amplitude grows from $O(\varepsilon^{1/3})$ to $O(\varepsilon^{1/5})$. Substituting equations (4.129a), (4.131), and (4.132) into equations (4.119) and comparing terms of order $\varepsilon^{-4/5}$ leads to the same first-order problem as before. The leading-order solutions are

$$\bar{\Phi}_0 = A \cos x \cosh(z + \bar{h}) \sin \tau \tag{4.133}$$

$$\bar{\eta}_0 = -A \cos x \sinh \bar{h} \cos \tau \tag{4.134}$$

In order to determine A, it is necessary to find a solvability condition for $\bar{\Phi}_4$ and $\bar{\eta}_4$. This means that terms of orders $\varepsilon^{-3/5}$, $\varepsilon^{-2/5}$, and $\varepsilon^{-1/5}$ from the comparison process must be evaluated to solve for $\bar{\Phi}_i$ and $\bar{\eta}_i$, $i = 1, 2, 3$. By using a symbolic manipulation computer package such as MATHEMATICA, the following results are obtained

$$\bar{\Phi}_1 = B\Im_1 \sin \tau + A^2(0.46 - 0.12\Im_2) \sin 2\tau - \frac{A^2}{8}\tau \tag{4.135a}$$

$$\bar{\Phi}_2 = C\Im_1 \sin\tau + AB(0.93 - 0.23\Im_2)\sin 2\tau - A^3(-0.59\Im_1 + 0.004\Im_3)\sin 3\tau$$
$$+ 0.01A^3\Im_3 \sin\tau - \frac{AB}{4}\tau + A\bar{h}_1 \cos x \sinh(z + \bar{h}_0)\sin\tau \tag{4.135b}$$

$$\bar{\Phi}_3 = D\Im_1 \sin\tau + \{[0.26A^4 - 0.23AC + 0.3\bar{h}_1 A^2 - 0.12B^2]\Im_2 + 0.004A^4\Im_4$$
$$+[1.02\bar{h}_1 A^2 + 0.46B^2 + 0.6A^2 + 0.93AC]\}\sin 2\tau + A^2B$$
$$(-1.76\Im_1 + 0.01\Im_3)\sin 3\tau + A^4(0.001\Im_4 + 0.15\Im_2 + 0.13)\sin 4\tau$$
$$+ \bar{h}_1[B\cos x \sinh(z + \bar{h}_0)\sin\tau - 0.23\cos 2x \sinh 2(z + \bar{h}_0)\sin 2\tau]$$
$$+ 0.03A^3B\Im_3 \sin\tau \tag{4.135c}$$

where B, C, and D are arbitrary constants and $\Im_n = \cos nx \cosh n(z + \bar{h}_0)$, $n = 1, 2, 3, 4$. The corresponding values of $\bar{\eta}_i$ are determined from equation (4.122).

Consider the $O(1)$ terms from equations (4.119a–d), (4.119f), and (4.122). This gives the following equations

$$\nabla^2\bar{\Phi}_4 = 0 \tag{4.136a}$$

$$\frac{\partial\bar{\Phi}_4}{\partial x} = \sin\tau, \qquad \text{on} \quad x = 0,\ \pi \tag{4.136b}$$

$$\frac{\partial\bar{\Phi}_4}{\partial z} = \bar{h}_1\frac{\partial^2\bar{\Phi}_2}{\partial z^2}, \qquad \text{on} \quad z = -\bar{h}_0 \tag{4.136c}$$

$$\frac{\partial\bar{\Phi}_4}{\partial z} + \tanh\bar{h}_0\frac{\partial^2\bar{\Phi}_4}{\partial\tau^2} = F(x,\tau), \qquad \text{on} \quad z = 0 \tag{4.136d}$$

$$\nabla\bar{\Phi}_4(x, z, \tau) = \nabla\bar{\Phi}_4(x, z, \tau + 2\pi) \tag{4.136e}$$

where $F(x, \tau)$ is a combination of $\bar{\Phi}_i$ and $\bar{\eta}_i$, and given by the following expression

$$F(x,\tau) = F_1(x,0,\tau) - \tanh\bar{h}_0\frac{\partial}{\partial\tau}F_2(x,0,\tau) + \bar{h}_1(\tanh\bar{h}_0 - 1)\frac{\partial}{\partial\tau}F_3(x,0,\tau)$$
$$-\tanh\bar{h}_0(\nu\varepsilon^{-4/5} + \bar{h}_1^2(\tanh^2\bar{h}_0 - 1)\frac{\partial^2}{\partial\tau^2}\bar{\Phi}_0(x,0,\tau) \tag{4.137}$$

The functions $F_i(x, 0, \tau)$ are given by the following expressions

$$
\begin{aligned}
F_1(x,0,\tau) = {} & \bar{\eta}_2 \frac{\partial^2 \bar{\Phi}_0}{\partial x \partial z} \frac{\partial \bar{\eta}_0}{\partial x} + \bar{\eta}_0 \bar{\eta}_1 \frac{\partial \bar{\Phi}_0}{\partial x} \frac{\partial \bar{\eta}_0}{\partial x} + \frac{1}{6} \bar{\eta}_0^3 \frac{\partial^2 \bar{\Phi}_0}{\partial x \partial z} \frac{\partial \bar{\eta}_0}{\partial x} + \bar{\eta}_1 \frac{\partial^2 \bar{\Phi}_0}{\partial x \partial z} \frac{\partial \bar{\eta}_1}{\partial x} \\
& + \bar{\eta}_1 \frac{\partial^2 \bar{\Phi}_1}{\partial x \partial z} \frac{\partial \bar{\eta}_0}{\partial x} + \frac{1}{2} \bar{\eta}_0^2 \frac{\partial \bar{\Phi}_0}{\partial x} \frac{\partial \bar{\eta}_1}{\partial x} + \frac{1}{2} \bar{\eta}_0^2 \frac{\partial^3 \bar{\Phi}_1}{\partial x \partial z^2} \frac{\partial \bar{\eta}_0}{\partial x} + \bar{\eta}_0 \frac{\partial^2 \bar{\Phi}_0}{\partial x \partial z} \frac{\partial \bar{\eta}_2}{\partial x} \\
& + \bar{\eta}_0 \frac{\partial^2 \bar{\Phi}_2}{\partial x \partial z} \frac{\partial \bar{\eta}_0}{\partial x} + \bar{\eta}_0 \frac{\partial^2 \bar{\Phi}_1}{\partial x \partial z} \frac{\partial \bar{\eta}_1}{\partial x} + \frac{\partial \bar{\Phi}_3}{\partial x} \frac{\partial \bar{\eta}_0}{\partial x} + \frac{\partial \bar{\Phi}_0}{\partial x} \frac{\partial \bar{\eta}_3}{\partial x} + \frac{\partial \bar{\Phi}_1}{\partial x} \frac{\partial \bar{\eta}_2}{\partial x} \\
& + \frac{\partial \bar{\Phi}_2}{\partial x} \frac{\partial \bar{\eta}_1}{\partial x} - \bar{\eta}_3 \bar{\Phi}_0 - \bar{\eta}_0 \bar{\eta}_2 \frac{\partial \bar{\Phi}_0}{\partial z} \frac{\partial \bar{\eta}_3}{\partial x} - \frac{1}{2} \bar{\eta}_1^2 \frac{\partial \bar{\Phi}_0}{\partial z} - \frac{1}{2} \bar{\eta}_0^2 \bar{\eta}_1 \bar{\Phi}_0 \\
& - \frac{1}{24} \bar{\eta}_0^4 \frac{\partial \bar{\Phi}_0}{\partial z} - \bar{\eta}_2 \frac{\partial^2 \bar{\Phi}_1}{\partial z^2} - \bar{\eta}_0 \bar{\eta}_1 \frac{\partial^3 \bar{\Phi}_1}{\partial z^3} \frac{\partial \bar{\eta}_3}{\partial x} - \frac{1}{6} \bar{\eta}_0^3 \frac{\partial^4 \bar{\Phi}_1}{\partial z^4} \\
& - \bar{\eta}_1 \frac{\partial^2 \bar{\Phi}_2}{\partial z^2} - \frac{1}{2} \bar{\eta}_0^3 \frac{\partial^3 \bar{\Phi}_2}{\partial z^3} - \bar{\eta}_0 \frac{\partial^2 \bar{\Phi}_3}{\partial z^2}
\end{aligned} \tag{4.138a}
$$

$$
\begin{aligned}
F_2(x,0,\tau) = {} & \bar{\eta}_3 \frac{\partial^2 \bar{\Phi}_0}{\partial \tau \partial z} + \bar{\eta}_0 \bar{\eta}_2 \frac{\partial \bar{\Phi}_0}{\partial \tau} + \frac{1}{2} \bar{\eta}_1^2 \frac{\partial \bar{\Phi}_0}{\partial \tau \partial z} + \frac{1}{2} \bar{\eta}_0^2 \bar{\eta}_1 \frac{\partial^2 \bar{\Phi}_0}{\partial \tau \partial z} + \frac{1}{24} \bar{\eta}_0^4 \frac{\partial \bar{\Phi}_0}{\partial \tau} \\
& + \bar{\eta}_2 \frac{\partial^2 \bar{\Phi}_1}{\partial \tau \partial z} + \bar{\eta}_0 \bar{\eta}_1 \frac{\partial^3 \bar{\Phi}_1}{\partial \tau \partial z^2} + \frac{1}{6} \bar{\eta}_0^3 \frac{\partial^4 \bar{\Phi}_1}{\partial \tau \partial z^3} + \bar{\eta}_1 \frac{\partial^2 \bar{\Phi}_2}{\partial \tau \partial z} + \frac{1}{2} \bar{\eta}_0^2 \frac{\partial^3 \bar{\Phi}_2}{\partial \tau \partial z^2} \\
& + \bar{\eta}_0 \frac{\partial^2 \bar{\Phi}_3}{\partial \tau \partial z} + \bar{\eta}_2 \frac{\partial \bar{\Phi}_0}{\partial x} \frac{\partial^2 \bar{\Phi}_0}{\partial x \partial z} + \bar{\eta}_0 \bar{\eta}_1 \left(\frac{\partial^2 \bar{\Phi}_0}{\partial x \partial z} \right)^2 + \bar{\eta}_0 \bar{\eta}_1 \left(\frac{\partial \bar{\Phi}_0}{\partial x} \right)^2 \\
& + \frac{2}{3} \bar{\eta}_0^3 \frac{\partial^2 \bar{\Phi}_0}{\partial x \partial z} \frac{\partial \bar{\Phi}_0}{\partial x} + \bar{\eta}_1 \frac{\partial \bar{\Phi}_1}{\partial x} \frac{\partial^2 \bar{\Phi}_0}{\partial x \partial z} + \bar{\eta}_1 \frac{\partial \bar{\Phi}_0}{\partial x} \frac{\partial^2 \bar{\Phi}_1}{\partial x \partial z} + \frac{1}{2} \bar{\eta}_0^2 \frac{\partial \bar{\Phi}_0}{\partial x} \frac{\partial \bar{\Phi}_1}{\partial x} \\
& + \bar{\eta}_0^2 \frac{\partial^2 \bar{\Phi}_0}{\partial x \partial z} \frac{\partial^2 \bar{\Phi}_1}{\partial x \partial z} + \frac{1}{2} \bar{\eta}_0^2 \frac{\partial \bar{\Phi}_0}{\partial x} \frac{\partial^3 \bar{\Phi}_1}{\partial x \partial z^3} + \bar{\eta}_0 \frac{\partial \bar{\Phi}_1}{\partial x} \frac{\partial^2 \bar{\Phi}_1}{\partial x \partial z} + \bar{\eta}_0 \frac{\partial^2 \bar{\Phi}_0}{\partial x \partial z} \frac{\partial \bar{\Phi}_2}{\partial x} \\
& + \bar{\eta}_0 \bar{\eta}_1 \left(\frac{\partial \bar{\Phi}_0}{\partial z} \right)^2 + \frac{2}{3} \bar{\eta}_0^3 \bar{\Phi}_0 \frac{\partial \bar{\Phi}_0}{\partial z} + \bar{\eta}_1 \bar{\Phi}_0 \frac{\partial \bar{\Phi}_1}{\partial z} + \bar{\eta}_1 \frac{\partial \bar{\Phi}_0}{\partial z} \frac{\partial^2 \bar{\Phi}_1}{\partial z^2} \\
& + \frac{1}{2} \bar{\eta}_0^2 \frac{\partial \bar{\Phi}_0}{\partial z} \frac{\partial \bar{\Phi}_1}{\partial z} + \bar{\eta}_0^2 \bar{\Phi}_0 \frac{\partial^2 \bar{\Phi}_1}{\partial z^2} + \frac{1}{2} \bar{\eta}_0^2 \frac{\partial \bar{\Phi}_0}{\partial z} \frac{\partial^3 \bar{\Phi}_1}{\partial z^3} + \bar{\eta}_0 \frac{\partial \bar{\Phi}_1}{\partial z} \frac{\partial^2 \bar{\Phi}_1}{\partial z^2} \\
& + \bar{\eta}_0 \bar{\Phi}_0 \frac{\partial \bar{\Phi}_2}{\partial z} + \bar{\eta}_0 \frac{\partial \bar{\Phi}_0}{\partial z} \frac{\partial^2 \bar{\Phi}_2}{\partial z^2} + \frac{\partial \bar{\Phi}_0}{\partial z} \frac{\partial \bar{\Phi}_3}{\partial z} + \frac{\partial \bar{\Phi}_1}{\partial z} \frac{\partial \bar{\Phi}_2}{\partial z}
\end{aligned} \tag{4.138b}
$$

$$
\begin{aligned}
F_3(x,0,\tau) = {} & \bar{\eta}_1 \frac{\partial^2 \bar{\Phi}_0}{\partial \tau \partial z} + \frac{1}{2} \bar{\eta}_0^2 \frac{\partial \bar{\Phi}_0}{\partial \tau} + \bar{\eta}_0 \frac{\partial \bar{\Phi}_1}{\partial \tau \partial z} + \frac{\partial \bar{\Phi}_2}{\partial \tau} + \bar{\eta}_0 \frac{\partial \bar{\Phi}_0}{\partial x} \frac{\partial^2 \bar{\Phi}_0}{\partial x \partial z} + \frac{\partial \bar{\Phi}_0}{\partial x} \frac{\partial \bar{\Phi}_1}{\partial x} \\
& + \bar{\eta}_0 \bar{\Phi}_0 \frac{\partial \bar{\Phi}_0}{\partial z} + \frac{\partial \bar{\Phi}_0}{\partial z} \frac{\partial \bar{\Phi}_1}{\partial z}
\end{aligned} \tag{4.138c}
$$

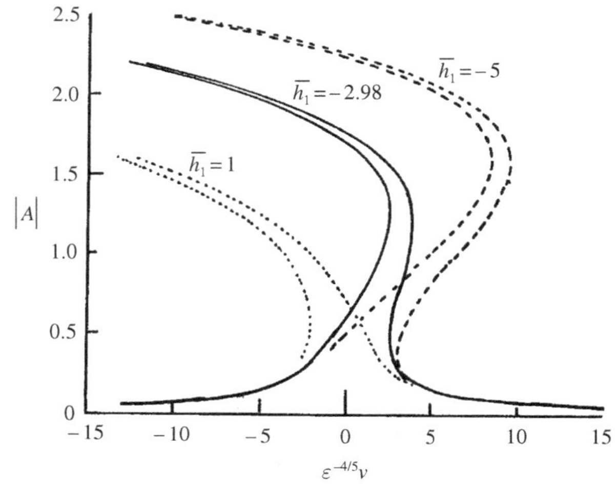

Figure 4.20 Amplitude–frequency response curves for three different values of fluid depth parameter perturbation. (Waterhouse, 1994)

The solvability condition for the system of equations governing $\bar{\Phi}_4$, described by equations (4.136), can be obtained using the Fredholm technique (Nayfeh, 1985). The resulting condition is

$$\frac{2}{\pi^2}\int_0^{2\pi}\int_0^{\pi} F(x,\tau)\cos x \sin\tau \, dt - 0.62\bar{h}_1 C = \frac{4}{\pi}\tanh\bar{h}_0 \tag{4.139}$$

where the coefficient C is obtained from $\bar{\Phi}_2$ given by equation (4.135b). Substituting the expressions $\bar{\Phi}_i$ and $\bar{\eta}_i$, $i=0,1,2,3$, into equation (4.137) and its components (4.138) gives

$$\varepsilon^{-4/5}\nu A\sinh\bar{h}_0 + A^3\bar{h}_1 H'(\bar{h}_0) + 1.80A^5 = \frac{4}{\pi}\tanh\bar{h}_0 \tag{4.140}$$

The dependence of the response amplitude $|A|$ on the detuning parameter $\varepsilon^{-4/5}\nu$ is shown in Figure 4.20 for three different values of $\bar{h}_1 = 1, -2.98$, and -5. One may notice that in the neighborhood of the critical depth, the large-amplitude response follows a soft spring characteristics. As $\bar{h}_1$ decreases to -2.98, the number of solutions switches from 3 to 5 and the hard spring characteristics begin to emerge. When $\bar{h}_1$ is very large and negative, the large-amplitude branches move off to infinity and the response follows hard spring characteristics with no switching.

Waterhouse (1994) obtained a uniformly valid response for all values of depths and detuning parameters by considering the dimensional expressions of the velocity potential and free-surface wave height

$$\Phi = \Omega^2 l\left[A_\varepsilon \cos x\cosh(z+\bar{h})\sin\tau + O(\varepsilon^{2/n})\right] \tag{4.141a}$$

$$\eta = l\left[-A_\varepsilon\cos x\sinh(\bar{h})\cos\tau + O(\varepsilon^{2/n})\right] \tag{4.141b}$$

where $A_\varepsilon = \varepsilon^{1/n} A$, and n takes the values 1, 2, or 3 depending on the detuning parameter range. The final response can be written independently of n, in the form

$$\nu A_\varepsilon \sinh \bar{h} + A_\varepsilon^3 H'(\bar{h}) + 1.80 A_\varepsilon^5 + O(\nu^2 A_\varepsilon) = \frac{4\varepsilon}{\pi} \tanh \bar{h}_0 \tag{4.142}$$

This equation encompasses the response equations given by (4.123), (4.127) and (4.140).

4.5 Conical tanks

The nonlinear analysis of liquid sloshing in inverted (V-shape) containers has been considered by Dokuchaev (1964), Lukovskii and Bilyk (1985), Bauer and Eidel (1988b), Lukovskii and Timokha (2000b), and Gavrilyuk, *et al.* (2001). Bauer and Eidel (1988b) expanded the nonlinear kinematic and dynamic free-surface conditions into Taylor series about the undisturbed free-surface level. Gavrilyuk, *et al.* (2001) employed a nonconformal transformation technique originally developed by Lukovskii (1975) and Lukovskii and Timokha (2000b). The following analysis is based on the work of Bauer and Eidel (1988b) and will be compared with the results of Gavrilyuk, *et al.* (2001). Gavrilyuk, *et al.* (2001) established a general modal system that saves its invariant structure with respect to tensor transformations. The tensor transformations map the circular conical domain into a circular cylinder such that the mean fluid depth remains its invariant value. Consider an inverted conical tank moving with velocity $\mathbf{V}_0$ and experiencing rotation with angular velocity vector $\boldsymbol{\omega}$, the fluid field equations are

$$\nabla^2 \Phi = 0 \qquad \text{inside the container} \tag{4.143a}$$

$$\frac{\partial \Phi}{\partial n} = \mathbf{V}_0 \cdot \mathbf{n} + \boldsymbol{\omega} \cdot (\mathbf{r} \times \mathbf{n}) \qquad \text{on the tank surface } S \tag{4.143b}$$

$$\frac{\partial \Phi}{\partial n} = \mathbf{V}_0 \cdot \mathbf{n} + \boldsymbol{\omega} \cdot (\mathbf{r} \times \mathbf{n}) - \frac{\partial \eta / \partial t}{|\nabla \eta|} \quad \text{on the free surface} \tag{4.143c}$$

$$\frac{\partial \Phi}{\partial t} + \frac{1}{2} (\nabla \Phi)^2 - \nabla \Phi \cdot (\mathbf{V}_0 + \boldsymbol{\omega} \times \mathbf{r}) + \Pi = 0 \ \text{ on the free surface} \tag{4.143d}$$

where $\mathbf{V}_0$ and $\boldsymbol{\omega}$ are the translational and angular velocity vectors of the tank with respect to the stationary coordinate frame $OXYZ$, $\mathbf{r}$ is the position vector of a fluid particle with respect to the moving frame $oxyz$, $\Pi = -\mathbf{g} \cdot \mathbf{r}'$ is the gravitational potential energy, $\mathbf{g}$ is the gravitational acceleration vector, and $\mathbf{V}'$ is the position vector of the fluid particle with respect to the stationary frame.

The nonconformal transformation is introduced by considering the original domain Q associated with the Cartesian coordinate frame $oxyz$, and the transformed domain $Q^* = (0, d) \times A$ with bottom A and height d as shown in Figure 4.21. The domain Q^* is described in terms of curvilinear coordinates (x_1, x_2, x_3). Consider the set of admissible smooth transformations mapping Q into Q^* as

$$x_1 = z, \qquad x_2 = x_2(x, y, z), \qquad x_3 = x_3(x, y, z) \tag{4.144a}$$

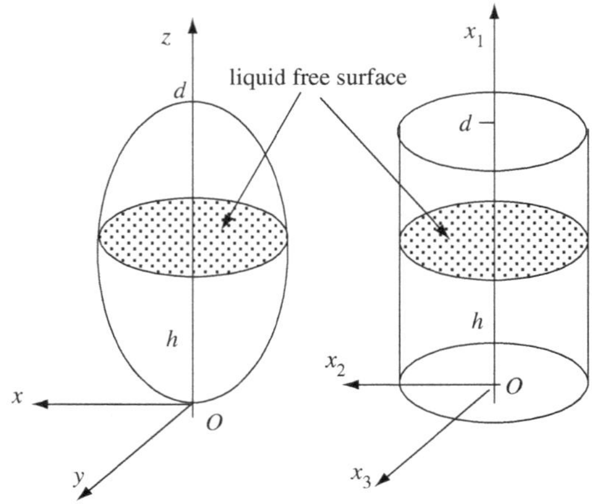

Figure 4.21 Admissible coordinate transformation.

$$z = x_1, \qquad x = x(x_1, x_2, x_3), \qquad y = y(x_1, x_2, x_3) \tag{4.144b}$$

Note that the mean free surface has a circular shape of radius $r_0 = h \tan \alpha$ such that

$$x_1 = z, \qquad x_2 = \tan\alpha, \qquad x = x_1 x_2 \cos x_3, \qquad y = x_1 x_2 \sin x_3 \tag{4.144c}$$

This transformation does not change the maximum height of the container cavity d and the mean static fluid depth. With reference to Figure 4.21, the plane Oxy is assumed tangent to the tank surface at the bottom at $z=0$ in Q, and the plane Ox_2x_3 coincides with the artificial bottom of Q^*. It is desirable to avoid singularities of transformations (4.144) on the tank surface S. However, this is not possible for conical and spherical tanks that do not possess a flat bottom at the origin O. Thus, the bottom vertex has to be transformed into an artificial bottom as shown in Figure 4.22. The admissible transformations are mathematically restricted to degenerating only in a limited number of isolated singular points on the surface S, where

$$J(x, y, z) = \frac{1}{J^*} = \infty; J * (x_1, x_2, x_3) = \begin{vmatrix} \dfrac{\partial x}{\partial x_1} & \dfrac{\partial x}{\partial x_2} & \dfrac{\partial x}{\partial x_3} \\ \dfrac{\partial y}{\partial x_1} & \dfrac{\partial y}{\partial x_2} & \dfrac{\partial y}{\partial x_3} \\ 1 & 0 & 0 \end{vmatrix} = 0 \tag{4.145}$$

The singular points are assumed to be bound away from the mean free surface $z = h$.

Applying a variational formulation (see, e.g., Luke, 1967) of the boundary value problem (4.143), we write

$$W(\Phi, \eta) = \int_{t_1}^{t_2} L\, dt \tag{4.146a}$$

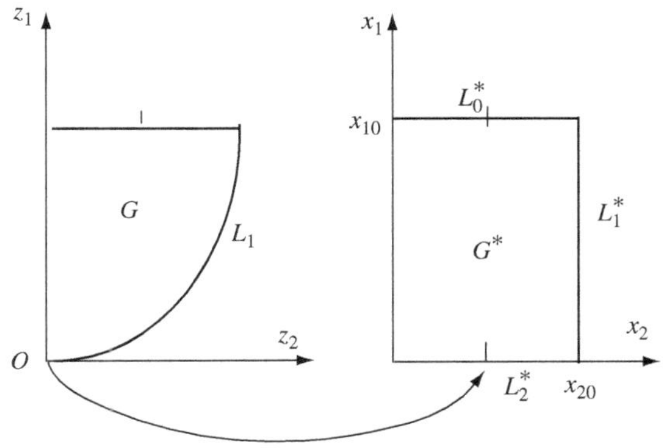

Figure 4.22 The original and transformed cross-sections $z=z_1$, $x=z_2\cos z_3$, $y=z_2\sin z_3$.

$$L=\int\limits_{Q(t)}(p-p_0)\,dQ=-\rho\int\limits_{Q(t)}\left[\frac{\partial\Phi}{\partial t}+\frac{1}{2}(\nabla\Phi)^2-\nabla\Phi\cdot(\mathbf{V}_0+\boldsymbol{\omega}\times\mathbf{r})+\Pi\right]dQ \qquad (4.146b)$$

within test functions

$$\delta\Phi|_{t_1,t_2}=0,\quad \delta\eta|_{t_1,t_2}=0 \qquad (4.147)$$

The pressure field is expressed using the Lagrange–Cauchy integral equation

$$\frac{\partial\Phi}{\partial t}+\frac{1}{2}(\nabla\Phi)^2-\nabla\Phi\cdot(\mathbf{V}_0+\omega\times\mathbf{r})+\Pi+\frac{p-p_0}{\rho}=0 \quad \text{in } Q \qquad (4.148)$$

The Lagrangian L is invariant relative to the admissible time-independent tensor transformation (4.144). Thus we write

$$L\equiv L^*=-\rho\int\limits_{Q(t)}\left[\frac{\partial\Phi^*}{\partial t}+\frac{1}{2}(\nabla^*\Phi^*)^2-\nabla^*\Phi^*\cdot(\mathbf{V}_0+\omega\times\mathbf{r})^*+\Pi^*\right]J^*\,dQ^* \qquad (4.149)$$

where $Q^*(t)$ is the transformation domain,

$$\Pi^*=\Pi(x(x_1,x_2,x_3)\ y(x_1,x_2,x_3),z(x_1,x_2,x_3),t)$$

$$\Phi^*=\Phi(x(x_1,x_2,x_3),y(x_1,x_2,x_3),z(x_1,x_2,x_3),t)$$

$$\nabla\Phi=\nabla^*\Phi^*=\left(g_{1,\,j}\frac{\partial\Phi^*}{\partial x_j},g_{2,\,j}\frac{\partial\Phi^*}{\partial x_j},g_{3,\,j}\frac{\partial\Phi^*}{\partial x_j}\right),\quad \text{with } g_{ij}=\frac{\partial\mathbf{r}}{\partial x_i}\frac{\partial\mathbf{r}}{\partial x_j} \qquad (4.150)$$

and $(\mathbf{V}_0+\boldsymbol{\omega}\times\mathbf{r})^*$ denotes the projection on unit vectors of a curvilinear coordinate system.

The equation of the free surface takes the normal form

$$\eta^*(x_1,x_2,x_3,t)=x_1-f^*(x_2,x_3,t) \qquad (4.151)$$

The nonlinear free-surface boundary condition is then transformed to an equivalent tensor form. The function f^* is expanded into the Fourier series

$$f^*(x_2, x_3, t) = h + \beta_0(t) + \sum_{i=1}^{\infty} \beta_i(t) F_i(x_2, x_3) \tag{4.152}$$

Similarly, the velocity potential $\Phi^*(x_1, x_2, x_3)$ is expressed in terms of the Fourier series

$$\Phi^* = \mathbf{V}_0 \cdot \mathbf{r} + \sum_{i=1}^{\infty} Z_n(t)\phi_n(x_1, x_2, x_3) \tag{4.153}$$

where $\beta_i(t)$ and $Z_n(t)$ are generalized coordinates. Introducing expansions (4.152) and (4.153) into the Lagrangian (4.149) and taking the variational to the functional (4.146a), gives the following set of modal nonlinear differential equations

$$\frac{dA_n}{dt} - \sum_k A_{nk} Z_k = 0, n = 1, 2, \ldots \tag{4.154a}$$

$$\sum_n \left(\frac{\partial A_n}{\partial \beta_i}\right) \frac{dZ_n}{dt} + \frac{1}{2} \sum_n \left(\frac{\partial A_{nk}}{\partial \beta_i}\right) Z_n Z_k + (\dot{V}_{01} - g_1) \frac{\partial l_1}{\partial \beta_i} + (\dot{V}_{02} - g_2) \frac{\partial l_2}{\partial \beta_i} + (\dot{V}_{03} - g_3) \frac{\partial l_3}{\partial \beta_i} = 0 \tag{4.154b}$$

where

$$A_n = \rho \int_D \left(\int_0^{f^*} \phi_n J^* \, dx_1 \right) dx_2 \, dx_3 \tag{4.155a}$$

$$A_{nk} = \rho \int_D \left(\int_0^{f^*} (\nabla^* \phi_n^*, \nabla^* \phi_k^*) J^* \, dx_1 \right) dx_2 \, dx_3 \tag{4.155b}$$

$$\frac{\partial l_1}{\partial \beta_i} = \rho \int_{\Sigma_0^*} F_i [x_1 J^*(x_1, x_2, x_3)]_{x_1 = f^*} \, dx_2 \, dx_3 \tag{4.156a}$$

$$\frac{\partial l_2}{\partial \beta_i} = \rho \int_{\Sigma_0^*} F_i [x(x_1, x_2, x_3) J^*(x_1, x_2, x_3)]_{x_1 = f^*} \, dx_2 \, dx_3 \tag{4.156b}$$

$$\frac{\partial l_3}{\partial \beta_i} = \rho \int_{\Sigma_0^*} F_i [y(x_1, x_2, x_3) J^*(x_1, x_2, x_3)]_{x_1 = f^*} \, dx_2 \, dx_3 \tag{4.156c}$$

Let r_0 denote the radius of the undisturbed free surface inside the circular conical container. The following scaling parameters are introduced

$$\mathbf{g} = \mathbf{g}/r_0, \ \bar{\dot{V}}_0 = \dot{\mathbf{V}}_0/r_0, \ \bar{\beta}_i(t) = \beta_i(t)/r_0, \ Z_i(t) = Z_i(t)/r_0 \tag{4.157}$$

Under lateral translation excitation of the tank we have

$$g_1 = -g, \ g_2 = g_3 = 0, \ \text{and} \quad V_{01} = 0 \tag{4.158}$$

According to Moiseev (1958) the amplitude of the lowest natural modes dominates the free-surface motion and forms a Duffing-like third-order asymptotic. Lukovskii (1990) introduced five modes approximation of the free surface and potential functions (4.152) and (4.153), which take the following form for conical tanks

$$\begin{aligned} f^*(x_2, x_3, t) &= h + f(x_2, x_3, t) = h + \beta_0(t) + p_0(t) f_0(x_2) \\ &\times \ [r_1(t) \sin x_3 + p_1(t) \cos x_3]\, f_1(x_2) \\ &+ [r_2(t) \sin 2x_3 + p_2(t) \cos 2x_3]\, f_2(x_2) \end{aligned} \tag{4.159}$$

$$\begin{aligned} \phi(x_1, x_2, x_3) &= P_0(t)\boldsymbol{\psi}_0(x_1, x_2) + [R_1(t) \sin x_3 + P_1(t) \cos x_3]\boldsymbol{\psi}_1(x_1, x_2) \\ &+ [R_2(t) \sin 2x_3 + P_2(t) \cos 2x_3]\boldsymbol{\psi}_2(x_1, x_2) \end{aligned} \tag{4.160}$$

Introducing the parameter ε that measures the ratio of the excitation amplitude and the radius of the free surface r_0

$$\max_{j=2,3} \left| \bar{\dot{V}}_{0j} \right| \sim \varepsilon \omega_{11}^2 \tag{4.161}$$

where $\omega_{11} = \sqrt{g\chi_{11}/r_0}$, and χ_{11} is the first eigenvalue (dimensionless) determined from the free vibration of the linear problem. The roots of different values of χ_{ij} are listed in Table 4.2. The ordering of the modal functions are

$$r_1 \sim R_1 \sim p_1 \sim P_1 \sim \varepsilon^{1/3}, \quad \text{and} \quad p_0 \sim P_0 \sim r_2 \sim R_2 \sim p_2 \sim P_2 \sim \varepsilon^{2/3} \tag{4.162}$$

The modal equations are asymptotically truncated by the terms $O(\varepsilon)$. Introducing equations (4.159) and (4.160) into equations (4.154) and keeping the ordering sequence (4.162) gives the following set of second-order nonlinear equations of the five modes

$$\begin{aligned} &\ddot{r}_1 + \omega_{11}^2 r_1 + D_1\left(r_1^2 \ddot{r}_1 + r_1 \dot{r}_1^2 + r_1 p_1 \ddot{p}_1 + r_1 \dot{p}_1^2\right) \\ &+ \ D_2\left(p_1^2 \ddot{r}_1 + 2p_1 \dot{r}_1 \dot{p}_1 - r_1 p_1 \ddot{p}_1 - 2r_1 \dot{p}_1^2\right) - D_3(p_2 \ddot{r}_1 - r_2 \ddot{p}_1 + \dot{r}_1 \dot{p}_2 - \dot{p}_1 \dot{r}_2) \\ &+ \ D_4(r_1 \ddot{p}_2 - p_1 \ddot{r}_2) + D_5(p_0 \ddot{r}_1 + \dot{r}_1 \dot{p}_0) D_6 r_1 \ddot{p}_0 \\ &+ \ \omega_{11}^2[\varsigma_1 r_1(r_1^2 + p_1^2) + \varsigma_2(p_1 r_2 - r_1 p_2) + \varsigma_3 r_1 p_0] + \Lambda \bar{\dot{V}}_{03} = 0 \end{aligned} \tag{4.163a}$$

Table 4.2 *Roots of different values of x_{ij}*

α°	χ 01	χ 11	χ 21
1	3.8228	1.8251	3.0323
3	3.8042	1.7925	2.9874
5	3.7844	1.7594	2.9414
7	3.7633	1.7258	2.8941
9	3.7407	1.6916	2.8457
11	3.7166	1.6570	2.7960
13	3.6909	1.6218	2.7451
15	3.6634	1.5862	2.6930
17	3.6340	1.5501	2.6397
19	3.6025	1.5135	2.5851
21	3.5689	1.4765	2.5293
23	3.5328	1.4390	2.4723
25	3.4943	1.4011	2.4140
27	3.4530	1.3627	2.3546
29	3.4088	1.3239	2.2939
31	3.3616	1.2848	2.2321
33	3.3110	1.2452	2.1691
35	3.2569	1.2052	2.1049
37	3.1991	1.1649	2.0396
39	3.1374	1.1242	1.9732
41	3.0715	1.0831	1.9056
43	3.0013	1.0417	1.8370
45	2.9266	1.0000	1.7674
47	2.8471	0.9580	1.6967
49	2.7628	0.9156	1.6250
51	2.6734	0.8730	1.5524
51	2.5789	0.8300	1.4738

Source: Gavrilyuk, *et al.* (2001).

$$\begin{aligned}
&\ddot{p}_1 + \omega_{11}^2 p_1 + D_1\left(p_1^2\ddot{p}_1 + p_1\dot{p}_1^2 + r_1 p_1 \ddot{r}_1 + p_1\dot{r}_1^2\right) \\
&\quad + \ D_2\left(r_1^2\ddot{p}_1 + 2r_1\dot{r}_1\dot{p}_1 - r_1p_1\ddot{r}_1 - 2p_1\dot{r}_1^2\right) + D_3(p_2\ddot{p}_1 + r_2\ddot{r}_1 + \dot{r}_1\dot{r}_2 + \dot{p}_1\dot{p}_2) \\
&\quad - \ D_4(p_1\ddot{p}_2 + r_1\ddot{r}_2) + D_5(p_0\ddot{p}_1 + \dot{p}_1\dot{p}_0) + D_6 p_1\ddot{p}_0 \\
&\quad + \ \omega_{11}^2[\varsigma_1 p_1(r_1^2 + p_1^2) + \varsigma_2(r_1r_2 + p_1p_2) + \varsigma_3 p_1 p_0] + \Lambda\dot{\bar{V}}_{02} = 0
\end{aligned} \tag{4.163b}$$

$$\ddot{r}_2 + \omega_{21}^2 r_2 - D_9(p_1\ddot{r}_1 + r_1\ddot{p}_1) - 2D_7\dot{r}_1\dot{p}_1 + 2\omega_{21}^2\varsigma_4 r_1 p_1 = 0 \tag{4.163c}$$

$$\ddot{p}_2 + \omega_{12}^2 p_2 + D_9\left(r_1\ddot{r}_1 - p_1\ddot{p}_1\right) + D_7\left(\dot{r}_1^2 - \dot{p}_1^2\right) - \omega_{12}^2\varsigma_4\left(r_1^2 - p_1^2\right) = 0 \tag{4.163d}$$

$$\ddot{p}_0 + \omega_{01}^2 p_0 + D_{10}\left(r_1\ddot{r}_1 + p_1\ddot{p}_1\right) + D_8\left(\dot{r}_1^2 + \dot{p}_1^2\right) + \omega_{01}^2\varsigma_5\left(r_1^2 + p_1^2\right) = 0 \tag{4.163e}$$

Table 4.3 *Coefficients of the nonlinear modal system*

α°	D_1	D_2	D_3	D_4	D_5	D_6	D_7	D_8	D_9	D_{10}
25	−0.1165	−0.4152	1.6116	−0.5428	2.0291	0.6715	0.6055	−0.3634	−1.2498	0.7113
27	−0.1927	−0.4287	1.6707	−0.5790	2.0773	0.7271	0.5954	−0.3428	−1.3445	0.8001
29	−0.2744	−0.4447	1.7330	−0.6163	2.1288	0.7833	0.5860	−0.3217	−1.4432	0.8964
31	−0.3623	−0.4635	1.7990	−0.6550	2.1838	0.8402	0.5773	−0.2999	−1.5465	1.0009
33	−0.4575	−0.4853	1.8691	−0.6953	2.2427	0.8980	0.5694	−0.2772	−1.6548	1.1146
35	−0.5609	−0.5106	1.9438	−0.7375	2.3060	0.9571	0.5623	−0.2536	−1.7689	1.2387
37	−0.6740	−0.5397	2.0236	−0.7818	2.3742	1.0177	0.5560	−0.2287	−1.8893	1.3742
39	−0.7984	−0.5731	2.1093	−0.8285	2.4478	1.0802	0.5506	−0.2026	−2.0170	1.5225
41	−0.9360	−0.6115	2.2015	−0.8780	2.5276	1.1450	0.5461	−0.1748	−2.1527	1.6851
43	−1.0891	−0.6556	2.3011	−0.9308	2.6145	1.2126	0.5426	−0.1454	−2.2978	1.8636
45	−1.2608	−0.7064	2.4092	−0.9872	2.709	1.2837	0.5402	−0.1139	−2.4532	2.0600
47	−1.4546	−0.7650	2.5270	−1.0480	2.8134	1.3588	0.5390	−0.0802	−2.6207	2.2763
49	−1.6750	−0.8330	2.6559	−1.1136	2.9281	1.4389	0.5392	−0.0440	−2.8019	2.5153
51	−1.9277	−0.9123	2.7976	−1.1851	3.0552	1.5248	0.5408	−0.0050	−2.9990	2.7797
53	−2.220	−1.0052	2.9543	−1.2633	3.1967	1.6179	0.5442	0.0372	−3.2146	3.073
α°	ς_1	ς_2	ς_3	ς_4	ς_5	λ	k_1			
25	−0.2541	0.6810	0.7638	0.4550	0.1446	1.3135	−0.1324			
27	−0.2981	0.7388	0.8314	0.4965	0.1604	1.2872	−0.1425			
29	−0.3469	0.7983	0.9007	0.5395	0.1774	1.2597	−0.1529			
31	−0.4009	0.8597	0.9717	0.5841	0.1957	1.2311	−0.1634			
33	−0.4607	0.9233	1.0448	0.6307	0.2155	1.2013	−0.1743			
35	−0.5272	0.9896	1.1202	0.6795	0.2370	1.1704	−0.1855			
37	−0.6012	1.0588	1.1984	0.7307	0.2603	1.1384	−0.1972			
39	−0.6840	1.1316	1.2797	0.7848	0.2857	1.1054	−0.2094			
41	−0.7768	1.2084	1.3647	0.8420	0.3136	1.0713	−0.2221			
43	−0.8815	1.2899	1.4539	0.9028	0.3440	1.0361	−0.2356			
45	−1.0000	1.3767	1.5482	0.9678	0.3774	1.0000	−0.2500			
47	−1.1350	1.4697	1.6483	1.0376	0.4142	0.9629	−0.2653			
49	−1.2897	1.5700	1.7552	1.1128	0.4548	0.9249	−0.2818			
51	−1.4683	1.6787	1.8703	1.1944	0.4996	0.8859	−0.2997			
53	−1.6760	1.7972	1.9951	1.2834	0.5493	0.8461	−0.3192			

Source: Gavrilyuk, *et al.* (2001).

where the coefficients Λ, D_i, and ς_i are listed in Gavrilyuk, *et al.* (2001). The dependence of these coefficients on the cone semi-angle, within the range $25^\circ < \alpha < 53^\circ$, are listed in Table 4.3.

Now consider sinusoidal excitation of the tank along the horizontal axis y, $V_{02}=0$, $V_{03}(t) = -\Omega\bar{X}\sin\Omega t$, where $\bar{X} = X_0/r_0$ is the ratio of the excitation amplitude to the radius r_0. The primary modal functional are approximated by the main periodic terms

$$r_1(t) = A\cos\Omega t + \bar{A}\sin\Omega t \tag{4.164a}$$

$$p_1(t) = \bar{B}\cos\Omega t + B\sin\Omega t \tag{4.164b}$$

where A, $\bar{A}$, B, and $\bar{B}$ are amplitudes to be determined. Substituting solutions (4.164) into equations (4.163c) through (4.163e) and collecting the primary harmonics, gives

$$p_0(t) = c_0 + c_1 \cos 2\Omega t + c_2 \sin 2\Omega t \tag{4.165a}$$

$$p_2(t) = s_0 + s_1 \cos 2\Omega t + s_2 \sin 2\Omega t \tag{4.165b}$$

$$r_2(t) = e_0 + e_1 \cos 2\Omega t + e_2 \sin 2\Omega t \tag{4.165c}$$

where

$$\begin{aligned} c_0 &= l_0(A^2 + \bar{A}^2 + B^2 + \bar{B}^2),\ \ c_1 = h_0(A^2 - \bar{A}^2 - B^2 + \bar{B}^2),\ \ c_2 = 2h_0(A\bar{A} + B\bar{B}) \\ s_0 &= l_2(A^2 + \bar{A}^2 - B^2 - \bar{B}^2),\ \ s_1 = h_2(A^2 - \bar{A}^2 + B^2 - \bar{B}^2),\ \ s_2 = 2h_2(A\bar{A} - B\bar{B}) \end{aligned} \tag{4.166a}$$

$$\begin{aligned} e_0 &= -2l_2(A\bar{B} + \bar{A}B),\ \ e_1 = 2h_2(\bar{A}B - A\bar{B}),\ \ e_2 = -2h_2(AB + \bar{A}\bar{B}) \\ h_0 &= \frac{D_{10} + D_8 - \varpi_{01}^2 \varsigma_5}{2(\varpi_{01}^2 - 4)},\ \ l_0 = \frac{D_{10} - D_8 - \varpi_{01}^2 \varsigma_5}{2\varpi_{01}^2},\ \ h_2 = \frac{D_9 + D_7 + \varpi_{21}^2 \varsigma_5}{2(\varpi_{21}^2 - 4)} \\ l_2 &= \frac{D_9 - D_7 + \varpi_{21}^2 \varsigma_5}{2\varpi_{21}^2},\ \ \varpi_{ij} = \omega_{ij}/\Omega \end{aligned} \tag{4.166b}$$

Substituting equations (4.164) and (4.165) into equations (4.163a) and (4.163b) the following set of algebraic equations are obtained for the unknown amplitudes A, $\bar{A}$, B, and $\bar{B}$

$$A\left[\varpi_{11}^2 - 1 - m_1\left(A^2 + \bar{A}^2 + \bar{B}^2\right) - m_2 B^2\right] + m_3 \bar{A} B \bar{B} = \bar{X}\Lambda \tag{4.167a}$$

$$\bar{A}\left[\varpi_{11}^2 - 1 - m_1\left(A^2 + \bar{A}^2 + B^2\right) - m_2 \bar{B}^2\right] + m_3 A B \bar{B} = 0 \tag{4.167b}$$

$$B\left[\varpi_{11}^2 - 1 - m_1\left(B^2 + \bar{A}^2 + \bar{B}^2\right) - m_2 A^2\right] + m_3 A \bar{A} \bar{B} = 0 \tag{4.167c}$$

$$\bar{B}\left[\varpi_{11}^2 - 1 - m_1\left(A^2 + B^2 + \bar{B}^2\right) - m_2 \bar{A}^2\right] + m_3 A \bar{A} B = 0 \tag{4.167d}$$

where

$$\begin{aligned} m_1 &= -D_5\left(\frac{h_0}{2} - l_0\right) + D_3\left(\frac{h_2}{2} - l_2\right) + 2D_6 h_0 + 2D_4 h_2 + \frac{D_1}{2} \\ &\quad - \varpi_{11}^2\left[\frac{3\varsigma_1}{4} - \varsigma_2\left(\frac{h_2}{2} + l_2\right) + \varsigma_3\left(\frac{h_0}{2} + l_0\right)\right] \\ m_2 &= D_3\left(\frac{3h_2}{2} + l_2\right) + D_5\left(\frac{h_0}{2} + l_0\right) - 2D_6 h_0 + 6D_4 h_2 - \frac{D_1}{2} + 2D_2 \\ &\quad - \varpi_{11}^2\left[\frac{\varsigma_1}{4} + \varsigma_2\left(l_2 - \frac{3h_2}{2}\right) + \varsigma_3\left(l_0 - \frac{h_0}{2}\right)\right] \\ m_3 &= m_2 - m_1 \end{aligned}$$

In view of the parameters listed in (4.166b), the coefficients m_1 are functions of excitation frequency and semi-cone angle. Since the damping is ignored in the analytical modeling, it is reasonable to drop the response along the x-axis component that is 90° out of phase with the excitation displacement in $r_1(t)$, that is, $\bar{A}=0$. The $\bar{B}$ is also dropped from the transverse displacement $p_1(t)$. In this case equations are reduced to the following two equations

$$A\left[\varpi_{11}^2 - 1 - m_1 A^2 - m_2 B^2\right] = \bar{X}\Lambda \tag{4.168a}$$

$$B\left[\varpi_{11}^2 - 1 - m_1 B^2 - m_2 A^2\right] = 0 \tag{4.168b}$$

These two equations have two possible solutions:

(1) The planar solution given by

$$B = 0, \text{ and } A\left[\varpi_{11}^2 - 1 - m_1 A^2\right] = \bar{X}\Lambda \tag{4.169}$$

(2) The rotational solution given by the solution of the two equations

$$A\left[\varpi_{11}^2 - 1 - m_1 A^2 - m_2 B^2\right] = \bar{X}\Lambda, \quad \text{and} \quad \varpi_{11}^2 - 1 - m_1 B^2 - m_2 A^2 = 0 \tag{4.170a, b}$$

Equation (4.170b) gives

$$B^2 = \frac{1}{m_1}\left(\varpi_{11}^2 - 1 - m_2 A^2\right) \tag{4.171a}$$

and A is given by the solution of the cubic equation

$$A\left[\left(\varpi_{11}^2 - 1\right) - (m_1 + m_2)A^2\right] = -\frac{m_1}{m_2 - m_1}\bar{X}\Lambda \tag{4.171b}$$

In order to estimate the amplitude–frequency response under lateral harmonic excitation, it is important to define the range of the semi-apex angle of the conical container over which the modal equations are applicable. For example, Gavrilyuk, *et al.* (2001) showed that $m_2 - m_1 = 0$ at $\alpha = 60°$. The sloshing characteristics belong to shallow fluid approximation as $\alpha \geq 60°$, and the amplitude of the rotating free-surface wave grows without limit. It was found that secondary internal resonance takes place for $\alpha \leq 25°$. Accordingly, the modal system equations (4.163) are only applicable over the angle range $25° < \alpha < 55°$. Figure 4.23 shows the amplitude–frequency responses of the amplitude A and the transverse amplitude B for $\alpha = 30°$, $\bar{X} = 0.02$, and $(m_1 + m_2) < 0$. The planar waves are indicated by branches P_1, P_2, and P_3, while branches R_1 and R_2 belong to rotating waves. The turning point T separates the stable P_1T sub-branch and unstable TP_2 sub-branch. The bifurcation point C gives rise to the rotating wave on the branch CR_2 as indicated in the nonzero transverse motion shown in Figure 4.23(b). With reference to Figure 4.23(a), the sub-branch P_4C refers to stable planar motion, and CP_3 is unstable. The point F separates unstable solution shown by the sub-branch R_1F and stable rotating motion shown by the sub-branch FR_2. Figure 4.24 shows the amplitude–frequency response curves of longitudinal and transverse motions for $\alpha = 45°$, that is, when $(m_1 + m_2) > 0$. Gavrilyuk, *et al.* (2001) conducted numerical simulation of the modal equations (4.177) for $\alpha = 30°$, under excitation amplitude $\bar{X} = 0.02$, excitation frequency ratio $\Omega/\omega_{11} = 0.9967$, $A = 0.35$, $B = 0.419$, and initial conditions: $r_1(0) = 0.35$, $p_2(0) = 0.122$, $p_0(0) = 0.469$, $\dot{p}_1(0) = 0.1494$, and $\dot{r}_2(0) = 0.316$.

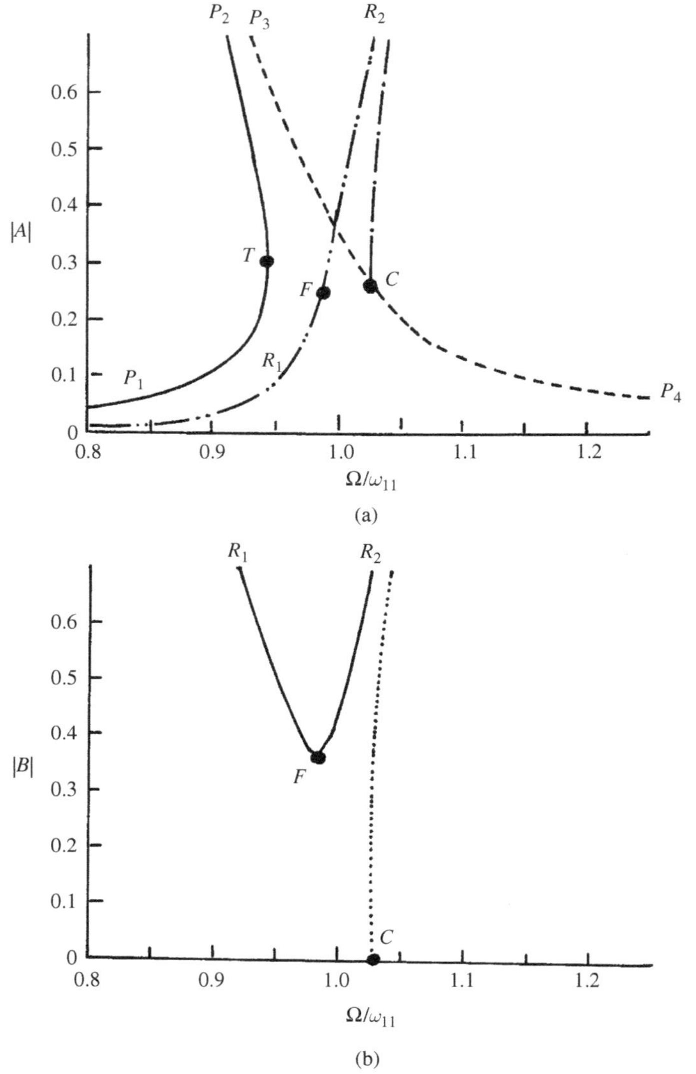

Figure 4.23 Amplitude–frequency responses for (a) in-phase amplitude, and (b) transverse amplitude of liquid free surface in a conical cone of semi angle $\alpha = 30°$. (Gavrilyuk, *et al.*, 2001)

4.6 Prolate spheroidal container

The nonlinear behavior of liquid surface in a prolate spheroidal container is not a simple task. In order to explore the nonlinear dynamic behavior of the liquid free surface inside a prolate spheroidal container one must conduct experimental forcing excitation in the neighborhood of resonance frequency. Analytically, Eidel (1989) estimated the backbone curves as a function of the liquid surface amplitude. He selected a coordinate system such that the container wall is specified as coordinate surfaces and the free liquid surface is described by a coordinate surface of the system. With reference to the prolate spheroid coordinates ξ, η, φ shown in Figure 1.14(a), Laplace's equation takes the form

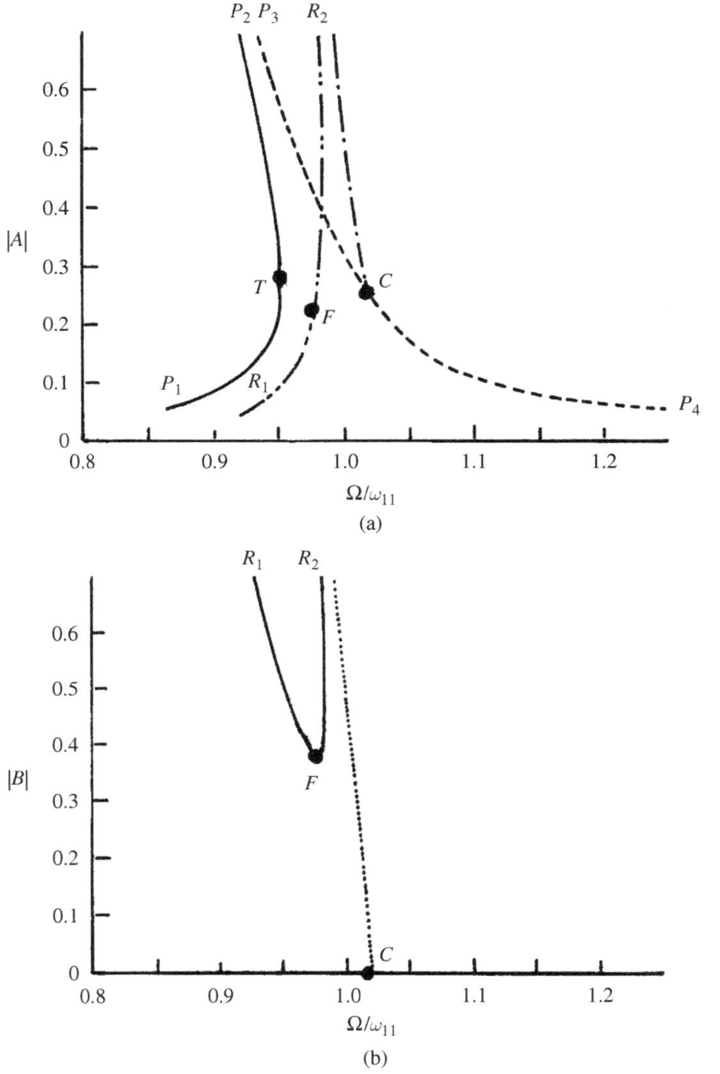

Figure 4.24 Amplitude–frequency responses for (a) in-phase amplitude, and (b) transverse amplitude of liquid free surface in a conical cone of semi angle $\alpha = 45^\circ$ (Gavrilyuk, *et al.*, 2001)

$$\sin\eta\frac{\partial}{\partial\xi}\left[\sinh\xi\frac{\partial\Phi}{\partial\xi}\right]+\sinh\xi\frac{\partial}{\partial\eta}\left[\sin\eta\frac{\partial\Phi}{\partial\eta}\right]+\frac{\left(\sinh^2\xi+\sin^2\eta\right)}{\sinh\xi\sin\eta}\frac{\partial^2\Phi}{\partial\varphi^2}=0 \tag{4.172}$$

Eidel (1989) considered different configurations shown in Figure 4.25. The boundary conditions depend on the configuration,

$$\frac{\partial\Phi}{\partial\xi}=0 \quad \text{at } \xi=\xi_0,\ \xi_1 \tag{4.173a}$$

where ξ_0 is the inside annular wall, and ξ_1 is the larger wall.

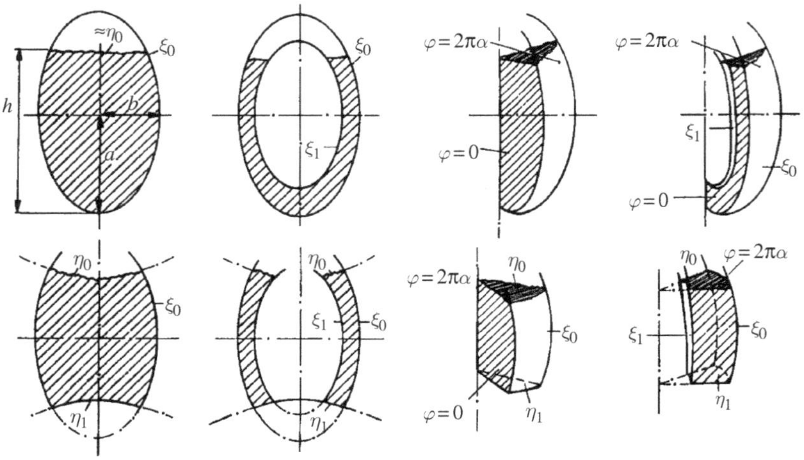

Figure 4.25 Different configurations of prolate spheroidal container. (Eidel, 1989)

$$\frac{\partial \Phi}{\partial \eta} = 0 \quad \text{at the container bottom } \eta = \eta_1 \tag{4.173b}$$

$$\frac{\partial \Phi}{\partial \varphi} = 0 \quad \text{at the container sector walls } \varphi = 0,\ 2\pi\alpha \tag{4.173c}$$

At the free surface the kinematic boundary condition is

$$\frac{\partial z}{\partial t} + \frac{1}{c^2(\sinh^2\xi + \sin^2\eta)}\left(\frac{\partial \Phi}{\partial \xi}\frac{\partial z}{\partial \xi} + \frac{\partial \Phi}{\partial \eta}\right) + \frac{1}{c^2\sinh^2\xi\sin^2\eta}\frac{\partial \Phi}{\partial \varphi}\frac{\partial z}{\partial \varphi} = 0, \quad \text{at } \eta = \eta_0 - z \tag{4.174}$$

where $a/2 = c$. The dynamic free-surface boundary condition is

$$\frac{\partial \Phi}{\partial t} + \frac{1}{2}\left\{\frac{1}{c^2(\sinh^2\xi + \sin^2\eta)}\left[\left(\frac{\partial \Phi}{\partial \xi}\right)^2 + \left(\frac{\partial \Phi}{\partial \eta}\right)^2\right] + \frac{1}{c^2\sinh^2\xi\sin^2\eta}\left(\frac{\partial \Phi}{\partial \varphi}\right)^2\right\} + gc\cosh\xi\cos\eta = 0 \quad \text{at } \eta = \eta_0 - z \tag{4.175}$$

Expanding conditions (4.174) and (4.175) into Taylor series in z at the undisturbed free-surface curvilinear $\eta \approx \eta_0$ up to third-order terms, gives

$$\frac{\partial z}{\partial t} + \frac{1}{c^2(\sinh^2\xi + \sin^2\eta)}\left[\frac{\partial \Phi}{\partial \eta} + \frac{2\sin\eta\cos\eta}{(\sinh^2\xi + \sin^2\eta)}\frac{\partial \Phi}{\partial \eta}z - \frac{\partial^2 \Phi}{\partial \eta^2}z + \frac{\partial \Phi}{\partial \xi}\frac{\partial z}{\partial \xi}\right] + \frac{1}{c^2\sinh^2\xi\sin^2\eta}\frac{\partial \Phi}{\partial \varphi}\frac{\partial z}{\partial \varphi} + \frac{1}{c^2(\sinh^2\xi + \sin^2\eta)^2} \times \left[2\sin^2\eta - 1 + \frac{4\sin^2\eta\cos^2\eta}{(\sinh^2\xi + \sin^2\eta)}\right]\frac{\partial \Phi}{\partial \eta}z^2 + \frac{1}{c^2(\sinh^2\xi + \sin^2\eta)}$$

$$\times\left[-\frac{2\sin\eta\cos\eta}{(\sinh^2\xi+\sin^2\eta)}\frac{\partial^2\Phi}{\partial\eta^2}z^2+\frac{1}{2}\frac{\partial^3\Phi}{\partial\eta^3}z^2+\frac{2\sin\eta\cos\eta}{(\sinh^2\xi+\sin^2\eta)}\frac{\partial\Phi}{\partial\xi}\frac{\partial z}{\partial\xi}z-\frac{\partial^2\Phi}{\partial\xi\partial\eta}\frac{\partial z}{\partial\xi}z\right]$$
$$+\frac{1}{c^2\sinh^2\xi\sin^2\eta}\left[2\frac{\cos\eta}{\sin\eta}\frac{\partial\Phi}{\partial\varphi}\frac{\partial z}{\partial\varphi}z-\frac{\partial^2\Phi}{\partial\varphi\partial\eta}\frac{\partial z}{\partial\varphi}z\right]=0\quad\text{at }\eta=\eta_0 \tag{4.176}$$

$$\underline{\frac{\partial\Phi}{\partial t}+cg\cosh\xi[z\sin\eta+\cos\eta]}+\frac{1}{2c^2(\sinh^2\xi+\sin^2\eta)}\left[\left(\frac{\partial\Phi}{\partial\xi}\right)^2+\left(\frac{\partial\Phi}{\partial\eta}\right)^2\right]$$
$$+\frac{1}{c^2\sinh^2\xi\sin^2\eta}\left(\frac{\partial\Phi}{\partial\varphi}\right)^2-\frac{1}{2}(cg\cosh\xi\cos\eta)z^2-z\frac{\partial^2\Phi}{\partial t\partial\eta}$$
$$+\frac{\sin\eta\cos\eta}{c^2(\sinh^2\xi+\sin^2\eta)^2}z\left[\left(\frac{\partial\Phi}{\partial\xi}\right)^2+\left(\frac{\partial\Phi}{\partial\eta}\right)^2\right]+\frac{\cos\eta}{c^2\sinh^2\xi\sin^3\eta}z\frac{\partial^2\Phi}{\partial\varphi^2}$$
$$-\frac{1}{c^2(\sinh^2\xi+\sin^2\eta)}z\left[\frac{\partial\Phi}{\partial\xi}\frac{\partial^2\Phi}{\partial\xi\partial\eta}+\frac{\partial\Phi}{\partial\eta}\frac{\partial^2\Phi}{\partial\eta^2}\right]-\frac{z}{c^2\sinh^2\xi\sin^2\eta}\frac{\partial\Phi}{\partial\varphi}\frac{\partial^2\Phi}{\partial\varphi\partial\eta}$$
$$-\frac{1}{6}(cg\cosh\xi\sin\eta)z^3+\frac{1}{2}z^2\frac{\partial^3\Phi}{\partial t\partial\eta^2}=0\quad\text{at }\eta=\eta_0 \tag{4.177}$$

The underlined terms constitute the linearized surface boundary conditions. Equations (4.176) and (4.177) are more accurate near $\eta_0=\pi/2$, and $\xi_0\ll 1$, which represent the case of a slender prolate spheroidal container as shown in Figure 1.14(a). The solution of Laplace's equation (4.172) that satisfies the wall boundary conditions may be written in the form

$$\Phi(\xi,\eta,\varphi,t)=\sum_{m=-\infty}^{\infty}\sum_{n=1}^{\infty}A_{mn}(t)L_{mn}(\xi)\bar{L}_{mn}(\eta)\,e^{im\varphi/2\alpha} \tag{4.178}$$

where $A_{mn}(t)=A_{-mn}(t)$ are time function generalized coordinates of the mode mn, and

$$L_{mn}(\xi)=P^{m/2\alpha}_{\lambda_{mn}}(\cosh\xi)Q'^{m/2\alpha}_{\lambda_{mn}}(\cosh\xi_0)-P'^{m/2\alpha}_{\lambda_{mn}}(\cosh\xi_0)Q'^{m/2\alpha}_{\lambda_{mn}}(\cosh\xi)$$
$$\bar{L}_{mn}(\eta)=P^{m/2\alpha}_{\lambda_{mn}}(\cos\eta)Q'^{m/2\alpha}_{\lambda_{mn}}(\cos\eta_1)-P'^{m/2\alpha}_{\lambda_{mn}}(\cos\eta_1)Q'^{m/2\alpha}_{\lambda_{mn}}(\cos\eta)$$

The functions $P^{m/2\alpha}_{\lambda_{mn}}$ and $Q^{m/2\alpha}_{\lambda_{mn}}$ are associated Legendre functions of the first and second kind, respectively, $\lambda_{mn}=-\frac{1}{2}-i\beta_{mn}$ are the roots of $L'_{mn}(\xi)=0$ for an annular container and are given by the solution of the determinant

$$\begin{vmatrix} P'^{m/2\alpha}_{\lambda}(\cosh\xi_0) & Q'^{m/2\alpha}_{\lambda}(\cosh\xi_0) \\ P'^{m/2\alpha}_{\lambda}(\cosh\xi_1) & Q'^{m/2\alpha}_{\lambda}(\cosh\xi_1) \end{vmatrix}=0 \tag{4.179}$$

For a nonannular container, λ_{mn} are given by the roots of the equation

$$P'^{m/2\alpha}_{\lambda}(\cosh\xi_0)=0 \tag{4.180}$$

The wave height of the free surface $z(\xi,\varphi,t)$ is determined from the solution

$$z(\xi, \varphi, t) = \sum_{m=-\infty}^{\infty} \sum_{n=1}^{\infty} a_{mn}(t) L_{mn}(\xi)\, e^{im\varphi/2\alpha} \tag{4.181}$$

where $a_{mn}(t)$ are time-dependent generalized coordinates of the free-surface elevation of mode mn. The coordinates $A_{mn}(t)$ and $a_{mn}(t)$ are determined by substituting equations (4.178) and (4.181) in the free-surface boundary conditions (4.176) and (4.177) and the left-hand sides are expanded in series of functions L_{mn} using the orthogonality property:

$$\int_{\xi_0}^{\xi_1} L_{mn}(\xi) L_{mk}(\xi) \sinh \xi \, d\xi = \begin{cases} 0 & n \neq k \\ \dfrac{\sinh \xi_1 L_{mk}(\xi_1)\left((\partial^2 L_{mk}(\xi_1))/(\partial \lambda_{mn} \partial \xi)\right) - \sinh \xi_0 L_{mk}(\xi_0)\left((\partial^2 L_{mk}(\xi_0)/(\partial \lambda_{mn} \partial \xi)\right)}{(2\lambda_{mn}+1)} & n = k \end{cases} \tag{4.182}$$

The result of this process is a set of an infinite number of nonlinear ordinary differential equations given in the form

$$D_{mn}^{(1)}(A_{jk}, a_{jk}, \dot{a}_{jk}; t) = 0, \quad m = 0, 1, 2, \ldots, \ n = 1, 2, 3, \ldots \tag{4.183}$$

$$D_{mn}^{(2)}(A_{jk}, \dot{A}_{jk}, \dot{a}_{jk}; t) = 0, \quad m = 0, 1, 2, \ldots, \ n = 1, 2, 3, \ldots \tag{4.184}$$

Considering only harmonic solutions of the form

$$a_{mn}(t) = Z_{mn} \cos \omega t, \ \ A_{mn}(t) = c^2 \omega_{m_0 b_0}^{(0)} Y_{mn} \sin \omega t \tag{4.185}$$

where $\omega_{m_0 b_0}^{(0)}$ is the linearized natural frequency of the mode $m_0 n_0$. Using the Ritz averaging method with the following conditions

$$\int_0^{2\pi/\omega} D_{mn}^{(1)}(A_{jk}, a_{jk}, \dot{a}_{jk}; t) \sin \omega \, dt = 0$$

$$\int_0^{2\pi/\omega} D_{mn}^{(2)}(A_{jk}, \dot{A}_{jk}, \dot{a}_{jk}; t) \cos \omega \, dt = 0 \tag{4.186}$$

with $m = 0, 1, 2, \ldots, n = 1, 2, 3, \ldots$, the following infinite set of nonlinear algebraic equations is obtained

$$-\frac{\omega}{\omega_{m_0 n_0}^{(0)}} Z_{mn} + \sum_{\nu=1}^{\infty} e^{(1)}(m, n, \nu) Y_{m\nu} + \frac{1}{4} \sum_{\substack{j_1+j_2+j_3=m \\ k_1, k_2, k_3}} [e^{(6)}(m, n, |j_1|, k_1, |j_2|, k_2, |j_3|, k_3)$$
$$- 2 \sin \eta_0 \cos \eta_0\, e^{(7)}(m, n, |j_1|, k_1, |j_2|, k_2, |j_3|, k_3)$$

$$
\begin{aligned}
&+\frac{1}{2}e^{(8)}(m,n,|j_1|,k_1,|j_2|,k_2,|j_3|,k_3)\\
&+2\sin\eta_0\cos\eta_0\,e^{(9)}(m,n,|j_1|,k_1,|j_2|,k_2,|j_3|,k_3)\\
&-e^{(10)}(m,n,|j_1|,k_1,|j_2|,k_2,|j_3|,k_3)\\
&-2j_1j_2\frac{\cos\eta_0}{\sin^3\eta_0}e^{(11)}(m,n,|j_1|,k_1,|j_2|,k_2,|j_3|,k_3)\\
&+\frac{j_1j_2}{\sin^2\eta_0}e^{(12)}(m,n,|j_1|,k_1,|j_2|,k_2,|j_3|,k_3)\Big]Y_{j_1k_1}Z_{j_2k_2}Z_{j_3k_3}=0,
\end{aligned}\tag{4.187}
$$

$$
\begin{aligned}
&\frac{\omega}{\omega^{(0)}_{m_0n_0}}\bar{L}_{mn}Y_{mn}+\frac{(g/c)\sin\eta_0}{\omega^{(0)}_{m_0n_0}}\sum_{\nu=1}^{\infty}e^{(13)}(m,n,\nu)X_{m\nu}\\
&+\sum_{\substack{j_1+j_2+j_3=m\\k_1,k_2,k_3}}\Bigg\{\frac{1}{4}\Big[\sin\eta_0\cos\eta_0 e^{(19)}(m,n,|j_1|,k_1,|j_2|,k_2,|j_3|,k_3)\\
&+e^{(20)}(m,n,|j_1|,k_1,|j_2|,k_2,|j_3|,k_3)-j_1j_2\frac{\cos\eta_0}{\sin^3\eta_0}\times e^{(21)}(m,n,|j_1|,k_1,|j_2|,k_2,|j_3|,k_3)\\
&-e^{(22)}(m,n,|j_1|,k_1,|j_2|,k_2,|j_3|,k_3)-e^{(23)}(m,n,|j_1|,k_1,|j_2|,k_2,|j_3|,k_3)\\
&-\frac{j_1j_2}{\sin^2\eta_0}e^{(24)}(m,n,|j_1|,k_1,|j_2|,k_2,|j_3|,k_3)\Big]Y_{j_1k_1}Y_{j_2k_2}Z_{j_3k_3}\\
&-\frac{1}{8}\frac{(g/c)\sin\eta_0}{\left(\omega^{(0)}_{m_0n_0}\right)^2}e^{(25)}(m,n,|j_1|,k_1,|j_2|,k_2,|j_3|,k_3)Z_{j_1k_1}Z_{j_2k_2}Z_{j_3k_3}\\
&+\frac{3}{8}\frac{\omega}{\omega^{(0)}_{m_0n_0}}e^{(26)}(m,n,|j_1|,k_1,|j_2|,k_2,|j_3|,k_3)Y_{j_1k_1}Z_{j_2k_2}Z_{j_3k_3}\Bigg\}=0
\end{aligned}\tag{4.188}
$$

where $e^{(l)}$ are real numbers and the summation with several indices stands for

$$
\sum_{\substack{j_1+j_2+j_3=m\\k_1,k_2,k_3}}a_{j_1k_1}a_{j_2k_2}a_{j_3k_3}=\sum_{j_1=-\infty}^{\infty}\sum_{k_1=1}^{\infty}\sum_{j_2=-\infty}^{\infty}\sum_{k_2=1}^{\infty}\sum_{j_3=-\infty}^{\infty}\sum_{k_3=1}^{\infty}a_{j_1k_1}a_{j_2k_2}a_{j_3k_3}\delta_{j_1+j_2+j_3,m}\tag{4.189}
$$

and the Kronecker delta is defined as follows

$$
\delta_{j_1+j_2+j_3,m}=\begin{cases}1 & j_1+j_2+j_3=m\\ 0 & j_1+j_2+j_3\neq m\end{cases}
$$

The amplitudes $|Z_{mn}|$ and $|Y_{mn}|$ decrease in their values as the mode mn increases and series (4.187) and (4.188) converge for a reasonable value of the indices m and n. Eidel (1989) numerically determined the first two ($n_0=1$, 2) natural frequencies of the asymmetric mode, $m_0=1$, for two complete prolate spheroidal containers of different slenderness ratios, $b/a=0.2$: $\xi_0=0.2027$; $b/a=0.7616$: $\xi_0=1$, and for different values of surface elevation. Figures 4.26 and 4.27 show the backbone curves of the two containers $b/a=0.2$, and $b/a=0.7616$, respectively. For the slender container, Figure 4.26 shows that for large filling heights, $\eta_0<\pi/2$, the softening spring characteristics are manifested, while for low values of liquid heights, $\eta_0<3\pi/5$, the liquid free surface exhibits hard spring characteristics. For the bulgy container, the behavior of the free surface is softening or hardening depending on the

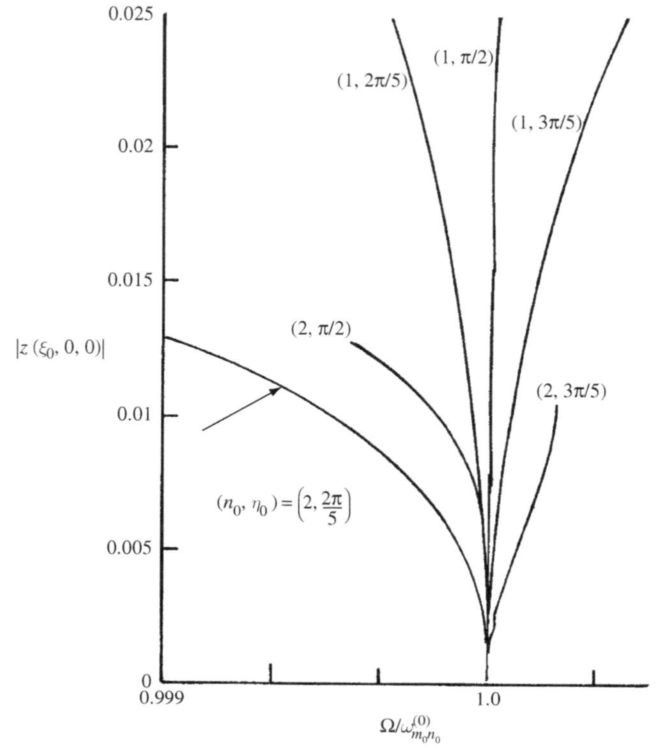

Figure 4.26 Backbone curves of asymmetric sloshing modes in prolate spheroidal container for minor-to-major semi-axis ratio $b/a = 0.2$, $m_0 = 1$. (Eidel, 1989)

liquid height and the mode number n_0. For $n_0 = 1$, the sloshing is hardening for all filling heights.

The previous sections considered the weakly nonlinear sloshing in selected tank geometries. To this end, the remainder of the chapter is devoted to related issues, namely, spatial resonance, nonlinear sloshing of magnetic fluids, and self-induced sloshing.

4.7 Spatial resonance

Spatial resonance describes a strong interaction between two classes of waves, which have related modal shapes but greatly different frequencies. This special type of free-liquid-surface instability was observed by Huntly (1972) and Huntly and Smith (1973). Acoustic fields can generate standing waves of large amplitude in the water (Lukovskii and Timokha, 2001). These waves have frequencies about two orders of magnitude lower than that of the acoustic field. Similar waves were generated by Huntly (1972) by exciting a beaker of water in its natural bell modes. Large amplitudes with very low frequency were observed to be generated. Waves with various mode patterns could also occur and some were maintained indefinitely. Moreover, different surface wave patterns could be generated by exciting higher-order bell modes. Mahony and Smith (1972) developed a model that describes the onset of low frequency waves in a system when high frequency waves are driven at a frequency close to resonance. It was

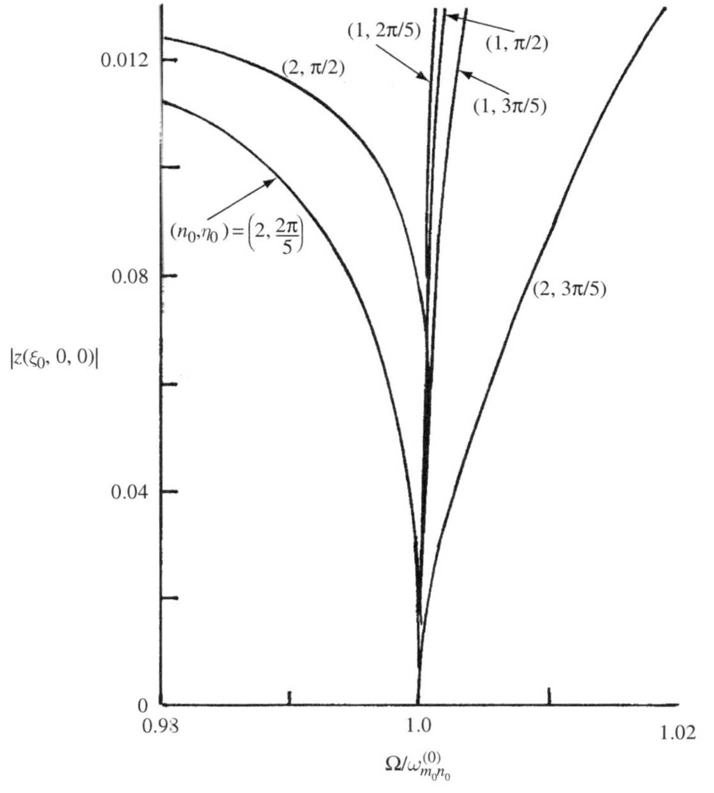

Figure 4.27 Backbone curves of asymmetric sloshing modes in prolate spheroidal container for minor-to-major semi-axis ratio $b/a = 0.7616$, $m_0 = 1$, $\xi_0 = 1$. (Eidel, 1989)

shown, experimentally, that under constant excitation amplitude and slowly varying frequency the steady-state amplitude of the low frequency wave exhibits hysteresis. Note that this type of resonance is not of the class parametric resonance since the phenomena are not sensitive to the frequency ratio of the two natural modes. Accordingly, these phenomena must be explained in terms of resonant nonlinear coupling.

Mahony and Smith suggested the following mechanism for such phenomena. Because the acoustic field is driven at an angular frequency ω close to resonance for a particular mode, whose spatial variation over the water surface is described by $X(x)$, this mode may dominate the acoustic field. Further, let there be present a small amount of standing surface wave whose spatial variation over the surface is given by $Y(x)\exp(i\sigma t)$. It is necessary that $X^2(x)$, when expanded by a modal decomposition, has a significant amount of the mode $Y(x)$ and that $X(x)Y(x)$ has a significant amount of the mode $X(x)$. The acoustic field drives a surface wave field of the form $X(x)\exp(i\omega t)$, but this has a small amplitude because its frequency is so much greater than the natural frequency of standing waves with modal form $X(x)$. The nonlinear surface conditions couple the two standing waves, so that very small amounts of the modes $X(x)\exp(i(\omega \pm \sigma)t)$ appear in the surface displacement. Now, if at least one value of $(\omega \pm \sigma)$ lies within the resonant bandwidth of the acoustic mode $X(x)$, the acoustic field will contain the modal response $X(x)\exp(i(\omega \pm \sigma)t)$. Moreover, the resonant amplification of the surface displacement at one or both of these frequencies may compensate for the very small surface response of the type $X(x)\exp(i(\omega \pm \sigma)t)$.

Nonlinear coupling between the acoustic modes $X(x)\exp(i\omega t)$ and $X(x)\exp(i(\omega \pm \sigma)t)$ and the assumed modal decomposition of $X^2(x)$, gives a very small forced excitation of the form $Y(x)\exp(i\sigma t)$ to the water surface. However, this is a natural mode of the surface wave system so there is a greatly enhanced response from such forced excitation. Thus, the presence of a surface wave mode may lead to the generation of the same mode through interaction with the acoustic field. Furthermore, there is a potential for instability but to confirm the instability it is necessary to show that the phase relationships will induce significant energy field into the standing wave.

In a related study, Lukovskii and Timokha (1991a, b) and Limarchenko, *et al.* (1992) developed nonlinear models of the dynamics of the free-surface separation between liquid and gas in the presence of a high-frequency acoustic field in the gas. They investigated liquid and gas oscillations during the acoustic action. They described bending of the capillary liquid surface under acoustic action (capillary–acoustic equilibrium shapes) and found wave number regions in which surface stabilization effects occur for negative critical loads and a swelling effect of the fluid surface. An acoustic mechanism was suggested for controlling the free fluid surface under conditions of near weightlessness.

4.8 Nonlinear sloshing of magnetic fluids

The linear lateral sloshing of magnetic fluids was discussed in Chapter 2 and the experimental results of Sawada, *et al.* (2002a, b) revealed the well-known nonlinear phenomenon of the fluid free surface near resonance. Sawada, *et al.* (1999) studied the lateral sloshing behavior of magnetic fluids in rectangular containers. Consider a rectangular container of length L, partially filled with a magnetic fluid to a depth h, subjected to sinusoidal lateral excitation $x(t) = X_0 \sin \Omega t$. The analysis is restricted to two dimensional flow with x, z placed at the undisturbed free surface. For irrotational, inviscid, and incompressible flow, a velocity potential function Φ exists which satisfies the continuity equation

$$\frac{\partial^2 \Phi}{\partial x^2} + \frac{\partial^2 \Phi}{\partial z^2} = 0 \tag{4.190}$$

The unsteady Bernoulli equation takes the form (Rosenweig, 1987)

$$\frac{p}{\rho} + \frac{1}{2}\left[\left(\frac{\partial \Phi}{\partial x}\right)^2 + \left(\frac{\partial \Phi}{\partial z}\right)^2\right] + \frac{\partial \Phi}{\partial t} + gz - \frac{\mu_0}{\rho}\int_0^H M\, dH = \Omega^2 X_0 x \sin \Omega t \tag{4.191}$$

where μ_0 is the magnetic permeability of vacuum, M is the magnetization, H (amp-turns/m) is the magnetic field. If the magnetization and magnetic field are assumed to be parallel, the magnetic term takes the form

$$\frac{\mu_0}{\rho}\int_0^H M\, dH = \frac{\mu_0}{2\rho}\chi_m H^2 \tag{4.192}$$

where χ_m is the susceptibility of the magnetic fluid. The magnetic field induction is $B = H\mu$ (weber/m^2, or newton/amper-meter). The magnetic field $H(z)$ may be approximated by the following expression

$$H(z) = H_0\, e^{-\alpha(z+h)} \tag{4.193}$$

where H_0 is the magnetic field intensity at the bottom of the tank ($z = -h$), and α is a constant determined from experimental measurements. The boundary conditions on the sidewalls and bottom of the tank require the normal component of the fluid vanishes, that is,

$$\left.\frac{\partial \Phi}{\partial x}\right|_{x=\pm L/2} = 0, \quad \text{and} \quad \left.\frac{\partial \Phi}{\partial z}\right|_{z=-h} = 0 \tag{4.194a, b}$$

The nonlinear kinematic and dynamic free-surface boundary conditions are, respectively,

$$\left[\frac{\partial \eta}{\partial t} + \frac{\partial \Phi}{\partial x}\frac{\partial \eta}{\partial x}\right]_{z=\eta} = \left.\frac{\partial \Phi}{\partial z}\right|_{z=\eta} \tag{4.195}$$

$$\begin{aligned} &\frac{\partial^2 \Phi}{\partial t^2} + 2\left(\frac{\partial \Phi}{\partial x}\frac{\partial^2 \Phi}{\partial x \partial t} + \frac{\partial \Phi}{\partial z}\frac{\partial^2 \Phi}{\partial z \partial t}\right) + \left\{\left(\frac{\partial \Phi}{\partial x}\right)^2 \frac{\partial^2 \Phi}{\partial x^2} + \left(\frac{\partial \Phi}{\partial z}\right)^2 \frac{\partial^2 \Phi}{\partial z^2}\right\} \\ &+ 2\frac{\partial \Phi}{\partial x}\frac{\partial \Phi}{\partial z}\frac{\partial^2 \Phi}{\partial x \partial z} \left[g + \frac{\mu_0}{\rho}\alpha \chi H_0^2\, e^{-2\alpha h}\right]\frac{\partial \Phi}{\partial z} \\ &- \Omega^3 X_0 x \cos \Omega t - \Omega^2 X_0 \frac{\partial \Phi}{\partial x} \sin \Omega t = 0 \text{ at } z = \eta \end{aligned} \tag{4.196}$$

Condition (4.195) is obtained from equation (4.191) by taking the time derivative of both sides at the free surface where the pressure is considered constant. The expression in the bracket represents the effective gravity due to the magnetic field. Equations (4.195) and (4.196) are linearized by expanding Φ, η, and p into power series of small parameter $\varepsilon = (X_0/L)^{1/3}$. It is convenient to write the expansions in the nondimensional form

$$\phi = \frac{\Phi}{L^2 \omega_1} = \varepsilon \phi_1 + \varepsilon^2 \phi_2 + \varepsilon^3 \phi_3 + \text{H.O.T.} \tag{4.197a}$$

$$\eta = \frac{\eta}{L} = \varepsilon \eta_1 + \varepsilon^2 \eta_2 + \varepsilon^3 \eta_3 + \text{H.O.T.} \tag{4.197b}$$

$$\bar{p} = \frac{p}{\rho g L} = -\bar{z} + \varepsilon p_1 + \varepsilon^2 p_2 + \varepsilon^3 p_3 + \text{H.O.T.} \tag{4.197c}$$

where H.O.T. stands for higher order terms, ω_1 is the first natural frequency of the free surface obtained from the linearized theory, and is given by the expression

$$\omega_n^2 = \frac{n\pi}{L}\left[g + \frac{\mu_0}{\rho}\alpha \chi H_0^2\, e^{-2\alpha h}\right] \tanh(n\pi h/L) \tag{4.198}$$

$$\bar{x} = x/L, \quad \bar{z} = \bar{z}/L, \quad \tau = \Omega t \tag{4.199}$$

Since the dynamic behavior of the fluid free surface is evaluated in the neighborhood of the first natural frequency, the following expansion is introduced

$$\frac{\Omega}{\omega_1} = 1 + \varepsilon\varpi_1 + \varepsilon^2\varpi_2 + \text{H.O.T.} \tag{4.200}$$

Transforming equations (4.195) and (4.196) into the nondimensional form, introducing expansions (4.197), and equating terms of equal powers of ε, gives a set of linear equations which are solved successively. The following are the linearized solutions of the velocity potential function and fluid elevation components up to third-order

$$\phi_1 = \frac{\eta_0}{\pi}\frac{\cosh \pi(\bar{z} + h/L)}{\sinh(\pi h/L)} \sin \pi\bar{x}\, \cos\tau \tag{4.201a}$$

$$\begin{aligned}\phi_2 = -\frac{\eta_0^2}{16\tanh(\pi h/L)}\Bigg\{ & 3\tanh(\pi h/L) + \frac{1}{\tanh(\pi h/L)} \\ & + 3\left(\frac{1}{\tanh^3(\pi h/L)} - \tanh(\pi h/L)\right)\frac{\cosh 2\pi(\bar{z}+h/L)}{\cosh(2\pi h/L)} \cos 2\pi\bar{x}\Bigg\} \cos 2\tau\end{aligned} \tag{4.201b}$$

$$\begin{aligned}\phi_3 = \frac{\pi\eta_0^3}{256\tanh(\pi h/L)}\Bigg\{ & -\left(1 + \frac{3}{\tanh^4(\pi h/L)} + \frac{11}{\tanh^2(\pi h/L)}\right. \\ & \left. - 15\tanh^2(\pi h/L)\right)\frac{\cosh 3\pi(\bar{z}+h/L)}{\cosh(3\pi h/L)} \sin 3\pi\bar{x}\, \cos\tau \\ & + \left(\frac{9}{\tanh^4(\pi h/L)} + \frac{62}{\tanh^2(\pi h/L)} - 31\right)\frac{\cosh \pi(\bar{z}+h/L)}{\cosh(\pi h/L)} \sin \pi\bar{x}\, \cos 3\tau \\ & + \left(\frac{9}{\tanh^6(\pi h/L)} + \frac{5}{\tanh^4(\pi h/L)} - \frac{53}{\tanh^2(\pi h/L)} + 39\right)\frac{\cosh 3\pi(\bar{z}+h/L)}{\cosh 3(\pi h/L)} \\ & \times \sin 3\pi\bar{x}\, \cos 3\tau\Bigg\} + \frac{4}{\pi^2}\sum_{n=2}^{\infty}\frac{(-1)^{n-1}\sin(2n-1)\pi\bar{x}}{(2n-1)^2\{(\omega_{2n-1}/\omega_1)^2 - 1\}} \\ & \times \frac{\cosh((2n-1)\pi(\bar{z}+h/L))}{\cosh((2n-1)(\pi h/L))} \cos\tau\end{aligned} \tag{4.201c}$$

$$\eta_1 = \eta_0 \sin \pi\bar{x} \sin\tau \tag{4.202a}$$

$$\eta_2 = \frac{\pi}{8}\eta_0^2\left\{\frac{3}{\tanh^3(\pi h/L)} - \frac{1}{\tanh(\pi h/L)}\right\} \cos 2\pi\bar{x}\, \cos 2\tau \tag{4.202b}$$

$$\eta_3 = \frac{\pi^2}{64}\eta_0^3\left(\frac{3}{\tanh^4(\pi h/L)} + \frac{12}{\tanh^2(\pi h/L)} - 1\right)\sin\pi\bar{x}\ \sin\tau$$

$$-\frac{\pi^2}{256}\eta_0^3\left(\frac{27}{\tanh^4(\pi h/L)} + \frac{51}{\tanh^2(\pi h/L)} - 15\tanh^2(\pi h/L) - 87\right)\sin 3\pi\bar{x}\ \sin\tau$$

$$+\frac{\pi^2}{256}\eta_0^3\left(\frac{9}{\tanh^4(\pi h/L)} + \frac{54}{\tanh^2(\pi h/L)} - 15\right)\sin\pi\bar{x}\ \sin 3\tau$$

$$+\frac{\pi^2}{256}\eta_0^3\left(\frac{81}{\tanh^6(\pi h/L)} - \frac{27}{\tanh^4(\pi h/L)} + \frac{27}{\tanh^2(\pi h/L)} - 9\right)\sin 3\pi\bar{x}\ \sin 3\tau$$

$$+\frac{2}{\pi}\tanh(\pi h/L)\sin\pi\bar{x}\ \sin\tau$$

$$+\frac{4\tanh(\pi h/L)}{\pi}\sum_{n=2}^{\infty}\frac{(-1)^{n-1}\sin(2n-1)\pi\bar{x}}{(2n-1)^2\{1-(\omega_{2n-1}/\omega_1)^2\}}\sin\tau \qquad (4.202c)$$

where η_0 is the amplitude of the free-surface elevation of the first-order approximation. This amplitude is obtained from the solution of the cubic equation

$$\varepsilon^3\frac{\pi^2}{64}\left(\frac{9}{\tanh^4(\pi h/L)} - \frac{14}{\tanh^2(\pi h/L)} - 3\right)\eta_0^3 - \varepsilon\left(\frac{\Omega}{\omega_1} - 1\right)\eta_0 - \frac{2X_0}{\pi L}\tanh(\pi h/L) = 0 \qquad (4.203)$$

The real roots of equation (4.203) are then introduced in equations (4.202) and (4.197b) to evaluate the maximum free-surface elevation η. The experimental results were obtained by Sawada, *et al.* (1999) for magnetic fluid of water-based 24% weight concentration of fine magnetic particles, Fe_3O_4 mixed with porous SiO_2 powder with a mean diameter of 0.9 μm. The magnetic fluid has kinematic viscosity $\nu = 4.2 \times 10^{-6}\,m^2/s$, density $\rho = 1240\ kg/m^3$, speed of sound 1440 m/s, fluid filling ratio, $h/L = 0.5$. A magnetic field was vertically applied. The dependence of η/h on the excitation frequency ratio, Ω/ω_E, where ω_E is the first resonant angular frequency for $B = 0$ is shown in Figure 4.28 for $B = 0$ mT, 20 mT, 40 mT, and 80 mT (mT = millitesla). It is seen that the first resonant frequency moves to a higher frequency as the magnetic field intensity increases.

Figure 4.29 shows the dependence of η/L on the excitation frequency ratio, Ω/ω_0, where ω_0 is the first natural frequency of the free surface as obtained from the linearized theory in the absence of the magnetic field. The experimental results are compared with those predicted by the linear and nonlinear analyses for $B = 0$ mT, 40 mT, and 80 mT. These figures reveal that the response possesses soft nonlinear characteristics. Furthermore, an increase of magnetic field induction resulted in a decrease of the value of the free-surface flow velocity. Due to the nonlinearity of fluid free-surface motion, the spatial velocity profile was asymmetric with respect to the central axis of the tank.

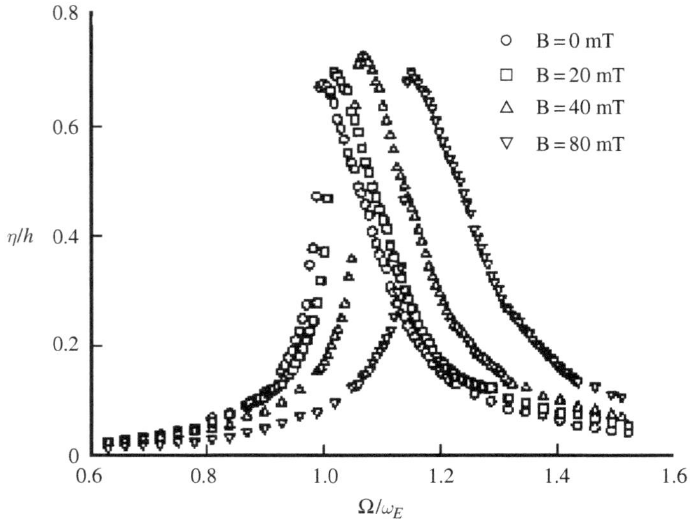

Figure 4.28 Amplitude–frequency response of the free surface of a magnetic fluid under different values of magnetic field induction $B = H\mu$. (Sawada, *et al.*, 1999)

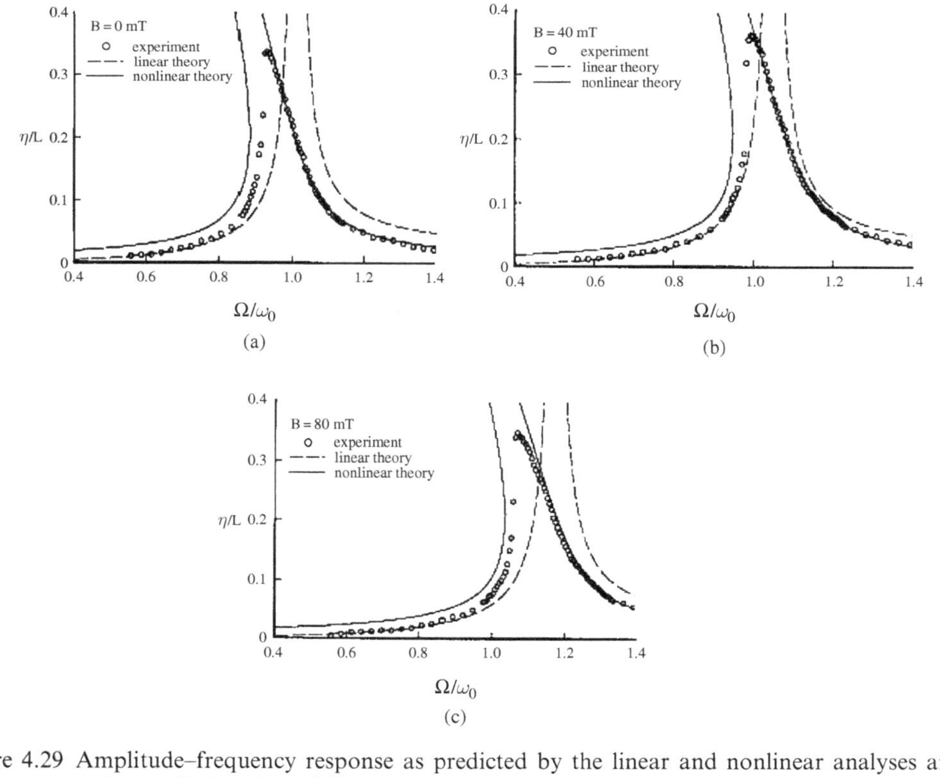

Figure 4.29 Amplitude–frequency response as predicted by the linear and nonlinear analyses and as measured experimentally for three different values of magnetic field induction parameter B. (Sawada, *et al.*, 1999)

4.9 Self-induced sloshing in nuclear reactors

4.9.1 Description of liquid dynamic problems in nuclear plants

Nuclear reactors employ pool type tanks for cooling purposes and to relieve the high-pressure that would result from the escaping steam during a loss-of-coolant accident (LOCA). In his retrospective, Abramson (2003) provided an assessment of the activities of the Southwest Research Institute, Texas, in liquid dynamics in different applications including nuclear power plants. These structures are required to maintain their design integrity under the influence of earthquake ground motion. The interaction of hydrodynamic pressure with the tank structure will be considered in Chapters 8 and 9. This section addresses the main experimental results from the study of patterns and mechanisms of liquid sloshing in suppression pools due to jet flow inside liquid tanks. Sloshing dynamics in pools subjected to earthquake ground motion was studied by Dong (1978). This type of sloshing is referred to as "self-induced sloshing" and is extensively studied by the Japanese.

Self-induced sloshing implies the interaction between a free surface and a submerged fluid jet. The free surface is thus excited by itself without any external excitation. The dominant frequency is close to the first or second natural frequencies of the liquid free surface. Two types of self-excited sloshing are observed in breeder reactors. The first type occurs at the natural frequency of sloshing and is known as coupled sloshing mode oscillation. The second type takes place in breeder reactors in a form of interaction between the sodium coolant free surface and a neighboring jet, or high-speed flow, and is referred to as self-induced sloshing. An upward plane jet impinging on the free surface can experience flutter (Madarame, *et al.*, 1993a, b). The jet flutter is accompanied by horizontal oscillations of the surface swell that forms on jet impingement. In a U-shaped double-tank system, self-induced U-tube oscillation is caused by coupled oscillation of two free surfaces. In this case, two separated surfaces move up and down alternatively with a remarkable variation in the flow pattern in the upstream tank (Fukaya, *et al.*, 1993, 1995, 1996, and Park, *et al.*, 2000).

This section does not deal with self-excitation of low-frequency liquid free-surface waves due to high-frequency vibration of its container. The reader may consult the work of Venediktov and Shibanov (1986) for this type of liquid motion.

In the liquid-metal fast breeder reactor (LMFBR), high-speed flow of the sodium coolant enters the reactor vessel to remove sufficient heat from the core (Inagaki, *et al.*, 1987). Reactor design criteria require that the free-surface flow remains stable under all operation conditions. However, high-speed flow under the free surface may cause undesirable surface phenomena such as breaking waves, gas entrainment, and self-induced sloshing. Yeh (1966), Okamoto, *et al.* (1991, 2000), Madarame, *et al.* (1992), and Fukaya *et al.* (1993) observed self-induced sloshing caused by a horizontal injected plane jet in different tank geometries. It was observed that pattern transformation of circulating flow with free-surface elevation caused first mode sloshing. Oscillation energy was suggested to be supplied by surface potential variations due to flow pattern transformation. Based on an analytical study by Madarame, *et al.* (1992), it was found that oscillation energy is supplied by pressure fluctuations caused by interaction between circulating flow and a sloshing motion.

The coupled sloshing mode type was studied by Aita and Gerbert (1986), Aita, *et al.* (1986), Fujita, *et al.* (1993, 1996), Nagakura and Kaneko (1995, 2000a, b), Koval'chuk and

Boyarshina (1998), and Kaneko, *et al.* (1999). Nagakura and Kaneko (1998a) analytically studied the coupled sloshing mode in a flexible plate weir. Nagakura and Kaneko (1998b, c) and Kaneko, *et al.* (1999) analyzed the case of a cylindrical weir, while Tanaka, *et al.* (1999, 2000) considered the case of a dual cylindrical weir. It was found that dual cylindrical shells oscillate in and out of phase with each other with an oval vibration mode under certain conditions. Hara and Suzuki (1992) analyzed the dynamic instability induced in a flexible weir, which separates a fluid into two volumes and subjects them to fluid discharge. Chiba, *et al.* (1997) experimentally studied the liquid oscillations in a thin cylindrical weir and observed three types of instability in the form of traveling waves, which travel either clockwise or counter-clockwise, or standing waves. They found that the velocity of the traveling waves is nearly constant and is independent of the flow rate and the falling height. Maschek, *et al.* (1992, 1998) studied the sloshing in liquid-metal fast-breeder reactors.

The dependency of the sloshing motion growth on geometrical and operational parameters was considered by Madarame, *et al.* (1992, 1993a, b), Okamoto, *et al.* (1992), Fukaya, *et al.* (1993), Takizawa and Kondo (1993, 1995), Someya, *et al.* (1995, 2000), Lu, *et al.* (1997), and Madarame and Ida (1998). Iida, *et al.* (1993) observed self-induced sloshing in a cylindrical tank with a round jet on the tank axis impinging on the free surface. Madarame, *et al.* (2002) conducted experiments to define the characteristics of liquid sloshing in cylindrical and rectangular tanks. It was found that the induced mode has one diametrical node, which is accompanied by the lateral motion of the swell. The mode is replaced by a lower one as the inlet–surface distance and velocity increase. The jet-flutter is coupled with the liquid sloshing and supplied energy to the liquid. Another form of flow-induced sloshing is due to overflow coupling with elastic oscillations.

Numerical simulations of self-induced sloshing were performed by Amano and Iwano (1991), Takizawa, *et al.* (1992), and Lu, *et al.* (1995). Takizawa and Kondo (1995) proposed numerical growth models using their previous simulation results (Takizawa and Kondo, 1993). Amano and Iwano (1991) simulated first mode sloshing using the vortex method and concluded that the phenomenon was excited by an unbalance of vortices formed over and under the jet near the inlet. Takizawa, *et al.* (1992) employed the physical components boundary fitted coordinate (PCBFC) method and showed that the free surface oscillates at certain values of inlet velocity and liquid level. It was assumed that oscillation energy is transferred from the kinetic energy of forced circulation by a nonlinear wave created by the secondary flow. Iida (2000) employed the three-dimensional finite element computer code SPLASH-ALE based on the arbitrary Lagrangian–Eulerian (ALE) method to analyze unstable phenomena caused by the interaction between flow and free surface in fast breeder reactors. Saeki *et al.* (2001) presented experimental and numerical investigations in an attempt to determine the excitation mechanism of self-induced sloshing due to a horizontal injected flat jet.

4.9.2 Experimental results

Rectangular tanks

Self-induced sloshing phenomena were experimentally studied by Okamoto, *et al.* (1991, 1993), Hu, *et al.* (1999), Saga, *et al.* (2000a, b), and Saeki, *et al.* (2001). Saeki, *et al.* (2001) used a

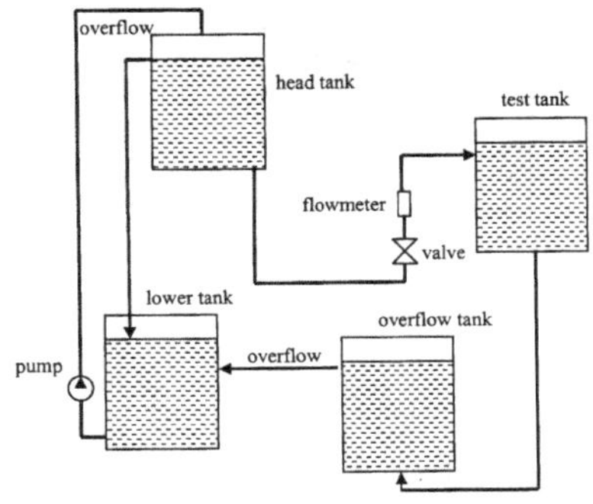

Figure 4.30 Schematic diagram of the test loop used by Saeki, *et al* (2001)

simple two-dimensional rectangular tank shown in Figure 4.30 to examine the dynamics of a liquid free surface with high-speed flow. Water is pumped up from the lower tank to the head tank. The water from the head tank is then supplied to the test tank. The inlet flow rate is controlled by a valve and measured by a floating-type flow meter. The cross-section of the overflow tank is ten times larger than that of the test tank so that the rate variation of the flow tank is about 5%. The breadth of the test tank, 100 mm, is much smaller than its width (1000 mm), such that the flow is considered to be two-dimensional. The tank has a rectangular inlet on the left-hand sidewall of length 600 mm and an outlet on the bottom of length 500 mm. The tank geometry can be changed by adding solid inserts, and the inlet width is either 100 mm or 20 mm. The inlet height from the bottom can be changed up to one of the three levels, 250 mm, 375 mm, and 450 mm. The distance of the outlet center, located at the tank bottom, from the tank left side is either 500 mm or 375 mm.

The characteristics of the free-surface flow and flow field are observed for each tank geometry by varying the inlet velocity, U_0, and mean water level, h. Three states of flow were reported and shown in Figure 4.31: (i) stable, (ii) oscillating, and (iii) reverse flow pattern. The reverse flow takes place when the water level is low and close to the inlet.

Self-induced sloshing was observed for a certain tank geometry, inlet velocity and water level. Figure 4.32 shows two different time history records for two different mean values of fluid depth. The first record shows that the oscillation amplitude reaches 80 mm. The second record shows a stable state. It was concluded that self-induced sloshing takes place only when one circulation is present on each side of the jet as shown in Figure 4.31. The dominant frequency of the free surface was measured and Figure 4.33 shows its dependence on the water mean level h for two different tank geometries. The figure also shows the theoretical frequency of the nth sloshing mode without circulation flow as documented in Chapter 1. The theoretical values are shown by solid curves.

The dependence of the dominant free-surface frequency on the inlet velocity is shown in Figure 4.34. It is seen that the theoretical values are independent of the inlet velocity for each inlet-surface distance $\Delta h = h - B = 0.4$ m, 0.3 m, and 0.2 m, where B is the inlet height from the tank bottom. The measured frequencies are given for two stages of self-induced sloshing: the first stage sloshing belongs to high velocity, while the second stage sloshing belongs to low inlet velocity. The dominant frequency of self-induced sloshing is considered to be dependent on the inlet velocity. Hara (1990) examined the frequency response of sloshing under lateral excitation by varying the under-surface flow velocity. Hara found the sloshing frequencies decrease with

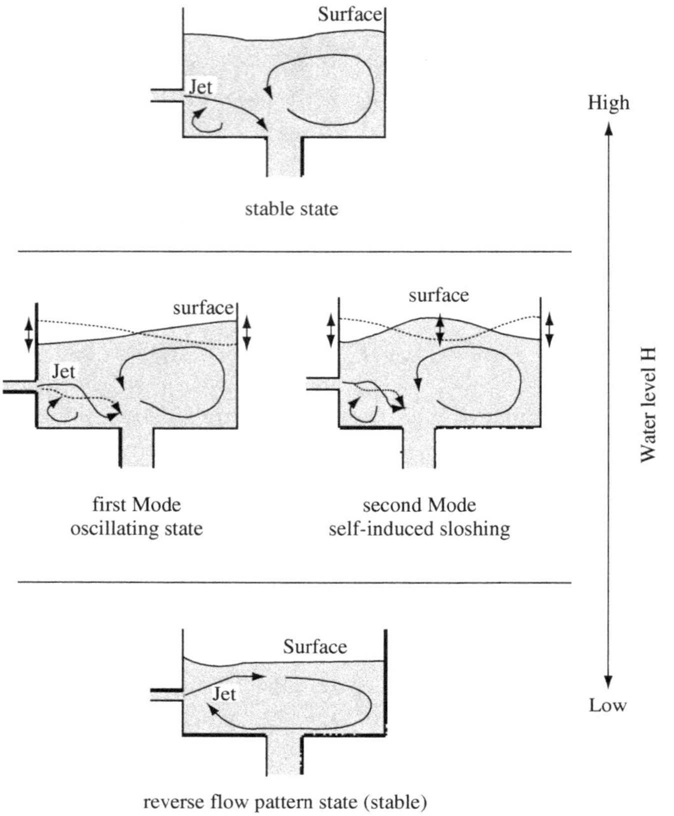

Figure 4.31 State of flow in the tank. (Saeki, *et al.*, 2001)

increasing under-surface flow velocity. The results of Hara (1990), however, appear to be in contradiction to the measured dominant frequency by Saeki, *et al.* (2001). The trend of the results of Figure 4.34 is a characteristic of self-induced sloshing and originates in the sloshing growth mechanism. The growth rate, γ, of the free-surface oscillation is defined as the natural logarithm of the ratio of two successive amplitudes, that is, $\gamma = \ln(A_{i+1}/A_i)$, which has the same definition of logarithmic decrement in measuring viscous damping.

Figure 4.35 shows the bifurcation diagram (or the excitation map) that separates the regions of first and second sloshing stages on a $(\Delta h, U_0)$ plane. Generally, the bifurcation diagram depends strongly on tank geometry.

In terms of the oscillation energy fed back to the sloshing motion, sloshing growth was analyzed quantitatively. Self-induced sloshing was found to be sustained by the feedback oscillation energy that was generated by the nonlinear interaction between sloshing motion and jet fluctuation. Furthermore, sloshing growth was primarily dependent on the spatial phase state of jet fluctuation.

Cylindrical tanks

Self-induced sloshing in a cylindrical tank with an upward round jet at the tank axis was studied by Madarame, *et al.* (2002) and Okamoto and Madarame (1998). With reference to the

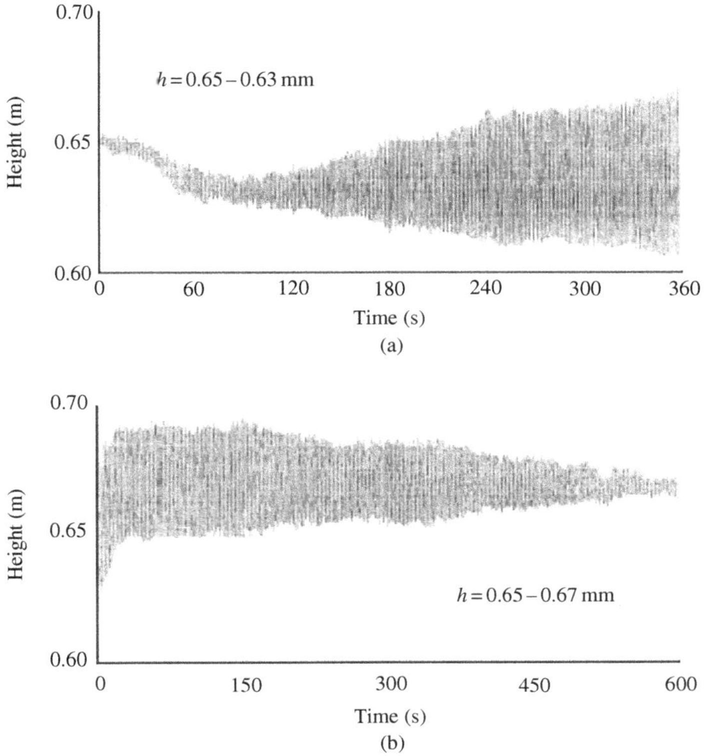

Figure 4.32 Time history records of free surface oscillations for two different mean values of liquid depth and for inlet flow velocity $U_0 = 0.67$ m/s, inlet width 100 mm, tank width 1000 mm, inlet height from bottom 250 mm, and outlet left distance 500 mm. (Saeki, *et al.*, 2001)

test tank shown in Figure 4.36, studied by Madarame, *et al.* (2002), the water is injected into the tank and forms a swell at the impingement point on the free surface. When the inlet velocity exceeds a certain value, the swell oscillates laterally in a diametral plane or rotates around the tank axis. The surrounding water sloshes synchronously with the swell. Some sketches of the oscillating water surface are shown in Figure 4.37(a–e). Figures 4.37(a) and (b) shows the oscillation in the direction perpendicular to a diametral plane, accompanied by sloshing of the surrounding water in the same direction. The sloshing mode was found to have only one nodal diameter and at least one circular node at the inner cylinder wall. Figure 4.37(a) depicts the first mode and Figure 4.37(b) is for the second mode. Figure 4.37(c) shows the case when the free-surface oscillation changes direction, either steadily or randomly, forming a spiral or more complex waves on the surrounding surface. Figure 4.37(d) shows a swell approximately 500 mm in height, which was observed to occur at the maximum flow rate. The swell and surrounding water surface rotates around the tank axis when an oscillation in a plane is superimposed on another oscillation of the same mode in the perpendicular direction. Figure 4.37(e) shows this case only for the first mode. The swell was observed always to lean towards the rising gradient of the sloshing surface during the rotation as well as during the uni-directional motion.

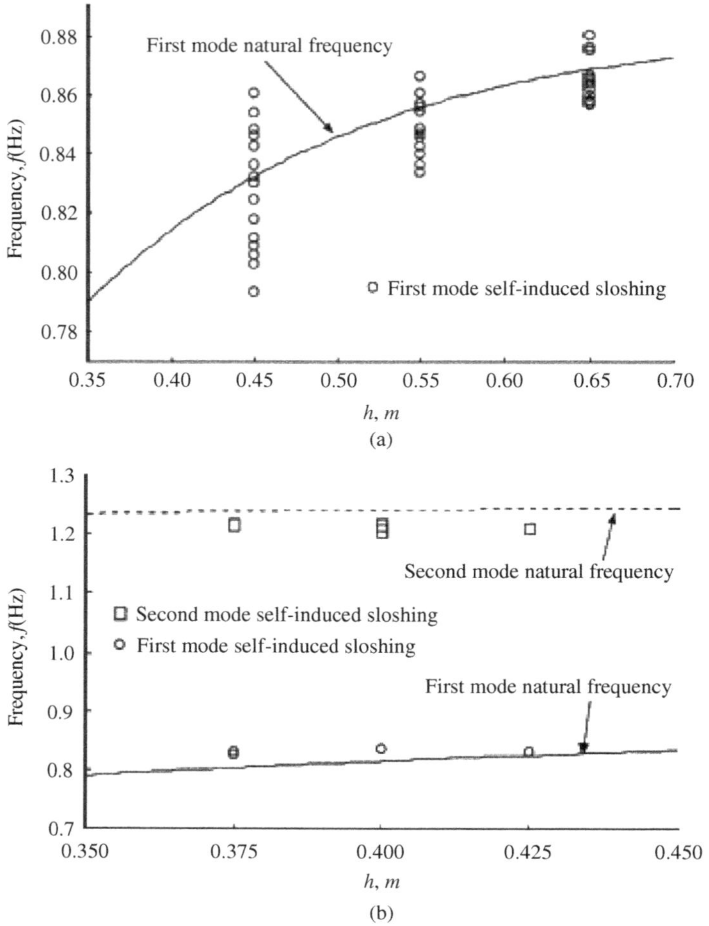

Figure 4.33 Dependence of the free-surface frequency on the fluid depth for two different tank geometries. (Saeki, *et al.*, 2001)

The frequency decomposition of the free-surface oscillations is shown by the power spectra (see Figure 4.38) for three different values of inlet velocity. It is seen that frequency spectrum has one or two dominant peaks at 1.2 Hz and 2.0 Hz for an inner cylinder diameter $D = 600$ mm. These frequencies are very close to the natural frequencies of the liquid. The spectrum with one spike at 2.0 Hz corresponds to the second mode of Figure 4.37(b), while the peak at 1.2 of Figure 4.38(c) corresponds to the first mode of Figure 4.37(a). Figure 4.38(b) corresponds to mixed mode interact shown in Figure 4.37(c). The oscillating regions were determined in terms of the inlet velocity, U_0, and the distance of the free surface from the inlet, Δh. Figure 4.39 shows a typical bifurcation diagram of three regions corresponding to first, second and third mode oscillations. The empty circles, •, refer to first mode oscillation with frequency 1.2 Hz, solid circles, •, refer to first and second mixed mode oscillations with frequencies 1.2 Hz, and 2.0 Hz, empty squares, □, refer to second mode oscillations with

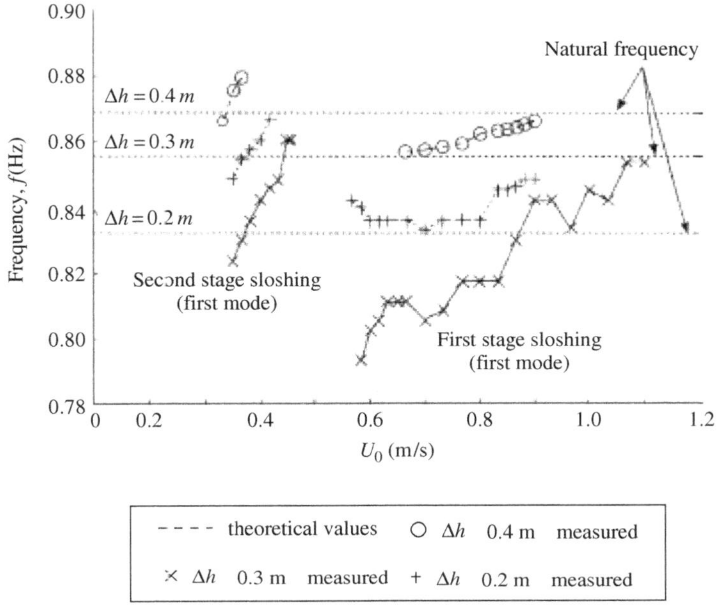

Figure 4.34 Dependence of the dominant free-surface frequency on the inlet velocity....... theoretical values independent of inlet velocitymeasured: ○ $\Delta h = 0.4$ m, x $\Delta h = 0.3$ m, + $\Delta h = 0.2$ m. (Saeki, *et al.*, 2001)

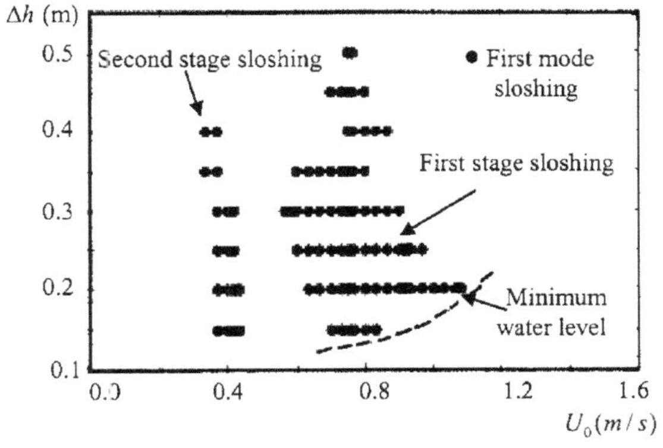

Figure 4.35 Bifurcation diagram obtained for a tank with inlet width 100 mm, tank width 1000 mm, inlet height from bottom 250 mm, and outlet left distance 500 mm. (Saeki, *et al.*, 2001)

frequency 2.0 Hz, solid squares, ■, refer to second and third mixed mode oscillations at 2.0 Hz and 2.5 Hz, respectively, delta symbols, Δ, refer to third mode oscillations with frequency 2.5 Hz. The dashes denote stable free surface, and small curved arrows denote rotational motion.

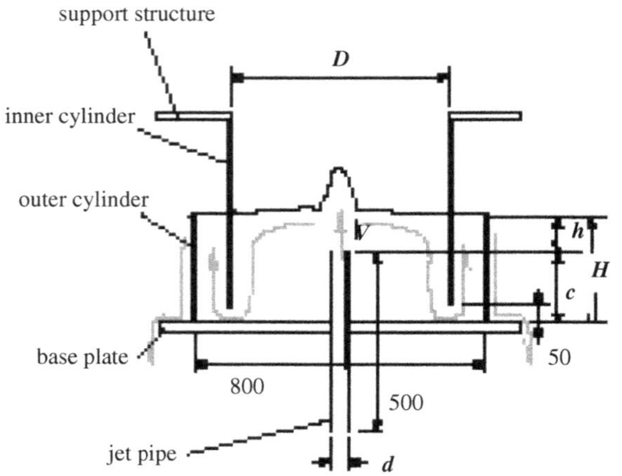

Figure 4.36 Test tank showing dimensions in mm. (Madarame, *et al.*, 2002)

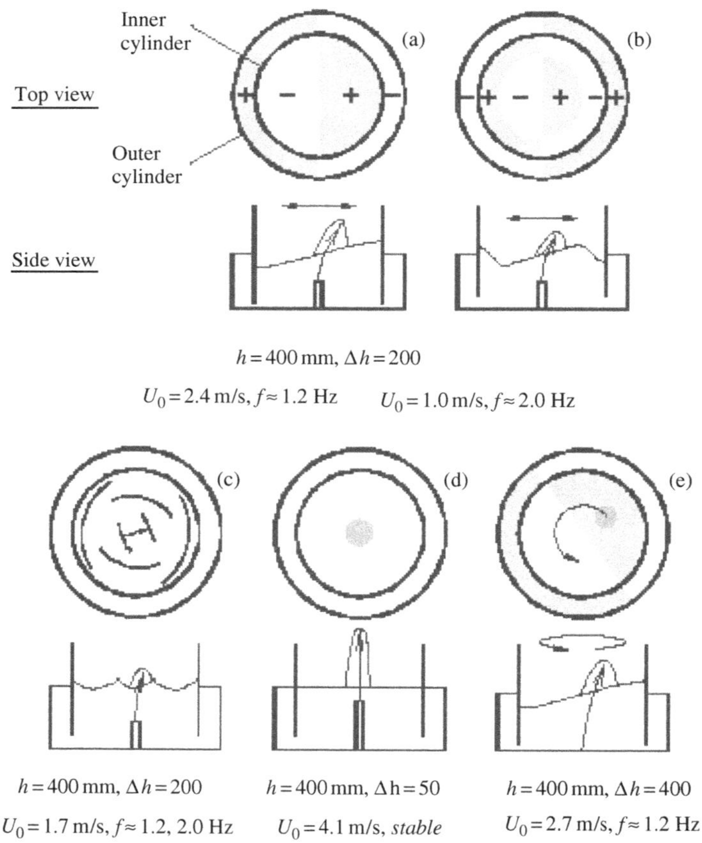

Figure 4.37 Possible surface patterns observed by Madarame, *et al.* (2002)

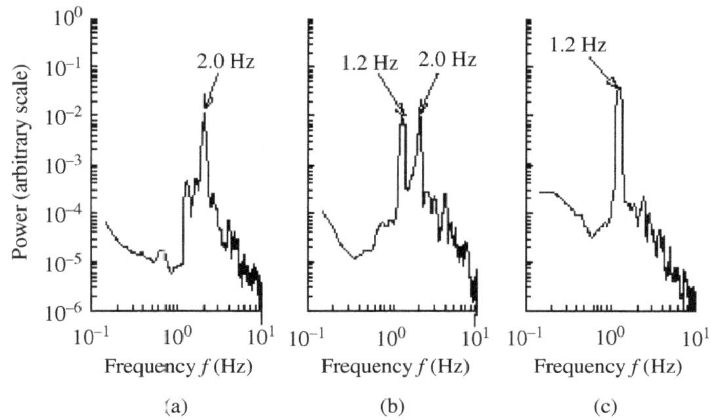

Figure 4.38 Typical frequency spectra for different values of inlet velocity, (a) $U_0 = 1.0$ m/s, (b) $U_0 = 1.7$ m/s, (c) $U_0 = 2.4$ m/s. (Madarame, *et al.*, 2002)

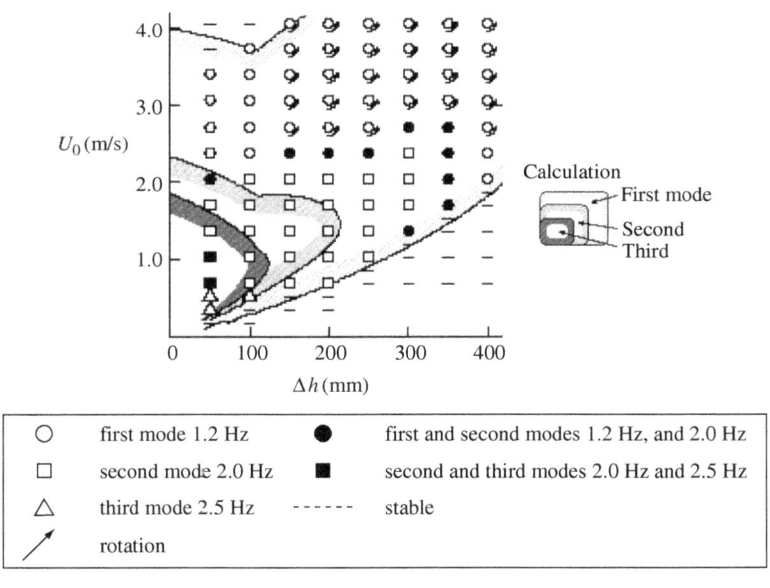

Figure 4.39 (Bifurcation diagram showing the regions of different oscillations) experiment. (Madarame, *et al.*, 2002)

4.10 Closing remarks

This chapter has introduced nonlinear surface wave phenomena in cylindrical and rectangular containers subjected to lateral excitations. In circular cylindrical tanks, rotary sloshing has received the attention of several studies motivated originally by experimental observations. The rotational motion is analogous to the dynamics of a spherical pendulum whose support point experiences lateral sinusoidal excitation. Some nonlinear theorems have been developed to predict not only rotary sloshing, but also other complex motions such as period doubling and chaos. Some experimental and analytical results of nonlinear liquid sloshing under random excitation with applications to aerospace and earthquake problems have been presented. In rectangular tanks, we focused our attention on nonlinear phenomena of standing longitudinal waves. In particular, attention was given to the switching of surface characteristics from hard to soft spring. Standing cross-waves at half the wave-maker frequency and their nonlinear interaction with longitudinal standing waves have not been addressed but the reader can refer to the work cited in Section 4.4. Nonlinear response analyses of conical and prolate spheroidal containers were treated based on nonconformal mapping and special coordinate frames. The main experimental results pertaining to the patterns and mechanism of self-induced sloshing in pool-type tanks of nuclear reactors have been reviewed. The next chapter will present the concept of linear and nonlinear equivalent mechanical models that emulate the dynamics of the free surface.

This chapter did not consider the nonlinear resonance of large-amplitude waves between two immiscible liquids of different densities and in stratified liquids. Boyarshina and Puchka (1985) and Boyarshina and Kholopova (1994) studied the nonlinear oscillations of such systems and showed that motion of the interface in the form of three-dimensional circular waves and experienced chaotic behavior. The influence of surface tension on the nonlinear free-surface waves was not addressed but the reader may refer to the work of Chen and Saffman (1979). Numerical simulation such as the boundary element method was employed to determine nonlinear surface waves by Ohyama and Fuji (1989), and Tanaka and Nakayama (1991). Tanaka and Fukuhara (1996) employed the boundary element method to determine the self-excited free surface sloshing. Other numerical simulations were employed by Wu, *et al.* (1998) and Yamamoto and Basoglu (1995). Chapter 7 will outline other numerical techniques for analyzing nonlinear sloshing problems.

This chapter did not address nonlinear sloshing under pitching or roll excitations. Under pitching excitation Rocca, *et al.* (2000) considered the problem under rotation about an axis parallel to the pitching axis. They (2002) also considered gravity waves in two different liquids.

Appendix

Orthogonality and recurrence relations

$$\int_0^{2\pi} \sin(m\theta)\sin(n\theta)\,d\theta = \begin{cases} \pi\delta_{mn} & \text{if } m \neq 0,\ n \neq 0 \\ 0 & \text{if } m \text{ or } n = 0 \end{cases} \tag{A4.1}$$

$$\int_0^{2\pi} \cos(m\theta)\cos(n\theta)\,d\theta = \begin{cases} \pi\delta_{mn} & \text{if } m \neq 0,\ n \neq 0 \\ 2\pi & \text{if } m \text{ or } n = 0 \end{cases} \tag{A4.2}$$

where the Kroneker $\delta_{mn} = 1$ if $m = n$ or $\delta_{mn} = 0$ if $m \neq 0$.

$$\int_0^{2\pi} \{\cos\theta,\ \sin\theta,\ \cos\theta\sin\theta,\ \cos^2\theta\sin\theta,\ \cos\theta\sin^2\theta,\ \cos^3\theta,\ \sin^3\theta\}\,d\theta = 0 \tag{A4.3}$$

$$\int_0^R rJ_0(\lambda_{0m}r)J_0(\lambda_{0n}r)\,dr = \begin{cases} 0 & m \neq n \\ \dfrac{R^2}{2}J_0^2(\lambda_{0n}R) & m = n \end{cases} \tag{A4.4}$$

$$\int_0^R rJ_2(\lambda_{2m}r)J_2(\lambda_{2n}r)\,dr = \begin{cases} 0 & m \neq n \\ \dfrac{\lambda_{2n}^2R^2 - 4}{2\lambda_{2n}^2}J_2^2(\lambda_{2n}R) & m = n \end{cases} \tag{A4.5}$$

$$\int_0^R rJ_1(\lambda_{1m}r)J_1(\lambda_{1n}r)\,dr = \begin{cases} 0 & m \neq n \\ \dfrac{\lambda_{1n}^2R^2 - 4}{2\lambda_{1n}^2}J_1^2(\lambda_{1n}R) & m = n \end{cases} \tag{A4.6}$$

$$\int_0^R r^{m+1}J_m(\lambda_{mn}r)\,dr = \frac{R^{m+1}}{\lambda_{mn}}J_{m+1}(\lambda_{mn}R) \tag{A4.7}$$

$$\frac{d}{dx}\left\{\frac{J_m(x)}{x^m}\right\} = -\frac{J_{m+1}(x)}{x^m}, \quad \text{for } m = 0, \quad \frac{dJ_0(x)}{dx} = -J_1(x) \tag{A4.8}$$

$$nJ_n(x) + x\frac{dJ_n(x)}{dx} = xJ_{n-1}(x) \tag{A4.9}$$

$$J_{n-1}(x) + J_{n+1}(x) = \frac{2n}{x}J_n(x) \tag{A4.10}$$

$$nJ_n(x) - x\frac{dJ_n(x)}{dx} = xJ_{n+1}(x) \tag{A4.11}$$

$$\frac{dJ_n(x)}{dx} = \frac{1}{2}[J_{n-1}(x) - J_{n+1}(x)] \tag{A4.12}$$

5
Equivalent mechanical models

5.1 Introduction

Generally, the liquid hydrodynamic pressure in rigid containers has two distinct components. One component is directly proportional to the acceleration of the tank and is caused by part of the fluid moving in unison with the tank. The second, known as "convective" pressure, experiences sloshing at the free surface. A realistic representation of the liquid dynamics inside closed containers can be approximated by an equivalent mechanical system. The equivalence is taken in the sense of equal resulting forces and moments acting on the tank wall. By properly accounting for the equivalent mechanical system representation of sloshing, the problem of overall dynamic system behavior can be formulated more simply. For linear planar liquid motion, one can develop equivalent mechanical models in the form of a series of mass-spring dashpot systems or a set of simple pendulums. For nonlinear sloshing phenomena, other equivalent models such as spherical or compound pendulum may be developed to emulate rotational and chaotic sloshing.

Graham (1951) developed an equivalent pendulum to represent the free-surface oscillations of a liquid in a stationary tank. Graham and Rodriguez (1952) introduced another model consisting of a sloshing point mass attached with springs to the tank wall at a specified depth and a fixed rigid mass. Pinson (1964) determined spring constants for liquid propellant in ellipsoidal tanks. Models in the form of mass-spring dashpot systems or a set of simple pendulums were considered by Ewart (1956), Bauer (1960a, 1961c, 1962b), Armstrong and Kachigan (1961), Abramson and Ransleben (1961c), Mooney, *et al.* (1964b). Bauer and Villenueva (1967b) developed a mechanical model that simulates the liquid dynamics in a rectangular container. Dodge, *et al.* (1994) introduced mechanical models for spinning spacecraft. Berry, *et al.* (1981) proposed analytical tools for simulating large amplitude liquid motion. Slosh models have been extensively used to examine the liquid behavior in various containers (Sandorff, 1960, Mixon, *et al.*, 1963, Mixon and Catherine, 1964a, b, Sumner, 1966, Candless and Walls, 1969, Pinson and Leonard, 1969, Huther, *et al.*, 1973, and Dodge, 1996). Equivalent moment of inertia of a liquid in cylindrical containers has been numerically estimated by Patrom (1985), and experimentally by Werner and Coldwell (1961). In multibody systems, liquid sloshing effects have been modeled using mechanical models by Ebert (1989), Enright and Wong (1994), and Rumold (1998, 2001).

The principals for constructing a mechanical model are based on the following conditions:

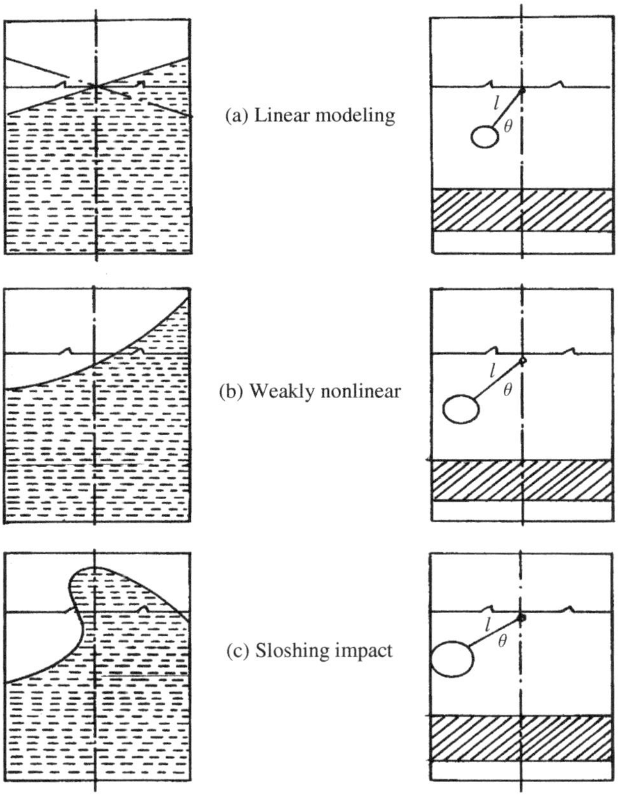

Figure 5.1 Regimes of free-liquid-surface motion and their equivalent mechanical modeling.

(1) The equivalent masses and moments of inertia must be preserved.
(2) The center of gravity must remain the same for small oscillations.
(3) The system must possess the same modes of oscillations and produce the same damping forces.
(4) The force and moment components under certain excitation must be equivalent to that produced by the actual system.

Equivalent mechanical models provide good representation of the liquid free-surface dynamics as long as the excitation frequency is remote from resonance. Near resonance, linear modeling isno longer valid and nonlinear representation must be considered. For example, for the first liquid sloshing mode, an equivalent pendulum may be used. Three dynamic regimes are possible, as shown in Figure 5.1:

(1) Small oscillations in which the fluid free surface remains planar without rotation of its nodal diameter (see Figure 5.1(a)). This regime can be described by an equivalent pendulum describing small oscillations, with $\sin\theta \approx \theta$.
(2) Relatively-large-amplitude oscillations in which the free liquid surface experiences nonplanar motion (see Figure 5.1(b)). This regime is described by a differential equation with weak nonlinearity and can be analyzed using the standard perturbation techniques. The equivalent mechanical model is a spherical pendulum describing relatively large motion such that $\sin\theta \approx \theta - \theta^3/3!$.
(3) Strongly nonlinear motion where the nonlinearity is mainly due to rapid velocity changes associated with hydrodynamic pressure impacts of the liquid motion close to the free surface (see Figure 5.1(c)).

The free-surface velocity changes are usually treated as being instantaneous (velocity jumps) giving various strong nonlinear features to the system behavior. The equivalent mechanical model of this regime is a pendulum describing impacts with the tank walls, as will be shown in Chapter 7.

This chapter presents the theory of linear and nonlinear equivalent mechanical models. Linear models will be developed for cylindrical and rectangular tanks. For other tanks, such as spherical, the modeling is usually developed experimentally. Nonlinear models that describe rotary sloshing will be analyzed using an equivalent spherical pendulum in addition to other mechanical elements. The theory of spherical pendulum will be presented to promote our understanding of rotary sloshing.

5.2 Spring-mass-dashpot modeling

Figure 5.2(a) shows a schematic diagram of an equivalent mechanical model consisting of a rigid mass m_o moving in unison with the tank, and a series of masses, m_n, representing the equivalent mass of each sloshing mode. Each modal mass, m_n, is restrained by a spring, K_n, and a dashpot, C_n. The equivalent model should satisfy the following conditions:

Fluid total mass

$$m_F = m_o + \sum_{n=1}^{\infty} m_n \tag{5.1a}$$

Mass moment of inertia about the y-axis that passes through the center of mass of the solidified liquid

$$I_f = I_o + m_o h_o^2 + \sum_{n=1}^{\infty} m_n h_n^2 \tag{5.1b}$$

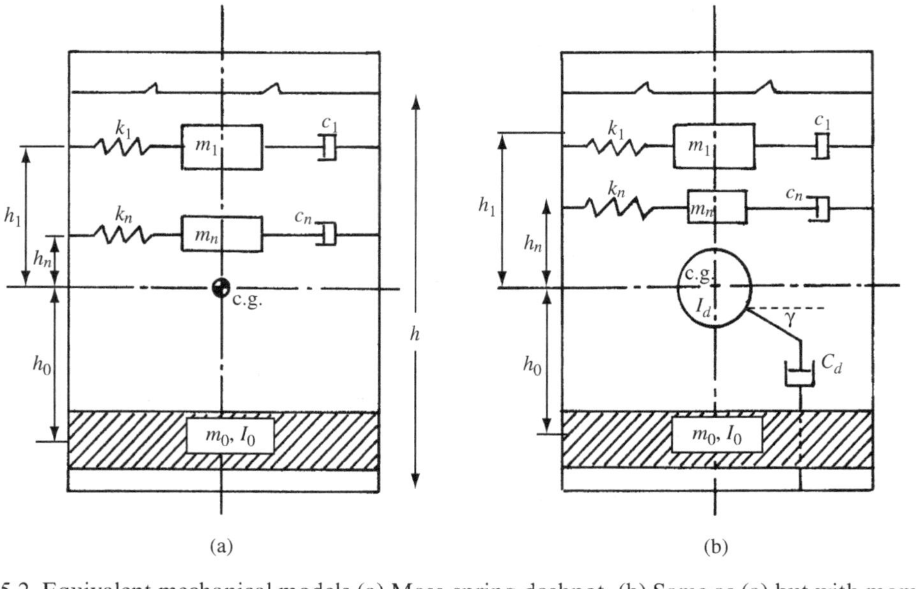

Figure 5.2 Equivalent mechanical models (a) Mass-spring-dashpot, (b) Same as (a) but with moment of inertia damping.

The *center of mass* should be preserved

$$m_o h_o - \sum_{n=1}^{\infty} m_n h_n = 0 \tag{5.1c}$$

The *spring constants* K_n can be determined from the definition of natural frequencies. For example, with reference to an upright cylindrical tank we have

$$\omega_n^2 = K_n / m_n = (g\xi_{1n}/R)\tanh(\xi_{1n}h/R) \tag{5.1d}$$

Let x_n be the displacement of the equivalent mass relative to the container wall, x the displacement of the container, ψ the rotational motion of the container about the y-axis through the center of gravity as shown in Figure 5.2(b). Bauer (1962b, 1964a) introduced a weightless disc with moment of inertia attached to the center of gravity together with a dashpot of damping coefficient C_d. Let the turning angle of the rigid disc relative to the bottom be γ. The equations of motion of the equivalent model may be derived using Lagrange's equation

$$\frac{d}{dt}\left(\frac{\partial}{\partial \dot{q}_i} L\right) - \frac{\partial}{\partial q_i} L = -\frac{\partial}{\partial \dot{q}_i}\Im + Q_i \tag{5.2}$$

where the Lagrangian $L = T - V$, q_i are the generalized coordinates, Q_i are the generalized forces, $\Im$ is the Rayleigh dissipation energy function, T and V are the kinetic and potential energies, respectively. The kinetic energy is

$$T = \frac{1}{2} m_o(\dot{x} - h_o\dot{\psi})^2 + \frac{1}{2} I_o \dot{\psi}^2 + \frac{1}{2}\sum_{n=1}^{\infty} m_n(\dot{x}_n + \dot{x} + h_n\dot{\psi})^2 + \frac{1}{2} I_d(\dot{\psi} + \dot{\gamma})^2 \tag{5.3a}$$

The potential energy is

$$V = \frac{1}{2} g\psi^2 m_o h_o - \frac{1}{2} g\psi^2 \sum_{n=1}^{\infty} m_n h_n - g\psi \sum_{n=1}^{\infty} m_n x_n + \frac{1}{2}\sum_{n=1}^{\infty} k_n x_n^2 \tag{5.3b}$$

The dissipation energy is

$$\Im = \frac{1}{2}\sum_{n=1}^{\infty} C_n \dot{x}_n^2 + \frac{1}{2} C_d \dot{\gamma}^2 = \sum_{n=1}^{\infty} m_n \omega_n \zeta_n \dot{x}_n^2 + \frac{1}{2} C_d \dot{\gamma}^2 \tag{5.3c}$$

The dashpot $C_n = 2m_n\omega_n\zeta_n$, and ζ_n is the damping factor of the equivalent dashpot. The generalized coordinates, q_i, and forces, Q_i, are given by the following vectors

$$\{q_i\} = \{x \ \ x_n \ \ \psi \ \ \gamma\}^T, \ \{Q_i\} = \{-F_x \ \ 0 \ \ M_y \ \ 0\} \tag{5.4}$$

Applying Lagrange's equation (5.2), using expressions (5.3) and (5.4), gives the following equations of motion

Force equation

$$m_o(\ddot{x} - h_o\ddot{\psi}) + \sum_{n=1}^{\infty} m_n(\ddot{x}_n + \ddot{x} + h_n\ddot{\psi}) = -F_x \tag{5.5a}$$

Slosh equation of nth mode

$$m_n(\ddot{x}+\ddot{x}_n+h_n\ddot{\psi})+K_nx_n+2m_n\omega_n\zeta_n\dot{x}_n-m_ng\psi=0 \tag{5.5b}$$

Moment equation

$$I_o\ddot{\psi}+m_oh_o(\ddot{x}-h_o\ddot{\psi})-g\sum_{n=1}^{\infty}m_nx_n + \sum_{n=1}^{\infty}m_nh_n(\ddot{x}_n+\ddot{x}+h_n\ddot{\psi})+I_d(\ddot{\psi}+\ddot{\gamma})=M_y \tag{5.5c}$$

Disc equation

$$I_d(\ddot{\psi}+\ddot{\gamma})+C_d\dot{\gamma}=0 \tag{5.5d}$$

Equations (5.5) fully describe the dynamics of the equivalent model in translation and pitching. From equation (5.5c) one can write the mass moment of inertia of a rigid body of fluid of equal mass as

$$I_{rigid}=I_0+m_0h_0^2+\sum_{n=1}^{\infty}m_nh_n^2+I_d$$

The following special cases are extracted from equations (5.5).

5.2.1 Lateral excitation of undamped models

For pure translational excitation, $x(t)=X_0\sin\Omega t$, $\Psi=\gamma=0$, and zero damping, the slosh equation takes the form

$$m_n\ddot{x}_n+K_nx_n=m_nX_0\Omega^2\sin\Omega t \tag{5.6}$$

The steady state solution of this equation is

$$x_n=\frac{\Omega^2}{\omega_n^2-\Omega^2}X_0\sin\Omega t \tag{5.7}$$

The force equation is

$$F_x=-m_o\ddot{x}-\sum_{n=1}^{\infty}m_n(\ddot{x}_n+\ddot{x})=X_0\Omega^2\sin\Omega t\left\{m_o+\sum_{n=1}^{\infty}m_n\left[\frac{\Omega^2}{\omega_n^2-\Omega^2}+1\right]\right\}$$
$$=m_FX_0\Omega^2\sin\Omega t\left\{1+\sum_{n=1}^{\infty}\frac{m_n}{m_F}\left(\frac{\Omega^2}{\omega_n^2-\Omega^2}\right)\right\} \tag{5.8}$$

The moment equation is

$$\begin{aligned}M_y&=m_oh_o\ddot{x}-g\sum_{n=1}^{\infty}m_nx_n+\sum_{n=1}^{\infty}m_nh_n(\ddot{x}_n+\ddot{x})\\&=-X_0\Omega^2\sin\Omega t\left\{m_oh_o+\sum_{n=1}^{\infty}\frac{gm_n}{(\omega_n^2-\Omega^2)}+\sum_{n=1}^{\infty}m_nh_n\left(\frac{\Omega^2}{\omega_n^2-\Omega^2}\right)\right\}\\&=-X_0\Omega^2\sin\Omega t\left\{m_oh_o+\sum_{n=1}^{\infty}m_n\left(x_n+\frac{g}{\omega_n^2}\right)+\sum_{n=1}^{\infty}m_n\left(x_n+\frac{g}{\omega_n^2}\right)\left(\frac{\Omega^2}{\omega_n^2-\Omega^2}\right)\right\}\end{aligned} \tag{5.9}$$

Equation (5.9) gives the moment due to pure translation taken about the center of mass of the solidified liquid.

5.2.2 Pitching excitation of undamped models

Under pitching excitation about the y-axis, $\psi = \Psi_0 \sin \Omega t$, the equations of motion are

$$m_n \ddot{x}_n + K_n x_n = m_n(g + h_n \Omega^2) \Psi_0 \sin \Omega t \tag{5.10}$$

The steady-state response is

$$x_n = \frac{(h_n \Omega^2 + g)}{\omega_n^2 - \Omega^2} \Psi_0 \sin \Omega t \tag{5.11}$$

The resulting force is

$$\begin{aligned} F_x &= -\left[m_o h_o \ddot{\psi} + \sum_{n=1}^{\infty} m_n (\ddot{x}_n + h_n \ddot{\psi}) \right] \\ &= \Psi_0 \Omega^2 \sin \Omega t \left\{ m_o h_o + \sum_{n=1}^{\infty} m_n (h_n \Omega^2 + g) \frac{\omega_n^2}{(\omega_n^2 - \Omega^2)} \right\} \end{aligned} \tag{5.12}$$

The moment about the y-axis is

$$\begin{aligned} M_y &= I_o \ddot{\psi} + m_o h_o^2 \ddot{\psi} - g \sum_{n=1}^{\infty} m_n x_n + \sum_{n=1}^{\infty} m_n h_n (\ddot{x}_n + h_n \ddot{\psi}) \\ &= \Psi_0 \Omega^2 \sin \Omega t \left\{ I_o + m_o h_o^2 + \sum_{n=1}^{\infty} m_n (h_n \Omega^2 + g)^2 \frac{\omega_n^2}{(\omega_n^2 - \Omega)^2} \right\} \end{aligned} \tag{5.13}$$

The values of the equivalent model parameters can be determined by comparing the values of the undamped model with the results of the ideal liquid presented in Chapter 2.

5.2.3 Model parameters for a circular upright cylinder

The hydrodynamic force due to combined translational and pitching excitations is

$$\begin{aligned} F_x = m_f X_0 \Omega^2 \sin \Omega t \left\{ 1 + \sum_{n=1}^{\infty} \frac{2R\Omega^2 \tanh(\xi_{1n} h/R)}{\xi_{1n} h (\xi_{1n}^2 - 1)(\omega_n^2 - \Omega^2)} \right\} + m_f \Psi_0 \Omega^2 \sin \Omega t \\ \times \sum_{n=1}^{\infty} \frac{2R \tanh(\xi_{1n} h/R) \Omega^2}{\xi_{1n} h (\zeta_{1n}^2 - 1)} \frac{\left[(h/2)\Omega^2 (1 - (4R/\xi_{in} h) \tanh(\xi_{1n} h/R)) + g\right]}{(\omega_n^2 - \Omega^2)} \end{aligned} \tag{5.14}$$

Comparing the first expression with equation (5.8), and the second expression with equation (5.12), the following equivalent parameters are obtained

$$\frac{m_n}{m_F} = \frac{2R}{\xi_{1n} h (\xi_{1n}^2 - 1)} \tanh(\xi_{1n} h/R) \tag{5.15a}$$

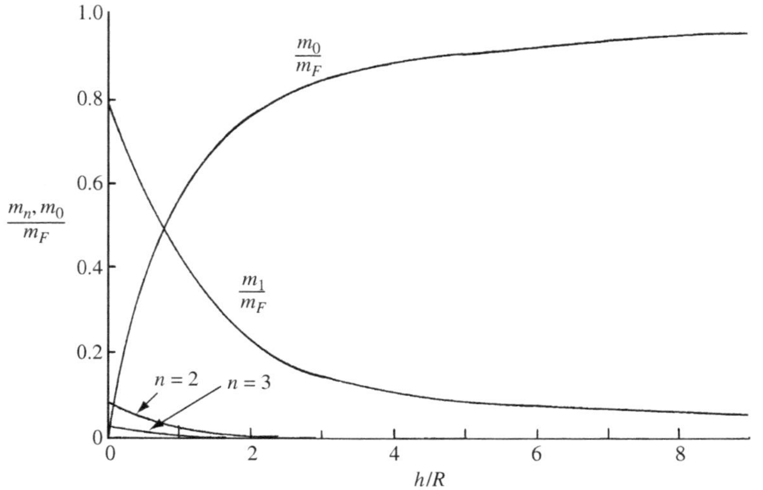

Figure 5.3 Ratios of mechanical model masses and fixed mass to the total fluid mass for a circular cylindrical tank.

$$\frac{m_o}{m_f} = 1 - \sum_{n=1}^{\infty} \frac{m_n}{m_f} = 1 - \sum_{n=1}^{\infty} \frac{2R}{\xi_{1n} h(\xi_{1n}^2 - 1)} \tanh(\xi_{1n} h/R) \tag{5.15b}$$

$$\frac{h_n}{h} = \frac{1}{2}\left[1 - \frac{4R\tanh(\xi_{1n}h/2R)}{\xi_{1n}h}\right] \tag{5.15c}$$

Figure 5.3 shows the dependence of the nonsloshing mass ratio, m_o/m_F, and modal slosh mass ratio, m_n/m_F, on the liquid-depth ratio, h/R, for the first three modes. One may observe that the size of slosh mass rapidly decreases for all modes exceeding the first.

From the definition of the center of mass (5.1c) and using relations (5.15), one can write

$$\frac{h_o}{h} = \frac{1}{1 - \sum\limits_{n=1}^{\infty} (2R/\xi_{1n}h(\xi_{1n}^2 - 1))\tanh(\xi_{1n}h/R)} \times \left[\frac{1}{2(h/R)^2} - \sum_{n=1}^{\infty} \frac{\xi_{1n}\tanh(\xi_{1n}h/R) + 4(R/(h\cosh(\xi_{1n}h/R))}{\xi_{1n}^2(\xi_{1n}^2 - 1)h/R}\right] \tag{5.15d}$$

The pitching moment caused by sloshing hydrodynamic forces is

$$M_y = \Psi_0 \Omega^2 \sin\Omega t \times \left\{ I_F + m_F \sum_{n=1}^{\infty} \frac{2Rh^2\Omega^2 \tanh(\xi_{1n}h/R)}{\xi_{1n}h(\xi_{1n}^2 - 1)} \frac{[1 - (4R/\xi_{in}h)\tanh(\xi_{1n}h/R)]^2}{4(\omega_n^2 - \Omega^2)} + m_F \sum_{n=1}^{\infty} \frac{2R\tanh(\xi_{1n}h/R)}{\xi_{1n}h(\xi_{1n}^2 - 1)} \left\{ \frac{(g/\Omega)^2 + gh[1 - (4R/\xi_{1n}h)\tanh(\xi_{1n}h/R)]}{(\omega_n^2 - \Omega^2)} \right\} \right\}$$

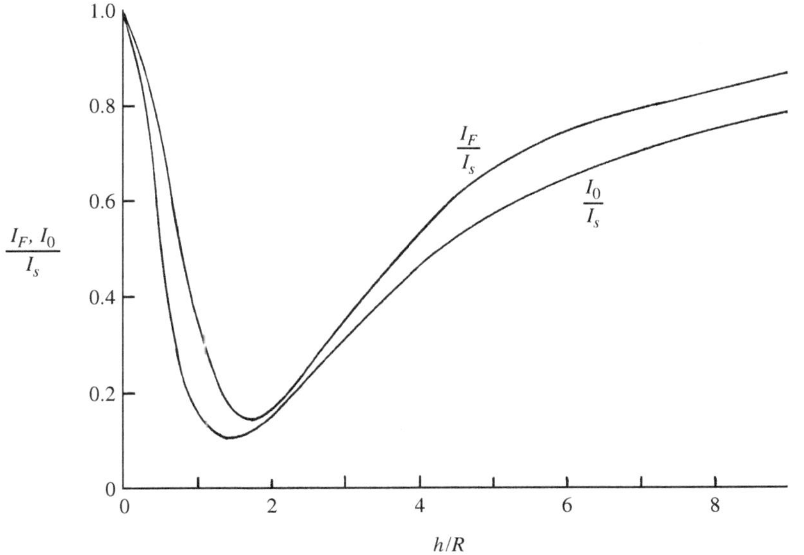

Figure 5.4 Ratios of fluid and fixed mass moments of inertia to the moment of inertia of solidified fluid in a circular cylindrical container.

$$+ m_F X_0 \Omega^2 \sin \Omega t \times \sum_{n=1}^{\infty} \frac{2R \tanh(\xi_{1n} h/R)}{\xi_{1n} h(\xi_{1n}^2 - 1)} \left\{ \frac{g + (h\Omega^2/2)[1 - (4R/\xi_{1n} h) \tanh(\xi_{1n} h/R)]}{(\omega_n^2 - \Omega^2)} \right\} \tag{5.16}$$

This moment is a combination of the moments due to pitching and translation, which should be equivalent to the sum of equations (5.9) and (5.13). Comparing this sum with equation (5.16), gives

$$I_F = I_o + m_o h_o^2 + \sum_{n=1}^{\infty} m_n h_n^2 \tag{5.17a}$$

$$= m_F \left\{ \frac{h^2}{12} + \frac{R^2}{4} - 8R^2 \sum_{n=1}^{\infty} \frac{1 - (2R/\xi_{1n} h) \tanh(\xi_{1n} h/2R)}{\xi_{1n}(\xi_{1n}^2 - 1)} \right\} \tag{5.17b}$$

where

$$m_F \left(\frac{h^2}{12} + \frac{R^2}{4} \right) = I_{\text{solid}}$$

represents the mass moment of inertia of the solidified liquid.

The moment I_F is the actual mass moment of inertia of the liquid about the y-axis through its center of mass. The actual moment of inertia can be obtained by determining the ratio, I_F/I_{solid}, of the actual liquid to rigid liquid moment of inertia for a cylindrical tank having an identical fluid height and fluid mass. Figure 5.4 shows the dependence of I_F/I_{solid} on the liquid depth h/R.

For a liquid depth close to $h/R=2$ the effective mass moment of inertia is about 18% of the value of the solidified liquid. The main reason of this drop is attributed to the fact that not all of the liquid participates in the pitching motion of the tank.

For the first slosh mode, the following equivalent parameters are used

$$
\begin{aligned}
K_1 &= m_F\left(\frac{g}{1.19h}\right)\left(\tanh 1.84\frac{h}{R}\right)^2 \\
m_1 &= m_F\left(\frac{R}{2.2h}\right)\tanh 1.84\frac{h}{R} \\
m_0 &= m_F - m_1 \\
h_1 &= \frac{h}{2} - \frac{R}{1.84}\tanh 1.84\frac{h}{R} \\
h_0 &= \frac{m_F}{m_0}\left[\frac{h}{2} - \frac{R^2}{2h}\right] - h_1\frac{m_1}{m_0}
\end{aligned} \tag{5.18}
$$

In many applications, one may obtain satisfactory results by considering only the fundamental mode.

5.2.4 Model parameters for a rectangular tank

Following the same steps of the previous section, the following parameters are obtained for a rectangular tank

$$
\begin{aligned}
\frac{m_n}{m_F} &= \frac{8}{\pi^3}\frac{\tanh((2n+1)\pi h/\ell)}{(2n+1)^3 h/\ell} \\
\frac{h_n}{h} &= \frac{1}{2} - \frac{\tanh((2n+1)\pi h/2\ell)}{(2n+1)\pi h/2\ell} \\
\frac{m_o}{m_F} &= 1 - \sum_{n=1}^{\infty}\frac{m_n}{m_F} \\
\frac{h_o}{h} &= -\frac{1}{(m_o/m_F)}\sum_{n=1}^{\infty}\frac{m_n}{m_F}\frac{h_n}{h} \\
\frac{hk_n}{m_F g} &= \frac{8\tanh^2((2n+1)\pi h/\ell)}{(2n+1)^2}
\end{aligned} \tag{5.19}
$$

$$
\frac{I_{o\,Y}}{m_F h^2} = \frac{I_{\text{solid}\,Y}}{m_F h^2}\left\{1 - \frac{4}{1+(h/\ell)^2} + \frac{768}{\pi^5(h/\ell)(1+(h/\ell)^2)}\sum_{n=1}^{\infty}\frac{\tanh((2n+1)\pi h/2\ell)}{(2n+1)^5}\right\}
- \frac{m_o}{m_F}\left(\frac{h_o}{h}\right)^2 - \sum_{n=1}^{\infty}\frac{m_n}{m_F}\left(\frac{h_n}{h}\right)^2
$$

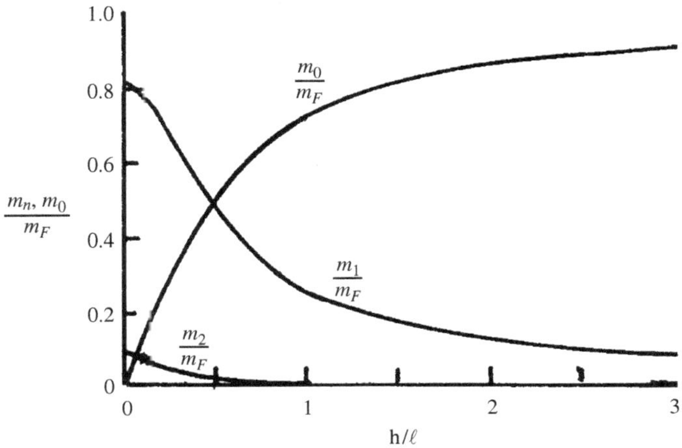

Figure 5.5 Mass ratios in a rectangular tank. (Graham and Rodriguez, 1952)

where $I_{\text{solid } Y}$ is the moment of inertia about the y-axis of the solidified fluid, I_{oY} is the moment of inertia of fixed mass.

The fluid moments of inertia about the y-axis axis, I_{FY}, and z-axis, I_{FZ} are, respectively

$$\frac{I_{FY}}{I_{\text{solid}Y}} = 1 - \frac{4}{1+(h/\ell)^2} + \frac{768}{\pi^5(h/\ell)(1+(h/\ell)^2)} \sum_{n=1}^{\infty} \frac{\tanh((2n+1)\pi h/2\ell)}{(2n+1)^5} \tag{5.20}$$

Figure 5.5 shows the dependence of the mass ratios, m_o/m_F, and m_i/m_F, $i = 1, 2$, on the fluid depth ratio h/ℓ. For very small depth ratio, that is, shallow tank, the modal sloshing mass ratios are dependent on the natural frequency of each mode as indicated in the logarithmic scale of Figure 5.6. Figure 5.7 shows the dependence of effective moment of inertia ratio, $I_{Fy}/I_{\text{solid } y}$, and the moment of inertia ratio of the fixed mass, $I_{oy}/I_{\text{solid } y}$, on the liquid depth ratio.

For a tank motion about the z-axis, the mechanical system may be composed of a fixed mass with moment of inertia about the z-axis, I_{oz}, and an infinite set of moving masses constrained to pivot about the z-axis. The moving masses possess moment of inertia I_{mnZ} given by the expression

$$\frac{I_{mn\,Z}}{I_{\text{solid } Z}} = \frac{768\left[(\ell/L)^2(2m+1)^2 - (2n+1)^2\right]^2}{\pi^6(1+(\ell/L)^2)(2m+1)^4(2n+1)^4\left[(\ell/L)^2(2m+1)^2 + (2n+1)^2\right]} \times \frac{\tanh\left((\pi h/\ell)\sqrt{\left[(\ell/L)^2(2m+1)^2 + (2n+1)^2\right]}\right)}{(\pi h/\ell)\sqrt{\left[(\ell/L)^2(2m+1)^2 + (2n+1)^2\right]}} \tag{5.21}$$

and the fixed mass moment of inertia is given by the expression

$$\frac{I_z}{I_{\text{solid } z}} = \frac{I_{Fz}}{I_{\text{solid } z}} - \sum_{m=1}^{\infty}\sum_{n=1}^{\infty} \frac{I_{mnz}}{I_{\text{solid } z}} \tag{5.22a}$$

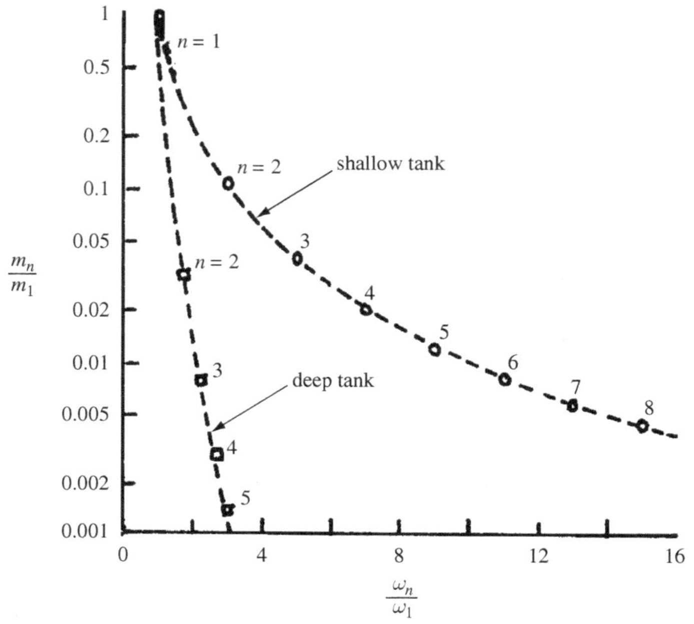

Figure 5.6 Modal masses as a function of their natural frequencies for shallow and deep rectangular tanks. (Graham and Rodriguez, 1952)

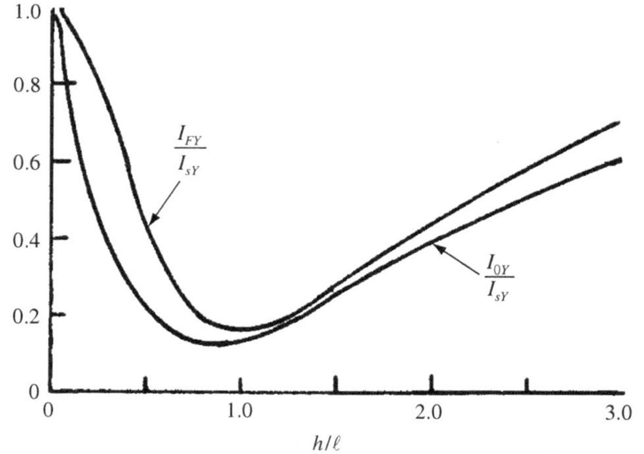

Figure 5.7 Mass moment of inertia of fluid, and fixed mass normalized to the solidified moment of inertia as a function of fluid depth ratio. (Graham and Rodriguez, 1952)

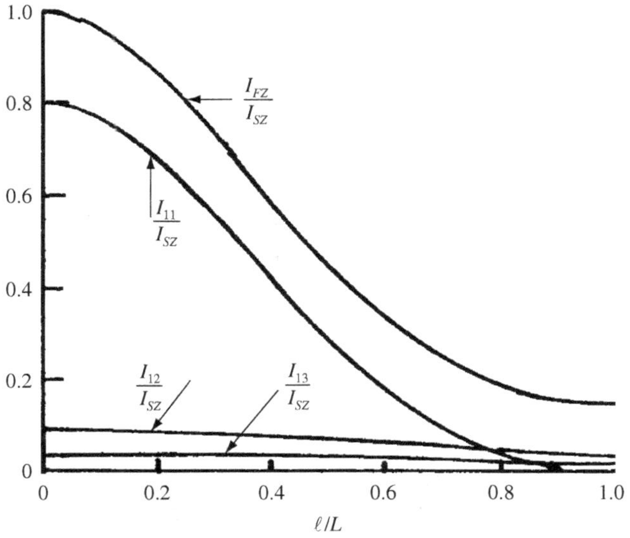

Figure 5.8 Mass moment of inertia of fixed mass and modal masses normalized to the solidified fluid inertia about the Z-axis in a rectangular tank. (Graham and Rodriguez, 1952)

where

$$\frac{I_{FZ}}{I_{\text{solid } Z}} = 1 - \frac{4}{1+(\ell/L)^2} + \frac{768}{\pi^5(\ell/L)(1+(\ell/L)^2)}\sum_{n=1}^{\infty}\frac{\tanh((2n+1)\pi\ell/2L)}{(2n+1)^5} \tag{5.22b}$$

Figure 5.8 shows the dependence of the effective fluid moment of inertia ratio, $I_{FZ}/I_{\text{solid } Z}$, and modal moment of inertia ratio, $I_{mnZ}/I_{\text{solid } Z}$, on the tank aspect ratio, ℓ/L. As the aspect ratio increases, the inertia ratio decreases for the first anti-symmetric mode. Higher modes reveal a slight drop as the aspect ratio increases.

5.3 Pendulum modeling

This modeling consists of a series of pendulums, each of mass, m_n, and length, ℓ_n, plus a rigid mass, m_o, as shown in Figure 5.9. The support point of the nth pendulum is at distance L_n below the undisturbed free surface and the rigid mass location is at distance L_o. The present analysis does not include energy dissipation. The mass preservation requires

$$m_F = m_o + \sum_{n=1}^{\infty} m_n \tag{5.23}$$

Equating the natural frequency of a simple pendulum, $\omega_n = \sqrt{g/\ell_n}$, to the natural frequency of the nth sloshing mode in cylindrical and rectangular tanks, yields the following expressions for the length of the nth pendulum for each tank, respectively:

For cylindrical tank:

$$\ell_n = \frac{R}{\xi_{1n}}\tanh(\xi_{1n}h/R) \tag{5.24a}$$

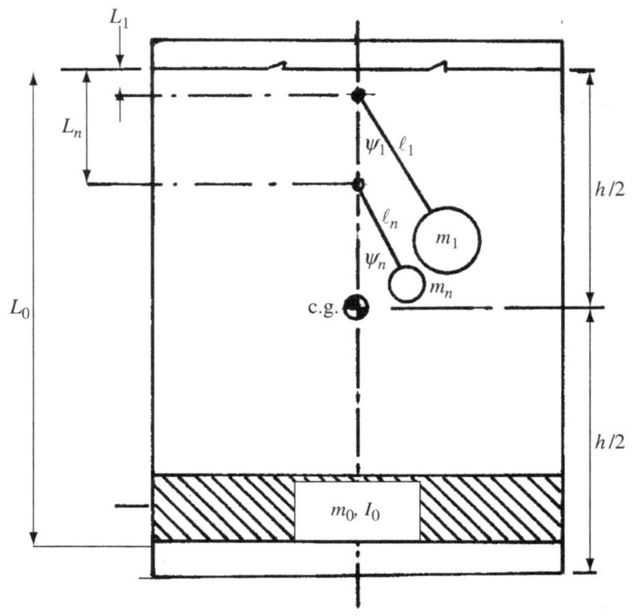

Figure 5.9 Pendulum equivalent model.

For rectangular tank:

$$\ell_{mn} = \frac{1}{\pi\sqrt{\dfrac{m^2}{\ell^2}+\dfrac{n^2}{L^2}}} \tanh\left(\pi h\sqrt{\frac{m^2}{\ell^2}+\frac{n^2}{L^2}}\right) \tag{5.24b}$$

where ℓ and L are the tank width and breadth, respectively, m and n are integers referring to the mode mn. Under combined horizontal, x, and pitching, ψ, excitations, one can develop the model equations of motion using Lagrange's equation. The kinetic energy is

$$T \approx \frac{1}{2}m_o\left[\dot{x}+\left(\frac{h}{2}-L_0\right)\dot{\psi}\right]^2+\frac{1}{2}I_0\,\dot{\psi}^2+\frac{1}{2}\sum_{n=1}^{\infty}m_n\left[\left(\frac{h}{2}-L_n-\ell_n\right)\dot{\psi}+\ell_n\dot{\psi}_n+\dot{x}\right]^2 \tag{5.25a}$$

The potential energy is

$$V = -m_0 g L_0\cos\psi - \sum_{n=1}^{\infty} m_n g[L_n\cos\psi+\ell_n\cos(\psi+\psi_n)] \tag{5.25b}$$

The generalized coordinates, q_i, and generalized forces, Q_i, are, respectively,

$$\{q_i\} = \{x,\ \psi,\ \psi_n\}^T, \quad \text{and} \quad \{Q_i\} = \{-F_z,\ M_y,\ 0\}^T \tag{5.25c}$$

The force equation is

$$-F_x = m_o\left[\ddot{x}+\left(\frac{h}{2}-L_0\right)\ddot{\psi}\right]+\sum_{n=1}^{\infty}m_n\left[\left(\frac{h}{2}-L_n-\ell_n\right)\ddot{\psi}+\ell_n\ddot{\psi}_n+\ddot{x}\right] \tag{5.26a}$$

The moment about the y-axis through the center of mass is

$$M_y = m_o\left[\ddot{x} + \left(\frac{h}{2} - L_0\right)\ddot{\psi}\right]\left(\frac{h}{2} - L_0\right) + I_o\ddot{\psi}$$

$$+ \sum_{n=1}^{\infty} m_n\left[\left(\frac{h}{2} - L_n - \ell_n\right)\ddot{\psi} + \ell_n\ddot{\psi}_n + \ddot{x}\right]\ell_n$$

$$+ \sum_{n=1}^{\infty} m_n g\left[L_n\psi + \ell_n(\psi + \psi_n)\right] \tag{5.26b}$$

The slosh equation is

$$\sum_{n=1}^{\infty} m_n\left[\left(\frac{h}{2} - L_n - \ell_n\right)\ddot{\psi} + \ell_n\ddot{\psi}_n + \ddot{x}\right]\ell_n + \sum_{n=1}^{\infty} m_n g\ell_n(\psi + \psi_n) = 0 \tag{5.26c}$$

The following special cases are considered:

5.3.1 Lateral excitation $x = X_0 \sin\Omega t,\ \psi = 0$

The slosh equation is

$$m_n\ell_n(\ell_n\ddot{\psi}_n + \ddot{x}) + m_n g\ell_n\psi_n = 0 \tag{5.27a}$$

The pendulum response is

$$\psi_n = \left[\frac{\Omega^2}{(\omega_n^2 - \Omega^2)}\right]\frac{X_0}{\ell_n}\sin\Omega t \tag{5.27b}$$

The resulting sloshing force and moment are, respectively,

$$F_x = m_F X_0\Omega^2\sin\Omega t\left\{1 + \sum_{n=1}^{\infty}\frac{m_n}{m_F}\frac{\Omega^2}{\omega_n^2 - \Omega^2}\right\} \tag{5.27c}$$

$$M_y = -X_0\Omega^2\sin\Omega t\left\{m_o\left(\frac{h}{2} - L_0\right) + \sum_{n=1}^{\infty} m_n\ell_n + \sum_{n=1}^{\infty} m_n\left(\ell_n - \frac{g}{\Omega^2}\right)\frac{\Omega^2}{\omega_n^2 - \Omega^2}\right\} \tag{5.27d}$$

5.3.2 Pitching excitation: $\psi = \Psi_0 \sin\Omega t,\ x = 0$

The slosh equation is

$$m_n\left[\left(\frac{h}{2} - L_n - \ell_n\right)\ddot{\psi} + \ell_n\ddot{\psi}_n\right]\ell_n + m_n g\ell_n(\psi + \psi_n) = 0 \tag{5.28a}$$

The response is

$$\psi_n = \frac{\Psi_0 \sin \Omega t}{(\omega_n^2 - \Omega^2)} \left[\frac{((h/2) - L_n - \ell_n)}{\ell_n} \Omega^2 - \omega_n^2 \right] \tag{5.28b}$$

The resulting slosh force and moment are, respectively,

$$\begin{aligned} F_x = & -\Omega^2 \Psi_0 \sin \Omega \Bigg[m_o L_o + \sum_{n=1}^{\infty} m_n ((h/2) - L_n - \ell_n) \\ & + \sum_{n=1}^{\infty} m_n \ell_n \left\{ \frac{((h/2) - L_n - \ell_n)}{\ell_n} - \frac{\omega_n^2}{\Omega^2} \right\} \frac{\Omega^2}{(\omega_n^2 - \Omega^2)} \Bigg] \end{aligned} \tag{5.28c}$$

$$\begin{aligned} M_y = & -\Omega^2 \Psi_0 \sin \Omega \Bigg[m_o \left(\frac{h}{2} - L_o \right)^2 + I_0 + \sum_{n=1}^{\infty} m_n \ell_n ((h/2) - L_n - \ell_n) \\ & + \sum_{n=1}^{\infty} m_n \ell_n^2 \left\{ \frac{((h/2) - L_n - \ell_n)}{\ell_n} - \frac{\omega_n^2}{\Omega^2} \right\} \frac{\Omega^2}{(\omega_n^2 - \Omega^2)} - m_o g \frac{L_o}{\Omega^2} \\ & - \sum_{n=1}^{\infty} m_n \ell_n (L_n + \ell_n) \frac{\omega_n^2}{\Omega^2} \Bigg] \end{aligned} \tag{5.28d}$$

Equations (5.28) are used to derive the model parameters by comparison with the ideal fluid equations derived in Chapter 2.

5.3.3 Mechanical parameters for a circular cylinder

A mode-by-mode comparison of the force and moment equations of the pendulum model and the ideal fluid equations gives the same values of m_n/m_F and m_o/m_F as in the case of mass-spring models in addition to relations (5.24). The remaining parameters for the circular upright cylinder are

$$L_n = -\frac{2R}{\xi_{1n} \sinh(\xi_{1n} h/R)} \tag{5.29a}$$

$$L_0 = -\left[\frac{h}{2} + \frac{1}{m_o} \sum_{n=1}^{\infty} m_n \left(\frac{h}{2} - L_n - \ell_n \right) \right] \tag{5.29b}$$

$$\begin{aligned} \frac{I_0}{m_F h^2} = & \frac{1}{12} + \frac{R^2}{4h^2} - \frac{m_0}{m_F} \left(\frac{h}{2} - L_0 \right)^2 - \sum_{n=1}^{\infty} \frac{m_n}{m_F} \left(\frac{h}{2} - L_n \right)^2 \\ & + \frac{R^2}{h^2} \left[\frac{2R}{h} \sum_{n=1}^{\infty} \frac{-4 + 5\cosh(\xi_{1n} h/R)}{\xi_{1n}^3 (\xi_{1n}^2 - 1) \sinh(\xi_{1n} h/R)} - \frac{3}{4} \right] \end{aligned} \tag{5.29c}$$

5.3.4 Mechanical parameters for a rectangular tank

Comparing the expression for hydrodynamic force due to lateral excitation of the tank to the total force expression due to the system of pendulums and rigid mass gives the following model parameters

$$\begin{aligned} m_n &= \frac{8\rho\ell L^2}{(2n-1)^3\pi^3}\tanh(2n-1)\frac{\pi h}{L} \\ m_0 &= \rho\ell Lh - \sum_{n=1}^{\infty} m_n \\ \ell_n &= \frac{L}{(2n-1)\pi}\coth(2n-1)\frac{\pi h}{L} \\ L_n &= -\left(\ell_n + \frac{h}{2} - \frac{2L}{(2n-1)\pi}\tanh(2n-1)\frac{\pi h}{L}\right) \end{aligned} \tag{5.30}$$

Similarly, the resulting moment about the center of mass yields the following constants. The mass moment of inertia of the rigid mass is

$$I_0 = I_F - m_0\left(L_0 - \frac{h}{2}\right)^2 - \sum_{n=1}^{\infty} m_n\left(\frac{h}{2} - L_n - \ell_n\right)^2 \tag{5.31a}$$

where I_F is the fluid mass moment of inertia about the center of mass and is given by the expression

$$I_F = \rho abh\left\{\frac{h^2}{12} + \frac{a^2}{16} - 2a^2\sum_{n=1}^{\infty}\frac{16}{(2n-1)^4\pi^4}\left[1 - \frac{2a}{(2n-1)\pi h}\tanh(2n-1)\frac{\pi h}{2a}\right]\right\} \tag{5.31b}$$

With reference to Figure 5.9, m_n is the nth pendulum mass, ℓ_n represents the nth pendulum length, L and ℓ are the tank width and breadth, respectively.

5.3.5 Spherical and oblate spheroidal containers

Based on the ideal fluid theory, the linear free and forced liquid-free-surface motions in spherical and oblate spheroidal containers are presented in Chapters 2 and 3. The development of analytical equivalent models of these containers is not a simple task. However, some attempts have been made to construct equivalent models based on experimental results. For example, Werner and Coldwell (1961), Sumner, *et al.* (1964), and Sumner (1965) experimentally determined the equivalent pendulum parameters for liquid sloshing in spheroidal and spherical tanks. Sumner (1965) used a spherical tank of radius 32 inch and an oblate-spheroidal tank of a circular diameter 32 inch and a minor diameter of 22.86 inch. Under horizontal sinusoidal excitation, $x(t) = X_0 \cos\Omega t$, at an excitation frequency much lower than its fundamental frequency ($\Omega \ll \omega$) the liquid oscillates at exactly the excitation frequency. The pendulum mass m_p was determined experimentally in terms of the measured horizontal sloshing force F_s, excitation amplitude X_0 and frequency Ω, that is,

$$m_p = \frac{F_s}{X_0}\left(\frac{1}{\Omega^2} - \frac{1}{\omega_e^2}\right) \tag{5.32}$$

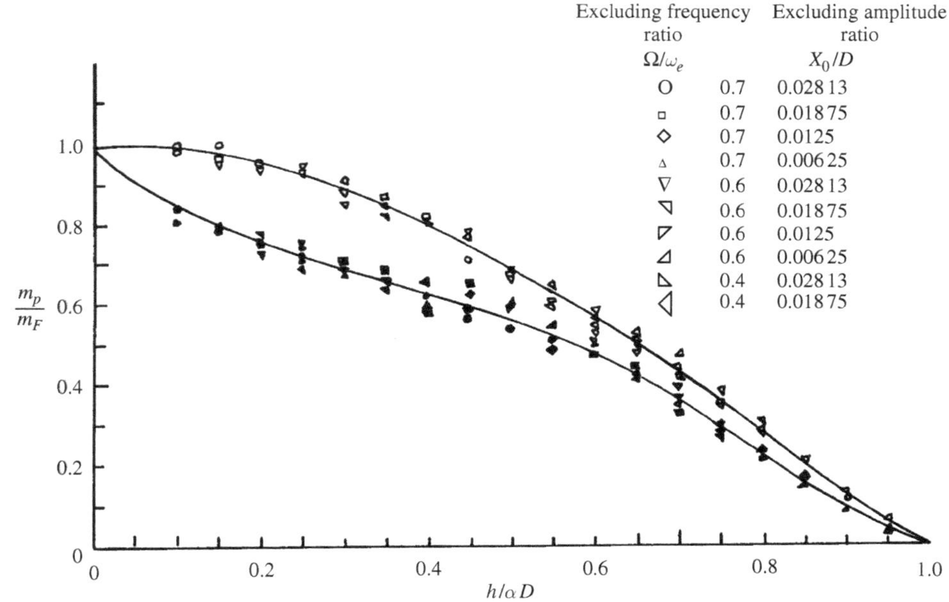

Figure 5.10 Dependence of sloshing mass to total liquid mass in partially filled tank. Open symbols denote oblate spheroid tank, and filled symbols denote spherical tank. (Sumner, 1965)

where the fundamental sloshing frequency ω_e was measured experimentally for the first several slosh cycles taken immediately after stopping the forced excitation test. The pendulum mass was measured for different liquid fillings of the tank. The dependence of the ratio of the sloshing mass to the total liquid mass, m_p/m_F, on the liquid-depth ratio, $h/\alpha D$, for the oblate-spheroidal and spherical tanks is shown in Figure 5.10, where $\alpha = d/D$, d is the minor diameter, D is the major diameter. It is seen that at low liquid-depth ratios, almost all of the liquid mass in each tank acts as the sloshing mass, while at the higher depth ratios, only a small portion of the liquid mass acts as the sloshing mass. The slosh mass ratio is greater for oblate-spheroidal tank at a given depth ratio.

The pendulum arm length, ℓ_p, for each tank was calculated directly from the pendulum fundamental frequency $\ell_p = g/\omega_e^2$. The ratio of pendulum arm length to tank height, $\ell_p/\alpha D$, is shown in Figure 5.11. The location of the hinge point of the pendulum arm, l_{op}, is obtained experimentally by measuring the sloshing moment about the center of the tank using the following expression

$$l_{op} = M/F_s \tag{5.33}$$

The ratio of hinge-point location to tank height, $l_{op}/\alpha D$, is shown in Figure 5.12. The hinge point is seen to move from near the top to the center of the oblate-spheroidal tank as the liquid-depth ratio increases from 0 to 1.0. The hinge-point location tends to remain at the center of the spherical tank regardless of the liquid-depth ratio.

McIver (1989) estimated sloshing frequencies for horizontal cylindrical and spherical containers filled to an arbitrary depth. For a horizontal circular container, he considered a two-dimensional sloshing model, and for spherical containers he used a three-dimensional model.

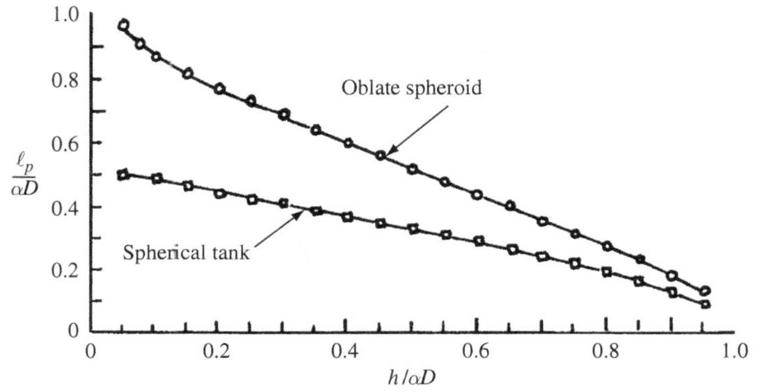

Figure 5.11 Dependence of pendulum arm length on fluid depth in oblate spheroidal and spherical tanks. (Sumner, 1965)

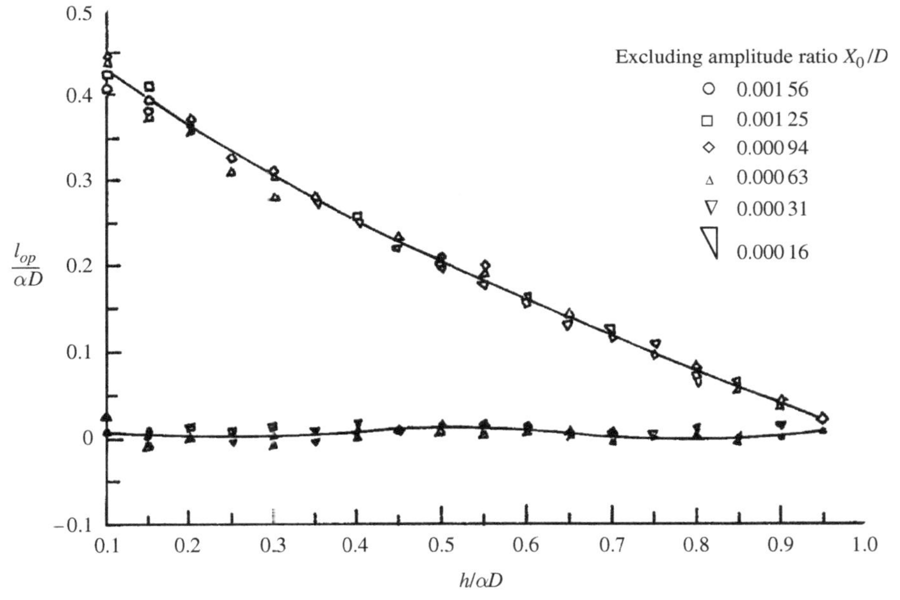

Figure 5.12 Dependence of pendulum hinge point location on fluid depth ratio. Open symbols denote oblate spheroid tank, and filled symbols denote spherical tank. (Sumner, 1965)

The linearized theory of water waves was used to determine the frequencies of free oscillations under gravitational influence by an arbitrary amount of fluid. The problem was formulated in terms of integral equations that were solved numerically for the eigenvalues. Bauer (1984d) developed a mass-spring-dashpot model to represent the translational oscillations of the liquid free surface. He derived the magnitude of the sloshing masses and their locations together with the location of the rigid mass.

5.4 Remarks on linear modeling

In view of the results obtained for mass-spring and pendulum equivalent models, one may establish a complete equivalence between the two systems. For example, the modal masses of the two models are the same and the spring stiffness is equivalent to the pendulum weight divided by the pendulum length. A correlation between the spring-mass position and pendulum mass position can also be established. The equivalence between the mass moment of inertia of the two cases can be obtained through some algebraic manipulations.

For large liquid heights, $h/R > 2.5$, the equivalent masses and their positions become almost independent of the tank filling but only depend on the container geometry. However, by increasing the natural frequency (by changing the tank geometry or by any other means) the disturbances penetrate less into the liquid and the equivalent masses for higher modes approach the free surface. On the other hand, for small liquid heights, the portion of the sloshing mass shifts toward the center of mass of the liquid volume.

The advantage of the equivalent modeling techniques is to provide a physical interpretation of the fluid free-surface motion by considering a certain portion of the liquid acting as a rigid mass, while the other sloshing portion is replaced by either a spring-mass system or a pendulum. This modeling can be incorporated into multibody dynamics packages so that the influence of liquid sloshing on the overall dynamics of the system can be studied.

5.5 Nonlinear modeling

The linear equivalent models are useful in studying sloshing dynamics as long as the excitation frequency is not close to the sloshing modal frequencies and the excitation amplitude is relatively small. As the excitation frequency approaches resonance, the liquid free surface experiences complex motions as demonstrated in Chapter 4. In this case, the linear equivalent model fails to emulate the nonlinear phenomena observed in the laboratory, and one must develop appropriate nonlinear equivalent models. The nonlinear rotational motion of the free surface can be understood by studying the stability of the spherical pendulum under forced excitation (see, for example, Berlot and Freed, 1956 and Miles, 1962a, 1984c, 1989). Bauer, *et al.* (1965) and Bauer (1966e, 1967) developed nonlinear mechanical models to describe the free-liquid-surface motion. Sayar and Baumgarten (1981) studied nonlinear fluid oscillation in spherical containers. They indicated that the pendulum model could not duplicate the nonlinear response of the fluid motion except at a particular fluid height. They also found that the mathematical prediction agrees favorably with experimental results after including cubic nonlinearity in the pendulum model. Liquid chaotic motion can also be explained by using the model of a swinging pendulum (Tritton, 1986a,b).

Kana, *et al.* (1985) and Unruh, *et al.* (1986) developed simple pendulum analytical models to represent liquid slosh in a Centaur rocket tank. However, it was found that at low liquid levels under harmonic excitation, the slosh response also contained a rotary-type wave motion in addition to the usual anti-symmetric slosh in the plane of excitation. Kana (1987, 1989) developed analytical modeling of a spherical pendulum combined with a simple pendulum to emulate the normal and rotary motions of the free surface for shallow liquid tanks. He showed that a compound pendulum could predict rotary sloshing in a scale model of a Centaur propellant tank. A portion of the liquid behaves as a spherical pendulum, which experiences rotary motion throughout a frequency range below, at, and above first mode resonance. The remainder of the fluid behaves as an ordinary linear pendulum. Phasing of the rotary motion was such that it subtracts from the effects of normal slosh below resonance and adds to the effects above resonance. It was also shown that the compound pendulum experiences bi-valued states and jump phenomena, in a range just above resonance. Another feature was that a significant magnitude cross-axis force exists for the rotary motion, which can only be predicted by the compound pendulum. Mechanical models for sloshing hydrodynamic impact will be discussed in Chapter 7.

5.5.1 Mechanical modeling of nonplanar sloshing

Bauer (1966e) developed an equivalent nonlinear mechanical model for the first sloshing mode. The model exhibits the same regions of softening response, unstable motion, and hardening rotary motion near the fundamental frequency. It also takes into account the influence of the vertical displacement of the center of gravity of the liquid. The model consists of a sliding mass point on some parabolic surface with an additional nonlinear spring attached to the mass as shown in Figure 5.13.

For an upright circular cylindrical tank subjected to harmonic excitation along the x-axis, $x(t) = X_0 \sin \Omega t$, the free-surface wave height is given by the expression

$$\eta(r, \theta, t) = \frac{\Omega^2 R}{g} X_0 \sin \Omega t \cos \theta \left[\frac{r}{R} + \sum \frac{2J_1(\xi_{1n} r/R)}{(\xi_{1n}^2 - 1) J_1(\xi_{1n})} \frac{\Omega^2}{\omega_n^2 - \Omega^2} \right] \tag{5.34}$$

Replacing the radial coordinate by the Bessel series $\frac{r}{R} = \sum \frac{2J_1(\xi_{1n} r/R)}{(\xi_{1n}^2 - 1) J_1(\xi_{1n})}$ and substituting into equation (5.34), gives

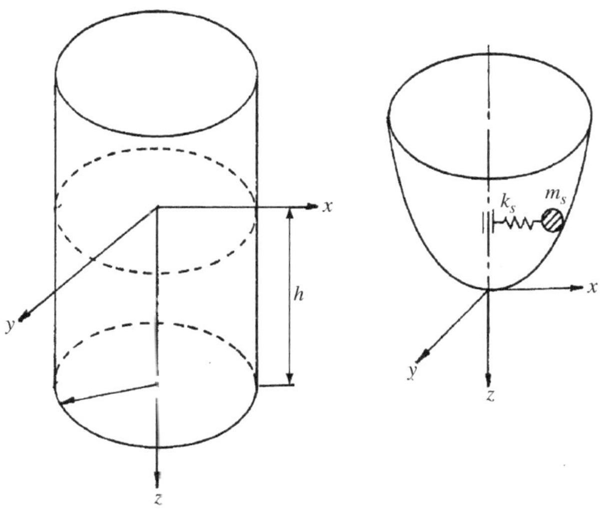

Figure 5.13 Nonlinear mechanical slosh model and coordinate system. (Bauer, 1966e)

$$\eta(r,\theta,t)=\frac{2\Omega^2 R}{g}X_0\sin\Omega t\cos\theta\sum\frac{\omega_n^2 J_1(\xi_{1n}r/R)}{(\xi_{1n}^2-1)J_1(\xi_{1n})(\omega_n^2-\Omega^2)} \tag{5.35}$$

Restricting the analysis to the first mode, equation (5.35) may be written in the form

$$\eta(r,\theta,t)=\eta_w\cos\theta\frac{J_1(\xi_{1n}r/R)}{J_1(\xi_{1n})} \tag{5.36}$$

where η_w is the wave height at the wall, that is,

$$\eta_w=\frac{2\Omega^2 R}{g}\frac{\omega_n^2 X_0\sin\Omega t}{(\xi_{1n}^2-1)(\omega_n^2-\Omega^2)} \tag{5.37}$$

It is not difficult to show that the radial and vertical displacements of the liquid center of gravity are determined from the elementary definition, measured from the center of gravity of the undisturbed liquid,

$$r_{\text{c.g}}=\frac{1}{\pi R^2 h}\int_0^R\int_0^{2\pi}\int_{-(h/2)-\eta}^{h/2} r^2\cos\theta\,dr\,d\theta\,dz=\frac{R}{h\xi_{11}^2}\eta_w \tag{5.38a}$$

$$z_{\text{c.g.}}=\frac{-1}{\pi R^2 h}\int_0^R\int_0^{2\pi}\int_{-(h/2)-\eta}^{h/2} rz\cos\theta\,dr\,d\theta\,dz=-\frac{(\xi_{11}^2-1)}{4h\xi_{11}^2}\eta_w^2 \tag{5.38b}$$

where the following integrals have been used

$$\int_0^R r^2 J_1(\xi_{11}r/R)\,dr=\frac{R^3}{\xi_{11}^2}J_1(\xi_{11})$$

$$\int_0^R r^2 J_1^2(\xi_{11}r/R)\,dr=\frac{R^2(\xi_{11}^2-1)}{2\xi_{11}^2}J_1^2(\xi_{11})$$

Eliminating η_w from relations (5.38) gives the following parabolic relation between the radial and vertical displacements of the liquid center of gravity,

$$z_{\text{c.g.}}=-Cr_{\text{c.g.}}^2/2R \tag{5.39}$$

where $C=(\xi^2{}_{11}-1)\xi^2{}_{11}(h/2R)$.

The essence of the analytical model is based on relation (5.39). The model is constructed such that the lower part of the liquid consists of a rigid mass that moves with the tank, while the sloshing mass is represented by a mass point whose motion is constrained along a paraboloid. The sloshing mass is allowed to exhibit rotary motion at a single frequency. A nonlinear spring providing a restoring force proportional to some power of the radial displacement was introduced and provides the necessary “hardening” for the proper description of the rotary motion. The idea of a high-power spring is very similar to that proposed by Hunt and Crossley (1975) and Pilipchuk and Ibrahim (1997) for the case of impact.

The sloshing mass of the liquid may be represented by a liquid volume of height, h_s, corresponding to the modal mass, m_s. The ratio of the first modal sloshing mass, m_s, to the total liquid mass, m_F, is equivalent to the height ratio, h_s/h, that is,

$$\frac{h_s}{h} = \frac{m_s}{m_F} = \frac{2\tanh(\xi_{11}h/R)}{\xi_{11}(\xi_{11}^2 - 1)h/R} \tag{5.40}$$

The displacement of the center of gravity of the sloshing part, z_s and r_s, is obtained by using equations (5.38),

$$z_s = -\frac{(\xi_{11}^2 - 1)}{4h\xi_{11}^2}\eta_w^2 = -\frac{(\xi_{11}^2 - 1)^2\eta_w^2}{8R\xi_{11}\tanh(\xi_{11}h/R)}$$

$$z_s = -C_s\frac{r_s^2}{2R}, \quad \text{with } C_s = \xi_{11}\tanh(\xi_{11}h/R) \tag{5.41}$$

With these parameters one can proceed to derive the equations of motion of the equivalent system using Lagrange's equation. The kinetic energy of the sloshing mass is

$$T = \frac{1}{2}m_s\left[(\dot{x}_s - \Omega X_0\cos\Omega t)^2 + \dot{y}_s^2 + \dot{z}_s^2\right] \tag{5.42a}$$

Expressing z_s in terms of x_s and y_s, using the constraint relationship

$$z_s = -\frac{C_z}{2R}r_s^2 = -\frac{C_z}{2R}\left(x_s^2 + y_s^2\right) \tag{5.42b}$$

The kinetic energy takes the form

$$T = \frac{1}{2}m_s\left[(\dot{x}_s - \Omega X_0\cos\Omega t)^2 + \dot{y}_s^2 + \frac{C_s^2}{R^2}(x_s\dot{x}_s + y_s\dot{y}_s)^2\right] \tag{5.42c}$$

The potential energy including a power spring of order, n, is

$$V = -m_s g z_s + \int_0^{r_s} k_s r_s^{2n-1} dr_s = \frac{m_s g C_s}{2R}\left(x_s^2 + y_s^2\right) + \frac{k_s}{2n}\left(x_s^2 + y_s^2\right)^n \tag{5.43}$$

Introducing a dashpot to account for the energy dissipation,

$$\Im = \frac{1}{2}c_s\left(\dot{x}_s^2 + \dot{y}_s^2 + \dot{z}_s^2\right) = m_s\omega_s\zeta_s\left[\dot{x}_s^2 + \dot{y}_s^2 + \frac{C_s^2}{R^2}(x_s\dot{x}_s + y_s\dot{y}_s)^2\right] \tag{5.44}$$

where $c_s = 2m_s\omega_s\zeta_s$ is the damping coefficient and ζ_s is the damping factor associated with the sloshing mass.

Introducing the following nondimensional parameters

$$X_s = \frac{x_s}{R}, \quad Y_s = \frac{y_s}{R}, \quad \alpha_s = \frac{k_s R^{2n-2}}{m_s\omega_s^2} \tag{5.45}$$

in equations (5.42c), (5.43) and (5.44), and applying Lagrange's equation (5.2) gives the following two nonlinear differential equations

$$\begin{aligned}&\ddot{X}_s + 2\omega_s\zeta_s\left[\dot{X}_s + C_s^2(X_s^2\dot{X}_s + X_sY_s\dot{Y}_s)\right] + C_s^2\left(X_s^2\ddot{X}_s + X_s\dot{X}_s^2 + X_sY_s\ddot{Y}_s + X_s\dot{Y}_s^2\right)\\&\quad + \omega_s^2\left[1 + \alpha_s\left(X_s^2 + Y_s^2\right)^{n-1}\right]X_s - \Omega^2\frac{X_0}{R}\cos\Omega t = 0\end{aligned} \tag{5.46a}$$

$$\begin{aligned}&\ddot{Y}_s + 2\omega_s\zeta_s\left[\dot{Y}_s + C_s^2(Y_s^2\dot{Y}_s + X_s\dot{X}_sY_s)\right] + C_s^2\left(Y_s^2\ddot{Y}_s + Y_s\dot{Y}_s^2 + X_s\ddot{X}_sY_s + Y_s\dot{X}_s^2\right)\\&\quad + \omega_s^2\left[1 + \alpha_s\left(X_s^2 + Y_s^2\right)^{n-1}\right]Y_s = 0\end{aligned} \tag{5.46b}$$

These equations include nonlinear inertia terms, due to the displacement of the center of gravity of the sloshing part represented by the mass m_s, and nonlinear stiffness due to the nonlinear spring.

Nonlinear planar modeling

The modeling of the planar motion can be derived from equations (5.46) by setting the coordinate Y_s to zero. Introducing the following solution into equation (5.46a)

$$X_s = A\cos(\Omega t + \psi) \tag{5.47}$$

where A is the response amplitude and ψ is the response phase angle. Assuming the nonlinear spring stiffness of cubic order, $n = 2$, and applying the Ritz averaging method, gives

$$\begin{aligned}\int_0^{2\pi}&\left\{\ddot{X}_s + 2\omega_s\zeta_s(\dot{X}_s + C_s^2X_s^2\dot{X}_s) + C_s^2\left(X_s^2\ddot{X}_s + X_s\dot{X}_s^2\right) + \omega_s^2(1 + \alpha_sX_s^3)X_s\right.\\&\quad\left. - \Omega^2\frac{X_0}{R}\cos\Omega t\right\}\cos(\Omega t)\,d(\Omega t) = 0\end{aligned}$$

The result after inserting (5.47) is

$$\left(1 - \nu^2 + \frac{3}{4}\alpha_sA^2 - \frac{1}{2}\nu^2C_s^2A^2\right)\cos\psi - 2\zeta_s\nu(1 + C_s^2A^2/4)\sin\psi - \frac{\nu^2X_0}{RA} = 0 \tag{5.48a}$$

and

$$\begin{aligned}\int_0^{2\pi}&\left\{\ddot{X}_s + 2\omega_s\zeta_s(\dot{X}_s + C_s^2X_s^2\dot{X}_s) + C_s^2\left(X_s^2\ddot{X}_s + X_s\dot{X}_s^2\right) + \omega_s^2(1 + \alpha_sX_s^3)X_s\right.\\&\quad\left. - \Omega^2\frac{X_0}{R}\cos\Omega t\right\}\sin(\Omega t)\,d(\Omega t) = 0\end{aligned}$$

The result of this integration after setting $\Omega/\omega_n = \nu$ is

$$\left(1 - \nu^2 + \frac{3}{4}\alpha_sA^2 - \frac{1}{2}\nu^2C_s^2A^2\right)\sin\psi + 2\zeta_s\nu(1 + C_s^2A^2/4)\cos\psi = 0 \tag{5.48b}$$

Equations (5.48) define the response amplitude and phase angle in terms of excitation and system parameters. If damping is ignored, one can solve for the cubic equation

$$\pm\left(1-\nu^2+\frac{3}{4}\alpha_s A^2-\frac{1}{2}\nu^2 C_s^2 A^2\right)A-\nu^2\frac{X_0}{R}=0$$

or

$$\nu^2=\frac{A\left(1+\frac{3}{4}\alpha_s A^2\right)}{A+\frac{1}{2}C_s^2 A^3\pm(X_0/R)} \tag{5.49}$$

Bauer made some attempts to obtain better approximation by including a third harmonic term in the assumed solution, $X_s=A\cos(\Omega t+\psi)+B\cos 3\Omega t$. His numerical solution resulted in a very small value of B (1% of the fundamental amplitude A over the tested frequency range.)

Nonlinear nonplanar modeling

The nonplanar modeling should include the out of plane motion, Y_s, and one should solve the coupled equations (5.46). In order to simplify the analysis one may consider the undamped case. Introducing the following solution

$$X_s=A\cos(\Omega t+\psi),\ Y_s=D\sin(\Omega t+\psi) \tag{5.50a, b}$$

and applying the Ritz averaging method, gives

$$\int_0^{2\pi}\left\{\ddot{X}_s+C_s^2\left(X_s^2\ddot{X}_s+X_s\dot{X}_s^2+X_sY_s\ddot{Y}_s+X_s\dot{Y}_s^2\right)\right.$$
$$\left.+\omega_s^2\left[1+\alpha_s\left(X_s^2+Y_s^2\right)\right]X_s-\Omega^2\frac{X_0}{R}\cos\Omega t\right\}\cos(\Omega t)\,d(\Omega t)=0$$
$$(1-\nu^2)A-\frac{1}{2}C_s^2\nu^2A(A^2-D^2)+\frac{\alpha_s}{4}A(3A^2+D^2)=\nu^2\frac{X_0}{R} \tag{5.51a}$$

$$\int_0^{2\pi}\{\ddot{Y}_s+C_s^2\left(Y_s^2\ddot{Y}_s+Y_s\dot{Y}_s^2+X_s\ddot{X}_sY_s+Y_s\dot{X}_s^2\right)$$
$$+\omega_s^2\left[1+\alpha_s\left(X_s^2+Y_s^2\right)\right]Y_s\}\sin(\Omega t)\,d(\Omega t)=0$$
$$(1-\nu^2)D+\frac{1}{2}C_s^2\nu^2D(A^2-D^2)+\frac{\alpha_s}{4}D(3D^2+A^2)=0 \tag{5.51b}$$

Equations (5.51) define the amplitude–frequency responses of the rotary motion. The spring constant, α_s, is usually obtained experimentally, and one may assume a value within the range $1/2\le\alpha_s\le 2/3$ for a cubic spring. For $\alpha_s=2/3$, Bauer obtained the response curves for planar motion given in Figure 5.14. This shows the results of the pendulum model represented by a dash–dot curve, the linearized fluid theory given by the dash curve, the present model

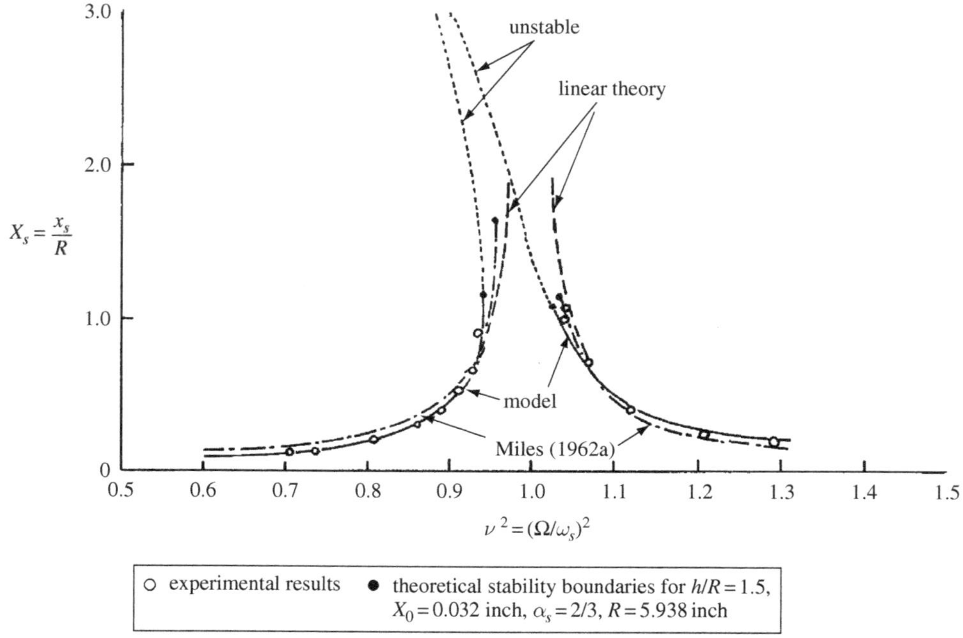

Figure 5.14 Dependence of response amplitude on excitation frequency and comparison with different methods: experimental. (Hutton, 1962), theoretical (Bauer, 1966e)

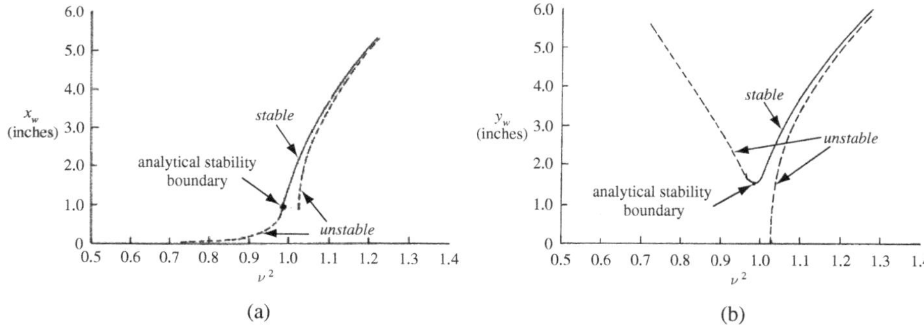

Figure 5.15 Nonplanar amplitude–frequency response (a) in the direction of excitation, (b) in the direction perpendicular to excitation, for $h/R = 1.5$, $X_0 = 0.032$ inch, $\alpha_s = 2/3$, $R = 5.938$ inch. (Bauer, 1966e)

(equations (5.48)), plotted by the solid curve, whereas the unstable response is represented by a dot curve. Figures 5.15(a) and (b) show the rotary response along and perpendicular to the excitation direction, respectively. These figures indicate that the planar motion essentially follows soft characteristics, while the nonplanar motion is essentially hard.

5.5.2 Dynamics of the spherical pendulum

The spherical (or conical) pendulum is an excellent model that exhibits the phenomenon of internal resonance since its natural frequencies in two orthogonal planes are identical. Its dynamical features resemble to a great extent the dynamics of liquid sloshing in an upright circular container. Several studies have been performed on the spherical pendulum under horizontal support excitation (see, e.g., Miles, 1962a, 1984c, 1984g, Tritton, 1986a, b, Kana, 1987, 1989, Irons, 1990, Bryant, 1993, and Kana and Fox, 1995). These studies were mainly initiated to provide some insight into the problem of rotary sloshing and its associated chaotic motion. There is no doubt that the modern theory of nonlinear dynamics inaugurated by the work of Poincaré and his successors, promoted our understanding of the complex phenomena of liquid free-surface motion and spherical pendulum. Distinctions between quasi-periodic and chaotic motions require the development of a detailed motion time history and its further analysis by means of the Poincaré space, Lyapunov exponents, and Fourier spectra.

There are different approaches to describe the equations of motion of a spherical pendulum in terms of different coordinate systems. These approaches model the pendulum in terms of two degrees of freedom. For example, Miles (1962a) described the equations of motion of the spherical pendulum in terms of the direction cosines. He (1984c) reformulated the problem in terms of Cartesian coordinates in order to study the pendulum dynamics using two control parameters, namely damping and frequency detuning parameters. Kana (1987, 1989), and Kana and Fox (1995) described the equations of motion in terms of the pitching angle about an axis orthogonal to the horizontal excitation and the rotational angle about the vertical axis. Kana (1987, 1989) and Kana and Fox (1995) also considered a compound pendulum (a spherical pendulum and a simple pendulum) to properly simulate low-depth liquid sloshing.

The equations of motion of a spherical pendulum excited along the x-axis by a harmonic excitation, $x(t) = X_0 \cos \Omega t$, are derived in terms of the Cartesian coordinates x, y, z. In order to derive the phase-space equations, Miles (1984c) used Hamilton's principle for the average of the Lagrangian

$$L = \frac{1}{2} m\left(\dot{x}^2 + \dot{y}^2 + \dot{z}^2\right) - mg(\ell - z) \tag{5.52}$$

where m is the pendulum mass and ℓ is the pendulum length. This Lagrangian is subject to the constraint equation

$$\ell^2 = (x - x(t))^2 + y^2 + z^2 \tag{5.53}$$

With this relation, the motion of the pendulum-bob can be fully described in terms of the x- and y-coordinates. Introducing the detuning parameter ν defined by the relationship

$$\nu = \frac{\Omega^2 - \omega_0^2}{\varepsilon^{2/3}\Omega^2} \tag{5.54}$$

where $\omega_0 = \sqrt{g/\ell}$ is the linearized natural frequency of the pendulum and $\varepsilon = X_0/\ell$ is a small parameter. The detuning parameter $\underset{\substack{\varepsilon \to 0 \\ \Omega \to \omega_0}}{\nu = O(1)}$. In this case, one may seek a solution of the form

$$x = \varepsilon^{1/3}\ell[p_1(\tau)\cos\theta + q_1(\tau)\sin\theta] \tag{5.55a}$$

$$y = \varepsilon^{1/3}\ell[p_2(\tau)\cos\theta + q_2(\tau)\sin\theta] \tag{5.55b}$$

where $\theta = \Omega t$, $\tau = \frac{\varepsilon^{2/3}}{2}\Omega t$, p_i and q_i, $i = 1, 2$, are slowly varying coordinates in a four-dimensional space. The trajectory of the pendulum-bob on the x, y horizontal plane is generally an ellipse with slowly varying axes. At an equilibrium point, the trajectory is a fixed ellipse. The relationship between the moving axes, $\hat{x}$, $\hat{y}$, and the fixed axes, x, y, is

$$\hat{x} = x\cos\varphi + y\sin\varphi \equiv \varepsilon^{1/3}\ell A\cos(\theta - \psi) \tag{5.56a}$$

$$\hat{y} = -x\sin\varphi + y\cos\varphi \equiv \varepsilon^{1/3}\ell B\sin(\theta - \psi) \tag{5.56b}$$

where φ is the slowly varying angle between the major $\hat{x}$-axis of the ellipse and the original x-axis, $\varepsilon^{1/3}\ell(A,B)$ are slowly varying semi-major and minor axes, and ψ is a slowly varying angle. Substituting equations (5.55) into equations (5.56) and equating coefficients of $\cos\theta$ and $\sin\theta$, gives

$$p_1 = A\cos\varphi\cos\psi + B\sin\varphi\sin\psi \tag{5.57a}$$

$$q_1 = A\cos\varphi\sin\psi - B\sin\varphi\cos\psi \tag{5.57b}$$

$$p_2 = A\sin\varphi\cos\psi - B\cos\varphi\sin\psi \tag{5.57c}$$

$$q_2 = A\sin\varphi\sin\psi + B\cos\varphi\cos\psi \tag{5.57d}$$

To this end, the average of the Lagrangian (5.52) with respect to θ is evaluated after substituting equations (5.55). Keeping the dominant terms in the limit, $\varepsilon \to 0$, and using the Hamilton principle gives the evolution equations in the Hamiltonian form

$$\frac{dp_i}{d\tau} = -\frac{\partial H}{\partial q_i}, \quad \frac{dq_i}{d\tau} = \frac{\partial H}{\partial p_i}, \quad (i = 1, 2) \tag{5.58a, b}$$

where

$$H = p_1 + \frac{1}{2}\nu E + \frac{1}{32}E^2 - \frac{3}{8}M^2 \tag{5.59a}$$

$$E = p_1^2 + q_1^2 + p_2^2 + q_2^2 \equiv \Re^2, \quad M = p_1 q_2 - p_2 q_1 \tag{5.59b, c}$$

where E is a measure of the pendulum energy, and M is a measure of the angular momentum about the vertical axis through $x = y = 0$ (generated by the y-component of the reaction force at the point of support), and $\Re$ is the hyper-radius in the (p_i, q_i) space and also a measure of the rms displacement, that is,

$$\sqrt{\frac{1}{2\pi}\int_0^{2\pi}(x^2 + y^2)\,d\theta} = \frac{\varepsilon^{1/2}}{\sqrt{2}}\ell\Re \tag{5.60}$$

Introducing linear viscous damping to account for energy dissipation, and writing the damping factor, ζ, in terms of the damping control parameter, α, through the relationship

$$\zeta = \frac{\varepsilon^{2/3}}{2}\left(\frac{\Omega}{\omega_0}\right)\alpha \tag{5.61}$$

where $\alpha = O(1)$ as $\varepsilon \to 0$, equations (5.58) take the following form in terms of the phase coordinates, (p_i, q_i), after introducing damping terms damping, $(-\alpha p_i, -\alpha q_i)$,

$$\dot{p}_1 = -\alpha p_1 - \left(\nu + \frac{1}{8}E\right)q_1 - \frac{3}{4}Mp_2 \tag{5.62a}$$

$$\dot{q}_1 = -\alpha q_1 - \left(\nu + \frac{1}{8}E\right)p_1 - \frac{3}{4}Mq_2 + 1 \tag{5.62b}$$

$$\dot{p}_2 = -\alpha p_2 - \left(\nu + \frac{1}{8}E\right)q_2 - \frac{3}{4}Mp_1 \tag{5.62c}$$

$$\dot{q}_2 = -\alpha q_2 - \left(\nu + \frac{1}{8}E\right)p_2 + \frac{3}{4}Mq_1 \tag{5.62d}$$

where a dot denotes differentiation with respect to the nondimensional time parameter τ. The advantage of writing the equations of motion in terms of the phase space rather than Cartesian coordinates is that they provide useful information about the system qualitative properties. For example, equations (5.62) are invariant under the reflection $(p_2, q_2) \to -(p_2, q_2)$ as a consequence of the Lagrangian (5.52) and the constraint condition (5.53) under the reflection $y \to -y$. Furthermore, the divergence of equations (5.62) is a constant, that is,

$$\Delta \equiv \frac{\partial \dot{p}_1}{\partial p_1} + \frac{\partial \dot{q}_1}{\partial q_1} + \frac{\partial \dot{p}_2}{\partial p_2} + \frac{\partial \dot{q}_2}{\partial q_2} = -4\alpha \tag{5.63}$$

This result means that an elemental volume in the phase space contracts like $\exp(-4\alpha\tau)$, and every trajectory ultimately must be confined to a limiting subspace of dimension less than four. Miles (1984c) showed that every trajectory ultimately must lie in $\Re \le 1/\alpha$.

The steady-state solutions (or fixed or singular points) are determined by setting the left-hand sides of equations (5.62) to zero and solving for (p_i, q_i). One may also express (p_i, q_i) in terms of E and M. First, multiplying equations (5.62) through by (p_1, q_1, p_2, q_2) and $(q_2, -p_2, -q_1, p_1)$, respectively, and summing the results, gives

$$q_1 = \alpha E, \quad p_2 = -2\alpha M \tag{5.64a, b}$$

Substituting (5.64) into equations (5.62a, d) and solving for p_1 and q_2, on the assumption that $\alpha \neq 0$, gives

$$p_1 = \frac{3}{2}M^2 - \left(\nu + \frac{1}{8}E\right)E, \quad q_2 = \left(\frac{1}{2}E - 2\nu\right)M \tag{5.65a, b}$$

Introducing equation (5.64) into equation (5.62c) and solving for M, gives two possible roots

$$M = 0, \quad M = \frac{5}{36}\left[E^2 + \frac{32}{5}\nu E - \frac{64}{5}\left(\nu^2 + \alpha^2\right)\right] \tag{5.66a, b}$$

Introducing equations (5.64) into equation (5.62b) and using either (5.66a) or (5.66b), gives the corresponding roots for E

$$E^3 + 16\nu E^2 + 64\left(\nu^2 + \alpha^2\right)E - 64 = 0 \tag{5.67a}$$

$$E^3 + 16\left(\alpha^2 - 3\nu^2\right)\nu E + 128\nu\left(\nu^2 + \alpha^2\right) - 24 = 0 \tag{5.67b}$$

The positive real roots of equation (5.67a) define the planar motion of the pendulum while those of equation (5.67b) define the nonplanar motion. The amplitude–frequency characteristics of each will be considered in the next two sections.

Planar motion

The algebraic equation (5.67a) is cubic in E and quadratic in the frequency parameter ν. Thus, it is easy to solve for ν in terms of the energy level E. The roots of ν are

$$\nu = -\frac{1}{8}E \mp \sqrt{\frac{1 - \alpha^2 E}{E}} \tag{5.68}$$

The corresponding values of (p_i, q_i) are

$$p_1 = \pm\sqrt{E(1 - \alpha^2 E)}, \quad q_1 = \alpha E, \quad p_2 = q_2 = 0 \tag{5.69a, b, c}$$

Figure 5.16 shows the dependence of the response amplitude, $\Re = \sqrt{E}$, on the frequency detuning parameter, ν for $\alpha = 1/4$. Branches I (extends to $\nu = -\infty$) and III_3 (extends to $\nu = \infty$) represent stable planar solutions. Branches II and III_2 represent unstable, planar solutions. The other branches, IV and V, are nonplanar motions and will be discussed later. The resonant peak on the left of the figure, on Branch III, is given by

$$E_{\max} = 1/\alpha^2 \quad \text{at } \nu = -1/(8\alpha^2) \tag{5.70a, b}$$

The different branches in Figure 5.16 are preserved for all values of damping parameter within the range $0 < \alpha < 0.433$.

The vertical tangency of the response curves is determined by setting the derivative of equation (5.68) to zero, which yields

$$E^3\left(1 - \alpha^2 E\right) = 16 \tag{5.71}$$

This equation has two positive real roots for sufficiently small damping. Substituting these roots into equation (5.68) with the upper sign yields the values of ν_1 and ν_3, at which vertical tangency occurs, for $\alpha^2 \ll 1$,

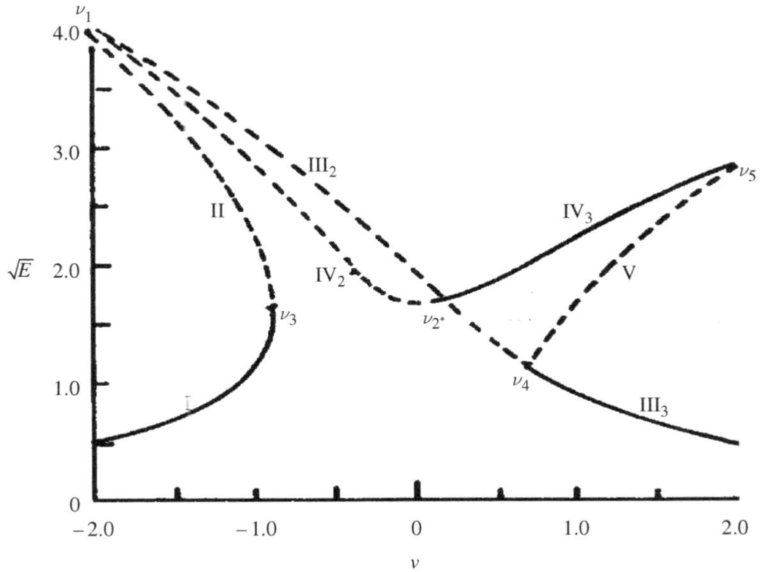

Figure 5.16 Resonance curves (loci of equilibrium points) for $\alpha = 0.25$. (Miles, 1984c)

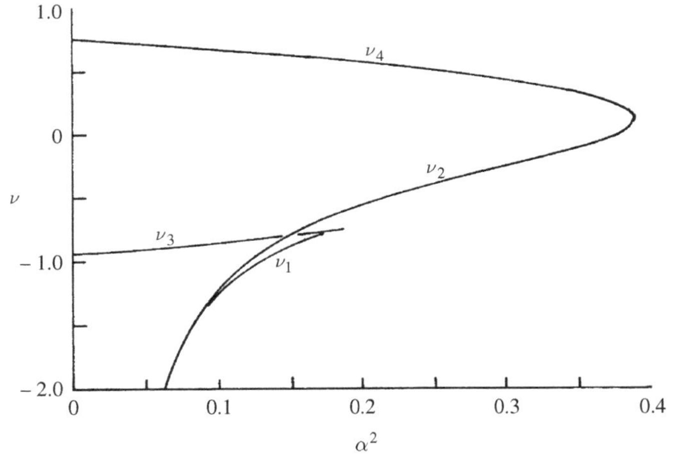

Figure 5.17 Bifurcation points on the frequency-damping diagram. (Miles, 1984c)

$$\nu_1 \approx -\frac{1}{8\alpha^2} - 2\alpha^4 \quad \text{and} \quad \nu_3 \approx -0.945 + 0.794\alpha^2 \tag{5.72a, b}$$

The coincidence of the positive real roots of equation (5.71) implies

$$\nu_1 = \nu_3 = -3/4 \quad \text{for } \alpha = \sqrt{3}/4 = 0.433 \tag{5.73}$$

Under this value of damping, branch II disappears and the planar resonance curve is a single value for all values of $\alpha > 0.433$. Figure 5.17 shows the bifurcation points on the ν, α^2 plane.

Nonplanar motion

The roots of equations (5.67b) and (5.66b) describe the nonplanar motion over the frequency range $\nu_2 < \nu < \nu_5$. There are two values of the detuning parameter, ν_2 and ν_4, at which M^2 changes sign and the nonplanar motion terminates on the planar resonance curve. There is another higher value of the frequency parameter, ν_5, below which curve (5.67b) has two roots. Thus, the nonplanar motion has either one branch over $\nu_2 < \nu < \nu_4$ (small portion of branch IV) or two branches over $\nu_4 < \nu < \nu_5$ (branches IV and V) as shown in Figure 5.16. The maximum and minimum values of the response amplitudes on branch IV are given by the values

$$E_{\max} \approx \frac{1}{2\alpha^2} + \frac{32}{9}\alpha^4 \text{ at } \nu \approx \frac{1}{8\alpha^2} - \frac{16}{9}\alpha^4 \tag{5.74a, b}$$

$$E_{\min} \approx 2.885 - 1.849\alpha^2 \text{ at } \nu \approx 0.462\alpha^2 \tag{5.75a, b}$$

By setting the left-hand side of equation (5.66b) to zero, that is, by setting $M^2 = 0$, and solving the resulting equation together with equation (5.67b) for E by eliminating ν from the two equations gives

$$E^3(1 - \alpha^2 E) = 16/9 \tag{5.76}$$

Equation (5.76) defines the common roots of equations (5.66b) and (5.67b), and it has two positive roots for very small values of the damping parameter α. Substituting these roots into equation (5.68) with the lower sign one obtains the bifurcation points ν_2 and ν_4 which are shown in Figure 5.17. For $\alpha^2 \ll 1$, these values are given by the following expressions

$$\nu_2 \approx -\frac{1}{8\alpha^2} + \frac{14}{9}\alpha^4, \quad \text{and} \quad \nu_4 \approx 0.757 - 0.795\alpha^2 \tag{5.77a, b}$$

The bifurcation point ν_5 is given by the vertical tangency of (5.67b) which is obtained by the larger root of the equation

$$\left(\nu^2 - \frac{1}{3}\alpha^2\right)^{3/2} = \nu(\nu^2 + \alpha^2) - \frac{3}{16} \tag{5.78}$$

For very small values of $\alpha^2 \ll 1$, the above equation gives the following approximate root

$$\nu_5 = -\frac{1}{8\alpha^2} + \frac{2}{9}\alpha^4 \tag{5.79}$$

Points ν_4 and ν_5 approach each other and coincide for $M = 0$ as the control parameter α increases such that

$$\nu_4 = \nu_5 = 0.47 \quad \text{and} \quad E = 1.433 \quad \text{for } \alpha = 0.526 \tag{5.80}$$

For $\alpha > 0.526$, branch V disappears.

When the two positive real roots of equation (5.76) coincide the two frequencies ν_2 and ν_4 coincide such that

$$\nu_2 = \nu_4 = 0.12 \quad \text{and} \quad E = 1.924 \quad \text{for } \alpha = 3^{5/6}/4 = 0.625 \tag{5.81}$$

Equation (5.76) has no real roots and the nonplanar motion disappears if $\alpha > 0.625$.

Note that the frequencies $\nu_{1,2,3,4,5}$ are all bifurcation points in Poincaré's sense. Such bifurcation requires a transition from stability to instability. The eigenvalues associated with the stability of fixed points equally vanish. Furthermore, there is a change in the number of fixed points. When the real parts of a pair of complex-conjugate roots of the stability eigenvalues become positive the fixed points experience Hopf-bifurcation which is indicated by point $\nu_2{}^*$ in Figure 5.16.

Miles (1984c) numerically integrated equations (5.62) and plotted the phase trajectory on the (p_1, p_2) plane, which represents the trajectory of the pendulum-bob on the horizontal plane. The trajectory for $\nu = 0.05$ is a simple limit cycle that projects as an oval on the (p_1, p_2) plane and is one frequency. The stability of this limit cycle suggests that the bifurcation is supercritical. Over the region $\nu = -0.04$ to -0.144, the trajectory possesses period doubling. A second period doubling first occurs as ν is decreased through the range -0.13 to -0.14 and the quarter-order subharmonics increase in strength down to $\nu = -0.144$. For $\nu = -0.144$ the phase trajectory involves period octupling (a third period doubling). For detuning parameters in the range $\nu = -0.145$ up to -0.20, the system experiences chaotic motion. The trajectories for $\nu = -0.21$ and -0.22 are periodic, with period doubling as ν increases from -0.22 to -0.21. Trajectories for $\nu = -0.23$ and -0.24 are chaotic and lie on a fractal attractor. Those of $-\nu = 0.25$ (0.01) then 0.30 (0.05) up to 0.45 are periodic, with period doubling at -0.25 as ν increases to -0.24 and at -0.40 and -0.45 as ν decreases to 0.50. Over the region of $-0.92 \leq \nu \leq -0.5$, Miles observed subintervals of both periodic and chaotic trajectories.

The spherical pendulum dynamic behavior resembles to a certain extent the behavior of a two-mode description of liquid sloshing in a circular upright tank that is subjected to a horizontal, planar harmonic excitation, provided that the liquid depth exceeds a certain critical value. If the depth is below that critical value, then the solution resembles that for a stretched string (Miles, 1984g) and may be nonplanar but not chaotic. This situation may be modeled by a combined pendulum as will be shown in Section 5.5.3.

Tritton (1986a, b) conducted an experimental investigation on a spherical pendulum consisting of a string with a rubber ball on its free end. The upper end of the string is fixed to a rod, which can be excited sinusoidally in a horizontal line by a crankshaft. For different ranges of the frequency ratio f/f_0, where f is the driving frequency and f_0 is the pendulum linear natural frequency, Tritton reported the following observations:

1.	$f/f_0 < 0.979$	Planar motion
2.	$0.979 < f/f_0 < 1.001$	Irregular fluctuations
2a.	$f/f_0 \approx 0.992$	Irregular fluctuations and periodicity
3.	$1.001 < f/f_0 < 1.008$	Semi-irregular fluctuations
3a.	$f/f_0 \approx 1.004$	Regular orbit
4.	$1.008 < f/f_0 < 1.024$	Regular orbit
5.	$1.024 < f/f_0 < 1.028$	Planar motion or regular orbit
6.	$1.028 < f/f_0$	Planar motion

A sequence of orbits projected on the horizontal plane at intervals of 15 periods of the driving period at frequency ratio $f/f_0 = 0.9875$ were found to be near ellipses, with varying degrees of eccentricity and angle relative to the excitation axis. Beyond a certain region of excitation parameters, these orbits break down into a complex motion that includes a slow change in shape and rotation with time. Tritton (1986b) compared his observations with those predicted by Miles (1984c) and found an agreement with Miles regarding irregular and semi-irregular fluctuations. The most unsatisfactory feature of the comparison was related to the phase relationship of the ball motion to the excitation. Detailed comparison was not possible since the detailed behavior of the pendulum was sensitive to the damping parameter α. Furthermore, observations might not distinguish between a truly chaotic motion and a complicated periodic one.

Irons (1990) also conducted an experiment to study the evolution of the spherical pendulum motion toward its asymptotic behavior in time. A range of forcing frequencies was found for which the sense of rotation of the pendulum about the vertical reversed intermittently during the pendulum oscillations. Bryant (1993) showed that there are broad frequency ranges in which the pendulum rotation is reversed for both periodic and chaotic oscillations.

Kana and Fox (1995) examined the validity of analytical and numerical results with experimental measurements for stability boundaries. Figure 5.18 shows the stability boundaries as computed by Miles (1984c) and those measured experimentally. The boundaries divide the excitation amplitude–frequency plane into the following regions:

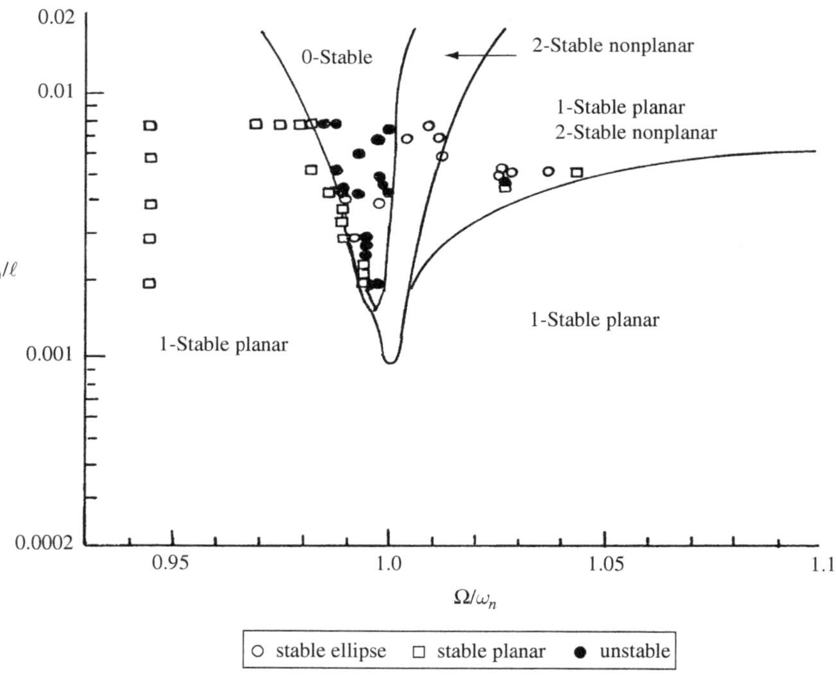

Figure 5.18 Experimental and analytical stability boundaries of spherical pendulum, $f_n = 0.433$ Hz, $\zeta = 0.003$. (Kana, 1987, 1989)

1-stable planar motion in the plane of excitation;

2-stable nonplanar motion characterized by closed orbits with out-of-plane components in which one response is counterclockwise and the other is clockwise;

1-stable planar/2-stable nonplanar motion in which the type of motion is determined by the initial conditions;

0-stable region characterized by complex motion where no stable solutions exist for excitation parameters covering this region. Unstable motion is represented by slowly shifting multiple periodic, quasi-periodic, or chaotic orbits.

For some regions of stable solutions, Miles' theory also predicted the existence of unstable solutions. The numerical simulation of the equations of motion of the pendulum revealed that the transition from quasi-periodic to chaotic motions can be carefully quantified in systems with very light damping.

5.5.3 Linear plus spherical pendulums

At relatively low-fill levels of tanks subjected to harmonic excitation, the liquid free-surface response may experience a rotary-type wave motion in addition to the usual asymmetric slosh in the plane of excitation. Kana (1987, 1989) developed a compound spherical and normal slosh equivalent shown in Figure 5.19. If the total weight of the liquid is W_1, then a fraction of that weight, say $\beta_1 W_1$, responds with spherical slosh, while another portion, $\beta_2 W_1$, responds with normal slosh (i.e. is, as an ordinary linear pendulum). Both pendulums have the same length, ℓ. The motion of the linear pendulum is described in terms of angle θ_2 and a phase lag $\gamma_0 - \varepsilon$ relative to the excitation displacement, where ε stands for a spatial phase difference

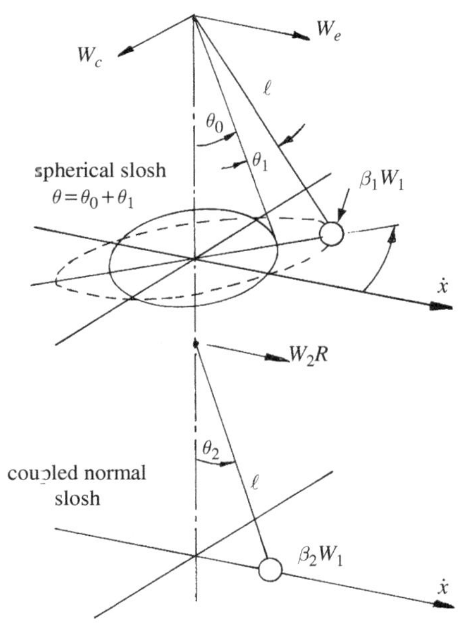

Figure 5.19 Compound spherical and linear pendulum system. (Kana, 1989)

observed in the experiments and to postulate a phase coupling between the two pendulums. The motion of the spherical pendulum is described by a steady-state angle θ_0 plus an oscillatory component, θ_1 along the excitation displacement, that is, $\theta = \theta_0 + \theta_1(t)$, and a precession angle φ about the vertical axis.

In terms of the angles θ and φ, the equations of motion of the spherical pendulum under support motion, $x(t) = X_0 \cos \Omega t$, are derived using Lagrange's equation. The kinetic energy is

$$T = \frac{1}{2} m \left[(\ell \dot{\varphi} \sin \theta - \dot{x} \sin \varphi)^2 + \left(\ell \dot{\theta} + \dot{x} \cos \varphi \cos \theta \right)^2 \right] \tag{5.82}$$

The potential energy is

$$V = mg\ell(1 - \cos \theta) \tag{5.83}$$

where $m = \beta_1 W_1/g$ is the spherical pendulum mass. Applying Lagrange's equation, and introducing damping, gives

$$\ddot{\theta} + \left(\omega_n^2 - \dot{\varphi}^2 \cos \theta\right) \sin \theta + 2\zeta_\theta \omega_n \dot{\theta} + \frac{\ddot{x}}{\ell} \cos \varphi \cos \theta = 0 \tag{5.84}$$

$$\ddot{\varphi} \sin \theta + 2\zeta_\varphi \omega_n \dot{\varphi} \sin \theta + 2\dot{\phi}\dot{\theta} \cos \theta - \frac{\ddot{x}}{\ell} \sin \varphi = 0 \tag{5.85}$$

The lower simple linear pendulum equation of motion is

$$\ddot{\theta}_2 + 2\zeta_\theta \omega_n \dot{\theta}_2 + \omega_n^2 \theta_2 = -\ddot{x}/\ell \tag{5.86}$$

where ζ_θ and ζ_φ are the damping ratios associated with θ and φ coordinates, respectively, and $\omega_n^2 = g/\ell$.

The simple pendulum equation (5.86) is a linear oscillator and has the following steady-state solution

$$\theta_2 = \frac{X_0}{\ell} \frac{(\Omega/\omega_n)^2}{\sqrt{\left(1 - (\Omega/\omega_n)^2\right)^2 + 4(\Omega/\omega_n)^2 \zeta_\theta^2}} \tag{5.87a}$$

and the phase angle

$$\gamma_0 = \tan^{-1} \left[\frac{2\zeta_\theta(\Omega/\omega_n)}{1 - (\Omega/\omega_n)^2} \right] \tag{5.87b}$$

An approximate solution of equations (5.84) and (5.85) may be obtained by using the method of harmonic balance for the steady-state deflection

$$\theta = \theta_0, \quad \varphi = +(\Omega t - \phi_0)\text{CCW}, \quad \varphi = -(\Omega t + \phi_0)\text{CW} \tag{5.88a, b, c}$$

where CCW and CW stand for counter-clockwise and clockwise, respectively, $+\varphi$ is measured counter-clockwise relative to the excitation plane, and ϕ_0 is a spatial lag angle measured clockwise relative to that plane.

Substituting equations (5.88) into equation (5.84) and performing harmonic balance gives

$$\tan\theta_0 - \left(\frac{\Omega}{\omega_n}\right)^2 \sin\theta_0 = \frac{1}{2}\left(\frac{\Omega}{\omega_n}\right)^2 \frac{X_0}{\ell}\cos\phi_0 \tag{5.89}$$

This expression is valid for either direction of rotation. It has solutions for the following conditions:

$$\frac{1}{2}\left(\frac{\Omega}{\omega_n}\right)^2 \frac{X_0}{\ell}\cos\phi_0 \geq 0, \quad \text{and} \quad \frac{\Omega}{\omega_n} \geq 0 \tag{5.90a}$$

$$\frac{1}{2}\left(\frac{\Omega}{\omega_n}\right)^2 \frac{X_0}{\ell}\cos\phi_0 \leq 0, \quad \text{and} \quad \frac{\Omega}{\omega_n} \geq 0 \tag{5.90b}$$

Performing harmonic balance for equation (5.85) using equations (5.88), gives

$$\zeta_\varphi = \pm\left(\frac{X_0}{4\ell}\frac{\Omega}{\omega_n}\right)\frac{\sin\phi_0}{\sin\theta_0} \tag{5.91}$$

where the positive sign belongs to counter-clockwise motion and the negative sign is for clockwise motion. Equation (5.91) implies the following location of ϕ_0,

(1) For CCW sin ϕ_0 is positive, which implies that ϕ_0 is in the 3rd or 4th quadrant,
(2) For CW sin ϕ_0 is negative, which implies that ϕ_0 is in the 1st or 2nd quadrant.

This means that either clockwise or counterclockwise motion can occur throughout the frequency range, and a double solution can exist for motion either way for Ω/ω_n.

The oscillatory component θ_1 can be studied by substituting $\theta = \theta_0 + \theta_1(t)$ into equation (5.85a) and using Taylor expansion, gives the linear differential equation

$$\ddot{\theta}_1 + 2\zeta_{1\theta}\varpi_n\dot{\theta}_1 + \varpi_n^2\theta_2 = \frac{X_0}{2\ell}\Omega^2\, e^{j(2\Omega t - \phi_0)}\cos\theta_0 \tag{5.92}$$

where $\varpi_n = 2\omega_n$ and $\zeta_{\theta 1} = \zeta_\theta/2$.

Equation (5.92) has the steady-state solution

$$\theta_1 = \frac{X_0}{8\ell}\frac{(\Omega/\omega_n)^2}{\sqrt{\left(1-(\Omega/\omega_n)^2\right)^2 + 4(\Omega/\omega_n)^2\zeta_{\theta 1}^2}}\cos\theta_0\, e^{j(2\Omega t - \phi_0 - \alpha_0)} \tag{5.93a}$$

The phase angle α_0 is given by the expression

$$\alpha_0 = \tan^{-1}\left[\frac{2\zeta_{\theta 1}(\Omega/\omega_n)}{1-(\Omega/\omega_n)^2}\right] \tag{5.93b}$$

The response of θ_1 occurs at twice the frequency of excitation. The complex notation j differs from the one used for the horizontal physical i-plane. The j-plane and i-plane are shown in

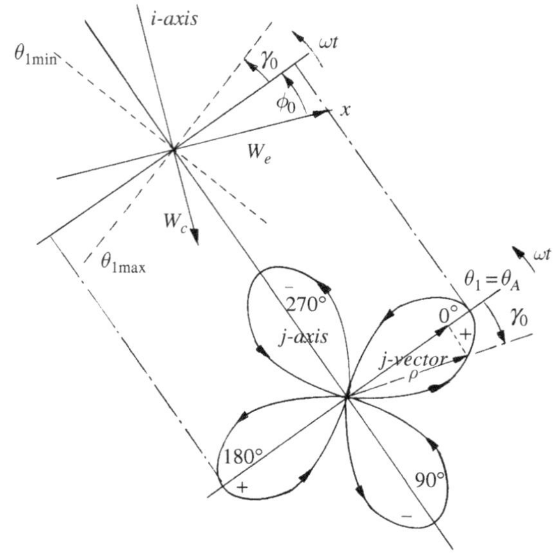

Figure 5.20 Physical and complex planes showing the in-line weight and cross-axis weight. (Kana, 1989)

Figure 5.20. The physical i-plane is shown in the horizontal plane at $\Omega t = \phi_0$. Indicated on the two axes of the i-plane are the in-line weight W_e and cross-axis weight W_c. Let the complex j-plane intersect orthogonally the i-plane at a line making $\Omega t = \phi_0$ as shown on the upper part of Figure 5.20. The j-plane contains the vector defining the oscillator response θ_1 given by equation (5.93a). Solution (5.93a) may be written as

$$\theta_1 = \Theta\, e^{-j\phi_0}\, e^{-j\alpha_0}\, e^{j2\Omega t} \tag{5.94}$$

The cos $2\Omega t$ component of θ_1 is represented by the cloverleaf shown in Figure 5.20. The vector ρ is traced out as its radius rotates counter-clockwise with frequency Ω. The interaction of the two planes is such that the projections onto the horizontal i-plane form a pair of oscillators fixed in the i-space, with one in the plane at $\Omega t = \phi_0 + \alpha_0$ and the other at $\Omega t = \phi_0 + \alpha_0 + \pi/2$.

The projection of the pendulum orbit onto the horizontal plane is defined by the complex physical i-space

$$W_s = W_e(\Omega) - iW_c(\Omega) \tag{5.95}$$

where the directions for the weight components are shown in Figure 5.19. The in-line weight for the nonrotating pendulum that oscillates in a plane at angle ψ relative to the x-axis can be written in the form

$$W_{et} = \beta_1 W_1 \ell(\ddot{\theta}/\ddot{x})\cos\theta \cos\psi + \beta_1 W_1(\ddot{x}/\ddot{x}) \tag{5.96a}$$

The cross-axis component is

$$W_{ct} = -\beta_1 W_1 \ell(\ddot{\theta}/\ddot{x})\cos\theta \sin\psi \tag{5.96b}$$

The effect of rotation at angular velocity Ω is determined by estimating the contributions of centrifugal and Coriolis accelerations. The in-line weight components are

$$W_{es} = -\beta_1 W_1 (\ell\dot{\varphi}^2/\ddot{x}) \sin\theta \cos\varphi - 2\beta_1 W_1 (\ell\Omega\dot{\theta}/\ddot{x}) \cos\theta \sin\psi \tag{5.97a}$$

$$W_{cs} = \beta_1 W_1 (\ell\dot{\varphi}^2/\ddot{x}) \sin\theta \cos\varphi - 2\beta_1 W_1 (\ell\Omega\dot{\theta}/\ddot{x}) \cos\theta \cos\psi \tag{5.97b}$$

Adding (5.96a) to (5.97a) and (5.96b) to (5.97b) gives the components of complex representation in the physical i-space, and after setting $\ddot{x}(t) = -\Omega^2 X_0 \cos\Omega t$, we write

$$W_\epsilon(\Omega)\cos\Omega t = \frac{\beta_1 W_1 \ell}{\Omega^2 X_0}\left(\dot{\varphi}^2 \sin\theta\cos\varphi + 2\Omega\dot{\theta}\cos\theta\sin\psi - \ddot{\theta}\cos\theta\cos\psi\right) + \beta_1 W_1 \cos\Omega t \tag{5.98a}$$

$$W_c(\Omega)\cos\Omega t = -\frac{\beta_1 W_1 \ell}{\Omega^2 X_0}\left(\dot{\varphi}^2 \sin\theta\sin\varphi - 2\Omega\dot{\theta}\cos\theta\cos\psi - \ddot{\theta}\cos\theta\sin\psi\right) \tag{5.98b}$$

The total weight may be written in the complex form

$$W_s(\Omega) = -\beta_1 W_1 \frac{\ell\dot{\varphi}^2}{\ddot{x}} \sin\theta\, e^{i\varphi} + 2\beta_1 W_1 (\ell\Omega\dot{\theta}/\ddot{x})\cos\theta\, e^{i(\psi+(\pi/2))} + \beta_1 W_1\left(1 + \frac{\ell\ddot{\theta}}{\ddot{x}} e^{i\psi}\cos\theta\right) \tag{5.99}$$

In writing equation (5.99), the relationship $\psi = \Omega t - (\gamma_0 + \phi_0) = \varphi - \gamma_0$ has been used. Substituting for $\ddot{x}(t) = -\Omega^2 X_0 \cos\Omega t$ and equation (5.93a), setting $\sin\theta \approx \sin\theta_0$, $\cos\theta \approx \cos\theta_0$, $\dot{\varphi} = \Omega$ and collecting the coefficients of $\cos\Omega t$ on both sides, gives

$$W_s(\Omega) = \beta_1 W_1 \left[1 + \frac{\ell}{X_0}(\sin\vartheta_0\cos\phi_0 - i\sin\theta_0\sin\phi_0) + \frac{(\Omega/\omega_n)^2\cos^2\theta_0}{\sqrt{\left(1-(\Omega/\omega_n)^2\right)^2 + 4(\Omega/\omega_n)^2\zeta_{\theta 1}^2}} \times \left(\cos 2(\gamma_0+\phi_0) - i\sin 2(\gamma_0+\phi_0)\right)\right] \tag{5.100}$$

Separating this equation into in-line axis, W_{CC}, (quadrature phase) and cross-axis, W_{CQ}, (co-phase) components, gives

$$W_{CC}(\Omega) = \beta_1 W_1 \left[\frac{\ell}{X_0}\sin\theta_0\sin\phi_0 + \frac{(\Omega/\omega_n)^2\cos^2\theta_0\sin 2(\gamma_0+\phi_0)}{\sqrt{\left(1-(\Omega/\omega_n)^2\right)^2 + 4(\Omega/\omega_n)^2\zeta_{\theta 1}^2}}\right] \tag{5.101a}$$

$$W_{CQ}(\Omega) = -\beta_1 W_1 \left[\frac{\ell}{X_0}\sin\theta_0\cos\phi_0 - 1 + \frac{(\Omega/\omega_n)^2\cos^2\theta_0\cos 2(\gamma_0+\phi_0)}{\sqrt{\left(1-(\Omega/\omega_n)^2\right)^2 + 4(\Omega/\omega_n)^2\zeta_{\theta 1}^2}}\right] \tag{5.101b}$$

where W_{CQ} is taken positive to the left.

The total reaction weights for the combined effects of the spherical pendulum and the linear pendulum can be written in terms of the real (in-line) and imaginary (not the cross-axis) components, respectively, as

$$W_R = W_1 - W_{CI} + \frac{\beta_2 W_1 (\Omega/\omega_n)^2 \cos\,\psi_0}{\sqrt{\left(1-(\Omega/\omega_n)^2\right)^2 + 4(\Omega/\omega_n)^2 \zeta_{\theta 2}^2}} \tag{5.102a}$$

$$W_I = W_{CR} + \frac{\beta_2 W_1 (\Omega/\omega_n)^2 \sin\,\psi_0}{\sqrt{\left(1-(\Omega/\omega_n)^2\right)^2 + 4(\Omega/\omega_n)^2 \zeta_2^2}} \tag{5.102b}$$

where

$$\tan\psi_0 = \frac{W_1 - W_{CR}}{W_R - W_1 + W_{CI}} \tag{5.102c}$$

for which W_1 is the total weight of liquid, W_{CI} and W_{CR} are the projection of the cross component of the spherical pendulum along the in-line and imaginary directions, respectively.

The experimental measurements of cross-weight were described in Chapter 4. It was observed that the steady state amplitude of rotary slosh increased as the excitation frequency increased until a frequency above resonance, $\Omega > \omega_n$ where the response dropped out to essentially a planar response at very low amplitude. Just before the dropout, the phase angle ϕ_0 of the CCW rotary motion was very near to 90°. These measurements were used to identify the spherical pendulum parameters. The pendulum length is determined using the definition of the first mode natural frequency. The solution of equations (5.89) and (5.90) gives the steady-state deflection θ_0. With an assumed value of circumferential damping, ζ_φ, the position lag angle, ϕ_0, was determined from equation (5.91). Figure 5.21(a), and (b) shows the dependence of θ_0 and ϕ_0 on the excitation frequency ratio Ω/ω_n. Equations (5.101) are used to determine the pendulum weight $\beta_1 W_1$. The estimated mass fraction β_1 is plotted against the excitation frequency ratio in Figure 5.22. Kana took into account the amplitudes distortion effects in estimating the mass fraction β_1.

The parameters of the lower linear pendulum were determined by substituting experimental weight data into equations (5.102) for a damping ratio $\zeta_{\theta 2} = 0.01$. The dependence of corresponding fraction weight β_2 on the frequency ratio is shown in Figure 5.23. The results show a significant variation with frequency. The algebraic sum of the two fractions $\beta_1 + \beta_2$ exceeds unity for excitation frequency ratio $\Omega/\omega_n \approx 0.89$. This observation should not signal any contradiction since the weight component of weight along each axis does not exceed the total weight W_1.

5.6 Closing remarks

Mechanical equivalent models serve as a tool to emulate linear and nonlinear liquid-free-surface dynamics under external excitations. The parameters of linear models for regular shape containers, such as circular cylindrical and rectangular tanks, are given in closed forms. For

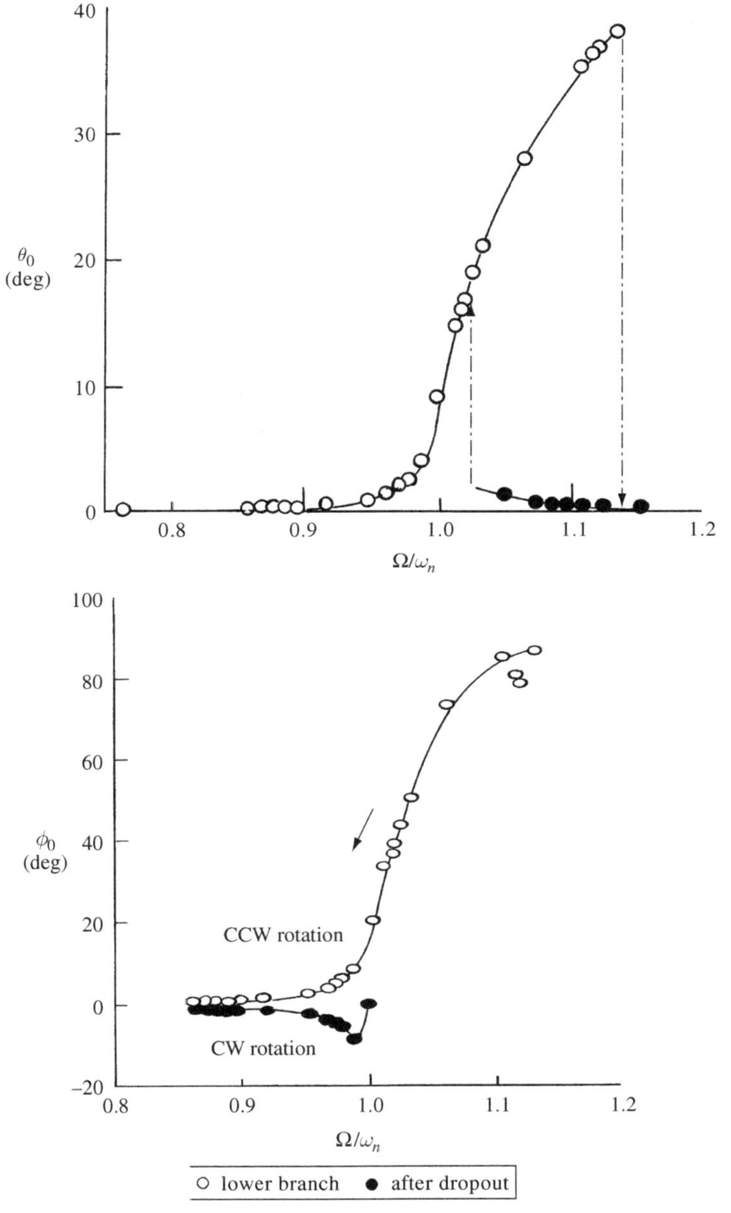

Figure 5.21(a) Dependence of the steady deflection angle on the excitation frequency ratio, (b) dependence of the position lag angle on excitation frequency. (Kana, 1989)

other geometries, the parameters are usually obtained experimentally. Linear models are valid as long as the excitation frequency is remote from the sloshing natural frequency. In the neighborhood of resonance, these models fail to represent complex phenomena such as rotary sloshing and chaotic motion. Nonlinear equivalent models, such as spherical and compound pendulums, exhibit similar phenomena to those observed for liquid free surface near

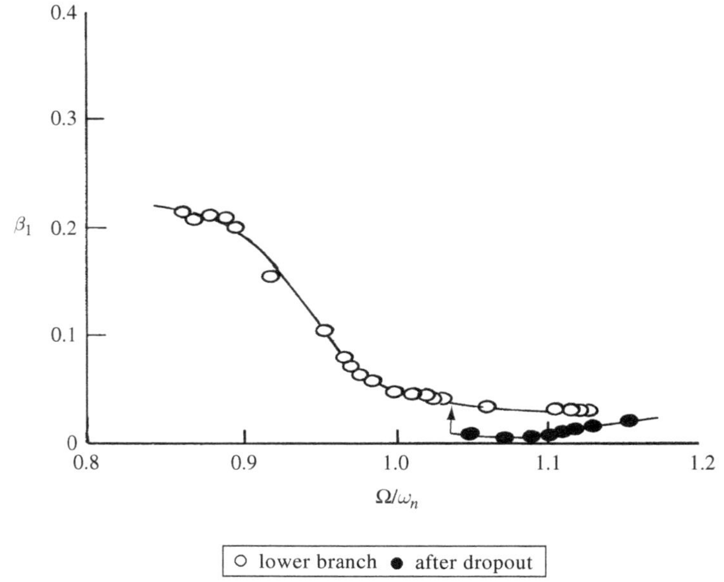

Figure 5.22 Dependence of spherical pendulum mass fraction on excitation frequency ratio. (Kana, 1989)

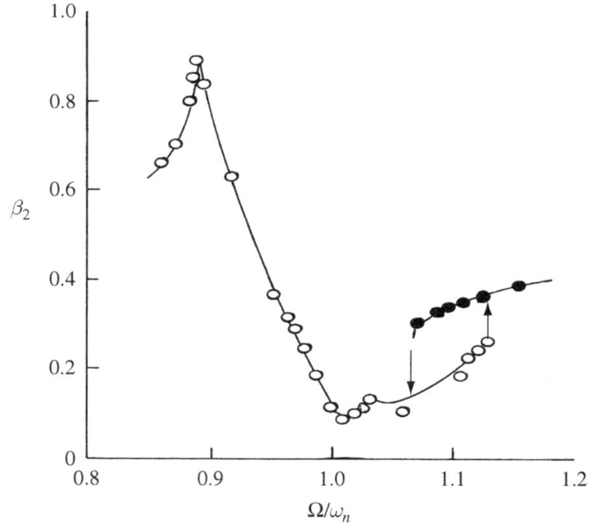

Figure 5.23 Dependence of linear pendulum mass fraction on excitation frequency. (Kana, 1989)

resonance. The nonlinearity of these models is weak and does not serve to simulate strong nonlinearity encountered in liquid-free-surface impact. Liquid sloshing impact modeling and analysis will be treated in Chapter 7. Under parametric excitation, the mass-spring-dashpot systems are not convenient because they cannot exhibit parametric resonance. Simple and spherical pendulums are convenient for simulating liquid sloshing under parametric excitation.

Related to mechanical equivalent modeling, is the simulation of liquid sloshing using a pendulum model and finite element (see, e.g., Aliabadi, *et al.*, 2003). Dodge and Kana (1966) determined the mass moment of inertia and damping of a liquid in baffled cylindrical containers.

This chapter did not address the techniques of building scale models for simulating liquid sloshing dynamics. The application of similitude theory and small models was considered by Abramson, Martin and Ransleben (1958), Abramson and Ransleben (1959a,b, 1960), Sandorff (1960), Abramson and Neville (1963), Abramson (1966a), and Eggleston (1968) discussed the similitude requirements and experimental techniques. Abramson and Ransleben (1960) presented satisfactory correlation of small-scale and full-scale data.

6

Parametric sloshing: Faraday waves

6.1 Introduction

Parametric sloshing refers to the motion of the liquid free surface due to an excitation perpendicular to the plane of the undisturbed free surface. Generally, parametric oscillation occurs in dynamical\ systems as a result of time-dependent variation of such parameters as inertia, damping, or stiffness. This variation may be due to the influence of externally applied forces or acceleration fields referred to as *parametric excitations*. For the case of parametric sloshing, the effective gravitational field becomes time dependent. Although parametric oscillation is regarded to be of a secondary interest, it can have catastrophic effects on mechanical systems near critical regions of parametric instability. For example, sloshing waves are generated when the liquid container is vertically excited at a frequency close to twice the natural frequency of the free surface. This type is referred to as Faraday waves since he (1831) was the first to observe and report them. Faraday waves are distinct from other waves with crests normal to a moving boundary (wave-maker) as cross-waves reported by Miles (1985a, 1988b, 1990a), Miles and Becker (1988), and Guthar and Wu (1991), see also Section 4.4. Hocking (1976, 1977, 1987b) observed surface waves produced by a vertically oscillating plate.

In 1831, Faraday observed the fluid inside a glass container oscillates at one-half of the vertical excitation frequency. Another similar series of experiments conducted by Mathiessen (1868, 1970) showed that the fluid oscillations are synchronous. The contradiction of the two observations led Lord Rayleigh (1883a,b, 1987) to make a further series of experiments with improved equipment and his observations supported Faraday's results. During that time, Mathieu (1868) formulated his equations, which helped Rayleigh to explain this phenomenon mathematically. The problem was investigated again by Lewis (1950), Taylor (1950), Benjamin and Ursell (1954), Konstantinov, *et al.* (1978), and Nevolin (1985) who explained mathematically the discrepancy between Faraday's and Rayleigh's observations and Matthiessen's findings. By setting $\ddot{Z}_0 = -\Omega^2 Z_0 \cos\Omega t$ in the free surface dynamic condition, and after linearization, their analysis led to a system of Mathieu equations in the form

$$\frac{d^2 A_{mn}}{dt^2} + 2\zeta_{mn}\omega_{mn}\frac{dA_{mn}}{dt} + \omega_{mn}^2\left(1 - \frac{\Omega^2 Z_0}{g}\cos\Omega t\right)A_{mn} = 0 \tag{6.1}$$

where A_{mn} is nondimensional wave height amplitude, ω_{mn} and ζ_{mn} are the natural frequency and damping ratio of mode mn, respectively, Ω is the excitation frequency, Z_0 is the excitation amplitude. Depending on the excitation amplitude, frequency, and damping ratio the solutions of equation (6.1) can be stable or unstable.

The analysis of parametric stability of differential equations with periodic coefficients of the type of equation (6.1) is outlined by McLachlan (1947) and Yakubovich and Starzhinskii (1975). The mechanics of parametric oscillation in various mechanical systems is well documented in Bolotin (1964), Schmidt (1975), Evan-Iwanowski (1976), and Nayfeh and Mook (1979). Under parametric random excitation the stochastic stability and response are presented by Ibrahim (1985).

In a sense, a single degree-of-freedom system with a nonlinear restoring force could be said to have a time-dependent stiffness because the stiffness is displacement dependent and the displacement varies with time. This relationship is implicit, however, and the problem is not classified as parametric. Alternatively, linear systems with time-dependent coefficients are similar in some ways to nonlinear systems and can be said to be classified between constant coefficient linear systems and nonlinear systems. This is most clearly seen when a linear periodic parametric system displays a half-order subharmonic response, an effect not found in a time-invariant linear system. Furthermore, the resulting response is orthogonal to the direction of the excitation, a feature that is opposite to the response of constant coefficient linear systems under forced excitation.

Some time-dependent variations in parameters apparently result from geometric changes, and these are always forces or torques that can do work on the system. Energy can flow into a system from external sources during resonance, which is dependent upon the frequency of the parameter variation and the natural frequencies of the system; the state so created is called *parametric resonance*. Thus, the dynamic system will experience parametric instability when the excitation frequency Ω is twice (or any integer multiple of) the system natural frequency ω_{mn}, that is,

$$\Omega = 2\omega_{mn} \quad \text{or} \quad \Omega = k\omega_{mn}, (k = 2,\ 3,\ \ldots) \tag{6.2}$$

The relationship $\Omega = 2\omega_{mn}$ is known as "principal parametric resonance" and differs from normal resonance, in which the forcing frequency and a natural frequency coincide.

When some average quantity of energy flows into a system, the amplitude of the system response increases. This behavior is referred to as *parametric instability*. The rate of increase in amplitude is generally exponential and thus potentially dangerous, more so in some ways than typical resonance in which the rate of increase is linear. Furthermore, damping effectively reduces the severity of typical resonance, but might reduce only the rate of increase during parametric resonance and thus have little or no effect on final amplitudes. Indeed, in some situations the introduction of damping can extend the instability region.

Because of some mathematical difficulties involved in analyzing tanks of various geometrical shapes, Kana (1966) conducted an experimental investigation of parametric excitation of liquid surface in 90° sector cylindrical and spherical tanks. The liquid response in both containers was found to be essentially similar to its behavior in a cylindrical tank especially the half frequency subharmonic response. The free-surface modes in a spherical tank are dependent on the fluid depth. It was anticipated that the first mode motion of the liquid in the 90° sector tank could exert a net torque about the longitudinal axis, if the proper phase relationships exist in the motion. Such a torque cannot occur in a nonsectored tank, or even in a sectored tank under lateral excitation. Woodward (1966), Lomen and Fontenot (1967), and Woodward and Bauer (1970) considered parabolic containers and annular sector cross-section tanks. They showed that the stability consideration would make the occurrence of harmonic and superharmonic responses very difficult.

The influence of parametric excitation on the discharge of liquid propellant from tanks of space vehicles has been experimentally investigated by Schoenhals, *et al.* (1967) and Nevolin (1983). A remarkable flow retardation was found when the acceleration amplitude level increased. On the other hand, the flow retardation was found to decrease as the excitation frequency increased for fixed acceleration amplitude. The problem of parametric excitation of a spinning liquid column was considered by Bauer (1989a) and will be considered in Chapters 11 and 12.

Different sources of parametric excitations are encountered in different applications. For example, under a micro-gravitational field, g-jitter acts as a parametric random excitation. Nontraditional sources of parametric excitation include electrostatic forces and convective temperature gradient. The parametric response of the interface between two dielectric liquids under an alternating electrostatic force was studied by Reynolds (1965), Devitt and Melcher (1965), Briskman and Shaidurov (1968), and Yih (1968a, b). Their studies showed that the stability of the interface required that the applied voltage must be high enough to suppress surface tension effects but lower than a certain analytically determined critical value. For voltages greater than this critical value, Reynolds (1965) indicated that the interface is unstable. Similar studies were carried out by Raco (1968) to determine the stability of a liquid jet in a time-dependent electric field. The effects of the frequency and the strength of the electric field on jet stability were determined from a stability analysis of the Mathieu equation. Gershuni and Zhukhovitskii (1963) discovered a parametric resonance called convective instability in a fluid body subject to a periodically varying temperature gradient. Later, Gershuni and Zhukhovitskii (1964) found that the modulation of the vertical temperature gradient had the same influence as the modulation of the angular velocity of rotation of the fluid as a rigid body. The stability of the equilibrium was found to depend not only on the mean temperature gradient, as in Rayleigh's problem (1892), but also on the amplitude and frequency modulation. The free-surface oscillations of a magnetic liquid were studied by Dodge and Garza (1972), Ohaba and Sudo (1995), Pursi and Malik (1995), and Sudo, *et al.* (1993, 1997).

Briskman, *et al.* (1976) provided further analysis of parametric excitation of liquids in communicating vessels. The global stability of liquid motion in ellipsoid containers was studied by Leonov and Morozov (1988). Bessem, *et al.* (1996) examined the stability of surface waves under two-frequency parametric excitation. The nonlinear phenomena associated with Faraday waves will be discussed in the next sections. Miles and Henderson (1990) presented a review of Faraday sloshing problems based the modern theory of bifurcation and chaos using normal mode expansion. Frandsen and Borthwick (2002, 2003) numerically simulated the nonlinear free surface motion under parametric excitation of storage containers using a two-dimensional finite difference algorithm.

The parametric excitation of internal waves of an ideal stratified fluid in closed containers was studied by Vladimirov (1981), Nesterov (1982), Sekerzhi-Zenkovich (1983), Kravtsov and Sekerzhi-Zenkovich (1993, 1996). Stratified fluids are characterized by having a gradual change in the density of the fluid such that the gradient of density with the fluid depth does not vanish.

This chapter presents the basic theory of liquid parametric sloshing under sinusoidal excitation. While the linear theory is only capable of defining the stability boundaries of the free surface, the nonlinear theory can predict the steady state response and uncover complex free-surface dynamic behavior such as quasi-periodic and chaotic motions, modal competition, and internal resonance. This chapter begins with a brief account of the linear theory of parametric sloshing. Section 6.3 will address the nonlinear theory of single mode excitation and the stability of fixed solutions. Modal competition and autoparametric interaction in cylindrical and rectangular tanks are treated in Sections 6.4–6.6. Section 6.7 describes some experimental investigation of random excitation tests. Surface disintegration is briefly reviewed in Section 6.8. The chapter is then closed by some remarks and conclusions.

6.2 Linear theory of parametric sloshing

The differential equation (6.1) of the free-surface amplitude under parametric excitation can be written in terms of two first-order differential equations in the matrix form

$$\dot{\mathbf{X}}(t) = \mathbf{A}(t)\mathbf{X}(t), \quad \text{with } \mathbf{A}(t+T) = \mathbf{A}(t) \tag{6.3}$$

where $\mathbf{X} = \{x_1, x_2\}^T$, $x_1 = A$, $\dot{x}_1 = x_2$, $\dot{x}_2 = -2\omega\zeta x_2 - \omega^2 x_1 + (\omega^2\Omega^2 Z_0/g)\, x_1 \cos\Omega t$. The subscript mn has been dropped. The matrix $\mathbf{A}(t)$ is usually real. Complex forms can arise if, for example, hysteretic damping or viscoelastic effects are considered. The solutions of system (6.3) are contained in the solution of the related matrix equation

$$\dot{\mathbf{Q}}(t) = \mathbf{A}(t)\mathbf{Q}(t) \tag{6.4}$$

If the solution of equation (6.4) is $\Phi(t)$, which is formed from independent solutions of equation (6.3), and if it can be assumed without losing generality that $\Phi(0) = \mathbf{I}$, the unit matrix, the form of the solution matrix after one period $\Phi(T)$ establishes whether or not the system is stable. The matrix $\Phi(T)$ is referred to as the monodromy matrix (Meirovitch, 1970) of equation (6.4). One of its several important properties is that it multiplies the solution at t, that is, $\Phi(t)$, to produce the solution at time $(t + T)$. In other words,

$$\Phi(t+T) = \Phi(t)\Phi(T) \tag{6.5}$$

This property holds for periods $2T$, $3T$, ..., so that

$$\Phi(t+nT) = \Phi(t)\Phi^n(T) \tag{6.6}$$

The powers of the monodromy matrix thus determine whether the solution will grow with successive periods. The solution will be stable provided the eigenvalues of $\Phi(T)$ lie within a unit circle on the complex plane. The eigenvalues λ_i are commonly referred to as the characteristic multipliers of the system. Alternatively, stability can be expressed in terms of characteristic exponents ρ_i of the system. These are defined by

$$\lambda_i = \exp(\rho_i) \tag{6.7}$$

In this case stability requires that the characteristic exponents have negative real parts. In fact, multipliers on the unit circle are also stable, except in certain solutions involving repeated multipliers.

Another useful property of $\Phi(t)$ is that it can be expressed in the form

$$\Phi(t) = \mathbf{P}(t)\exp(t\mathbf{R}) \quad \text{with } \mathbf{P}(t+T) = \mathbf{P}(t) \tag{6.8}$$

where $\mathbf{P}(t)$ is a periodic nonsingular matrix, and $\mathbf{R}$ is related to the monodromy matrix such that

$$\Phi(T) = \exp(T\mathbf{R}) \tag{6.9}$$

Here $\Phi(t)$ is the product of a periodic matrix $\mathbf{P}(t)$ times an exponential matrix $\exp(t\mathbf{R})$. It is a consequence of the basic work of Floquet (Cesari, 1963). Thus, the determination of the boundaries of the regions of instability is reduced to finding the conditions under which the given differential equation has periodic solutions with periods T, $2T$, $3T$, ... This is because the periodic motion is essentially the boundary case of vibrations with unboundedly increasing amplitudes. By dropping the subscript mn, equation (6.1) takes the form

$$\frac{d^2A}{dt^2} + 2\zeta\omega\frac{dA}{dt} + \omega^2(1 - 2\varepsilon\cos\Omega t)A = 0, \quad \varepsilon = \Omega^2 Z_0/(2g) \tag{6.10}$$

Considering the periodic solution with a period $2T$ in the form

$$A = \sum_{k=1,3,5}^{\infty}\left\{a_k \sin\left(\frac{k\Omega t}{2}\right) + b_k \cos\left(\frac{k\Omega t}{2}\right)\right\} \tag{6.11}$$

where the coefficients a_k and b_k are unknown and are related by substituting (6.11) into equation (6.10) and, after equating the coefficients of sin and cos yield the system of linear homogeneous coupled equations:
coefficients of $\sin(k\Omega t/2)$ are

$$\left(1 + \varepsilon - \frac{\Omega^2}{4\omega^2}\right)a_1 - \zeta\frac{\Omega}{\omega}b_1 - \varepsilon a_3 = 0 \tag{6.12a}$$

$$\cdots$$

$$\left(1 - \frac{k^4\Omega^2}{4\omega^2}\right)a_k - \zeta k\frac{\Omega}{\omega}b_k - \varepsilon(a_{k-2} + a_{k+2}) = 0 \tag{6.12b}$$

coefficients of $\cos(k\Omega t/2)$ are

$$\left(1 - \varepsilon - \frac{\Omega^2}{4\omega^2}\right)b_1 + \zeta\frac{\Omega}{\omega}a_1 - \varepsilon b_3 = 0 \tag{6.12c}$$

$$\cdots$$

$$\left(1 - \frac{k^4\Omega^2}{4\omega^2}\right)b_k + \zeta k\frac{\Omega}{\omega}a_k - \varepsilon(b_{k-2} + b_{k+2}) = 0 \tag{6.12d}$$

This system of equations is infinitely coupled. Approximate nontrivial solutions are obtained by truncating the number of equations and setting the determinant composed of the coefficients to zero. Restricting the analysis to $k = 1$, that is, subharmonic motion with period $2T$, the corresponding determinant takes the form

$$\begin{vmatrix} 1 + \varepsilon - \dfrac{\Omega^2}{2\omega^2} & -\zeta\dfrac{\Omega}{\omega} \\ \zeta\dfrac{\Omega}{\omega} & 1 - \varepsilon - \dfrac{\Omega^2}{2\omega^2} \end{vmatrix} = 0 \tag{6.13}$$

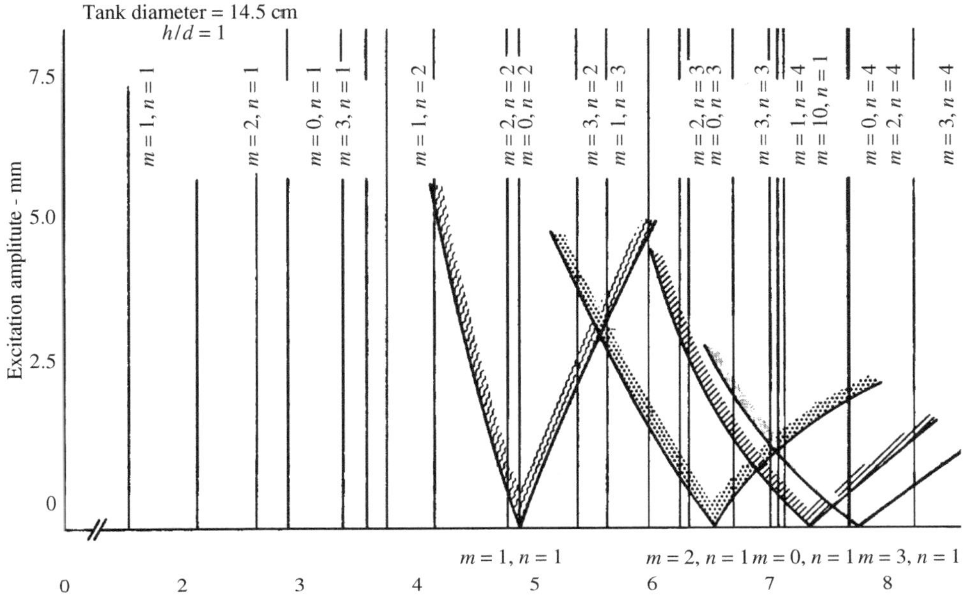

Figure 6.1 Stability boundaries of the first few sloshing modes. (Dodge, *et al.*, 1965)

This equation relates the excitation frequency Ω with the natural frequency, damping ratio of the free surface and the excitation amplitude. The boundaries of instability are given by the inequality

$$1 - \sqrt{\varepsilon^2 - \zeta_{mn}^2} < \left(\frac{\Omega}{2\omega_{mn}}\right)^2 < 1 + \sqrt{\varepsilon^2 - \zeta_{mn}^2} \tag{6.14}$$

It can be shown that if the free surface is unstable, the resulting motion could have frequency 1/2 times the excitation frequency. The analysis can be extended to include $k = 2, 3, \ldots$ which yields harmonic, 3/2-superharmonic, and $k/2$-superharmonic solutions. Experimentally, solutions with a 1/2-subharmonic are frequently observed. Dodge, *et al.* (1965) observed harmonic and double superharmonic liquid motions. The stability boundaries are usually given in the Ince–Mathieu chart. A typical chart is shown in Figure 6.1 where the ordinate represents the excitation amplitude parameter, Z_0, and the abscissa is excitation frequency. The chart includes the instability regions of the first few modes (1, 1), (2, 1), (0, 1), and (3, 1). If the excitation amplitude and frequency take any values in the shaded region, the fluid free-surface equilibrium state is unstable and oscillates with exponential growth amplitude.

Mathematical expressions for the stability boundaries are well documented (Abramowitz and Segun, 1968). Since the motion might be a half-frequency subharmonic, harmonic, or superharmonic, both Faraday and Matthiessen could be correct, however, the experimental results of Benjamin and Ursell only showed the half frequency subharmonic.

Bolotin (1956), Sorokin (1957), Woodward (1966), Rajappa (1970b), Goldberg (1972), Henstock and Sani (1974), and Hasegawa, *et al.* (1983) found the domains of instability could be slightly reduced if a linear viscous damping was introduced in equation (6.15). Brand and Nyborg (1965) carried out a series of experiments to measure the minimum value of the parametric excitation amplitude required to produce half-frequency surface waves. The measured values of excitation amplitude were found to be much greater than those predicted analytically. They attributed this difference to the lack of development in the previous theories of the damping coefficient of the free surface.

Kumar and Tuckerman (1994) considered the fluid viscosity and found that when viscous effects are included, the linearized evolution equations can no longer be reduced to a single ordinary differential equation. Furthermore, an eigenvalue problem involving partial-differential equations must be solved. The effect of viscosity was discussed by Henderson and Miles (1994), Cerda and Tirapegui (1997, 1998), and Henderson (1998). They found that the nature of the instability for slightly viscous fluids is fundamentally different from highly viscous fluids. In this case, the dynamics of the free surface in the presence of dominant viscous stresses may be described by a Mathieu equation. Decent (1997) considered the nonlinear damping of parametrically two-dimensional gravity waves. Decent and Craik (1997, 1999) and Foster and Craik (1997) studied the occurrence of limit cycles and sideband instability of Faraday waves.

Kravtsov and Sekerzh-Zenkovich (1993) examined the parametric excitation of a viscous two-layer fluid in a closed vessel. The stability of periodic waves of the free surface of deep liquids was determined by Zakharov (1968). Wright (1999) and Wright, *et al.* (2000) numerically studied the parametric excitation of two-dimensional standing periodic waves at the interface between two inviscid fluids. The fluid possesses different densities near normal and reduced gravity conditions. They used a boundary-integral method, applicable when the density of one fluid is negligible compared to that of the other fluid. They also used a vortex-sheet method for the more general case of arbitrary densities. The formation of complex interfacial structures, entrainment and ejection mixing of droplets from one fluid into the other, are possible nonlinear effects considered by Wright, *et al.* (2000).

Boyarshina, *et al.* (1987) studied the stability and critical forms for the fluid free under axial harmonic excitation. The differential equations of motion were obtained based on the Hamilton–Ostrogradsky principle (see Lukovskii, 1967). It was shown that under parametric resonance conditions, some complex spatial motions of the traveling wave type might propagate in a circumferential direction. Quantitatively, resonance mode equations and their coefficients are subject to the influence of nonresonant modes. However, the influence of instability of these modes on the dynamic characteristics of resonance modes is eliminated and the complexity of the problem is significantly diminished. When this is applied to the fluid motion in cylindrical or spherical containers, the procedure yields differential equations with two coupled modes having the same natural frequencies corresponding to eigenfunctions in the angular coordinate as demonstrated by Kubenko, *et al.* (1984). When the natural frequencies, corresponding to modes with different wave parameters, are close to one another, the procedure yields four equations involving the coupling and mutual influence of resonance modes.

6.3 Nonlinear parametric sloshing of a single mode

In parametric instability, the amplitude of a linear system grows without limit. As the amplitude increases, geometric nonlinearities emerge and the motion achieves bounded amplitude. For the case of parametric excitation of liquid free surface, the response exhibits other nonlinear phenomena such as nonplanar motion, rotational motion, chaotic motion and free-surface disintegration. Some of these phenomena are similar to those encountered in spherical pendulum motion under parametric excitation (see, e.g., Miles and Zhou, 1993). Skalak and Yarymovich (1962) considered the nonlinear analysis of deep-water two-dimensional waves in rectangular containers. For a cylindrical upright tank subjected to sinusoidal excitation, $\ddot{Z}_0(t)$, the fluid field equations take the form

$$\nabla^2\Phi = 0 \tag{6.15a}$$

$$\frac{\partial\Phi}{\partial r} = 0, \quad \text{at } r = R \tag{6.15b}$$

$$\frac{\partial\Phi}{\partial z} = 0, \quad \text{at } z = -h \tag{6.15c}$$

$$\frac{\partial\Phi}{\partial t} - \frac{1}{2}(\nabla\Phi\cdot\nabla\Phi) - (g + \ddot{Z}_0(t))\eta = 0, \quad \text{at } z = \eta \tag{6.15d}$$

$$\frac{\partial\eta}{\partial t} - \frac{\partial\Phi}{\partial r}\frac{\partial\eta}{\partial r} - \frac{1}{r^2}\frac{\partial\Phi}{\partial\theta}\frac{\partial\eta}{\partial\theta} + \frac{\partial\Phi}{\partial z} = 0, \quad \text{at } z = \eta \tag{6.15e}$$

The solution of the above boundary value problem may be obtained by adopting the asymptotic expansion analysis outlined in Section 4.2. The velocity potential function and wave height amplitude may be written in the form

$$\Phi(r,\theta,z,t) = \sum_{m=0}^{\infty}\sum_{n=1}^{\infty}\alpha_{mn}(t)J_m(\lambda_{mn}r)\cos m\theta\frac{\cosh[\lambda_{mn}(z+h)]}{\cosh\lambda_{mn}h} \tag{6.16}$$

$$\eta(r,\theta,t) = \sum_{m=0}^{\infty}\sum_{n=1}^{\infty}a_{mn}(t)J_m(\lambda_{mn}r)\cos m\theta \tag{6.17}$$

Following the analysis of Chapter 4, we obtain the following two differential equations for the generalized coordinates of the first anti-symmetric sloshing mode

$$\begin{aligned}\dot{\alpha}_{11} - (g - \Omega^2 Z_0\cos\Omega t)a_{11} + 0.104\,568\,2\lambda_{11}^2\dot{\alpha}_{11}a_{11}^2 \\ - 0.281\lambda_{11}^3\alpha_{11}^2 a_{11}\tanh\lambda_{11}h = 0\end{aligned} \tag{6.18}$$

$$\dot{a}_{11} + (\lambda_{11}\tanh\lambda_{11}h)\alpha_{11} - 0.122\ 515\lambda_{11}^3\alpha_{11}a_{11}^2\tanh\lambda_{11}h = 0 \tag{6.19}$$

where $\lambda_{11} = 1.841/R$.

Eliminating α_{11} and $\dot{\alpha}_{11}$ from equations (6.18) and (6.19) gives the following nonlinear differential equation of the first mode wave amplitude

$$\begin{aligned}&\ddot{a}_{11} + 2\omega_{11}\zeta_{11}\dot{a}_{11} + \omega_{11}^2\left(1 - \frac{\Omega^2 Z_0\cos\Omega t}{g}\right)a_{11} \\ &+ \lambda_{11}^2\left(0.227\,083a_{11}\ddot{a}_{11} + \left(0.245\,03 + \frac{0.281}{\tanh\lambda_{11}h}\right)\dot{a}_{11}^2\right)a_{11} = 0\end{aligned} \tag{6.20}$$

where the second term is introduced to account for energy dissipation. This nonlinear equation can be solved using one of the asymptotic perturbation techniques. First, we introduce the following nondimensional parameters

$$\begin{aligned}\tau=\omega_{11}t,\ x=a_{11}/R,\ C_1=0.220\,83\lambda_{11}^2R^2,\\ C_2=\left(0.245\,03+\frac{0.281}{\tanh\lambda_{11}h}\right)\lambda_{11}^2R^2,\ z=\Omega^2Z_0/g,\ \nu=\Omega/\omega_{11}\end{aligned}\tag{6.21}$$

Equation (6.20) takes the standard form

$$x''+x=\varepsilon\{-2\zeta x'+zx\cos\nu\tau-C_1x^2x''-C_2xx'^2\}\tag{6.22}$$

where a prime denotes differentiation with respect to τ, ε is a bookkeeping parameter, and the double subscript 11 has been dropped from the damping ratio.

The method of multiple scales is used and the solution may be written in the form

$$x(\tau;\varepsilon)=X_0(T_0,T_1)+\varepsilon X_1(T_0,T_1)+\varepsilon^2X_2(T_0,T_1)+\cdots\tag{6.23}$$

where $T_0=\tau$, $T_1=\varepsilon\tau$, $T_n=\varepsilon^n\tau$,

$$\begin{aligned}\frac{d}{d\tau}&=\frac{dT_0}{d\tau}\frac{\partial}{\partial T_0}+\frac{dT_1}{d\tau}\frac{\partial}{\partial T_1}+\cdots=D_0+\varepsilon D_1+\cdots\\ D_n&=\frac{\partial}{\partial T_n}\\ \frac{d^2}{d\tau^2}&=D_0^2+2\varepsilon D_0D_1+\varepsilon^2\left(D_1^2+2D_0D_2\right)+\cdots\end{aligned}\tag{6.24}$$

Different time scales are represented by T_n such that T_0 represents a fast scale, T_1 is the slower time scale, higher-order time scales represent even slower scales. The expansion is carried out up to first-order in ε, and thus we need to consider the two time scales T_0 and T_1. Substituting equation (6.23) into equation (6.22) and equating terms of equal power of ε gives

$$D_0^2X_0+X_0=0\tag{6.25a}$$

$$\begin{aligned}D_0^2X_1+X_1=&-2D_0D_1X_0+2\zeta D_0X_0+zX_0\cos\nu\tau-C_1X_0^2D_0^2X_0\\&-C_2X_0(D_0X_0)^2\end{aligned}\tag{6.25b}$$

The solution of equation (6.25a) may be written in the form

$$X_0=A(T_1)\,e^{iT_0}+\bar{A}(T_1)\,e^{-iT_0}\tag{6.26}$$

where over-bar denotes conjugate and $i=\sqrt{-1}$. Substituting this solution in equation (6.25b) and setting secular terms to zero, we obtain the solvability condition

$$-2iA'-2i\zeta A+\frac{z}{2}\bar{A}\,e^{2i\sigma T_1}+3C_1A^2\bar{A}-C_2A^2\bar{A}=0\tag{6.27}$$

where σ is the parametric detuning parameter defined by the relationship $\nu=2(1+\varepsilon\sigma)$ Introducing the complex polar transformation

$$A(T_1)=\frac{1}{2}a(T_1)\,e^{i\beta(T_1)}\tag{6.28}$$

into equation (6.27) and separating complex and real parts gives the two first-order differential equations

$$a' = -\zeta a + \frac{z}{4} a \sin(2\sigma T_1 - 2\beta) \tag{6.29a}$$

$$-a\beta' = \frac{z}{4} a \cos(2\sigma T_1 - 2\beta) + \frac{a^3}{8}(3C_1 - C_2) \tag{6.29b}$$

The steady state solution can be obtained by setting the left-hand sides of equations (6.29) to zero. In order to obtain the dependence of the steady state solution on the detuning parameter, σ, we set $\psi = 2\sigma T_1 - 2\beta$. It follows that equations (6.29) take the following form

$$a' = -\zeta a + \frac{z}{4} a \sin \psi \tag{6.30a}$$

$$a\psi' = 2\sigma a + \frac{z}{2} a \cos \psi + \frac{a^3}{4}(3C_1 - C_2) \tag{6.30b}$$

These equations form an invariant under the transformation

$$(a, \psi) \Leftrightarrow (-a, \psi) \tag{6.31}$$

The steady state solutions of these equations are given by the following roots

$$a = 0 \tag{6.32a}$$

$$a_{1,2}^2 = \frac{8}{(3C_1 - C_2)}\left[-\sigma \pm \sqrt{\frac{z^2}{16} - \zeta^2}\right], \quad \text{and} \quad \sin \psi = -4\frac{\zeta}{z} \tag{6.32b}$$

The zero solution represents the equilibrium position of the liquid free surface. The second solution is only valid for real positive roots. For positive real roots, $a_{1,2}^2 \geq 0$, the value of the detuning parameter σ must be negative provided $(3C_1 - C_2) > 0$, and $z \geq 4\zeta$. In this case, the dependence of the response amplitude on the detuning parameter bends to the left and exhibits soft spring characteristics. The value of $(3C_1 - C_2)$ is found to be negative for liquid depth ratio $h/R < 0.416$ and for real positive roots of (6.32b), the values of σ must be positive for which the free-surface amplitude experiences hard-spring behavior. Figure 6.2(a) and (b) shows the

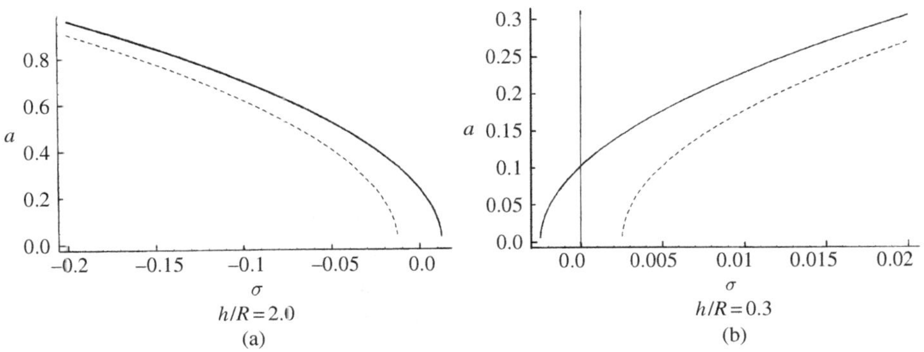

Figure 6.2 Dependence of the response amplitude on the detuning parameter for two different values of depth ratio: depth ratio (a) greater than the critical value, and (b) less that the critical value.

dependence of the response amplitude on the detuning parameter σ for liquid depth ratios $h/R = 0.3$ and $h/R = 2.0$.

It is not convenient to examine the stability of this solution in terms of equations (6.30). Instead, one may introduce the following Cartesian transformation

$$p = a\cos\frac{\psi}{2}, \quad q = a\sin\frac{\psi}{2} \tag{6.33}$$

Equations (6.30) take the form

$$p' = -\zeta p + \left(\frac{z}{4} - \sigma\right)q - \frac{1}{8}(3C_1 - C_2)q(p^2 + q^2) \tag{6.34a}$$

$$q' = -\zeta q + \left(\frac{z}{4} + \sigma\right)p + \frac{1}{8}(3C_1 - C_2)p(p^2 + q^2) \tag{6.34b}$$

These equations are invariant under the transformation

$$(p, q) \Leftrightarrow (-p, -q) \tag{6.35}$$

The Jacobian matrix of equations (6.34) associated with a fixed solution (p_0, q_0) is

$$\mathbf{J} = \begin{bmatrix} -\zeta - \dfrac{(3C_1 - C_2)}{4}p_0 q_0 & \left(\dfrac{z}{4} - \sigma\right) - \dfrac{(3C_1 - C_2)}{8}(p_0^2 + 3q_0^2) \\ \left(\dfrac{z}{4} + \sigma\right) - \dfrac{(3C_1 - C_2)}{8}(3p_0^2 + q_0^2) & -\zeta + \dfrac{(3C_1 - C_2)}{4}p_0 q_0 \end{bmatrix}$$

For the trivial solution, $(p_0, q_0) = (0, 0)$, the eigenvalues of the corresponding Jacobian are given by the solution of the quadratic equation

$$\lambda^2 + 2\zeta\lambda + \zeta^2 - \left(\frac{z^2}{16} - \sigma^2\right) = 0 \tag{6.36}$$

The sum of the two eigenvalues is -2ζ, which is negative. Thus, at least one of the two eigenvalues will always have a negative real part and this excludes the possibility of having a pair of purely imaginary eigenvalues. This implies that Hopf bifurcation cannot take place. On the other hand, one of the eigenvalues vanishes if the excitation level z reaches the following critical value

$$z^* = 4\sqrt{\zeta^2 + \sigma^2} \tag{6.37}$$

Obviously this excitation level defines the bifurcation point that separates the trivial solution (6.32a) from the nonzero solution (6.32b). This means that as the excitation level increases from zero, the undisturbed liquid free surface remains flat (i.e., stable) until the excitation level reaches a critical value given by equation (6.37) where the static bifurcation occurs. For $z > z^*$, the trivial fixed point is a saddle point, and the stable nontrivial fixed points given by (6.32b) emerge. This type of bifurcation is referred to as symmetry-breaking bifurcation because the static bifurcation at $z = z^*$ results in asymmetric solutions.

For $\sigma < 0$ and $(3C_1 - C_2) > 0$ the two roots given by (6.32b) are equal for

$$z = 4\zeta \tag{6.38}$$

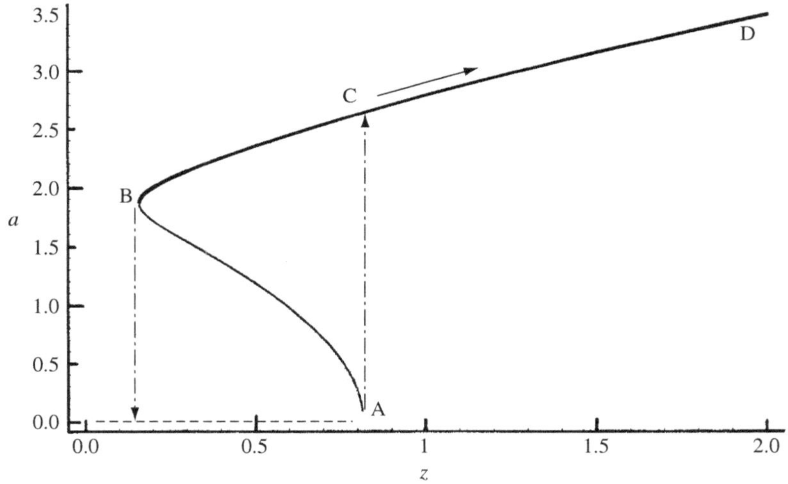

Figure 6.3 Dependence of response amplitude on excitation level.

and the corresponding fixed point is

$$a_{1,2}^* = \pm\sqrt{\frac{-8\sigma}{(3C_1 - C_2)}}, \quad \text{and} \quad \psi = -\pi/2 \tag{6.39a}$$

$$\begin{aligned} p_1^* &= +\sqrt{\frac{-4\sigma}{(3C_1 - C_2)}}, \quad q_1^* = -\sqrt{\frac{-4\sigma}{(3C_1 - C_2)}} \\ p_1^* &= -\sqrt{\frac{-4\sigma}{(3C_1 - C_2)}}, \quad q_1^* = \sqrt{\frac{-4\sigma}{(3C_1 - C_2)}} \end{aligned} \tag{6.39b}$$

The Jacobian of the first solution (p_1^*, q_1^*) is

$$\mathbf{J} = \begin{bmatrix} -\zeta - \sigma & \zeta + \sigma \\ \zeta - \sigma & -\zeta + \sigma \end{bmatrix} \tag{6.40}$$

This Jacobian has one zero eigenvalue. At the excitation level $z = 4\zeta$, the fixed point (p_1^*, q_1^*) experiences a saddle-node bifurcation at point B in Figure 6.3. This figure shows the dependence of the response amplitude on the excitation amplitude parameter z. As the excitation amplitude increases from zero value the liquid amplitude remains undisturbed until the excitation reaches the bifurcation value z^* at point A, due to a symmetry-breaking bifurcation. The liquid-free-surface amplitude will jump in a fast dynamic transition to point C, which is a form of a subcritical pitchfork bifurcation. As the excitation level increases above the bifurcation point z^*, the liquid amplitude increases monotonically along the solid curve CD. If, on the other hand, the excitation level begins to decrease the liquid-free-surface amplitude will decrease along the curve DCB until point B, at excitation level $z = 4\zeta$, at which the amplitude collapses to the static equilibrium position, where the tangent

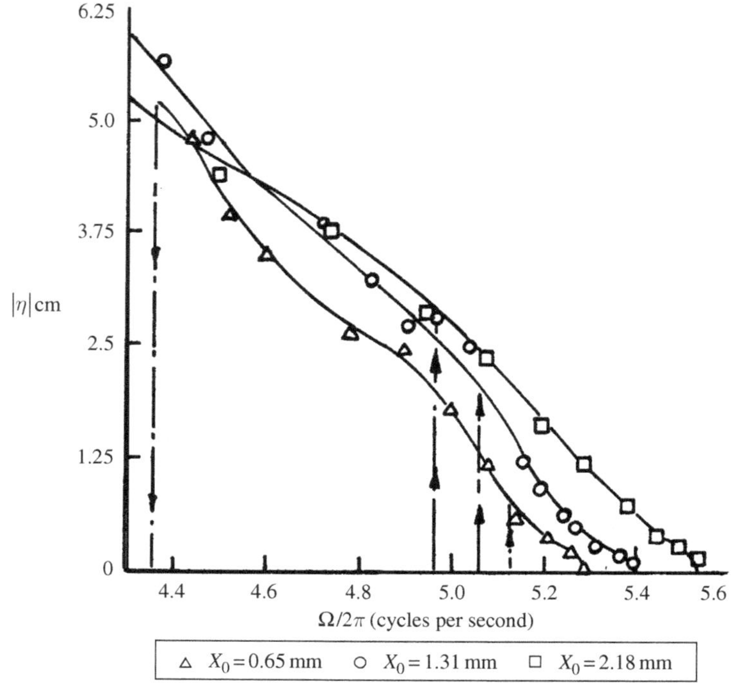

Figure 6.4 Experimental measurement of liquid response amplitude of sloshing mode (1, 1) as function of excitation frequency for 1/2 sub-harmonic response for different values of excitation amplitude. (Dodge, *et al.*, 1965)

$$\frac{d\sigma^2}{da^2} = (3C_1 - C_2)\left[\frac{(3C_1 - C_2)}{2}a^2 + 4\sigma\right] \tag{6.41}$$

vanishes at $z = 4\zeta$, at which the two roots (6.32b) are equal, and

$$a_{1,2}^2 = \frac{-8\sigma}{(3C_1 - C_2)} \tag{6.42}$$

Dodge, *et al.* (1965) conducted experimental investigations to measure the dependence of the liquid-free-surface height under three different excitation amplitudes as shown in Figure 6.4. The measured amplitude was taken at $r/R = 0.837$, and the wave height was estimated as being one-half the difference between the maximum and minimum liquid-free-surface excursions for depth ratio $h/R = 2$. The figure shows the points at which the response amplitude jumps and collapses, depending on whether the excitation frequency increases or decreases. Another important observation, that is common for both symmetric and asymmetric sloshing modes, is that the positive amplitude height is always greater than the negative amplitude of the free surface. This is only valid for nonlinear free-surface excitation.

6.4 Nonlinear modal competition

Under parametric excitation of frequency close to twice the natural frequency of a certain sloshing mode, the liquid free surface will oscillate exhibiting the shape of that mode. However, as the excitation amplitude increases, the input energy can stimulate other neighbor modes although their natural frequencies are not commensurable with the excited mode. The reason for this type of mixed mode interaction is that its instability regions overlap as the excitation amplitude increases, as indicated by the stability chart shown in Figure 6.1 For example, the instability regions of the first symmetrical mode (0, 1), the second and third anti-symmetric modes (2, 1) and (3, 1) overlap considerably. In the absence of internal resonance, Dodge (Abramson, 1966a) provided an excellent summary of the nonlinear response of liquid free surface in different tank geometries including rectangular and cylindrical tanks. The main feature of the amplitude–frequency response curves exhibits soft nonlinear characteristics. These studies considered deep fluid depths, which exceed the critical depth.

For a thin fluid layer, experimental measurements may exhibit complex surface motions, which are mainly chaotic. For example, under parametric harmonic excitation of two different sloshing modes, temporal chaos was experimentally observed in a thin fluid layer by Gollub and Meyer (1983) and Ciliberto and Gollub (1984, 1985a,b). They reported chaotic motions in a fluid layer in a circular tank with depth-to-radius ratio of $h/R = 0.16$ using surface-imaging and a laser-beam-deflection method. For such fluid depths, the sloshing natural frequency is strongly dependent on the fluid depth.

Gollub and Meyer (1983) found that the symmetric modes resulting from the initial instability undergo a secondary instability in which the circular symmetry is broken and the pattern precesses. The resulting flow is quasi-periodic because of the fast oscillation associated with the primary instability. This was particularly manifested when the first symmetric mode was parametrically excited. An oscillatory instability was observed to develop at a threshold amplitude at which the center of the circular pattern is displaced slightly from the tank, but precesses at a low frequency in a way that preserves the azimuthal symmetry in a time-average sense. At another threshold amplitude, azimuthal structure begins to appear that produces a chaotic modulation in the time dependence of the local fluid slope. At relatively higher excitation amplitudes, the free-surface pattern appeared disordered and the noise in the spectrum of the subharmonic amplitude was centered at the zero frequency.

Chaotic behavior may arise due to competition between two different spatial modes or patterns (see, e.g., Ciliberto and Gollub, 1984, 1985a,b). A region of mode competition emerges in which the fluid surface can be described as a superposition of the two modes with amplitudes having a slowly varying envelope. These slow variations can be either periodic or chaotic. Ciliberto and Gollub (1984) developed a phenomenological model in the form of the nonlinear Mathieu equation

$$\ddot{a}_m + \zeta_m \dot{a}_m + \left(\omega_m^2 - \psi_m Z_0 \cos \Omega t\right) a_m = \varsigma_m a_m^3 \tag{6.43}$$

where ς_m is the coefficient of the cubic term that limits the growth of the mode, and ψ_m is the gain coefficient of the parametric excitation term.

Figure 6.5 shows the digitized optical intensity fields formed by stable patterns of modes (4, 3) and (7, 2), where the first index indicates the number of angular maxima. The crosses are experimentally determined points on the stability boundaries (Ciliberto and Gollub, 1985a, b). Below the parabolic stability boundaries, the surface is essentially flat, except for a small circularly symmetric response at the driving frequency caused by the meniscus at the outer boundary of the fluid layer. At the intersection of the two stability boundaries, both modes oscillate simultaneously. Above the stability boundaries, the liquid surface oscillates at half the

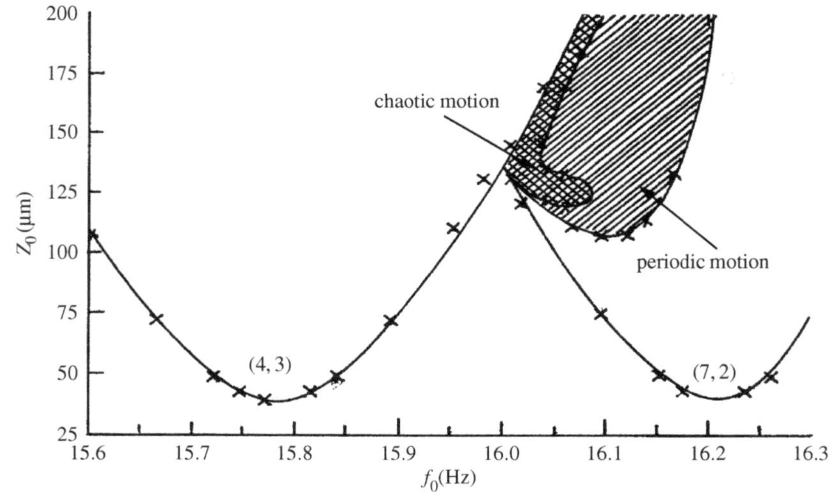

Figure 6.5 Stability boundaries of modes (4, 3) and (7, 2) including regions of chaotic and periodic motions. Crosses are experimental points. (Ciliberto and Gollub, 1984)

driving frequency in a single stable mode. The shaded areas are regions of mode competition, in which the surface can be described as a superposition of the (4, 3) and (7, 2) modes with amplitudes having a slowly varying envelope in addition to the fast oscillation at half the excitation frequency, $\Omega/2$. These slow variations can be either periodic or chaotic. At driving amplitudes higher than those shown in Figure 6.5, the surface can become chaotic even if the driving frequency is resonant, so that a single mode is dominant.

As one crosses from the region of slow periodic oscillations into the chaotic region, one finds a period-doubling bifurcation followed by a transition to chaos. A typical example is shown in Figure 6.6, where time history records are shown for three different driving amplitudes but a fixed driving frequency of 16.05 Hz. The time history records were obtained by measuring the light intensity $V(t)$ at a single point in the image as a function of time. The figure also shows the corresponding power spectra of the time series. It is seen that subharmonic bifurcation occurs at relatively higher excitation amplitudes ($Z_0 = 149$ μm and $Z_0 = 190$ μm). This is accompanied by a slight broadening of the peaks. Above $Z_0 = 180$ μm, the time history record is chaotic. Ciliberto and Gollub (1985a, b) indicated that at amplitudes higher than $Z_0 = 200$ μm, there was no mode competition and only the fourfold symmetric mode occurred.

Ciliberto and Gollub (1984) described the interaction between modes by assuming that the gain coefficients are coupled. For example, to describe the interaction of the fourth and seventh modes, $m = 4$, 7, the gain coefficients are expressed by the following expressions

$$\psi_4 = \bar{\psi}_4 + \beta_{47} a_7^2 \quad \text{and} \quad \psi_7 = \bar{\psi}_7 + \beta_{74} a_4^2 \tag{6.44}$$

where the coupling coefficient β_{47} is positive and β_{74} is negative. The origin of the sign difference is due to the observed phase difference between the two modes during the oscillations, such that a_7^0 leads a_4^0 by 90°. This implies that the seventh mode pumps energy to the fourth mode, while the fourth mode damps the seventh mode. The response is then expressed in terms of slowly varying amplitudes in the form

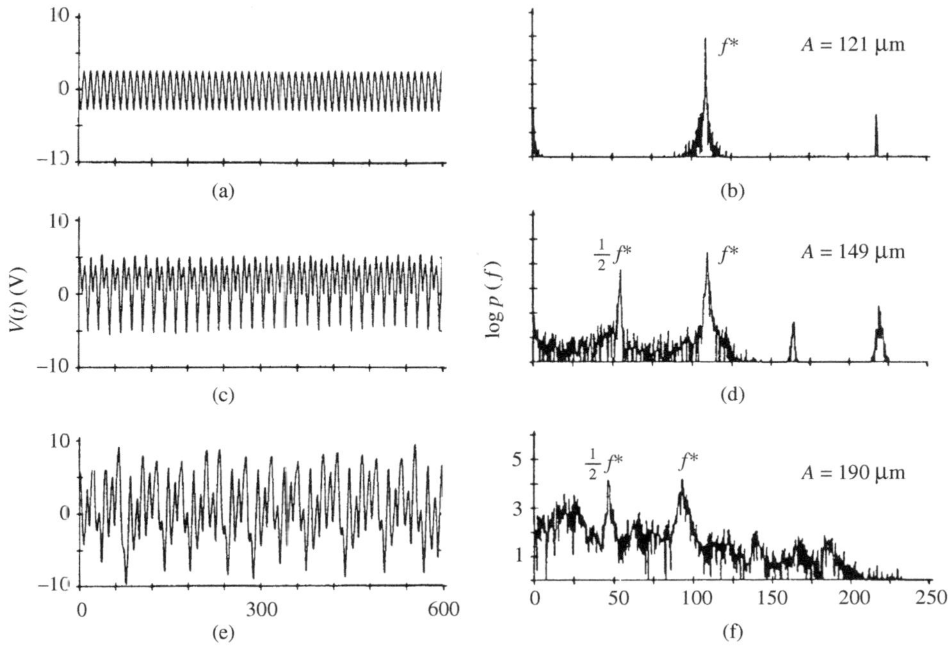

Figure 6.6 Time history records showing the transition from periodic to chaotic oscillations and corresponding power spectra under excitation frequency 16.05 Hz and three different excitation amplitudes. (Ciliberto and Gollub, 1984)

$$a_m(t) = C_m(t)\cos(\Omega t/2) + B_m(t)\sin(\Omega t/2) \tag{6.45}$$

The damping parameter ζ_m and the coefficient $\bar{\psi}_m$ are adjusted to fit the stability boundaries and the nonlinear coefficients ς_m are chosen to obtain the correct saturation amplitudes, all in a region where only a single mode is present. Substituting equations (6.45) and (6.44) into equation (6.43), keeping only terms oscillating at $\Omega/2$, and dropping $\ddot{B}_m$ and $\ddot{C}_m$ because the time scales of the fast oscillation and mode competition are very different, so that $\ddot{B}_m \ll 2\Omega\dot{C}_m$ and $\ddot{C}_m \ll 2\Omega\dot{B}_m$, the following four equations are obtained

$$\dot{C}_4 = -\frac{1}{2}\zeta_4 C_4 - \left[\frac{(\Omega^2 - 4\omega_4^2)}{4\Omega} - \frac{\bar{\psi}_4}{\Omega}Z_0 + \frac{3\varsigma_4}{4\Omega}(C_4^2 + B_4^2) - \frac{\beta_{47}A}{2\Omega}B_7^2\right]B_4 \tag{6.46a}$$

$$\dot{B}_4 = -\frac{1}{2}\zeta_4 B_4 - \left[\frac{(\Omega^2 - 4\omega_4^2)}{4\Omega} + \frac{\bar{\psi}_4}{\Omega}Z_0 + \frac{3\varsigma_4}{4\Omega}(C_4^2 + B_4^2) - \frac{\beta_{47}A}{2\Omega}C_7^2\right]C_4 \tag{6.46b}$$

$$\dot{C}_7 = -\frac{1}{2}\zeta_7 C_7 - \left[\frac{(\Omega^2 - 4\omega_7^2)}{4\Omega} - \frac{\bar{\psi}_7}{\Omega}Z_0 + \frac{3\varsigma_7}{4\Omega}(C_7^2 + B_7^2) - \frac{\beta_{74}A}{2\Omega}B_4^2\right]B_7 \tag{6.46c}$$

$$\dot{B}_7 = -\frac{1}{2}\zeta_7 B_7 + \left[\frac{(\Omega^2 - 4\omega_7^2)}{4\Omega} + \frac{\bar{\psi}_7}{\Omega}Z_0 + \frac{3\varsigma_7}{4\Omega}(C_7^2 + B_7^2) + \frac{\beta_{74}A}{2\Omega}C_4^2\right]C_7 \tag{6.46d}$$

These equations were numerically integrated and both periodic and chaotic oscillations were produced near the intersection of the stability boundaries. The two mode-coupling coefficients were adjusted to obtain an oscillatory domain similar in size to that found in the experiment shown in Figure 6.6. The values of the parameters used are $\zeta_4 = \zeta_7 = 0.4\ \text{s}^{-1}$, $(\bar{\psi}_4/\Omega) = 51.3\ \text{cm}^{-1}\text{s}^{-1}$, $(\bar{\psi}_7/\Omega) = 52.6\,\text{cm}^{-1}\text{s}^{-1}$, $(3\varsigma_4/4\Omega) = 1.0\,\text{s}^{-1}$, $(3\varsigma_7/4\Omega) = 0.1\text{s}^{-1}$, $(\beta_{47}/2\Omega) = 7.0\,\text{sm}^{-1}\text{s}^{-1}$, $(\beta_{74}/2\Omega) = -9.0\,\text{sm}^{-1}\text{s}^{-1}$, $\omega_4 = 49.59$ rad/s, $\omega_7 = 50.92$ rad/s.

The stability boundaries obtained by equations (6.46) with these parameters were found qualitatively similar to Figure 6.5. While the parabolic stability boundaries are in agreement with the measured ones, the shapes of the periodic and chaotic regimes are different from those measured experimentally.

Keolian, *et al.* (1981) and Keolian and Rudnick (1984), using liquid helium and water in thin annular troughs, observed both period-doubling and quasi-periodic motions apparently involving three modes. Meron and Procaccia (1986a, b) derived amplitude equations based on symmetry considerations to provide a theoretical basis for the experimental results. Miles (1993, 1994) and Yoshimatsu and Funakoshi (1998) addressed some features and patterns of Faraday waves. The disordered structure of the free-surface flow under relatively large harmonic excitation amplitude was also experimentally observed by Gollub and Meyer (1983). Their measurements showed a sequence of symmetry-breaking instabilities leading to a chaotic state.

Faraday waves were observed to be especially versatile (Bechhoefer, *et al.*, 1995, Bechhoefer, 1996, and Kurdolli and Gollub, 1996a, b). They exhibit the common patterns familiar from convection (stripes, squares, hexagons, spirals), as well as more exotic patterns. These patterns include triangles (Müller, 1993), quasi-patterns (Christiansen, *et al.*, 1992, Edwards and Fauve, 1994, and Kurdolli, *et al.*, 1998), super-lattice patterns (Kurdolli, *et al.*, 1998, Arbell and Fineberg, 1998, and Wagner, *et al.*, 1999), time-dependent rhombic patterns (Arbell and Fineberg, 2000a) and localized waves (Wagner, *et al.*, 1999 and Arbell and Fineberg, 2000b). Müller, *et al.* (1998) presented a review on Faraday wave pattern formation. Riecke (1990) analyzed the stable wave number kinks in parametrically excited standing waves.

It has been shown that waves, which are synchronous with the excitation, can be excited in thin layers of fluid vibrated at low frequency (see, e.g., Kumar, 1996, Müller, *et al.*, 1997, and Cerda and Tirapegui, 1997). They also occur in certain viscoelastic fluids (Wagner, *et al.*, 1999) and in fluids forced periodically, but with more than one frequency component (Edwards and Fuave, 1994, Beyer and Friedrich, 1995, and Bessem, *et al.*, 1996). For each case, it is possible to tune the forcing parameters in order to access the transition between subharmonic and harmonic response. At codimension-two[1] point, both instabilities set in simultaneously, but with different spatial wave numbers. Kumar (1996) showed that harmonic instability could only occur in a fairly shallow viscous liquid in a vessel of infinitely large horizontal dimensions.

Kalinichenko, *et al.* (1995, 2000) derived the conditions for harmonic instability of the free surface of a liquid of low viscosity in a rectangular vessel of finite horizontal dimensions. The velocity field of the liquid was divided into potential and eddy components, and the boundary layer method was employed. For a rectangular vessel of length 50 cm, width 4 cm, water depth of 15 cm, viscosity $\nu = 0.01\ \text{cm}^2/\text{s}$, the tank was excited at a frequency close to the second wave mode, $\omega_2 = 10.85$ rad/s. Figure 6.7 shows the analytical (shown by a solid curve) and experimental (shown by small circles) stability boundaries.

[1] A bifurcation that requires at least m control parameters to occur is called a codimension-m bifurcation.

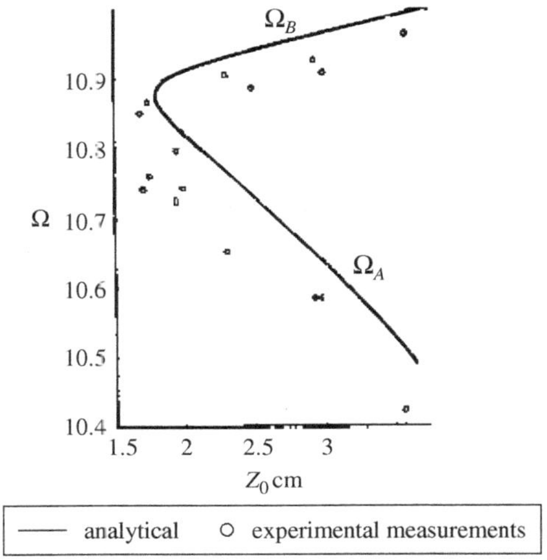

Figure 6.7 Harmonic instability boundaries of liquid free surface in a vertically oscillating rectangular container (l = 50 cm, L = 4 cm, h = 15 cm), fluid kinematic viscosity $\nu = 0.01\text{cm}^2/\text{s}$. The upper and lower branches define the excitation frequencies Ω_B, and Ω_A at which the liquid free-surface amplitude is zero. (Kalinichenko, *et al.*, 2000)

Kimura, *et al.* (1994b) and Kimura and Takahara (1997) examined the influence of higher-order radial modes on nonlinear free surface motion in circular cylindrical and rectangular tanks subjected to parametric and pitching excitations. Sudo and Hashimoto (1987) and Yamada, *et al.* (1987) determined the liquid response spectra due to horizontal and vertical excitations. Jacqmin and Duval (1988) and Jacqmin (1990) determined the instability conditions of fluid–fluid interface and fluid with a uniform density gradient due to an oscillatory parametric acceleration. Houghton (1968) studied the particle motion in vertically oscillating fluid. Craik (1986, 1995) and Craik and Armitage (1995) revisited the Faraday problem and were able to describe the wavelength selection and hysteresis observed experimentally. Jiang, *et al.* (1996) performed boundary-integral simulations of two-dimensional motion to determine the highly nonlinear motion of an interface with finite- and large-amplitude deformations.

Miles and Henderson (1990) reviewed the dynamics of the interface under the influence of a harmonic acceleration. Experimental studies conducted by Edwards and Fauve (1994), and Kurdolli and Gollub (1996a,b) indicated a harmonically forced system might sustain finite-amplitude waves over a broad range of frequencies and amplitudes of the container motion. The spatial form and temporal evolution of these waves were affected by the geometry of the container, the fluid's viscosity, and the behavior of the contact-angle or motion of the contact-line. Under most conditions, it was found that the waves have a square pattern and oscillate at half the frequency of the external excitation with modulated amplitude. Many other patterns were observed especially for polychromatic accelerations.

Zhang and Viñals (1997a,b) developed a weakly nonlinear analysis for the dynamics of small-amplitude surface waves on a semi-infinite weakly inviscid fluid layer. Kurdolli, *et al.* (1998) experimentally observed that two-frequency parametrically excited waves produce an intriguing "super-lattice" wave pattern near a codimension-two bifurcation point where both superhamonic and harmonic waves onset simultaneously, but with different spatial wave numbers. The super-lattice pattern is synchronous with forcing, spatially periodic on a large hexagonal lattice, and exhibits small-scale triangular structure. Silber and Proctor (1998) showed that similar patterns can exist and may be stable if the nonlinear coefficients of the bifurcation problem satisfy certain inequalities. Silber, *et al.* (2000) used the spatial and temporal symmetries to indicate that weakly damped harmonic waves may be critical to understanding the stabilization of such a pattern in the Faraday system.

Sloshing modal interaction in the presence of internal resonance is considered in the next sections. Note that internal resonance conditions may occur only for specific liquid depths depending on the tank cross-section geometry. The order of internal resonance is dictated by the order of nonlinear coupling. For example 1 : 1 is associated with cubic nonlinearity while 2 : 1 internal resonance is associated with quadratic nonlinearity.

6.5 Autoparametric interaction in cylindrical tanks

6.5.1 Historical overview

Complex free-liquid-surface motions owe their occurrence to the presence of nonlinear resonance conditions such as internal and parametric resonance conditions. Internal resonance implies the presence of a linear algebraic relationship among the natural frequencies of the interacted modes. These problems were studied by Miles (1976a, 1984a, b), Gollub and Simonelli (1989), Simonelli and Gollub (1989), Feng and Sethna (1989), Feng (1990), Funakoshi and Inoue (1991), and many others.

The modern theory of nonlinear dynamics has attracted several investigators to revisit Faraday's problem. Their studies dealt with weakly nonlinear phenomena associated with surface waves (see, for example, Keolian, *et al.*, 1981 and Kuz'ma and Kholopova, 1983). The most interesting phenomena, including chaotic behavior, are due to internal resonance, which occurs when the ratios of natural frequencies of two or more modes of motion are near small positive integers. Modal interaction is an avenue by which the free surface acquires strange attractors because the averaged equation for one mode (not coupled with other modes) does not have stabilized states of the chaotic attractor type. For coupled modes, when one mode is directly excited, the amplitude–frequency response of the excited mode has an additional region of instability besides the standard one. This new region arises from the zeroth solution instability of the second coupled mode, which is not directly excited. Krasnopolskaya and Shvets (1992) analyzed the oscillations of a free liquid surface in a cylindrical tank taking into account the effects of interaction with the energy source. They extended their work (1993) to investigate the behavior of parametric resonance and derived the equations of motion based on the works of Miles (1984a, c, d), Ciliberto and Gollub (1985b), Meron and Procaccia (1986b), Kamke and Umeki (1990), Miles and Henderson (1990), and Crawford and Knobloch (1991). The Lagrangian averaging procedure with respect to the motor shaft phase angle was used in order to obtain a system of equations for amplitudes of the dominant modes. The resulting equations were solved numerically. A control parameter, which characterizes the slope of the motor performance curve, was adopted as the bifurcation parameter. They reported a series of transitions between different limit cycles and chaotic motions when gradually increasing the control parameter. The results were given as phase space projections and power spectra. It was found that the transition from regular to chaotic motions adheres strictly to the Feigenbaum scenario or through intermittency. They showed that the interaction with the excitation mechanisms could generate chaos.

Miles (1984b, d) studied such wave phenomena when two modal frequencies are nearly equal (circular containers). Miles (1984a), Holmes (1986), Gu and Sethna (1987), and Gu, *et al.* (1988) also studied the case of parametric excitation when the modal frequencies are in the ratio of 1 : 2. Holmes (1986) qualitatively showed the existence of chaotic motions for certain parameter ranges close to 2 : 1 subharmonic resonances. Gu and Sethna (1987) studied periodic, almost periodic, and chaotic wave motions in a rectangular tank subjected to vertical sinusoidal excitation. The internal resonance condition 1 : 2 requires the fluid height to be relatively small, which causes excessive energy dissipation. Such energy dissipation suppresses nonlinear phenomena making it impossible to verify the analytical results experimentally. When the frequencies are nearly equal, the free oscillation in a nearly square container was discussed by Bridges (1987) for the case of standing waves. For the case of a nearly square container all nonsymmetric modes have nearly equal natural frequencies independent of the fluid depth. Silber and Knobloch (1989), Nagata (1989, 1991), and Feng and Sethna (1989) considered the case of 1 : 1 internal resonance and showed that the system is capable of exhibiting periodic and quasi-periodic standing and traveling waves. They were able to identify parameter values at which chaotic behavior can occur.

Weakly nonlinear models of Faraday resonance can be derived from the averaged Lagrangian approach proposed by Miles (1976a, 1984a) or from direct multiple-scale analysis (Gu and Sethna, 1987). These models were extended to include third-order viscous damping by Milner (1991), third order forcing by Miles (1993), and fifth-order frequency detuning effects by Decent (1995) and Decent and Craik (1995). These models describe the selection of patterns in a large system with many wavelengths (Miles, 1994) and include hysteresis effects (Craik and Armitage, 1995).

The pattern selection problem of the free surface was examined in terms of a symmetric-breaking bifurcation of the trivial fixed-point using the methods of equivalent bifurcation theory (Golubitsky, *et al.*, 1988). The application of the equivalent bifurcation theory to hydrodynamic pattern formulation problems has been reviewed by Crawford and Knobloch (1991).

The significant damping effect of contact-line hysteresis has been addressed by Miles (1967). A composite boundary condition on the wall was proposed by Hocking (1987a) and later improved by Miles (1990b) to represent the damping effect of the contact line. The complex local motion near an oscillating plate at high Reynolds number was shown experimentally by Ting and Perlin (1995). Contact-line hysteresis can also change the natural frequency. The amplitude equation for a single-mode Faraday wave was exploited by Jiang, *et al.* (1996) for periodic forcing with sideband noise to partially explain the modulation realized experimentally.

Nonlinear modal interaction was also studied for the case of spatio-temporal resonant triads in a horizontally unbounded fluid domain. The interaction takes place such that each critical wave number from linear analysis actually corresponds to a circle of critical wave vectors. In this situation it has been argued that resonant triads may play a central role in the Faraday's wave pattern selection problem as indicated by Binks and Van de Water (1997), Binks, *et al.* (1997), Zhang and Viñals, (1997b), Lifshitz and Petrich (1997) and Westera, Binks and De Water (2003). Resonant triads have three critical wave vectors that sum to zero, that is, $\mathbf{k}_3 = \mathbf{k}_1 \pm \mathbf{k}_2$, where $|\mathbf{k}_1| = |\mathbf{k}_2|$ is the wave number of one critical mode and $|\mathbf{k}_3|$ is the wave number of the other critical mode. Crawford, *et al.* (1990), Crawford, (1991a, b), Crawford, *et al.*, (1993), Silber and Skeldon (1999), and Silber, *et al.* (2000) formulated the bifurcation problem using a stroboscopic map and estimation of Floquet multipliers. It was shown that there is a fundamental difference in the pattern selection problems for subharmonic and harmonic instabilities near the codimension-two point. Many experimental (see, e.g., Binks and Van de Water, 1997 and Arbell and Fineberg, 1998) and theoretical studies (Lifshitz and Petrich, 1997 and Chen and Viñals, 1999) attributed the formation of exotic patterns near the codimension-two (or "bicritical") point to resonant triad interactions involving the critical or near-critical modes with different spatial wave numbers.

6.5.2 Lagrangian formulation

The analytical modeling of nonlinear coupled sloshing modes developed originally by Dodge, *et al.* (1965) is based on the assumption that the primary mode is the dominant mode and is coupled to second order symmetric modes. Miles (1984a) found that such equations involve some difficulty in the form of the nonlinear terms that couple the dominant mode with other modes. He found that "their equations of motion for the modal amplitudes violate the reciprocity conditions that are implicit in the underlying (Newtonian) mechanics." Instead, Miles (1984a) developed a Lagrangian formulation in which the potential energy is in the normal form. His formulation is convenient for studying the nonlinear Faraday waves in the presence of the internal resonance conditions 2 : 1 and 1 : 1. Other Lagrangian formulations of liquid waves were developed by Blak (1996) and Longuet-Higgins (2000). Nayfeh (1987) adopted the general fluid field equations and expanded the free-surface conditions and free-surface elevation into Taylor series in terms of linear mode shapes. The governing equation of the liquid free-surface elevation was in full agreement with Miles' Lagrangian formulation.

The Lagrangian is developed by including second- and fourth-order terms. The analysis leads to a Hamiltonian system for the slowly varying amplitudes. Following the analysis of "chaotic sloshing" in Chapter 4, the Lagrangian takes the form, after dividing by ρS,

$$L \equiv T - V = \frac{1}{2}a_n(\dot{\eta}_n^2 - \omega_n^2\eta_n^2) - \frac{1}{2}\ddot{Z}_0\eta_n\eta_n + \frac{1}{2}a_{\ell mn}\eta_\ell\dot{\eta}_m\dot{\eta}_n + \frac{1}{4}a_{j\ell mn}\eta_j\eta_\ell\dot{\eta}_m\dot{\eta}_n \tag{6.47}$$

where $a_n = (1/k_n)\coth k_n h \equiv g/\omega_n^2$, ω_n represents the natural frequency of the nth mode and a_n represents the length of an equivalent pendulum of the nth mode.

One can generate a discrete set of differential equations from equation (4.60) for any number of modes. In view of the fourth-order nonlinearity in the Lagrangian, the equations of motion contain terms of first- and third-order. One may introduce the following form of solution

$$\eta_n = \delta_{1n} l\left\{p(\tau)\cos\frac{\Omega}{2}t + q(\tau)\sin\frac{\Omega}{2}t\right\} + \frac{l^2}{a_1}\{A_n(\tau)\cos\Omega t + B_n(\tau)\sin\Omega t + C_n(\tau)\} \tag{6.48}$$

where $l = 2(1/k_1)\sqrt{\varepsilon/|A|}\tanh(k_1 h)$ is a reference length-scale, $\varepsilon = \omega_1^2 Z_0/g \ll 1$, $\tau = \varepsilon^2\Omega t/2$ is a nondimensional slow time parameter. Also, p_n, q_n, A_n, B_n, and C_n are slowly varying nondimensional amplitudes, and A is a parameter that determines the effect of nonlinearities to bring the surface wave height into a bounded value and is given by the expression

$$A = \frac{1}{2}\left\{ a_1 a_{1111} + a_{n11}^2 - \frac{1}{2} a_1 \frac{(4a_{11n} - a_{n11})^2}{4a_n - a_1} \right\} \tanh^4 k_1 h \tag{6.49}$$

where the parameters a_1 and a_{ijkl} are defined by relations (4.57). This parameter also determines the effect of finite amplitude on the resonant frequency according to the relationship

$$\Omega^2 = 4\omega_1^2 [1 - \varepsilon A(p^2 + q^2)] \tag{6.50}$$

Substituting equation (6.48) into the Lagrangian expression (6.47) and taking the average $\langle L \rangle$ over a 2π interval of Ωt while holding τ fixed, gives

$$\begin{aligned}\langle L \rangle = \frac{1}{2}\varepsilon g l^2 \Bigg[& \dot{p}q - p\dot{q} + \beta(p^2 + q^2) + p^2 q^2 \\ & + \frac{1}{\varepsilon}\left(\frac{l}{a_1}\right)^2 \left\{ \frac{1}{16} a_1 a_{1111} (p^2 + q^2)^2 - C_n^2 + \frac{1}{2\Omega_n}(A_n^2 + B_n^2) \right. \\ & + \left(a_{11n} - \frac{1}{4} a_{n11} \left[A_n(p^2 - q^2) + 2B_n pq \right] \right) \\ & \left. + \frac{1}{2} a_{n11} C_n (p^2 + q^2) \right\} \Bigg]\end{aligned} \tag{6.51}$$

where

$$\beta = \frac{\Omega^2 - 4\omega_1^2}{8\varepsilon\omega_1^2} \approx \frac{\Omega - 2\omega_1}{2\varepsilon\omega_1}, \quad \Omega_n = 4\left(\frac{\Omega}{2\omega_n}\right)^2 - 1 \approx \frac{4a_n - a_1}{a_1} \tag{6.52a, b}$$

A dot denotes differentiation with respect to τ. Requiring $\langle L \rangle$ to be stationary with respect to variations of p, q, A_n, B_n and C_n, yields

$$\{A_n, B_n\} = -\frac{1}{\Omega_n}\left(a_{11n} - \frac{1}{4} a_{n11} \right) \{p^2 - q^2, 2pq\} \tag{6.53a}$$

$$C_n = \frac{1}{4} a_{n11} (p^2 + q^2) \tag{6.53b}$$

In terms of p, and q, the average $\langle L \rangle$ can be written in the compact form

$$\langle L \rangle = \varepsilon g l^2 \left\{ \frac{1}{2}(\dot{p}_n q_n - p_n \dot{q}_n) + H(p, q) \right\} + O(\varepsilon) \tag{6.54}$$

where

$$H = \frac{1}{2}(1 + \beta)p^2 + \frac{1}{2}(\beta - 1)q^2 + \frac{1}{4}(p^2 + q^2)^2 \operatorname{sgn} A \tag{6.55}$$

The evolution equations are obtained using the canonical form

$$\dot{p} = -\frac{\partial H}{\partial q_n} = -(\beta - 1 + p^2 + q^2)q, \quad \dot{q} = \frac{\partial H}{\partial p} = (\beta + 1 + p^2 + q^2)p \tag{6.56a, b}$$

The fixed points are obtained by setting $\dot{p}=\dot{q}=0$,

$$p=q=0 \tag{6.57a}$$

$$p=0,\ \ q=\pm\sqrt{1-\beta},\quad \text{provided } \beta<1 \tag{6.57b}$$

$$p=\pm\sqrt{-1-\beta},\ \ q=0,\quad \text{provided } \beta<-1 \tag{6.57c}$$

Stability analysis of these fixed points reveals that the trivial solution (6.57a) is stable for $\beta^2>1$ and unstable for $\beta^2<1$. Solution (6.57b) is stable while (6.57c) is unstable. The stable fixed points are centers and the unstable ones are saddle points. These are summarized as follows:

(1) For $\beta>1$, we have center at $p=q=0$.
(2) For $-1<\beta<1$, we have saddle point at $p=q=0$, and two centers at $p=0$, $q=\pm\sqrt{1-\beta}$.
(3) For $\beta<-1$, we have three centers at $p=q=0$ and $p=0$, $q=\pm\sqrt{1-\beta}$ and two saddle points at $p=\pm\sqrt{-1-\beta}$ and $q=0$.

The motion relative to the basin vanishes at $p=q=0$, and is harmonic (within $(1+O\sqrt{\varepsilon})$) at $p=0$, $q=\pm\sqrt{1-\beta}$

Based on equation (6.48) and the fixed point (6.57b), the mean square displacement of the free surface is given by the expression

$$\begin{aligned}E[\eta^2]=\frac{1}{2}l^2q^2=\frac{1}{2}l^2(1-\beta)&=\frac{2\varepsilon\tanh^2(k_1h)}{Ak_1^2}(1-\beta)\\&=\frac{\tanh^2 k_1h}{Ak_1^2}\left\{1-\left(\frac{\Omega}{2\omega_1}\right)^2+2\varepsilon\right\}\end{aligned} \tag{6.58}$$

This relationship reveals singularity near $A=0$. It is not difficult to show that if one introduces damping, $0<(\zeta/\varepsilon)=\alpha<1$, equations (6.56) take the form

$$\dot{p}=-\alpha p-(\beta-1+p^2+q^2)q,\ \ \dot{q}=-\alpha q+(\beta+1+p^2+q^2)p \tag{6.59a, b}$$

These equations have the following fixed points

$$p=q=0 \tag{6.60a}$$

$$p+iq=\pm e^{i(\pi/2-\varphi)}\sqrt{\gamma-\beta}\quad \text{for } \beta<\gamma \tag{6.60b}$$

$$p+iq=\pm e^{i\varphi}\sqrt{-\gamma-\beta}\quad \text{for } \beta<-\gamma \tag{6.60c}$$

where $\cos 2\varphi=\gamma=\sqrt{1-\alpha^2}$, provided $\alpha<1$.

The stability analysis of this fixed point $p=q=0$ gives the following scenarios

(1) If $\beta>\gamma$, we have sink at $p=q=0$.
(2) If $-\gamma<\beta<\gamma$, we have a saddle point at $p=q=0$ and two sinks at (6.60b).
(3) If $\beta<-\gamma$, we have three sinks at $p=q=0$ and two saddle points at (6.60c).

Introducing the action-angle coordinates E and θ defined by the following coordinate transformation

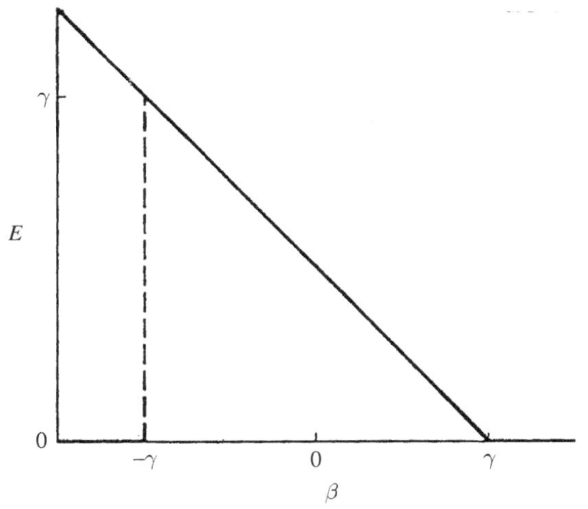

Figure 6.8 The dimensionless energy at the stable fixed points. The dash line indicates the jump from $E = 0$ as β is increased through $-\gamma$. (Miles, 1984a)

$$p = \sqrt{2E}\cos\theta, \quad q = \sqrt{2E}\sin\theta \tag{6.61a, b}$$

Substituting (6.61) into the Hamiltonian (6.55), gives

$$H = E^2 + E(\beta + \cos 2\theta) \tag{6.62}$$

Introducing damping, the new state equations take the form

$$\dot{E} = -\frac{\partial H}{\partial \theta} = 2E(\sin 2\theta - \alpha) \quad \dot{\theta} = \frac{\partial H}{\partial E} = 2E + \beta + \cos 2\theta \tag{6.63a, b}$$

where the action is a measure of the energy such that the total energy is

$$\hat{E} = \rho g l E\{1 + O(\varepsilon)\} \tag{6.64}$$

The dependence of the action E on the tuning parameter β at the stable fixed points is shown in Figure 6.8. If the motion starts from rest and β increases through $-\gamma$, E may be expected to jump from 0 to γ and then to decrease linearly to 0 at $\beta = \gamma$. Then E remains at 0 as β increases. If β decreases through γ, E may be expected to increase linearly from 0 to some value in excess of γ for $\beta < -\gamma$ but eventually to drop to 0 at some lower value of β.

Miles (1984a) extended the analysis to include the influence of 2 : 1 and 1 : 1 internal resonance conditions. In view of the complexity of the 1 : 1 internal resonance case, the analysis was restricted to the Hamiltonian of that condition. Nayfeh (1987) and Nayfeh and Nayfeh (1990) considered two cases of a 2 : 1 internal resonance condition in which each of the modes involved was parametrically excited.

6.5.3 Two-to-one internal resonance

Under parametric excitation, it may happen that the excited mode is in internal resonance with another sloshing mode. When the internal resonance is 1 : 1 the coupling is dictated by cubic nonlinearity. Another case examined by Miles (1984a), Nayfeh (1987), and Nayfeh and Nayfeh

(1990) involves 2:1 internal resonance, that is, $\omega_2 = 2\omega_1$, which is mainly governed by quadratic nonlinearity. The parametric excitation was studied for two different cases. The first is when the excitation frequency is twice the lowest natural frequency, that is, $\Omega = 2\omega_1$, while the second case is when it is close to twice the higher mode, that is, $\Omega = 2\omega_2$. The analysis is general for any container with an arbitrary cross-section. However, the analysis for an upright cylindrical tank will be presented.

Nayfeh (1987) expressed the wave height and velocity potential in terms of linear normal modes, that is,

$$\eta(x,y,\tau) = \sum a_m(\tau)\Psi_m(x,y) \tag{6.65a}$$

$$\Phi(x,y,z,\tau) = \sum \alpha_m(\tau)\Psi_m(x,y)\frac{\cosh(K_m(z+h)/l)}{\cosh(K_m h/l)} \tag{6.65b}$$

where l is a characteristic length of the container, $\tau = t\sqrt{g/l}$, K_m are wave numbers related to the natural frequency of the free surface, such that $\bar{\omega}_m^2 = K_m \tanh K_m h/l$, $\bar{\omega}_n = \omega_n/\sqrt{g/l}$. Their values depend on the tank cross-section geometry. The internal resonance condition $\bar{\omega}_2 = 2\bar{\omega}_1$ is satisfied if

$$K_2 \tanh(K_2 h/l) = 4K_1 \tanh(K_1 h/l) \tag{6.66}$$

The $\Psi_m(x,y)$ are the linear mode shapes governed by the partial differential equation

$$\nabla^2 \psi_m + k_m^2 \psi_m = 0 \tag{6.67}$$

subject to the boundary condition

$$\nabla\psi_m \cdot \mathbf{n} = 0 \quad \text{at the wetted wall} \tag{6.68a}$$

where $\mathbf{n}$ is the outward normal to wall. The shapes $\Psi_m(x,y)$ are orthonormalized, that is,

$$\int\int \psi_m \psi_n \, ds = \delta_{mn} S \tag{6.68b}$$

where δ_{mn} is the Kronecker delta.

The kinematic and dynamic free-surface conditions are expanded into Taylor series, that is,

$$\frac{\partial\eta}{\partial\tau} - \frac{\partial\Phi}{\partial z} - \frac{\partial^2\Phi}{\partial z^2}\eta + \nabla\Phi\cdot\nabla\eta + \cdots = 0, \quad \text{at } z=0 \tag{6.69a}$$

$$\frac{\partial\Phi}{\partial\tau} - \frac{\partial^2\Phi}{\partial\tau\partial z}\eta + (1+\ddot{z}_0(\tau))\eta + \frac{1}{2}(\nabla\Phi)^2 + \frac{1}{2}\left(\frac{\partial}{\partial z}\right)^2 + \cdots = 0, \quad \text{at } z=0 \tag{6.69b}$$

Substituting equations (6.65) into equations (6.69), multiplying the resulting equations by $\Psi_{m(x,y)}$, and using the orthonomality condition (6.68b) gives

$$\frac{\partial a_n}{\partial\tau} - \kappa_n\alpha_n + \sum (D_{nms} - K_s^2 C_{nms}) a_m\alpha_s = 0 \tag{6.70a}$$

$$\frac{\partial\alpha_n}{\partial\tau} + a_n + \ddot{z}_0(\tau)a_n + \sum \kappa_n C_{nms} a_n\frac{\partial\alpha_s}{\partial\tau} + \frac{1}{2}\sum (D_{nms} + \kappa_m\kappa_s C_{nms})\alpha_m\alpha_s = 0 \tag{6.70b}$$

where

$$\kappa_s = K_s \tanh K_s = \bar{\omega}_s^2 \tag{6.71a}$$

$$C_{nms} = \frac{1}{S}\int\int \Psi_n \Psi_m \Psi_s \, ds \tag{6.71b}$$

$$D_{nms} = \frac{1}{S}\int\int \Psi_n \nabla\Psi_m \cdot \nabla\Psi_s \, ds \tag{6.71c}$$

Integrating equation (6.71c) by parts using Green's theorem gives

$$D_{nms} = \frac{1}{2} C_{nms}\left(K_m^2 + K_s^2 - K_n^2\right) \tag{6.72}$$

Eliminating α_m from equations (6.70), using equation (6.72), and adding a weak linear damping term $2\zeta_n\dot{\eta}_n$ to account for energy dissipation gives the following nonlinear equation

$$\ddot{a}_n + 2\zeta_n\bar{\omega}_n\dot{a}_n + \bar{\omega}_n^2 a_n + \bar{\omega}_n^2 \ddot{z}_0(\tau)a_n + \sum \Gamma_{nms} a_m \ddot{a}_s + \sum \chi_{nms}\dot{a}_m\dot{a}_s = 0 \tag{6.73}$$

where a dot denotes differentiation with respect to the nondimensional time parameter $\tau = t\sqrt{g/l}$

$$\Gamma_{nms} = \frac{1}{2}\bar{\omega}_n^2 C_{nms}\left[2 + \left(K_m^4 - K_n^4 - K_s^4\right)/\bar{\omega}_n^2\bar{\omega}_s^2\right] \tag{6.74a}$$

$$\chi_{nms} = \frac{1}{2}\bar{\omega}_n^2 C_{nms}\left[1 + \frac{\left(K_m^4 + K_s^4 - K_n^4\right)}{2\bar{\omega}_m^2\bar{\omega}_s^2} + \frac{\left(K_m^4 - K_{ns}^4 - K_s^4\right)}{\bar{\omega}_n^2\bar{\omega}_s^2}\right] \tag{6.74b}$$

Also S denotes the container cross-section, and $z_0(\tau) = Z_0 \cos\bar{\Omega}\tau$.

The method of multiple scales is used to solve the wave height amplitude. A second-order expansion of equation (6.68) is introduced in the form

$$a_n = \varepsilon a_{n1}(T_0, T_1) + \varepsilon^2 a_{n2}(T_0, T_1) + \cdots \tag{6.75}$$

where $T_0 = \tau$ and $T_1 = \varepsilon\tau$, and the time derivatives in terms T_i are defined as follows

$$\frac{d}{d\tau} = D_0 + \varepsilon D_1 + \cdots, \quad \text{and} \quad \frac{d^2}{d\tau^2} = D_0^2 + 2\varepsilon D_0 D_1 + \cdots, \quad D_i = \frac{\partial}{\partial T_i} \tag{6.76}$$

The parametric excitation is ordered so that the effect of parametric resonance balances the effect of internal resonance. Set

$$\bar{\Omega}\bar{\omega}_n^2 \ddot{z}_0(\tau) = 2\varepsilon\xi_n \cos\bar{\Omega}\tau \tag{6.77}$$

Substituting equations (6.75) and (6.76) into equation (6.73) and equating coefficients of equal powers of ε, gives

Terms of order ε

$$D_0^2 a_{n1} + \bar{\omega}_n^2 a_{n1} = 0 \tag{6.78}$$

Terms of order ε^2

$$D_0^2 a_{n2} + \omega_n^2 a_{n2} = -2D_0 D_1 a_{n1} - 2\zeta_n D_0 a_{n1} + 2\Omega\xi_n a_{n1} \cos \Omega T_0$$
$$-\sum \Gamma_{nms} a_{m1} D_0^2 a_{s1} - \sum \chi_{nms} a_{m1} \left(D_0 a_{m1}\right)\left(D_0 a_{s1}\right) \tag{6.79}$$

The solution of equation (6.78) may be written in the form

$$a_{n1} = A_n(T_1)\, e^{i\omega_n T_0} + \bar{A}_n(T_1)\, e^{-i\omega_n T_0} \tag{6.80a}$$

where the over-bar denotes a complex conjugate. Alternatively, this solution can be written in the form

$$a_{n1} = \bar{a}_n(T_1) \cos(\bar{\omega}_n T_0 + \theta_n(T_1)) \tag{6.80b}$$

The response of the free surface will first be determined for the case when the sloshing mode of the lower frequency, ω_1, is parametrically excited. The second case considers the parametric excitation of the higher mode, ω_2. Both cases will be studied in the presence of the internal resonance condition $\omega_2 = 2\omega_1$. Here we omitted the over bar.

First mode parametric excitation

It is assumed that the parametric excitation frequency Ω is close to twice the natural frequency of the lower sloshing mode, that is, $2\omega_1$. The higher sloshing mode frequency, ω_2, is close to twice the lower mode frequency. These two conditions can be stated after introducing detuning parameters as:

$$\Omega = 2\omega_1 + \varepsilon\sigma_2,$$
$$\omega_2 = 2\omega_1 + \varepsilon\sigma_1 \tag{6.81a, b}$$

where ε is a small parameter and σ_i are detuning parameters.

The functions $A_n(T_1)$ can be determined by imposing the solvability condition that can be extracted by substituting solution (6.80a) into equation (6.79)

$$D_0^2 a_{12} + \omega_1^2 a_{12} = -2i\omega_1(A_1' + \zeta_1 A_1)\, e^{i\omega_1 T_0} + \Omega\xi_1 \bar{A}_1\, e^{i(\Omega-\omega_1)T_0}$$
$$+ 4\omega_1 \Lambda_1 A_2 \bar{A}_1\, e^{i(\omega_2-\omega_1)T_0} + \text{c.c.} + \text{NST} \tag{6.82a}$$

$$D_0^2 a_{22} + \omega_2^2 a_{22} = -2i\omega_2(A_2' + \zeta_2 A_2)\, e^{i\omega_2 T_0} + \Omega\xi_2 \bar{A}_2\, e^{i(\Omega-\omega_2)T_0}$$
$$+ 4\omega_2 \Lambda_2 A_1^2\, e^{2i\omega_1 T_0} + \text{c.c.} + \text{NST} \tag{6.82b}$$

where c.c. stands for complex conjugate of the preceding terms, NST refers to nonsecular terms, and

$$4\omega_1\Lambda_1 = \Gamma_{121}\omega_1^2 + \Gamma_{112}\omega_2^2 - \omega_1\omega_2(\chi_{112} + \chi_{121}),$$
$$4\omega_2\Lambda_2 = \omega_1^2(\Gamma_{211} + \chi_{211}) \tag{6.83a, b}$$

The resonance conditions (6.81a,b) can be rewritten in the form

$$\Omega T_0 = 2\omega_1 T_0 + \sigma_2 T_1, \ \omega_2 T_0 = 2\omega_1 T_0 + \sigma_1 T_1 \tag{6.84a, b}$$

Substituting these conditions into equations (6.83) and setting secular terms to zero gives the two solvability conditions

$$2i(A_1' + \zeta_1 A_1) - 4\Lambda_1 A_2 \bar{A}_1 e^{i\sigma_1} - 2\xi_1 \bar{A}_1 e^{i\sigma_2 T_1} = 0 \tag{6.85a}$$

$$2i(A_2' + \zeta_2 A_2) - 4\Lambda_2 A_1^2 e^{i\sigma_1 T_1} = 0 \tag{6.85b}$$

where a prime denotes differentiation with respect to the time scale T_1. One can write the complex function $A_n(T_1)$ in terms of polar coordinates, that is,

$$A_n(T_1) = \frac{1}{2} a_n(T_1) e^{i\theta_n(T_1)} \tag{6.86}$$

Introducing (6.86) into equations (6.85) and separating real and imaginary parts, gives

$$a_1' = -\zeta_1 a_1 + \Lambda_2 a_1 a_2 \sin(\theta_2 - 2\theta_1 + \sigma_1 T_1) + \xi_1 a_1 \sin(\sigma_2 T_1 - 2\theta_1) \tag{6.87a}$$

$$a_2' = -\zeta_2 a_2 - \Lambda_2 a_1^2 \sin(\theta_2 - 2\theta_1 + \sigma_1 T_1) \tag{6.87b}$$

$$a_1\theta_1' = -\Lambda_1 a_1 a_2 \cos(\theta_2 - 2\theta_1 + \sigma_1 T_1) - \xi_1 a_1 \cos(\sigma_2 T_1 - 2\theta_1) \tag{6.87c}$$

$$a_2\theta_2' = -\Lambda_2 a_1^2 \cos(\theta_2 - 2\theta_1 + \sigma_1 T_1) \tag{6.87d}$$

Let $\gamma_1 = (\theta_2 - 2\theta_1 + \sigma_1 T_1)$ and $\gamma_2 = (\sigma_2 T_1 - 2\theta_1)$. One can then write

$$\gamma_1' = (\theta_2' - 2\theta_1' + \sigma_1), \quad \text{and} \quad \gamma_2' = (\sigma_2 - 2\theta_1') \tag{6.88a, b}$$

The steady state solution of equations (6.87) can be determined by setting the left-hand sides to zero. It requires $\gamma_1' = \gamma_2' = 0$, and this yields

$$\theta_1' = \frac{1}{2}\sigma_2 \quad \text{and} \quad \theta_2' = \sigma_2 - \sigma_1 \tag{6.89a, b}$$

In this case, the following set of algebraic equations are obtained

$$-\zeta_1 a_1 + \Lambda_2 a_1 a_2 \sin\gamma_1 + \xi_1 a_1 \sin\gamma_2 = 0 \tag{6.90a}$$

$$\zeta_2 a_2 + \Lambda_2 a_1^2 \sin\gamma_1 = 0 \tag{6.90b}$$

$$\frac{1}{2}\sigma_2 a_1 + \Lambda_1 a_1 a_2 \cos\gamma_1 + \xi_1 a_1 \cos\gamma_2 = 0 \tag{6.90c}$$

$$(\sigma_2 - \sigma_1) a_2 + \Lambda_2 a_1^2 \cos\gamma_1 = 0 \tag{6.90d}$$

These equations have two different solutions:

(1) The trivial solution

$$a_1 = a_2 = 0 \tag{6.91}$$

(2) The nontrivial solution given by the following values

$$a_1^2 = S_1 \pm \sqrt{\left\{\frac{\xi_1^2\left[\zeta_2^2 + (\sigma_2 - \sigma_1)^2\right]}{\Lambda_1^2 \Lambda_2^2}\right\} - S_2^2} \tag{6.92a}$$

$$a_2 = \frac{|\Lambda_2| a_1^2}{\sqrt{\left[\zeta_2^2 + (\sigma_2 - \sigma_1)^2\right]}} \tag{6.92b}$$

where

$$S_1 = \frac{\frac{1}{2}\sigma_2(\sigma_2 - \sigma_1) - \zeta_1\zeta_2}{\Lambda_1\Lambda_2}, \quad S_2 = \frac{\frac{1}{2}\sigma_2\zeta_2 + \zeta_1(\sigma_2 - \sigma_1)}{\Lambda_1\Lambda_2}, \tag{6.93}$$

Careful inspection of (6.92a), reveals that if $S_1 < 0$, equation (6.92a) has one real root for all values of $\xi_1 \geq \sqrt{\zeta_1^2 + (\sigma_2^2/4)}$. On the other hand, if $S_1 > 0$, equation (6.92a) has one real root for all $\xi_1 > \sqrt{\zeta_1^2 + (\sigma_2^2/4)}$ and has two real roots for all $\Lambda_1\Lambda_2|S_2|/\sqrt{\zeta_2^2(\sigma_2 - \sigma_1)^2} \leq \xi_1 \leq \sqrt{\zeta_1^2 + (\sigma_2^2/4)}$. Thus, there are three possible steady-state solutions, namely, the zero solution and two finite-amplitude solutions. Nayfeh (1987) examined the stability of the trivial solution and the steady-state solutions by transforming the modulation equations (6.90) into Cartesian coordinates and the eigenvalues of the Jacobian of the system were then examined. The trivial solution was found to be stable if and only if

$$\xi_1 \leq \sqrt{\zeta_1^2 + (\sigma_2^2/4)}. \tag{6.94}$$

This means that as long as the liquid damping energy exceeds the input energy, the free surface remains undisturbed.

On the other hand, the steady-state nontrivial solution is stable under the following conditions:

$$\begin{aligned} &2(\zeta_1 + \zeta_2)\left[4\zeta_1\zeta_2 + \zeta_2^2 + (\sigma_2 - \sigma_1)^2 + 4\Lambda_1\Lambda_2 a_1^2\right] \\ &\quad - \left[2\zeta_1\left\{\zeta_2^2 + (\sigma_2 - \sigma_1)^2\right\} + 4(\zeta_1 + \zeta_2)\Lambda_1\Lambda_2 a_1^2\right] > 0 \end{aligned} \tag{6.95a}$$

$$\begin{aligned} &\left[2\zeta_1\left\{\zeta_2^2 + (\sigma_2 - \sigma_1)^2\right\} + 4(\zeta_1 + \zeta_2)\Lambda_1\Lambda_2 a_1^2\right] \\ &\times \left\{2(\zeta_1 + \zeta_2)\left[4\zeta_1\zeta_2 + \zeta_2^2 + (\sigma_2 - \sigma_1)^2 + 4\Lambda_1\Lambda_2 a_1^2\right]\right. \\ &\left. - \left[2\zeta_1\left\{\zeta_2^2 + (\sigma_2 - \sigma_1)^2\right\} + 4(\zeta_1 + \zeta_2)\Lambda_1\Lambda_2 a_1^2\right]\right\} > 0 \end{aligned} \tag{6.95b}$$

$$4\Lambda_1\Lambda_2 a_1^2\left[\Lambda_1\Lambda_2 a_1^2 + \zeta_1\zeta_2 - \frac{1}{2}\sigma_2(\sigma_2 - \sigma_1)\right] > 0 \tag{6.95c}$$

For a cylindrical tank, the internal resonance condition may take place between antisymmetric mode and axisymmetric sloshing modes of wave numbers

$$K_1 = 1.8412 \quad \text{and} \quad K_2 = 3.8317 \quad \text{for } h/R = 0.1523 \tag{6.96}$$

where R is the tank radius. The corresponding wave height components, based on the above analysis, are

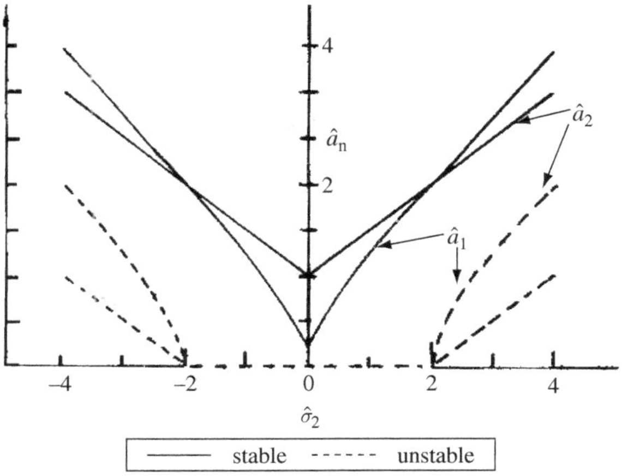

Figure 6.9 Amplitude–frequency response curves for the case of principal parametric resonance of the first mode, $\sigma_1 = 0$. (Nayfeh, 1987)

$$a_1 = \bar{a}_1 \cos\left(\frac{1}{2}\Omega t - \frac{1}{2}\gamma_1\right) + \cdots, \quad a_2 = \bar{a}_2 \cos(\Omega t + \gamma_1 - \gamma_2) + \cdots \tag{6.97a, b}$$

Figure 6.9 shows the dependence of the response amplitudes $\hat{a}_1 = \bar{a}_1\sqrt{\Lambda_1\Lambda_2}/\xi_1$ and $\hat{a}_2 = \bar{a}_2\Lambda_1/\xi_1$ on the external detuning parameter $\hat{\sigma}_2/\xi_1$, for exact internal detuning, that is, $\sigma_1 = 0$, and equal damping parameters, $\hat{\zeta}_n = \zeta_n/\xi_1 = 0.02$. Over the range of detuning parameters $-2.0 < \hat{\sigma}_2 < 2.0$, there exist unstable zero solutions and stable nontrivial solutions (periodic). Outside this range, we have a stable trivial solution, stable nontrivial solution (periodic) shown by solid curves, and unstable nontrivial solution (saddle point) shown by dash curves. Depending on the initial conditions, the stable trivial or nontrivial solutions can occur. Since $\hat{a}_n$ and $-\hat{a}_n$ are solutions, it follows that $\hat{\sigma}_2 = 2.0$ and -2.0 are sub-critical pitchfork bifurcation points. As $\hat{\sigma}_2$ increases past -2.0, the sink at the origin becomes a saddle point and the two saddles corresponding to $\hat{a}_n$ and $-\hat{a}_n$ shrink toward the sink at $\hat{\sigma}_2 = -2.0$. The same is applied if one decreases the frequency from above $\hat{\sigma}_2 = 2.0$. This implies that Hopf bifurcation does not take place in this case.

Figure 6.10 shows the dependence of the response amplitudes, $\hat{a}_1$ and $\hat{a}_2$, on the external detuning parameter, $\hat{\sigma}_2$, for nonperfect internal detuning parameter, $\hat{\sigma}_1 = \sigma_1/\xi_1 = 0.266$, damping parameters $\hat{\zeta}_n = \zeta_n/\xi_1 = 0.02$, and liquid depth ratio, $h/R = 0.13$. It is seen that the presence of internal detuning breaks the symmetry of the response curves. Careful inspection of these curves reveals that over the detuning range $0.14 < \hat{\sigma}_2 < 0.261$ (not shown in the figure), the fixed nontrivial curves lose their stability where the system Jacobian possesses a pair of complex conjugate eigenvalues, and thus Hopf bifurcation occurs. This means that the evolution equations have limit cycles with amplitudes proportional to $\sqrt{\hat{\sigma}_2 - 0.14}$ and $\sqrt{\hat{\sigma}_2 - 0.261}$.

Second mode excitation

This case deals with parametric excitation of the higher mode. The following resonance conditions, including detuning parameters, are considered

$$\Omega = 2\omega_2 + \varepsilon\sigma_2, \quad \omega_2 = 2\omega_1 + \varepsilon\sigma_1 \tag{6.98}$$

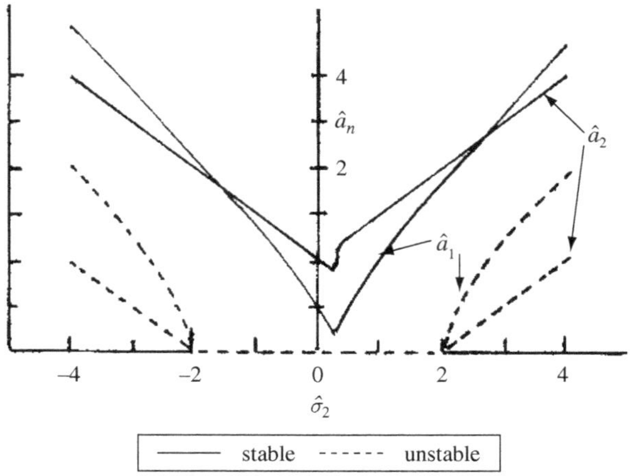

Figure 6.10 Amplitude–frequency response curves for the case of principal parametric resonance of the first mode, $\hat{\sigma}_1 = 0.266$. (Nayfeh, 1987)

Following the same analysis as in the previous case, the following solvability conditions are obtained

$$2i(A_1' + \zeta_1 A_1) - 4\Lambda_1 A_2 \bar{A}_1 e^{i\sigma_1} = 0 \tag{6.99a}$$

$$2i(A_2' + \zeta_2 A_2) - 4\Lambda_2 A_1^2 e^{-i\sigma_1 T_1} + 2\xi_2 \bar{A}_2 e^{i\sigma_2 T_1} = 0 \tag{6.99b}$$

Introducing the polar coordinate transformation (6.86) the following modulation equations are obtained

$$a_1' = -\zeta_1 a_1 + \Lambda_2 a_1 a_2 \sin(\theta_2 - 2\theta_1 + \sigma_1 T_1) \tag{6.100a}$$

$$a_2' = -\zeta_2 a_2 - \Lambda_2 a_1^2 \sin(\theta_2 - 2\theta_1 + \sigma_1 T_1) - \xi_2 a_2 \sin(\sigma_2 T_1 - 2\theta_2) \tag{6.100b}$$

$$a_1\theta_1' = -\Lambda_1 a_1 a_2 \cos(\theta_2 - 2\theta_1 + \sigma_1 T_1) \tag{6.100c}$$

$$a_2\theta_2' = -\Lambda_2 a_1^2 \cos(\theta_2 - 2\theta_1 + \sigma_1 T_1) - \xi_2 a_2 \cos(\sigma_2 T_1 - 2\theta_2) \tag{6.100d}$$

Let $\gamma_1 = (\theta_2 - 2\theta_1 + \sigma_1 T_1)$ and $\gamma_2 = (\sigma_2 T_1 - 2\theta_2)$. One can then write

$$\gamma_1' = (\theta_2' - 2\theta_1' + \sigma_1), \quad \text{and} \quad \gamma_2' = (\sigma_2 - 2\theta_2') \tag{6.101a, b}$$

The steady state solution of equations (6.100) is obtained by setting the left-hand sides to zero. The steady state is obtained by setting $\gamma_1' = \gamma_2' = 0$, and this implies

$$\theta_2' = \frac{1}{2}\sigma_2 \quad \text{and} \quad \theta_2' = \frac{1}{2}\left(\frac{1}{2}\sigma_2 - \sigma_1\right) \tag{6.102a, b}$$

In this case, the steady state solution can be estimated by solving the following set of algebraic equations

$$\zeta_1 a_1 - \Lambda_2 a_1 a_2 \sin\gamma_1 = 0 \tag{6.103a}$$

$$\zeta_2 a_2 + \Lambda_2 a_1^2 \sin\gamma_1 + \xi_2 a_2 \sin\gamma_2 = 0 \tag{6.103b}$$

$$\frac{1}{2}\left(\frac{1}{2}\sigma_2 + \sigma_1\right)a_1 + \Lambda_1 a_1 a_2 \cos\gamma_1 = 0 \tag{6.103c}$$

$$\frac{1}{2}\sigma_2 a_2 + \Lambda_2 a_1^2 \cos\gamma_1 - \xi_2 a_2 \cos\gamma_2 = 0 \tag{6.103d}$$

These equations have two different solutions:

(3) The trivial solution

$$a_1 = a_2 = 0 \tag{6.104}$$

(4) The nontrivial solution given by the following values

$$a_1^2 = -S_3 \pm \sqrt{(\xi_2 a_2/\Lambda_2)^2 - S_4^2} \tag{6.105a}$$

$$a_2 = \frac{1}{|\Lambda_2|}\sqrt{\left[\zeta_1^2 + \frac{1}{4}\left(\frac{1}{2}\sigma_2 + \sigma_1\right)^2\right]} \tag{6.105b}$$

where

$$S_3 = \frac{\left[\zeta_1\zeta_2 - \frac{1}{4}\sigma_2\left(\frac{1}{2}\sigma_2 + \sigma_1\right)\right]}{\Lambda_1\Lambda_2}, \quad S_2 = \frac{\left[\sigma_2\zeta_1 + \zeta_2\left(\frac{1}{2}\sigma_2 + \sigma_1\right)\right]}{2\Lambda_1\Lambda_2} \tag{6.106a,b}$$

If $S_3 > 0$, equation (6.105a) has no real roots when $\xi_2 < \sqrt{\zeta_2^2 + (\sigma_2/2)^2}$. It has one real root if $\xi_2 > \sqrt{\zeta_2^2 + (\sigma_2/2)^2}$. On the other hand, if $S_3 < 0$, equation (6.105a) has no real root when $\xi_2 < \sigma_2\zeta_1 + \zeta_2\left(\frac{1}{2}\sigma_2 + \sigma_1\right)\Big/\sqrt{4\zeta_1^2 + \left(\frac{1}{2}\sigma_2 + \sigma_1\right)^2}$, one root when $\xi_2 > \sqrt{\zeta_2^2 + (\sigma_2/2)^2}$, and two real roots when

$$\left\{\sigma_2\zeta_1 + \zeta_2\left(\frac{1}{2}\sigma_2 + \sigma_1\right)\Big/\sqrt{4\zeta_1^2 + \left(\frac{1}{2}\sigma_2 + \sigma_1\right)^2}\right\} < \xi_2 < \sqrt{\zeta_2^2 + (\sigma_2/2)^2} \tag{6.107}$$

The response behavior experiences a saturation phenomenon as revealed from equation (6.105b), where the response amplitude is independent of the excitation amplitude. This situation does not exist when the first mode is excited. Figure 6.11 shows the dependence of the response amplitudes $\hat{a}_1 = \bar{a}_1\sqrt{\Lambda_1\Lambda_2}$ and $\hat{a}_2 = \bar{a}_2|\Lambda_1|$, on the excitation amplitude, ξ_2, for $S_3 < 0$, $\varepsilon = 0.05$, $\hat{\zeta}_n = \zeta_n/\xi_2 = 0.02$, $\hat{\sigma}_1 = \sigma_1/\xi_2 = 0.02$, $\hat{\sigma}_2 = \sigma_2/\xi_2 = -0.2$. The solid curves (non-zero) refer to fixed stable solutions and are bounded by the excitation level defined by the range (6.107). The response exhibits saturation and jump phenomena and a Hopf bifurcation. Unstable solutions are indicated by dash curves and include trivial and nontrivial solutions.

Figure 6.12(a) shows amplitude–frequency response curves for $\xi_2 = 1.0$, $\hat{\zeta}_n = 0.02$ and zero internal tuning, $\hat{\sigma}_1 = \sigma_1/\xi_2 = 0$. Again, the solid parts of the response curves indicate stable fixed points, while the dash parts are unstable. Hopf bifurcation indicated on the left side is super-critical while that on the right is sub-critical. The symmetry shown in this figure is broken if the

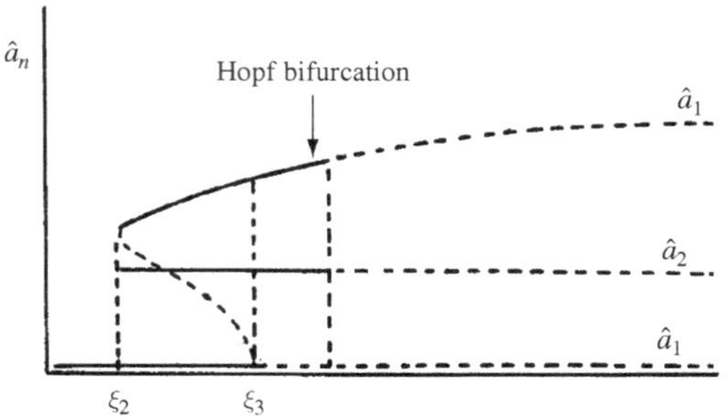

Figure 6.11 Dependence of response amplitude on excitation amplitude showing stable fixed-point (solid curves) and unstable solution (dash curves). (Nayfeh, 1987)

internal tuning differs from zero as shown in Figure 6.12(b) and (c). Over the frequency range bounded between the two points of Hopf bifurcation, the modulation equations possess limit-cycle solutions and they correspond to aperiodic motions of the liquid surface. Nayfeh and Nayfeh (1990) studied global bifurcation of the periodic solutions of the modulated equations and found that a cyclic-fold bifurcation occurs at a certain excitation frequency, causing a jump to a chaotic attractor. It extends over a large parameter region where it evolves continuously and smoothly while maintaining its structural characteristics.

6.6 Autoparametric interaction in rectangular tanks

6.6.1 Analytical modeling

In rectangular tanks, there is a critical fluid depth that separates two nonlinear regimes of the fluid free surface referred to as soft and hard spring characteristics. This depth was originally predicted by Tadjbakhsh and Keller (1960) and verified experimentally by Fultz (1962) and Hayama, *et al.* (1983) for waves produced by excitations in a direction parallel to the free surface. These observations motivated Gu (1986), Gu, *et al.* (1988), Viring, *et al.* (1988), Umeki and Kambe (1989), and Umeki (1991) to examine the role of the liquid critical depth in rectangular tanks subjected to vertical sinusoidal excitation, $z(t) = -Z_0 \cos \Omega t$. These studies included the influence of damping and surface tension. This section presents a general analytical formulation of the parametric excitation of liquid free surface in a rectangular tank. It allows the study of a single mode excitation or mixed-mode excitation where the interaction is governed by an internal resonance relationship. The nondimensional fluid field equations for this case are:

$$\nabla^2 \phi = 0 \tag{6.108a}$$

$$\frac{\partial \phi}{\partial x} = 0 \quad \text{at } x = 0,\ \chi \tag{6.108b}$$

$$\frac{\partial \phi}{\partial y} = 0 \quad \text{at } y = 0,\ 1 \tag{6.108c}$$

$$\frac{\partial \phi}{\partial z} = 0 \quad \text{at } z = -\xi \tag{6.108d}$$

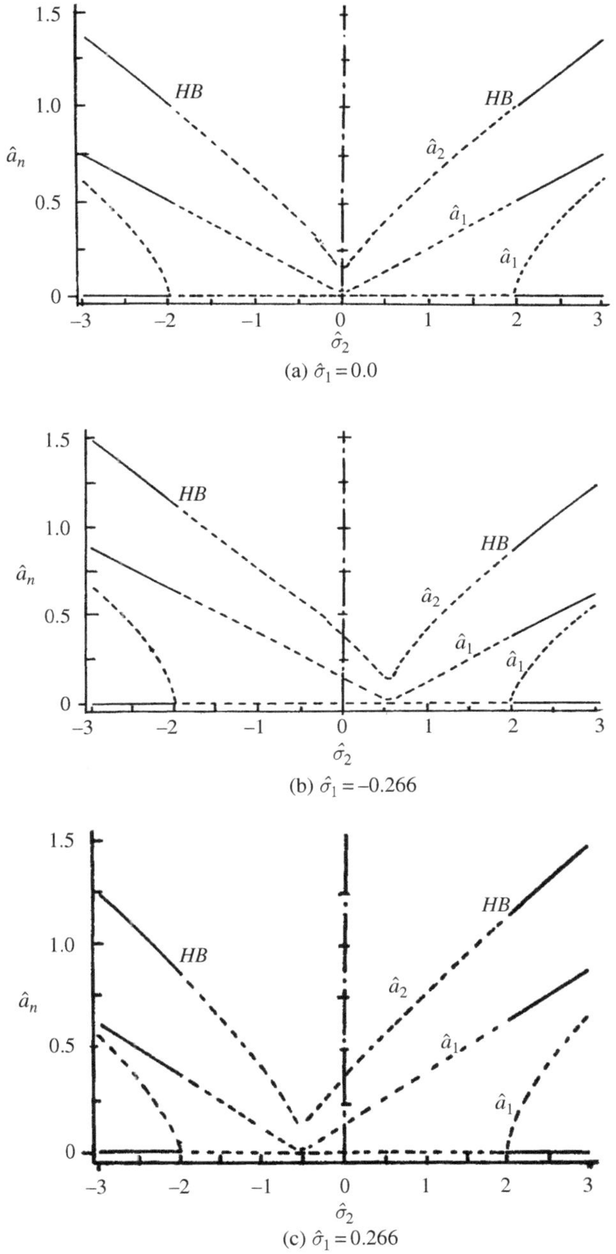

Figure 6.12 Amplitude–frequency response curves for different internal detuning parameters. (Nayfeh, 1987)

At $z = \varepsilon\eta(x, y, \tau)$, the dynamic boundary condition of the free surface is

$$\frac{\partial\phi}{\partial\tau} + \frac{1}{2}\varepsilon(\nabla\phi)^2 + (\Gamma + \varepsilon Z\cos\tau)\eta = \gamma\left\{\left(\frac{\partial^2\eta}{\partial x^2} + \frac{\partial^2\eta}{\partial y^2}\right)\right.$$
$$-\frac{\varepsilon^2}{2}\left(\frac{\partial^2\eta}{\partial x^2}\left(\frac{\partial\eta}{\partial y}\right)^2 + \frac{\partial^2\eta}{\partial y^2}\left(\frac{\partial\eta}{\partial x}\right)^2 + 3\frac{\partial^2\eta}{\partial x^2}\left(\frac{\partial\eta}{\partial x}\right)^2\right.$$
$$\left.\left. + 3\frac{\partial^2\eta}{\partial y^2}\left(\frac{\partial\eta}{\partial y}\right)^2 + 4\frac{\partial^2\eta}{\partial x\partial y}\frac{\partial\eta}{\partial x}\frac{\partial\eta}{\partial y}\right)\right\} + O(\varepsilon^2) \tag{6.108e}$$

and the kinematic boundary condition is

$$\frac{\partial\phi}{\partial z} = \frac{\partial\eta}{\partial\tau} + \varepsilon\left\{\frac{\partial\phi}{\partial x}\frac{\partial\eta}{\partial x} + \frac{\partial\phi}{\partial y}\frac{\partial\eta}{\partial y}\right\} \tag{6.108f}$$

where

$$\varepsilon Z = \frac{Z_0}{L},\ x = \frac{x^*}{L},\ y = \frac{y^*}{L},\ z = \frac{z^*}{L}\tau = \Omega t,\ \varepsilon\phi = \frac{\Phi}{\Omega L^2},$$
$$\varepsilon\eta = \frac{\eta^*}{L},\ \Gamma = \frac{g}{\Omega^2 L},\ \gamma = \frac{\sigma}{\rho\Omega^2 L^3},\ \chi = \frac{\ell}{L},\ \xi = \frac{h}{L} \tag{6.109}$$

Also ρ is the liquid density, σ is the surface tension, ℓ and L are the width and breadth of the tank, respectively, x^*, y^*, and z^* are the dimensional coordinate frame of axes attached to the tank at the free surface, η^* is the dimensional wave height of the free surface, and ε is a small parameter.

The condition of zero vertical velocity at the bottom of the tank, at $z = -\xi$, suggests that, see Lamb (1945, Section 257), the potential function may be written in the form

$$\phi(x, y, z) = S_{ij}(x, y)\frac{\cosh(K_{ij}(\xi + z))}{\cosh(K_{ij}\xi)} \tag{6.110}$$

In this case, Laplace's equation (6.108a) takes the form

$$\left[\frac{\partial^2}{\partial x^2} + \frac{\partial^2}{\partial y^2} + K_{ij}^2\right] S_{ij}(x, y) = 0 \tag{6.111}$$

$$S_{ij}(x, y) = \cos(i\pi x/\ell)\cos(j\pi y),\ K_{ij} = \pi\sqrt{\frac{i^2L^2}{\ell^2} + j^2} \tag{6.112}$$

The inner product of any two elements of S_{ij} is given by the following expression

$$\langle S_{ij}, S_{rs}\rangle \stackrel{\text{def}}{=} \frac{4}{\chi}2^{-(\delta_{or} + \delta_{os})}\int_0^{\chi}\int_0^1 S_{ij}S_{rs}\,dx\,dy, \quad \langle S_{ij}, S_{rs}\rangle = \delta_{ir}\delta_{js} \tag{6.113}$$

where δ denotes the Kronecker delta.

The solution of the boundary value problem is obtained by assuming the following asymptotic expansions

$$\phi = \phi(x, y, z, \mathbf{a}, \boldsymbol{\theta}, \tau, \varepsilon) = \sum_{n=0}^{N} \varepsilon^n \phi^{(n)}(x, y, z, \mathbf{a}, \boldsymbol{\theta}, \tau) \tag{6.114a}$$

$$\eta = \eta(x, y, \mathbf{a}, \boldsymbol{\theta}, \tau, \varepsilon) = \sum_{n=0}^{N} \varepsilon^n \eta^{(n)}(x, y, \mathbf{a}, \boldsymbol{\theta}, \tau) \tag{6.114b}$$

where $\mathbf{a}$ and $\boldsymbol{\theta}$ are infinite dimensional vector functions of τ, each component of which represents the amplitude or phase of a linear mode of motion, respectively. For the nonlinear case, these components satisfy the following autonomous equations

$$\dot{a}_{ij} = \sum_{n=1}^{N} \varepsilon^n A_{ij}^{(n)}(\mathbf{a}, \boldsymbol{\theta}),\ \dot{\theta}_{ij} = \sum_{n=1}^{N} \varepsilon^n B_{ij}^{(n)}(\mathbf{a}, \boldsymbol{\theta}), \quad i, j = 0,\ 1,\ 2,\ \ldots, \quad i = j \neq 0 \tag{6.115a,b}$$

Substituting (6.114a, b) into (6.108e, f), taking into account equations (6.115a, b), and equating equal powers of ε, gives:

Terms of $O(\varepsilon^0)$ give

$$\frac{\partial \phi^{(0)}}{\partial z} - \frac{\partial \eta^{(0)}}{\partial \tau} = 0 \tag{6.116a}$$

$$\frac{\partial \phi^{(0)}}{\partial \tau} + \Gamma \eta^{(0)} - \gamma \left(\frac{\partial^2 \eta^{(0)}}{\partial x^2} + \frac{\partial^2 \eta^{(0)}}{\partial y^2} \right) = 0 \tag{6.116b}$$

Terms of $O(\varepsilon)$ give

$$\frac{\partial \phi^{(1)}}{\partial z} - \frac{\partial \eta^{(1)}}{\partial \tau} = \frac{\partial \eta^{(0)}}{\partial a_{rs}} A_{rs}^{(1)} + \frac{\partial \eta^{(0)}}{\partial \theta_{rs}} B_{rs}^{(1)} + \frac{\partial \phi^{(0)}}{\partial x} \frac{\partial \eta^{(0)}}{\partial x} + \frac{\partial \phi^{(0)}}{\partial y} \frac{\partial \eta^{(0)}}{\partial y} - \frac{\partial^2 \phi^{(0)}}{\partial z^2} \eta^0 \tag{6.117a}$$

$$\frac{\partial \phi^{(1)}}{\partial \tau} + \Gamma \eta^{(1)} - \gamma \left(\frac{\partial^2 \eta^{(1)}}{\partial x^2} + \frac{\partial^2 \eta^{(1)}}{\partial y^2} \right) = -Z \cos(\tau) \eta^{(0)} - \left[\frac{\partial \phi^{(0)}}{\partial a_{rs}} A_{rs}^{(1)} + \frac{\partial \phi^{(0)}}{\partial \theta_{rs}} B_{rs}^{(1)} \right] - \frac{1}{2} \left[\left(\frac{\partial \phi^{(0)}}{\partial x} \right)^2 + \left(\frac{\partial \phi^{(0)}}{\partial y} \right)^2 + \left(\frac{\partial \phi^{(0)}}{\partial z} \right)^2 \right] - \frac{\partial^2 \phi}{\partial z \partial \tau} \eta^{(0)} \tag{6.117b}$$

Terms of $O(\varepsilon^2)$ give

$$\frac{\partial\phi^{(2)}}{\partial z}-\frac{\partial\eta^{(2)}}{\partial\tau}=\left(\frac{\partial\eta^{(0)}}{\partial a_{rs}}A_{rs}^{(2)}+\frac{\partial\eta^{(0)}}{\partial\theta_{rs}}B_{rs}^{(2)}\right)+\left(\frac{\partial\eta^{(1)}}{\partial a_{rs}}A_{rs}^{(1)}+\frac{\partial\eta^{(1)}}{\partial\theta_{rs}}B_{rs}^{(1)}\right)$$
$$+\frac{\partial\phi^{(0)}}{\partial x}\frac{\partial\eta^{(1)}}{\partial x}+\frac{\partial\phi^{(1)}}{\partial x}\frac{\partial\eta^{(0)}}{\partial x}+\frac{\partial\phi^{(0)}}{\partial y}\frac{\partial\eta^{(1)}}{\partial y}+\frac{\partial\phi^{(1)}}{\partial y}\frac{\partial\eta^{(0)}}{\partial y}$$
$$+\frac{\partial^2\phi^{(0)}}{\partial z\partial x}\eta^{(0)}\frac{\partial\eta^{(0)}}{\partial x}+\frac{\partial^2\phi^{(0)}}{\partial z\partial y}\eta^{(0)}\frac{\partial\eta^{(0)}}{\partial y}-\frac{\partial^2\phi^{(1)}}{\partial z^2}\eta^{(0)}$$
$$-2\frac{\partial^2\phi^{(0)}}{\partial z^2}\eta^{(1)}-\frac{1}{2}\frac{\partial^3\phi^{(0)}}{\partial z^3}\eta^{(0)}\eta^{(0)} \tag{6.118a}$$

$$\frac{\partial\phi^{(2)}}{\partial\tau}+\Gamma\eta^{(2)}-\gamma\left(\frac{\partial^2\eta^{(2)}}{\partial x^2}+\frac{\partial^2\eta^{(2)}}{\partial y^2}\right)=-Z\cos(\tau)\eta^{(1)}-\left[\frac{\partial\phi^{(0)}}{\partial a_{rs}}A_{rs}^{(2)}+\frac{\partial\phi^{(0)}}{\partial\theta_{rs}}B_{rs}^{(2)}\right]$$
$$-\left[\frac{\partial\phi^{(1)}}{\partial a_{rs}}A_{rs}^{(1)}+\frac{\partial\phi^{(1)}}{\partial\theta_{rs}}B_{rs}^{(1)}\right]-\left[\frac{\partial\phi^{(0)}}{\partial x}\frac{\partial\phi^{(1)}}{\partial x}+\frac{\partial\phi^{(0)}}{\partial y}\frac{\partial\phi^{(1)}}{\partial y}+\frac{\partial\phi^{(0)}}{\partial z}\frac{\partial\phi^{(1)}}{\partial z}\right.$$
$$\left.+\frac{\partial\phi^{(0)}}{\partial x}\frac{\partial^2\phi^{(1)}}{\partial z\partial x}\eta^{(0)}+\frac{\partial\phi^{(0)}}{\partial y}\frac{\partial^2\phi^{(1)}}{\partial z\partial y}\eta^{(0)}+\frac{\partial\phi^{(0)}}{\partial z}\frac{\partial^2\phi^{(1)}}{\partial z^2}\eta^{(0)}\right]$$
$$-\left[\frac{\partial^2\phi^{(1)}}{\partial\tau\partial z}\eta^{(0)}+2\frac{\partial^2\phi^{(1)}}{\partial\tau\partial z}\eta^{(1)}+\frac{1}{2}\frac{\partial^3\phi^{(1)}}{\partial^2 z\partial\tau}\eta^{(0)}\eta^{(1)}\right]$$
$$-\left[\eta^{(0)}\frac{\partial^2\phi^{(0)}}{\partial z\partial a_{rs}}A_{rs}^{(1)}+\eta^{(0)}\frac{\partial^2\phi^{(0)}}{\partial z\partial\theta_{rs}}B_{rs}^{(1)}\right]-\frac{\gamma}{2}\left[\frac{\partial^2\eta^{(0)}}{\partial x^2}\left(\frac{\partial\eta^{(0)}}{\partial y}\right)^2+\frac{\partial^2\eta^{(0)}}{\partial y^2}\left(\frac{\partial\eta^{(0)}}{\partial x}\right)^2\right.$$
$$\left.+3\frac{\partial^2\eta^{(0)}}{\partial x^2}\left(\frac{\partial\eta^{(0)}}{\partial x}\right)^2+3\frac{\partial^2\eta^{(0)}}{\partial y^2}\left(\frac{\partial\eta^{(0)}}{\partial y}\right)^2+4\frac{\partial\eta^{(0)}}{\partial x}\frac{\partial\eta^{(0)}}{\partial y}\frac{\partial^2\eta^{(0)}}{\partial x\partial y}\right] \tag{6.118b}$$

Let the potential function and wave-height be given by the following expansions

$$\phi=\sum_{n=0}^{N}\varepsilon^n\phi^{(n)}=\sum_{n=0}^{N}\sum_{ij=0}^{\infty}\varepsilon^n q_{ij}^{(n)}(\tau)S_{ij}(x,y)g_{ij}(z) \tag{6.119a}$$

$$\eta=\sum_{n=0}^{N}\varepsilon^n\eta^{(n)}=\sum_{n=0}^{N}\sum_{ij=0}^{\infty}\varepsilon^n p_{ij}^{(n)}(\tau)S_{ij}(x,y) \tag{6.119b}$$

where

$$g_{ij}(z)=\frac{\cosh(K_{ij}(\xi+z))}{\cosh(K_{ij}\xi)},\quad i,j=0,\ 1,\ 2,\ \ldots,\quad i=j\neq 0,\quad n=0,\ 1,\ 2,\ \ldots,\ N.$$

Substituting equations (6.119) into equations (6.116) through (6.118) and taking the projection using the orthogonality of the complete set of $S_{ij}(x,y)$ given by relation (6.113), gives:

Equations of zero-order

$$\frac{\partial p_{ij}^{(0)}}{\partial\tau}-\alpha_{ij}q_{ij}^{(0)}=0 \tag{6.120a}$$

$$\frac{\partial q_{ij}^{(0)}}{\partial \tau} + \frac{\nu_{ij}^2}{\alpha_{ij}} p_{ij}^{(0)} = 0 \tag{6.120b}$$

Equations of first-order

$$\frac{\partial p_{ij}^{(1)}}{\partial \tau} - \alpha_{ij} q_{ij}^{(1)} = -\left[\frac{\partial p_{ij}^{(0)}}{\partial a_{ij}} A_{ij}^{(1)} + \frac{\partial p_{ij}^{(0)}}{\partial \theta_{ij}} B_{ij}^{(1)}\right] - G_{uvrs}^{(1)ij} p_{uv}^{(0)} q_{rs}^{(0)} \tag{6.121a}$$

$$\begin{aligned}\frac{\partial q_{ij}^{(1)}}{\partial \tau} + \frac{\nu_{ij}^2}{\alpha_{ij}} p_{ij}^{(1)} = &- Z\cos(\tau) p_{ij}^{(0)} - \left[\frac{\partial q_{ij}^{(0)}}{\partial a_{ij}} A_{ij}^{(1)} + \frac{\partial q_{ij}^{(0)}}{\partial \theta_{ij}} B_{ij}^{(1)}\right] \\ &- \left[\frac{1}{2} G_{uvrs}^{(2)ij} q_{uv}^{(0)} q_{rs}^{(0)} + G_{uvrs}^{(3)ij} p_{uv}^{(0)} \frac{\partial q_{rs}^{(0)}}{\partial \tau}\right]\end{aligned} \tag{6.121b}$$

Equations of second-order

$$\begin{aligned}\frac{\partial p_{ij}^{(2)}}{\partial \tau} - \alpha_{ij} q_{ij}^{(2)} = &-\left[\frac{\partial p_{ij}^{(0)}}{\partial a_{ij}} A_{ij}^{(2)} + \frac{\partial p_{ij}^{(0)}}{\partial \theta_{ij}} B_{ij}^{(2)} + \frac{\partial p_{ij}^{(1)}}{\partial a_{ij}} A_{ij}^{(1)} + \frac{\partial p_{ij}^{(1)}}{\partial \theta_{ij}} B_{ij}^{(1)}\right] \\ &- G_{uvrs}^{(1)ij}\left(p_{uv}^{(1)} q_{rs}^{(0)} + p_{uv}^{(0)} q_{rs}^{(1)}\right) + H_{uvrsk\ell}^{(3)ij} p_{uv}^{(0)} q_{rs}^{(0)} q_{k\ell}^{(0)}\end{aligned} \tag{6.122a}$$

$$\begin{aligned}\frac{\partial q_{ij}^{(2)}}{\partial \tau} + \frac{\nu_{ij}^2}{\alpha_{ij}} p_{ij}^{(2)} = &- Z\cos(\tau) p_{ij}^{(1)} - \left[\frac{\partial q_{ij}^{(0)}}{\partial a_{ij}} A_{ij}^{(2)} + \frac{\partial q_{ij}^{(0)}}{\partial \theta_{ij}} B_{ij}^{(2)} + \frac{\partial q_{ij}^{(1)}}{\partial a_{ij}} A_{ij}^{(1)}\right] \\ &+ \frac{\partial q_{ij}^{(1)}}{\partial \theta_{ij}} B_{ij}^{(1)} - \left\{ G_{uvrs}^{(2)ij} q_{uv}^{(0)} q_{rs}^{(1)} + H_{uvrsk\ell}^{(2)ij} q_{uv}^{(0)} q_{rs}^{(0)} p_{k\ell}^{(0)} \right. \\ &+ G_{uvrs}^{(3)ij}\left(p_{uv}^{(0)} \frac{\partial q_{rs}^{(1)}}{\partial \tau} + p_{uv}^{(1)} \frac{\partial q_{rs}^{(0)}}{\partial \tau} + p_{uv}^{(0)} \frac{\partial q_{rs}^{(0)}}{\partial a_{rs}} A_{rs}^{(1)} + p_{uv}^{(0)} \frac{\partial q_{rs}^{(0)}}{\partial \theta_{rs}} B_{rs}^{(1)}\right) \\ &\left. + H_{uvrsk\ell}^{(3)ij} p_{uv}^{(0)} p_{rs}^{(0)} \frac{\partial q_{k\ell}^{(0)}}{\partial \tau} + H_{uvrsk\ell}^{(4)ij} p_{uv}^{(0)} p_{rs}^{(0)} p_{k\ell}^{(0)} \right\}\end{aligned} \tag{6.122b}$$

where repeated subscripts, except ij, denote summation. The coupling coefficients G and H are not zero when the equations involve spatial coupling with (u, v) modes. These coefficients are given by the following expressions:

$$\begin{aligned}G_{uvrs}^{(1)ij} &= \left\langle \frac{\partial S_{uv}}{\partial x}\frac{\partial S_{rs}}{\partial x}, S_{ij}\right\rangle + \left\langle \frac{\partial S_{uv}}{\partial y}\frac{\partial S_{rs}}{\partial y}, S_{ij}\right\rangle - K_{rs}^2\langle S_{uv}S_{rs}, S_{ij}\rangle \\ G_{uvrs}^{(2)ij} &= \left\langle \frac{\partial S_{uv}}{\partial x}\frac{\partial S_{rs}}{\partial x}, S_{ij}\right\rangle + \left\langle \frac{\partial S_{uv}}{\partial y}\frac{\partial S_{rs}}{\partial y}, S_{ij}\right\rangle - \alpha_{uv}\alpha_{rs}\langle S_{uv}S_{rs}, S_{ij}\rangle \\ G_{uvrs}^{(3)ij} &= \alpha_{rs}\langle S_{uv}S_{rs}, S_{ij}\rangle\end{aligned} \tag{6.123a}$$

$$H^{(1)ij}_{uvrsk\ell} = \frac{1}{2}K^2_{k\ell}\alpha_{k\ell}\langle S_{uv}S_{rs}S_{k\ell}, S_{ij}\rangle - \alpha_{k\ell}\left\{\left\langle S_{uv}\frac{\partial S_{rs}}{\partial x}\frac{\partial S_{k\ell}}{\partial x}, S_{ij}\right\rangle + \left\langle S_{uv}\frac{\partial S_{rs}}{\partial y}\frac{\partial S_{k\ell}}{\partial y}, S_{ij}\right\rangle\right\}$$

$$H^{(2)ij}_{uvrsk\ell} = K^2_{rs}\alpha_{uv}\langle S_{uv}S_{rs}S_{k\ell}, S_{ij}\rangle + \alpha_{rs}\left\{\left\langle \frac{\partial S_{uv}}{\partial x}\frac{\partial S_{rs}}{\partial x}S_{k\ell}, S_{ij}\right\rangle + \left\langle \frac{\partial S_{uv}}{\partial y}\frac{\partial S_{rs}}{\partial y}S_{k\ell}, S_{ij}\right\rangle\right\}$$

$$H^{(3)ij}_{uvrsk\ell} = \frac{1}{2}K^2_{k\ell}\langle S_{uv}S_{rs}S_{k\ell}, S_{ij}\rangle$$

$$\begin{aligned} H^{(4)ij}_{uvrsk\ell} = \frac{\gamma}{2}\Bigg[& \left\langle \frac{\partial^2 S_{uv}}{\partial x^2}\frac{\partial S_{rs}}{\partial y}\frac{\partial S_{k\ell}}{\partial y}, S_{ij}\right\rangle + \left\langle \frac{\partial^2 S_{uv}}{\partial y^2}\frac{\partial S_{rs}}{\partial x}\frac{\partial S_{k\ell}}{\partial x}, S_{ij}\right\rangle \\ & + 3\left\langle \frac{\partial^2 S_{uv}}{\partial x^2}\frac{\partial S_{rs}}{\partial x}\frac{\partial S_{k\ell}}{\partial x}, S_{ij}\right\rangle + 3\left\langle \frac{\partial^2 S_{uv}}{\partial y^2}\frac{\partial S_{rs}}{\partial y}\frac{\partial S_{k\ell}}{\partial y}, S_{ij}\right\rangle \\ & + 4\left\langle \frac{\partial S_{uv}}{\partial x}\frac{\partial S_{rs}}{\partial y}\frac{\partial^2 S_{k\ell}}{\partial x\partial y}, S_{ij}\right\rangle\Bigg] \end{aligned} \tag{6.123b}$$

where $\alpha_{uv} = K_{uv}\tanh(K_{uv}h/L)$ and $\nu_{ij} = \sqrt{(\Gamma + \gamma K^2_{ij})\alpha_{ij}}$.

This formulation enables one to study different types of surface motion such as single mode excitation and autoparametric resonance. These cases are considered in the following sections

6.6.2 Single mode excitation

In the absence of modal coupling, let (i, j) be the mode under consideration such that $\nu_{ij} \approx 1/2$, and $(i,j) \neq (u,v)$. Thus, the desired necessary conditions for any other mode not to interact or to be in resonance are that ν_{2i2j}, ν_{2i0}, and ν_{02j} must not be close to $2\nu_{ij}$. The deviation from exact parametric resonance is introduced by the detuning parameter λ defined by the relationship

$$\nu^2_{ij} = \left(\frac{1}{2}\right)^2 - \varepsilon\lambda \tag{6.124}$$

For $\varepsilon = 0$, equations (6.120) have the solution

$$p^{(0)}_{ij}(\tau) = a_{ij}\cos\left(\frac{\tau}{2} + \theta_{ij}\right) \tag{6.125a}$$

$$q^{(0)}_{ij}(\tau) = -\frac{a_{ij}}{2\alpha_{ij}}\sin\left(\frac{\tau}{2} + \theta_{ij}\right) \tag{6.125b}$$

where a_{ij} and θ_{ij} are two arbitrary constants representing the amplitude and phase, respectively. Note that modes other than ij mode have zero solution.

In order to evaluate the first-order steady-state solution of equations (6.121) we must set all coupled terms to zero, and equations (6.121) take the form after substituting the zero-order solution on the right-hand sides

$$\frac{\partial p^{(1)}_{ij}}{\partial \tau} = \alpha_{ij}q^{(1)}_{ij} - \left[\cos\left(\frac{\tau}{2} + \theta_{ij}\right)A^{(1)}_{ij} - a_{ij}\sin\left(\frac{\tau}{2} + \theta_{ij}\right)B^{(1)}_{ij}\right] \tag{6.126a}$$

$$\begin{aligned} \frac{\partial q^{(1)}_{ij}}{\partial \tau} = -\frac{1}{4\alpha_{ij}}p^{(1)}_{ij} - \Bigg[& (-(\lambda/\alpha_{ij}) + Z\cos\tau)a_{ij}\cos\left(\frac{\tau}{2} + \theta_{ij}\right) \\ & - \frac{1}{2a_{ij}}\left[\sin(\tau/2 + \theta_{ij})A^{(1)}_{ij} + a_{ij}\cos(\tau/2 + \theta_{ij})B^{(1)}_{ij}\right]\Bigg] \end{aligned} \tag{6.126b}$$

The method of averaging is used to determine periodic solutions of equations (6.126). The method is based on developing solvability conditions of equations (6.121) that define the values of the functions $A_{ij}^{(1)}$ and $B_{ij}^{(1)}$. The method is outlined in Hale (1969) and requires equations (6.126) to be written in the form

$$\begin{Bmatrix} \dot{x}_1 \\ \dot{x}_2 \end{Bmatrix} = [\boldsymbol{A}]\begin{Bmatrix} x_1 \\ x_2 \end{Bmatrix} + \begin{Bmatrix} f_1(\tau, \mathbf{x}) \\ f_2(\tau, \mathbf{x}) \end{Bmatrix} \tag{6.127}$$

where $\begin{Bmatrix} x_1 \\ x_2 \end{Bmatrix} = \begin{Bmatrix} p_{ij} \\ q_{ij} \end{Bmatrix}$, $[\boldsymbol{A}] = \begin{bmatrix} 0 & \alpha_{ij} \\ -1/4\alpha_{ij} & 0 \end{bmatrix}$, and the vector $\{f\}$ represents periodic functions whose elements are the bracketed expressions on the right-hand sides of equations (6.126). Equations (6.126) have periodic solutions of period T if the following condition is satisfied

$$\int_0^T \mathbf{y}(\tau)\mathbf{f}(\tau)\, d\tau = 0 \tag{6.128}$$

for all solutions of the adjoint equation

$$\mathbf{y} = \{\dot{y}_1 \quad \dot{y}_2\} = -\{y_1 \quad y_2\}[\boldsymbol{A}] \tag{6.129}$$

Equation (6.129) has the following solution

$$y_1 = a\cos\frac{\tau}{2} + b\sin\frac{\tau}{2} \quad \text{and} \quad y_2 = 2\alpha_{ij}\left(-a\sin\frac{\tau}{2} + b\cos\frac{\tau}{2}\right) \tag{6.130a,b}$$

This solution is periodic with period 4π. The solvability condition (6.128) is now applied and will yield the following values of the functions $A_{ij}^{(1)}$ and $B_{ij}^{(1)}$:

$$A_{ij}^{(1)} = \frac{\alpha_{ij}}{2} Z a_{ij} \sin 2\theta_{ij}, \quad B_{ij}^{(1)} = -\lambda + \frac{\alpha_{ij}}{2} Z \cos 2\theta_{ij} \tag{6.131a,b}$$

Substituting equations (6.131) into equations (6.126), gives

$$p_{ij}^{(1)} = \frac{\alpha_{ij}}{4} Z - a_{ij} \cos(\frac{3}{2}\tau + \theta_{ij}) \tag{6.132a}$$

$$q_{ij}^{(1)} = \frac{3}{8} a_{ij} Z \sin\left(\frac{3}{2}\tau + \theta_{ij}\right) - \frac{\lambda}{\alpha_{ij}} a_{ij} \sin\left(\frac{1}{2}\tau + \theta_{ij}\right) - \frac{Z}{2} a_{ij} \sin\left(\frac{1}{2}\tau - \theta_{ij}\right) \tag{6.132b}$$

The process continues to obtain second-order solutions of equations (6.118) after substituting equations (6.125), (6.131) and (6.132) and applying the averaging method to obtain the solvability condition. This yields the following values for the functions $A_{ij}^{(2)}$ and $B_{ij}^{(2)}$:

$$A_{ij}^{(2)} = 0, \quad B_{ij}^{(2)} = M_{ij} a_{ij}^2 + \frac{3}{4}\alpha_{ij} Z - \lambda^2 \tag{6.133a,b}$$

where

$$M_{ij} = -\left(\frac{1}{2} S_{ij}^{(1)} + \alpha_{ij} C_{ij}^{(2)}\right) \tag{6.134a}$$

$$S_{ij}^{(1)} = -\frac{G_{uvij}^{(1)ij}}{64\alpha_{ij}^3\nu_{uv}^2(\nu_{uv}^2-1)}\left[\left(-2\alpha_{uv}\alpha_{ij}G_{ijij}^{(1)uv}+3\alpha_{uv}G_{ijij}^{(1)uv}+4\alpha_{ij}G_{ijij}^{(1)uv}\right)\nu_{uv}^2\right.$$
$$\left.+4\alpha_{uv}\alpha_{ij}G_{ijij}^{(1)uv}-2\alpha_{uv}G_{ijij}^{(2)uv}\right]+\frac{G_{uvij}^{(1)ij}}{32\alpha_{ij}^3\alpha_{uv}(\nu_{uv}^2-1)}\left[4\alpha_{ij}\nu_{uv}^2G_{ijij}^{(1)uv}\right.$$
$$\left.+\alpha_{uv}G_{ijij}^{(2)uv}+2\alpha_{ij}\alpha_{uv}G_{ijij}^{(3)uv}\right]-\frac{1}{8\alpha_{ij}}H_{ijij}^{(1)ij} \tag{6.134b}$$

$$C_{ij}^{(2)} = -\frac{1}{128\alpha_{ij}^3(\nu_{uv}^2-1)}\left[G_{ijij}^{(3)uv}\left(\alpha_{uv}\alpha_{ij}\left(6-\frac{4}{\nu_{uv}^2}\right)G_{uvij}^{(3)ij}-4\alpha_{ij}G_{ijuv}^{(2)ij}+8\alpha_{ij}^2G_{ijuv}^{(3)ij}\right)\right.$$
$$\times G_{ijij}^{(2)uv}\left(\alpha_{ij}\left(-1+\frac{2}{\nu_{uv}^2}\right)G_{uvij}^{(3)ij}-2G_{ijuv}^{(2)ij}+4\alpha_{ij}G_{ijuv}^{(3)ij}\right)$$
$$\left.+G_{ijij}^{(1)uv}\left(4\alpha_{ij}G_{uvij}^{(3)ij}-8\nu_{uv}^2\frac{\alpha_{ij}}{\alpha_{uv}}G_{ijuv}^{(3)ij}\right)\right]$$
$$-\frac{1}{16\alpha_{ij}^2}H_{ijijij}^{(2)ij}-\frac{3}{4}H_{ijijij}^{(2)ij}+\frac{3}{16\alpha_{ij}}H_{ijijij}^{(3)ij} \tag{6.134c}$$

In order to construct the solutions of equations (6.115) up to second-order we substitute for the values of $A_{ij}^{1,2}$ and $B_{ij}^{1,2}$ as obtained by relations (6.131) and (6.133). Furthermore, a linear viscous damping, $\varepsilon\zeta_{ij}$, will be introduced, to account for energy dissipation. The new state equations (6.115) take the form

$$\dot{a}_{ij} = \varepsilon\left(-\zeta_{ij}+\frac{\alpha_{ij}}{2}Z\sin 2\theta_{ij}\right)a_{ij} \tag{6.135a}$$

$$\dot{\theta}_{ij} = \varepsilon\left(-\lambda+\frac{\alpha_{ij}}{2}Z\cos 2\theta_{ij}\right)+\varepsilon^2\left(M_{ij}a_{ij}^2+\frac{3}{4}\alpha_{ij}Z-\lambda^2\right) \tag{6.135b}$$

The steady-state solution of these equations is obtained by setting the left hand sides to zero. Squaring the resulting algebraic equations and adding gives

$$\varepsilon^2M^2a^4-2\varepsilon M\left[\varepsilon\left(\lambda^2-\frac{3}{4}\alpha Z\right)+\lambda\right]a^2+\left[\lambda+\varepsilon\left(\lambda^2-\frac{3}{4}\alpha Z\right)\right]^2$$
$$+\lambda^2-\frac{\alpha^2}{4}Z^2=0 \tag{6.136}$$

where the subscript ij has been dropped. The roots of this equation are

$$a_{1,2}^2 = \frac{\varepsilon(\lambda^2-3\alpha Z/4)+\lambda\pm\sqrt{\alpha^2Z^2/4-\zeta^2}}{\varepsilon M} \tag{6.137}$$

The corresponding phase angles are

$$\theta = \begin{cases} \mp\frac{1}{2}\tan^{-1}\left(\zeta/\sqrt{\alpha^2Z^2/4-\zeta^2}\right) \\ \pi\mp\frac{1}{2}\tan^{-1}\left(\zeta/\sqrt{\alpha^2Z^2/4-\zeta^2}\right) \end{cases} \tag{6.138}$$

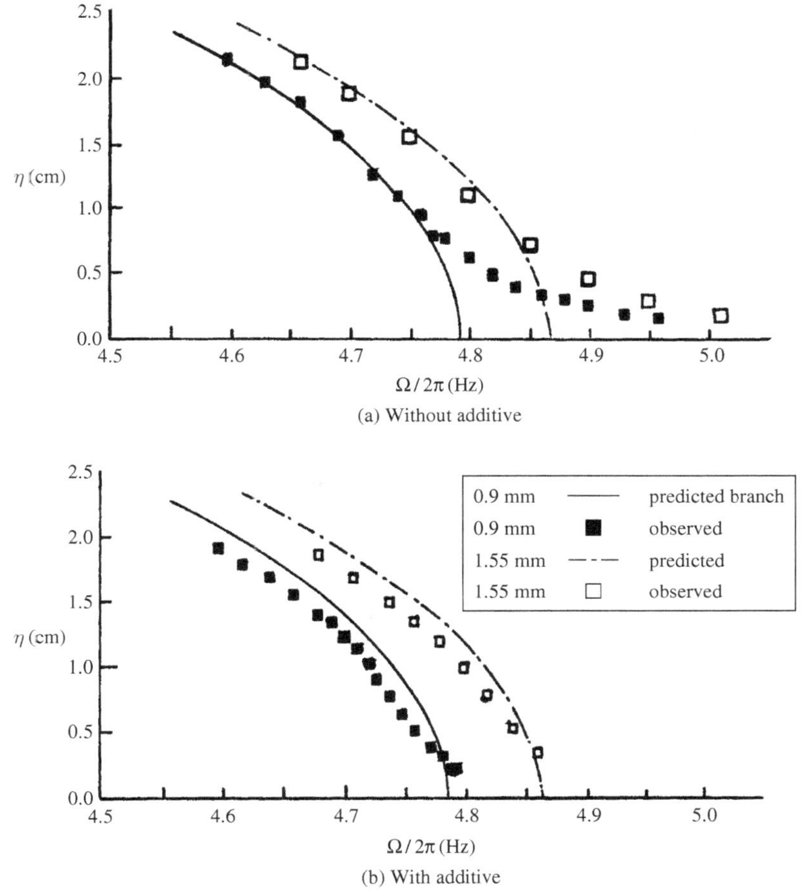

Figure 6.13 Amplitude–frequency response in a rectangular container under excitation amplitude. (Viring, *et al.*, 1988)

This analytical result was compared with experimental measurements conducted by Viring, *et al.* (1988). Figure 6.13(a) shows analytical and experimental amplitude–frequency response results of the sloshing mode (1, 1) for two excitation amplitudes as a function of excitation frequency. The results shown were taken for a rectangular container of width 17.78 cm, breadth 22.86 cm, and liquid depth of 11 cm. Only the stable branches of the analytical solution are shown by solid and dash-dot curves. It is interesting to observe that both experimental and predicted results have good agreement for higher values of wave amplitudes. At lower wave amplitudes, the experimental results reveal a "tailing" which is repeated for both increasing and decreasing excitation frequency at constant amplitude. The observed tailing effect is attributed to the influence of surface tension. Figure 6.13(b) shows a similar set of results after applying a surfactant (Kodak Photo-Flo 200 solution). It is seen that as the surface tension is reduced the tailing effect is eliminated. The inclusion of surface tension in the analysis of Gu, *et al.* (1988) revealed that the natural frequency of mode (1, 1) is reduced when the surface tension is reduced.

The dependence of the liquid wave amplitude on the excitation amplitude for constant excitation frequency is shown in Figure 6.14(a) for four different excitation frequencies after

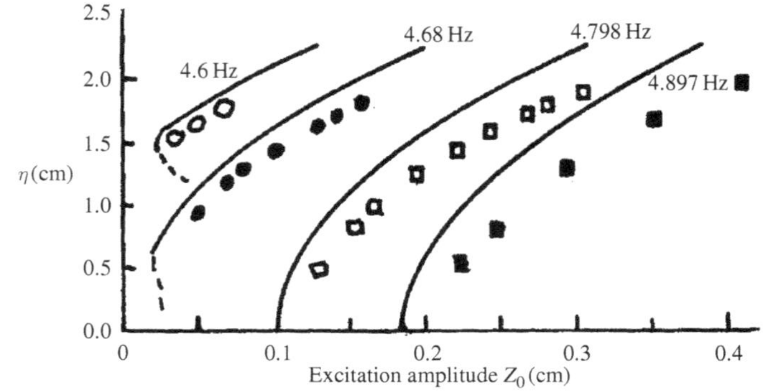

(a) response amplitude versus excitation amplitude for different excitation frequencies for fluid depth 11.43 cm for mode(1,1)

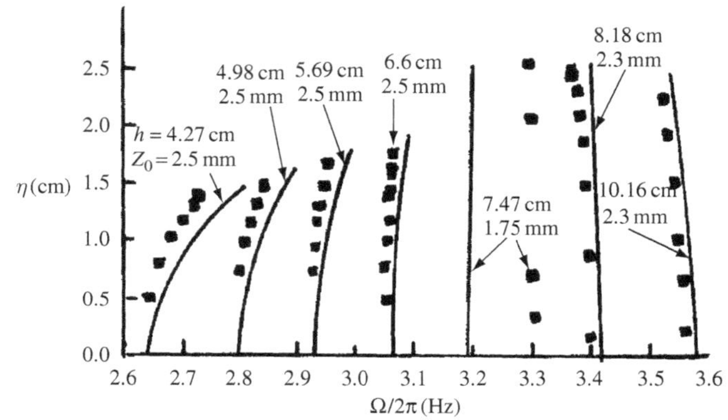

(b) amplitude–frequency response for different values of fluid depth and excitation amplitude for mode (0,1)

Figure 6.14 Dependence of fluid amplitude on excitation amplitude and frequency in a rectangular tank. (Viring, *et al.*, 1988)

applying a surfactant. The analytical results are seen to be higher than those measured experimentally and convergence of both results appears when the excitation frequency is close to twice the natural frequency of the first mode (4.692 Hz). For excitation frequencies less than twice the natural frequency, the analytical results reveal the occurrence of a saddle node bifurcation at critical excitation amplitude. The unstable branch is shown by a dash curve. The saddle node signifies the occurrence of a jump in the response amplitude and this point is dictated by the system damping ratio and excitation frequency.

The amplitude–frequency response is shown in Figure 6.14(b) for different values of liquid depth. It is known that below the critical depth, the liquid free surface behaves like a "hard" oscillator, that is, the amplitude increases with the excitation frequency. Above the critical depth, the "softening" characteristic takes place. For this case the predicted critical depth is 7.7 cm, while the experimental measurements suggest that it is between 6.6 cm and 7.47 cm.

6.6.3 One-to-one internal resonance

Gu and Sethna (1987) studied two-mode interaction in the neighborhood of 1:2 internal resonance. They showed the existence of periodic, almost periodic, and chaotic wave motions when the container is subjected to parametric excitation with its frequency twice the frequency of the lower mode. The 1:2 internal resonance was difficult to realize in a rectangular tank because it requires a relatively small liquid depth, which causes excessive energy dissipation. Bridges (1987), Feng and Sethna (1989), and Feng (1990) considered a nearly square container, $\ell \approx L$, in which all nonsymmetric modes have nearly equal natural frequencies independently of the fluid depth. Bridges (1987) considered the case of standing waves while Feng and Sethna (1989) showed that the system is capable of periodic standing waves in the form of mixed modes, traveling waves with steady rotations around "nodal points" of two linear modes, almost periodic motions in the two types of waves. Feng (1990) and Feng and Sethna (1989) obtained amplitude and phase equations of the two modes in the normal form

$$\dot{a}_1 = (-\zeta + \sin 2\theta_1)a_1 - M_3 a_1 a_2^2 \sin(2\theta_2 - 2\theta_1) \tag{6.139a}$$

$$\dot{\theta}_1 = -(\lambda - \beta) + \cos 2\theta_1 + M_1 a_1^2 + M_2 a_2^2 + M_3 a_2^2 \cos(2\theta_2 - 2\theta_1) \tag{6.139b}$$

$$\dot{a}_2 = (-\zeta + \sin 2\theta_2)a_2 - M_3 a_2 a_1^2 \sin(2\theta_1 - 2\theta_2) \tag{6.139c}$$

$$\dot{\theta}_2 = -(\lambda - \beta) + \cos 2\theta_2 + M_1 a_2^2 + M_2 a_1^2 + M_3 a_1^2 \cos(2\theta_1 - 2\theta_2) \tag{6.139d}$$

where

$$\begin{aligned} M_1 = -\omega_{mn}^2 \Bigg\{ & \frac{1}{128}\left[-\frac{9}{\tanh^6\left(\bar{h}\sqrt{1+(n/m)^2}\right)} \right. \\ & + \frac{24(1+(n/m)^4)}{(1+(n/m)^2)\tanh^4\left(\bar{h}\sqrt{1+(n/m)^2}\right)} \\ & \left. - \frac{5}{\tanh^2\left(\bar{h}\sqrt{1+(n/m)^2}\right)} + 46 \right] \\ & + \frac{1}{32\left[\sqrt{1+(n/m)^2}\tanh\left(\bar{h}\sqrt{1+(n/m)^2}\right)\right]^3} \\ & \times \left[\frac{\left(3\left[\sqrt{1+(n/m)^2}\tanh\left(\bar{h}\sqrt{1+(n/m)^2}\right)\right]^2 - 3 + (n/m)^2\right)^2}{\tanh(2\bar{h}) - 2\left[\sqrt{1+(n/m)^2}\tanh\left(\bar{h}\sqrt{1+(n/m)^2}\right)\right]} \right. \\ & \left. + \frac{\left(3\left[\sqrt{1+(n/m)^2}\tanh\left(\bar{h}\sqrt{1+(n/m)^2}\right)\right]^2 - 3(n/m)^2 + 1\right)^2}{(n/m)\tanh(2\bar{h}) - 2\left[\sqrt{1+(n/m)^2}\tanh\left(\bar{h}\sqrt{1+(n/m)^2}\right)\right]} \right] \Bigg\} \end{aligned} \tag{6.140a}$$

$$M_2 = -\omega_{mn}^2 \left\{ \frac{1}{16} \left[\frac{2}{\tanh^4\left(\bar{h}\sqrt{1+(n/m)^2}\right)} - \frac{2(1+(n/m)^4)}{(1+(n/m)^2)\tanh^4\left(\bar{h}\sqrt{1+(n/m)^2}\right)} \right. \right.$$

$$\left. - \frac{4}{\tanh^2\left(\bar{h}\sqrt{1}+(n/m)^2\right)} + 10 \right] + \frac{1}{16\left[\sqrt{1+(n/m)^2}\tanh\left(\bar{h}\sqrt{1+(n/m)^2}\right)\right]^3}$$

$$\times \left[\frac{\left(3\left[\sqrt{1+(n/m)^2}\tanh\left(\bar{h}\sqrt{1+(n/m)^2}\right)\right]^2 - 1 - (n/m)^2 - 4(n/m)\right)^2}{\sqrt{2}(1+(n/m))\tanh(\sqrt{2}\bar{h}(1+(n/m))) - 4\left[\sqrt{1+(n/m)^2}\tanh\left(\bar{h}\sqrt{1+(n/m)^2}\right)\right]} \right.$$

$$\left. + \frac{\left(3\left[\sqrt{1+(n/m)^2}\tanh\left(\bar{h}\sqrt{1+(n/m)^2}\right)\right]^2 - 1 - (n/m)^2 + 4(n/m)\right)^2}{\sqrt{2}(1+(n/m))\tanh\left(\sqrt{2}\bar{h}(1-(n/m))\right) - 4\left[\sqrt{1+(n/m)^2}\tanh\left(\bar{h}\sqrt{1+(n/m)^2}\right)\right]} \right]$$

$$+ \frac{1}{8\left[\sqrt{1+(n/m)^2}\tanh\left(\bar{h}\sqrt{1+(n/m)^2}\right)\right]^3}$$

$$\left. \times \frac{\left(3\left[\sqrt{1+(n/m)^2}\tanh\left(\bar{h}\sqrt{1+(n/m)^2}\right)\right]^2 - 1 - (n/m)^2\right)^2}{\sqrt{2}\sqrt{1+(n/m)^2}\tanh\left(\sqrt{2}\bar{h}\sqrt{1+(n/m)^2}\right) - 4\left[\sqrt{1+(n/m)^2}\tanh\left(\bar{h}\sqrt{1+(n/m)^2}\right)\right]} \right\} \tag{6.140b}$$

$$M_3 = -\omega_{mn}^2 \left\{ \frac{1}{16} \left[\frac{1}{\tanh^4\left(\bar{h}\sqrt{1+(n/m)^2}\right)} - \frac{(1+(n/m)^4)}{\left(1+(n/m)^2\right)\tanh^4\left(\bar{h}\sqrt{1+(n/m)^2}\right)} + \frac{2}{\tanh^2\left(\bar{h}\sqrt{1+(n/m)^2}\right)} + 1 \right] \right\} \tag{6.140c}$$

$\bar{h}=m\pi h/\ell$, $\zeta=2\zeta_{mn}\ell/Z_0$, $\varepsilon\lambda=((\Omega/2)-\omega_{nm})$, $\tau=\Omega t(\varepsilon Z_0/\ell)$, a dot denotes differentiation with respect to the nondimensional time parameter τ, and the parameter β measures the nearness to a square section container, $\ell\approx L$, and is given by the expression

$$\beta=\frac{\pi^2}{K_{mn}^2}\left(\frac{\ell-L}{\ell}\right)\left(\frac{n^2-m^2}{\varepsilon Z_0/\ell}\right)\left[1+\frac{\ell}{L}\left(\frac{K_{mn}^2-\omega_{mn}^4}{\omega_{mn}^2}\right)\right] \tag{6.140d}$$

Since this parameter measures the deviation of the aspect ratio of the tank, it also represents a detuning parameter for "internal resonance" of the system. This parameter is inversely proportional to the excitation amplitude. Equations (6.139) are in the normal form and involve six parameters ζ, λ, β, M_1, M_2, and M_3. The last three parameters M_1, M_2, and M_3 depend on the ratio of the mode numbers, n/m, and the liquid depth parameter $\bar{h}$. Feng and Sethna (1989) restricted the analysis to the values of M_1, M_2, and M_3 that result in surface waves. For $n\neq 0$, $\bar{h}>1$, and $n/m>0.13$, the following inequalities hold, independent of any specific mode number:

$$\begin{aligned} &M_1<0,\ M_3<0,\ M_1+M_2+M_3<0\\ &M_1-M_2-M_3>0,\ M_1+M_2-M_3>0 \end{aligned} \tag{6.141a}$$

$$0<p=\frac{-M_1-M_2+M_3}{M_1-M_2+M_3}<1,\quad \text{and}\quad q=\frac{M_1+M_2+M_3}{-M_1+M_2+M_3}>1 \tag{6.141b}$$

The restriction for $\bar{h}>1$ results in weak dissipation effects and will be unable to suppress nonlinear phenomena. The case for $n/m<0.13$ is excluded as it will involve large mode numbers. As the number of modes increases, the values of the natural frequencies of higher modes become close to each other.

The steady-state solutions of equations (6.139) include the trivial solution $a_1=a_2=0$, unimodal solution in which either $a_1=0$ or $a_2=0$, and the mixed mode solution. The stability of such solutions can be examined by writing equations (6.139) in terms of Cartesian coordinates through the following coordinate transformation:

$$x_1=a_1\cos\theta_1,\ y_1=a_1\sin\theta_1,\ x_2=a_2\cos\theta_2,\ y_2=a_2\sin\theta_2 \tag{6.142}$$

Equations (6.139) take the form

$$\begin{aligned}\dot{x}_1=&-\zeta x_1+(1+\lambda-\beta)y_1-y_1\left[M_1(x_1^2+y_1^2)+M_2(x_2^2+y_2^2)\right]\\ &-2M_3x_1x_2y_2+M_3y_1(x_2^2-y_2^2)\end{aligned} \tag{6.143a}$$

$$\dot{y}_1 = -\zeta y_1 + (1-\lambda+\beta)x_1 + x_1\left[M_1(x_1^2+y_1^2) + M_2(x_2^2+y_2^2)\right] + 2M_3 y_1 x_2 y_2 + M_3 x_1(x_2^2 - y_2^2) \tag{6.143b}$$

$$\dot{x}_2 = -\zeta x_2 + (1+\lambda+\beta)y_2 - y_2\left[M_2(x_1^2+y_1^2) + M_1(x_2^2+y_2^2)\right] - 2M_3 x_1 x_2 y_1 + M_3 y_2(x_1^2 - y_1^2) \tag{6.143c}$$

$$\dot{y}_2 = -\zeta y_2 + (1-\lambda-\beta)x_2 + x_2\left[M_2(x_1^2+y_1^2) + M_1(x_2^2+y_2^2)\right] + 2M_3 y_2 x_2 y_2 + M_3 x_2(x_1^2 - y_1^2) \tag{6.143d}$$

This system of equations exhibits the following symmetries:

(1) If $\beta = 0$, or if (x_1, y_1, x_2, y_2) is a solution, then $(x_1, y_1, -x_2, -y_2)$, $(-x_1, -y_1, -x_2, -y_2)$, and $(-x_1, -x_2, -y_2, -x_1, -y_1)$ are also solutions of equations (6.143).
(2) If $\beta \neq 0$, or if (x_1, y_1, x_2, y_2) is a solution, then $(-x_1, -y_1, x_2, y_2)$, $(x_1, y_1, -x_2, -y_2)$, and $(-x_1, -y_1, -x_2, -y_2)$, are also solutions of equations (6.143).
(3) If β is replaced by $-\beta$ and (x_2, y_2, x_1, y_1) is replaced by (x_1, y_1, x_2, y_2), the system is unchanged.

The stability of the zero solution of equations (6.143) can be examined by evaluating the eigenvalues μ of the Jacobian of equations (6.143) evaluated at this solution. They are given by the following roots

$$\mu_{1,2} = -\zeta \pm \sqrt{1-(\lambda+\beta)^2}, \quad \mu_{3,4} = -\zeta \pm \sqrt{1-(\lambda-\beta)^2} \tag{6.144a, b}$$

These eigenvalues vanish under the following values of detuning parameter λ

$$\lambda_1 = \beta + \sqrt{1-\zeta^2}, \quad \lambda_2 = \beta - \sqrt{1-\zeta^2}, \quad \lambda_3 = -\beta + \sqrt{1-\zeta^2}, \quad \text{and} \quad \lambda_4 = -\beta - \sqrt{1-\zeta^2} \tag{6.145a, b}$$

At one of these detuning parameters, the eigenvalues have a simple zero leading to codimension-one bifurcation on one-dimensional center manifold. In view of the above-mentioned symmetries, the resulting bifurcations are of the pitchfork type.

The mixed mode solution includes different phenomena. Solutions of equations (6.143) represent modal motions that take place at nearly the linear modal frequency with amplitude modulations at the small frequency. The nature of such waves is thus quasi-periodic. The amplitude-modulated motions often lead to chaotic behavior. The different types of physical phenomena depend on the values of the detuning parameters λ and β. As these parameters vary a transition of one type of motion to another type takes place at a bifurcation point. These bifurcations are local and codimension-one.

Case a: Unimodal Response of the Second Mode, the unimodal response of the second mode is examined, $a_1 = 0$, $\theta_1 = 0$, and

$$(a_2^2)_{1,2} = \left[(\lambda + \beta) \pm \sqrt{1 - \zeta^2}\right] \Big/ M_1, \quad \text{and} \quad \sin 2\theta_2 = \zeta \qquad (6.146a, b)$$

In terms of (6.145) these solutions are

$$(a_2)_1 = \{[(\lambda - \lambda_3]/M_1\}^{1/2}, \qquad (a_2)_2 = \{[(\lambda - \lambda_4]/M_1\}^{1/2} \qquad (6.147a, b)$$

These solutions are shown in Figure 6.15(a)–(c) for different values of aspect ratio parameter, β. The stability of these solutions is determined from the values of the eigenvalues of the Jacobian matrix, which are given by the roots of the equation

$$\begin{aligned}\mu^2 + 2\zeta\mu - \frac{1}{M_1^2}&\left\{\lambda(M_3 - M_2 + M_1) + \left(\pm\sqrt{1-\zeta^2} - \beta\right)(-M_3 + M_2 + M_1)\right\} \\ &\times \left\{\lambda(M_3 + M_2 - M_1) \pm \sqrt{1-\zeta^2}(-M_3 - M_2 + M_1) + \beta(M_3 + M_2 + M_1)\right\} \\ &= 0 \end{aligned} \qquad (6.148)$$

A zero eigenvalue occurs when the free expression in equation (6.148) vanishes. This condition occurs at the following two possible values of the detuning parameter λ:

$$\lambda_5 = p\left(\sqrt{1-\zeta^2} - \beta\right), \quad \text{and} \quad \lambda_6 = \sqrt{1-\zeta^2} - q\beta \qquad (6.149)$$

These bifurcations coalesce, that is, $\lambda_5 = \lambda_6$, for the following value of the aspect ratio

$$\beta^* = -\frac{M_1 - M_2 - M_3}{2M_3}\sqrt{1-\zeta^2} \qquad (6.150)$$

The corresponding detuning parameter is $\lambda = p(\sqrt{1-\zeta^2} - \beta^*)$. At this detuning parameter and the aspect ratio given by (6.150), we have codimension-two. Note that $\lambda_5 > \lambda_6$ if $\beta > \beta^*$, $\lambda_5 < \lambda_6$ if $\beta > \beta^*$, and $\lambda_6 < \lambda_3$ if $\beta > \sqrt{1-\zeta^2}$. These features are demonstrated in the bifurcation diagrams shown in Figure 6.15(a)–(c).

The second solution (6.147b) experiences a pitchfork bifurcation at

$$\lambda_7 = -\sqrt{1-\zeta^2} - q\beta \qquad (6.151)$$

Case b: Unimodal Response of the First Mode: The unimodal response of the first mode is, $a_2 = 0$,

$$(a_1^2)_1 = \frac{\lambda - \lambda_1}{M_1}, \quad (a_1^2)_2 = \frac{\lambda - \lambda_2}{M_1} \qquad (6.152a, b)$$

These solutions are also shown in Figure 6.15(a)–(c) and experience pitchfork bifurcations at

$$\lambda_8 = p\left(\beta + \sqrt{1-\zeta^2}\right), \quad \lambda_9 = p\left(\beta - \sqrt{1-\zeta^2}\right) \qquad (6.153a, b)$$

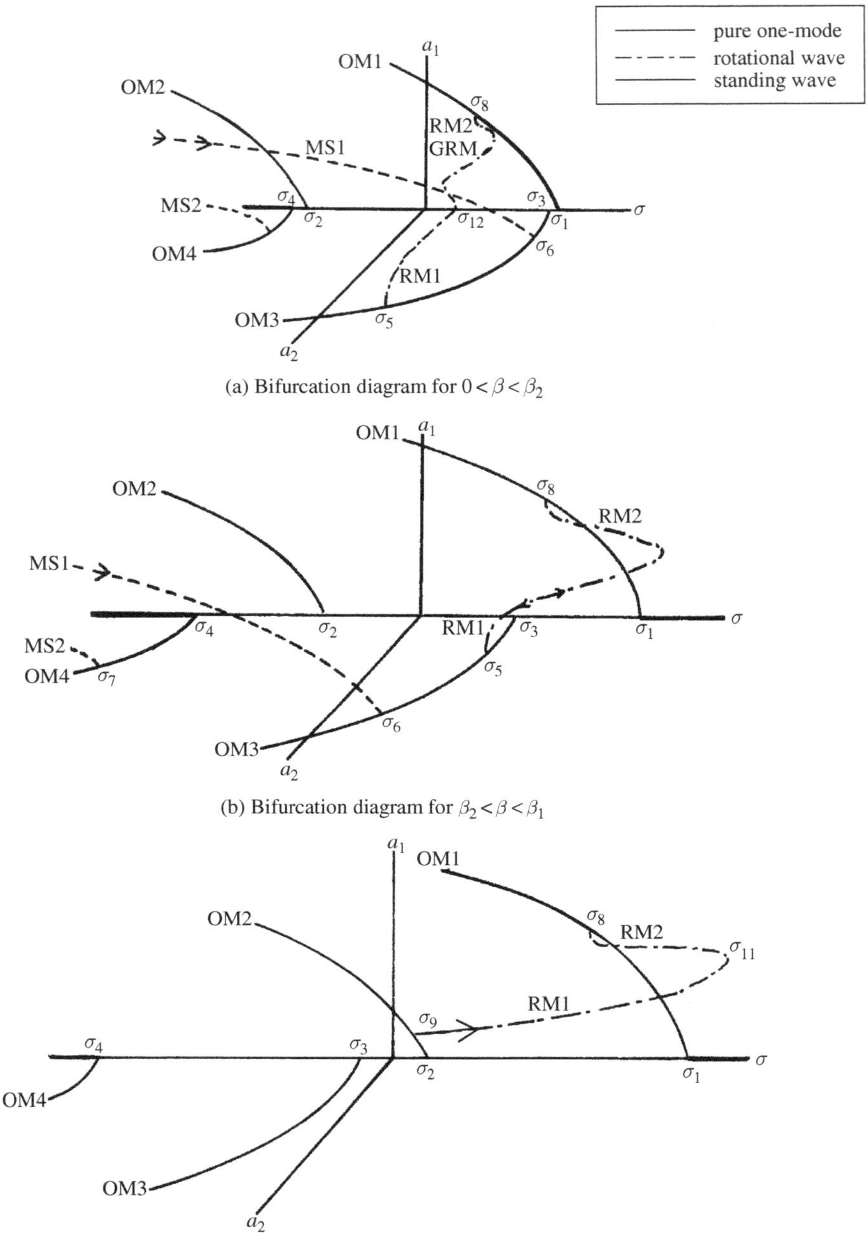

(a) Bifurcation diagram for $0<\beta<\beta_2$

(b) Bifurcation diagram for $\beta_2<\beta<\beta_1$

(c) Bifurcation diagram for $\beta>\beta_1$

Figure 6.15 Bifurcation diagrams for different ranges of the nearness to a square section parameter. (Feng and Sethna, 1989)

Case c: Mixed-mode Response: The above analysis is restricted to unimodal response characteristics away from the 1:1 internal resonance.In the neighborhood of 1:1 internal resonance mixed modal response is examined by considering the complete set of equations (6.139). The fixed solutions may be obtained by setting the left-hand sides of equations (6.139) to zero, which gives the following set of algebraic equations

$$\zeta - \sin 2\theta_1 + M_3 \sin(2\theta_2 - 2\theta_1)\, a_2^2 = 0 \tag{6.154a}$$

$$\beta - \lambda + M_1 a_1^2 + (M_2 + M_3 \cos(2\theta_2 - 2\theta_1))\, a_2^2 + \cos 2\theta_1 = 0 \tag{6.154b}$$

$$\zeta - \sin 2\theta_2 - M_3 \sin(2\theta_2 - 2\theta_1)\, a_1^2 = 0 \tag{6.154c}$$

$$-\beta - \lambda + M_1 a_2^2 + (M_2 + M_3 \cos(2\theta_1 - 2\theta_2))\, a_1^2 + \cos 2\theta_2 = 0 \tag{6.154d}$$

In view of the structure of these equations, it is difficult to extract a steady state solution. Alternatively, one may re-write these equations in the matrix form

$$\begin{bmatrix} 0 & 0 & 0 & -M_3 \sin(2\theta_2 - 2\theta_1) \\ -1 & 1 & M_1 & M_2 + M_3 \cos(2\theta_2 - 2\theta_1) \\ 0 & 0 & M_3 \sin(2\theta_2 - 2\theta_1) & 0 \\ -1 & -1 & M_2 + M_3 \cos(2\theta_1 - 2\theta_2) & M_1 \end{bmatrix} \begin{Bmatrix} \lambda \\ \beta \\ a_1^2 \\ a_2^2 \end{Bmatrix} = \begin{Bmatrix} \zeta - \sin 2\theta_1 \\ -\cos 2\theta_1 \\ \zeta - \sin 2\theta_2 \\ -\cos 2\theta_2 \end{Bmatrix} \tag{6.155}$$

A nontrivial solution for λ, β, a_1^2, and a_2^2 is guaranteed if the determinant of the coefficient matrix does not vanish, that is,

$$2M_3^2 \sin^2(2\theta_2 - 2\theta_1) \neq 0 \tag{6.156}$$

If $\sin(2\theta_2 - 2\theta_1) = 0$, then $\theta_1 = \theta_2$, and we have in-phase mixed mode motions. In this case, equations (6.139) are reduced to the following form after setting the left-hand sides to zero

$$\sin 2\theta_1 = \sin 2\theta_2 = \zeta \tag{6.157a}$$

$$-(\lambda - \beta) + \cos 2\theta_1 + M_1 a_1^2 + (M_2 + M_3) a_2^2 = 0 \tag{6.157b}$$

$$-(\lambda + \beta) + \cos 2\theta_2 + M_1 a_2^2 + (M_2 + M_3) a_1^2 = 0 \tag{6.157c}$$

It is not difficult to show that the steady state amplitude solutions are

$$a_1 = \sqrt{\frac{\lambda - \lambda_j}{M_1 + M_2 + M_3}}, \quad \theta_1 = \frac{1}{2}\sin^{-1}\zeta, \quad j = 6,\ 7 \tag{6.158a}$$

$$a_2 = \sqrt{\frac{\lambda - \left(\pm\sqrt{1-\zeta^2}\right) - q\beta}{M_1 + M_2 + M_3}}, \quad \theta_2 = \frac{1}{2}\sin^{-1}\zeta \tag{6.158b}$$

These solutions represent standing waves and bifurcate from the unimodal solutions at detuning parameters λ_6 and λ_7 as shown in Figure 6.15(a)–(c). For the case when $\sin(2\theta_2 - 2\theta_1) \neq 0$, one can solve the algebraic set of equations given by (6.155). The fixed solution is

$$\lambda = \frac{(-M_1 - M_2 + M_3)(\sin 2\theta_2 - \sin 2\theta_1)}{2M_3 \sin(2\theta_2 - 2\theta_1)} \tag{6.159a}$$

$$\beta = \frac{2\zeta(-M_1 + M_2 + M_3\cos(2\theta_2 - 2\theta_1)) + (M_1 - M_2 - M_3)(\sin 2\theta_2 + \sin 2\theta_1)}{2M_3 \sin(2\theta_2 - 2\theta_1)} \tag{6.159b}$$

$$a_1^2 = \frac{\zeta - \sin 2\theta_2}{2M_3 \sin(2\theta_2 - 2\theta_1)}, \quad a_2^2 = \frac{\sin 2\theta_1 - \zeta}{2M_3 \sin(2\theta_2 - 2\theta_1)} \tag{6.159c,d}$$

This solution is written in terms of the phase angles and with some algebraic manipulations, one can combine equations (6.159a, b) into a quartic equation in $(\theta_2 - \theta_1)$ whose solution can be used to determine the values of $2\theta_1$ and $2\theta_2$. The general mixed mode solutions emanate from the unimodal solutions at λ_5, λ_8, and λ_9. The general mixed mode solution passes through λ_{10}. The liquid surface wave height can be expressed as

$$\eta(x, y, t) = a_1 \sin(t + \theta_1)\cos(m\pi x)\cos(n\pi y) + a_2 \sin(t + \theta_2)\cos(n\pi x)\cos(m\pi y) \tag{6.160}$$

For standing waves, the nodal lines occur when $\eta = 0$ for all time. The coordinates of such points are

$$x_0 = \frac{2i-1}{2k}, \quad y_0 = \frac{2j-1}{2k}, \quad i, j = 1,\ 2,\ \ldots,\ k,\ k = m,\ n \tag{6.161}$$

The nodal curves can be determined by introducing polar coordinates at each point in (6.161), that is,

$$x = \frac{2i-1}{2m} + R_0\cos\varphi, \quad y = \frac{2j-1}{2m} + R_0\sin\varphi \tag{6.162}$$

The expressions for the nodal curves are then obtained by substituting (6.162) into (6.160) and setting $\eta = 0$. The resulting nodal curves move with time and thus represent the case of traveling waves. For very small values of R_0, the wave height can be expressed locally by the approximate expression

$$\eta(x,y,t) = m\pi R_0 a_2 \left[(-1)^i \frac{a_1}{a_2} \sin t \cos\varphi \cos\left(\frac{n\pi(2j-1)}{2m}\right) + (-1)^j \sin(t+\psi)\cos(n\pi x)\cos\left(\frac{n\pi(2j-1)}{2m}\right)\right] \tag{6.163}$$

where $\psi = \theta_2 - \theta_1$, and for $\eta = 0$, we have

$$\tan\varphi = (-1)^{i+j+1}\frac{a_1}{a_2}\frac{\cos\left(\frac{n\pi(2j-1)}{2m}\right)}{\cos\left(\frac{n\pi(2i-1)}{2m}\right)}\frac{\sin t}{\sin(t+\psi)} \tag{6.164a}$$

$$\dot{\varphi} = (-1)^{i+j+1}\frac{a_1}{a_2}\frac{\cos\left(\frac{n\pi(2j-1)}{2m}\right)}{\cos\left(\frac{n\pi(2i-1)}{2m}\right)}\frac{\sin\psi\cos^2\varphi}{\sin^2(t+\psi)} \tag{6.164b}$$

Thus, additional degenerate standing waves occur when φ is constant, and they occur when $\psi = 0$, that is, when $\theta_1 = \theta_2$. These standing waves are shown by the branches MS1 and MS2 in Figure 6.15(a)–(c). When $\psi = \pi/2$, and the amplitude ratio a_1/a_2 is constant, the resulting waves are of the traveling type and rotate about the nodal points with constant angular velocity. Such waves are represented by the degenerate branches RM1 and RM2.

6.6.4 Experimental results

Feng and Sethna (1989) conducted a series of experimental tests on two different containers that are nearly square (with cross-sections of 177.8 mm × 180.34 mm, and 177.8 mm × 190.50 mm. The fluid depth was not specified and tests were conducted for (1, 0) and (0, 1) sloshing modes. Figure 6.16 shows the measured and predicted amplitude frequency response curves for a tank with cross section of 177.8 mm × 180.34 mm under the condition $\sqrt{1-\zeta^2}(-M_1+M_2+M_3)/(2M_3) < \beta < \sqrt{1-\zeta^2}$, where the damping ratio was taken as four times that predicted by Miles (1967). The measurements were taken for fixed excitation amplitude of 1 mm, and excitation frequency that started from a relatively higher value and then decreased. One-mode motion corresponding to the curve OM1 was observed for the first mode and the other mode experienced a very small amplitude motion. As the excitation frequency was reduced further, the wave motion was observed to be a traveling mixed mode wave corresponding to curve RM1. With further reduction in frequency the second mode motion corresponding to curve OM3 was observed, and represented by triangle points Δ with small values of the first

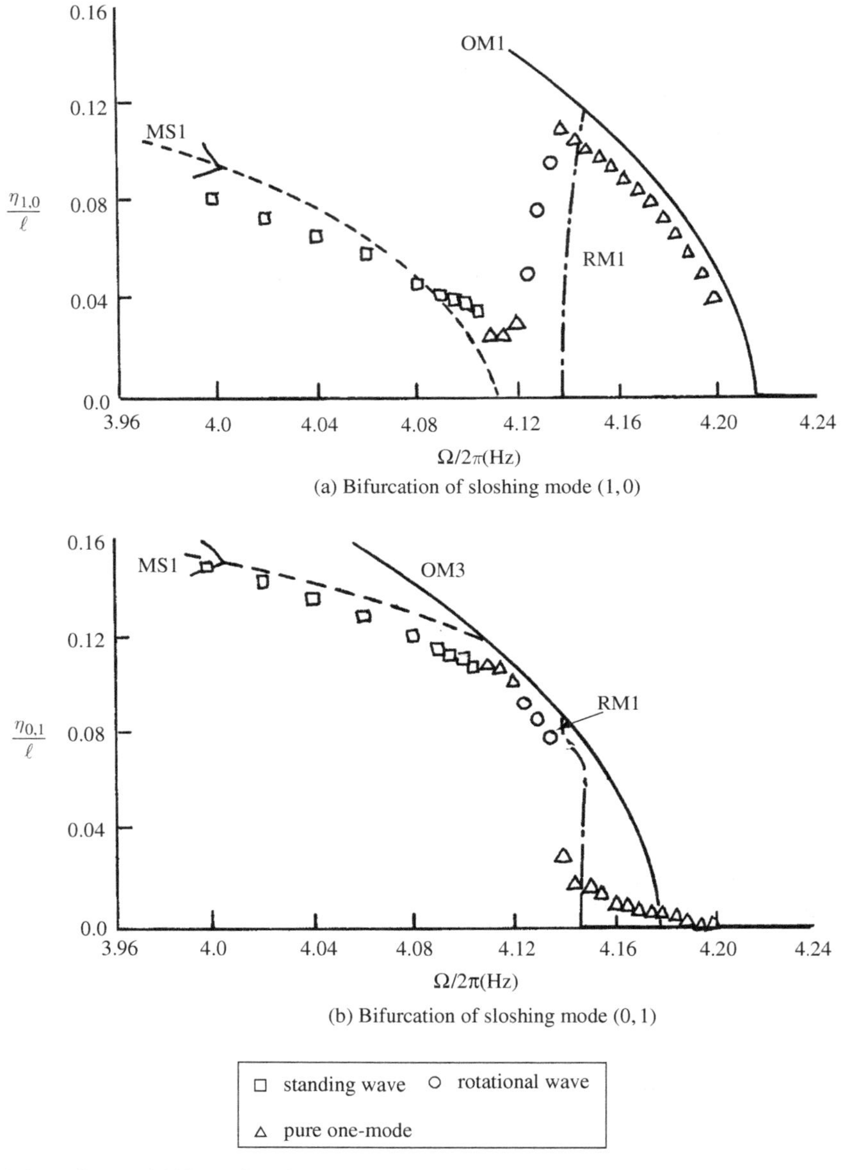

Figure 6.16 Experimental bifurcation for the case $\beta_2 < \beta < \beta_3$ for two different sloshing modes (Feng and Sethna, 1989)

mode. As the frequency was further decreased, mixed mode interaction was detected close to curve MS1 and is shown on the figure by square points □. Any further decrease in frequency gives a Hopf bifurcation of MS1 and a potential tendency for chaotic motion.

Figure 6.17 shows another set of response curves for a tank with a cross-section of 177.8 mm × 190.50 mm under excitation amplitude 1.25 mm. As the excitation frequency was decreased from a relatively high value, the first mode was observed to be close to OM1 and the second mode experienced very small amplitude. As the frequency further decreased, rotating

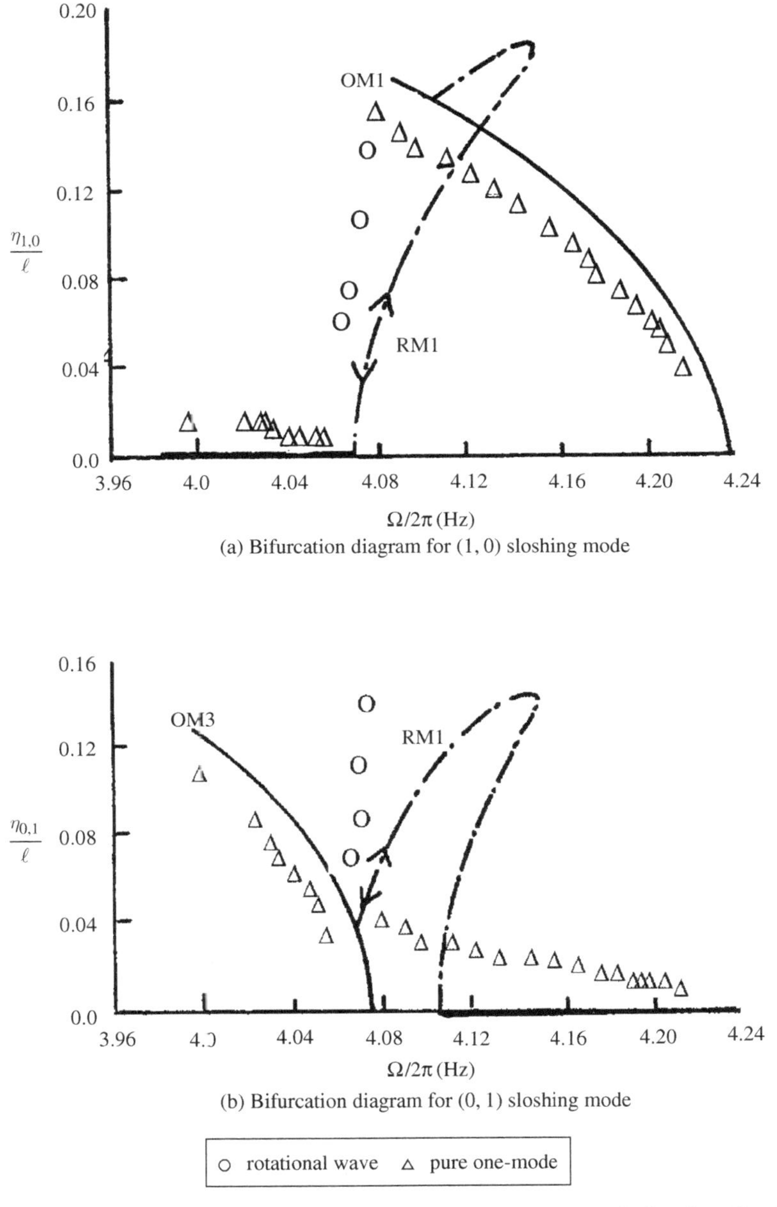

Figure 6.17 Experimental verification of bifurcation in a near square tank for $\beta_3 < \beta < \beta_1$. (Feng and Sethna, 1989)

traveling waves in one direction with amplitude modulation were observed. One may recall that the theory predicted the steady traveling waves indicated by the curve RM1 as well as Hopf bifurcation. Feng and Sethna (1989) reported that the rotating waves with varying amplitude had durations in the range of 50 to 150 seconds, depending on the excitation amplitude. After this period the motion became unimodal corresponding to the higher of the two frequencies OM1. This regime was followed by a rotational motion in the reverse direction that had the same duration as

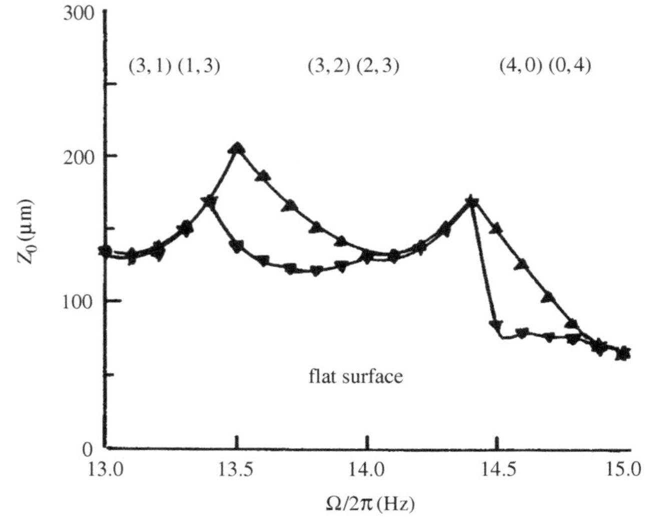

(a) Several resonances (asymmetric subcritical to the left, and supercritical to the right.

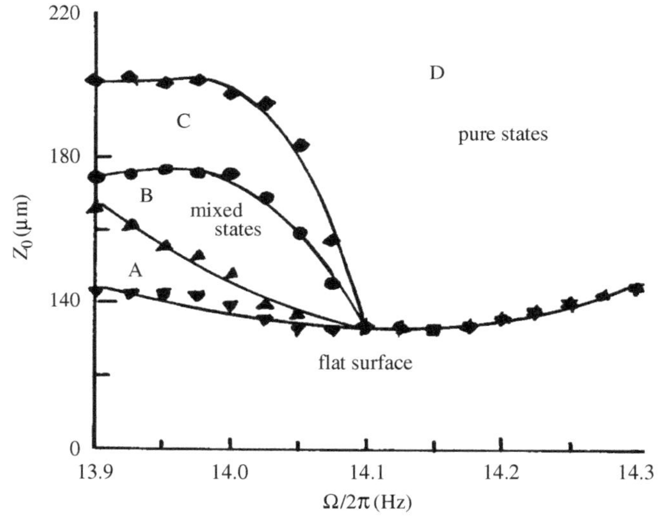

(b) Near (3, 2)–(2, 3) resonance showing three regions: flat surface, mixed states, and pure states. Region A and C are characterized by coexistence of different types of fixed points.

Figure 6.18 Free surface states in a square cell (6.17 × 6.17 cm). (Simonelli and Golub, 1989)

the previous rotational motion. Figure 6.17 shows only the average amplitude observed during this regime. Further reduction in the frequency resulted in unimodal motion corresponding to OM3.

Simonelli and Gollub (1989) experimentally studied the effects of symmetry and degeneracy of surface wave modal interactions. For example, they examined the interaction of two completely degenerate modes, namely (3, 2) and (2, 3) modes, in a square container of size 6.17 × 6.17 cm. Figure 6.18(a) shows the major resonance regions of the (3, 2) and (2, 3) modes and the neighboring modes in an experimental parameter space: excitation amplitude Z (μm) versus excitation frequency ($\Omega/2\pi$). The observed hysteresis takes place when the excitation amplitude is increasing and when it decreases. Figure 6.18(b) shows a detailed structure near (3, 2) and (2, 3) resonance. This figure reveals the following regions:

(1) The undisturbed (stable flat) free surface.
(2) Mixed states in region B characterized by equal amplitudes of the two modes.
(3) Pure state in region D depending on the initial conditions.
(4) Hysteretic intermediate regions A and C characterized by coexistence of different types of fixed points (flat or mixed in A, mixed or pure in C), which are realized for different initial conditions.

Figure 6.19 shows bifurcation diagrams, representing the dependence of the response amplitude, A_{32}, on the excitation amplitude, Z_0. It is obtained by increasing the excitation amplitude at fixed excitation frequency. Figure 6.19(a) belongs to negative detuning ($\Omega < 2\omega_{32} = 2\omega_{23}$) and Figure 6.19(b) is for positive detuning. The various regions indicated in Figure 6.18(b) are shown on the excitation axis. Solid curves indicate stable sinks, saddles are given by dash curves, while dot curves refer to sources.

Breaking the geometry symmetry occurs if the container cross-section is not perfectly square, and thus the two modes (3, 2) and (2, 3) become nonidentical. Simonelli and Gollub (1989) studied this case for tank cross-section 6.17 ×6.6 cm, which results in a difference of 1.5% between the two natural frequencies. Figure 6.20 shows the bifurcation diagram demonstrating the regions of different liquid motion regimes. For example, pure motion of (3, 2) takes place over excitation parameters defined in region B while region D will exhibit mode (2, 3). In region C the two modes coexist. In region A the flat surface and pure motion of mode (3, 2) coexist. It was reported that stable mixed mode motion does not take place, instead, time-dependent mixed states were found in region F. In region E both the flat surface state and time-dependent mixed states coexist. In region G, the pure (2, 3) mode motion and time-dependent mixed states coexist.

The time-dependent motion can be either periodic or chaotic. Periodic behavior prevails as the excitation amplitude increases. As the excitation amplitude, Z_0, increases very slowly toward the time-dependent region, a period relaxation oscillation takes place. It occurs in such a way that the liquid surface remains flat for a considerable time, then a large wave grows until it reaches its peak value, and decays. The time duration for the motion is shorter compared with the time of zero wave motions. A sample of the time history record and the corresponding configuration diagram are shown in Figure 6.21. This type of periodic motion can take place either for purely single mode or mixed modes. If the excitation amplitude is further increased, the free surface jumps discontinuously to a chaotic attractor centered on a mixed mode. The transition to chaos can be studied by observing the time history records and the corresponding configuration space for different values of excitation amplitude at excitation frequency $\Omega/2\pi = 13.75$ Hz, as shown in Figure 6.22. As the excitation amplitude is reduced, the bifurcations become more closely spaced as the sequence progresses. Below a certain excitation amplitude, the attractor becomes extremely asymmetric. Below that excitation level, shown in Figure 6.22 (i, j), the chaotic motion develops as shown in Figure 6.22 (k, l). This attractor persists until the excitation amplitude is reduced to a level where the free surface enters region E shown in Figure 6.20.

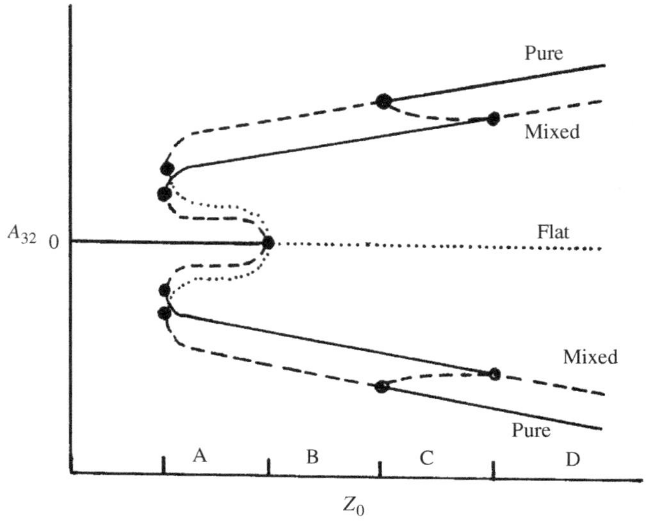

(a) Bifurcation diagram for increasing Z_0 with $f < 2f_{32} = 2f_{23}$

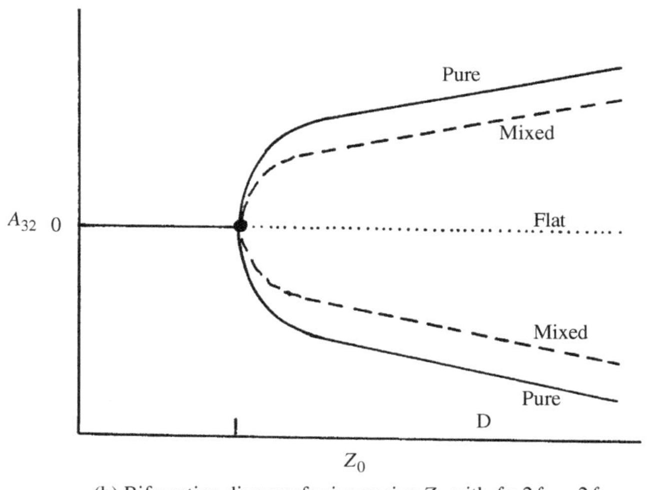

(b) Bifurcation diagram for increasing Z_0 with $f > 2f_{32} = 2f_{23}$

Figure 6.19 Bifurcation diagrams representing two-dimensional projections of the five- dimensional space spanned by the four mode amplitudes and excitation amplitude. (Simonelli and Golub, 1989)

6.7 Random parametric excitation

The stochastic stability of a liquid surface under random parametric excitation can be studied in terms of one of the stochastic modes of convergence. These modes include convergence in probability, convergence in the mean square, and almost sure convergence (see, e.g., Ibrahim, 1985). The linear stability analysis is based on the stochastic differential equation of the sloshing mode mn, that is,

$$A''_{mn} + 2\zeta_{mn} A'_{mn} + [1 + \xi''(\tau)] A_{mn} = 0 \tag{6.165}$$

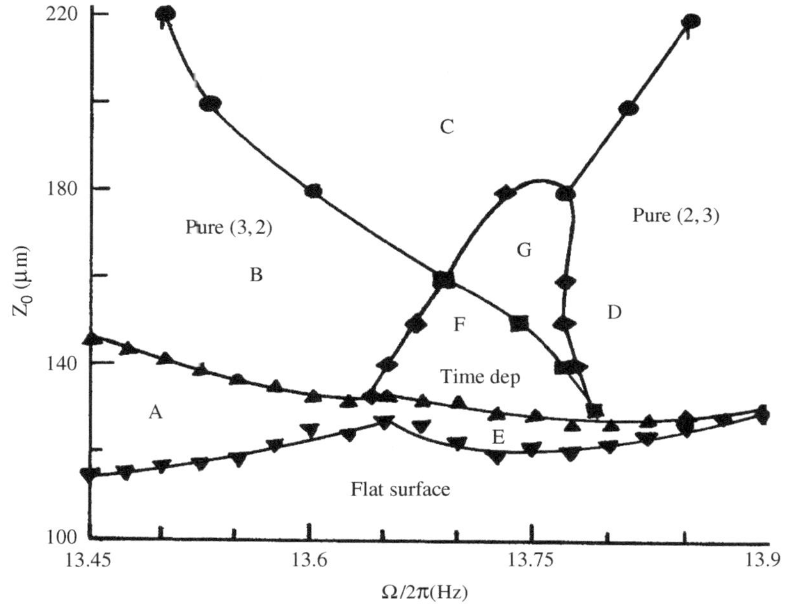

Figure 6.20 Experimental parameter space for the rectangular cell. (Simonelli and Golub, 1989)

where A_{mn} is a dimensionless free-liquid-surface amplitude of mode mn, a prime denotes differentiation with respect to the nondimensional time parameter $\tau = \omega_{mn}t$, ω_{mn} is the natural frequency of the sloshing mode mn, ζ_{mn} is the corresponding damping ratio, and $\xi''(\tau)$ is a dimensionless vertical wide band random acceleration of spectral density $2D$.

Fontenot and Lomen (1964), Fontenot, *et al.* (1965), Mitchell (1968), and Konstantinov and Lapshin (2001) determined the mean square stability condition of the response of equation (6.165), which is given by the inequality

$$D/2\zeta_{mn} < 1 \tag{6.166a}$$

On the other hand, the sample stability condition is

$$D/2\zeta_{mn} < 2 \tag{6.166b}$$

Dalzell (1967a) conducted an experimental investigation to measure the spectral density of the liquid-free-surface elevation under narrow- and wide-band random excitations. Different tests were conducted at different values of excitation level ranging from 0.09 to 0.18 g, rms. It was observed that such variation did not have significant influence on the response spectral density, implying a saturation feature. The liquid spectral density revealed also another component at the excitation frequency. Another important feature was that the rms acceleration level defining the onset of the 1/2-subharmonic response was not bracketed. This feature motivated Ibrahim and Heinrich (1988) to conduct another experiment, which revealed the occurrence of "on-off" intermittency as will be discussed later.

Under narrow-band random excitation centered on twice the frequency of the first axisymmetric mode, both harmonic and subharmonic responses were observed, but the acceleration defining the onset of such responses was not bracketed. As the excitation level decreases, the

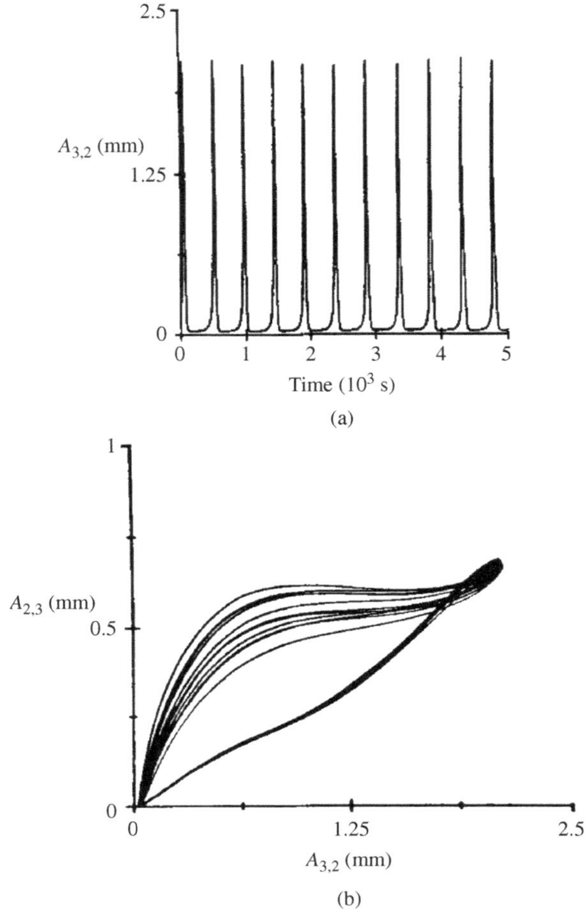

Figure 6.21 (a) Time history record and (b) configuration diagram of in-phase amplitudes of the periodic spiking process. (Simonelli and Golub, 1989)

response changes to predominantly harmonic. The wide-band excitation covers the first fifteen symmetric modes and two excitation levels were applied. The higher level exhibited subharmonic response. An abrupt transition between harmonic and subharmonic responses was observed when the excitation acceleration level was reduced. An important feature of the results showed that low-level harmonic response to random Gaussian excitation was nearly Gaussian. However, when large amplitude subharmonic response was excited, the probability distribution changed abruptly into a double-exponential distribution.

Dalzell (1967a) conducted a least-square fitting algorithm to develop a phenomenological probability distribution of the liquid-free-surface elevation. He found that the following double exponential distribution fits the experimental results

$$P(X) = \exp\left\{-\exp\left[-\left(\frac{\pi}{\sqrt{6}}X + \gamma\right)\right]\right\} \tag{6.167}$$

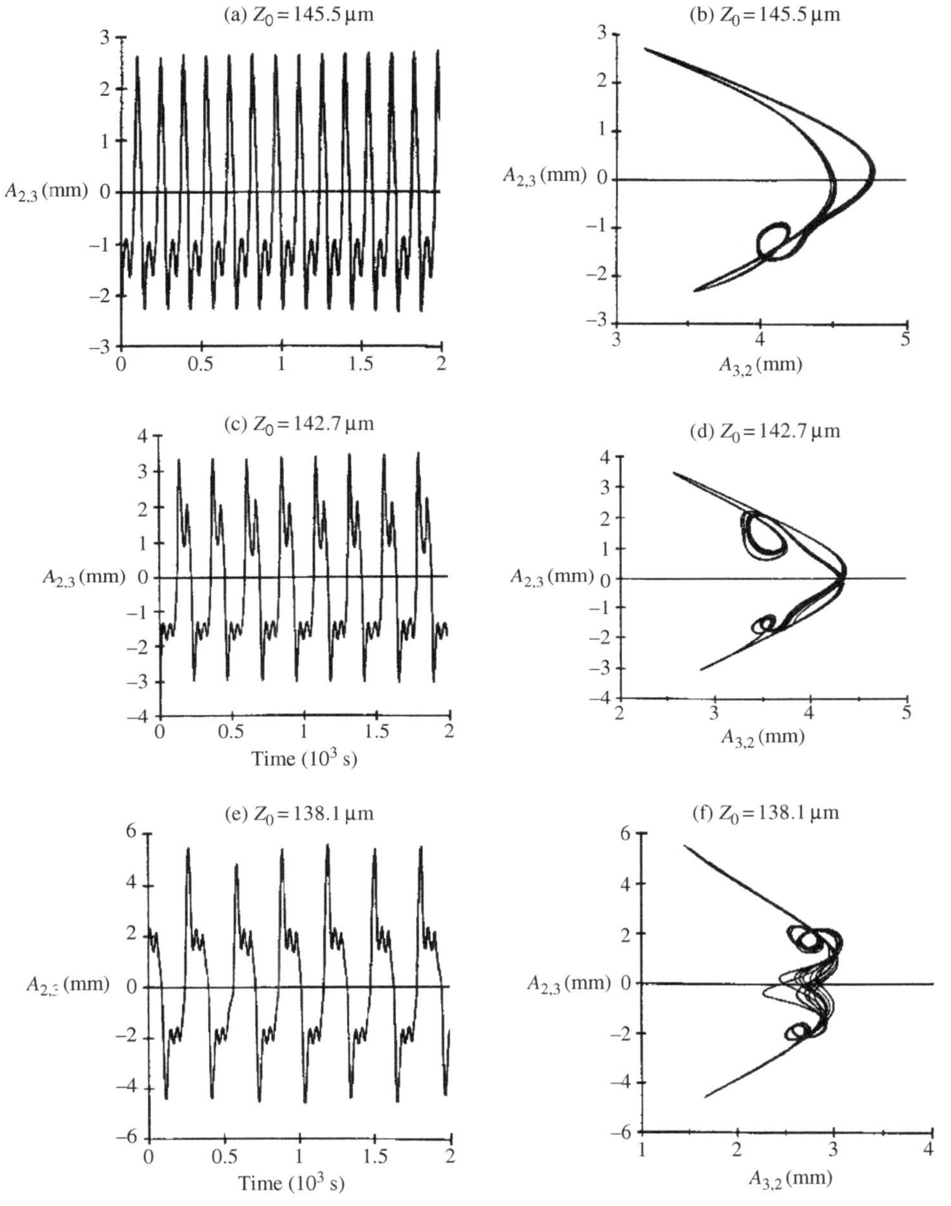

Figure 6.22 Sequence of time-series and configuration diagrams for decreasing excitation amplitude at excitation frequency $\Omega/2\pi = 13.75$ Hz. (Simonelli and Golub, 1989)

where γ is the Euler constant $= 0.577\,215\,665$, $X = (\eta - \bar{\eta})/\hat{\sigma}_\eta$, $\bar{\eta}$ is the mean value of the fluid elevation, and $\hat{\sigma}_\eta$ is the rms of the fluid elevation.

The nonlinear motion of the free liquid surface under random parametric excitation involves the estimation of stochastic stability and response statistics of the free surface (Dalzell, 1967a, Mitchell, 1968, Soundararajan (1983), Ibrahim and Soundararajan, 1983, 1985, and Ibrahim

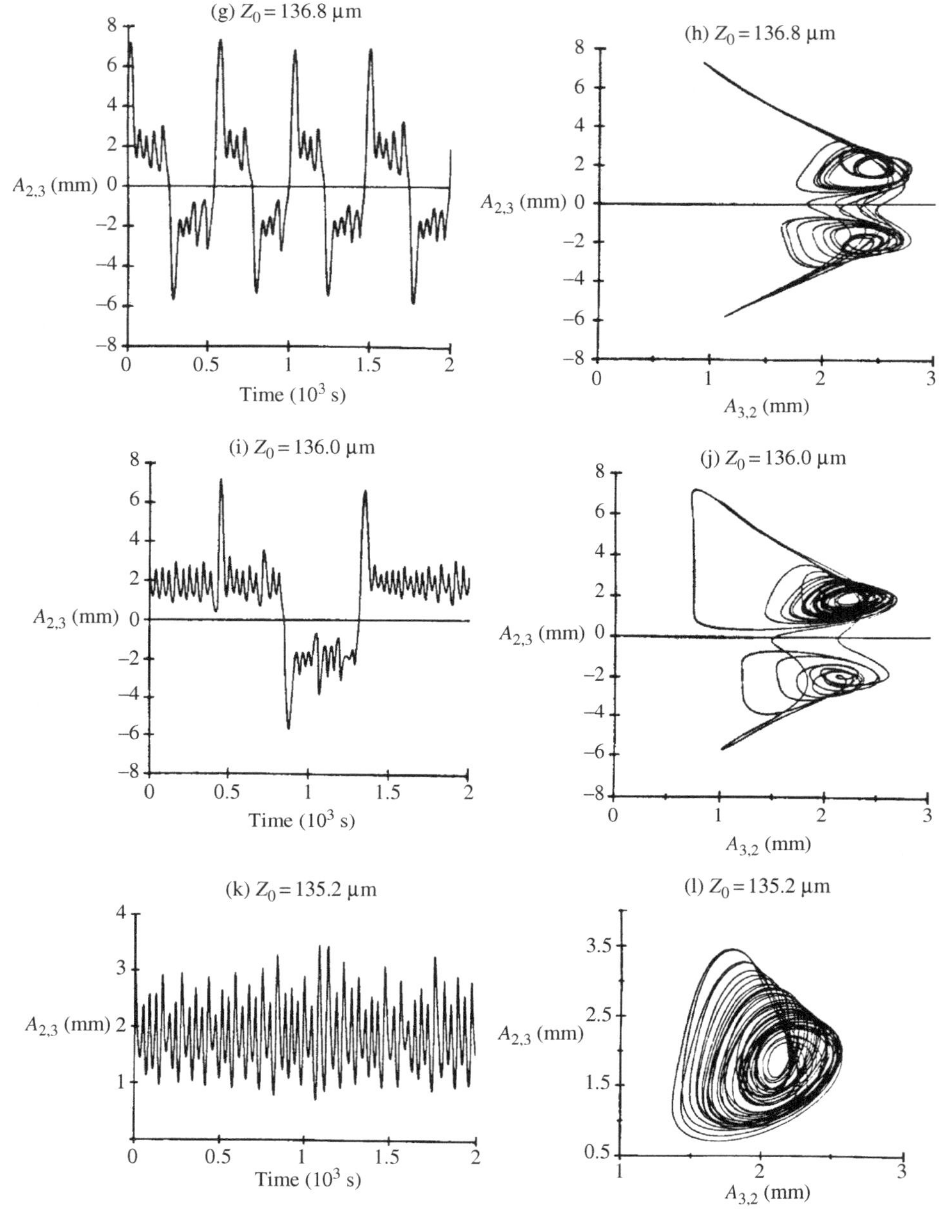

Figure 6.22 Continued.

and Heinrich, 1988). The free-liquid-surface height of a sloshing mode mn in a cylindrical container was found to be governed by the nonlinear differential equation

$$A''_{mn} + 2\zeta_{mn}A'_{mn} + [1 + \xi''(\tau)]A_{mn}(1 - K_1 A_{mn} - K_2 A^2_{mn}) + K_3 A'^2_{mn} + K_4 A_{mn} A''_{mn} + K_5 A_{mn} A'^2_{mn} + K_6 A^2_{mn} A''_{mn} = 0 \tag{6.168}$$

The last four terms in equation (6.168) represent quadratic (for symmetric modes) and cubic (for asymmetric modes) inertia nonlinearities. Equation (6.168) represents the nonlinear modeling of any mode mn and does not include nonlinear coupling with other sloshing modes. Douady, *et al.* (1989), and Douady (1990) examined the phase modulation of parametrically excited surface waves. Ibrahim and Heinrich (1988) observed the following regimes of the liquid-free-surface state:

(1) *Zero free-liquid-surface motion* is characterized by a delta Dirac function of the response probability density function. The free surface is always flat because the liquid damping force prevents any motion of the liquid. The excitation level for the first anti-symmetric and axisymmetric modes are given by the following ranges, respectively

$$0 < D/2\zeta_{11} < 1.55, \ 0 < D/2\zeta_{01} < 4.98 \tag{6.169a,b}$$

(2) *On-off intermittent motion* of the free liquid surface. This uncertain or intermittent motion takes place over the following ranges of excitation level for the two modes

$$1.55 < D/2\zeta_{11} < 1.82, \ 4.98 < D/2\zeta_{01} < 46.8 \tag{6.170a,b}$$

A corresponding regime, known as undeveloped sloshing, was predicted by Ibrahim and Soundararajan (1983). This regime is characterized differently by very small motion of the liquid free surface with an excitation level

$$2.0 < D/2\zeta_{mn} < 4.0 \tag{6.171}$$

Bosch and Van de Water (1993) and Bosch, *et al.* (1994) studied the spatio-temporal intermittency of liquid free surface under parametric excitation.

In deterministic bifurcation theory, the bifurcation point may be well defined as separating two different states of the system. However, if the control parameter experiences time variation, the bifurcation point may be subject to different scenarios (Haberman, 1979 and Erneux and Mandel, 1986) and one scenario is "*on-off intermittency*." In fluid mechanics (Townsond, 1976), the term on-off intermittency describes a flow alternating between long, regular laminar phases, interrupted by the shedding vortices. The mechanism of on-off intermittency is different from that reported by Pomeau and Manneville (1980) and Grebogi, *et al.* (1982, 1987). For example, the fixed point in the Pomeau and Manneville model corresponds to a periodic orbit in the continuous time system. Accordingly, intermittency is generally not of the on-off type for continuous systems. When a burst starts at the end of a laminar phase, this periodic motion instability is due to the modulus of at least one Floquet multiplier being greater than 1. This may occur in three different ways (Pomeau and Manneville, 1980): a real Floquet multiplier crosses the unit circle at $+1$ (intermittency of type I), or at -1 (type II), or two complex conjugate multipliers cross simultaneously (type III). To each of these three typical crossings there is one type of intermittency.

(3) *Partially developed random sloshing*. This regime is characterized by undeveloped sloshing where significant liquid-free-surface motion occurs for a certain time-period and then ceases for another period. At higher excitation levels, the time-period of liquid motion exceeds the period of zero motion. The excitation levels of this regime for the two modes are, respectively

$$1.82 < D/2\zeta_{11} < 14.05, \ 6.8 < D/2\zeta_{01} < 9.44 \tag{6.172a,b}$$

Heinrich (1986) and Ibrahim and Heinrich (1988) observed the development of circular motion of a central spike for first symmetric mode excitation. Occasionally, the spike is displaced from the center of the tank and precesses in such a way that preserves the azimuth symmetry in a time-average sense.

This motion was first reported by Gollub and Meyer (1983) who studied the amplitude dependence on the precession frequency under harmonic parametric excitation.

(4) *Fully developed sloshing* is characterized by continuous random liquid motion for all excitation levels exceeding the previous regime. When the first symmetric sloshing mode is excited, higher sloshing modes are excited as well. This is why the bandwidth of the symmetric mode excitation was narrowed in the test. The same observation of the interaction with other modes was reported by Dalzell (1967a).

With regard to the first anti-symmetric mode, another nonlinear phenomenon was observed, that is, the rotational motion of the nodal diameter. This same nonlinear phenomenon was also observed in the case of lateral excitation, as outlined in Chapter 4. It is known that such a phenomenon is created due to the coupling of the modes about two orthogonal axes of the circular container.

Figure 6.23(a) and (b) shows samples of the time history records of these different regimes for the two modes. One may also observe that the liquid-free-surface elevation does not have a zero mean response. In order to examine the stationarity of the response, several tests were conducted at the same excitation level and almost identical mean square responses were obtained for a test duration of 600 seconds.

Figure 6.24(a) and (b) shows the dependence of mean and mean square response of the first anti-symmetric sloshing mode on the excitation spectral density level $D/2\zeta_{11}$. The squares and crosses refer to the values as the excitation level increases then decreases, respectively. It is seen that the values of the mean and mean square responses remain zero up to a critical excitation level above which the free surface begins to lose stability. The uncertain sloshing region is included in the zero-response motion. Figure 6.25 shows the mean square response of the first axisymmetric sloshing mode. This figure demonstrates the uncertain region and another region where higher modes coexist with that mode.

Figure 6.26(a) and (b) shows the measured probability density function of the first anti-symmetric and axisymmetric sloshing modes for excitation levels $D/2\zeta_{11}=7.39$ and $D/2\zeta_{01}=9.43$, respectively. The dash curve represents the Gaussian probability density function. The diamond points refer to the measured probability density, while the solid curve is the double exponential proposed by Dalzell (1967a). It is seen that the double exponential curve gives a moderately good fit for most of the experimental results, except for the region that is close to the mean value of the liquid's elevation.

Note that the excitation spectral level was limited to a lower level within a narrow-band in order to avoid parametric excitation with other modes. Mixed mode interaction under random excitation has not been treated in the literature. The measured probability density function of the liquid response was found to be non-Gaussian for regions of large subharmonic motion with nonzero mean.

In view of the inertia nonlinearity of the equation of motion of one mode excitation, equation (6.168), there are a limited number of technical approaches to predicting the response statistics. The stochastic averaging carried out to second order of the deterministic part, is very powerful but tedious. Another alternative is the Monte Carlo simulation, which is also very useful in studying modal interaction.

6.8 Surface disintegration

Under relatively high-frequency parametric excitation, large-amplitude surface waves were observed by Yarymovich (1959) and Kana (1963a, b, 1966). This motion usually occurs in the

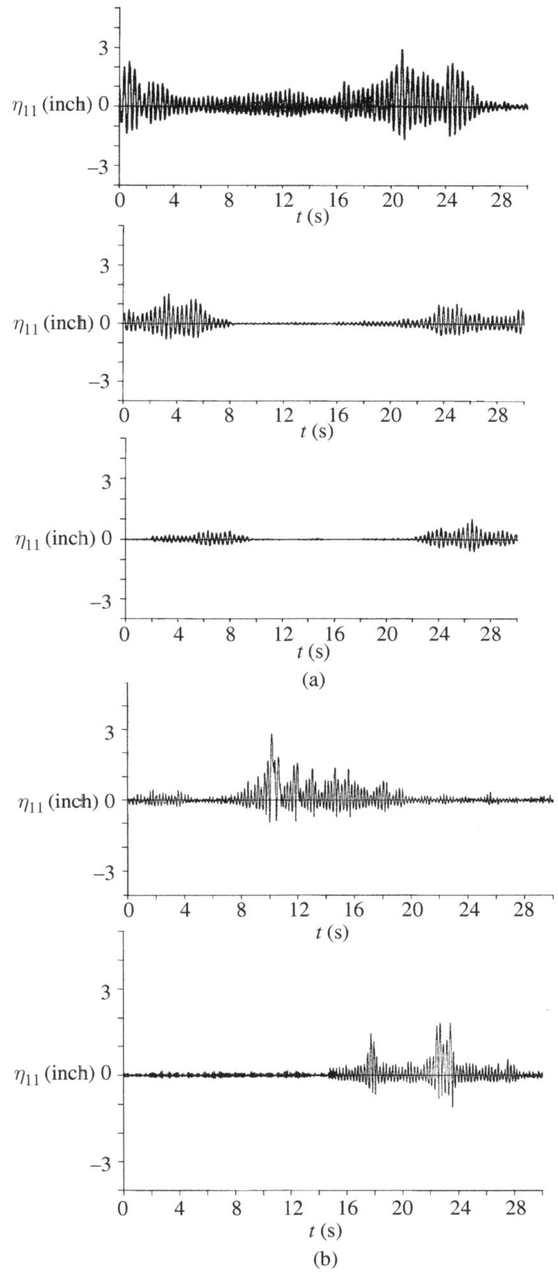

Figure 6.23 Fully and partially fully random sloshing regimes of (a) the first anti-symmetric mode, (b) the first symmetric mode. (Ibrahim and Heinrich, 1988)

form of typical spray-formed waves. Sometimes the surface motion becomes very violent at larger excitation levels with small vapor bubbles entrained in the liquid. The bubbles can become negatively buoyant and sink to the tank bottom. Bleich (1956b), Baird (1963), Ponder, *et al.* (1964), Abramson (1966a), Kana and Dodge (1966), and Gerlach (1968) reported

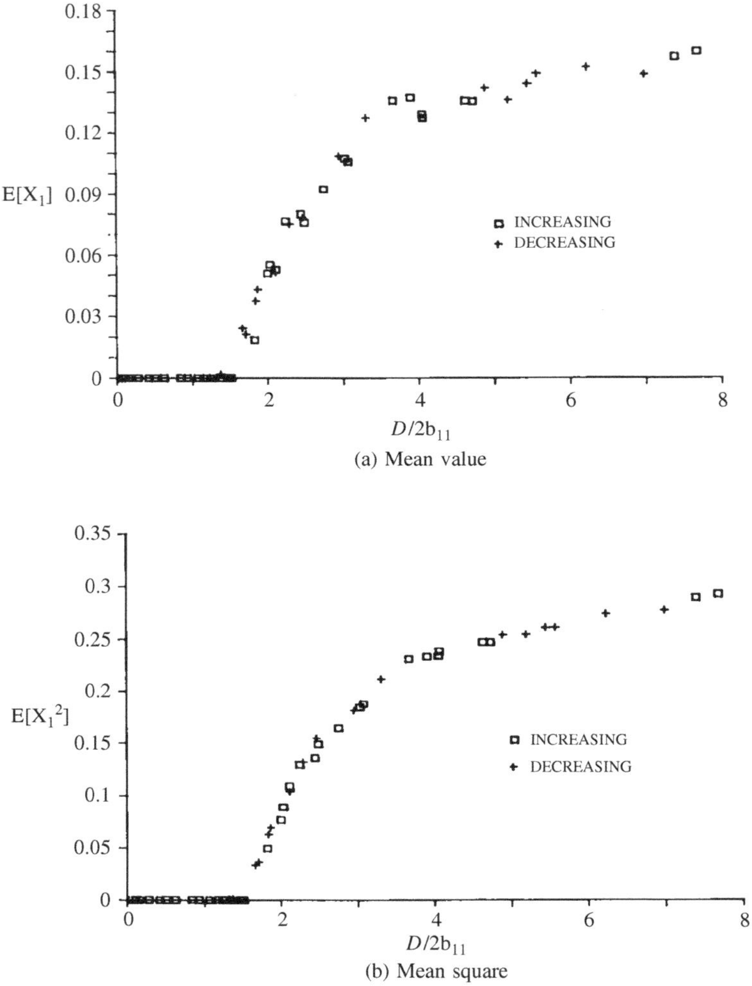

Figure 6.24 Dependence of mean and mean square response of the first anti-symmetric sloshing mode on excitation spectral density level. (Ibrahim and Heinrich, 1988)

analytical and experimental studies to examine the occurrence of surface disintegration with liquid-particles spray and the wave characteristics of spray-excited low-frequency waves. The surface disintegration and bubble formation in a vertically excited liquid column was theoretically and experimentally investigated by Buchanan, *et al.* (1962), Dodge (1963, 1968), Schoenhals and Overcamp (1965), Benjamin and Feir (1967), Schoenhals, *et al.* (1967), Hashimoto and Sudo (1980, 1982, 1984, 1985a, b, 1987, 1988), Sudo and Hashimoto (1987), and Sudo and Ohaba (1998). These studies showed that the smaller is the fluid height the larger is the minimum excitation amplitude required for the surface disintegration. It was also shown that the frequency of a spray-excited low-frequency wave is independent of the liquid height-to-diameter ratio. The parametric instability of a liquid-vapor interface was studied by Fauve, *et al.*, (1992).

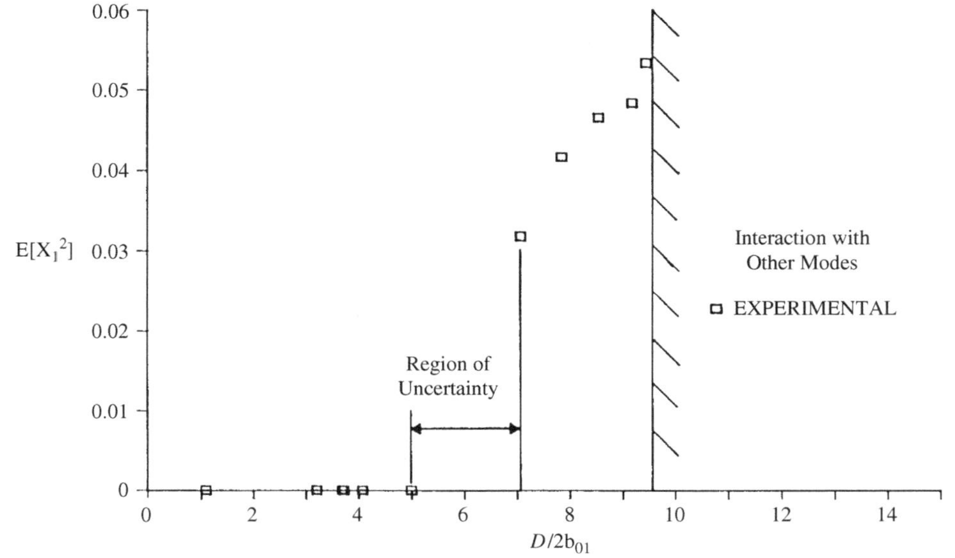

Figure 6.25 Dependence of the mean square of the first symmetric sloshing mode on the excitation spectral density parameter. (Ibrahim and Heinrich, 1988)

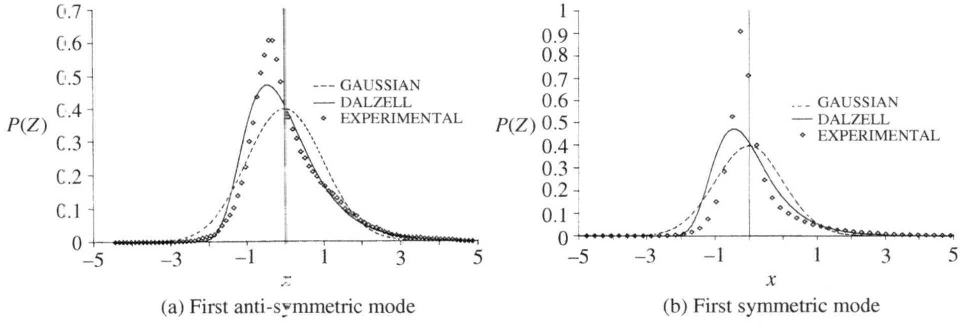

Figure 6.26 Comparison of measured and predicted of liquid free-surface response. (Ibrahim and Heinrich, 1988)

A series of experiments conducted by Crawford, *et al.* (1993) revealed that as the excitation amplitude increases, spatially periodic regular patterns with square hexagonal and triangular cells yield disordered and fluctuating states. Miles (1984e) indicated that under certain conditions soliton-like deformations develop adding to the diversity of the nonlinear motion. When the amplitude of oscillations exceeds a critical threshold value cusped interfacial waves, ejection of droplets, intense mixing, and temporal period tripling arise as reported by Mesquita, *et al.* (1992), Goodrich, *et al.* (1996, 1997), and Jiang, *et al.* (1998). Fauve, *et al.* (1992), Edwards and Fauve (1993), and Kumar and Bajaj (1995) indicated that two-dimensional striped patterns dominate as viscosity effects become important.

Wright, *et al.* (2000) provided interpretations pertaining to interface instability during vertical excitation of the free surface. They indicated that during one cycle of the fluid oscillation on either side of an interface, it is subjected to stabilizing and destabilizing acceleration directed from the light to the heavy fluid and vice versa. The interplay between the Rayleigh–Taylor instability[2] prevails during the first half-cycle of the vibration. The stabilizing influence of the fluid acceleration directed from the heavy to the light fluid, prevailing in the second half cycle of the vibration, is responsible for temporal asymmetry of the wave profiles during a complete cycle. This asymmetry should be contrasted with the symmetry of the unforced gravity or capillary waves. If the frequency of oscillations is small, the interface is subjected to destabilizing acceleration for a sufficiently long period. This allows the Rayleigh–Taylor instability to proceed into its nonlinear stages causing the development of spikes and bubbles. Lakiza (1995) and Kubenko and Lakiza (1996) studied experimentally the formation process of gas bubbles in cylindrical and spherical tanks undergoing the action of vibrations in different frequency domains and some resonance phenomena were observed.

6.9 Closing remarks

Parametric instability of liquid free surface has been the subject of extensive studies in many disciplines since the time of Faraday (1831). Faraday's waves and their related problems deserve an independent research monograph, and this chapter only provides a selected number of applications. The early research efforts focused on estimating instability boundaries in the presence and the absence of viscosity and surface tension forces. While the damping forces result in raising the threshold excitation amplitude, they do not bring the free-surface amplitude into a bounded value. It is the inherent free-surface wave amplitude nonlinearity that can limit the growth of the amplitude. Both nonlinear theory and experimental tests revealed that the liquid free surface exhibits "soft" or "hard" spring characteristics depending on the liquid depth ratio.

With the advent of the modern theory of nonlinear dynamics, many complex surface dynamic phenomena have been uncovered. These include mode competition, quasi-periodic motion, chaotic motion, and mixed mode motion in the presence of internal resonance. These phenomena have been studied in circular and rectangular tanks and the corresponding motion was identified in bifurcation diagrams under the variation of certain control parameters such as excitation amplitude and frequency.

Most of these complex phenomena should take place under random parametric excitation in terms of excitation bandwidth and spectral density function. Unlike the deterministic theory, random excitation tests revealed that the bifurcation of the free surface to 1/2-subharmonic response is not bracketed. Experimental measurements revealed that over a certain excitation an "on-off" intermittent motion takes place followed by a partial random motion of the free surface. In view of inertia nonlinearity, the stochastic averaging with second-order approximation should be applied to predict the response statistics and new response phenomena. One also may use Monte Carlo simulation for computing response statistics not only for single mode excitation but also for mixed mode coupling in the presence of internal resonance. A stochastic bifurcation study is also encouraged to define different regions of response motion

[2] Rayleigh–Taylor instability is generated when a heavy fluid sits above a lighter fluid in a gravitational field. The flow behavior is described in terms of bubbles of light fluid rising into the heavier medium and spikes of heavy fluid dropping into the lighter. More generally a material interface is said to be Rayleigh–Taylor unstable whenever the fluid acceleration has an opposite direction to the density gradient.

7
Dynamics of liquid sloshing impact

7.1 Introduction

An impulsive acceleration to a liquid can result in impact hydrodynamic pressure of the free surface on the tank walls. It can also occur during maneuvering or docking of spacecraft in an essentially low gravity field. Methods for estimating liquid impact and its hydrodynamic pressure are not well developed and are only identified by experimental studies (Pinson, 1963, Dalzell and Garza, 1964, and Stephens, 1965). It was found that when hydraulic jumps or traveling waves are present, extremely high impact pressures can occur on the tank walls. A hydraulic jump may occur in a liquid container undergoing oscillatory motion if the liquid height is relatively shallow and the excitation frequency is close to the natural frequency of the free surface.

Typical pressure–time history records under sloshing impact condition were reported by Cox, *et al.* (1980). Feng and Robertson (1971, 1972) studied the liquid propellant dynamics during docking. Rumyantsev (1969a) analyzed the collision of a body containing a viscous liquid with a rigid barrier. Hamlin, *et al.* (1986) indicated that the hydraulic jump could create localized high impact pressures on the container walls, which has a direct effect on the container dynamics and may result in structural damage. Verhagen and WijnGaarden (1965) reported that close to resonance a hydraulic jump was observed, which travels periodically back and forth between the container walls. This hydraulic jump is a nonlinear phenomenon, analogous to the shock wave appearing in one-dimensional gas flow under similar resonance conditions (Chester, 1964, 1968). Chester showed that the effect of dispersion is to introduce higher harmonics into the spectrum of liquid oscillations. The movement of hydraulic jumps affects the stability of oil tankers. Under these circumstances there is a strong coupling between the dynamics of the container structure and the movement of the hydraulic jump.

Hydraulic jumps have been modeled mathematically with a discontinuity in liquid surface elevation. Across the jump the mass and momentum of the liquid are conserved. However, Stoker (1957) stated that the conservation of the liquid energy does not hold across the jump because the liquid particles crossing the jump must either lose or gain energy. For example, the energy may be lost through the breaking of the waves. Furthermore, Stoker (1957) showed that the rate of change of energy across the jump is proportional to the cube of the difference between the liquid elevations on the two sides of the jump. The mechanics of liquid impact and associated hydrodynamic pressure were studied by Adler (1979), Korobkin and Pukhashov (1988), Cooker and Peregrine (1992, 1995), Chan (1994), Cooker (1994), Hattori, *et al.* (1994), Topliss (1994), and Zhang, Yue, and Tanizawa (1996).

Away from resonance, the hydraulic jump disappears and a solitary wave traveling back and forth between the tank walls is generated. Wehausen and Laitone (1960) showed that the solitary wave travels at a speed slightly higher than the critical speed corresponding to a Froude number of unity. To predict the solitary wave, a higher-order shallow-water theory should be used. When hydraulic jumps or traveling waves are present, extremely high impact pressures can occur on the tank walls in gasoline tankers and ship cargo tanks (see, e.g., Akita, 1967, Brathu, *et al.*, 1972, Filstead, 1972, Faltinsen, 1974, Faltinsen, *et al.*, 1974, Abramson, *et al.*, 1974, Bass, 1975, Faltinsen, 1978, Cox, *et al.*, 1980, Mikelis, *et al.*, 1984, Arai, 1986, and Arai, *et al.*, 1994).

By considering sea wave effects on ships, it is realistic that wave-induced ship motions can cause resonant fluid oscillations. This can lead to large local structural loads on the tank and has an important effect on the global ship motions. Vasta, *et al.* (1961) and Dalzell, *et al.* (1964) studied the stabilizing effect of liquid tanks on the roll motion of ships. Lee and Choi (1999) conducted experimental and numerical investigations of the sloshing impact in cargo tanks. They found hydraulic jumps formed when the excitation frequency was close to the natural frequency in the case of low filling depth. In the case of high filling depth, the flow resonates and hits the top of the tank, which results in large impact pressure.

The liquid sloshing can be more severe longitudinally than laterally if no transverse baffles are introduced. The longitudinal acceleration peaks are larger than the lateral ones. For some tank shapes, the liquid impact is probably more severe to the structure for longitudinal than for lateral sloshing. Ye and Birk (1994) measured the fluid pressure in horizontal partially filled cylindrical tanks with different values of length to diameter ratio ($L/2R$) when suddenly accelerated by impact along the longitudinal axis. Different types of pressure–time histories were obtained and revealed that the pressure profile changes with fill level. The peak pressure on the end of the tank was strongly affected by the fill level and the tank length-diameter ratio For a certain value, the pressure in the tank acted either like a water hammer[1] or like an accelerated fluid column, depending on the duration of the impact relative to the pressure wave transient time. For other values of height–diameter ratio, the pressure at the end of the tank was a function of liquid dynamic pressure. This causes the location of the maximum peak pressure to move from the impact end of the tank to the tank top. In some cases, the pressure on the tank top was two times larger than that at the tank impact end. The maximum pressure observed from tests was on the top of the tank.

The fluid sloshing impact causes a serious problem in underground radioactive waste storage tanks subjected to earthquakes. Some tanks were damaged at the roofs due to sloshing impact caused by strong earthquakes. The hydrodynamic pressure distribution of such impact loads is an important factor in studying the integrity and safety of tanks. Milgram (1969) studied the sloshing impact pressure in roofed liquid tanks. He carried out a series of experiments to distinguish some nonlinear sloshing phenomena in a reactor vessel of the pool type, which may cause damage to the vessel or inner structures. Test results for three different models using a long-period large-amplitude shaking table provided information on how the scales and the configurations affect the sloshing wave crest impact pressure.

The sloshing roof impact problem was studied by Kobayashi (1980), Minowa (1980, 1989), and Kurihara, *et al.* (1992). Minowa (1994) and Minowa, *et al.* (1994) conducted a series of shaking table tests of a rectangular tank to measure roof impact pressures, natural frequencies,

[1] Water hammer is a forceful slam, bang, or shudder that occurs in pipes when a sudden change in fluid velocity creates a significant change in fluid pressure.

and modes of bulging vibrations. Their measured results showed that the roof impact pressures possess great potential damage to tank as the pressure reaches as high as 30 psi, under 400 gal El-Centro seismic excitation. The liquid viscosity effects on sloshing response were found to be significant. Su, *et al.* (1982a,b) studied the nonlinear behavior and damping characteristics of liquid sloshing in partially filled prismatic tanks subjected to a large amplitude excitation.

The problem of hydrodynamic impact of blunt bodies with a flat free surface was first analyzed by Wagner (1932). Cointe and Armand (1987), Cointe (1989), Korobkin (1997), and Scolan and Korobkin (2001, 2003) extended the work of Wagner and developed two- and three-dimensional analytical solutions. During a drop test, the maximum body deceleration was found to be much higher than the gravitational acceleration. During the initial stage of impact, the flow region is divided into three zones: (i) outer zone, (ii) jet root zone, and (iii) jet region. It was shown that the flow description could be obtained by matching the three-dimensional Wagner solution in the main flow region (outer solution) with the two-dimensional solution in the jet root region (inner solution). The three-dimensional solution was used by Sconlan and Korobkin (2003) to evaluate the energy distribution during the impact process. It was shown that the energy is equally transmitted to the bulk of the fluid and to the spray jet provided that the velocity of the entering body is constant.

Some of the main problems encountered in the design of liquid propellant rockets include the motion of the center of mass, the vehicle attitude, and lateral bending of the vehicle structure. The sloshing technology developed for space applications is not applicable to road tankers, because emphasis has been placed on frequencies and total forces as they relate to control system requirements. Therefore, the effects of local peak impact pressure on structural requirements have not been studied to any extent. Further, excitation amplitudes considered in space applications are too small for road vehicle simulation.

This chapter presents the basic concepts of the theory of hydrodynamic slamming and impacting. It begins with the analogy of shock waves that take place in a gas column based on the work of Verhagen and WijnGaarden (1965). Analytical expressions of hydrodynamic impact forces and moments acting on the tank walls, due to sinusoidal pitching excitation, are given in closed approximate forms and compared with experimental measurements. Under lateral excitation, the modal interaction is presented based on the work of Faltinsen, *et al.* (2000). Experimental measurements of hydrodynamic pressure impacts due to sudden acceleration of the tank are displayed according to the experimental measurements of Ye and Birk (1994). A phenomenological modeling of impact forces and damping forces is proposed and used in the analytical modeling of elastic structures carrying liquid containers. The impact forces are replaced by a high power force, which covers both elastic and rigid impact cases. Three nontraditional analytical techniques, namely saw-tooth time transformation, nonsmooth coordinate transformation, and the Lie group transformations are briefly outlined. An application of a mechanical system simulating elastic structure carrying a liquid container allowing high power impact forces will be adopted. The response will be obtained using the method of multiple scales and the results will be compared with those obtained by nontraditional techniques. The analysis involves the simultaneous occurrence of parametric resonance and internal resonance conditions. The volume-of-fluid method based on finite difference solution of differential equations is outlined with application of surge excitation of ship tankers. A brief review of liquid sloshing in road tankers and their roll instability is presented.

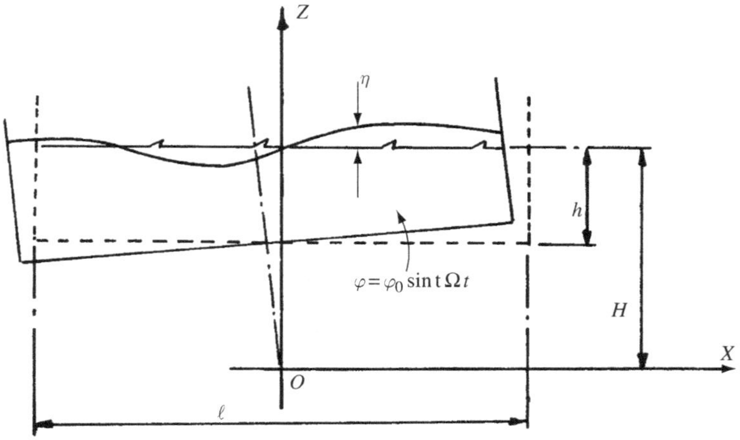

Figure 7.1 Rectangular container partially filled with liquid under pitching excitation.

7.2 Shock wave in a gas column analogy

Under sinusoidal excitation of a liquid tank, a hydrodynamic jump may be formed in the neighborhood of the resonance frequency. The jump usually moves back and forth between the walls of the tank. Verhagen and WijnGaarden (1965) employed a theory for one-dimensional gas flow to study the fluid oscillations experiencing a hydraulic jump in a rectangular tank. This theory was originally developed by Betchov (1958), Chu and Ying (1963), and Chester (1964). Figure 7.1 shows a schematic diagram of a rectangular container of width ℓ filled with liquid to a level h. The container is allowed to oscillate about axis Y, through O, with a small pitching angle $\varphi = \varphi_0 \sin \Omega t$. It is assumed that the breadth of the container in the Y-direction is large enough for the flow to be described as two-dimensional. The undisturbed liquid free surface is located at $Z = H$, while the surface elevation measured from this level is denoted by η During the tank oscillations, the bottom position is described by

$$z = H - h + \varphi_0 x \sin \Omega t \tag{7.1}$$

and the free surface is

$$z = H + \eta \tag{7.2}$$

Let λ be the surface level measured from the bottom of the container, such that

$$\lambda = h + \eta - \varphi_0 x \sin \Omega t \tag{7.3}$$

For $h/\ell \ll 1$, the motion of the liquid is governed by the "shallow-water theory" (see, e.g., Stoker 1957, Chapter 2 and Wehausen and Laitone, 1960). Thus, one can write the kinematic free-surface condition in the form

$$\frac{\partial \lambda}{\partial t} + u\frac{\partial \lambda}{\partial x} + \lambda \frac{\partial u}{\partial x} = 0 \tag{7.4}$$

where u is the fluid velocity in the x-direction, and the dynamic free-surface condition is

$$\frac{\partial u}{\partial t} + u\frac{\partial u}{\partial x} + g\left(\frac{\partial \lambda}{\partial x} + \varphi_0 \sin \Omega t\right) = 0 \tag{7.5}$$

In shallow-water approximation, the velocity, u, does not vary between the bottom and the surface. Thus, one writes

$$u = -\varphi_0 H\Omega \cos \Omega t \qquad \text{at } x = \pm \ell/2 \tag{7.6}$$

For very small values of φ_0, the linearized form of equations (7.4) and (7.5) is

$$\frac{\partial \lambda}{\partial t} + h\frac{\partial u}{\partial x} = 0, \quad \frac{\partial u}{\partial t} + g\left(\frac{\partial \lambda}{\partial x} + \varphi_0 \sin \Omega t\right) = 0 \tag{7.7a,b}$$

The solution of equations (7.7) satisfying condition (7.6) is

$$\lambda = h - \frac{\varphi_0 \ell \omega_0}{\pi\Omega} \frac{(1 + H\Omega^2/g)}{\cos(\pi\Omega/2\omega_0)} \sin \Omega t \sin(\pi\Omega x/\ell\omega_0) \tag{7.8}$$

where

$$\omega_0 = \frac{\pi}{\ell}\sqrt{gh} \tag{7.9}$$

It is seen that as $\Omega \to \omega_0$, the solution (7.8) gives $\lambda \to \infty$. However, experimental observation showed that as $\Omega \to \omega_0$, a hydraulic jump occurs. This implies that the linear theory fails to describe the actual behavior of the surface when $\Omega \to \omega_0$. This situation is analogous to the gas column dynamics under small amplitude oscillation by a piston. In terms of the gas density, ρ, and velocity, u, the continuity and momentum equations are given in the approximate forms

$$\frac{\partial \rho}{\partial t} + \rho_0\frac{\partial u}{\partial x} = 0, \quad \frac{\partial u}{\partial t} + \frac{a^2}{\rho_0}\frac{\partial \rho}{\partial x} = 0 \tag{7.10a,b}$$

where the subscript "0" denotes undisturbed quantity, and a is the speed of sound in the gas. If $u = 0$ at the fixed wall, $x = L$, and at the piston $u = X_0\Omega \cos \Omega t$, the solution of equations (7.10) is

$$u(x, t) = X_0\Omega \cos \Omega t \frac{\sin\{\Omega(L - x)/a\}}{\sin(\Omega L/a)} \tag{7.11}$$

This solution reveals that resonance occurs when $\Omega L/a = n\pi$, $n = 0, 1, 2, \ldots$ At resonance a shock wave is generated in the gas, which travels periodically back and forth in the column. This is similar to the motion of the hydraulic jump in liquid containers. This implies that the equations of shallow-water theory are equivalent to the equations of one-dimensional gas dynamics. Chu and Ying (1963) used the method of characteristics developed by Lin (1954) for traveling shock waves in tubes. The Chu–Ying–Lin method as applied to the hydraulic jump is described by introducing the following parameters

$$c = \sqrt{g\lambda}, \; c_0 = \sqrt{gh}, \; \varepsilon = \sqrt{\frac{\ell\varphi_0}{\pi h}} \tag{7.12a,b,c}$$

Equations (7.4) and (7.5) may be written in terms of the above parameters in the form

$$\left[\frac{\partial}{\partial t}+(u+c)\frac{\partial}{\partial x}\right]\left\{u+2c-\frac{\omega_0}{\Omega}\varepsilon^2 c_0\cos(\Omega t-\theta)\right\}=0 \tag{7.13a}$$

$$\left[\frac{\partial}{\partial t}+(u-c)\frac{\partial}{\partial x}\right]\left\{u-2c-\frac{\omega_0}{\Omega}\varepsilon^2 c_0\cos(\Omega t-\theta)\right\}=0 \tag{7.13b}$$

where the phase angle θ has been introduced. Define the characteristic coordinates α and β, such that along the C^+characteristics they are governed by the equation

$$\frac{\partial x}{\partial \alpha}=(u+c)\frac{\partial t}{\partial \alpha}, \quad \text{with} \quad \beta \quad \text{a constant} \tag{7.14a}$$

and along the C^-characteristics we have

$$\frac{\partial x}{\partial \beta}=(u-c)\frac{\partial t}{\partial \beta}, \quad \text{with} \quad \alpha \quad \text{a constant} \tag{7.14b}$$

It follows from equations (13) that along C^+

$$\frac{\partial}{\partial \alpha}\left\{u+2c-\frac{\omega_0}{\Omega}\varepsilon^2 c_0\cos(\Omega t-\theta)\right\}=0 \tag{7.15a}$$

and along C^-

$$\frac{\partial}{\partial \beta}\left\{u-2c-\frac{\omega_0}{\Omega}\varepsilon^2 c_0\cos(\Omega t-\theta)\right\}=0 \tag{7.15b}$$

Note that the Riemann invariants given by the expression in the braces in equations (7.15) assume a simple form involving u and c because the bottom is flat and its slope does not depend on x. In order to obtain a periodic solution of equations (7.14) and (7.15) involving hydraulic jumps, one has to use the method of characteristics outlined in Courant and Friedrichs (1948) for the gas-dynamics case. The procedure starts from the undisturbed conditions and constructs the evolution of the flow by the method of characteristics. The procedure might be continued until a quasi-stationary state is achieved in which a jump travels periodically back and forth.

Alternatively, one may start from a quasi-stationary state and attempt to construct an analytical solution. With reference to Figure 7.2, the paths of the jumps are indicated by the lines PQ, QR, RS, ..., etc. The solution has to satisfy equations (7.14) and (7.15) in the regions I, II, III, etc. The solution in each region has different energy and must be related to other solutions by the jump conditions. These conditions require the conservation of mass and momentum across the hydraulic jump. Across a shock the energy is preserved. The flow in region I must be repeated in region II due to the periodicity. Since the paths of the jumps are not known a priori, one may begin by assuming that these paths cannot be far from $dx/dt=\pm c_0$, as $\varphi_0\to 0$. The deviations of the jumps from these directions may conveniently be expressed as perturbations in terms of the characteristic coordinates described by Lin (1954). Thus, if the difference in level across the jump is $\Delta\lambda$, then the rate of energy loss is, Verhagen and WijnGaarden (1965),

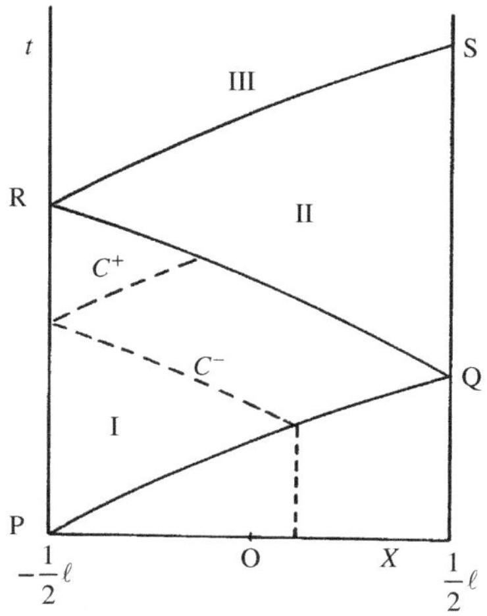

Figure 7.2 Paths of hydraulic jumps in $X - t$ plane. (Verhagen and WijnGaarden, 1965)

$$\frac{dE}{dt} \approx \frac{(\Delta\lambda)^3}{h^2} \ell h \Omega \rho g \tag{7.16}$$

The input power by the external force is

$$W \approx \Delta\lambda \ell^2 \varphi_0 \Omega \rho g \tag{7.17}$$

Equating the two expressions gives

$$\frac{\Delta\lambda}{h} \approx \sqrt{\frac{\varphi_0 \ell}{h}} \approx \varepsilon \tag{7.18}$$

This implies that the strength of the jump is of order ε. Verhagen and WijnGaarden (1965) followed a perturbation scheme for equations (7.14) and (7.15) in the form of asymptotic expansion series in terms of ε and obtained the following results for different regions:

In region I, the following parameters apply

$$u^{\mathrm{I}} = 4\varepsilon c_0 \sqrt{\frac{2}{3\pi}\left(1 + \frac{\pi^2 hH}{\ell^2}\right)} \sin\left(\frac{1}{2}\left(\omega_0 t - \theta - \frac{1}{2}\pi\right)\right) \sin\left(\frac{\pi(x + (\ell/2))}{2\ell}\right) + O(\varepsilon^2) \tag{7.19a}$$

$$\begin{aligned} c^{\mathrm{I}} = c_0 + \varepsilon \Bigg[& \frac{\ell(\Omega - \omega_0)}{\pi\varepsilon} + \frac{2c_0}{\pi}\sqrt{\frac{2}{3\pi}\left(1 + \frac{\pi^2 hH}{\ell^2}\right)} \sin\left(\frac{1}{2}\theta + \frac{1}{4\pi}\right) \\ & + 2c_0\sqrt{\frac{2}{3\pi}\left(1 + \frac{\pi^2 hH}{\ell^2}\right)} \cos\left(\frac{1}{2}\left(\omega_0 t - \theta - \frac{1}{2}\pi\right)\right) \cos\left(\frac{\pi(x + (\ell/2))}{2\ell}\right)\Bigg] \\ & + O(\varepsilon^2) \end{aligned} \tag{7.19b}$$

In region II, the following parameters apply

$$u^{\mathrm{II}} = 4\varepsilon c_0\sqrt{\frac{2}{3\pi}\left(1+\frac{\pi^2 hH}{\ell^2}\right)}\cos\left(\frac{1}{2}\left(\omega_0 t-\theta-\frac{1}{2}\pi\right)\right)\cos\left(\frac{\pi(x+(\ell/2))}{2\ell}\right) + O(\varepsilon^2) \tag{7.20a}$$

$$\begin{aligned} c^{\mathrm{II}} = c_0 + \varepsilon\Bigg[&\frac{\ell(\Omega-\omega_0)}{\pi\varepsilon} + \frac{2c_0}{\pi}\sqrt{\frac{2}{3\pi}\left(1+\frac{\pi^2\,hH}{\ell^2}\right)}\,\sin\left(\frac{1}{2}\,\theta+\frac{1}{4}\,\pi\right) \\ &+2c_0\,\sqrt{\frac{2}{3\pi}\left(1+\frac{\pi^2\,hH}{\ell^2}\right)}\,\sin\left(\frac{1}{2}\left(\omega_0 t-\theta-\frac{1}{2}\,\pi\right)\right)c\,\sin\left(\frac{\pi(x+(\ell/2))}{2\ell}\right)\Bigg] \\ &+ O(\varepsilon^2) \end{aligned} \tag{7.20b}$$

The path of the jump PQ, traveling from left to right, is

$$\begin{aligned} x(t,\Omega) = &-\frac{1}{2}\ell + c_0 t \\ &+\varepsilon\Bigg[\frac{\ell(\Omega-\omega_0)}{\pi\varepsilon}t+\frac{2}{\pi}\sqrt{\frac{2}{3\pi}\left(1+\frac{\pi^2 hH}{\ell^2}\right)}(c_0 t-\ell)\sin\left(\frac{1}{2}\theta+\frac{1}{4}\pi\right) \\ &-\frac{2\ell}{\pi}\sqrt{\frac{2}{3\pi}\left(1+\frac{\pi^2 hH}{\ell^2}\right)}\,\sin\left(\frac{1}{2}\left(\omega_0 t-\theta-\frac{1}{2}\pi\right)\right)\cos\left(\frac{1}{2}\omega_0 t\right)\Bigg] + O(\varepsilon^2) \end{aligned} \tag{7.21a}$$

On the other hand, the equation of the jump QR, traveling to the left, is

$$\begin{aligned} x(t,\Omega) = &\frac{3}{2}\ell - c_0 t \\ &-\varepsilon\Bigg[\frac{\ell(\Omega-\omega_0)}{\pi\varepsilon}t+\frac{2}{\pi}\sqrt{\frac{2}{3\pi}\left(1+\frac{\pi^2 hH}{\ell^2}\right)}(c_0 t-2\ell)\sin\left(\frac{1}{2}\theta+\frac{1}{4}\pi\right) \\ &+\frac{2\ell}{\pi}\sqrt{\frac{2}{3\pi}\left(1+\frac{\pi^2 hH}{\ell^2}\right)}\cos\left(\frac{1}{2}\left(\omega_0 t-\theta-\frac{1}{2}\pi\right)\right)\cos\left(\frac{1}{2}\omega_0 t\right)\Bigg] + O(\varepsilon^2) \end{aligned} \tag{7.21b}$$

In order to determine the phase angle, θ, between the jump and the container excitation one has to satisfy the condition of preserving the total volume of the water per unit length in the y-direction at all times, that is,

$$\int_{-\ell/2}^{\ell/2} \lambda\, dx = \ell h \tag{7.22a}$$

or by using relations (7.12), one writes

$$\int_{-\ell/2}^{\ell/2} c^2\, dx = \ell c_0^2 \tag{7.22b}$$

This integral can be evaluated by using equations (7.19b) and (7.20b). To obtain the contribution of terms of order ε, the relation between x and time t at the jump is needed only to zero order. This can be obtained from equation (7.21a) after setting t' to be the time at which the jump travels from the left to the right, that is,

$$x(t') = -\frac{1}{2}\ell + c_0 t' \tag{7.23}$$

Note that equation (7.19b) is used from $x = -\ell/2$ to $x = -(\ell/2) + c_0 t'$, and for the remainder of the integration interval, which represents the region in front of the jump one has to use equation (7.20b) by replacing t' by $t' + (2\ell/c_0)$. This process yields

$$\sin\left(\frac{1}{2}\theta + \frac{1}{2}\pi\right) = -\frac{\ell(\Omega - \omega_0)}{6\varepsilon c_0\sqrt{(2/3\pi)(1 + (\pi^2 hH/\ell^2))}} + O(\varepsilon) \tag{7.24}$$

This relation reveals that if $\Omega = \omega_0$, the phase angle is $\theta = -\pi/2$. This means that when the container is excited at $\Omega = \omega_0$, the jump is just passing the center of the container during its way from $x = \ell/2$ to $x = -\ell/2$. In order to determine the jump height, that is, the difference between the surface elevation behind and in front of the jump, the following relation is obtained from relations (7.3) and (7.12a, b)

$$c \approx c_0\left(1 + \frac{\eta}{2h}\right) \tag{7.25}$$

Using equations (7.19b), (7.20b), and (7.24) in equation (7.25), the wave elevations for regions I and II are, respectively,

$$\frac{\eta^{\mathrm{I}}}{h} \approx \frac{4}{3}\left(\frac{\Omega - \omega_0}{\omega}\right) + 4\varepsilon\sqrt{\frac{2}{3\pi}\left(1 + \frac{\pi^2 hH}{\ell^2}\right)}\cos\left(\frac{1}{2}\left(\omega_0 t - \theta - \frac{1}{2}\pi\right)\right)\cos\left(\frac{\pi(x + (\ell/2))}{2\ell}\right) \tag{7.26a}$$

$$\frac{\eta^{\mathrm{II}}}{h} \approx \frac{4}{3}\left(\frac{\Omega - \omega_0}{\omega}\right) + 4\varepsilon\sqrt{\frac{2}{3\pi}\left(1 + \frac{\pi^2 hH}{\ell^2}\right)}\sin\left(\frac{1}{2}\left(\omega_0 t - \theta - \frac{1}{2}\pi\right)\right)\sin\left(\frac{\pi(x + (\ell/2))}{2\ell}\right) \tag{7.26b}$$

Verhagen and WijnGaarden (1965) conducted a series of experimental tests to verify the above analytical results. For a rectangular container of width 1.2 m and liquid depth 0.09 cm, Figure 7.3 shows a comparison of analytical and experimental results of time history records taken at different positions along the tank $x/\ell = 0.0$, 0.333, 0.492, and 0.667 for excitation frequency $\Omega = \omega_0$. The analytical results, shown by solid curves, reveal some kind of saw tooth waveform, while the measured results show that the peaks of wave height are close to the analytical peaks. For $x/\ell = 0.492$, the wave peak reaches a value close to the liquid depth and twice the predicted value at half the period. The influence of excitation amplitude, φ_0, on the

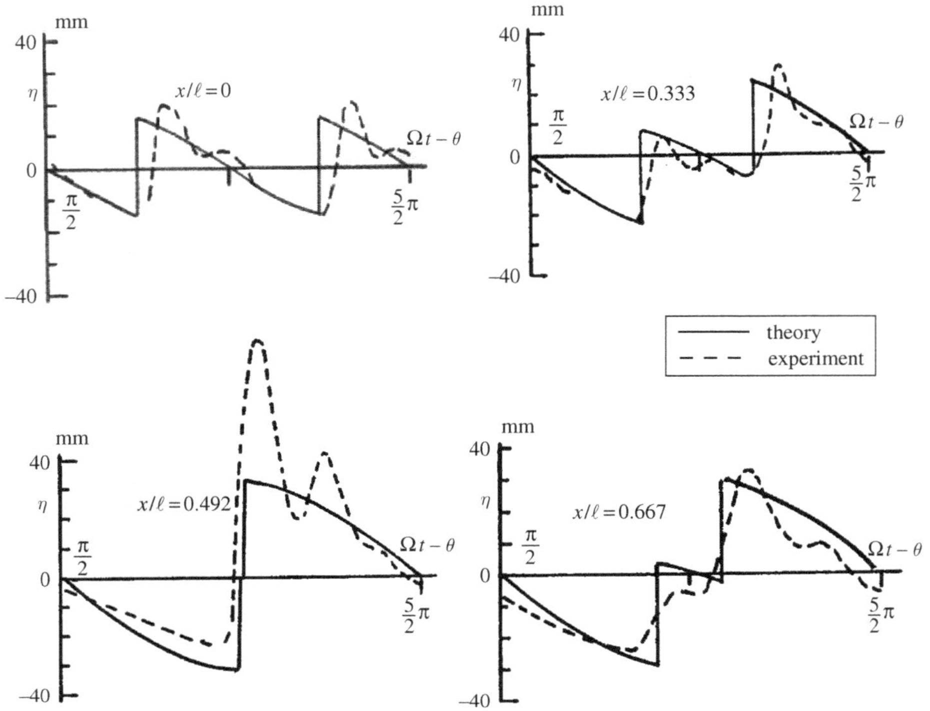

Figure 7.3 Time history records of surface elevation at different values of x, $\Omega = \omega_0$, $\varphi_0 = 2^\circ$. (Verhagen and WijnGaarden, 1965)

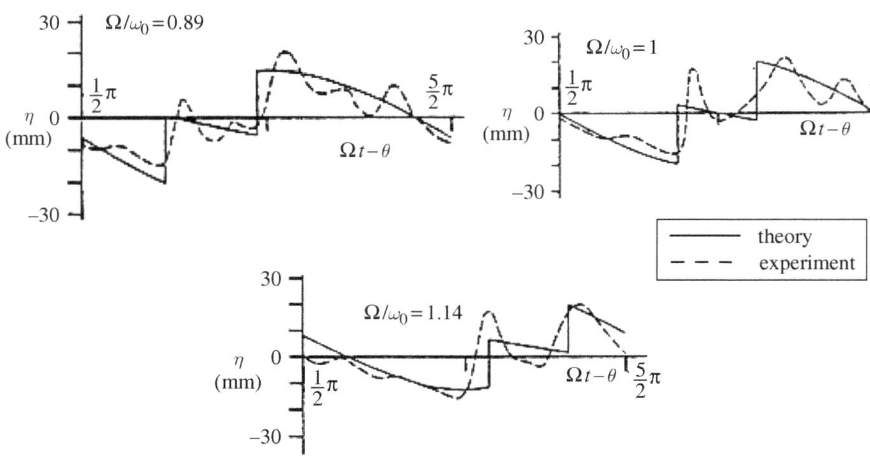

Figure 7.4 Time history records of surface elevation for different excitation frequencies and $\varphi_0 = 1^\circ$ at $x/\ell = 0.5$. (Verhagen and WijnGaarden, 1965)

surface elevation for different values of excitation frequency ratio $\Omega/\omega_0 = 0.89$, 1, and 1.14, at $x = \ell/4$ is shown in Figure 7.4.

The hydrodynamic moment about the y-axis through O, can be estimated in the neighborhood of resonance as follows

$$\frac{M(t)}{\rho g(\ell/2)^3} = -(2/\ell)^3 \int_{-\ell/2}^{\ell/2} \eta x\, dx \tag{7.27}$$

Using relations (7.21) and (7.26) the above integral can be evaluated. For $0 < t < \pi/\Omega$, the integral gives

$$\begin{aligned}\frac{M(t)}{\rho g(\ell/2)^3} = \frac{128\,\varepsilon h}{\pi^2 \ell}\sqrt{\frac{2}{3\pi}\left(1+\frac{\pi^2 hH}{\ell^2}\right)}\Bigg[\left(-1+\cos\frac{1}{2}\omega_0 t+\sin\frac{1}{2}\omega_0 t\right)\\ \times\cos\left(\frac{1}{2}\vartheta+\frac{1}{4}\pi\right)+\left(-\frac{1}{2}\omega_0 t+\frac{1}{4}\pi+\sin\frac{1}{2}\omega_0 t-\cos\frac{1}{2}\omega_0 t\right)\\ \times\sin\left(\frac{1}{2}\theta+\frac{1}{4}\pi\right)\Bigg]\end{aligned} \tag{7.28a}$$

For the other half of the period, $\pi/\Omega < t < 2\pi/\Omega$, the moment takes the form

$$\begin{aligned}\frac{M(t)}{\rho g(\ell/2)^3} = \frac{128\varepsilon h}{\pi^2 \ell}\sqrt{\frac{2}{3\pi}\left(1+\frac{\pi^2 hH}{\ell^2}\right)}\Bigg[\left(1+\cos\frac{1}{2}\omega_0 t-\sin\frac{1}{2}\omega_0 t\right)\\ \times\cos\left(\frac{1}{2}\theta+\frac{1}{4}\pi\right)+\left(\frac{1}{2}\omega_0 t-\frac{3}{4}\pi+\sin\frac{1}{2}\omega_0 t-\cos\frac{1}{2}\omega_0 t\right)\\ \times\sin\left(\frac{1}{2}\theta\ \frac{1}{4}\pi\right)\Bigg]\end{aligned} \tag{7.28b}$$

Away from resonance, the hydrodynamic moment can be determined from the linear theory according to equation (7.8), which gives

$$\frac{M(t)}{\rho g(\ell/2)^3} = 8\varphi_0(\omega_0/\pi\Omega)\sin(\Omega t-\theta)[(2\omega_0/\pi\Omega)\tan(\pi\Omega/2\omega_0)-1] \tag{7.28c}$$

The measured hydrodynamic moment about O was decomposed into different frequencies through the Fourier series

$$\frac{M(t)}{\rho g(\ell/2)^3} = \sum_{n=1}^{\infty} \overline{M}_n \sin\{n(\omega_0 t-\theta)-\psi_n\} \tag{7.29}$$

where ψ_n is the phase difference between the nth harmonics in the measured moment and the nth harmonic of the container excitation. Figure 7.5 shows the dependence of the measured values of the fundamental component $\overline{M}_1$ on the excitation frequency for different values of excitation amplitude $\varphi_0 = 1^\circ, 2^\circ, 3^\circ$, and 4°. Both measured and predicted plots reveal some deviations at resonance, which are attributed to the fact that damping forces are ignored in the analysis.

Lee, *et al.* (2002) employed Glimm's method to numerically simulate the liquid hydraulic jump and impact in a rolling tank similar to the one used by Verhagen and WijnGaarden. Figure 7.6, taken from Lee, *et al.* (2002), shows a comparison of a time history record of the

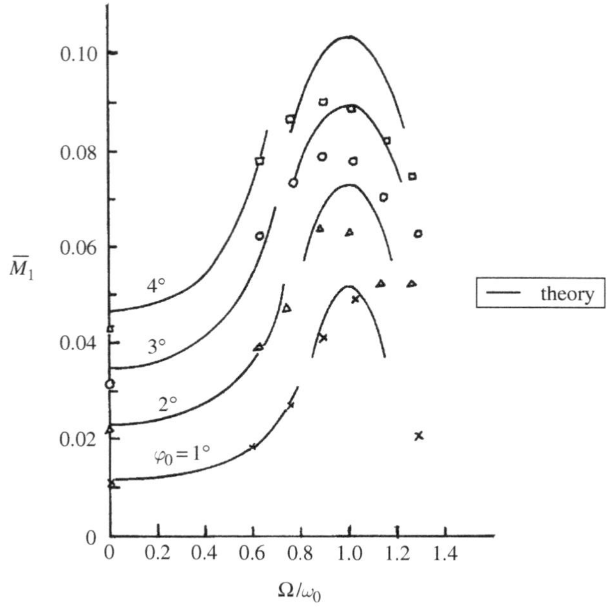

Figure 7.5 Dependence of the measured hydrodynamic moment acting on the container on the excitation frequency ratio for different excitation amplitudes. Symbols are experimental measurements. (Verhagen and WijnGaarden, 1965)

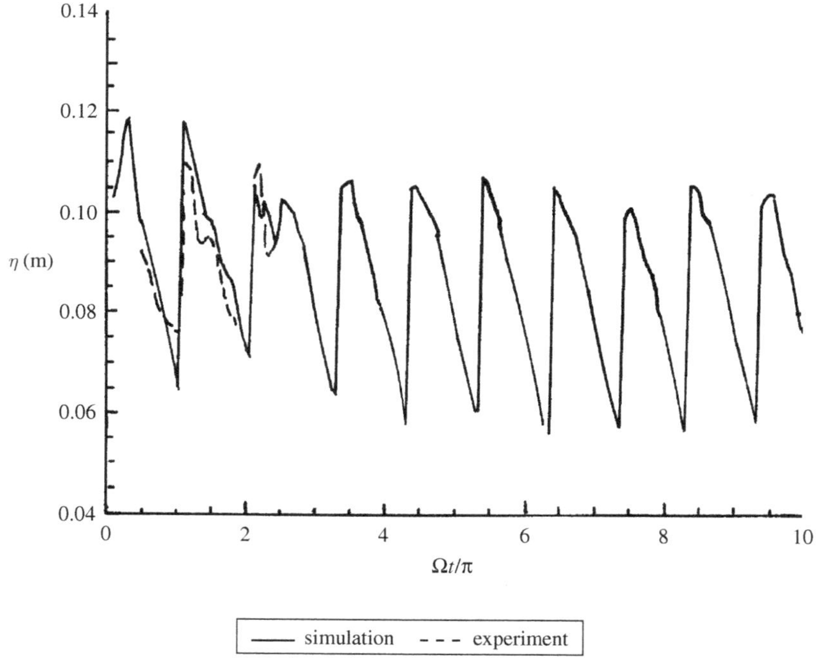

Figure 7.6 Time history record of surface elevation at the tank center for $\varphi_0 = 2°$ (Lee, *et al.*, 2002)

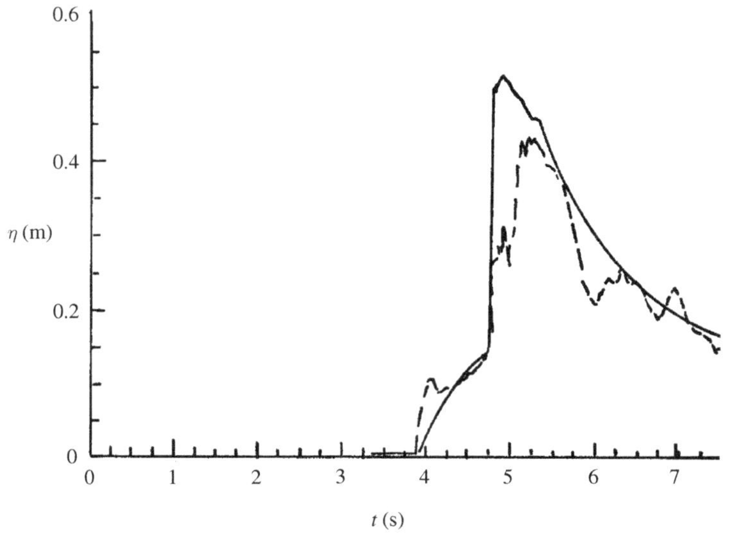

Figure 7.7 Development of jump at the left end of the tank. (Lee, *et al.*, 2002)

liquid surface elevation at the center of the tank under excitation amplitude $\varphi_0 = 2^\circ$. The experimental measurements are shown only for one period of oscillation. Both results show the nature of the hydraulic jump in the form of a saw-tooth waveform. The abrupt change in the water height was idealized mathematically with a discontinuity in the height. In reality, the water surface is always rounded on both sides of the jump due to the effects of surface tension and viscosity. Lee, *et al.* (2002) conducted a test to measure the dynamics of the water impact in a reservoir that was divided in two compartments by a flap. One compartment was full of water up to a depth of 0.6 m, while the other was empty. Upon lifting the flap, the water impacting on the tank bottom and wall was measured. The time history records at the left end of the tank as measured and numerically estimated are shown in Figure 7.7. Figure 7.8 shows a comparison of the impact pressure at a lower point of the left end of the tank wall. Both results show an abrupt jump in the pressure at $t \approx 4$ s, the time of impact with the wall. The numerical simulation gives a smooth pressure evolution while the measured results reveal two sharp spikes.

7.3 Lateral excitation of a rectangular tank

Under translational excitation of liquid containers, the motion of the free surface becomes complicated especially if the excitation frequency is close to the first sloshing mode. Other modes can be excited through nonlinear coupling of different modes. Faltinsen, *et al.* (2000, 2003) developed a general analytical model describing the free-surface motion and velocity potential in generalized Fourier series. They employed the variational principle to derive the general modal system. For a rectangular tank of width ℓ and liquid depth h, they expressed the liquid wave height by the following series

$$\eta(x, t) = \sum_{i=1}^{\infty} a_i(t) f_i(x) \tag{7.30}$$

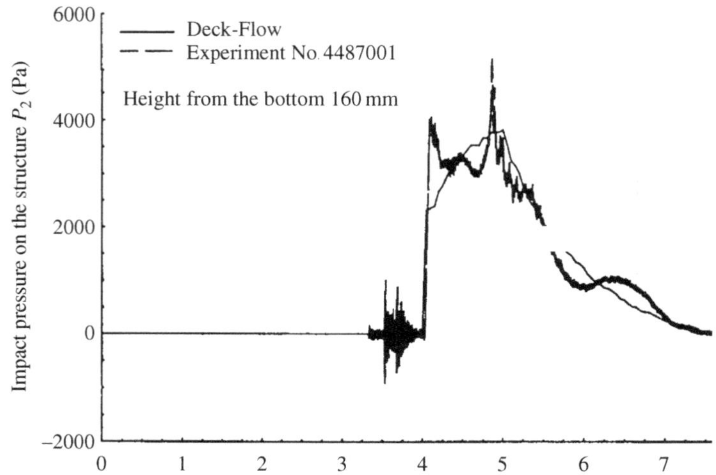

Figure 7.8 Development of impact pressure at a lower point of the tank left end. (Lee, *et al.*, 2000)

where a_i stand for the generalized coordinates and $f_i(x)$ are the mode shapes. The analysis of Faltinsen, *et al.* (2000) yields the following system of ordinary differential equations describing the modal oscillations of the liquid free surface under translational motion $V(_{0x}t) = \Omega\ X_0 \sin \Omega t$

$$\ddot{a}_1 + \zeta_1\dot{a}_1 + \omega_1^2 a_1 + c_1(\ddot{a}_1 a_2 + \dot{a}_1\dot{a}_2) + c_2(\ddot{a}_1 a_1^2 + \dot{a}_1^2 a_1) + c_3\ddot{a}_2 a_1 + P_1\dot{V}_{0x} = 0 \tag{7.31a}$$

$$\ddot{a}_2 + \zeta_2\dot{a}_2 + \omega_2^2 a_2 + c_4\ddot{a}_1 a_2 + c_5\dot{a}_1^2 = 0 \tag{7.31b}$$

$$\ddot{a}_3 + \zeta_3\dot{a}_3 + \omega_3^2 a_3 + c_6\ddot{a}_1 a_2 + c_7\ddot{a}_1 a_1^2 + c_8\ddot{a}_2 a_1 + c_9\dot{a}_1\dot{a}_2 + c_{10}\dot{a}_1^2 a_1 + P_3\dot{V}_{0x} = 0 \tag{7.31c}$$

The higher modes may be described by the linear uncoupled equations

$$\ddot{a}_i + \zeta_i\dot{a}_i + \omega_i^2 a_i + P_i\dot{V}_{0x} = 0 \tag{7.31d}$$

where $\omega_i^2 = ((ig\pi/\ell)\tanh(i\pi h/\ell))$, $\zeta_i\dot{a}_i$ is a linear damping term for mode i,

$$\begin{aligned}
P_{2i-1} &= -\frac{4}{\pi(2i-1)}\tanh((2i-1)\pi h/\ell)\\
c_1 &= \frac{\pi}{2\ell}\left[\frac{1}{\tanh(\pi h/\ell)} + \tanh(\pi h/\ell)\right]\\
c_2 &= \left(\frac{\pi}{2\ell}\right)^2\left[-1 + \frac{2}{\tanh(\pi h/\ell)\tanh(2\pi h/\ell)}\right]\\
c_3 &= \frac{\pi}{2\ell}\left[\tanh(\pi h/\ell) - \frac{1}{\tanh(2\pi h/\ell)}\right]\\
c_4 &= \frac{\pi}{\ell}\left[\tanh(2\pi h/\ell) - \frac{1}{\tanh(\pi h/\ell)}\right]\\
c_5 &= \frac{\pi}{2\ell}\left\{\tanh(2\pi h/\ell)\left[1 - \frac{1}{\tanh(\pi h/\ell)}\right] - \frac{2}{\tanh(\pi h/\ell)}\right\}
\end{aligned}$$

$$c_6 = \frac{3\pi}{2\ell}\left[\tanh(3\pi h/\ell) - \frac{1}{\tanh(\pi h/\ell)}\right]$$

$$c_8 = \frac{3\pi}{2\ell}\left[\tanh(3\pi h/\ell) - \frac{1}{\tanh(2\pi h/\ell)}\right]$$

$$c_7 = 6\left(\frac{\pi}{2\ell}\right)^2\left[\frac{3}{4} - \frac{\tanh(4\pi h/\ell)}{\tanh(\pi h/\ell)} - \tanh(3\pi h/\ell)\tanh(4\pi h/\ell) + \frac{1}{\tanh(\pi h/\ell)\tanh(2\pi h/\ell)} + \frac{\tanh(3\pi h/\ell)}{4\tanh(\pi h/\ell)}\right]$$

$$c_9 = \frac{3\pi}{2\ell}\left[-\frac{1}{\tanh(\pi h/\ell)} - \frac{1}{\tanh(2\pi h/\ell)}\left(1 + \frac{\tanh(3\pi h/\ell)}{\tanh(\pi h/\ell)} - \tanh(3\pi h/\ell)\tanh(\pi h/\ell)\right)\right]$$

$$c_{10} = 3\left(\frac{\pi}{2\ell}\right)^2\left[3 - 4\frac{\tanh(4\pi h/\ell)}{\tanh(\pi h/\ell)}(1 + \tanh(\pi h/\ell)\tanh(3\pi h/\ell) + \frac{3}{2\tanh(\pi h/\ell)\tanh(\pi h/\ell)} + 2\left(\frac{\tanh(3\pi h/\ell)}{\tanh(\pi h/\ell)} - \frac{\tanh(\pi h/\ell)}{\tanh(2\pi h/\ell)}\right)\right] \tag{7.32}$$

Equations (7.31a) and (7.31b) are coupled through nonlinear inertia terms. Equation (7.31c) is directly excited by the tank motion and nonlinearly coupled with the first two modes. To obtain the steady state periodic solution of equations (7.31) an asymptotic approximation technique will be employed. The solution of the first mode is assumed to be in the form

$$a_1(t) = A\cos\Omega t + O(A) \tag{7.33}$$

Substituting equation (7.33) into equation (7.31b), gives

$$a_2 = A^2\left(\frac{(c_4 - c_5)}{2}\frac{\Omega^2}{\omega_2^2} + \frac{c_4 + c_5}{2[(\omega_2^2/\Omega^2) - 4]}\cos 2\Omega t\right) + O(A^2) \tag{7.34}$$

The amplitude A can be determined by substituting equations (7.33) and (7.34) into equation (7.31a) by collecting harmonic terms of the lowest order. This process gives

$$F_h(r_1, r_2, \overline{A}) = (r_1^2 - 1)\overline{A} + \nu(r_2, \overline{h})\overline{A}^3 - \overline{P}_1\overline{X}_0 = 0 \tag{7.35}$$

where

$$\nu(r_2, \overline{h}) = \overline{c}_1\left(\frac{\overline{c}_4 - \overline{c}_5}{2r_2^2} + \frac{\overline{c}_4 + \overline{c}_5}{4(r_2^2 - 4)}\right) - \frac{1}{2}\,\overline{c}_2 - \overline{c}_3\frac{\overline{c}_4 + \overline{c}_5}{(r_2^2 - 4)} \tag{7.36}$$

The over-bar terms are nondimensional with respect to the length ℓ, and $r_i = \omega_1/\Omega$, where $i = 1, 2$. Expanding equation (7.35) into a Taylor series, gives

$$F_h(r_1, r_2, \overline{A}) = F_h(r_1, r_2 0) + \left.\frac{\partial F_h(r_1, r_2, \overline{A})}{\partial \overline{A}}\right|_{\overline{A}=0}\overline{A} + \frac{1}{2}\left.\frac{\partial^2 F_h(r_1, r_2, \overline{A})}{\partial \overline{A}^2}\right|_{\overline{A}=0}\overline{A}^2 + \frac{1}{6}\left.\frac{\partial^3 F_h(r_1, r_2, \overline{A})}{\partial \overline{A}^3}\right|_{\overline{A}=0}\overline{A}^3 + \left(\overline{A}^3\right) \tag{7.37}$$

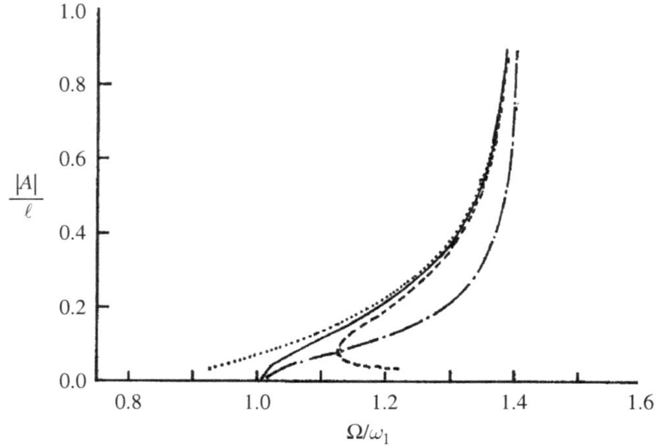

(a) $h/\ell = 0.173, \sqrt{\dfrac{\tanh(2\pi h/\ell)}{2\tanh(\pi h/\ell)}} = 0.9$

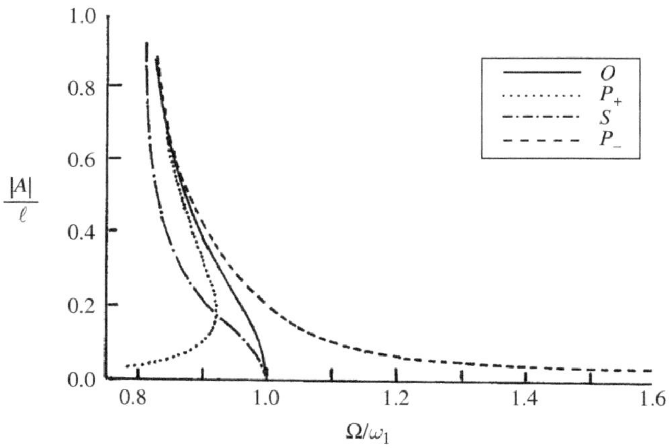

(b) $h/\ell = 1.0, \sqrt{\dfrac{\tanh(2\pi h/\ell)}{2\tanh(\pi h/\ell)}} = 0.7085$

Figure 7.9 Amplitude–frequency response for two different values of fluid depth ratio. (Faltinsen, *et al.*, 2000)

Equation (7.35) gives infinite solution at resonance, $r_1 = 1$. In this case, the condition for $\nu(\omega_2/\omega_1, \bar{h}) = 0$ provides the critical fluid height, $h/\ell = 0.3378$, which coincides with the critical depth obtained by Waterhouse (1994). This critical depth separates between the hard and soft liquid nonlinear characteristics. On the other hand, if Ω is close to ω_1, but $\Omega \neq \omega_1$, then equation $\nu(r_2, \bar{h}) = 0$ gives a different value of the critical depth that is dependent on Ω. Figure 7.9 shows the dependence of the response amplitude as determined from equation (7.35) on the excitation frequency for different values of the liquid depth and for fixed excitation amplitude $X_0/\ell = 0.0173$. Branch O represents the solution of equation (7.35) when $X_0 = 0$, that is, the backbone curve. It represents the dependence of the liquid natural frequency on the liquid amplitude. Branch S represents the set of all turning points of the branch P_+ or $P-$, depending on the liquid depth, for different excitation amplitudes. The turning points correspond to the vertical tangency of the curve. This can be obtained by differentiating equation (7.35) with respect to $\overline{A}$, which gives

$$\left(r_1^2 - 1\right) + 3\nu(r_2, \overline{h})\overline{A}^2 = 0 \tag{7.38}$$

Equation (7.38) reveals that branch S does not depend on the value of the excitation amplitude, but depends on the liquid depth ratio. The turning point separates between stable and unstable solutions. The branch above the turning point is unstable while the lower branch corresponds to the stable solution. In the absence of a turning point the branch represents a stable solution.

For periodic solutions there is a critical value of Ω/ω_1 for which the amplitude of the second mode tends to infinity, which can happen for small liquid depth. The condition for the secondary resonance is

$$\frac{\Omega}{\omega_1} \to \sqrt{\frac{\tanh(2\pi h/\ell)}{2\tanh(\pi h/\ell)}} \tag{7.39}$$

The above expression characterizes the applicability of the asymptotic approximation, and Figure 7.9 is given for two different values of frequency ratio.

In order to determine the hydrodynamic sloshing loads, the pressure distribution must be estimated first, that is,

$$p(x, z, t) = p_0 - \rho\left(\frac{\partial\Phi}{\partial t} + \frac{(\nabla\Phi)}{2} + gz - \Omega^2 x X_0 \cos(\Omega t) - \frac{1}{2}X_0^2\Omega^2\sin^2(\Omega t)\right) \tag{7.40}$$

$$\nabla\Phi = \sum_{i=1}^{N} i\frac{\pi}{\ell}f_i(a_i)\left\{-\sin\left(i\frac{\pi}{\ell}(x+\ell/2)\right)\frac{\cosh((i\pi/\ell)(z+h))}{\cosh((i\pi/\ell)h)},\ 0,\right.$$

$$\left.\cos\left(i\frac{\pi}{\ell}(x+\ell/2)\right)\frac{\sinh((i\pi/\ell)(z+h))}{\cosh((i\pi/\ell)h)}\right\} \tag{7.41}$$

$$\frac{\partial\Phi}{\partial t} = \sum_{i=1}^{N}\dot{f}_i(a_i)\cos\left(i\frac{\pi}{\ell}(x+\ell/2)\right)\frac{\cosh((i\pi/\ell)(z+h))}{\cosh((i\pi/\ell)h)} \tag{7.42}$$

where

$$f_1(a_i) = \frac{\dot{a}_1}{(\pi/\ell)\tanh(\pi h/\ell)} - \frac{\dot{a}_1 a_2}{2\tan^2(\pi h/\ell)} - \frac{a_1\dot{a}_2}{2\tan(\pi h/\ell)\tanh(2\pi h/\ell)}$$

$$+\frac{\pi}{2\ell}\left(-\frac{1}{2}+\frac{2}{\tan(\pi h/\ell)\tanh(2\pi h/\ell)}\right)\frac{a_1^2\dot{a}_1}{\tanh(\pi h/\ell)} \tag{7.43a}$$

$$f_2(a_i) = \frac{1}{4(\pi/2\ell)\tanh(2\pi h/\ell)}\left(\dot{a}_2 - \frac{2(\pi/2\ell)}{\tanh(\pi h/\ell)}a_1\dot{a}_1\right) \tag{7.43b}$$

$$f_3(a_i) = \frac{\dot{a}_3}{6(\pi/2\ell)\tanh(3\pi h/\ell)} - \frac{\dot{a}_1 a_2}{2\tanh(\pi h/\ell)\tanh(3\pi h/\ell)}$$

$$- \frac{a_1 \dot{a}_2}{2\tanh(2\pi h/\ell)\tanh(3\pi h/\ell)} + \frac{\pi}{2\ell}\left[\frac{3\tanh(2\pi h/\ell)}{2\tanh(3\pi h/\ell)}\right.$$

$$- \frac{\tanh(4\pi h/\ell)}{\tanh(\pi h/\ell)\tanh(3\pi h/\ell)} - \tanh(4\pi h/\ell)$$

$$+ \frac{1}{\tanh(\pi h/\ell)\tanh(2\pi h/\ell)\tanh(3\pi h/\ell)}$$

$$\left. + \frac{1}{\tanh(\pi h/\ell)\tanh(3\pi h/\ell)}\right]\dot{a}_1 a_1^2 \tag{7.43c}$$

$$f_i(a_i) = \frac{\ell}{i\pi\tanh(i\pi h/\ell)}\dot{a}_1, \quad i > 3 \tag{7.43d}$$

The hydrodynamic force is determined by integrating the pressure over the wetted area of tank sidewall and bottom. The following are the x- and z-force components

$$F_x/m_l = \Omega^2 X_0 \cos\Omega t + \frac{\ell}{\pi^2 h}\sum_{i=1}^{N}\ddot{a}_i(t)\frac{1}{i^2}\left[1 + (-1)^{i+1}\right] \tag{7.44a}$$

$$F_z/m_l = -g - \frac{1}{2h}\sum_{i=1}^{N}\left\{\ddot{a}_i(t)a_i(t) + \dot{a}_i^2\right\} \tag{7.44b}$$

where m_l is the total fluid mass. This analysis is based on the assumption that the water waves do not hit the ceiling. During liquid free-surface motion the x_c and z_c coordinates of the center of mass are

$$x_c = -\frac{1}{\pi h}\sum_{i=1}^{N}\frac{a_i}{i^2}\left(1 + (-1)^{i+1}\right), \quad z_c = -\frac{h}{2} + \frac{1}{4h}\sum_{i=1}^{N}a_i^2 \tag{7.45a,b}$$

Figure 7.10 shows the trajectory of the center of mass.

Faltinsen, *et al.* (2000) conducted a series of experimental tests on a rectangular tank partially filled with water in the neighborhood of the first mode natural frequency. Under translational (surge) impulse and sinusoidal excitations, Figure 7.11 shows measured and predicted time history records of the free-surface elevation at the left wall of the tank for $h/\ell = 0.289$, $X_0/\ell = 0.0116$, and $\Omega/\omega_1 = 1.1264$. It is seen that the experimental measurements reveal an impact with the tank ceiling. The observed disagreement between analytical and experimental results of the liquid wave height for the case of forcing impact is mainly attributed to the energy dissipation during impact. As the liquid hits the ceiling, a jet is formed and eventually the free surface overturns causing the liquid jet to hit the free surface. This also causes energy dissipation. An estimate of this energy loss was calculated by Faltinsen and Rognebakke (1999) based on the assumption that the kinetic and potential energies in the jet are dissipated. Figure 7.12 shows a comparison between predicted and measured time history records where the damping terms were included in equations (7.31) for $h/\ell = 0.289$,

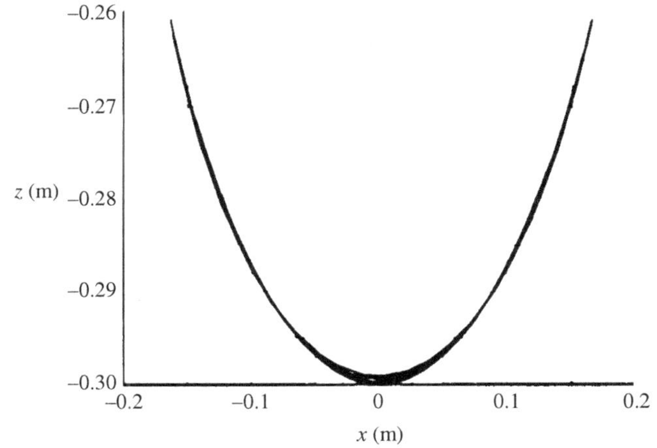

Figure 7.10 Trajectory of the center of mass. (Faltinsen, *et al.*, 2000)

$X_0/\ell = 0.0289$ and, $\Omega/\omega_1 = 1.235$. Figure 7.13 shows the contribution of the first three modes to the free-surface elevation as measured at the left wall of the tank for $h/\ell = 0.1734$, $\Omega/\omega_1 = 1.173$, and $X_0/\ell = 0.0231$. It is seen that the influence of the first two modes is reflected in the third mode since it contains components of their frequencies.

7.4 Impact due to sudden acceleration

The problem of liquid impact on tank bulkheads is a serious problem in liquid propellant rockets and space vehicles during a sudden thrust reversal. The influence of the liquid motion due to impulsive accelerations imposed on space vehicles was considered by Eide (1964). Thrust termination before virtually all the liquid propellants have been used may result in a possible rupture of a tank bulkhead or a structural breakup. In the case of hypergolic fuels, it can result in a fireball. Dalzell (Chapter 7.10 in Abramson, 1966a) presented an overview of three experimental studies of dome impact phenomenon conducted during the period 1957 to 1964. The objectives of these tests were to determine the pressure resulting from the impact of fuel on the tank head under sudden deceleration. These studies were reported by Epperson and Brown (1957), Epperson, *et al.* (1961), Cokonis, *et al.* (1963), Pinson (1963), Dalzell, *et al.* (1964), and Stephens (1965). The main results indicated that the pressure distribution over the upper bulkhead reached its maximum value at the center. Under vertical firing with hemispherical heads, the pressure near the edge of the dome averages about half of the pressures near the top (Young, 2002).

Sudden vehicle acceleration can result in high-pressure impact on the roof of liquid tanks. Most of the studies under sudden acceleration were mainly experimental. Ye and Birk (1994) measured the liquid pressure on a horizontal cylinder. A tank was subjected to a sudden acceleration induced by an impact hammer along the longitudinal axis. Figure 7.14 shows a series of time history records taken for different fill levels and at different locations. It is seen that for high fill levels, $h/D = 0.98$, where D is the tank diameter, the pressure spike is a water hammer effect, see Figure 7.14(d). On the other hand, for high fill levels and shorter tanks, the liquid acts like an accelerated fluid column and the pressure is proportional to the tank average

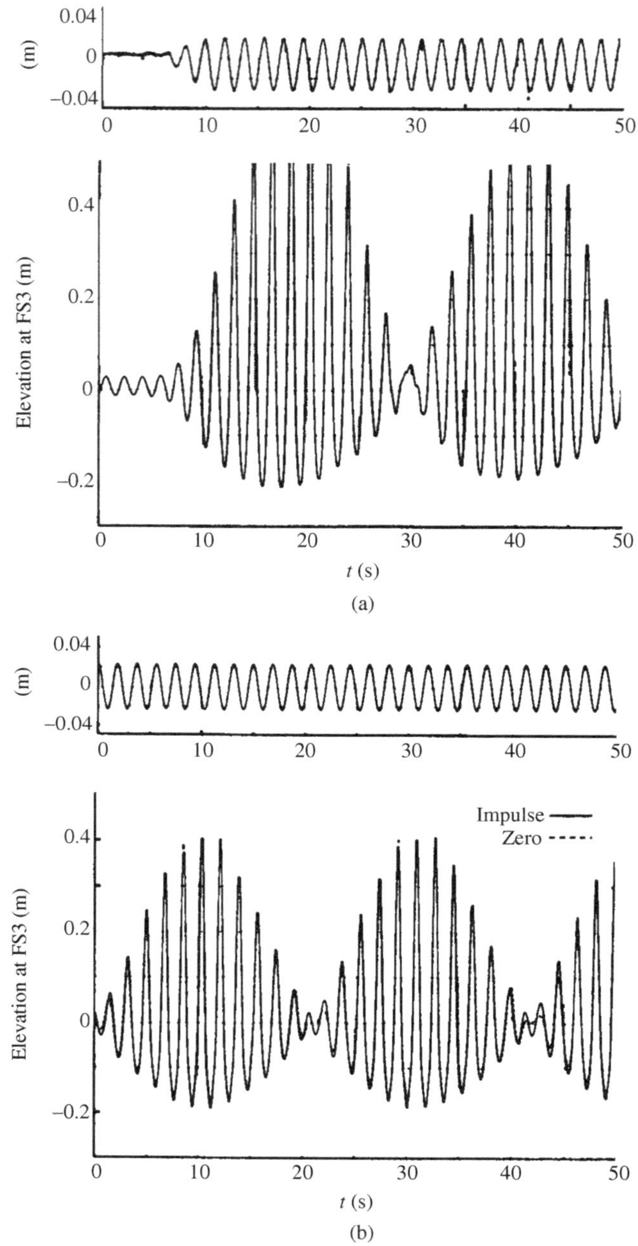

Figure 7.11 Measured and predicted time history records of free-surface elevation at the tank left wall. (Faltinsen, *et al.*, 2000)

acceleration. For low fill levels, $h/D < 0.95$, the liquid slosh dominates the pressure. The dependence of the liquid impact pressure on the impact velocity (represented by the Froude number $\mathrm{Fr} = U^2/\rho g$, where U is the post-impact velocity of the tank velocity) is shown in Figure 7.15(a) for different values of liquid fill level, and $L/D = 6$. The measurements were taken at the

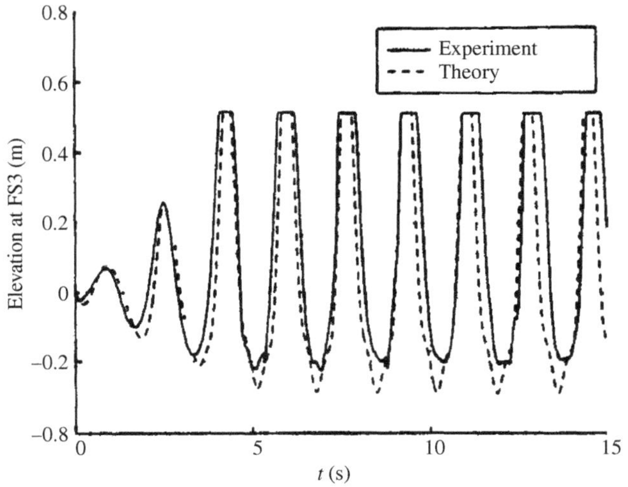

Figure 7.12 Measured and predicted time history of surface elevation after including damping. (Faltinsen, *et al.*, 2000)

impact end A. Figure 7.15(b) shows another set of pressure measurements for different values of length-to-diameter ratio. Note that the pressure is given by the nondimensional parameter $P^{**} = P_{\text{peak}}/(0.5\rho U^2)$, where ρ is the liquid density. In all cases, the impact pressure increases monotonically with the speed of the tank.

7.5 Modeling of hydrodynamic impact forces

In dealing with the dynamics of moving vehicles carrying liquid containers, it is important to allow for analytical modeling to represent the impact liquid forces on the container. Sloshing impact loading cannot be viewed as a single loading event since it can be repeated due to the inertia and restoring forces. This type of impact is known as vibro-impact (Babitsky, 1998). Systems involving vibro-impact of masses colliding with rigid or elastic barriers during their oscillation constitute a specific class of strongly nonlinear systems. The nonlinearity is mainly due to extremely rapid velocity changes during impacts. These changes are usually treated as being instantaneous (velocity jumps) and they lead to various strongly nonlinear features of system behavior. If the system is linear with constant coefficients and is subjected to impact loading it will experience nonlinear behavior. Liquid pressure impacts are a source of strong nonlinearity in a liquid tank system and belong to the broad class of nonsmooth dynamics.

When a particle impacts against a rigid barrier, its normal velocity changes direction. Such a finite change implies that the impact is regarded as instantaneous. Thus, the governing differential equations of motion should be supplemented with finite relations between the values of the generalized coordinates and velocities just before and after impact. These finite relations, which should be satisfied at the instants of impact, are in fact responsible for various specific strong nonlinear features of a vibro-impact systems' behavior. The time instants of impacts are not specified a priori but governed by the equations of vibro-impact systems. The nonlinearity becomes clear when trying to exclude the impact finite relations

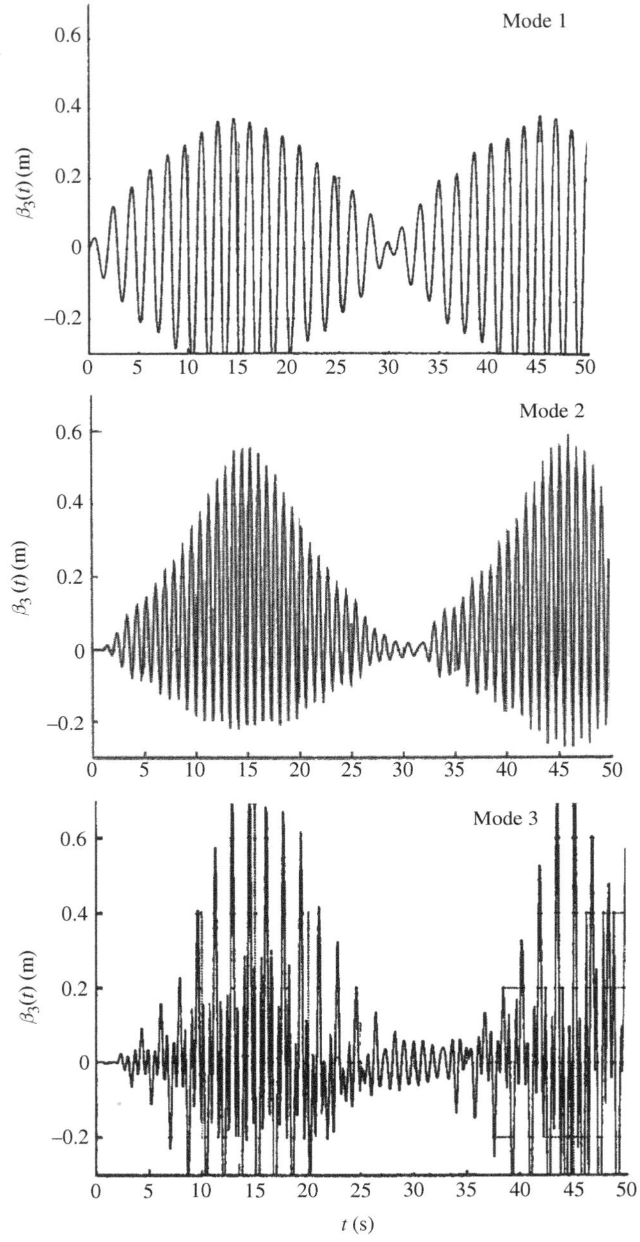

Figure 7.13 Contribution of the first three modes to the free-surface elevation at the tank left wall. (Faltinsen, *et al.*, 2000)

by using the Dirac delta-functions in the equations of motion. This can be easily done for the case of linear equations of motion between impacts. In this case, the equations of motion involve delta-function terms.

The vibro-impact problem of a pendulum in a vibrating container was analyzed by Babitsky (1966) and Buzhinskii and Stolbetsov (1987). One may model the liquid, which experiences

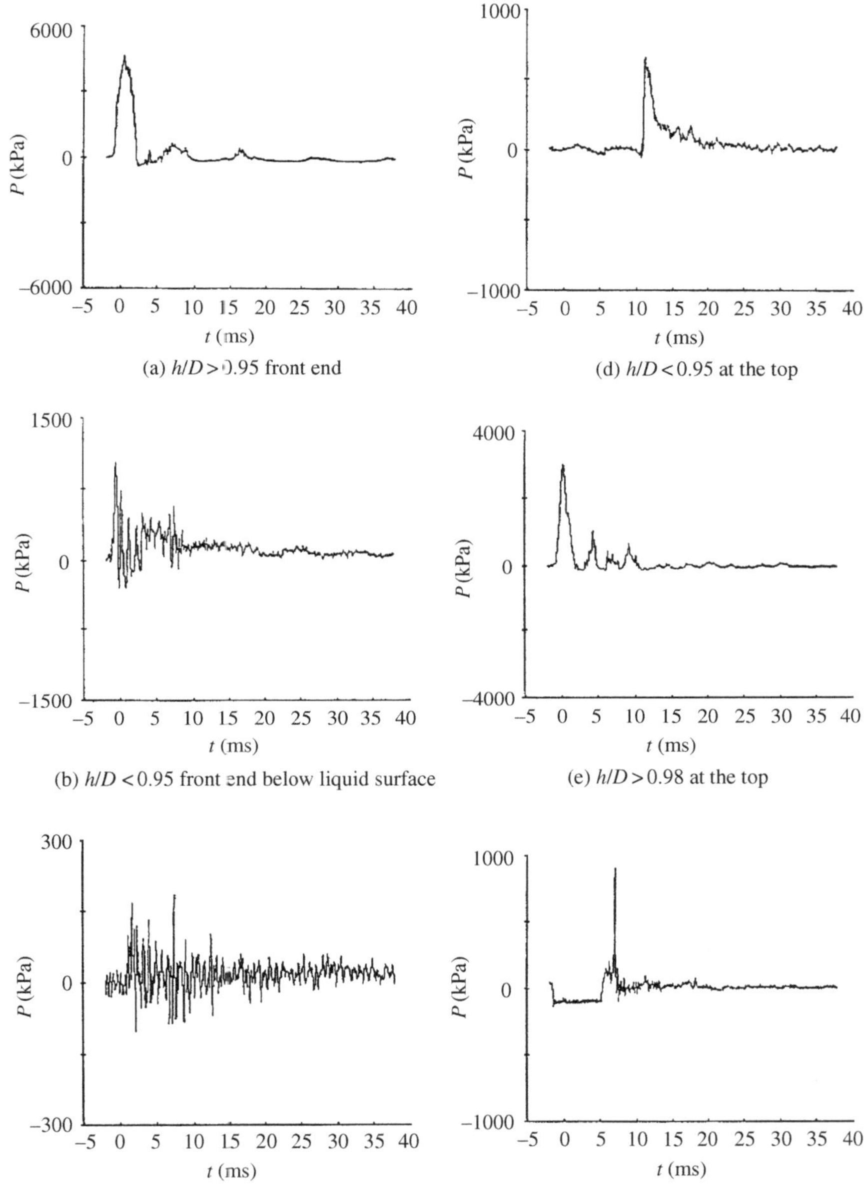

(a) $h/D > 0.95$ front end
(b) $h/D < 0.95$ front end below liquid surface
(c) $h/D < 0.8$ front end
(d) $h/D < 0.95$ at the top
(e) $h/D > 0.98$ at the top
(f) Full tank at rear end

Figure 7.14 Typical pressure time histories measured at different tank locations. (Ye and Birk, 1994)

impact motion, by a simple pendulum of mass m and length ℓ. The pendulum can reach the walls of the tank if its angle with the vertical axis is $\theta = \pm\theta_0$. Considering the pendulum and the tank walls as rigid bodies, one must introduce the constraint $|\theta| \leq \pm\theta_0$. From the point of view of analytical techniques in nonlinear mechanics such constraints essentially complicate the analysis because one must match solutions at points of interaction $\{t : \theta(t) = \pm\theta_0\}$, which is a priori unknown. Hence, it is useful to avoid operations with constraints. One can

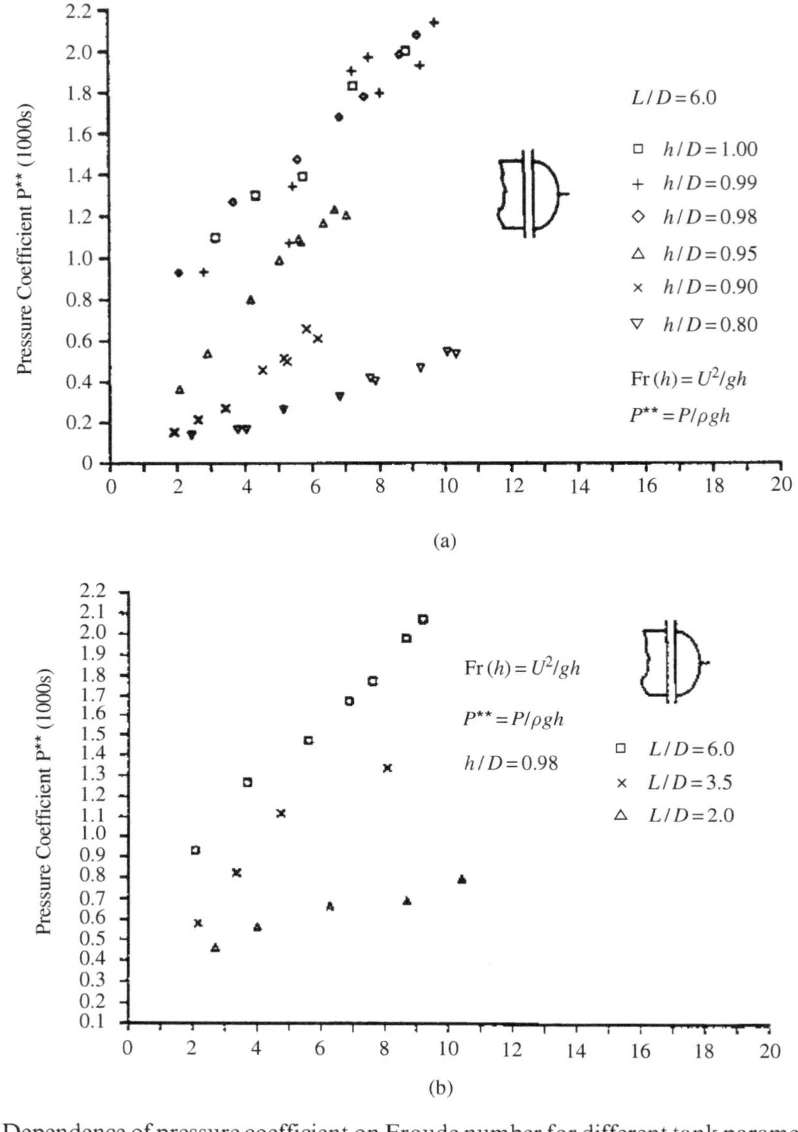

Figure 7.15 Dependence of pressure coefficient on Froude number for different tank parameters. (Ye and Birk, 1994)

phenomenologically describe the interaction between the pendulum and the tank walls with a special potential field of interaction, which is very weak in the region $|\theta| < \pm\,\theta_0$, but becomes fast-growing in the neighborhood of the points $|\theta| = \pm\theta_0$. For example, the desirable properties of the potential field can be provided by means of the following potential energy function (Hunt and Grossley, 1975 and Pilipchuk and Ibrahim, 1997),

$$\Pi_{\text{impact}} = \frac{b}{2n}\left(\frac{\theta}{\theta_0}\right)^{2n} \tag{7.46}$$

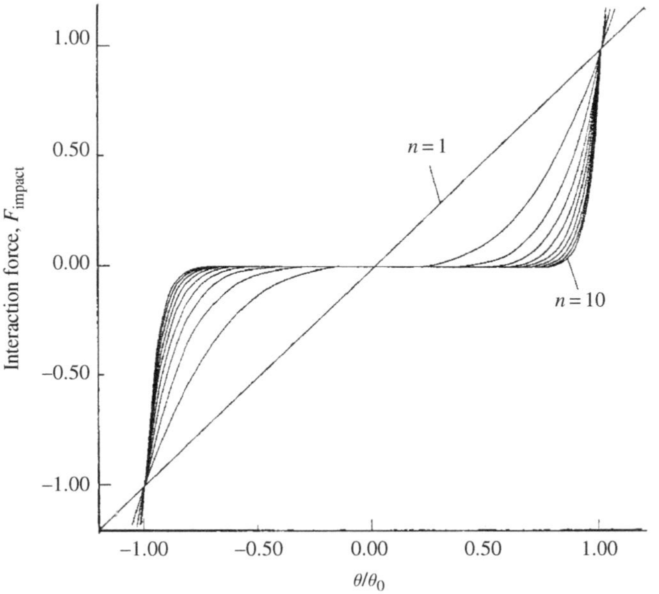

Figure 7.16 Variation of liquid impact force between the tank walls for different values of the exponent n.

where $n \gg 1$, is a positive integer, and b is a positive constant parameter.

The force of interaction is

$$F_{\text{impact}} = \frac{d}{d\theta}\left[\Pi_{\text{impact}}(\theta)\right] = b\left(\frac{\theta}{\theta_0}\right)^{2n-1} \tag{7.47}$$

One has a limit of absolutely rigid bodies interaction, if $n \to \infty$. For this case, the potential energy (7.46) takes the square well form. If the exponent $2n - 1$ is large and finite, then the interaction field is not absolutely localized at the points $\theta = \pm\theta_0$. This means that the tank walls and the pendulum mass are not rigid, but admit a small deformation about the points of contact $\theta = \pm\theta_0$. Thus, a finite value of n seems more realistic than the rigid body limit. Figure 7.16 shows the dependence of the impact force on the coordinate θ for different values of n. As n increases, the impact force at the walls approaches the δ function with some smoothness.

Suppose that the energy dissipation of the pendulum results from the pendulum interaction with the container walls. This means that the dissipation is spatially localized around the points $\theta = \pm\theta_0$. The localized dissipative force will be approximated by the expression

$$F_d = d\left(\frac{\theta}{\theta_0}\right)^{2p}\dot{\theta} \tag{7.48}$$

where d is a constant coefficient, $p \gg 1$ is a positive integer (generally $p \neq n$), and a dot denotes differentiation with respect to time t. Note that the constants b and d are determined experimentally. Relations (7.47) and (7.48) can be introduced into the analytical modeling of the equations of motions of systems involving vibro-impact. A typical case of simulating liquid sloshing impact will be presented in Section 7.7. The next section provides an overview of analytical methods that may be used for the vibro-impact problems.

7.6 Analytical approaches

A thorough description of various methods for the analysis of vibro-impact systems is well documented in Kobrinksy and Kobrinsky (1973) and Babitsky (1998). The following is a brief description of some selected methods.

7.6.1 Step-by-step integration method

This is known as the point-wise mapping method, where one has to match the solutions at points of impact.

7.6.2 Asymptotic approximation techniques

The asymptotic approximate methods were originally developed to treat weakly nonlinear dynamical systems (Nayfeh and Mook, 1979). These methods include perturbation techniques, asymptotic approximation methods, and averaging methods. Section 7.7 will consider the multiple scales method to analyze the interaction of liquid impact with an elastic support structure.

7.6.3 Nonsmooth coordinate transformation

This transformation was originally introduced by Zhuravlev (1977) and assumes rigid barriers and converts the vibro-impact system into an oscillator without barriers such that the equations of motion do not contain any impact terms. This technique has recently been used by Pilipchuk and Ibrahim (2000) to analyze the dynamic response of a system involving hydrodynamic sloshing impact. A brief demonstration of Zhuravlev's coordinate transformation using the example of a single-degree-of-freedom system that oscillates between two absolutely rigid barriers is introduced in this section. The main rationale of such a coordinate transformation is to convert the vibro-impact oscillator into an oscillator without barriers such that the corresponding equation of motion does not contain any *impact* terms. Consider a unit mass particle moving freely along the q-axis in a potential field with the potential energy $\Pi(q)$ under the condition $|q| \leq 1$. This condition can be provided by means of absolutely rigid walls at the points $q = \pm 1$. When the particle reaches the wall, the velocity changes its sign but preserves its magnitude. At the same time the coordinate $q(t)$ remains continuous with respect to time, t. The Lagrangian $L(q, \dot{q})$ of the system is given by the expression

$$L = \frac{1}{2}\dot{q}^2 - \Pi(q), \quad -1 \leq q \leq 1 \tag{7.49}$$

where a dot denotes differentiation with respect to time t. This is a mechanical system with constraints and hence no unique differential equation of motion can be written for the time domain $0 \leq t \leq \infty$. One should, in principal, consider different systems and match their solutions at the moments of impact, that is, the time instants of reflections with respect to the rigid walls. It is possible to obtain a unique differential equation of motion through a transformation to the coordinate x given by the expression

$$q = S(x) \tag{7.50}$$

where $S(x)$ is a periodic saw-tooth piece-wise linear function defined as

$$S(x) = \begin{cases} x & \text{if } -1 \leq x \leq 1 \\ 2 - x & \text{if } 1 \leq x \leq 3 \end{cases} \quad S(x + 4n) \overset{\forall x \in R}{=} S(x), \quad n = 1, 2, \ldots \tag{7.51}$$

Substituting transformation (7.50) into Lagrangian (7.49) gives

$$L = \frac{1}{2}[S'(x)\dot{x}]^2 - \Pi[S(x)] = \frac{1}{2}\dot{x}^2 - \Pi[S(x)], \quad (-\infty < x < \infty) \tag{7.52}$$

where a prime denotes differentiation with respect to x, and the condition

$$[S'(x)]^2 = 1 \tag{7.53}$$

has been used. Note that the condition of constraints in system (7.49) is automatically satisfied for any x due to $-1 \leq S(x) \leq 1$, and the Hamilton's principle, $\delta \int_{t_1}^{t_2} L\, dt = 0$, gives the following differential equation of motion with no constraints:

$$\ddot{x} + \frac{d\Pi[S(x)]}{dS(x)} S(x) = 0 \tag{7.54}$$

Recall that the function $S(x)$ belongs to a class of continuous but nonsmooth functions. Its first derivative, $S'(x)$, has bounded jumps at those points x for which $S(x) = \pm 1$. Physically equation (7.54) describes a particle moving in the periodic nonsmooth potential field as shown in Figure 7.17. It is seen that the transformed system (7.54) is written in terms of a smooth coordinate $x(t)$ without barriers. Note that the velocity, $\dot{q}(t)$, changes its signs on the borders of the interval $-1 \leq q \leq 1$, whereas the velocity of the transformed system, $\dot{x}(t)$, remains continuous (otherwise $\ddot{x}(t)$ would be a singular function in contradiction to equation (7.54). Furthermore, one will not deal anymore with the problem of matching different pieces of the solution at points $q = \pm 1$. The transformation (7.50) provides such matching automatically and gives a uniform expression for the solution that is more convenient for future manipulations. Finally, the most important advantage of the transformation is that it brings a special geometrical treatment of the vibro-impact motion in terms of the transformed system. Each impact event corresponds to a transmission (jump) into a new cell of the periodic potential field shown in Figure 7.17. Such visualization may reveal interesting physical properties when dealing with more complicated systems of many degrees of freedom, as will be illustrated in Section 7.7.

7.6.4 *Saw-tooth time-transformation (STTT) method*

This technique is based on a special transformation of time and gives explicit form of analytical solutions for nonlinearities of high power. The physical and mathematical principles of the STTT have been formulated by Pilipchuk and Ibrahim (1997). The method is demonstrated by considering the free undamped oscillation of the nonlinear system

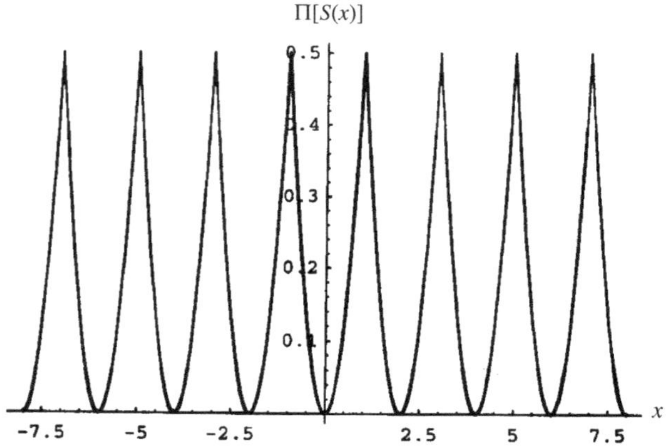

Figure 7.17 Potential energy of the harmonic oscillator between two absolutely rigid walls after the elimination of constraints. Each crossing of the peaks corresponds to an impact in terms of the original system. (Pilipchuk and Ibrahim, 2000)

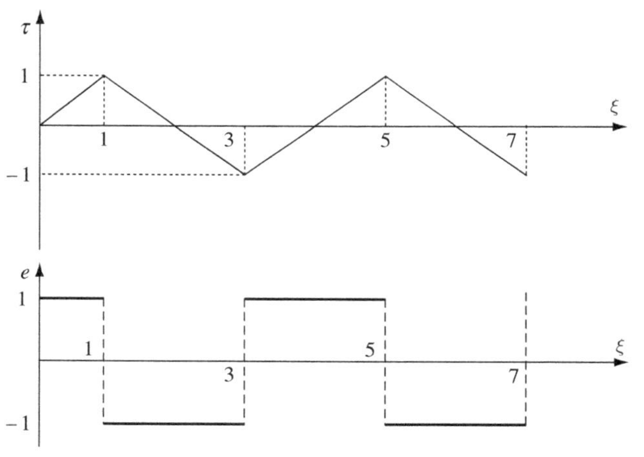

Figure 7.18 Saw-tooth functions sine and rectangular cosine.

$$\mathbf{m}\frac{d^2\mathbf{x}}{dt^2} + \mathbf{kx} + \beta\mathbf{N}(\mathbf{x}) = \mathbf{0} \tag{7.55}$$

where $\mathbf{m}$ is the system mass matrix, $\mathbf{k}$ is the linear stiffness matrix, $\mathbf{x}$ is the system generalized coordinates, $\mathbf{N}(\mathbf{x})$ is the vector of high power nonlinear terms that account for the strong impact nonlinearity, and β is a positive constant parameter.

The idea of STTT is similar to a great extent to the trigonometric generating functions $\{\sin\xi, \cos\xi\}$ frequently used in constructing solutions of linear and weakly nonlinear systems. Similarly, one can consider a pair of nonsmooth functions which have relatively simple forms and will be termed as the saw-tooth sine, $\tau(\xi)$, and the rectangular cosine, $e(\xi)$, which is the generalized derivative of $\tau(\xi)$ as shown in Figure 7.18. The functions $\{\tau(\xi), e(\xi)\}$ and $\{\sin\xi, \cos\xi\}$ describe the motions of the two simplest vibrating models, namely, the motion of a particle between two rigid barriers and a mass-spring oscillator, respectively.

A family of periodic solutions of equations (7.55) is proposed in the following form

$$\mathbf{x} = \mathbf{X}(\tau), \quad \tau = \tau(t/a) \tag{7.56}$$

where a is an unknown scaling factor, which is equal to a quarter of the period $T = 4a$ and must be defined for the autonomous case.

Thus, the solution is going to be constructed as a function of the saw-tooth function τ, which varies in the region $-1 \leq \tau \leq 1$. Note that equation (7.55) admits a group of transformations $\mathbf{x} \to -\mathbf{x}$. As a result, the solution can be constructed as an odd function: $\mathbf{X}(-\tau) \equiv -\mathbf{X}(\tau)$.

When substituting equation (7.56) into equation (7.55), one should take into account the following differentiation scheme of the expression (7.56). Due to the equality $e^2(t/a) = 1$, one can write

$$\frac{d\mathbf{x}}{dt} = \frac{1}{a}\frac{d\mathbf{X}}{d\tau}e, \qquad \frac{d^2\mathbf{x}}{dt^2} = \frac{1}{a^2}\frac{d^2\mathbf{X}}{d\tau^2} + \frac{1}{a^2}\frac{d\mathbf{X}}{d\tau}\frac{de(t/a)}{d(t/a)} \tag{7.57a,b}$$

The last term in equation (7.57) contains the series of Dirac delta functions

$$\frac{de(t/a)}{d(t/a)} = 2\sum_{j=-\infty}^{\infty}\left[\delta\left(\frac{t}{a} + 1 - 4j\right) - \delta\left(\frac{t}{a} - 1 - 4j\right)\right] \tag{7.58}$$

Note that, the delta functions (7.58) are 'localized' at points $\{t : \tau(t/a) = \pm 1\}$. It means that under the condition

$$\frac{d\mathbf{X}}{d\tau} = 0, \qquad \text{for } \tau = \pm 1 \tag{7.59}$$

all delta functions of the series will be eliminated, and as a result the second derivative in expressions (7.57a, b) becomes a continuous function. Substituting expressions (7.57a, b) into the equations of motion (7.55) one obtains the boundary value problem

$$\mathbf{m}\frac{d^2\mathbf{x}}{d\tau^2} = -a^2[\mathbf{kX} + \beta\mathbf{N}(\mathbf{X})] \tag{7.60a}$$

$$\left.\frac{d\mathbf{X}}{d\tau}\right|_{\tau=1} = 0, \quad \mathbf{X}(-\tau) \equiv -\mathbf{X}(\tau) \tag{7.60b,c}$$

The left-hand side of equation (7.60a) does not include the linear stiffness term, and by setting the right-hand side to zero, the left-hand side does not represent a harmonic oscillator as in the case of quasi-linear treatment. This means any generating solution for equation (7.60a) should be based on the solution of

$$\mathbf{m}\frac{d^2\mathbf{x}}{d\tau^2} = 0 \tag{7.61}$$

However, by setting the right-hand side to zero, the qualitative structure of the periodic motion of the system will be preserved. This property will be destroyed if the same argument is applied to the system equations of motion before performing STTT. The transformed generating equation (7.61) possesses a solution of the form

$$\mathbf{X}(\tau) = \mathbf{X}(0) + \mathbf{X}'(0)\tau \tag{7.62}$$

which is a periodic function of t. Now we seek a solution of the nonlinear boundary value problem (7.60) in form of a series of successive approximations:

$$\mathbf{X} = \mathbf{X}_0(\tau) + \varepsilon\mathbf{X}_1(\tau) + \varepsilon^2\mathbf{X}_2(\tau) + \cdots \tag{7.63a}$$

$$a^2 = \varepsilon h_0\left(1 + \varepsilon\gamma_1 + \varepsilon^2\gamma_2 + \cdots\right) \tag{7.63b}$$

where the formal parameter $\varepsilon = 1$ is introduced as bookkeeping to identify terms of different orders in the expansion. All terms of the first series are n-component columns, where n is the number the system degrees of freedom. These functions and the constants h_0, γ_1, γ_2, ... are determined using an iterative process.

This technique was used to analyze the response of systems involving liquid sloshing impact under support excitation by Pilipchuk and Ibrahim (1997). Their main results will be described in Section 7.7.

7.6.5 Lie group transformation

Basic concept

The quantitative theory of nonlinear vibration has been advanced by new developments in asymptotic expansion techniques originally developed for solving nonlinear differential equations. Most traditional methods are essentially based on perturbation methods or averaging techniques (Giancaglia 1972, and Nayfeh and Mook, 1979). The theory of Poincaré normal forms (Nayfeh, 1993), which is similar to averaging techniques, retains resonance terms, since all nonresonance terms are removed from the equations of motion by means of a special coordinate transformation. In this case, the method of Poincaré normal forms is qualified as the simplest possible form of the equations of motion. Alternatively, one can use Lie group operators, which can lead to the simplest form of the system equations of motion. The Lie group theory has become a powerful tool for studying differential equations among mathematicians and specialists, and needs to be adapted for dynamicists. Belinfante, *et al.* (1966) presented an overview of the mathematical structure of Lie groups and Lie algebras with applications to nonlinear differential equations. Hori (1966) used Lie series to construct an additional first integral in an autonomous Hamiltonian system. Zhuravlev (1986) developed an algorithm for the asymptotic integration of nonlinear differential equations as monomial Lie group transformations of the phase space into itself.

The method is essentially based on the work of Zhuravlev (1986) and Zhuravlev and Klimov (1988). An essential ingredient of the Lie group operators is the Hausdorff formula, which relates the Lie group operators of the original system and the new one, and the operator of coordinate transformation. Zhuravlev (1986) and Zhuravlev and Klimov (1988) made a conjecture that most averaging techniques reproduce this formula, each time implicitly, during the transformation process. There is no need to do this, since it is reasonable to start the transformation using Hausdorff's relationship. This section outlines the general scheme of transformations, including normal form coordinates and Lie group operators.

In terms of the principal coordinates q_k, a nonlinear dynamical system of n-degrees of freedom may be described by a set of $n+1$ autonomous differential equations written in the standard form

$$\ddot{q}_k + \omega_k^2 q_k = \varepsilon F_k(\dot{q}_1, \ldots, \dot{q}_{n+1}, \dot{q}_1, \ldots, q_{n+1}), \qquad k = 1, \ldots, n+1 \tag{7.64}$$

where a dot denotes differentiation with respect to time t, ε is a small parameter, and an external excitation has been replaced by the coordinate q_{n+1}. The functions F_k include all nonlinear terms and possibly parametric excitation terms, and ω_k are the normal mode frequencies.

The Poincaré normal form theory deals with sets of first-order differential equations written in terms of normal form coordinates (Nayfeh, 1993). In this case, it is convenient to transform the $(n+1)$ second-order differential equations (7.64) into $(n+1)$ first-order differential equations plus their conjugate set. This can be done by introducing the complex coordinates $y_k(t)$, (see, e.g., Landau and Lifshitz, 1959),

$$y_k = \dot{q}_k + i\omega_k q_k \tag{7.65a}$$

$$q_k = \frac{1}{2i\omega_k}(y_k - \bar{y}_k) \tag{7.65b}$$

$$\dot{q}_k = \frac{1}{2}(y_k + \bar{y}_k) \tag{7.65c}$$

The physical meaning of these coordinates can be understood by considering the linear case (obtained after setting $\varepsilon = 0$). The linear solution of equation (7.64) and its first derivative (the velocity) are, respectively,

$$q_k = A_k e^{i\omega_k t} + \bar{A}_k e^{-i\omega_k t}, \qquad i^2 = -1 \tag{7.66a}$$

$$\dot{q}_k = i\omega_k \left(A_k e^{i\omega_k t} - \bar{A}_k e^{-i\omega_k t}\right) \tag{7.66b}$$

where A_k and $\overline{A}_k$ are complex and an over-bar denotes conjugate.

The complex coordinates $y_k(t)$ and $\bar{y}_k(t)$ can be viewed as vectors rotating in the complex plane with angular velocities ω_k and $-\omega_k$, respectively. Thus, $y_k(t)$ and $\bar{y}_k(t)$ may be written in the form

$$y_k = \dot{q}_k + i\omega_k q_k = 2i\omega_k A_k e^{i\omega_k t}, \quad \text{and} \quad \bar{y}_k = -2i\omega_k A_k e^{-i\omega_k t} \tag{7.67a,b}$$

Introducing the transformation (7.65) into the equations of motion (7.64) gives

$$\begin{aligned}\ddot{q}_k + \omega_k^2 q_k &= \frac{d}{dt}(\dot{q}_k + i\omega_k q_k) - i\omega_k(\dot{q}_k + i\omega_k q_k) \\ &= \frac{dy_k}{dt} - i\omega_k y_k = \varepsilon F_k(y_1, \ldots, y_{n+1}; \bar{y}_1, \ldots, \bar{y}_{n+1})\end{aligned} \tag{7.68a}$$

or

$$\frac{dy_k}{dt} = i\omega_k y_k + \varepsilon F_k(y_1, \ldots, y_{n+1}; \bar{y}_1, \ldots, \bar{y}_{n+1}) \tag{7.68b}$$

$$\text{c. c. equations} \tag{7.68c}$$

where c. c. stands for complex conjugate of equations (7.68b), and

$$F_k(y_1, \ldots, y_{n+1}; \bar{y}_1, \ldots, \bar{y}_{n+1}) \\ = F_k(q_1, \ldots, q_{n+1}; \dot{q}_1, \ldots, \dot{q}_{n+1})|_{q_j=(1/2i\omega_j)(y_j-\bar{y}_j), \dot{q}_j=(1/2)(y_j+\bar{y}_j)} \tag{7.69}$$

These terms can be represented in the polynomial form

$$F_k = \sum_{|\sigma|=2,3,\ldots} F_k^{\sigma} y_1^{m_1} \cdots y_{n+1}^{m_{n+1}} \bar{y}_1^{l_1} \cdots \bar{y}_{n+1}^{l_{n+1}} \tag{7.70}$$

where the Taylor coefficients are defined by the partial differentiation of equation (7.69) as

$$F_k^{\sigma} = \frac{1}{\sigma!} \frac{\partial F_k}{\partial y_1^{m_1} \ldots \partial y_{n+1}^{m_{n+1}} \partial \bar{y}_1^{l_1} \ldots \partial \bar{y}_{n+1}^{l_{n+1}}} |_{y=0} \tag{7.71}$$

$\sigma = \{m_1, \ldots, m_{n+1}, l_1, \ldots, l_{n+1}\}$, $|\sigma| = m_1 + \cdots + m_{n+1} + l_1 + \cdots + l_{n+1}$, and $\sigma! = m_1! \ldots m_{n+1}! \, l_1 \ldots l_{n+1}!$ have been used. Equations (7.68b) correspond to the standard form, which is ready for analysis in terms of Lie group operators.

Lie group operators

To apply the theory of the Lie groups we rewrite equations (7.68b) in the form

$$\dot{\mathbf{y}} = A\mathbf{y}, \quad A = A_0 + \varepsilon A_1 \tag{7.72a,b}$$

where

$$\mathbf{y} = \{y_1, \ldots, y_{n+1}; \bar{y}_1, \ldots, \bar{y}_{n+1}\}^T$$

$$A_0 = \sum_{k=1}^{n+1} i\omega_k y_k \frac{\partial}{\partial y_k} + \text{c. c.},$$

$$A_1 = \sum_{k=1}^{n+1} F_k \frac{\partial}{\partial y_k} + \text{c. c.} \tag{7.73}$$

are operators of linear and nonlinear components of the system, respectively. In order to write the equations of motion in their simplest form, we introduce the coordinate transformation in the Lie series form

$$\mathbf{y} = e^{-\varepsilon U}\mathbf{z} = \mathbf{z} - \varepsilon U\mathbf{z} + \frac{\varepsilon^2}{2!} U^2 \mathbf{z} - \cdots \tag{7.74a}$$

$$\mathbf{y} = \{y_1, \ldots, y_{n+1}; \bar{y}_1, \ldots, \bar{y}_{n+1}\}^T \rightarrow \mathbf{z} = \{z_1, \ldots, z_{n+1}; \bar{z}_1, \ldots, \bar{z}_{n+1}\}^T \tag{7.74b}$$

where the operator of transformation U is represented in the power series form with respect to the small parameter ε

$$U = U_0 + \varepsilon U_1 + \cdots \tag{7.75}$$

The coefficients of this series are

$$U_j = \sum_{k=1}^{n+1} T_{j,\,k} \frac{\partial}{\partial z_k} + \text{c. c.} \tag{7.76}$$

where

$$T_{j,\,k} = T_{j,\,k}(z_1, \ldots, z_{n+1}; \bar{z}_1, \ldots, \bar{z}_{n+1});\ j = 0,\ 1, \ldots;\ k = 1, \ldots,\ n+1 \tag{7.77}$$

are unknown functions to be determined. One of the advantages of this process is that the inverted coordinate transformation to the form (7.74) can be easily written as

$$\mathbf{z} = e^{\varepsilon U}\mathbf{y} \tag{7.78}$$

where one simply replaces $\mathbf{z}$ with $\mathbf{y}$ in the operator of transformation, U.

If $\varepsilon = 0$, transformation (7.74) becomes identical, $\mathbf{y} = e^{(0)}\mathbf{z} = \mathbf{z}$. In this case, equation (7.72) has the simplest linear form and there is no need to transform the system. For $\varepsilon \neq 0$ transformation (7.74) converts the system (7.72) into the following one:

$$\dot{\mathbf{z}} = B\mathbf{z} \tag{7.79}$$

where the new operator B is given by the Hausdorff formula (Belinfante *et al.*, 1966)

$$B = A + \varepsilon[A, U] + \frac{\varepsilon^2}{2!}[[A, U], U] + \cdots \tag{7.80}$$

where $[A, U] = AU - UA$ is the commutator of operators A and U.

Substituting the power series expansions for A and U given by relations (7.72) and (7.74) into (7.80) gives

$$B = A_0 + \varepsilon(A_1 + [A_0, U_0]) + \varepsilon^2\left([A_0, U_1] + [A_1, U_0] + \frac{1}{2!}[[A_0, U_0], U_0]\right) + \cdots \tag{7.81}$$

A simple calculation gives

$$B = \sum_{k=1}^{n+1} \left\{ i\omega_k z_k + \varepsilon\left[F_k + (A_0 - i\omega_k)T_{0,\,k}\right]\right\} \frac{\partial}{\partial z_k} + \text{c. c.} + O(\varepsilon^2) \tag{7.82}$$

where terms of order ε^2 have been ignored, $F_k = F_k|_{y \to z}$, $A_0 = A_0|_{y \to z}$.

The above relationships show that a transformation of the system $\dot{\mathbf{y}} = A\mathbf{y} \to \dot{\mathbf{z}} = B\mathbf{z}$ can be considered in terms of operators $A \to B$. To bring the system into its normal form, one must eliminate as many nonlinear terms as possible from the transformed system such that the system dynamic characteristics are preserved. It follows from (7.82) that all nonlinear terms of order ε would be removed if

$$F_k + (A_0 - i\omega_k)T_{0,k} = 0, \quad \text{and} \quad \text{c. c.} \tag{7.83}$$

Representing the unknown functions in the polynomial form

$$T_{0,k} = \sum_{|\sigma|=2,3,\ldots} T_{0,k}^{\sigma} z_1^{m_1} \ldots z_{n+1}^{m_{n+1}}\ \bar{z}_1^{l_1} \cdots \bar{z}_{n+1}^{l_{n+1}} \tag{7.84}$$

and taking into account (7.70) one obtains the left hand sides of conditions (7.83) as

$$F_k + (A_0 - i\omega_k)T_{0,k} = \sum_{|\sigma|=2,3,\dots} \left(F_k^{\sigma} + i\Delta_k^{\sigma} T_{0,k}^{\sigma}\right) z_1^{m_1} \cdots z_{n+1}^{m_{n+1}} \bar{z}_1^{l_1} \cdots \bar{z}_{n+1}^{l_{n+1}} \quad \text{and} \quad \text{c. c.} \tag{7.85}$$

where $\Delta_{\sigma}^{k} = (m_1 - l_1 - \delta_{1,k})\omega_1 + \cdots + (m_{n+1} - l_{n+1} - \delta_{n+1,k})\omega_{n+1}$.

To reach the zeroth coefficient of the nonlinear form $z_1^{m_1} \cdots z_{n+1}^{m_{n+1}} \bar{z}_1^{l_1} \cdots \bar{z}_{n+1}^{l_{n+1}}$ one must put

$$T_{0,k}^{\sigma} = i\frac{F_k^{\sigma}}{\Delta_k^{\sigma}} \tag{7.86}$$

if only $\Delta_k^{\sigma} \neq 0$ If $\Delta_k^{\sigma} = 0$ for definite k and σ then the corresponding nonlinear term cannot be eliminated from the transformed equation since it is qualified as a resonance term.

Finally, the result of transformation is summarized as follows.

The Original Set:

$$\frac{dy_k}{dt} = i\omega_k y_k + \varepsilon \sum_{|\sigma|=2,3,\dots} F_k^{\sigma} y_1^{m_1} \cdots y_{n+1}^{m_{n+1}} \bar{y}_1^{l_1} \cdots \bar{y}_{n+1}^{l_{n+1}}, \quad k = 1,\ 2,\ 3 \tag{7.87}$$

The Transformation of Coordinates:

$$y_k = z_k - \varepsilon \sum_{\substack{|\sigma|=2,3,\dots \\ \Delta_k^{\sigma} \neq 0}} \frac{F_k^{\sigma}}{\Delta_k^{\sigma}} z_1^{m_1} \dots z_{n+1}^{m_{n+1}} \bar{z}_1^{l_1} \dots \bar{z}_{n+1}^{l_{n+1}} + O(\varepsilon^2) \tag{7.88}$$

The Transformed Set:

$$\dot{z}_k = i\omega_k z_k + \varepsilon \sum_{\substack{|\sigma|=2,3,\dots \\ \Delta_k^{\sigma} \neq 0}} F_k^{\sigma} z_1 m_1 \cdots z_{n+1}^{m_{n+1}} \bar{z}_1^{l_1} \dots \bar{z}_{n+1}^{l_{n+1}} + O(\varepsilon^2) \tag{7.89}$$

This is the normal form of the system, and the summation in this form is much simpler than the one in the original set (7.87). The summation in (7.89) contains only those terms that give rise to resonance while the first term on the right-hand side stands for the fast component of the motion. The fast component of the motion can be extracted by introducing the complex amplitudes $a_k(t)$ as

$$z_k = a_k\, e^{i\omega_k^t} \tag{7.90}$$

Substituting (7.90) into (7.89) and taking into account the resonance condition, $\Delta_k^{\sigma} = 0$, gives

$$\frac{da_k}{dt} = \varepsilon \sum_{\substack{|\sigma|=2,3,\dots \\ \Delta_k^{\sigma} = 0}} F_k^{\sigma} a_1^{m_1} \cdots a_{n+1}^{m_{n+1}} \bar{a}_1^{l_1} \cdots \bar{a}_{n+1}^{l_{n+1}} + O(\varepsilon^2) \tag{7.91}$$

Relationships (7.88)–(7.91) have been obtained under no concrete assumptions regarding the type of resonance. Pilipchuk and Ibrahim (1999) employed this technique for an elastic structure carrying a liquid container subjected to parametric support harmonic excitation.

7.7 Structure interaction with sloshing impact

Consider a liquid container supported by four massless rods of length L, which are restrained by four torsional springs of stiffness k at the base as shown in Figure 7.19. The base is subjected

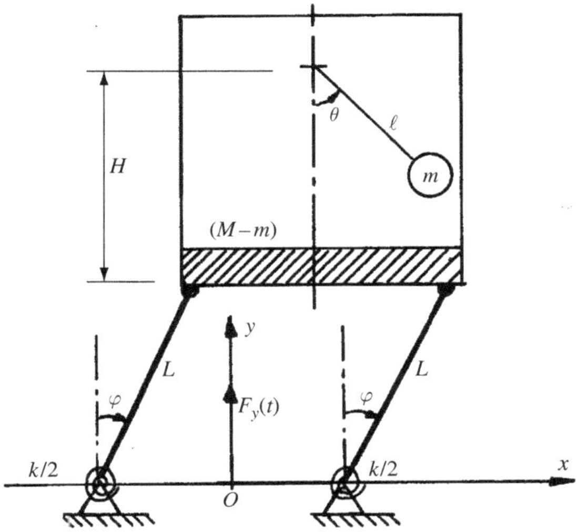

Figure 7.19 Schematic diagram of a liquid sloshing impact represented by a simple pendulum interacting with its support structure.

to vertical acceleration $F_y(t)$. Let M be the total mass of the container including liquid and m the equivalent sloshing mass of the first asymmetric mode of the liquid. The fluid sloshing mass, m, is modeled as a pendulum of length ℓ. The equivalent pendulum parameters for different types of container geometry are given in Chapter 5. Using Lagrange's equation with respect to the coordinates θ and φ, and introducing the approximations $\sin\theta \approx \theta - \theta^3/3!$ and $\cos\theta \approx 1 - \theta^2/2$, the equations of motion are (Sayad, 1999)

$$\ddot{\theta} + \frac{g}{\ell}\theta - \frac{L}{\ell}\ddot{\omega} = \frac{L}{2\ell}\ddot{\varphi}(\theta+\varphi)^2 + \frac{L}{\ell}\dot{\varphi}^2(\theta+\varphi) + \frac{F_y(t)}{\ell}\theta + \frac{g}{6\ell}\theta^3 - \frac{d}{m\ell^2}\left(\frac{\theta}{\theta_0}\right)^{2p}\dot{\theta} - \frac{b}{m\ell^2}\left(\frac{\theta}{\theta_0}\right)^{2n-1} \tag{7.92a}$$

$$\ddot{\varphi} + \left(\frac{k}{ML^2} - \frac{g}{L}\right)\varphi + \frac{m\ell}{ML}\ddot{\theta} = \frac{m\ell}{2ML}\ddot{\theta}(\theta+\varphi)^2 + ML(\theta+\varphi)\dot{\theta}^2 - \frac{g}{6L}\varphi^3 - \frac{F_y(t)}{L}\varphi \tag{7.92b}$$

Introducing the nondimensional parameters:

$$\Theta = \theta/\theta_0,\ \Phi = \varphi/\theta_0,\ \tau = \omega_l t,\ \omega_\ell^2 = g/\ell,\ \omega_L^2 = (k/ML^2) - g/L,$$
$$v = \omega_\ell/\omega_L,\ \mu = m/M,\ \lambda = \ell/L,\ f_y(\tau) = F_y(\tau/\omega_\ell)/\ell\omega_\ell^2\theta_0 \tag{7.93}$$

gives

$$\begin{pmatrix} 1 & 1/\lambda \\ \mu\lambda & 1 \end{pmatrix}\begin{Bmatrix} \Theta'' \\ \Phi'' \end{Bmatrix} + \begin{bmatrix} \mu\lambda^2 & 0 \\ 0 & v^2 \end{bmatrix}\begin{Bmatrix} \Theta \\ \Phi \end{Bmatrix} = \begin{Bmatrix} \mu\lambda^2 Q_{11} \\ Q_{22} \end{Bmatrix} \tag{7.94}$$

where a prime denotes differentiation with respect to τ. The right-hand functions, Q_{ii}, stand for nonlinear and excitation terms.

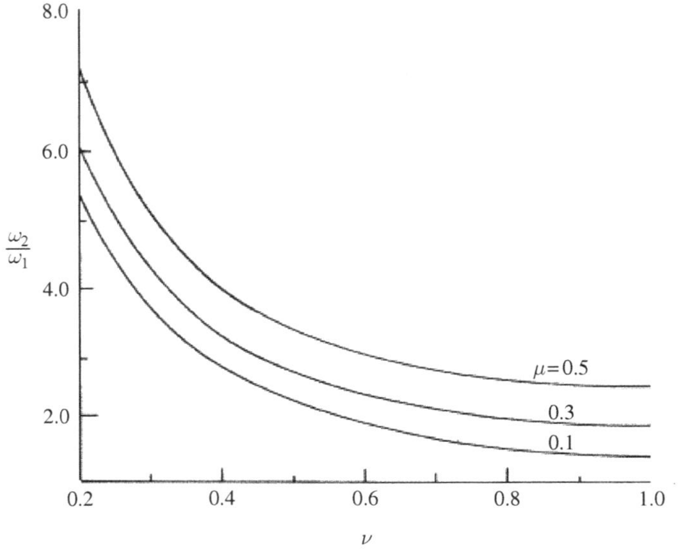

Figure 7.20 Dependence of normal mode frequency ratio on the local frequency ratio $\nu = \omega_\ell/\omega_L$ for three different values of mass ratio parameter $\mu = m/M$.

Equations (7.94) are linearly coupled through the mass matrix. They are also nonlinearly coupled through inertia nonlinearity. Coefficients b and d, and the exponents n and p are usually obtained experimentally. However, in the absence of experimental data, one may select $n = 3$ and $p = 2$, while the influence of the coefficients b and d will be examined in the present section. Both coordinates are parametrically excited. Higher-order exponents can be selected, but will result in more lengthy analysis. Any possible solution for equations (7.94) gives the total motion as a sum of responses in its characteristic modes of vibration. In order to derive the solution in terms of principal modes, a transformation to the principal coordinates $\mathbf{Y} = \{Y_1, Y_2\}^T$, where T denotes transpose, should be carried out by introducing the linear transformation into equations (7.92)

$$\begin{Bmatrix} \Theta \\ \Phi \end{Bmatrix} = [P]\begin{Bmatrix} Y_1 \\ Y_2 \end{Bmatrix}\begin{bmatrix} 1 & 1 \\ K_1 & K_2 \end{bmatrix}\begin{Bmatrix} Y_1 \\ Y_2 \end{Bmatrix} \tag{7.95}$$

where $[P]$ is the modal matrix with $K_{1,2} = \lambda(1 - \omega_{1,2}^2)/\omega_{1,2}^2$ and

$$\omega_{1,2}^2 = \frac{(1+\nu^2) \mp \sqrt{(1-\nu^2)^2 + 4\mu\nu^2}}{2(1-\mu)} \tag{7.96}$$

The dependence of the linear modal frequencies on the local frequency ratio, $v = \omega_\ell/\omega_L$, for fixed mass ratio parameter, $\mu = 0.5$, is illustrated in Figure 7.20. Now pre-multiplying equation (7.94) by $[P]^T$, gives

$$Y_1'' + \omega_1^2 Y_1 = \varepsilon\left\{-2\bar{\zeta}_1\omega_1 Y_1' + (\Psi_{11})_{\text{gn}} + (\Psi_{11})_{\text{impact}} + (\Psi_{11})_{\text{ex}}\right\} \tag{7.97a}$$

$$Y_2'' + \omega_2^2 Y_2 = -\varepsilon\frac{m_{11}}{m_{22}}\left\{-2\bar{\zeta}_2\omega_2 Y_2' + (\Psi_{22})_{\text{gn}} + (\Psi_{22})_{\text{impact}} + (\Psi_{22})_{\text{ex}}\right\} \tag{7.97b}$$

where a linear viscous damping $\zeta_i = \varepsilon\bar{\zeta}_i$, with $\varepsilon = \mu\lambda^2/m_{11}$, has been introduced to account for energy dissipation, ω_i are the normal mode frequencies, m_{ii} are constant coefficients. The right-hand sides of these equations include inertia and stiffness nonlinearities of cubic-order and are denoted by subscript "gn." They also include impact nonlinearities of quintic-order and are denoted by subscript "impact." The subscript "ex" stands for excitation terms. The impact and geometric nonlinear terms are treated as being of the same order of $O(\varepsilon)$ since the coefficients of the impact terms are of a higher order of magnitude than those coefficients of cubic terms. The functions Ψ_{ii} are listed in the appendix to this chapter.

In order to identify the critical nonlinear resonance conditions and to solve for the system response in the neighborhood of these conditions, the multiple scales method will be used. The solution of equations (7.97) may be expressed in a uniform expansion form

$$Y_i = Y_{i0}(T_0, T_1, T_2, \ldots) + \varepsilon Y_{i1}(T_0, T_1, T_2, \ldots) + \cdots \tag{7.98}$$

where $T_0 = \tau$, $T_1 = \varepsilon\tau$, $T_2 = \varepsilon^2\tau$, or $T_n = \varepsilon^n\tau$, $n = 0, 1, 2\ldots$ The time derivatives can be written in the form

$$\begin{aligned}\frac{d}{dt} &= \frac{\partial}{\partial T_0} + \varepsilon\frac{\partial}{\partial T_1} + \varepsilon^2\frac{\partial}{\partial T_2} + \cdots = D_0 + \varepsilon D_1 + \varepsilon^2 D_2 + \cdots\\ \text{where } D_n &= \frac{\partial}{\partial T_n}\frac{d^2}{dt^2} = D_0^2 + 2\varepsilon D_0 D_1 + \varepsilon^2(D_1^2 + 2D_0 D_2) + \cdots\end{aligned} \tag{7.99}$$

The expansion is carried out up to first-order in ε, and, thus, we need to deal with the two time scales T_0 and T_1. Substituting the solution (7.98) into equations (7.97), using the transformed time derivatives (7.99), gives

$$(D_0^2 + 2\varepsilon D_0 D_1 + \varepsilon^2(D_1^2 + 2D_0 D_2) + \cdots)Y_i + \omega_i^2 Y_i = \varepsilon\Psi_I \tag{7.100}$$

Equating the coefficients of equal powers of ε^0 and ε^1 (ε^n) gives a set of differential equations to be solved for Y_{i0}, and Y_{i1}. For equation (7.97a) the zero- and first-order equations in ε are, respectively

$$D_0^2 Y_{10} + \omega_1^2 Y_{10} = 0 \tag{7.101a}$$

$$D_0^2 Y_{11} + \omega_1^2 Y_{11} = -2D_0 D_1 Y_{10} - 2\bar{\zeta}_1\omega_1 D_0 Y_{10} + \Pi_{11}(Y_{ij}) \tag{7.101b}$$

For equation (7.97b) the zero- and first-order equations in ε are, respectively

$$D_0^2 Y_{20} + \omega_2^2 Y_{20} = 0 \tag{7.102a}$$

$$D_0^2 Y_{21} + \omega_2^2 Y_{21} = -2D_0 D_1 Y_{20} - 2\bar{\zeta}_2\omega_2 D_0 Y_{20} + \Pi_{22}(Y_{ij}) \tag{7.102b}$$

where Π_{ii} stands for nonlinear, damping, and excitation terms. The general solution of (7.101a) and (7.102a) can be written in the form

$$Y_{10} = A(T_1)\exp(i\omega_1 T_0) + \bar{A}(T_1)\exp(-i\omega_1 T_0) \tag{7.103a}$$

$$Y_{20} = B(T_1)\exp(i\omega_2 T_0) + \bar{B}(T_1)\exp(-i\omega_2 T_0) \tag{7.103b}$$

where an over-bar denotes conjugate, $i = \sqrt{-1}$, and $A(T_1)$ and $B(T_1)$ are functions of the time scale T_1 and will be determined later. Substituting solutions (7.103) into (7.101b) and (7.102b), gives

$$D_0^2 Y_{11} + \omega_1^2 Y_{11} = -2D_0 D_1\{A(T_1)\exp(i\omega_1 T_0) + \bar{A}(T_1)\exp(-i\omega_1 T_0)\} + \Psi_{11}(Y_{ij}) - 2i\omega_1^2\bar{\zeta}_1(A(T_1)\exp(i\omega_1 T_0) + \cdots \tag{7.104a}$$

$$D_0^2 Y_{21} + \omega_2^2 Y_{21} = -2D_0 D_1\{B(T_1)\exp(i\omega_2 T_0) + \bar{B}(T_1)\exp(-i\omega_2 T_0)\} + \frac{m_{11}}{m_{22}}\Psi_{22}(Y_{ij}) - 2i\omega_2^2\bar{\zeta}_2(B(T_1)\exp(i\omega_2 T_0) + \cdots \tag{7.104b}$$

The right-hand sides of these equations contain secular terms in Y_{i1}. In view of the system nonlinearities and excitation, these secular terms establish different types of resonance conditions. Under parametric excitation, two types of resonance conditions may arise. These are

(1) internal resonance condition of fourth-order

$$\omega_2 = 3\omega_1 \tag{7.105}$$

(2) parametric resonance conditions

(i) Principal parametric resonance of the first mode

$$\Omega_y = 2\omega_1 \tag{7.106a}$$

(ii) Principal parametric resonance of the second mode

$$\Omega_y = 2\omega_2 \tag{7.106b}$$

(iii) Combination parametric resonance of the summed type

$$\Omega_y = \omega_1 + \omega_2 \tag{7.106c}$$

Pilipchuk and Ibrahim (1997) employed a special saw-tooth time-transformation technique to analyze the strongly nonlinear periodic regimes of the system response described by equations (7.94). El-Sayad, *et al.* (1999) considered the same system and used the method of multiple scales to determine the response characteristics under parametric excitation and in the absence of internal resonance. When the first normal mode was parametrically excited the system exhibited hard nonlinear behavior and the impact loading reduced the response amplitude. On the other hand, when the second mode was parametrically excited, the impact loading resulted in a complex response behavior characterized by multiple steady-state solutions where the response switched from soft to hard nonlinear characteristics. Under combined parametric resonance the system behaves like a soft system in the absence of impact and as a hard system in the presence of impact.

The response characteristics corresponding to simultaneous occurrence of the internal resonance condition, $\omega_2 = 3\omega_1$, and one of the parametric resonance conditions are considered in the next three sections. In each case the solutions for the complex amplitudes A and B will be expressed in the complex polar form

$$A = \frac{a}{2}\exp(i\alpha), \quad B = \frac{b}{2}\exp(i\beta) \tag{7.107a,b}$$

7.7.1 First mode parametric excitation

Introducing the detuning parameters σ_y and σ_I defined by

$$\Omega_y = 2\omega_1 + \varepsilon\sigma_y, \quad \text{and} \quad \omega_2 = 3\omega_1 - \varepsilon\sigma_I \tag{7.108a,b}$$

and following the standard procedure of multiple scales gives the following set of first-order differential equations in the amplitudes a and b, and the new phase angles $\gamma_1 = \sigma_y T_1 - 2\alpha$, and $\gamma_2 = \sigma_I T_1 - \beta + 3\alpha$

$$\begin{aligned}\frac{\partial\gamma_1}{\partial T_1} = \sigma_y + \frac{2}{\omega_1}\Bigg\{ & G_{13}\frac{Y_0}{4}\sin\gamma_1 + \bar{G}_2 a^2 + \bar{G}_1 b^2 + \left(\frac{G_{118}\omega_2}{8}\sin\gamma_2 + \bar{G}_6\cos\gamma_2\right)ab \\ & + \frac{5C_{16}}{8}\left(\frac{1}{2}a^4 + 3b^4\right) - \bar{G}_4 a^2 b^2 - \bar{G}_7 a b^3 \sin\gamma_2 \\ & - \frac{3}{16}C_{15}(2\omega_1\sin\gamma_2 - 5\omega_2\cos\gamma_2)a^3 b\Bigg\}\end{aligned} \tag{7.109a}$$

$$\begin{aligned}\omega_1\frac{\partial a}{\partial T_1} = & -\frac{G_{13}}{4}Y_0 a\cos\gamma_1 - \omega_1^2\bar{\zeta}_1 a + \left(\bar{G}_6\sin\gamma_2 + \frac{G_{118}\omega_2}{8}\cos\gamma_2\right)a^2 b \\ & + \frac{\omega_1 C_{15}}{8}\left(3b^2 + \frac{1}{2}a^2\right)a^3 - \frac{3}{8}\omega_1 C_{15} a^4 b\cos\gamma_2 \\ & - \left(\bar{G}_7\cos\gamma_2 + \frac{15\omega_2}{16}C_{15}\sin\gamma_2\right)a^2 b^3 + \bar{G}_3 a b^4\end{aligned} \tag{7.109b}$$

$$\begin{aligned}\frac{\partial\gamma_2}{\partial T_1} = & \ \sigma_I - \frac{3}{4}\frac{G_{13}}{\omega_1}Y_0\sin\gamma_1 - \left(3\frac{\bar{G}_1}{\omega_1} + \frac{\bar{G}_9}{\omega_2}\right)b^2 - \left(3\frac{\bar{G}_2}{\omega_1} - \frac{\bar{G}_8}{\omega_2}\right)a^2 \\ & + \frac{\bar{G}_{11}}{\omega_2}\frac{a^3}{b}\cos\gamma_2 - \frac{3}{\omega_1}\left(\frac{G_{118}\omega_2}{8}\sin\gamma_2 + \bar{G}_6\cos\gamma_2\right)ab \\ & - \frac{15}{16}C_{16}\left(\frac{1}{\omega_1} - \frac{1}{\omega_2}\right)a^4 + \left(\frac{9}{8}C_{15} + \frac{\bar{G}_{12}}{\omega_2}\right)a^3 b\sin\gamma_2 \\ & + \left(3\frac{\bar{G}_4}{\omega_1} + \frac{\bar{G}_{10}}{\omega_2}\right)a^2 b^2 + \frac{3}{\omega_1}\left(\bar{G}_7\sin\gamma_2 - \frac{15}{16}C_{15}\omega_2\cos\gamma_2\right)ab^3 \\ & - \frac{5}{8}C_{16}\left(\frac{9}{\omega_1} - \frac{1}{2\omega_2}\right)b^4 + \frac{1}{32\omega_2}(\omega_1 C_{15}\sin\gamma_2 + 5C_{16}\cos\gamma_2)\frac{a^5}{b}\end{aligned} \tag{7.109c}$$

$$\begin{aligned}\omega_2\frac{\partial b}{\partial T_1} = & -\omega_2^2\bar{\zeta}_2 b - \bar{G}_{11}a^3\sin\gamma_2 + \frac{C_{15}}{16}(3\omega_1 a^4 + \omega_2 b^4)b \\ & + \bar{G}_{12}a^3 b^2\cos\gamma_2 + \frac{1}{32}C_{15}(\omega_1\cos\gamma_2 - 5\sin\gamma_2)a^5\end{aligned} \tag{7.109d}$$

These equations are integrated numerically for mass ratio $\mu = 0.2$, length ratio $\lambda = 0.2$, local frequency ratio $\nu = 0.5$, excitation amplitude ratio $Y_0 = 0.1$, impact coefficients $C_{16} = -0.5$, $C_{15} = -0.05$, and damping ratios $\zeta_1 = \zeta_2 = 0.1$. It is found that a stationary fixed solution cannot be obtained by setting the left-hand side to zero.

In the absence of impact the response is examined by dropping the fifth-order terms from equations (7.109). In this case, the amplitude and phase equations take the form

$$\frac{\partial \gamma_1}{\partial T_1} = \sigma_y + \frac{2}{\omega_1}\left\{ G_{13}\frac{Y_0}{4}\sin\gamma_1 + \bar{G}_1 b^2 + \bar{G}_2 a^2 + \left(\frac{G_{118}\omega_2}{8}\sin\gamma_2 + \bar{G}_6\cos\gamma_2\right)ab \right\} \tag{7.110a}$$

$$\omega_1 \frac{\partial a}{\partial T_1} = -G_{13}a\frac{Y_0}{4}\cos\gamma_1 - \omega_1^2\bar{\zeta}_1 a + \frac{G_{118}\omega_2}{8}a^2 b\cos\gamma_2 \tag{7.110b}$$

$$\frac{\partial \gamma_2}{\partial T_1} = \sigma_I - \frac{3}{\omega_1}\left\{ G_{13}\frac{Y_0}{4}\sin\gamma_1 + \frac{G_{118}\omega_2}{8}ab\sin\gamma_2 + \bar{G}_6 ab\cos\gamma_2 \right\} - \left(3\frac{\bar{G}_2}{\omega_1} - \frac{\bar{G}_6}{\omega_2}\right)a^2 - \left(3\frac{\bar{G}_1}{\omega_1} + \frac{\bar{G}_9}{\omega_2}\right)b^2 + \frac{\bar{G}_{11}}{\omega_2}\frac{a^3}{b}\cos\gamma_2 \tag{7.110c}$$

$$\omega_2 \frac{\partial b}{\partial T_1} = -\omega_2^2\bar{\zeta}_2 b - \bar{G}_{11}a^3\sin\gamma_2 \tag{7.110d}$$

It is found that in the presence of internal resonance the system responds in different ways depending on the internal detuning only at zero parametric detuning parameter. Figure 7.21 shows a sample of time history records for $\sigma_y = 0$ and $\sigma_I = 0$. The first mode oscillates about a mean value $\langle a \rangle \approx 0.2$ while the second mode fluctuates with occasional spikes about its zero equilibrium position $b = 0$. Figure 7.22 shows the unimodal amplitude–frequency response (in the absence of internal resonance taken from El-Sayad, *et al.* (1999) including one point in the presence of internal resonance at $\sigma_y = 0$ with $\sigma_I = 0$. Obviously, the first mode amplitude fluctuates around the unimodal parametric response. These fluctuations arise mainly due to internal resonance where an irregular energy distribution between the two modes takes place.

If the internal detuning parameter is selected to be remote from the exact internal resonance, the response may or may not achieve a steady state depending on initial conditions. For example, when $\sigma_I = -25$, Figure 7.23 shows a chaotic response, where the first mode oscillates about its unimodal parametric response, and the second mode oscillates about its zero equilibrium. If the initial conditions are selected to be very small, such as $a = b = 0.01$, the response amplitudes achieve fixed values $a = 0.22$ and $b = 0.1$. The second mode frequency spectrum is essentially wide band. As the internal detuning parameter σ_I increases the response becomes quasi-periodic. Numerical integration revealed that, within the range of $\sigma_I = 6 \rightarrow 17$, the response exhibits periodic solutions. This regime also oscillates about the unimodal response fixed point (as determined by the first mode parametric excitation in the absence of internal resonance). There are regions of internal detuning parameter within the ranges $\sigma_I = -20$ to $+6$, and 18 to 55 where the response looks random in behavior. Over these regions, the first mode involves few frequency components while the second mode has a wide frequency spectrum with several spikes. Furthermore, the second mode may or

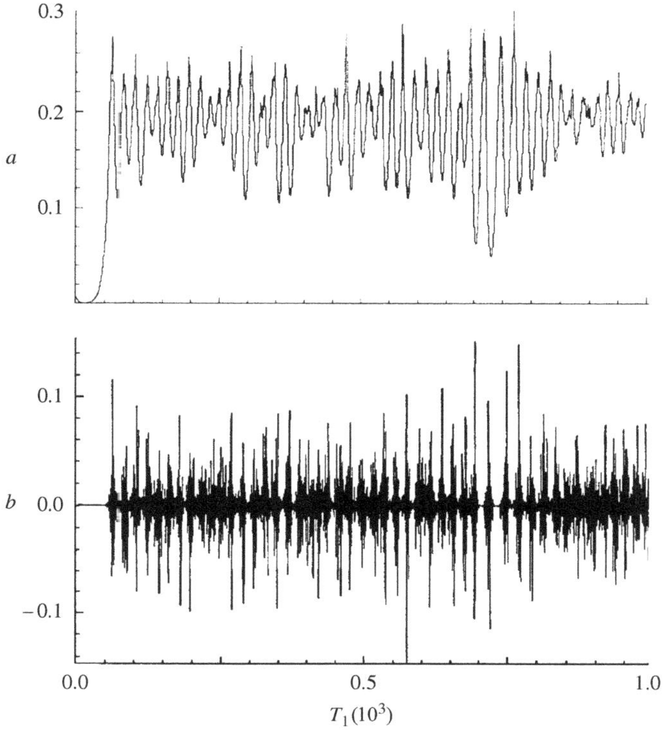

Figure 7.21 Time history records of response amplitudes under first mode parametric excitation $a(T_1 = 0) = 0.01$, $b(T_1 = 0) = 0.01$, for $Y_0 = 0.2$, $\mu = 0.2$, $\lambda = 0.2$, $\sigma_y = \sigma_I = 0$, $\zeta_1 = \zeta_2 = 0.1$. (Ibrahim and El-Sayad, 1999)

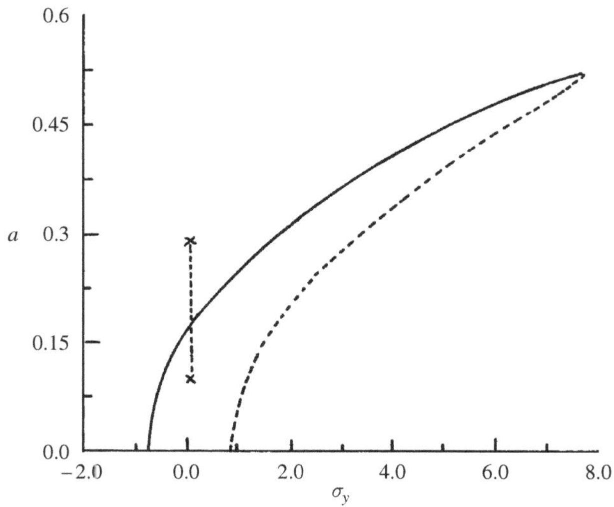

Figure 7.22 First mode amplitude–frequency response in the absence of internal resonance showing the system amplitude indicated by the vertical dashed line when the system is exactly internaly tuned, $\sigma_I = 0$ for $Y_0 = 0.2$, $\mu = 0.2$, $\lambda = 0.2$, $\zeta_1 = \zeta_2 = 0.1$. (Ibrahim and El-Sayad, 1999)

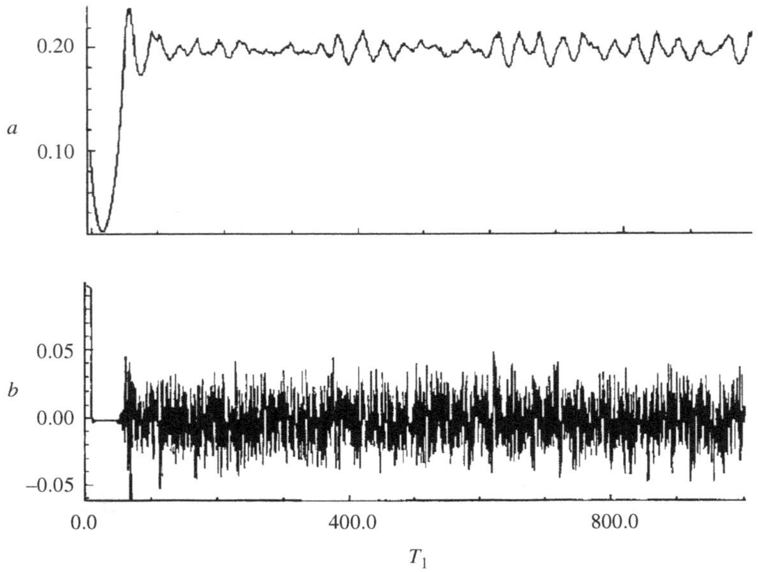

Figure 7.23 Time history records of response amplitudes for nonimpact case under first mode parametric excitation $a(T_1 = 0) = 0.01$, $b(T_1 = 0) = 0.01$, for $Y_0 = 0.2$, $\mu = 0.2$, $\lambda = 0.2$, $\sigma_y = 0$, $\sigma_I = -25$, $\zeta_1 = \zeta_2 = 0.1$. (Ibrahim and El-Sayad, 1999)

may not oscillate about its zero equilibrium position depending on the initial conditions. Figures 7.21 and 7.23 reveal occasional symmetric spikes indicating that the response processes are essentially non-Gaussian with zero skewness and kurtosis greater than 3. The skewness is a statistical parameter measuring the asymmetry of the process and is expressed in terms of third-order statistical moments. The kurtosis, on the other hand, measures the degree of flatness of the probability density curve about the mean value. The kurtosis is usually expressed in terms of the fourth-order moment.

For the impact case, the quintic terms should be included and equations (7.109) should be considered. The numerical integration yields nonstationary solutions in the form of chaotic motion as shown in Figure 7.24 for $\sigma_I = 0.0$. Comparing Figure 7.24 with Figure 7.21 reveals that the impact loading results in asymmetry of the second mode response indicating nonzero skewness. The second mode experiences a wide frequency spectrum and involves several spikes in both the time and frequency domains. The unimodal amplitude–frequency response under parametric excitation is shown by solid and dash curves in Figure 7.25 (taken from El-Sayad, *et al.*, 1999). Figure 7.25 also includes the first mode amplitude response in the presence of internal resonance $\sigma_I = 0.0$ and is indicated by the vertical dashed line at $\sigma_y = 0$. If the initial conditions are changed, the response takes a completely different scenario characterized by large asymmetric spikes on the positive side of the response amplitude. The spikes in the time domain of the second mode response are found to persist as the detuning parameter increases.

Further numerical integration revealed that within the internal detuning parameter range $\sigma_I = -15$ to -10, the response is periodic. The response amplitudes always oscillate about nonzero mean values as shown in Figure 7.26. For internal detuning $\sigma_I > -10$ the response is chaotic with a wide frequency spectrum in the second mode response.

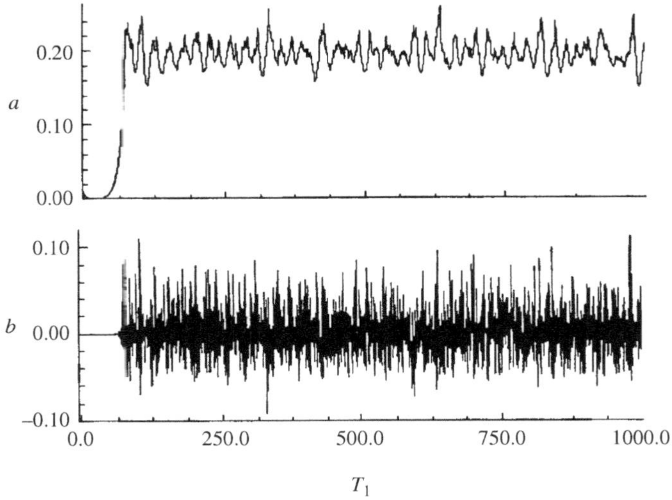

Figure 7.24 Time history records of response amplitudes for the impact case under first mode parametric excitation $a(T_1 = 0) = 0.1$, $b(T_1 = 0) = 0.1$, for $Y_0 = 0.2$, $\mu = 0.2$, $\lambda = 0.2$, $\sigma_y = 0$, $\sigma_I = 0$, $\zeta_1 = \zeta_2 = 0.1$, $C_{16} = -0.05$, $C_{15} = -0.05$. (Ibrahim and El-Sayad, 1999)

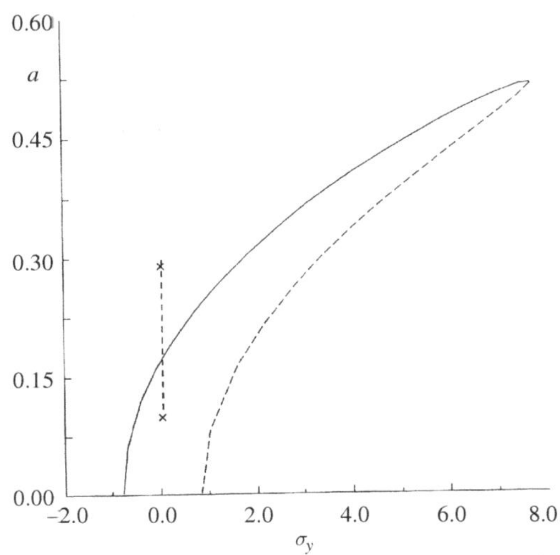

Figure 7.25 First mode amplitude–frequency response in the absence of internal resonance showing the response amplitude by a dash vertical line at $\sigma_y = 0.0$ when the system is internally tuned, $Y_0 = 0.2$, $\sigma_I = 0.0$, $\mu = 0.2$, $\lambda = 0.2$, $\zeta_1 = \zeta_2 = 0.1$. (Ibrahim and El-Sayad, 1999)

Pilipchuk and Ibrahim (1999) employed the theory of Lie group operators to analyze the dynamic behavior of the system described by equations (7.92). The system's equations of motion were reduced to their simplest (normal) form using operations with the linear differential operators according to Hausdorff's formula (Belinfante *et al.*, 1966). For different values of parametric excitation amplitude, the response time history records did not achieve

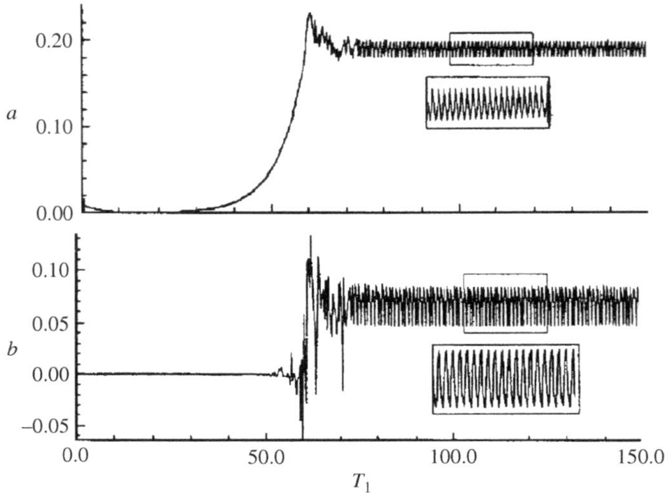

Figure 7.26 Time history records of response amplitudes for the impact case under first-mode parametric excitation $a(T_1 = 0) = 0.01$, $b(T_1 = 0) = 0.01$, for $Y_0 = 0.2$, $\mu = 0.2$, $\lambda = 0.2$, $\sigma_y = 0$, $\sigma_I = -10$, $\zeta_1 = \zeta_2 = 0.1$, $C_{16} = -0.05$, $C_{15} = -0.05$. (Ibrahim and El-Sayad, 1999)

any steady state. When the first mode is parametrically excited, it transfers energy to the second mode through nonlinear coupling. This energy transfer takes place when the two modes are in internal resonance. It is obvious that the simultaneous presence of internal and parametric resonances is the main source of the unsteady state of the system response.

7.7.2 Second mode parametric excitation

Introducing the detuning parameters σ_i such that $\Omega_1 = 2\omega_2 + \varepsilon\sigma_y$, and $\omega_2 T_0 = 3\omega_1 T_0 - \sigma_I T_1$. In terms of the new phase angles γ_i defined by $\gamma_1 = \sigma_y T_1 - 2\beta$, $\gamma_2 = \sigma_I T_1 - \beta + 3\alpha$, the following first-order differential equations are obtained

$$\frac{\partial \gamma_1}{\partial T_1} = \sigma_y + \frac{2}{\omega_2}\left\{ G_{22}\frac{Y_0}{4}\sin\gamma_1 + (\bar{G}_6 + \bar{G}_8)a^2 + \bar{G}_9 b^2 \right.$$
$$+ \bar{G}_{10}a^2b^2 + \frac{15}{16}C_{16}(a^4 + b^4)$$
$$\left. + \bar{G}_{11}\frac{a^3}{b}\cos\gamma_2 + \bar{G}_{12}a^3 b\sin\gamma_2 + \frac{1}{31}(\omega_1 C_{15}\sin\gamma_2 + 5C_{16}\cos\gamma_2)\frac{a^5}{b} \right\} \tag{7.111a}$$

$$\omega_1\frac{\partial a}{\partial T_1} = -\omega_1^2\bar{\zeta}_1 a + \bar{G}_3 ab^4 + \frac{3\omega_1}{8}C_{15}a^3b^2 + \frac{\omega_1}{16}C_{15}a^5 - \frac{3\omega_1}{8}C_{15}a^4 b\cos\gamma_2$$
$$+ \left(\frac{G_{118}\omega_2}{8}\cos\gamma_2 - \bar{G}_6\sin\gamma_2\right)a^2 b - \left(\bar{G}_7\cos\gamma_2 + \frac{15\omega_2}{16}C_{15}\sin\gamma_2\right)a^2b^3 \tag{7.111b}$$

$$\frac{\partial \gamma_2}{\partial T_1} = \sigma_I + \frac{Y_0}{4\omega_2} G_{22} \sin\gamma_1 + \left(\frac{\bar{G}_8}{\omega_2} - \frac{3\bar{G}_2}{\omega_1}\right) a^2 + \left(\frac{\bar{G}_9}{\omega_2} - \frac{3\bar{G}_1}{\omega_1}\right) b^2$$
$$- \frac{3}{\omega_1}\left(\bar{G}_6 \cos\gamma_2 + \frac{G_{118}\omega_2}{8} \sin\gamma_2\right) ab + \frac{\bar{G}_{11}}{\omega_2}\frac{a^3}{b} \cos\gamma_2$$
$$+ \left(\frac{\bar{G}_{10}}{\omega_2} + \frac{3\bar{G}_4}{\omega_1}\right) a^2 b^2 + \frac{15}{16} C_{16}\left(\frac{1}{\omega_2} - \frac{1}{\omega_1}\right) a^4 + \frac{15}{16} C_{16}\left(\frac{1}{3\omega_2} - \frac{6}{\omega_1}\right) b^4$$
$$- \frac{3}{\omega_1}\left(\frac{15}{16} C_{15}\omega_2 \cos\gamma_2 - \bar{G}_7 \sin\gamma_2\right) ab^3$$
$$+ \left(\frac{\bar{G}_{12}}{\omega_2} a^3 b \sin\gamma_2 + \frac{9}{8} C_{15}\right) \sin\gamma_2 a^3 b$$
$$+ \frac{1}{32\omega_2}(\omega_1 C_{15} \sin\gamma_2 + 5C_{16}\cos\gamma_2)\frac{a^5}{b} \qquad (7.111c)$$

$$\omega_2 \frac{\partial b}{\partial T_1} = -\left(\frac{G_{22}}{4} Y_0 \cos\gamma_1 + \omega_2^2 \zeta_2\right) b - \bar{G}_{11} a^3 \sin\gamma_2 + \frac{C_{15}}{16}(3\omega_1 a^4 + \omega_2 b^4) b$$
$$+ \bar{G}_{12} a^3 b^2 \cos\gamma_2 + \frac{C_{15}}{32}(\omega_1 \cos\gamma_2 - 5\sin\gamma_2) a^5 \qquad (7.111d)$$

Equations (7.111) define the response amplitudes and phase angles in the neighborhood of the second mode parametric resonance condition $\Omega_y = 2\omega_2$ and 3 : 1 internal resonance condition $\omega_2 = 3\omega_1$. The steady state response can be obtained by setting the left-hand side to zero since T1 is a slow time scale.

The nonimpact response can be examined by dropping the fifth-order terms from equations (7.111). The amplitude and phase equations take the form

$$\frac{\partial \gamma_1}{\partial T_1} = \sigma_y + \frac{2}{\omega_2}\left\{G_{22}\frac{Y_0}{4} \sin\gamma_1 + (\bar{G}_6 + \bar{G}_8) a^2 + \bar{G}_9 b^2 + \bar{G}_{11}\frac{a^3}{b} \cos\gamma_2\right\} \qquad (7.112a)$$

$$\omega_1 \frac{\partial a}{\partial T_1} = -\omega_1^2 \bar{\zeta}_1 a + \left(\frac{G_{118}\omega_2}{8} \cos\gamma_2 - \bar{G}_6 \sin\gamma_2\right) a^2 b \qquad (7.112b)$$

$$\frac{\partial \gamma_2}{\partial T_1} = \sigma_I + \frac{G_{22}}{4\omega_2} Y_0 \sin\gamma_1 + \left(\frac{\bar{G}_8}{\omega_2} - \frac{3}{\omega_1}\bar{G}_2\right) a^2 + \left(\frac{\bar{G}_9}{\omega_2} - \frac{3\bar{G}_1}{\omega_1}\right) b^2$$
$$- \frac{3}{\omega_1}\left(\frac{G_{118}\omega_2}{8} \sin\gamma_2 + \bar{G}_6 \cos\gamma_2\right) ab + \frac{\bar{G}_{11}}{\omega_2}\frac{a^3}{b} \cos\gamma_2 \qquad (7.112c)$$

$$\omega_2 \frac{\partial b}{\partial T_1} = -G_{22} b \frac{Y_C}{4} \cos\gamma_1 - \omega_2^2 \zeta_2 b - \bar{G}_{11} a^3 \sin\gamma_2 \qquad (7.112d)$$

Careful inspection of these equations for the steady state solution reveals that for the unimodal response of the second mode ($a = 0$, and $b \neq 0$), the resulting three algebraic equations (from 7.112a, c, d) are in two unknowns, namely b and γ_1. This means that the time derivative of γ_2 may not vanish. Note that if the left-hand sides are set to zero one may get more than one solution.

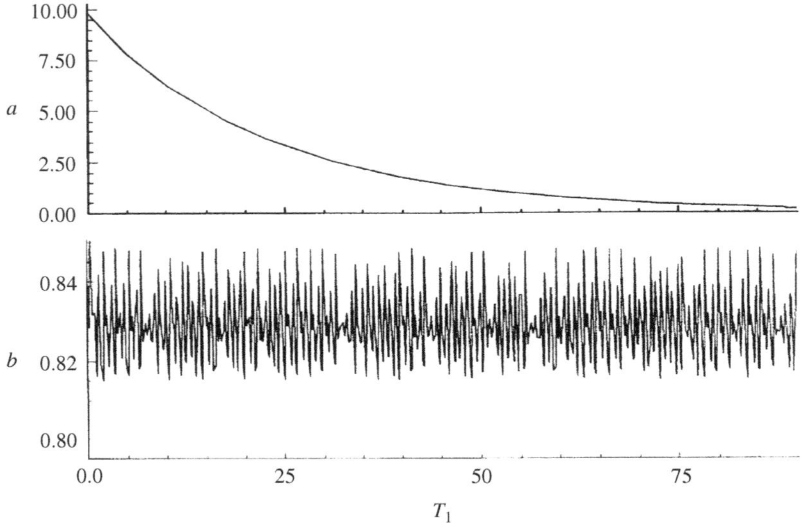

Figure 7.27 Time history records of response amplitudes for the nonimpact case under second-mode parametric excitation $a(T_1 = 0) = 0.1$, $b(T_1 = 0) = 0.086$, for $Y_0 = 0.2$, $\mu = 0.2$, $\lambda = 0.2$, $\sigma_y = -50$, $\sigma_I = 0.0$, $\zeta_1 = \zeta_2 = 0.1$. (Ibrahim and El-Sayad, 1999)

These solutions include the unimodal response, which can be obtained in the absence of internal resonance (see El-Sayad, *et al.*, 1999):

$$a = 0$$
$$b = 0, \quad \text{or} \quad b_{1,2}^2 = \frac{1}{4\bar{G}_9}\left\{2\omega_2\sigma_y \pm \sqrt{\bar{G}_{22}^2 Y_0^2 - 16\omega_2^4\bar{\zeta}_2^2}\right\} \tag{7.113}$$

The validity of these solutions is examined by numerically integrating equations (7.112) for mass ratio $\mu = 0.2$, length ratio $\lambda = 0.2$, local frequency ratio $\nu = 0.5$, excitation amplitude ratio $Y_0 = 0.1$, and damping ratios $\zeta_1 = \zeta_2 = 0.1$. Figure 7.27 shows that the first modal amplitude always reaches its zero equilibrium, while the second modal amplitude may or may not reach a fixed value depending on the initial conditions. When the second mode does not reach a steady state value, the first phase angle oscillates, while the second phase angle has a constant time rate. For a different set of initial conditions, the second-mode amplitude reaches a fixed point; however, its phase has a constant time rate, while the second phase possesses fixed steady state values.

The numerical integration always yields a unimodal response regardless of the value of internal detuning parameter and initial conditions. Furthermore, no energy transfer from the second mode to the first one is found to exist. For zero internal detuning parameter, $\sigma_I = 0$, the unimodal response is always a fixed point but takes two different values depending on the initial conditions over a parametric detuning parameter σ_y ranging from 0 to -40. For $\sigma_y < -40$, the response exhibits two possible different solutions one is fixed while the other is chaotic. For the chaotic response, equations (7.112) do not achieve a steady state for relatively large initial conditions while they achieve a fixed point for smaller initial conditions. It is also found that the internal detuning parameter does not have any appreciable effect on the response except when the internal detuning parameter assumes very large values. The initial

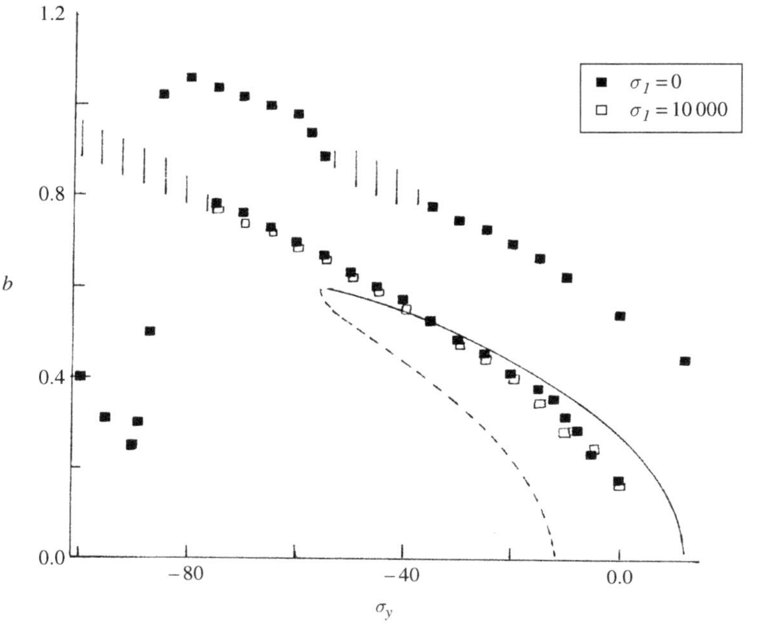

Figure 7.28 Second mode amplitude–frequency response nonimpact curve in the absence of internal resonance showing the response when the system is internally tuned vertical lines indicate unsteady amplitude. (Ibrahim and El-Sayad, 1999)

conditions can bring the response either to a fixed point or to periodic/quasi-periodic oscillations.

Figure 7.28 summarizes the different scenarios described earlier. It shows the amplitude–frequency response curves in the absence of internal resonance as indicated by the solid and dash curves (taken from El-Sayad, *et al.*, 1999). The solid square points show the response with zero internal detuning parameter and the empty squares are for $\sigma_I = 10\,000$. Over the larger portion of the parametric detuning parameter there are two different manifolds indicated by two solid squares. There is also a region showing one solid square (indicating a fixed solution) and vertical short lines indicating chaotic behavior.

In the presence of impact, equations (7.111) are considered. For the steady state solution of the second mode, the left-hand sides of equations (7.111a, c, d) are set to zero. There are three resulting algebraic equations with two unknowns, which implies that a steady state response is not possible at least for the second phase angle. Equations (7.111) are numerically integrated for mass ratio $\mu = 0.2$, length ratio $\lambda = 0.2$, local frequency ratio $\nu = 0.5$, excitation amplitude ratio $Y_0 = 0.1$, $C_{16} = -0.5$, $C_{15} = -0.05$, and damping ratio $\zeta_1 = \zeta_2 = 0.1$. The first mode amplitude always reaches its zero equilibrium position while the second mode amplitude reaches fixed solutions depending on the initial conditions. However, the first phase angle reaches a nonzero fixed value while the second phase angle increases linearly with time. This means that $d\gamma_2/dT_1 \neq 0$ and any analytical solution is not valid.

For zero internal detuning parameter, σ_I, and different values of parametric detuning parameter, σ_y, the response amplitude always reaches a fixed steady state value. For each parametric detuning parameter there are two different values of the second mode amplitude

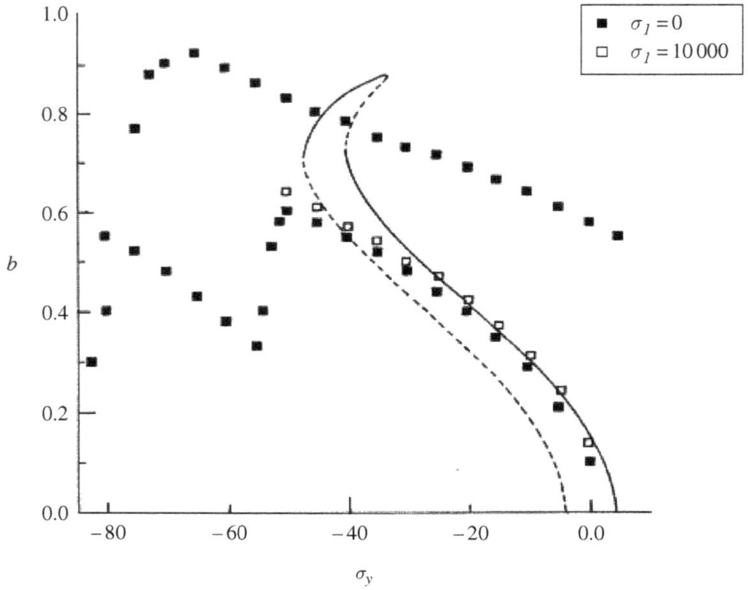

Figure 7.29 Second mode amplitude–frequency response impact curve in the absence of internal resonance showing the response when the system is internally tuned. (Ibrahim and El-Sayad, 1999)

corresponding to two different sets of initial conditions. It is important to note that the amplitude b reaches two different values for the same parametric detuning parameter and for zero internal detuning parameter. In the absence of internal resonance, the response approaches the value determined solely by second mode parametric excitation.

Figure 7.29 shows the amplitude–frequency response curves of the second mode excitation in the absence of internal resonance shown by solid and dash curves (taken from El-Sayad, *et al.*, 1999). The solid curves belong to stable manifolds while the dash ones are unstable. In the presence of internal resonance, $\sigma_I = 0$, the response is indicated by solid squares which increase with the negative values of the parametric detuning parameter up to one of the turning points $\sigma_y \approx -50$. For values $\sigma_y < -50$, the trend is changed. For very large values of $\sigma_I = 10\,000$, the response indicated by empty squares is very close to the solid curve (of second parametric excitation). The agreement holds for a range of parametric detuning parameter defined by the region $\sigma_y = 0$ to -30. For $\sigma_y < -30$, there is a deviation attributed to the effect of the time variation of phase angles.

The unimodal response of the second mode was also obtained by Pilipchuk and Ibrahim (1999), who employed the theory of Lie group operators. This means that the second mode, which is parametrically excited, does not transfer energy to the first mode.

7.7.3 Mixed mode parametric excitation

Introducing the detuning parameters σ_y and σ_I defined by $\Omega_y = \omega_1 + \omega_2 + \varepsilon\sigma_y$ and $\omega_2 = 3\omega_1 - \varepsilon\sigma_I$, and following the standard procedure of multiple scales gives the following set of first-order differential equations in the amplitudes a and b, and phase angles $\gamma_1 = \sigma_y T_1 - \alpha - \beta$, $\gamma_2 = \sigma_I T_1 - \beta + 3\alpha$,

$$\begin{aligned}\frac{\partial \gamma_1}{\partial T_1} &= \sigma_y + \frac{Y_0}{4}\left(\frac{G_{12}}{\omega_1}\frac{b}{a} + \frac{G_{22}}{4}\frac{a}{b}\right)\sin\gamma_1 + \left(\frac{\bar{G}_1}{\omega_1} - \frac{\bar{G}_9}{\omega_2}\right)b^2 + \left(\frac{\bar{G}_2}{\omega_1} + \frac{\bar{G}_6 + \bar{G}_8}{\omega_2}\right)a^2 \\ &+ \frac{1}{\omega_1}\left(\frac{G_{118}\omega_2}{8}\sin\gamma_2 + \bar{G}_6\cos\gamma_2\right)ab + \frac{\bar{G}_{11}}{\omega_2}\frac{a^3}{b}\cos\gamma_2 - \left(\frac{\bar{G}_4}{\omega_1} - \frac{\bar{G}_{10}}{\omega_2}\right)a^2b^2 \\ &+ \frac{5}{16}C_{16}\left(\frac{6}{\omega_1} + \frac{1}{\omega_2}\right)b^4 + \frac{5}{16}C_{16}\left(\frac{1}{\omega_1} + \frac{3}{\omega_2}\right)a^4 - \left(\frac{3}{8}C_{15} - \frac{G_{12}}{\omega_2}\right)a^3b\sin\gamma_2 \\ &- \frac{1}{\omega_1}\left(\bar{G}_7\sin\gamma_2 - \frac{15}{16}C_{15}\omega_2\cos\gamma_2\right)ab^3 \\ &+ \frac{1}{32\omega_2}(\omega_1 C_{15}\sin\gamma_2 + 5C_{16}\cos\gamma_2)\frac{a^5}{b} \end{aligned} \tag{7.114a}$$

$$\begin{aligned}\omega_1\frac{\partial a}{\partial T_1} &= -G_{12}b\frac{Y_0}{4}\cos\gamma_1 - \omega_1^2\bar{\zeta}_1 a + \frac{G_{118}\omega_2}{8}a^2b\cos\gamma_2 - \bar{G}_6a^2b\sin\gamma_2 \\ &- \frac{3\omega_1}{8}C_{15}a^4b\cos\gamma_2 + \frac{\omega_1 C_{15}}{8}\left(\frac{a^2}{2} + 3b^2\right)a^3 \\ &- \left(\bar{G}_7\cos\gamma_2 + \frac{15\omega_2}{16}C_{15}\sin\gamma_2\right)a^2b^3 + \bar{G}_3ab^4 \end{aligned} \tag{7.114b}$$

$$\begin{aligned}\frac{\partial \gamma_2}{\partial T_1} &= \sigma_I - \frac{1}{4}\left(\frac{3}{\omega_1}G_{12}\frac{b}{a} - \frac{G_{23}}{\omega_2}\frac{a}{b}\right)Y_0\sin\gamma_1 - \left(3\frac{\bar{G}_2}{\omega_1} - \frac{\bar{G}_8}{\omega_2}\right)a^2 \\ &- \left(3\frac{\bar{G}_1}{\omega_1} - \frac{\bar{G}_9}{\omega_2}\right)b^2 - \frac{3}{\omega_1}\left(\frac{G_{118}\omega_2}{8}\sin\gamma_2 + \bar{G}_6\cos\gamma_2\right)ab \\ &+ \frac{\bar{G}_{11}}{\omega_2}\cos\gamma_2\left(\frac{a^3}{b}\right) - \frac{15}{16}C_{16}\left(\frac{1}{\omega_1} - \frac{1}{\omega_2}\right)a^4 + \left(\frac{3}{8}C_{15} + \frac{\bar{G}_{12}}{\omega_2}\right)a^3b\sin\gamma_2 \\ &+ \left(3\frac{\bar{G}_4}{\omega_1} + \frac{\bar{G}_{10}}{\omega_2}\right)a^2b^2 + \frac{3}{\omega_1}\left(\bar{G}_7\sin\gamma_2 - \frac{15}{16}C_{15}\omega_2\cos\gamma_2\right)ab^3 \\ &- \frac{5}{8}C_{16}\left(\frac{9}{\omega_1} - \frac{1}{2\omega_2}\right)b^4 + \frac{1}{32\omega_2}(\omega_1 C_{15}\sin\gamma_2 + 5C_{16}\cos\gamma_2)\frac{a^5}{b} \end{aligned} \tag{7.114c}$$

$$\begin{aligned}\omega_2\frac{\partial b}{\partial T_1} &= -G_{23}a\frac{Y_0}{4}\cos\gamma_1 - \omega_2^2\bar{\zeta}_2 b - \bar{G}_{11}a^3\sin\gamma_2 + \frac{C_{15}}{16}(3\omega_1 a^4 + \omega_2 b^4)b \\ &+ \bar{G}_{12}a^3b^2\cos\gamma_2 + \frac{C_{15}}{32}(\omega_1\cos\gamma_2 - 5\sin\gamma_2)a^5 \end{aligned} \tag{7.114d}$$

These equations are numerically integrated for mass ratio $\mu = 0.2$, length ratio $\lambda = 0.2$, local frequency ratio $\nu = 0.5$, excitation amplitude ratio $Y_0 = 0.1$, and damping ratios $\zeta_1 = \zeta_2 = 0.1$.

First, we consider the nonimpact response by dropping the quintic terms from equations (7.114). For the steady state solution, the left-hand sides of the resulting equations are set to zero. Careful inspection of the algebraic equations reveals the possibility of two equilibrium positions which can only take place when the phase angles take the two specific values $\gamma_1 = \gamma_2 = 0$ or π. The corresponding steady state solutions are

$$a = -\frac{4\omega_2^2\bar{\zeta}_2}{G_{23}Y_0}b \tag{7.115}$$

and from (7.114b) one obtains

$$b = 0 \quad \text{or} \quad b = \pm \frac{1}{2\omega_2^2 \bar{\zeta}_2} \sqrt{\frac{|G_{12} G_{23} Y_0^2 - 4\omega_1^2 \omega_2^2 \bar{\zeta}_1 \bar{\zeta}_2| G_{23} Y_0}{2G_{118}}} \tag{7.116a,b}$$

Another solution is obtained from equation (7.114a, c) which gives (7.115) and

$$b = \pm \sqrt{\left(\sigma_y + \frac{\sigma_1}{3}\right) \bigg/ \left[\frac{4\bar{G}_9}{3\omega_2} - \frac{16\omega_2^4 \bar{\zeta}_2^2}{\bar{G}_{23}^2 Y_0^2}\left(\frac{1}{\omega_2}\left(\bar{G}_6 + \frac{\bar{G}_8}{3}\right) - \frac{16\omega_2 \bar{G}_{11} \bar{\zeta}_2}{3\bar{G}_{23} Y_0}\right)\right]} \tag{7.117}$$

Note that the zero solution, given in (7.116a), corresponds to the static equilibrium position, while the other two solutions (7.116b) and (7.117) can only exist under parametric excitation, that is, when $Y_0 \neq 0$. Obviously, the solution spirals out about each of the two unstable equilibrium points given by (7.116b) or (7.117) until its amplitude is sufficiently large when it is attracted to the other equilibrium point about which it then spirals. The jumping of the solution trajectory from the vicinity of one unstable equilibrium point to the other unstable point seems to take place in a random fashion. The alternation of the second mode between two states reflects the chaotic nature of the system similar to the case of a Lorenz attractor.

If $\gamma_1 \neq \gamma_2 \neq 0$, it is not possible to solve for the stationary response. In this case, the equations of motion are numerically integrated. Depending on the internal detuning parameter and initial conditions, the numerical integration gives different response regimes. For zero detuning parameters, the response amplitudes of the two modes experience quasi-periodic oscillations. If the initial conditions are changed, the response trajectories take another form of attractor of nearly periodic oscillations. In the absence of internal resonance, the amplitude–frequency curves give fixed points corresponding to steady state responses under purely parametric excitation. The presence of internal resonance obviously results in quasi-periodic motion and the energy is redistributed between the two modes.

For an internal detuning parameter, $\sigma_I = -30$, and exact parametric resonance, the response can be periodic as shown in Figure 7.30 or chaotic with irregular jumps between two unstable equilibria as shown in Figure 7.31a, depending on initial conditions. The phase of the first mode is periodic while it has constant time rate for the second mode. On the other hand, the phase of the switching sign of the amplitude b shown in Figure 7.31b reveals that the phase of the first mode assumes fixed values when the amplitude fluctuates around the positive value. When the first mode amplitude fluctuates around the negative value the corresponding phase increases monotonically with time. The spectra of Figure 7.31a reveal that the second mode amplitude has a wide frequency spectrum with a large number of closely packed peaks, which fall off rapidly as their frequency increases. The response is sensitive to initial conditions and different initial conditions can lead to an almost periodic response. The initial conditions have a significant effect on the response behavior.

Contrary to the nonimpact case, the impact loading brings the system response to a fixed stationary point at exactly zero detuning parameters $\sigma_I = 0$, and $\sigma_Y = 0$. The presence of internal resonance results in a redistribution of energy between the two modes. Over a small range of internal detuning parameter, $0 < \sigma_I < 1$, the system preserves a fixed point response. Away from this region, the response is essentially periodic over the two ranges of internal

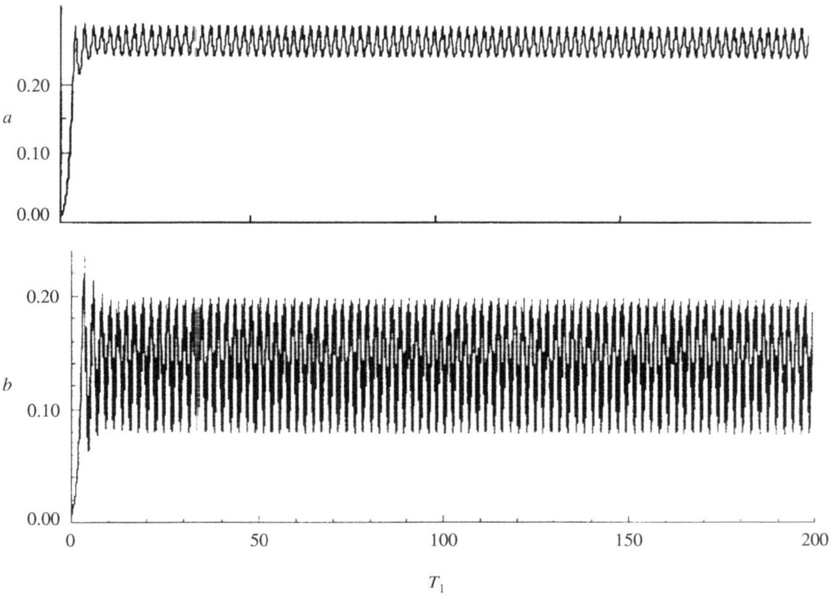

Figure 7.30 Time history records of response amplitudes for the nonimpact case under mixed-mode parametric excitation $a(T_1 = 0) = 0.1$, $b(T_1 = 0) = 0.086$, for $Y_0 = 0.2$, $\mu = 0.2$, $\lambda = 0.2$, $\sigma_y = 0$, $\sigma_I = -30$, $\zeta_1 = \zeta_2 = 0.1$. (Ibrahim and El-Sayad, 1999)

detuning parameter $-15 < \sigma_I < 0$, and $1 < \sigma_I < 15$. The response over these ranges possesses different attractors with different periods. Beyond these ranges, the response experiences complex characteristics where both quasi-periodic and switching phase characteristics can take place depending on initial conditions. Figure 7.32(a) and (b) show two sets of time history records corresponding to two different sets of initial conditions and for the same internal detuning parameter $\sigma_I = -15$. This trend continues for different values of σ_I. An extensive number of numerical integrations revealed the presence of three possible basic regimes: fixed point, quasi-periodic, and quasi-periodic/switching phase. Again each regime is governed by the initial conditions, internal detuning parameter, damping ratios and excitation amplitude.

Pilipchuk and Ibrahim (1999) employed the theory of Lie group operators to analyze the mixed mode response of the system described by equations (7.92). In the presence of internal resonance, the response does not achieve a steady state and is nearly quasi-periodic. The internal resonance is responsible for the exchange of energy between the two modes.

The fluid free-surface behavior characterized by relatively large wave height may be determined numerically. Some numerical algorithms have been developed and the next section will present an overview of some selected numerical codes with application to liquid sloshing in liquid cargo ships.

7.8 Numerical simulation of sloshing impact

7.8.1 Preliminaries

Numerical methods of solving free boundary problems associated with nonlinear liquid sloshing may be classified into three categories:

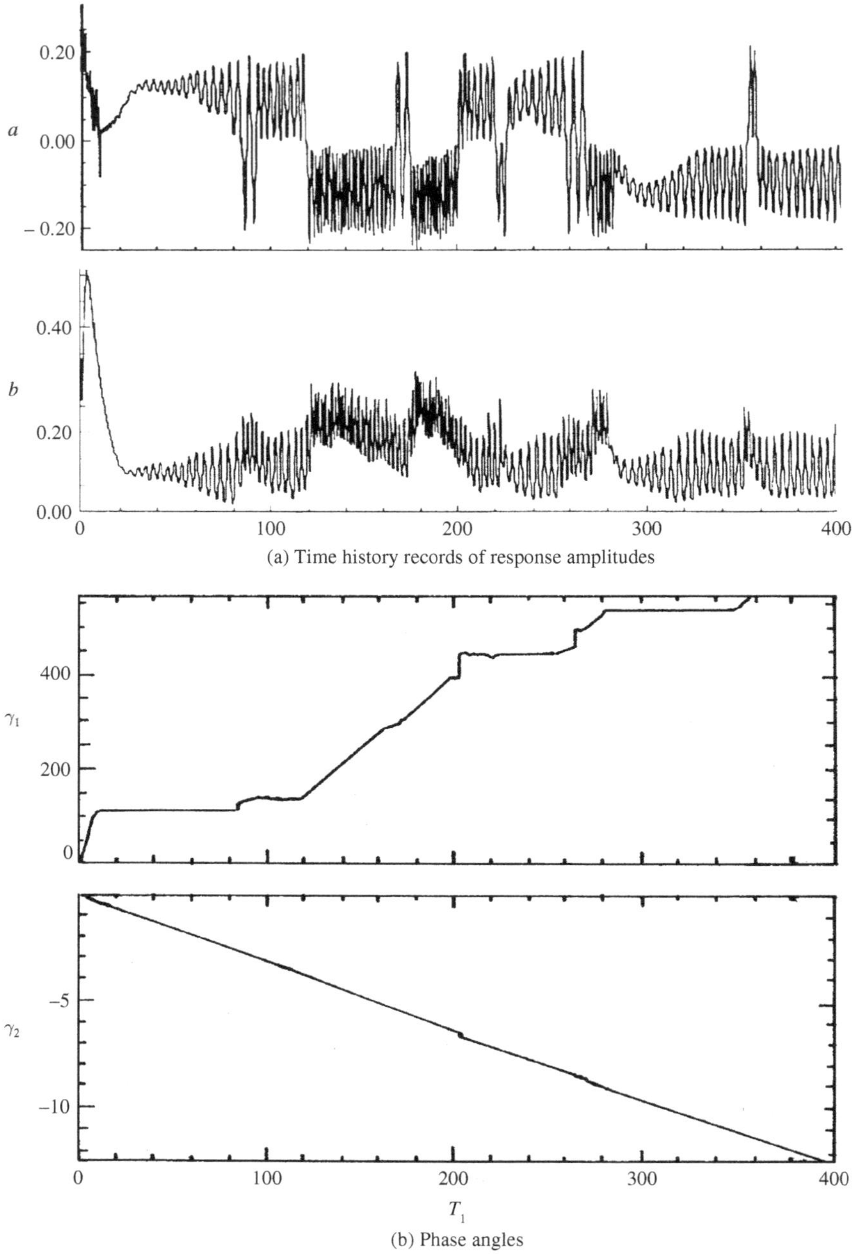

Figure 7.31 Time history records or response amplitudes and phase angles for the nonimpact case under mixed mode parametric excitation $a(T_1 = 0) = 0.2$, $b(T_1 = 0) = 0.2$, for $Y_0 = 0.2$, $\mu = 0.2$, $\lambda = 0.2$, $\sigma_y = 0$, $\sigma_I = -30$, $\zeta_1 = \zeta_2 = 0.1$. (Ibrahim and El-Sayad, 1999)

Lagrangian methods: These methods assume that any point of the free boundary moves with the fluid velocity. They include the line segments method (Nichols and Hirt, 1971, 1975 and Miyata, 1986), and the marker particles method (Harlow and Welch, 1965). These methods are very accurate but frequent remeshings are required. Okamoto and Kawahara (1990) and

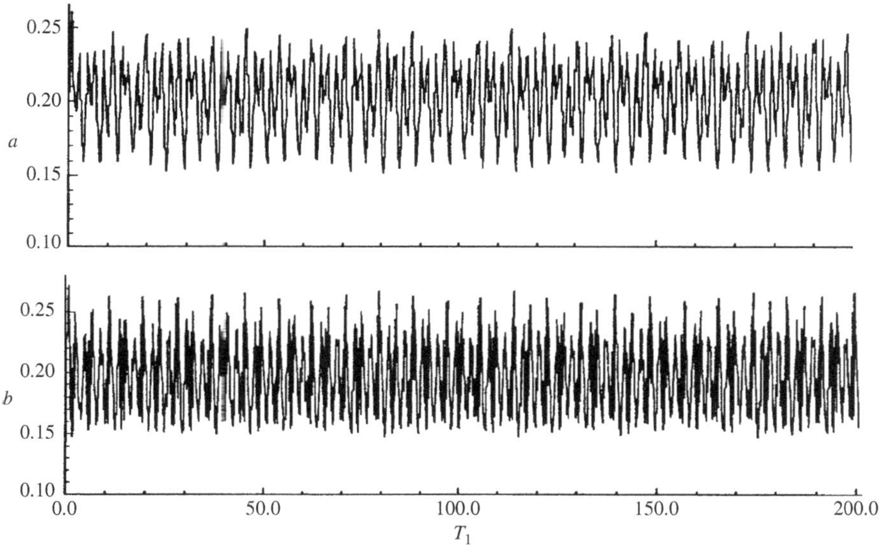

(a) Initial conditions $a(T_1 = 0) = 0.2$, $b(T_1 = 0) = 0.2$

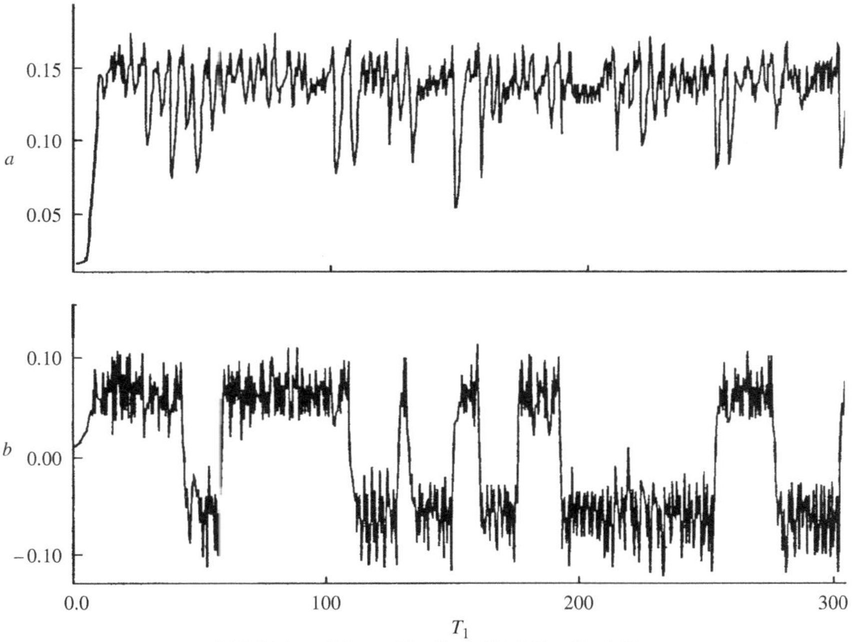

(b) Initial conditions $a(T_1 = 0) = 0.01$, $b(T_1 = 0) = 0.01$

Figure 7.32 Time history records of response amplitude for the impact case under mixed mode excitation for two different sets of initial conditions, $\sigma_I = -15$. (Ibrahim and El-Sayad, 1999)

Doğangün, *et al.* (1996) employed the Lagrangian finite element to study fluid dynamics in rectangular tanks.

Eulerian methods: The actual position of the points of the free boundary is localized in a fixed mesh. The fluid field equations are solved in a larger domain than that really occupied by the fluid. The interface between the dry and wet regions has to be distinguished by using the concept of volume-of-fluid function (Hirt and Nichols, 1981). In principle, these methods can be employed for general free and moving boundary problems; however, the identification of the interface needs a refined mesh in order to obtain accurate results. Ushima (1998) and Souli and Zolesio (2001) developed an arbitrary Lagrangian–Eulerian and free-surface methods for general applications in fluid mechanics.

Arbitrary Lagrangian–Eulerian method (ALE): these methods combine the advantages of the two previous groups. The computational domain is completely occupied by the fluid but it can be animated with its proper motion. At the free or moving boundary, the velocities of mesh points are related to the fluid velocities in order to satisfy the kinematic boundary condition.

Under large-amplitude excitation, the liquid behaves violently and numerical solutions become unstable. An improved numerical algorithm, which permits repeated liquid impacts, was developed by Su and Kang (1984a,b, 1986). Later, Su and Wang (1989, 1990) developed a three-dimensional finite difference scheme based on the volume-of-fluid (VOF) technique for large liquid amplitude motion in rectangular and cylindrical containers subjected to vertical and arbitrary excitations. The finite difference time-stepping-scheme method has been employed by Frandsen (2002, 04) and Franden and Borthwick (2002) to simulate free and forced sloshing motions in rectangular tanks. It was shown that for very steep waves, just before the waves overturn, and for large forcing frequency, a discrepancy in amplitude and phase was reported between the approximate forms and the numerical model. Su (1992) presented some numerical results for three-dimensional sloshing simulation. The results included the estimation of liquid total force and impact for vessels undergoing rapid movement. He observed that high sloshing waves are produced in the vicinity of surface centers. Under a step type displacement input, the deformations of sidewalls produced a heaving type sloshing mode.

Other numerical simulations of shallow liquid flows, such as finite difference and boundary element methods, were conducted to estimate hydrodynamic loads acting on the structures (Faltinsen, 1974, 1978, Lou, *et al.*, 1985, Lee, *et al.*, 1995, and Buchner, 1995). In applying these methods, three main problems were encountered: (1) Difficulty of accommodating the inherent discontinuity of the surface elevation on the two sides of the jump. (2) Difficulty of handling the propagation of the hydraulic jump. (3) Difficulty of predicting the magnitude and phase of the propagation velocity of the hydraulic jump. Dillingham (1981) adopted an algorithm originally developed by Glimm (1965), which was later modified by Chorin (1976, 1977). This algorithm has a unique feature in its capability to handle hyperbolic differential equations and discontinuities in the solution. Pantazopoulos and Adee (1987) and Pantazopoulos (1988) employed Glimm's method for two- and three-dimensional liquid containers. Yamaguchi (1996) employed a numerical algorithm for three-dimensional fluid dynamics with free-surface boundaries.

Park and Rezvani (1995) used a two-dimensional computational algorithm SOLA-VOF (SOLution Algorithm-Volume-Of-Fluid, originally developed by Nichols, *et al.*, 1980 and Hirt and Nichols, 1981) to determine the motion of waves in rectangular tanks subjected to pre-measured earthquake acceleration records. The numerical solution revealed a large liquid amplitude that reached the lid of the tank with secondary surges. Amano, *et al.* (1989) developed a three-dimensional analysis method using the boundary element method.

Hwang, *et al.* (1992) and Ferrant and Touze (2001) simulated sloshing waves in a three-dimensional tank using numerical pseudo-spectral methods. This method was found to be within 15% accuracy when compared with experimental measurements for the impulse case when the liquid surface collided with the container ceiling. Minowa and Kiyosumi (1997) conducted shaking table tests of an actual water tank and analyzed the sloshing impacts on the tank roof using nonlinear theories developed by Kurihara, *et al.* (1992) and Minowa, *et al.* (1994). They obtained a closed form expression for the roof impact pressure for cylindrical tanks. These studies were motivated by the bulging-type damage of water tanks caused by the 1995 Hyogokennabu earthquake.

Harlow and Welch (1965) and Welch, *et al.* (1965) developed and used the marker-and-cell (MAC) method. The method introduced massless markers that move with the liquid. Surfaces are defined as lying at the "boundary" between regions with and without marker particles. A mesh cell containing markers, but having a neighboring cell with no markers, is defined as containing a free surface. Harlow, *et al.* (1976) extended the marker particle method to three-dimensional computations. Some extended techniques were developed by Chan and Street (1970), Hirt, *et al.* (1975), and Miyata *et al.* (1980).

Chorin (1967) presented an extension of the MAC for calculating steady state solutions of the equations of motion of an incompressible fluid. Chorin introduced the use of an artificial density and sought a steady state solution for which the rates of change of this artificial density and of the velocities vanish. Viecelli (1969) adapted Chorin's method in the MAC scheme for solving the pressure and velocity fields simultaneously. The simplified marker-and-cell (SMAC) method presented by Amsden and Harlow (1971) was a significant revision of the MAC method and has been widely used. Johnson, *et al.* (1994) used the SMAC to simulate impacts of advancing fluid fronts with solid obstacles. The simulation takes into account the relatively rapid changes in both pressure and velocity in the simulation of impact.

The design of floating systems, storage and off-loading platforms, and floating storage units depends on the values of the forces acting on these systems. In general, the dynamic wave loads consist of two components. The first component constitutes the global loads, which are induced by the unsteady hydrodynamic pressure due to fluid oscillatory motions surrounding the hull. The second component represents the local or secondary loads, such as slamming and whipping, due to wave impacts. Ochi and Motter (1971), Kawakami, *et al.* (1977), and Wood *et al.* (2000) developed some simple formulas for estimating the slamming loads based on experimental results. Belik and Price (1982) and Belik, *et al.* (1988) divided the bottom slamming into impact and momentum slamming components. They found that the predicted magnitudes of the responses after a slam depend very much on the mathematical model adopted in estimating slamming loads. Molin, *et al.* (1996) found that the values of impact force as determined by conservation of momentum are almost identical to those predicted by conservation of kinetic energy.

Zhou, *et al.* (1999) used Glimm's (1965) method to numerically simulate green[2] water dynamics on deck. Ramos, *et al.* (2000) conducted experimental investigation and numerical simulation to evaluate the slamming forces. Faltinsen, *et al.* (2000) used a discrete modal system to calculate various kinematic and dynamic characteristics occurring due to interaction between the free

[2] Green water or green seas are defined as unbroken waves overtopping the bow, side, or stern of a vessel. Green water on the deck can cause damage to the sensitive equipment at the bow, such as fluid swivel, piping, and turret structure.

liquid surface and its container. They considered the example of ship tank dynamic response to wave-induced ship motion. The combined equations of motion for the global rigid ship motions and the equations describing sloshing dynamics gave complex fluid–structure interaction. Jones and Hulme (1987), Huang and Hsiung (1996), and Yilmaz, *et al.* (2003) showed that the flow of water over the deck edge resembles a suddenly released wall of water rather than a breaking wave. They simulated green water flow onto the deck by using dam breaking theory.

The method of volume-of-fluid (VOF) is based on the concept of a fractional volume-of-fluid, which is usually incorporated into the flow equations to track free fluid surface. The method is used to approximate free boundaries in finite difference simulations. For large surface deformations discontinuities exist in one or more variables. When the free boundary experiences large deformations one has to use Eulerian coordinates. The grid remains fixed in the Eulerian representation and the identity of individual fluid elements is not maintained. In the calculation it is necessary to compute the flow of fluid through the mesh. The convective flux calculation requires an averaging of the flow properties of all fluid elements that are found in a given mesh cell after some period of time. The averaging process, however, results in a smoothing of all variations in the flow quantities and thus smears surfaces of discontinuities. In this case, one has to introduce some special treatment that recognizes a discontinuity and avoids averaging across it. Among the problems encountered in the numerical treatment of free boundaries (Hirt and Nichols, 1981) are their discrete representation and their time evolution.

In a rectangular two-dimensional mesh of cells of width δx, and height δz one may define the vertical height, h, of the free boundary above the bottom of the mesh in each column of cells. The curve $h = f(x, t)$ is approximated by assigning values of h to discrete values of x. The time evolution of the fluid free surface, $h = f(x, t)$, is governed by the kinematic equation that the free surface must move with the fluid

$$w = \frac{\partial h}{\partial t} + u\frac{\partial h}{\partial x} \tag{7.118}$$

where u, and w are the fluid velocity components in the x- and z-directions, respectively. The free-surface boundary is represented by its distance, h, measured from the undisturbed position. The height function does not work well if the boundary slope, dh/dx, exceeds the mesh cell aspect ratio $\delta z/\delta x$, and does not work at all for multiple-valued surfaces having more than one z value for a given x value, such as in the case when the free surface takes the "S" shape. A generalization of the height function method is to use chains of short line segments, or points connected by line segments (Nichols and Hirt, 1971). Coordinates of each point are stored and accuracy is improved by limiting the distance between neighboring points to values less than the minimum mesh size δx or δz.

The evolution of a chain of line segments is obtained by moving each point with the local fluid velocity determined by interpolation in the surrounding mesh. When two surfaces intersect, or when a surface folds over on itself in the form of an "S" shape, segment chains must be reordered by adding or removing some chains. Nichols and Hirt (1975) extended the line segment method to three-dimensional surfaces.

The estimation of liquid sloshing loads in tanks of ships is important in the design of ship vessels such as liquefied natural gas (LNG) carriers and double-hull tankers. Shinkai, *et al.* (1994, 1995), Takemoto, *et al.* (1994), and Yamamoto, *et al.* (1995) found that ship motions of liquid vessels tend to be relatively large in comparison with those of large vessels. Casella, *et al.*

(1996) and Casella and Dogliani (1996) studied the sloshing-induced fatigue damage in moored floating-storage off-loading-unit tankers. Brosset (1995) and Park, *et al.* (1995) conducted numerical simulation of liquid sloshing in large LNG tankers and double hull tankers. Studies of reliability in ship design with liquid tanks include static and dynamic stability and structure integrity (Francescutto and Contento, 1994). Cariou and Casella (1999) presented a comparative study from a significant sample of 11 codes for two- and three-dimensional numerical simulations of liquid sloshing in ship tanks. The study confirmed that nonimpulsive phenomena are correctly simulated but impacts are still far more difficult to assess and need more development. For impulsive cases, quite large discrepancies were observed and three numerical codes did not produce results.

The interaction effect between liquid nonlinear sloshing dynamics and the floating tank motion under sea waves was numerically studied using a finite difference code by Chen and Chiang (2000). The study considered a simple two-dimensional rectangular tank with rigid walls under simultaneous actions of surge, heave, and pitch motions. It was found that in the absence of sea waves, the contribution of the sloshing load would enlarge the tank angular response as well as the rise of free surface. In addition, the sloshing effect delays the damping effect on angular displacement. On the other hand, under sea waves, the sloshing effect was found to decrease the tank dynamic motion and free-surface elevation.

7.8.2 The volume-of-fluid (VOF) method

The VOF method is developed on the basis that the fluid state is described in each cell of a mesh by one value of each dependent variable. The method introduces a function F whose value $F = 1$ corresponds to a cell full of fluid, and $F = 0$ to an empty cell. The average value of F in a cell would then represent the fractional volume of the cell occupied by fluid. Thus, the fractional volume-of-fluid method provides the same coarse interface information available to the marker particle method. Because F is a step function, its derivatives must be carefully computed. The time evolution of F is governed by the equation

$$\frac{\partial F}{\partial t} + u\frac{\partial F}{\partial x} + w\frac{\partial F}{\partial z} = 0 \tag{7.119}$$

Equation (7.119) states that the function F moves with the fluid. The VOF method is the basis of the SOLA-VM (SOLution Algorithm-Variable Mesh) VOF code. The SOLA-VM method uses Eulerian rectangular cells with variable sizes δx_i for the ith column and δz_j for the jth row. The fluid equations to be solved are the Navier–Stokes equations

$$\frac{\partial u}{\partial t} + u\frac{\partial u}{\partial x} + w\frac{\partial u}{\partial z} = -\frac{\partial p}{\partial x} + \nu\left[\frac{\partial^2 u}{\partial x^2} + \frac{\partial^2 u}{\partial z^2}\right] \tag{7.120a}$$

$$\frac{\partial w}{\partial t} + u\frac{\partial w}{\partial x} + w\frac{\partial w}{\partial z} = -\frac{\partial p}{\partial z} + g + \nu\left[\frac{\partial^2 w}{\partial x^2} + \frac{\partial^2 w}{\partial z^2}\right] \tag{7.120b}$$

The fluid density has been normalized to unity. Equations (7.120) must be supplemented with the continuity equation

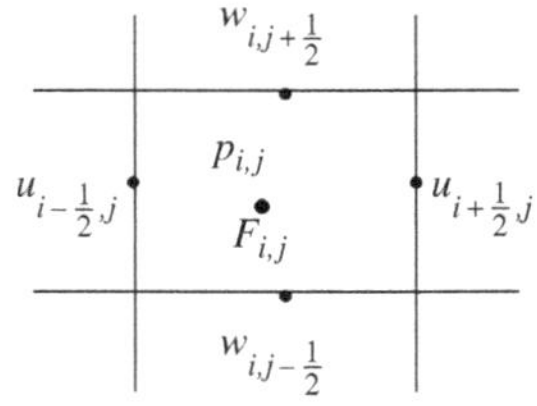

Figure 7.33 Schematic diagram showing a typical mesh cell and location of variables. (Hirt and Nichols, 1981)

$$\frac{\partial u}{\partial x} + \frac{\partial w}{\partial z} = 0 \tag{7.121}$$

Discrete values of the dependent variables, including the variable F, are located at the cell position shown in Figure 7.33. A free surface cell i, j is defined as a cell containing a nonzero value of F and having at least one neighboring cell, $(i \pm 1, j)$ or $(i, j \pm 1)$, that contains a zero value of F. Empty cells have zero values of F, and full or interior fluid cells have nonzero F. Explicit approximation of equations (7.120) is used to compute the first guess for new time-level velocities using the initial conditions or previous time-level values for all convective, pressure, and viscous accelerations. At the same time equation (7.121) has to be satisfied. The function F must be updated to give the new fluid configuration. At each step the boundary conditions must be imposed at all mesh and free-surface boundaries. A generic form for the finite-difference approximation of equations is

$$u_{i+1/2,j}^{n+1} = u_{i+1/2,j}^{n} + \delta t\left[-\frac{\left(p_{i+1,j}^{n+1} - p_{i,j}^{n+1}\right)}{\delta x_{i+1/2}} - \text{FUX} - \text{FUZ} + \text{VISX}\right] \tag{7.122a}$$

$$w_{i,j+1/2}^{n+1} = w_{i,j+1/2}^{n} + \delta t\left[-\frac{\left(p_{i,j+1}^{n+1} - p_{i,j}^{n+1}\right)}{\delta w_{j+1/2}} - \text{FWX} - \text{FWZ} + \text{VISW}\right] \tag{7.122b}$$

where $\delta x_{i+1/2} = (\delta x_i + \delta x_{i+1})/2$, $\delta w_{j+1/2} = (\delta w_j + \delta w_{j+1})/2$, FUX stands for the convective flux in the x-direction, and the same applies for the remaining expressions. These terms are evaluated using the old time level (n) values for velocities. In the marker-and-cell (MAC) method equations (7.120) and (7.121) are combined so that the convective flux terms are written in a divergence form, and FUX would be written, for example, as $\partial u^2/\partial x$, rather than $2u(\partial u/\partial x)$. The divergence form provides a simple way to ensure conservation of momentum in the difference approximation. This may be demonstrated by considering the control volume used for $\mathbf{u}_{(i+1/2,j)}$ shown by the dash lines in Figure 7.34. By using Green's theorem one can convert the integrated form of FUX over the control volume to boundary fluxes at its sides. The flux leaving one control volume will automatically be gained by the adjacent volume, and conservation during convection is thus guaranteed. Hirt and Nichols (1981) showed that conservation in a variable mesh does not automatically imply accuracy. The reason that the form of convective terms loses accuracy is that the control volumes are not centered about the positions where variables are located. Thus, the convective terms should be corrected to

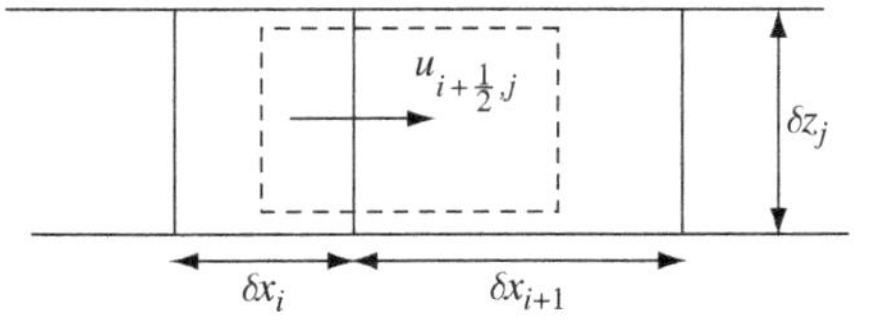

Figure 7.34 Schematic diagram of a control volume, showin by the dash rectangle, for constructing a finite-difference approximation. (Hirt and Nichols, 1981)

account for the difference in locations of the variables being updated and the centroids of their control volumes.

The stability advantages of the donor-cell method can be retained in a variable mesh with no reduction in accuracy, if the convective term $\mathbf{u} \cdot \nabla \mathbf{u}$ is used for the convective flux. It is possible to combine the donor-cell and centered-difference approximations into a single expression with a parameter, α, that controls the relative amount of each one. The general form at (i, 1/2, j) is (Hirt and Nichols, 1981),

$$\mathrm{FUX} = \frac{u_{i+1/2,\,j}}{\delta x_\alpha}[\delta x_{i+1}\mathrm{DUL} + \delta x_i \mathrm{DUR} + \alpha\,\mathrm{sgn}(u)(\delta x_{i+1}\mathrm{DUL} - \delta x_i \mathrm{DUR})] \tag{7.123}$$

where

$$\mathrm{DUL} = \frac{(u_{i+1/2,\,j} - u_{i-1/2,\,j})}{\delta x_i}$$

$$\mathrm{DUR} = \frac{(u_{i+3/2,\,j} - u_{i+1/2,\,j})}{\delta x_{i+1}}$$

$$\delta x_\alpha = \delta x_{i+1} + \delta x_i + \alpha\,\mathrm{sgn}(u)(\delta x_{i+1} - \delta x_i)]$$

sgn(u) refers to the sign of $u_{i+1/2,j}$. When $\alpha = 0$ this approximation reduces to a second-order accuracy referred to as centered-difference approximation. When $\alpha = 1$, the first-order donor-cell form is recovered. Thus, by using approximation (7.123) the formal accuracy is preserved. The basic idea of approximation (7.123) is to weight the upstream derivative of the quantity being fluxed more than the downstream value. The weighting factors are $1 + \alpha$ and $1 - \alpha$, for the up and downstream derivatives, respectively. The derivatives are also weighted by cell sizes in such a way that the correct order of approximation is maintained in a variable mesh. This type of approximation is used in SOLA-VOF for all convective flux terms appearing in equations (7.122). Viscous accelerations are approximated using standard centered approximations. Patrom (1987) employed the VOF method to simulate liquid sloshing in cylindrical containers.

7.8.3 Sloshing impact in ship tankers

The local pressure due to sloshing impact loads can cause a critical damage in the insulation box of liquid natural gas (LNG) carriers. Most ships with liquid cargo experience impulsive local loads on the cargo ceiling. Several studies have been conducted to examine the influence of liquid sloshing impact in LNG carriers (Filstead, 1972, Abramson, *et al.*, 1974, Faltinsen, *et al.*, 1974, Faltinsen, 1978, Cox, *et al.*, 1980, Bridges, 1982, Mikelis, *et al.*, 1984, Arai, 1986, and Arai, *et al.*, 1993, 1994). A moving boundary finite element approach to simulate nonlinear sloshing waves in ship tankers was developed and implemented by Ikegawa and Washizu (1973), Washizu, *et al.* (1977), Tokuda and Sakurai (1994), and Greaves, *et al.* (1997). At shallow filling, a hydraulic jump can be generated during a large excitation motion. In this case, the tank wall is exposed to large impact pressures, followed by secondary tails. Sometimes the fluid moves up to the tank ceiling, resulting in sharp pressure peaks on the top or in the corner. In relatively high filling, the impact pressure can be observed in a large area of the tank ceiling, the liquid runs up to the tank ceiling and moves along the top boundary. In this case, the impact area moves with the wave crest. If the liquid motion is in a standing mode, it will hit a large part of the ceiling. These phenomena are well described by Abramson, *et al.* (1974), Hamlin, *et al.* (1986), Kim, *et al.* (1994), and Kim (2001). When the liquid sloshing is violent, some local physical phenomena become difficult to analyze. For example, splash and wave breaking are typical phenomena in violent flow (Banner and Peregrine, 1993 and Banner and Tian, 1996). Any numerical simulation has to accommodate such flow motion. Park, *et al.* (1995) developed a practical method to predict sloshing loads in cargo ship tanks.

The VOF technique has been used by Celebi, *et al.* (1998) and Celebi and Akyildiz (2002) to simulate two-dimensional viscous liquid sloshing in moving rectangular baffled and un-baffled tanks. The excitation was assumed to be combined harmonic translational and pitching. Numerical simulation of axisymmetric surface flows was performed by Tome, *et al.* (2000). As the liquid free surface becomes violent under large excitation amplitude the numerical solution becomes unstable. This indicates that the VOF technique is limited up to the moment prior to the inception of the violent motion. Kim (2001) used a finite-difference scheme based on the Marker-and-Cell (MAC) method that assumes a single-valued surface profile. This scheme does not simulate the multi-valued surface profiles usually observed after the occurrence of impact. Kim (2001) considered a rectangular tank partially filled with an incompressible liquid and experienced forced excitation as shown in Figure 7.35. The moving coordinates x, y, z are attached to the center of the tank bottom. The tank rotates with angular velocity vector, $\mathbf{\Omega}$, and translational velocity vector, $\mathbf{U}$, with respect to the inertial frame X, Y, Z of origin G. Letting the liquid velocity with respect to the moving coordinates be $\mathbf{u}$, the continuity and Navier–Stokes equations are

$$\nabla \cdot \mathbf{u} = 0 \tag{7.124}$$

$$\frac{D\mathbf{u}}{Dt} = -\frac{1}{\rho}\nabla p + \nu\nabla^2\mathbf{u} + \mathbf{F} \tag{7.125}$$

where $\mathbf{F}$ is the external force vector and D/Dt stands for material derivative. The external force is the sum of the gravitational force, translational, and rotational inertia forces, that is,

$$\mathbf{F} = \mathbf{g} - \frac{d\mathbf{U}}{dt} - \frac{d\mathbf{\Omega}}{dt} \times (\mathbf{r} - \mathbf{R}) - 2\mathbf{\Omega} \times \frac{d(\mathbf{r} - \mathbf{R})}{dt} - \mathbf{\Omega} \times [\mathbf{\Omega} \times (\mathbf{r} - \mathbf{R})] \tag{7.126}$$

where $\mathbf{r}$ and $\mathbf{R}$ are position vectors of a fluid particle and the tank with respect to the inertial fixed frame.

The kinematic free-surface boundary condition is

$$\frac{D\mathbf{r}_f}{Dt} = \mathbf{u}_f \tag{7.127}$$

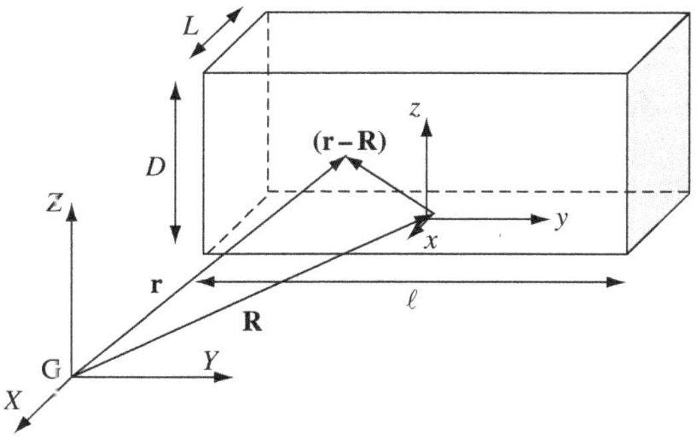

Figure 7.35 Schematic diagram of a rectangular tank showing the moving and fixed coordinate frames.

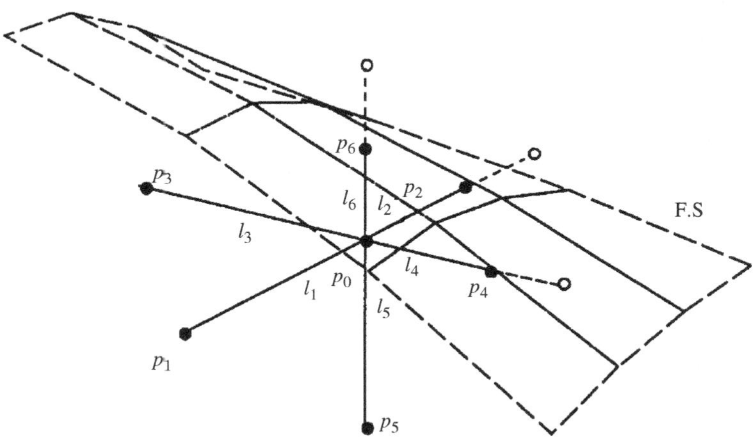

Figure 7.36 Schematic diagram of three-dimensional irregular-star method. (Kim, 2001)

The dynamic boundary condition is

$$p_f = p_{\text{atm}} \tag{7.128}$$

where the subscript f refers to the values on the free surface, and p_{atm} is the atmospheric or ullage[3] pressure inside the tank. The boundary conditions at the tank bottom and walls must be satisfied. The solution algorithm volume-of-fluid SOLA-VOF is implemented by discretizing the tank volume into Cartesian stagger grids as shown in Figure 7.36. The velocity components are defined at the centers of cell boundaries and pressure is computed at the center of the cell. The finite-difference approximation of the velocity at the $(n+1)th$ time step may be written in the form

$$\mathbf{u}_{ijk}^{(n+1)} = \mathbf{u}_{ijk}^{(n)} + \Delta t\left[-\frac{1}{\rho}(\nabla \mathrm{p})_{ijk}^{(n)} + \nu(\nabla^2\mathbf{u})_{ijk}^{(n)} + (\mathbf{f})_{ijk}^{(n+1)}\right] - \Delta t\left[(\mathbf{u}\cdot\nabla)_{ijk}^{(n)}\mathbf{u}_{ijk}^{(n)}\right] \tag{7.129}$$

[3] The space provided for thermal expansion of the liquid propellant and the accumulation of gases evolved from it. Ullage is usually referred to as the volume of a loaded tank of liquid propellant in excess of the volume of the propellant.

where ∇ is the discrete gradient. The last convection term is sensitive in the numerical solution and requires special treatment. Kim (2001) adopted the mixed form of the second-order central difference and the first-order up-winding difference. In the SOLA scheme, the pressure and velocity corrections are written in the form

$$\frac{\Delta p_{ijk}}{\rho} = -\frac{\gamma\left(\nabla \cdot \mathbf{u}_{ijk}^{(n+1)}\right)}{2\Delta t\left[(1/\Delta x)^2+(1/\Delta y)^2+(1/\Delta z)^2\right]} \tag{7.130}$$

$$\mathbf{u}_{ijk}^{(n+1)} = \mathbf{u}_{ijk}^{(n)} + \frac{\Delta p_{ijk}}{\rho}\frac{\Delta t}{\Delta(x, y, z)} \tag{7.131}$$

where Δt is the temporal segment, $\Delta(x, y, z)$ is the spatial segment, γ is the relaxation factor, usually over-relaxed at a full cell and under-relaxed at a partly filled cell. The velocity prediction and correction should be repeated until the pressure correction converges to the specified error criteria.

For the wall boundary condition, two cases may be considered. The first is when the viscosity effect on the tank boundary is significant and the no-slip condition must be imposed. The second, which is common for sloshing problems, is when the viscous effect is not significant and the boundary layer thickness is much less than the cell size. In this case, the free-slip condition is applied. Kim (2001) adopted the free-slip condition and applied the constant slope condition at the intersection of the free-surface with the wall. The dynamic free-surface condition (7.128) is imposed using the irregular-star method (Chan and Street, 1970 and Miyata, *et al.*, 1980). The irregular-star method assumes that the pressure at a cell near the free surface is interpolated from the pressure values at six neighboring points, including points on the free surface. At the cell where the free surface intersects, the pressure can be computed using Taylor expansion from the neighboring points. With reference to Figure 7.36, the pressure at the center of the cell near the free surface can be written as

$$p_0 = \sum_{l=1}^{6} C_l p_l \tag{7.132}$$

where the coefficient C_l depends on the distances $l_1, l_2, \ldots, l_6$. For example, the adjacent points 2, 4, and 6 shown in Figure 7.36 are on the free surface and $p_2 = p_4 = p_6 = p_{\text{atm}}$.

The kinematic free surface condition (7.127) defines the free-surface profile and for a single-valued surface, it is assumed that wave breaking does not occur. The wave elevation at the center of each cell column at the $(n+1)$th time step is

$$\eta_{ij}^{(n+1)} = \eta_{ij}^{(n)} + \Delta t\left[w - u\frac{\partial \eta}{\partial x} - v\frac{\partial \eta}{\partial y}\right]_{ij} \tag{7.133}$$

The local wave slopes are obtained with a proper mixture of central and up-winding differences. The fluid detachment is allowed when the vertical velocity is negative and a free surface exists at adjacent cells. The numerical algorithm checks for the following numerical stability conditions

$$\left(\Delta u_{ijk}, \Delta v_{ijk}, \Delta w_{ijk}\right)\Delta t < \Delta(x, y, z) \tag{7.134a}$$

$$v\Delta t < \frac{1}{2\Delta Z\left[(1/\Delta x)^2+(1/\Delta y)^2+(1/\Delta z)^2\right]} \tag{7.134b}$$

$$\left|\eta_{ij}^{(n+1)} - \eta_{ij}^{(n)}\right| < \Delta z \tag{7.134c}$$

Condition (7.134c) does not allow the free surface to change more than one cell height during one time segment. When the free surface hits the tank ceiling, the hydrodynamic pressure takes the form of a sharp impulse. If the ordinary wall boundary condition is imposed, the peak value critically depends on the spatial and temporal discretizations. For example, $u_{ijk} = w_{ijk} = 0$, and $v_{ijk} = V$ at the $(n)th$ time step can happen at tank corners. At the same step, when the free surface contacts the tank top, the pressure correction in equation (7.130) is written in the form, after setting $\partial w/\partial z \approx (w_{ijk} - w_{ijk-1})/\Delta z$,

$$\frac{\Delta p_{ijk}}{\rho} = -\frac{\gamma V}{2\Delta t\Delta z\left[(1/\Delta x)^2 + (1/\Delta y)^2 + (1/\Delta z)^2\right]} \tag{7.135}$$

When Δt or Δz becomes zero, equation (7.135) diverges to infinity and the solution becomes too sensitive on the spatial and temporal segments. Sometimes, it is hard to simulate the real physical phenomena.

Under sloshing impact, some physical phenomena, such as liquid compressibility, air cushion, and structural elasticity, play an important role in dampening the fluid motion during impact. The overall effect is a reduction of the impulsive pressure. These effects are difficult to consider in one formulation. The following condition is imposed on the fluid free surface as the fluid approaches the tank ceiling

$$F = \kappa\left[\frac{p_f - p_{\text{atm}}}{\rho}\right] - (1-\kappa)\frac{H_B}{\Delta t}w_f = 0 \tag{7.136}$$

where

$$\kappa = \begin{cases} (D-\eta)/H_B & D - H_B \le \eta < D \\ 1 & \eta \le D - H_B \\ 0 & \eta = D \end{cases} \tag{7.137}$$

for which D is the tank depth, and H_B is the size of the buffer zone in which this condition is applicable. Equation (7.136) is a linear combination of the flux wall condition and dynamic free-surface condition. Thus, the wall presence affects the fluid motion when the free surface is inside the buffer zone near the tank ceiling. This condition is imposed only during the upward motion of the free surface. The pressure correction in the buffer zone is computed using the Newton–Raphson method. The time history of impulsive pressure is not continuous because there are some intervals between the impacts on neighboring cells. In order to obtain more

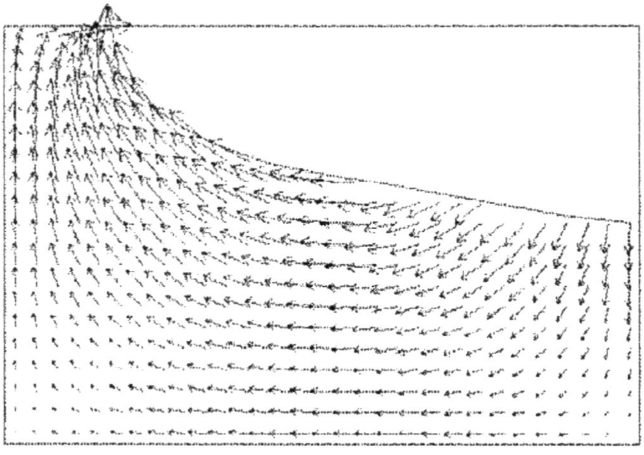

Figure 7.37 Computed velocity vectors and free surface elevation. (Kim, 2001)

continuous signals, the time averaging of the discrete signal is taken over several time steps, that is,

$$p_{\text{avg}}\left[\frac{1}{N}\sum_{n=1}^{N} t^{(n)}\right] = \frac{1}{N}\sum_{n=1}^{N} p^{(n)} \tag{7.138}$$

The averaging process results in a continuous signal and a drop in the peak value of impact pressure. For this reason N should be carefully selected when the magnitude of the peak is critical in the design process. Before an impact occurs, the fluid particles have a high vertical velocity and if the time segment is not very small the free surface can pass the buffer zone during one time segment. This situation should be avoided since equation (7.136) cannot be imposed properly. Accordingly, the following condition is introduced to select the time segment

$$\left|\eta_{ij}^{(n+1)} - \eta_{ij}^{(n)}\right| < H_B \tag{7.139}$$

Kim (2001) employed a numerical simulation scheme to a smooth tank without baffles (tank A) and another tank with one large vertical baffle and three horizontal baffles (tank B), each of width $\ell = 0.8$ m and height $D = 0.45$ m. In all computations, the combination of 25% central difference and 75% up-winding difference for the convection, $10^{-6}\,\text{m}^2/\text{s}$, term in equation (7.129) was adopted. For a tank 70% filled with fluid of kinematic viscosity under a surge excitation of amplitude $A_1 = 0.02$ m and of period $T_1 = 1.08$ s the computed fluid velocity vectors and elevation are shown in Figure 7.37 for tank A. Figure 7.38 shows the computed pattern for tank B under the same filling ratio and the same excitation period but with a surge amplitude of $A_1 = 0.04$ m based on $30 \times 1 \times 20$ meshes.

Figure 7.39 shows time history records of the computed pressure acting on the tank ceiling when impact occurs. The figure shows the effect of mesh size and the averaging step. The maximum peak of the measured pressure was found to be $9.07\ \text{kN/m}^2$. It is seen that when the mesh is coarse, the discretization provides significant numerical damping and predictably less

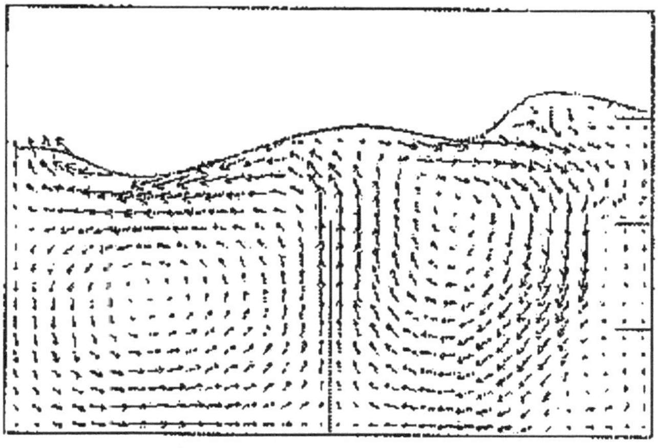

Figure 7.38 Computed velocity vectors and free surface elevation in a tank with horizontal and vertical baffles. (Kim, 2001)

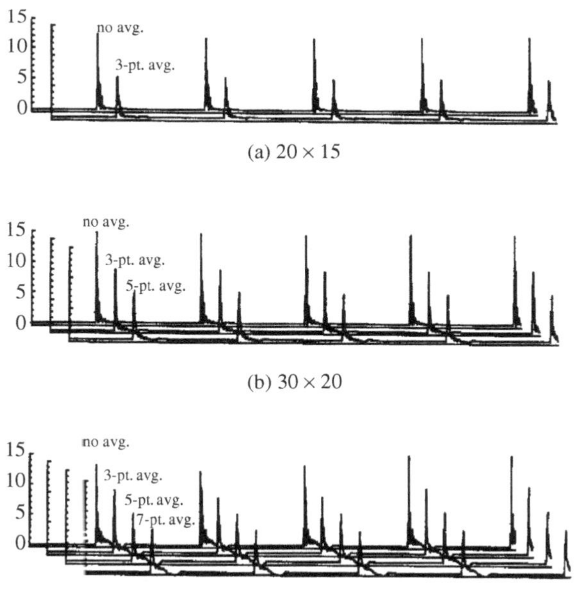

(c) 40 × 30

Figure 7.39 Time history records of pressure at the tank ceiling for different grids in a tank with 80% filling. Peak experimental measurement is 9.07 kN/m^2. (Kim, 2001)

impact pressure. Figure 7.39(a) shows that only 3-point averaging results in a dramatic change of the peak values. Generally, larger peaks and spiky signals can be observed when the mesh size becomes smaller. In a fine mesh computation, the secondary impulse due to impact in a neighboring cell follows quickly after the first impact. Thus, the time-averaged method is effective since the impulse profile becomes more continuous and peak values are closer to the

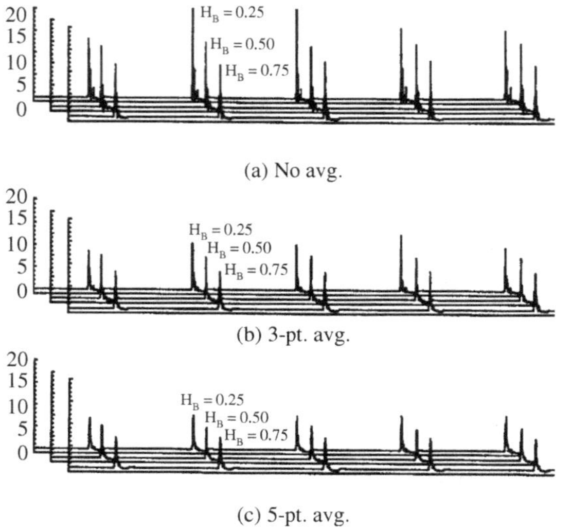

Figure 7.40 Time history records of pressure at the tank ceiling, showing the effect of the buffer zone in a tank with 70% filling. (Kim, 2001)

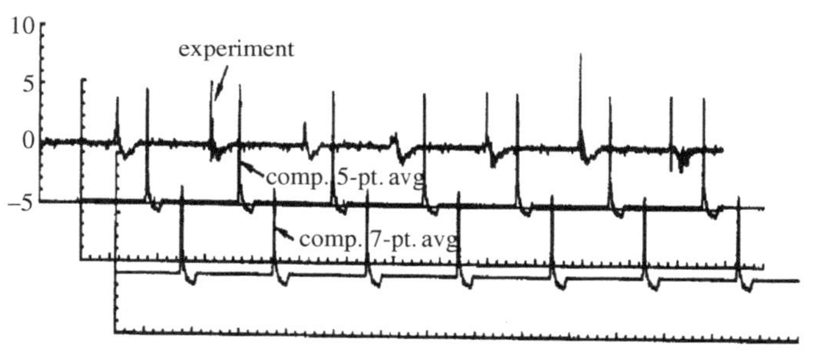

Figure 7.41 Time history records of computed and measured pressure at the tank ceiling, showing the effect of the buffer zone in a tank with 70% filling, with a 6.2 s time window. (Kim, 2001)

experimental measurements. Figure 7.40 shows the influence of the size of the buffer zone and it reveals that the impact pressure decreases as the buffer size increases. It was found that the mixed boundary condition (7.136) has a strong effect on the results when the free surface is close to the tank ceiling. Figure 7.41 shows experimental and computed time history records under surge amplitude $A_1 = 0.038$ m and period $T_1 = 0.98$ s. The computed record shows periodic impacts with the same magnitude, while the measured results show different peaks at each impact occurrence.

Figure 7.42 shows the free surface profiles at four different time steps under surge excitation of amplitude $A_1 = 0.02$ m and period $T_1 = 1.08$ s. Figure 7.43 shows time history records of wave elevation at two corners when tank B is under roll excitation. The flat crests indicate that the fluid contacts the tank ceiling. At the tank wall with horizontal members, additional flat signals are shown by dash curves. The horizontal members were found to cause a delay of the upward liquid motion.

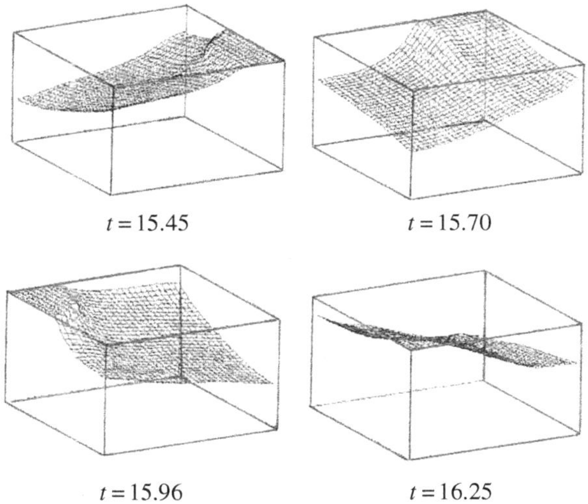

Figure 7.42 Free-surface profiles at four time steps for a tank with 70% filling. (Kim, 2001)

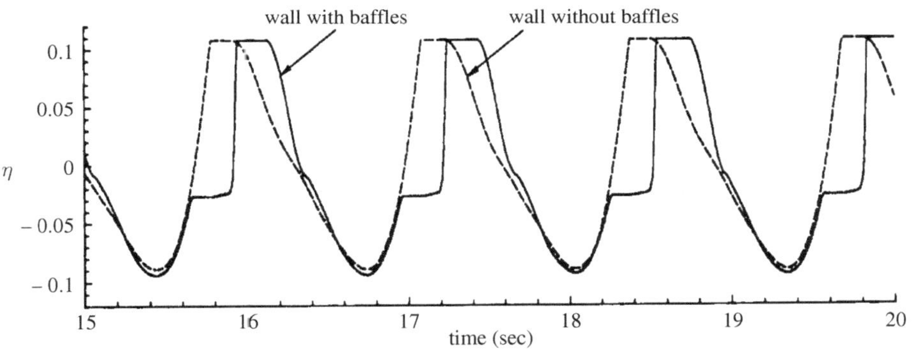

Figure 7.43 Time history records of wave elevation at two corners with and without horizontal baffles. (Kim, 2001)

7.9 Sloshing in road tankers

In road tankers, the free liquid surface may experience large excursions for even very small motions of the container. This is an undesirable feature, which may considerably endanger the stability and maneuvering quality of the vehicle. This problem is common for fuel or cargo tanks of automotive vehicles, railroad tank cars, fuel tanks of large ships and tankers. The study of liquid sloshing dynamics within a moving vehicle involves different types of modeling and analyses. Gustafson and Gustafson (1969) presented an extensive survey pertaining to the overturning problem of heavy vehicles. Isermann (1970) computed the overturning limit for some tank semi-trailers during steady state cornering. Bauer (1972, 1973, 1982b) and Popov, *et al.* (1992) provided a mathematical treatment for different container geometries partially filled with liquid. The analysis includes estimates for liquid natural frequencies, forces, and moments. Slibar and Troger (1975, 1976, 1977) analyzed the lateral wheel-load transfer ratio of a tractor semi-trailer system. The forces and moments from the liquid load for harmonic oscillation steering were introduced via a mechanical model approach. Each compartment includes one rigid and one moving mass. The moving masses are restrained by linear springs and dashpots. It is speculated that liquid-carrying vehicles have more accidents than other vehicles in part because of the large movement of the liquid cargo and related forces and moments. Liquid sloshing in a highway maneuver can lead to lateral and roll instabilities, decreased controllability/ maneuverability, and increased stress on tank structures. The dynamic analysis of articulated tank vehicles during turning was studied using equivalent mechanical models by Rakheja, *et al.* (1988) and Ranganathan, *et al.* (1989a, b).

The influence of large-amplitude liquid sloshing on the overturning and skidding stability of road tankers is very serious during dynamic maneuvers (Lidström, 1976, 1977). Standberg (1978) conducted experimental investigations to measure the liquid sloshing force in laterally oscillated model tanks with or without baffles and cross walls. The effects of liquid forces on overturning and skidding tendencies were evaluated from simplified computer vehicle models (no roll, no yaw). Standberg's study included comparisons with the effects from liquid motion during steady state cornering and with roll influence. The poorest lateral stability arises mainly due to a combination of two factors. The first is when the value of dynamic forces from liquid sloshing reaches twice the rigid load force. The second is when the liquid center of mass experiences large displacement, which can reach about 40 cm. The ratio between the center of mass height and effective width are often higher for heavy than for light tankers. This results in poor lateral stability and less capacity to perform escape maneuvers than for automobiles. Heavy vehicles often overturn during a less severe maneuver than during a skid. In fact, poor overturning stability contributes to the skidding tendency due to lateral load transfer and nonlinear tire characteristics. In conventional road tankers the free-liquid-surface motions can be more severe longitudinally than laterally if no transversal walls or baffles are introduced. Liquid impact loading is more severe to the structure for longitudinal than for lateral sloshing.

When the steering frequency is close to the natural frequency of the free liquid surface, the liquid mass will act like a pendulum swinging with the vehicle. This will result in a distinct increase in load transfer and the overturning risk factor. The overturning factor $\Re$ is defined by the expression

$$\Re = \left| 1 - \frac{F_l}{F_{lo}} \right| \tag{7.140}$$

where F_l is the instantaneous wheel load on the left side, and F_{lo} is the static wheel load on the left side. As $\Re \to 1$, the vehicle is close to inner wheel lift and overturning. The value of $\Re$ depends on the lateral acceleration whose critical value, known as the overturning limit, is reached when $\Re = 1$.

If the steering frequency is greater than the natural frequency, the liquid pendulum motion will lag the tank motion. Close to but above resonance, this may be especially dangerous if the risk factor with sloshing load is still close to its resonance level and thus will exceed the corresponding peak for rigid loads. Above resonance, the liquid will oscillate against the tank and stabilize the vehicle. Unfortunately, this stabilization is not effective in practice because the actual high steering frequencies are very rare. Furthermore, roll motions act like a low-pass filter suppressing high frequency motions in chassis before they reach the tank.

Large liquid displacement inside the tank will cause a larger overturning moment due to pure displacement of the center of mass and larger overturning forces when the relative motion of

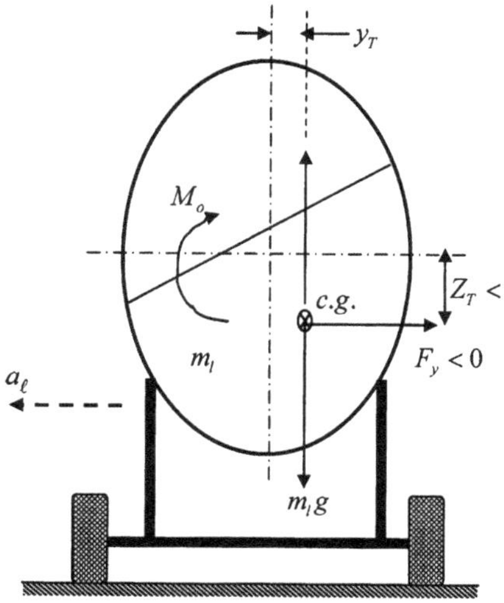

Figure 7.44 Schematic diagram of liquid displacement in a moving tanker (from Standberg, 1978).

the liquid is reversed by the tank wall. With reference to Figure 7.44, Standberg (1978) defines the side force coefficient (SC) by the expression

$$\mathrm{SC} = \frac{m_e a_\ell - F_y}{(m_e + m_l)g - F_z} \tag{7.141}$$

where m_e is the empty vehicle mass, m_l is the liquid mass, F_y is the resultant force to the left from the liquid load, F_z is the resultant upward force from the load after subtracting the weight, and a_ℓ is the lateral acceleration.

Lidström (1977) examined the distinct stabilization effect from cross walls under harmonic excitation and for a half filled tank. The overturning limits are compared for different tank partitions in Figure 7.45. Figure 7.46 shows the side force coefficient in terms of the oscillation frequency for different baffle orientations.

Consideration of tank cross-sectional shape, while maintaining the tank bottom at the same level, shows that the center height is higher for the circular horizontal tank than for the elliptic tank. The overturning limit minimum value and the worst oscillation frequency are found to be smaller for the elliptic tanks than for the circular horizontal ones.

7.10 Closing remarks

Liquid sloshing impact is a serious problem in several engineering applications such as liquid propellant rockets, road tankers, liquid storage tanks, and cargo ships. The hydrodynamic pressure variation is neither harmonic nor periodic, since the magnitude and duration of pressure peaks vary from cycle to cycle even though the tank experiences harmonic excitation. Having stated that, one may at least assume some kind of saw-tooth pressure time evolution in order to apply some analytical investigation. Over the last three decades, analytical and experimental investigations have been carried out to determine the free surface dynamics and hydrodynamic forces acting on the tank walls. The high-power impact-force modeling still needs to be verified experimentally, at least to determine the powers n, p, and the coefficients b and d of equations

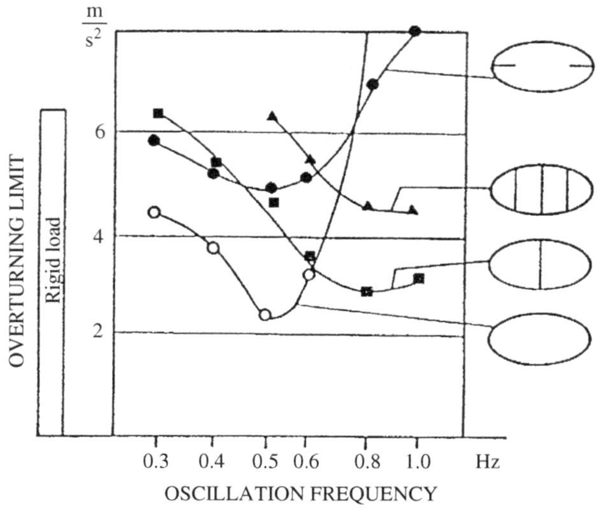

Figure 7.45 Dependence of overturning limits on the frequency for different tank baffles (from Standberg, 1978).

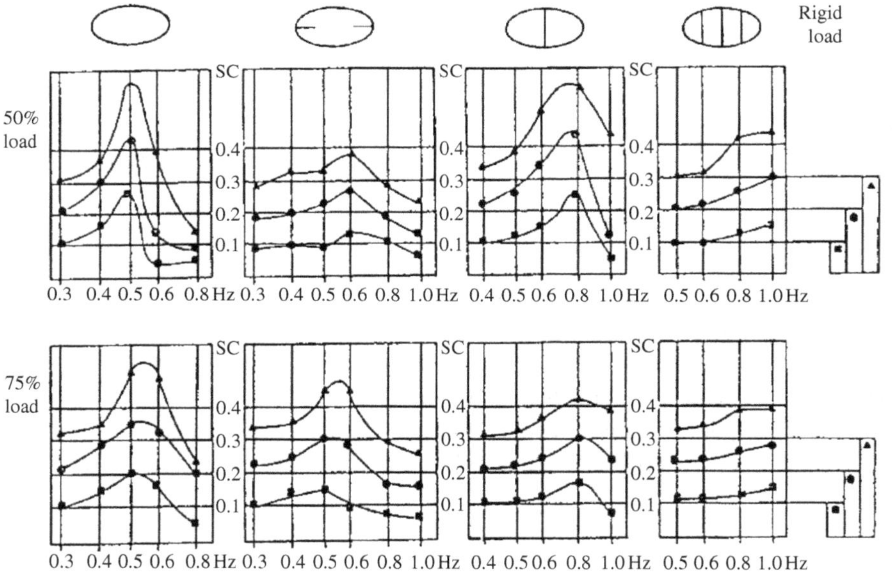

Figure 7.46 Side force coefficient (SC) as a function of the oscillation frequency for different baffle orientation (from Standberg, 1978).

(7.47) and (7.48). In the presence of simultaneous parametric and resonance conditions of an elastic structure carrying a liquid container, the system response was found to strongly depend on the initial conditions and internal detuning parameter. The response can be quasi-periodic or chaotic with irregular jumps between two unstable equilibrium positions. The phenomenological modeling presented in Section 7.5 should be implemented in multibody dynamics codes. This is an area open to possible future research.

In practice, the excitation is not harmonic or periodic, but random in nature. It is possible to consider random excitation of liquid containers and employ the nonsmooth coordinate transformation originally developed by Zhuravlev (1977) and the method of stochastic averaging as outlined by Dimentberg (1988). If one studies the random excitation of an elastic structure in the presence of resonance conditions, Zhuravlev coordinate transformation can also be used and the system response dynamic behavior can be determined using Monte Carlo simulation.

Appendix

Functions Ψ_{ii} *of equations* (7.54)

$$\begin{aligned}(\Psi_{11})_{gn} &= G_{18}Y_1^3 + G_{19}Y_2^3 + G_{110}Y_2^2Y_2'' + G_{111}Y_2^2Y_1'' + G_{112}Y_2Y_1Y_2'' \\ &\quad + G_{113}Y_2Y_1Y_1'' + G_{114}Y_1^2Y_2'' + G_{115}Y_2Y_2'^2 + G_{116}Y_1Y_2'^2 \\ &\quad + G_{117}Y_2Y_2'Y_1' + G_{118}Y_1Y_2'Y_1' + G_{119}Y_1Y_2^2 + G_{120}Y_1^2Y_2 \\ &\quad + G_{121}Y_2Y_1'^2 + G_{122}Y_1^2Y_1'' + G_{123}Y_1Y_1'^2\end{aligned} \tag{A7.1}$$

$$\begin{aligned}(\Psi_{11})_{impact} &= C_{16}Y_1^5 + C_{16}Y_2^5 + 5C_{16}Y_2Y_1^4 + 5C_{16}Y_2Y_2^4 \\ &\quad + 4C_{15}Y_1Y_1'Y_2^3 + 4C_{15}Y_1Y_2'Y_2^3 + 10C_{16}Y_1^2Y_2^3 \\ &\quad + 6C_{15}Y_1'Y_1^2Y_2^2 + 10C_{16}Y_1^3Y_2^2 \\ &\quad + 6C_{15}Y_2'Y_1^2Y_2^2 + 4C_{15}Y_2'Y_1^3Y_2 + 4C_{15}Y_1^3Y_2Y_1' + C_{15}Y_1^4Y_1' \\ &\quad + C_{15}Y_2^4Y_1' + C_{15}Y_2^4Y_2' + C_{15}Y_1^4Y_2'\end{aligned} \tag{A.7.2}$$

$$(\Psi_{11})_{ex} = G_{12}f_y(t)Y_2 + G_{13}f_y(t)Y_1 \tag{A7.3}$$

$$\begin{aligned}(\Psi_{22})_{gn} &= G_{28}Y_1^3 + G_{29}Y_2^3 + G_{210}Y_2^2Y_2'' + G_{211}Y_2^2Y_1'' + G_{212}Y_2Y_1Y_2'' \\ &\quad + G_{213}Y_2Y_1Y_1'' + G_{214}Y_1^2Y_2'' + G_{215}Y_2Y_2'^2 + G_{216}Y_1Y_2'^2 \\ &\quad + G_{217}Y_2Y_2'Y_1' + G_{218}Y_1Y_2'Y_1' + G_{219}Y_1Y_2^2 + G_{220}Y_1^2Y_2 \\ &\quad + G_{221}Y_2Y_1'^2 + G_{222}Y_1^2Y_1'' + G_{223}Y_1Y_1'^2\end{aligned} \tag{A7.4}$$

$$\begin{aligned}(\Psi_{22})_{impact} &= C_{16}Y_1^5 + C_{16}Y_2^5 + 5C_{16}Y_2Y_1^4 + 5C_{16}Y_2Y_2^4 + 4C_{15}Y_1Y_1'Y_2^3 \\ &\quad + 4C_{15}Y_1Y_2'Y_2^3 + 10C_{16}Y_1^2Y_2^3 + 6C_{15}Y_1'Y_1^2Y_2^2 + 10C_{16}Y_1^3Y_2^2 \\ &\quad + 6C_{15}Y_2'Y_1^2Y_2^2 + 4C_{15}Y_2'Y_1^3Y_2 + 4C_{15}Y_1'Y_1^3Y_2 \\ &\quad + C_{15}Y_1'Y_1^4 + C_{15}Y_1'Y_2^4 + C_{15}Y_2'Y_2^4 + C_{15}Y_2'Y_1^4\end{aligned} \tag{A7.5}$$

$$(\Psi_{22})_{ex} = G_{22}f_y(t)Y_2 + G_{23}f_y(t)Y_1 \tag{A7.6}$$

where

$$G_{12} = \frac{1}{g}\left(1 - \frac{K_1^2}{\mu\lambda}\right)$$
$$G_{13} = \frac{1}{g}\left(1 + \frac{K_1 K_2}{\mu\lambda}\right)$$
$$G_{18} = \frac{\theta_0^2}{6}\left(1 - \frac{K_1^4}{\mu\lambda^2}\right)$$
$$G_{111} = \frac{\theta_0^2 K_1}{\lambda}(1 + K_2)^2$$

$$G_{19} = \frac{\theta_0^2}{6}\left(1 - \frac{K_1 K_2^3}{\mu\lambda}\right)$$
$$G_{110} = \frac{\theta_0^2}{2\lambda}(K_1 + K_2)(1 + K_2)^2$$

$$G_{112} = \frac{\theta_0^2}{\lambda}(K_1 + K_2)(1 + K_2 + K_1 + K_1 K_2)$$

$$G_{113} = \frac{2\theta_0^2 K_1}{\lambda}(1 + K_2 + K_1 + K_1 K_2)$$
$$G_{114} = \frac{\theta_0^2}{2\lambda}(K_1 + K_2)(1 + K_1)^2$$

$$G_{115} = \frac{\theta_0^2}{\lambda}(1 + K_2)(K_1 + K_2^2)$$
$$G_{116} = \frac{\theta_0^2}{\lambda}(1 + K_1)(K_1 + K_2^2)$$
$$G_{117} = \frac{2\theta_0^2 K_1}{\lambda}(1 + K_2)^2$$

$$G_{118} = \frac{2\theta_0^2 K_1}{\lambda}(1 + K_1)(K_1 + K_2)$$
$$G_{119} = \frac{\theta_0^2}{2}\left(1 - \frac{K_1^2 K_2^2}{\mu\lambda}\right)$$
$$G_{120} = \frac{\theta_0^2}{2}\left(1 - \frac{K_1^3 K_2}{\mu\lambda}\right)$$

$$G_{121} = \frac{1}{2}G_{118}$$
$$G_{122} = \frac{\theta_0^2 K_1}{\lambda}(1 + K_1)^2$$
$$G_{123} = G_{122}$$

$$C_{15} = -\frac{d}{m\ell^2\omega_\ell}$$
$$C_{16} = -\frac{b}{m\ell^2\omega_\ell^2\theta_0}$$

$$G_{21} = \frac{1}{g\theta_0}\left(1 + \frac{K_2}{\mu\lambda^2}\right)$$
$$G_{22} = \frac{1}{g}\left(1 - \frac{K_1K_2}{\mu\lambda}\right)$$
$$G_{23} = \frac{1}{g}\left(1 - \frac{K_2^2}{\mu\lambda}\right)$$

$$G_{28} = \frac{\theta_0^2}{6}\left(1 - \frac{K_1^3K_2}{\mu\lambda}\right)$$
$$G_{29} = \frac{\theta_0^2}{6}\left(1 - \frac{K_2^4}{\mu\lambda}\right)$$
$$G_{210} = \frac{\theta_0^2K_2}{\lambda}(1 + K_2)^2$$

$$G_{211} = \frac{\theta_0^2}{2\lambda}(K_1 + K_2)(1 + K_2)^2$$
$$G_{212} = \frac{\theta_0^2K_2}{\lambda}(1 + K_1 + K_2 + K_1K_2)$$
$$G_{223} = G_{221}$$

$$G_{213} = \frac{\theta_0^2}{\lambda}(K_1 + K_2)(1 + K_1 + K_2 + K_1K_2)$$
$$G_{214} = \frac{\theta_0^2K_2}{\lambda}(1 + K_1)^2$$
$$G_{215} = G_{210}$$

$$G_{216} = \frac{\theta_0^2K_2}{\lambda}(1 + K_1)(1 + K_2)$$
$$G_{217} = 2G_{216}$$
$$G_{218} = 2G_{214}$$
$$G_{219} = \frac{\theta_0^2}{2}\left(1 - \frac{K_1K_2^3}{\mu\lambda}\right)$$

$$G_{220} = \frac{\theta_0^2}{2}\left(1 - \frac{K_1^2K_2^2}{\mu\lambda}\right)$$
$$G_{221} = \frac{\theta_0^2}{\lambda}(1 + K_2)(K_1^2 + K_2)$$
$$G_{222} = \frac{\theta_0^2}{2\lambda}(1 + K_1)^2(K_1 + K_2)$$

Part III

Sloshing – structure interaction

8
Linear interaction with elastic containers

8.1 Introduction

The problem of dynamic interaction of liquid sloshing with elastic structures may fall under one of the following categories:

(1) Interaction of liquid sloshing dynamics with the container elastic modes in breathing and bending. This type is addressed in this chapter and Chapter 9.
(2) Interaction of liquid sloshing dynamics with the supporting elastic structure. This type is treated in Chapter 10.
(3) Liquid interaction with immersed elastic structures. This class will not be addressed in this book and the reader may consult Chen (1987), Paidoussis (1998) and Dzyuba and Kubenko (2002).

This chapter presents the linear problem of liquid interaction with its elastic container. Two limiting cases may occur where interaction disappears. The first case deals with the excitation of liquid surface modes where significant elastic modes of the container are not participating. In this case, the analysis of liquid dynamics in a rigid container will provide a satisfactory description of the overall behavior. The second case deals with the excitation of the container elastic modes where significant liquid motion does not occur. In this case, the presence of liquid will contribute to the mass distributed to the tank walls, and the analysis can be carried out without considering any interaction with liquid sloshing dynamics.

The first step in studying the interaction of liquid dynamics with elastic tank dynamics is to consider the linear eigenvalue problem and response to external excitations. The coupling may take place between the liquid-free-surface dynamics and with either the tank bending oscillations or breathing modes (or shell modes). Breathing vibrations of the tank are essentially radial, such that both flexure and stretching of the wall occur while the longitudinal axis of the tank remains straight.

The combined liquid–structure system is very difficult to model and any analysis is usually based on some assumed simplifications. However, the interaction of the liquid sloshing dynamics with elastic deformations of the container must be considered in studying the overall system dynamics. Kana (in Abramson, 1966a, Chapter 9) presented an excellent overview of the interaction with elastic tank bending and breathing vibrations. The equations of motion of elastic shells partially filled with liquid were developed by Grigoliuk (1970) and Grigoliuk and Shkliarchuk (1970). Early attempts include the normal modal analysis of elastic containers, empty, partially, and completely filled with liquid (see, e.g., Reissner, 1956, 1957, Baron and Bleich, 1959, Mixon and Herr, 1962, Baron and Skalak, 1962, Fontenot and Lianis, 1963, Bhuta and Koval, 1964b, Natushkin and Rakhimov, 1964, Pengelley, 1968, Aganovic, 1981,

Kondo, 1981a, Ang, *et al.*, 1982, and Ohayon, 1989). Ganiev (1977) employed the principle of least action in the Hamilton–Ostrogradsky form developed by Moiseev and Rumyantsev (1968). He developed a set of nonlinear differential equations of motion, taking into account the interaction of elastic oscillations with the free-surface motion of the liquid. Moiseev (1959, 1960), Rand and Dimaggio (1967), Engin (1969a, b), Advani and Lee (1970), Boujot (1973), Morand and Ohayon (1975), Morand (1977), Gribkov (1982), and Guarino and Elger (1992) conducted different studies of modal analysis of fluid-filled spherical shells. The normal mode frequencies in a hydrodynamic system were estimated by Shklyarchuk (1965a,b) and Aganovic (1981) for symmetric and axisymmetric oscillations. The review articles by DiMaggio (1975, 1978, 1981, 1984) and Bagno and Guz (1997) provided overviews of different problems pertaining to the dynamic response of fluid-filled shells. Shveiko and Kuleshov (1973), Stillman (1973), Qinque and Lidu (1992), Gonçalves and Ramos (1996), Ergin (1997), Sivak (2001), and Ergin and Temarel (2002), determined the normal modes of free vibrations of partially filled horizontal shells. Huang (1991) analyzed the orthogonality of wet modes in shells partially filled with liquid.

This chapter presents the linear analysis of liquid sloshing interaction with elastic containers. It begins with the fundamental analytical formulation of kinematic relations and stress–strain relations with emphasis on the compatibility relations at the wetted container regions. The mathematics of interacting forces is formulated for a general motion of the liquid containers. The linear analysis deals mainly with the modal analysis of the coupled natural frequencies of liquid–structure systems. Specific cases, such as (i) rigid tank walls with an elastic bottom modeled as either a stretched membrane or an elastic plate, (ii) a rigid bottom with elastic walls experiencing either bending deformation or breathing vibrations, and (iii) interaction with a completely elastic container, are considered.

8.2 Basic problem of hydroelastic dynamics

This section presents a general analytical formulation of liquid sloshing interaction with elastic containers based on the work of Rapoport (1968), Mikishev and Rabinovich (1971), and Lukovskii, *et al.* (1989). It begins with the governing kinematic relations of the liquid free surface, container elastic deformation, and the compatibility relations of the two motions at the elastic wall. The hydroelastic interacting forces will then be analyzed within the framework of the linear theory.

8.2.1 Kinematic relations

Let XYZ and xyz refer to the stationary and moving coordinate frames, respectively. The frame xyz is attached to the undeformed container with the plane xy coinciding with the undisturbed free surface as shown in Figure 8.1. A fluid particle p on the wetted surface $S^{(t)}$, at time t, coincides with another particle of the elastic surface. After time Δt, the elastic surface will become $S^{(t+\Delta t)}$, and the fluid particle will assume position p_1, while the elastic particle will assume position p_2. The displacement of the fluid particle, pp_1, is defined by the vector $\mathbf{r}_1$, and the elastic deformation, pp_2, is given by $\mathbf{r}_2$. In terms of the fluid relative velocity, $\mathbf{v}_r$, and the relative velocity of the elastic displacement, $\mathbf{u}$, we write

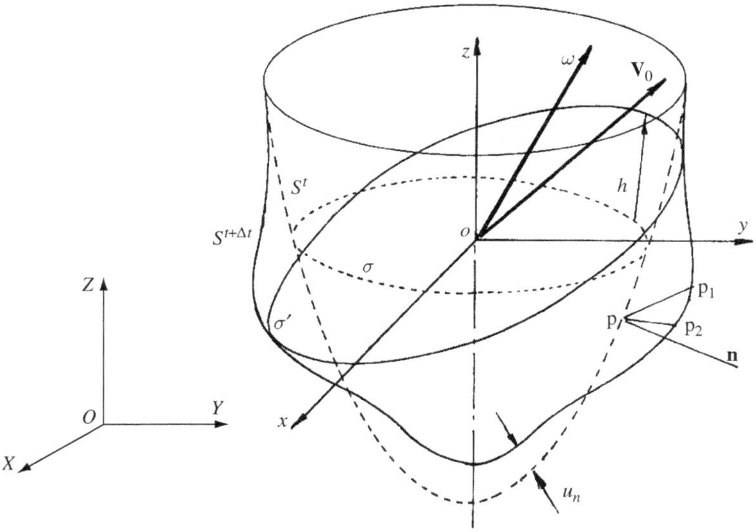

Figure 8.1 Elastic container interacting with liquid motion showing fixed and moving frames.

$$\mathbf{r}_1 = \mathbf{v}_r \Delta t, \qquad \mathbf{r}_2 = \frac{\partial \mathbf{u}}{\partial t} \Delta t \tag{8.1a,b}$$

Let $\mathbf{n}^{(t)}$ be a unit vector normal to the surface $S^{(t)}$. The projections of the fluid and elastic displacements on $\mathbf{n}^{(t)}$ are related by the relationship

$$\mathbf{v}_r \cdot \mathbf{n}^{(t)} \Delta t = \frac{\partial \mathbf{u}}{\partial t} \cdot \mathbf{n}^{(t)} \Delta t + \Delta \mathbf{r}_{1/2} \cdot \mathbf{n}^{(t)} \tag{8.2}$$

where $\Delta \mathbf{r}_{1/2} \cdot \mathbf{n}^{(t)}$ represents the projection of the vector $\Delta \mathbf{r}_{1/2}$ on the normal $\mathbf{n}^{(t)}$. Note that the length of the vector $\Delta \mathbf{r}_{1/2}$ is of order $(\Delta t)^2$, and dividing both sides of equation (8.2) by Δt gives the approximate relationship

$$\mathbf{v}_r \cdot \mathbf{n}^{(t)} \cong \frac{\partial \mathbf{u}}{\partial t} \cdot \mathbf{n}^{(t)} \quad \text{on } S^{(t)} \tag{8.3}$$

Defining the absolute velocity of the fluid particle as the gradient of the velocity potential, $\mathbf{V} = \nabla \Phi$, we write

$$\nabla \Phi \cdot \mathbf{n}^{(t)} = \mathbf{V}_0 \cdot \mathbf{n}^{(t)} + (\Omega \times \mathbf{r}) \cdot \mathbf{n}^{(t)} + \frac{\partial \mathbf{u}}{\partial t} \cdot \mathbf{n}^{(t)} \tag{8.4}$$

where $\mathbf{V}_0$ is the velocity of the origin of the moving coordinate frame xyz, Ω is its angular velocity vector, which is assumed very slow such that $\partial \Omega / \partial t$ is very small, and $\mathbf{r}$ is the position vector of the fluid particle with respect to the moving coordinate xyz.

With reference to Figure 8.2, let point A be at the fluid free surface $\sigma^{(t)}$. This point assumes position A_1 on the free surface $\sigma^{(t+\Delta t)}$ after time Δt. Let the projection of A meet the plane cross-section of the container at point A_0 such that the vector $\overrightarrow{A_0 A} = h(x, y, z, t)\mathbf{n}_0$. Let A_2 be the intersection of the normal $\mathbf{n}_0$ with the free surface $\sigma^{(t+\Delta t)}$. Thus, $\overrightarrow{A_0 A_2} = h(x, y, z, t + \Delta t)\mathbf{n}_0$, $\overrightarrow{AA_2} = [h'(x, y, z, t + \Delta t) - h(x, y, z, t)]\mathbf{n}_0$, and $\overrightarrow{AA_1} = \mathbf{v}_r \Delta t$ with an error of order $(\Delta t)^2$. The

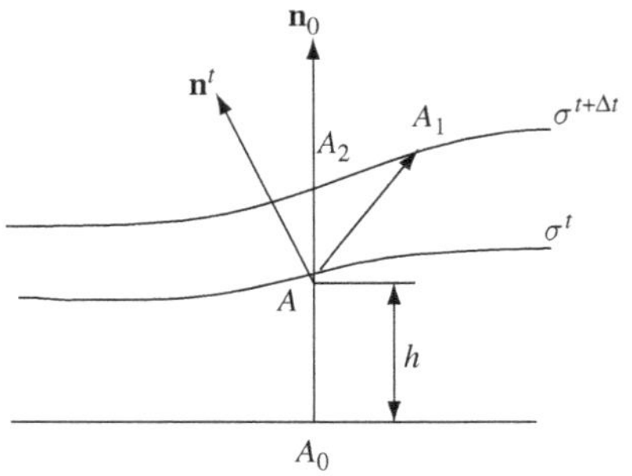

Figure 8.2 Kinematic of fluid motion and deformed free surface.

projection of $\overrightarrow{AA_1} = \mathbf{v}_r\Delta t$ on the normal $\mathbf{n}^{(t)}$ to the surface $\sigma^{(t)}$ must be equivalent to the projection of $\overrightarrow{AA_2}$ plus the projection of $\overrightarrow{A_2A_1}$ on $\mathbf{n}^{(t)}$, that is,

$$\mathbf{v}_r \cdot \mathbf{n}^{(t)}\Delta t = \frac{\partial h}{\partial t}\mathbf{n}_0 \cdot \mathbf{n}^{(t)}\Delta t + \overrightarrow{A_2A_1}\,\mathbf{n}^{(t)} \quad \text{on } \sigma^{(t)} \tag{8.5}$$

Note that $\overrightarrow{A_2A_1}\,\mathbf{n}^{(t)}$ is of order $(\Delta t)^2$. Dividing both sides of equation (8.5) by Δt. and taking the limits as $\Delta t \to 0$, gives

$$\mathbf{v}_r \cdot \mathbf{n}^{(t)} = \frac{\partial h}{\partial t}\mathbf{n}_0 \cdot \mathbf{n}^{(t)} \quad \text{on } \sigma^{(t)} \tag{8.6}$$

The projection of the fluid velocity on the normal $\mathbf{n}^{(t)}$ can be expressed in terms of the relative velocity of the fluid with respect to the container and the velocity of the container

$$\nabla\mathbf{\Phi} \cdot \mathbf{n}^{(t)} = \mathbf{V}_0 \cdot \mathbf{n}^{(t)} + (\Omega \times \mathbf{r}) \cdot \mathbf{n}^{(t)} + \frac{\partial h}{\partial t}\mathbf{n}_0 \cdot \mathbf{n}^{(t)} \quad \text{on } \sigma^{(t)} \tag{8.7a}$$

The velocity potential function $\mathbf{\Phi}$ can be split into a perturbed component, $\tilde{\mathbf{\Phi}}$, accounting for the motion of the fluid relative to the container plus another component, $\mathbf{V}_0(t) \cdot \mathbf{r} = V_{0x}(t)x + V_{0y}(t)y + V_{0z}(t)z$, representing the container motion, that is,

$$\Phi(x, y, z, t) = \tilde{\Phi}(x, y, z, t) + \mathbf{V}_0(t) \cdot \mathbf{r} \tag{8.8}$$

One can write Laplace's equations

$$\nabla^2\mathbf{\Phi} = 0 \quad \text{and} \quad \nabla^2\tilde{\mathbf{\Phi}} = 0 \tag{8.9a,b}$$

As stated earlier, the rotation of the moving coordinate is slow, and the elastic deformation of the container walls and fluid motion are small, that is, Ω, $\partial\mathbf{u}/\partial t$, and $\partial h/\partial t$ are small. This means that the unit vector $\mathbf{n}^{(t)}$ to the fluid surface $\sigma^{(t)}$ approximately coincides with the unit vector $\mathbf{n}^{(t)}$ to the surface $S^{(t)}$, and thus one can set $\mathbf{n}^{(t)} = \mathbf{n}$. In this case, the following compatibility conditions must be satisfied

$$\nabla\tilde{\Phi} \cdot \mathbf{n} = (\Omega \times \mathbf{r}) \cdot \mathbf{n} + \frac{\partial \mathbf{u}}{\partial t} \cdot \mathbf{n} \quad \text{on } S^{(t)} \tag{8.7b}$$

$$\nabla\Phi \cdot \mathbf{n} = (\Omega \times \mathbf{r}) \cdot \mathbf{n} + \frac{\partial h}{\partial t}\mathbf{n}_0 \cdot \mathbf{n} \quad \text{on } \sigma^{(t)} \tag{8.7c}$$

By using the vector cyclical property $(\mathbf{\Omega} \times \mathbf{r}) \cdot \mathbf{n} = (\mathbf{r} \times \mathbf{n}) \cdot \mathbf{\Omega}$, and

$$\nabla\tilde{\Phi} \cdot \mathbf{n} = \frac{\partial\tilde{\Phi}}{\partial x}\cos(n,x) + \frac{\partial\tilde{\Phi}}{\partial y}\cos(n,y) + \frac{\partial\tilde{\Phi}}{\partial z}\cos(n,z) = \frac{\partial\tilde{\Phi}}{\partial n} \tag{8.10a}$$

$$\frac{\partial \mathbf{u}}{\partial x} \cdot \mathbf{n} = \frac{\partial u_x}{\partial t}\cos(n,x) + \frac{\partial u_y}{\partial t}\cos(n,y) + \frac{\partial u_z}{\partial t}\cos(n,z) = \frac{\partial u_n}{\partial t} \tag{8.10b}$$

Conditions (8.7b, c) can be written in the form

$$\frac{\partial\tilde{\Phi}}{\partial n} = \mathbf{\Omega} \cdot (\mathbf{r} \times \mathbf{n}) + \frac{\partial u_n}{\partial t} \quad \text{on } S^{(t)} \tag{8.11a}$$

$$\frac{\partial\tilde{\Phi}}{\partial n} = \mathbf{\Omega} \cdot (\mathbf{r} \times \mathbf{n}) + \frac{\partial h}{\partial t} \quad \text{on } \sigma^{(t)} \tag{8.11b}$$

where u_n is the projection of the elastic deformation, $\mathbf{u}$, on the external normal to the wetted surface $S^{(t)}$.

Let the vector function $\mathbf{Q}(x, y, z)$ represent the solution of the boundary value problem

$$\nabla^2\mathbf{Q} = 0 \quad \text{in the fluid volume } V_f \tag{8.12a}$$

$$\frac{\partial\mathbf{Q}}{\partial n} = \mathbf{r} \times \mathbf{n} \quad \text{on the wet surface } S^{(t)} \tag{8.12b}$$

setting

$$\tilde{\Phi}(x,y,z,t) = \Theta(x,y,z,t) + \mathbf{\Omega}(t) \cdot \mathbf{Q}(x,y,z) \tag{8.13}$$

According to equations (8.9), (8.11), and (8.12), the function Θ should be determined from the following boundary value problem

$$\nabla^2\Theta = 0 \quad \text{in } V_f, \qquad \frac{\partial\Theta}{\partial n} = \frac{\partial u_n}{\partial t} \quad \text{on } S^{(t)}, \qquad \frac{\partial\Theta}{\partial n} = \frac{\partial h}{\partial t} \quad \text{on } \sigma \tag{8.14a,b,c}$$

Based on equations (8.8) and (8.13), the velocity potential Φ can be written in the form

$$\Phi(x,y,z,t) = \mathbf{V}_0(t) . \mathbf{r} + \boldsymbol{\omega}(t) \cdot \mathbf{Q}(x,y,z) + \Theta(x,y,z,t) \tag{8.15}$$

The boundary value problems described by equations (8.12) and (8.14) are called the Neumann problems, which must satisfy the following condition

$$\iint_{S+\sigma} (\mathbf{r} \times \mathbf{n})\, ds = 0, \qquad \iint_{S} \frac{\partial u_n}{\partial t}\, dS + \iint_{\sigma} \frac{\partial h}{\partial t}\, dS = 0 \tag{8.16a,b}$$

These kinematic relations are essential in developing the hydroelastic forces as will be shown in the next section.

8.2.2 Hydroelastic interacting forces

The pressure in the region occupied by the fluid is given by

$$p = -\rho\left[(\mathbf{a}_0 - \mathbf{g})\cdot\mathbf{r} + \frac{\partial\tilde{\Phi}}{\partial t}\right] = -\rho\left[(\mathbf{a}_0 - \mathbf{g})\cdot\mathbf{r} + \frac{\partial\Theta}{\partial t} + \frac{d\Omega}{dt}\cdot\mathbf{Q}\right] \tag{8.17}$$

where relation (8.15) has been used. Upon substituting the compatibility relations (8.11) one can write the pressure gradient in the form

$$\frac{\partial p}{\partial n} = -\rho\left[(\mathbf{a}_0 - \mathbf{g})\cdot\mathbf{n} + \frac{\mathrm{d}\Omega}{dt}\cdot(\mathbf{r}\times\mathbf{n}) + \frac{\partial^2 u_n}{\partial t^2}\right] \quad \text{on } S \tag{8.18a}$$

$$\frac{\partial p}{\partial n} = -\rho\left[(\mathbf{a}_0 - \mathbf{g})\cdot\mathbf{n} + \frac{\mathrm{d}\Omega}{dt}\cdot(\mathbf{r}\times\mathbf{n}) + \frac{\partial^2 h}{\partial t^2}\right] \quad \text{on } \sigma \tag{8.18b}$$

where $\mathbf{g}$ is the acceleration of gravity, $\mathbf{a}_0$ is the acceleration of the origin of the moving coordinate system. The term $-\rho\frac{d\Omega}{dt}\cdot\mathbf{Q}$ is the pressure contribution due to external moments acting on the container, and $-\rho\frac{\partial\Theta}{\partial t}$ determines the additional pressure due to the elastic wall vibration and free-surface oscillations. The potential function Θ can be written as the sum of two functions

$$\Theta(x, y, z, t) = \Theta_1(x, y, z, t) + \Theta_2(x, y, z, t) \tag{8.19}$$

where $\Theta_1(x, y, z, t)$ establishes the contribution of the elastic wall vibrations and $\Theta_2(x, y, z, t)$ defines the contribution of the free-surface oscillations. These functions are defined by the boundary-value problems

$$\nabla^2\Theta_1 = 0, \qquad \frac{\partial\Theta_1}{\partial n} = \frac{\partial u_n}{\partial t} \quad \text{on } S, \qquad \frac{\partial\Theta_1}{\partial n} = -\frac{\partial c}{\partial t} \quad \text{on } \sigma \tag{8.20a}$$

$$\nabla^2\Theta_2 = 0, \qquad \frac{\partial\Theta_2}{\partial n} = 0 \quad \text{on } S, \qquad \frac{\partial\Theta_2}{\partial n} = -\frac{\partial h}{\partial t} \quad \text{on } \sigma \tag{8.20b}$$

where c measures the contribution of the elastic deformation in changing the free-surface configuration, that is,

$$h(x, y, z, t) = h^o(x, y, z, t) - c(t) \tag{8.21}$$

The function $c(t)$ is defined such that the following condition is satisfied

$$\iint_S h^o\, dS = 0 \tag{8.22}$$

Furthermore, since the fluid is assumed incompressible, its volume is conserved at any time. Under small elastic deformations, $\mathbf{u}$, and small deviations of the free surface σ, the volumes will be equal if u_n and the surface deviation h are related by the condition

$$\iint_S u_n \, dS + \iint_\sigma h \, dS = 0 \tag{8.23}$$

Utilizing condition (8.23) in evaluating c, gives

$$c(t) = \frac{1}{S} \iint_S u_n \, ds \tag{8.24}$$

The pressure equation (8.17) takes the form

$$p = p_0 - \rho \left[(\mathbf{a}_0 - \mathbf{g}) \cdot (\mathbf{r} - \mathbf{r}_c) + \frac{d\Omega}{dt} \cdot \mathbf{Q} + \frac{\partial \Theta_1}{\partial t} + \frac{\partial \Theta_2}{\partial t} \right] \tag{8.25}$$

where $-\rho(\partial\Theta_1/\partial t)$ gives the pressure contribution due to the elastic wall vibrations, while $-\rho(\partial\Theta_2/\partial t)$ gives the contribution of the free-surface oscillations.

The elastic deformation vector $\mathbf{u}(x, y, z, t)$ is governed by the equilibrium equations of the elastic container by including the inertia forces. The body inertia force vector is defined by the expression

$$\mathbf{F} = F_x \mathbf{i} + F_y \mathbf{j} + F_z \mathbf{k} = -\rho \left\{ \mathbf{a}_0 - \mathbf{g} + \frac{d\Omega}{dt} \times \mathbf{r} + \frac{\partial^2 \mathbf{u}}{\partial t^2} \right\} \tag{8.26}$$

The governing equations of equilibrium of the elastic container are

$$\frac{\partial \sigma_x}{\partial x} + \frac{\partial \tau_{xy}}{\partial y} + \frac{\partial \tau_{xz}}{\partial z} + F_x = 0 \tag{8.27a}$$

$$\frac{\partial \tau_{xy}}{\partial x} + \frac{\partial \sigma_y}{\partial y} + \frac{\partial \tau_{yz}}{\partial z} + F_y = 0 \tag{8.27b}$$

$$\frac{\partial \tau_{xz}}{\partial x} + \frac{\partial \tau_{yz}}{\partial y} + \frac{\partial \sigma_z}{\partial z} + F_z = 0 \tag{8.27c}$$

These equations must satisfy the boundary conditions on the elastic wall surface S

$$\sigma_x \cos(n, x) + \tau_{xy} \cos(n, y) + \tau_{xz} \cos(n, z) = f_x \quad \text{on } S \tag{8.28a}$$

$$\tau_{xy} \cos(n, x) + \sigma_y \cos(n, y) + \tau_{yz} \cos(n, z) = f_y \quad \text{on } S \tag{8.28b}$$

$$\tau_{xz} \cos(n, x) + \tau_{yz} \cos(n, y) + \sigma_z \cos(n, z) = f_z \quad \text{on } S \tag{8.28c}$$

where σ_x, σ_y, and σ_z are the normal stresses in the elastic container walls, τ_{xy}, τ_{xz}, and τ_{yz} are the shear stresses, n is the direction of the external normal to the tank surface S. The surface force components f_x, f_y, and f_z represent the hydrodynamic pressure load components and can be represented vectorially as $\mathbf{f}(x, y, z, t) = p(x, y, z, t)\mathbf{n}$ on S; $\mathbf{n}$ is the external normal vector to S. The stress components in equations (8.27) together with the boundary conditions (8.28) are related to the elastic displacements through the relationships

$$\sigma_x = \frac{E}{(1+\nu)}\left(\frac{\partial u_x}{\partial x} + \frac{\nu}{1-2\nu}\nabla\cdot\mathbf{u}\right) \tag{8.29a}$$

$$\sigma_y = \frac{E}{(1+\nu)}\left(\frac{\partial u_y}{\partial y} + \frac{\nu}{1-2\nu}\nabla\cdot\mathbf{u}\right) \tag{8.29b}$$

$$\sigma_z = \frac{E}{(1+\nu)}\left(\frac{\partial u_z}{\partial z} + \frac{\nu}{1-2\nu}\nabla\cdot\mathbf{u}\right) \tag{8.29c}$$

$$\tau_{xy} = \frac{E}{2(1+\nu)}\left(\frac{\partial u_x}{\partial y} + \frac{\partial u_y}{\partial x}\right) \tag{8.29d}$$

$$\tau_{xz} = \frac{E}{2(1+\nu)}\left(\frac{\partial u_x}{\partial z} + \frac{\partial u_z}{\partial x}\right) \tag{8.29e}$$

$$\tau_{yz} = \frac{E}{2(1+\nu)}\left(\frac{\partial u_y}{\partial z} + \frac{\partial u_z}{\partial y}\right) \tag{8.29f}$$

where E is Young's modulus of elasticity, ν is Poisson's ratio. Substituting relations (8.29) into equations (8.27) one can write equations (8.27) in the vector form

$$\begin{aligned}\boldsymbol{\nabla}\left(\frac{\nu E(\boldsymbol{\nabla}\cdot\mathbf{u})}{(1-2\nu)(1+\nu)}\right) + 2\boldsymbol{\nabla}\cdot\left(\frac{E}{2(1+\nu)}\,\mathrm{grad}(\mathbf{u})\right)\\ + \boldsymbol{\nabla}\times\left(\frac{E}{2(1+\nu)}\boldsymbol{\nabla}\times\mathbf{u}\right) + \mathbf{F} = \mathbf{0}\end{aligned} \tag{8.30}$$

where $2\boldsymbol{\nabla}\cdot\left(\frac{E}{2(1+\nu)}\,\mathrm{grad}(\mathbf{u})\right)$ denotes a vector whose projections on the x, y, z axes are, respectively

$$2\boldsymbol{\nabla}\cdot\left(\frac{E}{2(1+\nu)}\,\mathrm{grad}\,u_x\right),$$

$$2\boldsymbol{\nabla}\cdot\left(\frac{E}{2(1+\nu)}\,\mathrm{grad}\,u_y\right),$$

$$2\boldsymbol{\nabla}\cdot\left(\frac{E}{2(1+\nu)}\,\mathrm{grad}\,u_z\right)$$

Substituting equation (8.26) into equations (8.30) and the boundary conditions (8.28) gives the boundary-value problem

$$\begin{aligned}\boldsymbol{\nabla}\left(\frac{\nu E(\boldsymbol{\nabla}\cdot\mathbf{u})}{(1-2\nu)(1+\nu)}\right) + 2\boldsymbol{\nabla}\cdot\left(\frac{E}{2(1+\nu)}\,\mathrm{grad}(\mathbf{u})\right) + \boldsymbol{\nabla}\times\left(\frac{E}{2(1+\nu)}\boldsymbol{\nabla}\times\mathbf{u}\right)\\ -\rho\frac{\partial^2\mathbf{u}}{\partial t^2} = \rho\left(\mathbf{a}_0 - \mathbf{g} + \frac{d\Omega}{dt}\times\mathbf{r}\right)\ \text{in}\ V\end{aligned} \tag{8.31}$$

$$\frac{\nu E}{(1+\nu)(1-2\nu)}\mathbf{n}(\nabla\cdot\mathbf{u})+\frac{E}{(1+\nu)}\frac{\partial\mathbf{u}}{\partial n}+\frac{E}{2(1+\nu)}\mathbf{n}\times(\nabla\times\mathbf{u})=p\mathbf{n}\quad\text{on } S \tag{8.32}$$

Equations (8.31) and (8 32) define the elastic displacement vector **u** for a given pressure $p(x, y, z, t)$ in the region occupied by the liquid in a container moving with translational acceleration $\mathbf{a}_0(t)$ and angular acceleration $d\Omega/dt$. To determine the hydrodynamic pressure one should use equation (8.18). The problem may be simplified by considering the interaction with different portions of the container such as the bottom or the walls. These cases are considered in the next sections.

8.3 Interaction with tank bottom

The interaction of the liquid surface motion with the tank elastic bottom, based on the assumption that tank sidewalls are rigid, may be examined by considering the fluid field and the elastic bottom equations. The coupling occurs in the boundary condition at the bottom where the fluid velocity normal to the bottom is the same as the velocity of the elastic bottom. The coupling also appears in the equation of the bottom by the hydrodynamic loading. The linear formulation gives the natural frequencies of the coupled boundary value problem in the form of an infinite order determinant. In most cases, the bottom is modeled as a stretched membrane or an elastic plate.

Liquid interaction with the tank elastic bottom was considered by Berry and Reissner (1958), Bhuta and Koval (1964a, b), Bhuta and Koval (1964a), Siekmann and Chang (1966, 1967, 1968), Tong (1966, 1967), Tsui and Small (1968), Chang (1969), Bauer, *et al.* (1971a, 1981a, b, 1995a), and Nagay and Takeuchi (1984). Bhuta and Koval (1964a) considered a thin and flat membrane tank bottom, which is governed by the circular membrane equation. Bhuta and Koval (1964b), Bauer and Siekmann (1969), and Chiba (1993a, 1994) also considered the case of a thin flat plate bottom, which is governed by the thin circular plate equation. These studies showed that the influence of the elastic bottom on the liquid response is more significant as the tank diameter increases and as the bottom thickness decreases. The normal mode frequencies of the liquid–membrane bottom system resulted in an infinite order determinant for the coupled natural frequencies. It was found that the bottom elasticity tends to lower the liquid surface natural frequencies below their values for the case of a rigid bottom. Siekmann and Chang (1966, 1967, 1968) calculated the natural frequencies of a liquid in a cylindrical tank with an elastic bottom including the surface tension. They found that the surface tension raises the natural frequencies. They confirmed the main conclusion of Bhuta and Koval that under a normal gravitational field the effect of an elastic bottom on the sloshing frequencies is small. Bauer (1981a, b) considered the coupled free vibration of liquid with an elastic bottom in a rectangular tank. Later, Bauer (1993a) estimated the natural frequencies of an elastic rectangular container partially filled with liquid.

Amabili (1997a) and Amabili and Dalpiaz (1998) used Ritz's method to study the bulging modes of an elastic bottom in circular and annular cylindrical containers partially filled with liquid. Cheung and Zhou (2000) and Zhou and Cheung (2000) analyzed the hydroelastic vibration of a rectangular container bottom plate and a vertical rectangular plate, respectively. Cheung and Zhou (2000) considered the bulging modes of a circular plate in contact with liquid whose free surface experiences sloshing motion. Some related studies were conducted by Amabili, *et al.* (1998a) who considered the effect of sidewall flexibility and ring stiffeners of the container. Espinosa and Gallego-Juarez (1984), Kwak (1991, 1997), Amabili (1996b), and Amabili and Kwak (1996, 1999) presented different studies to examine the hydroelastic vibration of circular plates in contact with an infinite liquid. Independent studies by Amabili and Dalpiaz (1998) and Cheung and Zhou (2002) found that the effect of free-surface waves on the bulging modes of the bottom plate is not significant, unless the plate is extremely flexible.

Nonlinear analysis was considered by Chiba (1996b) and Chiba and Abe (1999). Chiba (1997) studied the influence of flexible bottom on the response of liquid free surface to sinusoidal excitation. Bauer (1981b, 1995a) investigated the hydroelastic vibration of liquid in a rectangular and a circular cylindrical container, respectively. In both cases the free surface was covered by a flexible membrane or an elastic plate. It is known that if the free surface is partially covered by a structural element, its natural frequencies will increase. Bauer and Komatsu (2000) estimated the coupled natural frequencies of liquid free surface interacting with a flexible annular plate structure taking into account the effect of surface tension. It was found that flexible plates did not result in a significant increase of the coupled frequencies.

Chiba, *et al.* (2002) included the influence of surface tension in the linear hydroelastic free vibration analysis of a cylindrical container with membrane bottom. They found that coupling between membrane and liquid motion is manifested by a reduction in the membrane tension, which reduces the bulging (membrane) frequency, or by an increase in liquid surface tension. With an anchored edge condition, the uncoupled sloshing frequencies were found to be higher than those with a slipping edge condition. The difference between the two cases becomes larger with increasing surface tension. When the coupling between membrane and liquid motion becomes strong, coupled bulging and sloshing frequencies are of the same order.

8.3.1 Interaction with elastic bottom

The linear boundary-value problem of the combined free surface and elastic bottom is developed based on the assumption that the liquid is inviscid, incompressible, and initially irrotational. The elastic bottom may be assumed to be a membrane or an elastic plate as shown in Figure 8.3. The liquid field equations are

$$\frac{\partial^2\Phi}{\partial r^2}+\frac{1}{r}\frac{\partial\Phi}{\partial r}+\frac{1}{r^2}\frac{\partial^2\Phi}{\partial\theta^2}+\frac{\partial^2\Phi}{\partial z^2}=0 \tag{8.33a}$$

At the free surface, $z=h$, the dynamic and kinematic free-surface conditions are combined in one equation

$$\frac{\partial^2\Phi}{\partial t^2}+g\frac{\partial\Phi}{\partial z}=0 \tag{8.33b}$$

The liquid velocity normal to the wall vanishes, that is,

$$\left.\frac{\partial\Phi}{\partial r}\right|_{r=R}=0 \tag{8.33c}$$

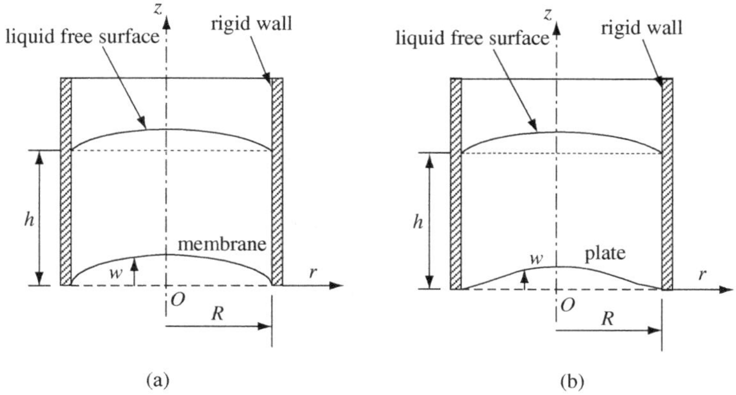

Figure 8.3 Rigid container with (a) membrane bottom, or (b) plate bottom

At the tank bottom the liquid dynamic pressure, p, is

$$p = -\rho \frac{\partial \Phi}{\partial t}\bigg|_{z=0} \tag{8.33d}$$

where ρ is the liquid density and g is the gravitational acceleration. Equations (8.33) are applied for the two models of the bottom.

Membrane bottom

If the bottom is represented by an elastic membrane, the following equation of motion is used

$$m^* \frac{\partial^2 w}{\partial t^2} = T\left(\frac{\partial^2 w}{\partial r^2} + \frac{1}{r}\frac{\partial w}{\partial r} + \frac{1}{r^2}\frac{\partial^2 w}{\partial \theta^2}\right) - p \tag{8.34a}$$

where m^* is the membrane mass per unit area, and T is the tensile force per unit length in the membrane. At the tank rigid sidewalls, the membrane deflection vanishes, that is,

$$w(r = R) = 0 \tag{8.34b}$$

The compatibility condition requires the fluid velocity in contact with the membrane to be the same as the velocity of the membrane, that is,

$$\frac{\partial w}{\partial t}\bigg|_{z=0} = \frac{\partial \Phi}{\partial z}\bigg|_{z=0}, \quad \text{at } z = 0 \tag{8.34c}$$

Equations (8.33) and (8.34) constitute the combined linear boundary-value problem of the coupled system. The solution of equation (8.33a) for the axisymmetric motion may be expressed in terms of Bessel functions of the first kind by the expression

$$\Phi_n(r, z) = A_n J_0(\lambda_n r)\left[e^{-\lambda_n z} + B_n\, e^{\lambda_n z}\right] e^{i\omega t}, \quad \lambda_n \neq 0 \tag{8.35a}$$

$$\Phi_0(r, z) = A_0[z + B_0]\, e^{i\omega t}, \quad \lambda_n = 0 \tag{8.35b}$$

where A_n, B_n, and λ_n are constants to be determined from the boundary conditions, ω is the coupled natural frequencies of the combined system. Applying the free-surface condition (8.33b) to the total potential function, $\Phi = \Phi_0 + \sum_{n=1} \Phi_n$, gives

$$B_n = \frac{(g\lambda_n + \omega^2)\, e^{-\lambda_n h}}{(g\lambda_n - \omega^2)\, e^{\lambda_n h}}, \quad \text{and} \quad B_0 = \frac{g}{\omega^2} - h \tag{8.36a,b}$$

Applying the boundary condition (8.33c) to equation (8.35a) gives $J_1(\lambda_n R) = 0$, which determines the values of λ_n. The case of $\lambda_n = 0$ requires separate treatment of the axisymmetric case.

The hydrodynamic pressure at $z = 0$ is determined using relation (8.33d) by substituting in equations (8.35)

$$p_n(r) = -i\rho\omega A_n J_0(\lambda_n r)[1 + B_n]\, e^{i\omega t}, \quad \lambda_n \neq 0 \tag{8.37a}$$

$$p_0 = -i\rho\omega A_n B_n\, e^{i\omega t}, \quad \lambda_n = 0 \tag{8.37b}$$

Let the solution of the membrane equation (8.34a) be represented in the form:

$$w(r,t) = W_n(r)\, e^{i\omega t}, \quad n \neq 0 \tag{8.38a}$$

$$w(r,t) = W_0(r)\, e^{i\omega t}, \quad n = 0 \tag{8.38b}$$

Substituting equations (8.37) and (8.38) into equation (8.34a) gives

$$\frac{d^2 W_n}{dr^2} + \frac{1}{r}\frac{dW_n}{dr} + \mu^2 W_n = -i\frac{\rho\omega}{T} A_n(1 + B_n) J_0(\lambda_n r) \quad n \neq 0 \tag{8.39a}$$

$$\frac{d^2 W_0}{dr^2} + \frac{1}{r}\frac{dW_0}{dr} + \mu^2 W_0 = -i\frac{\rho\omega}{T} A_0 B_0, \quad n = 0 \tag{8.39b}$$

where $\mu^2 = \dfrac{m^* \omega^2}{T}$. The solutions of these equations are

$$W_n(r) = \gamma_n \left[J_0(\lambda_n r) - \frac{J_0(\lambda_n R)}{J_0(\mu R)} J_0(\mu r) \right], \quad \text{where } \gamma_n = -i\frac{\rho\omega A_n(1 + B_n)}{T(\mu^2 - \lambda_n^2)} \tag{8.40a}$$

$$W_0(r) = \gamma_0 \left[1 - \frac{J_0(\mu r)}{J_0(\mu R)} \right], \quad \text{where } \gamma_0 = -i\frac{\rho\omega A_0 B_0}{T\mu^2} \tag{8.40b}$$

Equation (8.40a) is valid provided $\mu \neq \lambda_n$. To satisfy condition (8.34c), which requires the compatibility of the velocity of liquid and membrane surface, we substitute solutions (8.40) into equation (8.34c), which gives

$$i\omega\gamma_0 \left[1 - \frac{J_0(\mu r)}{J_0(\mu R)} \right] e^{i\omega t} + \sum_{n=1}^{\infty} i\omega\gamma_n \left[J_0(\lambda_n r) - \frac{J_0(\lambda_n R)}{J_0(\mu R)} J_0(\mu r) \right] e^{i\omega t}$$

$$= A_0\, e^{i\omega t} + \sum_{n=1}^{\infty} \lambda_n A_n (B_n - 1) J_0(\lambda_n r)\, e^{i\omega t} \tag{8.41}$$

In order to determine the natural frequency from equation (8.41) it is necessary to expand $J_0(\mu r)$ into a series of $J_0(\lambda_n r)$ and equate the coefficients of the corresponding terms on both sides. This procedure is valid since the functions $J_0(\lambda_n r)$ are orthogonal. The expansion of $J_0(\mu r)$ is (Whittaker and Watson, 1962):

$$J_0(\mu r) = \frac{2}{R\mu} J_1(\mu R) + \frac{2\mu J_1(\mu R)}{R} \sum_{p=1}^{\infty} \frac{J_0(\lambda_p r)}{(\mu^2 - \lambda_p^2) J_0(\lambda_p R)} \tag{8.42}$$

Substituting equation (8.42) into equation (8.41) gives

$$\Gamma_0 - \frac{2\Gamma_0 J_1(\mu R)}{R\mu J_0(\mu R)} - \frac{2\Gamma_1 \mu J_1(\mu R)}{R J_0(\mu R)} \sum_{p=1}^{\infty} \frac{1}{\left(\mu^2 - \lambda_p^2\right)} \frac{J_0(\lambda_p r)}{J_0(\lambda_p R)} + \sum_{1=1}^{\infty} \Gamma_n J_0(\lambda_p r)$$

$$- \frac{2J_1(\mu R)}{R\mu J_0(\mu R)} \sum_{n=1}^{\infty} \Gamma_n J_0(\lambda_p R) - \sum_{n=1}^{\infty} \Gamma_n \frac{J_0(\lambda_m R)}{J_0(\mu R)} \sum_{n=1}^{\infty} \frac{2\mu J_1(\mu R)}{R(\mu^2 - \lambda_p^2)} \frac{J_0(\lambda_p r)}{J_0(\lambda_p R)}$$

$$= A_0 + \sum_{n=1}^{\infty} \lambda_n A_n (B_n - 1) J_0(\lambda_n r) \tag{8.43}$$

where $\Gamma_0 = i\omega\gamma_0$, and $\Gamma_n = i\omega\gamma_n$.

The frequency equation is formed by setting the determinant of the coefficients of A_0, A_1, A_2, ... to zero

$$\begin{vmatrix} \alpha_{11} & \alpha_{12} & \alpha_{13} & \cdots \\ \alpha_{21} & \alpha_{22} & \alpha_{23} & \cdots \\ \alpha_{31} & \alpha_{32} & \alpha_{33} & \cdots \\ \cdots & \cdots & \cdots & \cdots \end{vmatrix} = 0 \tag{8.44}$$

where

$$\alpha_{11} = -\frac{2\rho\omega^2 B_0 J_1(\mu R)}{R\mu^3 T J_0(\mu R)} + \frac{\rho\omega^2 B_0}{T\mu^2} - 1$$

$$\alpha_{12} = -\frac{2\rho\omega^2(1 + B_1) J_0(\lambda_1 R) J_1(\mu R)}{R\mu T(\mu^2 - \lambda_1^2) J_0(\mu R)}$$

$$\alpha_{13} = -\frac{2\rho\omega^2(1 + B_2) J_0(\lambda_2 R) J_1(\mu R)}{R\mu T(\mu^2 - \lambda_2^2) J_0(\mu R)}$$

$$\alpha_{21} = -\frac{2\rho\omega^2 B_0 J_1(\mu R)}{R\mu T(\mu^2 - \lambda_1^2) J_0(\lambda_1 R) J_0(\mu R)}$$

$$\alpha_{22} = -\frac{2\mu\rho\omega^2(1 + B_1) J_1(\mu R)}{RT\left(\mu^2 - \lambda_1^2\right)^2 J_0(\mu R)} + \frac{\rho\omega^2(1 + B_1)}{T(\mu^2 - \lambda_1^2)} - \lambda_1(B_1 - 1)$$

$$\alpha_{23} = -\frac{2\mu\rho\omega^2(1+B_2)J_0(\lambda_2 R)J_1(\mu R)}{RT(\mu^2-\lambda_1^2)(\mu^2-\lambda_2^2)J_0(\mu R)J_0(\lambda_1 R)}$$

$$\alpha_{31} = -\frac{2\rho\omega^2 B_0 J_1(\mu R)}{R\mu T(\mu^2-\lambda_2^2)J_0(\lambda_2 R)J_0(\mu R)}$$

$$\alpha_{32} = -\frac{2\mu\rho\omega^2(1+B_1)J_0(\lambda_1 R)J_1(\mu R)}{RT(\mu^2-\lambda_1^2)(\mu^2-\lambda_2^2)J_0(\mu R)J_0(\lambda_2 R)}$$

$$\alpha_{33} = -\frac{2\mu\rho\omega^2(1+B_2)J_1(\mu R)}{RT(\mu^2-\lambda_2^2)^2 J_0(\mu R)} + \frac{\rho\omega^2(1+B_2)}{T(\mu^2-\lambda_2^2)} - \lambda_2(B_2-1) \tag{8.45}$$

Bhuta and Koval (1964a) studied two limiting cases. The first is when the membrane tension approaches infinity, the frequency determinant takes the form

$$J_0(\mu R)\prod_{n=1}^{\infty}\lambda_n(B_n-1)=0, \quad \text{or} \quad B_n=1 \tag{8.46}$$

This gives the natural frequency of the free surface in a rigid tank, that is,

$$\omega_n^2 = g\lambda_n \tanh \lambda_n h \tag{8.47}$$

The second is when the liquid density is zero and the determinant is reduced to

$$J_0\left(\sqrt{\frac{m^*}{T}}\omega_n R\right) = 0 \tag{8.48}$$

This equation gives the frequency of the membrane in the absence of liquid.

The above analysis is valid as long as $\mu \neq \lambda_n$. If $\mu = \lambda_n$, the frequencies are determined from the definition $\mu^2 = m^*\omega^2/T$, and the frequency is

$$\omega_n^2 = \frac{\lambda_n^2 T}{m^*} \tag{8.49}$$

This frequency expression is independent of the liquid properties and is not encountered in real problems.

Figure 8.4 shows the dependence of the natural frequencies of the first two modes on the tension in the membrane for liquid depth ratio $h/R = 0.1$, $R = 50$ inch, and liquid density $\rho = 0.935 \times 10^{-4}$ lb sec^2/in^4. It is seen that as the membrane tension increases the frequencies approach those of liquid free surface in a rigid tank.

Asymmetric case The solution of equation (8.33a) for the asymmetric case may be written in the form

$$\Phi(r,\theta,z) = A_{mn}J_m(\lambda_{mn}r)\left[e^{-\lambda_{mn}z} + B_{mn}\, e^{\lambda_{mn}z}\right]\cos m\theta\, e^{i\omega t} \tag{8.50}$$

Substituting equation (8.50) in the free-surface boundary condition (8.33b), gives

$$B_{mn} = \frac{(g\lambda_{mn}+\omega^2)\, e^{-\lambda_{mn}h}}{(g\lambda_{mn}-\omega^2)\, e^{\lambda_{mn}h}} \tag{8.51}$$

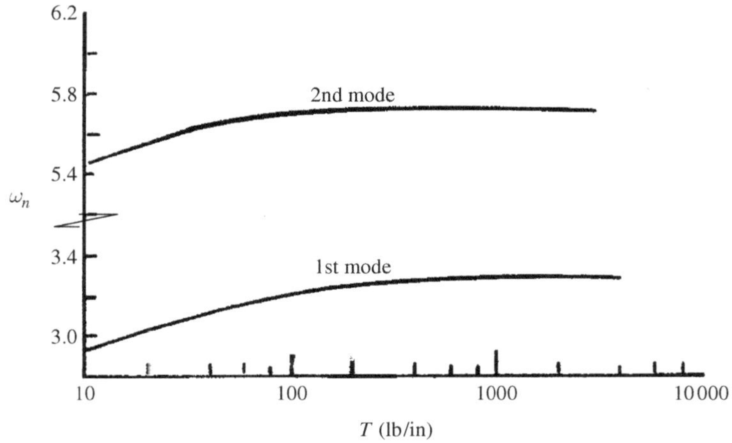

Figure 8.4 Dependence of the natural frequency of the first two modes on the membrane tension for $h/R = 0.1$. (Bhuta and Koval, 1964a)

Applying the boundary condition (8.33c), gives

$$\left.\frac{dJ_m(\lambda_{mn} r)}{dr}\right|_{r=R} = 0 \tag{8.52}$$

The hydrodynamic pressure is obtained using equation (8.33d)

$$p_{mn}(r, \theta, t) = -i\rho\omega A_{mn} J_m(\lambda_{mn} r)[1 + B_{mn}] \cos m\theta \, e^{i\omega t} \tag{8.53}$$

The membrane deflection may be assumed to be in the form

$$w(r, \theta, t) = W_{mn}(r) \cos m\theta \, e^{i\omega t} \tag{8.54}$$

Substituting equations (8.53) and (8.54) into equation (8.34a) gives

$$\frac{d^2 W_{mn}}{dr^2} + \frac{1}{r}\frac{dW_{mn}}{dr} + \left(\mu^2 - \frac{m^2}{r^2}\right) W_{mn} = -i\frac{\rho\omega}{T} A_{mn}(1 + B_{mn}) J_m(\lambda_{mn} r) \tag{8.55}$$

The solution of equation (8.55) satisfying the boundary condition (34b) is

$$W_{mn}(r) = \gamma'_{mn}\left[J_m(\lambda_{mn} r) - \frac{J_m(\lambda_{mn} R)}{J_m(\mu R)} J_m(\mu r)\right] \tag{8.56}$$

where

$$\gamma'_{mn} = -i\frac{\rho\omega A_{mn}(1 + B_{mn})}{T(\mu^2 - \lambda_{mn}^2)}, \quad \mu \neq \lambda_{mn} \tag{8.57}$$

Applying the velocity compatibility condition (8.34c), gives

$$\sum_{n=1}^{\infty} i\omega\gamma'_{mn}\left[J_m(\lambda_{mn}r) - \frac{J_m(\lambda_{mn}R)}{J_m(\mu R)}J_m(\mu r)\right]e^{i\omega t}$$

$$= \sum_{n=1}^{\infty} \lambda_{mn}A_{mn}(B_{mn}-1)J_m(\lambda_{mn}r)\,e^{i\omega t} \tag{8.58}$$

Introducing the asymptotic expansion,

$$J_m(\mu r) = \sum_{p=1}^{\infty} D_p J_m(\lambda_{mn}r) \tag{8.59}$$

where

$$D_p = \frac{2R\lambda_{mp}^2\{\lambda_{mp}J_m(\mu R)J_{m+1}(\lambda_{mp}R) - \mu J_m(\lambda_{mp}R)J_{m+1}(\mu R)\}}{(\lambda_{mn}^2R^2 - m^2)[J_m(\lambda_{mn}R)]^2\left(\lambda_{mp}^2 - \mu^2\right)} \tag{8.60}$$

Substituting equation (8.59) into equation (8.58) gives

$$\sum_{n=1}^{\infty}\Gamma'_{mn}J_m(\lambda_{mn}r) - \sum_{n=1}^{\infty}\Gamma'_{mn}\frac{J_m(\lambda_{mn}R)}{J_m(\mu R)}\sum_{p=1}^{\infty}D_pJ_m(\lambda_{mp}r)$$

$$= \sum_{n=1}^{\infty}\lambda_{mn}A_{mn}(B_{mn}-1)J_m(\lambda_{mn}r)\,e^{i\omega t} \tag{8.61}$$

where $\Gamma'_{mn} = i\omega\gamma'_{mn}$. Equating the determinant of the coefficients of A_{mn} to zero gives the frequency equation

$$\begin{vmatrix} \beta_{11} & \beta_{12} & \beta_{13} & \cdots \\ \beta_{21} & \beta_{22} & \beta_{23} & \cdots \\ \beta_{31} & \beta_{32} & \beta_{33} & \cdots \\ \cdots & \cdots & \cdots & \cdots \end{vmatrix} = 0 \tag{8.62}$$

where

$$\beta_{11} = -\frac{\rho\omega^2 D_1(1+B_{m1})J_m(\lambda_{m1}R)}{TJ_m(\mu R)(\mu^2-\lambda_{m1}^2)} + \frac{\rho\omega^2(1+B_{m1})}{T(\mu^2-\lambda_{m1}^2)} - \lambda_{m1}(B_{m1}-1)$$

$$\beta_{12} = -\frac{\rho\omega^2 D_1(1+B_{m2})J_m(\lambda_{m2}R)}{T(\mu^2-\lambda_{m2}^2)J_m(\mu R)}$$

$$\beta_{13} = -\frac{\rho\omega^2 D_1(1+B_{m3})J_m(\lambda_{m3}R)}{T(\mu^2-\lambda_{m3}^2)J_m(\mu R)}$$

$$\beta_{21} = -\frac{\rho\omega^2 D_2(1+B_{m1})J_m(\lambda_{m1}R)}{T(\mu^2-\lambda_{m1}^2)J_m(\mu R)}$$

$$\beta_{22} = -\frac{\rho\omega^2 D_2(1+B_{m2})J_m(\lambda_{m2}R)}{T(\mu^2-\lambda_{m2}^2)^2 J_m(\mu R)} + \frac{\rho\omega^2(1+B_{m2})}{T(\mu^2-\lambda_{m2}^2)} - \lambda_{m2}(B_{m2}-1)$$

$$\beta_{23} = -\frac{\rho\omega^2 D_2(1 + B_{m2})J_m(\lambda_{m3}R)}{T(\mu^2 - \lambda_{m3}^2)J_m(\mu R)}$$

$$\beta_{31} = -\frac{\rho\omega^2 D_3(1 + B_{m1})J_m(\lambda_{m1}R)}{T(\mu^2 - \lambda_{m1}^2)J_m(\mu R)}$$

$$\beta_{32} = -\frac{\rho\omega^2 D_3(1 + B_{m2})J_m(\lambda_{m2}R)}{T(\mu^2 - \lambda_{m2}^2)J_m(\mu R)}$$

$$\beta_{33} = -\frac{\rho\omega^2 D_3(1 + B_{m3})J_m(\lambda_{m3}R)}{T(\mu^2 - \lambda_{m3}^2)^2 J_m(\mu R)} + \frac{\rho\omega^2(1 + B_{m3})}{T(\mu^2 - \lambda_{m3}^2)} - \lambda_{m3}(B_{m3} - 1) \tag{8.63}$$

The natural frequencies estimated from equation (8.62) correspond to the asymmetric modes *mn*.

Elastic plate bottom

If the tank bottom is modeled by a thin elastic plate, Figure 8.3(b), the following equation of motion is considered

$$\rho_p h_p \frac{\partial^2 w}{\partial t^2} = -D_p\left(\frac{\partial^2}{\partial r^2} + \frac{1}{r}\frac{\partial}{\partial r} + \frac{1}{r^2}\frac{\partial^2}{\partial \theta^2}\right)^2 w - p \tag{8.64a}$$

where $D_p = Eh_p^3/[12(1-\nu^2)]$ is the flexural rigidity of the plate, E is Young's modulus, h_p is the plate thickness, ρ_p is the plate density. Note that the pressure in both equations (8.34a) and (8.64a) is negative since it is acting on the bottom in a direction opposite to the positive deflection of the bottom. For a clamped boundary of the plate, the following conditions are used

$$w(r = R) = 0, \qquad \frac{\partial w}{\partial r} = 0 \quad \text{at } r = R \tag{8.64b,c}$$

Condition (8.34c) is also imposed. Substituting equations (8.37a) and (8.38a) into equation (8.64a), gives

$$\frac{i\rho\omega A_n(1 + B_n)}{D_p} J_0(\lambda_n r) = \left(\frac{\partial^2}{\partial r^2} + \frac{1}{r}\frac{\partial}{\partial r} - \mu^2\right)\left(\frac{\partial^2}{\partial r^2} + \frac{1}{r}\frac{\partial}{\partial r} + \mu^2\right) W_n \tag{8.65a}$$

Similarly substituting equations (8.37b) and (8.38b) into equation (8.64a) gives

$$\frac{i\rho\omega A_0 B_0}{D_p} = \left(\frac{\partial^2}{\partial r^2} + \frac{1}{r}\frac{\partial}{\partial r} - \mu^2\right)\left(\frac{\partial^2}{\partial r^2} + \frac{1}{r}\frac{\partial}{\partial r} + \mu^2\right) W_0 \tag{8.65b}$$

where $\mu^4 = \rho_p h_p \omega^2 / D_p$.

The solution of equation (8.65a) that satisfies the boundary conditions (8.64) may be written in the form

$$W_n(r) = \gamma_n[C_n J_0(\mu r) + E_n I_0(\mu r) - J_0(\lambda_n r)], \quad \lambda_n \neq 0, \text{and } \mu \neq \lambda_n \tag{8.66}$$

where $C_n = J_0(\lambda_n R)I_1(\mu R)/\Delta, \qquad E_n = J_0(\lambda_n R)J_1(\mu R)/\Delta,$

$$\Delta = J_0(\mu R)I_1(\mu R) + I_0(\mu R)J_1(\mu R), \qquad \gamma_n = -i\frac{\rho\omega A_n(1+B_n)}{D(\mu^4 - \lambda_n^4)} \tag{8.67}$$

On the other hand, the solution of equation (8.65b) that satisfies the boundary conditions (8.64) may be written in the form

$$W_0(r) = \gamma_0'[C_0 J_0(\mu r) + E_0 I_0(\mu r) - 1] \tag{8.68}$$

where

$$C_0 = I_1(\mu R)/\Delta, \qquad E_0 = J_1(\mu R)/\Delta, \quad \text{and} \quad \gamma_0' = -i\frac{\rho\omega A_0 B_0}{D\mu^4} \tag{8.69}$$

Equating the velocities of the plate and of the liquid in contact with it, as required by condition (8.34c), gives

$$\begin{aligned} &i\omega\gamma_0'[C_0 J_0(\mu r) + E_0 I_0(\mu r) - 1]\, e^{i\omega t} \\ &\quad + \sum_{n=1}^{\infty} i\omega\gamma_n'[C_n J_0(\mu r) + E_n I_0(\mu r) - J_0(\lambda_n r)]\, e^{i\omega t} \\ &\quad = A_0\, e^{i\omega t} + \sum_{n=1}^{\infty} \lambda_n A_n (B_n - 1) J_0(\lambda_n r)\, e^{i\omega t} \end{aligned} \tag{8.70}$$

Following the same procedure for the case of membrane, the Bessel function $J_0(\mu r)$ is replaced by the expansion (8.42) and $I_0(\mu r)$ is expanded in series in terms of $J_0(\lambda_n r)$ as follows

$$I_0(\mu r) = \frac{2}{R\mu} I_1(\mu R) + \frac{2\mu I_1(\mu R)}{R} \sum_{p=1}^{\infty} \frac{J_0(\lambda_p r)}{(\mu^2 + \lambda_p^2) J_0(\lambda_p R)} \tag{8.71}$$

Substituting (8.42) and (8.71) into equation (8.70) gives

$$\begin{aligned} &2\omega^2 A_0 \Gamma_0 C_0 \frac{J_1(\mu R)}{R}\left[\frac{1}{\mu} + \mu \sum_{p=1}^{\infty} \frac{J_0(\lambda_p r)}{(\mu^2 - \lambda_p^2) J_0(\lambda_p R)}\right] \\ &\quad + 2\omega^2 A_0 \Gamma_0 E_0 \frac{I_1(\mu R)}{R}\left[\frac{1}{\mu} + \mu \sum_{p=1}^{\infty} \frac{J_0(\lambda_p r)}{(\mu^2 + \lambda_p^2) J_0(\lambda_p R)}\right] - \Gamma_0 \\ &\quad + \sum_{n=1}^{\infty} 2\omega^2 A_n \Gamma_n C_n \frac{J_1(\mu R)}{R}\left[\frac{1}{\mu} + \mu \sum_{p=1}^{\infty} \frac{J_0(\lambda_p r)}{(\mu^2 - \lambda_p^2) J_0(\lambda_p R)}\right] \\ &\quad + \sum_{n=1}^{\infty} 2\omega^2 A_n \Gamma_n E_n \frac{I_1(\mu R)}{R}\left[\frac{1}{\mu} + \mu \sum_{p=1}^{\infty} \frac{J_0(\lambda_p r)}{(\mu^2 + \lambda_p^2) J_0(\lambda_p R)}\right] \\ &\quad - \sum_{n=1}^{\infty} \omega^2 A_n \Gamma_n J_0(\lambda_n r) = -A_0 - \sum_{n=1}^{\infty} \lambda_n A_n (B_n - 1) J_0(\lambda_n r) \end{aligned} \tag{8.72}$$

where

$$\Gamma_0 = \rho B_0 / D\mu^4, \quad \text{and} \quad \Gamma_n = \rho(1 + B_n)/D(\mu^4 - \lambda_n^4) \tag{8.73}$$

Equating the coefficients of $J_0(\lambda_n r)$ gives a set of coupled equations. Setting the determinant of the coefficients A_0, A_1, ... to zero gives a frequency equation similar in form to equation (8.44) where the elements of the determinant are defined by the following expressions

$$\begin{aligned}
\alpha_{11} &= \frac{2\omega^2\Gamma_0}{\mu R}[C_0 J_1(\mu R) + E_0 I_1(\mu R)] - \omega^2\Gamma_0 + 1 \\
\alpha_{12} &= \frac{2\omega^2\Gamma_1}{\mu R}[C_1 J_1(\mu R) + E_1 I_1(\mu R)] \\
\alpha_{13} &= \frac{2\omega^2\Gamma_2}{\mu R}[C_2 J_1(\mu R) + E_2 I_1(\mu R)] \\
\alpha_{21} &= \frac{2\mu\omega^2\Gamma_0}{RJ_0(\lambda_1 R)}\left[\frac{C_0 J_1(\mu R)}{(\mu^2 - \lambda_1^2)} + \frac{E_0 I_1(\mu R)}{(\mu^2 + \lambda_1^2)}\right] \\
\alpha_{22} &= \frac{2\mu\omega^2\Gamma_1}{RJ_0(\lambda_1 R)}\left[\frac{C_1 J_1(\mu R)}{(\mu^2 - \lambda_1^2)} + \frac{E_1 I_1(\mu R)}{(\mu^2 + \lambda_1^2)}\right] - \Gamma_1\omega^2 + \lambda_1(B_1 - 1) \\
\alpha_{23} &= \frac{2\mu\omega^2\Gamma_2}{RJ_0(\lambda_1 R)}\left[\frac{C_2 J_1(\mu R)}{(\mu^2 - \lambda_1^2)} + \frac{E_2 I_1(\mu R)}{(\mu^2 + \lambda_1^2)}\right] \\
\alpha_{31} &= \frac{2\mu\omega^2\Gamma_0}{RJ_0(\lambda_2 R)}\left[\frac{C_0 J_1(\mu R)}{(\mu^2 - \lambda_2^2)} + \frac{E_0 I_1(\mu R)}{(\mu^2 + \lambda_2^2)}\right] \\
\alpha_{32} &= \frac{2\mu\omega^2\Gamma_1}{RJ_0(\lambda_2 R)}\left[\frac{C_1 J_1(\mu R)}{(\mu^2 - \lambda_2^2)} + \frac{E_1 I_1(\mu R)}{(\mu^2 + \lambda_2^2)}\right] \\
\alpha_{33} &= \frac{2\mu\omega^2\Gamma_2}{RJ_0(\lambda_2 R)}\left[\frac{C_2 J_1(\mu R)}{(\mu^2 - \lambda_2^2)} + \frac{E_2 I_1(\mu R)}{(\mu^2 + \lambda_2^2)}\right] - \omega^2\Gamma_2 + \lambda_2(B_2 - 1)
\end{aligned} \tag{8.74}$$

The frequency equation (8.44), with coefficients (8.74), has two limiting cases. If one sets the flexural rigidity of the plate $D \to \infty$, or the density of the liquid $\rho \to 0$, one obtains after some manipulations

$$\Delta \prod_{n=1}^{\infty} \lambda_n(B_n - 1) = 0 \tag{8.75}$$

Equation (8.75) implies that

$$\Delta = J_0(\mu R) I_1(\mu R) + I_0(\mu R) J_1(\mu R) = 0 \tag{8.76}$$

or $B_n = 1$, that is

$$\omega_n^2 = g\lambda_n \tanh \lambda_n h \tag{8.77}$$

Table 8.1 *Coupled frequencies* $h/R = 0.1$, $D = 279\,981$ lb inch, $h_p = 0.467$ inch

n	2×2	8×8	Rigid tank
1	3.280	3.280	3.288
2	–	5.724	5.726
3	–	7.766	7.770
4	–	9.459	9.459
5	–	10.865	10.865

Source: Bhuta and Koval, 1964b.

Table 8.2 *Coupled frequencies* $h/R = 0.05$, $D = 16\,021$ lb inch, $h_p = 0.26$ inch

n	2×2	8×8	Rigid tank
1	2.299	2.299	2.366
2	–	4.262	4.273
3	–	6.066	6.069
4	–	7.7339	7.740
5	–	9.278	9.278

Source: Bhuta and Koval, 1964b.

Table 8.3 *Coupled frequencies* $h/R = 0.1$, $D = 93\,326$ lb inch, $h_p = 0.467$ inch

n	2×2	8×8	Rigid tank
1	3.271	3.271	3.288
2	–	5.723	5.726
3	–	7.767	7.770
4	–	9.459	9.459
5	–	10.865	10.865

Source: Bhuta and Koval, 1964b.

Table 8.4 *Coupled frequencies* $h/R = 0.05$, $D = 48\,063$ lb inch, $h_p = 0.26$ inch

n	2×2	8×8	Rigid tank
1	2.342	2.342	2.366
2	–	4.269	4.272
3	–	6.068	6.068
4	–	7.740	7.740
5	–	9.278	9.278

Source: Bhuta and Koval, 1964b.

Equation (8.76) gives the frequency equation of axisymmetric vibrations of a clamped elastic circular plate in a vacuum. On the other hand, equation (8.77) gives the frequencies of the liquid free surface in a rigid circular upright cylindrical tank.

Bhuta and Koval (1964b) truncated the frequency equation (8.76), with coefficients (8.74), in stages at 2nd-, 3rd-, ... and 8th-order determinant and estimated the lower values of the coupled frequencies. The results of the numerical results are given in Tables 8.1–4 for different values of liquid depth ratio h/R, and different values of plate flexural rigidity, D. The effect of the elastic bottom is observed only for smaller values of liquid depth ratio and the plate flexural rigidity is decreased.

8.4 Interaction with tank walls

The interaction of liquid sloshing dynamics with the tank walls may take place between sloshing modes and bending modes; or breathing elastic modes. Bending deformation of the tanks does not involve stretching of the tank walls and the longitudinal axis of the tank experiences some type of deformation. On the other hand, breathing deformation involves flexure and stretching of the wall in the radial direction and the tank longitudinal axis remains straight. The two types of interaction are considered in the next two sections.

8.4.1 Interaction with bending modes

The modal analysis of liquid sloshing interaction with tank bending deformation involves the estimation of the system coupled natural frequencies. Within the framework of the linear theory of small oscillations, the liquid–elastic container coupling was studied by Merten and Stephenson (1952), Reissner (1956), Rabinovich (1956, 1959, 1964, 1980), Bauer (1958e, h, 1969b), Miles (1958a), Abramson, *et al.* (1962c), Lindholm, *et al.* (1963), and Bauer, Hsu, and Wang (1968). Reissner (1956) formulated the problem as an integral equation describing the tank wall displacement. He also outlined another approach based on a variational treatment. Miles (1958a) analyzed the coupled bending–liquid sloshing problem using Lagrangian formulation for potential sloshing in a bending cylindrical tank. He found that if the mass of the empty tank is small compared to that of the liquid, and the depth of the liquid is equal to the tank diameter, the presence of the free surface increases the bending frequency. Beal, *et al.* (1965) and Yangyi and Jingliang (1985) determined the effect of elastic tank inertia and bending stiffness on the axisymmetric modes of a partially filled cylindrical container. Chiba, *et al.* (1984a, b, 1985) analytically, numerically, and experimentally determined the natural frequencies of a clamped-free circular cylindrical shell partially filled with liquid. The case of free-free tanks filled with liquid was considered by Kreis and Klein (1991). Koleshov and Shveiko (1971) determined the asymmetric oscillations of cylindrical shells partially filled with liquid. Ohayon and Felippa (1990) and Genevaux and Lu (2000) examined the influence of container wall deformation on the liquid equations of motion.

Symmetric flow

A Lagrangian formulation for estimating the coupled natural frequencies of liquid in an elastic wall container is presented based on the work of Miles (1958a). Figure 8.5 shows a uniform

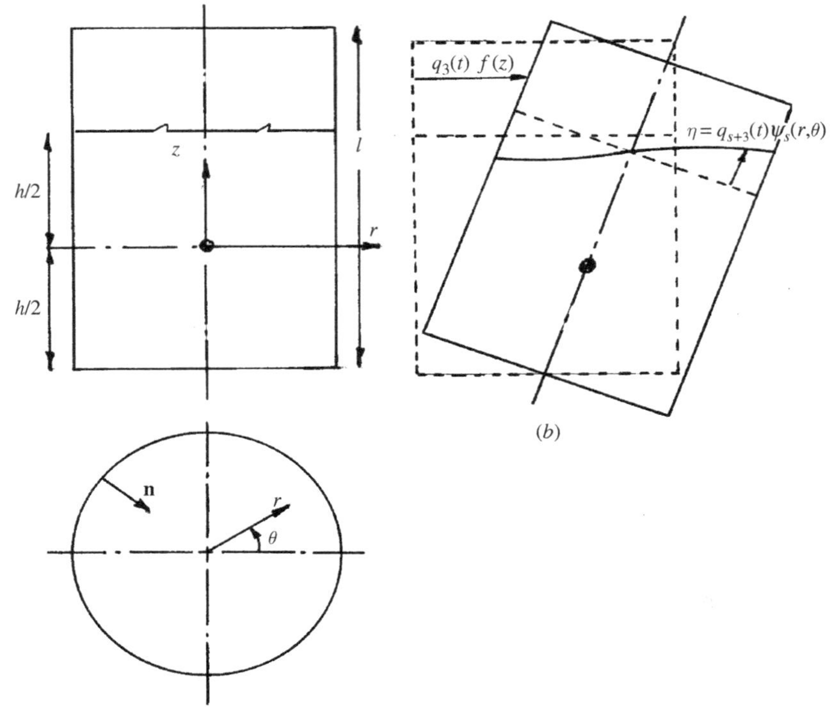

Figure 8.5 Elastic wall cylindrical container interacting with liquid free-surface motion. (Miles, 1958a)

thin-wall circular cylindrical tank partially filled with incompressible liquid of depth h with the coordinate frame (r, θ, z) located at the center of mass of the undisturbed tank. The liquid is assumed to be incompressible, frictionless, and initially irrotational. Let $q_1(t)$ be the tank center of mass translation along $\theta = 0$ and $q_2(t)$ be the rotation of the tank about the centroidal axis $|\theta| = \pi/2$, $q_3(t)f(z)$ be the bending displacement along $\theta = 0$ where simple inextension of beamlike bending of the tank walls is assumed. The sloshing displacements, $q_{s+3}(t)\boldsymbol{\Psi}_s(r, \theta)$, $s = 1, 2, \ldots$, of the free surface relative to the solidified free surface, $\eta_{2,3} = \eta_2 + \eta_3$, are shown in Figure 8.5(b). The total elevation of the free surface η is

$$\eta = \eta_{2,3} + \sum_{s=1}^{\infty} q_{s+3}(t)\boldsymbol{\Psi}_s(r, \theta) \tag{8.78}$$

The potential and kinetic energies of the system may be written in the general form

$$V = \frac{1}{2}\sum_i \sum_j k_{ij} q_i q_j, \qquad T = \frac{1}{2}\sum_i \sum_j m_{ij} \dot{q}_i \dot{q}_j \tag{8.79a,b}$$

Let $q_3^2 e_3$ denote the center of gravity elevation due to bending, where e_3 has dimension of 1/length. The potential energy of the system is

$$V = mgq_3^2e_3 + \frac{1}{2}\rho_f g \int\int_A (\eta - \eta_{2,3}^2)\, dA \tag{8.80}$$

where m is the total mass of the liquid, ρ_f is the liquid density, and A is the area of the free surface. Substituting equation (8.78) into equation (8.80) and comparing the result to equation (8.79a), gives

$$k_{33} = 2mge_3 \tag{8.81a}$$

$$k_{sj} = \rho_f g \int\int_A \Psi_{s+3}\Psi_{j+3}\, dA, \qquad s, j = 1, 2, \ldots \tag{8.81b}$$

$$k_{is} = \rho_f g \int\int_A \frac{\eta_i}{q_i}\Psi_{s+3}\, dA, \ i = 2, 3. \tag{8.81c}$$

The liquid kinetic energy may be calculated by using the well-known expression (see, e.g., Lamb, 1945)

$$T = -\frac{\rho_f}{2}\int\int_S \Phi\Phi_n\, ds \tag{8.82}$$

where Φ is the velocity potential function, S is the boundary area of the liquid, which includes the free surface and tank wetted walls and bottom, n is the inward normal to the boundary.

Based on the assumption of small displacement of the free surface, the potential function may be expressed in terms of the following series

$$\Phi(r, \theta, z, t) = \sum_i q_i(t)\phi_i(r, \theta, z) \tag{8.83}$$

Substituting equation (8.83) into equation (8.82) and comparing the result to equation (8.79b) gives

$$m_{ij} = -\rho_f \int\int\int_S \phi_i(\phi_j)_n ds = -\rho_f \int\int\int_S \phi_j(\phi_i)_n\, ds \tag{8.84}$$

The potential function is obtained by solving Laplace's equation

$$\Phi_{rr} + \frac{1}{r}\Phi_r + \frac{1}{r^2}\Phi_{\theta\theta} + \Phi_{zz} = 0 \tag{8.85}$$

The solution of this equation must satisfy the boundary conditions

$$\Phi_z|_{z=h/2} = \dot{\eta} = \dot{q}_2(t)r\cos\theta - \dot{q}_3(t)f'(h/2)r\cos\theta + \sum_{s=1}^{\infty} \dot{q}_{s+3}(t)\Psi_s(r,\theta) \tag{8.86a}$$

$$\Phi_z|_{z=-h/2} = \dot{q}_2(t)r\cos\theta - \dot{q}_3(t)f'(-h/2)r\cos\theta \tag{8.86b}$$

$$\Phi_r|_{r=R} = [\dot{q}_1(t) - \dot{q}_2(t)z + \dot{q}_3(t)f'(z)]\cos\theta \tag{8.86c}$$

It is not difficult to write these boundary conditions in terms of expansion (8.83).

The translational potential function can be written in terms of the elementary solution

$$\phi_1(r,\theta) = r\cos\theta \tag{8.87}$$

The rotational potential, ϕ_2, may be satisfied at $z=\pm h/2$, or $r=R$, by the elementary solutions $zr\cos\theta$, or $-zr\cos\theta$, respectively. However, it is convenient to write this potential in terms of cylindrical harmonics, that is,

$$\phi_2(r,\theta,z) = \left\{zr + \sum_{s=0}^{\infty} A_s I_1[(2s+1)(\pi r/h)]\sin[(2s+1)(\pi z/h)]\right\}\cos\theta \tag{8.88a}$$

$$\phi_2(r,\theta,z) = \left\{-zr + \sum_{s=0}^{\infty} B_s J_1(\xi_s r/R)\sinh(\xi_s z/R)\right\}\cos\theta \tag{8.88b}$$

where ξ_s are the roots of $J'_1(\xi_s)=0$, and the first few roots are $\xi_1=1.84$, $\xi_2=5.34$, $\xi_3=8.54$, $\xi_4=11.71,\ldots$ The coefficients A_s and B_s are determined by satisfying the boundary conditions at $r=R$ and $z=\pm h/2$, and utilizing the properties of Fourier and Fourier–Bessel series. This process yields

$$A_s = \frac{8(-1)^{s+1}h^2}{(2s+1)^3\pi^3 I_1'[(2s+1)\pi R/h]}, \quad B_s = \frac{4R^2}{\xi_s(\xi_s^2-1)J_1(\xi_s)\cosh(\xi_s h/2R)} \tag{8.89a,b}$$

A possible solution for ϕ_3 may be written in the form

$$\begin{aligned}\phi_3(r,\theta,z) = &\left[C_0 r + \sum_{s=1}^{\infty} C_s I_1(s\pi r/h)\cos[s\pi(z+h/2)/h]\right.\\ &+ \sum_{s=1}^{\infty} D_s J_1(\xi_s r/R)\{f'(h/2)\cosh[\xi_s(z+h/2)/R]\\ &\left.- f'(-h/2)\cosh[\xi_s(z-h/2)/R]\}\right]\cos\theta\end{aligned} \tag{8.90}$$

where the coefficients C_0, C_s, and D_s are obtained by satisfying the boundary conditions at $r=R$ and $z=\pm h/2$. This gives

$$C_0 = \frac{1}{h}\int_{-h/2}^{h/2} f(z)\,dz, \qquad C_s = \frac{hF_s}{\pi s I_1'(\pi s R/h)}, \qquad D_s = \frac{2R^2}{\xi_s(\xi_s^2-1)J_1(\xi_s)\sinh(\xi_s h/R)} \tag{8.91}$$

for $s \geq 1$, where $F_s = \frac{2}{h}\int_{-h/2}^{h/2} f(z)\cos[s\pi(z+h/2)/h]\,dz$.

The sloshing potential functions are similar to those of a rigid cylindrical tank, such that the normalized form is given by the expression

$$\boldsymbol{\Psi}_s(r,\theta) = \cos\theta \frac{J_1(\xi_s r/R)}{J_1(\xi_s)} \tag{8.92}$$

and the corresponding potentials are

$$\phi_{3+s}(r,\theta) = R\cos\theta \frac{J_1(\xi_s r/R)}{\xi_s J_1(\xi_s)} \frac{\cosh[\xi_s(z+h/2)/R]}{\cosh(\xi_s h/R)} \tag{8.93}$$

In order to determine the normal mode frequencies of the coupled system one needs to derive the equations of motion of free oscillations in the absence of external excitation. The kinetic and potential energies of this case are

$$T = \frac{1}{2} m_e \left(\bar{f}^2 + \frac{R^2}{2}\bar{f}'^2\right)\dot{q}_3^2 + \frac{1}{2} m_{33}\dot{q}_3^2 + \dot{q}_3 \sum_1^{\infty} m_{3,s+3}\dot{q}_{s+3} + \frac{1}{2}\sum_1^{\infty} m_{s+3,s+3}\dot{q}_{s+3}^2 \tag{8.94a}$$

$$V = \frac{1}{2} m_e \bar{f}^2 \omega_e^2 q_3^2 + \frac{1}{2} k_{33} q_3^2 + q_3 \sum_1^{\infty} k_{3,s+3} q_{s+3} + \frac{1}{2}\sum_1^{\infty} k_{s+3,s+3} q_{s+3}^2 \tag{8.94b}$$

where m_e is the mass of the empty tank, $\bar{f}^2 = \frac{1}{l}\int_0^l f^2(z)\,dz$, l is the tank length, ω_e is the fundamental natural frequency of the empty tank,

$$\begin{aligned} m_{33} = m\Bigg[& C_0^2 + \frac{1}{2}\sum_{s=1}^{\infty} \boldsymbol{\Psi}_s F_s^2 - \frac{2R^2}{h}\sum_{s=1}^{\infty} \frac{[f'(h/2)\chi_s(h/2) - f'(-h/2)\chi_s(-h/2)]}{(\xi_s^2-1)} \\ & + \frac{R^2 C_0}{4h}(f'(-h/2) - f'(h/2)) \\ & + \frac{h}{\pi^2}\sum_{s=1}^{\infty} \frac{(1-\boldsymbol{\Psi}_s)}{s^2}[(-1)^{s+1} f'(h/2) + f'(-h/2)] F_0 \frac{2R^3}{h} \\ & \sum_{s=1}^{\infty} \frac{\{[f'^2(h/2) + f'^2(-h/2)]\cosh(\xi_s h/R) - 2f'(h/2)f'(-h/2)\}}{\xi_s^3(\xi_s^2-1)\sinh(\xi_s h/R)}\Bigg] \end{aligned} \tag{8.95a}$$

$$\chi_s(h/2) = \sum_{p=0}^{\infty} \frac{(-1)^p F_p}{\left[\xi_s^2 + (p\pi R/h)^2\right]}$$

$$m_{3,s+3} = \frac{R}{h}\left[\chi_s(h/2) - \frac{R[f'(h/2)\cosh(\xi_s h/R) - f'(-h/2)]}{\xi_s^3 \sinh(\xi_s h/R)}\right] \tag{8.95b}$$

$$m_{s+3,s+3} = m\frac{R}{2h}\frac{(\xi_s^2-1)}{\xi_s^3}\coth(\xi_s h/R) \tag{8.95c}$$

$$k_{33}=mg\left\{\frac{R^2}{4h}[f'^2(h/2)-f'^2(h/2)]+\frac{1}{h}\int\limits_{-h/2}^{h/2}zf'^2(z)dz\right.$$
$$\left.-\frac{1}{2}\int\limits_{z_0}^{h/2}f^{e2}(z)dz+\frac{1}{2}\int\limits_{-h/2}^{z_0}f'^2(z)dz\right\} \tag{8.95d}$$

$$k_{3,s+3}=-mg\frac{R}{h\xi_s^2}f'(h/2) \tag{8.95e}$$

For the case of a cantilever-like tank of thickness $\hat{t}$, length l, density ρ_c and parabolic bending mode shape, $f(z)=[z+h/2]_2/l$, the following parameters are obtained

$$\bar{m}_{33}=\left(\frac{h}{l}\right)^4\left[\frac{1}{45}-\frac{1}{6}\left(\frac{2R}{h}\right)^2-\frac{7}{384}\left(\frac{2R}{h}\right)^4+\frac{16}{\pi^4}\sum_{s=1}^{\infty}\frac{\Psi_s}{s^4}\right.$$
$$\left.+2\left(\frac{2R}{h}\right)^3\sum_{s=1}^{\infty}\frac{\coth(\xi_s h/R)}{\xi_s^3(\xi_s^2-1)}\right] \tag{8.96a}$$

$$\bar{m}_{3,s+3}=\left(\frac{h}{l}\right)^2\left[\frac{1}{2\xi_s^2}\left(\frac{2R}{h}\right)-\left(\frac{2R}{h}\right)^2\frac{\coth(\xi_s h/R)}{\xi_s^3}+\frac{1}{4\xi_s^4}\left(\frac{2R}{h}\right)^3\right] \tag{8.96b}$$

$$\bar{m}_{s+3,s+3}=\frac{(\xi_s^2-1)}{4\xi_s^3}\left(\frac{2R}{h}\right)\coth(\xi_s h/R) \tag{8.96c}$$

$$\bar{m}_{\mathrm{e}}=\frac{2\hat{t}\rho_c}{h\rho_f}\left[\frac{l}{5R}+\frac{2R}{l}\right] \tag{8.96d}$$

$$\bar{k}_{33}=\frac{2g}{l}\left[\frac{h}{2l}\left(\frac{R}{l}\right)^2-\frac{4}{3}\left(\frac{h}{2l}\right)^3\right], \tag{8.96e}$$

$$\bar{k}_{3,s+3}=-\frac{2g}{l\xi_s^2}\left(\frac{R}{l}\right) \tag{8.96f}$$

Applying Lagrange's equation, the following equations of motion for bending and liquid free-surface motions are obtained

$$(\bar{m}_{\mathrm{e}}+\bar{m}_{33})\ddot{q}_3+\sum_{s=1}^{\infty}\bar{m}_{3,s+3}\ddot{q}_{s+3}+\left(\bar{m}_{\mathrm{e}}\omega_{\mathrm{e}}^2+\bar{k}_{33}\right)q_3+\sum_{s=1}^{\infty}\bar{k}_{3,s+3}q_{s+3}=0 \tag{8.97a}$$

$$\bar{m}_{3,s+3}\ddot{q}_3+\bar{m}_{s+3,s+3}\ddot{q}_{s+3}+\bar{k}_{3,s+3}q_3+\bar{k}_{s+3,s+3}q_{s+3}=0 \tag{8.97b}$$

where the over-bar inertia and stiffness terms in equations (8.97) are normalized with respect to the total mass of the liquid, m. For example, $\bar{m}_{\mathrm{e}}=(m_{\mathrm{e}}/m)[\bar{f}^2+R^2f'^2/2]$ represents the nondimensional effective mass of an empty tank. The frequency equation of equations (8.97) is

$$\omega^2 - \omega_3^2 - \sum_{s=1}^{\infty} \frac{\bar{m}_{3,s+3}^2}{\bar{m}_{s+3,s+3}(\bar{m}_e + \bar{m}_{33})} \frac{\left[\omega^2 - (\bar{k}_{3,s+3}/\bar{m}_{3,s+3})\right]^2}{\left(\omega^2 - \omega_{s+3}^2\right)} = 0 \tag{8.98}$$

where $\omega_3 = \sqrt{\dfrac{\bar{m}_e \omega_e^2 + \bar{k}_{33}}{\bar{m}_e + \bar{m}_{33}}}$ is the uncoupled bending frequency, $\omega_{s+3} = \sqrt{\bar{k}_{s+3,s+3}/\bar{m}_{s+3,s+3}}$ are the uncoupled sloshing frequencies.

The solution of the frequency equation (8.98) gives the natural frequencies of the coupled system. For the case of a small cylindrical tank, the liquid natural frequencies are relatively smaller than the coupled bending frequency. In this case, one can simplify the frequency equation to the following form

$$\omega^2 = \omega_3^2 \left[1 - \sum_{s=1}^{\infty} \frac{\bar{m}_{3,s+3}^2}{\bar{m}_{s+3,s+3}(\bar{m}_e + \bar{m}_{33})}\right]^{-1} \tag{8.99}$$

Note that the difference between the coupled, ω, and the uncoupled, ω_3, frequencies depends on the sign of the summation expression. All terms in the summation are positive except $\bar{m}_{3,s+3}$, which can assume a positive or negative value depending on the fluid depth h/R ratio. For a certain value of h/R, the inertia parameter $\bar{m}_{3,s+3} = 0$, and thus, the coupling becomes very weak. For $\bar{m}_{3,s+3} \neq 0$, the coupled bending frequency is always greater than the uncoupled frequency.

Lindholm, *et al.* (1963) conducted experiments on thin cylindrical shells with cantilever and simply supported ends to examine the validity of the numerical solution of equation (8.98). Their results are shown in Figure 8.6, which show the dependence of the resonant frequencies on the fluid depth ratio, h/l, for two cantilever tanks of length to diameter ratios $l/2R = 2.76$ and 5.03 with a capped liquid surface. The tank sizes were selected to lie in the region of maximum coupling effect. The solid curves are the theoretical frequency curves as predicted using equation (8.99). The influence of the mode shape on the predicted values for the case $l/2R = 2.76$ indicates that the measured frequencies fall between the two predicted results. The difference between the capped and uncapped resonant frequencies is shown in Figure 8.7 as the percentage $b[(\omega - \omega_3)/\omega_3] \times 100$ versus the liquid depth ratio. It is seen that the assumed mode shape has a significant effect on the accuracy of the predicted results. For the case of a simply supported tank, similar results are shown in Figures 8.8 and 8.9 based on a half-sine wave mode shape for the theoretical curves.

Axisymmetric flow

Bauer, Siekmann, and Wang (1968) followed a different approach based on bending theory as outlined in Section 8.2. The governing equation of the axisymmetric motion of liquid in cylindrical coordinates is

$$\Phi_{rr} + \frac{1}{r}\Phi_r + \Phi_{zz} = 0 \tag{8.100}$$

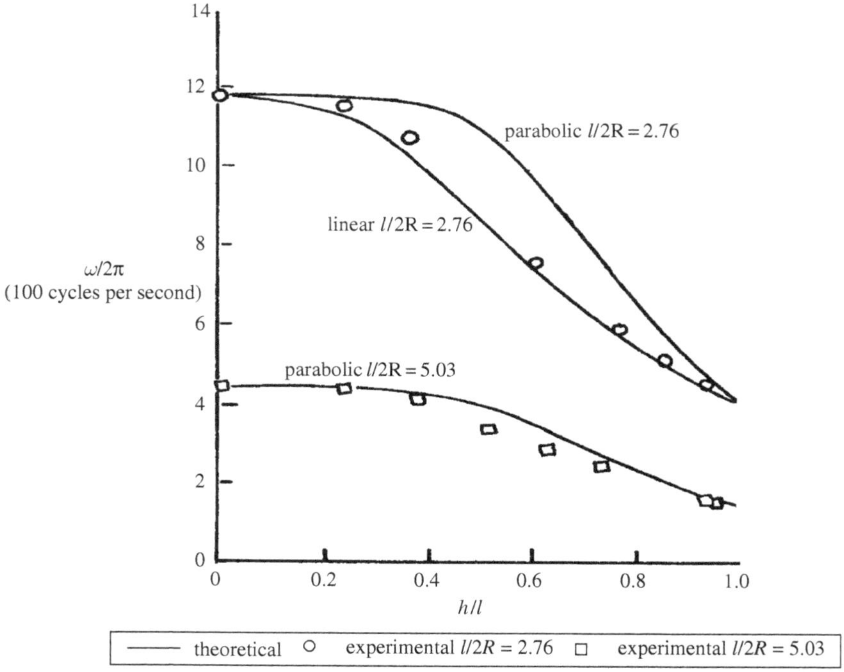

Figure 8.6 Dependence of the capped resonance bending frequency on the fluid depth ratio in cantilever tanks. (Lindholm, *et al.*, 1963)

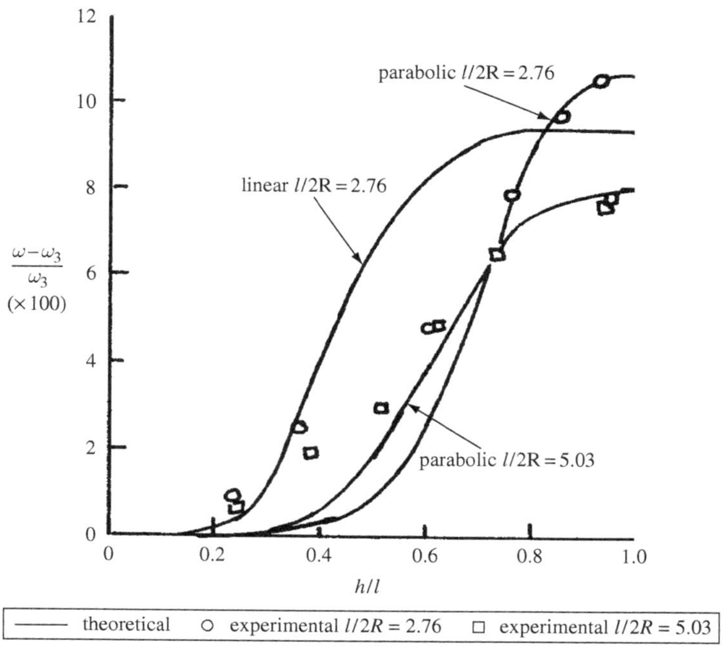

Figure 8.7 Percentage frequency increase versus fluid depth ratio in cantilever tanks. (Lindholm, *et al.*, 1963)

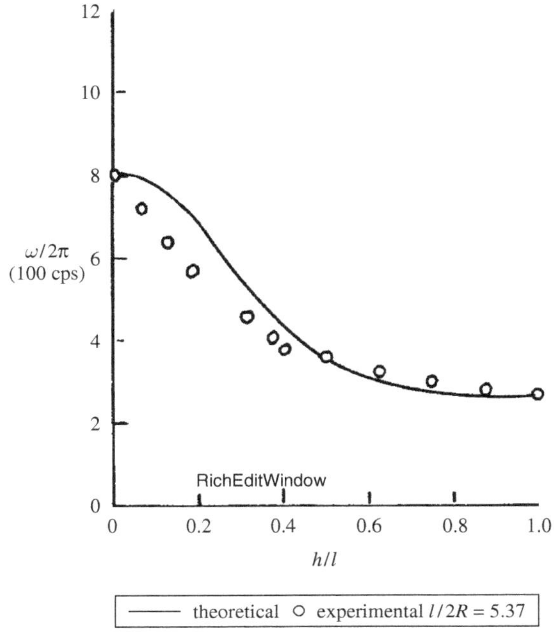

Figure 8.8 Dependence of capped natural frequency on fluid depth ratio for simply supported tanks. (Lindholm, *et al.*, 1963)

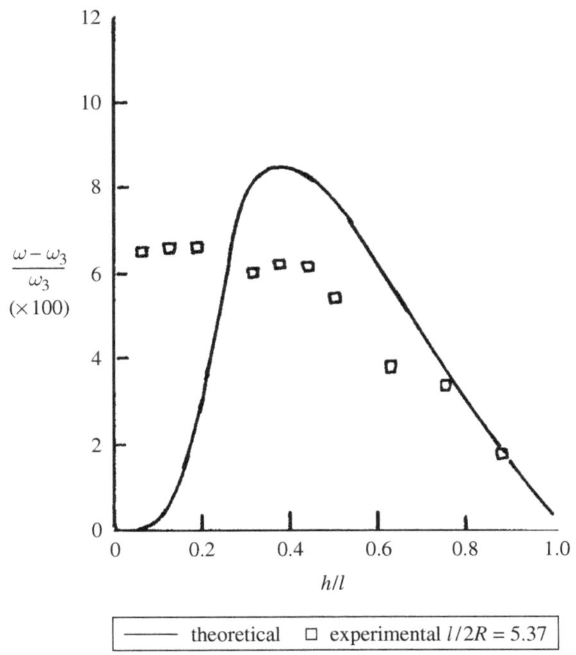

Figure 8.9 Percentage bending frequency increase versus fluid depth ratio in simply supported tanks. (Lindholm, *et al.*, 1963)

The solution of this equation must satisfy the boundary conditions at the rigid bottom and the free surface

$$\left.\frac{\partial\Phi}{\partial z}\right|_{z=0}, \qquad \left[\frac{\partial^2\Phi}{\partial t^2}+g\frac{\partial\Phi}{\partial z}\right]_{z=h}=0 \tag{8.101a,b}$$

The solution of equation (8.100) satisfying conditions (8.101) can be written in the form

$$\Phi(r,z,t)=e^{i\omega t}\sum_{n=1}^{\infty}A_n\cos(\varepsilon_n z)I_0(\varepsilon_n r) \tag{8.102}$$

where ε_n are the roots of the equation

$$\varepsilon\tan(\varepsilon h)=-(\omega^2/g) \tag{8.103}$$

The roots of equation (8.103) can be obtained numerically using, for example, Newton's method for specific values of fluid depth, h. The dynamic pressure at the tank wall is estimated from the pressure equation

$$p=-\rho_f\frac{\partial\Phi}{\partial t}=-i\omega\rho_f\,e^{i\omega t}\sum_{n=1}^{\infty}A_n\cos(\varepsilon_n z)I_0(\varepsilon_n r) \tag{8.104}$$

The equation of motion of the elastic cylinder is

$$\frac{\partial 4w}{\partial z^4}+\frac{E\hat{t}}{DR^2}w+\frac{\rho_c\hat{t}}{D}\frac{\partial^2 w}{\partial t^2}=-\left.\frac{\rho_f}{D}\frac{\partial\Phi}{\partial t}\right|_{r=R} \tag{8.105}$$

where $D=E\hat{t}^3/12(1-\nu^2)$ is the rigidity of the plate, w is the transverse displacement taken as positive along the outward normal and can be represented in the harmonic form

$$w(z,t)=W(z)\,e^{i\omega t} \tag{8.106}$$

Substituting (8.106) into equation (8.105) gives

$$\frac{d^4W}{dz^4}+4\lambda^4W=-\frac{i\omega\rho_f}{D}\sum_{n=1}^{\infty}A_n\cos(\varepsilon_n z)I_0(\varepsilon_n R) \tag{8.107}$$

where $4\lambda^4=\hat{t}/D\left(\frac{E}{R^2}-\rho_c\omega^2\right)$.

The general solution of equation (8.107) can be written in the form

$$W(z)=e^{\lambda z}(B_1\cos\lambda z+B_2\sin\lambda z)+e^{-\lambda z}(B_3\cos\lambda z+B_4\sin\lambda z)$$
$$-\frac{i\omega\rho_f}{D}\sum_{n=1}^{\infty}\frac{I_0(\varepsilon_n R)}{(\varepsilon_n^4+4\lambda^4)}\cos(\varepsilon_n z) \tag{8.108}$$

where B_j, $j=1, 2, 3, 4$, are constants determined by satisfying the boundary conditions. For a clamped-clamped cylindrical tank, the boundary conditions are

$$\left.W\right|_{z=0,\,l}=\left.\frac{dW}{dz}\right|_{z=0,\,l}=0 \tag{8.109}$$

The corresponding constants are

$$B_j = \frac{i\omega\rho_f}{D}\sum_{n=1}^{\infty}\frac{I_0(\varepsilon_n R)A_n}{\left(\varepsilon_n^4 + 4\lambda^4\right)}\bar{K}_{jn} \tag{8.110}$$

where

$$\begin{aligned}\bar{K}_{1n} = -\frac{1}{C_1}\Big\{ & e^{-\lambda l}\left[\cos\varepsilon_n l(\sin\lambda l - \cos\lambda l) - \frac{\varepsilon_n}{\lambda}\sin\varepsilon_n l\,\sin\lambda l\right] \\ & + e^{\lambda l}\left[\cos\varepsilon_n l(\sin\lambda l + \cos\lambda l) + \frac{\varepsilon_n}{\lambda}\sin\varepsilon_n l\sin\lambda l\right] \\ & + e^{-2\lambda l} - 2 - \sin 2\lambda l + \cos 2\lambda l\Big\}\end{aligned} \tag{8.111a}$$

$$\bar{K}_{2n} = -\frac{1}{C_2}\left[C_3\bar{K}_{1n} - 2\cos\varepsilon_n l\,\sin\lambda l + \frac{\varepsilon_n}{\lambda}\sin\lambda l(\sin\lambda l + \cos\lambda l)\right] \tag{8.111b}$$

$$\bar{K}_{3n} = (1 + \bar{K}_{1n}) \tag{8.111c}$$

$$\bar{K}_{4n} = (1 + 2\bar{K}_{1n} + \bar{K}_{2n}) \tag{8.111d}$$

with $C_1 = 4 - 2\cos 2\lambda l - e^{2\lambda l} - e^{-2\lambda l}$, $C_2 = e^{\lambda l}[2 + \sin 2\lambda l + \cos 2\lambda l] - e^{-\lambda l}$

$$C_3 = e^{-\lambda l} - e^{\lambda l}[\sin 2\lambda l + \cos 2\lambda l]$$

The functions $e^{\lambda z}\cos\lambda z$, $e^{\lambda z}\sin\lambda z$, $e^{-\lambda z}\cos\lambda z$, $e^{-\lambda z}\sin\lambda z$ may be expanded in Fourier cosine series

$$e^{\lambda z}\{\cos\lambda z,\ \sin\lambda z\} = \sum_{m=1}^{\infty}(a_m,\ b_m)\cos\varepsilon_m z \tag{8.112a}$$

$$e^{-\lambda z}\{\cos\lambda z,\ \sin\lambda z\} = \sum_{m=1}^{\infty}(c_m,\ c_m)\cos\varepsilon_m z \tag{8.112b}$$

where

$$\begin{aligned}a_m = \frac{2\varepsilon_m}{(2\varepsilon_m h + \sin 2\varepsilon_m h)}\Bigg[& \frac{\lambda\left[e^{\lambda h}\cos(\lambda+\varepsilon_m)h - 1\right] + (\lambda+\varepsilon_m)\,e^{\lambda h}\sin(\lambda+\varepsilon_m)h}{\lambda^2 + (\lambda+\varepsilon_m)^2} \\ & + \frac{\lambda\left[e^{\lambda h}\cos(\lambda-\varepsilon_m)h - 1\right] + (\lambda-\varepsilon_m)\,e^{\lambda h}\sin(\lambda-\varepsilon_m)h}{\lambda^2 + (\lambda-\varepsilon_m)^2}\Bigg]\end{aligned} \tag{8.113a}$$

$$\begin{aligned}b_m = \frac{2\varepsilon_m}{(2\varepsilon_m h + \sin 2\varepsilon_m h)}\Bigg[& \frac{\lambda\,e^{\lambda h}\sin(\lambda+\varepsilon_m)h - (\lambda+\varepsilon_m)\left[e^{\lambda h}\cos(\lambda+\varepsilon_m)h - 1\right]}{\lambda^2 + (\lambda+\varepsilon_m)^2} \\ & + \frac{\lambda\,e^{\lambda h}\sin(\lambda-\varepsilon_m)h - (\lambda-\varepsilon_m)\left[e^{\lambda h}\cos(\lambda-\varepsilon_m)h - 1\right]}{\lambda^2 + (\lambda-\varepsilon_m)^2}\Bigg]\end{aligned} \tag{8.113b}$$

$$c_m = \frac{2\varepsilon_m}{(2\varepsilon_m h + \sin 2\varepsilon_m h)}\left[\frac{(\lambda+\varepsilon_m)\lambda\, e^{-\lambda h}\sin(\lambda+\varepsilon_m)h - \lambda\left[e^{-\lambda h}\cos(\lambda+\varepsilon_m)h - 1\right]}{\lambda^2+(\lambda+\varepsilon_m)^2} + \frac{(\lambda-\varepsilon_m)\lambda\, e^{-\lambda h}\sin(\lambda-\varepsilon_m)h - \lambda\left[e^{-\lambda h}\cos(\lambda-\varepsilon_m)h - 1\right]}{\lambda^2+(\lambda+\varepsilon_m)^2}\right] \tag{8.113c}$$

$$d_m = \frac{2\varepsilon_m}{(2\varepsilon_m h + \sin 2\varepsilon_m h)}\left[\frac{\lambda\, e^{-\lambda h}\sin(\lambda+\varepsilon_m)h + (\lambda+\varepsilon_m)\left[e^{-\lambda h}\cos(\lambda+\varepsilon_m)h - 1\right]}{\lambda^2+(\lambda+\varepsilon_m)^2} + \frac{\lambda\, e^{-\lambda h}\sin(\lambda-\varepsilon_m)h + (\lambda-\varepsilon_m)\left[e^{-\lambda h}\cos(\lambda-\varepsilon_m)h - 1\right]}{\lambda^2+(\lambda+\varepsilon_m)^2}\right] \tag{8.113d}$$

Due to the elasticity of the tank wall, the radial velocity of the liquid at the wall must be identical to the radial velocity of the wall, that is,

$$\left.\frac{\partial \Phi}{\partial t}\right|_{r=R} = \frac{\partial w}{\partial r} \tag{8.114}$$

Substituting equations (8.102) and (8.106) in the compatibility equation (8.114) gives the following set of homogeneous algebraic equations

$$[\alpha_{mn}(\omega)]\{A_n\} = 0 \tag{8.115}$$

where

$$\alpha_{mn}(\omega) = \frac{\rho_f \omega^2}{D(\varepsilon_n^4 + 4\lambda^4)}\left(\bar{K}_{1n} a_m + \bar{K}_{2n} b_m + \bar{K}_{3n} c_m + \bar{K}_{4n} d_m\right) + \left\{\frac{\rho_f \omega^2}{D(\varepsilon_m^4 + 4\lambda^4)} - \varepsilon_m \frac{I_1(\varepsilon_m R)}{I_0(\varepsilon_m R)}\right\}\delta_{mn} \tag{8.116}$$

The frequency equation is obtained by setting the determinant of the coefficient matrix of equation (8.115) to zero, that is,

$$|\alpha_{mn}(\omega)| = 0 \tag{8.117}$$

This is an infinite-order determinant and can be truncated to any order. Bauer, Siekmann and Wang (1968) solved for the coupled frequencies for a tenth-order truncated determinant using an iterative scheme. Figure 8.10(a) shows the dependence of the natural frequency on the wall thickness ratio, $\hat{t}/R$, for different values of liquid depth. Figure 8.10(b) shows the dependence of the natural frequencies on the liquid height to tank length ratio, h/l, for different values of wall thickness. Both figures are obtained for $\rho_c g = 168\,\text{lb/ft}^3$, $\rho_f g = 65\,\text{lb/ft}^3$, $E = 1.44\times10^9\,\text{lb/ft}^2$, $\nu = 0.3$, and $R = 12\,\text{ft}$. It is seen that the natural frequency increases with increasing wall thickness and decreases with the liquid depth ratio.

8.4.2 Interaction with breathing modes

Historical account

The early work of the breathing vibration is believed to be due to Reissner (1956) who considered a nearly full-pressurized cylindrical shell containing a heavy liquid with a

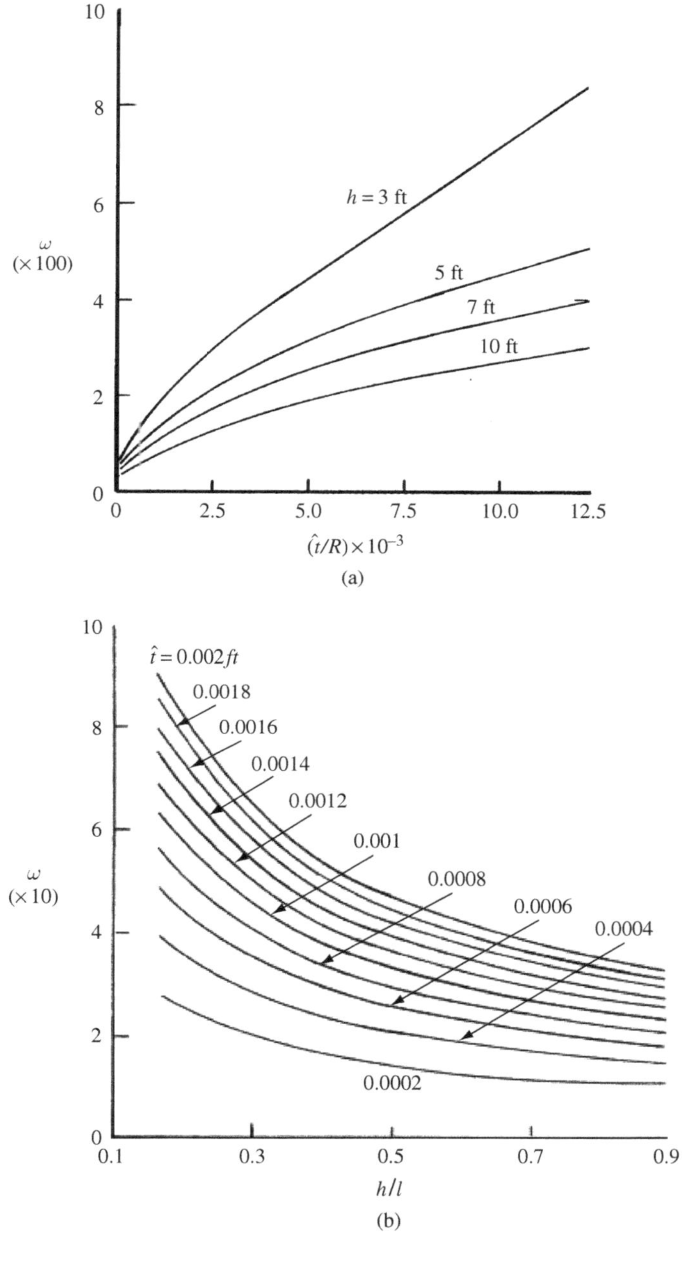

Figure 8.10 Dependence of the fundamental frequency of axisymmetric mode on (a) tank wall thickness ratio for different values of liquid depth, (b) liquid depth ratio for different values of tank wall thickness. (Bauer, *et al.*, 1968)

free surface. Reissner used a vibrating circular string equation to describe the shell motion. Fung, *et al.* (1957) and Berry and Reissner (1958) studied the breathing vibration of a pressurized shallow cylinder containing a compressible fluid where the influence of liquid elevation and free-surface motion were ignored. A shell is termed as shallow if the ratio of its height to its span is much less than 1. The frequency equation obtained by Berry and Reissner included a term for an apparent inertia of the shell walls due to the internal liquid. Mixon and Herr (1962) and Lindholm, *et al.*, (1962a) considered different assumptions for studying the breathing vibration of a nearly full cylindrical tank. Parkus (1982) and Markus (1988) documented the natural frequencies and associated modes of empty and liquid-filled cylindrical shells.

The breathing vibration of empty shells was extensively studied in the literature (Hu, 1964). Essentially, the work relies heavily on assumptions and simplifications involved in solving the equations of motion. These include Rayleigh's theories of in-extensional and extensional vibrations. For fundamental modes the in-extensional theory assumes the middle surface (situated at the mid-thickness of the shell wall) remains unstretched. The strain energy of this case is associated with bending stresses only with no membrane contribution. On the other hand, Rayleigh's theory of extensional vibrations assumes that the shell deformation consists mainly of stretching in the middle surface. In this theory, the membrane stresses are solely considered in the strain energy. Another two theories in common use are based on the Donnell equations for circular cylindrical shells and the Reissner equations for shallow shells. Kana (in Abramson, 1966a, Chapter 9) provided an excellent assessment of different shell theories and indicated that all theories predict the natural vibrational modes of a shell and the accuracy of the results depends on the degree to which the assumptions imposed in every case are satisfied.

For partially liquid-filled shells simply supported at the bottom and free at the top, Baron and Bleich (1959) and Baron and Skalak (1962) found that the coupled breathing modes are almost the same as those estimated in the absence of free-surface motion. Furthermore, additional resonances corresponding to liquid sloshing coupled modes occurred at very low frequencies, near the values they would have in a rigid container. The interaction of liquid motion with elastic tank breathing was considered by Chu (1963), Chu and Gonzales (1964), and Bauer (1964c). They found that the additional terms due to the free-surface effects are very small unless the excitation is in the vicinity of a frequency corresponding to a natural liquid mode for a similar rigid container. The influence of liquid free-surface motion on the breathing vibration of shells is negligible when the shell frequency is higher than the first several uncoupled sloshing frequencies. The influence of free-surface motion on breathing vibrations has been studied by Fontenot and Lianis (1963) for the case of a full, pressurized cylindrical shell. Rabinovich (1964) considered a partially full cylindrical shell to determine the hydrodynamic pressure using the Vlasov shell equation and a Lagrange–Cauchy integral. He obtained the natural frequencies and natural modes of the shell by expressing the modes as linear combinations of the empty shell modes. Natushkin and Rakhimov (1964) investigated the partially filled cylindrical shell with arbitrary end conditions, while Samoylov and Pavlov (1964) and Hwang (1965) considered the case of a liquid-filled hemispherical shell.

Further studies were conducted by Leroy (1963), Shmakov (1964), Palmer and Asher (1965), Triol (1965), Glaser (1967), Grigoliuk, *et al.* (1968), Siekmann and Chang (1971a), and Chen and Rosenberg (1975). They derived different forms for the frequency equation of the combined system. The linear formulation was adequate as long as the free liquid surface did not exceed a certain range above which nonlinearity became important. For a thin elastic tank wall, the fundamental axisymmetric coupled frequency was much smaller than the liquid frequency with rigid walls (Beam and Guist, 1967).

Saleme and Liber (1965) derived the equations of motion of an axially symmetrical shell filled with one or more nonmixable, nonviscous, compressible fluids under pressure. For the case of two fluids, there are two characteristic equations whose solutions are substituted into the continuity condition. They found that the frequencies are grouped into a low and a high range. The low-range frequencies are lower than the lowest, and the high-range frequencies are higher than the highest natural frequencies of the corresponding shell without fluid. For a given circumferential mode shape, there are, in general, several levels of the separation surface associated with the same natural frequency. Koval'chuk and Filin (2003a) determined the circumferential traveling waves in filled cylindrical shells.

Rumyantsev (1969b) examined the stability of an elastic body with a liquid filled cavity. The hydroelastic oscillations in cylindrical tanks were studied by Reifel (1964), Bauer, *et al.* (1967, 1972), Bauer (1970, 1971a, 1981b), Lakis and Païdoussis (1971), and Lakis and Sinno (1992) while shallow cylindrical tanks were considered by Gupta and Hutchinson (1988, 1989, 1991) and Gupta (1995a,b). The axisymmetric vibrations of cylindrical hemispherical and spherical liquid tanks were analyzed by Anisimov (1963), Samoylov and Pavlov (1964), Gossard, (1965), Hwang (1965), Balabukh (1966), Balabukh and Molchanov (1967), Balakirev (1967), Pozhalostin (1967), Rand and DiMaggio (1967), Coale and Nagano (1969), Engin (1969a, b), and Kana, *et al.* (1971). As the tank wall thickness increases the coupling effect diminishes and the liquid frequency approaches the case of a rigid tank.

Koga and Tsushima (1990) and Amabili and Dalpiaz (1995) carried out analyses and experiments to estimate the normal modes of breathing vibrations of horizontal circular shells partially filled with liquid. It was found that the first natural frequency in all filling levels is very close to that of a completely filled tank. It was concluded that a rather low level of liquid is enough to strongly reduce the first natural frequency.

Analysis

The linear analysis of breathing vibrations of a partially filled cylindrical shell is based on the assumption of ignoring inertia forces in the longitudinal and circumferential directions. This section is based on the work of Chu (1963) and Chu and Gonzales (1964) who adopted mode shapes of a partially filled shell identical to those of an empty shell. The dynamic shell equations based on Reissner's approximation for transverse vibrations were used. The derivation of the exact form of shell equations referred to as Donnell–Yu equations is well documented in Yu (1955). Thus, the approximate equations, referred to as Donnell–Yu–Reissner equations, describing the elastic displacements w, v, and u, along r, θ, z axes, respectively, are,

$$-\frac{\hat{t}^2}{12}\nabla^4 w-\frac{w}{R_c}+\frac{1}{R_c^2}\frac{\partial v}{\partial\theta}+\frac{\nu}{R_c}\frac{\partial u}{\partial z}=\frac{1-\nu^2}{E\hat{t}}\left[\rho_c\hat{t}\frac{\partial^2 w}{\partial t^2}-q\right] \tag{8.118a}$$

$$\nabla^4 v-\frac{2+\nu}{R_c^2}\frac{\partial^3 w}{\partial\theta\partial z^2}-\frac{1}{R_c^4}\frac{\partial^3 w}{\partial\theta^3}\approx 0 \tag{8.118b}$$

$$\nabla^4 u-\frac{\nu}{R_c}\frac{\partial^3 w}{\partial z^3}+\frac{1}{R_c^3}\frac{\partial^3 w}{\partial\theta^2\partial z}\approx 0 \tag{8.118c}$$

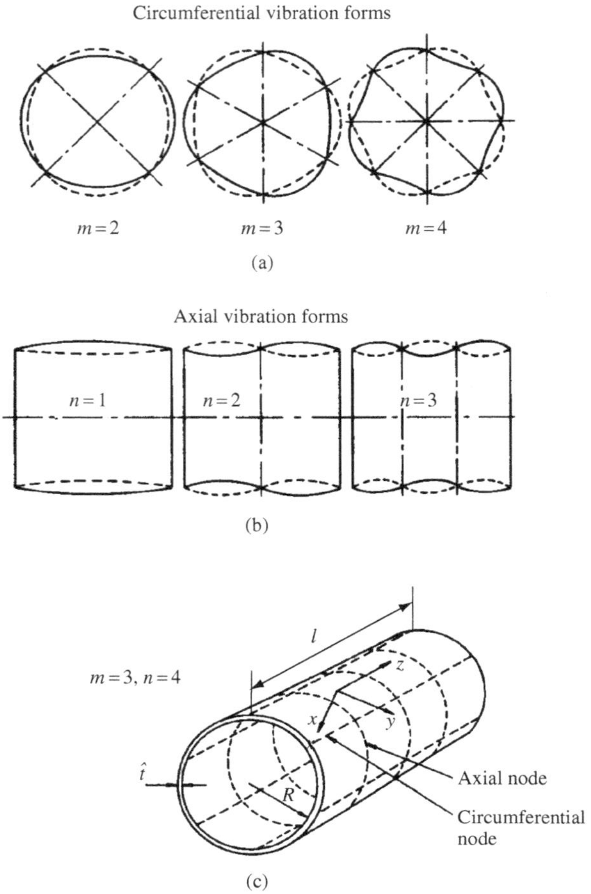

Figure 8.11 Forms of vibration of thin cylinders with freely supported ends. (Arnold and Warburton, 1949)

where q is the hydrodynamic loading and w is the displacement in the radial direction. The mode shapes of shells with simply supported end conditions were determined by Arnold and Warburton (1949) and the first few mode shapes are shown in Figure 8.11. The boundary conditions are satisfied by the following expressions of the three displacements

$$w = A\sin(n\pi z/l)\cos m\theta\cos\omega t \tag{8.119a}$$

$$v = B\sin(n\pi z/l)\sin m\theta\cos\omega t \tag{8.119b}$$

$$u = C\cos(n\pi z/l)\cos m\theta\cos\omega t \tag{8.119c}$$

where m denotes the number of circumferential stationary waves, and n specifies the number of axial stationary half waves.

Eliminating u and v from equation (8.118a) using equations (8.118b,c), gives

$$-\frac{E\hat{t}^2}{12(1-\nu^2)}\nabla^8 w + \frac{E\hat{t}}{R_c^2}\frac{\partial^4 w}{\partial z} = \nabla^4 q - \rho_c\hat{t}\frac{\partial^2}{\partial t^2}(\nabla^4 w) \tag{8.120}$$

where $\nabla^2 = \frac{\partial^2}{\partial z^2} + \frac{1}{R_c^2}\frac{\partial^2}{\partial \theta^2}$.

For breathing vibrations, one may express w in the form

$$w(\theta, z, t) = w_s = A_{mn}(t) Z_n(z) \cos m\theta \tag{8.121}$$

Regardless of the type of boundary conditions, upon substituting equation (8.121) into equation (8.120) the general form of the differential equation governing the generalized coordinate $A_{mn}(t)$ is

$$\ddot{A}_{mn}(t) + \omega_{mn}^2 A_{mn}(t) = \frac{q_{mn}(t)}{\rho_c \hat{t}} \tag{8.122}$$

where the generalized force is given by the expression

$$q_{mn}(t) = \frac{1}{\pi(1+\delta_{0m})} \frac{\int_{-\pi}^{\pi}\int_0^l q(t) Z_n(z) \cos m\theta \, dz \, d\theta}{\int_0^l (Z_n(z))^2 dz} \tag{8.123}$$

For simply–simply supported shells the function $Z_n(z)$ may be selected as $\sin(n\pi z/l)$, and the natural frequency in the absence of liquid ($q = 0$) is

$$\omega_{mn}^2 = \frac{E}{\rho_c R_c (1-\nu^2)} \left[\frac{m^4}{12} \frac{\hat{t}^2}{R_c^2} (1+\alpha^2)^2 + \frac{(1-\nu^2)\alpha^4}{(1+\alpha^2)^2} \right] \tag{8.124}$$

where $\alpha = \frac{n\pi R_c}{ml}$.

Equation (8.124) gives the natural frequencies of a simply supported empty shell and is identical to the result of Reissner. Chu (1963) considered the coupling natural frequency of the shell with the liquid free-surface motion. The mode of the breathing tank with liquid is assumed to be dominantly the same as that of the empty tank. In this case, the compatibility condition of the radial velocity of liquid and tank wall must be introduced, that is,

$$\left.\frac{\partial \Phi}{\partial r}\right|_{r=R} = -\frac{\partial w_s}{\partial t} = -\dot{A}_{mn} Z_n(z) \cos(m\theta) \tag{8.125}$$

Assuming a rigid bottom, one should impose the condition $(\partial\Phi/\partial z)|_{z=0} = 0$. In this case, the solution of Laplace's equation may be obtained using the method of separation of variables

$$\begin{aligned}\Phi = &- \dot{A}_{mn} R \sum_{k=0}^{\infty} D_{kn} \frac{I_m(k\pi r/h)}{I'_m(k\pi R/h)} \cos(k\pi z/h) \cos m\theta \\ &+ \delta_{0m} \dot{A}_{mn} \left[\sum_{k=0}^{\infty} R E_{k0} (-1)^k D_{km} \right] \\ &+ \dot{A}_{mn} \sum_{j=0}^{\infty} \frac{\sum_{k=0}^{\infty} (-1)^k D_{kn} R E_{kj}}{\cosh(\xi_{mj} h/R)} \frac{\Omega^2}{\left(\Omega^2 - \omega_{mj}^2\right)} \cos(\xi_{mj} z/R) J_m(\xi_{mj} r/R)\end{aligned} \tag{8.126}$$

where Ω is the excitation frequency, δ_{0m} is the Kronecker delta, and

$$D_{kn} = \frac{2}{h(1+\delta_{0m})}\int_0^h f_n(z)\cos(k\pi z/h)\,dz \tag{8.127a}$$

$$E_{kj} = \frac{2\xi_{mj}^2}{\left(\xi_{mj}^2 - m^2\right)J_m^2(\xi_{mj})}\int_0^a \frac{r}{R}\frac{I_m(k\pi r/h)J_m(\xi_{mj}r/R)}{(k\pi R/h)I'_m(k\pi R/h)}\frac{dr}{R}, \quad j>0 \tag{8.127b}$$

$$E_{kj} = \frac{2\xi_{mj}^2\left[(k\pi R/h)J_m(\xi_{mj})I_{m+1}(k\pi R/h) + \xi_{mj}I_m(k\pi R/h)J_{m+1}(\xi_{mj})\right]}{\left(\xi_{mj}^2 - m^2\right)J_m^2(\xi_{mj})(k\pi R/h)I'_m(k\pi R/h)\left[(k\pi R/h)^2 + \xi_{mj}^2\right]}, \quad k>1, \quad j>0 \tag{8.127c}$$

$$E_{0j} = \frac{2\xi_{mj}^3 J_{m+1}(\xi_{mj})}{\left(\xi_{mj}^2 - m^2\right)mJ_m2(\xi_{mj})\left[(k\pi R/h)^2 + \xi_{mj}^2\right]}, \quad j>0 \tag{8.127d}$$

$$E_{k0} = 2\int_0^a \frac{r}{R}\frac{I_m(k\pi r/h)}{(k\pi R/h)I'_m(k\pi R/h)}\frac{dr}{R} = \frac{2}{(k\pi R/h)} \tag{8.127e}$$

The linearized hydrodynamic pressure is

$$p = -q(t) = -\rho_f\frac{\partial\Phi}{\partial t}, \quad \text{for } 0<z<h \tag{8.128}$$

The generalized force given by equation (8.123) can be evaluated by substituting equations (8.128) and (8.126)

$$\begin{aligned} q_{mn} &= -\rho f R M^* \ddot{A}_{mn} - \rho_f R \ddot{A}_{mn}\sum_{j=1}^{\infty}\frac{\tilde{C}_{nj}(\omega_{mj}^2/\Omega^2)J_m(\xi_{mj})\sum_k(-1)^k E_{kj}D_{kn}}{\cosh(\xi_{mj}h/R)\left[1-(\omega_{mj}^2/\Omega^2)\right]} \\ &\equiv \bar{q}_{mn}(\varpi_{mn}^2)\ddot{A}_{mn} \end{aligned} \tag{8.129}$$

where

$$\begin{aligned} M^* &= \frac{1}{m}D_{0n}\tilde{D}_{0n} + \sum_{k=1}^{\infty}\frac{I_m(k\pi R/h)D_{kn}\tilde{D}_{kn}}{(k\pi R/h)I'_m(k\pi R/h)} \\ &\quad - \delta_{0m}\left[\sum_{k=0}^{\infty}E_{k0}(-1)^k D_{kn}\right]\left(\frac{2}{n\pi}\right)[1-(-1)^n]\tilde{E}_n \\ &\quad - \sum_{j=1}^{\infty}\frac{\tilde{C}_{nj}J_m(\xi_{mj})\sum_k(-1)^k E_{kj}D_{kn}}{\cosh(\xi_{mj}h/R)} \end{aligned} \tag{8.130a}$$

which represents the ratio of the added mass, due to the liquid, to the tank empty mass.

$$\tilde{C}_{nj} = \frac{\int_0^h \cos h(\xi_{mj}z/R) f_n(z)\, dz}{\int_0^l f_n^2(z)\, dz} \tag{8.130b}$$

$$\tilde{D}_{kn} = \frac{\int_0^h \cos(k\pi z/h) f_n(z)\, dz}{\int_0^l f_n^2(z)\, dz} \tag{8.130c}$$

$$\tilde{E}_n = \frac{\int_0^l f_n(z)\, dz}{\int_0^l f_n^2(z)\, dz} \tag{8.130d}$$

Substituting equation (8.129) into equation (8.122) gives the breathing frequency equation

$$-\varpi_{mn}^2 A_{mn} + \omega_{mn}^2 A_{mn} = -\varpi_{mn}^2 A_{mn} \frac{\bar{q}_{mn}(\varpi_{mn}^2)}{\rho_c \hat{t}} \tag{8.131}$$

where ϖ_{mn} is the breathing frequency in the presence of liquid, and ω_{mn} is the breathing frequency in the absence of liquid for breathing mode mn. Since the frequencies of the first few modes of liquid sloshing are much smaller than the shell coupled natural frequency, one may write the ratio of the natural breathing frequency, ϖ_{mn}, to the breathing natural frequency of the empty shell by the following approximate expression (by retaining the first expression in equation (8.129))

$$\frac{\varpi_{mn}}{\omega_{mn}} = \sqrt{1/\left[1 + M^* \frac{\rho_f R}{\rho_c \hat{t}}\right]} \tag{8.132}$$

For breathing vibration with fundamental longitudinal mode, $n=1$, the deviation of the actual mode shape from the assumed dominant shape, $w \propto \sin(\pi z/l)$ cos $m\theta$, see equation (8.119a), for all depths was found to be relatively small (Chu, 1963). Chu (1963) found that the theoretical frequency ratio is within 8% of the measured values for all liquid depths. For higher modes, the error may be as much as 20%, which is mainly attributed to the inaccuracy of the used mode shape. More accuracy may be achieved by using Galerkin's method with a finite expansion of the mode shape. In this case, the radial deflection of the shell may be represented by the expansion of orthogonal modes $f_n(z)$:

$$w(\theta, z, t) = \sum_{n=1}^{N} A_{mn}(t) f_n(z) \cos m\theta \tag{8.133}$$

where

$$\frac{\int_0^l f_j(z) f_k(z)\, dz}{\int_0^l f_j^2(z)\, dz} = \delta_{jk} \tag{8.134}$$

Galerkin's method yields

$$-\varpi_{mn}^2 A_{mn} + \omega_{mn}^2 A_{mn} = \frac{q_{mn}(\varpi_{mn}^2)}{\rho_c \hat{t}} \tag{8.135}$$

where

$$q_{mn}(t) = \frac{1}{\pi(1+\delta_{0m})} \frac{\int_{-\pi}^{\pi}\int_0^l q(t) f_n(z)\cos m\theta \, dz\, d\theta}{\int_0^l (f_n(z))^2\, dz}$$

$$= \rho_f R\varpi_{mn}^2 \sum_{n=1}^{N} A_{mn}\left\{\left[\frac{1}{m} D_{0n}\tilde{D}_{0n} + \sum_{k=1}^{\infty}\frac{I_m(k\pi R/h) D_{kn}\tilde{D}_{kn}}{(k\pi R/h) I'_m(k\pi R/h)}\right]\right.$$

$$\left. - \delta_{0m}\left[\sum_{k=0}^{\infty} E_{k0}(-1)^k D_{kn}\right]\frac{2}{n\pi}[1-(-1)^n]\tilde{E}_n - \sum_{j=1}^{\infty}\tilde{C}_{nj}\frac{\sum_{k=0}^{\infty}(-1)^k D_{kn}E_{kj}}{\cosh(\xi_{mj}h/R)} J_m(\xi_{mj})\right\}$$

$$+ \rho_f R\varpi_{mn}^2 \sum_{n=1}^{N} A_{mn} \sum_{j=1}^{\infty}\frac{\tilde{C}_{nj}\sum_{k=0}^{\infty}E_{kj}(-1)^k D_{kn}}{\cosh(\xi_{mj}h/R)} \frac{\left(\omega_{mj}^2/\varpi_{mn}^2\right) J_m(\xi_{mj})}{\left[1-\left(\omega_{mj}^2/\varpi_{mn}^2\right)\right]} \tag{8.136}$$

Chu and Gonzales (1964) included the effect of the mode shape distortion as given by equations (8.135) and (8.136) with $N = 10$ and with 50 terms of the first inner series in equation (8.19) to estimate the dependence of the breathing natural frequencies on the liquid depth ratio, h/l. The estimated and measured natural frequency ratio $\omega/\omega_{\text{empty}}$ as function of the liquid height–length ratio, h/l, is shown in Figures 8.12(a) and (b) for $n = 1$, $m = 4$, and $n = 2$, $m = 4$ modes, respectively, for shell wall thickness ratio $\hat{t}/R = 0.006\,06$ and slenderness ratio $R/l = 0.160\,87$. The absolute frequencies are compared with the experimental measurements for two modes in Figure 8.13. The agreement between the predicted and measured results is good except for the empty tank frequencies. This deviation was attributed to possible nonuniformity in the shell. In all cases, it appears that the effects of the free surface are negligible only when the shell frequency is higher than the first several uncoupled liquid sloshing frequencies.

8.5 Interaction with tank bottom and walls

The general case of hydrodynamic interaction with an elastic bottom and walls was considered by Bauer and Siekmann (1971a). Related to this problem are the hermetic cans composed of a circular cylindrical shell, joined to two circular end plates. Takahashi and Hirano (1970), Yuan and Dickinson (1992, 1994), and Huang and Soedel (1993) studied the free vibration of thin empty cans using different analytical approaches. Amabili (1997a,b,c, 2000a,b) studied the normal modes of fluid-filled cans and found that for the lower modes, the fluid pressure on the wetted surface is in phase with the structural acceleration, thus the fluid appears as an added mass. However, for symmetric modes, the fluid volume conservation plays an important role. Koval'chuk and Filin (2003b) considered the flexural vibrations of initially bent cylindrical shells.

Amabili and Garziera (2000, 2002) estimated the linear normal modes of circular cylindrical shells with nonuniform edge constraints using the Flügge theory of shells. They considered shells

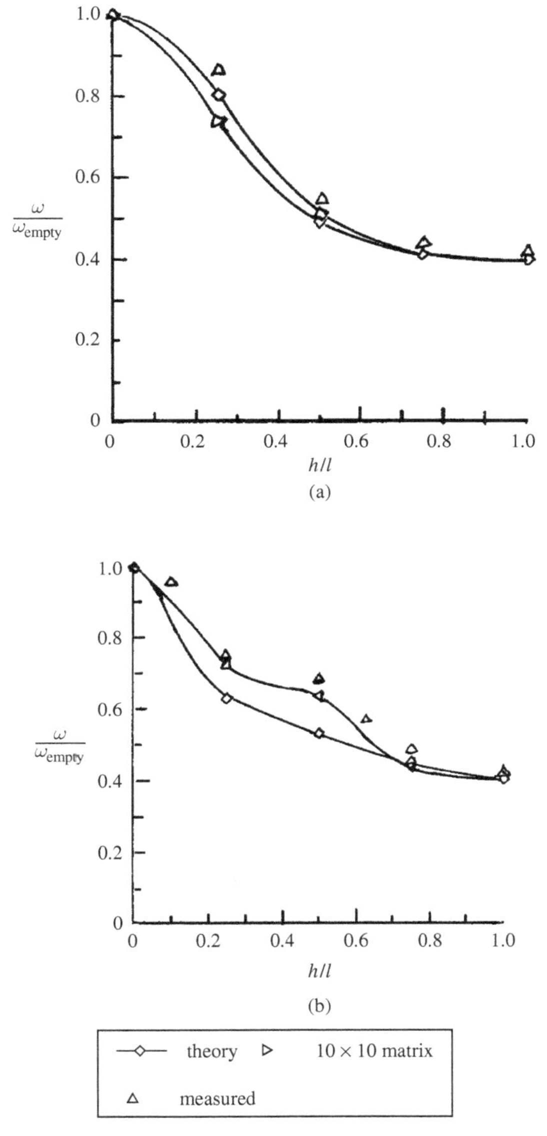

Figure 8.12 Estimated and measured natural frequency as function of fluid depth ratio of breathing mode $(m, n) =$ (a) 4.1, (b) 4.2. (Chu and Gonzales, 1964)

with either external or internal fluid flow or a combination. Their numerical results revealed that shells clamped at the upstream end and simply supported at the downstream end have a larger critical velocity than simply supported shells.

In addition to the basic assumptions that the liquid is inviscid and incompressible, a surface tension effect is introduced. With reference to Figure 8.14, the coordinate system r, θ, z is attached at the center of the container such that the plane r, θ coincides with the flat position of the undeformed bottom. The fluid field equations constitute the following boundary value problem:

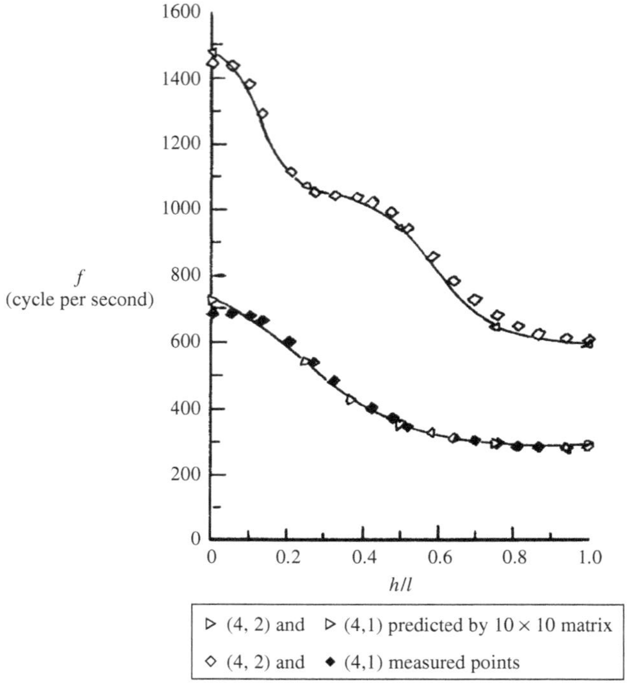

Figure 8.13 Comparison of predicted and measured breathing natural frequencies versus fluid depth. (Chu and Gonzales, 1964)

Continuity equation

$$\frac{\partial^2 \Phi}{\partial r^2} + \frac{1}{r}\frac{\partial \Phi}{\partial r} + \frac{1}{r^2}\frac{\partial^2 \Phi}{\partial \theta^2} + \frac{\partial^2 \Phi}{\partial z^2} = 0 \tag{8.137}$$

The combined free-surface condition including surface tension, σ, is

$$\frac{\partial^2 \Phi}{\partial t^2} + g\frac{\partial \Phi}{\partial z} + \frac{\sigma}{\rho_f}\frac{\partial^3 \Phi}{\partial z^3} = 0, \quad \text{at } z = h \tag{8.138}$$

At the wall, the fluid velocity must be the same as the radial velocity of the wall

$$\frac{\partial \Phi}{\partial r} = \frac{\partial u}{\partial t}, \quad \text{at } r = R \tag{8.139}$$

At the tank bottom, the compatibility condition gives

$$\frac{\partial \Phi}{\partial z} = \frac{\partial w_0}{\partial t}, \quad \text{at } z = 0 \tag{8.140}$$

where $w_0(r, v, t)$ is the transverse deflection of the bottom taken positive upward. The above equations are coupled with the elastic container equations. The analysis considers the two cases of elastic membrane bottom and elastic plate bottom.

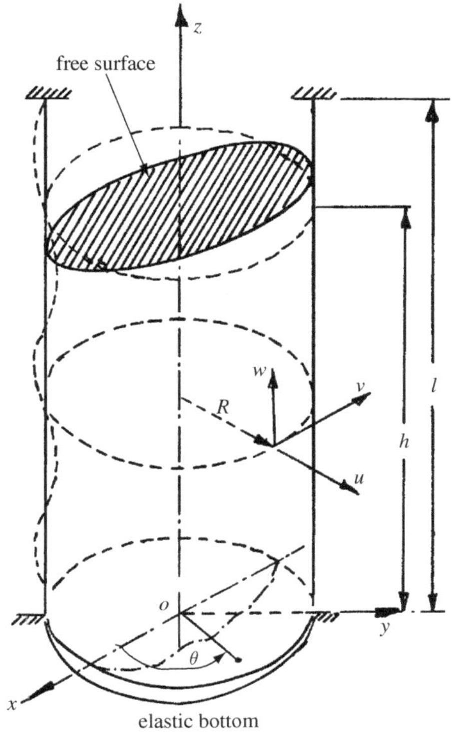

Figure 8.14 Elastic shell with elastic bottom partially filled with liquid showing the coordinate frame.

8.5.1 Shell with membrane bottom

The equations of motion of a stretched membrane bottom together with its boundary condition at $r = R$ are

$$\frac{\partial^2 w_0}{\partial r^2} + \frac{1}{r}\frac{\partial w_0}{\partial r} + \frac{1}{r^2}\frac{\partial^2 w_0}{\partial \theta^2} - \frac{\mu}{\tau}\frac{\partial^2 w_0}{\partial t^2} = -\frac{\rho_f}{\tau}\frac{\partial \Phi}{\partial t}\bigg|_{z=0} \tag{8.141a}$$

$$w_0(R, \vartheta, t) = 0 \tag{8.141b}$$

where μ is the mass per unit area of the membrane, τ is the tensile force per unit length of the membrane, ρ_f is the fluid density, and Φ is the fluid velocity potential function.

The shell wall equations of motion according to Donnell's approximation are based on the assumption that the changes in curvature and twist of the cylinder are the same as those of the flat plate. Furthermore, the effect of the transverse shear stress on the equilibrium of forces in the circumferential direction is negligible.

The equation of motion of the cylindrical wall in the radial direction, u, is

$$\begin{aligned} &\frac{\nu}{R}\frac{\partial w}{\partial z} + \frac{1}{R^2}\frac{\partial v}{\partial \theta} + \frac{u}{R^2} + \frac{\hat{t}^2}{12}\nabla^4 u + \frac{(1-\nu^2)\rho_w}{E}\frac{\partial^2 u}{\partial t^2} \\ &= \begin{cases} -((1-\nu^2)\rho_w/E\hat{t})(\partial\Phi/\partial t)\big|_{r=R} & \text{for } 0 \le z \le h \\ 0 & \text{for } h \le z \le 1 \end{cases} \end{aligned} \tag{8.142a}$$

where $\nabla^2 = \frac{\partial^2}{\partial z^2} + \frac{1}{R^2}\frac{\partial^2}{\partial \theta^2}$.

Equation of motion of the shell in the circumferential direction, v, is

$$\frac{(1+\nu)}{2R}\frac{\partial^2 w}{\partial z \partial \theta} + \frac{(1-\nu)}{2}\frac{\partial^2 v}{\partial z^2} + \frac{1}{R^2}\frac{\partial^2 v}{\partial \theta^2} + \frac{1}{R^2}\frac{\partial u}{\partial \theta} - \frac{(1-\nu^2)\rho_c}{E}\frac{\partial^2 v}{\partial t^2} = 0 \tag{8.142b}$$

The equation of motion of the shell in axial direction, w, is

$$\frac{\partial^2 w}{\partial z^2} + \frac{(1-\nu)}{2R^2}\frac{\partial^2 w}{\partial \theta^2} + \frac{(1+\nu)}{2R}\frac{\partial^2 v}{\partial z \partial \theta} + \frac{\nu}{R}\frac{\partial u}{\partial z} - \frac{(1-\nu^2)\rho_c}{E}\frac{\partial^2 w}{\partial t^2} = 0 \tag{8.142c}$$

These three equations are subject to the boundary conditions. For example, for clamped edges we have

$$u(\theta, 0, t) = 0, \quad \left.\frac{\partial u}{\partial z}\right|_{z=0} = 0, \quad u(\theta, l, t) = 0, \quad \left.\frac{\partial u}{\partial z}\right|_{z=l} = 0 \tag{8.143a–d}$$

$$v(\theta, 0, t) = 0, \quad v(\theta, l, t) = 0, w(\theta, 0, t) = 0, \quad w(\theta, l, t) = 0 \tag{8.144a–d}$$

These conditions are realized when rigid end rings are introduced in the shell. At the location of the free surface, $z = h$, the moments about the z-axis, the shear stresses in the transverse direction, the shear forces, the displacements and slopes in the radial, circumferential, and longitudinal directions, in the two domains stated in equation (8.142a), are equal. It is seen that the governing equations of liquid, elastic bottom, and shell wall are coupled. The solution of the governing equations must satisfy the respective boundary conditions. Bauer and Siekmann (1971a) introduced the following solution for the velocity potential function

$$\Phi(r, \theta, z, t) = e^{i\omega t}\left\{\phi_{00}^b + \sum_{n=1}^{\infty}\phi_{0n}^b + \sum_{m=1}^{\infty}\sum_{n=1}^{\infty}\phi_{mn}^b \cos m\theta + \sum_{n=1}^{\infty}\phi_{0n}^s + \sum_{m=1}^{\infty}\sum_{n=1}^{\infty}\phi_{mn}^s \cos m\theta\right\} \tag{8.144}$$

where superscripts b and s denote bottom and shell, respectively, and ω is the natural frequency of the coupled liquid–shell system to be determined. The index m denotes the circumferential wave number. Terms with subscript zero refer to potentials of the axisymmetric components. Any solution of equation (8.137) must satisfy the free surface condition (8.138), and the boundary conditions

$$\left.\frac{\partial \Phi^b}{\partial r}\right|_{r=R} = 0, \quad \left.\frac{\partial \Phi^s}{\partial z}\right|_{z=0} = 0 \tag{8.145a,b}$$

The following solutions are given by Bauer and Siekmann (1971a)

$$\phi_{00}^b = B_{00}\left(z + \frac{g}{\omega^2} - h\right) \tag{8.146a}$$

$$\phi_{0n}^b = B_{0n} J_0\left(\varepsilon_{0n}\frac{r}{R}\right)\left[\cosh\left(\varepsilon_{0n}\frac{z}{R}\right) - \frac{\left[\omega^2 - (\varepsilon_{0n}/R)\left(g + (\sigma/\rho)(\varepsilon_{0n}{}^2/R^2)\right)\tanh(\varepsilon_{0n}(h/R))\right]}{\left[\omega^2 \tanh(\varepsilon_{0n}(h/R)) - (\varepsilon_{0n}/R)(g + (\sigma/\rho)(\varepsilon_{0n}{}^2/R^2))\right]} \times \sinh\left(\varepsilon_{0n}\frac{z}{R}\right)\right] \tag{8.146b}$$

$$\phi_{mn}^b = B_{mn} J_m\left(\varepsilon_{mn}\frac{r}{R}\right)\left[\cosh\left(\varepsilon_{mn}\frac{z}{R}\right) - \frac{\left[\omega^2 - (\varepsilon_{mn}/R)\left(g + (\sigma/\rho)(\varepsilon_{mn}{}^2/R^2)\right)\tanh(\varepsilon_{mn}(h/R))\right]}{\left[\omega^2 \tanh(\varepsilon_{mn}(h/R)) - (\varepsilon_{mn}/R)(g + (\sigma/\rho)(\varepsilon_{mn}{}^2/R^2))\right]} \times \sinh\left(\varepsilon_{mn}\frac{z}{R}\right)\right] \tag{8.146c}$$

$$\phi_{0n}^s = A_{0n} I_0\left(\lambda_{0n}\frac{r}{R}\right)\cosh\left(\lambda_{0n}\frac{z}{R}\right) \tag{8.146d}$$

$$\phi_{mn}^s = A_{mn} I_m\left(\lambda_{mn}\frac{r}{R}\right)\cosh\left(\lambda_{mn}\frac{z}{R}\right) \tag{8.146e}$$

where ε_{mn} are the roots of $\frac{d}{dr} J_m(\varepsilon_{mn}(r/R))\big|_{r=R} = 0$, and λ_{mn} are the roots of the equation

$$\omega^2 = -\frac{\lambda_{mn}}{R}\left[g - \frac{\sigma}{\rho}\frac{\lambda_{mn}^2}{R^2}\right]\tanh\left(\lambda_{mn}\frac{h}{R}\right), \quad m = 0,\ 1,\ 2, \ldots, n = 1,\ 2,\ 3, \ldots \tag{8.147}$$

Equation (8.147) is obtained from the free-surface condition (8.138), J_m and I_m are Bessel functions and modified Bessel functions of the first kind, respectively, of order m.

The elastic deformation of the bottom can be expressed by the following expansion

$$w_0(r,\theta,t) = \left[\bar{W}_{00}(r) + \sum_{n=1}^{\infty} \bar{W}_{0n}(r) + \sum_{m=1}^{\infty}\sum_{n=1}^{\infty} \bar{W}_{mn}(r)\cos m\theta\right] e^{i\omega t} \tag{8.148}$$

The elastic shell displacements can also be expressed in the form

$$u(z,\theta,t) = \left[U_{00}(z) + \sum_{n=1}^{\infty} U_{0n}(z) + \sum_{m=1}^{\infty}\sum_{n=1}^{\infty} U_{mn}(z)\cos m\theta\right] e^{i\omega t} \tag{8.149a}$$

$$v(z,\theta,t) = e^{i\omega t}\sum_{m=1}^{\infty}\sum_{n=1}^{\infty} V_{mn}(z)\sin m\theta \tag{8.149b}$$

$$w(z,\theta,t) = \left[W_{00}(z) + \sum_{n=1}^{\infty} W_{0n}(z) + \sum_{m=1}^{\infty}\sum_{n=1}^{\infty} W_{mn}(z)\cos m\theta\right] e^{i\omega t} \tag{8.149c}$$

Substituting equations (8.148) and (8.144) into equation (8.141a) gives the following set of ordinary differential equations for the membrane bottom

$$\frac{d^2\bar{W}_{00}}{dr^2}+\frac{1}{r}\frac{d\bar{W}_{00}}{rdr}+\frac{1}{r^2}\frac{d^2\bar{W}_{00}}{d\theta^2}+\frac{\mu\omega^2}{\mathcal{T}}\bar{W}_{00}=-i\frac{\rho_f\omega}{\mathcal{T}}B_{00}\left(\frac{g}{\omega^2}-h\right) \tag{8.150a}$$

$$\frac{d^2\bar{W}_{0n}}{dr^2}+\frac{1}{r}\frac{d\bar{W}_{0n}}{dr}+\frac{\mu\omega^2}{\mathcal{T}}\bar{W}_{0n}=-i\frac{\rho_f\omega}{\mathcal{T}}\left\{A_{0n}I_0\left(\lambda_{0n}\frac{r}{R}\right)+B_{0n}J_0\left(\varepsilon_{0n}\frac{r}{R}\right)\right\} \tag{8.150b}$$

$$\frac{d^2\bar{W}_{mn}}{dr^2}+\frac{1}{r}\frac{d\bar{W}_{mn}}{dr}+\left(\frac{\mu\omega^2}{\mathcal{T}}-\frac{m^2}{r^2}\right)\bar{W}_{mn}=-i\frac{\rho_f\omega}{\mathcal{T}}\left\{A_{mn}I_m\left(\lambda_{mn}\frac{r}{R}\right)+B_{mn}J_m\left(\varepsilon_{mn}\frac{r}{R}\right)\right\} \tag{8.150c}$$

for which $m, n = 1, 2, 3, \ldots$ and the boundary conditions

$$\bar{W}_{00}(R)=0,\;\; \bar{W}_{0n}(R)=0,\;\; \bar{W}_{mn}(R)=0 \tag{8.151a,b,c}$$

For the shell elastic wall, equations (8.149) and (8.144) are substituted into equations (8.142) to obtain the following set of coupled ordinary differential equations

$$\frac{\nu}{R}\frac{dW_{00}}{dz}+\left[\frac{\hat{t}^2}{12}\frac{d^4}{dz^4}+\frac{1}{R^2}-\omega^2\frac{(1-\nu^2)\rho_w}{E}\right]U_{00}$$
$$=\begin{cases}-i\omega((1-\nu^2)\rho_w/E\hat{t})B_{00}(z+(g/\omega^2)-h) & \text{for } 0\le z\le h\\ 0 & \text{for } h\le z\le 1\end{cases} \tag{8.152 a}$$

$$\left(\frac{d^2}{dz^2}+\omega^2\frac{(1-\nu^2)\rho_w}{E}\right)W_{00}+\frac{\nu}{R}\frac{dU_{00}}{dz}=0 \tag{8.152b}$$

$$\frac{\nu}{R}\frac{dW_{0n}}{dz}+\left[\frac{\hat{t}^2}{12}\frac{d^4}{dz^4}+\frac{1}{R^2}-\omega^2\frac{(1-\nu^2)\rho_w}{E}\right]U_{0n}$$
$$=\begin{cases}P_n & \text{for } 0\le z\le h\\ 0 & \text{for } h\le z\le 1\end{cases} \tag{8.152c}$$

$$\left(\frac{d^2}{dz^2}+\omega^2\frac{(1-\nu^2)\rho_c}{E}\right)W_{0n}+\frac{\nu}{R}\frac{dU_{0n}}{dz}=0 \tag{8.152d}$$

$$\left[\frac{\hat{t}^2}{12}\left(\frac{d^4}{dz^4}-\frac{2m^2}{R^2}\frac{d^2}{dz^2}+\frac{m^4}{R^4}\right)+\frac{1}{R^2}-\omega^2\frac{(1-\nu^2)\rho_w}{E}\right]U_{mn}+\frac{\nu}{R}\frac{dW_{mn}}{dz}+\frac{m}{R^2}V_{mn}$$
$$=\begin{cases}P_{mn} & \text{for } 0\le z\le h\\ 0 & \text{for } h\le z\le 1\end{cases} \tag{8.152e}$$

$$\left[\frac{(1-\nu)}{2}\frac{d^2}{dz^2}+\omega^2\frac{(1-\nu^2)\rho_c}{E}-\frac{m^2}{R^2}\right]V_{mn}-\frac{m(1+\nu)}{2R}\frac{dW_{mn}}{dz}-\frac{m}{R^2}U_{mn}=0 \tag{8.152f}$$

$$\left[\frac{d^2}{dz^2}+\omega^2\frac{(1-\nu^2)\rho_c}{E}-\frac{m^2(1-\nu)}{2R^2}\right]W_{mn}+\frac{\nu}{R}\frac{dU_{mn}}{dz}+\frac{m(1+\nu)}{2R}\frac{dV_{mn}}{dz}=0 \tag{8.152g}$$

where

$$P_n=-i\omega\frac{\rho_f(1-\nu^2)}{E\hat{t}}\left\{A_{0n}I_0(\lambda_{0n})\cos\left(\lambda_{0n}\frac{z}{R}\right)+B_{0n}J_0(\varepsilon_{0n})\right.$$
$$\times\left[\cosh\left(\varepsilon_{0n}\frac{z}{R}\right)-\frac{\left[\omega^2-\frac{\varepsilon_{0n}}{R}\left(g+\frac{\sigma}{\rho}\frac{\varepsilon_{0n}^2}{R^2}\right)\tanh\left(\varepsilon_{0n}\frac{h}{R}\right)\right]}{\left[\omega^2\tanh\left(\varepsilon_{0n}\frac{h}{R}\right)-\frac{\varepsilon_{0n}}{R}\left(g+\frac{\sigma}{\rho}\frac{\varepsilon_{0n}^2}{R^2}\right)\right]}\right.$$
$$\left.\left.\times\sinh\left(\varepsilon_{0n}\frac{z}{R}\right)\right]\right\} \tag{8.153a}$$

$$P_{mn}=-i\omega\frac{\rho_f(1-\nu^2)}{E\hat{t}}\left\{A_{mn}I_m(\lambda_{mn})\cos\left(\lambda_{mn}\frac{z}{R}\right)+B_{mn}J_m(\varepsilon_{mn})\right.$$
$$\times\left[\cosh\left(\varepsilon_{mn}\frac{z}{R}\right)-\frac{\left[\omega^2-\frac{\varepsilon_{mn}}{R}\left(g+\frac{\sigma}{\rho}\frac{\varepsilon_{mn}^2}{R^2}\right)\tanh\left(\varepsilon_{mn}\frac{h}{R}\right)\right]}{\left[\omega^2\tanh\left(\varepsilon_{mn}\frac{h}{R}\right)-\frac{\varepsilon_{mn}}{R}\left(g+\frac{\sigma}{\rho}\frac{\varepsilon_{mn}^2}{R^2}\right)\right]}\right.$$
$$\left.\left.\times\sinh\left(\varepsilon_{mn}\frac{z}{R}\right)\right]\right\} \tag{8.153b}$$

The corresponding boundary conditions of equations (8.152) are

$$U_{00}(0)=\frac{d}{dz}U_{00}(z)\bigg|_{z=0}=W_{00}(0)=0 \tag{8.154a}$$

$$U_{0n}(0)=\frac{d}{dz}U_{0n}(z)\bigg|_{z=0}=W_{0n}(0)=0 \tag{8.154b}$$

$$U_{00}(l)=\frac{d}{dz}U_{00}(z)\bigg|_{z=l}=W_{00}(l)=0 \tag{8.154c}$$

$$U_{0n}(l)=\frac{d}{dz}U_{0n}(z)\bigg|_{z=l}=W_{0n}(l)=0 \tag{8.154d}$$

$$U_{mn}(0) = \frac{d}{dz}U_{mn}(z)\bigg|_{z=0} = V_{mn}(0) = W_{mn}(0) = 0 \tag{8.154e}$$

$$U_{mn}(l) = \frac{d}{dz}U_{mn}(z)\bigg|_{z=l} = V_{mn}(l) = W_{mn}(l) = 0 \tag{8.154f}$$

The displacements and their slopes of the shell wall below and above the free surface, $z = h$, must be equal.

The solutions of equations (8.150) subject to the boundary conditions (8.151) are:

$$\bar{W}_{00}(r) = i\frac{\rho_f}{\mu\omega}B_{00}\left(\frac{g}{\omega^2} - h\right)\left[\frac{J_0\left(\omega r\sqrt{\mu/\tau}\right)}{J_0\left(\omega R\sqrt{\mu/\tau}\right)} - 1\right] \tag{8.155a}$$

$$\begin{aligned}\bar{W}_{0n}(r) = {} & i\frac{\rho_f\omega}{\tau}\left\{\frac{A_{0n}I_0(\lambda_{0n})}{((\mu\omega^2/\tau) + (\lambda_{0n}^2/R^2))} + \frac{B_{0n}J_0(\varepsilon_{0n})}{((\mu\omega^2/\tau) - (\varepsilon_{0n}^2/R^2))}\right\}\frac{J_0\left(\omega r\sqrt{\mu/\tau}\right)}{J_0\left(\omega R\sqrt{\mu/\tau}\right)} \\ & - i\frac{\rho_f\omega}{\tau}\left\{\frac{A_{0n}I_0(\lambda_{0n}r/R)}{((\mu\omega^2/\tau) + (\lambda_{0n}^2/R^2))} + \frac{B_{0n}J_0(\varepsilon_{0n}r/R)}{((\mu\omega^2/\tau) - (\varepsilon_{0n}^2/R^-2))}\right\}\end{aligned} \tag{8.155b}$$

$$\begin{aligned}\bar{W}_{mn}(r) = {} & i\frac{\rho_f\omega}{\tau}\left[\frac{A_{mn}I_m(\lambda_{mn})}{((\mu\omega^2/\tau) + (\lambda_{0n}^2/R^2))} + \frac{B_{mn}J_m(\varepsilon_{mn})}{((\mu\omega^2/\tau) - (\varepsilon_{mn}^2/R^2))}\right]\frac{J_m\left(\omega r\sqrt{\mu/\tau}\right)}{J_m\left(\omega R\sqrt{\mu/\tau}\right)} \\ & - i\frac{\rho_f\omega}{\tau}\left[\frac{A_{mn}I_m(\lambda_{mn}r/R)}{((\mu\omega^2/\tau) + (\lambda_{mn}^2/R^2))} + \frac{B_{mn}J_m(\varepsilon_{mn}r/R)}{((\mu\omega^2/\tau) - (\varepsilon_{mn}^2/R^2))}\right]\end{aligned} \tag{8.155c}$$

The solutions of the shell equations (8.155) subject to the respective boundary conditions (8.154) are

$$U_{00} = \sum_{i=1}^{6}\begin{Bmatrix}A_{00i}\\ \tilde{A}_{00i}\end{Bmatrix}e^{\Lambda_i z} + \begin{Bmatrix}\alpha_{00}z + \beta_{00}\\ 0\end{Bmatrix}\quad\begin{matrix}\text{for } 0 \le z \le h\\ \text{for } h \le z \le l\end{matrix} \tag{8.156a}$$

$$W_{00} = \sum_{i=1}^{6}\begin{Bmatrix}C_{00i}\\ \tilde{C}_{00i}\end{Bmatrix}e^{\Lambda_i z} + \begin{Bmatrix}\gamma_{00}\\ 0\end{Bmatrix}\quad\begin{matrix}\text{for } 0 \le z \le h\\ \text{for } h \le z \le l\end{matrix} \tag{8.156b}$$

$$\begin{aligned}U_{0n} = {} & \sum_{i=1}^{6}\begin{Bmatrix}A_{0ni}\\ \tilde{A}_{0ni}\end{Bmatrix}e^{\Lambda_i z} \\ & + \begin{Bmatrix}\alpha_{0n}\cos\left(\lambda_{0n}\frac{z}{R}\right) + \gamma_{0n}\cosh\left(\varepsilon_{0n}\frac{z}{R}\right) + \delta_{0n}\sinh\left(\varepsilon_{0n}\frac{z}{R}\right)\\ 0\end{Bmatrix}\quad\begin{matrix}0 \le z \le h\\ h \le z \le l\end{matrix}\end{aligned} \tag{8.156c}$$

$$W_{0n} = \sum_{i=1}^{6} \begin{Bmatrix} C_{0ni} \\ \tilde{C}_{0ni} \end{Bmatrix} e^{\Lambda_i z} + \begin{cases} \overline{\beta}_{0n} \cos(\lambda_{0n}(z/R)) + \overline{\gamma}_{0n} \cosh(\varepsilon_{0n}(z/R)) + \overline{\delta}_{0n} \sinh(\varepsilon_{0n}(z/R)) & 0 \le z \le h \\ 0 & h \le z \le l \end{cases} \tag{8.156d}$$

$$U_{mn} = \sum_{i=1}^{6} \begin{Bmatrix} A_{mni} \\ \tilde{A}_{mni} \end{Bmatrix} e^{\Lambda_i^* z} + \begin{cases} \alpha_{mn} \cos(\lambda_{mn}(z/R)) + \gamma_{mn} \cosh(\varepsilon_{mn}(z/R)) + \delta_{mn} \sinh(\varepsilon_{mn}(z/R)) & 0 \le z \le h \\ 0 & h \le z \le l \end{cases} \tag{8.156e}$$

$$V_{mn} = \sum_{i=1}^{6} \begin{Bmatrix} A_{mni} \\ \tilde{A}_{mni} \end{Bmatrix} \xi e^{\Lambda_i^* z} + \begin{cases} \alpha_{mn} \cos(\lambda_{mn}(z/R)) + \gamma''_{mn} \cosh(\varepsilon_{mn}(z/R)) + \delta''_{mn} \sinh(\varepsilon_{mn}(z/R)) & 0 \le z \le h \\ 0 & h \le z \le l \end{cases} \tag{8.156f}$$

$$W_{mn} = \sum_{i=1}^{6} \begin{Bmatrix} A_{mni} \\ \tilde{A}_{mni} \end{Bmatrix} \eta i e^{\Lambda_i^* z} + \begin{cases} \beta_{mn} \cos(\lambda_{mn}(z/R)) + \gamma_{mn} \cosh(\varepsilon_{mn}(z/R)) + \delta_{mn} \sinh(\varepsilon_{mn}(z/R)) & 0 \le z \le h \\ 0 & h \le z \le l \end{cases} \tag{8.156g}$$

where

$$\begin{Bmatrix} C_{00i} \\ \tilde{C}_{00i} \end{Bmatrix} = -\frac{\nu}{R} \frac{\Lambda_i}{\Lambda_i^2 + \omega^2(1-\nu^2)(\rho/E)} \begin{Bmatrix} A_{00i} \\ \tilde{A}_{00i} \end{Bmatrix}, \quad i = 1,\ 2, \ldots, 6 \tag{8.157a}$$

$$\alpha_{00} = -i \frac{\rho_f \omega}{E\hat{t}} \frac{(1-\nu^2) B_{00}}{(1/R^2) - \omega^2(1-\nu^2)(\rho/E)} \tag{8.157b}$$

$$\beta_{00} = i \frac{\rho_f \omega}{E\hat{t}} \frac{(h + (g/\omega^2)) B_{00}}{(1/R^2) - \omega^2(1-\nu^2)(\rho/E)} \tag{8.157c}$$

$$\gamma_{00} = -\frac{\nu}{R}\frac{\alpha_{00}}{\omega^2(1-\nu^2)(\rho/E)} \tag{8.157d}$$

for which Λ_i is given by the roots of the cubic equation

$$\Lambda^6 + \omega^2(1-\nu^2)\frac{\rho}{E}\Lambda^4 + \frac{12}{\tilde{t}^2}\left[\frac{1}{R^2} - \omega^2(1-\nu^2)\frac{\rho}{E} - \left(\frac{\nu}{R}\right)^2\right]\Lambda^2$$

$$+\frac{12}{\tilde{t}^2}\omega^2(1-\nu^2)\frac{\rho}{E}\left[\frac{1}{R^2} - \omega^2(1-\nu^2)\frac{\rho}{E}\right] = 0 \tag{8.158}$$

for which Λ_i^*, $i = 1, 2, \ldots, 8$ is given by the roots of a biquartic equation. The constants A_{00i}, A_{0ni}, $\mathrm{A}_{mni}\tilde{\mathrm{A}}_{00i}$, $\tilde{\mathrm{A}}_{0ni}$, and $\tilde{\mathrm{A}}_{mni}$, $i = 1, 2, \ldots, 6$, are determined from the system non-homogeneous equations. Upon substituting the above results into the compatibility relations (8.139) and (8.140) the following equations are obtained

$$\sum_{n=1}^{\infty}\left.\frac{\partial\Phi_{0n}^s}{\partial r}\right|_{r=R} + \sum_{m=1}^{\infty}\sum_{n=1}^{\infty}\left.\frac{\partial\Phi_{mn}^s}{\partial r}\right|_{r=R}\cos m\theta$$

$$= i\omega\left[U_{00}(z) + \sum_{m=1}^{\infty}U_{0n}(z) + \sum_{m=1}^{\infty}\sum_{n=1}^{\infty}U_{mn}(z)\cos m\theta\right] \tag{8.159}$$

$$B_{00} + \sum_{n=1}^{\infty}\left.\frac{\partial\Phi_{0n}^b}{\partial z}\right|_{z=0} + \sum_{m=1}^{\infty}\sum_{n=1}^{\infty}\left.\frac{\partial\Phi_{mn}^b}{\partial z}\right|_{z=0}\cos m\theta$$

$$= i\omega\left[\bar{W}_{00}(r) + \sum_{n=1}^{\infty}W_{0n}(r) + \sum_{m=1}^{\infty}\sum_{n=1}^{\infty}\bar{W}_{mn}(r)\cos m\theta\right] \tag{8.160}$$

Equations (8.159) and (8.160) constitute an infinite number of algebraic equations and for a nontrivial solution the coefficient determinant must vanish. The roots of the resulting determinant give the coupled frequencies of the liquid–structure system. For the axisymmetric case, the system of equations is independent of the coordinate θ, that is, terms with $\cos m\theta$ will vanish from equations (8.159) and (8.160). In this case, equation (8.160) takes the form

$$\sum_{n=1}^{\infty}A_{0n}\frac{\lambda_{0n}}{R}I_0(\lambda_{0n})\cos\left(\lambda_{0n}\frac{z}{R}\right) = i\omega\left[\sum_{i=1}^{6}A_{00i}\,e^{\Lambda_i z} + \alpha_{00}z + \beta_{00}\right.$$

$$\left. + \sum_{n=1}^{\infty}\sum_{i=1}^{6}A_{0ni}\,e^{\Lambda_i z} + \alpha_{0n}\cos\left(\lambda_{0n}\frac{z}{R}\right) + \gamma_{0n}\cosh\left(\varepsilon_{0n}\frac{z}{R}\right) + \delta_{0n}\sinh\left(\varepsilon_{0n}\frac{z}{R}\right)\right] \tag{8.161}$$

Equation (8.160) takes the form

$$B_{00} - \sum_{n=1}^{\infty} B_{0n} \frac{(\omega^2 - \varpi_{0n}^2)(\varepsilon_{0n}/R)\tanh(\varepsilon_{0n}h/R)}{[\omega^2 \tanh^2(\varepsilon_{0n}h/R) - \varpi_{0n}^2]} J_0(\varepsilon_{0n}r/R)$$

$$= -\rho_f\omega \left[\frac{B_{00}}{\mu\omega}\left(\frac{g}{\omega^2} - h\right)\left\{ \frac{J_0\left(\omega r\sqrt{\mu/\tau}\right)}{J_0\left(\omega R\sqrt{\mu/\tau}\right)} - 1 \right\} \right.$$

$$+ \frac{\omega}{\tau}\sum_{n=1}^{\infty}\left\{ \left[\frac{A_{0n}I_0(\lambda_{0n})}{((\mu\omega^2/\tau) + (\lambda_{0n}^2/R^2))} + \frac{B_{0n}J_0(\varepsilon_{0n})}{((\mu\omega^2/\tau) + (\varepsilon_{0n}^2/R^2))} \right] \frac{J_0(\omega r\sqrt{\mu/\tau})}{J_0(\omega R\sqrt{\mu/\tau})} \right.$$

$$\left.\left. - \left[\frac{A_{0n}I_0(\lambda_{0n}r/R)}{((\mu\omega^2/\tau) + (\lambda_{0n}^2/R^2))} + \frac{B_{0n}J_0(\varepsilon_{0n}r/R)}{((\mu\omega^2/\tau) + (\varepsilon_{0n}^2/R^2))} \right] \right\} \right] \tag{8.162}$$

where

$$\varpi_{0n}^2 = \frac{\varepsilon_{0n}}{R}\left(g + \frac{\sigma\varepsilon_{0n}^2}{\rho_f R^2}\right)\tanh(\varepsilon_{0n}h/R) \tag{8.163}$$

This expression represents the square of the natural axisymmetric frequency of the liquid in a rigid circular tank including the effect of surface tension. The coefficients A_{00i}, α_{00}, and β_{00} depend on ω and B_{00}, while γ_{0n} and δ_{0n} depend on ω and B_{0n}. The coefficient A_{0ni} depends on ω and A_{0n}, while α_{00} depends on ω and A_{0n}. Similarly, Λ_i and χ depend on ω. For appropriate comparison of the coefficients, the functions of z on the right-hand side of equation (8.161) must be expanded into Fourier cosine series. Similarly, the functions of r on the right-hand side of equation (8.162) must be expanded into Bessel–Fourier series of $J_0(\varepsilon_{0n}r/R)$. Collecting the corresponding terms B_{00}, A_{0n}, and B_{0n} in equations (8.161) and (8.162), and setting their coefficient determinant to zero, gives the frequency equation of the coupled axisymmetric liquid–structure system. The infinite frequency equation is then truncated to any order, and solved numerically.

8.5.2 Shell with elastic plate bottom

The same fluid field equations and shell equations of the membrane case are applied to the case of an elastic plate bottom. Considering only those equations belonging to the bottom as an elastic plate, the equations of motion of elastic bottom modeled as a clamped plate are

$$\left[\frac{\partial^2}{\partial r^2} + \frac{1}{r}\frac{\partial}{\partial r} + \frac{1}{r^2}\frac{\partial^2}{\partial\theta^2}\right]^2 w_0 + \frac{\rho_p\hat{t}_p}{D_p}\frac{\partial^2 w_0}{\partial t^2} = \frac{\rho_f}{D_p}\frac{\partial\Phi}{\partial t}\bigg|_{z=0} \tag{8.164a}$$

$$w_0(R,\theta,t) = 0, \quad \frac{\partial w_0}{\partial r}\bigg|_{r=R} = 0 \tag{8.164b,c}$$

where $D_p = \frac{E_p \hat{t}_p^3}{12(1-\nu^2)}$ is the flexural rigidity of the plate, $\hat{t}_p$ is the plate thickness, ν is Poisson's ratio, and E_p is Young's modulus of the plate. The elastic deformation of the bottom can also be expressed by equation (8.148).

Substituting equations (8.148) and (8.144) into equation (8.164a) gives the following set of ordinary differential equations

$$\left\{\left[\frac{d^2}{dr^2}+\frac{1}{r}\frac{d}{dr}\right]^2-\frac{\rho_p \hat{t}_p}{D_p}\omega^2\right\}\bar{W}_{00} = i\frac{\rho_f}{D_p}\omega^2 B_{00}\left(\frac{g}{\omega^2}-h\right) \tag{8.165a}$$

$$\left\{\left[\frac{d^2}{dr^2}+\frac{1}{r}\frac{d}{dr}\right]^2-\frac{\rho_p \hat{t}_p}{D_p}\omega^2\right\}\bar{W}_{0n} = i\frac{\rho_f}{D_p}\omega^2\left[A_{0n}I_0\left(\lambda_{0n}\frac{r}{R}\right)+B_{0n}J_0\left(\varepsilon_{0n}\frac{r}{R}\right)\right] \tag{8.165b}$$

$$\left\{\left[\frac{d^2}{dr^2}+\frac{1}{r}\frac{d}{dr}-\frac{m^2}{r^2}\right]^2-\frac{\rho_p \hat{t}_p}{D_p}\omega^2\right\}\bar{W}_{mn} = i\frac{\rho_f}{D_p}\omega^2\left[A_{mn}I_m\left(\lambda_{mn}\frac{r}{R}\right)+B_{mn}J_m\left(\varepsilon_{mn}\frac{r}{R}\right)\right] \tag{8.165c}$$

with the boundary conditions

$$\bar{W}_{00}(R)=0,\ \ \bar{W}_{0n}(R)=0,\ \ \bar{W}_{mn}(R)=0 \tag{8.166a,b,c}$$

$$\frac{d}{dr}\bar{W}_{00}(r)\bigg|_{r=R}=0,\ \ \frac{d}{dr}\bar{W}_{0n}(r)\bigg|_{r=R}=0,\ \ \frac{d}{dr}\bar{W}_{mn}(r)\bigg|_{r=R}=0 \tag{8.167a,b,c}$$

The solutions of equations (8.165) subject to the boundary conditions (8.166) and (8.167) are:

$$\bar{W}_{00}(r) = i\frac{\rho_f}{\omega\rho_p\hat{t}_p\Delta_0}B_{00}\left(\frac{g}{\omega^2}-h\right)[I_1(\chi R)J_0(\chi r)+I_0(\chi r)J_1(\chi R)-\Delta_0] \tag{8.168a}$$

$$\begin{aligned}\bar{W}_{0n}(r) = i\frac{\rho_f\omega}{D_p\Delta_0}\Bigg[&\left\{\frac{A_{0n}[I_0(\lambda_{0n})I_1(\chi R)-((\lambda_{0n}/\chi R))I_1(\lambda_{0n})I_0(\chi R)]}{(\chi^4-(\lambda_{0n}^4/R^4))}\right.\\
&\times\left.\frac{B_{0n}J_0(\varepsilon_{0n})I_1(\chi R)}{(\chi^4-(\varepsilon_{0n}^4/R^4))}\right\}J_0(\chi r)\\
&+\left\{\frac{A_{0n}[(\lambda_{0n}/\chi R)J_0(\chi R)I_1(\lambda_{0n})+J_1(\chi R)I_0(\lambda_{0n})]}{(\chi^4-(\lambda_{0n}^4/R^4))}\right.\\
&+\left.\frac{B_{0n}J_0(\varepsilon_{0n})J_1(\chi R)}{(\chi^4-(\varepsilon_{0n}^4/R^4))}\right\}I_0(\chi r)\\
&-\frac{A_{0n}\Delta_0 I_0(\lambda_{0n}r/R)}{(\chi^4-(\lambda_{0n}^4/R^4))}-\frac{B_{0n}\Delta_0 J_0(\varepsilon_{0n}r/R)}{(\chi^4-(\varepsilon_{0n}^4/R^4))}\Bigg]\end{aligned} \tag{8.168b}$$

$$\bar{W}_{mn}(r) = i\frac{\rho_f\omega}{D_p\Delta_m}\left[\left\{\frac{A_{mn}[I_m(\lambda_{mn})I'_m(\chi R) - (\lambda_{mn}/\chi R)I'_m(\lambda_{mn})I_m(\chi R)]}{(\chi^4 - (\lambda_{mn}^4/R^4))}\right.\right.$$
$$\left.\times\frac{B_{mn}J_m(\varepsilon_{mn})I'_m(\chi R)}{(\chi^4 - (\varepsilon_{mn}^4/R^4))}\right\}J_m(\chi r)$$
$$+\left\{\frac{A_{mn}[(\lambda_{mn}/\chi R)J_m(\chi R)I'_m(\lambda_{mn}) + J'_m(\chi R)I_m(\lambda_{mn})]}{(\chi^4 - (\lambda_{mn}^4/R^4))}\right.$$
$$\left.+\frac{B_{mn}J_m(\varepsilon_{mn})J'_m(\chi R)}{(\chi^4 - (\varepsilon_{mn}^4/R^4))}\right\}I_m(\chi r)$$
$$\left.-\frac{A_{mn}\Delta_m I_m(\lambda_{mn}r/R)}{(\chi^4 - (\lambda_{mn}^4/R^4))} - \frac{B_{mn}\Delta_m J_m(\varepsilon_{mn}r/R)}{(\chi^4 - (\varepsilon_{mn}^4/R^4))}\right] \tag{8.168c}$$

where $\Delta_0 = I_0(\chi R)I_1(\chi R) + J_0(\chi R)I_1(\chi R)$, $\Delta_m = J_m(\chi R)I_m(\chi R) + J_m(\chi R)I_m(\chi R)$, and $\chi^4 = \rho_p\hat{t}_p\omega^2/D_p$. Equation (8.160) takes the form

$$B_{00} - \sum_{n=1}^{\infty} B_{0n}\frac{(\omega^2 - \varpi_{0n}^2)(\varepsilon_{0n}/R)\tanh(\varepsilon_{0n}h/R)}{[\omega^2\tanh^2(\varepsilon_{0n}h/R) - \varpi_{0n}^2]}J_0(\varepsilon_{0n}r/R)$$
$$= -\rho_f\omega\left[\frac{B_{00}}{\omega\rho_p\hat{t}_p\Delta_0}\left(\frac{g}{\omega^2} - h\right)\{I_1(\chi R)J_0(\chi r) + J_1(\chi R)I_0(\chi r) - \Delta_0\}\right.$$
$$+\frac{\omega}{D_p\Delta_0}\sum_{n=1}^{\infty}\left[\left[\left\{\frac{A_{0n}\left[I_0(\lambda_{0n})I_1(\chi R) - \frac{\lambda_{0n}}{\chi R}I_1(\lambda_{0n})I_0(\chi R)\right]}{(\chi^4 - (\lambda_{0n}^4/R^4))}\right.\right.\right.$$
$$\left.+\frac{B_{0n}J_0(\varepsilon_{0n})I_1(\chi R)}{(\chi^4 - (\varepsilon_{0n}^4/R^4))}\right\}J_0(\chi r)$$
$$+\left\{\frac{A_{0n}\left[\frac{\lambda_{0n}}{\chi R}J_0(\chi R)I_1(\lambda_{0n}) + J_1(\chi R)I_0(\lambda_{0n})\right]}{(\chi^4 - (\lambda_{0n}^4/R^4))}\right.$$
$$\left.+\frac{B_{0n}J_0(\varepsilon_{0n})J_1(\chi R)}{(\chi^4 - (\varepsilon_{0n}^4/R^4))}\right\}I_0(\chi r)$$
$$\left.\left.-\frac{A_{0n}\Delta_0 I_0(\lambda_{0n}r/R)}{(\chi^4 - (\lambda_{0n}^4/R^4))} - \frac{B_{0n}\Delta_0 J_0(\varepsilon_{0n}r/R)}{(\chi^4 - (\varepsilon_{0n}^4/R^4))}\right]\right] \tag{8.169}$$

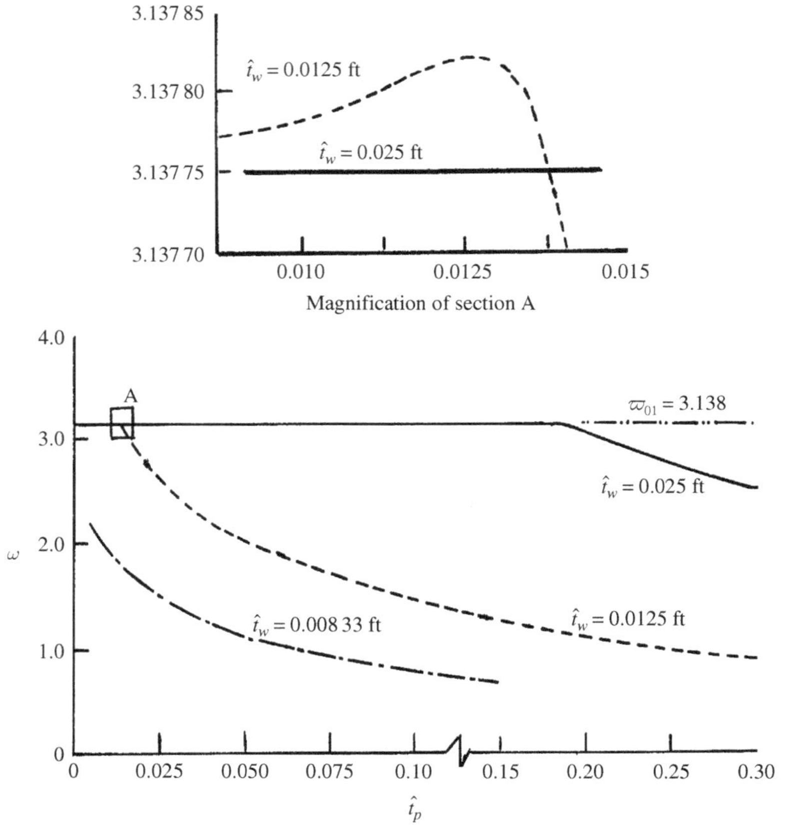

Figure 8.15 Dependence of coupled fundamental natural frequency on the bottom thickness for different values of shell wall thickness. (Bauer and Seikmann, 1971a, b)

The functions of r on the right-hand side of this equation are expanded into Bessel–Fourier series of $J_0(\varepsilon_{0n}r/R)$. Collecting the corresponding terms B_{00}, A_{0n}, and B_{0n} in equations (8.161) and (8.169) and setting their coefficient determinant to zero, gives the frequency equation of the coupled axisymmetric liquid–structure system. The infinite frequency equation is then truncated to any order, and solved numerically.

Bauer and Siekmann (1971a, b) numerically solved the truncated equations for the case of an aluminum container ($\rho g = 168$ lb/ft^3, $E = 14.4 \times 10^8$ lb/ft^2, $\nu = 1/3$) of diameter 24 ft and length 48 ft. The shell was clamped along the bottom edge and simply supported at the upper edge. The bottom was treated as a thin elastic plate. The container is partially filled with water, $\rho_f g = 65$ lb/ft^{3}for up to 6 ft depth. Figure 8.15 shows the dependence of the coupled frequency on the container bottom thickness $\hat{t}_p$ for different values of the shell wall thickness, $\hat{t}_w$. The straight line shown, ϖ_{01}, represents the fundamental axisymmetric natural frequency of the liquid in a rigid cylindrical tank. For thin walled cylindrical shells, the interaction between liquid and structure exhibits pronounced deviations from the frequency of the liquid in a rigid tank. With increasing wall thickness, the coupled frequency approaches the liquid frequency. As the bottom thickness increases, the coupled frequency approaches a fixed

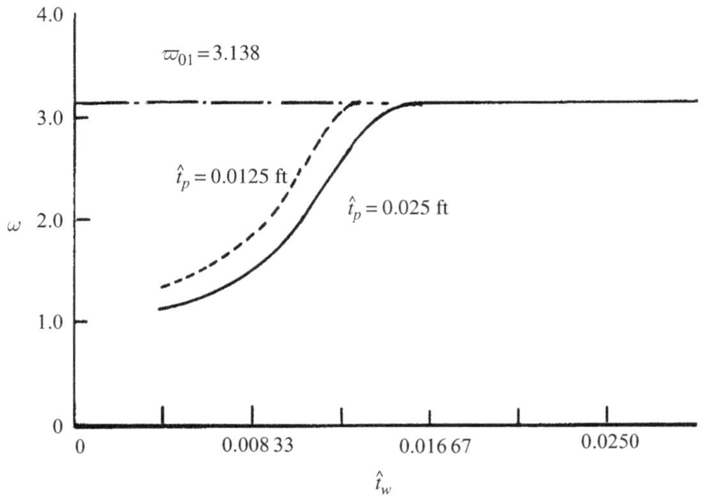

Figure 8.16 Dependence of coupled fundamental natural frequency on the wall thickness for two different values of bottom thickness. (Bauer and Seikmann, 1971a,b)

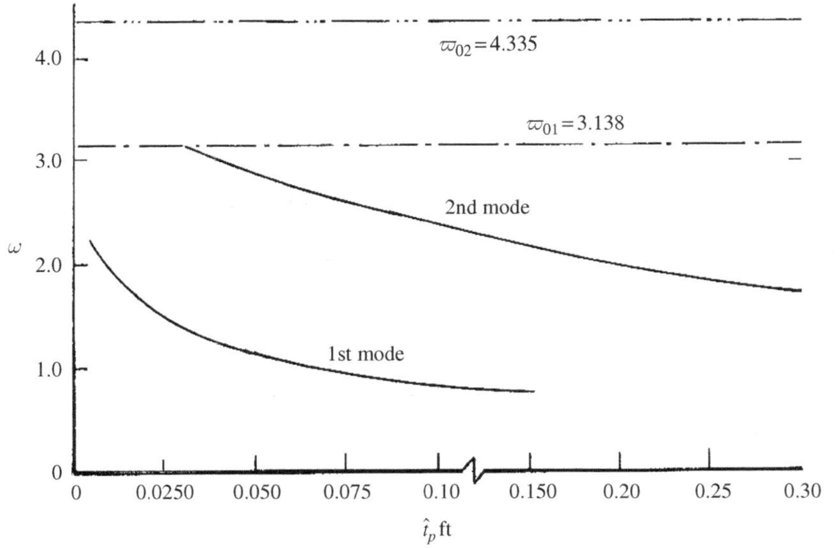

Figure 8.17 Dependence of the first two natural frequencies on the wall thickness for bottom thickness* $\hat{t}_p = 0.008\,33$ ft. (Bauer and Seikmann, 1971a, b)

value, which for thicker shell wall thickness results in values closer to ϖ_{01}. Generally, the bottom thickness results in a reduction in the coupled frequency for a given wall thickness.

Figure 8.16 shows the dependence of the coupled frequency on the container wall thickness for two different values of the bottom thickness, $\hat{t}_p$. As the wall thickness increases, the coupling effect is diminishing and the coupled frequency approaches the liquid frequency in a rigid tank. Figure 8.17 shows the dependence of the two lowest coupled frequencies on the bottom thickness for a tank wall thickness of 0.008 33 ft.

8.6 Closing remarks

The design of liquid containers involves the estimation of normal mode frequencies and mode shapes of the hydroelastic system. The coupling is significant when the elastic wetted surface experiences significant deformation that causes the fluid to be displaced. This deformation imposes different boundary conditions from those imposed for a rigid container. Different cases have been considered that include bending or breathing deformations. Special cases including (i) rigid tank walls with elastic bottom modeled as either a stretched membrane of elastic plate, (ii) rigid bottom with elastic walls experiencing either bending deformation or breathing vibrations, and (iii) interaction with completely elastic container are considered. The next chapter considers the nonlinear coupling problem that may arise due to the occurrence of nonlinear and parametric resonance conditions in the hydroelastic system.

The analysis presented in this chapter has been restricted to isotropic metal shells. However, other work has been reported in the literature pertaining to fiber-reinforced composite containers (Jain, 1974, Yamaki, *et al.*, 1984, and Shang and Lei, 1988). The modal analysis of liquid composite materials was based on approximate variational methods. Chen WQ and Ding (1999) determined the natural frequencies of fluid-filled isotropic cylindrical shells. Sharma, *et al.* (1996, 1998), Chen WQ, *et al.* (1997), Weiqiu, *et al.* (1997), and Zhang, *et al.* (2002) considered the modal analysis of multi-layered orthotropic fluid-filled cylindrical shells. They included transverse shear and rotary inertia. In addition, they used axial dependence of mode shapes in the form of a simple Fourier sine or cosine series. It was shown that the effect of the liquid depth on the natural frequency of the system is extremely significant for thin cylindrical containers and results in a reduction of the frequency parameter. Pal, *et al.* (2003) analyzed the coupled slosh dynamic problem of liquid-filled laminated composite containers. Biswal, *et al.* (2002, 2003) determined the natural frequencies of the coupled liquid-filled, laminated composite tanks, with baffles, using the finite element method. The natural frequencies were computed for different fiber orientations. It was found that the number of layers had negligible effect on the system frequencies. The influence of the annular plate baffle was found to increase the natural frequency. Toorani and Lakis (2001, 2003) studied the dynamic behavior of anisotropic cylindrical shells conveying fluid. The structural-acoustic finite element formation was used to determine the free vibration of baffled liquid-storage tanks by Cho, *et al.* (2001, 2002). The displacement-based finite element approach was used to determine the natural frequencies of a fluid–structure system by Olson and Bathe (1983).

The reader may notice that most of the work of interaction of shells with their contained liquid was restricted to circular cylindrical shells. Other geometries such as spherical and conical shells were considered by Kubenko and Kruk (1994), Kubenko and Lakiza (1999), and Kubenko and Savin (2002).

Another class of problems not treated in this chapter is the interaction of a compressible fluid contained in an elastic thin-walled cylinder. This class of problems deals with the wave propagation through the fluid. Lin and Morgan (1956) studied the dependence of the velocity of normal waves on the physical parameters of a compound waveguide within the framework of the Timoshenko model of cylindrical shells. The dispersion properties of normal waves in a cylinder containing a fluid were studied by Fuller and Fahy (1982) based on Donnell's theory of shells. Kumar (1971, 1972) studied the propagation of normal waves throughout a fluid-filled cylindrical shell using the elastic equations. Vollman and Dual (1997) and Vollman, *et al*

(1997) determined the dispersion spectra (relationship between the frequency and the propagation constant) for axisymmetric waves propagating in cylindrical shells both empty and filled with a fluid. The effects of elastic–liquid interaction in a compound waveguide were investigated by Grinchenko and Meleshko (1981), Grinchenko and Komissarova (1984, 1988), and Komissarova (1990) for a wide frequency range. Komissarova (2002) considered wave motions in thin-walled cylinders filled with a fluid and determined frequency ranges within which the interaction of partial subsystems is characterized by special properties. The lowest normal wave possessed specific properties and participates in the elastic–liquid interaction over a wide frequency range. Shul'ga and Shul'ga (2003) developed a numerical algorithm for solving the dispersion spectral problem of nonaxisymmetric waves propagating in containers made of anisotropic piezoelectric filled with a perfect compressible liquid. Bauer (1986b), Kayuk (1995), and Kayuk, *et al.* (1996) considered the fluid viscosity in estimating the coupled frequencies. Guz, *et al.* (2002) presented an extensive review of the Soviet studies pertaining to wave interaction of shells with a liquid.

9

Nonlinear interaction under external and parametric excitations

9.1 Introduction

The dynamic analysis of a cylindrical shell experiencing elastic deformation that is comparable to its wall thickness cannot be described within the framework of the linear theory. The same is applied if the liquid free-surface amplitude is relatively large. In both cases, nonlinear analysis should be carried out. The presence of nonlinearities may result in nonlinear resonance conditions that cause complex response characteristics. One of the main difficulties in nonlinear problems of shell–liquid systems is that the boundary conditions are essentially nonlinear. This is in addition to the fact that the strain state of an elastic shell and the shape of the liquid free surface are not known a priori. The treatment of the nonlinear interaction of a liquid–shell system is a nonclassical boundary-value problem and relies on mechanics of deformable solids, fluid dynamics, and nonlinear mechanics.

With reference to nonlinear vibrations of cylindrical shells *in vacuo*, the literature is very rich and reports some controversies regarding the influence of nonlinearities on the shell dynamic behavior. The main results have been reviewed by Vol'mir (1972, 1979), Leissa (1973), Evensen (1974), Kubenko, *et al.* (1984), Amiro and Prokopenko (1997), and Amabili, *et al.* (1998b). Some attempts have been made to reconcile the reported discrepancies (see, e.g., Dowell, 1998, Evensen, 1999, and Amabili, *et al.* 1999c). It is believed that Reissner (1955) made the first attempt to study the influence of large-amplitude vibration for simply supported shells. He used Donnell's shallow-shell theory for thin-walled shells and found that the nonlinearity could be either of the hardening or softening type, depending on the mode shape. Later, Chu (1961) and Nowinski (1963) found that the nonlinearity is of the hardening type.

Under external periodic excitation of circular cylindrical shells, it was observed experimentally that specific steady conditions of dynamic deformation in which the lines of nodes of the traditional standing wave moves in the azimuthal direction over time (see, e.g., Ganiev and Koval'chuk, 1980, Grigoliuk, 1980 and Kubenko, Koval'chuk, and Krasnopolskaya, 1984). Analytically, the dynamic deflection of the shell is usually represented as a single standing wave, considered in combination with an axisymmetric form with twice the number of harmonics along the shell surface generator. The presence of an axisymmetric form must be reflected by predominant inward deflection (Vol'mir, 1972). This approach does not allow the experimental observation of the rotation of the radial displacement around the longitudinal shell axis. Koval'chuk and Podchasov (1988) introduced a mixed space–time representation of the radial flexure, which allows the rotational waves. Evensen (1963, 1964, 1966, 1967) examined the nonlinear dynamics of rings and cylindrical shells and found that the nonlinearity was of the softening type for both rings and shells. Evensen (1963) proved that the work of

Nowinski (1963) did not guarantee zero transverse deflection at the ends of the shell. He also found that the analyses of Reissner (1955) and Chu (1961) do not satisfy the continuity of in-plane circumferential displacement. Olson (1965) experimentally observed softening nonlinear characteristics. This discrepancy motivated Evensen (1967) to introduce a different form of flexural mode shapes together with a companion mode. However, Evensen's assumed modes are not moment-free at the ends of the simply supported shell. In most cases, the assumed mode shapes are based on experimental observations. In large amplitude vibrations, the shell does not spend equal time intervals during its outwards and inwards displacements. The inwards maximum deflections measured from equilibrium are larger than the outwards ones. Thus, the mode expansion requires the addition of an axisymmetric mode term to asymmetric linear modes to account for the contraction of the shell.

Atluri (1972) noted some algebraic errors in the analysis of Dowell and Ventres (1968). He claimed that the nonlinearity is of the hardening type contrary to what was observed experimentally. The modal expansion of Dowell and Ventres (1968) and Atluri (1972) includes a sine term in the axial coordinate. Varadan, *et al.* (1989) indicated that the reported hardening nonlinearity is due to the choice of the axisymmetric term.

Matsuzaki and Kobayashi (1969a,b) observed softening type nonlinearities in their studies on simply supported shells and clamped shells. Dowell and Ventres (1968) found that as the length-to-radius ratio of the cylindrical shell becomes large, the behavior follows softening nonlinearity. It becomes of hard type for small length-to-radius ratios. Dowell (1998) commented that for cylindrical shells of intermediate length-to-width ratios, the results of Dowell (1967) and Dowell and Ventres (1968), on the one hand, and Evensen (1966, 1967) on the other, do not necessarily agree because the two studies were based on different assumptions for the axial variation of the axisymmetric mode. Evensen used an axial variation, while Dowell used the first axial linear eigenmode associated with the axisymmetric circumferential mode. For intermediate length-to-radius ratios, the two approaches may differ in the type of nonlinearity. To reconcile this discrepancy, Dowell (1998) suggested, based on the work of Amabili, *et al.* (1998b) that the most accurate approach would be to construct a higher-dimensional model based on an expansion in the linear eigenmodes.

Later, Ginsberg (1973, 1974) and Chen and Babcock (1975) made significant improvement approaches. For example, Ginsberg used Flugge–Lur'e–Byrne shell theory and then employed an asymptotic analysis to solve the nonlinear boundary conditions for a simply supported shell. Both softening and hardening nonlinearities were found, depending on some system parameters. Chen and Babcock (1975) used a perturbation technique to solve the nonlinear equations obtained by Donnell's shallow-shell theory. They did not select a particular deflection solution. Their analytical and experimental results were in good agreement and revealed softening nonlinearity.

Other studies were conducted by Radwan and Genin (1975, 1976) and Raju and Rao (1976) using Sanders' nonlinear theory of shells and the finite element method. However, these studies showed hardening nonlinear effects contrary to experimental observations for simply supported shells. Koval'chuck and Podchasov (1988) introduced a mode shape representation of the radial deflection to allow the nodal lines to travel in the circumferential direction. A mode shape expansion was introduced by Varadan, *et al.* (1989) who used Donnell's shallow-shell theory. They showed that the axisymmetric term gives rise to hardening effect. Chiba (1993b) conducted experimental tests to study large-amplitude vibrations of two cantilevered circular cylindrical shells made of polyester sheet. It was observed that almost all responses exhibit nonlinear softening behavior. For modes with the same axial wave number, the weakest degree of softening nonlinearity belongs to the mode of lowest natural frequency. Furthermore, shorter shells exhibit a larger softening nonlinearity than taller ones. Under parametric excitation, Popov (2003) studied the nonlinear vibration of shells in the neighborhood of parametric resonance.

In a series of studies Amabili, *et al.* (1999a,b, 2000a,b) studied the nonlinear dynamics and stability of simply supported circular cylindrical shells with liquid flow. They showed that the shell under the action of internal flow loses stability by divergence. When axisymmetric modes were allowed, the divergence is sub-critical, giving a softening-type behavior for small shell displacement, and becoming hardening for larger amplitudes. Amabili, *et al.* (2002) conducted an experimental investigation on two water-filled circular cylindrical shells made of steel subjected to harmonic excitation. The boundary conditions were approximated as simple supports. Experimental results showed a softening-type nonlinearity of about 4% for vibration amplitude twice the shell thickness.

It is important to match the frames of reference of the liquid and the shell in the analytical formulation of the matching conditions of the shell–liquid interface. Recall that the fluid field equations are described in terms of Euler variables, while the shell dynamics is described in terms of Lagrangian variables. Matching the two systems is not difficult if the shell displacements are small. In this case, one can neglect the spatial deformation of the boundaries and thus both Euler and Lagrange variables are identical. The difficulty arises in the presence of finite deflections for which Euler's technique does allow for automatic estimation of the coordinates of the deformed boundaries of the liquid–shell interface. This problem was addressed in Galiev (1977, 1981) and Ilgamov (1969, 1975).

The nonlinear analysis of shell–liquid systems has been documented in several research monographs and review articles. These include Lukovskii (1975, 1990), Grigoliuk and Gorshkov (1976), Bogoryad, *et al.* (1977), Bogoryad (1980), Ganiev and Koval'chuk (1980), Limarchenko (1987), Nikitin (1987), Limarchenko and Yasinskii (1996), and Amabili and Paidoussis (2003). It was established that, for cylindrical shells, nonlinear liquid formulation should be considered if the free-surface wave height exceeds 25% of the shell radius. For small wave heights, the linear liquid theory yields adequate results (Kubenko and Koval'chuk, 1987, 2000). In a series of papers, Koval'chuk and Pavolovskii (1992, 1993, 1994) examined the nonlinear interaction of shells with liquid under different forms of radial pressure.

This chapter considers the nonlinear interaction of liquid sloshing modes with its container elastic modes. The governing equations of motion of the shell and liquid are developed. The partial differential equations of the shell and liquid are discretized depending on the boundary conditions of special cases. Basically three groups of problems are considered in this chapter (i) modal analysis and nonlinear response to external excitation, (ii) parametric stability and nonlinear response under parametric excitation, and (iii) nonlinear free and forced excitation of nonlinear orthotropic shells.

The first group includes the free and forced vibrations of nonlinear shells partially filled with still and oscillating liquid. The case of shells completely filled with liquid involves the normal mode frequencies and mixed mode interactions in the presence of single and multiple internal resonances. The linear shell interaction with nonlinear liquid sloshing dynamics under external excitation is analyzed for the case of rigid and elastic shell bottom. The more general and complex case is the nonlinear shell interaction with the nonlinear sloshing dynamic. This case is studied for nonlinear modal analysis and forced autoparametric interaction. The second group covers the parametric excitation and stability of fundamental modes of shell–liquid systems. The case of shells completely filled with liquid is analyzed for parametric excitation of axisymmetric shell wall displacements. The third group addresses the free and forced vibration of orthotropic shells in the presence of one-to-one internal resonance.

An overview of research activities of storage liquid tanks subjected to earthquake excitations is presented in Section 9.8. Civil engineers studied the earthquake damages and failure modes of anchored tanks. The numerical techniques based on finite element and boundary element algorithms are discussed for liquid sloshing dynamics and sloshing–structure interaction.

9.2 General equations of motion

Consider a closed circular cylindrical shell of radius R, length l, and thickness $\hat{t}$. The shell is partially filled with inviscid and incompressible fluid to a depth h as shown in Figure 9.1. With reference to the cylindrical coordinates z, r, θ with the origin O placed at the center of one end, the displacements of a point in the middle surface of the shell are denoted as u, v, w in the axial, z, circumferential, θ, and radial, r, directions, respectively. The radial displacement, w, is taken to be positive towards the shell center. For thin wall shells, $\hat{t} \ll R$, one may use Donnell's nonlinear theory of shallow shells. This theory is based on neglecting in-plane inertia, transverse shear deformation, and rotary inertia. Furthermore, the theory gives accurate results only for modes of high circumferential wave number. According to Donnell's shallow-shell theory the governing equation for large deformation is (see, e.g., Vlasov, 1949, Mushtari and Galimov, 1957, Novozhilov, 1962, Evensen, 1967, Chen and Babcock, 1975, and Donnell, 1976)

$$D\nabla^4 w + c\hat{t}\dot{w} + \rho_c \hat{t}\ddot{w} = \frac{1}{R}\frac{\partial^2 F}{\partial z^2} + \frac{1}{R^2}\left(\frac{\partial^2 F}{\partial \theta^2}\frac{\partial^2 w}{\partial z^2} - 2\frac{\partial^2 F}{\partial z \partial \theta}\frac{\partial^2 w}{\partial z \partial \theta} + \frac{\partial^2 F}{\partial z^2}\frac{\partial^2 w}{\partial \theta^2}\right) + f(t) - p - N_z(t)\frac{\partial^2 w}{\partial z^2} \tag{9.1}$$

where $D = E\hat{t}^3/[12(1-\nu^2)]$, is the flexural rigidity, E is Young's modulus, ν is Poisson's ratio, ρ_c is the shell density, c is the damping coefficient, p is the internal hydrodynamic pressure acting on the wetted wall, $f(t)$ is an external excitation acting on the shell, and $N_z(t) = N_0 + N(t)$ is a dynamic excitation acting along the shell edges. A dot denotes differentiation with respect to time, and F is the in-plane stress function obtained by solving the equation

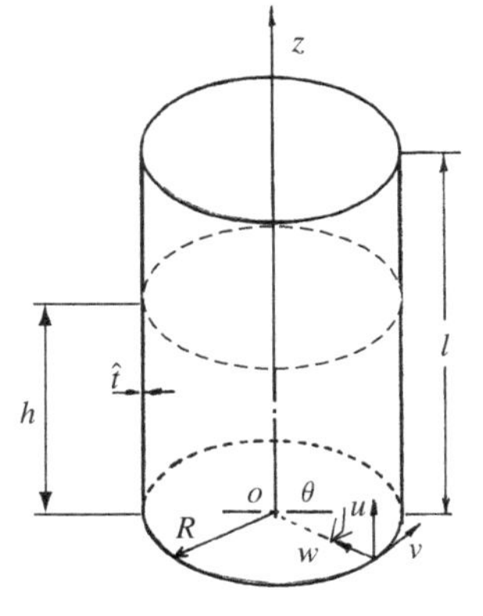

Figure 9.1 Elastic shell filled with liquid and the coordinate system.

$$\frac{1}{E\hat{t}}\nabla^4 F = \frac{1}{R}\frac{\partial^2 w}{\partial z^2} + \frac{1}{R^2}\left[\left(\frac{\partial^2 w}{\partial z \partial \theta}\right)^2 - \frac{\partial^2 w}{\partial z^2}\frac{\partial^2 w}{\partial \theta^2}\right] \tag{9.2}$$

where $\nabla^4 = \left[\frac{\partial^2}{\partial z^2} + \frac{1}{R^2}\frac{\partial^2}{\partial \theta^2}\right]^2$.

Later, Ogawa, *et al.* (1997) experimentally examined the fluid–structure-interaction behavior of the worm tank and observed that the acceleration in the lateral excitation is ten times larger than that in the longitudinal excitation. Gedikli and Ergüven (1999) considered the influence of baffles on storage tanks subjected to earthquakes.

The complete statement of the governing equations must include the fluid field equations. The hydrodynamic pressure, p, acting on the wetted lateral surface of the shell is determined from the Kelvin's equation

$$\left[\frac{\partial \Phi}{\partial t} + \frac{1}{2}(\nabla\Phi)^2 + \frac{p}{\rho} + gz\right]_S = 0 \tag{9.3}$$

where S is the shell-wetted surface. The boundary-value problem for evaluating the fluid velocity potential function, Φ, constitutes Laplace's equation

$$\frac{\partial^2 \Phi}{\partial r^2} + \frac{1}{r}\frac{\partial \Phi}{\partial r} + \frac{1}{r^2}\frac{\partial^2 \Phi}{\partial \theta^2} + \frac{\partial^2 \Phi}{\partial z^2} = 0 \tag{9.4}$$

The boundary condition at the wetted surface requires the compatibility condition

$$\left.\frac{\partial \Phi}{\partial r}\right|_{r=R} = -\frac{\partial w}{\partial t} - \nabla\Phi \cdot \nabla w \tag{9.5}$$

Since w is assumed to be positive towards the shell center, the negative sign is introduced. The dynamic and kinematic boundary conditions of the free surface are

$$\frac{\partial \Phi}{\partial t} + \frac{1}{2}(\nabla\Phi)^2 + gz = 0 \tag{9.6a}$$

$$\frac{\partial \eta}{\partial t} = \frac{\partial \Phi}{\partial z} - \nabla\eta \cdot \nabla\Phi \qquad \text{on the free surface} \tag{9.6b}$$

At the tank bottom we have

$$\left.\frac{\partial \Phi}{\partial z}\right|_{z=0} = 0 \qquad \text{for rigid bottom} \tag{9.7a}$$

$$\left.\frac{\partial \Phi}{\partial z}\right|_{z=0} = -\frac{\partial w}{\partial t} - \nabla\Phi \cdot \nabla w \qquad \text{for elastic bottom} \tag{9.7b}$$

Equations (9.1–9.7) establish the complete statement of the shell–liquid boundary-value problem. This formulation is general and some simplifications can be introduced to study the following special cases:

(1) Nonlinear shells completely filled with liquid.
(2) Linear shells interacting with nonlinear liquid sloshing.

(3) Rigid wall containers with linear elastic bottom interacting with nonlinear liquid sloshing.
(4) Nonlinear shells interacting with nonlinear liquid sloshing.

The above cases are studied for nonlinear modal analyses, forced excitation response, and stability and response to parametric excitation. These cases are considered in this chapter.

9.3 Shells partially filled with liquid

The nonlinear free vibration of partially filled shells deals with the estimation of normal mode frequencies of the combined system. Shkenev (1964) considered the case of nonlinear shells interacting with linear liquid sloshing dynamics. On the other hand, Grigorev (1966) treated the case of linear shells interacting with nonlinear sloshing dynamics. The case of cylindrical shells with spherical bottoms arbitrary filled with liquid (a shallow model) was studied by Petrenko (1969, 1970, 1978).

9.3.1 Free vibration of shells partially filled with still fluid

Kubenko, Koval'chuk, and Podchsov (1989) considered the free vibration of an elastic shell partially filled with liquid in the absence of free-surface sloshing. The deflection, $w(z, \theta, t)$, was assumed to be in the form

$$w(z,\theta,t) = [A(t)\cos n\theta + B(t)\sin n\theta]\sin(m\pi z/l) + C(t)\sin^2(m\pi z/l) \tag{9.8}$$

where A, and B, are generalized and companion coordinates, respectively. The generalized coordinate C accounts for the quasi-static deflection. Substituting equation (9.8) into equation (9.2) gives a partial differential equation for the stress function F. Equation (9.8) and the resulting solution for the stress function are then substituted in equation (9.1), after dropping the damping, external, and parametric excitations from equation (9.1), and applying Galerkin's method to give the following coupled nonlinear differential equations

$$\ddot{A} + \omega_1^2 A + \gamma_1(A^2+B^2)A + g_1(A^2+B^2)^2 A = 0 \tag{9.9a}$$

$$\ddot{B} + \omega_1^2 B + \gamma_1(A^2+B^2)B + g_1(A^2+B^2)^2 B = 0 \tag{9.9b}$$

where

$$\omega_1^2 = \frac{\omega_{nm}^{(0)2}}{1+m_{01}}, \quad \gamma_1 = \frac{\gamma_{01}}{1+m_{01}}, \quad g_1 = \frac{g_{01}}{1-m_{01}} \tag{9.10}$$

Frequencies $\omega_{nm}^{(0)}$ are the fundamental frequencies of the linear vibration of the empty shell, γ_{01} and g_{01} are constant parameters describing the nonlinear elastic properties of the shell, and $m_{01} \geq 0$ is the reduced mass due to the liquid. It is seen that the presence of the nonsloshing liquid decreases the fundamental frequencies of the shell. However, the qualitative nature of the shell geometric nonlinearity is not affected by the liquid mass. The solution of equations (9.9) may be obtained using the averaging method by introducing the following solution

$$A(t) = a(t)\cos(\omega_1 t + \vartheta_1(t)), \quad B(t) = b(t)\cos(\omega_1 t + \vartheta_2(t)) \tag{9.11a,b}$$

Following the standard averaging procedure, the averaged equations of amplitudes and phases give the first two integrals

$$a^2 + b^2 = c_1, \quad \Re(1-\Re)\left(1-\cos 2\bar{\theta}\right) = c_2 \tag{9.12a,b}$$

where $\Re = a^2/c_1$, $\bar{\theta} = \vartheta_2 - \vartheta_1$, and c_i are constants. The functions $\Re$ and $\bar{\theta}$ are obtained from the solution of the following equations

$$\frac{d\Re}{dt} = \frac{(\gamma_1 + g_1 c_1)c_1}{4\omega_1}\Re(1-\Re)\sin 2\bar{\theta} \tag{9.13a}$$

$$\frac{d\bar{\theta}}{dt} = \frac{(\gamma_1 + g_1 c_1)c_1}{8\omega_1}(2\Re - 1)\left(1-\cos 2\bar{\theta}\right) \tag{9.13b}$$

9.3.2 Influence of liquid free surface oscillations

In order to consider the influence of liquid free-surface oscillations, the coupled boundary-value problem of a shell–liquid system described by equations (9.1–9.7) should be considered, after dropping external and parametric excitations. The shell deflection and free-surface wave height may be represented by the following expansions:

$$w(z,\theta,t) = \sum_{n=0}^{\infty}\sum_{m=1}^{\infty}\left[A_1^{n,m}(t)\cos n\theta + A_2^{n,m}(t)\sin n\theta\right]\sin(m\pi z/l) \tag{9.14a}$$

$$\eta(r,\theta,t) = \sum_{n=0}^{\infty}\sum_{k=1}^{\infty}\left[a_1^{n,k}(t)\cos n\theta + a_2^{n,k}(t)\sin n\theta\right]J_n(\xi_{nk}r/R) \tag{9.14b}$$

Kubenko, *et al.* (1988), Kubenko, Podchsov (1989) and Koval'chuk and Kubenko (1991) obtained the following set of coupled ordinary nonlinear differential equations

$$\ddot{A}_i^{n,m} + \omega_{n,m}^2 A_i^{n,m} + \sum_{s,\, s\neq m} q_s^{n,m}\ddot{A}_i^{n,s} + \sum_{j=1}\delta_j^{n,m}a_i^{n,j} = \varepsilon F_i^{n,m}\left(A_i^{n,m}, a_i^{n,k}\right) \tag{9.15a}$$

$$\ddot{a}_i^{n,k} + \varpi_{n,k}^2 a_i^{n,k} + \sum_{\ell}\gamma_\ell^{n,k}A_i^{n,\ell} = \varepsilon G_i^{n,k}\left(A_i^{n,m}, a_i^{n,k}\right) \tag{9.15b}$$

with $n = 0, 1, 2, \ldots,\ m, s, k, \ell = 1, 2, \ldots,$ and $i = 1, 2$, where $\omega_{n,m}$ and $\varpi_{n,k}$ are the fundamental frequencies of the shell (with the effect of the apparent additional liquid mass) and the liquid free surface (in a rigid tank), respectively. The values $q_s^{n,m}$, $\delta_j^{n,m}$, and $\gamma_\ell^{n,k}$ are constants depending on the physical and geometrical parameters of the shell–liquid system. The right-hand functions $F_i^{n,m}$, and $G_i^{n,k}$ are nonlinear in the generalized coordinates $A_i^{n,m}$ and $a_i^{n,k}$, and ε is a small parameter.

Equations (9.15) include linear and nonlinear coupling of liquid and shell modes. In order to construct their integrals, they must be transformed into principal coordinates, that is, by eliminating the linear coupling on the left hand sides. Note that the damping was included in the previous formulation. Gorbunov (1965) conducted free vibration tests on cylindrical and spherical shells filled with liquid. It was found that with increasing vibration amplitudes, the logarithmic decrement did not maintain any regular trend. It was established that different structural factors contribute considerably to the dissipation energy. Sivak and Telalov (1991, 1992, 1998) showed that the damping capacity of longitudinally reinforced cylindrical shells partially filled with liquid depends on both the amplitudes of their flexural vibrations and the liquid free-surface motion.

9.3.3 Forced vibration of shells partially filled with still fluid

The simplest case of forced excitation of a shell–liquid system in the absence of liquid-free-surface oscillation is the interaction of one mode with its companion mode. This case was studied by Obraztsova (1976b), Shklyarchuk (1983), Kubenko, Koval'chuk, and Boyarshina (1992), and Koval'chuk, Pavlovskii, and Filin (1993). The modal equations (9.15) under sinusoidal excitation take the form

$$\ddot{A}+\omega_1^2A+\gamma_1(A^2+B^2)A+g_1(A^2+B^2)^2A=\frac{Q_{01}}{1+m_{01}}\cos\Omega t \tag{9.16a}$$

$$\ddot{B}+\omega_1^2B+\gamma_1(A^2+B^2)B+g_1(A^2+B^2)^2B=0 \tag{9.17b}$$

In the absence of cross coupling with the companion mode, that is, $B=0$, the solution of equation (9.16a) using the averaging method is obtained by substituting $A(t)=a(t)\cos(\omega_1 t+\theta_1(t))$. The following algebraic equation governing the amplitude–frequency unimodal response is obtained

$$\Omega^2=\omega_1^2+\frac{3}{4}\gamma_1 a_0^2\pm\frac{Q_{01}}{(1+m_{01})a_0} \tag{9.18}$$

where a_0 stands for unimodal response. Figure 9.2 shows the amplitude–frequency response in the absence of coupling with the companion mode. Solution (9.18) is stable outside the excitation frequency range $\Omega_1<\Omega<\Omega_2$, where

$$\Omega_1^2=\omega_1^2+\frac{9}{4}\sqrt[3]{\frac{9}{4}\frac{Q_{01}^2\gamma_1}{(1+m_{01})^2}},\quad \Omega_2^2=\omega_1^2+\frac{1}{4}\sqrt[3]{\frac{4Q_{01}^2\gamma_1}{(1+m_{01})^2}},\quad \gamma_1<0 \tag{9.19a, b}$$

The steady state mixed mode response, using the averaging method, is given by the following equations, after using equation (9.11),

$$\Omega^2=\omega_1^2+\gamma_1 a^2\pm\frac{1.5Q_{01}}{(1+m_{01})a} \tag{9.20a}$$

$$b^2=\frac{4}{3\gamma_1}\left[\Omega^2-\omega_1^2-\frac{\gamma_1}{4}a^2\right] \tag{9.20b}$$

Figure 9.2 shows the mixed amplitude–frequency response curves. Note that the interaction with the companion mode begins at excitation frequency Ω_2 and remains active down to frequency Ω_1. Above excitation frequency Ω_2, the first mode is governed by the unimodal

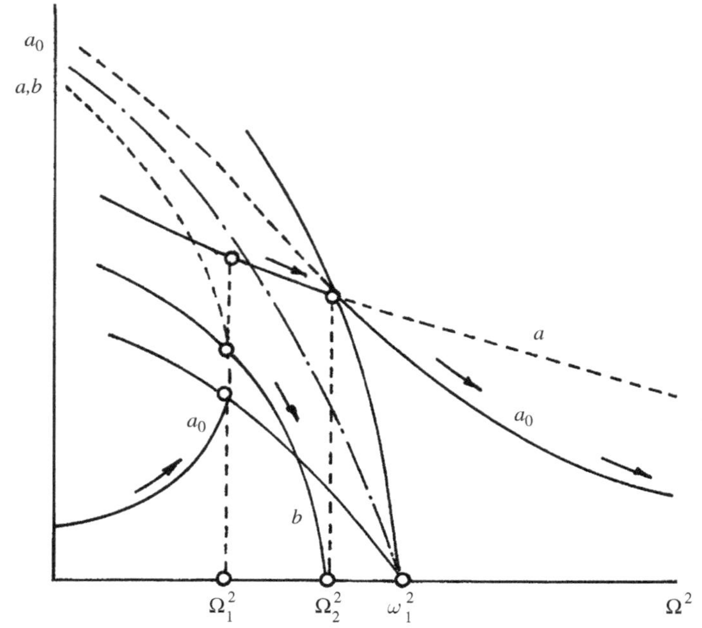

Figure 9.2 Typical plots of unimodal (a_0), and mixed-mode (a, b) amplitude–frequency responses of nonlinear shell partially filled with still liquid under forced excitation. (Kubenko and Koval'chuk, 2000)

response, and its coupled response is unstable. The response curves reveal that the shell–liquid system follows the behavior of soft nonlinear spring characteristics.

Kubenko, *et al.* (1988), Pavlovskii and Filin (1981), and Koval'chuk, *et al.* (2002) considered the influence of initial deflection of the shell on the forced response for different values of liquid filling depth. The fundamental frequencies (of the first bending mode and its companion mode) were found to deviate from each other. This difference results in a shift of the instability region of the steady state responses. Other cases studied in the literature include the forced excitation of a liquid contained between two coaxial cylindrical shells. Krasnopolskaya and Podchasov (1992a,b) considered the outer cylinder to be rigid, while the inner is an elastic shell with a specified time deformation corresponding to bending wave propagation in the circumferential direction. It was found that transversal (corrugated) waves with frequencies half that of the traveling wave frequency could be excited.

9.4 Shells filled with liquid

9.4.1 Historical overview

The study of nonlinear interaction between elastic shells and liquid-free-surface dynamics may be considered for three different cases: (a) linear elastic shell deformation interacting with nonlinear free-surface motion, (b) nonlinear shell deformation interacting with linear free-surface motion, and (c) nonlinear shell deformation interacting with nonlinear free-surface motion. Chu and Kana (1967) considered the interaction of linear elastic deformation interacting with nonlinear free-surface waves. The nonlinear interaction of liquid with elastic tank dynamics has been experimentally examined by Kana, *et al.* (1966), Pih and Wu (1969), Chu and Brown (1981), Liu and Ma (1982), Liu and Lam (1983), Sudo, *et al.* (1985), Chiba (1993d), Lu, *et al.* (1998), and Wunderlich and Seiler (2000). It was observed that laterally excited empty shells display essentially linear responses during breathing vibrations up to amplitudes as large as one-half of the wall thickness.

Analytical studies were presented by Kana (1967), Yu, *et al.* (1987), Gonçalves and Batista (1988), Bauer and Eidel (1988a), and Ortiz and Barhorst (1997). For shells that are completely or partially filled, some slight nonlinear softening was observed at larger amplitudes, where the liquid surface is in the vicinity of an axial node in the modal pattern of the breathing shell. If the liquid surface is not in the vicinity of an axial node, and particularly if it is in the vicinity of an axial antinode of the shell motion, then a significant nonlinear softening response occurs for the shell. A liquid-free-surface response in the form of high-frequency ripples is most pronounced when the mean surface level is at an antinode of the shell motion. Other cases were considered by Mololkov and Krauklis (1967). Some approximate and exact solutions of liquid oscillation in thin elastic containers were obtained by Pshenichnov (1969, 1971). Chen S.S., *et al.* (1994) developed velocity boundary conditions for simulating the free-surface liquid flow. Sakata, *et al.* (1986) studied the interaction of liquid sloshing with elastic walls of a rectangular tank. Soedel and Soedel (1994) studied the free and forced vibration for a plate supporting a sloshing liquid. Kuz'ma and Kholopova (1983) considered nonlinear oscillations of a cylindrical container containing a liquid under harmonic and random longitudinal perturbations.

Boyarshina (1984, 1988) analytically studied the nonlinear free and forced vibrations and stability of a circular cylindrical tank partially filled with a liquid. The nonlinearity is attributed to the interaction of free-surface waves and elastic flexural vibrations of the shell. Boyarshina (1984) examined the nonlinear interaction between liquid sloshing modes and a circular shell vibration in the neighborhood of three internal resonance cases. These are (i) the natural frequency of the shell is close to one of the sloshing natural frequencies, (ii) the natural frequency of the shell is twice the natural frequency of the sloshing modes, and (iii) the natural frequency of the shell is three times the natural frequency of the sloshing modes. Here the sloshing modes are characterized by a free-surface amplitude that is larger than the wall displacement, in contrast to "bulging modes" that are characteristic of the oscillating tank wall exciting the liquid. Amabili (1996a) and Amabili, *et al.* (1998b) reviewed some work pertaining to the nonlinear vibration of shells filled or surrounded by quiescent fluid.

Nayfeh and Raouf (1987) studied the nonlinear vibrations of circular cylindrical shells when the frequency of the axisymmetric mode is close to twice that of the asymmetric mode (known as internal or autoparametric resonance). McRobie, *et al.* (1999) and Popov, *et al.* (2001) considered the nonlinear modal interaction of a cylindrical shell near a 1:2 internal resonance. This type of autoparametric resonance can trigger both regular and chaotic dynamic characteristics. Amabili, *et al.* (2000a,b) investigated the dynamic response of a simply supported, fluid-filled circular cylindrical shell when the frequencies of any asymmetric mode, a companion mode, the first axisymmetric mode and the third axisymmetric mode have the ratio of 1:1:1:2.

Gonçalves and Batista (1988) considered simply supported circular cylindrical shells filled with inviscid fluid. They used Sanders' nonlinear theory of shells and a modal expansion that includes two terms in the radial direction (asymmetric and axisymmetric modes) and ten terms to describe the in-plane displacements. It was found numerically that the presence of a dense fluid gives more strongly softening results vis-à-vis those for the same shell in vacuum.

Chiba (1993c) conducted experimental tests to study large-amplitude vibrations of two vertical cantilevered circular cylindrical shells made of polyester sheet partially filled with water. It was observed that for bulging modes with the same axial wave number, the weakest degree of softening nonlinearity could be attributed to the mode having the minimum natural frequency. Such observation is also found for the same empty shells. It was also found that shorter tanks have a larger softening nonlinearity than taller tanks. The tank with a lower liquid height has a greater softening nonlinearity than the tank with a higher liquid level. Chiba (1993c) extended his experimental investigations to study the effect of a thin film floating on the free surface and the behavior of the free surface. He (1993e) also studied large-amplitude vibrations of two vertical clamped circular cylindrical shells, partially filled to different water levels. It was found that the responses displayed a generally softening nonlinearity. This section considers the shell interaction with linear liquid sloshing, multi-mode interaction, and auto-parametric resonance of shells filled with liquid.

9.4.2 Interaction with linear liquid sloshing

For an elastic shell completely filled with an inviscid and incompressible liquid, the liquid nonlinear effects are negligible and one can consider the linear fluid field equations. Ginsberg (1975), Gonçalves and Batista (1988), Selmane and Lakis (1997a,b), and Amabili, *et al.* 1998b) studied the case of the interaction of nonlinear shell deformation with linear fluid dynamics. The flow must satisfy Laplace' equation

$$\frac{\partial^2 \Phi}{\partial r^2} + \frac{1}{r}\frac{\partial \Phi}{\partial r} + \frac{1}{r^2}\frac{\partial^2 \Phi}{\partial \theta^2} + \frac{\partial^2 \Phi}{\partial z^2} = 0 \tag{9.21}$$

The boundary condition at the wetted surface requires the compatibility condition

$$\left.\frac{\partial \Phi}{\partial r}\right|_{r=R} = -\frac{\partial w}{\partial t} \tag{9.22}$$

For open shell ends, a zero pressure is assumed and one can write

$$\Phi|_{z=0} = \Phi|_{z=l} = 0 \tag{9.23}$$

A solution of equation (9.21) satisfying condition (9.23) may be given in the form (Amabili, *et al.* 1998b)

$$\Phi = \sum_{m=1}^{\infty}\sum_{n=0}^{\infty} [a_{mn}(t)\cos m\theta + b_{mn}(t)\sin m\theta][c_{mn} I_m(\lambda_n r) + d_{mn} K_m(\lambda_n r)]\sin \lambda_n z \tag{9.24}$$

where $I_m(\lambda_n r)$ and $K_m(\lambda_n r)$ are the modified Bessel functions of the first and second kind, respectively, of order n. In order to satisfy the compatibility condition (9.22), one should use the assumed mode shape

$$\begin{aligned} w(z,\theta,t) = {} & [A_{mn}(t)\cos m\theta + B_{mn}(t)\sin m\theta]\sin \lambda_n z \\ & + A_{0n}(t)[3\sin \lambda_n z - \sin 3\lambda_n z] \end{aligned} \tag{9.25}$$

Furthermore, the velocity potential function must be finite at $r = 0$. In this case, the solution for the velocity potential function is given in the form

$$\Phi = [\dot{A}_{mn}(t)\cos m\theta + \dot{B}_{mn}(t)\sin m\theta]\frac{I_m(\lambda_n r)\sin\lambda_n z}{\lambda_n I_m'(\lambda_n R)} + \dot{A}_{0n}(t)\left[\frac{3I_0(\lambda_n r)\sin\lambda_n z}{\lambda_n I_0'(\lambda_n R)} - \frac{I_0(3\lambda_n r)\sin 3\lambda_n z}{3\lambda_n I_0'(3\lambda_n R)}\right] \tag{9.26}$$

The hydrodynamic pressure is

$$p = \rho\left\{[\ddot{A}_{mn}(t)\cos m\theta + \ddot{B}_{mn}(t)\sin m\theta]\frac{I_m(\lambda_n r)\sin\lambda_n z}{\lambda_n I_m'(\lambda_n R)} + \ddot{A}_{m0}(t)\left[\frac{3I_0(\lambda_m r)\sin\lambda_m z}{\lambda_m I_0'(\lambda_m R)} - \frac{I_0(3\lambda_m r)\sin 3\lambda_m z}{3\lambda_m I_0'(3\lambda_m R)}\right]\right\} \tag{9.27}$$

The hydrodynamic pressure involves inertia effects given by asymmetric and axisymmetric terms. The inertia effects of the fluid are usually significant for axisymmetric modes, which have a direct effect on the nonlinear behavior of the shell. Applying Galerkin's method to equation (9.1), after dropping the parametric excitation term, using equation (9.25) and weighting functions as those associated with the generalized coordinates $A_{mn}(t)$, $B_{mn}(t)$, and $A_{0n}(t)$, gives the following ordinary nonlinear differential equations

$$\ddot{A}_{mn}(t) + 2\zeta_{mn}\omega_{mn}\dot{A}_{mn}(t) + \omega_{mn}^2 A_{mn}(t) + h_1 A_{mn}(t)A_{0n}(t) + h_2 A_{mn}^3(t) + h_2 A_{mn}(t)B_{mn}^2(t) + h_3 A_{mn}(t)A_{0n}^2(t) = f_{mn}\cos\Omega t \tag{9.28a}$$

$$\ddot{B}_{mn}(t) + 2\zeta_{mn}\omega_{mn}\dot{B}_{mn}(t) + \omega_{mn}^2 B_{mn}(t) + h_1 B_{mn}(t)A_{m0}(t) + h_2 B_{mn}^3(t) + h_2 B_{mn}(t)A_{mn}^2(t) + h_3 B_{mn}(t)A_{0n}^2(t) = 0 \tag{9.28b}$$

$$\ddot{A}_{0n}(t) + 2\zeta_{0n}\omega_{0n}\dot{A}_{0n}(t) + \omega_{0n}^2 A_{0n}(t) + k_1 A_{mn}^2(t) + k_1 B_{mn}^2(t) + k_2 A_{0n}(t)A_{mn}^2(t) + k_3 A_{0n}(t)B_{mn}^2(t) + k_3 A_{0n}^2(t) + k_4 A_{0n}^3(t) = 0 \tag{9.28c}$$

where h_i and k_i are coefficients depending on shell parameters, ω_{mn} is the linear natural frequency of the fluid–shell system having n axial half-waves and m nodal diameters, $\zeta_{mn} = c/(2\,\omega_{mn}\rho_v)$ is the corresponding damping factor, and ρ_v is the virtual density of the fluid–shell system given by the expression

$$\rho_v = \rho_c + \frac{\rho I_m(\lambda_n R)}{\lambda_n \hat{t} I_m'(\lambda_n R)} \tag{9.29a}$$

and $f_{mn} = f/(\rho_v\hat{t})$ is the amplitude of the external excitation, $\zeta_{0n} = c/(2\,\omega_{0n}\rho_{v0})$, with

$$\rho_{v0} = \rho_c + \frac{\rho}{\lambda_n\hat{t}}\left\{\frac{9I_m(\lambda_n R)}{I_m'(\lambda_n R)} + \frac{I_m(3\lambda_n R)}{3I_m'(3\lambda_n R)}\right\} \tag{9.29b}$$

Equations (9.28) can be solved numerically and analytically by using one of the asymptotic approximation techniques. Figure 9.3 shows the unimodal amplitude–frequency response of $\tilde{A}_{mn} = A_{mn}/\hat{t}$ in the absence of coupling with other modes for an external excitation amplitude $f_{mn} = 0.0012\hat{t}^2\rho_c\omega_{mn}^2$, damping ratio $\zeta_{mn} = 0.0005$, and $\omega_{mn} = 2\pi \times 564.2$ rad/s. The figure shows those results obtained by Chen and Babcock (1975), Ganapathi and

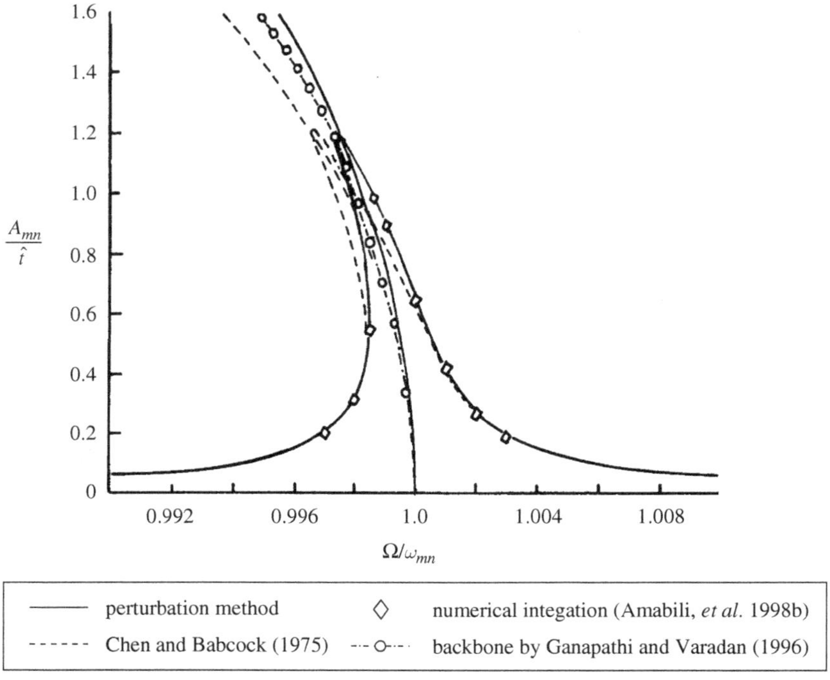

Figure 9.3 Amplitude–frequency response curves and backbone curves of the driven mode.

Varadan (1996), and Amabili, *et al.* (1998b). It is seen that in all cases the nonlinearity is of the softening type. Figure 9.4 shows the influence of coupling with the other mode $\tilde{B}_{mn}$, on the amplitude–frequency response. It is seen that some energy is transferred from the directly excited mode to the coupled mode through nonlinear coupling. Figure 9.5 shows the dependence of the backbone curves and the response amplitude for empty and water-filled shells on the excitation frequency for different nodal diameters m of the driven mode. It is seen that the liquid effect is to increase the degree of softening nonlinearity. The softening nonlinearity also increases with the number of nodal diameters m.

9.4.3 Free nonlinear multi-mode problem

The free nonlinear vibration problem of thin circular shells filled with incompressible liquids deals with the estimation of the free vibration of the shell–liquid system under initial conditions. The problem includes the estimation of normal mode frequencies and the energy distribution among the normal modes in the absence of energy dissipation sources. Kubenko, *et al.* (2003) studied the multi-mode periodic deformation of circular cylindrical shells filled with a liquid. They considered large deformations of the shell flexural modes. The system is essentially described by the shell Donnell's equations (9.1), (9.2) after dropping external and parametric excitations. The shell mode is approximated by the five-term expansion

$$\begin{aligned} w(z,\theta,t) = {} & [A_{m1}(t)\cos m_1\theta + B_{m1}(t)\sin m_1\theta]\sin(n\pi z/l) \\ & + [A_{m2}(t)\cos m_2\theta + B_{m2}(t)\sin m_2\theta]\sin(n\pi z/l) + D(t)\sin^4(n\pi z/l) \end{aligned} \tag{9.30}$$

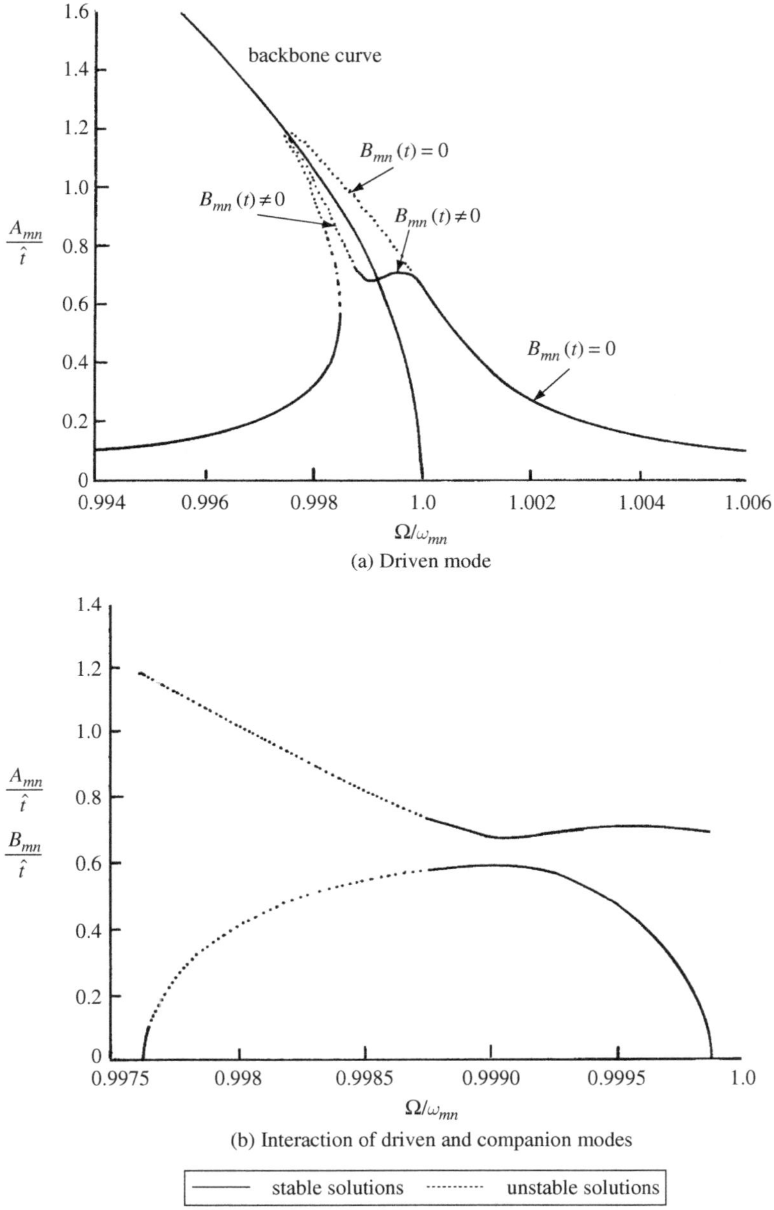

Figure 9.4 Amplitude–frequency response curves showing (a) unimodal and (b) mixed mode interaction. (Amabili, *et al.*, 1998b)

where $m_1 \neq m_2$, $A_{mi}(t)$ and $B_{mi}(t)$ are generalized coordinates and the companion (conjugate) modes, respectively. The generalized coordinate $D(t)$ is associated with the deformation of shells with large deformations that was experimentally observed to be of preferential inward buckling (Vol'mir, 1972 and Kubenko, *et al.*, 1984). Note that equation (9.30) must satisfy the boundary conditions for simply supported shell, that is,

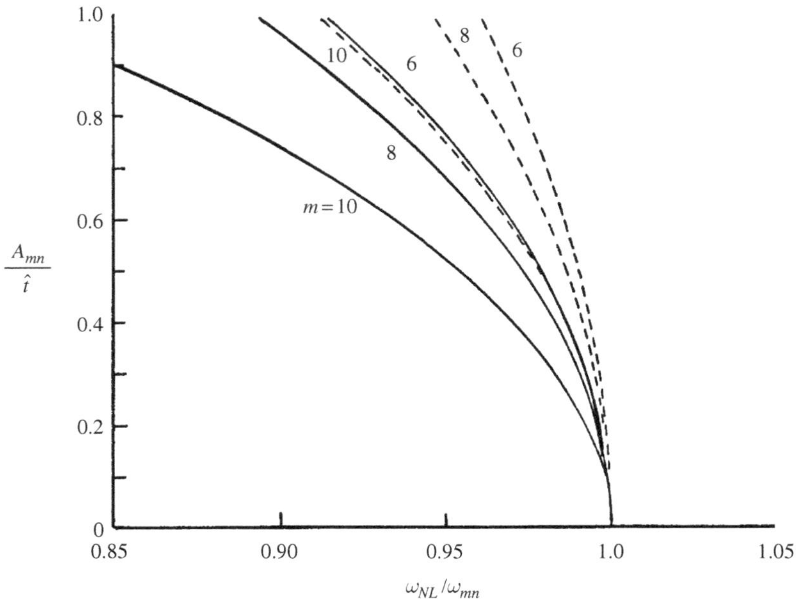

Figure 9.5 Dependence of the backbone curves of empty --- and water filled —— shells on the frequency ratio ω_{NL}/ω_{mn} for different numbers of nodal diameters. (Amabili, *et al.*, 1998b)

$$N_z = 0 \quad \text{at } z = 0,\ l, \quad \text{and} \quad v = 0 \quad \text{at } z = 0,\ l \tag{9.31a}$$

$$w = 0, M_z = -D\left(\frac{\partial^2 w}{\partial z^2} + \frac{\nu}{R^2}\frac{\partial^2 w}{\partial \theta^2}\right) = 0, \quad \text{at } z = 0,\ l \tag{9.31b}$$

where M_z is the bending moment per unit length. Substituting solution (9.30) into equation (9.2) gives the solution for the stress function F. By using equation (9.30), the solution of the linear fluid field equations (9.21)–(9.23) for the internal fluid is

$$\begin{aligned}\Phi = &-\left[\dot{A}_{m1}(t)\cos m_1\theta + \dot{B}_{m1}(t)\sin m_1\theta\right]\frac{lI_{m1}(\pi r/l)\sin(n\pi z/l)}{n\pi I'_{m1}(n\pi R/l)} \\ &-\left[\dot{A}_{m2}(t)\cos m_2\theta + \dot{B}_{m2}(t)\sin m_2\theta\right]\frac{lI_{m2}(n\pi r/l)\sin(n\pi z/l)}{n\pi I'_{m2}(n\pi R/l)} \\ &-\frac{\dot{D}(t)}{2}\sum_{k=1,3,5}\frac{192n^4}{(k\pi/l)[k^2-4n^2][k^2-16n^2]}\frac{I_0(k\pi r/l)}{k\pi I'_0(k\pi R/l)}\sin(k\pi z/l)\end{aligned} \tag{9.32}$$

The hydrodynamic pressure is

$$p = -\rho\dot{\Phi}|_{r=R} \tag{9.33}$$

Substituting equation (9.32) into equation (9.33), then substituting the result into equation (9.1) and applying Galerkin's method gives the five homogeneous differential equations

$$\begin{aligned}&\ddot{A}_{m1} + \omega_1^2 A_{m1} + k_{11}\left(A_{m1}^2 + B_{m1}^2\right)A_{m1} + k_{12}\left(A_{m2}^2 + B_{m2}^2\right)A_{m1} \\ &\quad + k_{13}A_{m1}D + k_{14}A_{m1}D^2 = 0\end{aligned} \tag{9.34a}$$

$$\ddot{B}_{m1} + \omega_1^2 B_{m1} + k_{11}\left(A_{m1}^2 + B_{m1}^2\right)B_{m1} + k_{12}\left(A_{m2}^2 + B_{m2}^2\right)B_{m1} + k_{13}B_{m1}D + k_{14}B_{m1}D^2 = 0 \tag{9.34b}$$

$$\ddot{A}_{m2} + \omega_2^2 A_{m2} + k_{21}\left(A_{m1}^2 + B_{m1}^2\right)A_{m2} + k_{22}\left(A_{m2}^2 + B_{m2}^2\right)A_{m2} + k_{23}A_{m2}D + k_{24}A_{m2}D^2 = 0 \tag{9.34c}$$

$$\ddot{B}_{m2} + \omega_2^2 B_{m2} + k_{21}\left(A_{m1}^2 + B_{m1}^2\right)B_{m2} + k_{22}\left(A_{m2}^2 + B_{m2}^2\right)B_{m2} + k_{23}B_{m2}D + k_{24}B_{m2}D^2 = 0 \tag{9.34d}$$

$$\ddot{D} + \omega_3^2 D + k_{31}\left(A_{m1}^2 + B_{m1}^2\right) + k_{32}\left(A_{m2}^2 + B_{m2}^2\right) + k_{33}\left(A_{m1}^2 + B_{m1}^2\right)D + k_{34}\left(A_{m2}^2 + B_{m2}^2\right)D = 0 \tag{9.34e}$$

where k_{ij} are constant coefficients depending on geometrical and physical parameters of the shell and waveform parameters, ω_j are the normal mode frequencies of the shell–liquid system and are given by the following expressions

$$\omega_1^2 = \frac{1}{\rho_c m_{01}}\left[\frac{E\hat{t}^2\left\{(n\pi/l)^2 + (m_1/R)^2\right\}^2}{12(1-\nu^2)} + \frac{E(n\pi/l)^4}{R^2\left\{(n\pi/l)^2 + (m_1/R)^2\right\}^2}\right] \tag{9.35a}$$

$$\omega_2^2 = \frac{1}{\rho_c m_{02}}\left[\frac{E\hat{t}^2\left\{(n\pi/l)^2 + (m_2/R)^2\right\}^2}{12(1-\nu^2)} + \frac{E(n\pi/l)^4}{R^2\left\{(n\pi/l)^2 + (m_2/R)^2\right\}^2}\right] \tag{9.35b}$$

$$\omega_3^2 = \frac{64}{35\rho_c m_{03}}\left[\frac{8E\hat{t}^2(n\pi/l)^4}{12(1-\nu^2)} + \frac{35E}{64R^2}\right] \tag{9.35c}$$

where $m_{0i} = 1 + \dfrac{\rho}{\rho_c}\dfrac{I_{mi}(n\pi R/l)}{I'_{mi}(n\pi R/l)}, \quad i = 1, 2,$

$$m_{03} = 1 + \frac{16\rho}{35\rho_c hl}\sum_{k=1,3,5}^{\infty}\frac{I_0(k\pi R/l)}{k\pi I_1(k\pi R/l)}\left[\frac{192n^4}{(k\pi/l)[k^2 - 4n^2][k^2 - 16n^2]}\right]^2 \tag{9.36}$$

For a cylindrical shell of Young's modulus, $E = 2 \times 10^{11}$ Pa, density, $\rho_c = 7.8 \times 10^3$ kg/m^3, radius, $R = 0.16$ m, Poisson's ratio, $\nu = 0.3$, thickness ratio, $\hat{t}/R = 3.125 \times 10^{-3}$, length-to-radius ratio, $l/R = 2.495$, and for $n = 1$, Kubenko, *et al.* (2003) estimated the natural frequencies of the empty shell (ω_s) and shell filled with water (ω_{sf}) according to equation (9.35a), and the values are listed in Table 9.1. It is seen that the shell–liquid natural frequencies are closely spaced. For example the natural frequencies of the modes 5 and 7 are close to each other for the shell filled with water. Also the modes 5 and 10 are in 1:2 internal resonance.

Equations (9.34) are nonlinearly coupled. The free vibration of the shell–liquid system in the neighborhood of internal resonance conditions under initial conditions results in an energy distribution among the interacting modes. The value of the fifth mode, D, can be replaced by its quasi-static solution by setting $\ddot{D} = 0$. The validity of this approach was addressed by

Table 9.1 *Natural frequencies of empty and filled shell*

m_1	2	3	4	5	6	7	8	9	10	11	12
$\omega_s/2\pi$	1430	756.1	461.7	325.9	277.8	288.1	335.1	405.1	490.2	587.5	695.6
$\omega_{sf}/2\pi$	326.6	203.2	140.1	108.7	100.0	110.6	136.1	172.4	217.6	270.8	331.6

Vol'mir (1972) and Kubenko, *et al.* (1989). As a first-order approximation, the resulting algebraic equation (9.34e) gives the following solution

$$\begin{aligned} D = & -\frac{1}{\omega_3^2}\left[k_{31}\left(A_{m1}^2+B_{m1}^2\right)+k_{32}\left(A_{m2}^2+B_{m2}^2\right)\right] \\ & +\frac{1}{\omega_3^4}\left[k_{31}k_{33}\left(A_{m1}^2+B_{m1}^2\right)^2+k_{32}k_{34}\left(A_{m2}^2+B_{m2}^2\right)^2\right. \\ & \left.+\left(k_{31}k_{34}+k_{32}k_{33}\right)\left(A_{m1}^2+B_{m1}^2\right)\left(A_{m2}^2+B_{m2}^2\right)\right] \end{aligned} \tag{9.37}$$

Introducing equation (9.37) into equations (9.34a–9.34d) and keeping terms up fifth-order, gives

$$\begin{aligned} & \ddot{A}_{m1}+\omega_1^2 A_{m1}+\gamma_1\left(A_{m1}^2+B_{m1}^2\right)A_{m1}+\delta_1\left(A_{m2}^2+B_{m2}^2\right)A_{m1} \\ & \quad +c_1\left(A_{m1}^2+B_{m1}^2\right)A_{m1}+d_1\left(A_{m1}^2+B_{m1}^2\right)\left(A_{m2}^2+B_{m2}^2\right)A_{m1} \\ & \quad +e_1\left(A_{m2}^2+B_{m2}^2\right)^2 A_{m1}=0 \end{aligned} \tag{9.38a}$$

$$\begin{aligned} & \ddot{B}_{m1}+\omega_1^2 B_{m1}+\gamma_1\left(A_{m1}^2+B_{m1}^2\right)B_{m1}+\delta_1\left(A_{m2}^2+B_{m2}^2\right)B_{m1} \\ & \quad +c_1\left(A_{m1}^2+B_{m1}^2\right)^2 B_{m1}+d_1\left(A_{m1}^2+B_{m1}^2\right)\left(A_{m2}^2+B_{m2}^2\right)B_{m1} \\ & \quad +e_1\left(A_{m2}^2+B_{m2}^2\right)^2 B_{m1}=0 \end{aligned} \tag{9.38b}$$

$$\begin{aligned} & \ddot{A}_{m2}+\omega_2^2 A_{m2}+\gamma_2\left(A_{m2}^2+B_{m2}^2\right)A_{m2}+\delta_2\left(A_{m1}^2+B_{m1}^2\right)A_{m2} \\ & \quad +c_2\left(A_{m2}^2+B_{m2}^2\right)A_{m2}+d_2\left(A_{m1}^2+B_{m1}^2\right)\left(A_{m2}^2+B_{m2}^2\right)A_{m2} \\ & \quad +e_2\left(A_{m1}^2+B_{m1}^2\right)^2 A_{m2}=0 \end{aligned} \tag{9.38c}$$

$$\begin{aligned} & \ddot{B}_{m2}+\omega_2^2 B_{m2}+\gamma_2\left(A_{m2}^2+B_{m2}^2\right)B_{m2}+\delta_2\left(A_{m1}^2+B_{m1}^2\right)B_{m2} \\ & \quad +c_2\left(A_{m2}^2+B_{m2}^2\right)^2 B_{m2}+d_2\left(A_{m1}^2+B_{m1}^2\right)\left(A_{m2}^2+B_{m2}^2\right)B_{m2} \\ & \quad +e_2\left(A_{m1}^2+B_{m1}^2\right)^2 B_{m2}=0 \end{aligned} \tag{9.38d}$$

where γ_i, δ_i, c_i, d_i, and e_i $(i=1, 2)$ are parameters that depend on the coefficients k_{ij} and ω_3^2.

In order to determine the periodic solutions of equations (9.38), one may introduce the following coordinate transformation

$$A_{m1}=A_1\cos\alpha_1,\ B_{m1}=A_1\sin\alpha_1,\ A_{m2}=A_2\cos\alpha_2,\ B_{m2}=A_2\sin\alpha_2 \tag{9.39}$$

where A_i and α_i, $i=1, 2$, are the amplitudes and phases of the shell oscillations. Substituting the coordinate transformation (9.39) into equations (9.38), gives

$$\ddot{A}_1+\left(\omega_1^2-\dot{\alpha}_1^2\right)A_1=-\gamma_1 A_1^3-\delta_1 A_1 A_2^2-c_1 A_1^5-d_1 A_1^3 A_2^2-e_1 A_1 A_2^4 \tag{9.40a}$$

$$\ddot{A}_2+(\omega_2^2-\dot{\alpha}_2^2)A_2=-\gamma_2 A_2^3-\delta_2 A_2 A_1^2-c_2 A_2^5-d_2 A_2^3 A_1^2-e_2 A_2 A_1^4 \tag{9.40b}$$

$$A_1\ddot{\alpha}_1+2\dot{A}_1\dot{\alpha}_1=0,\quad A_2\ddot{\alpha}_2+2\dot{A}_2\dot{\alpha}_2=0 \tag{9.40c, d}$$

The shell deflection given by equation (9.30) may be written in terms of the new coordinates in the form

$$\begin{aligned}W=&\,[A_1\cos(m_1\theta-\alpha_1)+A_2\cos(m_2\theta-\alpha_2)]\sin(n\pi z/l)\\&+\left\{-\frac{1}{\omega_3^2}(k_{31}A_1^2+k_{32}A_2^2)+\frac{1}{\omega_3^4}\left[k_{31}k_{33}A_1^4+k_{32}k_{34}A_2^4\right.\right.\\&\left.\left.+(k_{31}k_{34}+k_{32}k_{33})A_1^2A_2^2\right]\right\}\sin^4(n\pi z/l)\end{aligned} \tag{9.41}$$

Careful inspection of equations (9.40a,b) reveals the presence of secular terms that give rise to internal resonance conditions of the form

$$\omega_1=\omega_2,\quad \omega_1=\frac{1}{2}\omega_2,\quad \omega_1=2\omega_2 \tag{9.42}$$

Kubenko, *et al.* (2003) used the averaging method to determine the free periodic vibration of equations (9.40) in the neighborhood of the internal resonance condition $\omega_1=\omega_2/2$. It was shown that the period of free oscillations strongly depends on the initial amplitude $A_1(t=0)$. As $A_1(t=0)$ increases, the period increases for $\dot{\alpha}_i(t=0)=0$, and decreases for $\dot{\alpha}_i(t=0)\neq 0$.

9.4.4 Multiple internal resonances

Several studies of shell nonlinear interaction with a liquid have focused on simplified single-mode models. These studies are only valid when the natural frequencies are well spaced and not commensurable. The majority of cylindrical shells filled with liquid possesses a dense frequency spectrum and may be commensurable. In circular shells any asymmetric mode has a companion (orthogonal) mode possessing the same natural frequency. These two modes give rise to 1:1 internal resonance. In some geometry, it is possible that the first axisymmetric mode has almost the same frequency as the asymmetric mode. It may happen that the natural frequency of the third axisymmetric mode equals twice the frequency of the first axisymmetric mode. Amabili, *et al.* (2000c,d) considered the case of 1:1:1:2 internal resonances in a circular cylindrical tank filled with water. The nonlinear interaction between vibrations of a circular cylindrical shell and the free-surface waves was studied by Boyarshina (1984) for three different cases of internal resonance. The equations of motion of the first two anti-symmetric sloshing modes were coupled with the first two bending modes of the shell. The internal resonance conditions correspond to the cases when the first bending frequency is close to the first anti-symmetric sloshing mode frequency, or twice the sloshing frequency or three times, that is, 1:1, 1:2, and 1:3, respectively. The specific case of 1:2 internal resonance was carefully analyzed and showed that the interaction between the free surface and flexural shell vibrations took place in the form of energy exchange. Under external excitation of the shell, the response amplitude was found to be unstable over the frequency range where the fluid free surface oscillates.

Nayfeh and Raouf (1987) examined the nonlinear interaction in cylindrical shells in the neighborhood of 2:1 internal resonance when the frequency of the axisymmetric mode is close to twice that of the asymmetric mode.

The nonlinear forced excitation of a cylindrical shell filled with liquid was studied numerically and analytically in the neighborhood of multiple internal resonances by Amabili, *et al.* (2000c,d). The tank height was selected to be almost equivalent to its diameter to satisfy the internal resonance conditions. The mode expansion was modified to the following form

$$w(z,\theta,t) = \left[A_{m,1}(t)\cos m\theta + B_{m,1}(t)\sin m\theta\right]\sin(\pi z/l) + \sum_{n=1}^{2} A_{0,(2n-1)}(t)\sin(\lambda_{(2n-1)}z) \tag{9.43}$$

where $\lambda_{(2n-1)} = (2n-1)\pi/l$. By using the assumed mode shapes given by equation (9.43), the solution of the boundary value problem (9.21)–(9.23) for the internal fluid is

$$\Phi = \left[\dot{A}_{m1}(t)\cos m\theta + \dot{B}_{m1}(t)\sin m\theta\right]\frac{lI_m(\pi r/l)\sin(\pi z/l)}{\pi I'_m(\pi R/l)} + \dot{A}_{01}(t)\frac{lI_0(\pi r/l)\sin(\pi z/l)}{\pi I'_0(\pi R/l)} + \dot{A}_{03}(t)\frac{lI_0(3\pi r/l)\sin(3\pi z/l)}{3\pi I'_0(3\pi R/l)} \tag{9.44}$$

The hydrodynamic pressure is given by equation (9.33).

Substituting equation (9.43) into equation (9.2), gives a partial differential equation for the stress function F whose homogeneous, F_h, and particular integral, F_p, solutions are, respectively,

$$F_h = \frac{1}{2}\bar{N}_z R^2\theta^2 + \frac{1}{2}z^2\left\{\bar{N}_\theta + \frac{1}{2\pi Rl}\int_0^l\int_0^{2\pi}\left[\lambda_n^2 C_3(t)\sin\lambda_n z + 9C_{11}(t)\sin 3\lambda_n z\right]R\,d\theta\,dz\right\} - \bar{N}_{z\theta}zR\theta = \frac{1}{2}\bar{N}_z R^2\theta^2 + \frac{1}{2}z^2\left\{\bar{N}_\theta + \frac{1-(-1)^n}{l}\lambda_n[C_3(t)+3C_{11}(t)]\right\} - \bar{N}_{z\theta}zR\theta \tag{9.45a}$$

$$\begin{aligned}F_p = {} & C_1(t)\cos m\theta + C_2(t)\sin m\theta + C_3(t)\sin\lambda_n z + C_4(t)\cos m\theta\sin\lambda_n z \\ & + C_5(t)\sin m\theta\sin\lambda_n z + C_6(t)\cos 2\lambda_n z + C_7(t)\cos 2m\theta + C_8(t)\sin 2m\theta \\ & + C_9(t)\cos m\theta\cos 2\lambda_n z + C_{10}(t)\sin m\theta\sin 2\lambda_n z + C_{11}(t)\sin 3\lambda_n z \\ & + C_{12}(t)\cos m\theta\cos 4\lambda_n z + C_{13}(t)\sin m\theta\cos 4\lambda_n z\end{aligned} \tag{9.45b}$$

where $\bar{N}_z$, $\bar{N}_\theta$, and $\bar{N}_{z\theta}$ are the in-plane restraint stresses generated at the ends of the shell. The coefficients $C_i(t)$ are functions of the generalized coordinates $A_{mn}(t)$, $B_{mn}(t)$, and $A_{0n}(t)$, and shell parameters. Solution (9.45a) is not the most general homogeneous solution, but was selected by Amabili, *et al.* (1998b) on the basis that it satisfies the force per unit length in the axial and circumferential directions, and the shear force

$$N_z = \frac{1}{R^2}\frac{\partial^2 F}{\partial\theta^2}, \quad N_\theta = \frac{\partial^2 F}{\partial z^2}, \quad N_{z\theta} = \frac{1}{R}\frac{\partial^2 F}{\partial z\partial\theta} \tag{9.46}$$

Galerkin's method is used to generate four, second-order ordinary, nonlinear coupled differential equations for the amplitudes $A_{m,1}(t)$, $B_{m,1}(t)$, $A_{01}(t)$, and $A_{0,3}(t)$, together with the weighing functions $\cos(m\theta)\sin(\pi z/l)$, $\sin(m\theta)\sin(\pi z/l)$, $\sin(\pi z/l)$, and $\sin(3\pi z/l)$. The external excitation is assumed to be in the form

$$f(t) = f_{m,1}\cos(m\theta)\sin(\pi z/l)\cos(\Omega t)$$

and the following four equations are obtained

$$\begin{aligned}\ddot{A}_{m,1}(t) &+ 2\zeta_{m,1}\omega_{m,1}\dot{A}_{m,1}(t) + \omega_{m,1}^2 A_{m,1}(t) + h_1 A_{m,1}(t)A_{01}(t) + h_2 A_{m,1}^3(t)\\ &+ h_2 A_{m,1}(t)B_{m,1}^2(t) + h_3 A_{m,1}(t)A_{03}(t) + h_4 A_{m,1}(t)A_{01}^2(t)\\ &+ h_5 A_{m,1}(t)A_{0,3}^2(t) + h_6 A_{m,1}(t)A_{0,1}(t)A_{0,3}(t) = \frac{\pi l}{2m_1} f_{m1}\cos\Omega t\end{aligned} \tag{9.47a}$$

$$\begin{aligned}\ddot{B}_{m,1}(t) &+ 2\zeta_{m,1}\omega_{m,1}\dot{B}_{m,1}(t) + \omega_{m,1}^2 B_{m,1}(t) + h_1 B_{m,1}(t)A_{0,1}(t) + h_2 B_{mn}^3(t)\\ &+ h_2 B_{m,1}(t)A_{m,1}^2(t) + h_3 B_{m,1}(t)A_{03}(t) + h_4 B_{m,1}(t)A_{0,1}^2(t)\\ &+ h_5 B_{m,1}(t)A_{0,3}^2(t) + h_6 B_{m,1}(t)A_{0,1}(t)A_{0,3}(t) = 0\end{aligned} \tag{9.47b}$$

$$\begin{aligned}\ddot{A}_{01}(t) &+ 2\zeta_{01}\omega_{01}\dot{A}_{01}(t) + \omega_{01}^2 A_{01}(t) + k_1 A_{m,1}^2(t) + k_1 B_{m,1}^2(t)\\ &+ k_2 A_{01}(t)A_{m,1}^2(t) + k_2 A_{01}(t)B_{m,1}^2(t) + k_3 A_{0,3}(t)A_{m,1}^2\\ &+ k_3 A_{03}(t)B_{m,1}^2 = 0\end{aligned} \tag{9.47c}$$

$$\begin{aligned}\ddot{A}_{0,3}(t) &+ 2\zeta_{0,3}\omega_{0,3}\dot{A}_{0,3}(t) + \omega_{0,3}^2 A_{0,3}(t) + n_1 A_{m,1}^2(t) + n_1 B_{m,1}^2(t)\\ &+ n_2 A_{0,3}(t)A_{m,1}^2(t) + n_2 A_{0,3}(t)B_{m,1}^2(t) + n_3 A_{0,1}(t)A_{m,1}^2\\ &+ n_3 A_{0,1}(t)B_{m,1}^2 = 0\end{aligned} \tag{9.47d}$$

where

$$\begin{aligned}
m_1 &= \pi\rho_c\hat{t}l/2 + \frac{\rho l^2 I_m(/l)}{2I_m(\pi R/l)},\\
\omega_{m,1}^2 &= \frac{\pi l}{2m_1}\left[D\left(\frac{\pi^2}{l^2}+\frac{m^2}{R^2}\right)^2 + \frac{E\hat{t}\pi^4}{R^2 l^4(\pi^2/l^2)+(m^2/R^2)^2}\right]\\
\zeta_{m,1} &= \pi c\hat{t}l/(4\omega_{m,1}m_1),\\
\omega_{0,1}^2 &= \frac{\pi l}{m_{0,1}}\left[\frac{D\pi^4}{l^4}+\frac{E\hat{t}}{R^2}\right],\\
m_{0,1} &= \pi\rho_c\hat{t}l + \frac{\rho l^2 I_0(\pi R/l)}{I_0(\pi R/l)},\\
\zeta_{0,1} &= \pi c\hat{t}l/(2\omega_{0,1}m_{0,1}),\\
m_{0,3} &= \pi\rho_c\hat{t}l + \rho l^2 I_0(3\pi R/l)3I_0(3\pi R/l),\\
\omega_{0,3}^2 &= \frac{\pi l}{2m_{0,3}}\left[\frac{81D\pi^4}{l^4}+\frac{E\hat{t}}{R^2}\right],\\
\zeta_{0,3} &= \pi c\hat{t}l/(2\omega_{0,3}m_{0,3})
\end{aligned}$$

and h_i, k_i, and n_i are coefficients that depend on the geometry and material properties.

Note that the two modes $A_{m,1}$ and $B_{m,1}$ are orthogonal and in 1:1 internal resonance. Equations (9.47) were solved numerically for a cylindrical shell filled with water, shell length $l = 0.618$ m, radius $R = 0.3$ m, thickness $\hat{t} = 0.0006$ m, and material Young's modulus $E = 206 \times 10^9$ Pa, density $\rho_c = 7850\,\text{kg/m}^3$, Poisson's ratio $\varepsilon = 0.3$, and water density $\rho = 1000\,\text{kg/m}^3$. For the mode $m = 20$, $n = 1$, and damping ratio $\zeta_{m,1} = 0.01$, the natural frequencies of the water-filled shell were $\omega_{m,1} = 2\pi$ (323.9) rad/s, $\omega_{0,1} = 2\pi$ (324.0) rad/s, $\omega_{0,3} = 2\pi$ (663.6) rad/s.

Under excitation amplitude ratio $f_1 = 2\pi f_{m,1}/(2m_1\omega_{m,1}^2\hat{t}) = 0.01$, the amplitude–frequency response curves of the four modes are shown in Figure 9.6. Stable solutions are shown by solid curves while unstable solutions are given by dash curves. Over the frequency ratio range $0.98 \leq \Omega/\omega_{m,1} \leq 0.9965$ the solution is always unstable and direct integration by Amabili, *et al.* (2000d) revealed stable limit cycles with modulated amplitude of the harmonic response. For several values of the frequency ratio $\Omega/\omega_{m,1}$ there exist two or three stable solutions. The mode that is directly excited exhibits a second peak at the resonance of the third axisymmetric mode (3,0) due the 2:1 internal resonance. As the excitation frequency increases from below, the following regimes were reported: (i) initially the shell experiences a simple periodic response, (ii) when the excitation frequency reaches $0.97\omega_{m,1}$, the shell response exhibits a sudden increase in the amplitude if the system is perturbed due to internal resonance with the first axisymmetric mode, (iii) over the excitation frequency ratio $0.98 \leq \Omega/\omega_{m,1} \leq 0.9965$ there exists an amplitude modulated response as mentioned earlier, (iv) over the excitation frequency ratio $0.9965 \leq \Omega/\omega_{m,1} \leq 1.008$ a traveling wave superimposed to a stationary response was detected as a consequence of the internal resonance with the companion mode, (v) for $\Omega/\omega_{m,1} > 1.008$ a periodic response was obtained due to the internal resonance with the third axisymmetric mode.

The dependence of the response amplitudes on the excitation amplitude parameter f_1 at excitation frequency ratio $\Omega/\omega_{m,1} = 1.0$ is plotted in Figure 9.7. The saturation phenomenon is manifested for the directly excited mode. The amplitude of the first axisymmetric mode, which is in 1:1 resonance with the directly excited mode, increases after the beginning of the saturation regime. The companion mode does not participate along the saturation branch, while the third axisymmetric mode is simply dragged from the excited mode. The solution loses stability for $f_1 = 0.0057$ where branch "4" appears and the companion mode becomes active. For $f_1 = 0.0092$, branch "3" becomes stable and all response amplitudes are active but with different behavior especially for the axisymmetric modes.

9.4.5 Linear shell interaction with nonlinear liquid sloshing

The nonlinear behavior of elastic shells partially filled with liquid may also be created by large-amplitude motion of the liquid surface. The liquid motion in turn causes a nonlinear reaction force (apparent mass) on the shell wall. Both shell and liquid have linear governing equations, but the nonlinearities enter the problem through nonlinear boundary conditions. Chu and Kana (1967) developed a nonlinear treatment to study nonlinear transverse vibrations of liquid shells based on experimental tests conducted by Kana, *et al.* (1966). The experimental tests were carried out on a simply supported partially filled cylindrical shell with a rigid flat bottom and excited laterally. The excitation amplitude and frequency were adjusted to vibrate one of

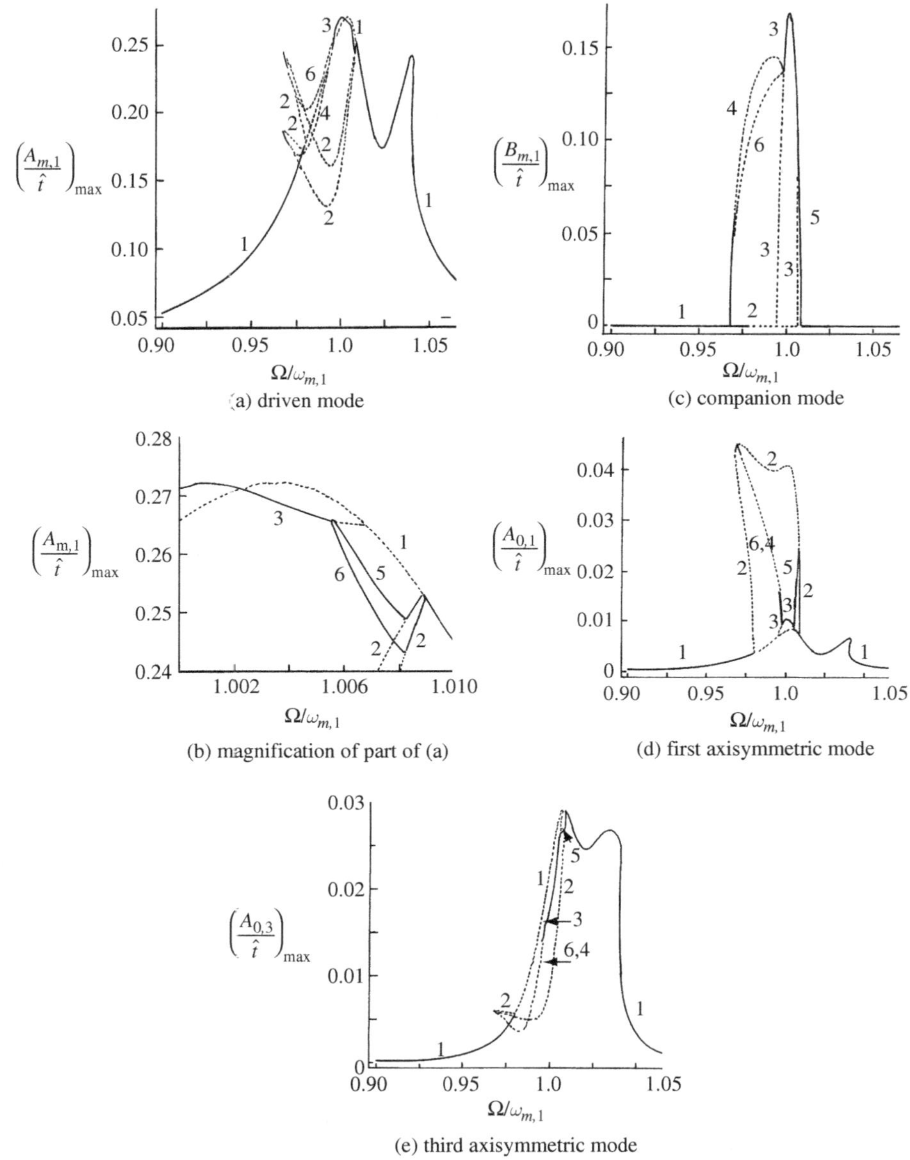

Figure 9.6 Amplitude–frequency response curves in 1:1:1:2 internal resonanceconditions under excitation level $f_1 = 0.01$, integer numbers refer to different solutions. (Amabili, *et al.*, 2000b)

its asymmetric modes as shown in Figure 9.8. The study considered shell vibration of four nodal diameters $m = 4$, and two half waves, $n = 2$. The shell amplitude–frequency response exhibits softening nonlinear characteristics with one region over which jump phenomena take place, and another region characterized by instability in which complicated nonlinear coupling occurs. The coupling was characterized by simultaneous occurrence of a low-frequency amplitude modulation of the high-frequency shell response and the excitation of the liquid

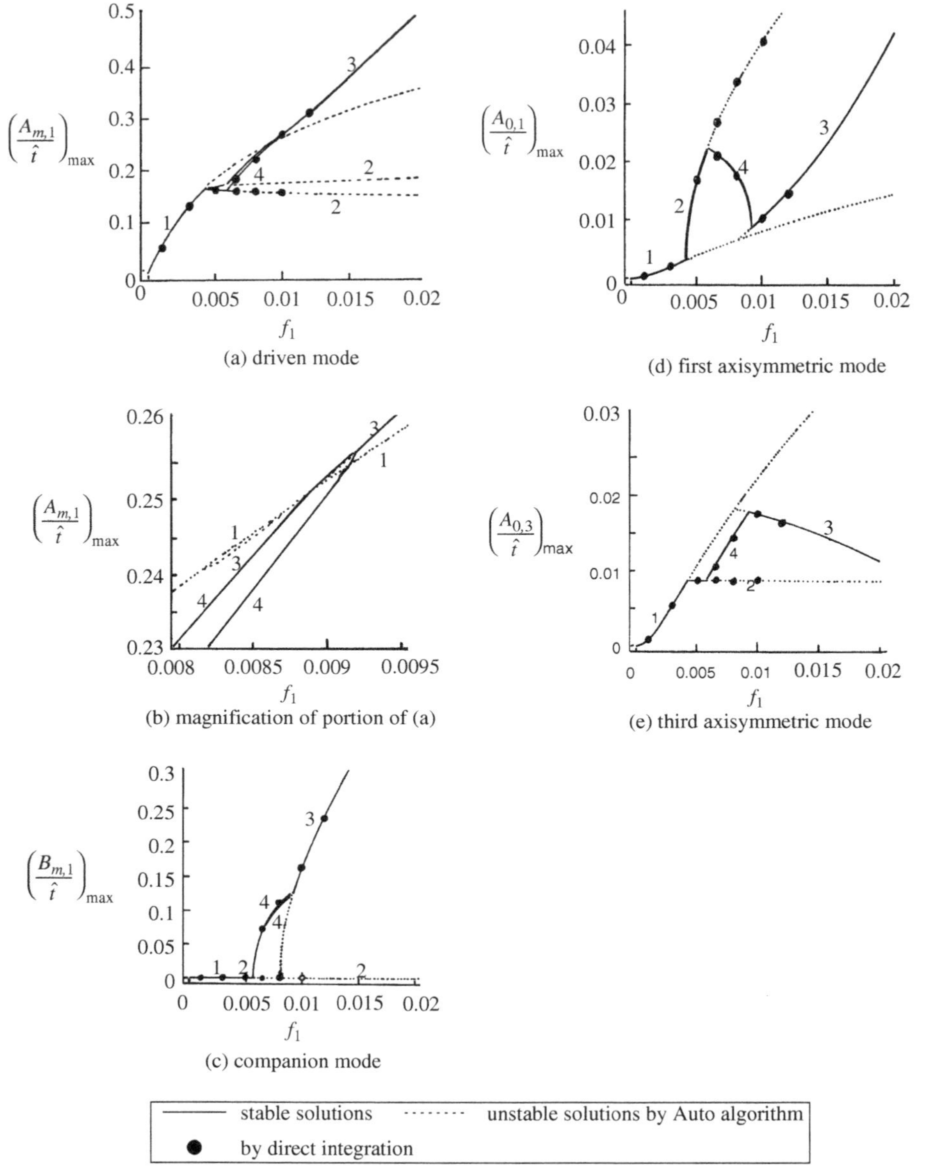

Figure 9.7 Dependence of response amplitudes on excitation amplitude in 1:1:1:2 internal resonance conditions integer numbers refer to different solutions. (Amabili, *et al.*, 2000b)

surface in one of its axisymmetric sloshing modes. The shell radial displacement can be expanded in a series of normal modes of the empty shell in the form

$$w(z, \theta, t) = \sum_{m=1}^{M} \sum_{n=1}^{N} A_{mn}(t) f_n(z) \cos m\theta \tag{9.48}$$

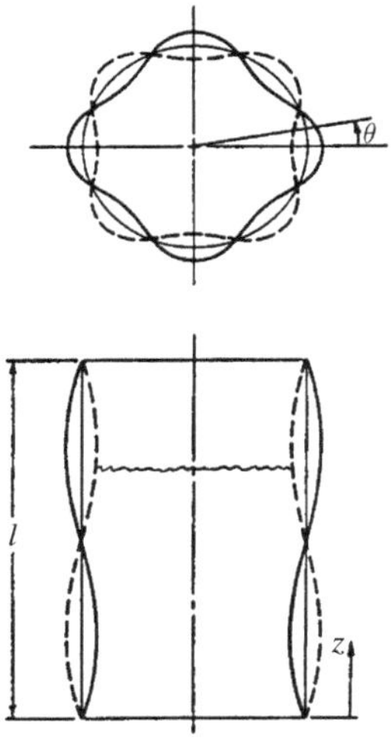

Figure 9.8 Schematic diagram of a partially full pin-ended shell showing its deformation under forced excitation.

where

$$\frac{\int_0^l f_n(z) f_m(z)\, dz}{\int_0^l f_n^2(z)\, dz} = \delta_{mn} \tag{9.49}$$

For simply supported end conditions the following mode shape may be used

$$f_n(z) = \sin(n\pi z/l) \tag{9.50}$$

The amplitude A_{mn} of the mth mode of order n is governed by equation (8.122) that is,

$$\ddot{A}_{mn}(t) + \omega_{mn}^2 A_{mn}(t) = \frac{q_{mn}(t)}{\rho_c \hat{t}} \tag{9.51}$$

where the generalized hydrodynamic loading is given by equation (8.123).

For a circular tank, the solution of Laplace's equation (9.21) that satisfies the normal velocity conditions on the wetted portion of the tank is

$$\Phi_{mn} = \phi(t) \cos m\theta J_m(\xi_{mn} r/R) \cosh(\xi_{mn} z/R) \tag{9.52}$$

where $J_m(\xi_{mn} r/R)$ is the Bessel function of the first kind of order m, and ξ_{mn} are the roots of $J_m\,(\xi_{mn}) = 0$. Under lateral excitation of the shell, a particular solution corresponding to an axial breathing mode shape $f_n(z)$ with circumferential mode shape $\cos m\theta$ is

$$\Phi_p = -R\dot{A}_{mn} \sum_{k=0}^{\infty} D_{kn} \frac{I_m(k\pi r/h) \cos(k\pi z/h) \cos m\theta}{(k\pi R/h) I_m'(k\pi r/h)} \tag{9.53}$$

where

$$D_{kn} = \frac{2}{h(1+\delta_{0k})} \int_0^h f_m(z) \cos(k\pi z/h)\, dz \tag{9.54}$$

At $z = h$, one may assume that the complementary velocity potential function $\Phi_c = -\Phi_p$. This is valid as long as the breathing frequency or the excitation frequency is much higher than the sloshing frequency. The kinematic free-surface boundary condition may be expanded in a Taylor series as:

$$\frac{\partial\eta}{\partial t} + \frac{\partial\eta}{\partial r}\left[\frac{\partial\Phi}{\partial r} + \eta\frac{\partial^2\Phi}{\partial z\partial r} + \cdots\right] + \frac{\partial\eta}{r\partial\theta}\left[\frac{\partial\Phi}{r\partial\theta} + \eta\frac{\partial^2\Phi}{r\partial\theta\partial z} + \cdots\right] - \left[\frac{\partial\Phi}{\partial z} + \eta\frac{\partial^2\eta}{\partial z^2} + \cdots\right] = 0 \tag{9.55}$$

Similarly, the dynamic free-surface boundary may be written in the form

$$\frac{\partial\Phi}{\partial t} + \frac{\partial^2\Phi}{\partial t\partial z}\eta + \cdots + \frac{1}{2}\left[\left(\frac{\partial\Phi}{\partial r}\right)^2 + \left(\frac{\partial\Phi}{\partial z}\right)^2 + \left(\frac{\partial\Phi}{r\partial\theta}\right)^2 + \cdots\right] + g\eta = 0 \tag{9.56}$$

Since the breathing frequency is much higher than the fundamental sloshing frequency, the dominant nonlinear terms belong to the dynamic pressure, and are apparently of second order. Thus, the free-surface amplitude is assumed to be of order $\varepsilon^{1/2}$ and the shell amplitude is of order ε. In this case, one may assume the axial fluid free-surface amplitude to be in the form

$$\eta = B_{00}(t) + \sum_{m_1}\sum_{k_1} B_{m_1k_1}(t) J_{m_1}(\xi_{m_1k_1} r/R) \cos m_1\theta \tag{9.57}$$

The velocity potential function of the shell may be assumed to be in the form

$$\Phi_{mn} = -R\dot{A}_{mn}\cos m\theta \sum_{k=0}^{\infty} D_{kn}\frac{I_m(k\pi r/h)\cos(k\pi z/h)}{(k\pi R/h) I'_m(k\pi r/h)} + \sum_{m_1}\sum_{k_1} C_{m_1k_1}(t) J_{m_1}(\xi_{m_1k_1} r/R)\frac{\cosh(\xi_{m_1k_1} z/R)\cos m_1\theta}{\cosh(\xi_{m_1k_1} h/R)} \tag{9.58}$$

One may consider $B_{m1k1}(t)$ to be of order $\varepsilon^{1/2}$ and A_{mn} of order ε. The ratio of the excitation frequency, Ω, to the sloshing frequency, ω_{mn}, is of order $\varepsilon^{1/2}$. The hydrodynamic loading acting on the shell consists of two components. The first is due to the unperturbed fluid motion, q^0_{mn}, and the other is due to the free-surface oscillations, q^a_{mn}. These are given by the following expressions

$$\frac{q^0_{mn}}{\rho_c\hat{t}} = \frac{2}{\pi l(1+\delta_{0m})}\frac{\rho R}{\rho_c\hat{t}} \times \int_{-\pi}^{\pi}\int_0^h\left[\frac{\partial\Phi}{\partial t} + \frac{1}{2}\left(\frac{\partial\Phi}{\partial r}\right)^2 + \frac{1}{2}\left(\frac{\partial\Phi}{\partial z}\right)^2 + \frac{1}{2}\left(\frac{\partial\Phi}{r\partial\theta}\right)^2 + O(\varepsilon^{3/2})\right] \times \sin(n\pi z/l)\cos m\theta\, dz\, d\theta \tag{9.59a}$$

$$\frac{q^a_{mn}}{\rho_c \hat{t}} = \frac{2}{\pi l(1+\delta_{0,n})} \frac{\rho R}{\rho_c \hat{t}} \int_{-\pi}^{\pi} \int_{h}^{h+\eta} \left[\frac{\partial \Phi}{\partial t} + O(\varepsilon)\right] \sin(n\pi z/l) \cos m\theta \, dz \, d\theta \tag{9.59b}$$

Chu and Kana (1967) studied a specific case in which the shell is 3/4 full of liquid, and the excitation frequency is in the neighborhood of m, $n = 4$, 2 breathing mode frequency. The liquid motion was assumed to be dominated by the first mode. Applying Galerkin's method to kimematic and dynamic conditions and the shell equation, gives the following three nonlinear-coupled ordinary differential equations

The shell equation is

$$[1 - S_1 \bar{B}_{01}(\tau) + S_2 \bar{B}_{81}(\tau)] \frac{d^2 \bar{A}_{42}(\tau)}{d\tau^2} + \frac{\bar{\omega}^2_{42}}{M^* \Omega^2} \bar{A}_{42}(\tau) = f_{42} \cos \tau \tag{9.60a}$$

The sloshing equations are

$$\frac{d^2 \bar{B}_{01}(\tau)}{d^2} + \frac{\bar{\omega}^2_{42} L_1}{M^* \Omega^2} \bar{B}_{01}(\tau) = L_2 \left(\frac{d\bar{A}_{42}(\tau)}{d}\right)^2 + L_3 \bar{A}_{42}(\tau) \frac{d^2 \bar{A}_{42}(\tau)}{d^2} \tag{9.60b}$$

$$\frac{d^2 \bar{B}_{81}(\tau)}{d\tau^2} + \frac{\bar{\omega}^2_{42} L_6}{M^* \Omega^2} \bar{B}_{81}(\tau) = -L_4 \left(\frac{d\bar{A}_{42}(\tau)}{d\tau}\right)^2 - L_5 \bar{A}_{42}(\tau) \frac{d^2 \bar{A}_{42}(\tau)}{d\tau^2} \tag{9.60c}$$

where M^* is the apparent mass of the (4,2) mode of the shell, defined by equation (8.130a) and $\tau = \Omega t$, $S_1 = 0.063\,88$, $S_2 = 0.024\,45$, $f_{42} = 0.2059 \dfrac{F_{42} \omega^2_{42}}{M^* \Omega^2}$, F_{42} is the excitation amplitude associated with the (4, 2) mode, $L_1 = 5.6284 \times 10^{-5}$, $L_2 = 0.545\,39$, $L_3 = 1.090\,78$, $L_4 = 5.479$, $L_5 = 10.958$. For the steady state solution, the shell amplitude response is assumed to be in the form

$$\bar{A}_{42} = Y \cos \tau \tag{9.61a}$$

Retaining only the $\varepsilon^{1/2}$ terms in $\bar{B}_{01}$ and $\bar{B}_{81}$, one writes

$$\bar{B}_{01} = -L_6 Y^2, \quad \text{and} \quad \bar{B}_{81} = 4 L_6 Y^2 \tag{9.61b, c}$$

Substituting equations (9.61) into equation (9.60a), gives a cubic equation of amplitude Y of the shell dominant mode in the form

$$a_3 Y^3 + a_1 Y + a_0 = 0 \tag{9.62}$$

where

$$a_3 \cong 0.7822 \times 10^3 \Omega M^* / \omega^2_{42},$$
$$a_1 = (\Omega^2 - (\omega^2_{42}/M^*))/\Omega^2,$$
$$a_0 = 0.2059 \omega^2_{42} / (M^* \Omega^2 F_{42})$$

The shell amplitude–frequency response curves for different excitation amplitudes are plotted in Figure 9.9 for $h/l = 0.75$, $R = 1.485$ inch, $l = 9.2$ inch, $\hat{t} = 0.009$ inch, $\rho = 1.938$ slug/ft^3, $\rho_c = 15.19$ slug/ft^3. The triangular empty points indicate results obtained using an analog

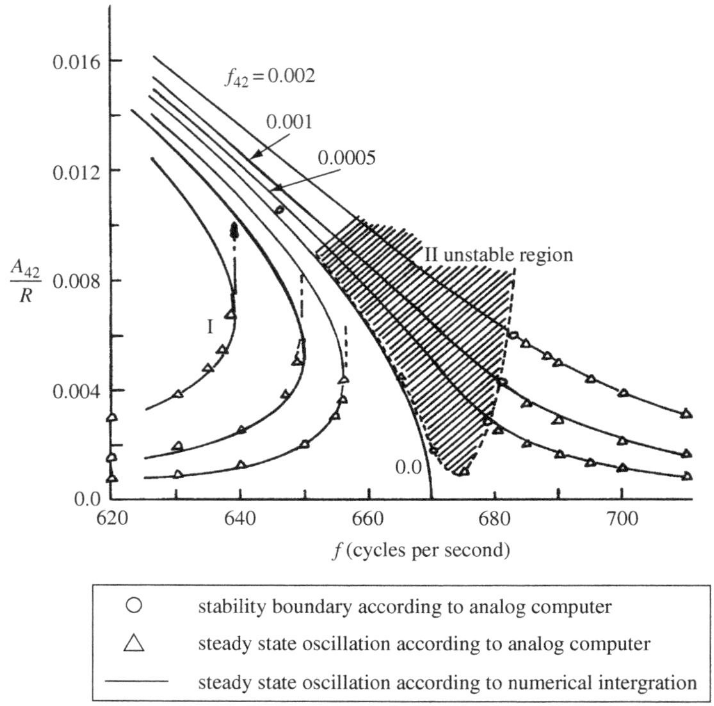

Figure 9.9 Amplitude–frequency response curves for different values of excitation amplitude level f_{42}. (Kana, *et al.*, 1966)

computer, while circular empty points indicate stability boundaries predicted by analog computer. Experimentally, it was observed that several liquid modes were excited and in order to have good quantitative comparison one should include higher sloshing modes and more terms in the expansion of the axial mode shape. The observed phenomena were attributed to the nonlinear liquid loading, which produces an amplitude-dependent apparent mass on the shell wall.

9.4.6 Nonlinear sloshing interaction with linear elastic bottom

The coupling of nonlinear free-surface motion with linear elastic bottom deformation is studied by considering the nonlinear free-surface dynamic condition. The analysis includes the nonlinear natural frequencies of the coupled system and the nonlinear response. Bhuta and Yeh (1965) analyzed the axisymmetric motion of the liquid free surface due to the outlet velocity fluctuations at the bottom. Their work motivated Bauer, Chang, and Wang (1971), Chiba (1989), Bauer and Chiba (2000a,b, 2001), and Chiba (1992, 1993a) to conduct experimental and analytical investigations to study the axisymmetric vibration of a liquid cylindrical tank with an elastic bottom when the bottom center is harmonically excited with a constant amplitude. Chiba (1994) examined the free axisymmetric vibration of an elastic bottom plate of a liquid-filled cylindrical tank supported on an elastic foundation. The influence of the in-plane force due to static deflection of the bottom plate was found to be significant for a thin-bottomed plate. The nonlinear axisymmetric free vibration of the coupled system was then considered using the Ritz method (Chiba, 1996b). Two distinct cases were considered, namely the bulging-type response in which plate motion is predominant, and the sloshing-type motion. For the first three modes, the natural frequencies were found to increase with the bottom deflection rms, indicating hard-spring nonlinearity. For the sloshing-type motion, the backbone curves for the first sloshing mode were essentially of soft-spring type. Chiba (1997) extended his previous work to examine the influence of the liquid depth on the free-surface response when the elastic bottom is excited, taking into account the effect of the bottom static deflection. It was found that the motion of the elastic bottom plate resulted in a free-surface peak at a fluid depth ratio $h/R=0.17$ and then a decrease with liquid height in the first axisymmetric mode. In the second and third modes, the maximum amplitude decreases with an increase in the order of vibration mode. Boyarshina (1994) and Kubenko and Savin (2000) modeled the bottom as a thin elastic plate clamped with the rigid cylinder wall. The fluid field equations are

$$\nabla^2 \Phi = 0 \tag{9.63a}$$

$$\left.\frac{\partial \Phi}{\partial r}\right|_{r=R} = 0, \tag{9.63b}$$

$$\left.\frac{\partial \Phi}{\partial z}\right|_{z=0} = -\left.\frac{\partial w}{\partial t}\right|_{z=0}, \tag{9.63c}$$

$$\left.\frac{\partial \Phi}{\partial z}\right|_{z=h} = \left.\frac{\partial \eta}{\partial t}\right|_{z=h} \tag{9.63d}$$

$$\frac{\partial \Phi}{\partial t} + \frac{1}{2}(\nabla \Phi)^2 + gz = 0|_{z=h} \tag{9.63e}$$

where w is the bending deflection of the bottom. The bottom deflection is governed by equation (8.64a) whose solution must satisfy the clamped boundary condition and may be approximated by the "single-mode" representation

$$w(r,\theta,t) = b(t)\,Y_{mn}(r)\cos[m\theta - \beta(t)] \tag{9.64}$$

where $b(t)$ is the generalized coordinate of the bottom deflection, $\beta(t)$ is the corresponding phase angle, and $Y_{mn}(r)$ is the normal mode shape of the plate given by the following expression

$$Y_{mn}(r) = J_m(k_{mn}r)I_m(k_{mn}R) - J_m(k_{mn}R)I_m(k_{mn}r) \tag{9.65}$$

where k_{mn} are the roots of the equation

$$J_{m+1}(k_{mn})I_m(k_{mn}) + J_m(k_{mn})I_{m+1}(k_{mn}) = 0 \tag{9.66}$$

The solution of the fluid field equation yields the following representation of the free-surface amplitude

$$\eta(r,\theta,t) = h + \eta_0(t)\cos[m\theta - \gamma(t)]\Re_{1n}(r) \tag{9.67}$$

where $\eta_0(t)$ is the generalized coordinate, $\gamma(t)$ is the corresponding phase angle, and $\Re_{1n}(r)$ is the eigenfunctions of the homogeneous boundary-value problem

$$\frac{d^2\Re}{dr^2}+\frac{1}{r}\frac{d\Re}{dr}+\left(\lambda_{1n}^2-\frac{n^2}{R^2}\right)\Re=0 \tag{9.68a}$$

$$\left.\frac{d\Re}{dr}\right|_{r=R}=0 \tag{9.68b}$$

in which λ_{1n} is the first eigenvalue corresponding to the sloshing mode $1n$.

Nonlinear modal analysis

Boyarshina (1994) employed Hamilton's principle and obtained the following set of differential equations of the combined liquid–bottom system

$$\ddot{\eta}_0+(\omega^2-\dot{\gamma}^2)\eta_0+\delta\left[\left(\dot{b}-b\dot{\beta}^2\right)\cos(\beta-\gamma)-b(\ddot{\beta}+2\beta)\sin(\beta-\gamma)\right]$$
$$=d_1\eta_0^2\ddot{\gamma}+d_2\eta_0^3\dot{\gamma}^2+d_3\dot{\eta}_0^2\eta_0 \tag{9.69a}$$

$$\eta_0\ddot{\gamma}+2\dot{\eta}_0\dot{\gamma}+\delta\left[b(\ddot{\beta}+2\beta)\cos(\beta-\gamma)-\left(\dot{b}-b\dot{\beta}^2\right)\sin(\beta-\gamma)\right]$$
$$=d_4\eta_0^2[\ddot{\gamma}\eta_0+\gamma\dot{\gamma}\dot{\eta}_0] \tag{9.69b}$$

$$\ddot{b}+(\varpi^2-\beta^2)b+\sigma\left(\dot{b}-b\dot{\beta}^2\right)+e\left[(\ddot{\eta}_0-\eta_0\dot{\gamma}^2)\cos(\gamma-\beta)\right.$$
$$\left.-(\eta_0\ddot{\gamma}+2\dot{\eta}_0\dot{\gamma})\sin(\gamma-\beta)\right]=0 \tag{9.69c}$$

$$\dot{b}\beta+2b\dot{\beta}+\sigma b(\ddot{\beta}+2\dot{\beta})+e[(2\dot{\eta}_0\dot{\gamma}+\eta_0\ddot{\gamma})\cos(\gamma-\beta)$$
$$-(\ddot{\eta}_0-\eta_0\dot{\gamma}^2)\sin(\gamma-\beta)]=0 \tag{9.69d}$$

where d_i, $i=1,\ldots,4$ are constant coefficients that depend on the liquid depth ratio h/R,

$$\omega^2=g\lambda_{1n}\tanh(\lambda_{1n}h)$$

$$\varpi^2=\frac{k^4E\hat{t}^2}{12\rho_b(1-\nu^2)}$$

$$\delta=-\frac{\lambda_{1n}\text{cotanh}(kh)}{k\,\text{cotanh}(\lambda_{1n})J_m(\lambda_{1n}R)}\int_0^R J_m(kr)r\,dr\int_0^R I_m(kr)J_m(\lambda_{1n}r)r\,dr$$

$$\sigma=\frac{J_m(kR)}{\rho_b\hat{t}k^2J_m^2(\lambda_{1n}R)}\left[\int_0^R I_m(kr)J_m(\lambda_{1n}r)r\,dr\right]^2$$

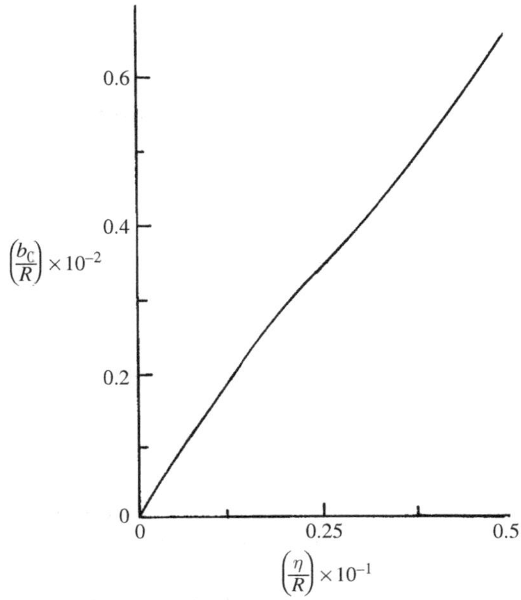

Figure 9.10 Dependence of bottom deformation of free-surface wave height. (Boyarshina, 1994)

$$e = -\frac{\text{cotanh}(kh)}{k\rho_b \hat{t} J_m(\lambda_{1n}R)} \int_0^R J_m(kr) r\, dr \int_0^R I_m(kr) J_m(\lambda_{1n} r) r\, dr$$

For a flexible bottom, the fluid free-surface motion is essentially the same as that when the bottom is rigid. However, the parameters of circular modes in the liquid depend on the parameters of traveling waves in the flexible bottom. This dependence is studied by considering the steady state solution of equations (9.69). One possible solution is

$$\dot{\eta}_0 = \eta^* = \text{const}, \qquad b = b_0 = \text{const}, \qquad \beta = \gamma = \omega_0 t \tag{9.70}$$

This solution describes the wave deformation of the bottom and the corresponding waveform of the liquid free surface. The fixed solution (9.70) is determined from the system equations

$$\eta^* \left[\omega^2 - (1 - d_2 \eta^{*2})\omega_0^2\right] - \delta b_0 \omega_0^2 = 0 \tag{9.71a}$$

$$b_0 \left[\varpi^2 - (1 + \sigma)\omega_0^2\right] - e\eta^* \omega_0^2 = 0 \tag{9.71b}$$

These equations give

$$\omega_0^2 = \frac{\omega^2 \eta^*}{\eta^* + \delta b_0 + d_2 \eta^{*3}} = \frac{\varpi^2 b_0}{b_0 + \sigma b_0 + e\eta^*} \tag{9.72}$$

Thus, the steady-state solution (9.70) characterizes the wave processes in the bottom and liquid free surface only for certain relations between specified initial values of the amplitudes b_0 and η^*. Figure 9.10 shows the dependence of the bottom amplitude ratio b_0/R on η^*/R for $E = 2 \times 10^{11}$ Pa, $\hat{t}/R = 5 \times 10^{-4}$, bottom density $\rho_b = 7.8 \times 10^3$ kg/m^3, liquid depth ratio $h/R = 2$, and sloshing mode $(m,n) = (1,1)$. The dependence of the system frequency parameter

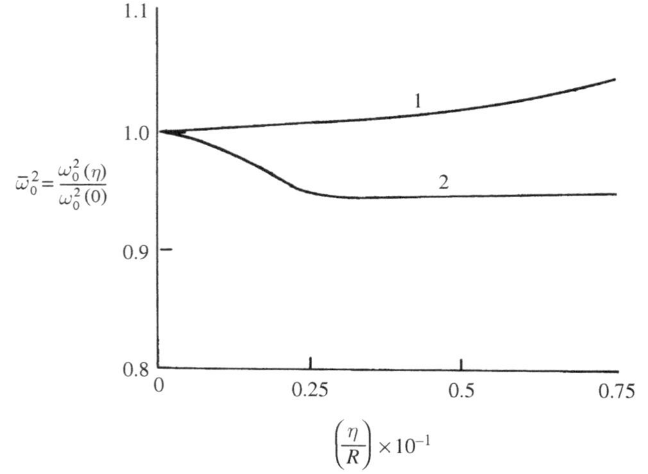

Figure 9.11 Dependence of the system natural frequency ratio on free-surface wave height: Curve 1 is for a rigid bottom, and curve 2 is for an elastic bottom. (Boyarshina, 1994)

$\bar{\omega}_0 = \omega_0(\eta^*)/\omega_0(\eta^* = 0)$ on the free-surface amplitude ratio η^*/R is shown in Figure 9.11 for the same parameters as those of Figure 9.10. Curve 1 is for a rigid bottom, and curve 2 for a flexible bottom. As the amplitude of a circular mode of the liquid free surface increases, its phase velocity ω_0/n can either decrease or increase for free oscillation of the entire system. In the case of a rigid bottom, this velocity increases as η^* increases. The flexibility of the bottom can even cause the direction of "rotation" of the free surface to change as η^* increases.

Another steady-state solution of equations (9.69) is obtained by setting

$$\gamma = \beta = \gamma_0 = \text{const} \tag{9.73}$$

This solution characterizes the planar wave motion of the free surface, and the bottom deformation exhibits standing wave mode with stationary nodes. Eliminating the phase angles from equation (9.69) and keeping terms up to cubic nonlinearity, gives

$$\ddot{\eta}_0 + \omega^2\left(1 + \frac{e\delta}{1+\sigma}\right)\eta_0 - \left(1 + \frac{e\delta}{1+\sigma}\right)\frac{\varpi^2}{1+\sigma}b = d_3\left(1 + \frac{e\delta}{1+\sigma}\right)\dot{\eta}_0^2\eta_0 - d_1\omega^2\eta_0^3 \tag{9.74a}$$

$$\ddot{b} + \frac{\varpi^2}{1+\sigma}\left[1 + \delta\left(1 + \frac{e\delta}{1+\sigma}\right)\right]b - \frac{e\omega^2}{1+\sigma}\left(1 + \frac{e\delta}{1+\sigma}\right)\eta_0$$
$$= d_1 e\omega^2\eta_0^3 - d_3 e\left(1 + \frac{e\delta}{1+\sigma}\right)\dot{\eta}_0^2\eta_0 \tag{9.74b}$$

Introducing the transformation to principal coordinates

$$\begin{Bmatrix} \eta_0 \\ b \end{Bmatrix} = \begin{bmatrix} 1 & 1 \\ n_1 & n_2 \end{bmatrix} \begin{Bmatrix} Y_1 \\ Y_2 \end{Bmatrix} \tag{9.75}$$

Equations (9.74) can be written in terms of the principal coordinates in the form

$$\ddot{Y}_1 + \omega_1^2 Y_1 = \frac{\varepsilon}{M_1}[F_1 + n_1 F_2] \tag{9.76a}$$

$$\ddot{Y}_2 + \omega_2^2 Y_2 = \frac{\varepsilon}{M_2}[F_1 + n_2 F_2] \tag{9.76b}$$

where

$$\omega_{1,2}^2 = \frac{\omega_{11}^2 + \varpi_{11}^2}{2} \mp \sqrt{\frac{(\omega_{11}^2 + \varpi_{11}^2)^2}{4} - e_1 e_2} \tag{9.77}$$

are the normal mode frequencies,

$$\omega_{11}^2 = \omega^2\left(1 + \frac{e\delta}{1+\sigma}\right), \qquad \omega_{22}^2 = \frac{\varpi^2}{1+\sigma}\left[1 + \delta\left(1 + \frac{e\delta}{1+\sigma}\right)\right]$$
$$M_1 = 1 - n_1^2, \qquad M_2 = 1 + n_2^2 \tag{9.78}$$

and F_1 and F_2 are nonlinear functions of principal coordinates.

Equations (9.76) may be solved using one of the asymptotic approximation techniques. Introducing the solution

$$Y_1 = A\cos(\varphi_1(t)), \qquad Y_2 = B\cos(\varphi_2(t)) \tag{9.79}$$

where A and B are constant amplitudes determined from the initial conditions. The angles φ_1 and φ_2 are defined by the first-order differential equations

$$\frac{d\varphi_1}{dt} = \omega_1 + \frac{1}{8M_1}\left[-3d_1 + d_3\left(1 - \frac{\delta e}{1+\sigma}\right)\right](1 - n_1 e)(A^2 + B^2) = \Omega_{11} \tag{9.80a}$$

$$\frac{d\varphi_2}{dt} = \omega_2 + \frac{1}{8M_2}\left[-3d_1 + d_3\left(1 - \frac{\delta e}{1+\sigma}\right)\right](1 - n_2 e)(AB + B^2) = \Omega_{22} \tag{9.80b}$$

Note that Ω_{ii} on the right-hand sides of equations (9.80) represent the nonlinear frequencies of the coupled system. The frequency Ω_{11} is close to the natural frequency of the liquid free surface in a container with a rigid bottom. The frequency Ω_{22} is close to the natural frequency of the deformable bottom in the absence of liquid. Figure 9.12 shows the dependence of the frequency Ω_{11} on the amplitude A for a given value of $B = 0.01$ and the same parameters as those of Figure 9.10. The upper curve 1 is the value of Ω_{11} for a rigid bottom, while curve 2 is for a flexible bottom.

Forcing response

Under periodic forcing excitation of the flexible bottom, resonance conditions may arise between the excitation frequency and the system normal mode frequencies. Let the radial excitation to the bottom be represented by the expression

$$F(r, \theta, t) = f(r, \theta)\cos\Omega t \tag{9.81}$$

where Ω is the excitation frequency, and the shape function $f(r,\theta)$ coincides with one of the normal modes of the bottom:

$$f(r, \theta) = f_0[J_m(k_{m1}r)I_m(k_{m1}R) - J_m(k_{m1}R)I_m(k_{m1}r)]\cos m\theta \tag{9.82}$$

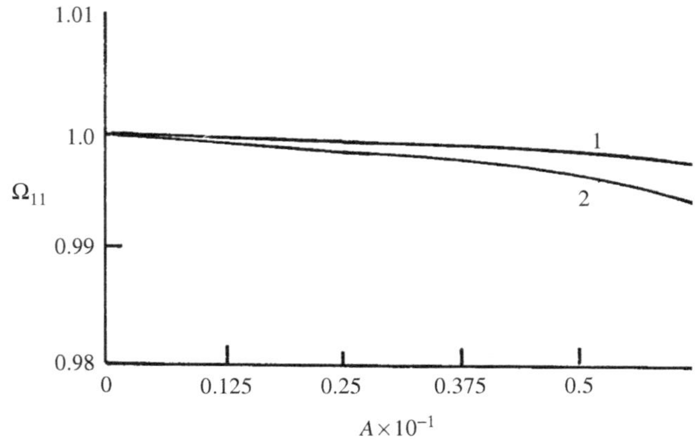

Figure 9.12 Dependence of the nonlinear frequency of the coupled system on the normal mode amplitude: Curve 1 is for a rigid bottom, and curve 2 is for an elastic bottom. (Boyarshina, 1994)

The same solutions of the bottom deformation, w, given by equation (9.64), and liquid wave height, η, given by equation (9.67), are adopted. When $n=1$, $\cos\theta=\pm 1$, $\theta=0$, π, 2π, ... The governing differential equations of the amplitudes η_0 and b are

$$\ddot{\eta}_0+\omega^2\left(1+\frac{e\delta}{1+\sigma}\right)\eta_0-\left(1+\frac{e\delta}{1+\sigma}\right)\frac{\varpi^2}{1+\sigma}b$$
$$=d_3\left(1+\frac{e\delta}{1+\sigma}\right)\dot{\eta}_0^2\eta_0-d_1\omega^2\eta_0^3-\frac{\delta f_0}{1+\sigma}\left[1+\frac{e\delta}{1+\sigma}+d_1\eta_0^2\right]\cos\Omega t \tag{9.83a}$$

$$\ddot{b}+\frac{\varpi^2}{1+\sigma}\left[1+\delta\left(1+\frac{e\delta}{1+\sigma}\right)\right]b-\frac{e\omega^2}{1+\sigma}\left(1+\frac{e\delta}{1+\sigma}\right)\eta_0$$
$$=d_1e\omega^2\eta_0^3-d_3e\left(1+\frac{e\delta}{1+\sigma}\right)\dot{\eta}_0^2\eta_0+\frac{f_0}{1+\sigma}\left[1+\delta\left(1+\frac{e\delta}{1+\sigma}\right)\right]\cos\Omega t$$
$$+d_1\frac{\delta f_0}{1+\sigma}\eta_0^2\cos\Omega t \tag{9.83b}$$

These equations can be written in terms of principal coordinates through transformation (9.75), such that the resulting equations take the form

$$\ddot{Y}_1+\omega_1^2Y_1=\frac{\varepsilon}{M_1}[\bar{F}_1+n_1\bar{F}_2] \tag{9.84a}$$

$$\ddot{Y}_2+\omega_2^2Y_2=\frac{\varepsilon}{M_2}[\bar{F}_1+n_2\bar{F}_2] \tag{9.84b}$$

The terms with over-bars on the right-hand sides of equations (9.84) include nonlinear terms in addition to excitation terms. The excitation terms give rise to secular terms that establish the following resonance conditions

$$\Omega=\omega_1,\ \Omega=\omega_2,\ \Omega=3\omega_1,\ \Omega=3\omega_2,\ \Omega=\omega_1+2\omega_2,\ \Omega=2\omega_1+\omega_2 \tag{9.85}$$

The present analysis is restricted to the excitation of the first normal mode under the primary resonance condition $\Omega=\omega_1$. Introducing the following solution

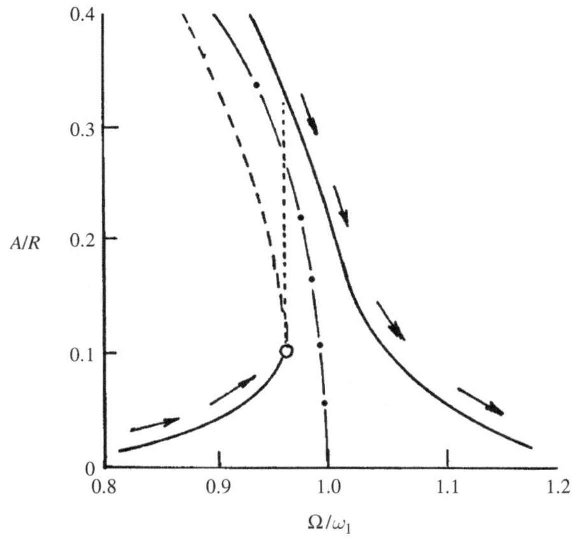

Figure 9.13 Amplitude–frequency response under sinusoidal excitation of shell elastic bottom. (Boyarshina, 1994)

$$Y_1 = A\cos(\omega_1 t + \varphi_1), \qquad Y_2 = B\cos(\omega_2 t + \varphi_2) + K\cos\Omega t \tag{9.86a,b}$$

where

$$K = \frac{f_0}{(\omega_2^2 - \Omega^2)(1+\sigma)}\left[e\delta\left(1 + \frac{e\delta}{1+\sigma}\right) - n_1\left\{1 + \delta\left(1 + \frac{e\delta}{1+\sigma}\right)\right\}\right] \tag{9.87}$$

The amplitude A and phase angle φ_1 are determined from the following equations

$$\frac{dA}{dt} = 0 \tag{9.88a}$$

$$\begin{aligned}\frac{d\varphi_1}{dt} &= \omega_1^2 - \Omega^2 - \frac{\omega_1^2}{8M_1}(1 - en_1)\left[3d_1 + d_3\left(1 + \frac{e\delta}{1+\sigma}\right)\right]A^2 \\ &\quad + \frac{f_0}{2M_1(1+\sigma)}\left[\delta\left(1 + \frac{e\delta}{1+\sigma}\right) - n_1\left\{1 + \delta\left(1 + \frac{e\delta}{1+\sigma}\right)\right\}\right] \\ &= 0\end{aligned} \tag{9.88b}$$

Equation (9.88b) establishes the amplitude–frequency relationship (A as a function of Ω). Figure 9.13 shows the dependence of the amplitude ratio A/R on the excitation frequency ratio Ω/ω_1. The dependence exhibits soft nonlinear characteristics. The figure shows the backbone curve represented by a dash-dot curve.

9.5 Nonlinear interaction with nonlinear sloshing

9.5.1 Governing equations of motion

The nonlinear interaction of a circular cylindrical shell with nonlinear sloshing dynamics under internal resonance conditions is considered in this section. This problem was originally considered by Boyarshina and Ganiev (1976) and Boyarshina (1984, 1988). The equations of motion of a thin shell partially filled with a liquid was derived using the Hamilton–Ostrogradsky principle

$$\int_{t_0}^{t_1}\left(\delta T+\sum_j \mathbf{F}_j\cdot\delta\mathbf{r}_j+\int_v \lambda\,\mathrm{div}\,\delta\mathbf{r}\,dv\right)dt=0 \tag{9.89}$$

where T is the kinetic energy of the system, $\mathbf{F}_j$ is the vector of the given applied force at the jth point of the system, $\mathbf{r}_j$ is the radius vector of this point, $\lambda=\lambda(\mathbf{r},\,t)$ is the Lagrange multiplier, which can be regarded as the hydrodynamic pressure, v is the volume of the fluid inside the container, t_0 and t_1 are two time instances.

The kinetic energy may be written in the form

$$T=\frac{\rho_c}{2}\int_{v_s}\dot{u}_s^2\,dv_s+\frac{\rho}{2}\int_v(\nabla\Phi)^2\,dv \tag{9.90a}$$

$$\sum_j \mathbf{F}_j\cdot\delta\mathbf{r}_j=-\delta\left\{\frac{1}{2}\int_{v_s}\left[\frac{E}{1-\nu^2}\left(\varepsilon_{11}^2+\varepsilon_{22}^2+2\nu\varepsilon_{11}\varepsilon_{22}\right)+G\varepsilon_{12}^2\right]dv_s\right\} \tag{9.90b}$$

$$\int_v \lambda\,\mathrm{div}\,\delta\mathbf{r}\,dv=-\rho\int_v\left[\frac{\partial\Phi}{\partial t}+gz+(\nabla\Phi)^2\right]\nabla\cdot\delta\mathbf{r}\,dv \tag{9.90c}$$

where v_s is the shell material volume, u_s is the middle surface displacement of the shell in bending, E is Young's modulus of elasticity, ε_{ij} are strain components expressed in terms of the shell middle surface displacements, G is the modulus of rigidity. Carrying out the variational of equation (9.89) gives the equations of motion and the boundary conditions.

The boundary value problem of the liquid is given in terms of the velocity potential function Φ in the form

$$\nabla^2\Phi=0 \quad \text{inside the liquid domain} \tag{9.91a}$$

$$\frac{\partial\Phi}{\partial\mathbf{n}}=u_n \quad \text{on the liquid free surface} \tag{9.91b}$$

$$\frac{\partial\Phi}{\partial r}=\dot{w} \quad \text{at the wetted shell surface} \tag{9.91c}$$

$$\frac{\partial\Phi}{\partial t}+\frac{1}{2}(\nabla\Phi)^2+gz=0 \quad \text{on the liquid free surface} \tag{9.91d}$$

where $\mathbf{n}$ is a unit vector along the outward normal of the free surface, u_n is the normal component of the liquid relative velocity with respect to the free surface, w is the shell deflection at the wetted surface.

Boyarshina (1984) showed that in the elastic shell–liquid system a rotational motion of the free surface of the liquid leads to the excitation of the elastic shell in the form of traveling waves in the azimuthal direction. The rotational motion and the associated traveling waves owe their origin to the nonlinear coupling between the coordinates describing the liquid orthogonal modes and the elastic vibrations of the shell. The interaction between these modes in the neighborhood of an internal resonance condition will be considered. Let the perturbed free surface of the liquid be represented by the two-parameter expansion

$$\eta(r,\theta,t)=\sum_{i=0}\sum_{s=1} a_{si}(t)\cos[i\theta-\alpha_{is}(t)]R_{si}(r) \tag{9.92}$$

where a_{si}(t) are the generalized coordinates, α_{is}(t) are the angular coordinates, and R_{si}(r) are the eigenfunctions of the homogeneous boundary-value problem

$$\frac{d^2R_{si}(r)}{dr^2}+\frac{1}{r}\frac{dR_{si}(r)}{dr}+\left(\lambda_{si}^2-\frac{i^2}{r^2}\right)R_{si}(r)=0 \tag{9.93a}$$

$$\left.\frac{dR_{si}(r)}{dr}\right|_{r=R}=0 \tag{9.93b}$$

where λ_{si} is the sth eigenvalues, or the roots of the resulting Bessel function that satisfies condition (9.93b).

For the case of the ends of a simply supported shell, the shell deflection may be approximated by the wave expansion

$$w(z,\theta,t)=\sum_m\sum_n b_{mn}(t)\cos[n\theta-\gamma_{mn}(t)]\sin[m\pi(z+h)/l] \tag{9.94}$$

where $b_{mn}(t)$ are the generalized coordinates and $\gamma_{mn}(t)$ are the angular coordinates of the shell deflection, $m=1,2,3,\ldots$, and $n=2,3,4,\ldots$

The liquid velocity potential function may be written as the sum of two components

$$\Phi=\Phi_l+\Phi_s \tag{9.95}$$

where Φ_l is due to the free-surface wave motion of the liquid, and Φ_s is due to the elastic deformation of the shell. These two components can be evaluated by using equations (9.92) and (9.94), and the following expressions are obtained

$$\begin{aligned}\Phi_l=&\sum_i \dot a_iA_{i0}\cos(i\theta-\alpha_i)+\sum_i a_i\dot\alpha_iA_{i0}\sin(i\theta-\alpha_i)\\&+\sum_i\sum_j \dot a_ia_jA_{ij}\cos(i\theta-\alpha_i)\cos(i\theta-\alpha_j)+\cdots\\&+\sum_i\sum_j\sum_k a_ia_ja_k\dot\alpha_iA_{ij}\sin(i\theta-\alpha_i)\cos(j\theta-\alpha_j)\cos(k\theta-\alpha_k)\end{aligned} \tag{9.96a}$$

for which $i,j,k=0,1,2,\ldots$

The subscript s has been dropped.

$$\Phi_s=\sum_m\sum_b\sum_k B_{mnk}\left[\dot b_{mn}\cos(n\theta-\gamma_{mn})+b_{mn}\dot\gamma_{mn}\sin(n\theta-\gamma_{mn})\right] \tag{9.96b}$$

where A_{i0}, A_{ij}, and A_{ijk} are harmonic functions depending on the spatial coordinates and they must satisfy the boundary condition of the liquid free surface. The functions B_{mnk} are evaluated from the shell boundary-value problem and are given by the expression

$$B_{mnk} = \frac{2}{n\pi}\frac{I_n(\pi kr/h)}{I'_n(\pi kR/h)}\cos\left[\frac{\pi k}{h}(z+h)\right]\int_0^h w(z,\theta,t)\cos\left[\frac{\pi k}{h}z\right]dz \tag{9.97}$$

A set of nonlinear, coupled, ordinary differential equations describing the time evolution of the generalized coordinates $a_{mn}(t)$ and $b_{mn}(t)$ can be obtained by applying Gaerkin's method, or another discrediting approach (Narimanov, 1957b and Narimanov, *et al.*, 1977), to the dynamical boundary condition (9.91d) and using equations (9.95) to (9.97), gives

$$\left(1+D_{mn}^{(1)}\right)\ddot{a}_{mn} + \left(\nu_{mn}^2 - \dot{\alpha}_{mn}^2\right)a_{mn} = -C_{mn}^{(1)}b_{mn}\cos(\gamma_{mn}-\alpha_{mn}) + \mu G_1^{mn} \tag{9.98a}$$

$$a_{mn}\ddot{\alpha}_{mn} + 2\dot{a}_{mn}\dot{\alpha}_{mn} = -C_{mn}^{(1)}b_{mn}\sin(\gamma_{mn}-\alpha_{mn}) + \mu G_2^{mn} \tag{9.98b}$$

$$\left(1+D_{mn}^{(2)}\right)\ddot{b}_{mn} + \left(\omega_{mn}^2 - \dot{\gamma}_{mn}^2\right)b_{mn} = -C_{mn}^{(2)}a_{mn}\cos(\gamma_{mn}-\alpha_{mn}) + \mu G_3^{mn} + f_1(t) \tag{9.98c}$$

$$b_{mn}\ddot{\gamma}_{mn} + 2\dot{b}_{mn}\dot{\gamma}_{mn} = -C_{mn}^{(2)}a_{mn}\sin(\gamma_{mn}-\alpha_{mn}) + \mu G_4^{mn} + f_2(t) \tag{9.98d}$$

where

$$\nu_{mn}^2 = g\lambda_{mn}\tanh(\lambda_{mn}h) \tag{9.99a}$$

are the square of the liquid natural frequency of sloshing mode mn, $\lambda_{mn}=\xi_{mn}/R$ are the roots of $dJ_m(\lambda_{mn}r)/dr|_{r=R}=0$,

$$\omega_{mn}^2 = \frac{1}{\rho_c}\left\{\frac{\hat{t}^2}{12}\frac{E}{(1-\varepsilon^2)}\left[\left(\frac{m\pi}{l}\right)^2+\left(\frac{n}{R}\right)^2\right] + \frac{E}{R^2}\frac{(m\pi/l)^4}{\left[(m\pi/l)^2+(n/R)^2\right]^2}\right\} \tag{9.99b}$$

are the shell square natural frequency of mode mn, $D_{mn}^{(i)}$ and $C_{mn}^{(i)}$, $i=1$, 2, are constants depending on the system parameters, G_j^{mn}, $j=1,2,\ldots,4$ are nonlinear functions of the generalized coordinates a_{mn}, α_{mn}, b_{mn}, and γ_{mn}, and their time derivatives, $f_1(t)$ and $f_2(t)$ are functions of the external excitation acting on the shell.

The right-hand sides of equations (9.98) include nonlinear terms that give rise to secular terms, which establish internal resonance conditions. The occurrence of internal resonance results in the interaction of the sloshing modes and elastic modes of the shell. Furthermore, direct excitation of one of these modes can lead to excitation of other coupled modes that are not directly excited. The following internal resonance condition was studied in details by Boyarshina (1988):

$$\omega_{mn} = 2\nu_{mn} \tag{9.100}$$

For the case of the first anti-symmetric sloshing mode $(1,1)$, one may set $a_{11}=a$, and $\alpha_{11}=\alpha$, while for the first bending mode of the shell $(1,2)$, one may set $b_{12}=b$, and $\gamma_{1,2}=\gamma$. The external excitation acting on the shell $F(t)$ was assumed to be in the form

$$F(t) = F_1 \cos \Omega t \cos(nr/R) \sin[m\pi(z+h)/l] + F_2 \sin \Omega t \sin(nr/R) \sin[m\pi(z+h)/l] \tag{9.101}$$

The corresponding equations of motion are

$$\ddot{a} + (\nu^2 - \dot{\alpha}^2)a = -e_{11}\omega^2 ab\cos(2\alpha-\gamma) + (d_1+e_{12})a^2\ddot{a} + (d_3+e_{13})a\dot{a}^2 + (d_2+e_{14})a^3\dot{\alpha}^2 \tag{9.102a}$$

$$\ddot{\alpha} = (d_5+e_{23})\dot{a}^2 - 2\dot{a}\dot{\alpha}/a - e_{12}\omega^2 b\sin(2\alpha-\gamma) + (d_2+e_{22})a\dot{a}\dot{\alpha} \tag{9.102b}$$

$$\ddot{b} + (\omega^2 - \dot{\gamma}^2)b = -e_{31}\nu^2 a^2\cos(2\alpha-\gamma) + e_{32}a^2\ddot{b} + e_{33}a^2 b\dot{\gamma}^2 + F_1\cos\gamma\cos\Omega t + F_2\sin\gamma\sin\Omega t \tag{9.102c}$$

$$\ddot{\gamma} = e_{43}a^2\ddot{\gamma} - 2\frac{\dot{b}}{b}\dot{\gamma} - e_{41}\nu^2\left(\frac{a^2}{b}\right)\sin(2\alpha-\gamma) + e_{42}\frac{a^2}{b}\dot{b}\dot{\gamma} + F_2\cos\gamma\sin\Omega t - F_1\sin\gamma\cos\Omega t \tag{9.102d}$$

where d_i and e_{ij} are constants depending on the system parameters. Note that the nonlinearity in the system equations of motion is of the inertia type. The following are special cases that may be considered.

9.5.2 Free nonlinear interaction

In the absence of excitation forces, $F_1 = F_2 = 0$, and damping, the following solution may be considered

$$a = a_0 = \text{const}, \qquad b = b_0 = \text{const}, \qquad \dot{\alpha} = \dot{\alpha}_0 = \text{const} \tag{9.103}$$

This solution corresponds to a circular wave of the liquid free surface with a constant amplitude, a_0, and phase velocity $\dot{\alpha}_0 R$. According to this solution and equation (9.102b) we have $\sin(2\alpha-\gamma) = 0$, which implies that $(2\alpha-\gamma) = k\pi$, $k = 0, 1, 2, \ldots$ This result leads to two solutions:

First solution

$$2\dot{\alpha} = \dot{\gamma} = \text{const} \tag{9.104}$$

Using this condition one may write the dependence of the frequencies and amplitudes of the liquid free surface and shell deformations, that is,

$$\dot{\alpha}_0^2 = \varpi_1^2(a_0, b_0) = \frac{\nu^2 + e_{11}\omega^2 b_0(-1)^k}{1+(d_2+e_{14})a_0^2}, \quad \dot{\gamma}_0^2 = \varpi_2^2(a_0, b_0) = \frac{e_{31}\nu^2 a_0 + \omega^2 b_0}{b_0(1+e_{33}a_0^2)} \tag{9.105}$$

These frequencies may satisfy the internal resonance condition

$$\varpi_2(a_0, b_0) = 2\varpi_1(a_0, b_0) \tag{9.106}$$

The fixed amplitude solution of b_0 may be written in terms of a_0 as

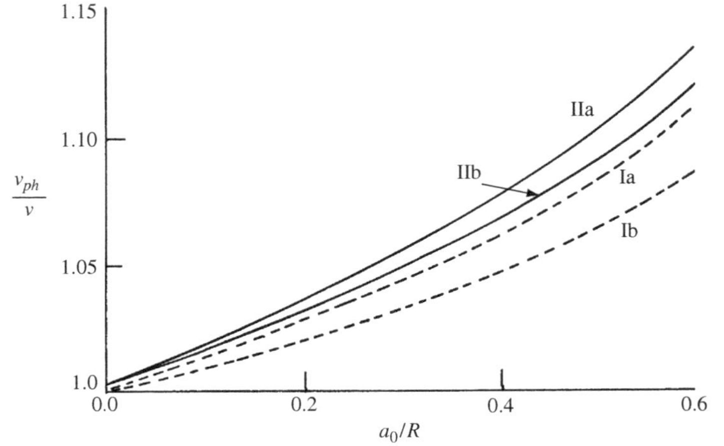

Figure 9.14 Dependence of the phase velocity ratio on the amplitude ratio. Curves I are for a rigid tank, curves II are for elastic shell, curves Ia and IIa are for $h/R=2$, and Ib and IIb are for $h/R=1$. (Boyarshina, 1988)

$$b_0 = \frac{a_0^2[e_{33} - (d_2 + e_{14})]}{8e_{11}(1 + e_{33}a_0^2)} \pm \sqrt{\frac{e_{11}e_{31}a_0^2 + \left[d_2^2/4 + e_{11}e_{31}(e_{33} + d_2 + e_{14})\right]a_0^4 + 4e_{11}e_{31}e_{33}(d_2 + e_{14})a_0^6}{16e_{11}^2(1 + e_{33}a_0^2)^2}} \tag{9.107}$$

This result states that any motion of the free surface of the liquid causes a corresponding traveling wave in the shell provided condition (9.106) is satisfied. The dependence of the phase velocity $v_{ph} = \varpi_1(a_0,b_0) = \varpi_2(a_0,b_0)/2$ on the fixed amplitude ratio a_0/R is shown in Figure 9.14 for the cases of the liquid in an elastic shell (solid curves) and a rigid tank (dash curves) for $\nu = 2\pi(7)$, $d_2 = -0.85$, $e_{11} = -0.1127$, and $e_{14} = 0.0048$. The curves II are estimated for $h/R = 2$, while curves I are obtained for $h/R = 1$. It is seen that the interaction of the liquid free surface with the elastic shell results in an increase of the phase velocity.

Second solution
Another possible solution resulting from the condition $(2\alpha - \gamma) = k\pi$ is

$$\alpha = \alpha_0 = \text{const} \qquad \gamma = \gamma_0 = \text{const} \tag{9.108}$$

Accordingly, the following equations of the amplitudes are obtained

$$\ddot{a} + \nu^2 a = -e_{11}\omega^2 ab(-1)^k - \nu^2(d_1 + e_{12})a^3 + (d_3 + e_{13})a\dot{a}^2 \tag{9.109a}$$

$$\ddot{b} + \omega^2 b = -e_{31}\nu^2 a^2(-1)^k + e_{33}a^2\ddot{b} + e_{32}\omega^2 a^2 b \tag{9.109b}$$

Note that the acceleration terms $\ddot{a}$ and $\ddot{b}$ involved in the nonlinear terms in equations (9.102a) and (9.102c) have been replaced by $\ddot{a} \approx -\nu^2 a$, and $\ddot{b} \approx -\omega^2 b$. The solutions of these two equations can be obtained by using the averaging approximation technique

$$\begin{aligned} a(t) &= A(t)\cos\Psi_1(t), \quad \Psi_1(t) = \nu t + \vartheta_1 \\ b(t) &= B(t)\cos\Psi_2(t), \quad \Psi_2(t) = \nu t + \vartheta_2 \end{aligned} \tag{9.110}$$

where $A(t)$, $B(t)$, $\Psi_1(t)$, and $\Psi_2(t)$ are slowly varying functions with time. Following the averaging technique, the amplitude and phase first-order differential equations are given in the form

$$\nu\dot{A} = \frac{1}{2}e_{11}\omega^2 AB\sin(2\vartheta_1 - \vartheta_2) \tag{9.111a}$$

$$\dot{\Psi}_1 = \nu + \frac{e_{11}}{4}\frac{\omega^2}{\nu}B\cos(2\vartheta_1 - \vartheta_2) + \frac{3}{8}\nu(d_1 + e_{12})A^2 - \frac{\nu}{8}(d_3 + e_{13})A^2 \tag{9.111b}$$

$$\omega\dot{B} = -\frac{1}{4}e_{31}\nu^2 A^2 \sin(2\vartheta_1 - \vartheta_2) \tag{9.111c}$$

$$B\dot{\Psi}_2 = \omega B + \frac{1}{4}A^2\left[e_{31}\frac{\nu^2}{\omega}\cos(2\vartheta_1 - \vartheta_2) + e_{32}\omega B\right] \tag{9.111d}$$

The first integral of these equations may be written in the form

$$\omega^2 e_{11}B + \frac{1}{2}\nu^2 e_{31}A^2 = \text{const} \tag{9.112}$$

This result reveals that the amplitudes of the wave motions of the shell and liquid are interacting in the form of energy exchange.

Setting $A = A_0 = \text{const}$ and $B = B_0 = \text{const}$, results in the stationary phase relationship $2\vartheta_1 - \vartheta_2 = k\pi$, and thus we get

$$\Psi_1 = \nu_1(A,B)t + \vartheta_{10}, \quad \text{and} \quad \Psi_2 = \omega_1(A,B)t + \vartheta_{20} \tag{9.113a,b}$$

where

$$\nu_1(A,B) = \nu + \frac{1}{4}(-1)^k e_{11}\frac{\omega^2}{\nu}B + \frac{3}{8}\nu(d_1 + e_{12})A^2 - \frac{1}{8}\nu(d_3 + e_{13})A^2 \tag{9.114a}$$

$$\omega_1(A,B) = \omega + \frac{1}{4}e_{32}\omega^2 A^2 + \frac{1}{4}(-1)^k\frac{\nu^2}{\omega}e_{31}\frac{A^2}{B} \tag{9.114b}$$

Equations (9.114) establish the frequencies of free oscillations of the shell and liquid system. The motion is of the "standing wave" type and the frequencies depend on the amplitudes A and B. The free oscillations are essentially governed by the initial perturbation of the system or the initial energy input.

9.5.3 Forced autoparametric interaction

Autoparametric interaction takes place when both internal resonance and external resonance conditions exist. In this section we will briefly discuss the relevant features of the problem of liquid–shell interaction under forced excitation.

Consider equal excitation amplitudes, $F_1 = F_2 = F_0$, then equations (9.102) will have a possible unimodal solution given by the well-known linear resonance solution

$$a = 0, \qquad b = b_0 = \frac{F_0}{\omega^2 - \Omega^2}, \qquad \gamma = \Omega t, \quad \text{and} \quad \alpha = 0 \tag{9.115}$$

Another steady-state solution is obtained by solving the coupled algebraic equations

$$\left(\omega^2-\Omega^2\right)b_0=-e_{31}\nu^2a_0^2+e_{33}a_0^2b_0\Omega^2+F_0 \tag{9.116a}$$

$$\nu^2-\frac{\Omega^2}{4}=-e_{11}\omega^2b_0+\frac{\Omega^2}{4}(e_{14}+d_2)a_0^2 \tag{9.116b}$$

with $\dot{\gamma}=\Omega$, and $\dot{\alpha}=\Omega/2$. Solving equations (9.116) gives

$$b_0=\frac{1}{\omega^2e_{11}}\left[\frac{\Omega^2}{4}(d_2+e_{14})a_0^2-\left(\nu^2-\frac{\Omega^2}{4}\right)\right] \tag{9.117a}$$

$$a_0^2=M\pm\sqrt{M^2-N} \tag{9.117b}$$

where

$$M=\frac{\frac{\Omega^2}{4}(d_2+e_{14})\left(\omega^2-\Omega^2\right)+\Omega^2\left(\nu^2-\frac{\Omega^2}{4}\right)e_{33}}{2\left[\frac{\Omega^2}{2}(d_2+e_{14})-e_{11}e_{31}\nu^2\omega^2\right]}$$

$$N=\frac{F_0e_{11}\omega^2+\left(\nu^2-\frac{\Omega^2}{4}\right)\left(\omega^2-\Omega^2\right)}{\left[\frac{\Omega^2}{2}e_{33}(d_2+e_{14})-e_{11}e_{31}\nu^2\omega^2\right]}$$

This mixed mode solution states that the mode which is directly excited transfers energy to the indirectly excited mode through nonlinear coupling due to the internal resonance condition. This solution corresponds to traveling waves in the shell and liquid free surface. The region of autoparametric interaction is governed by the liquid and shell parameters.

9.6 Interaction under parametric excitation

9.6.1 Historical overview

Parametric excitation of shell–liquid systems has been studied for two different cases. The first is when the liquid surface motion does not enter the basic problem and this has been studied extensively (see, e.g., Bleich, 1956a, Bublik and Merkulov, 1960, Shkenev, 1964, 1970, Bagdasaryan and Gnuni, 1966, Kana, *et al.*, 1966, 1968, Kana and Abramson, 1966, Grigorev, 1966, 1967, Kana and Gormley, 1967, Abramson and Kana, 1967, 1970, Kana and Craig, 1968, Kana and Chu, 1969, 1970, Koleshov and Shveiko, 1971, Liu and Belytchko, 1983, and Ulitin, 1986). The second is when the liquid surface is excited at a large amplitude in its first symmetric mode (see, e.g., Paidoussis and Denise, 1972, Manevich, *et al.*, 1976, and Timokha, 1992, 1997). For the first case, the dynamic response may follow one of the following scenarios:

(1) Direct linear harmonic response of axisymmetric modes occurs primarily as large pressure amplifications and is accompanied by very small wall deflections.
(2) Response of nonaxisymmetric modes appears in the form of only small pressure amplitudes but with rather large harmonic wall motions.

(3) Large symmetric harmonic pressure oscillations in the fluid accompanied by large-amplitude ½-subharmonic shell wall motions of nonaxisymmetric modes. This nonlinear response results from instabilities of the linear response and is usually observed in shells with little circumferential stiffening.

The shell–liquid dynamic behavior is complex in nature and the reported results indicate that most of the parameters considered have a significant influence on initial-state axisymmetric response, dynamic instability, and subsequent nonlinear responses. Druz and Magula (1967), Clary and Turner (1970), and Kolesinkov (1971) analytically and experimentally studied the parametric excitation of elastic liquid containers. Other studies were considered by Grigorev (1967), Obraztsova (1976a,b), Obraztsova and Shklyarchuk (1979), Pavlovskii and Filin (1979, 1985), Khandelwal and Nigam (1981), Zhu and Huang (1984), Chiba and Tani (1987), and Chiba, *et al.* (1987). It was found that the presence of the flexible base has the effect of reducing the regions of instability of the liquid surface. The dynamic stability of liquid shells was studied experimentally by Bublik and Merkulov (1960), Chiba (1986, 1993a,b, 1994, 1995, 1996a), Chiba, *et al.* (1989), and Uras (1989). The frequency of normal modes of a hydroelastic system was estimated by Aganovic (1981).

Okazaki, *et al.* (1993, 1995) studied the parametric excitation of a partially liquid-filled cylindrical shell and observed chaotic responses of the shell and liquid free surface. The sloshing and chaotic motion of the liquid free surface were found to appear in the lower frequency range than were the parametric resonance and chaotic motion of the shell walls. Parametric resonance in axisymmetric shells under horizontal excitation was examined by Takayangi, (1990). Obraztsova (1976a) and Obraztsova and Shklyarchuk (1979, 1980) studied the axisymmetric and nonaxisymmetric nonlinear parametric excitation of cylindrical shells filled with a liquid. They considered the cases of rigid and elastic shell bottom.

The hydroelastic vibration of coaxial cylindrical shells partially filled with liquid was studied by Au-Yang (1976), Chu, *et al.* (1979), Brown (1980), Chu and Brown (1981), Kobayashi and Chiba (1983), Paidoussis, *et al.* (1984, 1985), Fujita (1985), and Tani and Haiji (1986). These studies did not take into account the effects of the motion of the liquid free surface. Tani, *et al.* (1989c) considered the free linear vibration of clamped coaxial cylindrical shells partially filled with an inviscid liquid. It was found that the natural frequency decreases as the liquid depth increases. Under vertical excitation, Tani, *et al.* (1989b) determined the instability regions where parametric resonance takes place. They found that the principal instability resonance and the combination summed type resonance of two normal modes, which have the same circumferential wave number and different axial modes of vibration, are the two major instability sources. However, when the annular gap increased, it was difficult to observe the occurrence of the combination resonance of two separate vibrations dominated by the deflection of either the inner or outer cylinder. Yoshida and Miyoshi (1987, 1989), Okazaki, *et al.* (1990), and Okazaki and Tani (1991) studied the response of multi-walled coaxial cylindrical tanks under horizontal and vertical excitations using the finite element method. They included the static liquid pressure in their analysis but did not consider the effect of liquid sloshing dynamics.

Tani, *et al.* (1989a) studied the free vibration of a finite length rotating cylindrical shell partially filled with liquid. They found that the magnitude of the initial tension due to the centrifugal force depends on the rotation speed and liquid density and volume. The frequency was found to increase as the initial tension increases.

The effect of fluid compressibility on the parametric instability of cylindrical shells was considered by Shklyarchuk (1970a,b, 1973, 1983). Shklyarchuk treated this problem in two stages. The first stage considered the axisymmetric vibrations of a shell filled with a liquid by ignoring the gravitational waves. The second stage treated the perturbed nonaxisymmetric motions caused by periodic forces acting on the median surface due to changes in the hydrodynamic pressure. The inclusion of fluid compressibility resulted in a substantial reduction in the frequencies of the axisymmetric vibrations of the shell. Furthermore, the apparent additional mass resulted in some shift of the dynamic instability regions for the vibrations of axisymmetric and nonaxisymmetric deformations.

The nonlinear dynamic response of a shell–liquid system, in the case of still liquid, subjected to combined external and parametric excitations, was considered in some studies by Koval'chuk, Puchka, and Kholopova (1987), Koval'chuk (1990), and Koval'chuk, Pavlovskii, and Filin (1994, 2000). The influence of a small initial deflection on the nonlinear mixed mode parametric vibration of cylindrical shells partially filled with a (still) liquid was investigated by Pavlovskii and Filin (1985). Krasnopolskaya and Lavrov (1988) included the influence of liquid-free-surface oscillation under parametric excitation with limited intensity.

Boyarshina (1984) examined the nonlinear interaction between liquid sloshing modes and a circular shell vibration in the neighborhood of three internal resonance conditions. These are (i) the natural frequency of the shell is close to one of the sloshing natural frequencies, (ii) the natural frequency of the shell is twice the natural frequency of the sloshing modes, and (iii) a 3 : 1 internal resonance. Amabili (1996a) and Amabili, *et al.* (1998a,b, 1999a,b,c, 2000a,b) reviewed some work pertaining to the nonlinear vibration of shells filled or surrounded by quiescent fluid.

9.6.2 Parametric excitation of fundamental modes of a shell–liquid system

The case of parametric excitation of the fundamental mode and its companion (conjugate) mode is considered for the case of a nonlinear shell with still liquid. This case was studied by Koval'chuk (1990) and Koval'chuk, Pavlovskii, and Filin (2000). The modal equations of motion (9.9) up to cubic order under parametric excitation take the form

$$\ddot{A} + \left(\omega_1^2 - N\cos\Omega t\right)A + \gamma_1(A^2 + B^2)A = 0 \tag{9.118a}$$

$$\ddot{B} + \left(\omega_1^2 - N\cos\Omega t\right)B + \gamma_1(A^2 + B^2)B = 0 \tag{9.118b}$$

where $N = \dfrac{N_z(m\pi/l)^2}{\rho_c(1+m_{01})}$. The steady state response of equations (9.118) may be obtained using the averaging method by assuming the following solution:

$$A(t) = a(t)\cos\left(\frac{\Omega}{2}t + \vartheta_1(t)\right), \qquad B(t) = b(t)\cos\left(\frac{\Omega}{2}t + \vartheta_2(t)\right) \tag{9.119}$$

Following the standard averaging method, gives two possible solutions. The unimodal response is

$$a^2 = -\frac{4}{3\gamma_1}\left[\omega_1^2 - \frac{\Omega^2}{4} \pm \frac{N}{2}\right], \quad b = 0 \tag{9.120a}$$

or

$$b^2 = -\frac{4}{3\gamma_1}\left[\omega_1^2 - \frac{\Omega^2}{4} \pm \frac{N}{2}\right], \quad a = 0 \tag{9.120b}$$

On the other hand, the mixed mode response is

$$a^2 = \frac{1}{\gamma_1}\left[\frac{\Omega^2}{4} - \omega_1^2 \mp N\right], \qquad b^2 = \frac{1}{\gamma_1}\left[\frac{\Omega^2}{4} - \omega_1^2 \pm N\right] \tag{9.121a,b}$$

The domain of existence of this solution is bounded by the frequency band

$$\Omega < \Omega^* = 2\sqrt{\omega_1^2 - N} \tag{9.122}$$

9.6.3 Liquid-filled shell

Figure 9.15(a) shows a schematic diagram of a thin elastic cylindrical shell with a rigid flat bottom that is simply supported at both ends. The shell is completely filled with a liquid, which is assumed to be inviscid and compressible. It is assumed that the liquid-free-surface motion has a negligible effect on the coupled shell vibration. In developing the analytical modeling, Kana and Craig (1968) assumed the shell first responds in linear axisymmetric vibrational modes as the initial state. This assumption is based on experimental observations. Within certain ranges of excitation parameters, the response is followed by nonsymmetric perturbations superimposed on the initial state, which become unstable and the shell response subsequently includes a dominant nonaxisymmetric ½-subharmonic component displacement. This type of switching from the axisymmetric state to the asymmetric state can be described using one of the perturbational techniques by including additional nonlinear loadings that arise in passing from the initial state to the final state.

The final state of the shell displacements along z, θ, r are written respectively, as the sum of the initial and variational states, that is,

$$U = u_0 + u, \qquad V = v_0 + v, \qquad W = w_0 + w \tag{9.123a}$$

The velocity potential function, Φ, can also be written as the sum of initial, ϕ_0, and variational, $\tilde{\phi}$, components

$$\Phi = \phi_0 + \tilde{\phi} \tag{9.123b}$$

It is known that liquid pressure responses under longitudinal excitation are primarily sensitive to small axisymmetric shell wall displacements for a relatively incompressible liquid. Thus, one may expect that nonlinearity in the circumferential strain, ε_θ, would have the most pronounced effect on the liquid. Linear strain–displacement relations are used except for the strain, ε_θ, which includes a nonlinear term in the variational displacements, that is,

$$\varepsilon_\theta = \frac{1}{R}\frac{\partial V}{\partial \theta} - \frac{W}{R} + \frac{1}{2R^2}\left(\frac{\partial w}{\partial \theta}\right)^2 \tag{9.124}$$

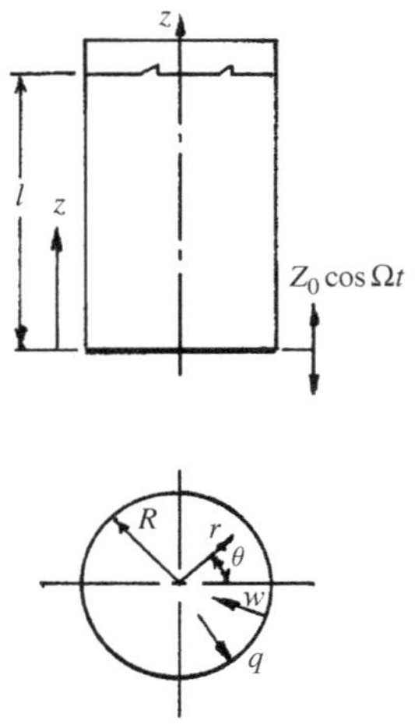

(a) Schematic diagram of elastic cylindrical shell partially filled with liquid.

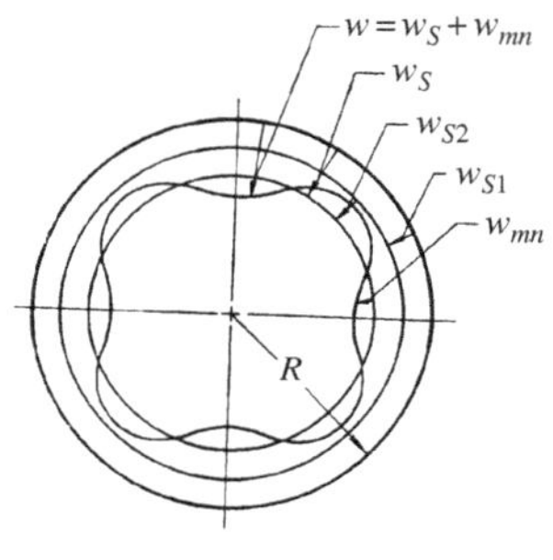

(b) Nonlinear variational shell displacements.

Figure 9.15 Schematic diagram of (a) a circular cylindrical shell partially filled with liquid and (b) the associated variational displacements.

The variational part of (9.124), $\tilde{\varepsilon}_\theta$, is

$$\tilde{\varepsilon}_\theta = \frac{1}{R}\frac{\partial v}{\partial \theta} - \frac{w}{R} + \frac{1}{2R^2}\left(\frac{\partial w}{\partial \theta}\right)^2 \tag{9.125}$$

The variational displacements are assumed to contain both symmetric and nonsymmetric (with subscript n) components. The symmetric components will be split into those induced by nonsymmetric displacements (with subscript $s1$), and those due to the change in internal symmetric pressure loading (with subscript $s2$). Thus one can write

$$u = u_{s1} + u_{s2} + u_n, \qquad v = v_n, \qquad w = w_{s1} + w_{s2} + w_n \tag{9.126}$$

where symmetric circumferential displacements are zero, and the radial displacements of the variational state are shown in Figure 9.15(b). Substituting equations (9.126) into equation (9.106) gives

$$\tilde{\varepsilon}_\theta = \frac{1}{R}\frac{\partial v_n}{\partial \theta} - \frac{1}{R}(w_n + w_{s2}) - \frac{1}{R}w_{s1} + \frac{1}{2R^2}\left(\frac{\partial w_n}{\partial \theta}\right)^2 \tag{9.127}$$

Evensen (1967) showed that the symmetric displacement, w_{s1}, is related to the nonsymmetric displacement, w_n, such that

$$-\frac{1}{R}w_{s1}+\frac{1}{2R^2}\left(\frac{\partial w_n}{\partial\theta}\right)^2=0 \tag{9.128}$$

The shell deformation is described by the membrane theory for all symmetric motions. Further, shell inertia in the axial and circumferential directions is neglected for both symmetric and nonsymmetric motions, while part of the radial shell inertia also is neglected for certain symmetric loading components. The total liquid pressure acting on the shell is

$$Q=(q_0+q_s)+q_n \tag{9.129}$$

Using Donnell's shell theory, the coupled shell equations are

$$\left(\frac{\partial^2}{\partial Z^2}+\frac{1-\nu}{2R^2}\frac{\partial^2}{\partial\theta^2}\right)U+\frac{1+\varepsilon}{2R}\frac{\partial^2 V}{\partial Z\partial\theta}-\frac{\nu}{R}\frac{\partial W}{\partial Z}+\underline{\frac{\varepsilon}{R}\frac{\partial w_{s1}}{\partial Z}}=0 \tag{9.130a}$$

$$\frac{1+\nu}{2}\frac{\partial^2 U}{\partial z\partial\theta}+R\frac{1-\nu}{2}\frac{\partial^2 V}{\partial z^2}+\frac{1}{R}\frac{\partial^2 V}{\partial\theta^2}-\frac{1}{R}\frac{\partial W}{\partial\theta}-\underline{\frac{(1-\nu^2)(q_0+q_s)}{E\hat{t}R}\frac{\partial w_n}{\partial\theta}}=0 \tag{9.130b}$$

$$\frac{\nu}{R}\frac{\partial U}{\partial z}+\frac{1}{R^2}\frac{\partial V}{\partial\theta}-\frac{W}{R^2}-\frac{\hat{t}^2}{12}\nabla^4 W \\ -\underline{\frac{(1-\nu^2)}{E\hat{t}}\left[\left(Q+m_c\frac{\partial^2 W}{\partial t^2}\right)-\frac{(q_0+q_s)}{R}\frac{\partial^2 w_n}{\partial\theta^2}\right]+\frac{w_{s1}}{R^2}}=0 \tag{9.130c}$$

The underlined expressions in equations (9.130) account for the loads that occur as a result of changes due to the fact that the system switches from the initial state to the final state. Thus, equations (9.130) represent the shell equations for the final state. These different states are considered in the next sections.

Initial state

The initial state of the shell–liquid system is mainly dominated by axisymmetric vibrational modes where the shell is only subjected to circumferential stresses as indicated by Bleich (1956a). In this case, one may write the shell equation from equation (9.130c) in the form

$$m_c\frac{\partial^2 w_0}{\partial t^2}+\frac{E\hat{t}}{R^2}w_0=-q_0 \tag{9.131}$$

For compressible fluid, the continuity equation satisfied by the velocity potential in the propagation of sound wave is (Thomson, 1965)

$$\frac{\partial^2\phi_0}{\partial r^2}+\frac{1}{r}\frac{\partial\phi_0}{\partial r}+\frac{\partial^2\phi_0}{\partial z^2}-\frac{1}{C_L^2}\frac{\partial^2\phi_0}{\partial t^2}=0 \tag{9.132}$$

where C_L is the sound speed in the liquid.

Equations (9.131) and (9.132) should be solved subject to the following boundary conditions:

The initial pressure loading acting on the wall shell, $r=R$, is given by the linearized Bernoulli equation

$$q_0 = -\rho \frac{\partial \phi_0}{\partial t} \tag{9.133a}$$

At the top of the tank, the zero ullage pressure requires

$$\left.\frac{\partial \phi_0}{\partial t}\right|_{z=l} = 0 \tag{9.133b}$$

At the bottom, the fluid must possess the same velocity as the shell bottom,

$$\left.\frac{\partial \phi_0}{\partial z}\right|_{z=0} = -\Omega z_0 \sin \Omega t \tag{9.133c}$$

At the tank wall, the following compatibility condition must be satisfied

$$\left.\frac{\partial \phi_0}{\partial r}\right|_{r=R} = -\frac{\partial w_0}{\partial t} \tag{9.133d}$$

The shell loading, q_0, for the initial membrane state is determined by solving equations (9.131) and (9.132) using the boundary conditions (9.133). For excitation frequency $\Omega \ll \omega_{00}$, where $\omega_{00}^2 = E\hat{t}/(R^2 m_c)$ is the natural frequency of the empty membrane shell and m_c is the mass per unit area of the shell, the solution for the velocity potential ϕ_0 that satisfies conditions (9.133a, b, d) is

$$\phi_0 = \left\{ C_0 \sin(\bar{\mu}_0(l-z)/R) I_0(\mu_0 r/R) + \sum_{i=1}^{\infty} C_i \sinh(\bar{\mu}_i(l-z)/R) J_0(\mu_i r/R) \right\} \times \sin \Omega t \tag{9.134}$$

where $\bar{\mu}_0^2 = \mu_0^2 + (\Omega R/C_L)^2$, $\bar{\mu}_i^2 = \mu_i^2 - (\Omega R/C_L)^2$, μ_0 is the single real root, and μ_i are the absolute values of the imaginary roots of the equation

$$\frac{I_0(\mu)}{\mu I_1(\mu)} = \frac{m_c}{\rho R}\left[\frac{\omega_{00}^2}{\Omega^2} - 1\right] \tag{9.135}$$

Equation (9.134) provides good approximation everywhere except at the bottom of the shell. To satisfy condition (9.133c), the following solution is obtained

$$\phi_0 = \frac{2 Z_0 \Omega R I_1(\mu_0)}{\bar{\mu}_0 \mu_0 \left[I_0^2(\mu_0) - I_1^2(\mu_0)\right]} \frac{I_0(\mu_0 r/R) \sin(\bar{\mu}_0(l-z)/R)}{\cos(\bar{\mu}_0 l/R)} \sin \Omega t \tag{9.136}$$

According to equation (9.131), for a shell with frequency ratio, $Et/(m_c R^2 \Omega^2) \gg 1$, equation (9.135) gives

$$\mu_0^2 = 2\frac{\rho \Omega^2 R^3}{E\hat{t}} \tag{9.137}$$

Since $\bar{\mu}_0^2 = \mu_0^2 + (\Omega R/C_L)^2$, and the speed of sound as $C_L^2 = \kappa/\rho$, where κ is the liquid compressibility index, one can use equation (9.137) to write

$$\bar{\mu}_0^2 = \rho \Omega^2 R^2 \left[2\frac{R}{E\hat{t}} + \frac{1}{\kappa}\right] \tag{9.138}$$

An equivalent sonic velocity may be defined as $C_e = 1/\sqrt{\rho(2(R/E\hat{t}) + (1/\kappa))}$, and one can write $\bar{\mu}_0 = \Omega R/C_e$. Using equations (9.136)–(9.138) in (9.133a) gives the initial state pressure loading

$$q_0 = -\frac{\rho l g g_z}{\alpha_e \cos \alpha_e} \sin[(\Omega/C_e)(l-x)] \cos \Omega t \tag{9.139}$$

where g is the gravitational acceleration, $g_z = \Omega^2 Z_0/g$ is an excitation acceleration parameter, $\alpha_e = \Omega l/C_e$ is a frequency parameter based on equivalent sonic velocity. Expression (9.139) amounts to that for a water-hammer wave of equivalent velocity C_e.

Nonlinear variational state

The nonlinear behavior of the system is determined by including the nonlinear terms in the shell equations of motion. Substituting equations (9.126) into equations (9.130) and separating the equations into symmetric and asymmetric displacements, gives

$$-w_{s1} + \frac{1}{2R}\left(\frac{\partial w_n}{\partial \theta}\right)^2 = 0 \tag{9.140a}$$

$$\frac{12(1-\varepsilon^2)}{R^2 \hat{t}^2} w_{s2} + \frac{1}{D}\left[q_s + m_c \frac{\partial^2}{\partial t^2}(w_{s1} + w_{s2})\right] = 0 \tag{9.140b}$$

$$D^4 \nabla^4 w_n + \frac{12(1-\varepsilon^2)}{R^2 \hat{t}^2} \frac{\partial^4 w_b}{\partial z^4} + \frac{1}{D} \nabla^4 \left[q_n - \frac{(q_0 + q_s)}{R} \frac{\partial^2 w_n}{\partial \theta^2} + m_c \frac{\partial^2 w_n}{\partial t^2}\right] = 0 \tag{9.140c}$$

$$\frac{\partial^2 \tilde{\phi}}{\partial r^2} + \frac{1}{r}\frac{\partial \tilde{\phi}}{\partial r} + \frac{1}{r^2}\frac{\partial^2 \tilde{\phi}}{\partial \theta^2} + \frac{\partial^2 \tilde{\phi}}{\partial z^2} - \frac{1}{C_L^2}\frac{\partial^2 \tilde{\phi}}{\partial t^2} = 0 \tag{9.140d}$$

Subject to the boundary conditions

$$q = -\rho \left.\frac{\partial \tilde{\phi}}{\partial t}\right|_{r=R}, \qquad \left.\frac{\partial \tilde{\phi}}{\partial t}\right|_{z=l} = 0 \tag{9.141a,b}$$

$$\left.\frac{\partial \tilde{\phi}}{\partial z}\right|_{z=0} = 0, \qquad \left.\frac{\partial \tilde{\phi}}{\partial r}\right|_{r=R} = -\frac{\partial w}{\partial t} \tag{9.141c,d}$$

The nonsymmetric part, w_{mn}, of the nonlinear displacement may be assumed to be in the form

$$w_{mn} = f_{mn}(t) \cos m\theta \sin k_n z \tag{9.142}$$

where $k_n = n\pi/l$. The symmetric nonlinear displacement, w_{s2}, may be determined by satisfying equation (9.128)

$$w_{s1} = \frac{m^2}{4R} f_{mn}^2(t) \sin^2 k_n z \tag{9.143}$$

Based on experimental observations (Kana and Craig, 1968), there is a change in symmetric fluid pressure that requires additional symmetric displacement, w_{s2}, as demonstrated in Figure 9.15(b). The nonlinear variational state may written in the form

$$w = w_{s1} + w_{s2} + w_{mn}, \quad q = q_{s1} + q_{s2} + q_{mn} \tag{9.144a, b}$$

In this case, $\tilde{\phi}_{s1}$ corresponding to w_{s1} is obtained from the solution of equation (9.140d) in the form

$$\tilde{\phi}_{s1} = \frac{m^2}{2} f_{mn}(t)\dot{f}_{mn}(t) \sum_{j=1}^{\infty} \left[\frac{\beta_{2ns}^2 + \beta_{00}^2}{(\gamma_{0j}^2 + \beta_{2ns}^2)(\gamma_{0j}^2 - \beta_{00}^2)} \right] \frac{\cosh \alpha_{0j} z}{\cosh \alpha_{0j} l} \frac{J_0(\gamma_{0j} r/R)}{J_0(\gamma_{0j})}$$
$$+ \frac{m^2}{4} f_{mn}(t)\dot{f}_{mn}(t) \left[\frac{I_0(\beta_{2ns} r/R)\cos(2k_n z)}{\beta_{2ns} I_1(\beta_{2ns})} + \frac{J_0(\beta_{00} r/R)}{\beta_{00} J_1(\beta_{00})} \right.$$
$$\left. - \frac{2\cos(\Omega z/C_L)}{\cos(\Omega l/C_L)} \left(\frac{\beta_{2ns}^2 + \beta_{00}^2}{\beta_{2ns}^2 \beta_{00}^2} \right) \right] \tag{9.145}$$

where $\beta_{2ns}^2 = R^2(4k_n^2 - (\Omega^2/C_L^2))$, $\beta_0^2 = (R^2\Omega^2)/C_L^2$, $\gamma_{0j}^2 = R^2\alpha\left(4k_{0j}^2 - (\Omega^2/C_L^2)\right)$, $J_0'(\gamma_{0j}) = 0$.

The corresponding pressure is obtained by using equation (9.141a)

$$q_{s1} = -\frac{m^2\rho}{4}\left[\dot{f}_{mn}^2 + f_{mn}(t)\ddot{f}_{mn}(t)\right]\left[\frac{I_0(\beta_{2ns})}{\beta_{2ns} I_1(\beta_{2ns})}\cos 2k_n z + \frac{J_0(\beta_{00})}{\beta_{00} J_1(\beta_{00})}\right.$$
$$\left. + \frac{2\cos(\Omega z/C_L)}{\cos(\Omega l/C_L)}\left(\frac{\beta_{2ns}^2 + \beta_{00}^2}{\beta_{2ns}^2 \beta_{00}^2}\right)\right] \tag{9.146}$$

The additional symmetric displacement, w_{s1}, may be expressed in terms of the natural modes of a membrane cylinder in the form

$$w_{s2} = f_{s2} \sum_{p=\text{odd}}^{N} a_p \cos(pk_1 z/2) \tag{9.147}$$

The corresponding $\tilde{\phi}_{s2}$ is obtained by solving equation (9.140d)

$$\tilde{\phi}_{s2} = -\dot{f}_{s2} R \sum_{p=\text{odd}}^{N} a_p \frac{I_0(\beta_{0p} r/R)}{\beta_{0p} I_1(\beta_{0p})} \cos(pk_1 z/2) \tag{9.148}$$

where $\beta_{0p}^2 = R^2\left[(pk_1/2)^2 - (\Omega_{s0p}^2/C_L^2)\right]$, $\Omega_{s0p}^2 = \dfrac{E\hat{t}}{R^2(m_c + m_p)}$ is the square of the natural frequency of full shell in pth symmetric mode, and m_p is the liquid apparent mass in pth symmetric mode. Further, β_{0p}^2 is given as the real roots of the equation

$$\frac{I_0(\beta_{0p})}{\beta_{0p} I_1(\beta_{0p})} = \frac{m_c}{\rho R}\left\{\frac{\Omega_{00}^2 R^2/C_L^2}{(pk_1 R/2)^2 - \beta_{0p}^2} - 1\right\} \tag{9.149}$$

Equations (9.145)–(9.148) satisfy equations (9.140) for the case of incompressible fluid. However, they are approximately valid in a lower frequency range. From equation (9.141a) the pressure loading corresponding to the additional symmetric motion is

$$q_{s2} = -\ddot{f}_{s2}\rho R \sum_{p=\text{odd}}^{N} a_p \frac{I_0(\beta_{0p})}{\beta_{0p} I_1(\beta_{0p})} \cos(pk_1 z/2) \tag{9.150}$$

To obtain approximate solutions of equations (9.140a–9.140c) with w_n given by equation (9.142), Galerkin's method is used with respect to θ by adopting the symmetric and $\cos m\theta$ form. From equation (9.140a) one may use the solution (9.143). This is followed by seeking a solution for equation (9.140b) by considering equations (9.143, 9.146–9.148, 9.150) and using Galerkin's method with respect to $\cos(pk_1z/2)$, which gives

$$\ddot{f}_{s2}a_p + \Omega_{s0p}^2 f_{s2}a_p + G_{2p}\left(\dot{f}_{mn}^2 + f_{mn}\ddot{f}_{mn}\right) = 0 \tag{9.151}$$

Applying Galerkin's method to equation (9.140c) and in view of equations (9.140, 9.143, 9.144, 9.146), one may write

$$\ddot{f}_{mn} + \Omega_{mn}^2 f_{mn} + f_{mn}G_3\cos\Omega t + f_{mn}\ddot{f}_{s2}\sum_{p=\text{odd}}^{N} a_p G_{4p} + G_5 f_{mn}\left(\dot{f}_{mn}^2 + f_{mn}\ddot{f}_{mn}\right) = 0 \tag{9.152}$$

where

$$G_{2p} = \frac{m^2\sin(p\pi/2)}{R(m_c+m_p)}\left[\frac{\rho p R}{\pi(16n^2-p^2)}\frac{I_0(\beta_{2ns})}{\beta_{2ns}I_1(\beta_{2ns})} - \frac{\rho R}{\pi p}\frac{J_0(\beta_{00})}{\beta_{00}J_1(\beta_{00})}\right.$$
$$\left. + \frac{2\pi\rho p R}{[4(\Omega l/C_L)^2 - p^2\pi^2]}\left(\frac{\beta_{2ns}^2+\beta_{00}^2}{\beta_{2ns}^2\beta_{00}^2}\right) + \frac{m_c}{p\pi}\frac{16n^2}{(16n^2-p^2)}\right]$$

$$G_3 = \frac{4\rho\pi^2m^2n^2lgg_z\cos\alpha_e(1-\cos\alpha_e)}{\alpha_e^2R(\alpha_e^2-4n^2\pi^2)(m_c+m_{mn})}$$

$$G_{4p} = \frac{32m^2n^2m_p\sin(p\pi/2)}{p\pi R(16n^2-p^2)(m_c+m_{mn})}$$

$$G_5 = \frac{m^4\rho}{4R(m_c+m_{mn})}\left[\frac{I_0(\beta_{2ns})}{2\beta_{2ns}I_1(\beta_{2ns})} - \frac{J_0(\beta_{00})}{\beta_{00}J_1(\beta_{00})}\right.$$
$$\left. - \frac{8\pi^2n^2\tan(\Omega l/C_L)}{(\Omega l/C_L)\left[(\Omega l/C_L)^2-4n^2\pi^2\right]}\left(\frac{\beta_{2ns}^2+\beta_{00}^2}{\beta_{2ns}^2\beta_{00}^2}\right)\right]$$

and m_{mn} is the liquid apparent mass in the mnth asymmetric mode.

A harmonic balance procedure is employed to obtain an approximate solution of equations (9.151) and (9.152). Introducing the harmonic and subharmonic solutions

$$f_{s2} = A_{s2}\cos\Omega t, \qquad f_{mn} = A_{mn}\sin\left[\frac{\Omega}{2}t+\beta\right]$$

Equation (9.151) requires $\beta = 0,\ \pm\pi,\ \pi/2$, and

$$A_{s2}a_p = -G_{2p}\frac{A_{mn}^2(\Omega^2/4)}{\left(\Omega_{0p}^2-\Omega^2\right)}\cos 2\beta.$$

Introducing these results in equation (9.152), gives for $\beta=0,\ \pm\pi$

$$\Omega_{mn}^2-\frac{\Omega^2}{4}-\frac{G_3}{2}=\frac{\Omega^2}{8}A_{mn}^2\left[G_5+\Omega^2\sum_{p=\mathrm{odd}}^{N}\frac{G_{2p}G_{4p}}{(\Omega_{s0p}^2-\Omega^2)}\right],\quad \beta=0,\ \pm\pi \tag{9.153a}$$

For $\beta=\pi/2$, the following solution is obtained

$$\Omega_{mn}^2-\frac{\Omega^2}{4}+\frac{G_3}{2}=\frac{\Omega^2}{8}A_{mn}^2\left[G_5+\Omega^2\sum_{p=\mathrm{odd}}^{N}\frac{G_{2p}G_{4p}}{(\Omega_{s0p}^2-\Omega^2)}\right],\quad \beta=\pi/2 \tag{9.153b}$$

The axisymmetric pressure components are evaluated from equation (9.129) at the center of tank bottom, with the result

$$\begin{aligned}Q\Big|_{\substack{r=0\\z=0}}=&-\frac{\rho lgg_z}{\alpha_e}\tan(\Omega l/C_e)\cos\Omega t-\frac{\Omega^2}{4}A_{mn}^2\cos 2\beta\\&\times\left\{\frac{m^2}{4R}\left[\frac{\rho R}{\beta_{2ms}I_1(\beta_{2ms})}+\frac{\rho R}{\beta_{00}J_1(\beta_{00})}-G_1\right]\right.\\&\left.-\Omega^2\sum_{p=\mathrm{odd}}^{N}\frac{\rho R}{\beta_{0p}I_1(\beta_{0p})}\frac{G_p}{(\Omega_{s0p}^2-\Omega^2)}\right\}\cos\Omega t\end{aligned} \tag{9.154}$$

Kana and Craig (1968) considered a tank-water system of length, $l=16$ inch, mean radius, $R=1.5$ inch, wall thickness, $\hat{t}=0.01$ inch, made of 4130 steel of Young's modulus, $E=30\times10^6$ psi, Poisson's ratio, $\varepsilon=0.3$, weight density, $\rho_c g_0=0.283\ \mathrm{lb/inch}^3$, water density, $\rho g_0=0.0361\ \mathrm{lb/inch}^3$, and sound speed in water, $C_L=56\ 000$ inch/sec. The theoretical (according to equation (9.149)) and experimental natural frequencies of the system for the axisymmetric modes, Ω_{s0p} in hertz, are for the first three modes:

- $\Omega_{s01}=442$ Hz (theoretical), 452 Hz (experimental)
- $\Omega_{s03}=1316$ Hz (theoretical), 1339 Hz (experimental)
- $\Omega_{s05}=2164$ Hz (theoretical), 2196 Hz (experimental)

The dependence of the shell displacement response and axisymmetric pressure response at the bottom on the excitation frequency are shown in Figure 9.16(a) and (b) for $m=4$ and $n=1$ mode. The analytical results are based on equations (9.153) and (9.154). The two figures include experimental measurements indicated by solid dots, •. The axisymmetric pressure acts as parametric excitation to the shell walls. Vertical arrows indicate experimental jump or collapse of the response. Figures 9.16 exhibits hard-type nonlinearity while for mode $m=5$ and $n=1$, the nonlinearity is switched to soft-type as shown in Figure 9.17.

9.7 Nonlinear interaction with orthotropic shells

9.7.1 Nonlinear free vibration

The nonlinear interaction of orthotropic shell deformation partially incompressible and inviscid fluid was considered by Koval'chuk and Kruk (2000) in the presence of one-to-one

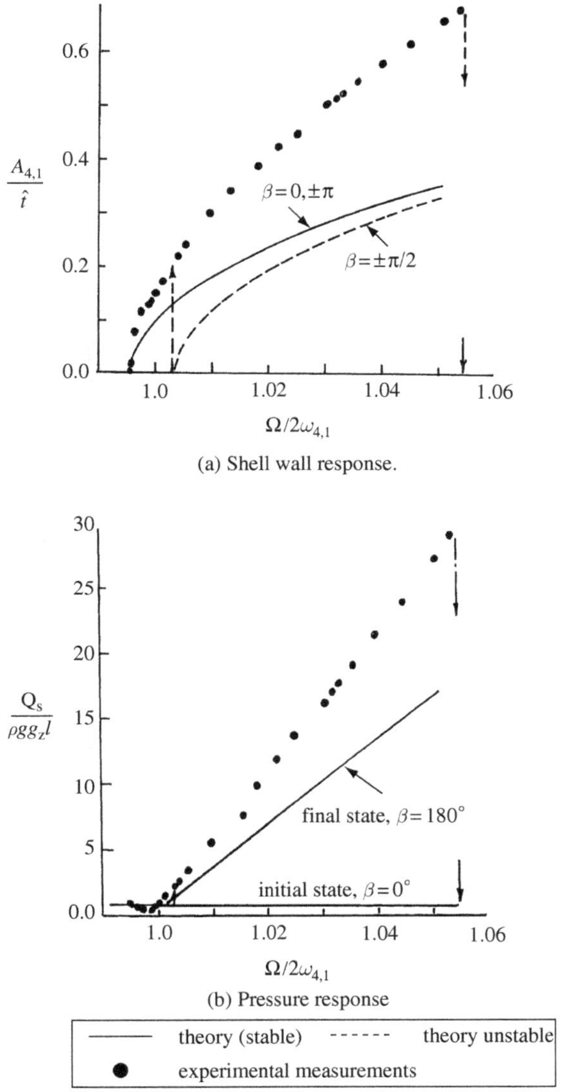

(a) Shell wall response.

(b) Pressure response

Figure 9 16 Parametric (a) amplitude–frequency, (b) pressure responses under parametric excitation of mode 4,1 of elastic shell partially filled with liquid, $g_z = \Omega^2 Z_0/g = 0.5$. (Kana and Craig, 1968)

internal resonance. The liquid free-surface oscillations are neglected. The dynamic equation of motion of an orthotropic shell, simply supported at both ends, is

$$D_1 \frac{\partial^4 w}{\partial z^4} + \frac{2}{R^2}(D_1\nu_2 + 2D_G)\frac{\partial^4 w}{\partial z^2 \partial \theta^2} + \frac{D_2}{R^4}\frac{\partial^4 w}{\partial \theta^4} = \frac{1}{R}\frac{\partial^2 F}{\partial z^2}$$
$$+ \frac{1}{R^2}\left(\frac{\partial^2 F}{\partial \theta^2}\frac{\partial^2 w}{\partial z^2} - 2\frac{\partial^2 F}{\partial z \partial \theta}\frac{\partial^2 w}{\partial z \partial \theta} + \frac{\partial^2 F}{\partial z^2}\frac{\partial^2 w}{\partial \theta^2}\right) - p - \rho_c \hat{t}\frac{\partial^2 w}{\partial t^2} \tag{9.155}$$

The stress function, F, is obtained by the solving the equation

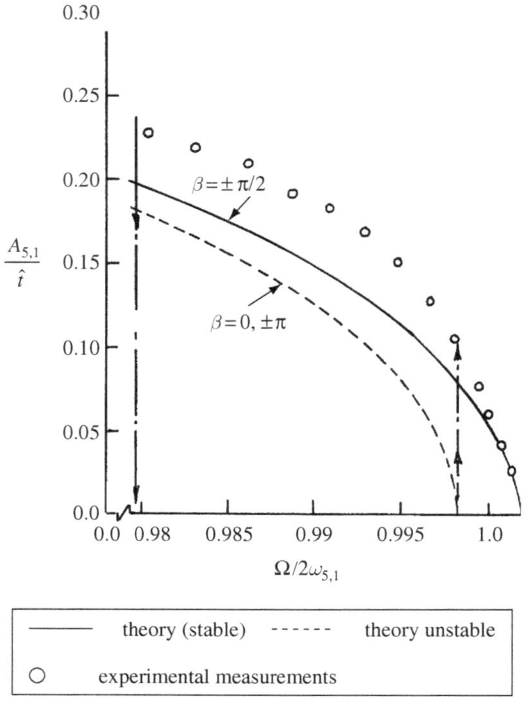

Figure 9.17 Parametric amplitude–frequency responses under parametric excitation of mode 5,1 of elastic shell partially filled with liquid $g_z = \Omega^2 Z_0/g = 0.5$. (Kana and Craig, 1968)

$$\frac{1}{E_2}\frac{\partial^4 F}{\partial z^4} + \frac{1}{R^2}\left[\frac{1}{G} - 2\frac{\nu_1}{E_1}\right]\frac{\partial^4 F}{\partial z^2 \partial \theta^2} + \frac{1}{E_1 R^4}\frac{\partial^4 F}{\partial \theta^4} = \frac{1}{R}\frac{\partial^2 w}{\partial z^2}$$
$$+ \frac{1}{R^2}\left[\left(\frac{\partial^2 w}{\partial z \partial \theta}\right)^2 - \frac{\partial^2 w}{\partial z^2}\frac{\partial^2 w}{\partial \theta^2}\right] \tag{9.156}$$

where $D_{1,2} = \dfrac{E_{1,2}\hat{t}^3}{12(1-\nu_1\nu_2)}$, $D_g = G\hat{t}^3/12$, E_1 and E_2 are elasticity moduli along z and θ coordinates, ν_1 and ν_2 are Poisson's ratios such that $E_1\nu_2 = E_2\nu_1$, and G is the shear modulus.

Consider two-mode interaction with wave numbers m_1 and m_2 along the longitudinal direction, and n_1 and n_2 along the circumferential direction of the shell. It is assumed that the natural frequencies of these two modes are nearly equal, that is, there is one-to-one internal resonance. In this case, the shell dynamic deflection may be expressed by the following modal summation

$$w(\theta, z, t) = [A_1(t)\cos n_1\theta + B_1(t)\sin n_1\theta]\sin(m_1\pi z/l)$$
$$+ [A_2(t)\cos n_2\theta + B_2(t)\sin n_2\theta]\sin(m_2\pi z/l) + c_1(t)\sin^2(m_1\pi z/l)$$
$$+ c_2(t)\sin^2(m_2\pi z/l) \tag{9.157}$$

where $A_i(t)$, $B_i(t)$, and $C_i(t)$ are the generalized coordinates. Introducing the following transformation

$$\begin{aligned} A_1(t) &= a(t)\cos\alpha(t), & B_1(t) &= a(t)\sin\alpha(t) \\ A_2(t) &= b(t)\cos\beta(t), & B_2(t) &= b(t)\sin\beta(t) \end{aligned} \tag{9.158}$$

Equation (9.157) takes the form

$$\begin{aligned} w(\theta, z, t) = {} & a(t)\cos(n_1\theta - \alpha(t))\sin(m_1\pi z/l) + b(t)\cos(n_2\theta - \beta(t))\sin(m_2\pi z/l) \\ & + c_1(t)\sin^2(m_1\pi z/l) + c_2(t)\sin^2(m_2\pi z/l) \end{aligned} \tag{9.159}$$

Substituting equation (9.159) into the stress function (9.156) one can solve for the stress function, F. The solution for F together with equation (9.159) are substituted in equation (9.155). Applying Galerkin's method, the following set of coupled nonlinear ordinary differential equations are obtained

$$\begin{aligned} & \ddot{a} + (\omega_{01}^2 - \dot{\alpha}^2)a + k_{11}a^3 + k_{12}ab^2 + k_{13}ac_1 + k_{14}ac_1^2 + k_{15}ac_2^2 \\ & \qquad = q_{11}\cos\alpha + q_{12}\sin\alpha \end{aligned} \tag{9.160a}$$

$$a\ddot{\alpha} + 2\dot{a}\dot{\alpha} = q_{12}\cos\alpha - q_{11}\sin\alpha \tag{9.160b}$$

$$\begin{aligned} & \ddot{b} + (\omega_{02}^2 - \dot{\beta}^2)b + k_{21}b^3 + k_{22}ba^2 + k_{23}bc_2 + k_{24}bc_2^2 + k_{25}bc_1^2 \\ & \qquad = q_{21}\cos\beta + q_{22}\sin\beta \end{aligned} \tag{9.160c}$$

$$b\ddot{\beta} + 2\dot{b}\dot{\beta} = q_{22}\cos\beta - q_{21}\sin\beta \tag{9.160d}$$

$$\ddot{c}_1 + \varpi_1^2 c_1 + k_{31}a^2 + k_{32}c_1a^2 + k_{33}c_1b^2 = q_{31} \tag{9.160e}$$

$$\ddot{c}_2 + \varpi_2^2 c_2 + k_{41}b^2 + k_{42}c_2b^2 + k_{43}c_2a^2 = q_{32} \tag{9.160f}$$

where ω_{0i} are the natural frequencies of the shell without fluid, given by the following expression

$$\begin{aligned} \omega_{0i}^2 = {} & \frac{1}{\rho_c\hat{t}}\left[D_1\left(\frac{m_i\pi}{l}\right)^4 + 2\frac{n_i^2}{R^2}\left(\frac{m_i\pi}{l}\right)^2\left\{\frac{G\hat{t}^3}{6} + \frac{E\nu_2\hat{t}^3}{12(1-\nu_1\nu_2)}\right\} + \frac{n_i^4}{R^4}D_2\right] \\ & - \frac{\left(\frac{m_i\pi}{l}\right)^4}{\rho_c R^2\left[\frac{(m_i\pi/l)^4}{E_2} + \left[\frac{1}{G} - \frac{2\nu_1}{E_1}\right](m_i\pi/l)^2(n_i/R)^2 + \frac{n_i^4}{R^4E_1}\right]}, \quad i = 1, 2 \end{aligned} \tag{9.161a}$$

$$\varpi_i^2 = \frac{16}{3}\frac{D_1(m_i\pi/l)^4}{\rho_c\hat{t}} + \frac{E_2}{3\rho_c R^2} \tag{9.161b}$$

$$\begin{Bmatrix} q_{11} \\ q_{12} \end{Bmatrix} = \frac{2}{\pi R l\rho_c\hat{t}}\int_0^l\int_0^{2\pi} p\sin(m_1\pi/l)\begin{Bmatrix}\cos n_1\theta \\ \sin n_1\theta\end{Bmatrix}R\,d\theta\,dz \tag{9.161c}$$

$$\begin{Bmatrix} q_{21} \\ q_{22} \end{Bmatrix} = \frac{2}{\pi R l \rho_c \hat{t}} \int_0^l \int_0^{2\pi} p \sin(m_2 \pi z/l) \begin{Bmatrix} \cos n_2\theta \\ \sin n_2\theta \end{Bmatrix} R\, d\theta\, dz \tag{9.161d}$$

$$\begin{Bmatrix} q_{31} \\ q_{32} \end{Bmatrix} = \frac{8}{3\pi R l \rho_c \hat{t}} \int_0^l \int_0^{2\pi} p \begin{Bmatrix} \sin^2(m_1 \pi z/l) \\ \sin^2(m_2 \pi z/l) \end{Bmatrix} R\, d\theta\, dz \tag{9.161e}$$

and k_{ij} are coefficients depending on the physical and geometrical parameters of the shell and wave numbers.

The hydrodynamic pressure, p, is determined from the equation

$$p = -\rho \left\{ \frac{\partial \Phi}{\partial t} - g(z-h) \right\}_{\substack{r=R \\ z \le h}} \tag{9.162}$$

where the fluid velocity function is obtained from the solution of Laplace's equation subject to the following boundary conditions

$$\left.\frac{\partial \Phi}{\partial z}\right|_{z=0} = 0, \qquad \left.\frac{\partial \Phi}{\partial z}\right|_{r=R} = \frac{\partial w}{\partial t}, \qquad \left.\frac{\partial \Phi}{\partial z}\right|_{z=h} = 0 \tag{9.163}$$

Note that the solution of Laplace's equation includes two waves identical to the two modes given in equation (9.159). Having solved for the velocity potential function, Φ, equation (9.162) is substituted into equations (9.161c–9.161e) and terms up to cubic orders are retained. This completes the estimation of the right-hand sides of equations (9.160). The functions $c_i(t)$ are determined from the quasi-static equations (9.160) based on the fact that $|c_i| \ll |a|$ and $|c_i| \ll |b|$. The following approximate solutions for $c_i(t)$ are obtained

$$c_1 \approx -\frac{k_{31}a^2}{\varpi_1^2 + k_{32}a^2 + k_{33}b^2}, \qquad c_2 \approx -\frac{k_{41}b^2}{\varpi_2^2 + k_{42}b^2 + k_{43}a^2} \tag{9.164a,b}$$

Introducing equations (9.164) into equations (9.160), the first four equations up to cubic order take the form

$$\ddot{a} + (\omega_1^2 - \dot{\alpha}^2)a + \gamma_1 a^3 + \delta ab^2 = 0 \tag{9.165a}$$

$$\ddot{b} + (\omega_2^2 - \dot{\beta}^2)b + \gamma_2 b^3 + \delta_2 ba^2 + k_{23}bc_2 = 0 \tag{9.165b}$$

$$[a\ddot{\alpha} + 2\dot{a}\dot{\alpha}](1 + m_{01}) = 0 \tag{9.165c}$$

$$\left[b\ddot{\beta} + 2\dot{b}\dot{\beta}\right](1 + m_{02}) = 0 \tag{9.165d}$$

where

$$\omega_i^2 = \frac{\omega_{0i}^2}{1+m_{0i}} \tag{9.166a}$$

$$m_{0i} = \frac{4\delta_{ni}}{\hat{t}lh}\left(\frac{\rho}{\rho_c}\right)\sum_{k=0}^{N}\frac{I_{ni}((k+1/2)\pi R/h)}{((k+1/2)\pi/h)I'_{ni}((k+1/2)\pi R/h)}$$
$$\times \int_0^h \cos((k+1/2)\pi z/h)\sin(m_i\pi z/l)\,dz \tag{9.166b}$$

$$\delta_{ni} = \begin{cases} 2 & \text{for } n_i = 0 \\ 1 & \text{for } n_i \neq 0 \end{cases}$$

$$\gamma_i = \frac{1}{(1+m_{0i})}\left\{\frac{1}{16}\left[E_1(m_i\pi/l)^4 + E_2(n_i/R)^4\right] + \frac{(n_i/R)^8}{48\rho_c^2R^2\varpi_i^2}\left[E_2 + \frac{8(n_i/R)^2}{\delta_i}\right]^2\right\} \tag{9.166c}$$

$$\delta_i = \frac{1}{16\rho_c(1+m_{i0})}\left\{\frac{[(m_1\pi/l)(n_2/R)-(m_2\pi/l)(n_1/R)]^4}{\Delta_A((m_1\pi/l)+(m_2\pi/l),(n_1/R)+(n_2/R))}\right.$$
$$+\frac{[(m_1\pi/l)(n_2/R)+(m_2\pi/l)(n_1/R)]^4}{\Delta_A((m_1\pi/l)-(m_2\pi/l),(n_1/R)+(n_2/R))}$$
$$+\frac{[(m_1\pi/l)(n_2/R)+(m_2\pi/l)(n_1/R)]^4}{\Delta_A((m_1\pi/l)+(m_2\pi/l),(n_1/R)-(n_2/R))}$$
$$\left.+\frac{[(m_1\pi/l)(n_2/R)-(m_2\pi/l)(n_1/R)]^4}{\Delta_A((m_1\pi/l)-(m_2\pi/l),(n_1/R)-(n_2/R))}\right\}$$

and the symbol Δ_A is given by the expression

$$\Delta_A(M,N) = \frac{M^4}{E_2} + \left(\frac{1}{G} - 2\frac{\nu_1}{E_1}\right)M^2N^2 + \frac{N^4}{E_1}$$

Koval'chuk and Kruk (2000) employed an extension of the method of averaging, based on the following coordinate transformation

$$a(t) = \sqrt{u_1(t) + v_1(t)\sin\psi_1(t)}$$

$$b(t) = \sqrt{u_2(t) + v_2(t)\sin\psi_2(t)}$$

$$\dot{a}(t) = \frac{\omega_1 v_1(t)}{a(t)}\cos\psi_1(t)$$

$$\dot{b}(t) = \frac{\omega_1 v_2(t)}{b(t)}\cos\psi_2(t)$$

$$\begin{aligned}\alpha(t) &= \varphi_1(t) + \arctan\left(\frac{u_1(t)\tan(\psi_1(t)/2) + v_1(t)}{\sqrt{u_1^2(t) - v_1^2(t)}}\right)\\ \dot{\alpha}(t) &= \frac{\omega_1\sqrt{u_1^2(t) - v_1^2(t)}}{a^2(t)}\\ \beta(t) &= \varphi_2(t) + \arctan\left(\frac{u_2(t)\tan(\psi_2(t)/2) + v_2(t)}{\sqrt{u_2^2(t) - v_2^2(t)}}\right)\\ \dot{\beta}(t) &= \frac{\omega_1\sqrt{u_2^2(t) - v_2^2(t)}}{b^2(t)}\\ \psi_i &= 2(\omega_i t + \vartheta_i(t))\end{aligned} \tag{9.167}$$

The time functions $u_i(t)$, $v_i(t)$, $\varphi_i(t)$, and $\vartheta_i(t)$ are determined from the averaged equations

$$\frac{du_i}{dt} = (-1)^i \frac{\delta_i}{2\omega_1} v_1 v_2 \sin 2(\vartheta_2 - \vartheta_1) \tag{9.168a}$$

$$\frac{dv_i}{dt} = (-1)^i \frac{\delta_i}{2\omega_1} u_i v_j \sin 2(\vartheta_2 - \vartheta_1) \tag{9.168b}$$

$$\frac{d\vartheta_i}{dt} = \frac{1}{4\omega_1}\left[3\gamma_i u_i + 2\delta_i u_j + 2\left(\omega_2^2 - \omega_1^2\right)\delta_{i,2} + \delta_i u_i \frac{v_j}{v_i}\cos 2(\vartheta_2 - \vartheta_1)\right] \tag{9.168c}$$

$$\frac{d\varphi_i}{dt} = -\frac{\sqrt{u_i^2(t) - v_i^2(t)}}{4\omega_1}\left[\gamma_i + \delta_i v_j\left(\frac{\cos 2(\vartheta_2 - \vartheta_1)}{v_i} - \frac{(-1)^i \sin 2(\vartheta_2 - \vartheta_1)}{u_i}\right)\right] \tag{9.168d}$$

where $i, j = 1, 2$, and $i \neq j$, $\delta_{i,2} = 1$ for $i = 2$, $\delta_{i,2} = 0$ for $i = 1$

The first integral of equations (9.168a, b) is

$$\begin{aligned}\frac{u_1(t)}{\delta_1} + \frac{u_2(t)}{\delta_2} &= \frac{1}{\omega_1^2}\left\{\frac{1}{\delta_1}\left[\dot{a}^2(0) + a^2(0)\left(\omega_1^2 + \dot{\alpha}^2(0)\right)\right]\right.\\ &\quad\left. + \frac{1}{\delta_2}\left[\dot{b}^2(0) + b^2(0)\left(\omega_1^2 + \dot{\beta}^2(0)\right)\right]\right\} = C_0\end{aligned} \tag{9.169}$$

The input energy to the system depends on the initial conditions. This energy is shared between the two amplitudes $u_1(t)$ and $u_2(t)$. Another two integrals can be obtained for $u_i(t)$ and $v_i(t)$ from equations (9.14) in the form

$$u_1^2 - v_1^2 = \frac{\dot{\alpha}^2(0)a^4(0)}{\omega_1^2} = C_1, \quad \text{and} \quad u_2^2 - v_2^2 = \frac{\dot{\beta}^2(0)b^4(0)}{\omega_1^2} = C_2 \tag{9.170a,b}$$

In contrast to the integral (9.169), integrals (9.170) describe the energy interchange between the amplitude functions $u_i(t)$ and $v_i(t)$ of complementary modes, that is, modes with wave numbers m_1 and n_2, and modes with numbers m_2 and n_1.

Equations (9.168) can be simplified in the form

$$\frac{du_1}{dt} = -\frac{\delta_1}{2\omega_1}\sqrt{\left[\left(\frac{u_1}{\delta_1} - C_0\right)^2 \delta_2^2 - C_2\right](u_1^2 - C_1)}\,\sin 2(\vartheta_2 - \vartheta_1) \tag{9.171a}$$

$$2\frac{d(\vartheta_2 - \vartheta_1)}{dt} = G - Nu_1 - \frac{\delta_1}{2\omega_1}\cos 2(\vartheta_2 - \vartheta_1) \times \frac{u_1\left[(u_1/\delta_1 - C_0)^2\delta_2^2 - C_2\right] - \delta_2/\delta_1\,(u_1^2 - C_1)(C_0\delta_2 - \delta_2/\delta_1 u_1)}{\sqrt{\left[(u_1/\delta_1 - C_0)^2\delta_2^2 - C_2\right](u_1^2 - C_1)}} \tag{9.171b}$$

where

$$G = \frac{1}{2\omega_1}\left[(3\gamma_2 - 2\delta_1)C_0\delta_2 + 2(\omega_2^2 - \omega_1^2)\right] \tag{9.172a}$$

$$N = \frac{1}{2\omega_1}\left[3(\gamma_1 + \gamma_2\delta_2/\delta_1) - 4\delta_2\right] \tag{9.172b}$$

The following more general integral is obtained

$$Gu_1 - \frac{Nu_1}{2} - \frac{\delta_1}{2\omega_1}\sqrt{\left[\left(\frac{u_1}{\delta_1} - C_0\right)^2 \delta_2^2 - C_2\right](u_1^2 - C_1)}\,\cos 2(\vartheta_2 - \vartheta_1) = C_3 \tag{9.173}$$

The constant C_3 depends on the initial values of amplitudes and phases of the shell two modes and is given by the expression

$$C_3 = Gu_1(0) - N\frac{u_1^2(0)}{2} - \frac{\delta_1}{2\omega_1^3}\Big\{a(0)\dot{a}(0)b(0)\dot{b}(0) + \omega_1^2\left[(\dot{a}^2(0) - u_1(0))(\dot{b}^2(0) - u_2(0))\right]\Big\} \tag{9.174a}$$

$$u_1(0) = \frac{1}{2\omega_1^2}\left[\dot{a}^2(0) + a^2(0)(\omega_1^2 + \dot{\alpha}^2(0))\right] \tag{9.174b}$$

$$u_2(0) = \frac{1}{2\omega_1^2}\left[\dot{b}^2(0) + b^2(0)(\omega_1^2 + \dot{\beta}^2(0))\right] \tag{9.174c}$$

The constants of motion given by equations (9.169), (9.170), and (9.173) describe different characteristics of a shell–liquid system. Each equation reveals the redistribution of the initial energy between the two bending modes as a result of a given initial energy imparted to one of the modes. This energy sharing owes its origin to the nonlinear coupling of the two modes when they are in one-to-one internal resonance $\omega_1 \approx \omega_2$. Koval'chuk and Kruk (2000) determined the dependence of the first two fundamental frequencies for the following system parameters

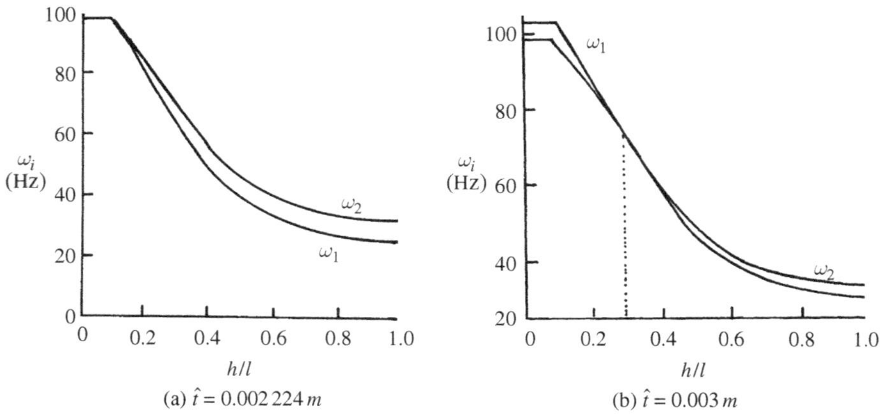

Figure 9.18 Dependence of nonlinear normal mode frequencies of orthotropic shell with still liquid on the liquid depth ratio for (a) modes $m_1, n_1 = 1, 3$ and $1, 5$, (b) $m_1, n_1 = 1, 3$ and $1, 4$. (Koval'chuk and Kruk, 2000)

$$E_1 = 2.12 \times 10^9 \text{ Pa}, \ E_2 = 1.23 \times 10^9 \text{ Pa}, G = 0.21 \times 10^9 \text{ Pa},$$
$$\nu_1 = 0.19, l = 0.48 \text{ m}, \ R = 0.16 \text{ m}, \ \rho_c = 1.65 \times 10^3 \text{ kg/m}^3, \ \rho = 1000 \text{ kg/m}^3$$

In the absence of fluid, the two fundamental frequencies of modes $(m_1, n_1 = 1, 3)$ and $(m_1, n_1 = 1, 5)$ are almost equal, that is, $\omega_{01}/2\pi = \omega_{02}/2\pi = 98.36$ Hz, for shell thickness $\hat{t} = 0.002\ 224$ m. The presence of liquid causes some detuning between the two frequencies, and the dependence of ω_1 and ω_2 on the liquid depth ratio, $\bar{h} = h/l$ is shown in Figure 9.18(a). For shell thickness $\hat{t} = 0.003$ m, Figure 9.18(b) shows the dependence of the modal frequencies on the liquid depth ratio for mode numbers $(m_1, n_1 = 1, 3)$ and $(m_1, n_1 = 1, 4)$. It is seen that the frequencies of the two modes have identical values at liquid depth ratio $\bar{h} = h/l = 0.35$. For given nonzero initial conditions of the amplitudes and phase rates, the time integration of equations (9.168) was found to reveal that the response amplitudes increase with the liquid depth while phase rates decrease with the liquid depth. Koval'chuk and Kholopova (1997), and Koval'chuk, *et al.* (1997) reported similar results for elastic shell–liquid systems.

9.7.2 Excitation of nonlinear sloshing interacting with linear orthotropic shells

The forced excitation of orthotropic shell partially filled with liquid was examined by Koval'chuk and Kruk (2002). The shell is described by the linearized equation (9.155) subjected to sinusoidal excitation. The linear shell equation takes the form

$$\begin{aligned} D_1 \frac{\partial^4 w}{\partial z^4} + \frac{2}{R^2}(D_1\nu_2 + 2D_G)\frac{\partial^4 w}{\partial z^2 \partial \theta^2} + \frac{D_2}{R^4}\frac{\partial^4 w}{\partial \theta^4} \\ = \frac{1}{R}\frac{\partial^2 F}{\partial z^2} + f_0(R\theta, z)\cos\Omega t - p - \rho_c \hat{t}\frac{\partial^2 w}{\partial t^2} \end{aligned} \tag{9.175}$$

The corresponding stress function is

$$\frac{1}{E_2}\frac{\partial^4 F}{\partial z^4} + \frac{1}{R^2}\left[\frac{1}{G} - 2\frac{\nu_1}{E_1}\right]\frac{\partial^4 F}{\partial z^2 \partial \theta^2} + \frac{1}{E_1 R^4}\frac{\partial^4 F}{\partial \theta^4} = \frac{1}{R}\frac{\partial^2 w}{\partial z^2} \tag{9.176}$$

where $f_0(\theta, z)\cos\Omega$t is the transverse sinusoidal pressure acting on the shell.

The hydrodynamic pressure, p, is first obtained by solving the fluid field equations described by equations (9.3) to (9.7). The linearized form of boundary condition (9.7b) is used. The free-surface wave height is approximated by the three-mode approximation

$$\eta(r,\theta,t) = [a_1(t)\cos n\theta + a_2 \sin n\theta] J_n(\lambda_{n1} r) + a_0(t) J_0(\lambda_{01} r) \tag{9.177}$$

The excitation frequency is assumed to be close to the fundamental shell natural frequency ω_{11}, that is, $\Omega \approx \omega_{11}$. Considering the interaction of the first mode along the longitudinal direction, and, the circumferential direction together with its conjugate, the shell deflection may be represented by the following approximate expression

$$w(\theta,z,t) = [A_1(t)\cos n\theta + A_2(t)\sin n\theta]\sin(m\pi z/l), \quad n, m = 1, 1 \tag{9.178}$$

Having solved for the velocity potential function of the liquid, and evaluating the hydrodynamic pressure, the shell equation (9.175) is discretized using Galerkin's method to yield the following two coupled nonlinear differential equations

$$\ddot{A}_i + \omega_{ii}^2 A_i - q a_i = F_i \cos\Omega t + \Psi_i(A_i, a_k) \quad k = 0, 1, 2, \quad i = 1, 2 \tag{9.179}$$

where ω_{ii} are defined by equation (9.166a) by replacing the double subscript by one single subscript,

$$q = \frac{2\rho g J_n(\lambda_{nj} R) N_{nm}}{\rho_c (1 + m_{01}) l\hat{t} \cosh(\lambda_{nj} h)}$$

$$N_{nm} = \int_0^h \cosh(\lambda_{nj} z)\sin(\pi z/l)\, dz$$

$$\begin{Bmatrix} F_1 \\ F_2 \end{Bmatrix} = \frac{2}{\pi R l \rho_c \hat{t}(1 + m_{01})} \int_0^l \int_0^{2\pi} f_0(\theta, z)\sin(\pi z/l) \begin{Bmatrix} \cos n\theta \\ \sin n\theta \end{Bmatrix} R\, d\theta\, dz, \ \Psi_i(A_i, a_k)$$

are nonlinear functions of the generalized coordinates of shell deflection and liquid wave height.

Equations (9.179) are coupled with the equations of liquid wave height motion

$$\ddot{a}_i + \varpi_i^2 a_i - \bar{\gamma}_i \ddot{A}_i = H_i(A_i, a_i) \tag{9.180}$$

where H_i are nonlinear functions of the generalized coordinates of shell deflection and liquid wave height, $\varpi_1^2 = \varpi_2^2 = g\lambda_{11}\tanh(\lambda_{11} h)$, $\varpi_0^2 = g\lambda_{01}\tanh(\lambda_{01} h)$, and

$$\bar{\gamma}_1 = \bar{\gamma}_2 = \frac{2\lambda_{11}^2 R N_{nm}}{\cosh(\lambda_{11} h)(\lambda_{11}^2 R^2 - n^2) J_1(\lambda_{11} R)}, \quad \bar{\gamma}_0 = 0$$

Equations (9.179) and (9.180) are linearly and nonlinearly coupled. In order to remove the linear coupling, the following coordinate transformation is introduced

$$a_1 = u_1 + v_1, \ a_2 = u_2 + v_2, \ A_1 = r_1 u_1 + r_2 v_1, \ A_2 = r_1 u_2 + r_2 v_2 \tag{9.181}$$

where the modal fractions r_i are given by the following expressions

$$r_1 = \frac{\omega_1^2 - (\varpi_1^2 + \bar{\gamma}_1 q)}{\gamma_1 \omega_{11}^2}, \quad r_2 = \frac{\omega_2^2 - (\varpi_1^2 + \bar{\gamma}_1 q)}{\gamma_1 \omega_{11}^2} \tag{9.182}$$

The normal mode coupled frequencies, $\omega_{1,2}$, of the shell-liquid system are

$$\omega_{1,2}^2 = \frac{\omega_{11}^2 + \varpi_1^2}{2} \pm \sqrt{\frac{\left(\omega_{11}^2 - \varpi_1^2\right)}{2} + \bar{\gamma}_1 q \omega_{11}^2} \tag{9.183}$$

Note that the root with the upper sign is close to the first mode frequency of the shell, that is, $\omega_1 \approx \omega_{11}$, while the root with lower sign corresponds to the first sloshing mode frequency, that is, $\omega_2 \approx \varpi_1$. In terms of the normal coordinates, equations (9.179) and (9.180) take the form

$$\ddot{u}_1 + \omega_1^2 u_1 = \frac{1}{1 + \bar{\gamma}_1 \omega_{11}^2 r_1^2 / q}\left[H_1 + \bar{\gamma}_1 (F_1 \cos\Omega t + \Psi_1)\left(\frac{r_1 \omega_{11}^2}{q} + 1\right)\right] \tag{9.184a}$$

$$\ddot{u}_2 + \omega_1^2 u_2 = \frac{1}{1 + \bar{\gamma}_1 \omega_{11}^2 r_1^2 / q}\left[H_2 + \bar{\gamma}_1 (F_2 \cos\Omega t + \Psi_2)\left(\frac{r_1 \omega_{11}^2}{q} + 1\right)\right] \tag{9.184b}$$

$$v_1 + \omega_1^2 v_1 = \frac{1}{1 + \bar{\gamma}_1 \omega_{11}^2 r_2^2 / q}\left[H_1 + \bar{\gamma}_1 (F_1 \cos\Omega t + \Psi_1)\left(\frac{r_1 \omega_{11}^2}{q} + 1\right)\right] \tag{9.184c}$$

$$v_2 + \omega_1^2 v_2 = \frac{1}{1 + \bar{\gamma}_1 \omega_{11}^2 r_2^2 / q}\left[H_2 + \bar{\gamma}_1 (F_2 \cos\Omega t + \Psi_2)\left(\frac{r_1 \omega_{11}^2}{q} + 1\right)\right] \tag{9.184d}$$

The functions H_i includes terms up to third-order of a_1 and a_2, and second-order of a_0. Since $\Omega \approx \omega_{11} \approx \omega_1$, the following detuning parameter is introduced

$$\sigma = \omega_1^2 - \Omega^2 \tag{9.185}$$

The spatial external pressure excitation can be written in a form compatible with the shell first mode as

$$f_0(\theta, z)\cos\Omega t = F_0 \cos\theta \sin \pi z / l \tag{9.186}$$

The construction of periodic solutions of nonautonomous coupled systems of differential equation was outlined by Boyarshina, *et al.* (1998). Since the first mode (u_1, v_1) is externally excited, the second mode (u_2, v_2) and a_0 will maintain their initial values. Introducing the solution for the first mode

$$u_1(t) = a(t)\cos(\Omega t + \varphi_1(t)), \quad v_1(t) = b(t)\cos(\Omega t + \varphi_2(t)) \tag{9.187}$$

Following the standard procedure of the averaging method, gives the following four first-order differential equations

$$\frac{da}{dt} = \frac{\delta_1 a b^2}{8\Omega}\sin 2(\varphi_1 - \varphi_1) - \frac{\bar{F}}{2\Omega}\sin\varphi_1 \tag{9.188a}$$

$$\frac{db}{dt} = -\frac{\delta_1 a^2 b}{8\Omega}\sin 2(\varphi_1 - \varphi_1) \tag{9.188b}$$

$$a\frac{d\varphi_1}{dt} = \frac{\sigma a}{2\Omega} + \frac{3\delta_2}{8\Omega}a^3 + \frac{\delta_3}{2\Omega}b^2 a + \frac{\delta_1}{8\Omega}b^2 a \cos 2(\varphi_1 - \varphi_1) - \frac{\bar{F}}{2\Omega}\cos\varphi_1 \tag{9.188c}$$

$$b\frac{d\varphi_2}{dt} = \frac{\sigma a}{2\Omega} + \frac{3\delta_2}{8\Omega}b^3 + \frac{\delta_3}{2\Omega}a^2 b + \frac{\delta_1}{8\Omega}a^2 b\cos 2(\varphi_1 - \varphi_1) \tag{9.188d}$$

where

$$\delta_1 = \frac{2\omega_1^2(k_{10} - 2k_{20})}{1 + \bar{\gamma}_1 r_1^2 \omega_{11}^2/q}, \quad \delta_2 = \frac{2\omega_1^2 k_{10}}{3(1 + \bar{\gamma}_1 r_1^2 \omega_{11}^2/q)}, \quad \delta_3 = \frac{\omega_1^2 k_{20}}{1 + \bar{\gamma}_1 r_1^2 \omega_{11}^2/q},$$

$$\bar{F} = \frac{F_0 \bar{\gamma}_1 (1 + r_1 \omega_{11}^2/q)}{\rho_c \hat{t}(1 + m_{01})(1 + \bar{\gamma}_1 r_1^2 \omega_{11}^2/q)} \tag{9.189}$$

and the constants k_{i0}, $i = 1, 2$ depend on the hydrodynamic forces.

The steady-state solution of equations (9.188) is obtained by setting the left-hand sides to zero. One possible fixed solution is

$$\Omega^2 - \omega_1^2 = \frac{3}{4}\delta_2 a^2 \pm \frac{\bar{F}}{a}, \quad b = 0, \quad \cos\varphi_1 = \pm 1 \tag{9.190}$$

This solution represents plane wave motions of the free-liquid-surface and unimodal shell deformation provided zero initial conditions for a_2, a_0, and A_2. It is stable over the frequency range

$$\omega_1^2 + \frac{3\omega_1^2 k_{10}}{2(1 + \bar{\gamma}_1 r_1^2 \omega_{11}^2/q)} a^2 < \Omega^2 < \omega_1^2 + \frac{\omega_1^2(k_{10} + 2k_{20})}{2(1 + \bar{\gamma}_1 r_1^2 \omega_{11}^2/q)} \tag{9.191}$$

Figure 9.19 shows the dependence of the response amplitude ratio, ($\bar{a} = a/(\sqrt{2}R)$), on the excitation frequency ratio, ($\bar{\Omega}^2 = \Omega^2/\omega_1^2$), according to the solution (9.191) for excitation amplitude, $F_0 = 10$ kg/m s, shell parameters: $E_1 = 2.12 \times 10^9$ Pa, $E_2 = 1.23 \times 10^9$ Pa, $G = 0.9 \times 10^9$ Pa, $\rho_c = 1650$ kg/m^3, $l/R = 4$, $\hat{t}/R = 0.5 \times 10^{-3}$, fluid depth ratio, $h/R = 3$, and fluid density, $\rho = 1000$ kg/m^3. The solid branches are the stable solution while the dash curves are unstable. Curves AA_1 and BB_1 represent stable unimodal responses where only one coordinate, u_1, of the coupled system is excited. Curves KK_1 and KK_2 separate the stable and unstable solutions.

Another solution is

$$\begin{aligned}
&\Omega^2 - \omega_1^2 = \frac{2\omega_1^2 k_{20}}{(1 + \bar{\gamma}_1 r_1^2 \omega_{11}^2/q)} a^2 \pm \frac{\bar{F} k_{10}}{2a(k_{10} - 2k_{20})} \\
&b^2 = -\frac{4}{3\delta 2}\left[\sigma + \frac{\omega_1^2(2k_{20} - k_{10}/2)}{(1 + \bar{\gamma}_1 r_1^2 \omega_{11}^2/q)} a^2\right] \\
&\sin 2(\varphi_1 - \varphi_2) = 0, \quad \sin\varphi_1 = 0
\end{aligned} \tag{9.192}$$

This solution characterizes more complex modes of the shell–liquid system and describes spatial liquid free-surface waves together with mixed mode shell deformation. Figure 9.20 shows the dependence of the resultant amplitude ratio, $\bar{A}_0 = \sqrt{(a^2 + b^2)/2R^2}$, on the excitation frequency ratio, $\bar{\Omega}^2 = \Omega^2/\omega_1^2$, for the same parameters of Figure 9.19. This case represents intensive interaction of the shell wave deformation and the liquid-free-surface motion for excitation frequencies that exceed the critical excitation corresponding to stable system response, that is, $\bar{\Omega} > \bar{\Omega}_c$. Since solutions (9.190) and (9.192) are the only two steady-state

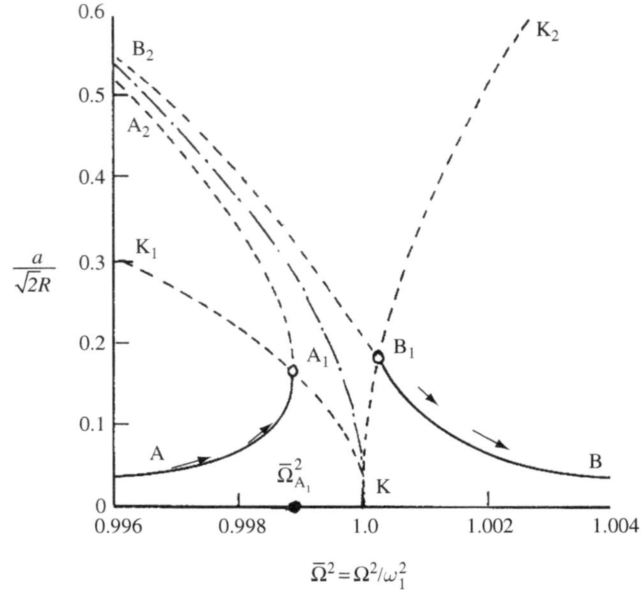

Figure 9.19 Unimodal amplitude–frequency response of nonlinear sloshing interaction with linear orthotropic shell. (Koval'chuk and Kruk, 2002).

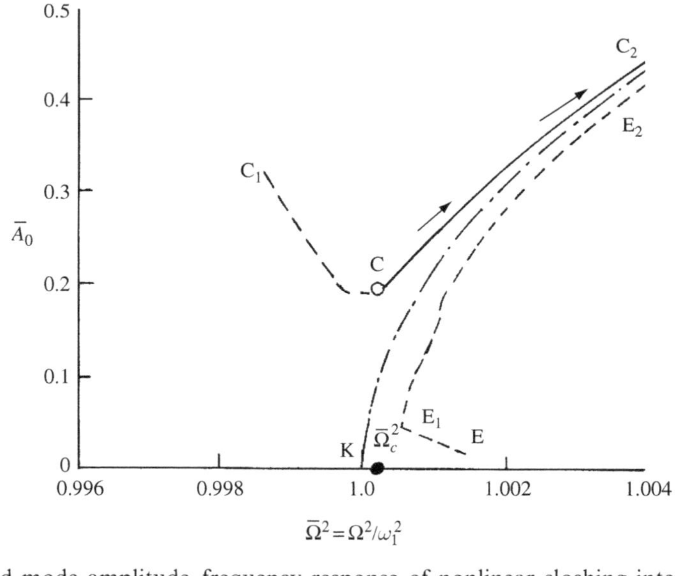

Figure 9.20 Mixed mode amplitude–frequency response of nonlinear sloshing interaction with linear orthotropic shell. (Koval'chuk and Kruk, 2002).

solutions of the coupled system, it follows that periodic solutions are impossible over the frequency range $\bar{\Omega}_{A_1} \leq \bar{\Omega} \leq \bar{\Omega}_c$.

The previous sections considered analytical approaches for different cases of a nonlinear shell–liquid system. However, analytical approaches are limited for cylindrical and rectangular containers with walls parallel to the longitudinal axis. For other geometries, numerical and experimental approaches should be considered. The next two sections are devoted to the modes of failure of storage liquid tanks and numerical algorithms currently employed for their analysis.

9.8 Storage liquid tanks

Storage liquid containers are usually mounted on the ground in two different ways: unanchored and anchored. Large-sized unanchored tanks with flat bases usually experience different kinds of damage under the action of ground motions. The best known damages are the elephant foot bulge, which takes the form of buckling at the bottom part of the tank, and cracks at the corner of the bottom plate-shell. Both classes are related to uplifting of the bottom plate and thus involve strong nonlinearity due to the associated large displacement and the separation between the bottom plate and foundation for unanchored tanks. The failure and damage of liquid tanks under earthquake excitations have been studied by many civil engineers (see, for example, Shibata, *et al.*, 1965, 1986, Shepherd, 1969, Hanson, 1973, Clough, 1977, Clough and Niwa, 1979, Fujita, 1981, Shih, 1981, Haroun and Housner, 1981a,b, 1982, Niwa and Clough, 1982, Haroun, 1983, 1991, Haroun and Mourad, 1990, and Isaacson and Subbiach, 1991). Hatano and Konno (1966) numerically estimated the hydrodynamic pressure on an arch dam during earthquakes.

Sakai (1985) and Shimizu (1983, 1990) presented a review of seismic studies of cylindrical liquid storage tanks. Shimizu (1990) classified the main interests of seismic studies of storage tanks into the following categories:

(1) The correlation between sloshing dynamics and the characteristics of long-period earthquake ground motions.
(2) The interaction between tanks' elastic dynamics and liquid sloshing under vertical ground motion.
(3) The interaction between liquid tanks and soil.
(4) The motions of rocking and uplifting of tanks.

Earthquake damage and failure modes caused by different earthquakes were analyzed by Kobayashi (1986). The following regimes were reported:

(1) Much damage and failures caused by liquid sloshing were observed in roof and shell walls for tanks of over 5000 m^3 capacity.
(2) Much damage and failures caused by liquid sloshing were observed in lower parts of tanks of less than 5000 m^3 capacity.
(3) Much damage and failures caused by inertia forces and overturning moments were observed around corner joints of the shell wall plates and bottom plates.

The behavior of anchored tanks, for which the base is not allowed to be separated from the foundation, has been studied by Veletsos and Yang (1977), Fischer (1979), Balendra, *et al.* (1982a,b), Ma, *et al.* (1982), and Haroun (1983). Veletsos (1974) adopted a simple procedure for evaluating the hydrodynamic forces induced in flexible liquid-filled tanks. The procedure was based on the assumption that the tank vibrates in a prescribed single mode and remains circular during vibrations. Veletsos and Yang (1977) presented simplified formulas to obtain the fundamental natural frequencies of liquid-filled shells using the Rayleigh–Ritz energy method. Haroun and Housner (1981a,b) studied the dynamic behavior of deformable anchored cylindrical tanks. Ishida and Kobayashi (1985), Auli, *et al.* (1985), Zui and Shinke (1985), Manos and Clough (1982, 1985), Manos and Talaslidis (1986), Sakai and Isoe (1989), Ishida and Mieda (1989), Sakai, *et al.* (1989), Uras and Liu (1990), Yi and Natsiavas (1990), Malhotra (1995), Shrimali and Jangid (2002), and Nachtogall, *et al.* (2003) conducted different computational and experimental studies to determine the uplifting effects on liquid containers. These studies were carried out based on the static tilt developed originally by Clough (1979), Clough and Niwa (1979), and Clough, *et al.* (1979). Manos (1989, 1990) studied the behavior of a cylindrical liquid tank subjected to lateral earthquake loading using tilt-test conditions. Rammerstofer, *et al.* (1989) estimated the maximum loads of nonlinear motion of uplifting liquid storage tanks by using the response spectrum method together with an iterative procedure. Nash, *et al.* (1989) considered the large-amplitude motion of the liquid free-surface and its effect on shell dynamics.

Note that the strong nonlinearity, associated with the unanchored tanks, results in a reduction in the effective beam type stiffness of the tank. Accordingly, the tank predominant frequency response affects the fluid pressures through the fluid–structure interaction. Furthermore, large amplitude sloshing can be induced by the long-period components of ground motion. Shih (1981) and Natsiavas (1987) indicated that shell flexibility effects might be neglected when evaluating seismic loads for tall unanchored tanks. Natsiavas (1989) considered two simplified models for studying the dynamic response of tall, unanchored, fluid-filled tanks, based on neglecting the shell flexibility. It was found that the sloshing effects on the overall tank dynamics were negligible for both models. Lau and Zeng (1992) developed a time-history analysis procedure to evaluate the nonlinear dynamic uplift response of unanchored cylindrical liquid tanks to horizontal earthquake support excitation. Contrary to previous results, the time-history analysis of Lau and Zeng (1992) revealed that the base uplift has a significant effect on the hydrodynamic forces generated from the vibrating liquid, particularly the convective component. Furthermore, the flexibility of the tank system was found to be important.

Natsiavas and Babcock (1988) developed analytical models of unanchored liquid tanks and analyzed the buckling behavior statically using the maximum value of pressure induced by dynamic and static loadings. Chiba, *et al.* (1986), Chiba and Tani, (1987), Liu and Uras (1989a,b), and Fujita, *et al.* (1990) included the dynamic interaction of liquid and structure, and the modal coupling in both axial and circumferential directions. Experimental results of Liu and Uras (1989a,b) did not reveal the higher-order buckling loads, which were attributed mainly to the presence of cylindrical shell imperfections in the actual experimental tests. Later, Uras, *et al.* (1990), Tsukimori, *et al.*, (1991), and Amabili (2003) were motivated to develop the equations of motion of cylindrical liquid shells taking into account the influence of imperfections. The finite element (FE) formulation of Tsukimori, *et al.* (1991) included the additional higher-order modal coupling, which depends mainly on the imperfection pattern.

Ogawa, *et al.* (1996) studied the sloshing behavior of worm tanks (tanks with a series of incomplete circular cross-sections). For this type of tank, the fluid–structure interaction and the sloshing dynamics can take place simultaneously. It was found that the natural frequency of the worm tank could be calculated using the formula of the frequency of a cylindrical tank or a rectangular tank for the lower modes. Ogawa, *et al.* (1996) estimated the surface behavior and impact mode using the rectangular tank results. Under seismic excitation along the tank short side, the tank slip occurred and the amplitude of the response in the seismic direction was twice that of the long side excitation. Later, Ogawa, *et al.* (1997) experimentally examined the fluid–structure-interaction behavior of the worm tank and observed that the acceleration in the lateral excitation is ten times larger than that in the longitudinal excitation.

If the tank is supported near its top and becomes separated from its foundation, then a three-dimensional FE analysis should be used. Chen W. W. (1991) used the 3-D FE analysis of a sodium-filled tank subjected to seismic loading to study the liquid sloshing effects on the tank structure. For the fluid depth–diameter ratio considered in the study, it was found that the sloshing motion participation was negligible.

The liquid–structure interaction of a reactor vessel subjected to a horizontal excitation was studied using combined computational schemes. The FE method was used for the structure and the boundary element method for the liquid (Chang, *et al.*, 1987, Yasiro, *et al.*, 1987, Kondo, 1989, and Kondo, *et al.*, 1989).

The problem of recovery and purification of fuel discharged from the fast reactors involves the dynamic response of tanks containing two liquids (Burris, *et al.*, 1987 and Bandyopadhyay, 1991). Tang, *et al.* (1991) showed that the sloshing motion in a tank containing two liquids is quite different from that containing only one liquid, and the sloshing wave height computed based on one liquid may lead to some errors. Tang (1992, 1993), Tang and Chang (1993a,b), and Veletsos and Shivakumar (1993) analytically and numerically studied the dynamic responses of both rigid and flexible tanks containing two liquids subjected to lateral and rocking base motions. They determined the hydrodynamic pressure, base shear, and base moments. Veletsos and Tang (1990) and Tang (1994) studied the effect of the soil–structure interaction on the dynamic response of tanks containing two liquids. His results indicated that the soil–structure interaction reduces the dynamic response.

9.9 Numerical techniques

9.9.1 Numerical simulation of liquid sloshing

Analytical approaches used to predict the dynamic response of liquid-filled tanks could not handle special types of tank geometry. Numerical methods are useful when the container walls are not vertical and straight or when the liquid wave heights are large. Three computational methods are in common use: (1) the finite difference (FD, see, e.g., Harlow and Welch, 1965, 1966)), (2) the finite element (FE, see, e.g., Zienkiewicz and Bettes (1978), Hughes, *et al.*, 1981, Barton and Parker, 1987, Babu, *et al.*, 1996, Wu, Ma, and Taylor, 1998, and Zhang, *et al.*, 2001), and (3) the boundary element (BE, see, e.g., Nakayama and Washizu, 1981). Under the assumption of small displacements, an inviscid fluid can be modeled by a degenerate solid finite element with no shear resistance. This approach is referred to as the "mock" fluid element or pseudo-elastic element (Kalinowski, 1975, Hamdi and Ousset, 1978, and Shi, 1987). A combination of analytical and finite element formulation was developed by Kochupillai *et al.*, 2002. The BE method is convenient for studying liquid sloshing behavior because both the FD and FE methods require very long computational time and a large amount of input data. Shiojiri and Hagiwara (1990) applied the BE method to compute the two-dimensional nonlinear sloshing in containers with inclined walls. Under sinusoidal excitation with frequency near resonance, the FE method results agreed fairly with the experimental measurements for a container with two inclined opposite walls.

The new developments of FE and BE methods have promoted the study of liquid sloshing in regular and arbitrary tank geometries. Luke (1967) developed a variational approach for the free-liquid-surface motion. It is based on establishing a mathematical formulation of the superlative, which is usually taken by minimizing the integral of some potential function of the system, and its gradient. Other variational formulations were developed by Bogoryad (1962), Shmakov (1964), Shklyarchuk (1966), Limarchenko (1978a,b, 1980, 1981, 1983a,b), Liu and Uras (1988), and Lui, *et al.* (1991). The precise form of the potential function is determined by the distribution of the various "sources" (or scalar potential functions) in space and by the boundary conditions. The solution of the boundary value problem, using the Green's function, establishes the variational formulation of the liquid sloshing modes for which the discretization by finite elements can be introduced. This approach is well documented in Morand and Ohayon (1995).

A pseudo-variational principle was originally developed by Ikegawa (1974). Nakayama and Washizu (1980, 1981) and Washizu *et al.* (1984) applied this principle to two-dimensional problems, including rectangular tanks experiencing transnational or pitching motions. There are, at least, three different methods available to implement the FE approach to handling liquid motion in solids. These are:

(1) The surface source distribution method, which results in a two-dimensional grid.
(2) Subdivision of the total liquid into small volumes, which produces a three-dimensional mesh.
(3) Representation of liquid behavior by a series of generalized functions weighted by unspecified coefficients. Bermudez and Rodriguez (1994) computed the vibration modes of a fluid–solid system using FE method.

Guyan *et al.* (1968) introduced a FE formulation for the axisymmetric modes of a hemispherical tank. Their approach utilized the simple source distribution representation on the free liquid surface. Luk (1969) and Khabbaz (1971) extended the analysis to develop a nonaxisymmetric FE by dividing the liquid into triangular annular elements and, therefore, subdividing the entire volume of liquid. Tokuda and Sakurai (1994) used exciting structural FE methods to determine the fluid free-surface motion. A numerical method based on Moiseev's theory (1958) for two-dimensional sloshing in a general tank shape was developed by Solaas and Faltinsen (1997) to determine the velocity potential function.

Ikegawa and Washizu (1973) and Ikegawa (1974) originally developed a pseudo-variational principle. Washizu, *et al.* (1977), Nakayama and Washizu (1980), and Washizu *et al.* (1984) implemented the principle for two-dimensional problems with application to rectangular tanks experiencing transnational or pitching motion. Takahara, *et al.* (1993a,b, 1994a,b, c) and Pawell (1997) conducted different studies on the liquid sloshing interaction with cylindrical tanks subjected to pitching excitation.

9.9.2 Numerical simulation of sloshing–structure interaction

Analytical techniques used to describe liquid–structure interaction are based on simplified assumptions and restricted to certain regular container geometry. In order to better describe this type of problem, computational algorithms are very useful and powerful, especially with the advent of high-speed computer machines. Haroun and Tayel (1984, 1985a, b, c) studied different cases of the axisymmetric vibrations of tanks subjected to seismic excitations. Natsiavas (1987 , 1988 , 1989) and Natsiavas and Babcock (1988) developed analytical models of unanchored liquid tanks and analyzed their dynamic characteristics under ground excitation. Peek (1986, 1988) and Peek and Jennings (1988) analyzed the dynamic response of unanchored liquid storage tanks to seismic lateral excitations. Fu (1993) used triangular elements to study the lateral sloshing dynamics in axisymmetric tanks. Wu, *et al.* (1998) developed a numerical simulation of sloshing waves in a three-dimensional tank based on FE methods.

Siekmann and Schilling (1974a,b) discretized the boundary of the fluid domain in developing boundary integrals to determine the natural frequencies of the free liquid surface in a container of arbitrary geometry. Morand and Ohayon (1995) noted that the *n*th approximate eigenvalue is higher than the exact eigenvalue. A paradoxical result by Morand (1977) was that the *n*th frequency of the liquid in a large tank is higher than the *n*th frequency in a smaller tank. Numerical and experimental investigations were carried out by Chiba, *et al.* (1984a, b, 1985) to determine the free vibration of circular cylindrical shells partially filled with liquid. Their studies focused mainly on the normal mode shapes of the coupled system. Tanaka and Nakayama (1991) and Chen, H. H., *et al.* (1994) employed the boundary element method to determine the liquid sloshing dynamics in three-dimensional containers.

Finite-difference methods were also used by Anisimov (1968) and Morand and Ohayon (1975, 1989) to study the symmetric vibrations of liquid shells. Boujot (1973), Berger, *et al.* (1975), Ohayon and Morand (1995), Schotte and Ohayon (1999, 2001, 2002), and Schotte (2001) estimated numerically the natural frequencies of elastic tanks partially filled with liquid. A series representation was used by Housner (1980) to study the hydroelastic vibration characteristics of partially filled shells.

Civil and earthquake engineers treated the dynamics of ground storage tanks via FE techniques. In the standard procedure of the FE method, the unknown function was approximated by trial functions, which did not satisfy the continuum equations exactly either in the domain or, in general, on the boundaries. The unknown nodal values were determined by approximately satisfying both the differential equations and the boundary conditions in an integrated mean sense. Various FE treatments of earthquake-loaded-liquid filled elastic tanks have recently been developed by Edwards (1969), Veletsos (1974), Veletsos and Yang (1977), Veletsos and Turner (1979), Brown (1980, 1982a, b, c), Veletsos and Kumar (1984), Fischer and Rammerstofer (1982, 1984, 1999), Veletsos and Tang (1986), and Rammerstofer *et al.* (1990, 1991). Shaaban and Nash (1975), Balendra and Nash (1978, 1980), Haroun (1980), Balendra, *et al.* (1982a, b), and Yi and Natsiavas (1990) used FE approaches to determine the natural frequencies and associated mode shapes of a coupled liquid–elastic tank subjected to arbitrary base excitation. A number of investigators (Hsiung and Weingarten, 1973, Shaaban and Nash, 1975, Brown and Hsu, 1978, and Brown and Chu, 1983) solved the liquid region using the BE technique and the elastic structure using the FE method. The FE displacement method was used by Hori, *et al.* (1994) to investigate the two-dimensional coupling between liquid and elastic structures.

Most of the reported FE techniques for ground storage tanks did not consider large displacement fluid flow (Wilson and Khalvati, 1983). For a cylindrical tank under two-dimensional horizontal earthquake excitation, no direction can be determined in which a horizontal earthquake is particularly dangerous. If a real earthquake excitation is considered, a dominant direction of excitation is hard to find. Scharf (1990) indicated that, by using a simple model and time integration, the maximum pressure due to two-dimensional excitation does not necessarily occur at an angle corresponding to the angle of maximum acceleration. Despite the fact that vertical excitation does not produce overturning moments, its influence on structural behavior with respect to stresses and stability may become very significant. The coupled oscillations of a liquid and a cylindrical shell subjected to vertical excitations were studied for small motions by Kondo (1981b) using a variational principle for the linear behavior of liquids and solids. This analysis led to a Rayleigh quotient that gives natural frequencies and mode shapes. Yamamoto (1981), Felippa (1991), and Felippa and Ohayon (1990) introduced mixed variational FE formulations to analyze the problem of elasto-acoustic interaction with liquid sloshing.

Note that it may be difficult to find water impact loads and a local structural response using these numerical techniques. One reason for these difficulties is that liquid impact often requires a very fine discretization in time and space. Hydroelasticity may also have to be considered.

9.10 Closing remarks

This chapter adopted typical problems of nonlinear interaction of elastic and composite thin shells with liquid sloshing dynamics. The interaction was considered under the influence of external and parametric excitations in the presence and absence of internal resonance conditions. In the presence of internal resonance, the energy sharing and redistribution among the interacting modes resulted in complex dynamic behavior.

This chapter did not address in detail the problem of storage liquid tank failure modes under the effect of earthquakes. Instead, it presented a review of the main research activities and results reported in the open literature by civil engineers. An overview of numerical techniques used to simulate liquid sloshing dynamics and sloshing–structure interaction has been presented. The subject of this chapter deserves an independent research monograph together with the linear shell–liquid interaction considered in the previous chapter.

This chapter did not address other nonlinear phenomena in shell–liquid systems such as chaos associated with nonlinear resonance. A few attempts were made by Krasnopolskaya and Podchasov (1993). Also the nonstationary response characteristics of nonlinear shells partially filled with liquid was not treated. The basic theory of nonstationary interaction was documented by Kubenko (1979). Koval'chuk and Podchasov (1996, 1999, 2001), Koval'chuk, Puchka, and Kholopova (1999), and Koval'chuk and Kholopova (2001) considered the problem of nonstationary response nonlinear shell–liquid systems under forced and parametric excitations as the system passes through resonance. Under polyharmonic excitation, Kubenko and Lakiza (1998) estimated the response of shell–liquid–gas systems near resonance. The influence of shell material properties was not addressed and the literature reported few studies on that topic (see, e.g., Prokopalo, *et al.* 1980, Lakis and Laveau, 1991, Sivak, 1998, and Pavlovskii, 2000). The influence of liquid viscosity on the forced response of a fluid-filled elastic shell was considered by Su (1983). One may notice, from the literature, that no attempts have been made to study the shell–liquid nonlinear interaction under random excitations. This is an open area for future research.

10

Interaction with support structures and tuned sloshing absorbers

10.1 Introduction

The interaction between the free-liquid-surface dynamics and supported elastic-structure dynamics, based on the assumption that the liquid container is rigid, is considered in this chapter. If the base of the supporting structure moves, the fluid container experiences motion in a certain trajectory governed by the excitation and the liquid response. The free-liquid-surface motion exerts hydrodynamic forces that are fed back to the supporting structure. Examples of liquid sloshing interaction with elastic support structures include elevated water towers and multistory buildings with tuned sloshing absorbers.

The dynamics of elevated water towers under seismic excitation was examined by Ifrim and Bratu (1969), Sonobe and Nishikawa (1969), Van Erp (1969), Shepherd (1972), Yang (1976), and Niwa (1978). The nonlinear interaction in elevated water towers subjected to vertical sinusoidal ground motion was examined in the neighborhood of internal resonance by Ibrahim and Barr (1975a,b), Ibrahim (1976), and Ibrahim *et al.* (1988). The interaction is critical when the liquid sloshing modes are coupled with the support structure dynamics through inertia nonlinearity. The nonlinear coupling may give rise to the occurrence of internal resonance conditions among the interacting modes. Internal resonance implies the presence of a linear relationship between the normal mode frequencies of the interacting modes (i.e., $\sum_{j=1}^{n} k_j\omega_j = 0$, where k_j are integers, and ω_j are the natural frequencies of the coupled modes, the number $K = \sum_{j=1}^{n} |k_j|$ is known as the order of internal resonance). This type of coupling is referred to as *autoparametric interaction* when an externally excited mode can act as a parametric excitation to other modes. The problem of internal resonances in nonlinearly coupled oscillators is of interest in connection with the redistribution of energy among the various natural modes. This energy sharing is usually brought about by resonant interactions among the natural modes of the system. The coupling among these modes plays a crucial role in such interactions. In a straightforward perturbation theory, internal resonances lead to the problem of small divisors.

Under principal internal resonance conditions (i.e., when one of the normal mode frequencies is twice another mode frequency), the system possesses a steady-state response (Ibrahim and Barr, 1975a). Ibrahim and Barr (1975b) found that under the summed or difference internal resonance conditions (i.e., one of the normal mode frequencies equals the sum or difference of another two mode frequencies) the system does not achieve a constant steady-state response.

Nonstationary responses involving violent system motion, which can lead to collapse of the system, were reported in the neighborhood of multiple internal resonances by Ibrahim (1976).

Multiple internal resonances may occur when two or more sloshing modes interact with the vertical and horizontal motions of the structure. In the neighborhood of the summed internal resonance and one-to-one internal resonance, the structure and free liquid surface simultaneously oscillate with a continuous increase in their amplitudes. This growth could lead to structural failure if the shaker excitation is not stopped. In the presence of one-to-two and one-to-one internal resonance conditions, experimental observations showed a steady-state response over a frequency range defined by the regions of instability. The regions of instability were indicated by a collapse in the response amplitudes. Another type of instability, manifested by a jump in amplitudes, was caused by a weak energy flow between the fluid modes and structure modes for a few cycles. Within a short period of time, the system achieves a steady state response.

In Hamiltonian autonomous systems, the simultaneous presence of several internal resonance relationships may result in an instability caused by certain nonlinear terms in the system equations of motion. Stability analysis of such systems was considered for different cases of internal resonance conditions (see, for example, Kunitsyn, 1971, Khazina, 1974, Goltser and Kunitsyn, 1975, Aa and Sanders, 1979, and Kunitsyn and Matveyev, 1991). Kunitsyn and Matveyev (1991) formulated the normal form of a system that contains the first nonlinear terms for an arbitrary number of noninteracting resonances of odd-order. They classified internal resonance conditions as weak and strong. Weak resonance preserves the stability of the system, while strong resonance results in system instability. The interaction of several weak internal resonances linked by more than one common frequency can result in system instability. The stability of the steady state responses of dynamic systems with multiple odd- and even-order internal resonances was considered by Kunitsyn and Markeev (1979) and Kunitsyn and Perezhogin (1985), respectively. These studies revealed that the problem of stability of systems with fourth-order internal resonance is more complicated than for third-order internal resonance. The stability of the equilibrium position of multi-dimensional Hamiltonian systems was determined for multiple independent and interacting resonance conditions. Interacting resonances imply that one or more frequencies are common in two internal resonance conditions. Kunitsyn and Tuyakbayev (1992) found that if among resonance conditions there exists at least one strong resonance, then the trivial solution of the system is unstable. On the other hand, if all independent resonances are weak then the trivial solution is stable (Kunitsyn and Muratov, 1993). Zhuravlev (1992) considered different types of oscillation shapes in the configuration and manifold spaces in the presence of multiple internal resonances. In the absence of perturbation, a subspace exists in which every trajectory is a closed curve. These trajectories become unstable under infinitesimally small perturbations.

Under horizontal periodic motion, Ibrahim and Li (1988), Ikeda and Nakagawa (1995, 1997), Ikeda (1997, 2003), and Ikeda and Murakami (1999) considered the nonlinear interaction of liquid sloshing in cylindrical and rectangular tanks with an elastic structure. It was shown that the response frequency curves experience a change from soft to hard response characteristics as the water depth decreases. Under vertical sinusoidal excitation of an elastic structure carrying a rigid rectangular tank, Ikeda (1997) determined the response of the coupled system when the structure natural frequency is about twice the liquid sloshing frequency. As the excitation frequency approached the structure natural frequency the free liquid surface was excited through autoparametric resonance and energy was transferred from the structure to the free liquid surface. An asymptotic expansion of the wave height and velocity potential of the liquid coupled with structural dynamics was developed by Limarchenko and Yasinski (1996) and Lukovskii and Timokha (1995) for simplified models of spacecraft.

In the absence of internal resonance, Haroun and Elliathy (1985a, b) and Haroun, *et al.* (1989, 1991) combined a finite element model of a tower with a mechanical model of an elevated vessel. They included the hydrodynamic forces due to liquid sloshing and its interaction with the motion of the supporting tower. They found that the fundamental sloshing mode combined with the lateral translation and the global rotation of the top of the supporting tower yield maximum values for the shearing force and overturning moment on the tower. The flexibility of the tank wall would definitely increase these maximum forces, though such an effect can be neglected in small capacity tanks. Kareem and Sun (1987) studied the stochastic response of structures with liquid tanks in the absence of internal resonance.

Soundararajan and Ibrahim (1988) examined more realistic cases such as simultaneous random horizontal and vertical ground excitations in the presence of 1:3 internal resonance. They used Gaussian and non-Gaussian closure schemes to determine the system response statistics. It was found that both Gaussian and non-Gaussian solutions deviate appreciably from the linear solution as the system approaches internal resonance but they converge when the system is detuned away from the exact internal resonance. The autoparametric interaction was identified by an irregular energy exchange between the two modes.

Among the community of civil engineers, the concept of a tuned liquid sloshing absorber is used to suppress the motion of multistorey buildings under the action of earthquakes. They coined the terms "tuned liquid damper" although the suppression effect owes its efficacy to the purely conservative coupled system if energy dissipation sources are ignored. The concept of linear vibration absorbers first introduced by the undamped Frahm's absorber will be discussed in Section 10.2. Section 10.3 deals with tuned liquid sloshing absorbers, which are classified as tuned liquid dampers and liquid column absorbers. The nonlinear analysis of liquid sloshing absorbers under horizontal and vertical ground harmonic excitations is presented in Section 10.4 in the neighborhood of parametric and internal resonance conditions. Section 10.5 considers the same absorbers under random excitations using Gaussian and non-Gaussian closure schemes and Monte Carlo simulation. Sections 10.6 and 10.7 present the concept of autoparametric sloshing absorbers of structures that are allowed to oscillate horizontally and vertically. Section 10.8 introduces the concept of ship roll stabilization using liquid tanks.

10.2 Basic concept of linear vibration absorbers

When a mechanical system experiences severe vibration at a particular frequency, one can introduce an auxiliary system that can absorb energy from the main system. These tuned vibration absorbers (also known as Frahm's absorbers) have different forms and are designed such that their own frequency equals the excitation frequency (Snowdon, 1968 and Hunt, 1979). The basic idea of undamped and damped tuned vibration absorbers, also known as tuned mass dampers, is shown in Figure 10.1. The tuned liquid dampers (TLDs) are based on exciting maximum liquid sloshing motion by tuning the liquid frequency to the system frequency. For example, the reduction of rolling of ships by using auxiliary wing tanks connected by pipes was realized a long time ago. The use of anti-rolling tanks in the German luxury liners *Bremen* and *Europa* reduced the maximum roll from 15° to 5° (Reed, 1961). Currently, TLDs have been proposed for suppressing the vibration of tall buildings, of long

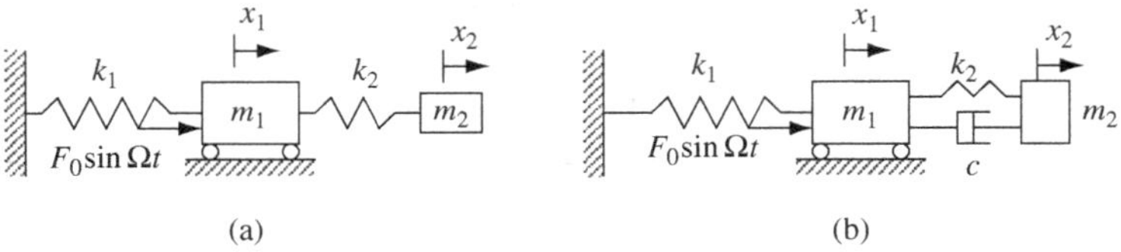

Figure 10.1 Schematic diagrams of undamped and damped vibration absorbers.

span bridges, and of offshore structures subjected to wind and earthquakes (Lee and Reddy, 1982, Sun, 1994, Xue, 1999, and Yoneda, 1989). They were proposed to suppress the pitching motion of structures by Xue, *et al.* (2000). Passive, semi-active, and active control systems for seismic protection of structures have been reviewed by Soong and Constantinou (1994) and Symans and Constantinou (1999).

Figure 10.1(a) shows a schematic representation of a structure of mass m_1 and stiffness k_1 acted upon by a harmonic excitation $F_0 \sin \Omega t$. The absorber consists of a comparatively small vibratory system of mass m_2 and stiffness k_2 attached to the main system mass m_1. The local natural frequency $\sqrt{k_2/m_2}$ of the absorber is chosen to be equal to the disturbing force frequency Ω. It will be shown that the main mass m_1 does not vibrate at all, and that the absorber mass m_2 vibrates in such a way that its spring force is at all instants equal and opposite to $F_0 \sin \Omega t$. Thus, there is no net force acting on m_1 and therefore it does not vibrate. The equations of motion are:

$$m_1\ddot{x}_1 + (k_1 + k_2)x_1 - k_2x_2 = F_0 \sin \Omega t \tag{10.1a}$$

$$m_2\ddot{x}_2 + k_2x_2 - k_2x_1 = 0 \tag{10.1b}$$

Consider only the steady state solution of the system of the form

$$x_{1p}(t) = X_1 \sin \Omega t, \qquad x_{2p}(t) = X_2 \sin \Omega t \tag{10.2a,b}$$

Introducing the following parameters

$$\omega_{11} = \sqrt{k_1/m_1}, \qquad \omega_{22} = \sqrt{k_2/m_2}, \qquad X_0 = F_0/k_1$$

$$r_1 = \Omega/\omega_{11}, \qquad r_2 = \Omega/\omega_{22}, \qquad r_{21} = \omega_{22}/\omega_{11}, \qquad \mu = m_2/m_1$$

$$k_2/k_1 = (k_2/m_2)(m_2/m_1)(m_1/k_1) = \omega_{22}^2\mu/\omega_{11}^2 = \mu r_{21}^2 \tag{10.3}$$

Substituting (10.2) into equations (10.1), gives the following solution

$$\frac{X_1}{X_0} = \frac{(1 - r_2^2)}{r_2^4 r_{21}^2 - r_2^2\left[1 + r_{21}^2(1 + \mu)\right] + 1} \tag{10.4a}$$

$$\frac{X_2}{X_0} = \frac{1}{r_2^4 r_{21}^2 - r_2^2\left[1 + r_{21}^2(1 + \mu)\right] + 1} \tag{10.4b}$$

Recall that $r_2 = \Omega/\omega_{22}$, we see that equation (10.4a) reveals that the amplitude X_1 of the main mass vanishes when the numerator $(1 - r_2^2)$ is zero, and this occurs when the excitation

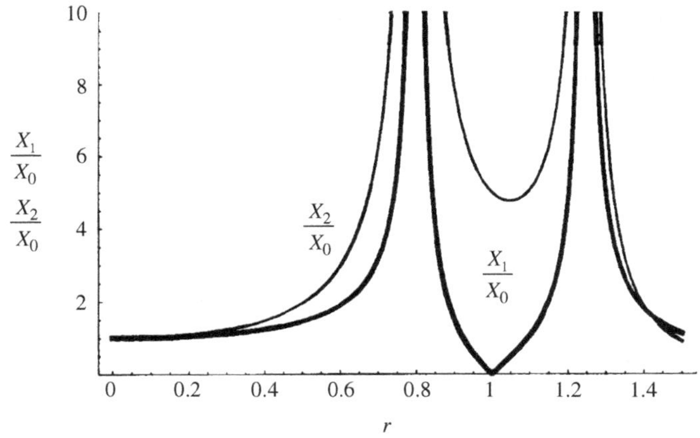

Figure 10.2 Amplitude–frequency response curves of the main mass (thick curves) and the absorber (thin curves),

frequency equals the local frequency of the absorber, that is, when $\Omega = \omega_{22}$. Thus, when $r_2 = 1$, equation (10.4b) gives

$$\frac{X_2}{X_0} = \frac{-1}{r_{21}^2 \mu} \quad \text{or} \quad X_2 = \frac{-X_0}{r_{21}^2 \mu} = -X_0 \frac{k_1}{k_2} = -\frac{F_0}{k_2} \tag{10.5}$$

With the main mass standing still and the mass of the absorber having the motion $-(F_0/k_2)$ sin Ωt, the force in the absorber spring varies as $-F_0$ sin Ωt, which is actually equal and opposite to the external force. These relations are true for any value of the ratio $r_1 = \Omega/\omega_{11}$. There is not much reason for the addition of an absorber unless the original system is in resonance or at least near it. Resonance occurs if the denominators of equations (10.4) vanish. The value of r_2 and hence Ω at which resonance occurs can be found by setting the denominator to zero, that is,

$$r_2^4 r_{21}^2 - r_2^2\left[1 + r_{21}^2(1+\mu)\right] + 1 = 0 \tag{10.6}$$

Thus, for two values of the external frequency Ω both denominators of (10.4) become zero and consequently X_1 and X_2 become infinitely large. These two frequencies are the resonant or natural frequencies of the system.

For a given system, equation (10.6) has two roots for r_2^2. The two roots depend on the values of the physical constants k_i and m_i of the system. The two resonant frequencies can be determined for any given values of r_{21} and μ. For the case $r_{21} = 1$, equation (10.6) becomes

$$r_2^4 - (2+\mu) r_2^2 + 1 = 0 \tag{10.7}$$

which has the two roots

$$\left(r_2^2\right)_{1,2} = \left(1 + \frac{\mu}{2}\right) \mp \sqrt{\left(1 + \frac{\mu}{2}\right)^2 - 1} = \left(1 + \frac{\mu}{2}\right) \mp \sqrt{\mu + \frac{\mu^2}{2}} \tag{10.8}$$

It is useful to understand the effectiveness of the absorber by plotting the amplitude–frequency response curves. Consider for the sake of demonstration the case when the mass ratio $\mu = 0.2$ and $\omega_{11} = \omega_{22}$, that is, $r_1 = r_2 = r$. Figure 10.2 shows the amplitude ratio, (X_1/X_0), of the main mass and the absorber, (X_2/X_0), versus the frequency ratio $r_1 = r_2 = r$. For $\mu = 0.2$, the two natural frequency ratios of the system are 0.8 and 1.25.

It is seen that the main mass amplitude ratio $X_1/X_0 = 1$ for $r = 0$ and increases as r increases until it reaches ∞ at resonance $r = 0.8$. At $r = 0.8$ the denominators of equations (10.4) are zero then become negative as r increases beyond 0.8. This means that both amplitude ratios, X_1/X_0 and X_2/X_0, are negative until $r = 1$, at which the main mass displacement vanishes but the absorber mass displacement remains negative. Note that the absolute values are shown in the figure. At the second resonance $r = 1.25$, the denominator changes sign once more with negative displacement of the main mass. The change of sign of the response curve means a change of 180° in the phase angle. This absorber is effective at only one frequency, $r = 1$, and if Ω varies appreciably, the absorber may cause serious trouble because the range between the two resonant frequencies is rather small.

If a dashpot is inserted between the main mass and the absorber mass the equations of motion take the form

$$m_1\ddot{x}_1 + (k_1 + k_2)x_1 - k_2x_2 + c(\dot{x}_1 - \dot{x}_2) = F_0 \sin \Omega t \tag{10.9a}$$

$$m_2\ddot{x}_2 + k_2x_2 - k_2x_1 - c(\dot{x}_1 - \dot{x}_2) = 0 \tag{10.9b}$$

By using complex algebra, it is not difficult to show that the steady state response of the main mass is

$$\frac{X_1}{X_0} = \sqrt{\frac{(2\zeta r_1 r_{21})^2 + (r_1^2 - r_{21}^2)^2}{(2\zeta r_1 r_{21})^2[r_1^2(1+\mu) - 1]^2 + [\mu r_1^2 r_{21}^2 - (r_1^2 - 1)(r_1^2 - r_{21}^2)]^2}} \tag{10.10}$$

Obviously the presence of damping will not bring complete suppression to the main mass at the conditions, $r_1 = r_{21}$, or $r_2 = 1$.

Figure 10.3 shows the amplitude–frequency response of the main system and reveals that excessive damping of the absorber results in poor suppression of the motion of the main

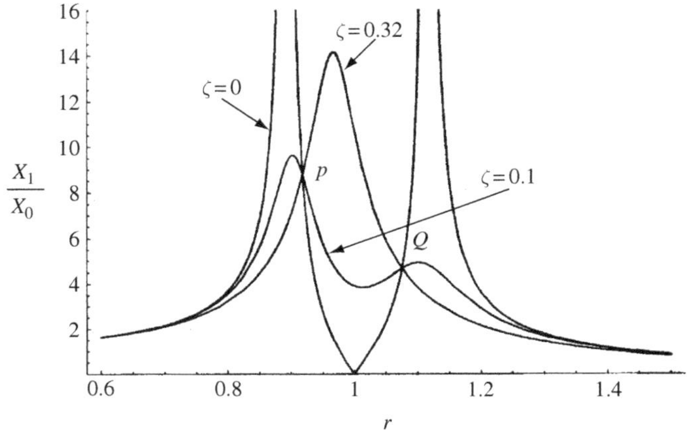

Figure 10.3 Amplitude–frequency response curves of the main mass for three different values of absorber damping ratio, $\mu = 0.05$, $r_{21} = 1$.

system However, the presence of a reasonable amount of damping brings the amplitude to a bounded value at the two resonance peaks shown for the case of zero damping. There is an optimum value of damping at which energy dissipation reaches its maximum value. It is interesting to note that all response curves intersect at two points, P and Q. This means that all curves pass through these two points regardless of the value of the damping ratio. By equating the amplitude response for $\zeta = 0$ and $\zeta = \infty$, one obtains the following values of the excitation frequency for the points P and Q for $r_{21} = 1$.

$$r_1^2 = 1 \pm \sqrt{\frac{\mu}{2 + \mu}} \tag{10.11}$$

Optimum tuning can be achieved by selecting the damping ratio of the absorber to yield the peak response amplitude at points P and Q. This can be achieved by differentiating the response amplitude (10.1) with respect to r_1 and equating the result to zero.

10.3 Tuned liquid sloshing absorbers

According to the equivalent mechanical models outlined in Chapter 5, a liquid in a container can be replaced by either a mass–spring system or a pendulum system. In both cases, the liquid acts as a vibration absorber as outlined in the previous section. Basically, there are two types of tuned liquid absorbers in common use by civil engineers. The first is referred to as the tuned liquid sloshing absorber (TLSA) or the tuned liquid damper (TLD) and the second is the tuned liquid column damper (TLCD), both are schematically shown in Figure 10.4. Most of the work of liquid sloshing absorbers has been motivated by civil engineers dealing with the control of structural vibration. For example liquid sloshing absorbers have been installed in several buildings in Japan such as the Yokohama Marine Tower, the Sin Yokohama Prince Hotel, and the new control tower at the Narita Airport.

10.3.1 Tuned liquid dampers

It is known that a shallow liquid level in a container experiences traveling sloshing waves. On the other hand, deeper liquid levels exhibit a standing sloshing wave in the fundamental mode. As demonstrated in Chapter 4, for large wave amplitudes, there is a critical depth, above which the liquid waves possess soft nonlinear spring characteristics and below that level, the

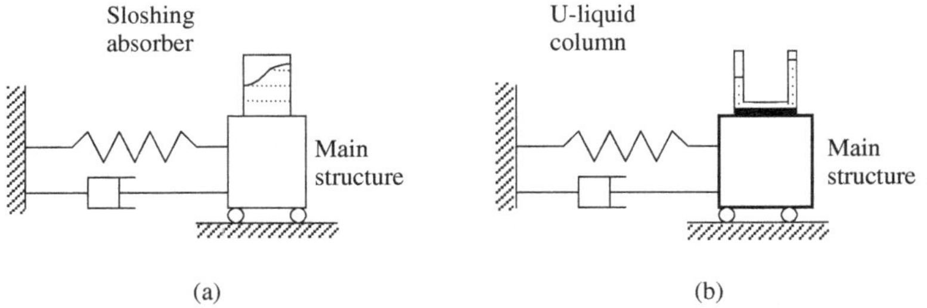

Figure 10.4 (a) Tuned liquid damper (TLD), (b) Tuned liquid column damper.

nonlinearity is of the hard type. Furthermore, standing waves are associated with poor energy dissipation (see, e.g., Kaneko and Yoshida, 1999). The effectiveness of a TLD has been studied analytically, numerically, and experimentally. Most of the analytical studies did not follow the coupled differential equations of the liquid interaction with the main structure. Instead, the effective liquid force was represented in the structure equation of motion. In some few cases, the coupling was considered numerically (Yamamoto and Kawahara, 1999). This section summarizes the main results of TLDs reported in the literature.

Liquid sloshing in a circular tank was shown to reduce wind-induced vibrations at Nagasaki Airport Tower and Yokohama Marine Tower to about half of the uncontrolled values (Fujii, *et al.*, 1990, Tamura, 1990, Ueda, *et al.*, 1992, and Wakahara, 1993). Tuned liquid dampers have found many applications such as: (i) satellites (Amieux and Dureigne, 1972 and Anzai, *et al.*, 1981), (ii) marine vessels (Matsuura, *et al.*, 1986), (iii) towers and tall buildings (Tamura, *et al.*, 1988, 1992, Fujii, *et al.*, 1988, Wakahara, *et al.*, 1991, and Chen, *et al.*, 1995), and (iv) pylons of cable-stayed bridges (Hagiuda, 1989, Yoneda, *et al.*, 1991, and Steel Structure and Civil Engineering Division, 1991). Liquid sloshing absorbers were used to control the libration motion of satellites (Schneider and Likins, 1973 and Alfriend and Spencer, 1981). These devices are characterized by their effectiveness for small- amplitude vibrations, and their ability to tune to the natural frequency of the structural system. An optimum design of such devices can be achieved by adjusting the damping factor. The damping factor can be adjusted by using rough tank walls (Fujino, *et al.*, 1988a,b), adding surface contamination (Fujino, *et al.*, 1990), using submerged nets (Noji, *et al.*, 1988, 1990), or inserting poles (Nakagaki, *et al.*, 1990). The design and performance of tuned sloshing water dampers were studied by Fujino, *et al.* (1992), Fujino and Abe (1993), and Fediw, *et al.* (1995).

Different analytical and computational models of shallow- and deep-water types of TLDs were developed by Modi and Welt (1987, 1988), Modi, *et al.* (1988), Welt and Modi (1989a,b, 1992a,b), Chaiseri, *et al.* (1989), Saoka, *et al.* (1989), Fujino, *et al.* (1989, 1991), Kaneko (1992), Lepelletier and Raichlen (1988), Williams and Wang (1992), Kaneko and Ishikawa (1999), and Kaneko and Yoshida (1999). The purpose of these models was to determine the energy dissipation and the effectiveness of each TLD. Fujino and Sun (1993), Koh, *et al.* (1994,1995), and Shankar and Balendra (2002) proposed the use of multiple liquid dampers tuned to several natural frequencies of a multi-degree-of-freedom structure. Kaneko and Mizota (1996) presented an analytical model describing the effectiveness of a deep water type cylindrical TLD with a submerged net for suppressing the horizontal vibration of structural systems. Specifically, they analytically and experimentally studied the effect of hydraulic resistance produced by a submerged net and the liquid depth ratio. In the presence of the net, the system response was reduced to less than 40%. Miyata, *et al.* (1988, 1989), and Fujino, *et al.* (1988a,b, 1989) conducted experiments to determine the effectiveness of tuned liquid dampers. Under large-amplitude excitation, Yalla and Kareem (1999) developed analytical models of TLD experiencing sloshing-slamming.

Sun, *et al.* (1989), Fujino, *et al.* (1992), and Yu (1997) proposed a nonlinear model for a TLD using a rectangular tank filled with shallow liquid. Their analytical formulation was based on continuous free-surface motion without breaking of waves. Sun and Fujino (1994) and Sun *et al.* (1995) modified the analytical model to accommodate breaking waves by introducing two coefficients into the fluid equation of motion. The first coefficient accounts for the damping coefficient representing the increase in liquid damping due to breaking waves. The second coefficient takes into account the frequency shift and represents the phase velocity shift of liquid motion due to breaking waves. Reed, *et al.* (1998a,b) examined the efficacy of TLD systems numerically and experimentally. They used the random choice method for solving the fully nonlinear shallow-water-wave equations and the results revealed the occurrence of breaking waves. Gardarsson, *et al.* (2001) studied the behavior of sloped-bottom tuned liquid dampers.

The efficiency of TLDs in suppressing the wind-induced response of an airport tower was measured experimentally by Tamura, *et al.* (1996). Modi and Munshi (1998) presented a parametric study designed to enhance the energy dissipation efficiency of a rectangular liquid damper through the introduction of a two-dimensional obstacle. The study aimed at optimum size and location of the obstacle inside the tank. In the presence of an obstacle, the energy dissipation could increase by 60%.

Kareem (1993) and Modi, *et al.* (1995) presented literature reviews of the application of TLD devices for the suppression of wind-induced instabilities. Xu, *et al.* (1992a,b), Seto (1996), Seto and Modi (1997), Modi and Seto (1997, 1998), and Popplewell, *et al.* (2002) studied the effectiveness of rectangular nutation dampers in controlling vortex resonance and the galloping type of wind-induced instabilities encountered in industrial aerodynamic problems. They found that liquid sloshing provides an efficient mechanism for energy dissipation. For a given amount of liquid, energy dissipation can be improved through the optimal partitioning of a rectangular damper in the direction of wave motion. The optimal properties of the TLD for suppressing the vortex-excited vibration of tall building were analytically and experimentally studied by Chang and Gu (1999). They conducted a series of wind tunnel experiments on a scaled-down building model equipped with rectangular TLDs of various sizes. It was found that the optimal frequency of the TLD should be within the range of 0.9 to 1.0 of the building model frequency. Anderson, *et al.* (2000a,b, 2001) and Troung *et al.* (2003) conducted a series of numerical and experimental studies of the standing-wave-type sloshing absorber including the use of vertical and horizontal cylindrical containers. They also considered the influence of floating or immersed dumbbell-type controllers consisting of two plates separated by rigid rods.

The feasibility of using active control of the TLD to reduce the vibration of large structures was studied by Chang, *et al.* (1998) and Lou, *et al.* (1994). They proposed a simple mechanism to actively control the tuning of the system by adjusting the length of the liquid tank with rotatable baffles driven by a stepping motor. Their control strategy is based on detecting the frequency content and two-state control to establish the way for positioning the baffles. Numerical simulation techniques including large sloshing amplitude were employed for tuned liquid dampers by Tam (1998) and Zhoa and Fujino (1993).

10.3.2 Liquid column vibration absorbers

Another type of liquid vibration absorber is the tuned liquid column damper (TLCD), or liquid column vibration absorber (LCVA). It takes the shape of a U-shaped tube of uniform rectangular cross-section containing liquid as shown schematically in Figure 10.4(b). Watkins (1991) introduced an intermediate horizontal tube with different cross-sectional area to each end of the vertical columns. When the structure experiences vibration, its energy is transferred to the TLCD liquid. The motion of the main structure is suppressed by the TLCD through the gravitational restoring force acting on the displaced liquid, and the energy is then dissipated by the viscous interaction between the liquid and its rigid tube. The effectiveness of TLCD under random excitation was studied by Won *et al.* (1996, 1997). Watanabe (1969) and Kagawa, *et al.* (1989, 1990) discussed the use of U-shaped tubes in marine vessels as anti-roll measures. Alfriend (1974) outlined the use of a partially filled viscous ring nutation on a spinning satellite. Sakai, Takaeda, and Tamaki (1989, 1991a, b) and Rakesh (1997) showed that this type is effective in reducing the vibrations of civil engineering structures.

Kagawa, *et al.* (1989) derived the natural frequency of liquid in the U-type damper using the energy method. Shyu and Kuo (1996) determined the natural frequencies of the entire system from the frequency-response curves (for the forced vibration problem) instead of solving the free vibration problem of the coupled system. The dynamic characteristics of the coupled structure–liquid damper were studied by Yalla and Kareem (2001a,b). They found that when the normal mode frequencies approach each other, the beating phenomenon took place, indicating an exchange of energy between the two modes. Wu and Hsieh (2002) found that one of the two natural frequencies of the coupled system is close to the uncoupled natural frequency of the liquid in the U-tube and the other is close to the frequency of the structure.

The influence of liquid damping and its natural frequency on LCVA systems has received extensive studies. For example, Kwok (1984), Kwok, *et al.* (1991), and Samali, *et al.* (1992a, b, c, d) conducted analytical and experimental investigations to examine the viscous interaction between the liquid and its rigid boundaries. Watkins and Hitchcock (1992) indicated that for a LCVA to successfully mitigate structural vibrations it is necessary to select the correct combination of LCVA natural frequency, liquid damping ratio, and its mass. Hitchcock, *et al.* (1997a, b) conducted experimental tests on unidirectional and bi-directional liquid column absorbers to determine the factors affecting the characteristics of LCVAs. The natural frequency of both LCVAs was found to depend on the ratio of the cross-sectional area of the vertical columns to the cross-sectional area of the horizontal column, and the lengths of all columns. Gao and Kwok (1997) studied the effectiveness of U-shaped and V-shaped TLCDs in controlling structural vibration, and estimated the optimum parameters of TLCDs for maximum reduction of peak structural response to harmonic excitation in a wide frequency range. Their results showed that the mass ratio and structural damping ratio govern the optimum parameters. Other related studies were conducted by Balendra, *et al.* (1994, 1998, 1999, 2001), Chang and Hsu (1998), Gao, *et al.* (1999), and Yalla and Kareem (2000).

Semi-active and active controls of TLCD systems were proposed and studied by Kareem (1994), Haroun, *et al.* (1994a, b, 1996), Yalla, *et al.* (1998), and Yalla and Kareem (1999, 2000, 2001a, b). Different control strategies were proposed such as full scale feedback, fuzzy control logic, and observer-based feedback.

10.4 Analytical modeling of liquid sloshing absorbers

An accurate study of liquid sloshing absorbers should include linear and nonlinear interaction between liquid sloshing dynamics and structural dynamics. Figure 10.5(a) and (b) represents elastic structures carrying rigid circular cylindrical and rectangular containers, respectively. The containers are assumed to be rigid and partially filled with an incompressible liquid. The two systems have been studied by Li (1985), Ibrahim and Li (1988), and Ikeda and Nakagawa (1997).

First, consider the general case of a circular container partially filled with liquid and supported by an elastic structure. The structure is subjected to prescribed vertical and horizontal base motions $y_b(t)$ and $x_b(t)$, respectively. Under ground motion the fluid container

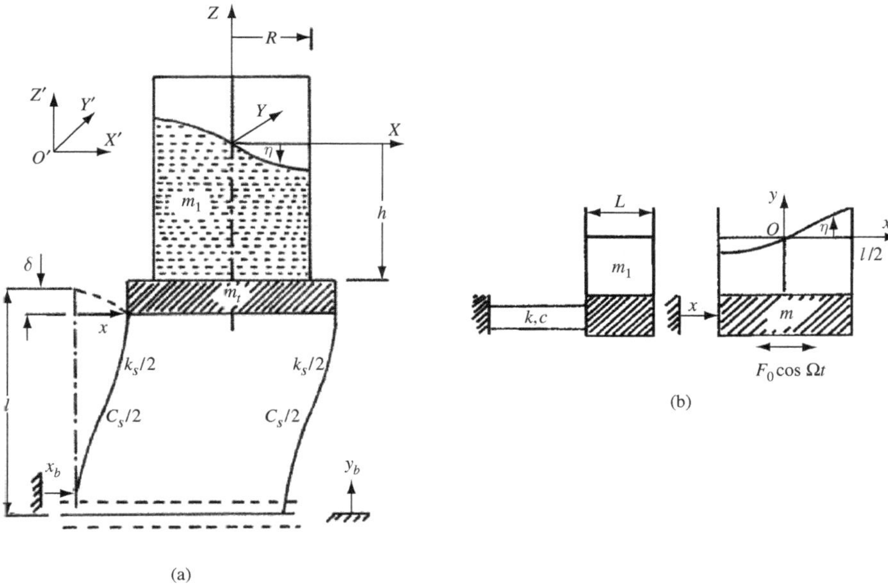

Figure 10.5 Schematic diagrams of liquid containers supported on elastic structures. (a) Circular cylindrical tank, (b) rectangular container.

moves along some trajectory. It is convenient to refer the fluid motion to a moving coordinate system, $OXYZ$, in which the OXY plane coincides with the undisturbed free surface and the OXZ plane is the trajectory plane. The moving frame coincides with the inertial frame $O'X'Y'Z'$ when the tank is at rest. Let $\mathbf{V}_0$ be the velocity of the origin O relative to O', where

$$\mathbf{V}_0 = \dot{X}_0\mathbf{i}_x + \dot{Z}_0\mathbf{i}_z = \dot{X}_0 \cos\theta\mathbf{i}_r - \dot{X}_0 \sin\theta\mathbf{i}_\theta + \dot{Z}_0\mathbf{i}_z \tag{10.12}$$

where $\mathbf{i}_x$ and $\mathbf{i}_z$ are unit vectors in the X_0 and Z_0 directions, respectively, and the unit vectors $\mathbf{i}_r$, $\mathbf{i}_\theta$ and $\mathbf{i}_z$ are in the r, θ, and Z directions. Following the same assumptions as those used in Chapter 1, the velocity potential function satisfies Laplace's equation of continuity

$$\nabla^2\Phi = 0 \tag{10.13}$$

The potential function Φ can be split into a disturbance function, $\tilde{\Phi}$, which accounts for the fluid motion relative to the tank, and a potential function, Φ_0, which defines the tank motion, that is,

$$\Phi = \tilde{\Phi} + \Phi_0 \tag{10.14}$$

The function Φ_0 can be determined by integrating equation (10.12), that is,

$$\Phi_0 = \dot{X}_0 r\cos\theta + \dot{Z}_0 Z + \frac{1}{2}\int \left(\dot{X}_0^2 + \dot{Z}_0^2\right) dt \tag{10.15}$$

The Euler pressure equation for unsteady flow as referred to the moving coordinates is

$$\frac{P}{\rho} + \frac{\partial\tilde{\Phi}}{\partial t} + \frac{1}{2}(\nabla\tilde{\Phi}\cdot\nabla\tilde{\Phi}) + (g + \ddot{Z}_0)Z + \ddot{X}_0 r\cos\theta = 0 \tag{10.16}$$

where P is the fluid pressure and ρ is the fluid density. The solution of equation (10.13) must satisfy the boundary conditions. At the wall and bottom the velocity component normal to the boundaries must vanish, that is,

$$\left.\frac{\partial\tilde{\Phi}}{\partial r}\right|_{r=R} = 0, \quad \left.\frac{\partial\tilde{\Phi}}{\partial z}\right|_{z=-h} = 0 \tag{10.17a,b}$$

At the free surface, $Z=\eta(r, \theta, t)$, the dynamic boundary condition is obtained from equation (10.16) by setting the pressure to zero, thus

$$\frac{\partial\tilde{\Phi}}{\partial t} + \frac{1}{2}(\nabla\tilde{\Phi}\cdot\nabla\tilde{\Phi}) + (g+\ddot{Z}_0)Z + \ddot{X}_0 r\cos\theta = 0 \tag{10.18}$$

At the free surface the kinematic condition requires that the vertical velocity of a fluid particle located on the free surface equals the vertical velocity of the free surface itself, that is,

$$\frac{\partial\tilde{\Phi}}{\partial z} = \left.\frac{\partial\eta}{\partial t} + \frac{\partial\tilde{\Phi}}{\partial r}\frac{\partial\eta}{\partial r} + \frac{1}{r^2}\frac{\partial\tilde{\Phi}}{\partial\theta}\frac{\partial\eta}{\partial\theta}\right|_{z=\eta} \tag{10.19}$$

A solution of equation (10.13) that satisfies conditions (10.17) is

$$\tilde{\Phi}(r,\theta,Z,t) = \sum_{m=0}^{\infty}\sum_{n=1}^{\infty}\alpha_{mn}(t)J_m(\lambda_{mn}r)\cos m\theta\frac{\cosh[\lambda_{mn}(Z+h)]}{\cosh[\lambda_{mn}h]} \tag{10.20}$$

where $\lambda_{mn}=\xi_{mn}/R$ are the roots of $J'_m(\xi_{mn})=0$. The free-surface elevation can be written in the form

$$\eta(r,\theta,t) = \sum_{m=0}^{\infty}\sum_{n=1}^{\infty}a_{mn}(t)J_m(\lambda_{mn}r)\cos m\theta \tag{10.21}$$

The coefficients $\alpha_{mn}(t)$ and $a_{mn}(t)$ are time dependent and are obtained by satisfying the free-surface equations (10.18) and (10.19). The acceleration terms, and $\ddot{X}_0$ and $\ddot{Z}_0$, in equation (10.18), are dependent on the excitation and the mechanical characteristics of the system. Their values and those of $\alpha_{mn}(t)$ and $a_{mn}(t)$ are not known explicitly since they are coupled with the equation of motion of the elastic structure.

The equation of motion of the structure is derived by applying Lagrange's equation. The fluid hydrodynamic forces are considered as external forces. In order to determine the kinetic and potential energies of the system, the amount of vertical drop, δ, of the tank as a result of its lateral deflection, x, needs to be known. An approximate expression is based on assuming that the static deflection curve of a beam fixed at both ends is $\delta=(3x^2/5l)$, where l is the beam length of the support structure. Applying Lagrange's equation for the lateral displacement coordinate, x, and ignoring the gravitational term due to vertical drop, gives

$$m_t\left\{\ddot{x}+\ddot{x}_b-\frac{6}{5l}x\ddot{y}_b+\frac{36}{25l^2}x(x\ddot{x}+\dot{x}^2)\right\}+k_s x$$

$$= \int_{-h}^{0}\int_{0}^{2\pi} P(r,\theta,z,t)|_{r=R}R\cos\theta\, d\theta\, dz \tag{10.22}$$

The acceleration terms in equation (10.18) can now be written in terms of the coordinate x and base acceleration components $\ddot{x}_b$ and $\ddot{y}_b$ as

$$\ddot{Z}_b = \ddot{y}_b - \ddot{\delta} = \ddot{y}_b - \frac{6}{5l}(x\ddot{x} + \dot{x}^2) \tag{10.23a}$$

$$\ddot{X}_0 = \ddot{x} + \ddot{x}_b - \frac{6}{5l}x\ddot{y}_b + \frac{26}{25l^2}x(x\ddot{x} + \dot{x}^2) \tag{10.23b}$$

Using expressions (10.23) in equation (10.16) and writing equation (10.22) in terms of $\tilde{\Phi}$ gives

$$\begin{aligned} m\ddot{x} + m\ddot{x}_b + k_s x - \frac{6}{5l}mx\ddot{y}_b + \frac{26}{25l^2}mx(x\ddot{x} + \dot{x}^2) \\ = \rho \int_{-h}^{0} \int_{0}^{2\pi} \left\{ \frac{\partial \tilde{\Phi}}{\partial t} - \frac{1}{2}(\nabla\tilde{\Phi} \cdot \nabla\tilde{\Phi}) \right\} R\cos\theta \, d\theta \, dZ \end{aligned} \tag{10.24}$$

where m_t is the mass of the top structure, m is the total mass of the structure plus liquid, and k_s is the stiffness of the support structure. The analysis proceeds by expanding the free-surface conditions (10.18) and (10.19) in a Taylor series about $\eta=0$. The Fourier–Bessel series expansion is then used after introducing expressions (10.20), (10.21) and (10.23). Considering only the first anti-symmetric sloshing (1,1), equations (10.18), (10.19), and (10.24) can be written in their respective forms:

$$\begin{aligned} \dot{\alpha}_{11} - \left[g + \ddot{y}_b - \frac{6}{5l}(x\ddot{x} + \dot{x}^2) \right] a_{11} + D_1\dot{\alpha}_{11}a_{11}^2 - D_2\alpha_{11}^2 a_{11} \\ - D_3\left[\ddot{x} + \ddot{x}_b - \frac{6}{5l}x\ddot{y}_b + \frac{26}{25l^2}x(x\ddot{x} + \dot{x}^2) \right] = 0 \end{aligned} \tag{10.25}$$

$$\dot{a}_{11} + D_4\alpha_{11} - D_5\alpha_{11}a_{11}^2 = 0 \tag{10.26}$$

$$m\ddot{x} + m\ddot{x}_b + k_s x - \frac{6}{5l}mx\ddot{y}_b + \frac{26}{25l^2}mx(x\ddot{x} + \dot{x}^2) - D_6\dot{\alpha}_{11} = 0 \tag{10.27}$$

where $D_1 = 0.104\,568\,2\lambda_{11}^2$, $D_2 = 0.281\,025\,6\lambda_{11}^2 \tanh\lambda_{11}h$, $D_3 = (2R\tanh\lambda_{11}h)/((\xi_{11}^2 - 1)$, $J_1(\xi_{11}))$, $D_4 = \lambda_{11}\tanh\lambda_{11}h$, $D_5 = 0.122\,515\lambda_{11}^3 \tanh\lambda_{11}h$, $D_6 = (\pi\rho R/\lambda_{11})J_1(\xi_{11})\tanh\lambda_{11}h$.

Eliminating α_{11} and $\dot{\alpha}_{11}$ reduces equations (10.25)–(10.27) into the two equations of motion

$$\begin{aligned} \begin{bmatrix} 1 & m_{12} \\ m_{21} & 1 \end{bmatrix} \begin{Bmatrix} \ddot{x} \\ \ddot{a}_{11} \end{Bmatrix} + \begin{bmatrix} k_s/m & 0 \\ 0 & \lambda_{11}g\tanh\lambda_{11}h \end{bmatrix} \begin{Bmatrix} x \\ a_{11} \end{Bmatrix} \\ = -\ddot{x}_b(t) \begin{Bmatrix} 1 \\ m_{21} \end{Bmatrix} + \frac{\ddot{y}_b(t)}{l} \begin{bmatrix} 1.2 & 0 \\ 1.2m_{21} & -\xi_{11}l/R \end{bmatrix} + \frac{1}{l^2} \begin{Bmatrix} \Psi_1 \\ \Psi_2 \end{Bmatrix} \end{aligned} \tag{10.28}$$

where $m_{12} = \frac{m_l J_1(\xi_{11})}{m(h/R)\xi_{11}^2}$, $m_{21} = \frac{2R\tanh\lambda_{11}h}{(\xi_{11}^2 - 1)J_1(\xi_{11})}$, m_l is the total mass of the liquid, Ψ_i, $i=1$, 2, are nonlinear inertia terms given by the following expressions

$$\Psi_1 = 1.44x(x\ddot{x} + \dot{x}^2) - 0.141\,65m_{12}\xi_{11}^2(l/R)^2 a_{11}(a_{11}\ddot{a}_{11} + 2\dot{a}_{11}^2) \tag{10.29a}$$

$$\begin{aligned}\Psi_1 &= 1.2[1.2m_{21}x - \xi_{11}(l/R)a_{11}](x\ddot{x} + \dot{x}^2) \\ &\quad - \xi_{11}^2(l/R)^2 a_{11}[0.738\,62\dot{a}_{11}^2 + 0.03709a_{11}\ddot{a}_{11}]\end{aligned} \tag{10.29b}$$

It is convenient to transform equations (10.28) into the principal coordinates $\mathbf{y} = \{y_1\ y_2\}^T$ through the linear transformation

$$\begin{Bmatrix} x \\ a_{11} \end{Bmatrix} = [\mathbf{Q}]\begin{Bmatrix} y_1 \\ y_2 \end{Bmatrix} \tag{10.30}$$

where the modal matrix $[\mathbf{Q}] = \begin{bmatrix} 1 & 1 \\ n_1 & n_2 \end{bmatrix}$, $n_i = \frac{(k_s/m)m_{21}}{\lambda_{11}g\tanh\lambda_{11}h - (1 - m_{12}m_{21})w_i^2}$, ω_i are the normal mode frequencies given by the following expressions

$$\omega_{1,2}^2 = \frac{(k_s/m) + \lambda_{11}g\tanh\lambda_{11}h \mp \sqrt{\begin{array}{c}((k_s/m) + \lambda_{11}g\tanh\lambda_{11}h)^2 \\ -4(k_s/m)\lambda_{11}g\tanh\lambda_{11}h(1 - m_{12}m_{21})\end{array}}}{2(1 - m_{12}m_{21})} \tag{10.31}$$

Writing equations (10.28) in terms of the principal coordinates, pre-multiplying both sides by $[\mathbf{Q}]^{-1}\,[\mathbf{M}]^{-1}$, and inserting a diagonal damping matrix to account for energy dissipation, gives

$$\begin{aligned}&\begin{bmatrix} 1 & 0 \\ 0 & 1 \end{bmatrix}\begin{Bmatrix} \ddot{y}_1 \\ \ddot{y}_2 \end{Bmatrix} + \begin{bmatrix} 2\tilde{\zeta}_1\omega_1 & 0 \\ 0 & 2\tilde{\zeta}_2\omega_2 \end{bmatrix}\begin{Bmatrix} \dot{y}_1 \\ \dot{y}_2 \end{Bmatrix} + \begin{bmatrix} \omega_1^2 & 0 \\ 0 & \omega_2^2 \end{bmatrix}\begin{Bmatrix} y_1 \\ y_2 \end{Bmatrix} \\ &= -\ddot{x}_b(t)[\mathbf{Q}]^{-1}[\mathbf{M}]^{-1}\begin{Bmatrix} 1 \\ m_{21} \end{Bmatrix} \\ &\quad + \frac{\ddot{y}_b(t)}{l}[\mathbf{Q}]^{-1}[\mathbf{M}]^{-1}\begin{bmatrix} 1.2 & 0 \\ 1.2m_{21} & -\xi_{11}l/R \end{bmatrix}[\mathbf{Q}]\begin{Bmatrix} y_1 \\ y_2 \end{Bmatrix} + \begin{Bmatrix} \bar{\Psi}_1 \\ \bar{\Psi}_2 \end{Bmatrix}\end{aligned} \tag{10.32}$$

where $[\mathbf{M}] = \begin{bmatrix} 1 & m_{12} \\ m_{21} & 1 \end{bmatrix}$, and $\bar{\Psi}_i$ are new inertia nonlinear terms in terms of the principle coordinates.

The system dynamic characteristics are studied under deterministic then random vertical and horizontal excitations in the following sections.

10.4.1 Vertical ground harmonic excitation

Consider the case of harmonic vertical excitation, $y_b(t) = y_0\cos\Omega t$, $x_b(t) = 0$, where y_0 and Ω are the excitation amplitude and frequency, respectively. It is convenient to introduce the following nondimensional parameters, Gau (1985),

$$\tau = \omega_1 t, r = \omega_2/\omega_1, r_f = \Omega/\omega_1, Y_i = y_i/y_0, 2\tilde{\zeta}_i = \varepsilon\zeta_i, \varepsilon = \ddot{y}_{b\max}/(l\omega_1^2) \ll 1 \tag{10.33}$$

Equations (10.32) can be written in the following standard form

$$\begin{aligned} Y_1'' + Y_1 &= \varepsilon\{-\cos r_f\tau(K_1 Y_1 + K_2 Y_2) + \zeta_1 Y_1'\} \\ &\quad + \varepsilon^2\{K_3 Y_1^2 Y_1'' + K_4 Y_1 Y_2 Y_1'' + K_5 Y_2^2 Y_1'' + K_6 Y_1^2 Y_2'' + K_7 Y_1 Y_2 Y_2'' \\ &\qquad + K_8 Y_2^2 Y_2'' + K_9 Y_1 Y_1'^2 + K_{10} Y_1 Y_1' Y_2' + K_{11} Y_1 Y_2'^2 + K_{12} Y_2 Y_1'^2 \\ &\qquad + K_{13} Y_2 Y_1' Y_2' + K_{14} Y_2 Y_2'^2\} \end{aligned} \tag{10.34a}$$

$$\begin{aligned} Y_2'' + r^2 Y_2 &= \varepsilon\{-\cos r_f\tau(L_1 Y_1 + L_2 Y_2) + r\zeta_2 Y_2'\} \\ &\quad + \varepsilon^2\{L_3 Y_1^2 Y_1'' + L_4 Y_1 Y_2 Y_1'' + L_5 Y_2^2 Y_1'' + L_6 Y_1^2 Y_2'' + L_7 Y_1 Y_2 Y_2'' \\ &\qquad + L_8 Y_2^2 Y_2'' + L_9 Y_1 Y_1'^2 + L_{10} Y_1 Y_1' Y_2' + L_{11} Y_1 Y_2'^2 + L_{12} Y_2 Y_1'^2 \\ &\qquad + L_{13} Y_2 Y_1' Y_2' + L_{14} Y_2 Y_2'^2\} \end{aligned} \tag{10.34b}$$

where a prime denotes differentiation with respect to τ, K_i and L_i are functions of the system parameters and are listed in Ibrahim, *et al.* (1988). It is assumed that damping terms and parametric excitation are of order ε. The solution of equations (10.34) is obtained using the Struble method of asymptotic approximation (Struble, 1962). Attention is given to the effect of nonlinearity on the dynamic parametric response in the absence of internal resonance. The solution of equations (10.34) may be written in the form

$$Y_1(\tau) = A(\tau)\cos[r_1\tau + \psi_1(\tau)] + \varepsilon a_1(\tau) + \varepsilon^2 a_2(\tau) + \cdots \tag{10.35a}$$

$$Y_2(\tau) = B(\tau)\cos[r_2\tau + \psi_2(\tau)] + \varepsilon b_1(\tau) + \varepsilon^2 b_2(\tau) + \cdots \tag{10.35b}$$

where the amplitudes, $A(\tau)$ and $B(\tau)$, and phase angles, $\psi_i(\tau)$, are slowly varying functions of the nondimensional time τ. These functions exhibit only long-period fluctuations. The additional perturbations, a_i and b_i, represent higher harmonics and are referred to as short-period perturbations, r_1 and r_2 are unknowns to be determined.

Substituting expansions (10.35) into equations (10.34) and equating coefficients of equal powers in ε on both sides, results in a set of differential equations. Those equations generated from terms of order zero in ε are called the variational equations, while other terms associated with coefficients ε, ε^2, ... constitute the perturbational equations. At each step of the iteration process the variational equations are associated with fundamental harmonic terms with arguments $r_i\tau + \psi_i(\tau)$, $i = 1, 2$, and the perturbation equations with the remaining nonresonant terms. All terms, which appear as resonant (or secular) in the perturbation equations, are transferred to the variational equations. Since $A(\tau)$, $B(\tau)$, and $\psi_i(\tau)$ are slowly varying functions of time, terms involving A'', B'', ψ_i'', $A'\psi_1'$, and $B'\psi_2'$ are neglected. At the same time, terms such as A', B', and ψ_i' on the right-hand sides will also be ignored because they are of order ε.

The fundamental variational equations are

$$-2r_1 A\,\psi_1' + (1 - r_1^2)A = 0, \quad -2r_1 A' = 0 \tag{10.36a,b}$$

$$-2r_2 B\,\psi_2' + (r^2 - r_2^2)B = 0, \quad -2r_2 B' = 0 \tag{10.36c,d}$$

The first-order perturbational equations (terms of order ε) are obtained after expressing all higher-order trigonometric functions in terms of the sum and difference of their arguments

$$\begin{aligned} a_1'' + a_1 = {} & \zeta_1 r_1 A \sin(r_1 \tau + \psi_1) - \frac{1}{2} K_1 A \cos[(r_f - r_1)\tau - \psi_1] \\ & - \frac{1}{2} K_1 A \cos[(r_f + r_1)\tau + \psi_1] - \frac{1}{2} K_2 B \cos[(r_f - r_2)\tau - \psi_2] \\ & - \frac{1}{2} K_2 B \cos[(r_f + r_2)\tau + \psi_2] \end{aligned} \tag{10.37a}$$

$$\begin{aligned} b_1'' + b_1 = {} & \zeta_2 r r_2 B \sin(r_2 \tau + \psi_2) - \frac{1}{2} L_1 A \cos[(r_f - r_1)\tau - \psi_1] \\ & - \frac{1}{2} L_1 A \cos[(r_f + r_1)\tau + \psi_1] - \frac{1}{2} L_2 B \cos[(r_f - r_2)\tau - \psi_2] \\ & - \frac{1}{2} L_2 B \cos[(r_f + r_2)\tau + \psi_1] \end{aligned} \tag{10.37b}$$

The first term on the right side of each of these equations is obviously a resonant term. Other terms may give rise to the following parametric resonance conditions

$$r_f = 2r_1, \; r_f = 2r_2, \; r_f = |r_1 \pm r_2| \tag{10.38a,b,c}$$

The second-order perturbation equations (coefficients of ε^2) can also be obtained. These equations are not listed here; however, secular terms that give rise to the above parametric resonance conditions will be identified. Obviously, the nonlinear parametric response of the system cannot be determined by restricting the analysis to first-order perturbation equations. Second-order terms must be included in order to predict the system response as demonstrated in the following sections.

First mode parametric excitation: $r_f = 2r_1$

Extracting the secular terms corresponding to this condition from first- and second-order perturbation equations, and transferring them to the variational equations (10.36), gives

$$-r_f A \, \psi_1' = -\left(1 - \frac{r_f^2}{4}\right) A - \frac{1}{2} \varepsilon K_1 A \cos 2\psi_1 + \frac{1}{16} \varepsilon^2 r_f^2 (-3K_3 + K_9) A^3 \tag{10.39a}$$

$$-r_f A' = \frac{1}{2} \varepsilon \zeta_1 r_f A - \frac{1}{2} \varepsilon K_1 A \sin 2\psi_1 \tag{10.39b}$$

The steady-state response is obtained by setting the left-hand sides of equations (10.39) to zero. The solutions of the resulting algebraic equations are

$$A = 0, \quad \text{or} \quad A_{1,2}^2 = \frac{-\left(1 - (r_f^2/4)\right) \pm \sqrt{(\varepsilon^2 K_1^2/2) - \zeta_1^2 (r_f^2/4)}}{\varepsilon^2 (r_f^2/16)(K_9 - 3K_3)} \tag{10.40a,b}$$

The trivial solution, $A = 0$, establishes the equilibrium position. The other solution has two branches, which may overhang to the right or to the left depending on the sign of the denominator. For the case of system parameters $l/R = 5.0$, $h/R = 2.0$, $m_l/m = 0.7$, and $\varepsilon = 0.008$, the amplitude–frequency response is shown in Figure 10.6 for three different values

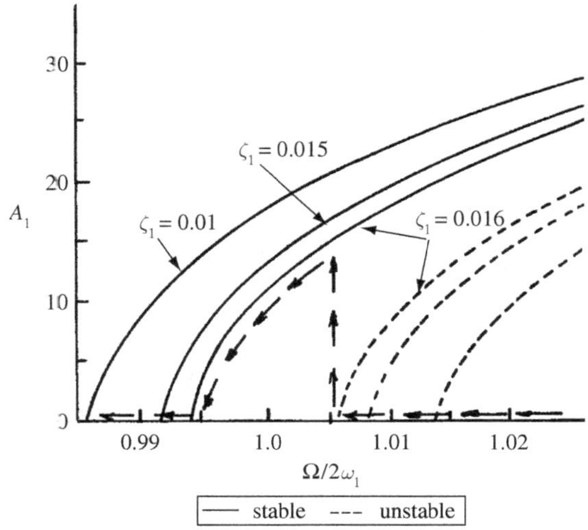

Figure 10.6 Amplitude–frequency response curves under first mode parametric excitation for different values of damping ratio ζ_1, $\varepsilon = 0.008$. (Ibrahim *et al.*, 1998)

of damping ratio $\zeta_1 = 0.01, 0.015$ and 0.016. It is seen that the response curves bend to the right indicating that the first mode excitation is of hard spring characteristics. Note that in the first mode both the liquid and structure move in phase. The two solutions given by (10.40b) will meet at an excitation frequency given by the expression

$$\frac{r_f}{2} = \frac{\Omega}{2\omega_1} = \frac{\varepsilon K_1}{2\zeta_1} \tag{10.41}$$

This excitation frequency determines the position of the turning or saddle point, and above this frequency ratio the response amplitude collapses to the static equilibrium position.

Second mode parametric excitation: $r_f = 2r_2$

Following the same procedure as in the previous case, the variational equations are

$$-r_f B\psi_2' = -\left(r^2 - \frac{r_f^2}{4}\right)B - \frac{1}{2}\varepsilon L_2 B\cos 2\psi_2 + \frac{1}{16}\varepsilon^2 r_f^2(-3L_8 + L_{14})B^3 \tag{10.42a}$$

$$-r_f B' = \frac{1}{2}\varepsilon\zeta_2 r r_f B - \frac{1}{2}\varepsilon L_2 B\sin 2\psi_2 \tag{10.42b}$$

The steady state response is given by the following roots

$$B = 0,\ B_{1,2}^2 = 4r^2\left\{\frac{1 - (r_f/2r)^2 \pm \sqrt{(\varepsilon^2 L_2^2/4r^2) - (\zeta_2^2/r^2)(r_f/2r)^2}}{\varepsilon^2 (r_f/r)^2 (L_{14} - L_8)}\right\} \tag{10.43a, b}$$

The amplitude–frequency response is shown in Figure 10.7 for various values of damping ratio ζ_2. It is seen that the system response follows soft nonlinear characteristics. The dotted

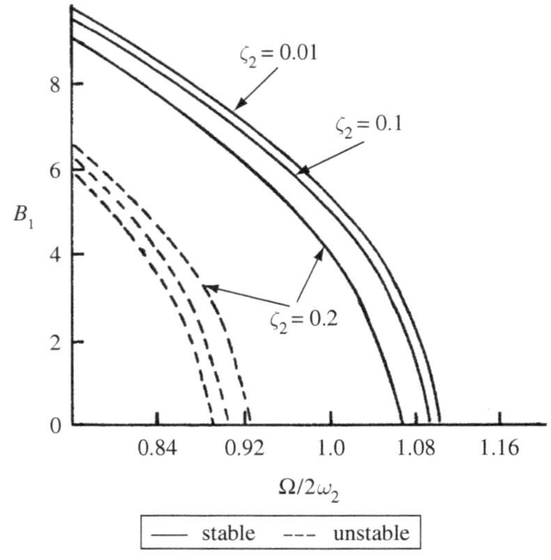

Figure 10.7 Amplitude–frequency response curves under second-mode parametric excitation for different values of damping ratio ζ_2, $\varepsilon = 0.005$. (Ibrahim *et al.*, 1988)

branches are the unstable response. Both stable and unstable branches meet at the turning point at an excitation frequency given by the expression

$$\frac{\Omega}{2\omega_f} = \frac{\varepsilon L_2}{2\zeta_2} \tag{10.44}$$

At this frequency the amplitude drops if the excitation frequency is decreasing.

Mixed mode excitation: $r_f = r_1 + r_2$

The variational equations of this case are

$$\begin{aligned} r_1 \psi_1' = & -\frac{1}{2}(r_1^2 - 1) + \frac{1}{4}\varepsilon K_2 \frac{B}{A}\cos(\psi_1 + \psi_2) \\ & + \frac{1}{2}\varepsilon^2 \left\{ \left(\frac{3}{4}K_3 - \frac{1}{4}K_9\right) A^2 + \frac{1}{2}(K_5 + K_7 r_2^2 - K_{11} r_2^2) B^2 \right\} \end{aligned} \tag{10.45a}$$

$$A' = \frac{1}{2}\varepsilon\zeta_1 A + \frac{1}{4}\varepsilon K_2 B \sin(\psi_1 + \psi_2) \tag{10.45b}$$

$$\begin{aligned} r_2 \psi_2' = & -\frac{1}{2}(r_2^2 - r^2) + \frac{1}{4}\varepsilon L_1 \frac{A}{r_2 B}\cos(\psi_1 + \psi_2) \\ & + \frac{1}{2}\varepsilon^2 \left\{ \left(\frac{3}{2}L_8 - \frac{1}{2}L_{14}\right) r_2 B^2 + (L_4 + L_6 r_2^2 - L_{12}) \frac{A^2}{r_2} \right\} \end{aligned} \tag{10.45c}$$

$$B' = \frac{1}{4}\varepsilon r \zeta_2 B + \frac{1}{4}\varepsilon \frac{L_1}{r_2} A \sin(\psi_1 + \psi_2) \tag{10.45d}$$

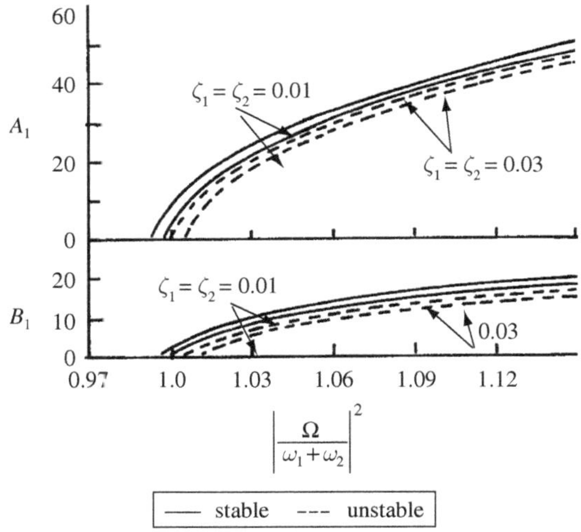

Figure 10.8 Amplitude–frequency response curves under combination parametric resonance for two different values of damping ratio ζ_2, $\varepsilon = 0.005$. (Ibrahim *et al.*, 1988)

The steady state solution is obtained by setting the left sides of equations (10.45) to zero. However, the resulting four algebraic equations will have only three unknowns. Instead, one combines equations (10.45a) and (10.45c) into one equation by setting $\gamma = \psi_1 + \psi_2$, and $\gamma' = \psi_1' + \psi_2'$. In this case one will have only three equations in three unknowns, namely, A, B, and γ. The steady state response is obtained numerically as shown in Figure 10.8 for two different values of damping ratios $\zeta_1 = \zeta_2 = 0.01$ and 0.03, and for $r_1 = 1$, and $r_2 = 2$. The general trend of the amplitude–frequency curves possesses hard nonlinear characteristics.

Attempts were made to determine the system response under parametric resonance of the difference type, $r_f = |r_1 - r_2|$. However, both numerical integration and algebraic solutions revealed that the system was always stable, and under any set of initial conditions the response converges to the static equilibrium state. This state is known as the "static attractor." Schmidt and Weidenhammer (1961), Tondl (1965), and Mettler (1968) showed that parametric instability of the difference type occurs if the value of $K_2 L_1$ is negative, which leads to stability boundaries. In the present case, this value is positive and parametric instability is not likely to occur.

10.4.2 Horizontal ground harmonic excitation

Consider the system response to a prescribed horizontal harmonic excitation $x_b\ (t) = X_{b0} \sin \Omega t$, $y_b\ (t) = 0$ in the neighborhood of fourth-order internal resonance. In this case, the equations of motion (10.34) take the form

$$
\begin{aligned}
Y_1'' + Y_1 &= \varepsilon^2(-2\zeta_1 Y_1' + f_1 \sin r_f \tau) \\
&+ \varepsilon^2\{K_3 Y_1^2 Y_1'' + K_4 Y_1 Y_2 Y_1'' + K_5 Y_2^2 Y_1'' + K_6 Y_1^2 Y_2'' + K_7 Y_1 Y_2 Y_2'' \\
&+ K_8 Y_2^2 Y_2'' + K_9 Y_1 Y_1'^2 + K_{10} Y_1 Y_1' Y_2' + K_{11} Y_1 Y_2'^2 + K_{12} Y_2 Y_1'^2 \\
&+ K_{13} Y_2 Y_1' Y_2' + K_{14} Y_2 Y_2'^2\} \qquad (10.46a)
\end{aligned}
$$

$$Y_2'' + r^2 Y_2 = \varepsilon^2\left(-2r\zeta_2 Y_2' + f_2 \sin r_f \tau\right)$$
$$+ \varepsilon^2\{L_3 Y_1^2 Y_1'' + L_4 Y_1 Y_2 Y_1'' + L_5 Y_2^2 Y_1'' + L_6 Y_1^2 Y_2'' + L_7 Y_1 Y_2 Y_2''$$
$$+ L_8 Y_2^2 Y_2'' + L_9 Y_1 Y_1'^2 + L_{10} Y_1 Y_1' Y_2' + L_{11} Y_1 Y_2'^2 + L_{12} Y_2 Y_1'^2$$
$$+ L_{13} Y_2 Y_1' Y_2' + L_{14} Y_2 Y_2'^2\} \tag{10.46b}$$

where a prime denotes differentiation with respect to $\tau = \omega_1 t$, $Y_i = y_i/X_0$, $\varepsilon = X_0/l$,
$f_1 = \dfrac{X_{b0}}{\varepsilon^2 X_0} \dfrac{[m_{21}(1 - n_1 - m_{21}) + n_2]}{(n_2 - n_1)(1 - m_{12}m_{21})}$
$f_2 = \dfrac{X_{b0}}{\varepsilon^2 X_0} \dfrac{(1 - m_{21}^2)}{(n_2 - n_1)(1 - m_{12}m_{21})}$ $r_f = (\Omega/\omega_1)$ and X_0 is a reference length taken as the amplitude of the structure deflection when the liquid is frozen, that is,

$$X_0 = \frac{r_0^2 X_{b0}}{\sqrt{(1 - r_f^2)^2 + (2r_f\zeta_1)^2}} \tag{10.47}$$

For a small- but finite-amplitude response, the equations of motion (10.46) are solved using the multiple scales method. The analysis is restricted to the response characteristics in the neighborhood of autoparametric resonance conditions (internal and external resonance conditions). Internal resonance conditions arise basically from nonlinear coupling of inertia terms. According to this method, the solution of equations (10.46) is written in the form

$$Y_1(\tau) = Y_{10}(T_0, T_2) + \varepsilon^2 Y_{12}(T_0, T_2) + \varepsilon^3 Y_{13}(T_0, T_2) + \cdots \tag{10.48a}$$

$$Y_2(\tau) = Y_{20}(T_0, T_2) + \varepsilon^2 Y_{22}(T_0, T_2) + \varepsilon^3 Y_{23}(T_0, T_2) + \cdots \tag{10.48b}$$

where $T_n = \varepsilon^n \tau$. Since the nonlinearities in equations (10.46) are of order ε^2 this solution does not include terms of order ε, T_0 is regarded as a fast-time scale and T_2 is a slow scale, which is associated with modulations in the amplitudes and phases resulting from the system nonlinearities. According to this expansion the nonlinear response consists of the fundamental component represented by the first term and perturbational components of orders ε^2 and higher. The time derivatives in the system equations of motion are transformed according to the following operators

$$\frac{d}{d\tau} = D_0 + \varepsilon^2 D_2 + \cdots, \quad \frac{d^2}{d\tau^2} = D_0^2 + \varepsilon^2(2D_0D_1) + \cdots \tag{10.49a,b}$$

where $D_i = \frac{\partial}{\partial T_i}$. Substituting equations (10.48) and (10.49) into equations (10.46), and equating terms of equal powers of ε, gives

Fundamental equations (zero-order of ε)

$$D_0^2 Y_{10} + Y_{10} = 0, \ D_0^2 Y_{20} + r^2 Y_{20} = 0 \tag{10.50a,b}$$

Second-order perturbational equations (terms of order ε^2)

$$
\begin{aligned}
D_0^2 Y_{12} + Y_{12} = & -2D_0 D_2 Y_{10} - 2\zeta_1 D_0 Y_{10} + f_1 \sin r_f T_0 + K_3 Y_{10}^2 D_0^2 Y_{10} \\
& + K_4 Y_{10} Y_{20} D_0^2 Y_{10} + K_5 Y_{20}^2 D_0^2 Y_{10} + K_6 Y_{10}^2 D_0^2 Y_{20} \\
& + K_7 Y_{10} Y_{20} D_0^2 Y_{20} + K_8 Y_{20}^2 D_0^2 Y_{20} \\
& + K_9 Y_{10}(D_0 Y_{10})^2 + K_{11} Y_{10}(D_0 Y_{20})^2 + K_{10} Y_{10} D_0 Y_{10} D_0 Y_{20} \\
& + K_{12} Y_{20}(D_0 Y_{20})^2 + K_{13} Y_{20} D_0 Y_{10} D_0 Y_{20} + K_{14} Y_{20}(D_0 Y_{10})^2
\end{aligned} \tag{10.51a}
$$

$$
\begin{aligned}
D_0^2 Y_{22} + r^2 Y_{22} = & -2D_0 D_2 Y_{20} - 2r\zeta_2 D_0 Y_{20} + f_2 \sin r_f T_0 + L_3 Y_{10}^2 D_0^2 Y_{10} \\
& + L_4 Y_{10} Y_{20} D_0^2 Y_{10} + L_5 Y_{20}^2 D_0^2 Y_{10} + L_6 Y_{10}^2 D_0^2 Y_{20} \\
& + L_7 Y_{10} Y_{20} D_0^2 Y_{20} + L_8 Y_{20}^2 D_0^2 Y_{20} \\
& + L_9 Y_{10}(D_0 Y_{10})^2 + L_{11} Y_{10}(D_0 Y_{20})^2 + L_{10} Y_{10} D_0 Y_{10} D_0 Y_{20} \\
& + L_{12} Y_{20}(D_0 Y_{20})^2 + L_{13} Y_{20} D_0 Y_{10} D_0 Y_{20} + L_{14} Y_{20}(D_0 Y_{10})^2
\end{aligned} \tag{10.51b}
$$

The solution of equations (10.50) may be written in the form

$$
Y_{10} = A_1(T_2)\, e^{iT_0} + \mathrm{CC}, \quad Y_{20} = A_2(T_2)\, e^{irT_0} + \mathrm{CC} \tag{10.52a,b}
$$

where CC stands for complex conjugate of the preceding terms, and the amplitudes A_i are functions of T_2. Substituting solutions (10.52) into equations (10.51), gives the following equations

$$
\begin{aligned}
D_0^2 Y_{12} + Y_{12} = & \frac{1}{2} f_1\, e^{ir_f T_0} \\
& + e^{iT_0}\big[-2i(A_1' + \zeta_1 A_1) - 2(K_5 + r^2 K_7 - r^2 K_{11}) A_1 A_2 \bar{A}_2 \\
& \qquad + (K_9 - 3K_3) A_1^2 \bar{A}_1\big] \\
& + e^{irT_0}\big[-2k_4 A_1 A_2 \bar{A}_1 - 2r^2 K_6 A_1 A_2 \bar{A}_1 - 3r^2 K_8 A_2^2 \bar{A}_2 \\
& \qquad + r^2 K_{12} A_2^2 \bar{A}_2 + 2K_{14} A_1 A_2 \bar{A}_1\big] - r^2 e^{i3rT_0}(K_8 + K_{12}) A_2^3 \\
& - e^{iT_0(1+2r)}\big[K_5 + r^2(K_7 + K_{11}) + rK_{13}\big] A_1 A_2^2 \\
& - e^{i3T_0}(K_3 + K_9) A_1^3 \\
& - e^{iT_0(2r-1)}\big[K_5 + r^2(K_7 + K_{11}) + rK_{13}\big] \bar{A}_1 A_2^2 \\
& - e^{iT_0(2+r)}\big[K_4 + r^2 K_6 - rK_{10} + K_{14}\big] A_1^2 A_2 \\
& - e^{iT_0(r-2)}\big[K_4 + r^2 K_6 - rK_{10} + K_{14}\big] \bar{A}_1^2 A_2 + \mathrm{CC}
\end{aligned} \tag{10.53a}
$$

$$
\begin{aligned}
D_0^2 Y_{22} + r^2 Y_{12} = {} & \frac{1}{2} f_2 \, e^{irfT_0} \\
& + e^{irT_0}[-2i(rA_2' + r\zeta_2 A_2) - 2(L_4 + r^2 L_6 - L_{14}) A_1 A_2 \bar{A}_1 \\
& \qquad + r^2 (L_{12} - 3L_8) A_2^2 \bar{A}_2] \\
& + e^{iT_0}[(L_9 - 3L_3) A_1^2 \bar{A}_1 - 2(L_5 + r^2 L_7 - r^2 L_{11}) A_1 A_2 \bar{A}_2] \\
& - e^{i3T_0}(L_3 + L_9) A_1^3 - r^2 \, e^{i3rT_0}[L_8 + L_{12}] A_2^3 \\
& - e^{iT_0(2r+1)} \left[L_5 + r^2 (L_7 + L_{11}) + rL_{13}\right] \bar{A}_1 A_2^2 \\
& - e^{iT_0(2r-1)} \left[L_5 + r^2 (L_7 + L_{11}) + rL_{13}\right] \bar{A}_1 A_2^2 \\
& - e^{iT_0(2+r)} \left[L_4 + r^2 L_6 + rL_{10} + L_{14}\right] A_1^2 A_2 \\
& - e^{iT_0(r-2)} \left[L_4 + r^2 L_6 + rL_{10} + L_{14}\right] \bar{A}_1^2 A_2 + \mathrm{CC}
\end{aligned}
\qquad (10.53\text{b})
$$

where a prime denotes differentiation with respect to T_2, and an over-bar denotes the complex conjugate. Equations (10.53) contain secular terms, which give rise to the following resonance conditions:

External resonance conditions:

$$\Omega = \omega_1, \quad \Omega = \omega_2 \qquad (10.54\text{a,b})$$

Internal resonance condition:

$$\omega_2 = 3\omega_1 \qquad (10.54\text{c})$$

The effect of the departure from exact internal resonance is examined by introducing a detuning parameter σ_{I} defined by the relation

$$r = \frac{\omega_2}{\omega_1} = 3 + \varepsilon^2 \sigma_I \qquad (10.55)$$

First mode excitation: $\Omega = \omega_1$

The response of the system is determined when the first mode is externally excited. An external detuning parameter σ_2 is introduced such that

$$r_f = \frac{\Omega}{\omega_1} = 1 + \varepsilon^2 \sigma_2 \qquad (10.56)$$

Eliminating secular terms that lead to resonance conditions (10.54a) and (10.54c) from equations (10.53), yields the two solvability conditions

$$
\begin{aligned}
& \frac{1}{2} f_1 \, e^{iT_2} - 2i(A_1' + \zeta_1 A_1) + (K_9 - 3K_3) A_1^2 \bar{A}_1 \\
& - 2(K_5 + r^2 K_7 - r^2 K_{11}) A_1 A_2 \bar{A}_2 \\
& + e^{i\sigma_I T_2}[rK_{10} - k_4 - r^2 K_6 - K_{14}] \bar{A}_1^2 A_2 = 0
\end{aligned}
\qquad (10.57\text{a})
$$

$$e^{i\sigma_I T_2}[-2i(rA_2' + r\zeta_2 A_2) - 2(L_4 + r^2 L_6 - L_{14})A_1 A_2 \bar{A}_1 + r^2(L_{12} - 3L_8)A_2^2 \bar{A}_2] - (L_3 + L_9)A_1^3 = 0 \quad (10.57b)$$

Introducing the polar coordinate transformation

$$A_j(T_2) = \frac{1}{2} a_j(T_2)\, e^{i\theta_j(T_2)}, \; j = 1, \; 2 \quad (10.58)$$

into equations (10.57), separating real and imaginary parts, and setting the phase angles

$$\beta_1 = \sigma_I T_2 + \theta_2 - 3\theta_1, \; \beta_2 = \sigma_2 T_2 - \theta_1 \quad (10.59a,b)$$

gives

$$2a_1' = -2\zeta_1 a_1 + 2C_{11} a_1^2 a_2 \sin\beta_1 + f_1 \sin\beta_2 \quad (10.60a)$$

$$ra_2' = r\zeta_2 a_2 + C_{21} a_1^3 \sin\beta_1 \quad (10.60b)$$

$$\beta_1' = 3\beta_2' - 3\sigma_2 + \sigma_I + \frac{C_{21} a_1^3}{r a_2} \cos\beta_1 + \frac{1}{r}\left[C_{22} a_2^2 + C_{23} a_1^2\right] \quad (10.60c)$$

$$\beta_2' = \sigma_2 + C_{11} a_1 a_2 \cos\beta_1 + C_{12} a_1^2 + \frac{f_1}{2a_1} \cos\beta_2 \quad (10.60d)$$

where $C_{11} = \frac{1}{8}(-r^2 K_6 + rK_{10} - K_4 - K_{12})$, $C_{12} = \frac{1}{8}(K_9 - 3K_3)$, $C_{13} = \frac{1}{4}(r^2 K_{11} - K_5 - r^2 K_7)$, $C_{21} = \frac{1}{8}(L_3 + L_9)$, $C_{22} = (r^2/8)(3L_8 - L_{12})$, $C_{23} = \frac{1}{4}(L_4 + r^2 L_6 - L_{14})$.

The transient and steady state responses are found to be sensitive to initial conditions and the external detuning parameter. Figures 10.9 and 10.10 show the time history records for $\sigma = -120$, and for two different sets of initial conditions. It is seen that the system possesses two different stable steady-state responses for the same excitation level and damping factors. Figure 10.11 shows the amplitude–frequency response curves for damping factors $\zeta_1 = \zeta_2 = 0.001$, and forcing parameter $f_1 = 0.075$. Points H and D correspond to the two steady state responses indicated in Figures 10.9 and 10.10, respectively. It is observed that this nonlinear interaction is characterized by a large amplitude of the first mode, which is directly excited, and a relatively small amplitude for the second mode. This means that the suppression effect, known for autoparametric systems with quadratic nonlinearity, does not take place in the present case because the excitation frequency is remote from the second normal mode frequency.

Figure 10.12 shows a set of response curves in the frequency domain for various levels of forcing excitation. For a very small excitation level ($f_1 = 0.0125$) the system response depicts linear resonance behavior. As the excitation level increases, the nonlinear effects become significant and the system response follows the same features as those of soft nonlinearity. For higher excitation amplitudes ($f_1 = 0.075$ and 0.1) complex response characteristics occur in the neighborhood of external resonance condition as shown in Figures 10.11 and 10.13.

The stability of the steady state responses is examined and those stable solutions are shown by solid curves while dot curves are unstable responses. The response curves can be divided into five groups (see Figure 10.13). Each group occupies a certain region of the external

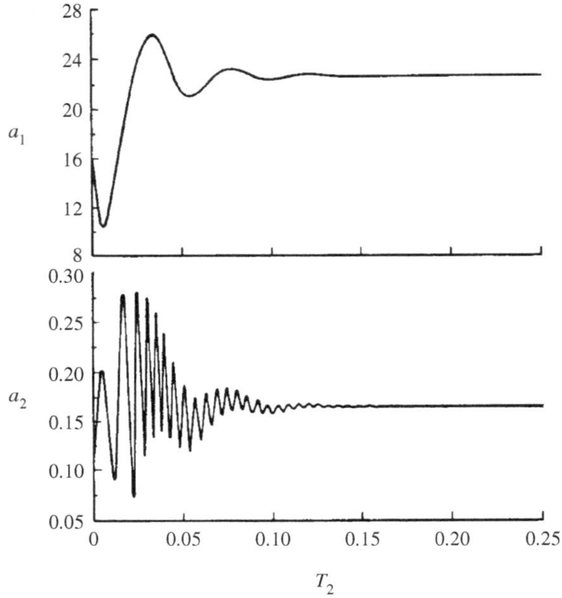

Figure 10.9 Time history response records under first mode excitation, $\sigma_1 = 0.0$, $\sigma_2 = -120$, $\hat{\zeta}_1 = \hat{\zeta}_2 = 0.001$, $\varepsilon = 0.005$, $\hat{f}_1 = 0.075$, with initial conditions: $a_1 = 16.5$, $a_2 = 0.132$, $\beta_1 = 0.4$, $\beta_2 = 0.772$. (Ibrahim and Li, 1988)

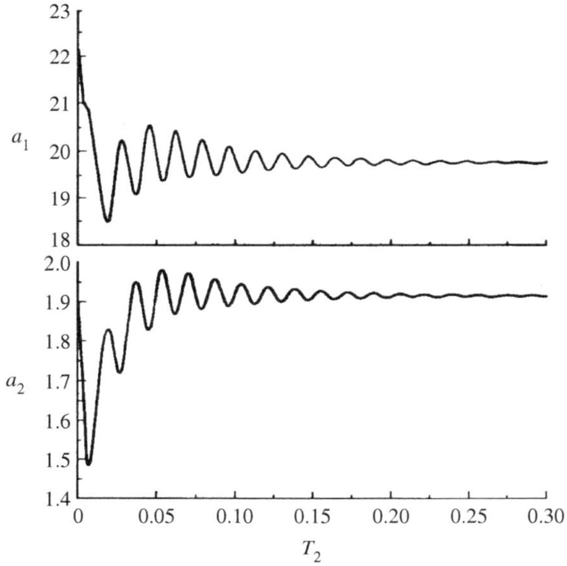

Figure 10.10 Time history response records under first mode excitation, $\sigma_1 = 0.0$, $\sigma_2 = -120$, $\hat{\zeta}_1 = \hat{\zeta}_2 = 0.001$, $\varepsilon = 0.005$, $\hat{f}_1 = 0.075$, with different initial conditions $a_1 = 22.5$, $a_2 = 1.91$, $\beta_1 = 0.01$, $\beta_2 = 0.01$. (Ibrahim and Li, 1988)

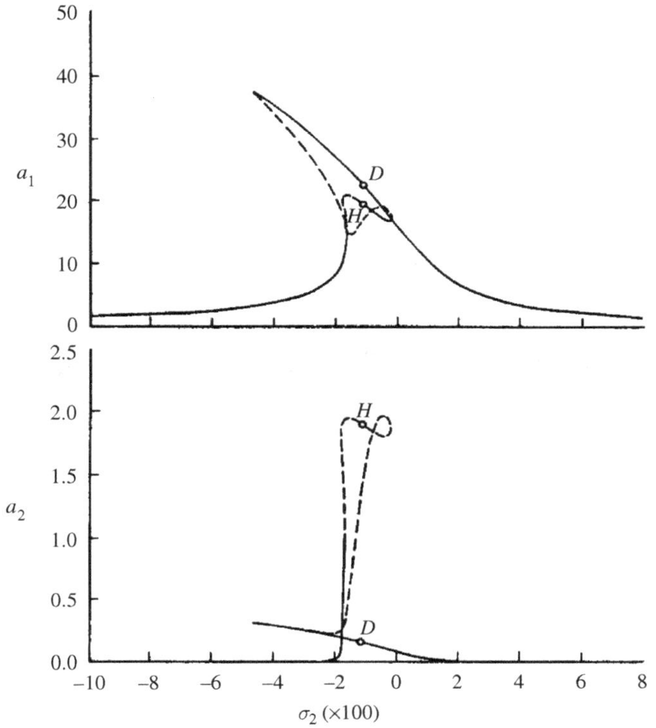

Figure 10.11 Amplitude–frequency response curves under first mode excitation $\hat{\zeta}_1 = \hat{\zeta}_2 = 0.001$, $\varepsilon = 0.005$, $\hat{f}_1 = 0.075$. (Ibrahim and Li, 1988)

detuning parameter σ_2. These groups include the following regions: region (I) bounded by the curve ab: $-1000 < \sigma_2 < -20.54$ defines a unimodal response for which $a_1 \neq 0$, $\beta_1 \neq 0$, and $a_2 = \beta_2 \approx 0$, this solution is always stable, region (II) bounded by the curve bc: $-306 < \sigma_2 < -209.54$ defines unstable mixed mode response, region (III) given by the curve cd: $-306 < \sigma_2 < 21$ defines stable mixed mode response, region (IV) represented by the curve deff′g: $-820 < \sigma_2 < 21$ defines unstable mixed mode response, and region (V) given by the curve gg′h: $-820 < \sigma_2 < 0$ defines stable response. It is seen that the stable and unstable response curves ff′g and gg′ are very close to each other only for the second amplitude and first phase angle. However, they are distinct for the first mode amplitude and second phase angle.

Second mode excitation: $\Omega = \omega_2$

Introducing the external detuning parameter, σ_2,

$$\frac{\Omega}{\omega_1} = r + \varepsilon^2 \sigma_2 \tag{10.61}$$

and extracting all secular terms from equations (10.61) gives the solvability conditions

$$\begin{aligned} &-2i(A_1' + \zeta_1 A_1) + (K_9 - 3K_3)A_1^2 \bar{A}_1 - 2(K_5 + r^2 K_7 - r^2 K_{11}) \\ &\times A_1 A_2 \bar{A}_2 + e^{i\sigma_I T_2}[rK_{10} - k_4 - r^2 K_6 - K_{14}]\bar{A}_1^2 A_2 = 0 \end{aligned} \tag{10.62a}$$

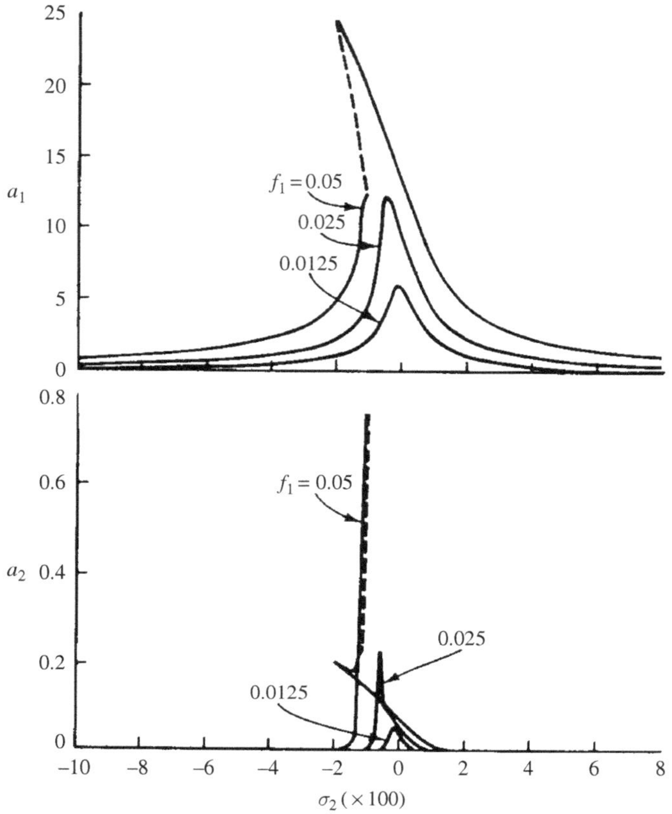

Figure 10.12 Amplitude–frequency response curves under first mode excitation for different excitation levels; $\hat{\zeta}_1 = \hat{\zeta}_2 = 0.001$, $\varepsilon = 0.005$. (Ibrahim and Li, 1988)

$$\frac{1}{2} f_2 \, e^{i\sigma_2 T_2} - [2i(rA_2' + r\zeta_2 A_2) + 2(L_4 + r^2 L_6 - L_{14}) A_1 A_2 \bar{A}_1 + r^2 (3L_8 - L_{12}) A_2^2 \bar{A}_2] - e^{i\sigma_I T_2} (L_3 + L_9) A_1^3 = 0 \tag{10.62b}$$

Following the same procedure as that of the previous case, expressing the amplitudes A_j in terms of the polar coordinates (10.58) and separating equations (10.62) into real and imaginary parts gives the following four equations

$$a_1' = -\zeta_1 a_1 + C_{11} a_1^2 a_2 \sin \beta_1 \tag{10.63a}$$

$$r a_2' = \frac{1}{2} f_2 \sin \beta_2 - r \zeta_2 a_2 + C_{21} a_1^3 \sin \beta_1 \tag{10.63b}$$

$$\beta_1' = \sigma_I + \sigma_2 - \beta_2' + 3[C_{12} a_1^2 + C_{13} a_2^2 + C_{11} a_1 a_2 \cos \beta_1 \tag{10.63c}$$

$$r\beta_2' = r\sigma_2 + \frac{1}{a_2} \left[\frac{1}{2} f_2 \cos \beta_2 - C_{21} a_1^3 \cos \beta_1 \right] - C_{22} a_2^2 - C_{23} a_1^2 \tag{10.63d}$$

where $\beta_1 = \sigma_I T_2 + \theta_2 - 3\theta_1$, $\beta_2 = \sigma_2 T_2 - \theta_2$.

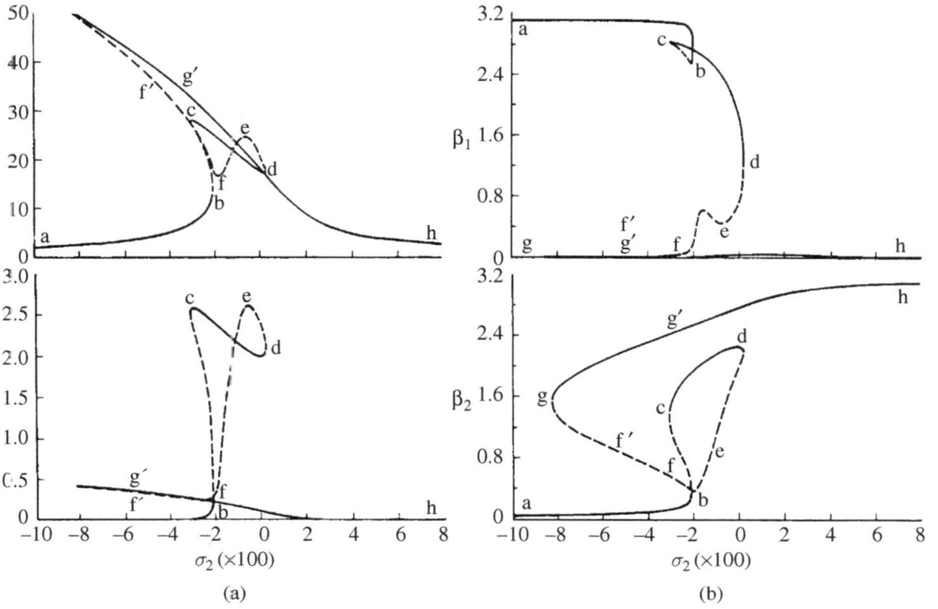

Figure 10.13 (a) amplitude–frequency response curves and (b) phase angle-frequency curves under first mode excitation $\hat{\zeta}_1 = \hat{\zeta}_2 = 0.001$, $\varepsilon = 0.005$, $\hat{f}_1 = 0.1$. (Ibrahim and Li, 1988)

Inspecting the resulting algebraic equations, obtained by setting the left-hand sides to zero, reveals two possible solutions. These are: (a) the unimodal response, $a_1 = 0$ and $a_2 \neq 0$, and (b) the mixed mode response $a_1 \neq 0$ and $a_2 \neq 0$.

When the second mode is externally excited, the nonlinear interaction with the first mode may be lost if the phase angle β_1 vanishes. Under this condition, the steady state response is obtained by solving the two algebraic equations

$$\frac{1}{2} f_2 \sin \beta_2 - r\zeta_2 a_2 = 0 \tag{10.64a}$$

$$r\sigma_2 a_2 + \frac{1}{2} f_2 \cos \beta_2 - C_{22} a_2^3 = 0 \tag{10.64b}$$

Eliminating β_2, gives the cubic equation in a_2^2

$$C_{22}^2 a_2^6 - 2C_{22} r\sigma_2 a_2^4 + r^2(\zeta_2^2 + \sigma_2^2) a_2^2 - \frac{1}{4} f_2^2 = 0 \tag{10.65}$$

Figure 10.14 shows the amplitude–frequency response of the second mode, which exhibits hard spring characteristics.

Now the mixed mode interaction under second mode excitation is obtained by solving the four algebraic equations obtained from equations (10.63). Figures 10.15–10.17 show the amplitude–frequency responses of the two modes for three different levels of external excitation $\hat{f}_2 = 1.5, 5, 10$, respectively. It is seen that the autoparametric absorbing effect, which takes place in systems with quadratic nonlinearity, does not occur in the case of cubic nonlinearity. The second normal mode amplitude is noticed to be higher than the unimodal response level. In

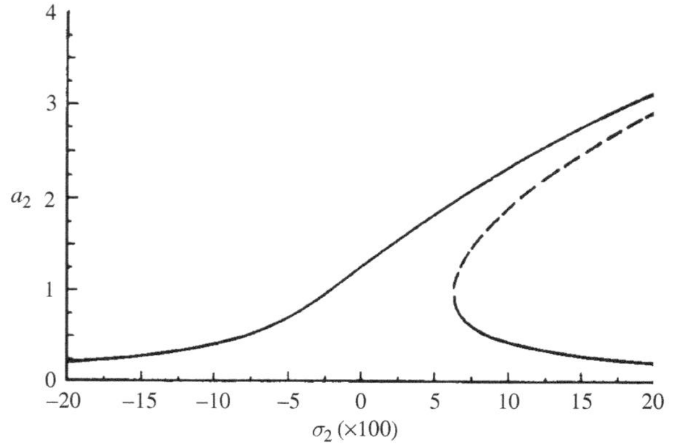

Figure 10.14 Amplitude–frequency response curves under second mode excitation; $\hat{\zeta}_1 = \hat{\zeta}_2 = 0.001$, $\varepsilon = 0.005$, $\hat{f}_2 = 1.0$. (Ibrahim and Li, 1988)

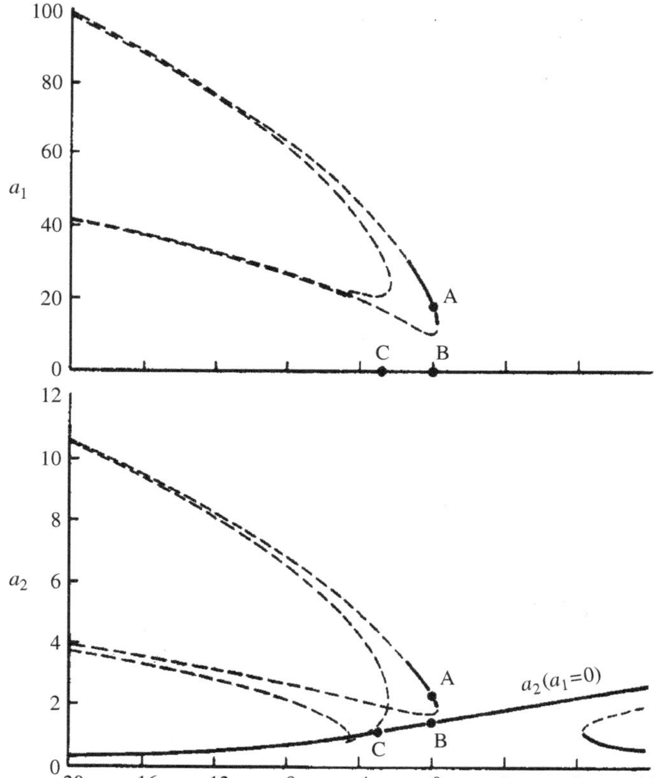

Figure 10.15 Mixed mode response curves under second mode excitation; $\sigma_1 = 0$, $\varepsilon = 0.02$, $\hat{\zeta}_1 = \hat{\zeta}_2 = 0.01$, $\hat{f}_2 = 1.5$. (Ibrahim and Li, 1988)

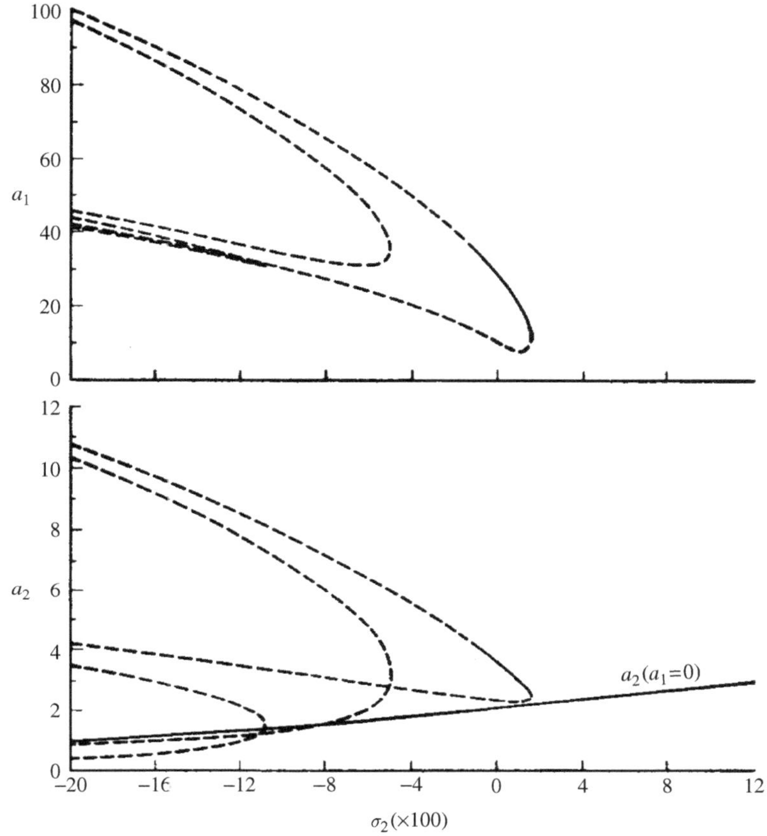

Figure 10.16 Mixed mode response curves under second mode excitation; $\sigma_1 = 0$, $\varepsilon = 0.02$, $\hat{\zeta}_1 = \hat{\zeta}_2 = 0.01$, $\hat{f}_2 = 5.0$. (Ibrahim and Li, 1988)

addition, the nonlinear interaction results in multi-solutions indicated by three response curves shown in Figures 10.16 and 10.17. The effect of increasing the excitation level is shown in Figure 10.18. For a small excitation level ($\hat{f}_2 = 1.0$) the nonlinear interaction is unstable and the system response possesses a unimodal response.

Numerical integration of equations (10.63) revealed that, depending on the initial conditions, the system might reach either nontrivial solutions for both modes (corresponding to point A in Figure 10.15, or a unimodal response (indicated by point B). It was found that the stable response of the two-mode interaction exists over a limited range of the external detuning parameter, σ_2: for the case of Figure 10.15, it is $-140 < \sigma_2 < 24$. Figure 10.19 shows the boundaries between the unimodal response and mixed mode response in the excitation–frequency plane. The shown two boundaries belong to two different values of damping ratios.

10.5 Random excitation

Under random excitation that covers a wide band frequency, it is important to design the absorber to suppress the motion of structures. Koh, *et al.* (1995) considered the case of

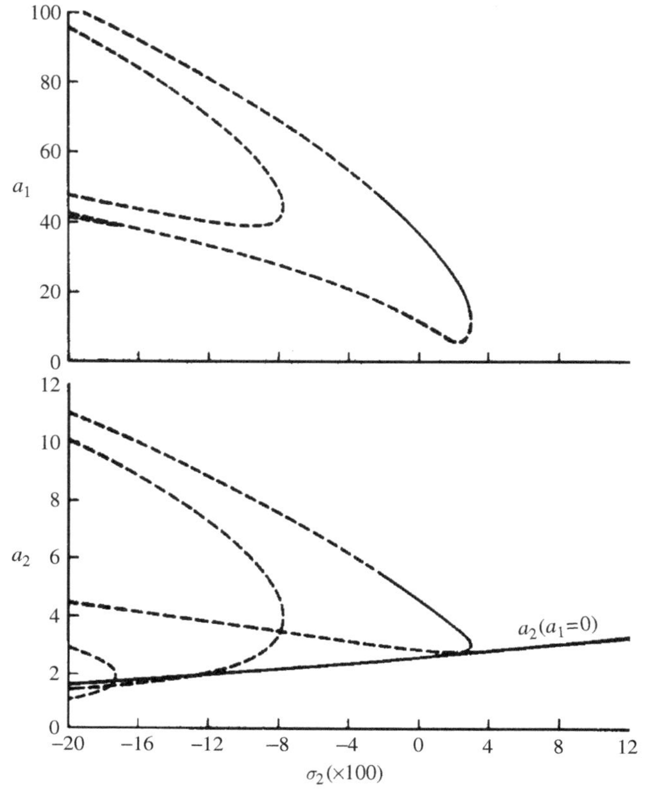

Figure 10.17 Mixed mode response curves under second mode excitation; $\sigma_1 = 0$, $\varepsilon = 0.02$, $\hat{\zeta}_1 = \hat{\zeta}_2 = 0.01$, $\hat{f}_2 = 10.0$. (Ibrahim and Li, 1988)

arbitrary excitation while Ibrahim and Soundararajan (1988) and Ikeda and Ibrahim (2002) considered random ground excitations. This section deals with the random excitation of such systems.

10.5.1 Horizontal and vertical random excitations

The response statistics of the system is considered under random vertical and horizontal excitations. The equations of motion in terms of the normal coordinates are written using the following nondimensional parameters

$$\tau = \omega_1 t, \quad Y_i = y_i/X_0, \quad \varepsilon = X_0/l, \quad \xi_1''(\tau) = \frac{\ddot{y}_b}{\omega_1^2 X_0}, \quad \xi_2''(\tau) = \frac{\ddot{x}_b}{\omega_1^2 X_0}$$

where X_0 is a reference displacement taken as the root mean square of the system response when it is reduced to a single degree-of-freedom. In this case, the liquid is frozen and the parametric excitation is removed. In other words $E[X_0^2]$ is the mean square response of the linear oscillator

$$\ddot{X}_0 + 2\zeta_s \omega_{11} \dot{X}_0 + \omega_{11}^2 X_0 = -\ddot{x}_b(t) \tag{10.66}$$

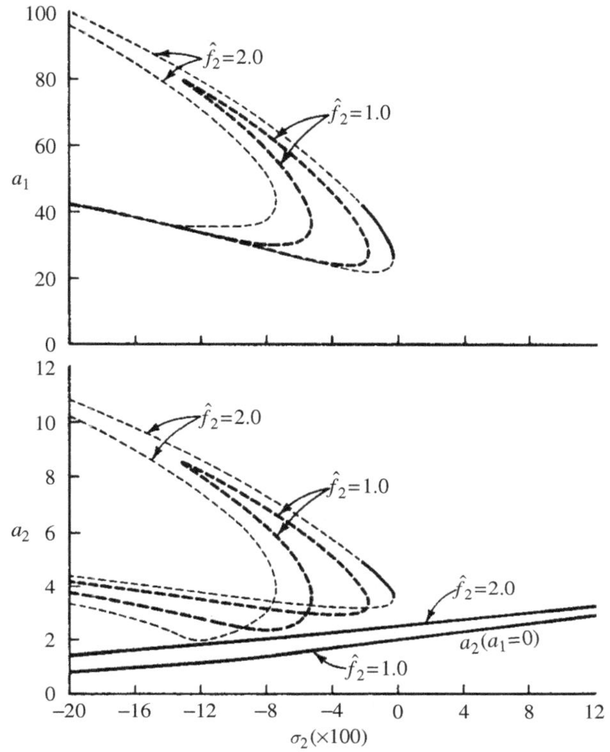

Figure 10.18 Mixed mode response curves under second mode excitation for different excitation levels; $\sigma_1 = 0$, $\varepsilon = 0.01$, $\hat{\zeta}_1 = \hat{\zeta}_2 = 0.01$. (Ibrahim and Li, 1988)

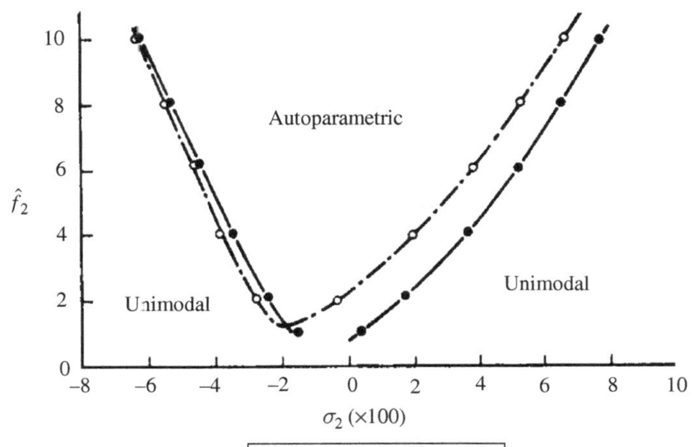

Figure 10.19 Boundaries of unimodal and autoparametric interaction for two sets of damping ratios; $\varepsilon = 0.01$. (Ibrahim and Li, 1988)

The present form of the equations of motion cannot be written in the Markov vector form due to the presence of nonlinear acceleration terms. However, these terms can be removed by successive elimination. Terms up to third order are retained, and the equations of motion are given in the nondimensional form

$$Y_1'' = -2\zeta_1 Y_1' - Y_1 + \varepsilon(B_{11}Y_1 + B_{12}Y_2)\xi_1''(\tau) + s_1\left[1 + \varepsilon^2 f_1(\mathbf{Y})\right]\xi_2''(\tau) + \varepsilon^2\phi_1(\mathbf{Y}, \mathbf{Y}') \tag{10.66a}$$

$$Y_2'' = -2\zeta_2 r Y_1' - r^2 Y_1 + \varepsilon(B_{21}Y_1 + B_{22}Y_2)\xi_1''(\tau) + s_2\left[1 + \varepsilon^2 f_2(\mathbf{Y})\right]\xi_2''(\tau) + \varepsilon^2\phi_2(\mathbf{Y}, \mathbf{Y}') \tag{10.66b}$$

where

$$r = \omega_2/\omega_1$$
$$B_{11} = (l/R)[\psi_{11}C_0 + \psi_{12}(C_1 + C_6 n_1)]$$
$$B_{12} = (l/R)[\psi_{11}C_0 + \psi_{12}(C_1 + C_6 n_2)]$$
$$B_{21} = (l/R)[\psi_{21}C_0 + \psi_{22}(C_1 + C_6 n_1)]$$
$$B_{22} = (l/R)[\psi_{21}C_0 + \psi_{22}(C_1 + C_6 n_2)]$$
$$\psi_{11} = \frac{n_2 + m_{21}}{\det[Q]\det[M]}$$
$$\psi_{12} = \frac{-n_2 m_{12} - 1}{\det[Q]\det[M]}$$
$$\psi_{21} = \frac{-n_1 - m_{21}}{\det[Q]\det[M]},$$
$$\psi_{22} = \frac{1 + n_1 m_{12}}{\det[Q]\det[M]}$$
$$s_1 = -\psi_{11} - \psi_{12}m_{21}, \quad s_2 = -\psi_{21} - \psi_{22}m_{21}$$

The functions f_i are nonlinear functions of the displacement vector $\mathbf{Y}$, while the functions ϕ_i are nonlinear of $\mathbf{Y}$ and its time derivative. The nondimensional excitations $\xi_1''(\tau)$ and $\xi_2''(\tau)$ are Gaussian wide-band random processes with zero means. The correlation time of these processes is assumed to be very small when compared with any characteristic time of the system. In this case one can write equations (10.66) in the form of the Stratonovich stochastic differential equation

$$\mathbf{X}' = \mathbf{f}(\mathbf{X}, \tau) + \mathbf{G}(\mathbf{X}, \tau)\boldsymbol{\xi}''(\tau) \tag{10.67}$$

through the coordinate transformation

$$\{X_1, X_2, X_3, X_4\} = \{Y_1, Y_2, Y_1', Y_2'\} \tag{10.68}$$

where $\mathbf{f}(\mathbf{X}, \tau)$ is the drift vector whose elements are linear and nonlinear functions of the state coordinates $\mathbf{X}$, and $\mathbf{G}(\mathbf{X}, \tau)$ is the diffusion matrix whose elements are also linear and nonlinear. Since the excitations, $\xi_i''(\tau)$, are assumed to be wide band, they can be replaced by independent white noise processes, $W_i(\tau)$, with zero means and delta autocorrelation functions

$$R_{W_i}(\tau) = E[W_i(\tau)W_i(\tau + \Delta\tau)] = 2D_i\delta(\Delta\tau) \tag{10.69}$$

where $2D_i$ is the spectral density of the white noise $W_i(\tau)$. Upon replacing $W_i(\tau)$ by the formal derivative of the Brownian motion process, $W_i(\tau) = \sigma_i dB_i(\tau)/d\tau$, where $\sigma_i^2 = 2D_i$, equation (10.67) may be written as a stochastic differential equation of the Itô type:

$$dX_i = \left\{ f_i(X, \tau) + \frac{1}{2}\sum_{k=1}^{4}\sum_{j=1}^{4} G_{kj}\frac{\partial G_{ij}}{\partial X_k} \right\} d\tau + \sum_{j=1}^{4} G_{ij}\, dB_j(\tau) \tag{10.70}$$

where the double summation expression in equation (10.70) is known as the Wong–Zakai correction term, which is the result of replacing the physical wide band random process $\xi_i''(\tau)$ by the white noise process $W_i(\tau)$. Upon adopting the above procedure, the Itô equations of system (10.66) are

$$\begin{aligned}
X_1' &= X_3, \quad X_2' = X_4 \\
X_3' &= -2\zeta_1 X_3 - X_1 + \varepsilon W_1(\tau)[B_{11}X_1 + B_{12}X_2] \\
&\quad + s_1 W_2(\tau)[1 + \varepsilon^2 f_1(\mathbf{X})] + \varepsilon^2 \hat{\phi}_1(\mathbf{X}) \\
X_4' &= -2\zeta_2 r X_4 - r^2 X_2 + \varepsilon W_1(\tau)[B_{21}X_1 + B_{22}X_2] \\
&\quad + s_2 W_2(\tau)[1 + \varepsilon^2 f_2(\mathbf{X})] + \varepsilon^2 \hat{\phi}_2(\mathbf{X})
\end{aligned} \tag{10.71}$$

where

$$\begin{aligned}
f_1(\mathbf{X}) &= (g_1 + g_2 n_1 + g_3 n_1^3)X_1^2 + (g_1 + g_2 n_2 + g_3 n_1 n_2^2)X_2^2 \\
&\quad + (2g_1 + g_2 n_1 + g_2 n_2 + 2g_3 n_1^2 n_2)X_1 X_2 \\
f_2(\mathbf{X}) &= (h_1 + h_2 n_1 + h_3 n_1^2 n_2)X_1^2 + (h_1 + h_2 n_2 + h_3 n_2^3)X_2^2 \\
&\quad + (2h_1 + h_2 n_1 + h_2 n_2 + 2h_3 n_1 n_2^2)X_1 X_2 \\
\hat{\phi}_1(\mathbf{X}) &= (-2\zeta_1 X_3 - X_1)[(g_1 + g_2 n_1 + g_3 n_1^3)X_1^2 \\
&\qquad + (2g_1 + g_2 n_1 + g_2 n_2 + 2g_3 n_1^2 n_2)X_1 X_2 \\
&\qquad + (g_1 + g_2 n_2 + g_3 n_2^3)X_2^2] \\
&\quad + (-2\zeta_2 r X_4 - r^2 X_2)[(g_1 + g_2 n_1 + g_3 n_1^2 n_2)X_1^2 \\
&\qquad + (2g_1 + g_2 n_1 + g_2 n_2 + 2g_3 n_1 n_2^2)X_1 X_2 + (g_1 + g_2 n_2 + g_3 n_2^3)X_2^2] \\
&\quad + \left[(g_1 + g_2 n_1 + g_4 n_1^3)X_1 + (g_1 + g_2 n_2 + g_4 n_1^2 n_2)X_2\right]X_3^2 \\
&\quad + \left[(g_1 + g_2 n_1 + g_4 n_1 n_2^2)X_1 + (g_1 + g_2 n_2 + g_4 n_2^3)X_2\right]X_4^2 \\
&\quad + 2\left[(g_1 + g_2 n_1 + g_4 n_1^2 n_2)X_1 + (g_1 + g_2 n_2 + g_4 n_1 n_2^2)X_2\right]X_3 X_4
\end{aligned}$$

$$
\begin{aligned}
\hat{\phi}_2(\mathbf{X}) = {} & (-2\zeta_1 X_3 - X_1)[(h_1 + h_2 n_1 + h_3 n_1^3)X_1^2 \\
& + (2h_1 + h_2 n_1 + h_2 n_2 + 2h_3 n_1^2 n_2)X_1 X_2 \\
& + (h_1 + h_2 n_2 + h_3 n_1 n_2^2)X_2^2] + (-2\zeta_2 r X_4 - r^2 X_2) \\
& \times [(h_1 + h_2 n_1 + h_3 n_1^2 n_2)X_1^2 + (2h_1 + h_2 n_1 + h_2 n_2 + 2h_3 n_1 n_2^2)X_1 X_2 \\
& + (h_1 + h_2 n_2 + h_3 n_2^3)X_2^2] + [(h_1 + h_2 n_1 + h_4 n_1^3)X_1 \\
& + (h_1 + h_2 n_2 + h_4 n_1^2 n_2)X_2]X_3^2 \\
& + [(h_1 + h_2 n_1 + h_4 n_1 n_2^2)X_1 + (h_1 + h_2 n_2 + h_4 n_2^3)X_2]X_4^2 \\
& + 2[(h_1 + h_2 n_1 + h_4 n_1^2 n_2)X_1 + (h_1 + h_2 n_2 + h_4 n_1 n_2^2)X_2]X_3 X_4
\end{aligned} \tag{10.72}
$$

$$
\begin{aligned}
g_1 &= -\psi_{11} C_0^2 - \psi_{12} C_3 \\
g_2 &= \psi_{12} C_7 \\
g_3 &= \psi_{11} C_2 + \psi_{12} C_5, \\
g_4 &= 2\psi_{11} C_2 + \psi_{12} C_4 \\
h_1 &= -\psi_{21} C_0^2 - \psi_{22} C_3 \\
h_2 &= \psi_{22} C_7 \\
h_3 &= \psi_{21} C_2 + \psi_{22} C_5, \\
h_4 &= 2\psi_{21} C_2 + \psi_{22} C_4
\end{aligned}
$$

The evolution of the joint probability density of the response coordinates, $p(\mathbf{X}, \tau)$, can be described by the Fokker–Planck equation

$$
\frac{\partial p(\mathbf{X},\tau)}{\partial \tau} = -\sum_{i=1}^{4} \frac{\partial}{\partial X_i}[a_i(\mathbf{X},\tau)p(\mathbf{X},\tau)] + \frac{1}{2}\sum_{i}^{4}\sum_{j}^{4} \frac{\partial^2}{\partial X_i \partial X_j}\left[b_{ij}(\mathbf{X},\tau)p(\mathbf{X},\tau)\right] \tag{10.73}
$$

where the drift $a_i(\mathbf{X}, \tau)$ and diffusion $b_{ij}(\mathbf{X}, \tau)$ coefficients are the incremental moments defined by

$$
a_i(\mathbf{X},\tau) = \lim_{\Delta\tau \to 0} \frac{1}{\Delta\tau} E[X_i(\tau + \Delta\tau) - X_i(\tau)] \tag{10.74a}
$$

$$
b_{ij}(\mathbf{X},\tau) = \lim_{\Delta\tau \to 0} \frac{1}{\Delta\tau} E[\{X_i(\tau + \Delta\tau) - X_i(\tau)\}\{X_j(\tau + \Delta\tau) - X_j(\tau)\}] \tag{10.74b}
$$

The general dynamic moment equation of the system response can be derived by multiplying both sides of the Fokker–Planck equation by the scalar function $X_1^i X_2^j X_3^k X_4^\ell$ and integrating over the entire state space $-\infty < \mathbf{X} < \infty$. This procedure results in the general first-order differential moment equation of order n, where $n = i + j + k + \ell$:

$$
\begin{aligned}
\frac{dm_{ijk\ell}}{d\tau} = {} & im_{i-1,j,k+1,\ell} + jm_{i,j-1,k,\ell+1} \\
& + k[-2\zeta_1 m_{i,j,k,\ell} - m_{i+1,j,k-1,\ell} + b_1 m_{i+3,j,k-1,\ell} + b_2 m_{i,j+3,k-1,\ell} \\
& \quad + b_3 m_{i+2,j+1,k-1,\ell} + b_4 m_{i+1,j+2,k-1,\ell+1} + b_5 m_{i+2,j,k,\ell} + b_6 m_{i+1,j,k+1,\ell} \\
& \quad + b_7 m_{i+2,j,k-1,\ell} + b_8 m_{i+1,j,k-1,\ell+2} + b_9 m_{i,j+2,k,\ell} + b_{10} m_{i,j+1,k+1,\ell} \\
& \quad + b_{11} m_{i,j+2,k-1,\ell} + b_{12} m_{i,j+1,k-1,\ell+2} + b_{13} m_{i+1,j,k,\ell-2} \\
& \quad + b_{14} m_{i-1,j+1,k-1,\ell+1} + b_{15} m_{i,j+1,k,\ell+1} + b_{16} m_{i+1,j+1,k,\ell}] \\
& + \ell[-2\zeta_2 r m_{i,j,k,\ell} - r^2 m_{i+1,j,k,\ell-1} + q_1 m_{i+3,j,k,\ell-1} + q_2 m_{i,j+3,k,\ell-1} \\
& \quad + q_3 m_{i+2,j+1,k,\ell-1} + q_4 m_{i+1,j+2,k,\ell-1} + q_5 m_{i+2,j,k+1,\ell-1} \\
& \quad + q_6 m_{i+1,j,k+2,\ell-1} + q_7 m_{i+2,j,k,\ell} + q_8 m_{i+1,j,k,\ell+1} + q_9 m_{i,j+2,k+1,\ell-1} \\
& \quad + q_{10} m_{i,j+1,k+2,\ell-1} + q_{11} m_{i,j+2,k,\ell} + q_{12} m_{i,j+1,k,\ell+1} + q_{13} m_{i+1,j,k+1,\ell} \\
& \quad + q_{14} m_{i+1,j+1,k,\ell} + q_{15} m_{i,j+1,k+1,\ell} + q_{16} m_{i+1,j+1,k+1,\ell-1}] + k(k-1) \\
& \times [\varepsilon^2 D_1 (B_{11}^2 m_{i+2,j,k-2,\ell} + 2B_{11}B_{12} m_{i+1,j+1,k-2,\ell} + B_{12}^2 m_{i,j+2,k-2,\ell}) \\
& \quad + s_1^2 D_2 m_{i,j,k-2,\ell} + \varepsilon^2 D_2 (f_{11} m_{i+2,j,k-2,\ell} + f_{12} m_{i+1,j+1,k-2,\ell} + f_{22} m_{i,j+2,k-2,\ell})] \\
& + k\ell[D_1 (B_{11}B_{21} m_{i+2,j,k-1,\ell-1} + (B_{11}B_{22} \\
& \quad + B_{12}B_{21}) m_{i+1,j+1,k-1,\ell-1} + B_{12}B_{22} m_{i,j+2,k-1,\ell-1}) + s_1 s_2 D_2 m_{i,j,k-1,\ell-1} \\
& \quad + \varepsilon^2 D_2 (e_{11} m_{i+2,j,k-1,\ell-1} + e_{12} m_{i+1,j+1,k-1,\ell-1} + e_{22} m_{i,j+2,k-1,\ell-1})] \\
& + \ell(\ell-1) \\
& \times [\varepsilon^2 D_1 (B_{21}^2 m_{i+2,j,k,\ell-2} + 2B_{21}B_{22} m_{i+1,j+1,k,\ell-2} + B_{22}^2 m_{i,j+2,k,\ell-2}) \\
& \quad + s_2^2 D_2 m_{i,j,k,\ell-2} + \varepsilon^2 D_2 (v_{11} m_{i+2,j,k,\ell-2} + v_{21} m_{i+1,j+1,k,\ell-2} \\
& \quad + v_{22} m_{i,j+2,k,\ell-2})]
\end{aligned}
\tag{10.75}
$$

where $m_{ijk\ell} = E[X_1^i X_2^j X_3^k X_4^\ell]$. The constants b_i, q_i, ... depend on the coefficients of equations (10.72). It can be observed that a moment equation of order n contains moments of order $n+1$ and $n+2$; thus equation (10.75) constitutes a set of infinite coupled equations. These equations can be truncated by using Gaussian or non-Gaussian closure schemes (see, e.g., Ibrahim, 1985).

Gaussian closure solution

The Gaussian closure solution is determined by generating the dynamic moment equations up to the second order from equation (10.75). Four equations are obtained for first-order moments and 10 equations for second-order moments. These equations contain higher-order moments up to fourth order. Based on the assumption that the nonlinear coupling is weak, the infinite hierarchy may be closed by assuming the response to be "nearly" Gaussian distributed, and thus third- and higher-order cumulants κ_i, $i \geq 3$, can be set to zero. In terms of joint moments, the third- and fourth-order cumulants can be expressed as follows:

$$
\kappa_3[X_i X_j X_k] = E[X_i X_j X_k] - \sum^{3} E[X_i]E[X_j X_k] + 2E[X_i]E[X_j]E[X_k] = 0 \tag{10.76a}
$$

$$\kappa_4[X_iX_jX_kX_\ell] = E[X_iX_jX_kX_\ell] - \sum^{4} E[X_i]E[X_jX_kX_\ell] - \sum^{4} E[X_iX_j]E[X_kX_\ell] + 2\sum^{6} E[X_i]E[X_j]E[X_kX_\ell] - 6E[X_i]E[X_j]E[X_k]E[X_\ell] = 0 \quad (10.76b)$$

where the number above the summation sign denotes the number of terms obtained in the expansion of the term following the summation without allowing permutation of indices. It is evident from equations (10.76) that any third- or fourth-order moment can be expressed in terms of first- and second-order moments. This procedure closes the infinite hierarchy and the 14 differential equations become consistent.

These equations are then solved numerically and Figures 10.20 and 10.21 show the time evolutions of the displacement mean square responses for two different values of the internal detuning parameter $r = \omega_2/\omega_1 = 0.31$ and 0.33, respectively. In both cases, after a sufficient time, $\tau > 800$, the response fluctuates between two limits, indicating that the system does not achieve a stationary response. It is seen that the degree of fluctuations is very high as the system approaches the critical internal resonance condition $r = 0.33$. For $r = 0.31$, that is, when the system is remote from the exact internal resonance, the degree of autoparametric interaction is less pronounced. The autoparametric interaction takes the form of energy exchange between the two modes during the steady state period. The detailed sections of the time histories reveal that the mean square of the first mode oscillates at a frequency three times the frequency of the second mode.

Non-Gaussian closure solution

If the system response is not Gaussian, which is true for nonlinear systems subjected to Gaussian excitation, third- and higher-order cumulants do not vanish. In fact, the deviation of the response from normality can be determined if these cumulants are known. Higher-order cumulants contribute very little to the Edgeworth expansion of the probability density, especially if the distribution does not deviate significantly from Gaussian (Ibrahim, 1985). Inclusion of third- and fourth-order cumulants alone has been shown to be sufficient to meet convergence requirements (Wu and Lin, 1984 and Ibrahim, *et al.*, 1985). It should also be noted that the inclusion of any higher-order cumulant increases the degree of nonlinearity of the moment differential equations and this will require more computational time.

Equation (10.75) is used to generate dynamic moment equations up to order 4. These equations are coupled through fifth- and sixth-order moment terms and can be closed by setting cumulants of order 5 and 6 to zero, as follows:

$$\begin{aligned}\kappa_5[X_iX_jX_kX_\ell X_m] &= E[X_iX_jX_kX_\ell X_m] - \sum^{5} E[X_i]E[X_jX_kX_\ell X_m] \\ &\quad + 2\sum^{10} E[X_i]E[X_j]E[X_kX_\ell X_m] + 2\sum^{15} E[X_iX_j]E[X_kX_\ell]E[X_m] \\ &\quad - 6\sum^{10} E[X_iX_j]E[X_k]E[X_\ell]E[X_m] - \sum^{10} E[X_iX_j]E[X_kX_\ell X_m] \\ &\quad + 24E[X_i]E[X_j]E[X_k]E[X_\ell]E[X_m] = 0 \end{aligned} \quad (10.77a)$$

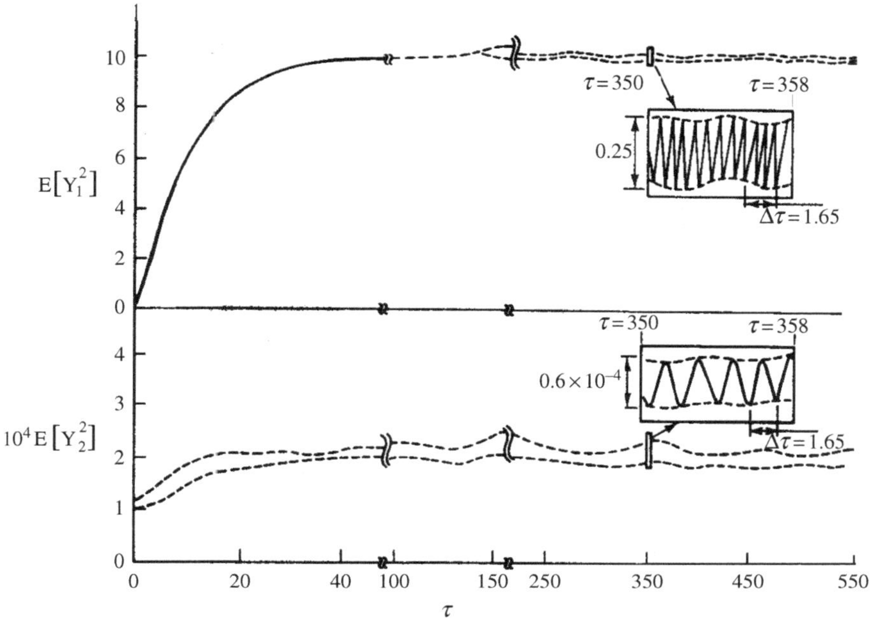

Figure 10.20 Gaussian time history records of the normal mode mean square displacements for $r = 0.31$, $\varepsilon = 0.02$, $\zeta_{1,2} = 0.02$. (Soundararajan and Ibrahim, 1988)

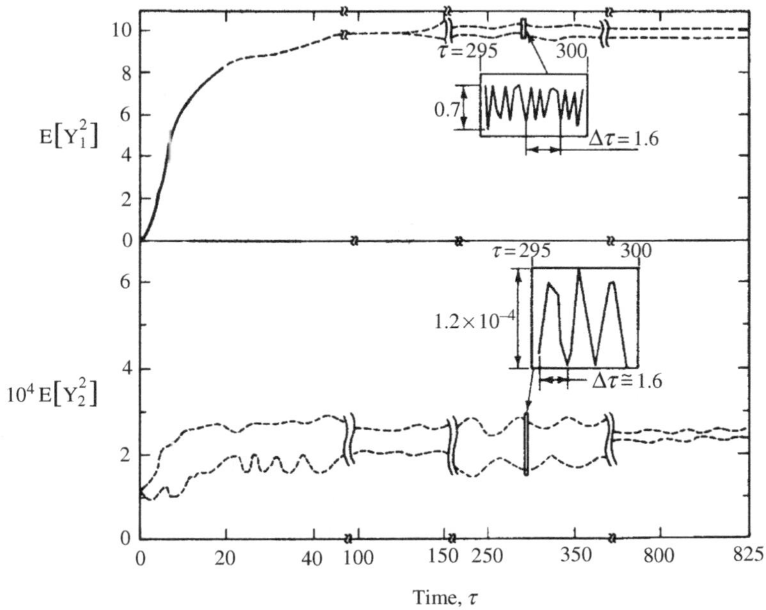

Figure 10.21 Gaussian time history records of the normal mode mean square displacements for $r = 0.33$, $\varepsilon = 0.02$, $\zeta_{1,2} = 0.02$. (Soundararajan and Ibrahim, 1988)

$$\kappa_6[X_iX_jX_kX_\ell X_mX_n] = E[X_iX_jX_kX_\ell X_mX_n] - \sum^{6} E[X_i]E[X_jX_kX_\ell X_mX_n]$$
$$+ 2\sum^{15} E[X_i]E[X_j]E[X_kX_\ell X_mX_n] - \sum^{15} E[X_iX_j]E[X_kX_\ell X_mX_n]$$
$$+ 2\sum^{60} E[X_i]E[X_jX_k]E[X_\ell X_mX_n] - \sum^{10} E[X_iX_jX_k]E[X_\ell X_mX_n]$$
$$- 6\sum^{20} E[X_iX_jX_k]E[X_\ell]E[X_m]E[X_n]$$
$$+ 2\sum^{15} E[X_iX_j]E[X_kX_\ell]E[X_mX_n]$$
$$+ 24\sum^{15} E[X_iX_j]E[X_k]E[X_\ell]E[X_m]E[X_n]$$
$$- 6\sum^{45} E[X_iX_j]E[X_kX_\ell]E[X_m]E[X_n]$$
$$- 120E[X_i]E[X_j]E[X_k]E[X_\ell]E[X_m]E[X_n] = 0 \qquad (10.77b)$$

These relations establish the first-order non-Gaussian closure and thus one can express fifth- and sixth-order moments in terms of fourth- and lower-order moments. The total number of differential equations is 69. These equations consist of four equations of first-order moments, 10 equations for second-order, 20 equations for third-order, and 35 equations for fourth-order. The closed differential equations are solved numerically and Figures 10.22 and 10.23 show the time evolution of the mean square responses for two different values of internal resonance ratios $r = \omega_2/\omega_1 = 0.31$ and 0.33, respectively. In both cases, the mean square responses achieve a stationary state. At $r = 0$, the second mode exhibits a fluctuation between $\tau = 100$ and 250, and then becomes strictly stationary. At $r = 0.31$ this fluctuation disappears.

Figures 10.24–10.30 demonstrate the influence of internal detuning parameter on the mean square responses for various system parameters. The mean square responses are normalized with respect to the linear solution of the moment equations. The upper and lower limits of the mean squares as obtained by Gaussian closure are represented by dash curves. For very small values of the coupling parameter ε, the Gaussian and non-Gaussian closure solutions do not depart significantly from the linear solution. This can be observed in Figures 10.24 and 10.25 for $\varepsilon = 0.005$, and damping ratios $\zeta_1 = \zeta_2 = 0.02$ and 0.05, respectively. Figures 10.26–10.28 reveal three different regions of response regimes; (i) autoparametric interaction, (ii) unstable solution, and (iii) linear motion. The Gaussian and non-Gaussian solutions approach the linear solution when the system is remote from the internal resonance, see, for example, Figure 10.27.

The region of autoparametric resonance, region (i), is restricted between some limits bracketing the internal resonance $r = 0.33$. Below a lower limit $r = 0.33 - O(\varepsilon)$ and above an upper limit $r = 0.33 + O(\varepsilon)$, regions (ii), the numerical integration of both solutions does not converge but grows monotonically with time. These limits define the unstable regions, which are mainly governed by the system damping ratios and the nonlinear coupling parameter. Outside these

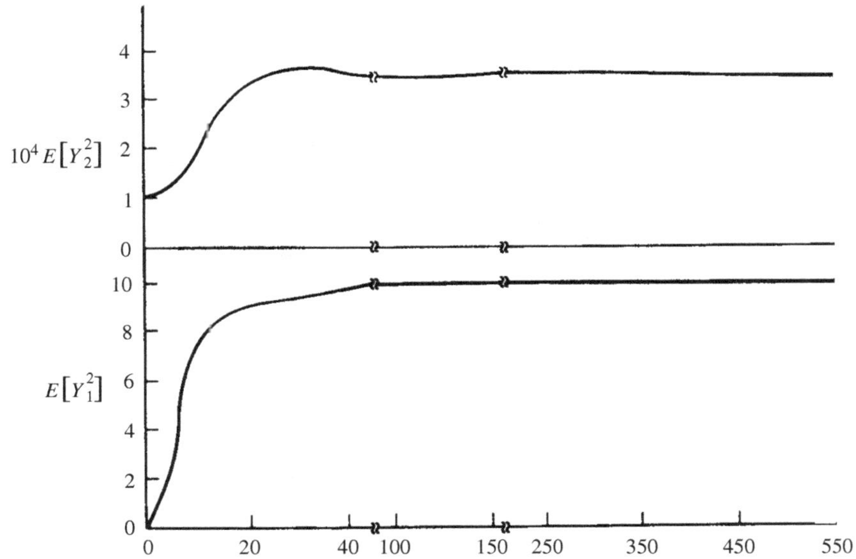

Figure 10.22 Non-Gaussian time history of the normal mode mean square displacements for $r = 0.31$, $\varepsilon = 0.02$, $\zeta_{1,2} = 0.02$. (Soundararajan and Ibrahim, 1988)

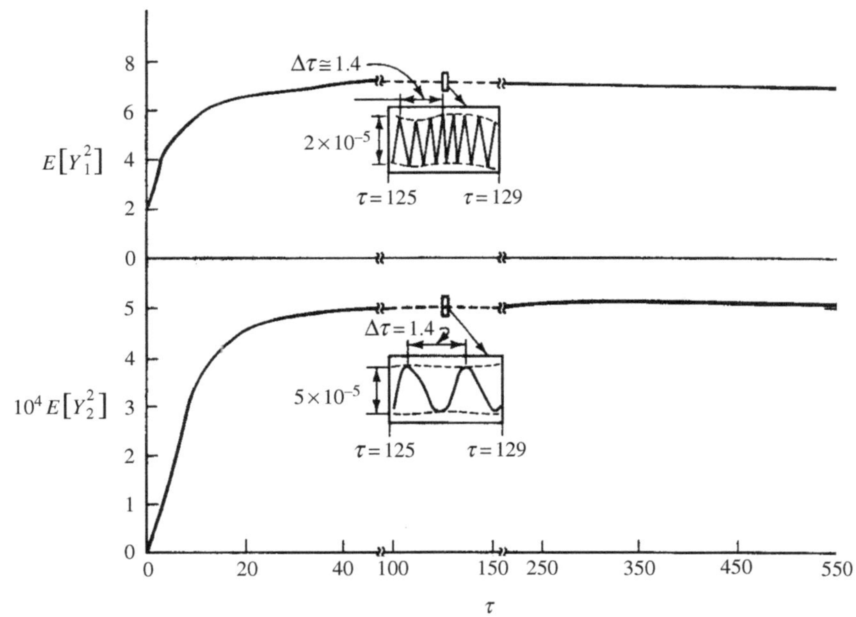

Figure 10.23 Non-Gaussian time history of the normal mode mean square displacements for $r = 0.33$, $\varepsilon = 0.02$, $\zeta_{1,2} = 0.02$. (Soundararajan and Ibrahim, 1988)

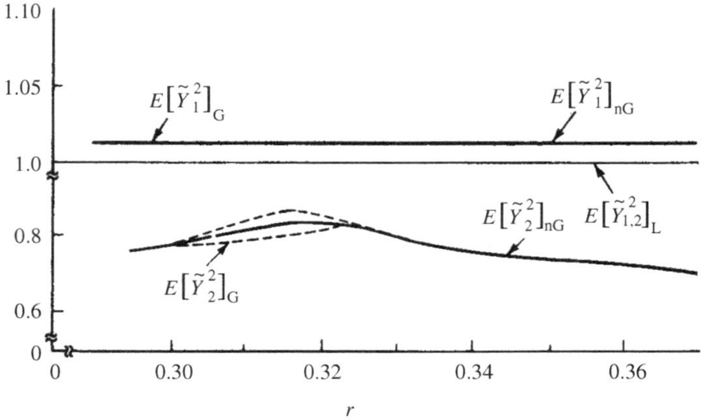

Figure 10.24 Dependence of normalized Gaussian and non-Gaussian mean square displacements on the frequency ratio $r=\omega_2/\omega_1$, $\varepsilon=0.005$, $\zeta_{1,2}=0.02$. (Soundararajan and Ibrahim, 1988)

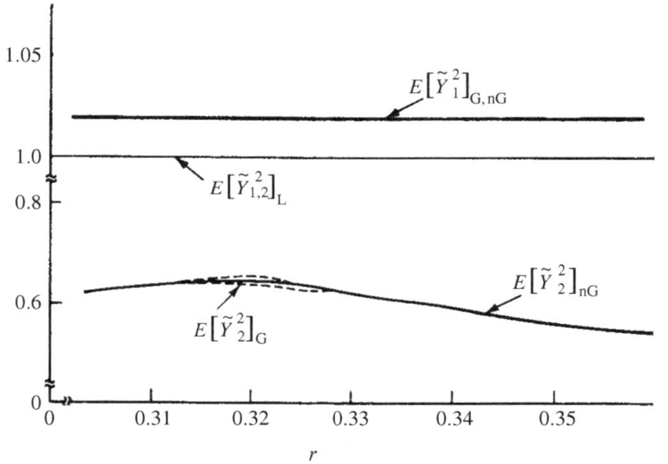

Figure 10.25 Dependence of normalized Gaussian and non-Gaussian mean square displacements on the frequency ratio $r=\omega_2/\omega_1$, $\varepsilon=0.005$, $\zeta_{1,2}=0.05$. (Soundararajan and Ibrahim, 1988)

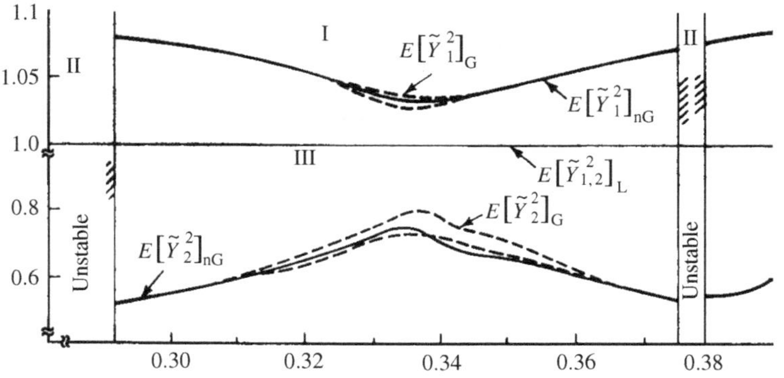

Figure 10.26 Dependence of normalized Gaussian and non-Gaussian mean square displacements on the frequency ratio $r=\omega_2/\omega_1$, $\varepsilon=0.01$, $\zeta_{1,2}=0.02$. (Soundararajan and Ibrahim, 1988)

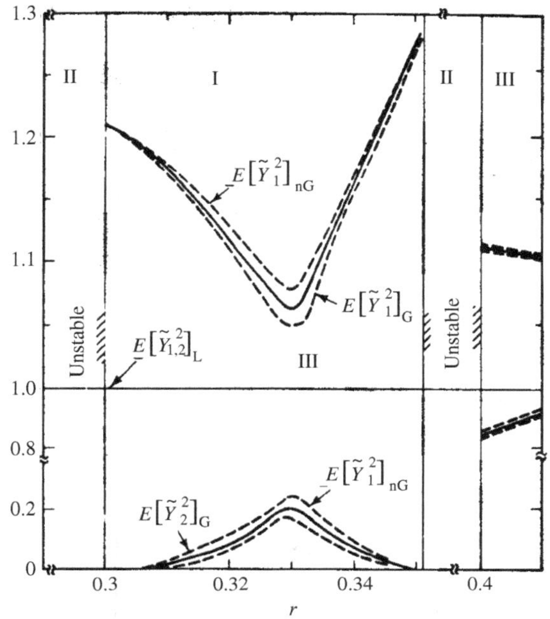

Figure 10.27 Dependence of normalized Gaussian and non-Gaussian mean square displacements on the frequency ratio $r = \omega_2/\omega_1$, $\varepsilon = 0.02$, $\zeta_{1,2} = 0.02$. (Soundararajan and Ibrahim, 1988)

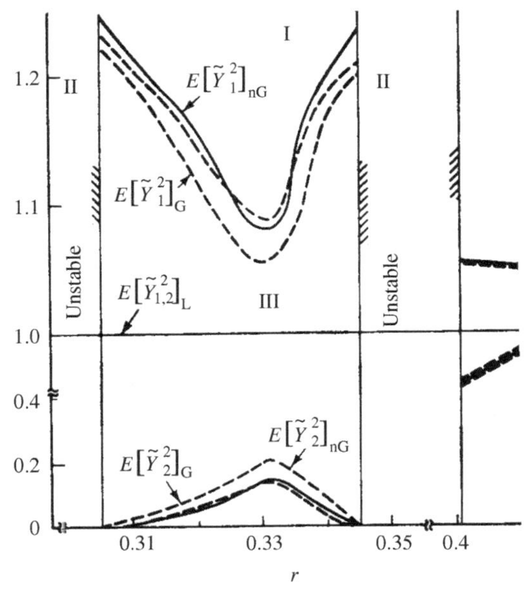

Figure 10.28 Dependence of normalized Gaussian and non-Gaussian mean square displacements on the frequency ratio $r = \omega_1 \varepsilon = 0.025$, $\zeta_{1,2} = 0.02$. (Soundararajan and Ibrahim, 1988)

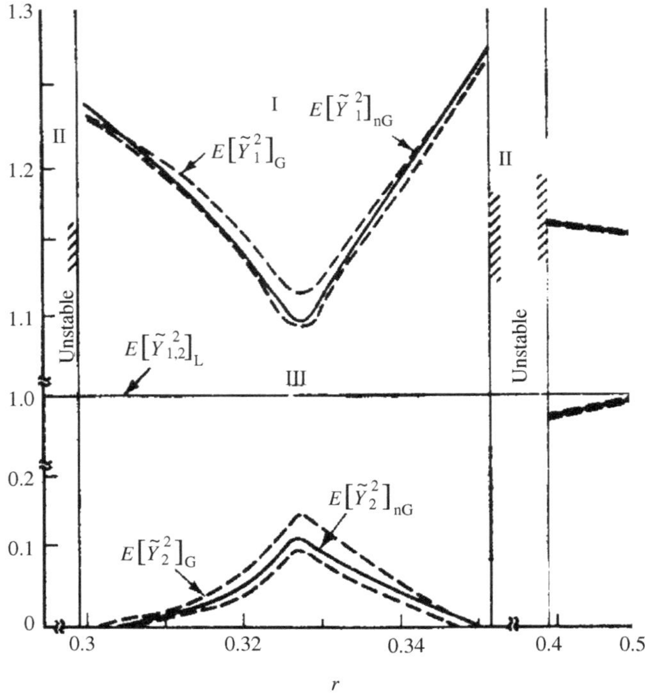

Figure 10.29 Dependence of normalized Gaussian and non-Gaussian mean square displacements on the frequency ratio $r=\omega_2/\omega_1$, $\varepsilon=0.02$, $\zeta_{1,2}=0.025$. (Soundararajan and Ibrahim, 1988)

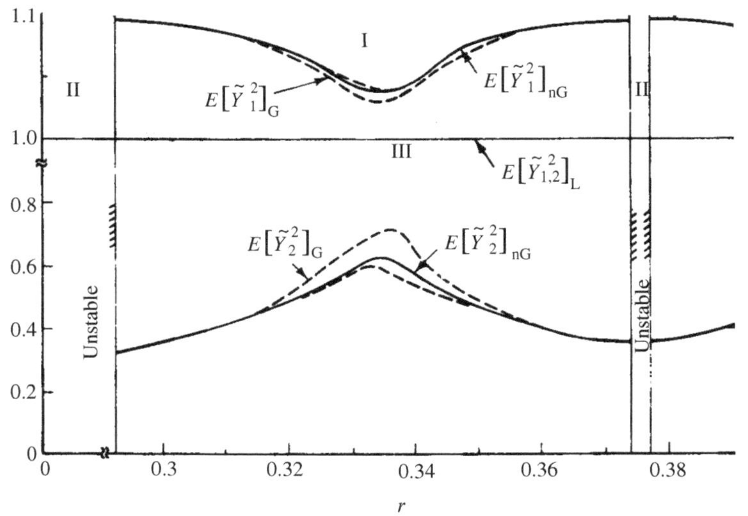

Figure 10.30 Dependence of normalized Gaussian and non-Gaussian mean square displacements on the frequency ratio $r=\omega_2/\omega_1$, $\varepsilon=0.01$, $\zeta_{1,2}=0.05$. (Soundararajan and Ibrahim, 1988)

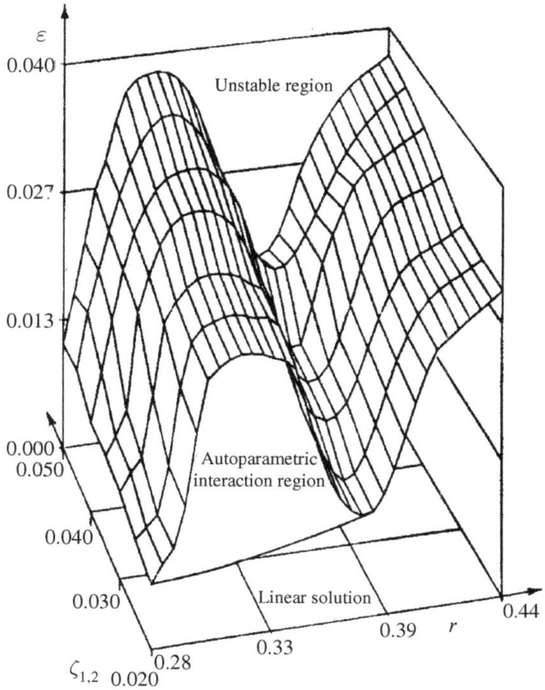

Figure 10.31 Stability boundaries for Gaussian closure solution ($D_1 = 0.06$, $D_2 = 0.4$). (Soundararajan and Ibrahim, 1988)

regions, (region iii), the system response approaches the linear equilibrium state governed by the horizontal acceleration. The boundary of instability regions is plotted on the ε, ζ, r space as shown in Figure 10.31. The instability region is found to increase with the system nonlinearity parameter, ε, and decrease with damping ratios.

10.5.2 Vertical random excitation

The case of vertical random excitation was considered by Ikeda and Ibrahim (2002). Figure 10.32 shows a rigid circular cylindrical tank supported on an elastic structure represented by a mass m, spring stiffness k, and dashpot with damping coefficient c. The tank is partially filled with liquid to a depth h. A moving rectangular coordinate frame (x, y, z) and a circular cylindrical coordinate (r, θ, z), are fixed to the liquid undisturbed free surface. The xy plane coincides with the equilibrium position of the liquid surface. The vertical displacement of the mass m, measured from its equilibrium position when the tank is partially filled is Z_0. The vertical displacement of the liquid surface is $\eta\ (r, \theta, t)$ measured from the undisturbed free surface. In terms of the velocity potential function, $\Phi(r, \theta, z, t)$, the liquid motion inside the tank is governed by the continuity equation (Laplace's equation):

$$\frac{\partial^2 \Phi}{\partial r^2} + \frac{1}{r}\frac{\partial \Phi}{\partial r} + \frac{1}{r^2}\frac{\partial^2 \Phi}{\partial \theta^2} + \frac{\partial^2 \Phi}{\partial z^2} = 0. \tag{10.78}$$

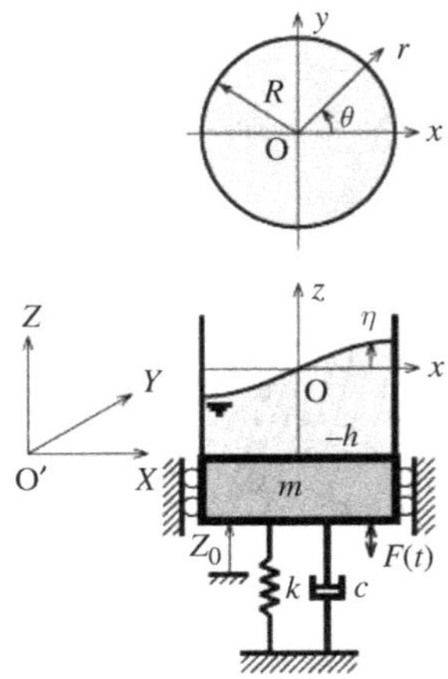

Figure 10.32 Schematic diagram of a partially filled cylindrical tank supported on elastic structure.

The hydrodynamic pressure $P(r, \theta, z, t)$ is determined from Bernoulli's equation

$$\frac{\partial \Phi}{\partial t} + \frac{1}{2}\left\{\left(\frac{\partial \Phi}{\partial r}\right)^2 + \frac{1}{r^2}\left(\frac{\partial \Phi}{\partial \theta}\right)^2 + \left(\frac{\partial \Phi}{\partial z}\right)^2\right\} + gz + \frac{P}{\rho} = -\ddot{Z}_0 z, \tag{10.79}$$

where a dot denotes differentiation with respect to time t.

The equation of motion of the structure subjected to vertical random excitation, $F(t)$, is

$$m\ddot{Z}_0 + c\dot{Z}_0 + kZ_0 = F_l + F(t), \tag{10.80}$$

where F_l is the fluid hydrodynamic force acting in the z-direction on the bottom of the tank, and m_l is the liquid mass. The fluid force F_l is given by integrating the fluid pressure at the bottom of the tank as follows:

$$F_l = -\int_0^{2\pi}\int_0^R r\,P(r, \theta, z, t)\big|_{z=-h}\,dr\,d\theta, \tag{10.81}$$

where $P(r, \theta, z, t)$ is determined from equation (10.79).

The random excitation, $F(t)$, is assumed to be generated from the following linear shaping filter

$$\frac{d^2F}{dt^2} + \gamma_0\frac{dF}{dt} + \Omega_0^2 F = \Omega_0\,W(t) \tag{10.82}$$

where γ_0 is the filter bandwidth, Ω_0 is the filter center frequency, $W(t)$ is a zero-mean Gaussian white noise possessing constant power spectral density D.

The following parameters are introduced

$$z_C = Z_0/R, \quad \bar{\eta} = \eta/R, \quad \bar{r} = r/R, \quad \bar{z} = z/R, \quad \bar{h} = h/R, \quad \mu_1 = m/M,$$

$$\mu_2 = m_l/(\pi M\bar{h}), \quad \phi = \Phi/(R^2\omega_{11}), \quad \bar{k} = k/(M\omega_{11}^2), \quad \zeta = c/(M\omega_{11}),$$

$$f = F/(MR\omega_{11}^2), \quad w(\tau) = W(t)/(MR\omega_{11}^3), \quad f_l = F_l/(MR\omega_{11}^2), \quad p = P/(\rho R^2\omega_{11}^2),$$

$$\tau = \omega_{11}t, \quad \xi_{mn} = \lambda_{mn}R, \quad \Omega = \Omega_0/\omega_{11}, \quad \gamma = \gamma_0/\omega_{11}, \quad M = m + m_l,$$

$$\omega_{11} = \sqrt{g\lambda_{11}\tanh(\lambda_{11}h)}, \quad \psi_{11} = \xi_{11}\tanh(\xi_{11}h/R), \quad \bar{\omega}_s = \sqrt{k/M}/\omega_{11} \tag{10.83}$$

where λ_{mn} is the nth positive root of $(d/dr)\{\mathrm{J}_m(\lambda r)\}\,|_{r=R} = 0$. The subscript (m, n) refers to m nodal lines and the nth-order to a nodal concentric circle. In terms of the new parameters (10.83), equations (10.78)–(10.80), and (10.82) are:

$$\frac{\partial^2\phi}{\partial\bar{r}^2} + \frac{1}{\bar{r}}\frac{\partial\phi}{\partial\bar{r}} + \frac{1}{\bar{r}^2}\frac{\partial^2\phi}{\partial\theta^2} + \frac{\partial^2\phi}{\partial\bar{z}^2} = 0 \tag{10.84}$$

$$\frac{\partial\phi}{\partial\tau} + \frac{1}{2}\left\{\left(\frac{\partial\phi}{\partial\bar{r}}\right)^2 + \frac{1}{\bar{r}^2}\left(\frac{\partial\phi}{\partial\theta}\right)^2 + \left(\frac{\partial\phi}{\partial\bar{z}}\right)^2\right\} + \frac{1}{\psi_{11}}\bar{z} + p = -\frac{d^2z_0}{d\tau^2}\bar{z} \tag{10.85}$$

$$\mu_1\frac{d^2z_0}{d\tau^2} + \zeta\frac{dz_0}{d\tau} + \bar{k}z_0 = f_l + f(\tau) + \frac{\pi\mu_2\bar{h}}{\psi_{11}}, \tag{10.86}$$

$$\frac{d^2f}{d\tau^2} + \gamma\frac{df}{d\tau} + \Omega^2 f = \Omega w(\tau), \tag{10.87}$$

where

$$f_l = -\mu_2\int_0^{2\pi}\int_0^1 \bar{r}p(\bar{r},\theta,\bar{z},\tau)|_{\bar{z}=-\bar{h}}\,d\bar{r}\,d\theta, \tag{10.88}$$

and $w(\tau)$ is a zero-mean stationary Gaussian white noise process with variance σ_w^2 and constant power spectral density intensity S_0. The fluid field equations are completed by including the boundary conditions at the tank walls and bottom together with the kinematic and dynamic free-surface boundary condition. The dynamic boundary condition is coupled with the structure equation of motion (10.86) and its filter equation (10.87).

The solution of Laplace's equation (10.84) that satisfies the boundary conditions may be written in the form

$$\phi(\bar{r},\theta,\bar{z},\tau) = a_c(\tau) + \sum_{m=0}^{\infty}\sum_{n=1}^{\infty}\{a_{mn}(\tau)\cos m\theta + b_{mn}(\tau)\sin m\theta\} \times \mathrm{J}_m(\xi_{mn}\bar{r})\frac{\cosh\{\xi_{mn}(\bar{z}+\bar{h})\}}{\cosh(\xi_{mn}\bar{h})} \tag{10.89}$$

The free-surface elevation may also be written in the form

$$\bar{\eta}(\bar{r},\theta,\tau) = \sum_{m=0}^{\infty}\sum_{n=1}^{\infty}\{c_{mn}(\tau)\cos m\theta + d_{mn}(\tau)\sin m\theta\}J_m(\xi_{mn}\bar{r}) \tag{10.90}$$

where a_c, a_{mn}, b_{mn}, c_{mn}, and d_{mn} are the generalized coordinates. The term $a_c(\tau)$, in equation (10.89), is very important because it is associated with the fluid force.

In the present analysis, the interaction of the three sloshing modes (0, 1), (1, 1) and (2, 1) with the structure dynamics is considered. The first anti-symmetric sloshing mode, (1, 1), has two different modal amplitudes c_{11} and d_{11} whose nodal diameters are perpendicular to each other. In the nonlinear analysis, these modes are coupled.

A small parameter, ε, is introduced and the following orders of the amplitudes and the damping coefficient ζ are adopted

$$a_{11}, b_{11}, c_{11}, d_{11}, z_0; \zeta \approx O(\varepsilon^{1/3}) \tag{10.91a}$$

$$a_c,\ a_{01},\ c_{01},\ a_{21},\ b_{21},\ c_{21},\ d_{21} \approx O(\varepsilon^{2/3}) \tag{10.91b}$$

$$a_{31}, b_{31}, c_{31}, d_{31} \approx O(\varepsilon^{3/3}) \tag{10.91c}$$

The orders of all other amplitudes, which are not shown in relations (10.91), are assumed to be smaller than $O(\varepsilon)$. The procedure begins by considering $\bar{\eta}$ to be small, and expanding the kinematic and dynamic boundary conditions about $\bar{\eta}=0$, and substituting equations (10.89) and (10.90). By expanding the result into a Fourier–Bessel series in terms of $J_0(\xi_{01}\bar{r})$, $J_m(\xi_{m1}\bar{r})\cos m\theta$ and $J_m(\xi_{m1}\bar{r})\sin m\theta$ $(m=1,2)$, and equating the constant terms and the coefficients of $J_0(\xi_{01}\bar{r})$, $J_m(\xi_{m1}\bar{r})\cos m\theta$ and $J_m(\xi_{m1}\bar{r})\sin m\theta$ on both sides of these equations, gives the modal equations as follows.

The equation of motion of the structure:

$$Q_1\ddot{z}_0 + \zeta\dot{z}_0 + \bar{k}z_0 + Q_2\dot{a}_c + G_1(a_{11}, b_{11}) = f(\tau) \tag{10.92a}$$

Sloshing modal equations of the first anti-symmetric mode:

$$\dot{a}_{11} + ((1/\psi_{11}) + \ddot{z}_0)\,c_{11} + G_2(\dot{a}_{i1}, \dot{b}_{j1}, a_{i1}, b_{j1}, c_{i1}, d_{j1}) = 0 \tag{10.92b}$$

$$\dot{b}_{11} + ((1/\psi_{11}) + \ddot{z}_0)\,d_{11} + G_3(\dot{a}_{i1}, \dot{b}_{j1}, a_{i1}, b_{j1}, c_{i1}, d_{j1}) = 0 \tag{10.92c}$$

$$\dot{c}_{11} - \psi_{11}a_{11} + G_4(a_{i1}, b_{j1}, c_{i1}, d_{j1}) = 0 \tag{10.92d}$$

$$\dot{d}_{11} - \psi_{11}b_{11} + G_5(a_{i1}, b_{j1}, c_{i1}, d_{j1}) = 0 \tag{10.92e}$$

$$\dot{a}_c + G_6(\dot{a}_{11}, \dot{b}_{11}, \dot{z}_0, a_{11}, b_{11}, c_{11}, d_{11}) = 0 \tag{10.92f}$$

Sloshing modal equations of the first symmetric mode:

$$\dot{a}_{01} + ((1/\psi_{11}) + \ddot{z}_0)\,c_{01} + G_7(\dot{a}_{11}, \dot{b}_{11}, a_{11}, b_{11}, c_{11}, d_{11}) = 0 \tag{10.92g}$$

$$\dot{c}_{01} - \psi_{01} a_{01} + G_8(a_{11}, b_{11}, c_{11}, d_{11}) = 0 \tag{10.92h}$$

Sloshing modal equations of the second anti-symmetric mode:

$$\dot{a}_{21} + ((1/\psi_{11}) + \ddot{z}_0)\, c_{21} + G_9(\dot{a}_{11}, \dot{b}_{11}, a_{11}, b_{11}, c_{11}, d_{11}) = 0 \tag{10.92i}$$

$$\dot{b}_{21} + ((1/\psi_{11}) + \ddot{z}_0)\, d_{21} + G_{10}(\dot{a}_{11}, \dot{b}_{11}, a_{11}, b_{11}, c_{11}, d_{11}) = 0 \tag{10.92j}$$

$$\dot{c}_{21} - \psi_{21} a_{21} + G_{11}(a_{11}, b_{11}, c_{11}, d_{11}) = 0 \tag{10.92k}$$

$$\dot{d}_{21} - \psi_{21} b_{21} + G_{12}(a_{11}, b_{11}, c_{11}, d_{11}) = 0 \tag{10.92l}$$

where a dot denotes differentiation with respect to the nondimensional time parameter τ, $\psi_{m1} = \xi_{m1} \tanh(\xi_{m1}\bar{h})$, $(m = 0, 1, 2)$, $Q_1 = \mu_1 + \pi\mu_2\bar{h}$, $Q_2 = -\pi\mu_2$, and G_k $(k = 1, 2, \ldots, 12)$ represent the nonlinear terms. Equation (10.92a) is obtained from the structure equation of motion (10.86). Eliminating a_c from equations (10.92a) and (10.92f), gives

$$Q_1\ddot{z}_0 + \zeta\dot{z}_0 + \bar{k} z_0 + G_{13}(\dot{a}_{11}, \dot{b}_{11}, a_{11}, b_{11}, c_{11}, d_{11}) = f(\tau), \tag{10.93}$$

where G_{13} contains nonlinear terms. Eliminating the variables a_{11}, b_{11}, a_{01}, a_{21} and b_{21} from equations (10.92) and (10.93) and considering the modal ordering assumption (10.91), gives: The structure equation of motion:

$$Q_1\ddot{z}_0 + \zeta\dot{z}_0 + \bar{k} z_0 + H_1(\ddot{z}_0, \dot{c}_{11}, \dot{d}_{11}, c_{11}, d_{11}) = f(\tau) \tag{10.94a}$$

Equations of motion of first anti-symmetric sloshing mode:

$$\ddot{c}_{11} + 2\zeta_{11}\dot{c}_{11} + (1 + \psi_{11}\ddot{z}_0)c_{11} + H_2(\dot{c}_{i1}, \dot{d}_{j1}, c_{i1}, d_{j1}) = 0 \tag{10.94b}$$

$$\ddot{d}_{11} + 2\zeta_{11}\dot{d}_{11} + (1 + \psi_{11}\ddot{z}_0)d_{11} + H_3(\dot{c}_{i1}, \dot{d}_{j1}, c_{i1}, d_{j1}) = 0 \tag{10.94c}$$

Equation of motion of first symmetric sloshing mode:

$$\ddot{c}_{01} + 2\zeta_{01}\omega_{01}\dot{c}_{01} + \omega_{01}^2(1 + \psi_{11}\ddot{z}_0)c_{01} + H_4(\ddot{z}_0, \dot{c}_{11}, \dot{d}_{11}, c_{11}, d_{11}) = 0 \tag{10.94d}$$

Equations of motion of second anti-symmetric sloshing mode:

$$\ddot{c}_{21} + 2\zeta_{21}\omega_{21}\dot{c}_{21} + \omega_{21}^2(1 + \psi_{11}\ddot{z}_0)c_{21} + H_5(\ddot{z}_0, \dot{c}_{11}, \dot{d}_{11}, c_{11}, d_{11}) = 0 \tag{10.94e}$$

$$\ddot{d}_{21} + 2\zeta_{21}\omega_{21}\dot{d}_{21} + \omega_{21}^2(1 + \psi_{11}\ddot{z}_0)d_{21} + H_6(\ddot{z}_0, \dot{c}_{11}, \dot{d}_{11}, c_{11}, d_{11}) = 0 \tag{10.94f}$$

where $\omega_{mn}^2 = \psi_{mn}/\psi_{11}$, and $H_l(l = 1, \ldots, 6)$ represent the nonlinear terms. Viscous damping terms are added in equations (10.94b)–(10.94f) in order to account for energy dissipation, ζ_{mn}, is the damping ratio corresponding to the sloshing mode (m, n). Equations (10.94) involve quadratic and cubic inertia nonlinear coupling terms that give rise to the occurrence of internal resonance. From the nonlinear coupling terms in equations (10.94), it is not difficult to show that principal internal resonance can take place if the natural frequency of the structure $\bar{\omega}_s$ is close to twice the natural frequency of the first anti-symmetric sloshing mode.

Consider first the mean square response of the structure with the liquid as a frozen block. This single freedom response is useful in verifying the accuracy of Monte Carlo simulation and will be used to normalize the response of the coupled system.

$$Q_1\ddot{z}_0 + \zeta\dot{z}_0 + \bar{k}z_0 = f(\tau) \tag{10.95}$$

where $f(\tau)$ is defined by equation (10.88). The autocorrelation function $R_{z_0}(\tau')$ of $z_0(\tau)$ is

$$R_{z_0}(\tau') = \int_{-\infty}^{\infty} |H(j\omega)|^2 S_w(\omega) e^{j\omega\tau'}\, d\omega \tag{10.96}$$

where $H(j\omega)$ is the frequency response function of $z_0(\tau)$ to $w(\tau)$ in a complex form, which can be obtained from the transfer function

$$H(s) = \frac{\Omega}{s^2 + \gamma s + \Omega^2} \cdot \frac{1}{Q_1 s^2 + \zeta s + \bar{k}} \tag{10.97}$$

$S_w(\omega) = S_0$ (= constant) is the power spectral density of the white noise. Substituting $\tau' = 0$ into equation (10.96), gives the mean square response $E[z_0^2]$

$$E[z_0^2] = R_{z_0}(0) = S_0 \int_{-\infty}^{\infty} |H(j\omega)|^2 d\omega$$

$$= \frac{\left(\pi S_0\left\{(\zeta+\gamma Q_1)(Q_1\Omega^2+\zeta\gamma+\bar{k}) - Q_1(\gamma\bar{k}+\zeta\Omega^2)\right\}\right)}{\left(\bar{k}\left\{(\zeta+\gamma Q_1)(Q_1\Omega^2+\zeta\gamma+\bar{k})(\gamma k+\zeta\Omega^2) - \bar{k}\Omega^2(\zeta+\gamma Q_1)^2 - Q_1(\gamma\bar{k}+\zeta\Omega^2)^2\right\}\right)} \tag{10.98}$$

This solution is used to verify Monte Carlo simulation. The white noise $w(\tau)$ is generated from a random number series with zero mean value and variance $\sigma_w^2 = 2\pi S_0/\Delta\tau$ based on the Box and Muller (1958) method. In the simulation, the power spectral density $S_w(\omega)$ is defined by

$$S_w(\omega) = \begin{cases} S_0 & (-\omega_N \le \omega \le \omega_N), \\ 0 & (-\infty < \omega < -\omega_N, \omega_N < \omega < \infty) \end{cases}$$

where ω_N is the Nyquist frequency, which is given by $\omega_N = 2\pi/(2\Delta\tau) = \pi/\Delta\tau$, with $\Delta\tau = 0.25$.

The nonlinear modal interaction of the liquid free-surface motion and structure dynamics is considered using Monte Carlo simulation. The numerical simulation of equations (10.94) is carried out when the natural frequency, ω_s, of the structure is close to twice the natural frequency of the sloshing mode (1, 1), that is, $\omega_s\left(\equiv \sqrt{k/M}\right) \approx 2\omega_{11}$. Figure 10.33 shows a sample of excitation and response time history records for fluid depth ratio $\bar{h} = 1.2$, structure mass ratio $\mu_1 = 0.87$, liquid mass ratio $\mu_2 = 0.034$, structure stiffness parameter $\bar{k} = 4.0$, structure damping parameter $\zeta = 0.03$, excitation spectral density $S_0 = 2.0 \times 10^{-7}$, filter bandwidth $\gamma = 0.1$, filter center frequency ratio $\Omega = 2.0$, and fluid damping ratios $\zeta_{01} = \zeta_{11} = \zeta_{21} = 0.01$. The liquid elevation $\bar{\eta}(\bar{r},\theta)$ is estimated with reference to the position $(\bar{r},\theta) = (1, 0)$. The displacement $z_0(\tau)$ exhibits a narrow-band random process, while the time histories of c_{11} and d_{11} reflect irregular rotational motion of the nodal diameter. The rotation of the nodal diameter is measured by the angle β as defined in Figure 10.33(b). This angle is estimated by considering the (1, 1) sloshing mode and expressing the liquid elevation as follows:

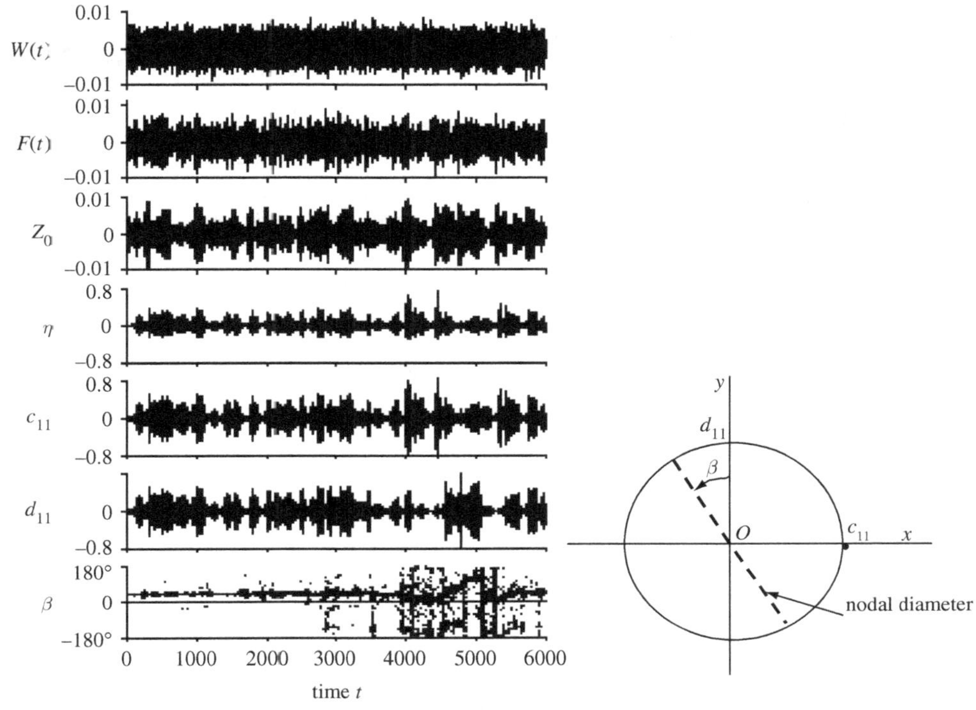

Figure 10.33 Time history response records of the coupled system for $h/R = 1.2$, $\nu_1 = 0.87$, $\nu_2 = 0.034$, $\bar{\kappa} = 4.0$, $\varepsilon = 0.03$, $\zeta_{01} = \zeta_{11} = \zeta_{21} = 0.01$, $S_0 = 2.0 \times 10^{-7}$, $\gamma = 0.1$, $\Omega = 2.0$, and $(r, \theta) = (1, 0)$. (Ikeda and Ibrahim, 2002)

$$\bar{\eta}(\bar{r}, \theta) = (c_{11}\cos\theta + d_{11}\sin\theta)J_1(\xi_{11}\bar{r}) = A\cos(\theta - \beta)J_1(\xi_{11}\bar{r}) \tag{10.99}$$

where $A = \sqrt{c_{11}^2 + d_{11}^2}$ and $\beta = \tan^{-1}(d_{11}/c_{11})$. The numerical simulation reveals that the angle β is irregular as shown by the discretized points in Figure 10.33.

Figures 10.34(a)–(c) shows three sets of time history records whose time axes are enlarged for the same parameters of Figure 10.33. Each figure also includes the corresponding trajectories on the (c_{11}, d_{11})-plane for three time intervals (a) $1375 \le \tau \le 1500$, (b) $4500 \le \tau \le 4625$ and (c) $5000 \le \tau \le 5125$, respectively. It is seen that β is close to $+135^\circ$ or -45° in Figure 10.34(a). This means that the nodal diameter maintains its orientation and does not rotate. On the other hand, Figures 10.34(b) and (c) reveals that β continuously varies with time. This means that the nodal diameter rotates counterclockwise then clockwise then switches to counterclockwise and so on in an irregular manner.

Figure 10.35 shows the response curves for the mean square response, $E[z_0^2]$, of the structure and the liquid mean square responses, $E[c_{11}^2]$ and $E[d_{11}^2]$, of the sloshing mode (1,1). The symbols • and ∘ represent the numerical results, which are calculated by Monte Carlo simulations using 100 different sets of random number series, while the dash-dot curve represents the theoretical result obtained from equation (10.98). In Figure 10.35 and subsequent figures, the statistics are estimated from 100 different sets of time histories over the time duration from

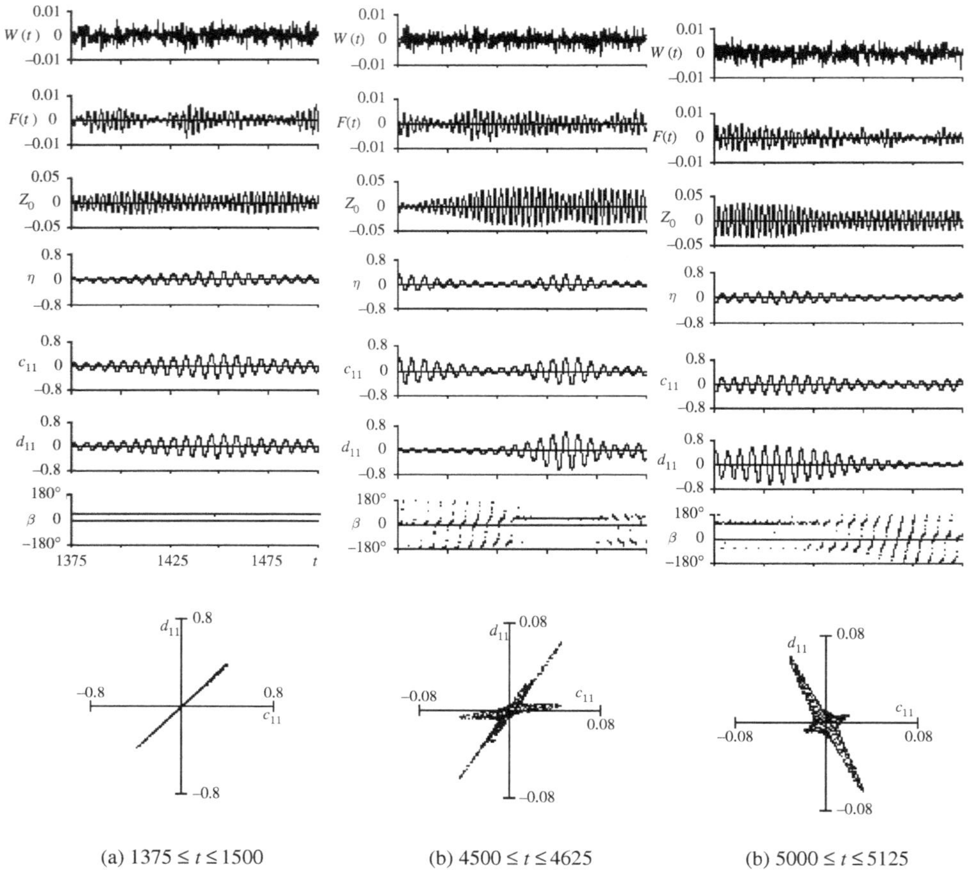

Figure 10.34 Magnifications of time histories of Figure 10.49 for different time intervals and the corresponding trajectories of (c_{11}, d_{11}). (Ikeda and Ibrahim, 2002)

$t = 1000$ to $t = 6000$ in order to eliminate the transient responses. Figures 35(a)–(c) shows the dependence of the mean square responses on the center frequency Ω of a narrow-band random excitation for three different values of bandwidth $\gamma = 0.1$, 0.2, and 0.3, respectively, and for the same parameters of Figure 10.33. Comparing these results with those of the uncoupled system reveals that the mean square response, $E[z_0^2]$, in the coupled system drops over a finite range of Ω, where the liquid-free-surface motion interacts with the structure through nonlinear coupling. As the filter bandwidth, γ, increases, the peak of $E[z_0^2]$, shown by solid circles, decreases and is associated with a shrinking in the range of Ω, over which the interaction with liquid motion takes place.

Figure 10.36 shows the dependence of the normalized mean square of the structure $E[z_0^2]_c/E[z_0^2]_f$ on the center frequency for various values of bandwidth, γ, where the subscript "c" denotes coupling, and "f" denotes frozen. This normalized representation provides direct information regarding the degree of nonlinear coupling and the energy transfer between the structure and liquid surface motions. It can be seen that the amount of the energy transfer from the structure motion to the liquid surface motion becomes predominant as γ decreases.

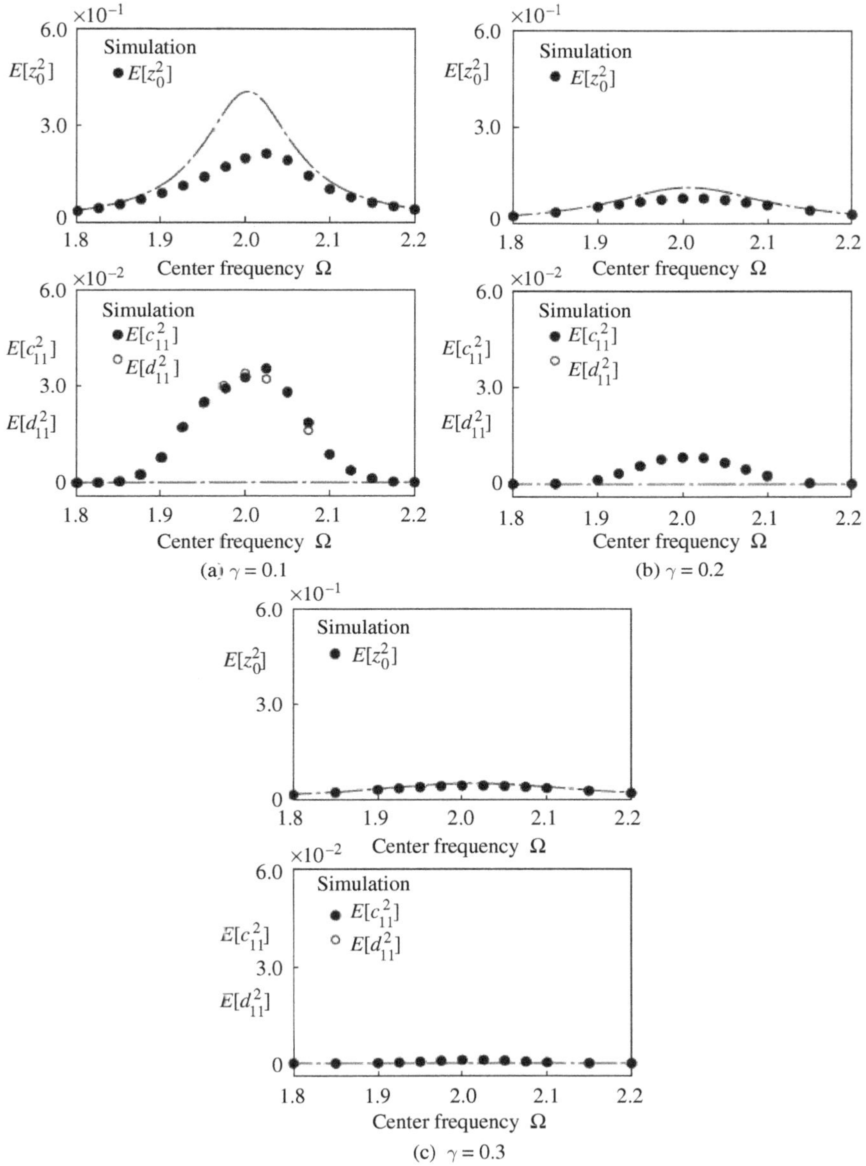

Figure 10.35 Dependence of mean square responses of the structure $E[z_0^2]$ and fluid elevation components $E[c_{11}^2]$ and $E[d_{11}^2]$ on the center frequency Ω for three different values of the bandwidth γ and the same parameters as Figure 10.49. (Ikeda and Ibrahim, 2002)

Figure 10.37(a) shows the dependence of the mean square response of the liquid amplitudes c_{11} and d_{11} on the excitation spectral density S_0. This figure is magnified in Figure 10.37(b) and both figures reveal that the mean square value of the liquid elevation gradually increases as the intensity S_0 increases, and that $E[c_{11}^2]$ is nearly equal to $E[d_{11}^2]$. The stability boundary of liquid sloshing is defined in this analysis by taking the value of the mean square level $A_{ms} = 2.0 \times 10^{-4}$ that corresponds to $S_0 \approx 0.15 \times 10^{-7}$. This threshold mean square level refers to either mode c_{11} or d_{11}.

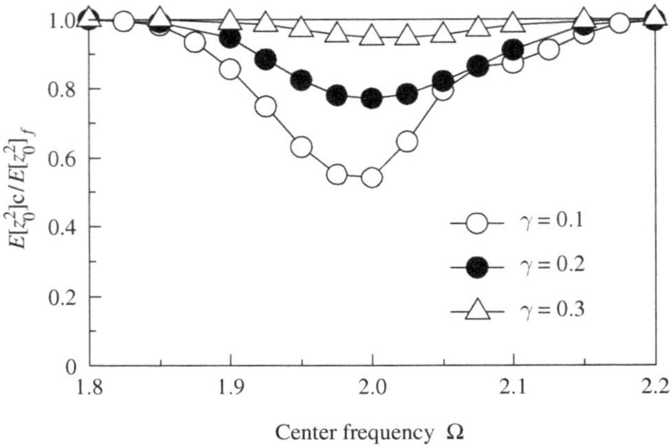

Figure 10.36 Normalized of the structure mean square response with response to its mean square with frozen liquid for different values of bandwidth γ, $h/R = 1.2$, $\nu_1 = 0.87$, $\nu_2 = 0.034$, $k = 4.0$, $c = 0.03$, $\zeta_{01} = \zeta_{11} = \zeta_{21} = 0.01$. (Ikeda and Ibrahim, 2002)

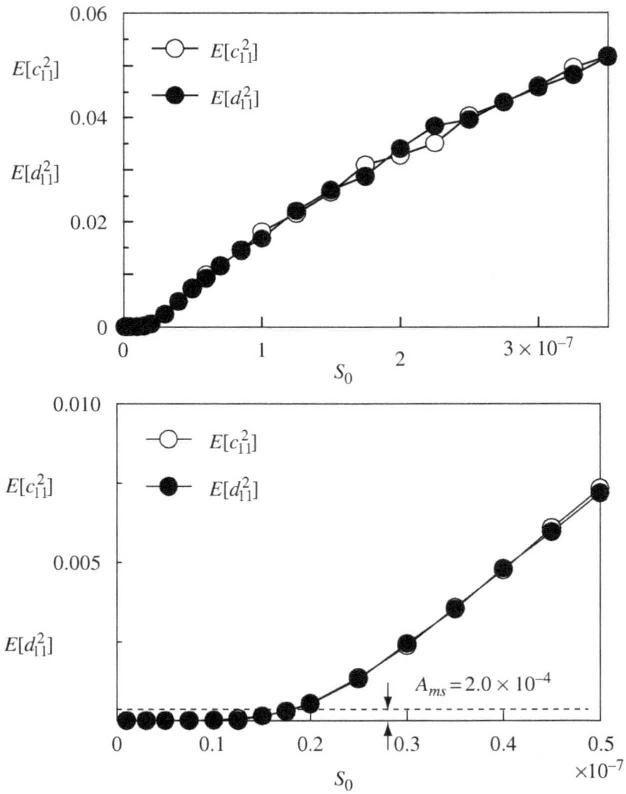

Figure 10.37 (a) Dependence of the mean square response of the liquid free-surface elevation components on the excitation spectral density level, and (b) magnification of a portion of (a). (Ikeda and Ibrahim, 2002)

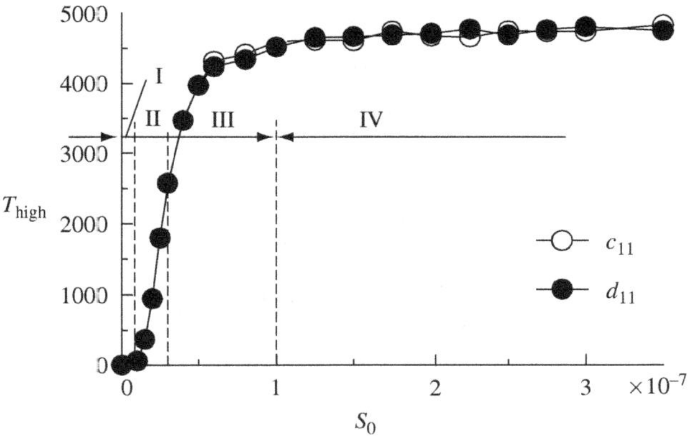

Figure 10.38 Averaged time length of different regimes of liquid height: region I zero liquid motion, II uncertain (intermittent) liquid motion, III partially developed liquid random motion, and IV fully developed random motion; $h/R = 1.2$, $\nu_1 = 0.87$, $\nu_2 = 0.034$, $k = 4.0$, $c = 0.03$, $\zeta_{01} = \zeta_{11} = \zeta_{21} = 0.01$, $\Omega = 2.0$. (Ikeda and Ibrahim, 2002)

Figure 10.38 shows the averaged time lengths of different regimes of liquid motion for the same values of parameters as those in Figure 10.37(a). These time history records reflect the way in which the energy is redistributed among the fluid and structure modes due to the presence of internal resonances and excitation levels in addition to the physical properties of the system. The value $A_{ms} = 2.0 \times 10^{-4}$ corresponds to the value $A_0 = 0.02$ of the amplitude of c_{11}, or d_{11}, since the relation $A_0 = \sqrt{2A_{ms}}$ is satisfied. The time length T_{high} of high liquid level is defined when the amplitudes of c_{11} or d_{11} exceed the value A_0. From Figures 10.37(b) and 10.38, four kinds of region are identified depending on S_0 as reported in Ibrahim and Heinrich (1988). Region I corresponds to $S_0 \leq 0.15 \times 10^{-7}$ (zero liquid motion), where the liquid free surface is stable. Region II corresponds to $0.15 \times 10^{-7} \leq S_0 \leq 0.3 \times 10^{-7}$ (uncertain zero motion, or on-off intermittency), where the mean square values of c_{11} and d_{11} are very small, that is, T_{high} is close to zero, but the liquid sloshing occurs intermittently. In region II, T_{high} is less than 50% of the whole time interval as shown in Figure 10.38. Region III, $0.3 \times 10^{-7} \leq S_0 \leq 1.0 \times 10^{-7}$, corresponds to partially developed motion, that is, a large amplitude of sloshing occurs over a finite time interval and then decays for the next time interval. Region IV, $S_0 \geq 1.0 \times 10^{-7}$, corresponds to fully developed motion and is characterized by continuous random liquid motion. In region IV the time length of high liquid level is more than 90% as shown in Figure 10.38.

10.6 Autoparametric sloshing absorber

For a time, civil engineers (Chen, 1961 and Newmark and Rosenbluetch, 1971) indicated that ground motions might be idealized ordinarily as horizontal, that is, the vertical component could be neglected for the calculations of hydrodynamic pressures without introducing significant error. Yet, another study by Chopra (1967) has yielded surprisingly high values of the hydrodynamic pressures due to vertical ground motion. The idea of an autoparametric sloshing absorber is similar to the autoparametric absorber developed by Haxton and Barr (1972) in which the absorber is only nonlinear coupled to the structure, and through this nonlinear coupling the absorbing effect takes place when the structure parametrically excites the absorber.

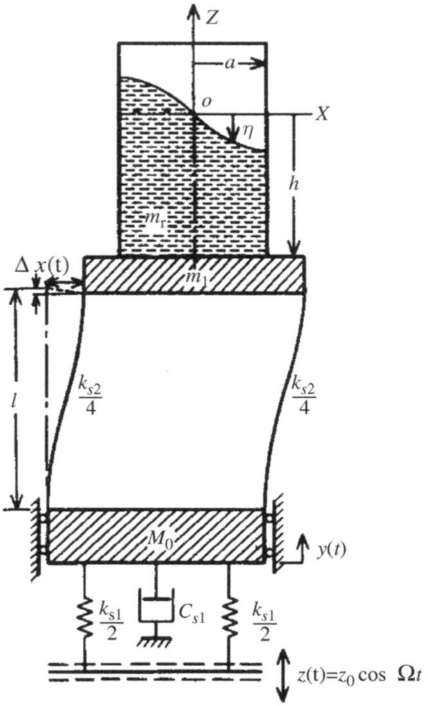

Figure 10.39 Schematic diagram of a circular cylindrical container supported on two elastic structures

The autoparametric sloshing absorber is shown schematically in Figure 10.39, which consists of three subsystems: (a) a container partially filled with liquid, (b) a spring-mass system carrying the liquid container, and (c) a damped spring-mass system carrying the other two systems. The interaction of the fluid dynamics with the dynamics of the structure systems is studied when the base is subjected to a vertical sinusoidal excitation $Z_0 \cos\Omega t$. Under this type of excitation, the fluid container moves up and down and, under conditions of autoparametric resonance, moves laterally as well. Accordingly, the free surface of the liquid exhibits normal sloshing, and hydrodynamic forces are fed back to the structure. The analytical model follows the same procedure as that of the previous section.

The equation of motion of the liquid first anti-symmetric sloshing mode amplitude is:

$$\begin{aligned} A''_{11} + 2\varepsilon\zeta_l A'_{11} + A_{11} + C_0 X' = & -\varepsilon Y''(A_{11} - C_1 X) \\ & - \varepsilon^2\{A_{11}(C_2 A_{11} A''_{11} + C_3 A'^2_{11}) \\ & - C_4 A_{11}(XX'' + X'^2) + C_5 X(XX'' + X'^2)\} \end{aligned} \tag{10.100}$$

The system equation of motion in the lateral direction, X, is

$$\begin{aligned} X'' + 2\varepsilon r_{s2}\zeta_{s2} X' + r_{s2}^2 X + K_0 A''_{11} = & \varepsilon K_1 X Y'' \\ & - \varepsilon^2\{K_1^2 X(XX'' + X'^2) \\ & + K_2 A_{11}(A_{11} A''_{11} + 2A''^2_{11})\} \end{aligned} \tag{10.101}$$

The system equation of motion in the vertical direction is

$$Y'' + 2\varepsilon r_{s1}\zeta_{s1} Y' + r_{s1}^2 Y = \varepsilon\{L_2(XX'' + X'^2) + L_1 A_{11}'^2 + f \cos \sigma\tau\} \tag{10.102}$$

where

$$\tau = \omega_{11} t, \quad \varepsilon = Z_0/R, \quad \begin{Bmatrix} A_{11} \\ Y \\ X \end{Bmatrix} = \frac{\lambda_{11}}{\varepsilon} \tanh \lambda_{11} h \begin{Bmatrix} a_{11} \\ y \\ x \end{Bmatrix},$$

$$\begin{Bmatrix} \sigma \\ r_{s1} \\ r_{s2} \end{Bmatrix} = \frac{1}{\omega_{11}} \begin{Bmatrix} \Omega \\ \omega_{s1} \\ \omega_{s2} \end{Bmatrix}, \quad f = \frac{r_{s1}^2 \xi_{11}}{\varepsilon} \tanh \lambda_{11} h,$$

$$\omega_{11}^2 = g\lambda_{11} \tanh \lambda_{11} h, \quad \omega_{s1}^2 = \frac{K_{s1}}{M_0 + m}, \quad \omega_{s2}^2 = \frac{K_{s2}}{m}, \quad \text{and}$$

$$\lambda_{11} = \frac{\xi_{11}}{R} \text{ is the root of } J_1'(\xi_{11}) = 0$$

It is convenient to transform the linear undamped homogeneous equations (10.100)–(10.102) into the principal coordinates Y_i, through the linear transformation

$$\begin{Bmatrix} A_{11} \\ X \\ Y \end{Bmatrix} = \mathbf{R} \begin{Bmatrix} Y_1 \\ Y_2 \\ Y_3 \end{Bmatrix} \tag{10.103}$$

where

$$\mathbf{R} = \begin{bmatrix} 1 & 1 & 0 \\ n_1 & n_2 & 0 \\ 0 & 0 & 1 \end{bmatrix} \text{ is the modal matrix, } \quad n_1 = \frac{1 - r_1^2}{C_0 r_1^2}, \quad n_2 = \frac{1 - r_2^2}{C_0 r_2^2}, \quad \text{and}$$

$$r_{1,2}^2 = \frac{1 + r_{s2}^2 \pm \sqrt{(1 + r_{s2}^2)^2 - 4(1 - C_0 K_0) r_{s2}^2}}{2(1 - C_0 K_0)} \tag{10.104}$$

The problem can be drastically simplified if cubic terms are removed from equations (10.100)–(10.102) because in the present analysis the system behavior is mainly governed by quadratic nonlinear coupling, which gives rise to internal resonance of order 3. Figure 10.40 shows the dependence of the normal mode frequencies and their combination on the upper structure dynamic property r_{s2}. Equations (10.100)–(10.102) can be written in terms of the principal coordinates in the form

$$Y_1'' + S_1^2 \nu^2 Y_1 = \varepsilon\{\varepsilon^{-1}(S_1^2 \nu^2 - r_1^2) Y_1 - 2r_1\zeta_1 Y_1' - C_6 Y_1 Y_3'' - C_7 Y_2 Y_3''\} \tag{10.105}$$

$$Y_2'' + S_2^2 \nu^2 Y_2 = \varepsilon\{\varepsilon^{-1}(S_2^2 \nu^2 - r_2^2) Y_2 - 2r_2\zeta_2 Y_2' + K_3 Y_1 Y_3'' + K_4 Y_2 Y_3''\} \tag{10.106}$$

$$\begin{aligned} Y_3'' + S_3^2 \nu^2 Y_3 = \varepsilon\{&\varepsilon^{-1}(S_3^2 \nu^2 - r_3^2) Y_3 - 2r_3\zeta_3 Y_3' + L_3 Y_1'^2 + L_4 Y_1' Y_2' + L_5 Y_2'^2 \\ &+ L_6 Y_1 Y_1'' + L_7(Y_1 Y_2'' + Y_1'' Y_2) + L_8 Y_2 Y_2'' + f \cos n\nu\tau\} \end{aligned} \tag{10.107}$$

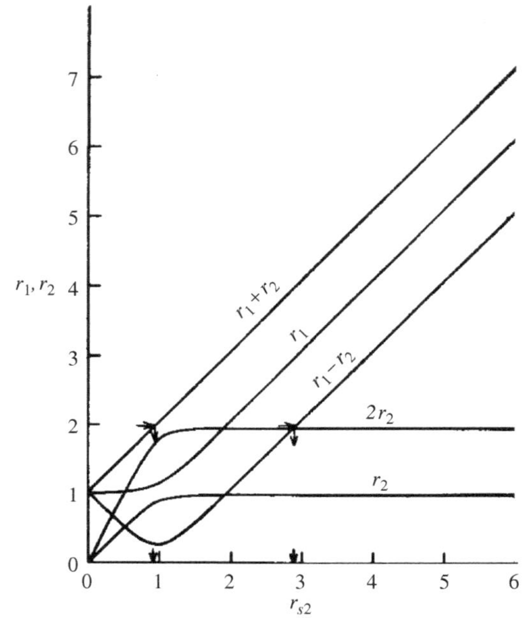

Figure 10.40 Dependence of normal mode frequencies and their combinations on the frequency of the upper structure

where the forcing frequency, σ, has been replaced by $n\nu$, where n is a positive real number and ν is a frequency ratio in the neighborhood of one of the principal mode frequency ratios, and S_i are positive real numbers such that $|S_i^2\nu^2 - r_i^2| < \varepsilon$.

Introducing the approximate asymptotic expansion as outlined by Struble (1962)

$$Y_1(\tau) = A(\tau)\cos(r_1\tau + \varphi(\tau)) + \varepsilon a_1 + \cdots \tag{10.108a}$$

$$Y_2(\tau) = \mathrm{B}(\tau)\cos(\mathrm{r}_2\tau + \theta(\tau)) + \varepsilon \mathrm{b}_1 + \cdots \tag{10.108b}$$

$$Y_3(\tau) = C(\tau)\cos(r_3\tau + \gamma(\tau)) + \varepsilon c_1 + \cdots \tag{10.108c}$$

Substituting expressions (10.108) into equations (10.105)–(10.107) gives the following equations up to first-order in ε:

$$\begin{aligned}
&\left[A'' - A(r_1+\varphi')^2\right]\cos(r_1\tau+\varphi) + [A\varphi'' + 2A'(r_1+\varphi')]\sin(r_1\tau+\varphi) \\
&+ \varepsilon a_1'' + \cdots + S_1^2\nu^2[A\cos(r_1\tau+\varphi) + \varepsilon a_1 + \cdots] \\
=\ & \varepsilon\{\varepsilon^{-1}(S_1^2\nu^2 - r_1^2)[A\cos(r_1\tau+\varphi) + \varepsilon a_1] - 2r_1\zeta_1\left[-r_1 A\sin(r_1\tau+\varphi) + \varepsilon a_1'\right] \\
&+ C_6 r_3^2 AC\cos(r_1\tau+\varphi)\cos(r_3\tau+\gamma) - C_7 r_3^2 BC\cos(r_2\tau+\theta)\cos(r_3\tau+\gamma)\}
\end{aligned} \tag{10.109}$$

$$\left[B'' - B(r_2+\theta')^2\right]\cos(r_2\tau+\theta) + [B\theta'' + 2B'(r_2+\theta')]\sin(r_2\tau+\theta) + \varepsilon b_1'' + \cdots$$
$$+ S_2^2\nu^2[B\cos(r_2\tau+\theta) + \varepsilon b_1 + \cdots]$$
$$= \varepsilon\{\varepsilon^{-1}(S_2^2\nu^2 - r_2^2)[B\cos(r_2\tau+\theta) + \varepsilon b_1] - 2r_2\zeta_2\left[-r_2 B\sin(r_2\tau+\theta) + \varepsilon b_1'\right]$$
$$- K_3 r_3^2 AC\cos(r_1\tau+\varphi)\cos(r_3\tau+\gamma) - K_4 r_3^2 BC\cos(r_2\tau+\theta)\cos(r_3\tau+\gamma)\} \tag{10.110}$$

$$\left[C'' - C(r_3+\gamma')^2\right]\cos(r_3\tau+\gamma) + [C\gamma'' + 2C'(r_3+\gamma')]\sin(r_3\tau+\gamma)$$
$$+ \varepsilon c_1'' + \cdots + S_3^2\nu^2[C\cos(r_3\tau+\gamma) + \varepsilon c_1 + \cdots]$$
$$= \varepsilon\{\varepsilon^{-1}(S_3^2\nu^2 - r_3^2)[C\cos(r_3\tau+\gamma) + \varepsilon c_1] - 2r_3\zeta_3\left[-r_3 C\sin(r_3\tau+\gamma) + \varepsilon c_1'\right]$$
$$+ L_3 r_1^2 A^2\sin^2(r_1\tau+\varphi) + L_4 r_1 r_2 AB\sin(r_1\tau+\varphi)\sin(r_2\tau+\theta)$$
$$+ L_5 r_2^2 B^2\sin^2(r_2\tau+\theta) - L_6 r_1^2 A^2\cos^2(r_1\tau+\varphi) - L_7\left(r_1^2 + r_2^2\right)$$
$$AB\cos(r_1\tau+\varphi)\cos(r_2\tau+\theta) - L_8 r_2^2 B^2\cos^2(r_2\tau+\theta)\} \tag{10.111}$$

The fundamental variational equations are obtained by equating terms of zero-order in ε, in equations (10.109)–(10.111). At the same time extracting terms of first-order in ε, gives the first-order-perturbational equations. The resulting perturbational equations contain secular terms that give rise to the following internal resonance conditions:

$$r_3 = 2r_1, \quad r_3 = 2r_2, \quad \text{and} \quad r_3 = |r_1 \pm r_2| \tag{10.112a,b,c}$$

In addition, the third equation gives also the external resonance condition

$$r_3 = n\nu \tag{10.112d}$$

The internal resonance conditions (10.112a, b, c) are realized by examining Figure 10.40. Thus, if the value of r_3 lies on one of the curves then one can define the value of r_{s2} for which the relevant internal resonance occurs.

10.6.1 Summed type internal resonance: $r_3 = r_1 + r_2$, $r_3 = n\nu$

Transferring all secular terms corresponding to these resonance conditions to the fundamental variational equations, gives

$$-2r_1 A\varphi' = \varepsilon\left\{\varepsilon^{-1}(S_1^2\nu^2 - r_1^2)A + \frac{1}{2}C_7 r_3^2 BC\cos(\gamma - \varphi - \theta)\right\} \tag{10.113a}$$

$$-2r_1 A' = \varepsilon\left\{2r_1^2\zeta_1 A - \frac{1}{2}C_7 r_3^2 BC\sin(\gamma - \varphi - \theta)\right\} \tag{10.113b}$$

$$-2r_2 B\theta' = \varepsilon\left\{\varepsilon^{-1}(S_2^2\nu^2 - r_2^2)B - \frac{1}{2}K_3 r_3^2 AC\cos(\gamma - \varphi - \theta)\right\} \tag{10.113c}$$

$$-2r_2 B' = \varepsilon\left\{2r_2^2\zeta_2 B + \frac{1}{2}K_3 r_3^2 AC\sin(\gamma - \varphi - \theta)\right\} \tag{10.113d}$$

$$-2r_3C\gamma' = \varepsilon\left\{\varepsilon^{-1}(S_3^2\nu^2 - r_3^2)C - \frac{1}{2}L_{47}r_3^2AB\cos(\gamma-\varphi-\theta) + f\cos\gamma\right\} \quad (10.113e)$$

$$-2r_3C' = \varepsilon\left\{2r_3^2\zeta_3C - \frac{1}{2}L_{47}r_3^2AB\sin(\gamma-\varphi-\theta) + f\sin\gamma\right\} \quad (10.113f)$$

where $L_{47} = [L_4r_1r_2 + L_7(r_1^2 + r_2^2)]/r_3^2$. Introducing the following variables

$$\begin{Bmatrix} A \\ B \\ C \end{Bmatrix} = \frac{\sqrt{f}}{r_3|L_{47}|}\begin{Bmatrix} b_1 \\ b_2 \\ b_3 \end{Bmatrix}, \quad \tau = \frac{2T}{\varepsilon\sqrt{f|L_{47}|}}, \quad \psi = \gamma - \varphi - \theta$$

$$\eta_i = \frac{2\zeta_i r_i}{\sqrt{f|L_{47}|}}, \quad \Gamma = \frac{r_3^2 - \nu^2}{\varepsilon r_3\sqrt{f|L_{47}|}}, \quad S_i = r_i/r_3 \quad (10.114)$$

Equations (10.113) take the new form

$$-b_1\dot{\varphi} = -S_1\Gamma b_1 + \frac{1}{2}\frac{C_7}{S_1|L_{47}|}b_2b_3\cos\psi \quad (10.115a)$$

$$-\dot{b}_1 = \eta_1b_1 - \frac{1}{2}\frac{C_7}{S_1|L_{47}|}b_2b_3\sin\psi \quad (10.115b)$$

$$-b_2\dot{\theta} = -S_2\Gamma b_2 - \frac{1}{2}\frac{K_3}{S_2|L_{47}|}b_1b_3\cos\psi \quad (10.115c)$$

$$-\dot{b}_2 = \eta_2b_2 + \frac{1}{2}\frac{K_3}{S_2|L_{47}|}b_1b_3\sin\psi \quad (10.115d)$$

$$-b_3\dot{\gamma} = -\Gamma b_3 - \frac{1}{2}\frac{L_{47}}{|L_{47}|}b_1b_2\cos\psi + \cos\gamma \quad (10.115e)$$

$$-\dot{b}_3 = \eta_3b_3 - \frac{1}{2}\frac{L_{47}}{|L_{47}|}b_1b_2\sin\psi + \sin\gamma \quad (10.115f)$$

where a dot denotes differentiation with respect to the new time parameter, T.

The steady state solution of equations (10.115) is obtained by setting the left-hand sides to zero. Careful inspection of the resulting algebraic equations reveals that there are five unknowns in six equations. Alternatively, one can combine the phase equations such that $\dot{\psi} = \dot{\gamma} - \dot{\varphi} - \dot{\theta}$. In this case, one obtains the new set of five equations in five unknowns:

$$b_3\dot{\psi} = \Gamma(1 - S_1 - S_2)b_3 + \frac{1}{2|L_{47}|}\left[L_{47}b_1b_2 + \left(\frac{C_7}{S_1}\frac{b_2}{b_1} - \frac{K_3}{S_2}\frac{b_1}{b_2}\right)b_3^2\right]\cos\psi - \cos\gamma \quad (10.116a)$$

$$-\dot{b}_1 = \eta_1b_1 - \frac{1}{2}\frac{C_7}{S_1|L_{47}|}b_2b_3\sin\psi \quad (10.116b)$$

$$-\dot{b}_2 = \eta_2 b_2 + \frac{1}{2}\frac{K_3}{S_2|L_{47}|} b_1 b_3 \sin\psi \tag{10.116c}$$

$$-\dot{b}_3 = \eta_3 b_3 - \frac{1}{2}\frac{L_{47}}{|L_{47}|} b_1 b_2 \sin\psi + \sin\gamma \tag{10.116d}$$

$$-b_3\dot{\gamma} = -\Gamma b_3 - \frac{1}{2}\frac{L_{47}}{|L_{47}|} b_1 b_2 \cos\psi + \cos\gamma \tag{10.116e}$$

It is not possible to obtain a closed-form analytical solution of the algebraic equations of system (10.116). Instead, a numerical integration is carried out. The integration of equations (10.115) results in a steady state solution for the three amplitudes and the phase angle γ; however, the phase angles φ and θ exhibit unsteady time history records whose degree of unsteadiness decreases as $n \to 1$. The system possesses a complete steady state at exact resonance $n = 1$. Figures 10.41 (a)–(b) shows samples of time history records of the three amplitudes for two different values of detuning parameter $n = 1.04$ and 1.0, respectively. Figure 10.42 shows the amplitude–frequency response of the three amplitudes in which the first two normal modes interact nonlinearly with the third mode over a limited range of excitation frequency. Outside this range, the response follows the behavior of a linear single-degree-of-freedom system.

In the absence of damping of the first two modes, that is, $\eta_1 = \eta^2 = 0$, one can obtain the following closed form steady state solutions

$$b_2^2 = 4\frac{S_1^2}{C_7}\left\{|L_{47}|\Gamma^2 \pm \sqrt{-\frac{K_3 C_7}{(2S_1 S_2)^2} - \Gamma^2\eta_3^2 L_{47}^2}\right\} \tag{10.117a}$$

$$b_1^2 = -\frac{C_7}{K_3}(S_2/S_1)^2 b_2^2 \tag{10.117b}$$

$$b_3 = -4\frac{(S_1 S_2 \Gamma)^2}{C_7 K_3} L_{47}^2 \tag{10.117c}$$

In addition, we have the unimodal response, $b_1 = b_2 = 0$, and

$$b_3 = \frac{1}{\sqrt{\Gamma^2 + \eta_3^2}} \tag{10.118}$$

which is the normal resonance solution.

The amplitude–frequency response curves are shown in Figure 10.43 for a liquid depth ratio $h/R = 1.0$, and excitation parameter $\varepsilon = 0.0105$. Points of vertical tangents define the boundaries of the region of parametric instability (turning or saddle points). Two kinds of vertical tangents exist on each side of the resonance frequency, $n = 1$. The first occurs at the forcing frequency point where $b_1 = b_2 = 0$ at which the main mass amplitude, b_3, of solutions (10.117c) and (10.118) are equal (point F) and the values of b_1 and b_2 jump from point J to point K on b_1 and K′ on b_2. As the forcing frequency increases the responses of b_1, b_2, and b_3 follow the curves KLM, K′L′M′, and FE′ GV, respectively, with peaks in b_1 and b_2 near M and M′, at which

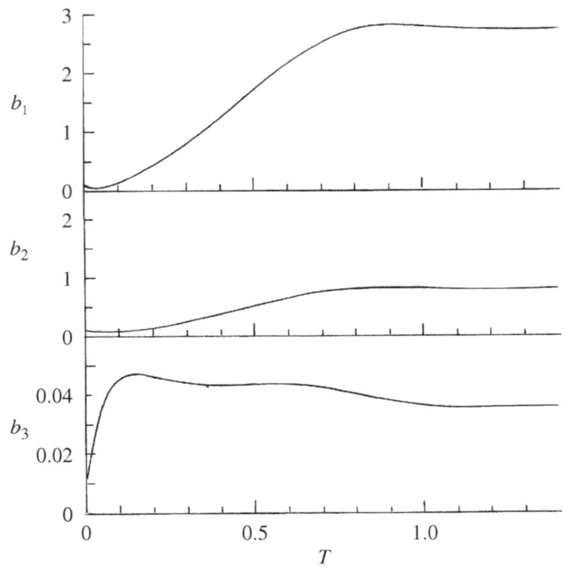

(a) $n=1.04$, $\varepsilon=0.021$, $h/R=1.0$, $l/R=2.55$, $\bar{\zeta}_1=0.005$, $\bar{\zeta}_2=0.0085$, $\bar{\zeta}_3=0.05$.

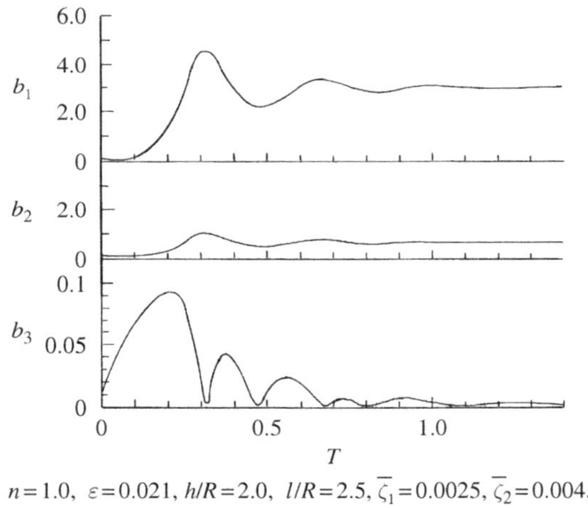

(b) $n=1.0$, $\varepsilon=0.021$, $h/R=2.0$, $l/R=2.5$, $\bar{\zeta}_1=0.0025$, $\bar{\zeta}_2=0.004$, $\bar{\zeta}_3=0.05$.

Figure 10.41 Numerical simulation of the system response in the neighborhood of summed internal resonance $r_3=r_1+r_2$. (Ibrahim and Barr, 1975b)

$|db_1/dn|=|db_2/dn|=\infty$ (the second kind of vertical tangents), and the values of b_1 and b_2 drop to point H. At this frequency another collapse occurs in b_3, which brings its response down to point V′ at which the system follows the response of a single degree of freedom.

When the frequency decreases from point H, the amplitudes b_1, b_2, and b_3 follow the paths HJ′LNH′, HJ′L′N′H′, and YGE′UU′X. It can be noticed that the main mass response, b_3, is suppressed through autoparametric action and meets the frequency axis at $n=1.0$, and instead of the linear resonance curve FEG energy is transferred to the first two modes, which exhibit large peaks on either side of $n=1.0$. If the exact internal resonance $r_3=r_1+r_2$ is not met

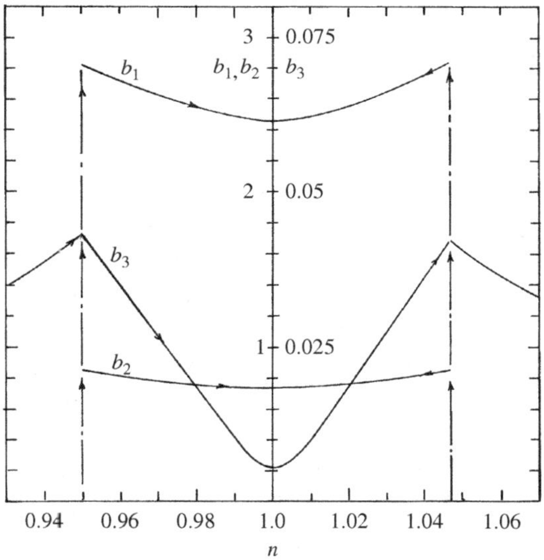

Figure 10.42 Numerical simulation of three-mode interaction in the neighborhood of combination internal resonance $r_3 = r_1 + r_2$. $\varepsilon = 0.021$, $h/R = 1.0$, $l/R = 2.55$, $\bar{\zeta}_1 = 0.005$, $\bar{\zeta}_2 = 0.0085$, $\bar{\zeta}_3 = 0.05$. (Ibrahim and Barr, 1975b)

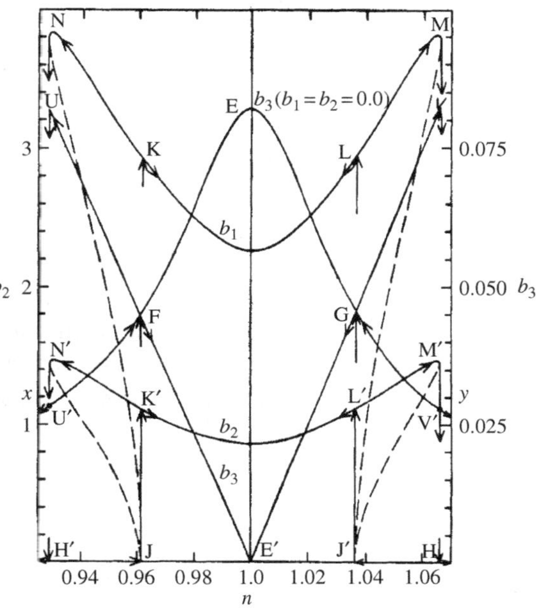

Figure 10.43 Numerical simulation of three-mode interaction in the neighborhood of combination internal resonance $r_3 = r_1 + r_2$, $\varepsilon = 0.0105$, $h/R = 2.0$, $l/r = 1.0$, $\bar{\zeta}_1 = \bar{\zeta}_2 = 0.0$, $\bar{\zeta}_3 = 0.025$, dotted curves indicate unstable solution. (Ibrahim and Barr, 1975b)

precisely, the effect of imperfect tuning is to shift the diagram to the left or to the right of $n = 1.0$ and the symmetry will be broken.

The case of difference type internal resonance $r_3 = r_1 - r_2$, $r_3 = n\nu$, yields to a set of equations and solutions similar to the summed type resonance case. Experimental tests of the system require a strong disturbance to initiate autparametric interaction

According to experimental observations and time history records reported by Ibrahim (1974), and Ibrahim and Barr (1975b), the system possesses a quasi-steady state response, and an energy exchange between the liquid and the structure takes place in a form of beating with constant maximum amplitude. It was observed that the system did not respond to any parametric excitation – except at the exact resonance – unless an initial lateral disturbance was applied. At nearly exact resonance, a considerable time elapses before the liquid free surface can be observed to oscillate with very small amplitude. The resulting small sloshing force causes the structure to oscillate laterally. This energy flow continues with a build up in amplitude until the damping of each mode and nonlinear effects control this growth and the system follows the response of Figure 10.44. The envelopes of the liquid and upper structure responses oscillate with a long period of 1.5 s at a constant amplitude. The transient time before the system achieves the quasi-steady state is about 40 seconds. If the internal resonance is distorted, that is, $r_3 \neq r_1 + r_2$, by reducing or increasing the length of the leaf spring, the system possesses a complete steady-state response as shown in Figure 10.45.

Figures 10.46(a)–(c) shows the amplitude-frequency response curves for two different values of excitation amplitude, $\varepsilon = 0.002$ and 0.0026, and liquid depth ratio $h/R = 2.0$. Both the liquid and upper structure amplitude curves exhibit instability at an excitation frequency less than ω_{s1} (i.e., $n < 1.0$), indicated by the jump and collapse amplitudes. On the right side of $n = 1$, the liquid amplitude, a_{11}, and the structure lateral displacement, X, decrease gradually as n increases until they cease at an excitation frequency slightly higher than $2\omega_{11}$. Within the frequency range bounded by $2\omega_{11}[1 - O(\varepsilon)] < \Omega < 2\omega_{11}$, the liquid free surface oscillates about a nodal diameter perpendicular to the tank oscillations. When the excitation frequency, Ω, approaches $2\omega_{11}$, the liquid amplitude decreases rapidly and then ceases. With a slight increase in Ω, the liquid free surface reorients itself to oscillate about a nodal diameter perpendicular to the first one; eventually the nodal diameter begins to rotate around the tank centerline with a new increase in the wave-height. This rotation sometimes reverses its direction and sometimes continues in the same direction without achieving constant amplitude in both cases. The observed asymmetry in the a_{11} and X curves is attributed to two sources. These are the nonlinear soft characteristics of the liquid and the rotation of the liquid nodal diameter. The main mass response, y, shows the characteristics of the autoparametric vibration absorber. The absorber effect does not occur exactly at $n = 1$ because it is difficult to adjust the system to possess an exact internal resonance condition.

10.6.2 *Principal type internal resonance:* $r_3 = 2r_2$, $r_3 = n\nu$

The variational equations after including secular terms corresponding to this type of resonance are

$$-b_2\dot{\theta} = -\frac{1}{2}\Gamma b_2 - \frac{K_4}{|L_{58}|} b_2 b_3 \cos(\gamma - 2\theta) \tag{10.119a}$$

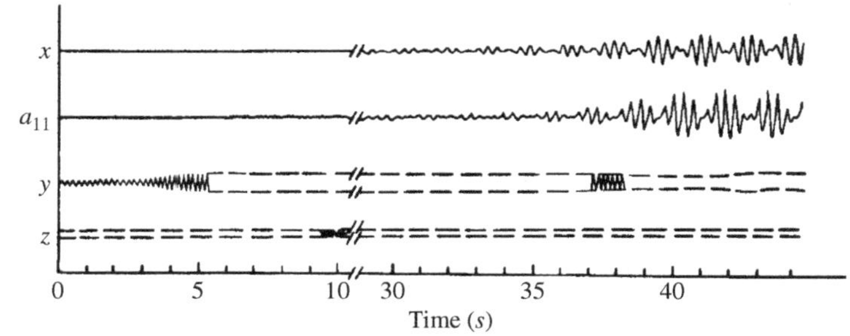

Figure 10.44 Measured time history records of transient and quasi-steady state response of the structure lateral x and vertical y displacements, liquid displacement a_{11}, and excitation z, under autoparametric resonance $r_3 = n\nu$, $r_3 = r_1 + r_2$. (Ibrahim and Barr, 1975b)

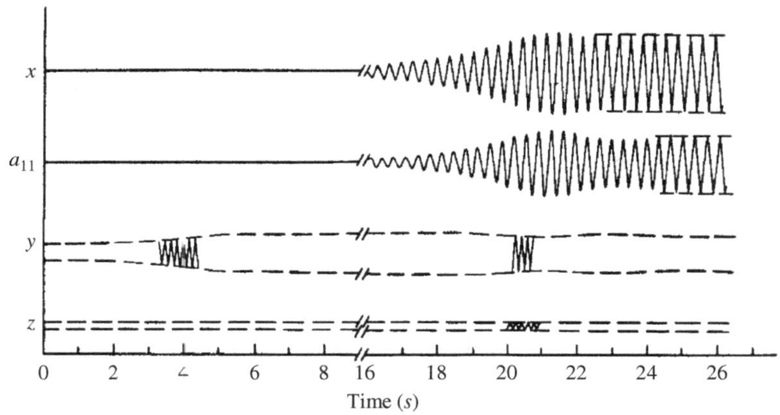

Figure 10.45 Measured time history records of three-mode interaction with distorted internal resonance $r_3 - (r_1 + r_2) \neq 0$ of the structure lateral x and vertical y displacements, liquid displacement a_{11}, and excitation z. (Ibrahim and Barr, 1975b)

$$- \dot{b}_2 = \eta_2 b_2 + \frac{K_4}{|L_{58}|} b_2 b_3 \sin(\gamma - 2\theta) \tag{10.119b}$$

$$- b_3 \dot{\gamma} = -\Gamma b_3 - \frac{1}{8} \frac{L_{58}}{|L_{58}|} b_2^2 \cos(\gamma - 2\theta) + \cos\gamma \tag{10.119c}$$

$$- \dot{b}_3 = \eta_3 b_3 - \frac{L_{58}}{|L_{58}|} b_2^2 \cos(\gamma - 2\theta) + \sin\gamma \tag{10.119d}$$

where $L_{58} = L_5 + L_8$. The steady state solution is

$$b_3 = \frac{|L_{58}|}{2K_4} \sqrt{\frac{\Gamma^2}{4} + \eta_2^2} \tag{10.120a}$$

$$b_2^2 = \frac{4}{K_4} \left\{ L_{58}(\Gamma^2 - 2\eta_2\eta_3) \pm |L_{58}| \sqrt{4(K_4/L_{58})^2 - \Gamma^2(2\eta_2 + \eta_3)^2} \right\} \tag{10.120b}$$

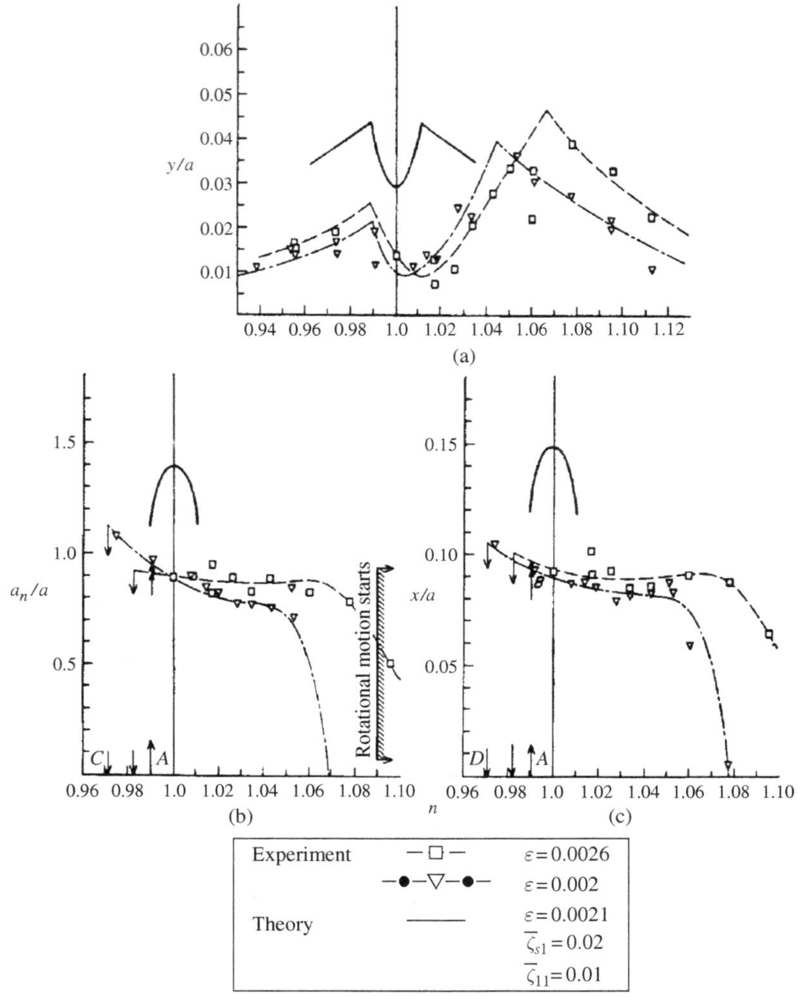

Figure 10.46 Experimental measurements of amplitude–frequency responses under summed internal resonance $r_3 = r_1 + r_2$ of the (a) main mass, (b) liquid free surface, and (c) upper structure; $h/R = 2.0$, $l/R = 2.1$, $\bar{\zeta}_{s1} = 0.0092$, $\bar{\zeta}_{11} = 0.0282$, $\bar{\zeta}_{s2} = 0.005$. (Ibrahim and Barr, 1975b)

There is another unimodal solution given by

$$b_2 = 0, \quad \text{and} \quad b_3 = \frac{1}{\sqrt{\Gamma^2 + \eta_3^2}} \tag{10.121}$$

Figure 10.47 shows the amplitude–frequency response curves of both mixed mode interaction and unimodal response. The unimodal response takes place outside the frequency range governed by the autoparametric interaction over which the top liquid– structure system acts as an autoparametric vibration absorber to the main mass motion.

Figure 10.48 shows the experimental measurements of the amplitude–frequency response curves for two different values of the main system damping parameter, $\zeta_{s1} = 0.0282$ and 0.0444

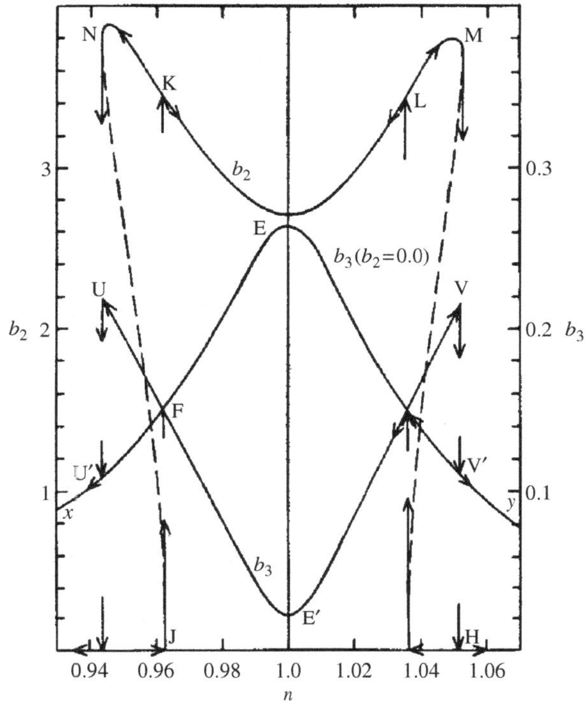

Figure 10.47 Amplitude–frequency response of two-mode interaction under principal internal resonance $r_3 = 2r_2$, $h/R = 1.0$, $l/R = 2.55$, $\varepsilon = 0.021$, $\bar{\zeta}_1 = 0.005$, $\bar{\zeta}_3 = 0.025$. Dash curves indicate unstable solution. (Ibrahim and Barr, 1975b)

and two excitation amplitude levels, $\varepsilon = 0.0042$ and 0.009. For higher damping ratio, $\zeta_{s1} = 0.0444$, the system did not respond for excitation level $\varepsilon =$ 0.0052 and the main mass motion follows normal resonance curve of single degree of freedom.

Some remarks

In all cases of autoparametric interaction, there are two possible kinds of steady-state solution: one gives mixed mode interaction in the vicinity of $n = 1 \pm O(\varepsilon)$, and the other gives unimodal linear response of the main mass. Note that the liquid container is mounted in such a way that longitudinal and lateral motions are possible, and the liquid free surface can undergo the following types of motions:

(1) Suppose that, in the first instance, the tank receives purely relative vertical harmonic excitation at a frequency below twice the liquid natural frequency. The behavior of the free surface is governed mainly by the Mathieu stability chart, and, as long as the excitation amplitude and frequency are not in one of the instability regions, the free-surface displacement remains identically zero and the system follows the linear unimodal solution.
(2) As the amplitude and frequency of the main mass enter the instability region, the free surface starts to move back and forth. This sloshing motion exerts a lateral force, which causes lateral oscillation of the container with its structure system. This in turn acts as a lateral excitation to the liquid and energy exchange between the two systems takes place until either a steady-state condition is achieved or the

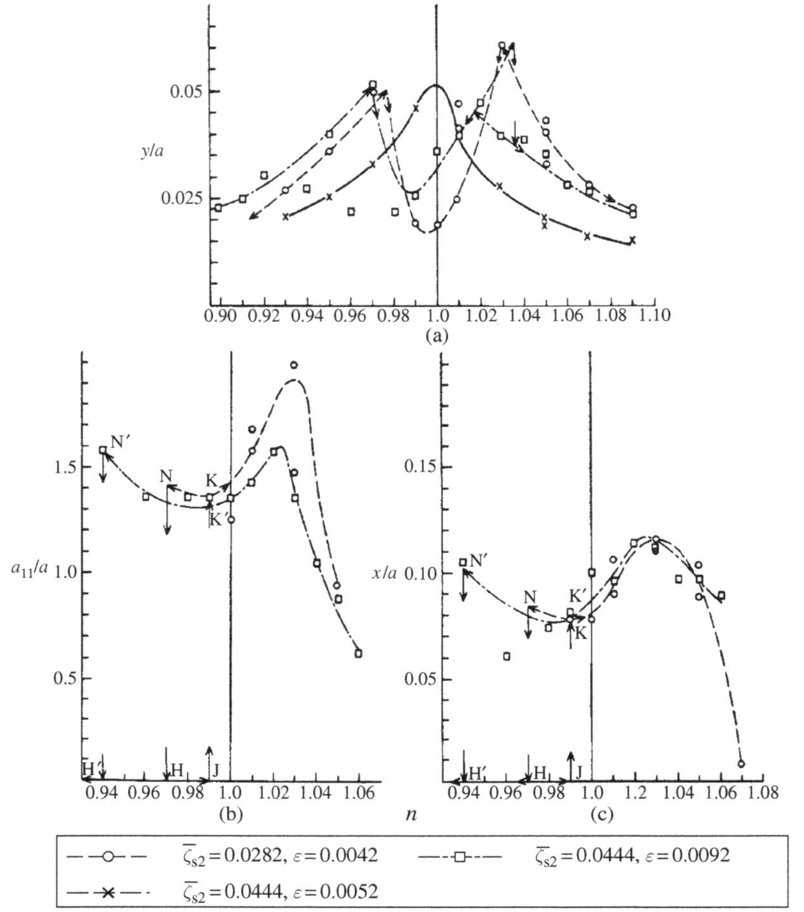

Figure 10.48 Experimental measurements of amplitude–frequency responses under principal internal resonance $r_3=2r_2$ of the (a) main mass, (b) liquid free surface, and (c) upper structure; $h/R=2.0$, $l/R=2.23$, $\bar{\zeta}_{s1}=0.005$, $\bar{\zeta}_{11}=0.0092$, $(x=a_{11}=0.0)$. (Ibrahim and Barr, 1975b)

resulting motion becomes of the beating type. In the steady-state case, the liquid motion has a constant peak wave height and a single nodal diameter perpendicular to the direction of excitation. As the tank lateral excitation frequency increases, the liquid wave height increases. Then, at a frequency a little less than the natural frequency, the nodal diameter begins to rotate and the peak wave height varies. This unstable behavior takes place up to a frequency a little above the liquid natural frequency, where once again the unsteady state liquid motion has a constant peak wave height and a single nodal diameter perpendicular to the tank lateral motion. Such unstable sloshing will occur only when the peak acceleration of a fluid particle on the free surface becomes equal to the acceleration due to gravity as reported by Eulitz and Glazer (1961). This complex motion can be attributed to a strong coupling between the two orthogonal directions, lateral and transverse, where the sloshing frequencies are the same in both directions, causing unstable motion to occur rapidly. According to the previous considerations, it is anticipated that the picture of the analytical response curves shown in the previous figures will eventually be distorted and shifted to the right side of $n=1$ by virtue of the liquid rotational motion at an excitation frequency above twice the sloshing frequency.

10.7 Nonlinear sloshing absorber in a rectangular tank

Liquid sloshing dynamics in rectangular tanks is different from that in cylindrical containers. However, in both cases, the liquid sloshing forces can suppress the motion of the elastic support structure. Ikeda and Nakagawa (1997) and Ikeda and Murakami (2001) analytically and experimentally studied the nonlinear sloshing absorber in a rectangular and cylindrical containers. Their system is shown schematically in Figure 10.5(a) and has a cantilever beam carrying a mass m, which in turn carries a rigid rectangular container. The cantilever beam has a stiffness k and damping coefficient c. The end mass is subjected to a harmonic excitation $F_0 \cos \Omega t$. The tank width is l, breadth L, and partially filled with liquid to a depth h. The moving coordinates, Oxy, are placed such that the axis x coincides with the undisturbed liquid free surface. Let the displacements of the structure and liquid free surface be x_1 and η, respectively.

The equation of motion of the structure is

$$m\ddot{x}_1 + c\dot{x}_1 + kx_1 = F_l + F_0 \cos \Omega t \tag{10.122}$$

where F_ℓ is the hydrodynamic sloshing force exerted on the side walls of the tank in the direction of x, and is given by integrating the hydrodynamic pressure over the tank projected area, that is,

$$F_l = L\left\{\int_{-h}^{\eta_1} p(l/2, y, t)\, dy - \int_{-h}^{\eta_2} p(-l/2, y, t)\, dy\right\} \tag{10.123}$$

where η_1 and η_2 are the liquid surface elevations at the two sides of the container, and $p(l/2,y,t)$ and $p(-l/2,y,t)$ are the liquid pressures at $x = \pm\, l/2$. The value of the hydrodynamic pressure is estimated from the equation

$$\frac{\partial\tilde{\Phi}}{\partial t} + \frac{1}{2}\left[\left(\frac{\partial\tilde{\Phi}}{\partial x}\right)^2 + \left(\frac{\partial\tilde{\Phi}}{\partial y}\right)^2\right] + gy + \frac{p}{\rho} - \frac{1}{2}\dot{x}_1^2 + \ddot{x}_1 x = 0 \tag{10.124}$$

Introducing the nondimensional parameters

$$X_1 = x_1/l, \quad X = x/l, \quad Y = y/l, \quad H = h/l, \quad D = \rho L l^2/(m + m_1), \quad \tilde{\eta} = \eta/l,$$

$$\phi = \tilde{\Phi}/(l^2\omega_1), \quad \omega_1^2 = (\pi g/l)\tanh(\pi h/l), \quad \mu = m/(m + m_l),$$

$$K = k/[(m + m_l)\omega_1^2], \quad C = c/[(m + m_l)\omega_1], \quad f_l = F_l/[(m + m_l)l\omega_1^2],$$

$$f_0 = F_0/[(m + m_l)l\omega_1^2], \quad P = p/[(m + m_l)l^2\omega_1^2], \quad \nu = \Omega/\omega_1, \quad \tau = \omega_1 t \tag{10.125}$$

where m_l is the mass of the liquid.

The solution of Laplace's equation must satisfy the boundary conditions

$$\left.\frac{\partial\phi}{\partial x}\right|_{X=\pm 1/2} = 0, \quad \left.\frac{\partial\phi}{\partial Y}\right|_{Y=-H} = 0 \tag{10.126a, b}$$

The kinematic and dynamic free-surface boundary conditions are, respectively

$$\frac{\partial \tilde{\eta}}{\partial \tau} = \frac{\partial \phi}{\partial Y} - \frac{\partial \phi}{\partial X}\frac{\partial \tilde{\eta}}{\partial X} \quad \text{at } Y = \tilde{\eta} \tag{10.126c}$$

$$\frac{\partial \phi}{\partial \tau} + \frac{1}{2}\left\{\left(\frac{\partial \phi}{\partial X}\right)^2 + \left(\frac{\partial \phi}{\partial Y}\right)^2\right\} + \frac{Y}{\pi \tanh(\pi H)} - \frac{1}{2}\left(\frac{\partial X_1}{\partial \tau}\right)^2$$
$$- \frac{\partial^2 X_1}{\partial \tau^2} X = 0, \quad \text{at } Y = \tilde{\eta} \tag{10.126d}$$

The solution of the velocity potential function and wave height in terms of asymmetric and symmetric sloshing modes may be written in the form

$$\phi(X, Y, \tau) = \sum_{n=1}^{\infty}\left[a_{2n-1}(\tau)\sin(\lambda_{2n-1}X)\frac{\cosh\{\lambda_{2n-1}(Y+H)\}}{\cosh(\lambda_{2n-1}H)}\right.$$
$$\left. + a_{2n}(\tau)\cos(\lambda_{2n}X)\frac{\cosh\{\lambda_{2n}(Y+H)\}}{\cosh(\lambda_{2n}H)}\right] \tag{10.127a}$$

$$\eta(X, \tau) = \sum_{n=1}^{\infty}[b_{2n-1}(\tau)\sin(\lambda_{2n-1}X) + b_{2n}(\tau)\cos(\lambda_{2n}X)] \tag{10.127b}$$

where λ_{2n-1} and λ_{2n} are the roots satisfying the characteristic equation $\sin\lambda = 0$, that is,

$$\lambda_{2n-1} = (2n-1)\pi, \qquad \lambda_{2n} = 2n\pi, \quad n = 1, 2, \ldots \tag{10.128}$$

Expressing X in condition (10.126d) into a series of the eigenfunctions of equations (10.127), gives

$$X = \sum_{n=1}^{\infty}\{S_{2n-1}\sin(\lambda_{2n-1}X) + S_{2n}\sin(\lambda_{2n}X)\} \tag{10.129}$$

where

$$S_{2n-1} = \frac{4(-1)^{n+1}}{(2n-1)^2\pi^2}, \qquad S_{2n} = 0 \tag{10.130}$$

The order of magnitude of the generalized coordinates a_i, b_i, and X_1 is assumed to be as follows

$$a_1, \quad b_1, \quad X_1 \sim O(\varepsilon^{1/3}), \qquad a_2, \quad b_2, \quad C \sim O(\varepsilon^{2/3}) \tag{10.131}$$

Upon expanding equations (10.126c) and (10.126d) in Taylor series about $\tilde{\eta} = 0$, substituting equations (10.105), and equating the coefficients of $\sin\lambda_1 x$ and $\cos\lambda_2 x$ on both sides, gives

$$\dot{a}_1 + \zeta a_1 + Q_1 b_1 + Q_2\ddot{X}_1 + Q_3\dot{a}_1 b_2 + Q_4\dot{a}_1 b_1^2 + Q_5\dot{a}_2 b_1 + Q_6 a_1 a_2 + Q_7 a_1^2 b_1 = 0 \tag{10.132a}$$

$$\dot{a}_2 + Q_8 b_2 + Q_9\dot{a}_1 b_1 + Q_{10}a_1^2 = 0 \tag{10.132b}$$

$$\dot{b}_1 + Q_{11}a_1 + Q_{12}a_1b_2 + Q_{13}a_2b_1 + Q_{14}a_1b_1^2 = 0 \tag{10.132c}$$

$$\dot{b}_2 + Q_{15}a_2 + Q_{16}a_1\dot{b}_1 = 0 \tag{10.132d}$$

Evaluating the integral of the hydrodynamic force (10.123) and substituting in equation (10.122), gives the structure equation of motion

$$Q_{17}\ddot{X}_1 + C\dot{X}_1 + KX_1 + Q_{18}\dot{a}_1 + \zeta Q_{19}a_1 + Q_{20}b_2\dot{a}_1 + Q_{21}b_1^2\dot{a}_1$$
$$+ Q_{22}b_1\dot{a}_2Q_{13}b_2\ddot{X}_1 + Q_{24}b_1\dot{X}_1^2 + Q_{25}a_1a_2 + Q_{26}b_1b_2 + Q_{27}a_1^2b_1 = f_0\cos\nu\tau \tag{10.132e}$$

where a linear viscous damping is introduced into the liquid amplitude equation (10.132a). The coefficients Q_i are listed in Ikeda and Nakagawa (1997). The solution of equations (10.110) may be assumed to be in the form

$$a_1 = u_1\cos\nu\tau - v_1\sin\nu\tau + w_1\cos 3\nu\tau - z_1\sin 3\nu\tau \tag{10.133a}$$

$$b_1 = u_2\cos\nu\tau - v_2\sin\nu\tau + w_2\cos 3\nu\tau - z_2\sin 3\nu\tau \tag{10.133b}$$

$$X_1 = u_3\cos\nu\tau - v_3\sin\nu\tau + w_3\cos 3\nu\tau - z_3\sin 3\nu\tau \tag{10.133c}$$

$$a_2 = e_1\cos 2\nu\tau - f_1\sin 2\nu\tau \tag{10.133d}$$

$$b_2 = e_2\cos 2\nu\tau - f_2\sin 2\nu\tau + r_0 \tag{10.133e}$$

where the amplitudes u_i, v_i, w_i, z_i, e_i and f_i are slowly varying with time and their orders of magnitude are in proportion to those of *ai*, *bi*, and X_1. Every time the amplitudes are differentiated with respect to time their orders are assumed to decrease by $O(\varepsilon^{2/3})$ as follows:

$$u_i,\ vv_i \sim O(\varepsilon^{1/3}),\quad \dot{u}_i,\ \dot{v}_i \sim O(\varepsilon^{3/3}),\quad \ddot{u}_i,\ \ddot{v}_i(\varepsilon^{5/3})$$
$$w_i,\ z_i,\ e_j, f_j,\ r_0 \sim O(\varepsilon^{2/3}),\quad \dot{w}_i,\ \dot{z}_i,\ \dot{e}_{j,j},\ \dot{r}_0 \sim O(\varepsilon^{4/3}),\quad \ddot{w}_i,\ \ddot{z}_i,\ \ddot{e}_j, \ddot{f}_j,\ \ddot{r}_0 \sim O(\varepsilon^{6/3}) \tag{10.134}$$

Substituting equations (10.133) into equations (10.132) and setting the coefficients of the terms with frequencies ν, 2ν, and 3ν, and the constant term to zero, subject to the orders of magnitude shown (10.134), gives

$$\dot{u}_1 - 2Q_2\nu\dot{v}_3 = -\zeta u_1 + \nu v_1 - Q_1u_2 + Q_2\nu^2u_3 + g_1(u_i, v_i, e_i, f_i, r_0) \tag{10.135a}$$

$$\dot{v}_1 + 2Q_2\nu\dot{u}_3 = -\nu u_1 - \zeta v_1 - Q_1v_2 + Q_2\nu^2v_3 + g_2(u_i, v_i, e_i, f_i, r_0) \tag{10.135b}$$

$$\begin{aligned}\dot{u}_2 &= -Q_{11}u_1 + \nu v_2 + g_3(u_i, v_i, e_i, f_i, r_0)\\ \dot{v}_2 &= -\nu u_2 - Q_{11}v_1 + g_4(u_i, v_i, e_i, f_i, r_0)\end{aligned} \tag{10.135c,d}$$

$$Q_{18}\dot{u}_1 - 2Q_{17}\nu\dot{v}_3 = -\zeta Q_{19}u_1 + Q_{18}\nu v_1 - (K - Q_{17}\nu^2)u_3 + C\nu v_3 + f_0$$
$$+ g_5(u_i, v_i, e_i, f_i, r_0) \tag{10.135e}$$

$$Q_{18}\dot{v}_1 + 2Q_{17}\nu\dot{u}_3 = -Q_{18}\nu u_1 - \zeta Q_{19}v_1 - C\nu u_3 - (K - Q_{17}\nu^2)v_3 + f_0 + g_6(u_i, v_i, e_i, f_i, r_0) \tag{10.135f}$$

$$\begin{aligned} -3\nu z_1 + Q_1 w_2 - 9Q_2\nu^2 w_3 &= g_7(u_i, v_i, e_i, f_i) \\ 3\nu w_1 + Q_1 z_2 - 9Q_2\nu^2 z_3 &= g_8(u_i, v_i, e_i, f_i) \end{aligned} \tag{10.135g, h}$$

$$\begin{aligned} Q_{11}w_1 - 3\nu z_2 &= g_9(u_i, v_i, e_i, f_i) \\ Q_{11}z_1 + 3\nu w_2 &= g_{10}(u_i, v_i, e_i, f_i) \end{aligned} \tag{10.135i, j}$$

$$-3Q_{18}\nu z_1 + (K - 9Q_{17}\nu^2)w_3 = g_{11}(u_i, v_i, e_i, f_i) \tag{10.135k}$$

$$3Q_{18}\nu w_1 + (K - 9Q_{17}\nu^2)z_3 = g_{12}(u_i, v_i, e_i, f_i) \tag{10.135l}$$

$$\begin{aligned} e_1 &= g_{13}(u_i, v_i, e_i, f_i) \\ f_1 &= g_{14}(u_i, v_i, e_i, f_i) \\ e_2 &= g_{15}(u_i, v_i, e_i, f_i) \end{aligned} \tag{10.135m, n, o}$$

$$\begin{aligned} f_2 &= g_{16}(u_i, v_i, e_i, f_i) \\ r_0 &= g_{17}(u_i, v_i) \end{aligned} \tag{10.135p, q}$$

The explicit expressions of the nonlinear functions g_i are listed in Ikeda and Nakagawa (1997). For the case of frozen liquid the system behaves as a single degree of freedom governed by the equation

$$Q_{17}\ddot{X}_1 + C\dot{X}_1 + KX_1 = f_0 \cos \nu\tau \tag{10.136}$$

This equation has the steady-state solution, $a_i = b_i = 0$,

$$|X_1| = \frac{f_0}{\sqrt{(k - Q_{17}\nu^2)^2 + C^2\nu^2}} \tag{10.137}$$

The liquid–structure interaction, when the main system natural frequency is close to the liquid first mode asymmetric sloshing frequency, ω_1, was obtained by solving equations (10.135). The equations were numerically solved for the steady state by setting $\dot{u}_i = \dot{v}_i = 0$. The solution was obtained for the system parameters: mass ratio of the main mass to the total mass, $\mu = 0.93$, main system frequency ratio, $K = 1.0$, main system damping parameter, $C = 0.013$, liquid damping parameter, $\zeta = 0.024$, liquid depth ratio, $H = 0.6$, ratio of the liquid mass to the system total mass, $D = 0.1$, excitation force amplitude ratio, $f_0 = 0.0015$, excitation frequency ratio, $\nu = 1.06$, and at $X = 0.15$. Figure 10.49 shows the amplitude–frequency curves, where $X_1 = \sqrt{u_3^2 + v_3^2}$, $b_1 = \sqrt{u_2^2 + v_2^2}$, $b_2 = \sqrt{e_2^2 + f_2^2}$, and $-r_0$ is the constant component. The solid and dash curves represent the stable and unstable solutions, respectively. The unimodal response of the system, estimated according to equation (10.137), is shown in Figure 10.49(a) by the dash-dot curve. It is seen that at the resonance frequency, $\nu \approx 1.0$, the liquid interaction imparts suppression effect to the motion of the main system. The solid

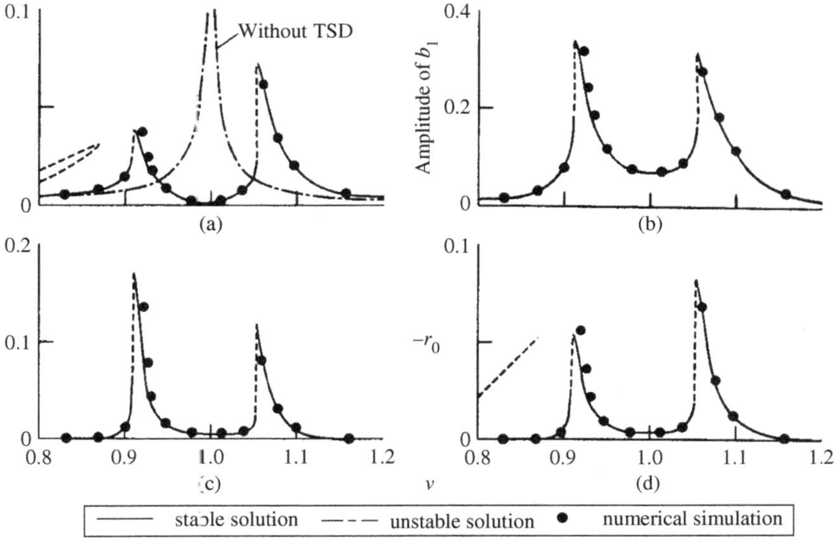

Figure 10.49 Amplitude–frequency response curves for $h/l=0.6$, $D=0.1$, $C=0.013$, $\mu=0.94$, $K=1.0$, $f_0=0.0015$, $X=0.15$ (a) amplitude of X_1 associated with frequency ν, (b) amplitude of b_1 associated with frequency ν, (c) amplitude of b_2 associated with frequency 2ν, (d) amplitude of the constant component r_0 in b_2. (Ikeda and Nakagawa, 1997)

points, •, are obtained from the FFT of the time history solutions of equations (10.132). The amplitude–frequency curves for two different values of liquid depth ratio $H=0.33$ and 0.2, revealed that the response curves change their trend from softening to hardening characteristics. It was shown that the nonlinear coupling contributes to time history records, whose frequency content includes super-summed and differential harmonic oscillations.

Figure 10.50 show a comparison between analytical and experimental resonance curves of the structure and liquid wave height, for a structure mass $m=6.08$ kg, spring stiffness $k=1789$ N/m, structure damping coefficient $c=1.109$ N s/m, tank length $l=0.1$ m, tank width $L=0.06$ m, liquid depth $h=0.06$ m, liquid mass $m_l=0.367$ kg, excitation amplitude $f_0=0.205$ N, and liquid natural frequency 2.725 Hz. The solid and dash curves are analytical results, the dash-dot curve is the unimodal response from equation (10.137). The experimental results are shown by empty circles, o, the experimental results in the absence of interaction with the liquid free surface are shown by solid points, •. The arrows indicate the occurrence of jump and collapse phenomena. Figures 10.51 and 10.52 show other amplitude–frequency response results of the structure for two different values of liquid depth $h=0.03$ m, and 0.02 m, respectively. Each figure is for different structure and tank parameters. The two figures reveal the switching from soft to hard spring characteristics as the water depth is decreased.

10.8 Ship roll stabilization using liquid tanks

The idea of liquid sloshing absorber systems described in Section 10.3 for the roll stabilization of ships has been realized for a long time. Froude (1874) is believed to have been the first to use anti-roll water tanks to reduce the ship roll motion. Later, Watts (1883, 1885) described in detail the mechanism by which a roll damping moment is created by the wave action of the

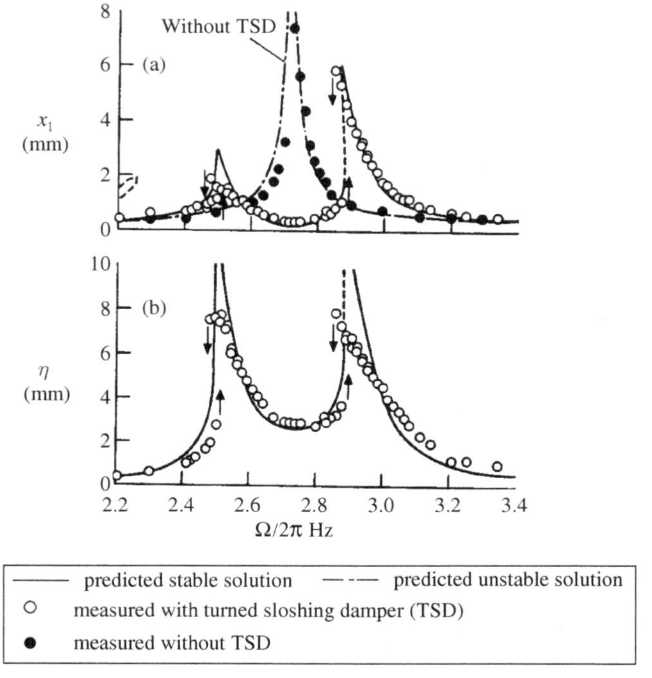

Figure 10.50 Comparison of predicted and measured amplitude-frequency response curves for $h = 60$ mm. (Ikeda and Nakagawa, 1997)

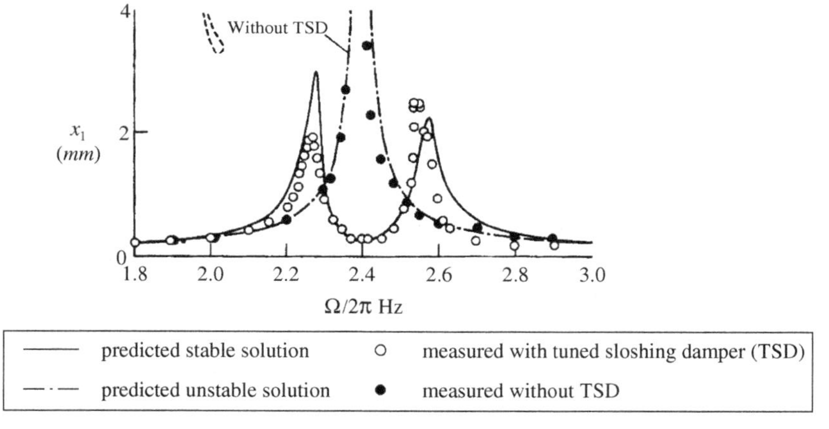

Figure 10.51 Comparison of predicted and measured amplitude-frequency response curves of x_1 for $h = 30$ mm. (Ikeda and Nakagawa, 1997)

liquid in a rectangular tank placed aboard a ship. Frahm (1911) introduced an anti-roll tank in the form of a U-tube shown in Figure 10.53(a) and (b) (taken from Den Hartog, 1985). For these tanks, a moment counteracting the rolling motion is generated by the oscillating water. Figure 10.53(a) shows the old version of Frahm's anti-rolling tanks. The two tanks are half filled with water and communicate through a water pipe below and through an air pipe with a valve V above. Figure 10.53(b) shows the modern construction in which the lower connecting

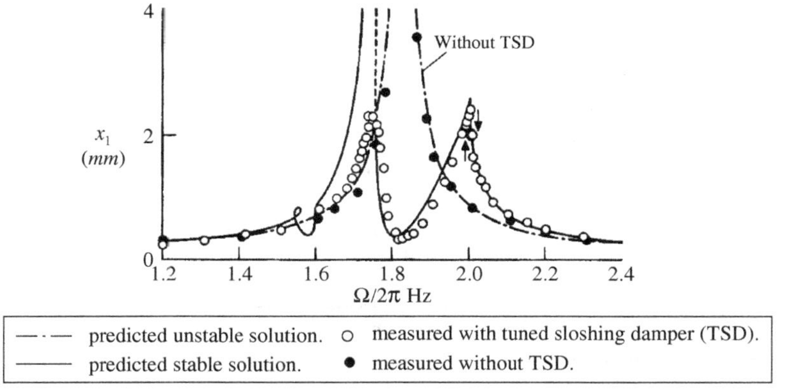

Figure 10.52 Comparison of predicted and measured amplitude-frequency response curves of x_l for $h = 20$ mm. (Ikeda and Nakagawa, 1997)

Figure 10.53 Frahm's anti-roll ship tanks: (a) Two tanks each half filled with water (old type), (b) Modern "blister" construction of Frahm's anti-rolling tanks. (Den Hartog, 1985)

pipe between the tanks is omitted and replaced by the open ocean. The "blisters" extended along two-thirds of the length of the ship and are subdivided into three or more compartments by vertical partitions.

The roll stabilization is only effective if the roll frequency is larger than or equal to the natural frequency of the water oscillation. Chadwick and Klotter (1954) and Vasta, *et al.* (1961) proposed the use of liquid tanks as passive anti-roll stabilization systems. The use of anti-rolling tanks in the German luxury liners *Bremen* and *Europa* reduced the maximum roll from 15° to 5° (Reed, 1961). Van Den Bosch and Vugts (1964, 1966), Stigter (1966), Vugts (1969), and Webster, *et al.* (1988) studied the performance of various types of anti-rolls tanks. Field and Martin (1976) presented an analysis of the influence of different parameters of anti-roll tanks using a U-tube. Lewson (1976) developed a mathematical model to optimize the design of free-surface passive tanks. Barr and Ankudinov (1977) presented an assessment of a number of analytical methods for predicting the roll motion and its reduction using anti-roll tanks. The effectiveness of flow obstructions inside the tank in the ship roll stability was examined by Lee Vassalos (1996).

The equations of motion of the liquid in the tank and the tank–ship system were developed by Stigter (1966) and Plank, *et al.* (1972). They compared their analytical results with those obtained experimentally. Abdel Gawad, *et al.* (2001) modeled the ship motion by a single

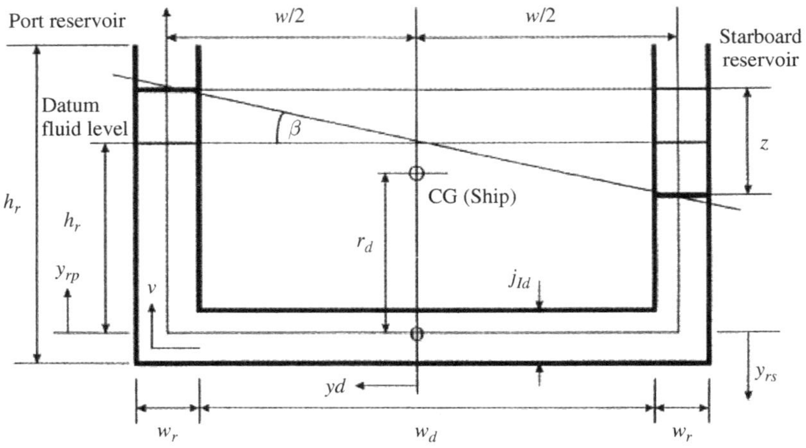

Figure 10.54 Schematic diagram of U-tube tank and its geometric parameters. (Abdel Gawad, *et al.*, 2001)

degree-of-freedom in roll. Numerical simulations of U-tanks were performed by Yamagushi and Shinkai (1995), Zhong, *et al.* (1998), Van Daalan, *et al.* (1999, 2000), Kleefsman (2000), Souto, *et al.* (2001), and Iglesias, *et al.* (2004).

Weng (1992) and Bass (1998) conducted experimental and analytical investigations to evaluate the behavior of anti-roll liquid tanks. The numerical models were used in combination with experimental tests with the purpose of selecting the optimum shape of the model. Iglesias, *et al.* (2004) used a numerical meshless method to simulate large free-surface liquid displacements.

Passive anti-roll liquid containers may not always provide satisfactory ship roll stabilization. In this case, active controlled-passive tanks may be introduced as proposed by Minrosky (1935) and Bell and Walker (1966). The control is effected by valves in the water channel between the two tanks in the U-shape case. At the same time, the control is effected by valves in the air channel. Bell and Walker proposed an active tank system with a propeller continuously driven in one direction to save power. Webster (1967) presented a detailed study of the control of pump-activated U-tube tanks. He indicated that tank control can be best selected on the basis of minimizing the response of the ship to an impulsive roll velocity.

10.8.1 Ship in pure rolling motion

Abdel Gawad, *et al.* (2001) presented a linear analysis to optimize the performance of a U-tube passive tank over a wide range of excitation frequency. With reference to the schematic diagram shown in Figure 10.54, Abdel Gawad, *et al.* derived the equations of motion of a ship exposed to beam waves and a U-tube tank in pure roll oscillations. With reference to the origin O located at the midpoint of the connecting duct and the axis y running along the duct and up the reservoirs of the U-tube, the liquid velocity v is taken along the positive y direction. The motion of a unit mass of liquid in the tank is governed by the one-dimensional Euler's equation

$$\frac{\partial v}{\partial t} + v\frac{\partial v}{\partial y} = Y - \frac{1}{\rho}\frac{\partial P}{\partial t} \tag{10.138}$$

where Y is the external force per unit mass, ρ is the density of the tank liquid. The cross-sections of the duct and reservoirs are assumed to be uniform such that

$$\frac{\partial v}{\partial y} = 0 \tag{10.139}$$

everywhere except at the junctions between the duct and the reservoirs. Neglecting the corner effects, equation (10.138) may be reduced to the form

$$\frac{\partial v}{\partial t} = Y - \frac{1}{\rho}\frac{\partial P}{\partial t} \tag{10.140}$$

Integrating along the tank y-axis, gives the flow velocity and consequently the liquid displacement, which determines the angel β. The details of this process are given by Lloyd (1989). The equation of motion of the tank angle, β, is

$$a\ddot{\beta} + b\dot{\beta} + c\beta + \left(d\ddot{\phi} + e\phi\right) = 0 \tag{10.141}$$

where ϕ is the ship roll angle, $a = Qw_r\left(\frac{w}{2h_d} + \frac{h_r}{w_r}\right)$, $b = Qqw_r\left(\frac{w}{2h_d^2} + \frac{h_r}{w_r^2}\right)$, $c = Q$, $d = Q(r_d + h_r)$, $e = Qg$, $Q = \frac{1}{2}\rho w_r w^2 x_t$, x_t is the tank length (in the fore–aft direction), q is the damping coefficient in the tank, g is the gravitational acceleration, and all other parameters are defined in Figure 10.54.

The equation of motion of the ship roll angle, ϕ, under roll excitation moment is

$$I\ddot{\phi} + \varsigma\dot{\phi} + C\phi + (d\ddot{\beta} + e\beta) = M \sin \Omega t \tag{10.142}$$

where M is the amplitude of the excitation roll moment, Ω is the wave encounter frequency, I is the ship roll inertia, ς is the ship roll damping, and C is the ship restoring moment coefficient. Note that the expression $\left(d\ddot{\beta} + e\beta\right)$ represents the roll stabilization moment from the liquid inside the tank. The liquid in the tank natural frequency, ω_t, and the natural frequency of the ship, ω_s, are, respectively,

$$\omega_t = \sqrt{\frac{c}{a}} = \sqrt{\frac{2gh_d}{w_t w + 2h_r h_d}}, \qquad \omega_s = \sqrt{\frac{C}{I}} \tag{10.143a,b}$$

Note that equations (10.141) and (10.142) are linearly coupled dynamically and statically. Their form is typical of linear vibration absorber equations described in Section 10.2. The only difference is the type of coupling. The steady state solution of equations (10.141) and (10.142) may be written in the form

$$\beta(t) = \beta_0\, e^{i\Omega t}, \quad \phi(t) = \phi_0\, e^{i\Omega t} \tag{10.144a,b}$$

where the response amplitudes (including phase angles), β_0 and ϕ_0, are obtained by substituting (10.144) into equations (10.141) and (10.142), to give

$$\frac{\beta_0}{M/C} = \frac{(\omega_t/\omega_s)^2 - (d/a)(\Omega/\omega_s)^2}{D} \tag{10.145a}$$

$$\frac{\phi_0}{M/C} = \frac{\sqrt{\left[(\omega_t/\omega_s)^2 - (\Omega/\omega_s)^2\right]^2 + ((b/a\omega_s)(\Omega/\omega_s))^2}}{D} \tag{10.145b}$$

where

$$D = \left\{ \left\{ \left[\left(\frac{\omega_t}{\omega_s}\right)^2 - \left(\frac{d}{a}\right)\left(\frac{\Omega}{\omega_s}\right)^2 \right]^2 \left(\frac{a}{I}\right) - \left[\left(\frac{\omega_t}{\omega_s}\right)^2 - \left(\frac{\Omega}{\omega_s}\right)^2 \right] \left[1 - \left(\frac{\Omega}{\omega_s}\right)^2 \right] \right. \right.$$

$$\left. + \left(\frac{b}{a\omega_s}\frac{\Omega}{\omega_s}\right)\left(\frac{\varsigma}{a\omega_s}\frac{\Omega}{\omega_s}\frac{a}{I}\right) \right\}^2$$

$$\left. + \left\{ \left[\left(\frac{\omega_t}{\omega_s}\right)^2 - \left(\frac{\Omega}{\omega_s}\right)^2 \right] \left(\frac{\varsigma}{a\omega_s}\frac{\Omega}{\omega_s}\frac{a}{I}\right) + \left[1 - \left(\frac{\Omega}{\omega_s}\right)^2 \right] \left(\frac{b}{a\omega_s}\frac{\Omega}{\omega_s}\right) \right\}^2 \right\}^{1/2}$$

In the absence of damping, equation (10.145b) reveals that the ship roll motion, ϕ, will be suppressed in the excitation frequency Ω that is identical to the liquid natural frequency, ω_t. Figure 10.55(a) and (b) shows the ship and liquid responses in the frequency domain for different values of damping parameter, $b/2a\omega_s$. There is an optimum value of damping at which energy dissipation reaches its maximum values. As indicated in Section 10.2, all response curves intersect at two points regardless of the value of the damping ratio. Optimum tuning can be achieved by selecting the damping parameter, $b/2a\omega_s$, to yield the peak response amplitudes that equal the response at frequency ratio $\Omega/\omega_s = 1.0$. The optimum damping parameter was found to be $b/2a\omega_s = 0.17$, and the corresponding response is shown by the solid curves.

10.8.2 Influence of other ship motions

Youssef, *et al.* (2002) considered the ship roll stabilization by passive anti-roll tanks by taking into account all six degrees of freedom of ship motion. However, they used one freedom modeling to describe the roll stabilization. An improved model of the passive tank–liquid motion describing the ship six degrees of freedom was developed by Youssef, *et al.* (2003). With reference to the body ship axes shown in Figure 10.56, the translational equations of motion of the ship along surge, x, and sway, y, and heave, z, body axes are, respectively, see, e.g., Bhattacharyya (1978) and Fossen (1994),

$$m\left[\dot{u} - v\dot{\psi} + w\dot{\theta} - x_G\left(\dot{\theta}^2 + \dot{\psi}^2\right) + y_G\left(\dot{\phi}\dot{\theta} - \ddot{\psi}\right) + z_G\left(\dot{\phi}\dot{\psi} + \ddot{\theta}\right)\right] = X_H + X_T \tag{10.146a}$$

$$m\left[\dot{v} - w\dot{\phi} + u\dot{\psi} - y_G\left(\dot{\psi}^2 + \dot{\phi}^2\right) + z_G\left(\dot{\psi}\dot{\theta} - \ddot{\phi}\right) + x_G\left(\dot{\theta}\dot{\phi} + \ddot{\psi}\right)\right] = Y_H + Y_T \tag{10.146b}$$

$$m\left[\dot{w} - u\dot{\theta} + v\dot{\phi} - z_G\left(\dot{\phi}^2 + \dot{\theta}^2\right) + x_G\left(\dot{\psi}\dot{\phi} - \ddot{\theta}\right) + y_G\left(\dot{\psi}\dot{\theta} + \ddot{\phi}\right)\right] = Z_H + Z_T \tag{10.146c}$$

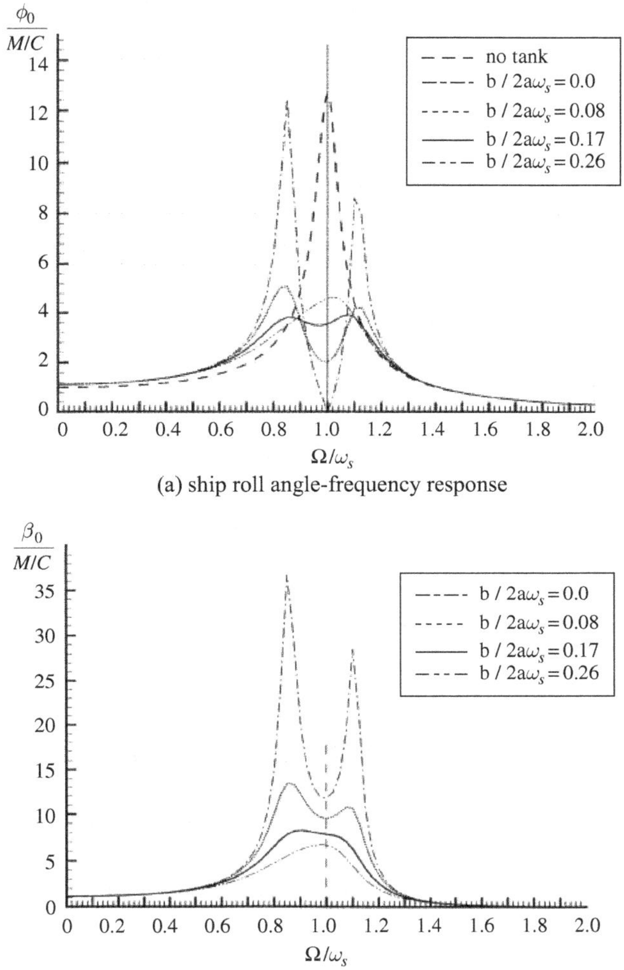

(a) ship roll angle-frequency response

(b) Liquid angle amplitude-frequency response

Figure 10.55 Ship (a) and liquid (b) amplitude–frequency responses for different values of damping parameter. (Gawad, *et al.*, 2001)

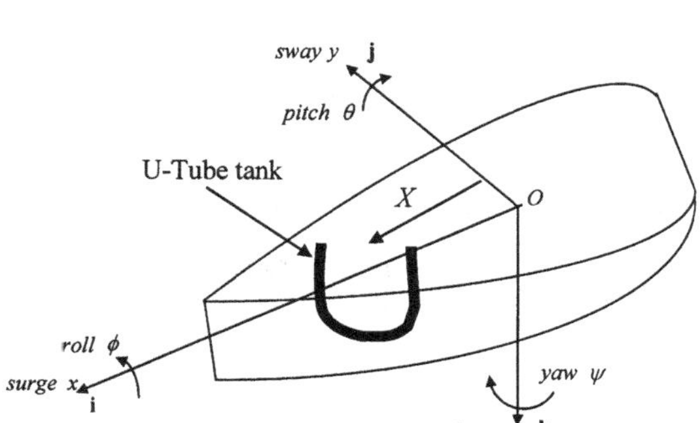

Figure 10.56 Schematic diagram of a ship showing the body axes and rotational motion and the location of a U-tube tank

The rotational equations of motion about the x, y, and z body axes are, respectively,

$$\begin{aligned} I_{xx}\ddot{\phi} + (I_{zz} - I_{yy})\dot{\theta}\dot{\psi} - I_{xz}(\ddot{\psi} + \dot{\phi}\dot{\theta}) + I_{yz}(\dot{\psi}^2 - \dot{\theta}^2) - I_{xy}(\ddot{\theta} - \dot{\phi}\dot{\psi}) \\ + m\left[y_G(\dot{w} - u\dot{\theta} + v\dot{\phi}) - z_G(\dot{v} - w\dot{\phi} + u\dot{\psi})\right] = M_{xH} + M_{xT} \end{aligned} \tag{10.147a}$$

$$\begin{aligned} I_{yy}\ddot{\phi} + (I_{xx} - I_{zz})\dot{\theta}\dot{\psi} - I_{xy}(\ddot{\psi} + \dot{\phi}\dot{\theta}) + I_{zx}(\dot{\psi}^2 - \dot{\theta}^2) - I_{yz}(\ddot{\theta} - \dot{\phi}\dot{\psi}) \\ + m\left[z_G(\dot{u} - v\dot{\psi} + w\dot{\theta}) - x_G(\dot{w} - u\dot{\theta} + v\dot{\phi})\right] = M_{yH} + M_{yT} \end{aligned} \tag{10.147b}$$

$$\begin{aligned} I_{zz}\ddot{\phi} + (I_{yy} - I_{xx})\dot{\theta}\dot{\phi} - I_{yz}(\ddot{\theta} + \dot{\psi}\dot{\phi}) + I_{xy}(\dot{\theta}^2 - \dot{\phi}^2) - I_{zx}(\ddot{\phi} - \dot{\psi}\dot{\theta}) \\ + m\left[x_G(\dot{v} - w\dot{\phi} + u\dot{\psi}) - y_G(\dot{u} - v\dot{\psi} + w\dot{\theta})\right] = M_{zH} + M_{zT} \end{aligned} \tag{10.147c}$$

where m is the mass of the ship, I_{xx}, I_{yy}, and I_{zz} are the ship mass moments of inertia about the x, y, and z body axes, respectively, x_G, y_G, and z_G represent the position of the ship center of gravity alone, u, v, and w are components of the linear velocity of the body origin O along the x-, y-, and z-axes, respectively. The angular rotations ϕ, θ, and ψ represent the roll, pitch, and yaw about the x-, y-, *and z-axes, respectively,* X_H, Y_H, *and* Z_H *are the components of hydrodynamic forces, while* X_T, Y_T, *and* Z_T *are the force components acting by the tank–liquid motion on the ship. The corresponding moment components due to hydrodynamic forces and tank–liquid motion are* M_{xH}, M_{yH}, M_{zH}, *and* M_{xT}, M_{yT}, M_{zT}, *respectively.*

The hydrodynamic force components X_H, Y_H, and Z_H, and hydrodynamic moment components were estimated by using the Large-Amplitude Program (LAMP) code.[1] This code is capable of calculating simultaneously and interactively in the time domain the forces and moments exerted by the tank system on the ship and the motions of both of the ship and the moving liquid in the tanks.

The forces and moments exerted on the ship by the sloshing liquid inside the tank (shown in Figure 10.57) on the ship are, see Youssef, *et al.* (2003),

$$\begin{aligned} X_T = -\rho A_1[C_1(g\sin\theta + \dot{u} + \dot{\theta}w - \dot{\psi}v - (\dot{\theta}^2 + \dot{\psi}^2)X) + 2B(\ddot{\psi} - \dot{\phi}\dot{\theta})Z \\ + C_2(\ddot{\theta} + \dot{\phi}\dot{\psi}) + (2B\dot{\psi} - 4\dot{\theta}Z)\dot{Z}] \end{aligned} \tag{10.148a}$$

$$\begin{aligned} Y_T = -\rho A_1[C_1(-g\sin\phi\cos\theta + \dot{v} + \dot{\psi}\dot{u} - \dot{\phi}w + (\ddot{\psi} + \dot{\phi}\dot{\theta})X) + 4\dot{\phi}Z\dot{Z} \\ - C_2(\ddot{\phi} - \dot{\theta}\dot{\psi}) + 2B(\dot{\phi}^2 + \dot{\psi}^2)Z - 2B\ddot{Z}] \end{aligned} \tag{10.148b}$$

$$\begin{aligned} Z_T = -\rho A_1[C_1(-g\cos\phi\cos\theta + \dot{w} + \dot{\phi}v - \dot{\theta}u - (\ddot{\theta} - \dot{\phi}\dot{\psi})X) - 4B\dot{\phi}\dot{Z} - 2\dot{Z}^2 \\ - C_2(\dot{\phi}^2 + \dot{\theta}^2) - 2B(\ddot{\phi} + \dot{\theta}\dot{\psi})Z - 2Z\ddot{Z}] \end{aligned} \tag{10.148c}$$

[1] Science Applications International Corporation, 1998, large-amplitude motions program (LAMP) for ship motions and wave loads predictions, Version 2.9.2 User's Guide, Annapolis, Maryland.

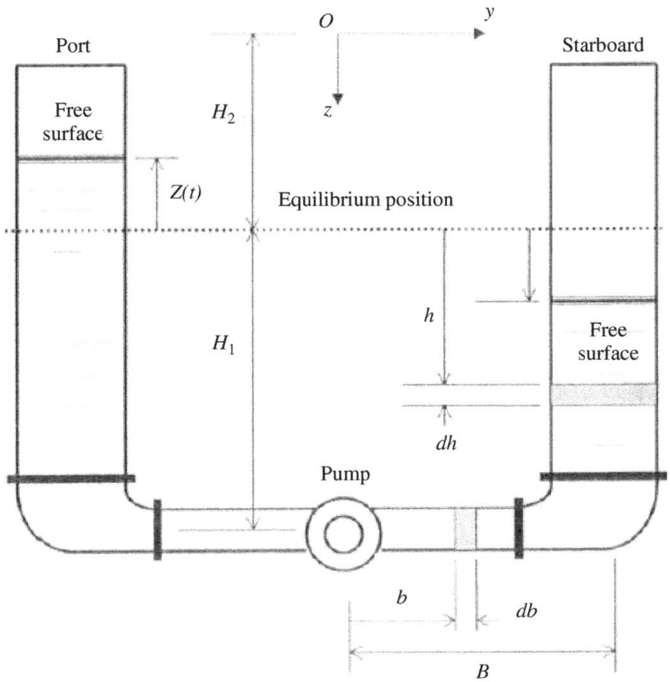

Figure 10.57 Schematic diagram of the U-tube tank system used by Youssef, *et al.* (2003).

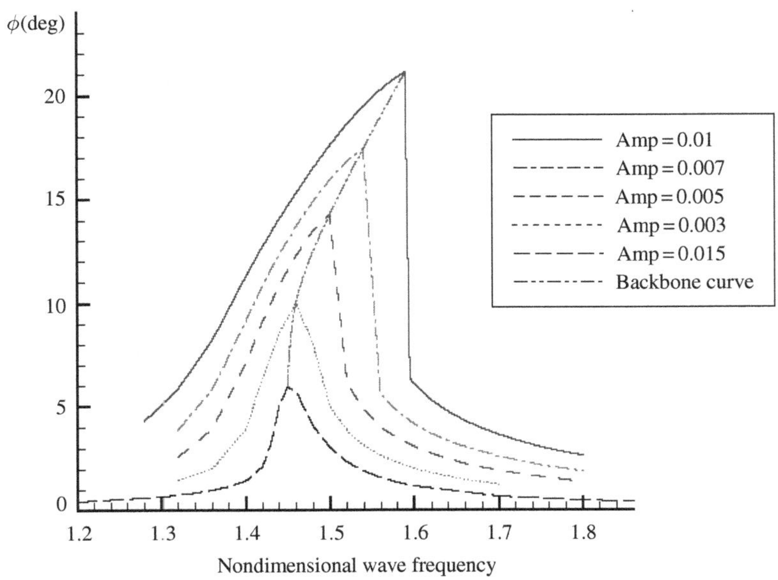

Figure 10.58 Ship roll angle-frequency response in the absence of anti-roll liquid tank for different values of beam wave amplitude. (Youssef, *et al.*, 2003)

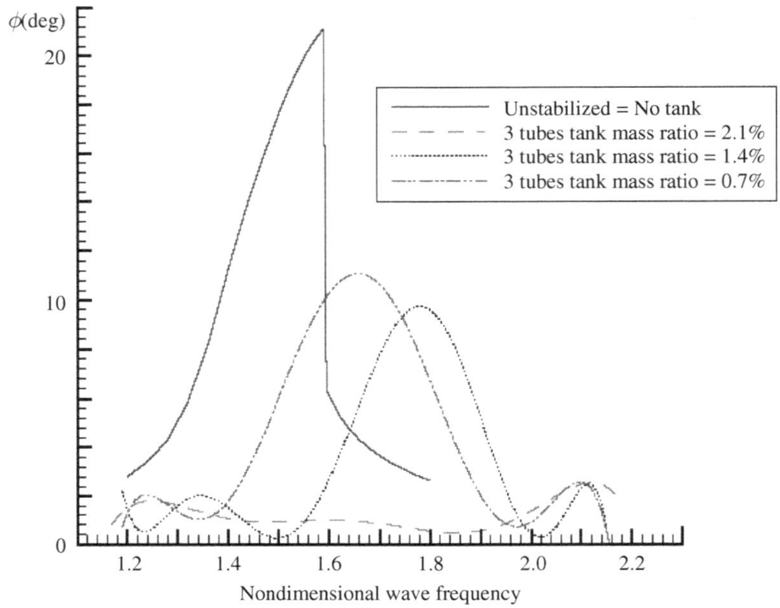

Figure 10.59 Ship roll angle-frequency response in the absence and presence of anti-roll liquid tank for different values of tank mass ratio under a beam wave amplitude of 0.01. (Youssef, *et al.*, 2003)

$$
\begin{aligned}
M_{xT} = & -\rho A_1 \Big[2(B+\tilde{B})H_3\ddot{Z} - 4H_2\dot{\phi}Z\dot{Z} - 2BZ(\dot{w} + \dot{\phi}v - \dot{\theta}u - H_2(\dot{\theta}^2 - \dot{\psi}^2) \\
& - X(\ddot{\theta} - \dot{\phi}\dot{\psi}) + g\cos\phi\cos\theta) + \frac{2}{3}\Big\{ B^2\tilde{B} + 3H_3^2(\tilde{B} + 3H_2) \\
& \qquad + 3H_2^2H_3 + H_3^3 - 3H_2Z^2 \Big\}(\ddot{\phi} - \dot{\theta}\dot{\psi}) \\
& + 2B^2H_3(\ddot{\phi} + \dot{\theta}\dot{\psi}) - (2H_2H_3 + H_3^2 + 2\tilde{B}H_3 - Z^2)(\dot{v} - \dot{\phi}w + \dot{\psi}u) \\
& + Z(\ddot{\psi} + \dot{\phi}\dot{\theta}) - g\sin\phi\cos\theta \Big]
\end{aligned}
\tag{10.149a}
$$

$$
\begin{aligned}
M_{yT} = & -\rho A_1 [2XZ\ddot{Z} + 4(\tilde{B}X\dot{\phi} - H_2Z\dot{\theta} + BH_3\dot{\psi}) \\
& - 2X(\tilde{B} + H_3)(\dot{w} - \dot{\theta}u - \dot{\phi}v - X(\dot{\theta} - \dot{\phi}\dot{\psi}) - g\cos\phi\cos\theta) \\
& + \frac{2}{3}\{B^2\tilde{B} + 3H_3^2(\tilde{B} + 3H_2) + 3H_2^2H_3 + H_3^3 - 3H_2Z^2\}(\ddot{\theta} + \dot{\phi}\dot{\psi}) \\
& + 2BH_2Z(\ddot{\psi} - \dot{\phi}\dot{\theta}) + 2BXZ(\ddot{\phi} + \dot{\theta}\dot{\psi}) + 2X\dot{Z}^2 \\
& + (2H_2H_3 + 2\tilde{B}H_3 + H_3^2 - Z^2)(\dot{u} - \dot{\psi}v + \dot{\theta}w) \\
& + X(\dot{\phi}^2 - \dot{\psi}^2) + g\sin\theta]
\end{aligned}
\tag{10.149b}
$$

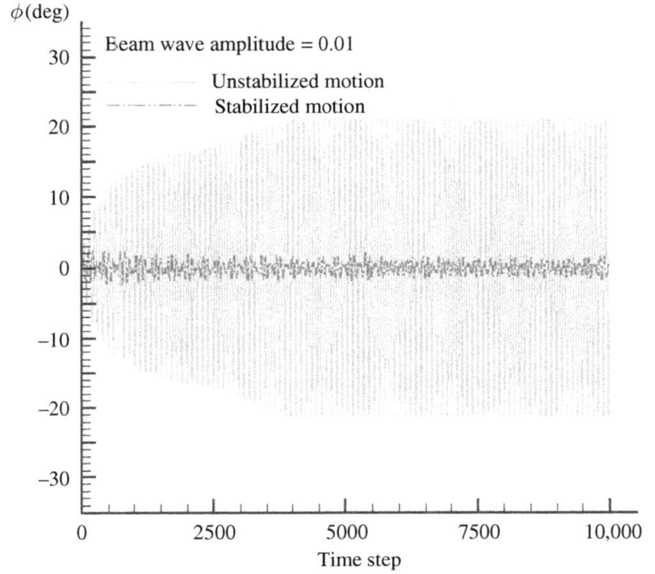

(a) Ship roll angle

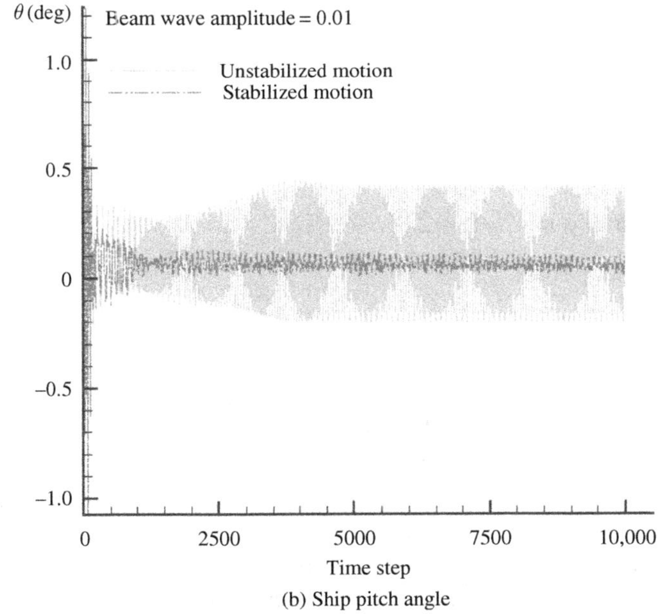

(b) Ship pitch angle

Figure 10.60 Time history records of ship roll and pitch angles in the absence and presence anti-roll tank. (Youssef, *et al.*, 2003)

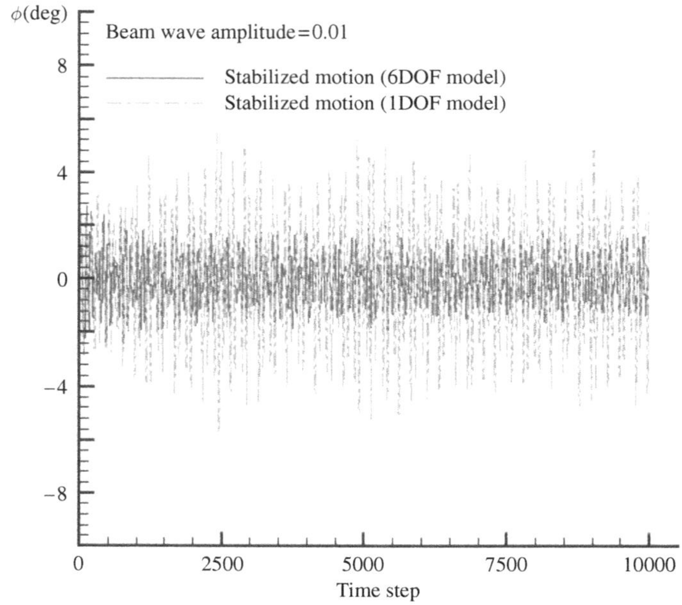

Figure 10.61 Time history records in the presence of anti-roll liquid tank system as estimated from 6DOF and 1DOF models in the nonlinear range in beam sea wave. (Youssef, *et al.*, 2003)

$$\begin{aligned} M_{zT} = -\rho A_1 \big[& 2X\tilde{B}\ddot{Z} + 4\big(XZ\dot{\phi} - BH_3\dot{\theta}\big)\dot{Z} + 2BZ\big(\dot{u} - \dot{\psi}v + \dot{\theta}w + H_2(\ddot{\theta} + \dot{\phi}\dot{\psi}) \\ & + X(\dot{\phi}^2 - \dot{\theta}^2) + g\sin\theta\big) - X\big(2H_2H_3 + 2\tilde{B}H_3 + H_3^2 - Z^2\big)\big(\ddot{\phi} - \dot{\theta}\dot{\psi}\big) \\ & + 2B^2H_2\big(\ddot{\psi} - \dot{\phi}\dot{\theta}\big) + 2X\big(\tilde{B} + H_3\big)\big(\dot{v} + \dot{\psi}u - \dot{\phi}w\big) + X\big(\ddot{\psi} + \dot{\phi}\dot{\theta}\big) \\ & - g\sin\phi\cos\theta\big] \end{aligned} \tag{10.149c}$$

where $R = A_1/A_2$, A_1 and A_2 are the cross-sectional areas of each tank and connecting pipe, respectively, $H_3 = H_1 + H_2$, $\tilde{B} = B/R$, and B, H_1, H_2, and Z are defined in Figure 10.57. Also, $C_1 = 2(H_1 + B/R)$, $C_2 = (2H_3B/R) + H_3^2 - H_2^2 - Z^2$, and X is the location of the tank from the plane *yoz* of the ship as shown in Figure 10.56.

For a ship with a linear roll natural frequency of 1.45, and different values of beam wave amplitude, the roll amplitude–frequency response in the absence of liquid tanks is shown in Figure 10.58. The frequency axis represents the nondimensional encounter-wave frequency. For relatively large wave amplitude, the response is nonlinear of hard type characteristics. In the presence of a liquid–tank system (of natural frequency 1.59) with three tubes mounted along the ship length, the roll angle is significantly reduced as shown in Figure 10.59, which shows three response corves corresponding to three different values of tank/ship mass ratio. It is seen that best performance is achieved for the tank/ship mass ratio of 2.1%. The three tubes are arranged in the ship such that one is placed at the mid-ship section, then two tubes are placed 5 m fore and aft of the mid-ship section. The total length of the ship is about 135 m.

In the nonlinear range of roll motion in a beam wave with an amplitude of 0.01, the time history records of ship roll and pitch motions in the absence and presence of the liquid–tank system are shown in Figure 10.60(a) and (b). It is seen that the roll motion is suppressed from

21° to 2° as a result of the presence of the liquid–tank system. The passive tank has a similar on the pitch motion. Figure 10.61 shows a comparison between time history records as obtained using single degree-of-freedom and six degrees-of-freedom models in the nonlinear range under wave amplitude of 0.01. It is seen that the 6DOF model yields a reduction of roll angle to 2° compared to 4° obtained using the 1DOF model. The 1DOF model results were obtained by Youssef, *et al* (2002).

10.9 Closing remarks

Liquid sloshing hydrodynamic forces are exploited to suppress the dynamic motion of multi-storey buildings subjected to earthquakes and wind forces. Civil engineers utilized the liquid containers and restricted their analyses to linear interaction. Their analyses and experimens did not take into account the nonlinear interaction between the liquid free surface and support structures. For specific liquid–structure design, the nonlinear interaction can give rise to the occurrence of internal resonance and makes the liquid container act as an autoparametric vibration absorber. The nonlinear interaction under horizontal and vertical excitations has been treated in this chapter. The treatment considered both harmonic and random excitations. While liquid tanks can act as vibration absorbers in multistorey buildings, these buildings need, at the same time, isolation from the ground motion. The need for the isolators is because, at the moment of earthquake occurrence, the building response may be severe and there may not be enough time for the liquid tank to absorb the building's motion.

Part IV

Rotating fluid and low gravity sloshing

11
Dynamics of rotating fluids

11.1 Introduction

The problem of fluid dynamic behavior in spinning tanks is encountered in the study of stability and control of rockets, space vehicles, liquid-cooled gas turbines, centrifuges, and oceans. Some spacecraft vehicles are designed to spin in order to gain gyroscopic stiffness during the transfer from low earth orbit. This spin helps to control liquid propellant location in its container. Another class of problems deals with the dynamics and stability of rotating rigid bodies as applied to the evolution of celestial bodies and astronavigation control. Stabilization is achieved when the spacecraft spins about its axis of minimum moment of inertia. A satellite that spins about its axis of minimum moment of inertia may experience instability in the presence of energy dissipation. This is similar to a spinning top on a rough surface and as a result of friction the top's nutation angle increases as it seeks to conserve angular momentum.

The theory of rotating fluids is well documented in Lamb (1945), Morgan (1951), Howard (1963), and Greenspan (1968). Usually, the fluid motion is characterized by certain types of boundary layers and dissipative behavior in addition to complicated viscous layers controlled processes for spin-up and spin-down (Wilde and Vanyo, 1993). Lighthill (1966) presented a survey on the dynamics of rotating fluid. The results always have to be distinguished by the magnitude of the ratio of spin speed and gravity, that is, by the parameter $\Omega^2 R/g$, where Ω is the spin speed, R is the tank radius, and g is the gravitational acceleration. If the free surface motion is mainly governed by the predominant axial acceleration (gravity) the surface exhibits a paraboloidal geometry. On the other hand, if $\Omega^2 R/g \gg 1$ the geometry of the liquid free surface is dominated by the centrifugal acceleration, yielding a free surface of an approximately circular cylindrical geometry.

The dynamic characteristics of rigid bodies containing liquid cavities involve estimation of stability and bifurcation of steady rotation. The spin axis is taken as the axis of symmetry. A well-known stability condition for the stability of a spinning top in the sleeping position is (Chetayev, 1957),

$$I_{zz}^2 \Omega^2 > 4mgZ_{cg} I_{xx} \tag{11.1}$$

where I_{zz} and I_{xx} are the mass moments of inertia of the top about the spin and transverse axes through a fixed point, respectively, m is the top mass, Z_{cg} is the distance of the center of gravity from the fixed point. Rumyantsev (1960) considered a top with an annular cavity in the form of a torus that does not intersect with the spin axis and obtained similar stability condition. The

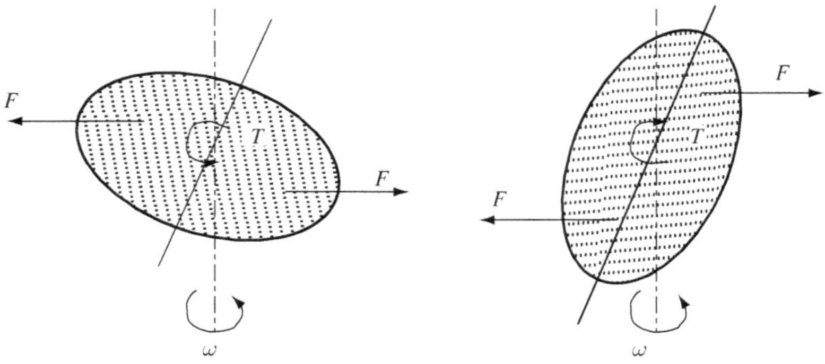

Figure 11.1 (a) Stabilizing and (b) destabilizing effects of rotating fluids.

stability of a spinning axisymmetric rocket with dissipative internal mass motion was considered by Yam, *et al.* (1997),

In the absence of viscosity, a spherical shell cannot communicate any of its rotational motion to the interior fluid, which remains unaffected. Consideration of boundary layer effects is required to determine the nature of the forced flow. The small convective circulation established by viscous boundary layers is the most important process controlling spin-up.

The physics of rotating fluids has direct application to fuel explosives in artillery shells with a large ratio of height to diameter ratio. They possess a large angular velocity and the tank geometry may be unfavorable to dynamic stability of the complete shell–fluid system. Pfeiffer (1977) presented an assessment of the problems of contained rotating fluids with reference to aerospace application. Magnus (1971) demonstrated the stability of rotating fluids in terms of the induced torque on the spinning bodies shown in Figure 11.1. Consider, for example, a fluid rotating about the axis of symmetry of an ellipsoidal cavity. If the axis of rotation is inclined, the fluid tends initially to continue its former motion around the old attitude of the spin axis.

The motion of the fluid itself in spinning bodies is very complicated and can be characterized by three types of wave patterns:

(1) The geostrophic flow (or mode) occurring when the Coriolis force and pressure gradient are in exact balance over many periods of revolution. In geostrophic flow, a column of fluid of constant height moves about the interior of the container as a unit, always maintaining a constant length and exists only when the container has closed constant height contours.
(2) Inertia waves characterized by real eigenvalues, ω, over the nondimensional range $-2 \leq \omega/\Omega \leq +2$, where Ω is the angular velocity of the container.
(3) Rossby waves (modes) characterized by low frequency inertial waves, replacing geostrophic flow when the container does not have constant height contours as in the case of a sliced cylinder whose bottom surface is not perpendicular to the cylinder axis. The ratio of the convective acceleration to the Coriolis acceleration is usually measured by the Rossby number, ε,

$$\varepsilon = \frac{U}{\Omega L} \tag{11.2}$$

where U is the relative velocity of a particular motion and L is a characteristic length. This number provides an overall estimate of the relative importance of nonlinear terms.

In all cases, the fluid motion is characterized by certain types of boundary layers and dissipative behavior as well as complicated viscous layers controlled processes for spin-up and spin-down.

The stability of rigid bodies with cavities filled with liquid is believed to have been first considered by Stokes (1949a,b), who introduced the definition of the "equivalent body" representing the fluid by an equivalent moment of inertia. The problem was considered earlier by Helmholtz (1860), Lübeck (1873), and Lamb (1945). Zhukovskii (1936) proved that the irrotational motion of the liquid in the cavity is determined by the rotation of the body and does not depend on its forward motion. The motion of the body proceeds as if the liquid had been replaced by an equivalent mass. A spin-stabilized top can become dynamically unstable if it is filled with a liquid (Kelvin, 1877). Kelvin (1877, 1880, 1882) described some experiments in which a liquid-filled sphere became violently unstable when it was slightly deformed into a prolate form. He also observed that a top is stable if the cavity is oblate. The mathematical treatment of the homogeneous vortex motion of a liquid in an ellipsoid cavity was considered by Greenhill (1880), Slutskii (1895), Hough (1895), Poincaré (1910), and Basset (1914). For example, Hough (1895) obtained stability conditions of the small oscillations about the state of uniform rotation about the principal axis of inertia of the rigid body and its fluid. The stability of rigid bodies with a liquid-filled cavity using Lyapunov's method was treated by Rumyantsev (1964b, 1968).

According to Stewartson (1959), the stability of the top containing a cylindrical liquid-filled cavity can be predicted if the liquid is in rigid-body rotation. However, for lightly viscous fluids, a relatively long time is required for the liquid to reach full spin. During the transition period, the top might become dynamically unstable even though it might be stable at its final state if the liquid attained rigid-body rotation.

There are two methods for studying the stability of the steady state motion of these systems. The first is the spectrum analysis (Sobolev, 1960, Zhu, 1988, and Liu, 1992), and the second is based on Lyapunov's method (Pozharitskiy, 1962, 1964, Pozharitskiy and Rumyantsev, 1963, Rumyantsev, 1964a,b, Moiseev and Rumyantsev, 1968, Li, 1983a, b, and Zhu, 1984, 1992). Rumyantsev (1964a,b) and Moiseev and Rumyantsev (1968) developed three theorems dealing with the stability of the steady rotation of a solid body with a liquid-filled cavity possessing a conservative component of the angular momentum. If a rigid body and its contained fluid do not rotate with the same speed, then the kinetic energy will be dissipated due to friction caused by the relative motion between the fluid and the body. The final motion must again be a rigid body rotation and for this motion to be stable, the system must rotate around one of the principal axes of inertia. Rumyantsev (1964a,b) used the sum of kinetic and potential energies as Lyapunov function. He introduced what is known as the motive potential, W, defined by the expression,

$$W = \frac{(M_\zeta^{(e)})^2}{2I_\zeta} + \Pi \tag{11.3}$$

where Π and I_ζ are the potential energy and moment of inertia about axis ζ, respectively, $M_\zeta^{(e)}$ is the angular momentum of the steady state rotation $\Omega^{(e)}$.

Theorem I If the expression of motive potential W has an isolated minimum $W^{(e)}$, then $\Omega^{(e)}$ is stable.

Theorem II If the liquid is viscous and the steady rotation $\Omega^{(e)}$ is isolated and if, in the neighborhood of $\Omega^{(e)}$, W has some values less than the value $W^{(e)}$ at $\Omega^{(e)}$ then the unperturbed rotation $\Omega^{(e)}$ is unstable.

Theorem III If the liquid is viscous and there exist no other steady motions satisfying the condition $M_\zeta = M_\zeta^{(e)}$ near $\Omega^{(e)}$ and, if in the neighborhood of $\Omega^{(e)}$, W has some values less than the value $W^{(e)}$ at $\Omega^{(e)}$ then the unperturbed rotation $\Omega^{(e)}$ is unstable.

The literature covers a wide spectrum of problems dealing with the dynamics and stability of spinning containers with fluid–filled cavities. The results are characterized by paradoxes and contradictions. For this reason, this chapter includes historical reviews of two groups of research activities. The first group deals with dynamical problems of bodies with cavities or containers filled completely with fluid. The first group is presented in Section 11.2. The second group addresses partially filled containers with liquid free surface and is treated in Section 11.3 for two- and three-dimensional problems. The stability theories of the two groups will first be briefly described. The parametric excitation of spinning containers partially filled with liquid is analyzed in Section 11.4. Sections 11.5 and 11.6 present the inertia waves in rotating containers and the influence of Rossby waves on inertia waves of the free surface in high spin-speed containers, respectively. The influence of rotating containers on the free-surface shape under spin up and spin down in a microgravity environment is considered in Section 2.7.

11.2 Fluid-filled spinning containers

11.2.1 Historical background

The dynamics of bodies with containers filled with liquid was the subject of early studies, which go back to Kelvin (1877, 1880), Lamb (1945), Zhukovskii (1948, 1949), and Chetayev (1957). The stability of rotating bodies having different principal moments of inertia as well as bodies rotating about a single stationary point in a gravitational field is the fundamental problem in this group. For example, Lamb (1945) obtained a criterion for the stability which requires that $I_{zz}/I_{xx} < 1$ or $I_{zz}/I_{xx} < 3$. The system is unstable within the range $1 < I_{zz}/I_{xx} < 3$, where I_{xx}, I_{zz} are the semi-axes of the spheroid that has the I_{zz}-axis as its spin axis.

The presence of fluid gives rise to no new physical phenomena. The stability conditions for the case of an ideal liquid have an invariable form, however, inertial properties involved in these conditions are changed at the cost of additional masses and moments of inertia. The conditions for stable regular precession of a symmetrical rigid body with a fixed point have been obtained by Karapetyan (1972) for bodies having an ellipsoid cavity completely filled with an ideal liquid. The stability conditions of rotations of a symmetrical rigid body, containing an ellipsoidal cavity filled with an ideal fluid, were examined for the case of motion on a smooth horizontal plane and a plane with sliding friction by Markeyev (1985a, b), Karapetyan and Prokomina (2000), and Rudenko (2002). Karapetyan (2001) considered the symmetrical rigid body, containing a spherical cavity filled with an ideal fluid. Grigoryeva (1999, 2000) indicated that the ellipsoidal equilibrium configurations of a rotating liquid (for motions with uniform deformations) can exist, but only when the corresponding velocity field of the liquid particles is superimposed to the rigid-body rotation and internal uniformly vortex motions.

Fluid viscosity causes the rotation to be stable only about the axis of maximum moment of inertia. These results were found to be in contradiction to those obtained by Ishlinskiy and Temochenko (1960), Temochenko (1969), Ishlinskiy, *et al.* (1986), and Crandall and Mroszczyk (1986). With reference to the stability of a body filled with ideal fluid and suspended on a string, it was indicated that there exist one or two ranges of unstable rotational velocities depending on the cavity shape. For the case of a flexible rotor with distributed parameters supported on two rigid bearings, it was shown that for the whole range of speeds higher than the first critical speed, the system is unstable due to the presence of fluid. The experimental results of Malashenko (1960) and Desyatov (1986) showed that the observed intensive oscillations of high rotational speeds of liquid-filled containers suspended by a string could not be analytically predicted.

Several studies revealed that the liquid payloads enhance the spin decay of projectiles (see, e.g., Rumyantsev, 1954, 1955, 1959a,b, 1960, 1966, Görtler, 1957, Sobolev, 1960, Miles and Troesch, 1961, Winch, 1962, Kolesnikov, 1962, Kuipers, 1964, Wedemeyer, 1964, Skalak and Conly, 1964, Levleva, 1964, 1966, Ackerberg, 1965, Kumok and Novgorodtseva, 1965, Wichterle and Wicterle, 1970, and Kitchens, 1980). It can produce flight dynamic instabilities as a result of resonance between the projectile nutational motion and inertial oscillations in the rotating liquid.

The analysis of liquid dynamics in a spinning container involves vorticity and Coriolis acceleration terms in the governing equations of motion. Fluid motion takes place during the transient periods of spin-up or spin-down. When the container achieves steady spin, the fluid rotates as a rigid body with secondary motions. Parks (1979) proposed a conjecture that viscosity will destabilize spinning spheroids with $I_{zz}/I_{xx} < 3$. With reference to spin-stabilized spacecraft, Hill and Baumgarten (1992) proposed a linear feedback control to stabilize and control the spacecraft. For large liquid volume, its dynamic state affects the stability and response of space vehicles. Ebert (1984), Harrison and Murphy (1987), Flugrad and Obermaier (1992), and Manasseh (1993) treated different aspects of this problem. For example, Or (1992) and Or, *et al.* (1994) attributed the observed instability in the Perigee Assist Module to be due to the asymmetric gas or liquid motions in the motor chamber. Harrison and Murphy (1987) and Pocha (1986) reported that the nutation angle grew exponentially. Flugrad and Obermaier (1992) developed a mathematical model to simulate the effects of liquid on a spin-stabilized satellite.

11.2.2 Fluid-filled spinning cylinder

The unsteady flow of a liquid within a cylinder of finite length suddenly started to spin about its axis was analyzed by Hocking and Michael (1959), Wedemeyer (1964), and Hocking (1965). Wedemeyer found that a secondary flow caused by the end walls of the container has a strong effect on the generation of spin in the liquid. In the vicinity of the end walls, the fluid motion is characterized by a boundary-layer flow, which can be either laminar or turbulent. The fluid within the boundary layers rotates faster than that at a large distance from the end walls. This portion is moved radially outward by the centrifugal forces. This motion creates a secondary motion within the spinning liquid. Due to the secondary flow, the transport of angular momentum from the walls to the interior is accomplished by convection rather than diffusion.

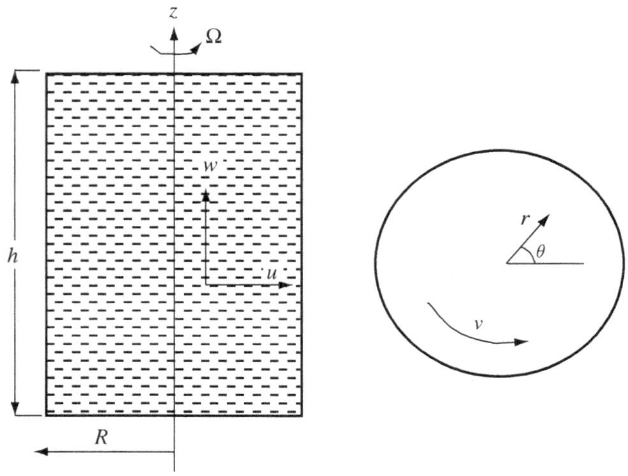

Figure 11.2 Axial section and cross-section of spinning cylinder.

Figure 11.2 shows a cylinder of height h, radius R, rotating about the z-axis with angular velocity Ω. With reference to the cylindrical coordinate frame, r, θ, z, and fluid velocity components along these axes, respectively, u, v, and w, the Navier–Stokes equations for the flow having rotational symmetry, that is, $\partial/\partial\theta = 0$, are

$$\frac{\partial u}{\partial t} + u\frac{\partial u}{\partial r} + w\frac{\partial u}{\partial z} - \frac{v^2}{r} + \frac{1}{\rho}\frac{\partial p}{\partial r} = \nu\left[\frac{\partial^2 u}{\partial r^2} + \frac{\partial}{\partial r}(u/r) + \frac{\partial^2 u}{\partial z^2}\right] \tag{11.4a}$$

$$\frac{\partial v}{\partial t} + u\left(\frac{\partial v}{\partial r} + \frac{v}{r}\right) + w\frac{\partial v}{\partial z} = \nu\left[\frac{\partial^2 v}{\partial r^2} + \frac{\partial}{\partial r}(v/r) + \frac{\partial^2 v}{\partial z^2}\right] \tag{11.4b}$$

$$\frac{\partial w}{\partial t} + u\frac{\partial w}{\partial r} + w\frac{\partial w}{\partial z} + \frac{1}{\rho}\frac{\partial p}{\partial z} = \nu\left[\frac{\partial^2 w}{\partial r^2} + \frac{1}{r}\frac{\partial w}{\partial r} + \frac{\partial^2 w}{\partial z^2}\right] \tag{11.4c}$$

and the continuity equation is

$$\frac{\partial(ru)}{\partial r} + \frac{\partial(rw)}{\partial z} = 0 \tag{11.4d}$$

The boundary conditions are

$$\begin{aligned} &u = w = 0, \quad v = \Omega r \quad \text{at } z = 0 \text{ and } h \\ &u = w = 0, \quad v = R\Omega \quad \text{at } r = R, \quad \text{with } u = v = 0 \text{ at } r = 0 \end{aligned} \tag{11.5}$$

In the vicinity of the end walls, $z = 0$ and h, the circumferential component v must depend on z. The fluid particles at the cylinder ends rotate with the velocity of the walls and are subject to centrifugal forces. These forces drive the fluid particles close to the end walls. This motion forms a secondary flow with velocity components u and w. Wedemeyer (1964) divided the entire flow region into two parts: the *boundary-layer region* close to the end walls, and the rest of the flow referred to as the *core flow*.

Let the velocity components in the core flow be u_0, v_0, and w_0, and the pressure p_0. On the assumption that the secondary motion is very slow in the core flow, equation (11.4a) can be applied to the core flow and takes the form

$$\frac{1}{\rho}\frac{\partial p_0}{\partial r}=\frac{v_0^2}{r} \tag{11.6a}$$

Accordingly, v_0 is independent of z, and equation (11.4b) may be written for the core flow in the form

$$\frac{\partial v_0}{\partial t}+u_0\left(\frac{\partial v_0}{\partial r}+\frac{v_0}{r}\right)=\nu\left[\frac{\partial^2 v_0}{\partial r^2}+\frac{\partial}{\partial r}(v_0/r)\right] \tag{11.6b}$$

Since v_0 is independent of z, equation (11.6a) implies that u_0 is also independent of z.

The flow within the spinning cylinder must be symmetric with respect to the middle plane, $z=h/2$, the analysis will be restricted to the boundary layer at the bottom end, $z=0$. The boundary-layer equations can be obtained from the Navier–Stokes equations (11.4) by replacing the radial pressure gradient within the boundary layer by the pressure gradient of the outer flow of the core according to equation (11.4a), $(1/\rho)(\partial p/\partial r)=v_0^2/r$. Although the boundary-layer flow is unsteady, it can be treated as being quasi-steady by neglecting the local acceleration terms, $\partial v/\partial t$, etc. After the cylinder starts to spin, the local acceleration terms are very small compared to the convective terms, and the boundary-layer equations take the form

$$u\frac{\partial u}{\partial r}+w\frac{\partial u}{\partial z}-\frac{v^2}{r}+\frac{v_0^2}{r}=\nu\frac{\partial^2 u}{\partial z^2} \tag{11.7a}$$

$$u\left(\frac{\partial v}{\partial r}+\frac{v}{r}\right)+w\frac{\partial v}{\partial z}=\nu\frac{\partial^2 v}{\partial z^2} \tag{11.7b}$$

$$\frac{\partial(ru)}{\partial r}+\frac{\partial(rw)}{\partial z}=0 \tag{11.7c}$$

The boundary conditions are

$$\begin{aligned}&u=w=0, \quad v=\Omega r \quad \text{at } z=0,\\ &u=0, \quad v=v_0(r,t) \quad \text{at } z=\infty,\ r=0.\end{aligned} \tag{11.8}$$

Equations (11.7) have been the subject of extensive studies in the literature using the integral method (von Karman, 1921), and linearized solutions (Ludwieg, 1951 and Squire, 1958). Wedemeyer (1964) used the momentum-integral method and expressed the radial mass-flow within the boundary layer by the expression

$$m(r)=2\pi\rho r\int_0^{\delta}u(r,z)\,dz \tag{11.9}$$

where δ is the boundary-layer thickness.

Imposing the condition that the total radial mass-flow (within the two boundary layers and the core flow) be zero, yields

$$2\pi\rho r\left[2\int_0^\delta u(r,z)\,dz + hu_0(r)\right] = 0 \quad \longrightarrow \quad -\frac{1}{2}hu_0(r) = \int_0^\delta u(r,z)\,dz \tag{11.10}$$

When the radial mass-flow distribution, $m(r)$, is determined, for a given distribution of $v_0(r)$, the radial velocity in the core flow, $u_0(r)$, can be found. Thus, one has the functional dependence

$$u_0(r) = F(v_0(r)) \tag{11.11}$$

At the beginning of the fluid motion, the circumferential component of the core flow, v_0, is zero. In this case, the boundary-layer problem reduces to the problem of the rotating disk flow, which was solved by Cochran (1934) in the form

$$\int_0^\infty u\,dz = 0.443r\Omega\sqrt{\frac{\nu}{\Omega}} \tag{11.12}$$

Substituting equation (11.12) in equation (11.10), gives

$$-\frac{1}{2}hu_0(r) = 0.443r\Omega\sqrt{\frac{\nu}{\Omega}} \tag{11.13}$$

When the fluid reaches the state of rigid rotation, $v_0 = r\Omega$, the boundary-layer equations (11.7) will have the trivial solution $v = r\Omega$, $u = w = 0$, and hence $u_0 = 0$.

The simplest possible approximation for general distribution of v_0 is to assume that

$$-\frac{1}{2}hu_0(r) = 0.443(r\Omega - v_0)\sqrt{\frac{\nu}{\Omega}} \tag{11.14}$$

This is a linear interpolation between the two extreme cases. This approximation can be examined by assuming the core flow to be almost a rigid-body rotation with angular velocity Ω, such that

$$v_0 = r\Omega + v_0', \quad u = u', \quad v = \Omega + v', \quad w = w', \quad \text{provided } v_0' \ll r\Omega \tag{11.15}$$

Substituting relations (11.15) into equations (11.7a,b) and neglecting higher-order terms in the primed quantities, gives the linearized equations

$$\nu\frac{\partial^2 u'}{\partial z^2} + 2\Omega(v' - v_0') = 0, \qquad \nu\frac{\partial^2 v'}{\partial z^2} - 2\Omega u' = 0 \tag{11.16}$$

subject to the boundary conditions

$$v' = 0, \quad u' = 0, \quad \text{at } z = 0; \quad v' = v_0', u' = 0 \quad \text{at } z = \infty \tag{11.17}$$

Equations (11.16) are satisfied by the following solution

$$u' = -v'_0 e^{-z\sqrt{\Omega/\nu}} \sin\left(\sqrt{\frac{\Omega}{\nu}} z\right), \quad v' = v'_0 \left[1 - e^{-z\sqrt{\Omega/\nu}} \cos\left(\sqrt{\frac{\Omega}{\nu}} z\right)\right] \tag{11.18a, b}$$

The radial component in the core flow, u_0, is determined by substituting (11.18a) into equation (11.10) and using (15), that is,

$$-\frac{1}{2} h u_0(r) = \int_0^\infty u' dz = -\frac{1}{2} v'_0 \sqrt{\frac{\nu}{\Omega}} = \frac{1}{2} (r\Omega - v_0) \sqrt{\frac{\nu}{\Omega}} \tag{11.19}$$

Comparing this result with the approximate solution (11.14) reveals an error of 13% in the approximate solution (11.14).

In order to solve for $v_0(r, t)$, relation (11.14) is introduced in equation (11.6b) to give

$$\frac{\partial v_0}{\partial t} + k(v_0 - r\Omega)\left(\frac{\partial v_0}{\partial r} + \frac{v_0}{r}\right) = \nu \left[\frac{\partial^2 v_0}{\partial r^2} + \frac{\partial}{\partial r}(v_0/r)\right] \tag{11.20a}$$

where $k = 0.443\left(2R/(h\sqrt{\mathrm{Re}})\right)$, $\mathrm{Re} = R^2\Omega/\nu$ is the Reynolds number. The initial and boundary conditions are

$$v_0 = 0 \quad \text{for } t < 0, \qquad v_0 = R\Omega \quad \text{for } r = R \text{ and } t \geq 0 \tag{11.20b}$$

Introducing the nondimensional parameters $v^* = v_0/R\Omega$, and $r^* = r/R$, and multiplying equation (11.20a) by $1/R\Omega^2$, equation (11.20a) takes the new form, for the case of constant spin, Ω, at $t = 0$,

$$\frac{\partial v^*}{\partial \Omega t} + k(v^* - r^*)\left(\frac{\partial v^*}{\partial r^*} + \frac{v^*}{r^*}\right) = \frac{1}{\mathrm{Re}} \left[\frac{\partial^2 v^*}{\partial r^{*2}} + \frac{\partial}{\partial r^*}(v^*/r^*)\right] \tag{11.21a}$$

The boundary conditions take the new form

$$v^* = 1 \quad \text{for } r^* = 1 \text{ and } \Omega t \geq 0 \tag{11.21b}$$

If $k\,\mathrm{Re} \gg 1$, then one can ignore viscous terms on the right-hand side of equation (11.21a) except for small times when the gradient $\partial v^*/\partial r^*$ is large near the wall, $r^* = 1$. For $\Omega t \gg 0$ and $k\,\mathrm{Re} \gg 1$, one obtains the solution of the inviscid equation in the form

$$v^* = \left[r^* e^{2k\Omega t} - \frac{1}{r^*}\right] \Big/ \left[e^{2k\Omega t} - 1\right] \quad \text{for } r^* \geq e^{-k\Omega t} \tag{11.22a}$$

$$v^* = 0 \quad \text{for } r^* \leq e^{-k\Omega t} \tag{11.22b}$$

Figure 11.3 shows the profile of the core flow velocity v^* for different values of time parameter $k\Omega t$. It is seen that as the time progresses the core flow velocity has the linear gradient from the tank center to the tank wall where $v^* = 1$.

The solution (11.22) satisfies the complete equation (11.21a) including viscous terms except at the point $r^* = e^{-k\Omega t}$, where the first derivative is discontinuous. Note that the inviscid equation is of first-order in the derivatives and hence only the solution function has to be continuous. On the other hand, the complete equation involves second-order derivatives and

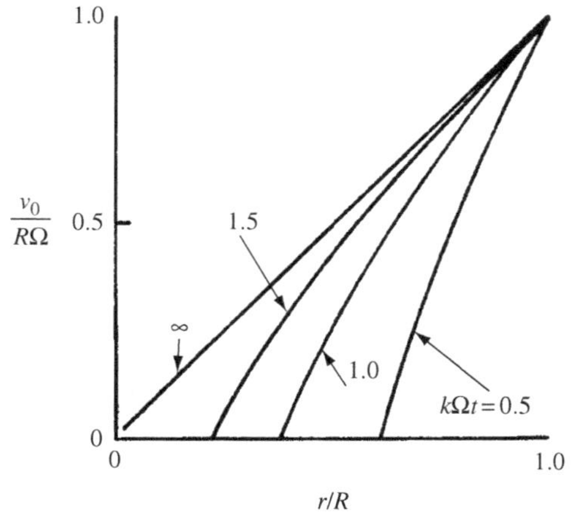

Figure 11.3 The core velocity profiles for different values of time parameter $k\Omega t$. (Wedemeyer, 1964)

the solution must have continuous derivatives. Thus, the role of viscosity is to smooth the corner at $r^* = e^{-k\Omega t}$. From equation (11.14) the radial velocity component is

$$u_0(r) = -k(r\Omega - v_0) \tag{11.23}$$

From the continuity equation (11.4d) and the condition that the flow must be symmetric at the mid-plane, $z = h/2$, one obtains the longitudinal velocity

$$w_0 = -\frac{[z - (h/2)]}{r}\frac{\partial r u_0}{\partial r} \tag{11.24}$$

Two different regions separated by a cylindrical front, $r = R\,e^{-k\Omega t}$, that moves towards the tank center, are identified as follows:

Region 1: $0 < r/R < e^{-k\Omega t}$ is characterized by no rotation of the fluid particles, that is, $v_0 = 0$.

Region 2: $e^{-k\Omega t} < r/R < 1$ is distinguished by rotational motion of the fluid particles with velocity $v_0 = R\Omega v^*$ as given by equation (11.22).

In region 1, the particles ahead of the cylindrical front will remain ahead of it until they hit one of the boundary layers at the end of the walls, $z = 0$ or $z = h$. This can be determined by computing the trajectories of the particles in region 1. For $v_0 = 0$, equation (11.23) gives

$$u_0(r) = \frac{dr}{dt} = -kr\Omega \quad \rightarrow \quad r = r_0\,e^{-k\Omega t} \tag{11.25a}$$

From equation (11.24) one can write

$$w_0 = \frac{dz}{dt} = 2k[z - (h/2)]\Omega \quad \rightarrow \quad z = \frac{h}{2} + \left(z_0 - \frac{h}{2}\right)\left(\frac{r_0}{r}\right)^2 \tag{11.25b}$$

where (r_0, z_0) is the particle position at $t = 0$.

The solution given by (11.25) describes the fluid particle trajectories in region 1. After the cylinder starts at $t = 0$, the fluid particles move along hyperbolas given by equations (11.25) until they hit the boundary layers at one of the end walls $z = 0$ or $z = h$. Inside the boundary layer, the flow direction changes rapidly, and the particles acquire rotational motion and are thrown radially outwards until they emerge from the boundary layer at some distance behind the moving front, $r = R\,e^{-\kappa\Omega t}$, with a rotational component v_0.

The above analysis is only valid for laminar flow for which the Reynolds number does not exceed 3×10^5. For turbulent boundary-lay flow, Wedemeyer (1964) obtained the following expression for the radial component of the core flow:

$$u_C(r) = -0.035\left(\frac{2R}{h}\right)(\mathrm{Re})^{-1/5}\frac{(r\Omega - v_0)^{8/5}}{(R\Omega)^{3/5}} \tag{11.26}$$

Wedemeyer (1964) provided some experimental results to support the analytical prediction. The axial spin decay of liquid-filled shells fired from a gun was measured following its departure from the gun. The decrease of the shell angular velocity is caused by absorption of angular momentum in liquid and by the torque due to air resistance. The contribution of air resistance was measured by observing the spin decrease of the empty shell. The difference was plotted in Figure (11.4)(a) and (b) for two different values of Reynolds number. Both experimental measurements are compared with those predicted analytically with and without secondary flow.

Gershuni and Zhukhovitskii (1964) studied the stability of the modulated motion of a viscous fluid in a rotating container whose angular velocity varies periodically with time. The governing equation of the fluid perturbed velocity involved periodic coefficients. Contrary to the case of stationary rotation, the modulation resulted in a destabilizing effect. Beig (1984) studied the slosh dynamics of fluid-filled containers on three-axis spin stabilized spacecraft.

11.3 Dynamics of partially filled spinning containers

11.3.1 Historical background

The dynamics of rotating fluids and their stability depends on the speed of rotation, liquid viscosity, and surface tension. Under a low gravitational field the surface tension is important in governing the profile of the free surface. Under a regular gravitational field, Miles and Troesch (1961), Conly (1963), and Skalak and Conly (1964) studied the modes and stability of small oscillations of an incompressible, inviscid fluid in a partially filled cylindrical container rotating about its vertical axis, in a state of steady spin. When the frequency of the free surface is less than twice the rotational speed, the motion is governed by an elliptic equation. For this case, the free-surface motions do not penetrate deeply into the fluid. On the other hand, when the free-surface frequency is less than the rotational speed, the fluid motion is governed by a hyperbolic equation. The effects of Coriolis acceleration are manifested in generating radial nodes and in frequency splitting for a given nodal wave pattern. The free-surface waves propagating in the direction of rotation have a different frequency from the waves propagating in the opposite direction. Under vertical periodic excitation, the excitation amplitude required to cause unstable motion at the subharmonic frequencies is larger than that for a nonrotating container.

Some nonlinear models were developed to describe the dynamics of bodies partially filled with liquid by Narimanov, *et al.* (1977), Limarchenko (1987, 1991), Lukovskii (1990), and Limarchenko, *et al.* (1992). They considered different geometries of the body such as a

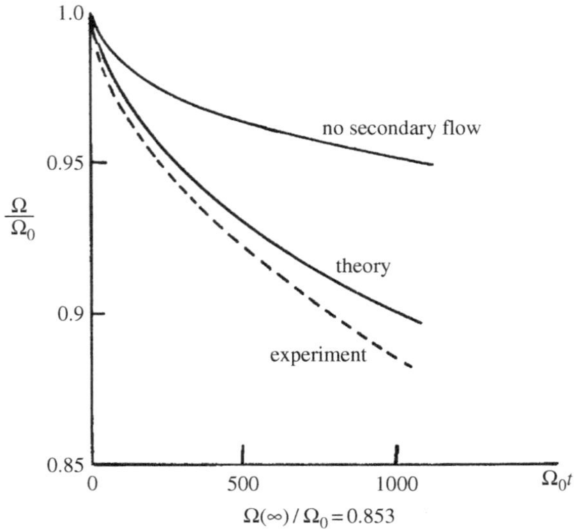

(a)

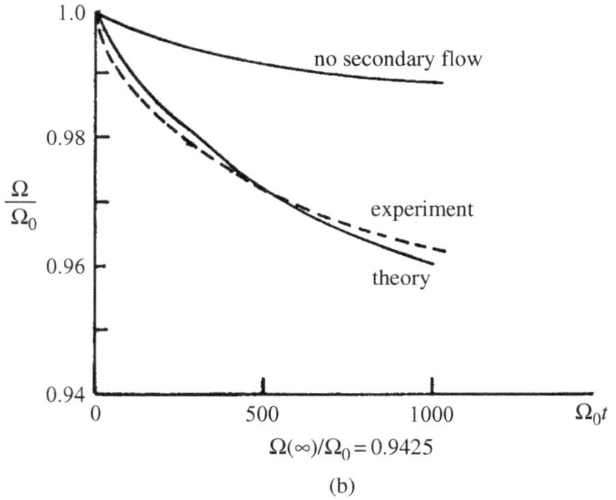

(b)

Figure 11.4 (a) Spin decay for laminar boundary-layer flow, $\mathrm{Re} = 1.76 \times 10^5$, $h/2R = 2.68$, (b) spin decay for turbulent boundary-layer flow, $\mathrm{Re} = 6.1 \times 10^5$, $h/2R = 2.68$. (Wedemeyer, 1964)

cylinder, coaxial cylinder, rectangular parallelepiped, sphere, and cone. The stability of the steady rotation of a rigid body with a cavity containing an ideal fluid was re-examined by Vladimirov and Ilin (1999) who obtained similar results to those reported earlier by Moiseev and Rumyantsev (1968). They found that the system is stable provided it rotates around the principal axis of its moment of inertia tensor, which corresponds to the maximum value of the moment of inertia. Limarchenko (1995) studied the unsteady rotational motion of a cylindrical tank partially filled with viscous fluid.

El-Raheb and Wagner (1981) considered the motion of liquid free surface in a spinning spherical tank using a finite element model. They assumed constant vorticity over the entire fluid volume and small-amplitude free-surface motion. It was found that the vorticity generated from spinning has a strong spatial dependence. The results were valid for tanks over half full or in a low gravitational field. For the case of less than half full tanks, vorticity cannot be held constant and the liquid free surface exhibits instability. Homicz and Gerber (1987) and Schulkes and Cuvelier (1991) numerically solved the Navier–Stokes equations using a finite element algorithm to estimate the free oscillations and the rotation-induced internal modes for a rotating viscous fluid with surface tension. It was found that when the angular velocity of the container is large, the solution results in mixed internal-free-surface oscillation modes. Weihs and Dodge (1991) analyzed the free-surface sloshing and internal wave motions in a rotating nonaxisymmetric, partially filled wedge-shaped tank under zero gravity. The liquid was assumed to be inviscid and incompressible. Furthermore, surface tension and contact angle effects were neglected. Their results revealed sloshing waves of many kinds and internal waves with frequencies less than twice the spinning frequency. Chen and Pletcher (1993) and Chen *et al.* (1994) developed a model of the free-surface dynamics in a half-filled spherical tank in a spin-stabilized satellite. They used a coupled strongly implicit procedure (CSIP) to solve the incompressible Navier–Stokes equations for the three-dimensional surface flow. Agrawal (1993) considered the fluid boundary layer effects in partially filled containers on the fluid natural frequency and nutation time constants of a spinning spacecraft. Note the nutation angle was proportional to $\exp(t/\tau)$, where t is the time and τ is a time constant determined from experimental drop test data.

Moiseev and Rumyantsev (1968) and Rumyantsev, *et al.* (1979) derived the equations of motion of fluid-containing rigid bodies. The derivation is based on the principle of least action in the Hamilton form. According to this principle, for the absolute motion (i.e., measured with respect to an inertial reference frame $O'X'Y'Z'$) of any mechanical system with ideal geometric constraints, one can write

$$\int_{t_0}^{t_1} \left(\delta T + \sum_j \mathbf{F}_\nu \cdot \delta \mathbf{r}'_j \right) dt = 0 \tag{11.27}$$

where T is the kinetic energy of the system, $\mathbf{F}_j$ is the vector of the given applied forces at the jth point of the system, $\mathbf{r}'_j$ is the radius vector of this point with the origin O', and t_0 and t_1 are the initial and current times. Let v be the volume of the fluid inside the body. The fluid is assumed incompressible satisfying the condition div $\delta \boldsymbol{r} = 0$. This condition holds at each point of the fluid, $\boldsymbol{r}$, for any time t. The principle of least action and constraint can be represented in the form

$$\int_{t_0}^{t_1} \left(\delta T + \sum_j \mathbf{F}_j \cdot \delta \mathbf{r}'_j + \int_v \lambda \operatorname{div} \delta \mathbf{r} \, dv \right) dt = 0 \tag{11.28}$$

where $\lambda = \lambda(\vec{r}, t)$ is the indeterminate Lagrange multiplier, which can be regarded as a "constraint reaction." Since the new integral term should have work units, the Lagrange multiplier λ is identical to the hydrodynamic pressure. The differential equations of motion are obtained from (11.28) in the general vector form. All further derivations took into account the geometrical details of the problem, including possible spatial symmetries, boundary conditions, etc. Implementing this procedure, Moiseev and Rumyantsev (1968) considered different cases of the dynamics and stability of rigid bodies containing a fluid. The stability problems were investigated by means of Lyapunov's theorems. They also included the influence of viscosity and free-surface tension phenomena.

Narimanov (1957a,b,c) derived the differential equations of motion of a gyroscope partially filled with a liquid. As a result, the free surface of the liquid was considered to be close to the cylindrical shape. The differential equation for liquid motion is given in the form

$$\frac{\partial \mathbf{V}_{\text{rel}}}{\partial t} + 2\Omega \times \mathbf{V}_{\text{rel}} + \frac{d\Omega}{dt} \times \mathbf{r} + \Omega \times (\Omega \times \mathbf{r}) = -\frac{1}{\rho}\nabla p \tag{11.29}$$

where $\mathbf{V}_{\text{rel}}$ is the relative velocity of the fluid particles, $\mathbf{r}$ is a position vector of the particles with respect to an immovable point on the gyroscope, Ω is the angular velocity vector of the gyroscope, and p is the pressure. Finally, an infinite set of ordinary differential equations of motion for modal coordinates of the free-surface shape and the perturbed (due to the liquid) gyroscope motion were obtained. Frasier and Scott (1971) determined the stability of a liquid-filled gyroscope.

Ganiev and Kholopova (1975), Boyarshina and Ganiev (1976), and Boyarshina and Koval'chuk (1990) developed the differential equations of motion of absolutely rigid and elastic containers containing a liquid in terms of the Euler aircraft angles. Their derivations were based on Hamilton's principle. The projections of the instantaneous angular velocity on the axis of the moving coordinate system were approximated by second-degree polynomials with respect to the Euler angles. A nonlinear boundary value problem was formulated for the fluid velocity potential by Lukovskii (1975). For the case of cylindrical tanks, the analysis of the wave shape characteristics of the fluid surface yields multidimensional nonlinear oscillatory systems with internal resonance, resulting from geometrically identical fluid surface shapes with different nodal line orientations. The related eigenfrequencies were either the same or very close to each other. This kind of internal resonance could be predicted, since an unperturbed stationary state of the system admitted a rotation group around the axial symmetry of axes. Implicitly, this group property was employed by Boyarshina and Koval'chuk (1990), who considered a "mixed" spatio-temporal mode shape for the fluid free-surface perturbed motion in the form of circular waves traveling in the circumferential direction.

Poznyak (1994), in an unpublished review article, provided an extensive overview of paradoxes in the dynamics of rotating containers partially filled with liquid. He summarized the paradoxes into the following groups: (i) The independence of the critical rotor speed of the fluid filling level; (ii) the destabilizing effect of external damping forces; (iii) the stabilizing effect of liquid viscous forces.

Based on analytical and experimental studies on centrifugal models with flexible shafts and containers partially filled with liquid, it was reported that oscillations occurred only beyond the first critical speed and existed in the range of speeds dependent on container filling level (see, e.g., Sokolov, 1954 and Yepishev, 1959). The oscillations have the form of cyclic precession with a frequency different from the rotor spin, however, its speed is near to the critical rotor speed and increasing with the speed. The conditions of these vibrations and their intensities are only slightly affected by the fluid viscosity. These oscillations do not take place when the container is empty or completely filled. Kollmann (1962) conducted experiments on a rotor system with a vertical symmetric flexible shaft and with a container filled with liquid at its center. For different filling levels, the fluid viscosity varies over a wide range. Kollmann made similar observations to those reported by Yepishev (1959).

Stewartson (1959) examined the stability of a spinning top, with its support point below the center of gravity (sleeping gyroscope), which had a cavity partially filled with inviscid fluid.

The analysis resulted in several instability regions. The main region of instability boundary exhibited fluid wave motion with a speed close to the nutation frequency of the empty top. The analysis of Rumyantsev (1962) did not reveal any instability similar to that predicted by Stewartson (1959). However, Kuipers (1964) showed that in the absence of damping, instability existed at speeds exceeding the critical speed depending on the fluid filling level. When viscous damping forces are introduced, Kuipers (1964, 1966) showed that all rotational speeds become unstable independent of the values of damping forces. This result appears to be as paradoxical as that for rotor systems stabilized by external damping. Kuipers (1966) compared the experimental results of Kollmann (1962) with his earlier results (Kuipers, 1964, 1966) and found a good agreement of the two instability results in the absence of damping. Wolf (1968), Lichtenberg (1982), and Wohlsbruck (1985) determined the instability boundaries of an undamped rotor partially filled with ideal liquid. They showed that the instability region does not propagate beyond the double critical rotor speed. Zhang, *et al.* (1996) determined the stability boundaries of a rotor filled or partially filled with liquid.

The stability of a rigid container partially filled with ideal fluid supported by anisotropic elastically damping bearings was studied by Daich (1971). It was shown that the tank is unstable over the entire range of rotational speed. One portion of the speed region is characterized by strong instability governed by bearings stiffness and the remaining region with weak instability governed by the destabilizing effect of external damping. Daich and Bar (1973) included the fluid viscosity, which was found to transform the weak instability region into an asymptotic stability region. They also found fluid viscosity diminishes the strong unstable region. Hendricks and Morton (1979), Hendricks and Klauber (1984), Hendricks (1986), and Saito and Someya (1980) obtained similar results to those reported by Daich and Bar (1973). They analyzed a circular whirling motion of the rotor and gave the viscosity correction by means of a procedure introduced by Greenspan (1968), where not all the boundary conditions are satisfied. Daich, *et al.* (1976) and Bermann, *et al.* (1985) considered the nonlinear dynamic behavior of rigid rotating containers partially filled with viscous fluid. They solved for the amplitude of vibration in the main unstable region. Stationary rotation stability conditions of a cylinder partially filled with viscous fluid together with Hopf bifurcation were determined by Derendyayev and Sanalov (1982) and Derendyayev (1988). Holm-Christensen and Träger (1991) directly used the full Navier–Stokes equations and solved them numerically. The stability of a spinning liquid-filled spacecraft was investigated by Bao and Pascal (1997) by using the Liapunov direct method. Three types of characteristic regimes were considered; (i) the free motion of inviscid fluid, (ii) slosh motion, and (iii) nonslosh motion. These three cases were solved using the finite element method or the boundary element method. It was demonstrated that the viscosity of the fluid has a dissipative effect at large Reynolds number, while the slosh motion causes destabilizing effect. The nonslosh model did not affect the stability criterion.

Tao and Zhang (2002a,b) studied the dynamic stability of a rotor partially filled with a viscous liquid. The equations of motion were formulated in terms of the dynamic forces acting on the rotor. They compared their results with those obtained by Hendricks and Morton (1979) and Holm-Christensen and Träger (1991). They found that the lower spin region is stable, and the smaller the external damping the larger the stable region. However, there was no stable region when the Reynolds number was beyond 2.5×10^5.

The governing equations of motion of an incompressible viscous fluid constitute the conservation of mass and conservation of momentum (Navier–Stokes equation), that is,

$$\nabla \cdot \mathbf{q} = 0 \tag{11.30}$$

$$\begin{aligned} &\frac{\partial \mathbf{q}}{\partial t} + \frac{1}{2}\nabla(\mathbf{q}\cdot\mathbf{q}) + (\nabla \times \mathbf{q}) \times \mathbf{q} + 2\Omega \times \mathbf{q} + \frac{\partial \Omega}{\partial t} \times \mathbf{r} + \Omega \times (\Omega \times \mathbf{r}) \\ &= -\frac{1}{\rho}\nabla p + \mathbf{F} - \nu\nabla \times (\nabla \times \mathbf{q}) \end{aligned} \tag{11.31}$$

where $\mathbf{q}$ is the fluid particle velocity vector measured in a coordinate frame rotating with a constant angular velocity $\Omega = \Omega\mathbf{k}$, $\mathbf{r}$ is the position vector, p is the pressure, ρ is the fluid density, $\mathbf{F}$ is the body force per unit mass. In order to complete the boundary value problem, equations (11.30) and (11.31) must be supplemented by the boundary conditions of the container walls. Note that in view of the viscosity effect given by the expression $\nu\nabla \times (\nabla \times \mathbf{q})$, and for a rigid impermeable surface, the viscous fluid must move with the boundary, since no slipping or penetration can occur. The boundary condition then requires that the relative velocities in body-fixed axes must vanish at the container walls.

Baines (1967) and Bauer (1971a) studied the dynamic behavior of an incompressible and inviscid liquid with a free surface in a uniformly fast rotating infinitely-long elastic circular cylinder. For a given filling, he obtained two possible wave frequencies, one greater than the rotational speed and the other smaller. For decreasing liquid mass, these two frequencies approach each other. Under translational forced excitation of the container, the liquid force component in the spin direction does not exhibit any resonance, while that in the opposite direction creates two singularities in the magnification function.

11.3.2 Inviscid fluid in partially filled upright cylinder

Two-dimensional case

It is known from the basic principles of fluid mechanics that if a liquid container is rotating about its vertical axis, the motion is transmitted by shear stresses to the liquid. The liquid after a short time interval will possess the same angular velocity throughout the container and the free surface takes a parabolic surface AOB, shown in Figure 11.5, given by the equation $z = \Omega^2 r^2/(2g)$, where r is the radius of a particle on the free surface. Miles (1959a, b, c, 1963,1964) showed that the disturbed motion of the fluid free surface could not remain irrotational due to a vertical gradient of velocity that necessarily generates vorticity. With reference to shallow water in circular tanks, it was observed that the tank rotation results in a splitting of the pairs of natural frequencies that would be of equal magnitude and opposite sign in the absence of rotation. Miles and Ball (1963) considered the free-surface oscillations in a rotating paraboloid.

If, on the other hand, a closed cylinder is partially filled with incompressible liquid and its angular velocity gradually increases, the liquid will rise up and will not spill over since the tank is closed. Instead, the liquid will form an annular ring similar to that shown in Figure 11.6. This case was considered by Sun (1960) and Miles and Troesch (1961). If the free surface at radius $r = r_0$ is disturbed, one may write the equations of small motion in the form

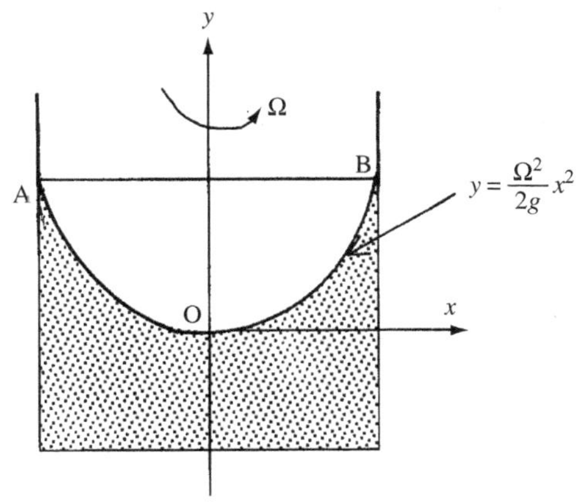

Figure 11.5 Rotating liquid in a cylindrical tank.

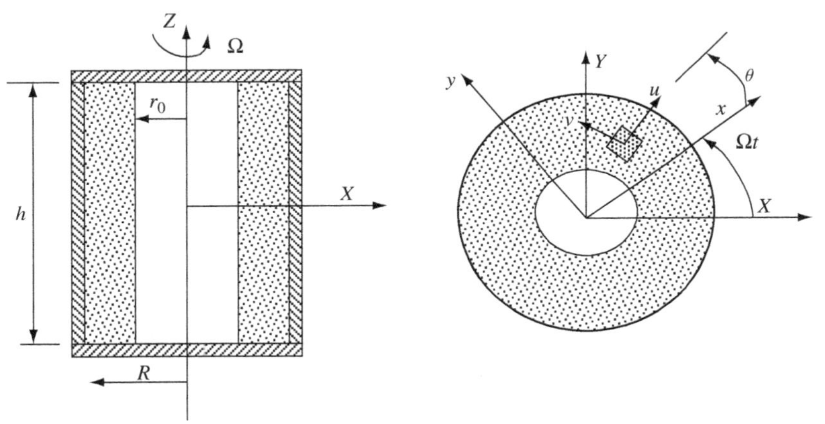

Figure 11.6 Rotating liquid forming an annular in a closed cylinder showing fixed coordinate and rotating frames.

$$\frac{\partial u}{\partial t} - 2\Omega \text{v} - \Omega^2 r = -\frac{1}{\rho}\frac{\partial P}{\partial r} \tag{11.32a}$$

$$\frac{\partial v}{\partial t} + 2\Omega u = -\frac{1}{\rho r}\frac{\partial P}{\partial \theta} \tag{11.32b}$$

The continuity equation is

$$\frac{\partial (ur)}{\partial r} + \frac{\partial v}{\partial \theta} = 0 \tag{11.33}$$

where P is the total pressure, which is composed of stationary pressure caused by the centripetal acceleration and additional pressure $p(r, \theta, t)$ due to small movements of the liquid, that is,

$$P = \frac{\rho\Omega^2}{2}\left(r^2 - r_0^2\right) + p(r, \theta, t) \tag{11.34}$$

The solution for the velocity components and pressure must satisfy the following boundary conditions:

At the cylinder wall, $r = R$, the velocity must vanish

$$u|_{r=R} = 0 \tag{11.35a}$$

On the free surface, $r = r_0 + \eta\ (\theta, t)$, we have the following kinematic condition

$$u|_{r=r_0+\eta} = \frac{\partial \eta}{\partial t} \tag{11.35b}$$

and the total pressure vanishes,

$$P|_{r=r_0+\eta} = 0, \quad \text{or} \quad p(r, \theta, t) + \rho\Omega^2 r_0 \eta \approx 0 \tag{11.35c}$$

where η is assumed to be small in the second condition of (11.35c) and a higher order term in η^2 was dropped. The second condition of (11.35c) can also be written in the form

$$\frac{\partial p(r, \theta, t)}{\partial t} + \rho\Omega^2 r_0 u \approx 0 \tag{11.35d}$$

A harmonic solution for u, v, p, and η can be written in the form

$$u(r, \theta, t) = U(r)\, e^{i(n\theta - \omega t)}, \quad v(r, \theta, t) = V(r)\, e^{i(n\theta - \omega t)} \tag{11.36a, b}$$

$$p(r, \theta, t) = p(r)\, e^{i(n\theta - \omega t)}, \quad \eta(\theta, t) = \eta\, e^{i(n\theta - \omega t)} \tag{11.36c, d}$$

where ω is an unknown angular velocity in the rotating coordinate frame $x, y, n = 1, 2, 3, \ldots$ The case $n = 0$ corresponds to no surface waves.

Substituting solutions (11.36) into equations (11.33a, b), and expressing $U(r)$ and $V(r)$ in terms of $p(r)$, gives

$$U(r) = \frac{i}{\rho(4\Omega^2 - \omega^2)} \left[\omega \frac{\partial p(r)}{\partial r} - \frac{2n\omega}{r} p(r) \right] \tag{11.37a}$$

$$V(r) = \frac{\omega}{\rho(4\Omega^2 - \omega^2)} \left[\omega \frac{\partial p(r)}{\partial r} - \frac{2n\omega}{r} p(r) \right] + \frac{1}{r} \frac{\partial p(r)}{\partial r} \tag{11.37b}$$

Substituting equations (11.37) in the continuity equation (11.33c) gives the differential equation in $p(r)$

$$r^2 \frac{\partial^2 p(r)}{\partial r^2} + r \frac{\partial p(r)}{\partial r} - n^2 p(r) = 0 \tag{11.38}$$

The solution of this equation may be written in the form

$$p(r) = A_1 r^n + A_2 r^{-n} \tag{11.39}$$

where A_1 and A_2 are constants to be determined from the boundary conditions (11.35a,d), after using equation (11.37a) at the corresponding boundaries. Equating the determinant of A_1 and A_2 from these two conditions to zero, gives the frequency equation

$$\gamma_n \omega^2 - 2\Omega\omega - n\Omega^2 = 0 \tag{11.40}$$

where $\gamma_n = [1 + (r_0/R)^{2n}]/[1 - (r_0/R)^{2n}]$.

The case of $r_0/R = 0$, $\gamma = 1$, corresponds to the case of a completely filled cylinder, while for $r_0/R = 1$, $\gamma = \infty$ belongs to an empty container. For values of r_0/R between 0 and 1, the following two real roots for the natural frequency are obtained

$$\omega_{1,2} = \frac{\Omega}{\gamma_n}\left[1 \pm \sqrt{1 + n\gamma_n}\right] \tag{11.41}$$

The first root, with the plus sign, characterizes surface-wave motion of an ellipse in the rotating coordinate frame, x, y. The root with the negative sign specifies surface waves in the opposite direction. The first root proved to be more probable than the other as the motion is dictated by the Coriolis acceleration as given by the middle term of equation (11.40). The free surface given by equation (11.36d) takes the form

$$\eta(\theta, t) = \eta_n^+ \cos(n\theta - \omega_n^+ t) + \eta_n^- \cos(n\theta - \omega_n^- t) \tag{11.42}$$

where η_r^- and η_n^- are wave amplitudes determined from initial perturbations. Of particular interest is the case of the first mode, $n = 1$, for which the natural frequencies take the form

$$\omega_{1,2} = \frac{\Omega}{\gamma}\left[1 \pm \sqrt{1 + \gamma}\right] = \Omega \frac{\left[1 - (r_0/R)^2\right]}{\left[1 + (r_0/R)^2\right]}\left[1 \pm \sqrt{1 + \left\{\frac{1 + (r_0/R)^2}{1 - (r_0/R)^2}\right\}}\right] \tag{11.43}$$

The free-surface equation of this case characterizes the deviation from the stationary cylindrical surface, $r = r_0$, given by the following equation

$$\eta_1(\theta, t) = \eta_1^+ \cos(n\theta - \omega_1^+ t) + \eta_1^- \cos(n\theta - \omega_1^- t) \tag{11.44}$$

This motion may be regarded as a result of mutual movements of two cylinders with the same radius, $r = r_0$. The center of the first cylinder, O_1, is stationary and defines the stationary position of the free surface. The center of the other surface, O_2, specifies the perturbed condition, which is displaced along a circular path with radius η_n^+ and oscillates with frequency $\omega_1^{+,-}$ as shown in Figure 11.7. This means the motion of the center O_2 measured with respect to the moving frame, x, y, is given by the following components

$$x = \eta^{+,-} \cos\omega_1^{+,-} t, \quad y = \eta^{+,-} \sin\omega_1^{+,-} t \tag{11.45}$$

The frequencies measured from the fixed observer in the frame X, Y are $\Lambda_{1,2} = \Omega + \omega_{1,2}$, are given for the first mode in the form

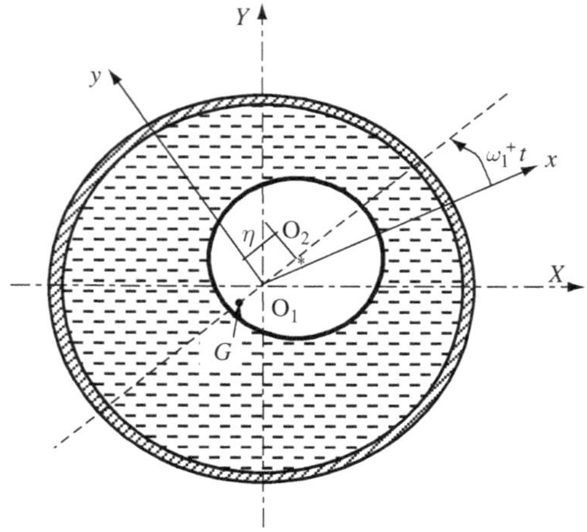

Figure 11.7 Perturbed free-surface motion around the original tank axis O_1

$$\Lambda_{1,2} = \Omega\left\{1 + \frac{1}{\gamma}\left[1 \pm \sqrt{1+\gamma}\right]\right\}$$
$$= \Omega\left\{1 + \frac{\left[1-(b/a)^2\right]}{\left[1+(b/a)^2\right]}\left[1 \pm \sqrt{1 + \left\{\frac{1+(b/a)^2}{1-(b/a)^2}\right\}}\right]\right\} \tag{11.46}$$

These frequencies are positive, one is greater than the rotor angular velocity (fast wave) and the other is slower than the rotor speed (slow wave). Note that the free oscillations of the first mode results in an unbalanced fluid configuration that generates radial forces rotating with frequencies $\omega_{1,2}$ in the moving frame and with frequencies $\Lambda_{1,2}$ in the fixed frame.

Three-dimensional case

Miles and Troesch (1961) considered three-dimensional oscillations of the same system as that of the previous section by adding the effect of the fluid depth along the Z-axis. It was found that due to the Coriolis force, there exist modes with the radial velocity component vanishing inside the fluid (nodal cylinders). The vanishing of the radial velocity for a given nodal pattern corresponds to two frequencies that differ in magnitude as well as sign. Miles and Troesch (1961) introduced the acceleration potential χ, whose gradient gives the acceleration vector, that is,

$$\chi = \frac{p}{\rho} - g\frac{\Omega^2 r^2}{2g} + \text{const} \tag{11.47}$$

The equations of motion of a small motion in terms of cylindrical coordinate, r, θ, z, are (Poincaré, 1885),

$$\frac{\partial u}{\partial t} - 2\Omega v = -\frac{\partial \chi}{\partial r}, \qquad \frac{\partial v}{\partial t} + 2\Omega u = -\frac{\partial \chi}{r\partial \theta}, \qquad \frac{\partial w}{\partial t} = -\frac{\partial \chi}{\partial z} \tag{11.48a, b, c}$$

where u, v, w are the relative velocity components along r, θ, z respectively. Alternatively, equations (11.31) can be written in the vector form, with $\mathbf{q} = u\mathbf{e}_r + v\mathbf{e}_\theta + w\mathbf{e}_z$,

$$\frac{\partial \mathbf{q}}{\partial t} + 2\Omega \times \mathbf{q} = -\nabla \chi \tag{11.49}$$

An immediate consequence of the Coriolis acceleration in equation (11.49) is that the velocity field cannot remain irrotational unless it is independent of z, since by taking the curl of equation (11.49), one writes

$$\frac{\partial}{\partial t}(\nabla \times \mathbf{q}) = 2\Omega \frac{\partial \mathbf{q}}{\partial z} \tag{11.50}$$

In order to derive a differential equation governing the acceleration potential χ, we take the divergence of both sides of equation (11.49) and use the continuity equation $\nabla \cdot \mathbf{q} = 0$. This gives

$$\left[\nabla \cdot \left(\frac{\partial \mathbf{q}}{\partial t} + 2\Omega \times \mathbf{q}\right)\right] = -[\nabla^2 \chi] \quad \rightarrow \quad \nabla \cdot (2\Omega \times \mathbf{q}) = -[\nabla^2 \chi] \tag{11.51}$$

Differentiating both sides of equation (11.51) with respect to time gives

$$\frac{\partial^2}{\partial t^2}[\nabla \cdot (2\Omega \times \mathbf{q})] = -\frac{\partial^2}{\partial t^2}[\nabla^2 \chi] \quad \rightarrow \quad -2\Omega \cdot \frac{\partial^2}{\partial t^2}[\nabla(\times \mathbf{q})] = -\nabla^2\left(\frac{\partial^2}{\partial t^2}\chi\right)$$

Using equation (11.50) the above result becomes

$$-4\Omega\Omega \cdot \frac{\partial^2}{\partial z \partial t}\mathbf{q} = -\nabla^2\left(\frac{\partial^2}{\partial t^2}\chi\right) \quad \rightarrow \quad \nabla^2\left(\frac{\partial^2}{\partial t^2}\chi\right) + 4\Omega^2 \frac{\partial^2}{\partial z^2}\mathbf{q} = 0 \tag{11.52}$$

The solutions of equations (11.48) and (11.52) must satisfy the boundary conditions:

(1) At the tank bottom and top surfaces

$$u|_{r=R} = 0, \quad w|_{z=0} = w|_{z=h} = 0 \tag{11.53a, b}$$

(2) At the free surface, $r = r_0 + \eta\ (\theta, z, t)$, the kinematic and dynamic boundary conditions are

$$u|_{r=r_0+\eta} = \frac{\partial \eta}{\partial t}, \quad \text{and} \quad p|_{r=r_0+\eta} = 0 \tag{11.54a, b}$$

Introducing an oscillatory solution to equations (11.48), (11.52) and (11.53b) in the form

$$u(r, \theta, z, t) = \left[\frac{df(r)}{dr} + \frac{\varpi m}{r} f(r)\right] e^{i(\omega t + m\theta)} \cos(n\pi z/h) \tag{11.55a}$$

$$v(r, \theta, z, t) = i\left[\varpi \frac{df(r)}{dr} + \frac{m}{r} f(r)\right] e^{i(\omega t + m\theta)} \cos(n\pi z/h) \tag{11.55b}$$

$$w(r, \theta, z, t) = (\varpi^2 - 1)\left(\frac{n\pi}{f}\right) f(r)\, e^{i(\omega t + m\theta)} \sin(n\pi z/h) \tag{11.55c}$$

$$\frac{1}{\rho} p(r, \theta, z, t) = \frac{1}{2}\Omega^2 (r^2 - r_0^2) + i\omega(\varpi^2 - 1) f(r)\, e^{i(\omega t + m\theta)} \cos(n\pi z/h) \tag{11.55d}$$

where ω stands for the angular frequency of the free liquid surface in the rotating frame and $\varpi = \frac{2\Omega}{\omega}$. n refers to the number of z-nodes, m stands for θ-nodes. The radial nodes are denoted by the number j. Note that the angular frequency of the free fluid surface for a fixed observer is $\omega - m\Omega$, m and n are integers. The function $f(r)$ must satisfy the modified Bessel equation

$$\frac{d^2f(r)}{dr^2} + \frac{1}{r}\frac{df(r)}{dr} - \left(k^2 + \frac{m^2}{r^2}\right)f(r) = 0 \tag{11.56}$$

where $k^2 = \left(1 - {}^2\right)(n\pi/h)^2$. Condition (11.53a) may be written in the form after using (11.55a)

$$R\frac{df(r)}{dr}\bigg|_{r=R} + m\varpi f(R) = 0 \tag{11.57}$$

The kinematic free-surface condition (11.54b) at $r = r_0 + \eta\,(\theta, z, t)$ is

$$\Omega^2 r_0\eta - i\Omega\left(1 - \varpi^2\right)f(r_0)\,e^{i(\omega t + m\theta)}\cos(n\pi z/h) = 0 \tag{11.58a}$$

Eliminating η from equation (11.58) by using (11.54a), gives

$$\varpi^2 r_0\frac{df}{dr}\bigg|_{r=r_0} + \left[4\left(1 - \varpi^2\right) + m\varpi^3\right] f(r_0) = 0 \tag{11.58b}$$

The natural frequencies of the free surface are obtained by estimating the frequency ratio ϖ for which the Bessel equation (11.56) has nonzero solutions that satisfy the boundary conditions (11.57) and (11.58b). The simple case is the two-dimensional motion, that is, $n = 0$, that is, $k = 0$, for which equation (11.56) possesses the following solution

$$f(r) = Ar^m + \frac{B}{r^m} \tag{11.59}$$

Substituting equation (11.13) in conditions (11.57) and (11.58b) gives two algebraic equations in A and B. The determinant of the coefficients yields the frequency equation

$$m\varpi^2 - 4\varpi - 4\left[\frac{(R/r_0)^{2m} + 1}{(R/r_0)^{2m} - 1}\right] = 0 \tag{11.60}$$

The roots of this equation are

$$\varpi = \frac{2}{m}\left\{1 \pm \sqrt{1 + m\left[\frac{(R/r_0)^{2m} + 1}{(R/r_0)^{2m} - 1}\right]}\right\} \tag{11.61}$$

The frequency of the free surface is

$$\omega = \frac{m\Omega}{\left\{1 \pm \sqrt{1 + m\left[\frac{(R/r_0)^{2m} + 1}{(R/r_0)^{2m} - 1}\right]}\right\}} \tag{11.62}$$

The absolute frequency is, $\omega_{abs} = \omega - m\Omega$, that is,

$$\omega_{abs} = m\Omega \left[\frac{\mp \sqrt{1 + m \left[\dfrac{(R/r_0)^{2m} + 1}{(R/r_0)^{2m} - 1} \right]}}{\left\{ 1 \pm \sqrt{1 + m \left[\dfrac{(R/r_0)^{2m} + 1}{(R/r_0)^{2m} - 1} \right]} \right\}} \right] \tag{11.63}$$

The more general case is the three-dimensional solution corresponding to $n \neq 0$, or $k \neq 0$. The solution can be expressed in terms of the Bessel functions for the following cases

$$f(r) = AI_m(kr) + BK_m(kr) \quad \text{for } \varpi^2 < 1 \tag{11.64a}$$

or

$$f(r) = AJ_m(\bar{k}r) + BY_m(\bar{k}r) \quad \text{for } \varpi^2 > 1 \tag{11.64b}$$

where I_m and K_m are modified Bessel functions of the first and second kind respectively, J_m and Y_m are Bessel functions of the first and second kind, respectively, $k = (n\pi/h)\sqrt{1 - \varpi^2}$ and $\bar{k} = (n\pi/h)\sqrt{\varpi^2 - 1}$. Introducing equation (11.64a) or (11.64b) into the boundary conditions (11.57) and (11.58b) and equating the determinant of their coefficients to zero gives the frequency equations for $\varpi^2 < 1$ and $\varpi^2 > 1$, respectively,

$$\begin{vmatrix} kRI'_m(kR) + m\varpi I_m(kR) & kRK'_m(kR) + m\varpi K_m(kR) \\ \varpi^2 kr_0 I'_m(kr_0) & \varpi^2 kr_0 K'_m(kr_0) \\ \quad + [4(1-\varpi^2) + m\varpi^3] I_m(kr_0) & \quad + [4(1-\varpi^2) + m\varpi^3] K_m(kr_0) \end{vmatrix} = 0 \quad \text{for } \varpi^2 < 1 \tag{11.65a}$$

$$\begin{vmatrix} \bar{k}RJ'_m(\bar{k}R) + m\varpi J_m(\bar{k}R) & \bar{k}RY'_m(\bar{k}R) + m\varpi Y_m(\bar{k}R) \\ \varpi^2 \bar{k}r_0 J'_m(\bar{k}r_0) & \varpi^2 \bar{k}r_0 Y'_m(\bar{k}r_0) \\ \quad + [4(1-\varpi^2) + m\varpi^3] J_m(\bar{k}r_0) & \quad + [4(1-\varpi^2) + m\varpi^3] Y_m(\bar{k}r_0) \end{vmatrix} = 0 \quad \text{for } \varpi^2 > 1 \tag{11.65b}$$

The most significant case is for $m = 1$, and the above frequency equations take the form

$$\begin{vmatrix} kRI_0(kR) + (\varpi - 1)I_m(kR) & -kRK_0(kR) + (\varpi - 1)K_1(kR) \\ \varpi^2 kr_0 I_0(kr_0) & -\varpi^2 kr_0 K_0(kr_0) \\ \quad + [4 - 5\varpi^2 + \varpi^3] I_1(kr_0) & \quad + [4 - 5\varpi^2 + \varpi^3] K_1(kr_0) \end{vmatrix} = 0 \tag{11.66a}$$

$$\text{for } \varpi^2 < 1$$

$$\begin{vmatrix} \bar{k}RJ_0(\bar{k}R) + (\varpi - 1)J_m(\bar{k}R) & -\bar{k}RY_0(\bar{k}R) + (\varpi - 1)Y_1(\bar{k}R) \\ \varpi^2\bar{k}r_0J_0(\bar{k}r_0) & -\varpi^2\bar{k}r_0Y_0(\bar{k}r_0) \\ \quad + [4 - 5\varpi^2 + \varpi^3]J_1(\bar{k}r_0) & \quad + [4 - 5\varpi^2 + \varpi^3]Y_1(\bar{k}r_0) \end{vmatrix} = 0 \tag{11.66b}$$

$$\text{for } \varpi^2 > 1$$

Equations (11.66) are solved numerically for given values of R, r_0, and h. However, if the solution $f(r)$ does not vanish for $r_0 \leq r \leq R$, Miles and Troesch (1961) outlined another representation by introducing the following transformation

$$y = k^2r^2, \quad \text{and} \quad F(y) = r\frac{df(r)/dr}{f(r)} \tag{11.67}$$

Introducing (11.67) into equation (11.56) gives the Ricatti equation

$$2y\frac{dF(y)}{dy} + F^2(y) = m^2 + y \tag{11.68}$$

and conditions (11.57) and (11.58b) become

$$F[\alpha^2(1 - \varpi^2)] + m\varpi = 0, \quad \varpi^2 F[\beta^2(1 - \varpi^2)] + 4(1 - \varpi^2) + m\varpi^3 = 0 \tag{11.69a, b}$$

where $\alpha = n\pi\ R/h$, and $\beta = n\pi\ r_0/h$.

For given values of m, α, and β, and assuming a guessing value for ϖ, say μ_0, which can be obtained from the two-dimensional case, the first-order differential equation (11.68) is integrated numerically from the value of $y_0 = \beta^2(1 - \varpi^2)$, which gives from equation (11.69b)

$$F(y_0) = -\left[4(1 - \varpi_0^2) + m\varpi_0^3\right]/\varpi_0^2 \tag{11.70a}$$

to the value $y_1 = \alpha^2(1 - \varpi^2)$, and determine from equation (11.69a)

$$\tilde{\varpi} = -F(y_1)/m \tag{11.70b}$$

The iteration process continues until $\tilde{\varpi}$ agrees with ϖ_0.

For tanks with $k\ (R - r_0) \gg 1$, and $\varpi < 1$, the sloshing is confined to the neighborhood of the free surface. Equation (11.66a) takes the following form after using the asymptotic expansion of I_1 and K_1

$$\begin{vmatrix} (kR + \varpi)\, e^{2kR} & -kR + \varpi \\ \varpi^2kr_0I_0(kr_0) & -\varpi^2kr_0K_0(kr_0) \\ + [4 - 5\varpi^2 + \varpi^3]I_1(kr_0) & + [4 - 5\varpi^2 + \varpi^3]K_1(kr_0) \end{vmatrix} = 0 \tag{11.71}$$

and thus the leading term of (11.71) is

$$-\varpi^2kr_0K_0(kr_0) + [4 - 5\varpi^2 + \varpi^3]K_1(kr_0) = 0 \tag{11.72}$$

For small values of $h/2nr_0$, the Bessel functions K_1 and K_0 may be replaced by their asymptotic expansion and the following result is obtained

$$\varpi \approx \pm 2\sqrt{\frac{h}{n\pi r_0}}, \quad \text{for } h/2nr_0 < 1 \tag{11.73}$$

For extreme values of $h/2nr_0 \to \infty$, that is, when $n = 0$, the value of ϖ approaches the two-dimensional solution. For $\varpi > 1$, r_0/R is small, and $h/2nr_0$ is large, an approximate solution can be obtained from equation (11.66b), with the following leading term

$$\left\{\bar{k}RJ_0(\bar{k}R) + (\varpi - 1)J_m(\bar{k}R)\right\}[4 - 5\varpi^2 + \varpi^3] = 0 \tag{11.74}$$

This equation has two solutions. One of the solutions is obtained from the expression in the brackets

$$\varpi = 2\left(\sqrt{2} + 1\right) \tag{11.75}$$

This solution represents the asymptotic value for $h/2nr_0 \to \infty$, $r_0/R \to 0$, and is in agreement with the two-dimensional case given by equation (11.61). The other solution is given by the first expression of equation (11.74), that is,

$$\bar{k}RJ_0(\bar{k}R) + (\varpi - 1)J_m(\bar{k}R) = 0 \tag{11.76}$$

Figure 11.8 shows the dependence of the eigenvalue ϖ on the depth ratio $h/2nr_0$ for different values of radii ratio r_0/R. The figure shows the approximate solution obtained from equation (11.76) by small circles. The radial nodes can occur for free-surface oscillations on a rotating liquid. The two frequencies for a given nodal pattern differ in magnitude and in sign due to the presence of Coriolis acceleration. The eigenvalue ϖ decreases monotonically with the number of z-nodes (i.e., n), and thus the frequency ω increases with n.

11.3.3 Free oscillations of spinning viscous liquid

The free oscillations of liquid free surface in a rotating cylindrical container was considered by Bauer (1993b), Kimura, *et al.* (1994a), and Bauer and Eidel (1997b). The fluid is assumed to be incompressible and viscous. Let the rotational speed about the container z-axis be $\boldsymbol{\Omega} = \Omega\mathbf{k}$ such that $\Omega^2 R/g$ is small. The undisturbed free liquid surface is described by the equation

$$z = h + \eta_0(r), \quad \eta_0(r) = \frac{\Omega^2}{2g}\left(r^2 - \frac{R^2}{2}\right) \tag{11.77}$$

The Navier–Stokes equations (11.4) for axisymmetric motion, $\partial/\partial\theta = 0$, take the form

$$\frac{\partial u}{\partial t} + u\frac{\partial u}{\partial r} + w\frac{\partial u}{\partial z} - \frac{v^2}{r} + \frac{1}{\rho}\frac{\partial p}{\partial r} = \nu\left[\frac{\partial^2 u}{\partial r^2} + \frac{\partial}{\partial r}(u/r) + \frac{\partial^2 u}{\partial z^2}\right] \tag{11.78a}$$

$$\frac{\partial v}{\partial t} + u\left(\frac{\partial v}{\partial r} + \frac{v}{r}\right) + w\frac{\partial v}{\partial z} = \nu\left[\frac{\partial^2 v}{\partial r^2} + \frac{\partial}{\partial r}(v/r) + \frac{\partial^2 v}{\partial z^2}\right] \tag{11.78b}$$

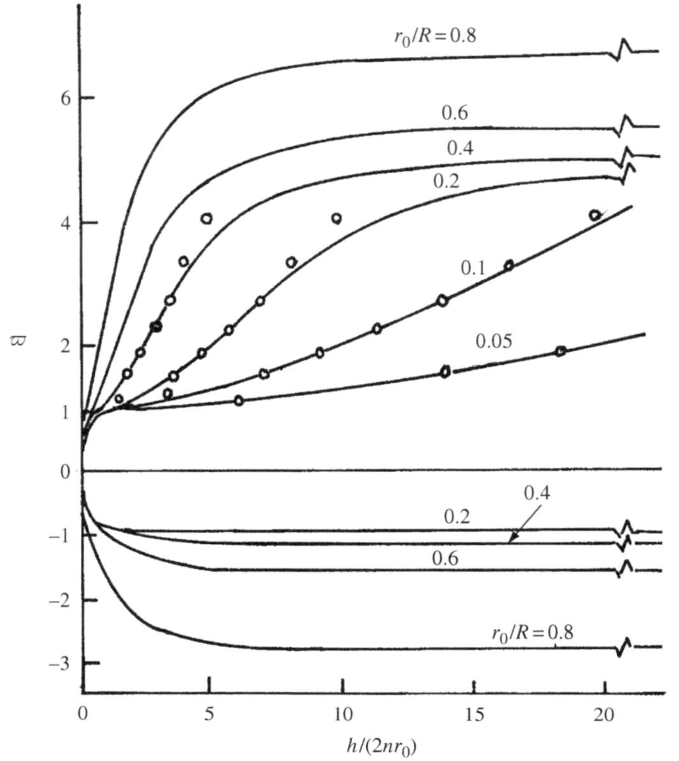

Figure 11.8 Dependence of the eigenvalues ϖ on the depth ratio $h/(2nr_0)$ for different values of radii ratio r_0/R. (Miles and Troesch, 1961)

$$\frac{\partial w}{\partial t}+u\frac{\partial w}{\partial r}+w\frac{\partial w}{\partial z}+\frac{1}{\rho}\frac{\partial p}{\partial z}=\nu\left[\frac{\partial^2 w}{\partial r^2}+\frac{1}{r}\frac{\partial w}{\partial r}+\frac{\partial^2 w}{\partial z^2}\right]-g \tag{11.78c}$$

The circumferential velocity component v may be written in terms of the spinning speed in the form

$$v=\Omega r+V(r,z,t) \tag{11.79}$$

The pressure distribution may be written in the form

$$p(r,z,t)=p_0+\bar{p}(r,z,t)+\frac{\rho}{2}\Omega^2 r^2-\rho g(z-h) \tag{11.80}$$

For small displacement of the free liquid surface and small velocities, equations (11.78) take the form, after using equations (11.79) and (11.80)

$$\frac{\partial u}{\partial t}-2\Omega V+\frac{1}{\rho}\frac{\partial p}{\partial r}=\nu\left[\frac{\partial^2 u}{\partial r^2}+\frac{\partial}{\partial r}(u/r)+\frac{\partial^2 u}{\partial z^2}\right] \tag{11.81a}$$

$$\frac{\partial V}{\partial t} + 2\Omega u = \nu\left[\frac{\partial^2 V}{\partial r^2} + \frac{\partial}{\partial r}(V/r) + \frac{\partial^2 V}{\partial z^2}\right] \tag{11.81b}$$

$$\frac{\partial w}{\partial t} + \frac{1}{\rho}\frac{\partial \bar{p}}{\partial z} = \nu\left[\frac{\partial^2 w}{\partial r^2} + \frac{1}{r}\frac{\partial w}{\partial r} + \frac{\partial^2 w}{\partial z^2}\right] \tag{11.81c}$$

These equations have to be solved together with the continuity equation (11.4d) subject to the following the boundary conditions

$$u = V = w = 0, \quad \text{at } z = 0 \tag{11.82a}$$

$$\frac{d\eta}{dt} = w \tag{11.82b}$$

$$\bar{p} - 2\mu\frac{\partial w}{\partial z} + \sigma\left[\frac{\partial^2 \eta}{\partial r^2} + \frac{1}{r}\frac{\partial \eta}{\partial r} - \frac{\rho g}{\sigma}\eta\right] = \text{const} \quad \text{at } z = h \tag{11.82c}$$

where μ is the liquid dynamic viscosity, σ is the liquid surface tension, and $\eta(r, t)$ is the surface elevation above $\eta_0(r)$. If the contact line is slipping at the side wall of the container, $r = R$, the two conditions given by equations (11.82b, c) can be combined to give the following equation

$$\frac{\partial \bar{p}}{\partial t} - 2\mu\frac{\partial^2 w}{\partial z \partial t} + \sigma\left[\frac{\partial^2 w}{\partial r^2} + \frac{1}{r}\frac{\partial w}{\partial r} - \frac{\rho g}{\sigma}w\right] = |_{z=h} 0 \tag{11.83}$$

If the contact line is anchored at the container wall the condition $\eta(R, t) = 0$ at the container wall must be satisfied.

Anchored contact-line

With reference to Figure 11.9, at the container wall, $r = R$, the following conditions have to be satisfied

$$u = V = 0, \quad w \neq 0, \quad \text{at } r = R \tag{11.84a}$$

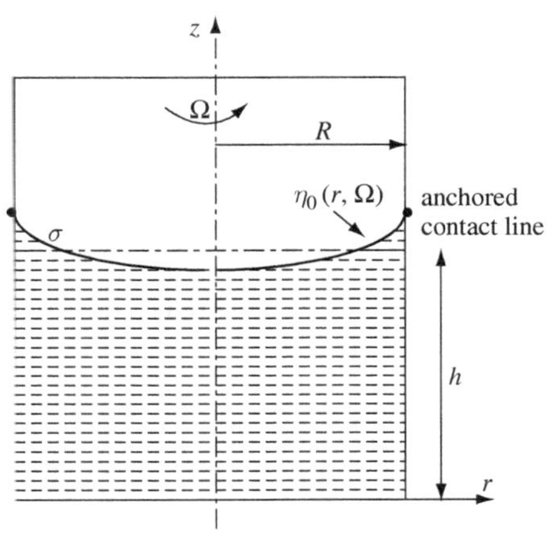

Figure 11.9 Rotating liquid with anchored contact line.

However, for an anchored contact line we have

$$\eta(R,t)=0 \tag{11.84b}$$

In this case, a paraboloid shape for the undisturbed free-liquid-surface position, that is, η_0, should be satisfied at $z=h+\eta_0\,(r)$. This would mean that the kinematic condition

$$\frac{d\eta}{dt}=w-u\frac{d\eta_0(r)}{dr} \tag{11.85}$$

could be developed. In addition, the shear stress must vanish at the free surface, that is,

$$\tau_{rz}=\mu\left[\frac{\partial w}{\partial r}+\frac{\partial u}{\partial z}\right]_{z=h}=0, \quad \text{and} \quad \tau_{\theta z}=\mu\frac{\partial v}{\partial z}\bigg|_{z=h}=0 \tag{11.86a, b}$$

The solution of the boundary value problem described by equations (11.81)–(11.84) may be assumed to be in the form

$$u(r,z,t)=e^{st}\sum_{n=1}^{\infty}U_n(z)J_1(\xi_n r/R), \quad V(r,z,t)=e^{st}\sum_{n=1}^{\infty}V_n(z)J_1(\xi_n r/R) \tag{11.87a,b}$$

$$w(r,z,t)=e^{st}\sum_{n=1}^{\infty}W_n(z)J_0(\xi_n r/R), \quad \bar{p}(r,z,t)=e^{st}\sum_{n=1}^{\infty}P_n(z)J_0(\xi_n r/R) \tag{11.87c,d}$$

where $s=\zeta+i\omega$ is the complex frequency to be determined, and in view of conditions (11.84a), ξ_n are the roots of $J_1\,(\xi_n)=0$.

Substituting equations (11.87) into equations (11.81) and using the recurrence relations of Bessel functions, gives

$$\frac{d^2U_n}{dz^2}-\varpi_n^2U_n+\frac{2\Omega}{\nu}V_n=\frac{\xi_n}{\mu R}P_n \tag{11.88a}$$

$$\frac{d^2V_n}{dz^2}-\varpi_n^2V_n+\frac{2\Omega}{\nu}U_n=0 \tag{11.88b}$$

$$\frac{d^2W_n}{dz^2}-\varpi_n^2W_n=\frac{1}{\mu}\frac{dP_n}{dz} \tag{11.88c}$$

Also substituting (11.87) in the continuity equation (11.4d), gives

$$\frac{\zeta_n}{R}U_n+\frac{dW_n}{dz}=0 \tag{11.89}$$

where $\varpi_n^2=\dfrac{\xi_n^2}{R^2}+\dfrac{s}{\nu}$. In order to determine the functions $U_n\,(z)$, $V_n\,(z)$, $W_n\,(z)$, and $P_n\,(z)$, we set each as $e^{\lambda z}$ and establish the characteristic equation from equations (11.88) and (11.89)

$$\left(\Lambda^2-\bar{\varpi}_n^2\right)^2\left(\Lambda^2-\xi_n^2\right)+4\bar{\Omega}_0^2\Lambda^2=0 \tag{11.90}$$

where $\Lambda = R\lambda$, $\bar{\varpi}_2^2 = \xi_n^2 + \frac{sR^2}{\nu}$, and $\bar{\Omega}_0 = \frac{\Omega R^2}{\nu}$. There are three roots of the characteristic equation (11.90). Accordingly, the functions $U_n(z)$, $V_n(z)$, $W_n(z)$, and $P_n(z)$ may be written in the form

$$U_n(z) = \sum_{j=1}^{3} \{A_{nj}\cosh(\Lambda_j z/R) + B_{nj}\sinh(\Lambda_j z/R)\} \tag{11.91a}$$

$$V_n(z) = \sum_{j=1}^{3} \frac{2\bar{\Omega}_0}{\left(\Lambda_j^2 - \bar{\varpi}_n^2\right)} \{A_{nj}\cosh(\Lambda_j z/R) + B_{nj}\sinh(\Lambda_j z/R)\} \tag{11.91b}$$

$$W_n(z) = -\sum_{j=1}^{3} \frac{\xi_n}{\Lambda_j} \{B_{nj}\cosh(\Lambda_j z/R) + A_{nj}\sinh(\Lambda_j z/R)\} \tag{11.91c}$$

$$P_n(z) = -\frac{\mu}{R}\sum_{j=1}^{3} \frac{\xi_n\left(\Lambda_j^2 - \bar{\varpi}_n^2\right)}{\Lambda_j^2} \{A_{nj}\cosh(\Lambda_j z/R) + B_{nj}\sinh(\Lambda_j z/R)\} \tag{11.91d}$$

where the coefficients A_{nj} and B_{nj} are obtained by satisfying the boundary condition (11.82a) at the container bottom. This gives

$$A_{n1} + A_{n2} + A_{n3} = 0 \tag{11.92a}$$

$$\frac{A_{n1}}{(\Lambda_1^2 - \bar{\varpi}_{n2})} + \frac{A_{n2}}{(\Lambda_2^2 - \bar{\varpi}_{n2})} + \frac{A_{n3}}{(\Lambda_3^2 - \bar{\varpi}_{n2})} = 0 \tag{11.92b}$$

$$\frac{A_{n1}}{\Lambda_1} + \frac{A_{n2}}{\Lambda_2} + \frac{A_{n3}}{\Lambda_3} = 0 \tag{11.92c}$$

with $n = 1, 2, \ldots$ The vanishing shear stresses given by equations (11.86) at the free surface gives

$$\sum_{j=1}^{3} \frac{\left(\Lambda_j^2 - \xi_n^2\right)}{\Lambda_j} \{A_{nj}\sinh(\Lambda_j z/R) + B_{nj}\cosh(\Lambda_j z/R)\} = 0 \tag{11.93a}$$

$$\sum_{j=1}^{3} \frac{\Lambda_j}{\left(\Lambda_j^2 - \bar{\varpi}_n^2\right)} \{A_{nj}\sinh(\Lambda_j z/R) + B_{nj}\cosh(\Lambda_j z/R)\} = 0 \tag{11.93b}$$

For the anchored contact line at the container wall the kinematic condition (11.82a) gives

$$\eta(r, t) = -e^{st}\sum_{n=1}^{\infty} \frac{\nu\xi_n}{SR^2} J_0(\xi_n r/R) \sum_{j=1}^{3} \{\bar{B}_{nj}\cosh(\Lambda_j h/R) + \bar{A}_{nj}\sinh(\Lambda_j h/R)\} \tag{11.94}$$

Introducing (11.91c, d) into the dynamic free-surface condition (11.82b), gives

$$\frac{d^2\bar{\eta}}{dr^2} + \frac{1}{r}\frac{d\bar{\eta}}{dr} - \frac{\rho g}{\sigma}\bar{\eta} = \frac{\mu}{\sigma R}\sum_{n=1}^{\infty}\xi_n J_0(\xi_n r/R)\sum_{j=1}^{3}\frac{\left(\Lambda_j^2 - \bar{\varpi}_n^2\right)}{\Lambda_j^2}$$
$$\times\left\{A_{nj}\cosh(\Lambda_j h/R) + B_{nj}\sinh(\Lambda_j h/R)\right\}$$
$$-\frac{2\mu}{\sigma R}\sum_{n=1}^{\infty}\xi_n J_0(\xi_n r/R)\sum_{j=1}^{3}\{A_{nj}\cosh(\Lambda_j h/R)$$
$$+ B_{nj}\sinh(\Lambda_j h/R)\} - P_0\frac{\rho g}{\sigma} \tag{11.95}$$

This equation possesses the following solution

$$\bar{\eta}(r) = A_0 I_0\left(\sqrt{\frac{\rho g}{\sigma}}r\right) - \frac{\rho\nu^2}{\sigma R}\sum_{n=1}^{\infty}\frac{\xi_n J_0(\xi_n r/R)}{\xi_n^2 + \frac{\rho g R^2}{\sigma}}\sum_{j=1}^{3}\frac{\left(\Lambda_j^2 - \bar{\varpi}_n^2\right)}{\Lambda_j^2}$$
$$\times\frac{R^2}{\nu}\{A_{nj}\cosh(\Lambda_j h/R) + B_{nj}\sinh(\Lambda_j h/R)\} \tag{11.96}$$
$$+\frac{2\rho\nu^2}{\sigma R}\sum_{n=1}^{\infty}\frac{\xi_n J_0(\xi_n r/R)}{\xi_n^2 + \frac{\rho g R^2}{\sigma}}\sum_{j=1}^{3}\frac{R^2}{\nu}\{A_{nj}\cosh(\Lambda_j h/R) + B_{nj}\sinh(\Lambda_j h/R)\} + P_0$$

The anchored contact line condition (11.84b) gives

$$A_0 I_0\left(\sqrt{\frac{\rho g}{\sigma}}r\right) - \frac{\rho\nu^2}{\sigma R}\sum_{n=1}^{\infty}\frac{\xi_n J_0(\xi_n r/R)}{\xi_n^2 + \frac{\rho g R^2}{\sigma}}\sum_{j=1}^{3}\frac{\left(\Lambda_j^2 - \bar{\varpi}_n^2\right)}{\Lambda_j^2}$$
$$\times\frac{R^2}{\nu}\{A_{nj}\cosh(\Lambda_j h/R) + B_{nj}\sinh(\Lambda_j h/R)\}$$
$$+\frac{2\rho\nu^2}{\sigma R}\sum_{n=1}^{\infty}\frac{\xi_n J_0(\xi_n r/R)}{\xi_n^2 + \frac{\rho g R^2}{\sigma}}\sum_{j=1}^{3}\frac{R^2}{\nu}\{A_{nj}\cosh(\Lambda_j h/R) + B_{nj}\sinh(\Lambda_j h/R)\} + P_0 = 0 \tag{11.97}$$

The modified Bessel function is expanded in the Dini series

$$I_0\left(\sqrt{\frac{\rho g}{\sigma}}r\right) = \frac{2}{R}\sqrt{\frac{\sigma}{\rho g}}I_1(R\sqrt{\frac{\rho g}{\sigma}}) + 2R\sqrt{\frac{\rho g}{\sigma}}\sum_{n=1}^{\infty}\frac{I_1\left(R\sqrt{\frac{\rho g}{\sigma}}\right)J_0(\xi_n r/R)}{\left(\xi_n^2 + \frac{\rho g R^2}{\sigma}\right)J_0(\xi_n)} \tag{11.98}$$

Introducing this expansion into equation (11.96) and comparing the result with the one obtained from the kinematic condition (11.94), the following conditions are obtained

$$2A_0 I_1\left(R\sqrt{\frac{\rho g}{\sigma}}\right) + R\sqrt{\frac{\rho g}{\sigma}}P_0 = 0 \tag{11.99a}$$

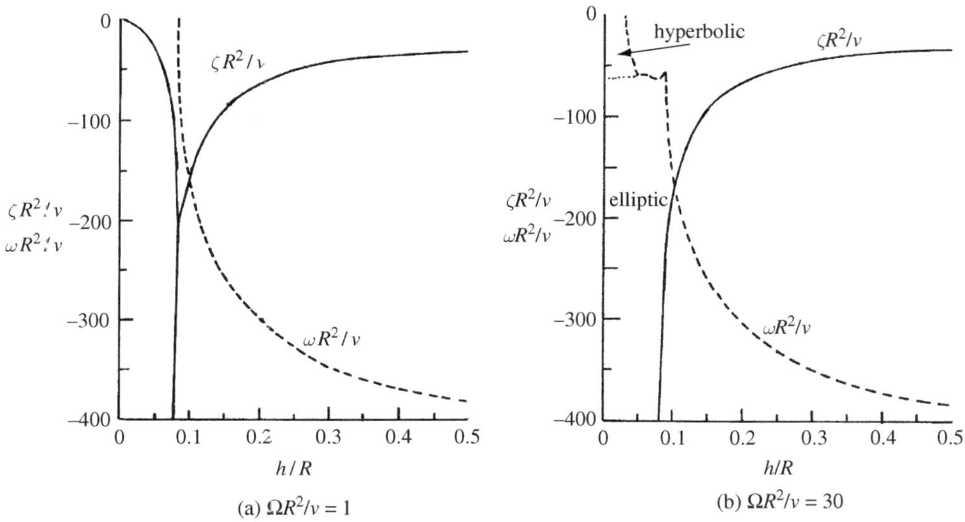

Figure 11.10 Dependence of the complex natural frequency components on the fluid depth for an anchored contact line, two different values of spin speed parameter, and for axisymmetric mode $n = 1$. (Bauer and Eidel, 1997b)

$$2R\sqrt{\frac{\rho g}{\sigma}}\,\frac{I_1\left(R\sqrt{\frac{\rho g}{\sigma}}\right)}{J_0(\xi_n)}$$
$$-\frac{\rho\nu^2\xi_n}{\sigma R}\sum_{j=1}^{3}\frac{\left(\Lambda_j^2-\bar{\varpi}_n^2\right)}{\Lambda_j^2}\times\frac{R^2}{\nu}\{A_{nj}\cosh(\Lambda_j h/R)+B_{nj}\sinh(\Lambda_j h/R)\}$$
$$+\frac{2\rho\nu^2\xi_n}{\sigma R}\sum_{j=1}^{3}\frac{R^2}{\nu}\{A_{nj}\cosh(\Lambda_j h/R)+B_{nj}\sinh(\Lambda_j h/R)\}$$
$$+\frac{(\xi_n^2+(\rho g R^2/\sigma))\xi_n}{sR^2/\nu}\sum_{j=1}^{3}\frac{R^2}{\nu\Lambda_j}\{A_{nj}\sinh(\Lambda_j h/R)+B_{nj}\cosh(\Lambda_j h/R)\}=0 \qquad (11.99b)$$

Equations (11.92), (11.93), (11.97), and (11.99) constitute a set of $(6n+2)$ homogeneous algebraic equations in A_{nj}, B_{nj}, A_0, and P_0. The determinant of the coefficients gives the damped frequency equation that may be solved numerically. Figures11.10(a) and (b) shows the dependence of the complex natural frequency components of the first axisymmetric mode, $(\zeta R^2/\nu)+i(\omega R^2/\nu)$, on the fluid depth for two different values of spinning speed $\bar{\Omega}_0 = \Omega R^2/\nu = 1$ and 30, respectively. The figures are obtained for surface tension parameter $\sigma^* = \sigma R/(\rho\nu^2) = 10^3$, and gravity parameter $\bar{g} = gR^3/\nu^2 = 10^4$. For a small depth ratio, $h/R < 0.1$, the free surface only performs an aperiodic motion if it is disturbed. For a fluid depth ratio exceeding 0.1, the fluid exhibits damped oscillations. Figure 11.10(b), reveals that for speed ratio $\bar{\Omega}_0 = 30$, the region of aperiodic motion is decreased, and at $\omega R^2/\nu \approx 60$ there is a bifurcation from the hyperbolic region below $\omega R^2/\nu < 60$ to the elliptic region > 60. Within the hyperbolic region the damping is very large and plays no important role in the oscillation of the rotating liquid. This indicates that for natural frequencies $\omega < 2\Omega$, that is, within the hyperbolic range, no appreciable response of the liquid free surface takes place.

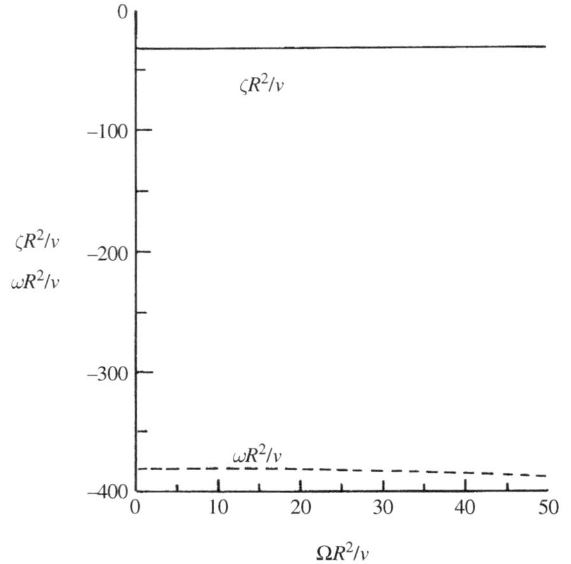

Figure 11.11 Dependence of the complex natural frequency components on the spin speed parameter, for an anchored contact line axisymmetric mode $n=1$, $h/R=0.5$. (Bauer and Eidel, 1997b)

Figure 11.11 shows the dependence of the natural frequency components on the spin speed parameter $\bar{\Omega}_0$. An increase in the spin speed results in a decrease of the natural frequency component $(\omega R^2/\nu)$ while there is no appreciable variation in the damping component $\zeta R^2/\nu$. Bauer and Eidel (1997b) found that the case of an anchored contact line exhibits a larger decay than that of a slipping contact line. Furthermore, they found that in the anchored case, there is a stronger change of the velocity distribution than in the slipping case, resulting in a stronger damping for the anchored case.

Adhesion condition

The case of axisymmetric free oscillations of liquid free surface in a slowly spinning cylindrical container was considered by Bauer and Eidel (2002b). The adhesion condition of the free liquid surface at the container wall condition (11.84a) is

$$u=V=w=0, \quad \text{at } r=R \tag{11.100}$$

The governing equations of motion and other boundary conditions are the same as in the previous case of an anchored contact line. Introducing the stream function Ψ (r, z, t) and the circulation function Θ (r, z, t) such that

$$u=\frac{1}{r}\frac{\partial\Psi}{\partial z}, \qquad v=\frac{1}{r}\Theta=\Omega r+V, \quad \text{and} \quad w=-\frac{1}{r}\frac{\partial\Psi}{\partial r} \tag{11.101}$$

the governing equations of motion can be written in terms of Ψ and Θ in the form

$$\left(\frac{\partial^2}{\partial r^2}-\frac{1}{r}\frac{\partial}{\partial r}+\frac{\partial^2}{\partial z^2}\right)\left[\left(\frac{\partial^2}{\partial r^2}-\frac{1}{r}\frac{\partial}{\partial r}+\frac{\partial^2}{\partial z^2}\right)\Psi-\frac{1}{\nu}\frac{\partial\Psi}{\partial t}\right]+\frac{2\Omega}{\nu}\frac{\partial\Theta}{\partial z}=0 \tag{11.102}$$

$$\left(\frac{\partial^2}{\partial r^2}-\frac{1}{r}\frac{\partial}{\partial r}+\frac{\partial^2}{\partial z^2}\right)\Theta-\frac{1}{\nu}\frac{\partial\Theta}{\partial t}-\frac{2\Omega}{\nu}\frac{\partial\Psi}{\partial z}=0 \tag{11.103}$$

where $\Theta=\Omega r^2+\Theta_0$, such that $V=\Theta_0/r$. The boundary conditions in terms of the functions Ψ and Θ_0 are

$$\left.\frac{\partial\Psi}{\partial z}\right|_{r=R}=0,\qquad \left.\frac{\partial\Psi}{\partial r}\right|_{r=R}=0,\qquad \Theta_0|_{r=R}=0,\qquad \left.\frac{\partial\Psi}{\partial r}\right|_{z=0}=0 \tag{11.104}$$

The shear stress conditions are

$$\frac{\partial^2\Psi}{\partial r^2}-\frac{1}{r}\frac{\partial\Psi}{\partial r}+\frac{\partial^2\Psi}{\partial z^2}=\Big|_{z=h}0,\quad\text{and}\quad \frac{\partial\Theta_0}{\partial z}\Big|_{z=h}=0 \tag{11.105a, b}$$

In addition to the combined free-surface condition (11.83). The solution of equations (11.102) and (11.103) may be assumed to be in the form

$$\Psi(r,z,t)=e^{st}\sinh(kz)R(r),\qquad \Theta_0(r,z,t)=e^{st}\cosh(kz)Q(r) \tag{11.106a, b}$$

Introducing equations (11.106) into equations (11.102) and (11.103), gives

$$\left(\frac{d^2}{dr^2}-\frac{1}{r}\frac{d}{dr}+k^2\right)\left[\left(\frac{d^2}{dr^2}-\frac{1}{r}\frac{d}{dr}+k^2\right)R-\frac{s}{\nu}R\right]+\frac{2k\Omega}{\nu}Q=0 \tag{11.107}$$

$$\left(\frac{d^2}{dr^2}-\frac{1}{r}\frac{d}{dr}+k^2\right)Q-\frac{s}{\nu}Q-\frac{2k\Omega}{\nu}R=0 \tag{11.108}$$

Applying the operation of divergence to equations (11.81) gives the equation for the pressure distribution

$$\left(\frac{\partial^2}{\partial r^2}+\frac{1}{r}\frac{\partial}{\partial r}+\frac{\partial^2}{\partial z^2}\right)\bar{p}-\frac{2}{r}\rho\Omega\left(\frac{\partial Q}{\partial r}\right)e^{st}\cosh(kz)=0 \tag{11.109}$$

Equations (11.107) and (11.108) are satisfied by the following solutions

$$R(r)=ARrJ_1(\lambda r),\quad\text{and}\quad Q(r)=BrJ_1(\lambda r) \tag{11.110a, b}$$

where λ is determined by substituting solutions (11.110) into equations (11.107) and (11.108),

$$\left(k^2-\lambda^2\right)\left(k^2-\lambda^2-\frac{s}{\nu}\right)AR+\frac{2k\Omega}{\nu}B=0 \tag{11.111a}$$

$$-\frac{2k\Omega}{\nu}AR+\left(k^2-\lambda^2-\frac{s}{\nu}\right)B=0 \tag{11.111b}$$

The characteristic equation of λ is obtained from equations (11.111) in the form

$$\left(K^2-\Lambda^2\right)\left(K^2-\Lambda^2-S\right)^2+4K^2\bar{\Omega}_0^2=0 \tag{11.112}$$

where $K = kR$, $\Lambda = \lambda R$, $S = sR^2/\nu$, and $\bar{\Omega}_0 = \Omega R^2/\nu$. The characteristic equation (11.112) is bi-cubic and has three roots, λ_j, $j = 1, 2, 3$. The solutions (11.110) may be written in the form

$$R(r) = \sum_{j=1}^{\infty} A_j R^2 \frac{r}{R} J_1(\Lambda_j r/R), \qquad Q(r) = \sum_{j=1}^{\infty} \frac{2K\bar{\Omega}_0}{(K^2 - \Lambda^2 - S)} A_j R \frac{r}{R} J_1(\Lambda_j r/R) \quad \text{(11.113a, b)}$$

where B_j has been written in terms of A_j using equation (11.111b). The boundary conditions (11.104) give

$$\sum_{j=1}^{\infty} A_j J_1(\Lambda_j) = 0, \qquad \sum_{j=1}^{\infty} A_j \Lambda_j J_0(\Lambda_j) = 0, \qquad \sum_{j=1}^{\infty} \frac{A_j J_1(\Lambda_j)}{(K^2 - \Lambda^2 - S)} = 0 \qquad \text{(11.114a, b, c)}$$

The determinant of the coefficients of equations (11.114) gives the characteristic equation of $K_n = k_n R$, which is a function of $\bar{\Omega}_0$ and S

$$\begin{vmatrix} J_1(\Lambda_1) & J_1(\Lambda_2) & J_1(\Lambda_3) \\ \Lambda_1 J_0(\Lambda_1) & \Lambda_2 J_0(\Lambda_2) & \Lambda_3 J_0(\Lambda_3) \\ \dfrac{J_1(\Lambda_1)}{(K^2 - \Lambda_1^2 - S)} & \dfrac{J_1(\Lambda_2)}{(K^2 - \Lambda_2^2 - S)} & \dfrac{J_1(\Lambda_3)}{(K^2 - \Lambda_3^2 - S)} \end{vmatrix} = 0 \qquad \text{(11.115)}$$

The amplitude ratios are

$$\frac{A_2}{A_1} = \frac{\Lambda_1 J_0(\Lambda_1) J_1(\Lambda_3) - \Lambda_3 J_0(\Lambda_3) J_1(\Lambda_1)}{\Lambda_3 J_0(\Lambda_3) J_1(\Lambda_2) - \Lambda_2 J_0(\Lambda_2) J_1(\Lambda_3)},$$
$$\frac{A_3}{A_1} = \frac{\Lambda_2 J_0(\Lambda_2) J_1(\Lambda_1) - \Lambda_1 J_0(\Lambda_1) J_1(\Lambda_2)}{\Lambda_3 J_0(\Lambda_3) J_1(\Lambda_2) - \Lambda_2 J_0(\Lambda_2) J_1(\Lambda_3)} \qquad \text{(11.116b, c)}$$

The velocity distribution is then given in the form

$$u(r, z, t) = e^{st} \sum_{n=1}^{\infty} \sum_{j=1}^{3} A_{nj} J_1(\Lambda_{nj} r/R) \cosh(K_n z/R) \qquad \text{(11.117a)}$$

$$v(r, z, t) = \Omega r + 2\, e^{st} \sum_{n=1}^{\infty} \sum_{j=1}^{3} \frac{A_{nj} K_n \bar{\Omega}_0 J_1(\Lambda_{nj} r/R)}{\left(K_n^2 - \Lambda_{nj}^2 - S\right)} \cosh(K_n z/R) \qquad \text{(11.117b)}$$

$$w(r, z, t) = -e^{st} \sum_{n=1}^{\infty} \sum_{j=1}^{3} A_{nj} \Lambda_{nj} J_0(\Lambda_{nj} r/R) \sinh(K_n z/R) \qquad \text{(11.117c)}$$

The shear stress conditions give

$$\sum_{n=1}^{\infty} \sum_{j=1}^{3} A_{nj} \left(\Lambda_{nj}^2 + K_n^2\right) J_1(\Lambda_{nj} r/R) \sinh(K_n h/R) = 0 \qquad \text{(11.118a)}$$

$$\sum_{n=1}^{\infty} \sum_{j=1}^{3} \frac{A_{nj} K_n^2 J_1(\Lambda_{nj} r/R)}{\left(K_n^2 - \Lambda_{nj}^2 - S\right)} \sinh(K_n h/R) = 0 \qquad \text{(11.118b)}$$

where A_{nj} are unknown constants. The pressure distribution (11.109) and the combined free-liquid-surface condition (11.83) give, respectively,

$$\bar{p}(r,z,t) = 4\frac{\mu\bar{\Omega}_0^2}{R}\sum_{n=1}^{\infty}\sum_{j=1}^{3}\frac{A_{nj}\Lambda_{nj}K_n J_0(\Lambda_{nj}r/R)}{\left(K_n^2-\Lambda_{nj}^2-S\right)\left(K_n^2-\Lambda_{nj}^2\right)}\sinh(K_n z/R) \tag{11.119}$$

and

$$\sum_{n=1}^{\infty}\sum_{j=1}^{3}A_{nj}\Lambda_{nj}J_0\left(\frac{\Lambda_{nj}r}{R}\right)\left\{\left[\frac{4\bar{\Omega}_0^2}{\left(K_n^2-\Lambda_{nj}^2-S\right)\left(K_n^2-\Lambda_{nj}^2\right)}+2\right]\right.$$
$$\left.\times SK_n\cosh\left(\frac{K_n h}{R}\right)+\left(\frac{\sigma R}{\rho\nu^2}\Lambda_{nj}^2+\frac{gR^3}{\nu^2}\right)\sinh\left(\frac{K_n h}{R}\right)\right\}=0 \tag{11.120}$$

Equations (11.118a) and (11.120) depend on the radial coordinate r. These equations can be satisfied at discrete locations in the range $0\le r/R\le 1$, such that $r/R=m/M$, for $m=0, 1, 2, \ldots, (M-1)$. This discretization results in $n=1, 2, \ldots, N=M-1$ algebraic equations whose coefficient determinant together with equation (11.112) gives an approximate estimation of the damped frequency equation. The discretized form of equations (11.118) and (11.120) is

$$\sum_{n=1}^{M-1}\sum_{j=1}^{3}A_{nj}\left(\Lambda_{nj}^2+K_n^2\right)J_1(\Lambda_{nj}m/M)\sinh(K_n h/R)=0 \tag{11.121a}$$

$$\sum_{n=1}^{M-1}\sum_{j=1}^{3}\frac{A_{nj}K_n^2 J_1(\Lambda_{nj}m/M)}{\left(K_n^2-\Lambda_{nj}^2-S\right)}\sinh(K_n h/R)=0 \tag{11.121b}$$

$$\sum_{n=1}^{\infty}\sum_{j=1}^{3}A_{nj}\Lambda_{nj}J_0\left(\frac{\Lambda_{nj}m}{M}\right)\left\{\left[\frac{4\bar{\Omega}_0^2}{\left(K_n^2-\Lambda_{nj}^2-S\right)\left(K_n^2-\Lambda_{nj}^2\right)}+2\right]\right.$$
$$\left.SK_n\cosh\left(\frac{K_n h}{R}\right)+\left(\frac{\sigma R}{\rho\nu^2}\Lambda_{nj}^2+\frac{gR^3}{\nu^2}\right)\sinh\left(\frac{K_n h}{R}\right)\right\}=0 \tag{11.121c}$$

for $m=0, 1, 2, \ldots, (M-1)$.

The effect of the spin speed parameter $\Omega^2 R/g$ on the undisturbed spinning free liquid surface is shown in Figure 11.12 for surface tension parameter $\sigma R/(\rho\nu^2)=10^3$, gravity parameter $gR^3/\nu^2=10^4$, and Bond number $\mathsf{Bo}=\rho g R^2/\sigma=10$. It is seen that all curves are intersecting at the location $r=R/\sqrt{2}$. As the spin-speed parameter increases the parabolic free surface is manifested.

The dependence of the damped natural frequency components $(\zeta R^2/\nu)+i(\omega R^2/\nu)$ on the liquid depth ratio h/R is shown in Figure 11.13 for the first sloshing mode, $n=1$, discretization of $M=21$ divisions, and two different values of spin-speed parameter $\Omega^2\ R/g=5$ and 30. With increasing liquid depth ratio, the natural frequency component $(\omega R^2/\nu)$ increases up to depth ratio $h/R=0.5$ above which the value reaches asymptotic value. On the other hand, the decay component, $(\zeta R^2/\nu)$, increases with depth ratio up to $h/R=0.2$ above which it decreases.

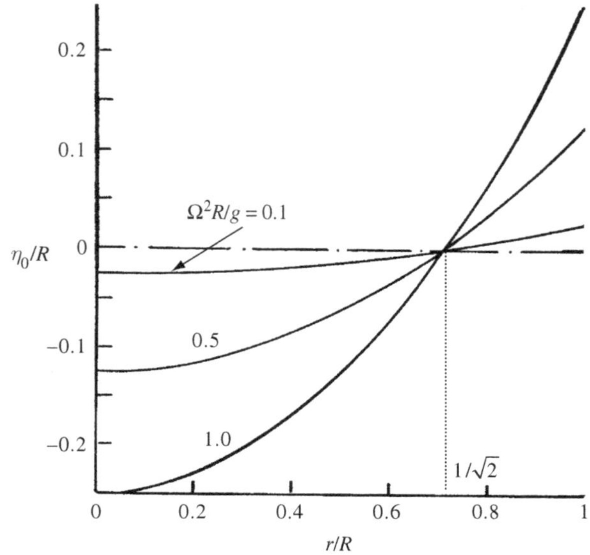

Figure 11.12 Undisturbed spinning equilibrium position of the free liquid surface for different values of spin-speed parameter. (Bauer and Eidel, 2002b)

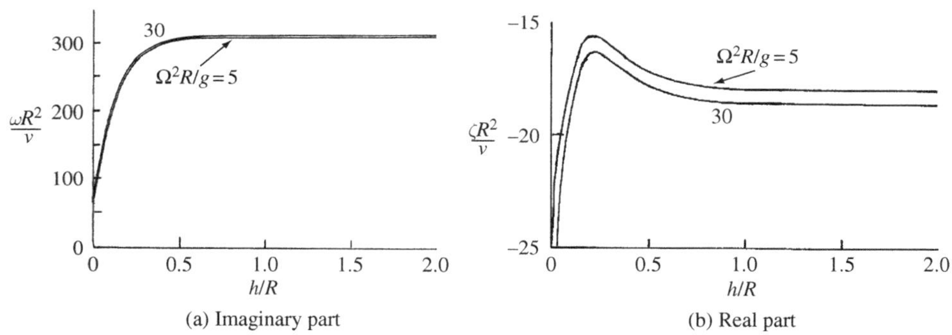

Figure 11.13 Dependence of the complex natural frequency components on the fluid depth for adhesion free-surface condition, two different values of spin-speed parameter, for axisymmetric mode $n=1$, $\rho g R^2/\sigma = 10$, and $\sigma R/(\rho \nu^2) = 1000$. (Bauer and Eidel, 2002b)

Note that the decay component increases as the spin speed increases due to the associated increase of wall area. With increasing spin speed, the increase of damping makes the liquid "stiff", which results in a faster decay of any disturbance to the free surface. The dependence of the damped natural frequency components on the surface tension parameter, $\sigma R/(\rho \nu^2)$, is shown in Figure 11.14 for the first axisymmetric mode and spin-speed parameter $\Omega^2 R/g = 30$. The surface tension causes an increase in both components of the natural frequency. Figure 11.15 shows the dependence of the damped frequency components on the Bond number, $\mathrm{Bo} = \rho g R^2/\sigma$. With an increase of Bond number, both components of the natural frequency increase. The effect of spin-speed parameter on the damped natural frequency components is shown in Figure 11.16. It is seen that the imaginary component increases while the decay component decreases with the increase of the spin speed.

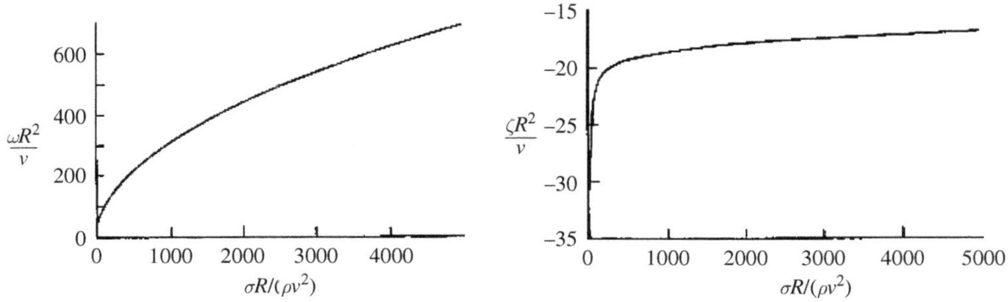

(a) Imaginary part (b) Real part

Figure 11.14 Dependence of the complex natural frequency components on the surface tension parameter for adhesion free-surface condition, spin-speed parameter 30, for axisymmetric mode $n = 1$, $\rho g R^2/\sigma = 10$, and $h/R = 2$. (Bauer and Eidel, 2002b)

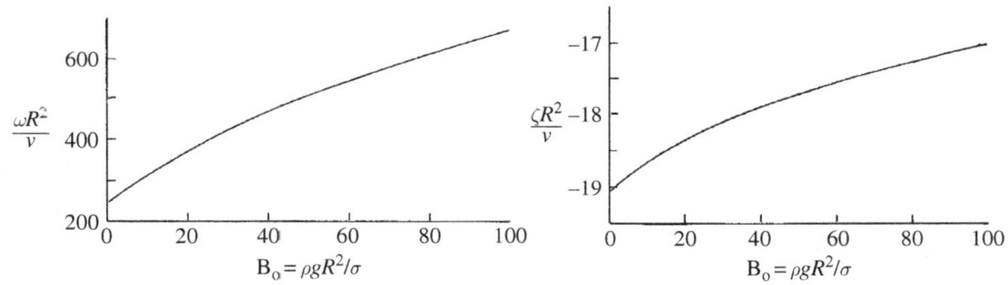

(a) Imaginary part (b) Real part

Figure 11.15 Dependence of the complex natural frequency components on the Bond number for adhesion free-surface condition, spin speed parameter 30, for axisymmetric mode $n = 1$, $\sigma R/(\rho \nu^2) = 1000$, and $h/R = 2$. (Bauer and Eidel, 2002b)

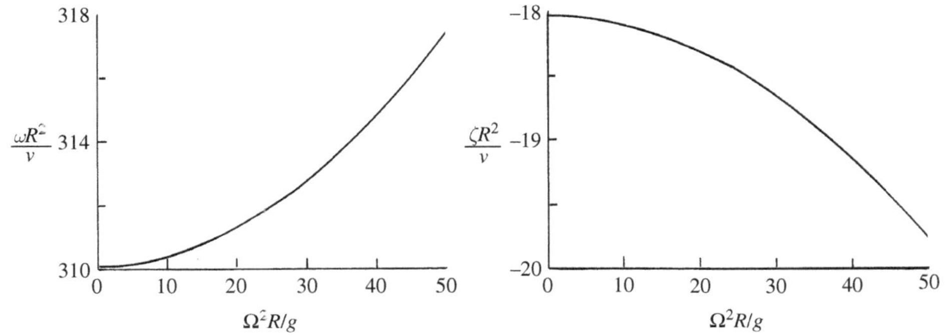

(a) Imaginary part (b) Real part

Figure 11.16 Dependence of the complex natural frequency components on the spin-speed parameter for adhesion free-surface condition, for axisymmetric mode $n = 1$, $\rho g R^2/\sigma = 10$, $h/R = 2$, and $\sigma R/(\rho \nu^2) = 1000$. (Bauer and Eidel, 2002b)

11.4 Parametric excitation of a spinning liquid

Skalak and Conly (1964) extended the work of Miles (1959c) to study the stability of the rotating fluid under parametric excitation, $\varepsilon g \sin \Omega_p t$, where ε is a positive number, g is the acceleration of gravity, and Ω_p is the parametric excitation frequency. Ignoring the convective acceleration, the equation of motion of the liquid in a moving coordinate frame is

$$\frac{\partial \mathbf{q}}{\partial t} + 2\Omega \times \mathbf{q} + \Omega \times (\Omega \times \mathbf{r}) = -\boldsymbol{\nabla}\left[\frac{p}{\rho} + gz(1 + \varepsilon \sin \Omega_p t)\right] \tag{11.122}$$

where $\mathbf{q}$ is the relative velocity of the liquid motion relative to a coordinate frame moving vertically with the container and rotating with angular velocity $\Omega = (0, 0, \Omega\mathbf{k})$, $\mathbf{r}$ is the position vector. The acceleration potential χ, whose gradient gives the acceleration vector, is introduced, and

$$\chi = \frac{p}{\rho} - gz_0 + gz(1 + \varepsilon \sin \Omega_p t) \tag{11.123}$$

where

$$z_0 = \frac{\Omega^2}{2g}\left(r^2 - \frac{1}{2}R^2\right) \tag{11.124}$$

Note that the plane free surface for the system at rest is selected as $z = 0$, such that $z = z_0$ is the free surface under rigid-body rotation. Assuming a harmonic time factor $e^{i\omega t}$ in $\mathbf{q}$ and χ, equation (11.122) may be written in the form

$$i\omega\mathbf{q} + 2\Omega \times \mathbf{q} = -\boldsymbol{\nabla}\chi \tag{11.125}$$

where ω represents the frequency of the free-liquid-surface waves. Alternatively, equation (11.125) may be written in terms of the velocity components u, v, and w, along r, θ, and z, respectively

$$u = \frac{i}{\omega(1 - \varpi^2)}\left(\frac{\partial \chi}{\partial r} - i\frac{\varpi}{r}\frac{\partial \chi}{\partial \theta}\right) \tag{11.126a}$$

$$v = \frac{i}{\omega(1 - \varpi^2)}\left(\frac{1}{r}\frac{\partial \chi}{\partial \theta} - i\varpi\frac{\partial \chi}{\partial r}\right) \tag{11.126b}$$

$$w = \frac{i}{\Omega}\frac{\partial \chi}{\partial r} \tag{11.126c}$$

where $\varpi = \dfrac{2\Omega}{\omega}$. Substituting equations (11.125) in the continuity equation, $\nabla \cdot \mathbf{q} = 0$, gives

$$\nabla^2\chi - \varpi^2\frac{\partial^2 \chi}{\partial z^2} = 0 \tag{11.127}$$

In the absence of parametric excitation, equation (11.127) is identical to the one derived by Miles (1959c). The solution of equation (11.127) must satisfy the boundary conditions. On the free surface the pressure is zero. Skalak and Conly (1964) introduced the effect of damping in an approximate manner. It was postulated that the pressure at the surface is proportional to the vertical velocity of the surface and always opposes the motion. Since the slope of the free surface is $\Omega^2 r/g$, the velocity of the free surface normal to itself, w_n, is

$$w_n = \frac{w - u(\Omega^2 r/g)}{\sqrt{1 + (\Omega^2 r/g)^2}} \tag{11.128}$$

The vertical velocity of the surface is

$$w_v = w - u(\Omega^2 r/g) \tag{11.129}$$

Replacing the pressure p in equation (11.123) by $\rho \nu w_v$, where ν is a pseudo-viscosity per unit length, the dynamical free-surface condition on $z = z_0 + \eta$ is

$$\chi = \nu\left[w - u(\Omega^2 r/g)\right] + \varepsilon g z_0 \sin \Omega_p t + g\eta(1 + \varepsilon \sin \Omega_p t) \tag{11.130}$$

where $\eta(r, t)$ is the vertical displacement of the free surface. The kinematic free-surface condition on $z = z_0 + \eta$ is

$$\frac{\partial \eta}{\partial t} = w - u\left(\frac{\Omega^2 r}{g}\right) \tag{11.131}$$

Assuming a harmonic factor $e^{i\omega t}$ for η, equations (11.130) and (11.131) give the following combined free-surface condition

$$\chi = \varepsilon g z_0 \sin \Omega_p t + \left[w - u(\Omega^2 r/g)\right]\left\{\frac{g\eta}{i\omega}(1 + \varepsilon \sin \Omega_p t) + \nu\right\} \tag{11.132}$$

Since η is assumed to be small, equation (11.132) will be applied on $z = z_0$, rather than on $z = z_0 + \eta$. The other boundary conditions require the velocity components normal to the container walls and bottom to vanish. Two distinct cases will be studied depending on the tank spin speed Ω. The first case is when the spin speed parameter $\varpi = 2\Omega/\omega < 1$, the elliptic case, while the second case is when $\varpi = \frac{2\Omega}{\omega} > 1$, the hyperbolic case.

For slow spin speed, $\varpi = 2\Omega/\omega < 1$, equation (11.127) is an elliptic partial differential equation The solution of this equation may be derived using the method of separation of variables through the parabolic coordinate transformation

$$\begin{aligned} \varsigma_1 &= \sqrt{\sqrt{\left[r^2 + \frac{(z+\hbar)^2}{(1-\varpi^2)}\right]} + \frac{(z+\hbar)}{\sqrt{1-\varpi^2}}}, \\ \varsigma_2 &= \sqrt{\sqrt{\left[r^2 + \frac{(z+\hbar)^2}{(1-\varpi^2)}\right]} - \frac{(z+\hbar)}{\sqrt{1-\varpi^2}}} \end{aligned} \tag{11.133}$$

where $\hbar$ is a constant that locates the parabolic coordinates with respect to the origin of z and will be selected such that the parabolic free surface coincides with a surface $\varsigma_2 = $ const. For $\hbar = 0$, the coordinates ς_1 and ς_2 are shown in Figure 11.17. In terms of the parabolic coordinates equation (11.127) takes the form

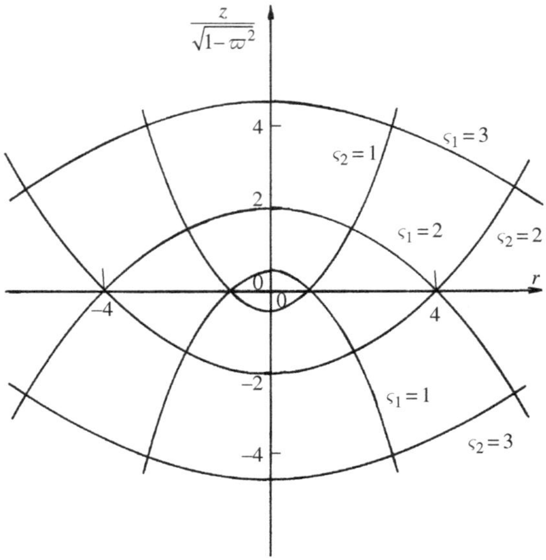

Figure 11.17 Parabolic coordinate transformation for the elliptic case. (Skalak and Conly, 1964)

$$\frac{\partial}{\partial\varsigma_1}\left(\frac{\varsigma_1\varsigma_2}{\sin\theta}\frac{\partial\chi}{\partial\varsigma_1}\right)+\frac{\partial}{\partial\varsigma_2}\left(\frac{\varsigma_1\varsigma_2}{\sin\theta}\frac{\partial\chi}{\partial\varsigma_2}\right)+\frac{\partial}{\partial(\cos\theta)}\frac{(\varsigma_1^2+\varsigma_2^2)\sin\theta}{\varsigma_1\varsigma_2}\frac{\partial\chi}{\partial(\cos\theta)}=0 \tag{11.134}$$

Introducing in equation (11.134) $\chi(\varsigma_1,\varsigma_2,\theta)=\chi_1(\varsigma_1)\,\chi_2(\varsigma_2)\,e^{i(\omega t-m\theta)}$, where m is an integer, and following the method of separation of variables, gives the two ordinary differential equations

$$\frac{1}{\varsigma_1}\frac{d}{d\varsigma_1}\left(\varsigma_1\frac{\partial\chi_1}{\partial\varsigma_1}\right)+\left(k^2-\frac{m^2}{\varsigma_1^2}\right)\chi_1=0,\qquad \frac{1}{\varsigma_2}\frac{d}{d\varsigma_2}\left(\varsigma_2\frac{\partial\chi_2}{\partial\varsigma_2}\right)-\left(k^2-\frac{m^2}{\varsigma_2^2}\right)\chi_2=0 \quad \text{(11.135a, b)}$$

where k is a separation constant and represents the wave number. The general solution of equation (11.135a) is a linear combination of the Bessel functions $J_m(k\varsigma_1)$, and $Y_m(k\varsigma_1)$. Since $Y_m(k\varsigma_1)$ is singular at $\varsigma_1=0$, it will be dropped. Equation (11.135b) is also satisfied by a linear combination of the modified Bessel functions $I_m(k\varsigma_2)$ and $K_m(k\varsigma_2)$, but $I_m(k\varsigma_2)$ will not be used because solutions that decay with the fluid depth are anticipated. Thus, the solution of equation (11.134) is

$$\chi(\varsigma_1,\varsigma_2,\theta)=A_m J_m(k\varsigma_1)K_2(k\varsigma_2)\,e^{i(\omega t+m\theta)} \tag{11.136}$$

where A_m is a complex amplitude.

Consider the free surface to be given by $\varsigma_2=\text{const}=\varsigma_{2s}$, the inverse of transformation (11.133) is

$$r=\varsigma_1\varsigma_2,\quad\text{and}\quad z=\frac{1}{2}(\varsigma_1^2-\varsigma_2^2)\sqrt{1-\varpi^2}-\hbar \quad \text{(11.137a, b)}$$

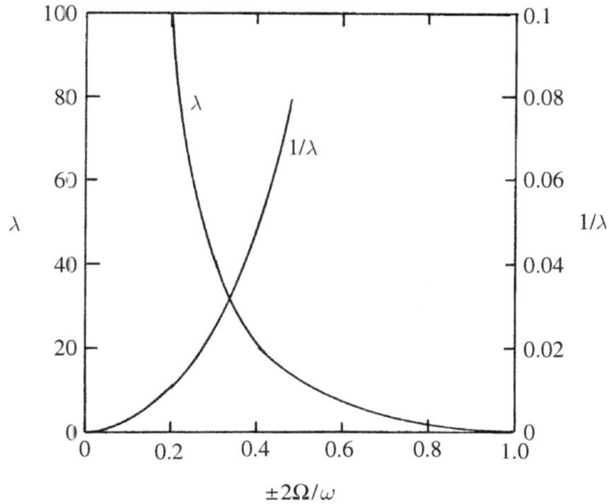

Figure 11.18 Dependence of the wave number $\lambda = kg/\Omega(1 - 4\Omega^2/\omega^2)^{1/4}$ on the frequency ratio for $m = 0$. (Skalak and Conly, 1964)

Equating the free surface equation (11.124) to equation (11.137b), and setting $\varsigma_2 = \varsigma_{2s}$, gives

$$\left(\varsigma_1^2 - \varsigma_{2s}^2\right)\sqrt{1-\varpi^2} - 2\hbar = \frac{\Omega^2}{g}\left(\varsigma_1^2\varsigma_{2s}^2 - \frac{1}{2}R^2\right) \tag{11.138}$$

This equation requires

$$\varsigma_{2s}^2 = \sqrt{\frac{g}{\Omega^2}\sqrt{1-\varpi^2}}, \quad \text{and} \quad \hbar = \frac{\Omega^2 R^2}{4g} - \frac{g}{2\Omega^2}\left(1-\varpi^2\right) \tag{11.139a, b}$$

The combined free-surface condition (11.132) may be written in terms of the parabolic coordinates on $\varsigma_2 = \varsigma_{2s}$ using equations (11.125), (11.133), and (11.139), as

$$\chi = \frac{\left[\frac{g}{i\omega}(1 + \varepsilon \sin \Omega_p t) + \nu\right]}{i\omega\varsigma_{2s}\sqrt{1-\varpi^2}}\left[\frac{\partial \chi}{\partial \varsigma_2} + \frac{m\varpi\chi}{\varsigma_{2s}}\right] + \frac{\varepsilon}{2}\Omega^2\left(\varsigma_1^2\varsigma_{2s}^2 - \frac{R^2}{2}\right)\sin \Omega_p t, \quad \text{on } \varsigma_2 = \varsigma_{2s} \tag{11.140}$$

In order to estimate the free vibration of the free-surface we drop the parametric excitation and set $\nu = 0$. Substituting equation (11.136) into equation (11.140), gives the frequency equation for free undamped surface oscillations

$$k\varsigma_{2s}K_m'(k\varsigma_{2s}) + \left[4\left(\frac{1}{\varpi^2} - 1\right) + m\varpi\right]K_m(k\varsigma_{2s}) = 0 \tag{11.141}$$

For $m = 0$, equation (11.141) gives the frequency of axisymmetric waves in terms of the wave number k. Figure 11.18 shows the dependence of the wave number parameter λ on the frequency ratio $2\Omega/\omega$, where

$$\lambda = \frac{kg}{\Omega}\left(1 - \frac{4\Omega^2}{\omega^2}\right)^{1/4} \tag{11.142}$$

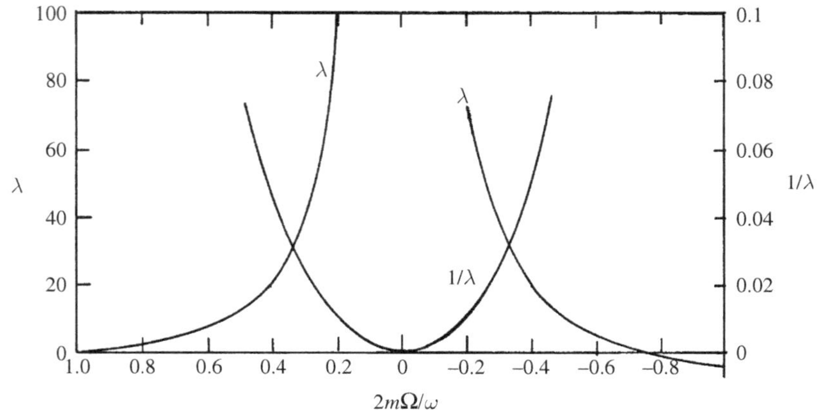

Figure 11.19 Dependence of the wave number $\lambda = kg/\Omega(1 - 4\Omega^2/\omega^2)^{1/4}$ on the frequency ratio for the $m = \pm 2$ elliptic case. (Skalak and Conly, 1964)

For $m = \pm M$, where M is a positive integer, two different values of ω are obtained from equation (11.141) depending on whether m is positive or negative. This means that the surface waves propagating in the direction of spin represented by $e^{i(\omega t - M\theta)}$ have a different frequency than waves of the same shape traveling in the opposite direction, represented by $e^{i(\omega t + M\theta)}$. Miles (1959a, b) called this phenomenon frequency splitting and it is due to the rotation. Figure 11.19 shows the roots of equation (11.141) for the case of $m = \pm 2$.

Under parametric excitation, the combined free-surface condition (11.140) establishes the response of the liquid free surface. Under the condition $\Omega_p = \omega$, the solution may be written as the Fourier–Bessel expansion

$$\chi = \sum_{m=-\infty}^{\infty} \sum_{n=1}^{\infty} A_{mn} J_m(k_{mn}\varsigma_1) K_m(k_{mn}\varsigma_2)\, e^{i(\omega t + m\theta)} \tag{11.143}$$

where k_{mn} are obtained by satisfying the condition at the side walls, ς_{1w} and ς_{2w}, and are given by the roots of the following equation

$$k\varsigma_{1w} J'_m(k\varsigma_{1w}) + \frac{2\Omega}{\Omega} m J_m(k\varsigma_{1w}) = 0 \tag{11.144}$$

In solving equation (11.140), we substitute equation (11.143) into equation (11.140) and drop the term $\varepsilon \sin \Omega_p t$ from the first expression in view of its smallness to 1. On the other, we retain and express $\varepsilon \sin \Omega_p t$ by $\varepsilon\, e^{i\Omega_p t}$ in the second expression because it represents the excitation function. Furthermore, the forcing coefficient is expanded into the Fourier–Bessel series

$$\left(\varsigma_1^2 \varsigma_{2s}^2 - \frac{R^2}{2}\right) = 4\varsigma_{2s}^2 \sum_{n=1}^{\infty} \frac{J_0(k_{0n}\varsigma_1)}{k_{0n}^2 J_0(k_{0n}\varsigma_{1w})} \tag{11.145}$$

Substituting equations (11.143) and (11.145) into equation (11.140) and equating coefficients of $J_m(k_{mn}\varsigma_1)\, e^{im\theta}$ gives the amplitude A_{0n}, since the forcing term (11.145) contains only $m = 0$,

$$A_{0n} = \frac{2\varepsilon\Omega^2 \varsigma_{2s}^2}{\left[k_{0n}^2 J_0(k_{0n}\varsigma_{1w})\right]\left[K_0(k_{0n}\varsigma_{2s}) + \dfrac{((g/i\omega)+\nu)k_{0n}K_1(k_{0n}\varsigma_{2s})}{i\omega\xi_{2s}\sqrt{1-\varpi^2}}\right]} \tag{11.146}$$

This result shows that every axisymmetric mode shape is excited at every excitation frequency. Note that the amplitude is bounded at resonance due to the presence of the pseudo viscosity parameter ν.

The half-subharmonic response is examined by setting $\omega = \Omega_p/2$. In this case the term $\varepsilon \sin \Omega_p t$ in the first expression is retained because the half-subharmonic response is created by this parametric excitation. The stability of one mode shape is studied by adopting the approximate form of χ

$$\chi = \left\{(A_1 + iA_2)\, e^{i(\omega t + m\theta)} + (A_3 + iA_4)\, e^{i(\omega t - m\theta)}\right\} J_m(k\varsigma_1) K_m(k\varsigma_2) \tag{11.147}$$

where A_i, $i = 1, 2, 3, 4$ are real, m is a positive integer, and k is the root of equation (11.144). Equation (11.147) represents the first term in the general periodic waveform. The procedure to be followed is to substitute equation (11.147) into equation (11.140), the resulting trigonometric functions are expanded into terms containing one of the products $\sin\omega t \sin m\theta$, $\sin\omega t \cos m\theta$, $\cos\omega t \sin m\theta$, $\cos\omega t \cos m\theta$ or higher harmonics of ωt. The higher harmonics are dropped since equation (11.147) represents the first term of a Fourier series. Equating coefficients of each of the four trigonometric products to zero gives the following four equations in the unknown amplitudes A_i

$$\nu B_1 A_1 + \left(K_m(k\varsigma_{2s}) + \frac{g}{\omega} B_1\right) A_2 - \left(\frac{\varepsilon g}{2\omega}\right) B_2 A_3 = 0 \tag{11.148a}$$

$$\left(K_m(k\varsigma_{2s}) + \frac{g}{\omega} B_1\right) A_1 - \nu B_1 A_2 - \left(\frac{\varepsilon g}{2\omega}\right) B_2 A_4 = 0 \tag{11.148b}$$

$$\left(\frac{\varepsilon g}{2\omega}\right) B_1 A_1 - \nu B_2 A_3 - \left(K_m(k\varsigma_{2s}) + \frac{g}{\omega} B_2\right) A_4 = 0 \tag{11.148c}$$

$$\left(\frac{\varepsilon g}{2\omega}\right) B_1 A_2 - \left(K_m(k\varsigma_{2s}) + \frac{g}{\omega} B_2\right) A_3 + \nu B_2 A_4 = 0 \tag{11.148d}$$

where

$$B_1 = \frac{\left[kK_m'(k\varsigma_{2s}) + \left(\dfrac{m\varpi}{\varsigma_{2s}}\right) K_m(k\varsigma_{2s})\right]}{(\omega\varsigma_{2s})\sqrt{1-\varpi^2}} \tag{11.149a}$$

$$B_2 = \frac{\left[kK_m'(k\varsigma_{2s}) - \left(\dfrac{m\varpi}{\varsigma_{2s}}\right) K_m(k\varsigma_{2s})\right]}{(\omega\varsigma_{2s})\sqrt{1-\varpi^2}} \tag{11.149b}$$

For nontrivial solution of equations (11.148) the determinant of the coefficients of the A_i must vanish. The resulting equation establishes the boundaries of stable and unstable parabolic free surface in the (ε, Ω_p) plane. For axisymmetric waves, $m = 0$, $B_1 = B_2$, the determinant gives

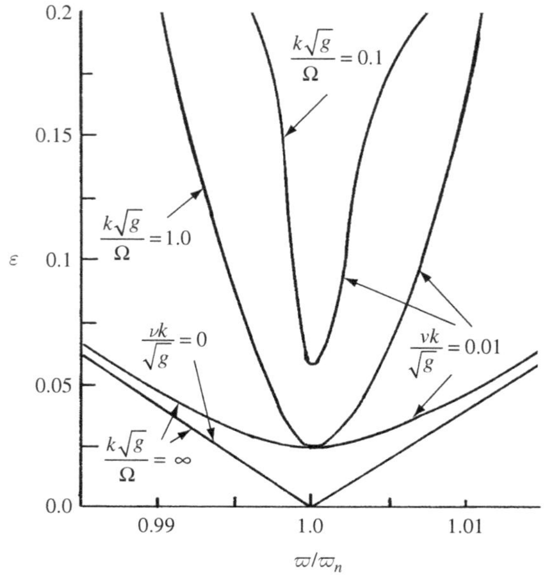

Figure 11.20 Stability boundaries for elliptic case, $m = 0$, for different values of damping parameter and spin speed parameter. (Skalak and Conly, 1964)

$$\varepsilon = \frac{2\omega}{gB_1}\sqrt{\left(K_m(k\varsigma_{2s}) + \frac{gB_1}{\omega}\right)^2 + \nu^2 B_1^2} \tag{11.150}$$

Figure 11.20 shows the stability boundaries on the $(\varepsilon, \varpi/\varpi_n)$ plane, for different values of spin speed parameter, $k\sqrt{g}/\Omega$, and the damping parameter, $\nu k/\sqrt{g}$, where ϖ_n is the value of ϖ at the natural frequency corresponding to the given value of k. The nonrotating case is shown by the curves corresponding to $k\sqrt{g}/\Omega = \infty$. It is clear that the spin speed has the effect of shrinking the unstable region, and the spin has a stabilizing effect on the free surface.

For asymmetric sloshing modes, $m \neq 0$, the first mode, $m = 1$, is considered for zero damping, that is, for $\nu = 0$. For half-subharmonics, the determinant gives the following relationship

$$\varepsilon = \frac{2\omega}{g\sqrt{B_1 B_2}}\sqrt{\left(K_m(k\varsigma_{2s}) + \frac{gB_1}{\omega}\right)\left(K_m(k\varsigma_{2s}) + \frac{gB_2}{\omega}\right)} \tag{11.151}$$

Figure 11.21 shows the stability boundaries for $m = 1$ on the $(\varepsilon, \varpi/\varpi_n)$, where ϖ_n is taken as the mean value of the frequency of the free vibration frequencies of forward and backward moving waves. The intersections of the curves for nonzero spin speed, that is, for $k\sqrt{g}/\Omega < \infty$ with the frequency axis, correspond to the free-surface frequencies.

For high-spin speed, $\varpi > 1$, equation (11.127) is hyperbolic since there are solutions that do not decay with depth. Skalak and Conly (1964) obtained special solutions of equation (11.127), which possesses the following conical surfaces

$$z = \pm r\sqrt{\varpi^2 - 1} + c \tag{11.152}$$

where c is a constant.

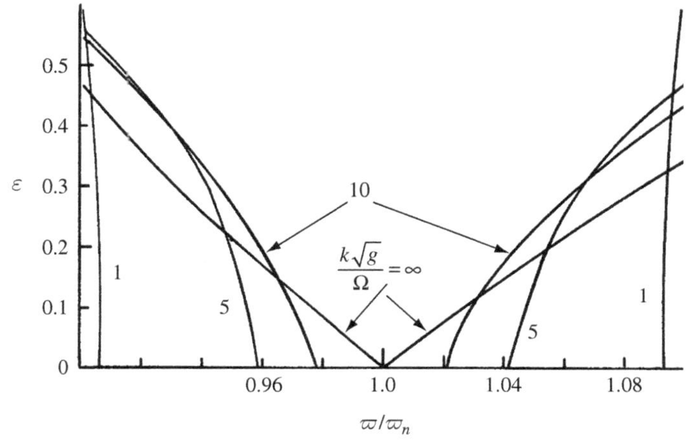

Figure 11.21 Stability boundaries for elliptic case, $m=1$, for different values of spin-speed parameter. (Skalak and Conly, 1964)

11.5 Inertia waves in a rotating fluid

Consider the problem of a linear inviscid incompressible fluid in a cylindrical tank of radius R, which rotates with a uniform angular velocity Ω about its longitudinal z-axis. The fluid free surface, $z=0$, is at rest, and under a uniform rotation Ω it takes the paraboloidal form, $z=z_0(r)=\Omega^2(r^2-R^2/2)/(2g)$. Note that viscous forces are required in order for the liquid to achieve this state of equilibrium. However, it is assumed that the time required for these forces to have an appreciable effect is large compared with the characteristic time of the inviscid motion, which is taken as the period of free oscillations.

The fluid dynamic behavior is governed by the continuity and Euler equations

$$\nabla \cdot \mathbf{q}=0 \tag{11.153a}$$

$$\frac{\partial \mathbf{q}}{\partial t}+2\boldsymbol{\Omega}\times\mathbf{q}=-\nabla P \tag{11.153b}$$

where $P=[(P_z-Pz_0)/\rho]+g(z-z_0)$. At the free surface of equilibrium we have $P_z=P_{z0}$, and $z=z_0$. An immediate consequence of the Coriolis acceleration in equation (11.153b) is that the velocity field cannot remain irrotational unless it is independent of z, since taking the curl of equation (11.153b), and recalling that Ω is directed along the z-axis, gives

$$\frac{\partial}{\partial t}(\nabla\times\mathbf{q})=2\Omega\frac{\partial \mathbf{q}}{\partial z} \tag{11.154}$$

Taking the divergence of equation (11.153b), and using equations (11.153a), gives

$$\left[\nabla\cdot\left(\frac{\partial \mathbf{q}}{\partial t}+2\Omega\times\mathbf{q}\right)\right]_{tt}=-\left[\nabla^2 P\right]_{tt} \quad\rightarrow\quad -2\Omega\cdot(\nabla\times\mathbf{q})_{tt}=-\left[\nabla^2 P\right]_{tt}$$

Using equation (11.154), gives

$$\nabla^2 P_{tt} + 4\Omega^2 P_{zz} = 0 \tag{11.155}$$

Introducing the wave-like solution

$$\mathbf{q} = \mathbf{Q}\, e^{i\omega t}, \quad P = \Psi\, e^{i\omega t} \tag{11.156a, b}$$

Equations (11.153a) and (11.155) take the form

$$\nabla \cdot \mathbf{Q} = 0, \quad \nabla^2 \Psi - 4\frac{\Omega^2}{\omega^2}\Psi_{zz} = 0 \tag{11.157a, b}$$

The solution given by equations (11.156) must satisfy the boundary conditions at the container wall whose outward unit normal vector is $\mathbf{n}$, and at the free surface, that is,

$$\mathbf{Q} \cdot \mathbf{n} = 0 \tag{11.158}$$

To establish the free-surface condition, let η denote the displacement of the surface relative to $z = z_0$, that is, $z = z_0 + \eta$, and thus $p = g\eta$. The kinematic condition requires the vertical acceleration to be given by the condition

$$\frac{\partial^2 \eta}{\partial t^2} = \frac{\partial w}{\partial t} = -\frac{\partial P}{\partial z} \quad \text{at } z = z_0 + \eta \tag{11.159}$$

The boundary condition at the equilibrium position is written in the form

$$\frac{\partial^2 P}{\partial t^2} + g\frac{\partial P}{\partial z} = 0 \quad \text{at } z = z_0 \tag{11.160}$$

The eigenvalue problem (11.157) admits solutions in closed domains for real values of $\lambda = \omega/\Omega$, $\lambda < 2$ for which equation (11.157b) is hyperbolic in its spatial dependence. A governing equation of hyperbolic type means that discontinuities can occur in the fluid across characteristic surfaces, which are in this case given by the cones, Greenspan (1968),

$$\sqrt{x^2 + y^2} \pm \frac{\lambda z}{\sqrt{4 - \lambda^2}} = \text{const} \tag{11.161}$$

These cones were demonstrated experimentally by Görtler (1957), and Oser (1958). Their experiments were conducted by placing a small disk in the middle of a cylindrical container filled with incompressible liquid. The disk was allowed to oscillate at frequency ω in the direction normal to its own surface and parallel to the axis of the cylinder. When the cylinder was not rotating, the fluid motion behaved quite ordinaryly in the appearance of potential flow. When the cylinder was allowed to rotate uniformly at an angular velocity Ω, extraordinary effects were observed as long as $\omega \leq 2\Omega$. The pattern of the flow of free shear layers in the fluid is reminiscent of Mach cones in compressible aerodynamics. As ω increases towards 2Ω, the characteristic cones become more nearly horizontal planes. Beyond the critical value, the wave system disappears entirely and the flow resembles a potential motion once again. Mathematically, a changeover from hyperbolic to elliptic character must occur at the precise value $\omega = 2\Omega$. Miles (1959c) studied in detail the natural frequencies for the free-surface oscillations. He found that the effect of rotation is to split the pairs of frequencies (that would have been of equal magnitude and opposite sign) for the nonrotating liquid.

11.6 Periodic breakdown of free surface

Stewartson (1959) showed that the liquid-filled top could be made unstable at almost any filling ratio by making the motion to have a large angle. In other words, the axis of the top departs significantly from the vertical axis. However, other peculiar phenomena have been reported by McEwan (1970) and Scott (1973, 1975), who observed two additional interesting phenomena associated with the "large angle" motion. The first is the failure of the axis of the hollow cylindrical core of the liquid to remain coincident with the axis of rotation of the right circular cylindrical rotor as the latter gyrated through large angles. The second is the resonant amplitude growth rate of the liquid-filled gyroscope at these angles as shown in Figure 11.22. These phenomena may have some relevance to the stability of spin-stabilized liquid-payload projectiles. The stability of these projectiles at large angles does not conform to the small angle stability theory of Stewartson (1959)

With reference to Figure 11.22, the axes X, Y, Z are fixed relative to the cylinder, which rotates with angular velocity components Ω_X, Ω_Y, and Ω_Z about X, Y, and Z, respectively. The moving axes x, y, z are moving with the bulk motion of the fluid, and their angular velocity vector Ω has components Ω_x, Ω_y, and Ω_z about x, y and z, respectively. When the spinning cylinder is gyrating, the Euler equation takes the form

$$\frac{\partial \mathbf{q}}{\partial t} + 2\boldsymbol{\Omega} \times \mathbf{q} = -\boldsymbol{\nabla}\left(\frac{p'}{\rho} + \frac{1}{2}\mathbf{q} \cdot \mathbf{q}\right) + \mathbf{r} \times \frac{\partial \boldsymbol{\Omega}}{\partial t} + \mathbf{q} \times (\boldsymbol{\nabla} \times \mathbf{q}) \tag{11.162}$$

where the gravity effect has been neglected, and $\mathbf{q}$ is the fluid velocity vector measured with respect to the moving frame x, y, z, and

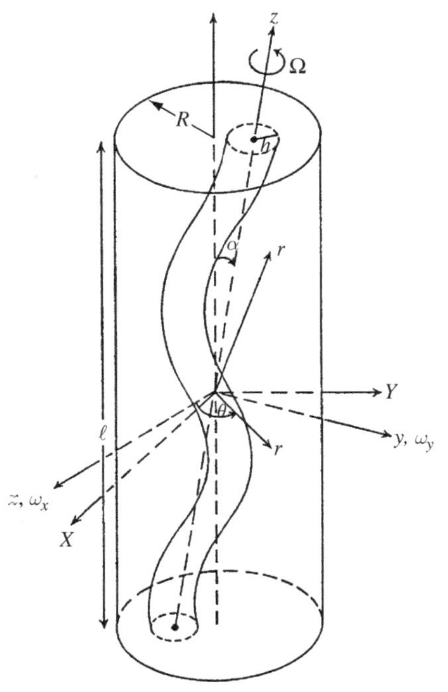

Figure 11.22 Partially filled, spinning cylinder gyrating through a large angle α. (Scott, 1975)

$$\frac{p'}{\rho} = \frac{p}{\rho} + \frac{1}{2}\Omega^2 b^2 - \frac{1}{2}(\mathbf{\Omega} \times \mathbf{r}) \cdot (\mathbf{\Omega} \times \mathbf{r}) \tag{11.163}$$

Scott (1975) introduced a system of Rossby waves, that is, very low frequency inertial waves (in the rotating x, y, z frame) and constructed nonlinear solutions for cylinders using series expansion in the angle α, the angle between the fluid axis of rotation and the vertical Z-axis. The method of multiple time scales is used and the solution may be given the form

$$\mathbf{q}(\mathbf{r}, t) = \mathbf{q}_0(\mathbf{r}, t) + \alpha \mathbf{q}_1(\mathbf{r}, T_1, T_2, \ldots) + \alpha^2 \mathbf{q}_2(\mathbf{r}, T_1, T_2, \ldots) + \cdots \tag{11.164a}$$

$$p'(\mathbf{r}, t) = p_0(\mathbf{r}, t) + \alpha p_1(\mathbf{r}, T_1, T_2, \ldots) + \alpha^2 p_2(\mathbf{r}, T_1, T_2, \ldots) + \cdots \tag{11.164b}$$

where $\mathbf{q}_0(\mathbf{r}, t)$ and $p_0(\mathbf{r}, t)$ belong to the inertial flow, $\mathbf{q}_j(\mathbf{r}, T_1, T_2, \ldots)$, and $p_j(\mathbf{r}, T_1, T_2, \ldots)$, $j \geq 1$ belong to Rossby waves, and $T_n = \alpha^n t$ are the time scales.

The velocity and pressure of the inertial waves may be represented by the expansions

$$\mathbf{q}_0(\mathbf{r}, t) = \sum_m A_m(t) \mathbf{Q}_m(\mathbf{r}) = \sum_m e^{i\omega_m t} \mathbf{Q}_m(\mathbf{r}) \tag{11.165a}$$

$$p_0(\mathbf{r}, t) = \sum_m A_m(t) P_m(\mathbf{r}) = \sum_m e^{i\omega_m t} P_m(\mathbf{r}) \tag{11.165b}$$

where the components $\mathbf{Q}_m(\mathbf{r})$ satisfy the equations

$$[i\lambda_m \Omega + 2\Omega \mathbf{k} \times] \mathbf{Q}_m(\mathbf{r}) = -\nabla P_m(\mathbf{r}) \tag{11.166a}$$

$$\nabla \cdot \mathbf{Q}_m(\mathbf{r}) = 0 \tag{11.166b}$$

On the wetted surface, we have the condition

$$\mathbf{Q}_m(\mathbf{r}) \cdot \mathbf{n} = 0 \tag{11.167}$$

Greenspan (1968) showed that the velocity vector $\mathbf{Q}_m$ is expressed in terms of the pressure by inverting equation (11.166a)[1]

$$\mathbf{Q}_m = \frac{1}{1 - (\lambda_m/2)^2} \left[\frac{1}{2} \mathbf{k} \times \nabla P_m(\mathbf{r}) - \frac{i\lambda_m}{4} \nabla P_m(\mathbf{r}) - \frac{1}{i\lambda_m} \mathbf{k}(\mathbf{k} \nabla P_m(\mathbf{r})) \right] \tag{11.168}$$

[1] The inversion of equation (11.166a) is obtained by writing the functions $\mathbf{Q}$ and ∇P_m in terms of their components, that is, $\mathbf{Q} = Q\mathbf{i} + Q\mathbf{j} + Q\mathbf{k}$, $\nabla P_m = \nabla_1 P_m \mathbf{i} + \nabla_2 P_m \mathbf{j} + \nabla_2 P_m \mathbf{k}$, such that $\mathbf{k} \times \mathbf{Q} = -Q_2 \mathbf{i} + Q_1 \mathbf{j}$. In this case, one can write equation (11.166a) in terms of its components as (after setting = 1)

$$i\lambda Q_1 - 2Q_2 = -\nabla_1 P_m, i\lambda Q_2 + 2Q_1 = -\nabla_2 P_m, i\lambda Q_3 = -\nabla_3 P_m$$

These equations can be solved for Q_i as

$$Q_1 = (1/(\lambda^2 - 4))[i\lambda \nabla_1 P_m + 2\nabla_2 P_m], Q_2 = (1/(\lambda^2 - 4))[-2\nabla_1 P_m + i\lambda \nabla_2 P_m], Q_3 = -(1/(i\lambda))\nabla_3 P_m$$

Summing up the three components, gives

$$\mathbf{Q} = -(1/(i\lambda))\mathbf{k}(\mathbf{k} \cdot \nabla P) + (i\lambda/(\lambda^2 - 4))\left[\nabla P - (\mathbf{k}/|\mathbf{k}|^2)(\mathbf{k} \cdot \nabla P)\right] - (2/(\lambda^2 - 4))\mathbf{k} \cdot \nabla P$$

This expression is essentially the same as equation (11.168).

where $\lambda_m = \omega_m/\Omega$, and $P_m(\mathbf{r})$ is obtained by solving equation (11.157b) which is written in the form

$$\nabla^2 P_m(\mathbf{r}) - \frac{4}{\lambda_m^2}\frac{\partial^2 P_m(\mathbf{r})}{\partial z^2} = 0 \tag{11.169}$$

The solution of this equation may be written in the form

$$P_m(\mathbf{r}) = \left\{ AJ_1\left(\xi_j r\sqrt{(2/\lambda_m)^2 - 1}\right) + BY_1\left(\xi_j r\sqrt{(2/\lambda_m)^2 - 1}\right)\right\} e^{i\theta}\cos\left(\xi_j(z + \frac{\ell}{2})\right) \tag{11.170}$$

where ℓ is the cylinder height, A and B are constants to be determined from the boundary conditions, $\xi_j = \pi(2j+1)/\ell$. The eigenvalues λ_m for a cylindrical annulus of liquid of inner and outer radii b and R are determined from the condition

$$\begin{aligned}&\left[\xi_j\beta J_1'(\xi_j\beta R) + \frac{2}{\lambda_m R}J_1(\xi_j\beta R)\right]\left\{\xi_j\beta Y_1'(\xi_j\beta b) + \left[\frac{2}{\lambda_m b} + \frac{4-\lambda_m^2}{b}\right]Y_1(\xi_j\beta b)\right\} \\ &\quad - \left[\xi_j\beta Y_1'(\xi_j\beta R) + \frac{2}{\lambda_m R}Y_1(\xi_j\beta R)\right]\left\{\xi_j\beta J_1'(\xi_j\beta b) + \left[\frac{2}{\lambda_m b} + \frac{4-\lambda_m^2}{b}\right]J_1(\xi_j\beta b)\right\} = 0\end{aligned} \tag{11.171}$$

where $\beta = \sqrt{(4/\lambda_m^2) - 1}$, and r is the radial cylindrical coordinate.

Equation (11.171) indicates that λ_m depends critically on the value of $\ell/[2R(2j+1)] = \pi/(2R\xi_j)$, where $(2j+1)$ is the number of half cosine waveforms that can be fitted into the cylinder height. Since this equation assumes that the liquid is spinning about the central axis of the cylinder, that is, $\alpha = 0$, Scott (1975) introduced the effect of a nonzero value of α on the number of waves by setting

$$\xi_j = \xi_{j0} + \alpha\xi_{j1} + \alpha^2\xi_{j2} + \cdots, \quad \text{where } \xi_{j0} = \pi(1+2j)/\ell \tag{11.172a}$$

$$\mathbf{Q}_m = \mathbf{Q}_{m0}(\mathbf{r}, \xi_{j0}) + \alpha\xi_{j1}\frac{\partial\mathbf{Q}_{m0}(\mathbf{r}, \xi_{j0})}{\partial\xi_{j0}} + \cdots \tag{11.172b}$$

$$\Psi_m = \Psi_{m0}(\mathbf{r}, \xi_{j0}) + \alpha\xi_{j1}\frac{\partial\Psi_{m0}(\mathbf{r}, \xi_{j0})}{\partial\xi_{j0}} + \cdots \tag{11.172c}$$

Following the asymptotic expansion technique up to second-order for equations (11.162) through (11.167), Scott (1975) obtained the following first-order approximation for $\mathbf{Q}_{m0}(\mathbf{r}, \xi_0)$

$$\left\{\mathbf{Q}_{m0}\cdot\mathbf{i}_r - r\frac{\partial(\mathbf{Q}_{m0}\cdot\mathbf{k})}{\partial z}\right\}_{z=\ell/2} = \left[\xi_1\frac{\partial(\mathbf{Q}_{m0}\cdot\mathbf{k})}{\partial\xi_0}\right]_{z=\ell/2} \tag{11.172d}$$

One can obtain a first order approximation of the generalized coordinate $A_m(t)$ of equations (11.165) as a function of the inertial and Rossby velocity components, that is, $\mathbf{Q}_m$ and $\mathbf{q}_1$. The formula revealed the frequency shift effect on the interaction between the inertial mode given $\mathbf{q}_0(\mathbf{r}, t)$ and the assumed Rossby mode $\mathbf{q}_1$. The Rossby mode was found to be independent of z, that is,

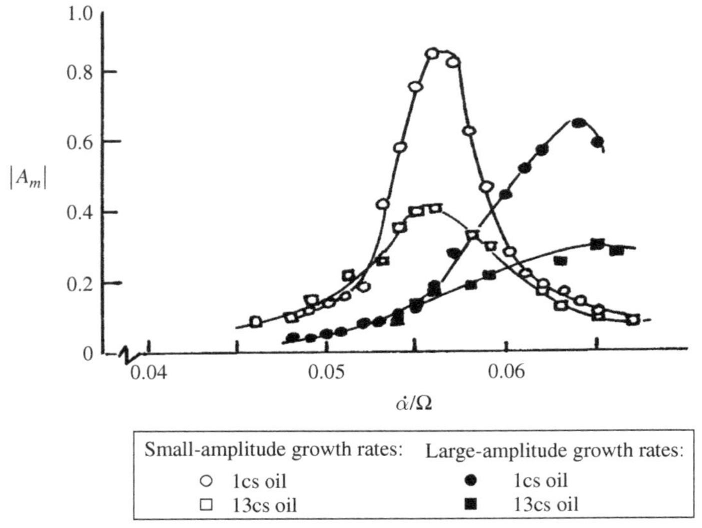

Figure 11.23 Dependence of amplitude growth rate on gyroscopic nutational frequency for 77% filled gyroscope, $\ell = 7.48$ inch, $R = 1.25$ inch, $\Omega = 5000$ r.p.m. (Scott, 1975)

$$\frac{\partial p_1}{\partial z} = 0 \quad \rightarrow \quad \frac{\partial \mathbf{q}_1}{\partial z} = 0 \tag{11.173}$$

This implies that the low frequency motion is geostrophic, that is, the motion characterized by $\mathbf{q}_1$ is columnar. It was concluded that the Rossby-mode effect was to allow the angle α to vary. On physical grounds, however, it was not possible to explain how this variation in α could affect the wave motion. Thus, by dismissing the Rossby wave effect, equation (11.172d) may be used to determine the effect of the nonzero α on the inertial modes. For $r = R$, the first term in (11.172d) vanishes and we have

$$\xi_{j1} = -R\left[\frac{\partial(\mathbf{Q}_{m0}\cdot\mathbf{k})}{\partial z}\right]_{z=\ell/2} \Bigg/ \left[\frac{\partial(\mathbf{Q}_{m0}\cdot\mathbf{k})}{\partial \xi_{j0}}\right]_{z=\ell/2} \tag{11.174}$$

Using equations (11.168) and (11.170), equation (11.174) is reduced to the simple expression

$$\xi_{j1} = -R\xi_{j0}/\ell \tag{11.175}$$

In this case the value of $\xi_j = \xi_{j0} + \alpha\xi_{j1} = \xi_{j0}(1 - R\alpha/\ell) = \pi(1 + 2j)/\ell$ can be used in equation (11.171) to the inertial wave frequency.

Figures 11.23 and 11.24, taken from the experimental results of Scott (1975), show the dependence of the gyroscope amplitude growth on the gyroscope nutational frequency for 77% and 100% filling, respectively. For a partially filled cavity, Figure 11.23 shows two different growth-rate curves having peaks at a gyroscope frequency of 0.0565. Both curves were generated by letting the amplitude of the gyroscope motion increase from the sleeping position. Another growth-rate peaking at a gyroscope frequency of 0.064 was generated by giving the gyroscopic motion initial amplitude of 3°. According to the results of Stewartson (1959), the inertial wave frequency is 0.065 which slightly exceeds the experimental value of 0.064.

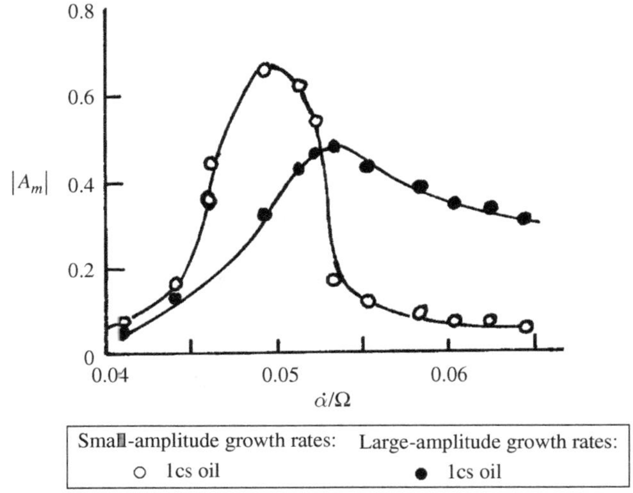

Figure 11.24 Dependence of amplitude growth rate on gyroscopic nutational frequency for 77% filled gyroscope, $\ell = 7.917$ inch, $R = 1.24$ inch, $\Omega = 5000$ r.p.m. (Scott, 1975)

Figure 11.24 shows two different growth-rate curves for a completely filled cavity. The peak of the curve was obtained by letting the amplitude grow from the sleeping position occurring at a gyroscopic frequency of 0.049. By giving the gyroscope initial amplitude of about 3°, the growth-rate curve reaches its peak at a gyroscopic frequency of 0.053. Again, Stewartson's results (1959) give an inertial wave frequency of 0.054. It was concluded that, up to second-order approximation, the Rossby modes excited by the geometry do not have any effect on the inertial modes.

11.7 Rotating liquids in microgravity

As stated earlier, the main problem of rotating fluids is the stability and motion of liquid-filled projectiles and liquid propellant in spacecraft vehicles. Under low gravity, the equilibrium shape of the free surface for the unsteady state rotating fluids is governed by the balance of capillary, bulk pressure, viscous, centrifugal, and gravity forces. Surface tension plays an important role in the liquid free-surface dynamics under microgravity. Blackhear and Eide (1964) determined the shape of the free surface under low gravitational field and centrifugal force. The shape of the free surface of rotating fluid at zero gravity was analyzed by Slobozhanin (1966) and Bauer (1982d, 1991c, j). Bauer (1994e) and Bauer and Eidel (1997b, 1999a) estimated the natural damped frequencies and axial response of a sloshing rotating viscous liquid in a cylindrical container. Bauer and Eidel (2002b) considered the same problem and imposed the adhesive condition at the cylindrical wall. They found that with increasing liquid height ratio both the natural frequencies and decay were increased. By increasing the surface tension the natural frequencies increase but the decay was decreased. Bauer and Siekmann (1971b) and Raake and Siekmann (1991) determined the shape of a rotating fluid consisting of a gas bubble enclosed in a liquid globe. Cutshall, *et al.* (1996) and Deffenbaugh, *et al.* (1998) developed analytical modeling of the low-gravity motion of liquid in spinning spacecraft tanks. Dodge, *et al.* (1991, 1999, 2000) designed and analyzed a spinning slosh rig. They estimated the boundary layer of viscous energy dissipation in spinning spacecraft tanks.

Leslie (1985) conducted a series of experimental measurements on rotating free-surface shapes in the low-gravity environment of a free-falling aircraft (KC-135). The apparatus consisted of a transparent cylinder partially filled with ethanol of a surface tension of 2.28×10^{-2} –N/m with gas at 20° C, and almost zero contact angle. The cylinder was mounted on a turntable rotating about the cylinder's axis. Leslie measured and numerically computed the shape of the free surface at various ratios of centrifugal force to surface tension force in 2, 4, and 6.3 cm deep cylinders. Hung and Leslie (1988) and Hung, Tsao, *et al.* (1988c) estimated the shape of the rotating free surface when it intersects both the top and bottom walls of the tank under different levels of gravity forces and different rotating speeds. Hung, Tsao, *et al.* (1988b, c, 1989a, b, c) numerically simulated the Navier–Stokes equations to estimate the time-dependent dynamical behavior of surface tension on partially-filled rotating fluids in both low and microgravity environments.

Hung, Tsao, *et al.* (1989a, b, c) considered a circular cylindrical tank partially filled with a Newtonian fluid of constant density ρ and kinematic viscosity ν. The cylinder rotates about its axis with time-dependent angular velocity, $\Omega(t)$. For the case of axial symmetric flow, that is, independent of the angular coordinate, θ, the fluid field equations are given by the continuity and momentum equations

$$\frac{1}{r}\frac{\partial(ru)}{\partial r} + \frac{\partial(w)}{\partial z} = 0 \tag{11.176a}$$

$$\frac{\partial u}{\partial t} + u\frac{\partial u}{\partial r} + w\frac{\partial u}{\partial z} - \frac{v^2}{r} = -\frac{1}{\rho}\frac{\partial p}{\partial r} + \nu\left(\nabla^2 u - \frac{u}{r^2}\right) \tag{11.176b}$$

$$\frac{\partial v}{\partial t} + u\frac{\partial v}{\partial r} + w\frac{\partial v}{\partial z} + \frac{uv}{r} = \nu\left(\nabla^2 v - \frac{v}{r^2}\right) \tag{11.176c}$$

$$\frac{\partial w}{\partial t} + u\frac{\partial w}{\partial r} + w\frac{\partial w}{\partial z} = -\frac{1}{\rho}\frac{\partial p}{\partial z} - g + \nu\nabla^2 w \tag{11.176d}$$

where u, v, and w are the fluid velocity components along the cylindrical coordinates r, θ, and z, respectively, $\nabla^2 = \frac{1}{r}\frac{\partial}{\partial r}\left(r\frac{\partial}{\partial r}\right) + \frac{\partial^2}{\partial z^2}$. Let the initial profile of the interface between the liquid and gas (at $t = 0$) and its value at any time be given by the expressions, respectively,

$$\eta_0(r, z, 0) = 0, \quad \text{and} \quad \eta(r, z, t) = 0 \tag{11.177a, b}$$

In order to solve the fluid field equations, the following boundary conditions, based on Leslie's (1985) measurements, must be imposed:

(1) At the container wall, both the normal and tangential velocity components must vanish. A constant angle of contact was assumed when the free surface intersects the container wall.
(2) Along the interface, the kinematic surface boundary condition is

$$\frac{\partial \eta}{\partial t} + u\frac{\partial \eta}{\partial r} + w\frac{\partial \eta}{\partial z} = 0 \tag{11.178}$$

(3) At the interface, the stress must be continuous along the normal and tangential directions of the interface. Along the tangential to the interface the following condition must hold

$$[\tau \cdot \mathbf{n} - (\mathbf{n} \cdot \tau.\mathbf{n})]_{\text{liquid}} = [\tau \cdot \mathbf{n} - (\mathbf{n} \cdot \tau.\mathbf{n})]_{\text{gas}} \tag{11.179}$$

where $\tau_{ij} = \mu\left(\frac{\partial u_i}{\partial x_j} + \frac{\partial u_j}{\partial x_i} - \frac{2}{3}\frac{\partial u_k}{\partial x_k}\delta_{ij}\right) + \zeta\frac{\partial u_k}{\partial x_k}\delta_{ij}$ is the viscous stress tensor, μ is the dynamic viscous coefficient of the first kind, ζ is the viscous coefficient of the second kind, $\mathbf{n}$ is the unit vector normal to the interface, δ_{ij} is the Dirac delta function. The normal component to the interface is governed by the condition

$$P_G - P_L - (\mathbf{n} \cdot \tau.\mathbf{n})_{\text{gas}} + (\mathbf{n} \cdot \tau.\mathbf{n})_{\text{liquid}} = -\frac{\sigma}{r}\frac{d}{dr}\left[\frac{r(dz/dr)}{\sqrt{1+(dz/dr)^2}}\right] \tag{11.180}$$

where P_G and P_L are the gas and liquid pressures at the interface, respectively, and σ is the surface tension. For the special case of a steady-state rotating fluid, P_G and P_L are determined from the Navier–Stokes equations in the form

$$P_G = P_{0G} + \frac{1}{2}\rho_G\Omega^2 r^2 - \rho_G g z, \qquad P_L = P_{0L} + \frac{1}{2}\rho_L\Omega^2 r^2 - \rho_L g z \tag{11.181a, b}$$

Substituting equations (11.181) into equation (11.180), and neglecting the normal component of the viscous stress at both sides of the interface, gives

$$P_0 r + \frac{1}{2}\rho\Omega^2 r^3 - \rho g z r = -\sigma\frac{d}{dr}\left[\frac{r(dz/dr)}{\sqrt{1+(dz/dr)^2}}\right] \tag{11.182}$$

where $P_0 = P_{0G} - P_{0L}$, and $\rho = \rho_G - \rho_L$. Equation (11.182) must be satisfied at the intersection of the fluid free surface and the tank wall, which requires

$$\left.\frac{dz}{dr}\right|_{r=R} = \cot\theta \tag{11.183}$$

where θ is the contact angle between the vertical wall and the tangent to the fluid surface at the tank wall as shown in Figure 11.25(a). Integrating equation (11.182) gives

$$\frac{1}{2}P_0 r^2 + \frac{1}{8}\rho\Omega^2 r^4 - \rho g\int rz\,dr = -\frac{\sigma r(dz/dr)}{\sqrt{1+(dz/dr)^2}} + C \tag{11.184}$$

where C is the constant of integration determined by evaluating equation (11.184) at $r = R$ and satisfying condition (11.180), that is,

$$C = \frac{1}{2}P_0 R^2 + \frac{1}{8}\rho\Omega^2 R^4 - \rho g\int rz\,dr|_{r=R} + \sigma R\cos\theta \tag{11.185}$$

The value of P_0 can be evaluated at $r = 0$, the bottom of the free surface, where $dz/dr \to 0$, that is,

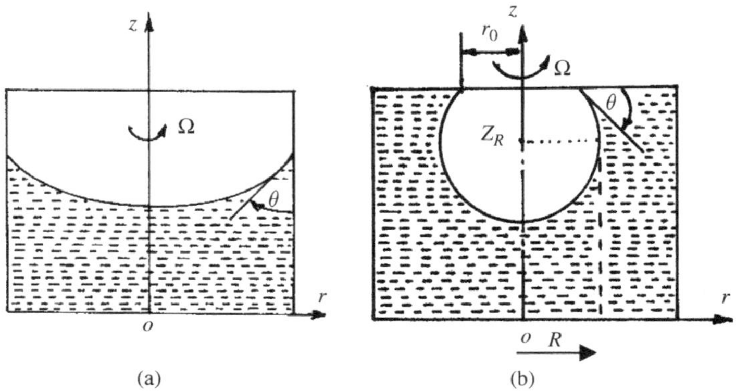

Figure 11.25 (a) Free surface intersection with the sidewall of a spinning cylinder, (b) formation of the bubble intersecting the upper wall of a rotating cylinder.

$$P_0 = -2\left[\frac{1}{8}\rho\Omega^2 R^2 - \frac{\rho g}{R^2}\int_0^R rz\,dr + \frac{\sigma}{R}\cos\theta\right] \tag{11.186}$$

Substituting equations (11.185), and (11.186) into equation (11.184) gives

$$-\frac{\sigma r(dz/dr)}{\sqrt{1+(dz/dr)^2}} = -\left[\frac{1}{8}\rho\Omega^2 R^2 - \frac{\rho g}{R^2}\int_0^R rz\,dr - \frac{\sigma}{R}\cos\theta\right](r^2 - R^2)$$
$$+\frac{1}{8}\rho\Omega^2(r^4 - R^4) - \rho g\int_r^R rz\,dr - \sigma R\cos\theta \tag{11.187}$$

Solving for the slope

$$dz/dr = \frac{\psi}{\sqrt{1-\psi^2}} \tag{11.188}$$

where

$$\psi = \frac{(dz/dr)}{\sqrt{1+(dz/dr)^2}}$$
$$= \frac{R}{r}\left\{\left(1-\frac{r^2}{R^2}\right)\left[\mathsf{We} - \frac{\mathsf{Bo}}{R^3}\int_0^R rz\,dr - \cos\theta\right] - \mathsf{We}\left(1-\frac{r^4}{R^4}\right) + \frac{\mathsf{Bo}}{R^3}\int_r^R rz\,dr\right.$$
$$\left. - \cos\theta\right\} \tag{11.189}$$

$$\mathsf{We} = \frac{\rho\Omega^2 R^2}{8\sigma} = \frac{\text{Centrifugal Force}}{\text{Surface Tension Force}} = \text{Weber Number} \tag{11.190a}$$

$$\mathrm{Bo} = \frac{\rho g R^2}{\sigma} = \frac{\text{Gravitational Force}}{\text{Surface Tension Force}} = \text{Bond Number} \tag{11.190b}$$

The profile of the interface is given by the following integration

$$z = \int_r^R \frac{\psi}{\sqrt{1-\psi^2}}\, dr \tag{11.191}$$

The above analysis gives the steady state fluid free-surface shape and it is valid as long as the free surface intersects only the vertical wall of the cylinder. If the tank is given high rotating speed, the free surface will intersect the roof of the tank at radius r_0 as shown in Figure 11.25(b), and the free surface may attain a maximum radius, $R_0 \leq R$, for which $dz/dr \to -\infty$. In this case, the angle of contact θ is measured between the horizontal surface of the roof and the tangent of the surface at intersection. The profile of the free surface is given by the expression

$$z = \int_{r_0}^{r} \frac{\psi}{\sqrt{1-\psi^2}}\, dr \tag{11.192}$$

where

$$\psi = \frac{R_0}{r}\left\{\left(\frac{r_0^2 - r^2}{R_0^2 - r_0^2}\right)\left[1 + \frac{r_0}{R_0}\sin\theta + \mathrm{We}\left(1 - \frac{r_0^4}{R_0^4}\right) - \frac{\mathrm{Bo}}{R_0^3}\int_{r_0}^{R_0} rz\, dr\right]\right.$$
$$\left. + \mathrm{We}\left(\frac{r^4 - r_0^4}{R_0^4}\right) - \frac{\mathrm{Bo}}{R_0^3}\int_{r_0}^{r} rz\, dr + \frac{r_0}{R_0}\cos\theta\right\} \tag{11.193}$$

and $\mathrm{We} = \frac{\rho\Omega^2 R_0^2}{8\sigma}$, $\mathrm{Bo} = \frac{\rho g R_0^2}{\sigma}$. The following cases were considered by Hung, Tsao, *et al.* (1989a, b, c).

11.7.1 Nonrotating free surface shape

The equilibrium shapes of the free surface governed by a balance of capillary and gravity forces for different liquids were estimated for nonrotating fluid. It was shown that the greater the coefficient of surface tension, the smaller the radius of the free surface intersecting the tank roof. As the gravity level decreases the radius of the free surface at the tank roof is reduced. In the limit, when the gravity level vanishes, the free-surface shape switches from paraboloid to ellipsoid. Figure 11.26 shows the evolution of a nonrotating free surface in low and microgravity environments for a container of radius 3 cm, and height 4 cm.

11.7.2 Rotating free surface shape

Leslie (1985) measured the free-surface shapes of rotating fluid in the low gravity environment of a free-falling aircraft (KC-135) in a cylinder of radius 3 cm and height 2 cm. The container was partially filled with ethanol and the volume of the empty space in the container was 30 cm^3.

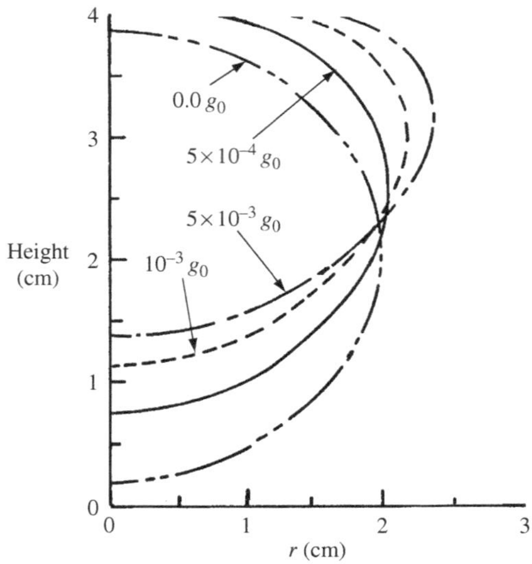

Figure 11.26 Equilibrium of nonrotating shapes of bubble between air and ethanol ($\sigma = 0.0228\ N/m$) in different values of gravitational field. (Hung, Tsao, *et al.*, 1989a, b, c)

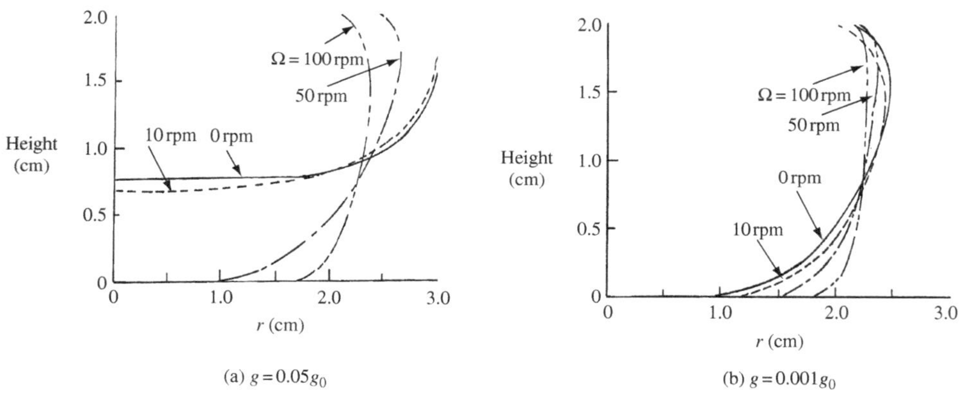

Figure 11.27 Free-surface patterns of rotating ethanol between ($\sigma = 0.0228\ N/m$) for different values of spinning speed. (Hung, Tsao, *et al.*, 1989a, b, c)

Figure 11.27(a) and (b) shows the shapes of rotating equilibrium free surface for different values of speed under gravitational levels of $0.05g$ and $0.001\,g$, respectively. The influence of the gravitational level has a similar trend to that of the nonrotating case. It is seen that as the rotating speed increases, the location of the free-surface intersection with the tank wall changes from the sidewall to the tank roof. As the speed is further increased, the free surface intersects with the bottom and the top of the tank. The process continues until the free surface becomes annulus.

11.7.3 Time-dependent rotation and gravitational field

Hung, Tsao *et al.* (1989a, b, c) numerically integrated the fluid field equations (11.176) subject to the initial and boundary conditions (11.177–11.180) for the unsteady case, which involves

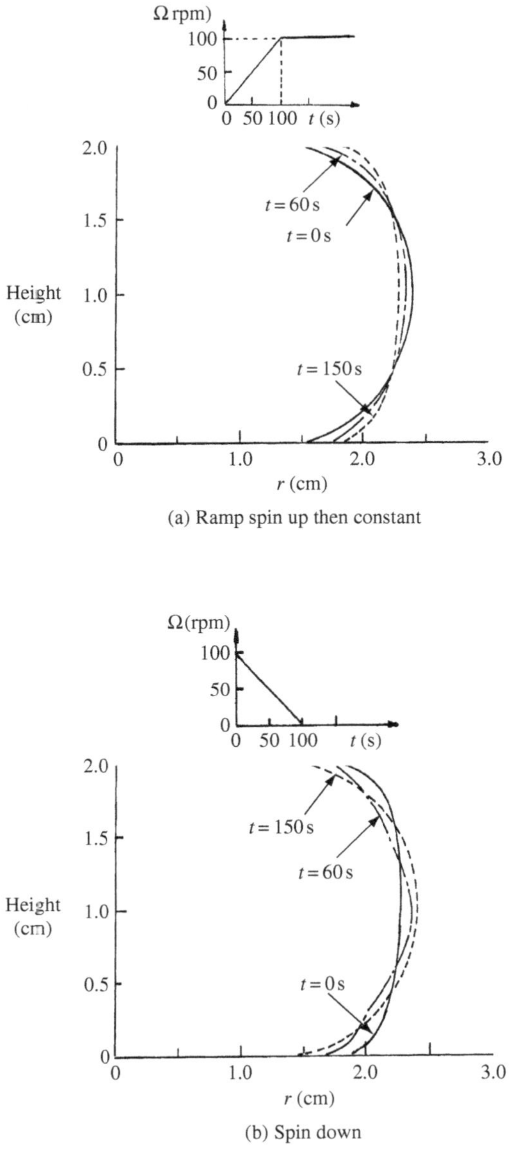

Figure 11.28 Free-surface patterns of rotation at different time intervals, $g = 0g_0$, for (a) spin-up and (b) spin-down. (Hung, Tsao, *et al.* (1989a, b, c)

time dependence of the rotating speed and gravitational field. They used a staggered grid for the velocity components developed originally by Harlow and Welch (1965) based on a finite difference scheme.

Figure 11.28(a) and 11.28(b) shows the free-surface profiles at different time steps for linear spin-up and spin-down, respectively. Both figures were estimated for a zero gravitational field. For every time step, the free surface deviates from the corresponding steady state cases. Figures 11.29(a) and (b) show the free-surface profiles for linear spin-up and spin-down, respectively

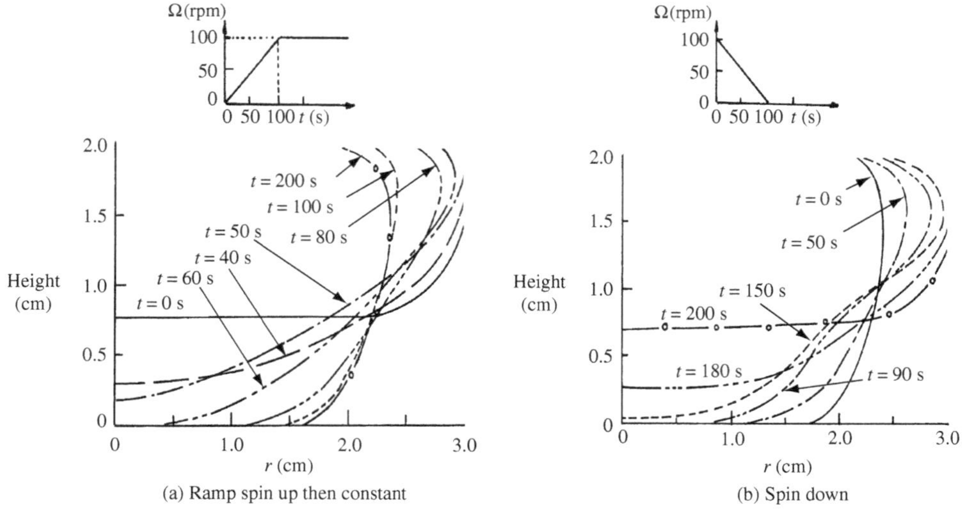

Figure 11.29 Free-surface patterns of rotation at different time intervals, $g = 0.05g_0$, for (a) spin-up and (b) spin-down. (Hung, Tsao, *et al.*, 1989a, b, c)

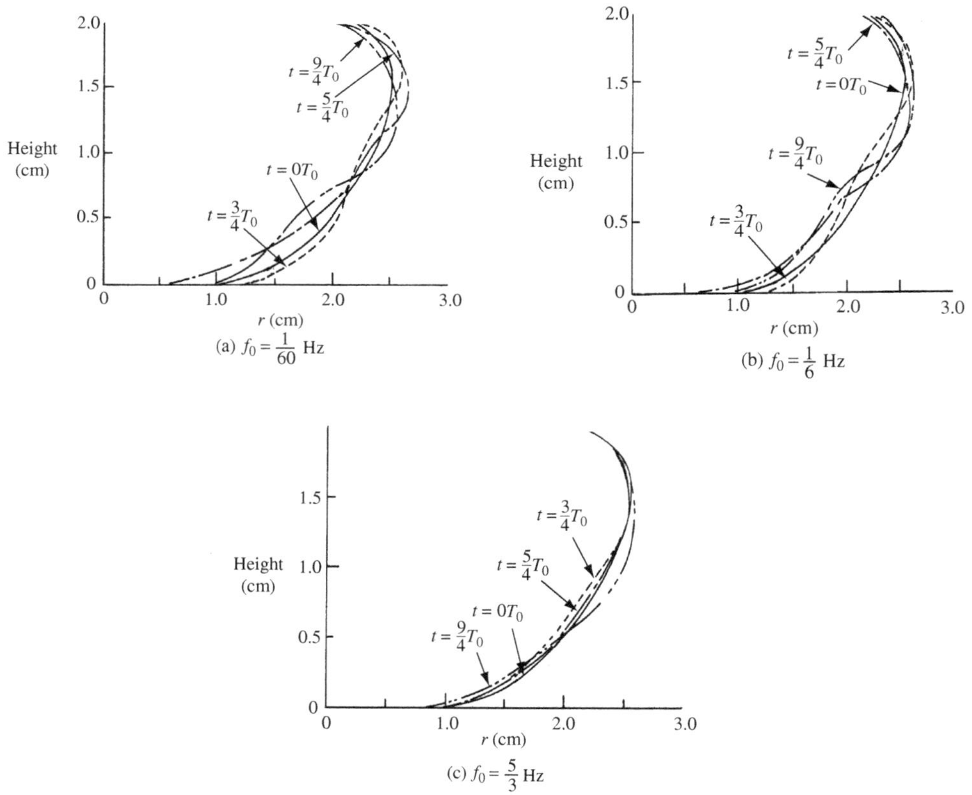

Figure 11.30 Free-surface patterns of rotating at different time intervals under sinusoidal gravitational field $g = 0.001[1 + 0.5 \sin 2\pi f_0 t]\, g_0$, and constant spin speed 1/min, for three different values of gravity oscillation frequency (a) $f_0 = \frac{1}{60}$ Hz (b) $f_0 = \frac{1}{6}$ Hz, and (c) $f_0 = \frac{5}{3}$ Hz. (Hung, Tsao, *et al.*, 1989a, b, c)

for $0.05g$. Again, the deviations of the free surface profiles at each time step from those estimated for steady state rotations are not significant. Under a sinusoidal gravitational environment given by the expression $g = g_0 [1 + 0.5 \sin (2\pi f_0 t)]$, where g_0 is the mean value, and f is the oscillation frequency in Hz, the free-surface profiles were estimated for different values of tank speed, and different gravitational frequencies. Figure 11.30(a–c) shows the free-surface profiles at different time steps for tank speed of 1 r.p.m, and three different gravitational frequencies 1/60 Hz, 1/6 Hz, and 5/3 Hz, respectively, and a mean gravitational level of $0.001g_0$. It is seen that the amplitude of oscillation of the free surface is larger for a lower frequency of gravity environment than that for a higher frequency. The point of intersection of the free surface with the bottom of the tank moves back and forth over a wider range than the motion of the point of intersection with the top of the tank. As the speed increases to a higher value, 100 r.p.m., the effect of gravity oscillations becomes negligible and the free surface does not experience any oscillations and the tank rotating speed increases.

11.8 Closing remarks

The stability of rotating liquid containers has been studied extensively in the former Soviet Union. These studies have been documented in a research monograph by Moiseev and Rumnyantsev (1968). In the West, the studies focused on the stability of rockets and space vehicles. In particular, the fluid free-surface orientation in rotating containers under microgravity environment has been analyzed. The spinning and unavoidable wobbling and precession of spacecraft vehicles cause the contained liquid to oscillate, and thus generate dynamic forces, which can destabilize the spacecraft. Note that the kinetic energy of the spinning spacecraft is dissipated by virtue of the liquid motion and this destabilizes the vehicle and causes it to tumble. Dodge (1999), Dodge and Abramson (2000), and Dodge and Green (2000) reviewed and developed analytical approaches to determine the energy dissipation due to unsteady viscous boundary layers. Their analysis includes the damping of the vortex motion and the rate of energy dissipation caused by the vortex motions. The next chapter is mainly devoted to different problems dealing with the physics and behavior of fluids in microgravity.

This chapter did not address the nonlinear phenomena of the free surface in horizontal rotating cylinders including the effect of surface tension (Gans and Yalisover, 1981, Aitta, 1991, Thoroddsen and Mahadevan, 1997, and Ashmore, *et al*, 2003). Baines (1967) considered the forced oscillations of an enclosed rotating fluid. Other problems include the dynamic simulation of spin-stabilized spacecraft with sloshing fluid containers (Hill, *et al.*, 1988, Slafer and Challoner, 1988, and Tan, 1994).

12

Microgravity sloshing dynamics

12.1 Introduction

Regular gravitational potential has a stabilizing effect in that it brings the liquid volume toward the bottom of its container. When the body force diminishes, the liquid volume assumes any location inside its container in an unpredictable manner. The problems of liquid sloshing dynamics under microgravity are different from those encountered under a regular gravitational field. These problems include liquid reorientation and the difficulty of moving and handling, since the body force is almost negligible. Under microgravity, surface tension forces become predominant. The Bond number, given by the ratio of the gravitational to capillary forces, plays a major role in the free-liquid-surface characteristics. For very small values of the Bond number, capillary forces predominate and the free surface will not be flat, but will rise around the vertical walls of a container.

The problems of liquid behavior at a low and zero gravity include the mechanics and thermodynamics of capillary systems, heat transfer in cryogenic tanks and mechanisms of energy transport, capillary hydrostatics, low-gravity sloshing, and fluid handling. The early work on free-liquid-surface behavior under a low gravity field considered different problems of the surface–vapor interface (Petrash and Otto, 1962, 1964, Shuleikin, 1962, Petrash, *et al.* 1962, 1963, Petrash and Nelson, 1963, Neu and Good, 1963, Paynter, 1964a,b, Zenkevich, 1965, Abramson, 1966a, Seebold *et al.*, 1967 and Hastings and Rutherford, 1968). Fluid physics under microgravity has been the subject of NASA biennial conferences since 1992 in the USA. Some research monographs have been devoted to microgravity fluid physics (see, e.g., Moiseev, 1968, Babskii, *et al.*, 1976a, Walter, 1987, Myshkis, *et al.*, 1987, 1992, and Monti, 2001). Some approximate methods and unresolved problems pertaining to zero gravity fluid mechanics were discussed by Babskii, *et al.* (1976b, 1980). Bauer (1993d) presented some analytical approaches for microgravity fluid dynamics. The free-fall experiments are very useful in exploring the physical phenomena of liquid free surface (Benedict, 1959, 1960). Flight test experiments are valuable for understanding of liquid free-surface behavior under microgravity (Swalley, *et al.*, 1965, Ryan and Buchanan, 1966, Platt, 1967, Slabinski, 1977, Vreeburg, 1986, 1992, 1999, Vreeburg and Vogels, 1986, Hung, Lee and Tsao, 1990, Chato, *et al.*, 1998, and Vreeburg and Veldman, 2001).

In view of the diversity of microgravity fluid physics, this chapter will address selected topics of fundamental importance. It begins with the mechanics of a free liquid surface under microgravity, the stability of static and dynamic free liquid surface, contact line, and contact angle. The modal analysis includes the estimation of free-surface natural frequencies and associate surface shapes under micro- and zero-gravity environments. The problem of forcing sloshing for slipping and anchored contact line will be addressed for simple cases. The influence of g-jitter and liquid handling are discussed briefly. Capillary systems including Marangoni flow and liquid bridges will be treated for static and dynamic cases. The chapter also includes reviews of the problem of thermocapillary fluid flow and liquid bridges, sloshing of cryogenics, and hydroelastic oscillations under microgravity.

12.2 Kinetics and geometry of liquid free surface

12.2.1 Surface tension and bond number

Under a regular gravitational field, it is customary to assume that the pressure along the free surface is constant corresponding to the pressure in the gap above the fluid volume. This is an approximation, but in reality the free surface experiences unbalanced cohesive forces acting on the liquid molecules at the free surface. These molecules are subjected to a force toward the interior. The unbalanced force along the surface results in a tensile force acting in the plane of the surface along any line in the surface. The intensity of the molecular attraction per unit length along any line in the surface is known as the surface tension σ (newtons/meter). Under a low gravitational field, cohesive forces on the surface are manifested and surface-tension effects cannot be ignored. The pressure along the edge of the liquid is different from the pressure outside the liquid volume. To establish the effect of this pressure differential, consider an element, Δx, of the surface shown in Figure 12.1. Let the surface tension and slope at the left side of the element be σ and $\partial\eta/\partial x$, respectively. Let the surface tension and slope at the right side of the element, up to the first-order of the Taylor series, be $\sigma + (\partial\sigma/\partial x)\Delta x$ and $\partial\eta/\partial x + (\partial^2\eta/\partial x^2)\Delta x$, respectively. Let the gas pressure just above the surface be P_g and just below the surface be $P(x, t)$. The summation of forces along the vertical gives

$$\left[P(x,t) - P_g\right]\Delta x + \left(\sigma + \frac{\partial\sigma}{\partial x}\Delta x\right)\left(\frac{\partial\eta}{\partial x} + \frac{\partial^2\eta}{\partial x^2}\Delta x\right) - \sigma\frac{\partial\eta}{\partial x} = 0$$

Simplifying and considering the case of constant surface tension, that is, $\partial\sigma/\partial x = 0$, gives

$$P(x,t) = P_g - \sigma\frac{\partial^2\eta}{\partial x^2} \tag{12.1}$$

If P_g is constant, then the time derivative of equation (12.1) takes the form

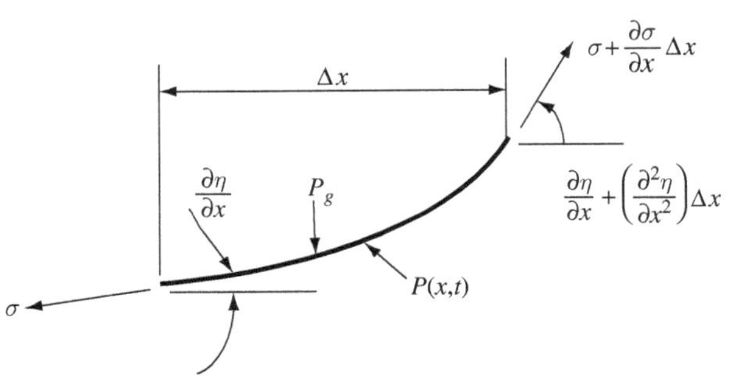

Figure 12.1 Surface forces acting on an element of liquid surface.

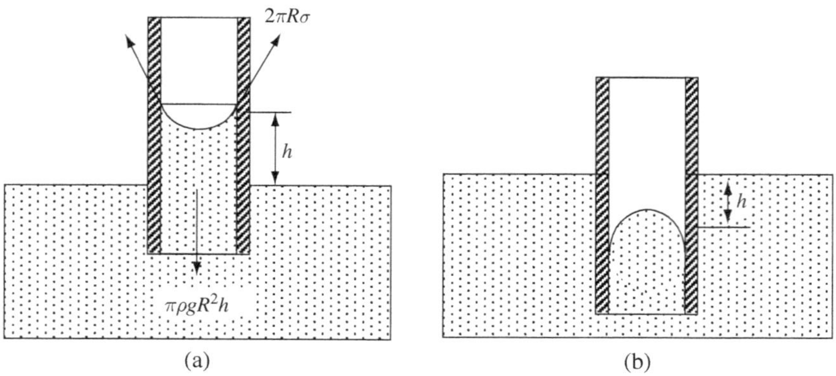

Figure 12.2 Effect of surface tension in small tubes: (a) Adhesion causes the liquid to rise (wet), (b) cohesion causes the liquid to be depressed.

$$\frac{\partial P(x,t)}{\partial t} = -\sigma \frac{\partial^2}{\partial x^2}\left(\frac{\partial \eta}{\partial t}\right) = -\sigma \frac{\partial^3 \Phi(x,0,t)}{\partial x^2 \partial z} \tag{12.2}$$

where $\partial\eta/\partial t$ has been replaced by the gradient of the potential function, $\partial\Phi/\partial z$, for potential flow.

The surface tension results in the rise or fall of a liquid in a capillary tube as shown in Figure 12.2. For the case of Figure 12.2(a), there is an adhesion between the tube wall and liquid molecules, which is strong enough to overcome the cohesion of the molecules, pulling up the wall to height h, and thus the liquid is said to *wet* the solid surface. On the other hand, if the adhesion is weak compared to the cohesion, the liquid will not wet the surface and the level is depressed as shown in Figure 12.2(b). The liquid in this case is said to be *nonwetting*. The surface tension is relatively an invariant property of the interface. However, its value is affected by the temperature. Generally, it decreases with the temperature and vanishes at a critical point. Early attempts to determine the surface tension from experimental measurements were made by Porter (1933).

With reference to the wet liquid, Figure 12.2(a), it may be assumed that the free surface is spherical for which the surface tension at the intersection with the wall has zero angle. The height h may be estimated from the static equilibrium between points 1 and 2 as $P_2 - P_1 = \rho g h$. The pressure force, $\pi R^2(P_2 - P_1) = \pi R^2 \rho g h$ must be in balance by the surface tension force, $\sigma 2\pi R \cos\vartheta_s$, where ϑ_s is the static contact angle measured from the liquid side of the contact line. Thus, the height is obtained from the equilibrium condition

$$\sigma 2\pi R \cos\vartheta_s = \pi R^2 \rho g h \rightarrow \frac{h}{R} = 2\frac{\sigma}{\rho g R^2}\cos\vartheta_s = \frac{2\cos\vartheta_s}{\mathsf{Bo}} \tag{12.3}$$

where $\mathsf{Bo} = \rho g R^2/\sigma$ is referred to as the Bond number, which measures the ratio of gravitational to surface tension forces. For values of $\mathsf{Bo} \gg 1$, the gravitational force predominates, and for $\mathsf{Bo} \ll 1$, the surface tension force predominates and the fluid free surface climbs on the container walls. Netter and Eckhardt (1981), Eckhardt and Netter (1982), Hung, Tsao, Hong, and Leslie (1990), Snyder (1999, 2001), and Passerone, *et al.* (2001) reported some experimental results from studying the dynamic behavior of fluid in a surface tension tank under microgravity conditions.

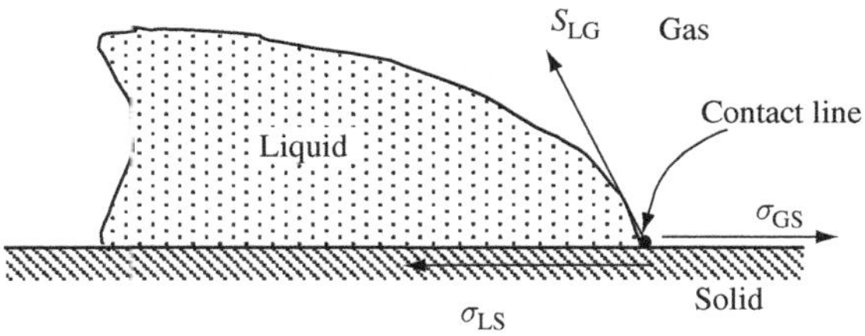

Figure 12.3 Surface tension between gas, liquid, and solid surface in static equilibrium.

12.2.2 Static and dynamic contact-angle

With reference to Figure 12.3, Young (1805) derived an expression for the static contact angle ϑ_s. He assumed the three material boundaries, gas/liquid, liquid/solid, and solid/gas, each possesses a constant surface tension, σ_{LG}, σ_{LS}, and σ_{GS}, respectively. The static equilibrium along the horizontal requires

$$\sigma_{LG} \cos \vartheta_s = \sigma_{GS} - \sigma_{LS} \tag{12.4}$$

The above equation is referred to as Young's equation. The measured values of $(\sigma_{GS} - \sigma_{LS})$ were often obtained by substituting the measured values of σ_{LS} and ϑ_s in equation (12.4).

In order to balance the vertical component of surface tension forces at the contact line, one must admit the existence of a reaction force in the solid, $-\sigma_{LG} \sin \vartheta_s$. However, this condition cannot be satisfied, since the solid is being implicitly modeled as a rigid body. The question then arises, what restricts this force to having only a vertical component? Gauss (1829) provided a rigorous theoretical basis for equation (12.4) indicating that the static contact angle is a unique property of the material. However, this usually disagrees with experimental observations.

The angle of contact of a liquid drop on a flat horizontal surface can be measured directly by drawing on an enlarged photograph a line tangent to the gas/fluid interface at the contact line. The second method was indirectly proposed by Padday (1969) who used the solution for the shape of the meniscus along the measured height of the apex of the drop and its volume. Padday (1971) provided a comprehensive treatment of the theory of surface tension including contact-line and contact angle. There are several other methods for determining the static contact angle described by Dussan (1976, 1979). The static menisci mechanics was studied by Concus (1968).

Dussan (1979) studied the contact-line behavior in terms of the relationship between the contact-angle and the relative motion of the contact line. She gave a thorough discussion of static and dynamic contact angles for uni-directional relative velocities. Her results were representative for experiments with small Reynolds number, small Weber number, and small Froude number. Dussan's work essentially dealt with uni-directional flow. It was shown that the contact-angle, ϑ_c, is a function of the contact-line velocity only, that is, $\vartheta_c = f(U)$ as shown in Figure 12.4, where ϑ_R represents the receding angle defined when $U \to 0^-$ where the contact point moves toward the liquid, and ϑ_a represents the advancing angle defined as $U \to 0^+$, where the contact point moves away from the liquid. There is a critical speed, $U_c > 0$, below which a transition may occur and the motion of the contact line appears to be unsteady and spasmatic (referred to as "stick-slip"). Figure 12.4 shows two data points reported exactly at $U = 0$, denoted by ϑ'_a and ϑ'_R. These angles were experimentally observed by Elliott and

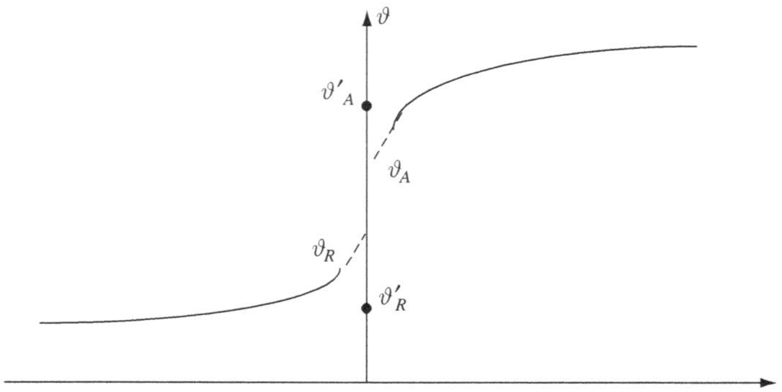

Figure 12.4 Dependence of dynamic contact angle ϑ_1 on the speed of the contact line.

Riddiford (1967) when the contact line was initially static and the contact angle was allowed to vary until the contact line was observed to move. Obviously, there exists an interval $[\vartheta_R, \vartheta_a]$ for which the contact line does not appear to move, a situation similar to the static and kinetic coefficients of friction. This nonuniqueness in the static contact angle is often referred to as *contact-angle hysteresis.* The contact-line boundary condition is nonlinear due to the occurrence of hysteresis, at low-Reynolds-number uni-directional steady motion (Perlin and Schultz, 1996, 2000). Figure 12.4 reveals a feature similar to the friction coefficient dependence on the relative sliding velocity, known as the *differential inclusions.* Differential inclusions can be regarded as differential equations that consist of set-valued or multi-valued terms. Accordingly, the existence and uniqueness of solutions are no longer guaranteed, except in a few cases (Filippov, 1964 and Dupont, 1992).

The presence of the inherent liquid viscosity creates some difficulties in the solution of the fluid flow field even with a (uni-directional) moving contact line. Dussan and Davis (1974) showed that a nonintegrable-stress singularity occurs at the contact-line if a moving contact-line is forced to obey the no-slip boundary condition. A popular model, which avoids the singularity that allows the contact point to move along the solid surface, was originated by Navier. He postulated that the resistant stress at a fixed solid surface is proportional to the relative velocity or $\beta_0 V = \mu \partial V / \partial n$, where β_0 is a constant, V is the velocity parallel to the fixed solid surface, n is the normal to the solid surface, and μ is the liquid dynamic viscosity. Goldstein (1938) provided a systematic description of this boundary condition.

The boundary conditions at the oscillating contact line for high Reynolds number were examined experimentally on a vertically oscillating plate by Ting and Perlin (1995). Oscillatory contact-line boundaries are important in wave and other flow interactions with a solid boundary. Benjamin and Scott (1979) and Graham-Eagle (1983, 1984) investigated waves propagating in a narrow open-channel with pinned-end (fixed contact-line position) edge conditions. Hocking (1987a) demonstrated the importance of surface tension effects at a contact line and showed that the damping of gravity capillary waves at a rigid boundary is primarily due to capillary effects. Several other studies demonstrated the importance of capillary effects in wave motion and damping (see, e.g., Miles, 1967, 1990b, 1991a, 1992,

Joo, *et al.*, 1990, Hocking and Mahdmina, 1991, and Cocciaro, *et al.*, 1991, 1993). The effect of flow rate on the dynamic contact angle was studied by Coney and Masica (1969).

Based on the relationship of uni-directional contact-lines given by Dussan (1979), Young and Davis (1987) proposed four possible relationships between contact-angle and contact-line velocity for an oscillating contact-line boundary. These relationships deal with (a) contact-angle hysteresis; (b) fixed contact line; (c) fixed contact angle; and (d) smooth contact-angle variation (no contact-angle hysteresis). Using these relationships and a force balance at the contact line, they presented a solution to the oscillatory contact line in the creeping-flow limit. They considered a small motion of the plate so that the inertial effect was small. Accordingly, the contact-line behavior is governed by a relationship similar to that of Dussan (1979). They found that the contact-angle hysteresis and steepening of contact angle with increasing contact-line speed resulted in dissipation effects. They also found that the contact-line motion tends to lag behind the plate motion due to inertia effects.

Hocking (1987a) used an oscillatory contact-line boundary condition to calculate the waves generated by a vertically oscillating vertical plate. He included capillary effects at the contact line. Miles (1990b) considered the effect of viscosity and a nonzero initial free-surface meniscus. A slip length, l_s, was adopted to replace the no-slip boundary condition along the entire plate. Based on a phenomenological hypothesis, he introduced the slip boundary condition (equivalent to Navier's slip model with $l_s = \mu/\beta_0$)

$$V - V_{\text{solid}} = l_s \frac{\partial V}{\partial n} \tag{12.5}$$

where V is the vertical velocity along the plate ($\partial\eta/\partial t$ at the contact line). Miles assumed l_s to be function of position along the plate surface, which vanishes at a distance given by the viscous length scale, $l_\nu = \sqrt{2\nu/\omega}$, away from the contact line, where ω is the frequency of the contact line oscillations. This means that the flow obeys the no-slip boundary condition at some distance, l_s, away from the contact line. According to Miles (1990b, 1991b) the slip length is not constant in the oscillatory case. Ting and Perlin (1995) indicated that the edge condition at an oscillatory boundary (triad junction) is not well developed. However, they provided some qualitative description pertaining to the contact-line movement changes as the amplitude of the plate changes, shown in Figure 12.5. The dependence of the mean slip coefficient on the oscillation amplitude is shown in Figure 12.6. Hocking (1987a) and Henderson and Miles (1990) indicated that an increase of the viscosity magnitude of up to three times is required to match numerical results with the measured dissipation. Perlin and Schultz (1996) proposed that the so-called "effective viscosity" is the result of contact-line-generated dissipation.

Jiang, *et al.* (1996) performed boundary-integral simulations of two-dimensional motion to the contact-line effects on Faraday waves. They discussed the competition between viscous effects, which tend to decrease frequency, and contact-line effects, which tend to increase frequency. They measured the threshold forcing amplitude for the fundamental mode and obtained a stability Ince chart for three slightly different fluids and two tank-aspect ratios.

Under low gravity, the meniscus will have large curvature and some conditions must be applied to the slope of the interface at the tank wall. Three possibilities for the surface slope at the tank wall are:

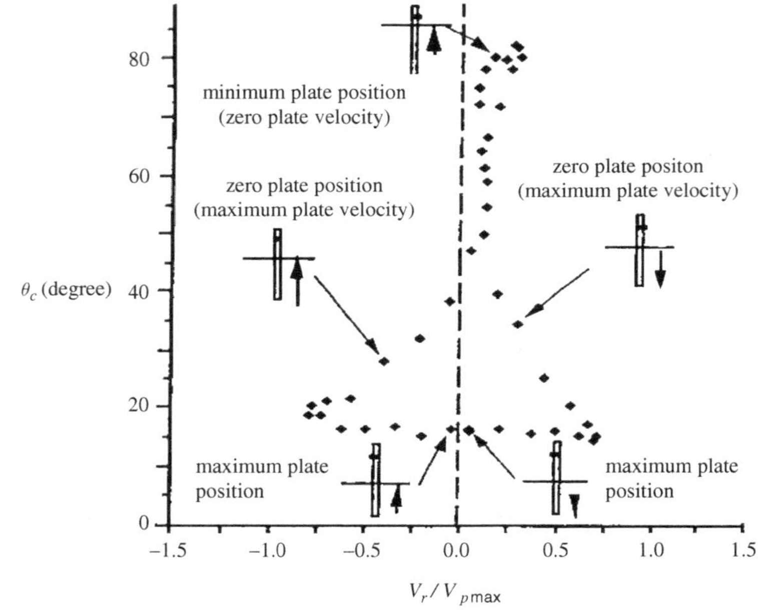

Figure 12.5 Dependence of the contact angle on the nondimensional relative velocity. (Ting and Perlin, 1995)

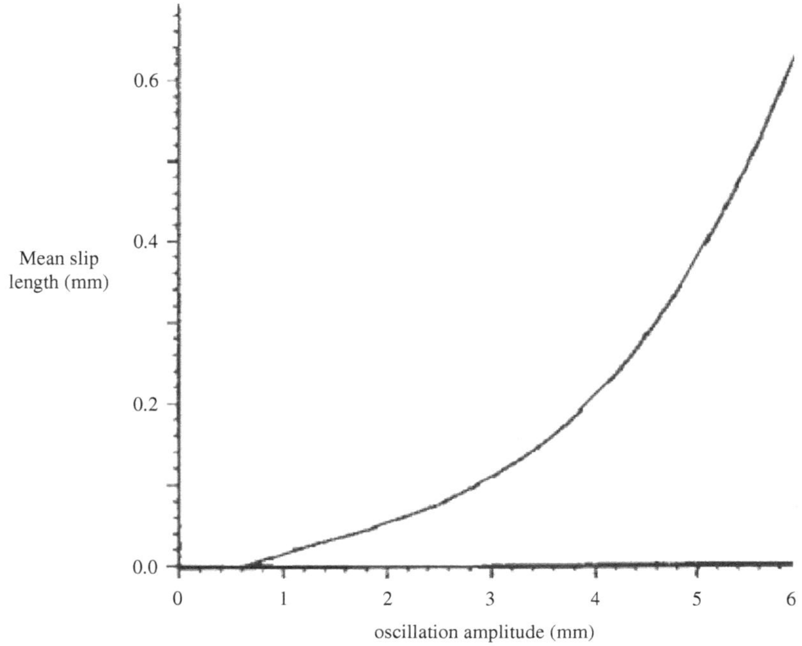

Figure 12.6 Dependence of mean slip length on stroke amplitude. (Ting and Perlin, 1995)

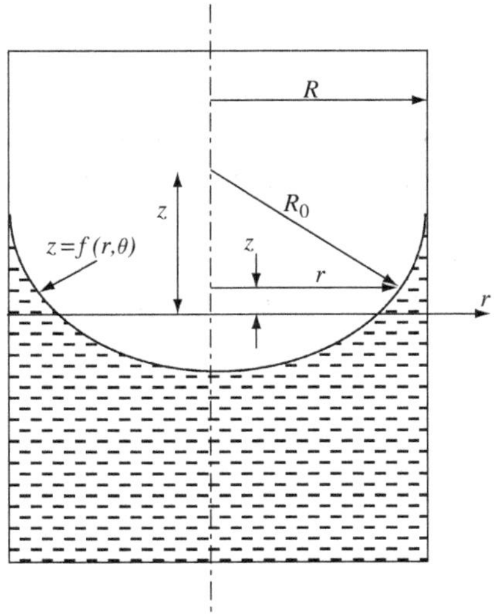

Figure 12.7 Configuration of special vase of spherical shape of undisturbed free liquid surface in a cylindrical tank.

(1) The slope remains constant, for which dynamic contact-angle hysteresis is absent (free-edge condition),
(2) the edge of the interface remains fixed (stuck-edge condition),
(3) some intermediate condition prevails.

The third scenario was considered by Reynolds and Satterlee (Chapter 11 in Abramson, 1996a). It was assumed that the contact angle measured in liquid having an undisturbed surface is zero, which is typical of several liquid–tank systems. However, it is possible that the angle at which the moving wave meets the wall is not the same as the static contact angle, that is, hysteresis.

The most important force is the capillary force usually measured in terms of Bond number. The characteristics of capillary waves were examined by Siegel (1961), Siegert *et al.* (1964, 1965), Habip (1965), Koval (1968), and Koval and Bhutta (1968). The theory of capillary surface waves is well documented by Finn (1974, 1978, 1979, 1986, 1999) and Myshkis, *et al.* (1987). For an axisymmetric meniscus, the curvature κ was given by the following expression, (Chu, 1964c and Satterlee and Reynolds, 1964),

$$\kappa = \frac{1}{r}\frac{\partial}{\partial r}\left(\frac{rf_r}{\sqrt{1+f_r^2+(1/r^2)f_\theta^2}}\right) + \frac{1}{r^2}\frac{\partial}{\partial \theta}\left(\frac{f_\theta}{\sqrt{1+f_r^2+(1/r^2)f_\theta^2}}\right) \tag{12.6}$$

where a subscript denotes differentiation with respect to the subscripted variable, and $z=f(r,\theta)$ is the undisturbed free-surface height of the meniscus, see Figure 12.7. The curvature κ has an important role in establishing the free-surface boundary condition. Assuming inviscid fluid, the associated normal pressure across the free surface experiences discontinuity by an amount proportional to the product of the interfacial tension and the mean surface curvature is given by equation (12.1), that is,

$$P_g - P = \sigma\kappa \tag{12.7}$$

Jahsman (1961) determined the equilibrium shape of the free liquid surface in a cylindrical tank. Fung (1965), and Hastings and Toole (1968) studied the dynamic response of the free liquid surface experiencing sudden weightless field. Christiansen, *et al.* (1992) studied three states of capillary waves: quasi-crystals, hexagons, and radial waves. Veldman and Vogels (1984) and Utsumi (1988, 1989, 1997, 1998) considered different cases of liquid sloshing in axisymmetric containers in a low gravitational field. The shape of the liquid surface for low Bond number was numerically determined by Wohlen, *et al.* (1975). Edge constraint effects on the capillary waves were examined by Benjamin and Scott (1979). Utsumi and Kondo (1987) determined the static configuration of a liquid surface in a spherical container. They derived the differential equation that governs the liquid surface equilibrium using the principle of virtual work instead of using a Young–Laplace equation.

Chao, *et al.* (1992) numerically studied the lateral g-jitter effects on the free-surface deformation using a finite-difference scheme. It was assumed that the influence of hysteresis could be represented by the relationship

$$\left.\frac{\partial \eta}{\partial r}\right|_{r=R} = C_1 \eta \tag{12.8}$$

If C_1 is a constant then relation (12.8) does not explain the damping or energy dissipation caused by the hysteresis. Rather than introducing an arbitrary functional relationship, hysteresis will be neglected in the analysis, and the contact-line is assumed to slide easily along the tank walls (the free-edge condition). In other words $\frac{\partial \eta}{\partial r}\big|_{r=R} = 0$, and in this case the contact-angle ϑ_c is defined by the expression

$$\cot \vartheta_c = \left.\frac{\partial \zeta}{\partial r} \middle/ \sqrt{1 + \left(\frac{\partial \zeta}{r \partial \vartheta}\right)}\right|_{r=R} \tag{12.9}$$

This means that the static equilibrium free-surface shape should be defined a priori. For very low values of Bond number, the equilibrium interface can be assumed to be spherical (Satterlee and Chin, 1965) and can be expressed by the shape function $f(r) = R - \sqrt{R^2 - r^2}$ for $\mathrm{Bo} \ll 1$. As Bo increases, the interface becomes flatter and a modified shape was suggested by Satterlee and Chin (1965) for values of Bond number in the range $10 < \mathrm{Bo} < 100$. Further studies on the nature of the contact-angle were given by Ngan and Dussan (1982).

The vibration of rotating menisci was studied by Chun, Ehmann, *et al.* (1987). Bauer and Eidel (1989c) found that small-amplitude liquid oscillations have a pronounced influence, for liquids with small contact-angles, on the dynamics of the containers. Dodge, *et al.* (1991) analytically studied the small-amplitude low gravity sloshing in axisymmetric containers. Hung, Lee, and Leslie (1991a,b,c) showed that resettlement of the liquid can be achieved by rotating the tank. The static meniscus configuration and stability of a free surface in propellant tanks under reduced and zero-gravity were studied by Siekman, *et al.* (1981), Slobozhanin, *et al.* (1999a,b and Slobozhanin and Alexander (2000).

12.2.3 Kinematics of spherical surface

The free-surface shape can be determined analytically for simple tank geometries. If the tank wall is a shell of revolution, the free-surface shape is spherical. The tank geometry, fluid volume, and contact angle between tank wall and free surface should be known in order to determine the shape of the free surface. Chin and Gallaghar (1964) and Babenko and Yurev (1980) analyzed the shape of the surface of a capillary liquid in a cylindrical container of arbitrary cross-section. If the coordinate frame is placed at $z = h$, where h is the liquid height, Bauer and Eidel (1990) assumed the spherical surface of the free surface in a cylindrical container of radius R. With reference to Figure 12.7, the shape of the undisturbed free liquid surface may be written in the form

$$r^2 + (Z - z)^2 = R_0^2, \quad z = Z \mp \sqrt{R_0^2 - r^2} \tag{12.10}$$

The condition of constant volume of the liquid requires

$$\pi R^2 h = \int_{-h}^{z} \int_0^{2\pi} \int_0^R r\, dr\, d\theta\, dz = \pi R^2 h + 2\pi \int_0^R r z(r)\, dr \tag{12.11}$$

Introducing relation (12.10) into (12.11) and carrying out the integration gives

$$Z = \frac{2}{3R^2} \left[R_0^3 - (R_0^2 - R^2)^{3/2} \right] \tag{12.12}$$

From the geometry, we have $R_0 = R/\cos\vartheta_c$, and introducing (12.12) into equation (12.10) gives

$$r^2 + \left[z - \frac{2R}{3\cos^3\vartheta_c} (1 - \sin^3\vartheta_c) \right]^2 = \frac{R^2}{\cos^2\vartheta_c} = R_0^2 \tag{12.13}$$

This surface is described by

$$z_0 = f_0(r) = \frac{R}{\cos\vartheta_c} \left[\frac{2(1 - \sin^3\vartheta_c)}{3\cos^2\vartheta_c} - \sqrt{1 - (r\cos\vartheta_c/R)^2} \right] \tag{12.14}$$

It is not difficult to show that for $\vartheta_c = \pi/2$, the right-hand side of equation (12.14) vanishes. From equation (12.14), the elevation at $r = 0$ is

$$z_0 = f_0(0) = \frac{R}{\cos\vartheta_c} \left[\frac{2(1 - \sin^3\vartheta_c)}{3\cos^2\vartheta_c} - 1 \right] \tag{12.15}$$

At $r = R$, the elevation is

$$z_0 = f_0(R) = \frac{R}{3\cos^3\vartheta_c} (2 - 2\sin^3\vartheta_c - 3\cos^3\vartheta_c \sin\theta_c) \tag{12.16}$$

For a cylindrical container of general cross-section in zero gravity the surface change can be discontinuous or "nearly discontinuous," leading to large shifts of the liquid mass arising from small changes in contact-angle. The height of a capillary surface was determined by Concus and Finn (1976). Concus and Finn (1974a, b, 1987, 1990a, b, 1991b, 1992) and Finn (1983, 1984) were able to develop some analytical studies dealing with the design basis of the container cross-sections. For particular cylindrical cross-sections, a discontinuous kind of change can be realized as the contact angle crosses a critical value. Above that critical value, there exists an equilibrium configuration of liquid that covers the base of the cylindrical container. For contact angles smaller than the critical value, no such equilibrium configuration can exist. Concus and Finn (1974a,b, 1990a, 1991b) studied the behavior of the interface for a near-rhombus section where the behavior can be nearly discontinuous.

Concus, *et al.* (1999) reported experimental results dealing with the interface configuration experiments performed aboard the Space Shuttle and the *Mir* Space Station. The "exotic" containers used are rotationally symmetric and have the property that for a given contact angle and liquid volume a continuum of distinct rotationally symmetric equilibrium configurations can appear, all of which have the same mechanical energy. The symmetric equilibrium configurations were found to be unstable and deformations that are not rotationally symmetric can be shown mathematically to yield configurations with lower energy. It was shown experimentally that distinct locally stable configurations could form and have different dynamic characteristics, confirming analytical and numerical results. This phenomenon of asymmetric local energy was found to occur if exotic conditions are not entirely met. Other recent research activities have included the mechanics of contact line and the associated boundary conditions. The contact line is the intersection of two distinct fluids with a solid.

12.3 Modal analysis

12.3.1 Modal analysis under microgravity

The fluid field equations under microgravity must include the influence of surface tension. Consider an upright circular cylindrical container partially filled with an inviscid fluid. Let S be the interface of the undisturbed free surface, which makes a contact angle ϑ_c, and S' is the disturbed surface. For an incompressible and irrotational flow the continuity equation is

$$\nabla^2\Phi = 0 \tag{12.17}$$

where Φ is the velocity potential function. The solution of this equation must satisfy the boundary conditions. The velocity normal to the tank wall and bottom must vanish, that is,

$$\left.\frac{\partial\Phi}{\partial r}\right|_{r=R} = 0, \quad \left.\frac{\partial\Phi}{\partial z}\right|_{z=-h} = 0 \tag{12.18a,b}$$

At the disturbed free surface, the unsteady Bernoulli's equation may be written in the form

$$\frac{P}{\rho} + \frac{1}{2}\left[\left(\frac{\partial\Phi}{\partial r}\right)^2 + \left(\frac{\partial\Phi}{r\partial\theta}\right)^2 + \left(\frac{\partial\Phi}{\partial z}\right)^2\right] + g\eta(r,\theta,t) + \frac{\partial\Phi}{\partial t} = \psi(t) \tag{12.19}$$

where P is the pressure just inside the interface, ρ is the fluid density, and $\eta(r,\theta,t) = \eta_0(r) + \tilde{\eta}(r,\theta,t)$ is the total fluid free-surface elevation consisting of the sum of the undisturbed component, $\eta_0(r)$, plus the disturbed displacement, $\tilde{\eta}(r,\theta,t)$, and $\psi(t)$ is an arbitrary function of time. The pressure P is related to the pressure just outside the surface, P_g, through relation (12.7), which is given in terms of the surface curvature (12.6).

The kinematic free-surface boundary condition is

$$\frac{\partial \eta}{\partial t} = \frac{\partial \Phi}{\partial z} - \frac{\partial \Phi}{\partial r}\frac{\partial \eta}{\partial r} - \frac{1}{r^2}\frac{\partial \Phi}{\partial \theta}\frac{\partial \eta}{\partial \theta} \tag{12.20}$$

The arbitrary function, $\Psi(t)$, can be replaced by the equilibrium liquid pressure divided by the density, P_0/ρ, at the vertex of the equilibrium free surface. By using the surface pressure equilibrium, $p_g - P = \sigma\kappa$, one can write equation (12.19) in the form

$$\frac{P_g - P_0}{\rho} - \frac{\sigma}{\rho}\kappa + \frac{1}{2}\left[\left(\frac{\partial \Phi}{\partial r}\right)^2 + \left(\frac{\partial \Phi}{r\partial \theta}\right)^2 + \left(\frac{\partial \Phi}{\partial z}\right)^2\right] + g\eta(r,\theta,t) + \frac{\partial \Phi}{\partial t} = 0 \tag{12.21}$$

The linear modal analysis of the fluid free surface is carried out by solving equation (12.17) subject to the linearized boundary conditions. If the velocity potential function is of order ε, where ε is a small parameter, the free-surface elevation may be expressed in the form

$$\eta(r,\theta,t) = \eta_0(r) + \varepsilon\tilde{\eta}(r,\theta,t) \tag{12.22}$$

Satterlee and Reynolds (1964) expressed the velocity potential function and its derivatives on the disturbed free surface in a Taylor's series expansion about their values on the undisturbed meniscus. This in effect allows the single boundary condition to be transferred from S' to S. The linearized dynamic free-surface boundary condition takes the form

$$\frac{\sigma}{\rho}\left\{\frac{1}{r}\frac{\partial}{\partial r}\left[\frac{r\tilde{\eta}_r}{\sqrt{1+\eta_{0r}^2}}\right] + \frac{1}{r^2}\frac{\partial}{\partial \theta}\left[\frac{\tilde{\eta}_\theta}{\sqrt{1+\eta_{0r}^2}}\right]\right\} + g\tilde{\eta} + \frac{\partial \Phi}{\partial t} = 0 \tag{12.23}$$

The linearized kinematic free-surface boundary is

$$\frac{\partial \tilde{\eta}}{\partial t} = \frac{\partial \Phi}{\partial z} - \frac{\partial \Phi}{\partial r}\frac{\partial \eta_0}{\partial r} \tag{12.24}$$

It is also assumed that the perturbation in the slope $\partial\tilde{\eta}/\partial r$ at the contact line is given by, to order ε,

$$\left.\frac{\partial \tilde{\eta}}{\partial r}\right|_{r=R} = \gamma\tilde{\eta} \tag{12.25}$$

where γ is a parameter that accounts for the effects of contact angle hysteresis. In order to express the fluid field equations in terms of nondimensional parameters that include Bond's number, the following parameters are introduced

$$\Re = r/R, \quad Z = z/R, \quad f = \eta_0/R, \quad \phi(\Re,\theta,Z)\sin\omega t = \frac{\Phi(r,\theta,z,t)}{\sqrt{gR^3}},$$

$$\mathrm{Bo} = \frac{\rho g R^2}{\sigma}\,H(\Re,\theta)\cos\omega t = \frac{\tilde{\eta}}{\sqrt{Rg/\omega^2}}, \quad \varpi = \frac{\rho R^3\omega^2}{\sigma}, \quad \Gamma = R\gamma, \quad L = h/R \tag{12.26}$$

The free-surface dynamic and kinematic boundary conditions are, respectively,

$$\frac{1}{\Re}\frac{\partial}{\partial\Re}\left[\frac{\Re H_{\Re}}{\sqrt{1+f_{\Re}^2}}\right]+\frac{1}{\Re^2}\frac{\partial}{\partial\theta}\left[\frac{H_{\theta}}{\sqrt{1+f_{\Re}^2}}\right]+(\mathsf{Bo})H+\varpi^2\phi=0 \tag{12.27}$$

$$H=\frac{\partial\phi}{\partial Z}-\frac{\partial f}{\partial\Re}\frac{\partial\phi}{\partial\Re} \tag{12.28}$$

The contact angle condition (12.25) is

$$\left.\frac{\partial H}{\partial\Re}\right|_{\Re=1}=\Gamma H \tag{12.29}$$

Based on the assumption that the meniscus shape, f, is flat, that is, $H=\partial\phi/\partial Z$, the solution of the Laplace equation, $\nabla^2\phi=0$, may be written in the form

$$\phi(\Re,\theta,Z)=J_m(\xi_{mn}\Re)\cos m\theta\cosh[\xi_{mn}(L+Z)] \tag{12.30}$$

where ξ_{mn} are the roots of $J'_m(\xi_{mn})=0$. In view of the kinematic condition, $H=\partial\phi/\partial Z$, and $f_{\Re}^2=0$, condition (12.27) takes the form

$$\frac{1}{\Re}\frac{\partial}{\partial\Re}(\Re\phi_{Z\Re})+\frac{1}{\Re^2}\phi_{Z\theta\theta}+(\mathsf{Bo})\phi_Z+\varpi^2\phi=0 \tag{12.31}$$

Substituting equation (12.30) into (12.31), one obtains the natural frequency of the free surface

$$\varpi_{mn}^2=\rho R^3\omega_{mn}^2/\sigma=\left[m^2\xi_{mn}^2+(\mathsf{Bo})\xi_{mn}\right]\tanh(\xi_{mn}L) \tag{12.32}$$

This solution is based on the assumption that the contact angle does not change and $\Gamma=0$. Lateral sloshing in cylindrical tanks with hemispherical bottoms was studied by Concus, *et al.* (1967). Barnyak (1971a,b) and Labus (1969) employed approximate methods to determine the natural frequencies of an ideal liquid under a weak gravitational field. Bauer (1987c, 1991b) and Bauer and Kamerer (1992, 1993) determined the natural frequencies of spinning and nonspinning liquid layers under a microgravity field.

12.3.2 Modal analysis under zero gravity

Although the zero gravity condition is not realized, it can be adopted as the limiting case because the residual gravitational field is very small. The modal analysis in a rectangular container under zero gravity was considered by Bauer and Eidel (1989c). Bauer and Eidel (1989c, 1990) and Bauer (1995d) considered the zero gravity case and the influence of contact angle. The lack of gravity makes the surface tension the dominant force and yields an equilibrium position of the free surface in the form of a spherical surface whose radius depends on the angle of contact, ϑ_c. The equilibrium position of the free surface, $\eta_0(r)$, is determined from the integration of Laplace–Young equation and Young's condition on the contact line. The associated normal pressure across the free surface experiences discontinuity by an amount proportional to the product of the interfacial tension and the mean surface curvature, assuming inviscid fluid, and we write

$$P_g - P = \sigma_{LG}\left[\frac{1}{R_1} + \frac{1}{R_2}\right] = \text{const} \tag{12.33}$$

where R_1 and R_2 are the principal radii of curvature. With reference to equation (12.4), the static equilibrium along the tank wall requires

$$\sigma_{LG}\cos\vartheta_c = \sigma_{GS} - \sigma_{LS} \tag{12.34}$$

where σ_{LG}, σ_{LS}, and σ_{GS} are the surface tensions betweens between the gas/liquid, liquid/solid, and solid/gas, respectively. The mean curvature (12.6) in terms of the liquid free surface $\eta(r, \theta)$ is

$$\frac{1}{R_1} + \frac{1}{R_2} = \frac{1}{r}\frac{\partial}{\partial r}\left\{\frac{r\eta_r}{\sqrt{1 + \eta_r^2 + (\eta_\theta^2/r^2)}}\right\} + \frac{1}{r^2}\frac{\partial}{\partial\theta}\left\{\frac{\eta_\theta}{\sqrt{1 + \eta_r^2 + (\eta_\theta^2/r^2)}}\right\} \tag{12.35}$$

If the coordinate frame is placed at $z=h$, where h is the liquid height, the shape of the undisturbed free liquid surface under zero gravity conditions is given in terms of the contact angle

$$r^2 + \left[\eta - \frac{2R}{3\cos^3\vartheta_c}\left(1 - \sin^3\vartheta_c\right)\right]^2 = \frac{R^2}{\cos^2\vartheta_c} = R_s^2 \tag{12.36}$$

where $R_s = R/\cos\vartheta_c$ is the radius of curvature of the undisturbed free surface. This surface is described by

$$\eta_0 = \frac{R}{\cos\vartheta_c}\left[\frac{2(1 - \sin^3\vartheta_c)}{3\cos^2\vartheta_c} - \sqrt{1 - (r\cos\vartheta_c/R)^2}\right] \tag{12.37}$$

The free surface oscillates about the equilibrium position, $\eta_0(r)$, under free sloshing. The liquid free-surface displacement measured from the surface $\eta_0(r)$ is $\tilde{\eta}(r,\theta,t)$, such that $\eta(r,\theta,t) = \eta_0(r) + \tilde{\eta}(r,\theta,t)$. The linearized mean curvature (12.35) takes the form

$$\frac{1}{R_1} + \frac{1}{R_2} = -\frac{2}{R_s} - \frac{1}{r}\frac{\partial}{\partial r}\left\{r\tilde{\eta}_r\left(1 - (r/R_s)^2\right)^{3/2}\right\} - \frac{1}{r^2}\tilde{\eta}_{\theta\theta}\sqrt{1 - (r/R_s)^2} \tag{12.38}$$

The fluid dynamic behavior is governed by equations (12.17) and (12.18) and the free-surface boundary conditions.

The dynamic free-surface boundary condition is

$$\frac{\sigma}{\rho}\left(\frac{1}{R_1} + \frac{1}{R_2}\right) + \frac{\partial\Phi}{\partial t} = \frac{2\sigma\cos\vartheta_c}{\rho R} \tag{12.39}$$

The kinematic boundary condition is

$$\frac{\partial\tilde{\eta}}{\partial t} = \frac{\partial\Phi}{\partial z} - \frac{\partial\Phi}{\partial r}\left(\frac{\partial\tilde{\eta}}{\partial r} + \frac{\partial\eta_0}{\partial r}\right) - \frac{1}{r^2}\frac{\partial\Phi}{\partial\theta}\frac{\partial\tilde{\eta}}{\partial\theta} \tag{12.40a}$$

Expanding this condition into a Taylor series about $\eta = 0$, gives the linearized form for small $\tilde{\eta}$,

$$\frac{\partial\Phi}{\partial z} - \frac{\partial\tilde{\eta}}{\partial t} - \frac{\partial\Phi}{\partial r}\frac{\partial\eta_0}{\partial r} + \sum_{j=1}^{\infty}\frac{\eta_0^j}{j!}\frac{\partial^{j+1}\Phi}{\partial z^{j+1}} - \frac{\partial\eta_0}{\partial r}\sum_{j=1}^{\infty}\frac{\eta_0^j}{j!}\frac{\partial^{j+1}\Phi}{\partial r\partial z^j} = 0 \tag{12.40b}$$

The contact angle at $r=R$ is ϑ_c and requires the following condition

$$\left.\frac{\partial\tilde{\eta}}{\partial r}\right|_{r=R}=0 \tag{12.41}$$

In addition, the fluid volume must be preserved, that is,

$$\int_0^{2\pi}\int_0^R \tilde{\eta}(r,\theta,t)r\,dr\,d\theta=0 \tag{12.42}$$

The solution of the Laplace equation (12.17) that satisfies conditions (12.18a, b) for this case may be written in the form

$$\Phi(r,\theta,z,t)=\sum_{m=-\infty}^{\infty}\sum_{n=1}^{\infty}A_{mn}(t)J_{|m|}(\xi_{|m|n}r/R)\frac{\cosh[\xi_{|m|n}(z+h)/R]}{\cosh[\xi_{|m|n}h/R]}e^{im\theta} \tag{12.43}$$

The free-surface displacement that satisfies conditions (12.41) and (12.42) may be written in the form

$$\tilde{\eta}(r,\theta,t)=\sum_{m=-\infty}^{\infty}\sum_{n=1}^{\infty}a_{mn}(t)J_{|m|}(\xi_{|m|n}r/R)e^{im\theta} \tag{12.44}$$

Introducing equations (12.43) and (12.44) into the dynamic free-surface conditions (12.39) and (12.40b), gives the eigenvalue problem

$$\left[\beta_{k\ell}^{(m)}\right]\left[\gamma_{k\ell}^{(m)}\right]\begin{Bmatrix}\mathbf{A}\\ \mathbf{a}\end{Bmatrix}=\chi\left[\alpha_{k\ell}^{(\mathrm{m})}\right]\begin{Bmatrix}\mathbf{A}\\ \mathbf{a}\end{Bmatrix} \tag{12.45}$$

where $\chi=\dfrac{\omega^2\rho R^3}{\sigma}$,

$$\alpha_{k\ell}^{(m)}=\frac{2\xi_{|m|k}^2}{\left[\xi_{|m|k}^2-m^2\right]J_{|m|}^2(\xi_{|m|k})}\int_0^1\left[\cosh(\xi_{|m|\ell}\eta_0(\bar{r})/R)+\tanh(\xi_{|m|\ell}h/R)\sinh(\xi_{|m|\ell}\eta_0(\bar{r})/R)\right]\times\bar{r}J_{|m|}(\xi_{|m|k}\bar{r})J_{|m|}(\xi_{|m|\ell}\bar{r})\,d\bar{r} \tag{12.46a}$$

$$\beta_{k\ell}^{(m)}=\frac{2\xi_{|m|k}^2}{\left[\xi_{|m|k}^2-m^2\right]J_{|m|}^2(\xi_{|m|k})}\int_0^1\left\{\xi_{|m|k}\xi_{|m|\ell}\left(1-\bar{r}^2\cos^2\vartheta_c\right)^{3/2}J'_{|m|}(\xi_{|m|k}\bar{r})J_{|m|}(\xi_{|m|\ell}\bar{r})\times\frac{m^2}{\bar{r}}\sqrt{1-\bar{r}^2\cos^2\vartheta_c}J_{|m|}(\xi_{|m|k}\bar{r})J_{|m|}(\xi_{|m|\ell}\bar{r})\right\}d\bar{r} \tag{12.46b}$$

$$\begin{aligned}\gamma_{k\ell}^{(m)}&=\frac{2\xi_{|m|k}^2}{[\xi_{|m|k}^2-m^2]J_{|m|}^2(\xi_{|m|k})}\\&\quad\times\int_0^1\left\{\left[\sinh(\xi_{|m|\ell}\eta_0(\bar{r})/R)+\tanh(\xi_{|m|\ell}h/R)\cosh(\xi_{|m|\ell}\eta_0(\bar{r})/R)\right]\right.\\&\quad\times\bar{r}J_{|m|}(\xi_{|m|k}\bar{r})J_{|m|}(\xi_{|m|\ell}\bar{r})-\left[\cosh(\xi_{|m|\ell}\eta_0(\bar{r})/R)+\tanh(\xi_{|m|\ell}h/R)\right.\\&\quad\left.\left.\times\sinh(\xi_{|m|\ell}\eta_0(\bar{r})/R)\right]\times\frac{\bar{r}^2\cos^2\vartheta_c}{\sqrt{1-\bar{r}^2\cos^2\vartheta_c}}J_{|m|}(\xi_{|m|\ell}\bar{r})J_{|m|}(\xi_{|m|k}\bar{r})\right\}d\bar{r},\quad \bar{r}=r/R\end{aligned} \tag{12.46c}$$

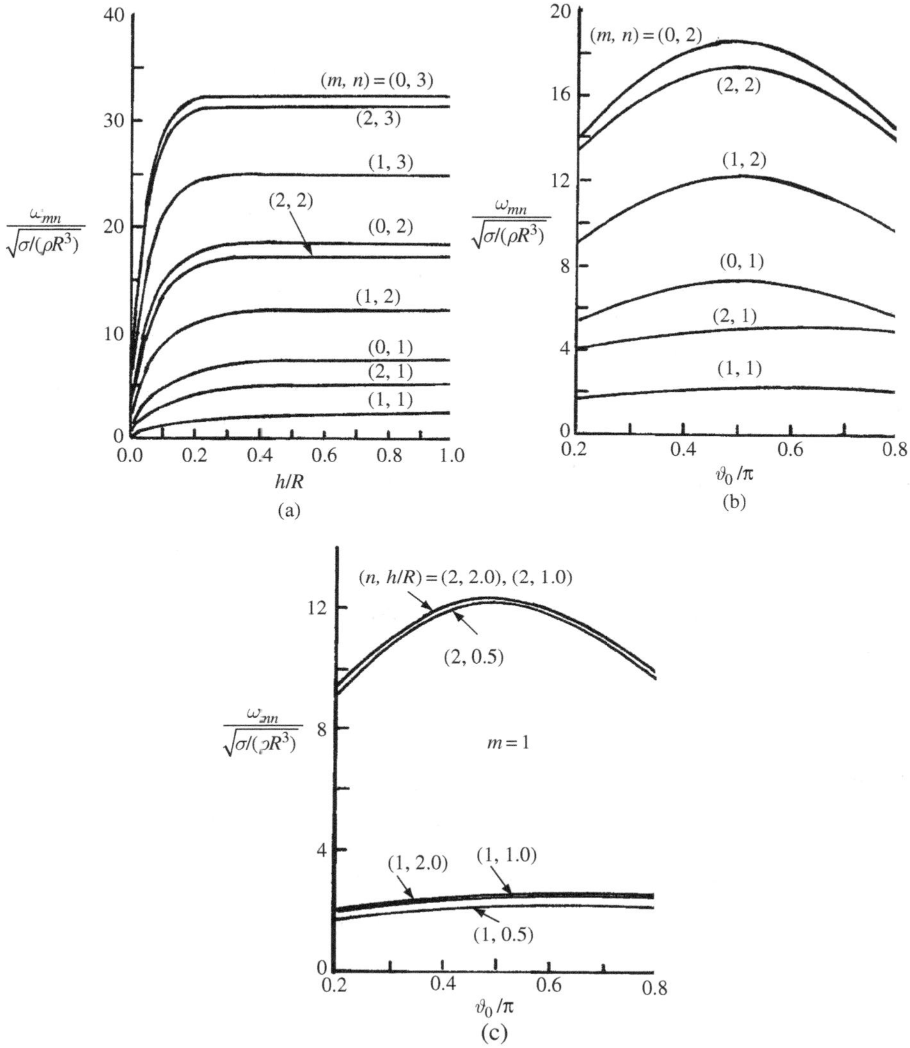

Figure 12.8 Dependence of liquid natural frequencies in zero gravity on (a) liquid depth ratio for contact angle $\vartheta_0 = 0.5\pi$, (b) contact angle for liquid depth $h/R = 0.5$, and (c) on contact angle for $m = 1$, and different values of liquid depth ratio and n. (Bauer and Eidel, 1990)

For the special case of a flat free surface, that is, $\vartheta_c = \pi/2$, the above coefficients take the values $\alpha_{k\ell}^{(m)} = \delta_{k\ell}$, $\beta_{k\ell}^{(m)} = \xi_{|m|k}^2 \delta_{k\ell}$, $\gamma_{k\ell}^{(m)} = \xi_{|m|k} \tanh\left(\xi_{|m|k} h/R\right) \delta_{k\ell}$, where δ_{kl} is the Kronecker delta. Introducing these special values in equation (12.45), one can obtain the well-known expression for the natural frequency

$$\omega_{mn}^2 = \frac{\sigma \xi_{mn}^3}{\rho R^3} \tanh\left(\xi_{mn} h/R\right), \quad \text{for } \vartheta_c = \pi/2 \tag{12.47}$$

Figure 12.8(a) shows the dependence of the natural frequency ratio $\omega_{mn}/\sqrt{\sigma/(\rho R^3)}$ on the fluid depth ratio, h/R, for several sloshing modes for contact angle $\vartheta_c = \pi/2$. The dependence of the natural frequency ratio on the contact angle ϑ_c is shown in Figure 12.8(b) for fluid depth $h/R = 0.5$ and for different modes. It is seen that the natural frequency of the axisymmetric mode, $m = 0$, reaches its maximum value at nearly $\vartheta_c = \pi/2$ above which it decreases again. Figure 12.8(c) shows the same dependence but for $m = 1$ and different combinations of fluid depth ratio and n.

12.3.3 Experimental modal analysis

The main objective of microgravity experimental modal analysis is to measure the natural frequencies, mode shapes, and damping coefficient of the free liquid surface for low values of Bond number (see, e,g., Maulard and Jourdin, 1966, Salzman, *et al.*, 1967, 1968, 1973, Salzmann and Masica, 1969, and Bauer, 1995c). For a cylindrical tank, the Bond number is $\mathsf{Bo} = gR^2/(\sigma/\rho)$, where g is the equivalent gravitational acceleration (cm/s^2), R is the tank radius, σ is the surface tension (dynes/cm or 10^{-5} N/cm), and ρ is the liquid density. The maximum Bond number during a free-fall drop was 0.002, while the majority of free-fall drops resulted in Bond numbers close to 0.001. The experimental results of Salzman and Masica (1967), using the 5 seconds free-fall facility, revealed that the value of the centerline liquid depth depends on the magnitude of the Bond number and liquid volume. For large liquid depth, $h/R > 2$, and zero static contact-angle, they obtained the following empirical relationship for the liquid first mode natural frequency ω_1

$$\omega_1^2 = (2.6 + 1.84\,\mathsf{Bo})\frac{\sigma}{\rho R^3} \tag{12.48}$$

where the constant 2.6 represents the capillary contribution to the lateral natural frequency. Equation (12.48) reveals that capillary effects begin to appear for Bond numbers less than 20. The dependence of the nondimensional natural frequency $\Omega_1^2 = \omega_1^2 R^3/\beta$, where $\beta = \sigma/\rho$, on the Bond number parameter ($\mathsf{Bo} + 1.4$) is plotted on a log-log graph as shown in Figure 12.9.

For shallow liquid depth $h/R < 2$ and zero Bond number, the following relation was obtained

$$\omega_1^2 = \frac{2.6\beta}{R^3}\tanh\frac{2h}{R} \tag{12.49}$$

Salzman and Masica (1967) measured the dependence of the liquid natural frequency on the fluid depth and Bond number in spherical tanks. The fundamental sloshing mode shape depends on Bond number. In the fundamental mode, the vertex of the liquid surface remains at the centerline of the cylinder and maximum displacement occurs at the cylinder wall. Free lateral sloshing was observed to occur on a thin layer of liquid rather than on an unwetted surface.

The influence of surface tension of the free liquid surface on the natural frequencies was estimated by Schilling and Siekmann (1981) and Bauer (1991h). Bauer (1982c) determined the natural frequencies of an inviscid liquid in an annular cylindrical container. Bauer (1984b) numerically determined the natural frequency of a viscous liquid in a rigid container. Schulkes and Cuvelier (1991) computed the normal modes of a rotating viscous liquid with a capillary free boundary. Albright (1977) developed mathematical and computational algorithms to study the stability of axisymmetric annular free surfaces.

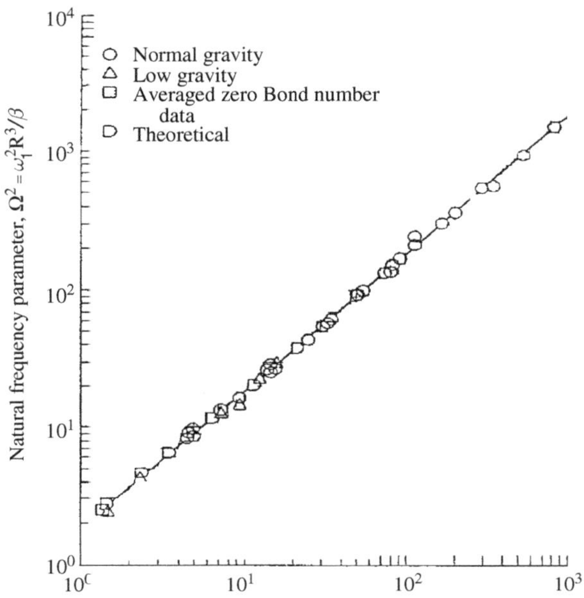

Figure 12.9 Dependence of liquid natural frequency on the Bond number for liquid depth ratio > 2. (Salzman, *et al.*, 1967)

The damping coefficient c was obtained by Clark and Stephens (1967) and Dodge and Garza (1967a,b), in the form

$$\frac{c}{(c)_{\mathrm{Bo}=0}} = \frac{1}{35.7} K_d (2.6 + 1.8\,\mathrm{Bo})^{1/4} \tag{12.50}$$

where $(c)_{\mathrm{Bo}=0} = \dfrac{28.1}{2\pi}\sqrt{\dfrac{\mu}{\rho R^2}}\left(2.6\dfrac{\sigma}{\rho R^3}\right)^{1/4}$ is the damping coefficient at zero Bond number, K_d is an explicit function of Bond number, and μ is the liquid viscosity. The measured results showed the normalized damping coefficient $c/(c)_{\mathrm{Bo}=0}$ remained constant for all Bond numbers below 100. The decrease in the natural frequency compensates for the increase in the damping coefficient. It was concluded that for identical radii and liquids, the damping coefficient c is relatively independent of acceleration in the Bond number range of 0 to 100. Bauer (1984a, 1988c 1989b) and Bauer and Eidel (1990) analytically estimated the linear motion of immiscible liquids in cylindrical containers under low gravitational field.

Yeh (1967) extended the variational formulation of Satterlee and Reynolds (1964) and obtained an additional integral, which emerged as a result of the application of Green's theorem. Yeh obtained the liquid eigenvalues and mode shapes for the free surface using the Ritz method. Chu (1970) developed a semi-numerical approach for an arbitrary axisymmetric tank subjected to both pitching and translational oscillations and estimated the resulting sloshing forces and moments. Concus, *et al.* (1968, 1969) and Coney and Salzman (1971) considered the lateral sloshing in hemispherically bottom and spheroidal containers under low gravity conditions. Dodge and Garza (1967a) used relation (12.49) to develop an equivalent mechanical model to describe the sloshing characteristics (natural frequencies, forces and moments exerted on the container). They found that the amount of liquid taking part in low-gravity sloshing is less than that for high-gravity sloshing for the same tank size and same liquid volume. Note that more of the liquid is in contact with the walls under low-gravity conditions, thus more of the fluid must follow the motion of the tank. In other words, more of the liquid must be assigned to be a rigidly attached mass in the mechanical model.

12.4 Sloshing with slipping and anchored contact lines

The behavior of liquid sloshing in microgravity was studied on several occasions by Gilbert, *et al.* (1984), Gilbert (1985), and Gerrits, *et al.* (1999). The liquid-free-surface contact line may be anchored at the container wall or may slip up and down during the oscillation of the free surface as shown in Figure 12.10. Each case may take place depending on the magnitude of the external excitation. Bauer (1992d) and Bauer and Eidel (1997a, c) analyzed the axisymmetric and asymmetric oscillations of viscous liquid in cylindrical containers. They were not able to satisfy the adherence condition at the cylindrical wall completely. Only the normal velocity condition could be satisfied, while circumferential and axial velocities could not be made to vanish at the wall. This means that for slipping at the cylindrical wall only the internal damping and adherence at the tank bottom contribute to the damping. A large damping, however, can be attributed to the adherence at the wall. To account for this damping, the adherence condition was introduced by assuming an anchored contact line at the cylindrical wall. Utsumi (1988, 1995) determined the response of the liquid-surface displacement under forced excitation using a variational approach. Bauer and Eidel (1997a) showed that the axial velocity was considerably reduced in comparison with the slipping liquid.

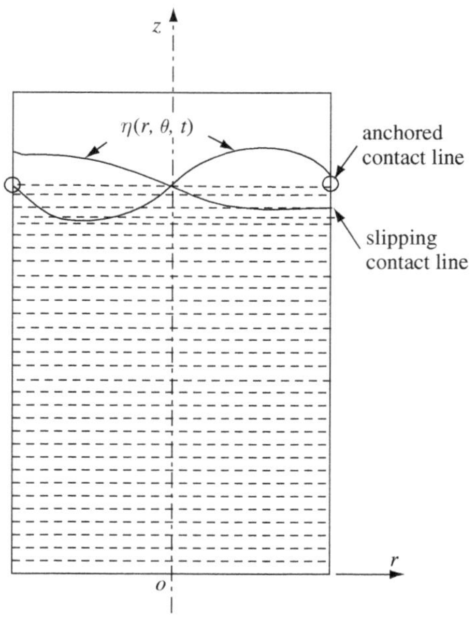

Figure 12.10 Cylindrical container partially filled with viscous liquid showing the anchored and slipping contact lines and the coordinate system.

Consider a circular cylinder of radius R partially filled with a viscous incompressible fluid of depth h. The container is subjected to translational harmonic excitation $x(t) = X_0 e^{i\Omega t}$. The continuity equation in cylindrical coordinates is

$$\frac{\partial u}{\partial r} + \frac{u}{r} + \frac{\partial v}{r\partial \theta} + \frac{\partial w}{\partial z} = 0 \tag{12.51}$$

The linearized Navier–Stokes equations for isothermal fluid are written in the form

$$\frac{\partial u}{\partial t} + \frac{1}{\rho}\frac{\partial p}{\partial z} = \nu\left[\frac{\partial^2 u}{\partial r^2} + \frac{\partial u}{r\partial r} - \frac{u}{r^2} + \frac{\partial^2 u}{r^2\partial \theta^2} - \frac{2}{r^2}\frac{\partial v}{\partial \theta} + \frac{\partial^2 u}{\partial z^2}\right] - \Omega^2 X_0 e^{i\Omega t}\cos\theta \tag{12.52a}$$

$$\frac{\partial v}{\partial t} + \frac{1}{\rho}\frac{\partial p}{r\partial \theta} = \nu\left[\frac{\partial^2 v}{\partial r^2} + \frac{\partial v}{r\partial r} - \frac{v}{r^2} + \frac{\partial^2 v}{r^2\partial \theta^2} + \frac{2}{r^2}\frac{\partial u}{\partial \theta} + \frac{\partial^2 v}{\partial z^2}\right] + \Omega^2 X_0 e^{i\Omega t}\sin\theta \tag{12.52b}$$

$$\frac{\partial w}{\partial t} + \frac{1\partial p}{\rho\partial z} = \nu\left[\frac{\partial^2 w}{\partial r^2} + \frac{1}{r}\frac{\partial w}{\partial r} + \frac{1}{r^2}\frac{\partial^2 w}{\partial \theta^2} + \frac{\partial^2 w}{\partial z^2}\right] - g \tag{12.52c}$$

Equations (12.51) and (12.52) together with the boundary conditions establish the boundary-value problem. Two cases will be considered, the modal analysis of viscous fluid sloshing and the forced excitation response of the fluid free surface.

12.4.1 Modal analysis

The modal analysis of free oscillations of viscous fluids in the absence of the external excitation is studied by setting $x(t) = 0$ in equations (12.52). Assuming the following solution of equations (12.52)

$$u(r, \theta, z, t) = \sum_{m=0}^{\infty} U_m(r, z)\, e^{im\theta}\, e^{\varpi t} \tag{12.53a}$$

$$v(r, \theta, z, t) = \sum_{m=0}^{\infty} V_m(r, z)\, e^{im\theta}\, e^{\varpi t} \tag{12.53b}$$

$$w(r, \theta, z, t) = \sum_{m=0}^{\infty} W_m(r, z)\, e^{im\theta}\, e^{\varpi t} \tag{12.53c}$$

$$p(r, \theta, z, t) = p_0 - \rho g z + \rho \sum_{m=0}^{\infty} P_m(r, z)\, e^{im\theta}\, e^{\varpi t} \tag{12.53d}$$

where m is an integer and $\varpi = \zeta + i\omega$ is a complex frequency. Substituting equations (12.53) into equations (12.51) and (12.52) gives

$$\frac{\partial U_m}{\partial r} + \frac{U_m}{r} + \frac{im}{r} V_m + \frac{\partial W_m}{\partial z} = 0 \tag{12.54}$$

$$\frac{\partial^2 U_m}{\partial r^2} + \frac{1}{r}\frac{\partial U_m}{\partial r} - \left[\frac{(1+m^2)}{r^2} + \frac{\varpi}{\nu}\right] U_m - \frac{2im}{r^2} V_m + \frac{\partial^2 U_m}{\partial z^2} = \frac{1}{\nu}\frac{\partial P_m}{\partial r} \tag{12.55a}$$

$$\frac{\partial^2 V_m}{\partial r^2} + \frac{1}{r}\frac{\partial V_m}{\partial r} - \left[\frac{(1+m^2)}{r^2} + \frac{\varpi}{\nu}\right] V_m - \frac{2im}{r^2} U_m + \frac{\partial^2 V_m}{\partial z^2} = \frac{im}{\nu r} P_m \tag{12.55b}$$

$$\frac{\partial^2 W_m}{\partial r^2} + \frac{1}{r}\frac{\partial W_m}{\partial r} - \left[\frac{m^2}{r^2} + \frac{\varpi}{\nu}\right] W_m + \frac{\partial^2 W_m}{\partial z^2} = \frac{1}{\nu}\frac{\partial P_m}{\partial r} \tag{12.55c}$$

Introducing the complex transformation

$$\Phi_m = U_m + iV_m, \qquad \Psi_m = U_m - iV_m \tag{12.56a,b}$$

$$U_m = \frac{1}{2}(\Phi_m + \Psi_m), \qquad V_m = \frac{i}{2}(\Psi_m - \Phi_m) \tag{12.56c,d}$$

Adding and subtracting equations (12.55a) and (12.55b) gives after using (12.56a, b)

$$\frac{\partial^2 \Phi_m}{\partial r^2} + \frac{1}{r}\frac{\partial \Phi_m}{\partial r} - \left[\frac{(m+1)^2}{r^2} + \frac{\varpi}{\nu}\right] \Phi_m + \frac{\partial^2 \Phi_m}{\partial z^2} = \frac{1}{\nu}\left[\frac{\partial}{\partial r} - \frac{m}{r}\right] P_m \tag{12.57a}$$

$$\frac{\partial^2 \Psi_m}{\partial r^2} + \frac{1}{r}\frac{\partial \Psi_m}{\partial r} - \left[\frac{(m-1)^2}{r^2} + \frac{\varpi}{\nu}\right] \Psi_m + \frac{\partial^2 \Psi_m}{\partial z^2} = \frac{1}{\nu}\left[\frac{\partial}{\partial r} + \frac{m}{r}\right] P_m \tag{12.57b}$$

The continuity equation (12.54) can be written in terms of the complex functions Φ_m and Ψ_m in the form

$$\frac{\partial \Phi_m}{\partial r} + \frac{1+m}{r}\Phi_m + \frac{\partial \Psi_m}{\partial r} - \frac{(m-1)}{r}\Psi_m = -2\frac{\partial W_m}{\partial z} \tag{12.58}$$

The solution of equations (12.57), (12.58) and (12.55c) that satisfies the boundary condition at the wall, $u(r = R, \theta, z, t) = 0$, may be written in the form

$$\Phi_{mn}(r, z) = A_{mn}(z) J_{m+1}(\xi_{mn} r/R) \tag{12.59a}$$

$$\Psi_{mn}(r, z) = -A_{mn}(z) J_{m-1}(\xi_{mn} r/R) \tag{12.59b}$$

$$W_{mn}(r, z) = C_{mn}(z) J_m(\xi_{mn} r/R) \tag{12.59c}$$

$$\Phi_{mn}(r, z) = D_{mn}(z) J_m(\xi_{mn} r/R) \tag{12.59d}$$

where ξ_{mn} are the roots of $J'_m(\xi_{mn} r/R)|_{r=R} = 0$, $m = 0, 1, 2, \ldots$ and $n = 1, 2, \ldots$, and the coefficients $A_{mn}(z)$, $C_{mn}(z)$, and $D_{mn}(z)$ must satisfy the following coupled ordinary differential equations obtained by substituting equations (12.59) into equations (12.57) and (12.58)

$$\frac{d^2 A_{mn}}{dz^2} - \left(\frac{\xi_{mn}^2}{R^2} + \frac{\varpi}{\nu}\right) A_{mn} = -\frac{\xi_{mn}}{\nu R} D_{mn} \tag{12.60a}$$

$$\frac{d^2C_{mn}}{dz^2} - \left(\frac{\xi_{mn}^2}{R^2} + \frac{\varpi}{\nu}\right)C_{mn} = -\frac{1}{\nu}\frac{dD_{mn}}{dz} \tag{12.60b}$$

$$\frac{dC_{mn}}{dz} = -\frac{\xi_{mn}}{R}A_{mn} \tag{12.60c}$$

It is not difficult to solve the coupled equations (12.60) whose characteristic function has the roots, $\pm\xi_{mn}/R$ and $\pm\sqrt{\xi_{mn}^2 + (sR^2/\nu)}/R$. Using the following recurrence relations of Bessel functions

$$\frac{1}{2}\left[J_{m-1}(\xi_{mn}r/R) - J_{m+1}(\xi_{mn}r/R)\right] = \frac{dJ_m(\xi_{mn}r/R)}{dr}$$

$$J_{m-1}(\xi_{mn}r/R) + J_{m+1}(\xi_{mn}r/R) = \frac{2m}{\xi_{mn}r/R}J_m(\xi_{mn}r/R)$$

and introducing the nondimensional parameters $S_{mn} = \varpi_{mn}R^2/\nu$ and $\tau = t\nu/R^2$, one obtains the following velocity components

$$\begin{aligned} u(r,\theta,z,\tau) = -&\sum_{m=0}^{\infty}\sum_{n=1}^{\infty} J'_m(\xi_{mn}r/R)\, e^{im\theta}\, e^{S_{mn}\tau} \\ &\times \Bigg\{A_{1mn}\cosh(\xi_{mn}z/R) \\ &+ A_{2mn}\sinh(\xi_{mn}z/R) + A_{3mn}\cosh\left(\frac{z}{R}\sqrt{\xi_{mn}^2 + S_{mn}}\right) \\ &+ A_{4mn}\sinh\left(\frac{z}{R}\sqrt{\xi_{mn}^2 + S_{mn}}\right)\Bigg\} \end{aligned} \tag{12.61a}$$

$$\begin{aligned} v(r,\theta,z,\tau) = -i&\sum_{m=0}^{\infty}\sum_{n=1}^{\infty}\frac{mJ_m(\xi_{mn}r/R)}{\xi_{mn}r/R}\, e^{im\theta}\, e^{S_{mn}\tau} \\ &\times \Bigg\{A_{1mn}\cosh(\xi_{mn}z/R) + A_{2mn}\sinh(\xi_{mn}z/R) \\ &+ A_{3mn}\cosh\left(\frac{z}{R}\sqrt{\xi_{mn}^2 + S_{mn}}\right) + A_{4mn}\sinh\left(\frac{z}{R}\sqrt{\xi_{mn}^2 + S_{mn}}\right)\Bigg\} \end{aligned} \tag{12.61b}$$

$$\begin{aligned} w(r,\theta,z,\tau) = -&\sum_{m=0}^{\infty}\sum_{n=1}^{\infty} J_m(\xi_{mn}r/R)\, e^{im\theta}\, e^{S_{mn}\tau} \\ &\times \Bigg\{A_{1mn}\sinh(\xi_{mn}z/R) + A_{2mn}\cosh(\xi_{mn}z/R) \\ &+ \frac{\xi_{mn}}{\sqrt{\xi_{mn}^2 + S_{mn}}}\Bigg[A_{3mn}\sinh\left(\frac{z}{R}\sqrt{\xi_{mn}^2 + S_{mn}}\right) \\ &+ A_{4mn}\cosh\left(\frac{z}{R}\sqrt{\xi_{mn}^2 + S_{mn}}\right)\Bigg]\Bigg\} \end{aligned} \tag{12.61c}$$

The pressure distribution is given in the form

$$p(r,\theta,z,\tau) = p_0 - \rho g z + \frac{\mu}{R}\sum_{m=0}^{\infty}\sum_{n=1}^{\infty}\frac{S_{mn}}{\xi_{mn}} J_m\left(\frac{\xi_{mn} r}{R}\right) e^{im\theta}\, e^{S_{mn}\tau} \times \left\{A_{1mn}\cosh\left(\frac{\xi_{mn} z}{R}\right) + A_{2mn}\sinh\left(\frac{\xi_{mn} z}{R}\right)\right\} \tag{12.62}$$

The integration constants, A_{imn}, $(i=1, 2, 3, 4)$, are determined from the following boundary conditions:

At the tank bottom

$$u|_{z=-h} = v|_{z=-h} = w|_{z=-h} = 0 \tag{12.63}$$

The shear stress condition at the liquid free surface

$$\tau_{zr}|_{z=0} = \mu\left[\frac{\partial u}{\partial z} + \frac{\partial w}{\partial r}\right]_{z=0} = \tau_{z\theta}|_{z=0} = \mu\left[\frac{\partial v}{\partial z} + \frac{1}{r}\frac{\partial w}{\partial \theta}\right]_{z=0} = 0 \tag{12.64}$$

The kinematic free-surface boundary condition is

$$w = \left.\frac{\partial \eta}{\partial t}\right|_{z=0} \tag{12.65}$$

The dynamic free-surface condition is

$$p - 2\mu\frac{\partial w}{\partial z} + \sigma\left[\frac{\partial^2 \eta}{\partial r^2} + \frac{1}{r}\frac{\partial \eta}{\partial r} + \frac{1}{r^2}\frac{\partial^2 \eta}{\partial \theta^2} - \frac{\rho g}{\sigma}\eta\right] = \text{const} \quad \text{at } z=0 \tag{12.66}$$

The solution of the free-surface oscillations of the fluid depends on whether the contact line is allowed to slip or be anchored. The two cases are considered in the following two sections.

Slipping contact line

When the contact line is allowed to slip at the container wall, $r = R$, the combined free-surface condition is obtained by taking the time derivative of equations (12.56) and using condition (12.65) to give

$$\varpi\bar{p} - 2\varpi\mu\frac{\partial w}{\partial z} + \sigma\left[\frac{\partial^2 w}{\partial r^2} + \frac{1}{r}\frac{\partial w}{\partial r} + \frac{1}{r^2}\frac{\partial^2 w}{\partial \theta^2} - \frac{\rho g}{\sigma}w\right] = 0 \tag{12.67}$$

The free-surface displacement is obtained by integrating equation (12.61c) and using condition (12.65)

$$\eta(r,\theta,\tau) = \sum_{m=0}^{\infty}\sum_{n=1}^{\infty}\eta_{mn}\, e^{im\theta}\, e^{S_{mn}\tau} J_m(\xi_{mn} r/R) \tag{12.68}$$

Substituting for the velocity w and the pressure in the combined free-surface boundary condition (12.67), gives

$$\left(S_{mn}^2+2\xi_{mn}^2 S_{mn}\right)A_{1mn}+\frac{\sigma R}{\rho\nu^2}\xi_{mn}\left(\xi_{mn}^2+\frac{\rho g R^2}{\sigma}\right)A_{2mn}$$

$$+2\xi_{mn}^2 S_{mn}A_{3mn}\frac{\sigma R}{\rho\nu^2}\frac{\xi_{mn}^2\left(\xi_{mn}^2+\left(\rho g R^2/\sigma\right)\right)}{\sqrt{\xi_{mn}^2+S_{mn}}}A_{4mn}=0 \tag{12.69}$$

The boundary conditions at the bottom (12.63) give

$$A_{1mn}\cosh\left(\frac{\xi_{mn}h}{R}\right)-A_{2mn}\sinh\left(\frac{\xi_{mn}h}{R}\right)+A_{3mn}\cosh\left(\frac{h}{R}\sqrt{\xi_{mn}^2+S_{mn}}\right)$$

$$-A_{4mn}\sinh\left(\frac{h}{R}\sqrt{\xi_{mn}^2+S_{mn}}\right)=0 \tag{12.70a}$$

$$A_{1mn}\sinh\left(\frac{\xi_{mn}h}{R}\right)-A_{2mn}\cosh\left(\frac{\xi_{mn}h}{R}\right)$$

$$+\frac{\xi_{mn}}{\sqrt{\xi_{mn}^2+S_{mn}}}\left[A_{3mn}\sinh\left(\frac{h}{R}\sqrt{\xi_{mn}^2+S_{mn}}\right)-A_{4mn}\cosh\left(\frac{h}{R}\sqrt{\xi_{mn}^2+S_{mn}}\right)\right]=0 \tag{12.70b}$$

Substituting in the shear stress conditions (12.64) at the free surface gives

$$2\xi_{mn}\sqrt{\xi_{mn}^2+S_{mn}}A_{2mn}+\left(2\xi_{mn}^2+S_{mn}\right)A_{4mn}=0 \tag{12.71}$$

Equations (12.69)–(12.71) are four homogeneous algebraic equations in the four unknowns A_{imn}, $i=1, 2, 3, 4$. The determinant of the coefficients establishes the frequency equation for the complex frequency, $\varpi_{mn}=\zeta_{mn}+i\omega_{mn}$. The resulting frequency equation is valid for zero gravity by setting $g=0$ as long as the contact angle, ϑ_c, of the contact line of the free surface is close to $\pi/2$.

Figure 12.11(a) and (b) shows the dependence of the real and imaginary parts of the liquid natural frequency, $\varpi R^2/\nu=(\zeta+i\omega)R^2/\nu$, on the fluid depth ratio, h/R, for the asymmetric modes, $m, n=1, 1$ and $1, 2$, respectively, surface tension parameter $\sigma^*=\sigma R/(\rho\nu^2)=1000$, and gravity parameter $\bar{g}=gR^3/\nu^2=10\,000$. For each figure the natural frequency of the frictionless fluid, $\bar{\omega}_0=\omega_0 R^2/\nu$, is shown to reflect the contribution of the liquid viscosity. The real part, ζ, represents the decay of the free oscillations due to liquid viscosity and the imaginary part, ω, is the frequency of free oscillations. In contrast to the frictionless liquid, it is seen that the viscous liquid experiences an aperiodic motion over a small range of fluid depth ratio, $0\le h/R\le 0.15$. With increasing mode number, n, that is, the radial modes, the aperiodic region decreases, indicating that a higher radial mode oscillates for lower liquid-depth ratio. As the depth ratio increases the natural frequency of the frictionless liquid approaches the oscillation frequency of the viscous liquid from above. Figure 12.12 shows the boundaries that separate between aperiodic and oscillatory motion regimes on the $(\sigma_{cr}^*, h/R)$ plane for different values of gravity parameter, $\bar{g}$, for $m, n=1, 1$. σ_{cr}^* is the critical value of the surface tension parameter at which the liquid ceases to oscillate. This is similar to a great extent to the critical damping ratio in linear viscous oscillators. It is seen that with decreasing gravity parameter, the aperiodic region expands for relatively small values of σ^*.

Bauer and Eidel (1997a) estimated the natural frequency of the axisymmetric oscillations, where $\partial/\partial\theta=0$ and $m=0$. The first axisymmetric mode, $n=1$, is shown in Figure 12.13 (a) and (b)

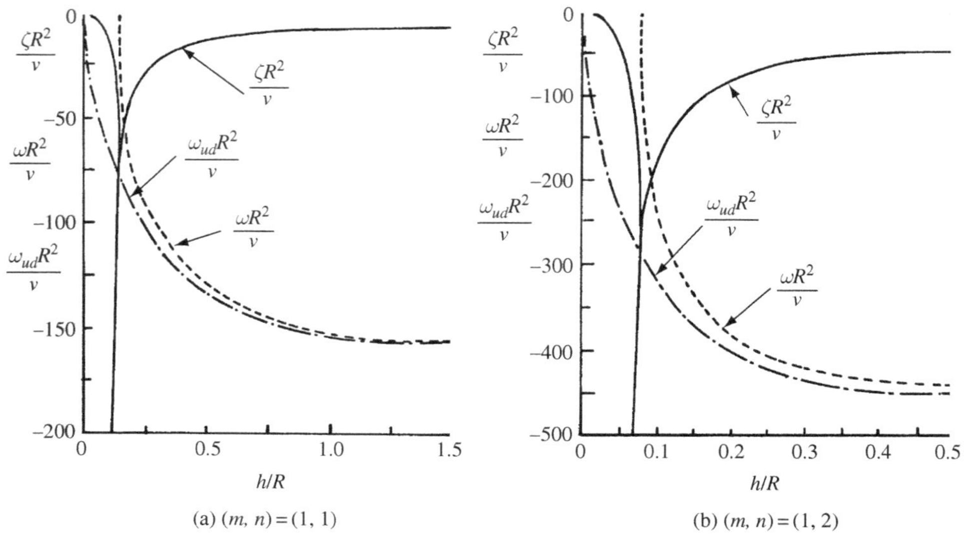

Figure 12.11 Dependence of asymmetric complex frequency components and undamped frequency on fluid depth for the case of slipping contact line, two different modes, $\sigma^* = 1000$, and $\bar{g} = 10\,000$. (Bauer and Eidel, 1997b)

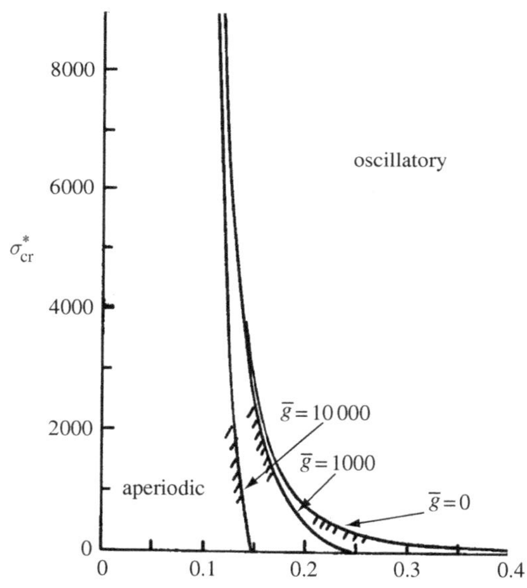

Figure 12.12 Boundaries separating aperiodic and oscillatory regimes for different values of gravity parameter and asymmetric mode $(m, n) = (1, 1)$ with slipping contact line. (Bauer and Eidel, 1997b)

for two different values of surface tension parameter $\sigma^* = 1$ and 10 000, and two gravity parameters $\bar{g} = 10\,000$ and 1000, respectively. Figure 12.14 shows the dependence of the liquid natural frequency components for a zero gravity condition. The dependence of the natural frequency components on the gravity parameter $\bar{g}$ is shown in Figure 12.15 for a zero surface

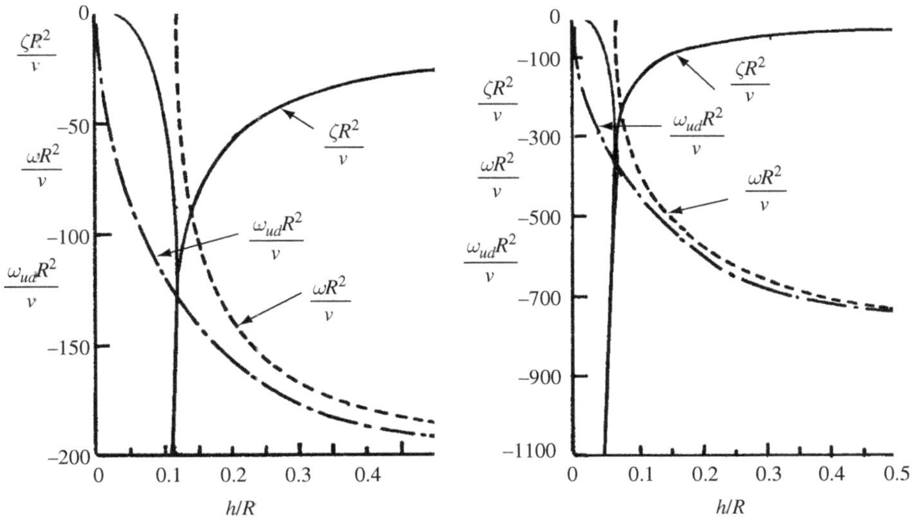

(a) $\sigma^* = 1$, $\bar{g} = 10\,000$ (b) $\sigma^* = 10\,000$, $\bar{g} = 1000$

Figure 12.13 Dependence of axisymmetric ($n = 1$) complex frequency components and undamped frequency on fluid depth for the case of slipping contact line, two different values of surface tension and gravity parameters. (Bauer and Eidel, 1997a)

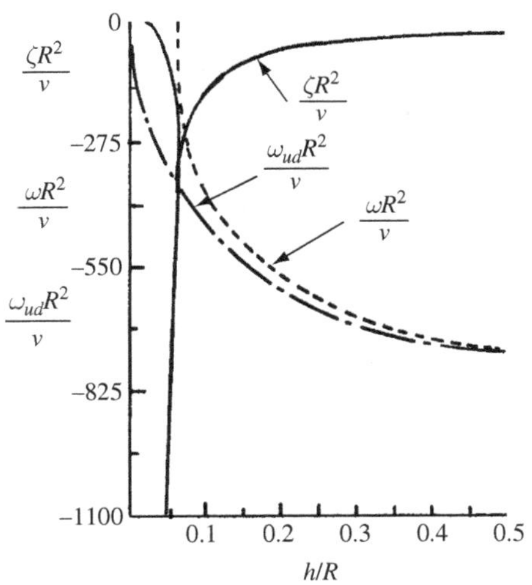

Figure 12.14 Dependence of axisymmetric ($n = 1$) complex frequency components and undamped frequency on fluid depth for the case of slipping contact line, $\sigma^* = 1000$, $\bar{g} = 0$. (Bauer and Eidel, 1997a)

tension parameter and for fluid depth ratio $h/R = 1$. With an increase of gravity, the motion of the liquid decays faster at a larger oscillation frequency, which is smaller but closer to the natural frequency of the frictionless liquid. The effect of surface tension on the liquid natural frequency components is shown in Figure 12.16 for zero gravity parameter and fluid depth ratio $h/R = 0.5$. The increase of surface tension results in a slightly stronger decay of the liquid motion and a strong increase of oscillation frequency. Figure 12.17 shows the boundaries that

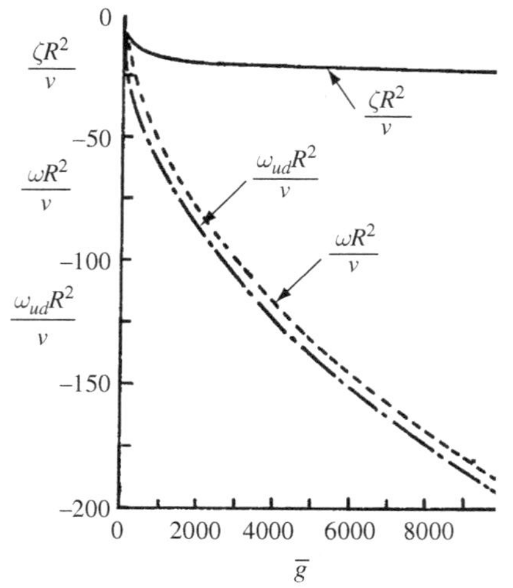

Figure 12.15 Dependence of axisymmetric ($n=1$) complex frequency components and undamped frequency on gravity parameter $\bar{g}=gR^3/\nu^2$ for the case of slipping contact line, $\sigma^*=0$, $h/R=1.0$. (Bauer and Eidel, 1997a)

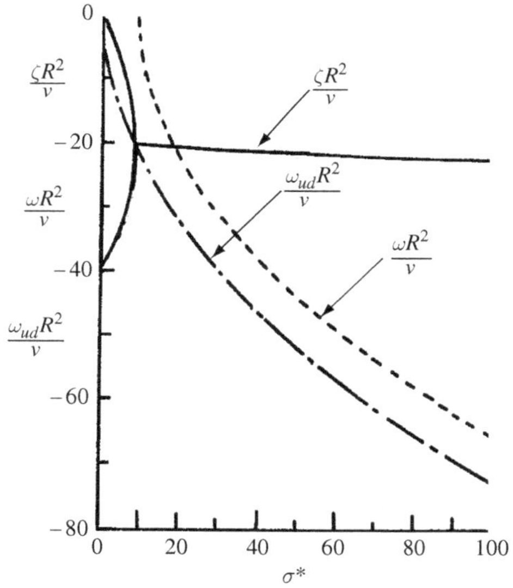

Figure 12.16 Dependence of axisymmetric ($n=1$) complex frequency components and undamped frequency on surface tension parameter $\sigma^*=\sigma R^3/(\rho\nu^2)$ for the case of slipping contact line, $\vartheta_0=\pi/2$, $h/R=0.5$. (Bauer and Eidel, 1997a)

separate aperiodic from oscillatory motion on $(\sigma^*_{\text{cr}}, h/R)$ plane for different values of gravity parameter for m, $n=0, 1$.

Anchored contact-line

For strong surface tension and a sharp rim of the container wall the contact line of the free surface remains anchored to the rim. The free-surface displacement

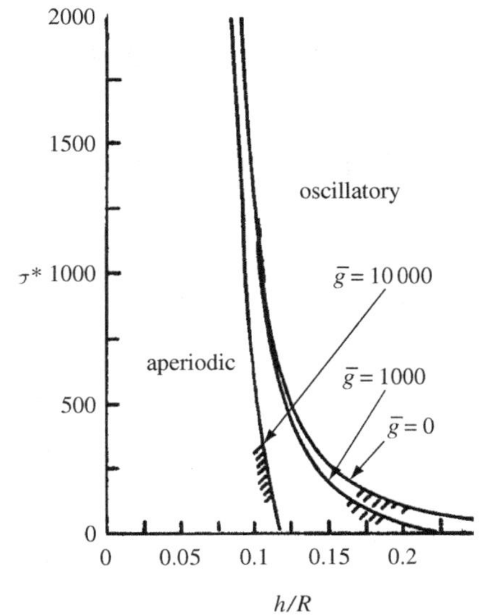

Figure 12.17 Boundaries separating aperiodic and oscillatory regimes for different values of gravity parameter and mode for axisymmetric mode, $n = 1$, with anchored contact line. (Bauer and Eidel, 1997a)

$$\eta(r, \theta, \tau) = \sum_{m=0}^{\infty} \sum_{n=1}^{\infty} \eta_{mn} e^{im\theta} e^{S_{mn}\tau} J_m(\xi_{mn} r/R) = \sum_{m=0}^{\infty} \bar{\eta}_m e^{im\theta} e^{\varpi\tau} \tag{12.72}$$

is substituted in the kinematic free-surface boundary condition (12.65) and by using equation (12.61c) at $z = 0$, gives

$$\varpi\eta_{mn} = -A_{2mn} - \frac{\xi_{mn}}{\sqrt{\xi_{mn}^2 + S_{mn}}} A_{4mn} \tag{12.73}$$

Substituting equations (12.61c), (12.62), and (12.72) into the dynamic boundary condition (12.66) gives the nonhomogeneous ordinary differential equation

$$\frac{d^2\bar{\eta}_m}{dr^2} + \frac{1}{r}\frac{d\bar{\eta}_m}{dr} - \left(\frac{m^2}{r^2} + \frac{\rho g}{\sigma}\right)\bar{\eta}_{mn} = -\frac{\mu}{\sigma R}\sum_{n=1}^{\infty}\frac{S_{mn}}{\xi_{mn}} A_{1mn} J_m(\xi_{mn} r/R)$$
$$- \frac{2\mu}{\sigma R}\sum_{n=1}^{\infty} \xi_{mn}(A_{1mn} + A_{3mn}) J_m(\xi_{mn} r/R) \tag{12.74}$$

The solution of equation (12.74) is

$$\bar{\eta}_m(r) = \bar{A}_r I_m\left(\frac{r}{R}\sqrt{\frac{\rho g R^2}{\sigma}}\right) + \frac{\rho\nu^2}{\sigma R}\sum_{n=1}^{\infty}\frac{S_{mn}\bar{A}_{1mn} J_m(\xi_{mn} r/R)}{\xi_{mn}\left(\xi_{mn}^2 + (\rho g R^2/\sigma)\right)}$$
$$+ \frac{2\rho\nu^2}{\sigma R}\sum_{n=1}^{\infty}\frac{\xi_{mn}(\bar{A}_{1mn} + \bar{A}_{3mn}) J_m(\xi_{mn} r/R)}{\left(\xi_{mn}^2 + (\rho g R^2/\sigma)\right)} \tag{12.75}$$

where $\bar{A}_{jmn} = A_{jmn}R^2/\nu$, $j = 1, 3$, and $I_m(x) = i^{-m}J_m(ix)$ is the modified Bessel function of the first kind of order m. The anchored edge condition requires $\eta = 0$ at $r = R$, and this converts equation (12.75) into the form

$$\bar{A}_m I_m\left(\sqrt{\frac{\rho g R^2}{\sigma}}\right) + \frac{\rho\nu^2}{\sigma R}\sum_{n=1}^{\infty}\frac{S_{mn}\bar{A}_{1mn}J_m(\xi_{mn})}{\xi_{mn}(\xi_{mn}^2 + (\rho g R^2/\sigma))}$$

$$+ \frac{2\rho\nu^2}{\sigma R}\sum_{n=1}^{\infty}\frac{\xi_{mn}(\bar{A}_{1mn} + \bar{A}_{3mn})J_m(\xi_{mn})}{(\xi_{mn}^2 + (\rho g R^2/\sigma))} = 0 \tag{12.76}$$

Expanding I_m into the Dini series

$$I_m\left(\frac{r}{R}\sqrt{\frac{\rho g R^2}{\sigma}}\right) = \sum_{n=1}^{\infty}\beta_{mn}J_m(\xi_{mn}r/R) \tag{12.77}$$

where $\beta_{mn} = \frac{2\xi_{mn}^2\sqrt{\rho g R^2/\sigma}\,I_m'\left(\sqrt{\rho g R^2/\sigma}\right)}{\left(\xi_{mn}^2+(\rho g R^2/\sigma)\right)\left(\xi_{mn}^2-m^2\right)J_m(\xi_{mn})}$ and comparing coefficients with those of equation (12.72) gives

$$\bar{A}_m\beta_{mn} + \frac{\rho\nu^2}{\sigma R}\frac{S_{mn}\bar{A}_{1mn}}{\xi_{mn}(\xi_{mn}^2 + (\rho g R^2/\sigma))} + \frac{2\rho\nu^2}{\sigma R}\frac{\xi_{mn}(\bar{A}_{1mn} + \bar{A}_{3mn})}{(\xi_{mn}^2 + (\rho g R^2/\sigma))}$$

$$+ \frac{1}{S_{mn}}\left(\bar{A}_{2mn} + \frac{\xi_{mn}}{(\xi_{mn}^2 + S_{mn})}\bar{A}_{4mn}\right) = 0 \tag{12.78}$$

The eigenvalue problem is now described by the $4n+1$ homogeneous algebraic equations, namely (12.70a)–(12.71), (12.76), and (12.78), in the unknowns $\bar{A}_m$, and $\bar{A}_{imn}$, $i = 1, 2, 3, 4$. The frequency equation is obtained by setting the truncated determinant of coefficients $\bar{A}_m$, and $\bar{A}_{imn}$ to zero.

For the case of zero gravity, $g = 0$, and contact angle close to $\pi/2$, the natural frequencies are estimated based on replacing $\bar{A}_m I_m$ by $\bar{A}_m(r/R)^m$, which can be expanded in the Dini series

$$\left(\frac{r}{R}\right)^m = \sum_{n=1}^{\infty}\frac{2\xi_{mn}J_{m+1}(\xi_{mn})}{(\xi_{mn}^2 - m^2)J_m^2(\xi_{mn})}J_m(\xi_{mn}r/R) \tag{12.79}$$

Figure 12.18 (a) and (b) shows the dependence of the natural frequency components on the fluid depth for two different modes $(m, n) = (1, 1)$ and $(1, 2)$, respectively, and for surface tension parameter $\sigma^* = \sigma R/(\rho\nu^2) = 1000$, and gravity parameter $\bar{g} = gR^3/\nu^2 = 10\,000$. It is seen that the oscillation frequency is larger than for the slipping case. On the other hand, the decay magnitude is decreased. The undamped natural frequency of a frictionless fluid approaches from above the damped oscillation frequency as the liquid depth ratio increases. The range of aperiodic motion takes place over a small range of fluid depth ratio $0 \le h/R \le 0.12$.

Figures 12.19 shows the boundaries separating aperiodic and oscillatory motion on the $(\sigma_{cr}^*, h/R)$ chart for different values of gravity parameter and for $m, n = 1, 1$. The figure exhibits the same features for the slipping case.

Bauer and Eidel (1997a) estimated the natural frequency of the axisymmetric oscillations, where, $m = 0$, and their results for the first axisymmetric mode, $n = 1$, are shown in

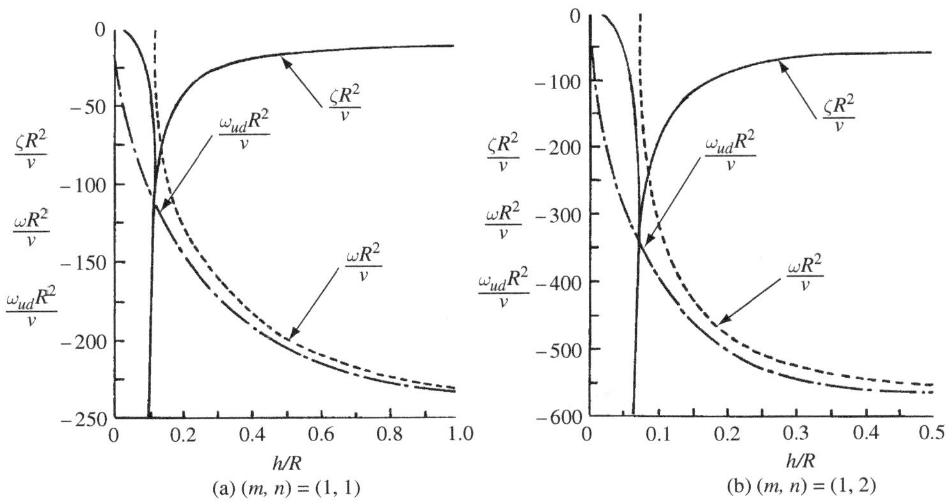

Figure 12.18 Dependence of asymmetric complex frequency components and undamped frequency on liquid-depth ratio for the case of anchored contact line, $\sigma^* = 1000$, $\bar{g} = 10\,000$ for two different modes. (Bauer and Eidel, 1997b)

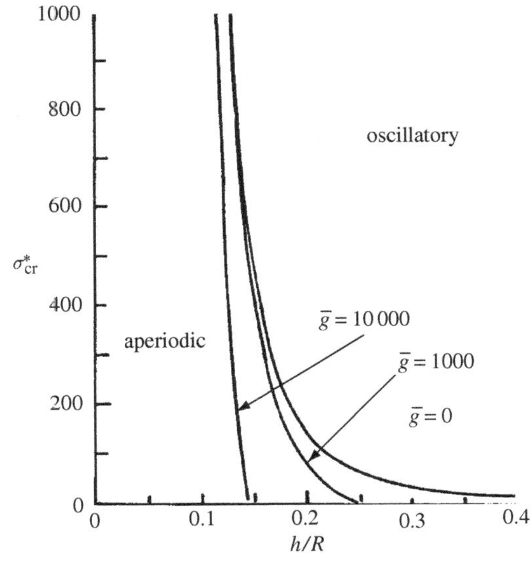

Figure 12.19 Boundaries separating aperiodic and oscillatory regimes for different values of gravity parameter asymmetric mode (1,1), with anchored contact line. (Bauer and Eidel, 1997b)

Figure 12.20(a) and (b) for two different values of surface tension parameter $\sigma^* = 1$ and $10\,000$, respectively, and gravity parameters $\bar{g} = 10\,000$ for both cases.

12.5 Forced excitation

The liquid behavior under micro- and zero-gravity environments received extensive studies. For example, Schwartz (1961) and Toole and Hastings (1968) made some attempts to study the

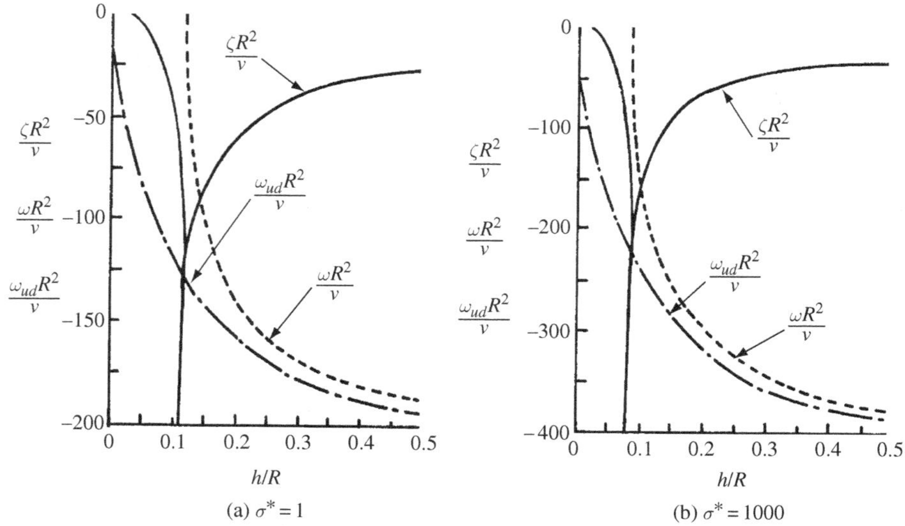

Figure 12.20 Dependence of axisymmetric complex frequency components and undamped frequency on liquid depth ratio for the case of anchored contact line, $n = 1$ and two different values of surface tension parameter, $\bar{g} = 10\ 000$. (Bauer and Eidel, 1997a)

liquid behavior under zero gravity and sudden reduction in axial acceleration. Hung and Pan (1994a) studied the asymmetric sloshing modes, while Moiseev and Chernousko (1965), Hung, Shyu, and Lee (1991, 1992a,b), Hung and Shyu (1992b,c, 1995), and Miles (1996) considered the free-surface response under impulsive thrust. Shen, *et al.* (1993) studied the forced capillary-gravity waves in a circular basin. In order to determine the response of the liquid free surface to sinusoidal lateral excitation, equations (12.51) and (12.52) are considered. The pressure distribution for case $m = 1$ is given by the expression

$$p(r,\theta,z,t) = p_0 - \rho g z + \rho P(r,z)\cos\theta\, e^{\varpi t} - \rho X_0 \Omega^2 r \cos\theta\, e^{i\Omega t} \tag{12.80}$$

Introducing the transformation (12.56) and dropping the subscript, m, equations (12.51) and (12.52) take the form

The continuity equation is

$$\frac{\partial \Phi}{\partial r} + \frac{2}{r}\Phi + \frac{\partial \Psi}{\partial r} = -2\frac{\partial W}{\partial z} \tag{12.81}$$

The Stokes equations are

$$\frac{\partial^2 \Phi}{\partial r^2} + \frac{1}{r}\frac{\partial \Phi}{\partial r} - \left[\frac{4}{r^2} + \frac{i\Omega}{\nu}\right]\Phi + \frac{\partial^2 \Phi}{\partial z^2} = \frac{1}{\nu}\left[\frac{\partial}{\partial r} - \frac{1}{r}\right]P \tag{12.82a}$$

$$\frac{\partial^2 \Psi}{\partial r^2} + \frac{1}{r}\frac{\partial \Psi}{\partial r} - \frac{i\Omega}{\nu}\Psi + \frac{\partial^2 \Psi}{\partial z^2} = \frac{1}{\nu}\left[\frac{\partial}{\partial r} + \frac{1}{r}\right]P \tag{12.82b}$$

$$\frac{\partial^2 W}{\partial r^2} + \frac{1}{r}\frac{\partial W}{\partial r} - \left[\frac{1}{r^2} + \frac{i\Omega}{\nu}\right]W + \frac{\partial^2 W}{\partial z^2} = \frac{1}{\nu}\frac{\partial P}{\partial z} \tag{12.82c}$$

The solution of these equations, which satisfies the boundary condition $u=0$ at $r=R$ may be written in the form

$$\Phi(r,z) = A_{in}J_2(\xi_{1n}r/R),\ \ \Psi(r,z) = -A_{in}J_0(\xi_{1n}r/R)$$

$$W(r,z) = C_{in}J_1(\xi_{1n}r/R),\ \ P(r,z) = D_{in}J_1(\xi_{1n}r/R) \tag{12.83a–d}$$

where ξ_{1n} are the roots of $J_1'\ (\xi_{1n})=0$. Using the recurrence relations $J_0 - J_2 = 2J_1'$, and $J_0 + J_2 = 2RJ_1/\xi_{1n}r$, the velocity distributions based on definitions (12.53) and replacing ϖ by $i\Omega$, may be written in the form

$$u(r,\theta,z,t) = -\sum_{n=1}^{\infty} J_1'(\xi_{1n}r/R)\cos\theta\, e^{i\Omega t}\left\{A_{1n}\cosh(\xi_{1n}z/R) + A_{2n}\sinh(\xi_{1n}z/R)\right.$$
$$+ A_{3n}\cosh\left(\frac{z}{R}\sqrt{\xi_{1n}^2 + \frac{i\Omega R^2}{\nu}}\right)$$
$$\left.+ A_{4n}\left(\frac{z}{R}\sqrt{\xi_{1n}^2 + \frac{i\Omega R^2}{\nu}}\right)\right\} \tag{12.84a}$$

$$v(r,\theta,z,t) = \sum_{n=1}^{\infty} \frac{J_1(\xi_{1n}r/R)}{\xi_{1n}r/R}\sin\theta\, e^{i\Omega t}\left\{A_{1n}\cosh(\xi_{1n}z/R) + A_{2n}\sinh(\xi_{1n}z/R)\right.$$
$$+ A_{3n}\cosh\left(\frac{z}{R}\sqrt{\xi_{1n}^2 + \frac{i\Omega R^2}{\nu}}\right)$$
$$\left.+ A_{4n}\sinh\left(\frac{z}{R}\sqrt{\xi_{1n}^2 + \frac{i\Omega R^2}{\nu}}\right)\right\} \tag{12.84b}$$

$$w(r,\theta,z,t) = -\sum_{n=1}^{\infty} J_1(\xi_{1n}r/R)\cos\theta\, e^{i\Omega t}$$
$$\times\left\{A_{1n}\sinh(\xi_{1n}z/R) + A_{2n}\cosh(\xi_{1n}z/R) + \frac{\xi_{1n}}{\sqrt{\xi_{1n}^2 + i\Omega R^2/\nu}}\right.$$
$$\left.\left[A_{3n}\sinh\left(\frac{z}{R}\sqrt{\xi_n^2 + \frac{i\Omega R^2}{\nu}}\right) + A_{4n}\cosh\left(\frac{z}{R}\sqrt{\xi_{1n}^2 - \frac{i\Omega R^2}{\nu}}\right)\right]\right\} \tag{12.84c}$$

The pressure distribution is given in the form

$$p(r,\theta,z,\tau) = p_0 - \rho g z - \rho X_0\Omega^2 r\cos\theta\, e^{i\Omega t}$$
$$+ \frac{\mu}{R}\sum_{n=1}^{\infty}\frac{i\Omega R^2}{\nu\xi_{1n}} J_1\left(\frac{\xi_{1n}r}{R}\right)\cos\theta\, e^{i\Omega t}\times\left\{A_{1n}\cosh\left(\frac{\xi_{1n}z}{R}\right) + A_{2n}\sinh\left(\frac{\xi_{1n}z}{R}\right)\right\} \tag{12.85}$$

The integration constants are obtained by satisfying the boundary conditions at the bottom (12.63) and the vanishing shear stresses at the free surface (12.64). The following algebraic equations are obtained

$$A_{1n}\cosh\left(\frac{\xi_{1n}h}{R}\right) - A_{2n}\sinh\left(\frac{\xi_{1n}h}{R}\right) + A_{3n}\cosh\left(\frac{h}{R}\sqrt{\xi_{1n}^2 + \frac{i\Omega R^2}{\nu}}\right)$$
$$- A_{4n}\sinh\left(\frac{h}{R}\sqrt{\xi_{1n}^2 + \frac{i\Omega R^2}{\nu}}\right) = 0 \tag{12.86a}$$

$$A_{1n}\sinh\left(\frac{\xi_{1n}h}{R}\right) - A_{2n}\cosh\left(\frac{\xi_{1n}h}{R}\right) + \frac{\xi_{1n}}{\sqrt{\xi_{1n}^2 + i\Omega R^2/\nu}}$$
$$\times \left[A_{31n}\sinh\left(\frac{h}{R}\sqrt{\xi_{1n}^2 + \frac{i\Omega R^2}{\nu}}\right) - A_{4n}\cosh\left(\frac{h}{R}\sqrt{\xi_{1n}^2 + \frac{i\Omega R^2}{\nu}}\right)\right] = 0 \tag{12.86b}$$

$$2\xi_{1n}\sqrt{\xi_{1n}^2 + \frac{i\Omega R^2}{\nu}}A_{2n} + \left(2\xi_{1n}^2 + \frac{i\Omega R^2}{\nu}\right)A_{4n} = 0 \tag{12.86c}$$

The response of the free surface is considered for two conditions of the contact line in the next sections.

12.5.1 Slipping contact line

The response of the free-surface displacement for the case of a slipping contact line may be estimated by integrating the kinematic free-surface condition (12.65) to give

$$\eta(r,\theta,t) = e^{i\Omega t}\cos\theta\sum_{n=1}^{\infty}\eta_{1n}J_m(\xi_{1n}r/R) \tag{12.87}$$

where η_{1n} is a function of the constants A_{in}. The combined free-surface boundary condition takes the form

$$i\Omega p - 2i\Omega\mu\frac{\partial w}{\partial z} + \sigma\left[\frac{\partial^2 w}{\partial r^2} + \frac{1}{r}\frac{\partial w}{\partial r} + \frac{1}{r^2}\frac{\partial^2 w}{\partial \theta^2} - \frac{\rho g}{\sigma}w\right] = 0 \tag{12.88}$$

Substituting the velocity components and pressure distribution (12.84) and (12.85) in the combined free-surface condition (12.88), gives

$$(\bar{\Omega}^2 + 2\xi_{mn}^2\bar{\Omega})A_{1n} + \frac{\sigma R}{\rho\nu^2}\xi_{mn}\left(\xi_{mn}^2 + \frac{\rho g R^2}{\sigma}\right)A_{2n}$$
$$+ 2\xi_{mn}^2\bar{\Omega}A_{3n}\frac{\sigma R}{\rho\nu^2}\frac{\xi_{1n}^2(\xi_{1n}^2 + (\rho g R^2/\sigma))}{\sqrt{\xi_{1n}^2 + \bar{\Omega}}}A_{4n}$$
$$= -\frac{2\bar{\Omega}^3\xi_{1n}\nu X_0}{(\xi_{1n}^2 - 1)R^2 J_1(\xi_{1n})} \tag{12.89}$$

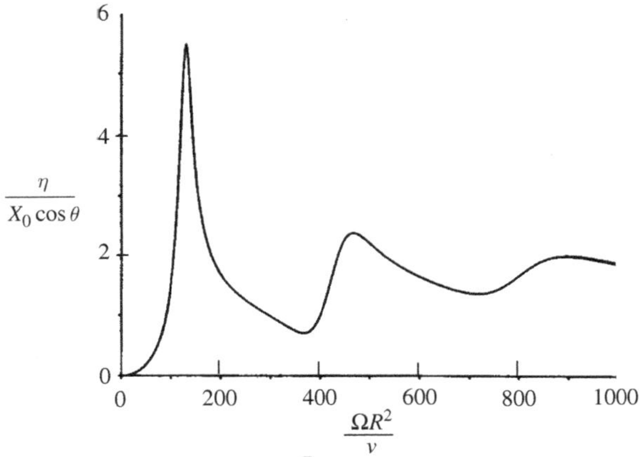

(a) $\sigma^* = \frac{\sigma R}{\rho \nu^2} = 1000$, $\bar{g} = gR^3/\nu^2 = 10\,000$

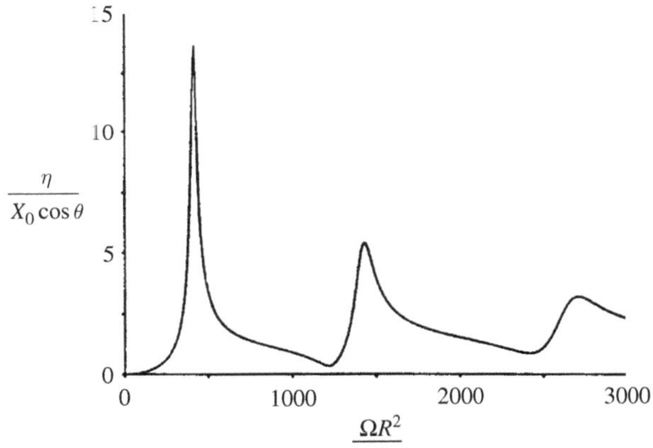

(b) $\sigma^* = \frac{\sigma R}{\rho \nu^2} = 10\,000$, $\bar{g} = gR^3/\nu^2 = 100\,000$

Figure 12.21 Dependence of the liquid free-surface displacement on the excitation frequency at $r = R$ for the case of slipping contact line, $h/R = 0.5$, and different surface tension and gravity parameters. (Bauer and Eidel, 1997c)

where $\bar{\Omega} = i\Omega R^2/\nu$. In developing equation (12.89) the following Dini expansion has been used

$$\frac{r}{R} = 2\sum_{n=1}^{\infty} \frac{J_1(\xi_{1n}r/R)}{(\xi_{1n}^2 - 1)J_1(\xi_{1n})} \tag{12.90}$$

The constants of integration A_{in} are obtained by solving the nonhomogeneous algebraic equations (12.86) and (12.89). Having obtained these constants one can estimate the velocity and pressure distributions and the free-surface displacement.

Figures 12.21(a) and (b) show the dependence of the liquid free-surface displacement ratio, $\eta/X_0 \cos\theta$, on the excitation frequency parameter, $\Omega R^2/\nu$, for two different sets of surface tension parameter and gravity parameter $(\sigma^*, \bar{g}) = (1000, 10\,000)$, and $(10\,000, 100\,000)$, respectively. Both figures are obtained for liquid depth ratio $h/R = 0.5$ and the liquid free displacement is estimated at $r=R$. It is seen that the first asymmetric mode experiences larger displacement at resonance than the other two next highest modes.

12.5.2 Anchored contact-line

For an anchored contact line, one may write the free-surface displacement (12.87) in the form

$$\eta(r,\theta,t)=e^{i\Omega t}\cos\theta\sum_{n=1}^{\infty}\eta_{1n}J_m(\xi_{1n}r/R)=e^{i\Omega t}\cos\theta\bar{\eta}_1(r) \tag{12.91}$$

Substituting (12.91) and (12.84c) into the kinematic boundary condition (12.65) gives

$$i\Omega\eta_{1n}=-A_{2n}-\frac{\xi_{1n}}{\sqrt{\xi_{1n}^2+\bar{\Omega}}}A_{4n} \tag{12.92}$$

Substituting the same in the dynamic condition (12.66) gives

$$\begin{aligned}\frac{d^2\bar{\eta}_1}{dr^2}+\frac{1}{r}\frac{d\bar{\eta}_1}{dr}-\left(\frac{1}{r^2}+\frac{\rho g}{\sigma}\right)\bar{\eta}_1 &= -\frac{\mu}{\sigma R}\sum_{n=1}^{\infty}\frac{\bar{\Omega}}{\xi_{1n}}A_{1n}J_1(\xi_{1n}r/R)\\ &\quad-\frac{2\mu}{\sigma R}\sum_{n=1}^{\infty}\xi_{1n}(A_{1n}+A_{3n})J_1(\xi_{1n}r/R)+\frac{\rho X_0\Omega^2}{\sigma}r\end{aligned} \tag{12.93}$$

The solution of this equation may be written in the form

$$\begin{aligned}\bar{\eta}_1(r)=\bar{A}_1I_1\left(\frac{r}{R}\sqrt{\frac{\rho gR^2}{\sigma}}\right)&+\frac{\rho\nu^2}{\sigma R}\sum_{n=1}^{\infty}\frac{\bar{\Omega}\bar{A}_{1n}J_1(\xi_{1n}r/R)}{\xi_{1n}(\xi_{1n}^2+\rho gR^2/\sigma)}\\ &+\frac{2\rho\nu^2}{\sigma R}\sum_{n=1}^{\infty}\frac{\xi_{1n}(\bar{A}_{1n}+\bar{A}_{3n})J_1(\xi_{1n}r/R)}{(\xi_{1n}^2+\rho gR^2/\sigma)}-\frac{\Omega^2R}{g}X_0\left(\frac{r}{R}\right)\end{aligned} \tag{12.94}$$

Imposing the anchored edge condition $\eta=0$ at $r=R$ gives

$$\begin{aligned}\bar{A}_1I_1\left(\sqrt{\frac{\rho gR^2}{\sigma}}\right)&+\frac{\rho\nu^2}{\sigma R}\sum_{n=1}^{\infty}\frac{\bar{\Omega}\bar{A}_{1n}J_1(\xi_{1n})}{\xi_{1n}(\xi_{1n}^2+\rho gR^2/\sigma)}\\ &+\frac{2\rho\nu^2}{\sigma R}\sum_{n=1}^{\infty}\frac{\xi_{1n}(\bar{A}_{1n}+\bar{A}_{3n})J_1(\xi_{1n})}{(\xi_{1n}^2+\rho gR^2/\sigma)}=\frac{\Omega^2R}{g}X_0\end{aligned} \tag{12.95}$$

Expanding I_1 into a Bessel–Fourier series

$$I_1\left(\frac{r}{R}\sqrt{\frac{\rho gR^2}{\sigma}}\right)=\sum_{n=1}^{\infty}a_nJ_1(\xi_{1n}r/R) \tag{12.96}$$

where

$$a_n=\frac{2\xi_{1n}^2\sqrt{\rho gR^2/\sigma}I_1'\left(\sqrt{\rho gR^2/\sigma}\right)}{\left(\xi_{1n}^2+(\rho gR^2/\sigma)\right)\left(\xi_{1n}^2-1\right)J_1(\xi_{1n})}$$

and comparing equation (12.94) with the result of the kinematic equation (12.92) gives

$$\bar{A}_1 a_n + \frac{\rho\nu^2}{\sigma R}\frac{\bar{\Omega}\bar{A}_{1n}}{\xi_{1n}(\xi_{1n}^2 + (\rho g R^2/\sigma))} + \frac{2\rho\nu^2}{\sigma R}\frac{\xi_{1n}(\bar{A}_{1n} + \bar{A}_{3n})}{(\xi_{mn}^2 + (\rho g R^2/\sigma))}$$
$$+ \frac{1}{\bar{\Omega}}\left(\bar{A}_{2n} + \frac{\xi_{1n}}{\sqrt{\xi_{mn}^2 + \bar{\Omega}}}\bar{A}_{4n}\right) = \frac{2\Omega^2 R}{(\xi_{1n}^2 - 1)J_1(\xi_{1n})g}X_0 \tag{12.97}$$

Equations (12.86a–c), (12.95), and (12.97) constitute $4n+1$ nonhomogeneous algebraic equations in the unknowns $\bar{A}_1$, $\bar{A}_{in}$, $i = 1, 2, 3, 4$, which have to be truncated. The estimation of $\bar{A}_1$, $\bar{A}_{in}$ makes it possible to determine the velocity and pressure distributions and the free-surface displacement under lateral sinusoidal excitation.

For the case of zero gravity, $g = 0$, the combined free-surface condition (12.93) takes the form

$$\frac{d^2\bar{\eta}_1}{dr^2} + \frac{1}{r}\frac{d\bar{\eta}_1}{dr} - \frac{1}{r^2}\bar{\eta}_1 = -\frac{\mu}{\sigma R}\sum_{n=1}^{\infty}\frac{\bar{\Omega}}{\xi_{1n}}A_{1n}J_1(\xi_{1n}r/R)$$
$$-\frac{2\mu}{\sigma R}\sum_{n=1}^{\infty}\xi_{1n}(A_{1n} + A_{3n})J_1(\xi_{1n}r/R) + \frac{\rho X_0\Omega^2}{\sigma}r \tag{12.98}$$

The solution of this equation is

$$\bar{\eta}_1(r) = \bar{A}_1\frac{r}{R} + \frac{\rho\nu^2}{\sigma R}\sum_{n=1}^{\infty}\frac{\bar{\Omega}\bar{A}_{1n}J_1(\xi_{1n}r/R)}{\xi_{1n}^3} + \frac{2\rho\nu^2}{\sigma R}\sum_{n=1}^{\infty}\frac{(\bar{A}_{1n} + \bar{A}_{3n})J_1(\xi_{1n}r/R)}{\xi_{1n}}$$
$$+ \frac{\rho\bar{\Omega}^2 X_0 R^3}{8\sigma}\left(\frac{r}{R}\right)^3 \tag{12.99}$$

The anchored end condition $\bar{\eta}_1 = 0$ at $r = R$ gives

$$\bar{A}_1 + \frac{\rho\nu^2}{\sigma R}\sum_{n=1}^{\infty}\frac{\bar{\Omega}J_1(\xi_{1n})}{\xi_{1n}^3}\bar{A}_{1n} + \frac{2\rho\nu^2}{\sigma R}\sum_{n=1}^{\infty}\frac{J_1(\xi_{1n})}{\xi_{1n}}(\bar{A}_{1n} + \bar{A}_{3n}) = -\frac{\rho}{\sigma}\Omega^2 R^3 X_0 \tag{12.100}$$

Expanding r/R and $(r/R)^3$ in equation (12.99) into Dini series as given by series (12.79) and (12.90) and comparing coefficients of $J_1(\xi_{1n}r/R)$ gives

$$\frac{2\bar{A}_1}{(\xi_{1n}^2 - 1)J_1(\xi_{1n})} + \frac{\rho\nu^2\bar{\Omega}}{\sigma R\xi_{1n}^3}\bar{A}_{1n} + \frac{2\rho\nu^2}{\sigma R\xi_{1n}}(\bar{A}_{1n} + \bar{A}_{3n}) + \frac{1}{\bar{\Omega}}\left[\bar{A}_{2n} + \frac{\xi_{1n}\bar{A}_{4n}}{\sqrt{\xi_{1n} + \bar{\Omega}}}\right]$$
$$= -2\rho\Omega^2 R^3 X_0\frac{[\xi_{1n}J_2(\xi_{1n}) - 2J_3(\xi_{1n})]}{\sigma(\xi_{1n}^2 - 1)J_1^2(\xi_{1n})} \tag{12.101}$$

Equations (12.86a–c), (12.100) and (12.101) constitute $4n+1$ nonhomogeneous equations in the unknowns $\bar{A}_1$, $\bar{A}_{in}$, $i = 1, 2, 3, 4$.

Figures 12.22(a) and (b) show the dependence of the liquid free-surface displacement parameter $\eta/X_0\cos\theta$ on the excitation frequency parameter $\Omega R^2/\nu$ for two different sets of

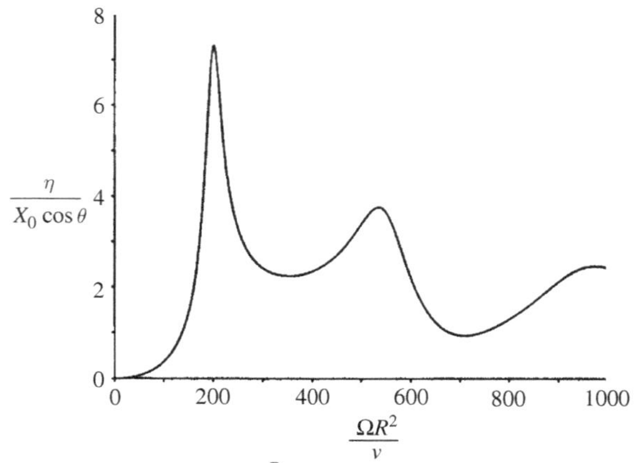

(a) $\sigma^* = \frac{\sigma R}{\rho\nu^2} = 1000$, $\bar{g} = gR^3/\nu^2 = 10\,000$

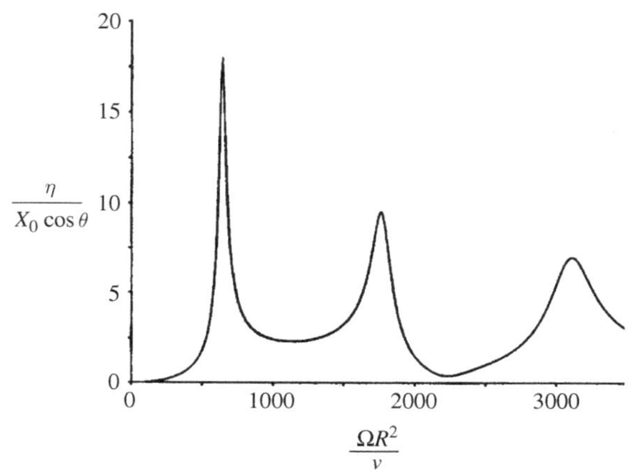

(b) $\sigma^* = \frac{\sigma R}{\rho\nu^2} = 10\,000$, $\bar{g} = gR^3/\nu^2 = 100\,000$

Figure 12.22 Dependence of the liquid free-surface displacement on the excitation frequency at $r = R$ for the case of anchored contact line, $h/R = 0.5$, and different surface tension and gravity parameters. (Bauer and Eidel, 1997c)

surface tension parameter and gravity parameter $(\sigma^*, \bar{g}) = (1000, 10\,000)$, and $(10\,000, 100\,000)$, respectively. Both figures are obtained for liquid depth ratio $h/R = 0.5$ and the liquid free displacement is estimated at $r = R/2$. It is seen that the first asymmetric mode experiences larger displacement at resonance than the other two next highest modes.

12.6 G-jitter modeling and effects

During space missions, microgravity experiments revealed significant levels of residual accelerations referred to as g-jitter (Walter, 1987, Nelson, 1991, Thomson, *et al.*, 1996, Savino and Monti, 1996, 1999, 2001, DeLombard, *et al.* 1997, and Savino, 1997, 2000). The residual acceleration field can be decomposed into a quasi-steady (or systematic component) and a fluctuating component known as g-jitter. Typical values of the quasi-steady component $\|g_s\|$ are around $10^{-6}g_E$ (g_E is the gravitational acceleration on the Earth's surface). The fluctuating contribution, $g(t)$, is random in nature and has a characteristic frequency of 1 Hz or higher. Alexander (1990) provided experimental results pertaining to the sensitivity of liquid behavior to residual acceleration at a low gravitational field.

The influence of g-jitter on fluid flow ranges from order of magnitude estimates to detailed numerical calculations (see, e.g., Alexander 1990 and Monti, 1990b). These include modeling the acceleration field by some simple analytic functions in which the acceleration is typically decomposed into steady and time-dependent components. The time-dependent component was considered periodic by Gershuni and Zhukhovitskii (1976, 1986), Kamotani, *et al.* (1981), Alexander, *et al.* (1991), Kamotani and Platt (1992), Farooq and Homsy (1994), Gershuni and Lyubimov (1998), and Grassia and Homsy (1998a,b). Alexander, *et al.* (1997) addressed the consequences of isolated pulses of short duration. Statistical description of the residual acceleration field on board spacecraft was considered by Zhang, *et al.* (1993) and Thomson, *et al.* (1995, 1996). Jacqmin and Duval (1988) studied the instabilities of the interface between two immiscible viscous fluids caused by oscillatory accelerations. Jacqmin (1990) extended this work to a fluid layer with a uniform density gradient and explicitly considered components of the gravity vector parallel and normal to the density gradient.

The effects of g-jitter on the capillary surface motion were studied by Chao, *et al.* (1992). The current state-of-knowledge of g-jitter problems in microgravity was documented by Nelson (1991). Each spatial Fourier mode of the free-surface displacement in a slightly viscous fluid satisfies the equation of a damped parametric harmonic oscillator in the linear regime. Zhang, *et al.* (1993) described the free-surface motion by a linear differential equation with a random coefficient and examined the stochastic stability boundary of the equilibrium state of the form

$$\frac{d^2\hat{\eta}}{dt^2} + 4\nu k^2 \frac{d\hat{\eta}}{dt} + k[g_0 - g_z(t)]\hat{\eta} + \frac{\sigma}{\rho} k^3 \hat{\eta} = 0 \tag{12.102}$$

where $\hat{\eta} = \hat{\eta}(\mathbf{k}, t)$ is the two-dimensional Fourier transform of the interface displacement $\eta(x, y, t)$, ν is the fluid kinematic viscosity, k is the characteristic wave number, g_0 is a constant background gravitational field, and $g_z(t)$ is the fluctuating component of the residual acceleration in the z-direction. Equation (12.102) can be written in the form

$$\frac{d^2\hat{\eta}}{dt^2} + \gamma \frac{d\hat{\eta}}{dt} + [\omega_0^2 - \xi(t)]\hat{\eta} = 0 \tag{12.103}$$

where $\gamma = 4\nu k^2$, ${\omega_0}^2 = g_0 k + \sigma k^3/\rho$, and $\xi(t) = -kg_z(t)$. Zhang, *et al.* (1993) modeled the residual acceleration field, $\xi(t)$, as a zero-mean Gaussian narrow-band random noise defined by the expression

$$\xi(t) = G_1(t)\cos\Omega t + G_2(t)\sin\Omega t \tag{12.104}$$

where $G_1(t)$ and $G_2(t)$ are two independent stationary Gaussian processes possessing zero-means and the following correlation function

$$R_G(t - t') = <\xi^2> \delta_{ij}\, e^{-|t-t'|/\tau}, \quad i, j = 1,\ 2 \tag{12.105}$$

where τ is a correlation time. The correlation function of $\xi(t)$ is

$$R_{\xi}(t-t') = <\xi^2> e^{-|t-t'|/\tau} \cos[\Omega(t-t')] \tag{12.106}$$

The corresponding power spectral density function is

$$S_g(\omega) = \frac{1}{2\pi} <g^2> \left[\frac{1}{1+\tau^2(\Omega+\omega)^2} + \frac{1}{1+\tau^2(\Omega-\omega)^2}\right] \tag{12.107}$$

Analytical results for the stability of the response second moment were presented in the limits of low-frequency oscillations and near the region of subharmonic parametric resonance. Casademunt, *et al.* (1993) extended the work of Zhang, *et al.* (1993) and included an additive term for the lateral acceleration component. For a given spectrum of g-jitter, they found a band of unstable modes at low frequencies.

Hung, Lee, and Leslie (1993a) considered Coriolis force, angular acceleration, and centrifugal, viscous, and surface tension forces in their numerical algorithm. They numerically solved the liquid sloshing dynamics using the method of marker-and cell. Hung and Pan (1995a,b) examined three different cases of liquid responses in a spinning spacecraft: (i) when gravity gradient is the dominant driving acceleration, (ii) when g-jitter is the dominant acceleration, and (iii) when g-jitter and gravity gradient accelerations are comparable. Their numerical simulation showed that large-amplitude fluctuating fluid forces and torques could be exerted on a spacecraft. Other studies pertaining to the influence of g-jitter include Hung, Lee and Leslie (1990a,b, 1992a), Hung, Lee and Wu (1990), Hung, Lee and Shyu (1991), Hung, Long, and Pan (1994), Hung and Pan (1994b), and Hung, Pan, and Leslie (1994a,b).

Bauer (1984c) indicated that both oscillatory thermo-capillary flow and g-jitter generated flow are related to free-surface motion and they might interact with each other. Both may also have a detrimental effect on the material processing. The susceptibility of melts to g-jitter disturbance not only enhances the displacement of the free surface, but also changes the steady heat transfer during the material processing to an irregular one, which may be harmful to the crystal growth process (Kamotani and Ostrach, 1987). The g-jitter is generated from a number of sources outlined by Ostrach (1982).

Spradley, *et al.* (1975), Kamotani, *et al.* (1981), and Danabasoglu and Biringen (1988) studied different problems dealing with g-jitter induced convection in a completely enclosed container. However, in microgravity containerless materials processing, the dynamics of free surface can no longer be neglected. In the linearized study of free surface, Curvelier (1985), and Curvelier and Schulkes (1990) showed that the free-surface natural frequency is dependent on the liquid viscosity. However, in Curvelier's study, the contact angle was assumed to be fixed at 90°, no other contact-line conditions; nonlinear effects were considered. Veldman and Vogels (1984) showed that, using the Marker-and-Cell dimensional calculation of the axisymmetrical sloshing in a low-gravity condition, the liquid free-surface natural frequencies are dependent on the contact angle. However, no variation of the dynamic contact angle was allowed during sloshing in their calculations.

Chao, *et al.* (1992) studied the effects of the lateral g-jitter on the free-surface deformation of liquid in open containers. They incorporated a hysteresis parameter into the SOLA-SURF FE scheme in order to allow for different contact-line conditions. It was shown that, for all conditions of the contact-line, the g-jitter effect is most intense only in the free-surface layer then the effect diminishes very fast as the liquid depth increases. As the Ohnesorge number increases the free-surface deformation decreases.

Drolet and Viñals (1998), and Volfson and Viñals (2001) discussed the flow-induced oscillation in an otherwise quiescent fluid by the random vibration of a solid boundary. The velocity boundary was assumed to be a narrow-band stochastic process. The variance of the velocity field decays exponentially away from the wall with a characteristic decay length given by the Stokes layer thickness, $\delta_s = \sqrt{2\nu/\Omega}$, where Ω is the angular frequency of the boundary.

12.7 Liquid handling

A major problem in the design of orbital propellant transfer is vapor pull-through caused by large-amplitude deformation of the liquid surface during outflow. Pull-through results in a portion of the propellant in the tank being unstable, either for transfer to another tank or for engine supply. During liquid drainage, the free liquid surface experiences distortion (Bhuta and Koval, 1965, Gluck and Gille, 1965, Gluck, *et al.*, 1966, Gold, *et al.*, 1967, Easton and Catton, 1970, and Amkhin, *et al.*, 1975). This is a serious problem in systems incorporating turbo-pumps, because of the potential danger of pump cavitations and destruction if gas is introduced into the system. Saad and Oliver (1964) and Bhuta and Koval (1965) analytically examined the interaction between the surface deformation and the average flow rate. Andracchio and Abdalla (1968) conducted an experiment to study the fluid flow in a baffled spherical container in a microgravity field. Lubin and Springer (1967) established a momentum balance at the outlet and were able to relate the parameters of the problem with experimental data. They obtained an empirical relationship between critical fluid height and outflow rate. Experimental investigations were conducted in a low gravitational field by Nussle, *et al.* (1965), Derdul, *et al.* (1966), Gluck, *et al.* (1966), and Grubb and Petrash (1967) using a drop tower. Most of the data on pull-through depths have been found to correlate quite well with Froude numbers for the range of conditions most suitable to the transfer of propellants in large vehicles. At low Bond numbers, the experimental results of Gluck, *et al.* (1966) indicated that either the Bond number or Weber number must be included as another parameter. Easton and Catton (1970) considered the full nonlinear dynamic and kinematic boundary conditions in addition to the influence of surface tension. They established the range of applicability of the linear theory. Lacovic, *et al.* (1968), Dodge and Bowles (1984), Defelice (1987), Dodge (1990, 1991), and Guadagnoli (1991) presented different studies of fluid management technology and control of liquid sloshing.

Bauer (1971b) and Hung, Tsao, and Hong (1988a) analyzed the migration of a large gas-bubble under a microgravitational field. Ward, *et al.* (1992) outlined three possible interfacial configurations of a vapor-fluid system in an isothermal cylindrical container subjected to a reduced gravitational intensity. The first one is the "single interface" which is normally observed under the regular gravitational field. The curvature of the interface would be altered at the reduced gravitational field. If the gravitational intensity is sufficiently reduced, this configuration becomes metastable, and as a result the system makes a transition out of this configuration based on the Bond number criterion. The second takes place in the form of a "two-interface" configuration in which a portion of the liquid phase would be above the vapor phase and a portion below. The third possibility is for the system to adopt the "bubble-configuration" in which a vapor volume is present and is surrounded by the liquid phase.

Ward, *et al.* (1992) conducted thermodynamic analysis to predict the equilibrium configuration for the liquid when subjected to near zero-gravitational intensity. They found that when the system is at zero gravity, the thermodynamic theory leads to certain values of the contact angle less than 35.83°, the single-interface configuration is metastable and either the two-interface or the bubble-configuration is the stable configuration.

McCaughan and Bedir (1992) studied the influence of the capillary number (Ohnesorge number) $\mu\nu/(\sigma L)$, where L is the fluid layer mean depth, which measures the ability of the free surface to deform, on the linear stability of a single layer of fluid with double diffusion. When viewed in the space of the Marangoni temperature and concentration numbers, the base conduction solution is stable when both of these numbers are negative. Usually, increasing the capillary number will result in a decrease of the size of the stable region. A nonzero capillary number also has a strong effect on the value of critical wave number, which in turn influences the nature of the new flow that will develop when the base conduction solution loses stability.

The influence of surface tension and viscosity on Taylor instability was analyzed by Bellman and Pennington (1954). Schultz, *et al.*, (1998) found that the presence of surface tension prevents the formation of cusped waves. Wright, *et al.* (2000) indicated that the presence of higher harmonics in the case of strong parametric excitation is associated with the formation of paired waves traveling within each period. The constructive interference of these waves yields momentary geometrical peaks with high curvature. When the curvature of the interface becomes large, sudden local acceleration occurs due to the effects of surface tension. In the case of a free surface, but not an interface, weak viscous effects at high Reynolds numbers may be accounted for by means of boundary-layers analysis. However, there are difficulties associated with the seaming of the boundary layers at free-surface stagnation points. In this case, direct numerical simulation based on the Navier–Stokes equations is convenient for studying large-amplitude motions.

Kalinichenko *et al.* (1995) experimentally studied the surface waves under Faraday resonance excitation of a fluid layer. Wave breakdown and onset of unstable oscillation modes were considered. In view of highly developed resonance oscillations, the spouting of droplets from the free surface of the oscillating liquid may occur (Sorokin, 1957). In low or zero gravity cases, the dynamics of drops presents practical and important problems, which are also reasonable in certain fields of chemical engineering (Savino *et al.*, 2003). In mathematical terms, one must consider the three-dimensional boundary value problem with free moving boundaries. In the case of sufficiently large droplets, the surface tension may appear to be relatively low so that the droplets will perform very large vibration amplitude. Some experiments clearly show this qualitative feature of droplets dynamics.

Azuma and Yoshinara (1999) experimentally studied three-dimensional large-amplitude oscillations of a mercury drop under electrical excitation in low gravity using a drop tower. The large-amplitude mode shapes were obtained. Multi-lobed (from three to six lobes) and polyhedral oscillations were observed as well as axisymmetric oscillation patterns. Zinchenko *et al.* (1999) numerically studied interacting deformable drops by means of a boundary-integral algorithm. The code was applied to study relative motion of two bubbles or droplets under gravity for moderately high Bond numbers. Interfacial patterns and waves were determined by Velarde and Nepomnyashchy (2001).

Under a microgravity field, the fluid can be reoriented inside the tank and may be located close to the other end or can be distributed around the walls. Concus (1963, 1964), Concus, *et al.* (1965) and Levy, *et al.* (1964) examined the stability of free liquid surface in an inverted tank. Madsen and Easton (1970) used numerical analysis to predict the fluid flow under low gravity since the weak nonlinear theory fails to predict such motion. Dodge and Garza (1968, 1970a) carried out a simulated low-gravity sloshing in tanks of different geometries. Scholl, *et al.* (1971) studied the anomalous behavior of liquid propellant in an Apollo lunar module propellant gage including the order–disorder transitions in capillary ripples (Tufillaro, *et al.*, 1989). Experiments were carried out by Satterlee (1962), Masica *et al.* (1964a,b), Hollister and Satterlee (1965), Masica and Salzman (1965, 1969), Masica (1967), and Salzman, *et al.* (1973) to examine the stabilization of liquid. It was found that the insertion of ring baffles drastically changed the reorientation flow profiles (see, e.g., Levy, *et al.*, 1964, Hung, Lee, and Shyu, 1990, Hung and Shyu, 1991a,b,c, 1992a,c, 1996, and Hung, 1995b). The problem of stabilization of the liquid profile needs more investigation of the possibility of stabilization by mechanically imposing a random field. Concus and Finn (1991a) studied capillary surfaces in nontraditional containers. Hurwitz and Melcher (1968) used the dielectrophoretic to control liquid sloshing in a low gravitational field.

The stabilization of drops by applying a high-frequency vibration has attracted some research of stabilization of the Rayleigh–Taylor instability. For example Wolf (1968) found two strange stationary interfacial profiles occurring in a horizontally shaking tank. These profiles significantly differ from the capillary meniscus. Under microgravity conditions, similar phenomena were reported by Ganiev, *et al.* (1976, 1977) and Bezdenezhik, *et al.* (1984). Vibration positioning and governing the time-dependent convective flows, due to a high-frequency vibrating container, were examined by Lyubimov *et al.* (1981, 2002), Lyubimov and Cherepanov (1986, 1989), Bezdenezhik, *et al.* (1991), and Ivanova, *et al.* (1997, 1998, 2001a,b). These studies included experimental and computational investigations on fluids with two layers of close densities.

12.8 Capillary systems

12.8.1 Marangoni flow

Marangoni (1871) proposed a model for surface liquid phenomena, which presume that if there exists a difference of surface tension along a free liquid surface, it will flow towards the region of higher surface tension. Since then, surface tension gradient driven flow along an interface separating two fluids has been referred to as the Marangoni effect. Levich and Krylov (1969) discussed some facets of surface tension driven phenomena. A historical review of Marangoni flows was presented by Castagnolo and Monti (2001) based on a web page by Toshio Funada (http://nctd35.denshi.numazu-ct.ac.jp/dr-th2/dr-book/node4.html).

Benard (1901) observed a hexagonal cell pattern of convections in a horizontal spermaceti layer of thickness 0.5–1 mm, vertically bounded below by a heated metal plate and above by a free surface. Lord Rayleigh (1916) considered convections caused by buoyancy. He showed, using linear stability theory, that for a horizontal fluid layer with two free boundaries, a quiescent state of the layer becomes unstable. Furthermore, cellular convections can occur if the Rayleigh number (i.e., the temperature difference imposed on the layer) exceeds a critical value, at which the corresponding critical wave number gives the cell size. The qualitative agreement in the cell size between his results and Benard's experiments stimulated further studies by Chandrasekhar (1961). Block (1956) found, however, that such a hexagonal cell pattern actually occurs when a liquid layer is very shallow and even when it is cooled from below.

Pearson (1958) showed theoretically that the surface tension gradient due to variations of temperature or of concentration (the Marangoni effect) could be responsible for convections in a liquid layer with a free surface. For the lower boundary of the layer, the heat flux is prescribed by (i) conducting (isothermal) condition and by (ii) insulating condition. The results in case (i) revealed that the convections observed by Benard were certainly due to the surface tension gradient rather than due to the buoyancy. This is attributed to the fact that the Marangoni number estimated from Benard's data was larger than the critical value. On the basis of Pearson's model, Scriven and Sternling (1960, 1964) studied the surface deformation effects on the steady Marangoni convections. They showed that an up-flow occurs beneath the depression of the free surface and a down-flow occurs beneath the elevation. This flow pattern is the same as that found by Benard in his experiments, but is converse to that in buoyancy-driven convections shown by Jeffries (1951).

Sternling and Scriven (1959) studied the Marangoni instability due to mass transfer. They adopted two immiscible liquid layers with semi-infinite depth, at the interface of which both the surface shear and dilational viscosities were also taken into account. Their results included various factors pertaining to instability, such as viscosities of bulk fluids, diffusivities of compositions, and the direction of the mass transfer across the interface.

The Rayleigh–Benard convections were theoretically and experimentally studied. For example, Smith (1966) studied the Marangoni instability for two immiscible horizontal fluid layers with finite depth whose interface is deformable. He showed that the depth ratio affects the onset of convection and the resulting flow pattern, where the gravity waves have a stabilizing effect on the convection. Koschmieder (1967) conducted a series of experiments for a shallow layer of silicone oil on a circular plate heated from below. Concentric circular rolls appeared and then broke into hexagonal cells. Hoard, *et al.* (1970) found that the cell pattern in a layer of silicone oil is concentric circular rolls when the layer is covered by a glass lid. It is hexagons when it is in contact with a thin air layer. Note that the experiments were made by taking into account the theoretical results for buoyancy-driven convections by Schluter, Lortz and Busse (1965) and Krishnamurti (1968). The experimental results conducted by Palmer and Berg (1971) on a layer of silicone oil, heated from below and bounded above by a very thin air layer, confirmed the linear stability theory for convections due to the Marangoni effect.

In the 1970s, numerous studies were devoted to examining the Marangoni effect in low-gravity environments. The dramatic experiments for the onset of convections were made in Apollo 14 where the acceleration due to gravity was evaluated at $10^{-6} \times g$ as reported by Grodzka and Bannister (1972). It was confirmed that the surface tension gradient alone could drive cellular convections of visible magnitude. The observed convections were affected, however, by the radial thermal conduction, which occurred along the bottom plate and in the sidewall of the circular container. With an improved apparatus, the experiments were made in a weaker gravity field of $10^{-8} \times g$ in Apollo 17 as reported by Grodzka and Bannister (1972). Polygonal cells were observed and their size was found to be exactly the same order of magnitude as that predicted by Pearson's theory. Bauer (1982f,g,h, 1983c, 1984f,g, 1985c) considered different problems of Marangoni-convection in liquid systems under zero gravitational field and a fixed or time-dependent temperature field. Monti (1990a) discussed the oscillatory regimes in Marangoni flows. Castagnolo and Monti (2001) presented a literature survey of research results pertaining to thermal Marangoni flows. Marangoni instability of a thin film heated from below was studied by Kaliadasis, *et al.* (2003)

12.8.2 Special forms

Special forms of fluid capillary systems under a microgravity environment are classified into drops, jets, and liquid bridges. A *drop* is a single-connected blob of liquid completely surrounded by another fluid (isolated drop) or touching just a solid surface in a single-connected solid–liquid interface (a captive drop). A *jet* is formed by issuing a liquid mass through a hole in a solid and then through the ambient fluid. Alexander (1997) presented an overview of experimental and analytical studies of drops and jets. A *liquid bridge* is a liquid mass extended between two solid supports and held together by surface tension and capillary forces in a weightless environment, and in case of disturbances the surface tension acts as the restoring force. It is assumed that the supports are planar, circular, and coaxial discs. The liquid is assumed to be anchored to the edges of the discs and is surrounded by a gas or an immiscible fluid. The pinning condition of the liquid at the ends is crucial to avoid the intricate problem of having a free-moving contact line and its associated problems of wetting and spreading (Concus and Finn, 1998, and Martinez and Perales, 2001). Other geometrical configurations, such as unbounded liquid bridges were considered by Vogel (1982). Other forms known as rectangular "boats" were experimentally investigated and used in orbit flights (Chun, 1980a, and Schwabe, *et al.*, 1990, 1991, 1992).

The main research activities of liquid bridges include, (see, e.g., Martinez and Perales, 2001): (i) static and dynamic stability problems in the absence of thermal effects, (ii) Marangoni convection flows due to convection-induced by temperature gradients, and (iii) unidirectional solidification due to phase change processes. Experimental studies on liquid bridges were carried out in different test facilities such as Plateau (1863) tanks, millimetric bridges, short-time microgravity platforms, sounding rockets platform, and orbiting platforms. In Plateau tanks, the liquid bridge is established inside an isodense immiscible liquid. The Plateau tank is the best technique for creating large capillary systems >1 cm on ground. In millimetric bridges, very small liquid bridges are established in the air under a magnifier. In short-time microgravity platforms, the liquid bridge is formed at the start of a parabolic flight of an aircraft for a one-minute duration. Sounding-rockets platforms can last between 6 to 15 minutes in microgravity environment. The study of the mechanical behavior of axisymmetric liquid bridges involves the study of equilibrium shapes and their stability limits, frequency response and resonance, rotation of bridges, the decay of the motion, and breakage. An experiment on the German D-2 Spacemission was performed under the title "Liquid Column Resonances" (LICORE) in order to measure the natural frequencies and the response of the liquid to axial excitation (Bauer, 1990i, 1992i and Bauer, *et al.*, 1994). Bontozoglou and Hanratty (1990) studied the capillary-gravity Kelvin–Helmholtz waves in the neighborhood of resonance.

12.8.3 Static stability of liquid bridges

The static stability of liquid bridges includes the equilibrium state and stability limits. Rayleigh (1892, 1945) analytically showed that the maximum length, l, to diameter, $2R$, ratio, $\Re$, of a cylindrical liquid bridge, $\Re = l/2R = \pi$, can be formed in the absence of gravity. Above that length ratio the liquid bridge would collapse. The stability and ultimate collapse of such bridges have been extensively studied because of their application in a low gravitational environment. However, in the presence of gravity, the cylindrical shape of an axisymmetric bridge tends to deform and reduce the maximum length ratio $\Re$. Later, Lamb (1945) found that an infinitely long column with a wave motion in the axial direction of wave number k exhibits stability for $kR > 1$, while it is unstable for $kR < 1$. This means that the axisymmetric moving column disintegrates and tends to form drops. The axial wavelength, λ_z, exhibiting the most pronounced instability is $\lambda_z = 4.508(2R)$. For a circular cylindrical liquid column, exhibiting only circumferential surface waves, Lord Rayleigh (1879) and Lamb (1945) treated the stability problem for $k = 0$ and $\lim_{x=0} [xI'_m(x)/I_m(x)] = m$, where $I_m(x)$ is the modified Bessel function of order m. Bauer (1983b) examined the surface and interface oscillations of freely floating spheres of immiscible viscous liquids. Carruthers and Grasso (1972) studied the behavior of floating liquid zones in a simulated zero-gravity environment.

In contrast to a floating globule, where only a unique equilibrium condition such as the sphere is possible, there are an infinite number of equilibrium conditions that are possible in a floating zone. These conditions depend upon the volume and the contact angle at the end disc of the liquid column. For large enough end discs, the contact angle can be preserved under any possible free motion of the liquid boundary on the end discs. This would yield the boundary condition $dr/dz = \pm \cot \vartheta_c$ at the end discs. Another case would be that the end discs are completely covered by the liquid, in which case the boundary condition of such an "anchored" condition will be $r = R$, where R is the radius of the disc. There is a series of stable shapes of the liquid bridge depending on the volume of the liquid and the ratio $\Re$. Padday (1976) demonstrated these shapes for a nonrotating floating liquid zone, where with successively decreasing volume the meniscus shape changes from an outward nodoid to sphere, outward unduloid, cylinder, inward unduloid, catenoid, and finally to an inward nodoid. The ratio $\Re$ at which instability sets in and at which the liquid bridge breaks down ranges from infinity for a spherical meniscus to π for a cylinder and 0.663 for a catenoid. For the inward nodoid, this ratio varies over the range $0 < \Re < 0.663$, while for an inward unduloid it is $0.663 < \Re < \pi$. These results were obtained with an anchored boundary for a perfectly wetted disc.

The stability and possible shapes of axisymmetric bridges were analytically studied by Sanz and Martinez (1983), Nicolás (1992), Langbein (1992), Langbein, *et al.* (1995), Lowry and Steen (1995), Nicolás, *et al.* (1997, 1998), and Patel, *et al.* (2002). Alexander, *et al.* (2000) determined the stability limits of nonaxisymmetric liquid bridges. Sankaran and Saville (1993), Marston, *et al.* (1996), and Morse, *et al.* (1996) conducted experiments in a reduced gravity environment in which the bridge is surrounded by a density-matched immiscible fluid or in a space-borne microgravity environment. Mahajan, *et al.* (1998, 1999a, b) reported the results of an experiments in which magnetic levitation was used to simulate a low-gravity environment and create quasi-cylindrical liquid columns of a paramagnetic fluid in air. The paramagnetic fluid was composed of manganese chloride tetrahydrate dissolved in water. The idea of the magnetic compensation of gravity is based on creating a magnetic field, H, such that the resulting force per unit volume is almost equal to the gravitational force per unit volume, that is,

$$\frac{1}{2}\chi \nabla H^2_{\text{comp}} \approx \rho g \tag{12.108}$$

where χ is the material volumetric magnetic susceptibility, and H_{comp} is the magnetic field whose gradient just compensates gravity. The corresponding Bond number including the effect of the magnetic field is given by the expression

$$\mathsf{Bo} = \frac{(\rho g - \chi H_x \partial_z H_x) d^2}{4\sigma} \tag{12.109}$$

where d is the bridge diameter, and $\chi H_x \partial_z H_x$ is the z-component of the magnetic force. The stability of the liquid bridges subjected to both axial and lateral gravity was studied as a function of the bond number and volume of the supported liquid. The measured stability boundary was compared with the predicted boundary under lateral gravity by Laverón-Simavilla and Perales (1995), who obtained the following asymptotic form

$$\Re = \frac{l}{2R} = \pi \left[1 - \frac{2}{4} \pi^2 (\mathsf{Bo})^2 \right] \tag{12.110}$$

Mahajan, *et al.* (1999a) reported disagreement between their measurements and the predicted boundary (12.110) and attributed the differences to be in part due to wetting of the sides of the supporting rods. Later, Mahajan, *et al.* (1999b) experimentally examined the stability of cylindrical bridges of liquid crystal octylcyanobiphenyl in an immiscible liquid bath. Liquid crystal exhibits internal order intermediate between an isotropic liquid and a solid (DeGennes and Prost, 1994).

12.8.4 Dynamic stability of liquid bridges

The dynamics of liquid bridges deals with the estimation of the natural frequencies, transient response, periodic or steady states, and the return to equilibrium or breaking. Lord Rayleigh (1879, 1945), Lamb (1945), Bauer (1983a, 1984b,e 1986a, 1989a,d,e, 1991a), Higuera, *et al.* (1994), and Higuera and Nicolás (1997) performed linear analyses for different geometries of liquid columns to determine their natural frequencies. The natural frequencies and stability of circular cylindrical immiscible liquid systems were documented for various liquid bridges by Bauer (1989b). For a frictionless single liquid column of infinite length the natural frequencies are given by the formula

$$\omega^2_{m\ l\to\infty} = \frac{\sigma}{\rho R^3} (m^2 + K^2 R^2 - 1) K R \frac{I'_m(KR)}{I_m(KR)} \tag{12.111}$$

where K is the wave number, I_m is the modified Bessel function of the first kind and mth order, a prime denotes derivative with respect to the radial coordinate r. It is seen that the axisymmetric mode, $m = 0$, experiences instability for $KR < 1$.

For a frictionless single column of finite length, l, the natural frequencies are

$$\omega^2_{mn} = \frac{\sigma}{\rho R^3} \left(m^2 + \frac{n^2 \pi^2 R^2}{l^2} - 1 \right) \frac{n\pi R}{l} \frac{I'_m(n\pi R/l)}{I_m(n\pi R/l)} \tag{12.112}$$

It is seen that the axisymmetric mode, $m = 0$, is unstable if $n\pi < l/R$. If the liquid column of finite length is spinning with angular velocity Ω_0 the frequency equation for natural frequencies below twice the frequency of spin, that is, $\omega < 2\Omega_0$ (hyperbolic case) is, (Bauer, 1982a),

$$\frac{\Theta^2}{n^2\pi^2}\left(\frac{l}{R}\right)^2 J_m(\Theta)+\frac{1}{4}\left[\frac{\sigma}{\rho R^3\Omega_0^2}\left(m^2+\frac{n^2\pi^2R^2}{l^2}-1\right)-1\right]\left(1+\frac{\Theta^2(l^2/R^2)}{n^2\pi^2}\right)$$

$$\times\left[m\sqrt{1+\frac{\chi^2(l^2/R^2)}{n^2\pi^2}}J_m(\Theta)+\chi J'_m(\Theta)\right]=0 \tag{12.113}$$

where $\Theta=(n\pi R/l)\sqrt{(4\Omega_0^2/\omega_0^2)-1}$. The frequency equation for natural frequencies above twice the frequency of spin, that is, $\omega>2\Omega_0$ (elliptic case) is

$$\frac{\Theta^2}{n^2\pi^2}\left(\frac{l}{R}\right)^2 I_m(\Theta)-\frac{1}{4}\left[\frac{\sigma}{\rho R^3\Omega_0^2}(m^2+\frac{n^2\pi^2R^2}{l^2}-1)-1\right]\left(1-\frac{\Theta^2(l^2/R^2)}{n^2\pi^2}\right)$$

$$\times\left[m\sqrt{1-\frac{\chi^2(l^2/R^2)}{n^2\pi^2}}I_m(\Theta)+\chi I'_m(\Theta)\right]=0 \tag{12.114}$$

For this case, $\Theta=(n\pi R/l)\sqrt{1-(4\Omega_0^2/\omega_0^2)}$. The rotating column becomes unstable if

$$\Omega_0^2\geq\frac{\sigma}{\rho R^3}\left(m^2+\frac{n^2\pi^2R^2}{l^2}-1\right) \tag{12.115}$$

In terms of the Bond number, $\mathsf{Bo}=\rho R\Omega_0^2R^2/\sigma$, this condition takes the form

$$\mathsf{Bo}\geq\left(m^2+\frac{n^2\pi^2R^2}{l^2}-1\right) \tag{12.116}$$

Bauer (1989b,c,f), and Bauer and Eidel (1998c) documented expressions for the natural frequencies of finite and infinite liquid columns, rotating and nonrotating, for the following cases:

(1) Two immiscible liquids of radii R_1 and R_2.
(2) A cylindrical bubble in an annular cylindrical column.
(3) Two immiscible liquids, the inner has radius R_2 while the outer has radius $R_1\to\infty$.
(4) Infinite annular liquid column around a rigid center core.
(5) Two immiscible liquids completely filling a cylindrical container of infinite length.
(6) Coupled frequencies of hydroelastic liquid system where a liquid layer surrounds an infinite long elastic shell (Bauer, 1987a, b).

The case of complex natural frequencies of an incompressible and viscous infinite long liquid column was numerically estimated by Bauer (1984b, 1989b). He found that the axisymmetric mode, $m=0$, exhibits instability for $KR<1$, while all other modes are stable. For an incompressible visco-elastic liquid of infinite length, Bauer (1986a, 1989b) numerically estimated the complex natural frequencies. If the liquid is deformed elastically within small time intervals, time-wise decaying shear stresses remain after the deformation and fade away after a sufficiently long time. The time decay of the stresses is called the Maxwell relaxation time. During the oscillations of such a visco-elastic liquid, for which the period is large compared to the relaxation time, the liquid behaves approximately like a viscous liquid. On the other hand, for a period smaller than the relaxation time, it behaves like an elastic solid. Bauer (1986c) studied the induced free-surface oscillations in a freely floating visco-elastic liquid sphere due to an oscillatory temperature gradient. The free oscillations of spherical and conical systems were determined by Bauer and Eidel (1986, 1987a,b), Bauer and Offenbach (1986), and Bauer (1992f, l, 1995b). The case of an amphora liquid column was considered by Bauer and Eidel (1991b,c, 1993c, 1994).

Transient problems depend on the initial conditions, and particular attention was given to the evolution during the breakage of the column beyond the stability limit (Meseguer, 1983). Sankaran and Saville (1993) imposed an axial electric field to stabilize the liquid bridge and by a sudden change of the electric field they monitored the collapse of the bridge. Monti and Savino (1995) studied the effect of an unsteady thermal boundary condition on Marangoni flow in liquid bridges.

Under harmonic excitations, the dynamic characteristics of liquid bridges include the bridge periodic oscillations under axial or lateral support excitations. Sanz (1985), Sanz and Diez (1989), Martinez (1987), Perales and Meseguer (1992), Nicolás and Vega (1996), and Morse, *et al.* (1996) studied the bridge resonances under different conditions. The natural damped frequencies of an infinitely long column of immiscible viscous liquids were estimated by Bauer (1984b). The eigenfrequencies (in terms of the Weber number $\sqrt{\rho\omega^2 R^3/\sigma}$) of a cylindrical volume of liquid bridge were calculated by Sanz (1985) as a function of the bridge slenderness ratio, $l/2R$. Perales and Meseguer (1992) estimated the response of a cylindrical liquid column to forced excitation imposed to the supporting discs with particular attention to the in-phase or out-of-phase cases. Langbein, *et al.* (1995) studied the resonance frequencies of an axially driven bridge as a function of the bridge's length-to-diameter ratio under zero Bond number. Marr-Lyon, *et al.* (1997) showed that the presence of an outer bath significantly affects the boundary layer term in the force equation, resulting in a very different oscillatory behavior from a bridge in equilibrium with its vapor. This analytical observation was verified experimentally by Asaki and Marston (1995). The dynamic stability of liquid bridges was experimentally studied by Mahajan, *et al.* (2000), who formed a paramagnetic liquid bridge between the tips of two collinear rods in a strong magnetic field gradient, then they introduced a sudden change of the magnet current. This current change created a change in the Bond number, which resulted in a deformation and ultimate collapse of the liquid bridge. The collapse time was found to be independent of the bridge length when other parameters were kept constant. Under either axial or lateral oscillations of the magnetic field, Mahajan, *et al.* (2002) found that the frequency of the first resonance peak reaches its maximum value when the total body force vanishes.

The analysis of these problems is usually governed by the general Navier–Stokes equations for the liquid bulk and Young–Laplace capillary equation for the interface. One may introduce various terms to account for their effects such as zero velocity for studying static shapes, or sinusoidal motion for periodic vibrations. One also may introduce boundary conditions such as perfect anchorage of the interface to the edges of the support discs, or residual acceleration. Martinez, Haynes and Langbein (1987), and Meseguer, Slobozhanin, and Perales (1995) presented analytical modelings and experimental results of the statics and dynamics of liquid bridges. Based on the assumption that the fluid is incompressible, except for the buoyancy term (Boussinesq model), the boundary value problem of liquid bridges was described by Martinez and Perales (2001). The equilibrium shapes of nonaxisymmetric liquid bridges were determined by Laverón-Simavilla and Perales (1995).

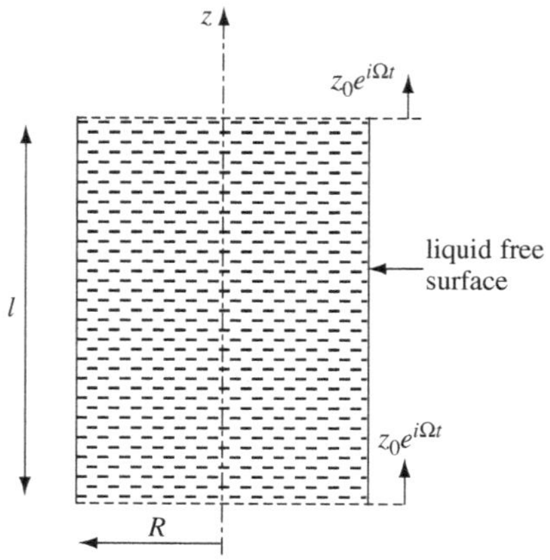

Figure 12.23 Geometry of a liquid bridge between two plates a distance l apart under zero gravity.

12.8.5 Axial excitation of liquid bridges

Consider a circular cylindrical liquid bridge of length l and radius R placed between two horizontal plates a distance l apart in a zero gravity environment as shown in Figure 12.23. The liquid is assumed to be viscous and incompressible. Let the walls be excited along the z-axis by the excitation $z(t) = z_0 e^{i\Omega t}$. In this case, the response of the free surface is restricted to the radial, u, and axial, w, velocity components. The continuity equation is

$$\frac{\partial u}{\partial r} + \frac{u}{r} + \frac{\partial w}{\partial z} = 0 \tag{12.117}$$

The linearized Navier–Stokes equations for isothermal fluid are written in the component form

$$\frac{\partial u}{\partial t} + \frac{1}{\rho}\frac{\partial p}{\partial r} = \nu\left[\frac{\partial^2 u}{\partial r^2} + \frac{1}{r}\frac{\partial u}{\partial r} - \frac{u}{r^2} + \frac{\partial^2 u}{\partial z^2}\right] \tag{12.118a}$$

$$\frac{\partial w}{\partial t} + \frac{1}{\rho}\frac{\partial p}{\partial z} = \nu\left[\frac{\partial^2 w}{\partial r^2} + \frac{1}{r}\frac{\partial w}{\partial r} + \frac{\partial^2 w}{\partial z^2}\right] - z_0\Omega^2 e^{i\Omega t} \tag{12.118b}$$

It is assumed that at the top and bottom, the liquid is anchored (stuck-edge). This imposes the boundary condition

$$u|_{z=\pm l/2} = w|_{z=\pm l/2} = 0 \tag{12.119}$$

At the free surface, $r = R$, the dynamic boundary condition is

$$p - 2\mu\frac{\partial u}{\partial r} + \frac{\sigma}{R^2}\left[\eta + R^2\frac{\partial^2 \eta}{\partial z^2}\right] = \frac{\sigma}{R} \tag{12.120a}$$

The kinematic boundary condition is

$$u|_{r=R} = \frac{\partial \eta}{\partial t} \tag{12.120b}$$

where $\eta = \eta(z, t)$ is the free-surface displacement from the equilibrium position, $r = R$. The shear stress τ_{rz} vanishes at the liquid surface, which implies

$$\left(\frac{\partial w}{\partial r} + \frac{\partial u}{\partial z}\right)_{r=R} = 0 \tag{12.121a}$$

The Laplace equation for the pressure is obtained from equations (12.117) and (12.118)

$$\frac{\partial^2 p}{\partial r^2} + \frac{1}{r}\frac{\partial p}{\partial r} + \frac{\partial^2 p}{\partial z^2} = 0 \tag{12.121b}$$

Bauer (1990i) obtained a solution of the boundary value problem described by equations (12.117)–(12.121). For the axisymmetric modes, the pressure and velocity distributions were assumed to be in the form

$$\begin{aligned} p(r,z,t) &= P(r,z)\,e^{i\Omega t}, \quad u(r,z,t) = U(r,z)\,e^{i\Omega t} \\ w(r,z,t) &= [W(r,z) + i\Omega z_0]\,e^{i\Omega t}, \quad \eta(z,t) = H(z)\,e^{i\Omega t} \end{aligned} \tag{12.122}$$

Introducing (12.122) into equation (12.117), gives the continuity equation in the form

$$\frac{\partial U}{\partial r} + \frac{U}{r} + \frac{\partial W}{\partial z} = 0 \tag{12.123a}$$

The Navier–Stokes equations (12.118) take the form

$$\frac{i\Omega}{\nu} + \frac{1}{\mu}\frac{\partial P}{\partial r} = \frac{\partial^2 U}{\partial r^2} + \frac{1}{r}\frac{\partial U}{\partial r} - \frac{U}{r^2} + \frac{\partial^2 U}{\partial z^2} \tag{12.123b}$$

$$\frac{i\Omega}{\nu}W + \frac{1}{\mu}\frac{\partial P}{\partial z} = \frac{\partial^2 W}{\partial r^2} + \frac{1}{r}\frac{\partial W}{\partial r} + \frac{\partial^2 W}{\partial z^2} \tag{12.123c}$$

The boundary conditions (12.119) take the form

$$U|_{z=\pm l/2} = 0, \quad W(r, z = \pm l/2) = -i\Omega z_0 \tag{12.124a, b}$$

The shear stress condition (12.121a) at the free surface takes the form

$$\left(\frac{\partial W}{\partial r} + \frac{\partial U}{\partial z}\right)_{r=R} = 0 \tag{12.124c}$$

Taking the time derivative of both sides of the dynamic condition (12.120a) and using the kinematic condition (12.120b) gives

$$i\Omega P - 2i\mu\Omega\frac{\partial U}{\partial r} + \frac{\sigma}{R^2}\left[U + R^2\frac{\partial^2 U}{\partial z^2}\right] = 0 \tag{12.124d}$$

If the radial adhesion condition, $U|_{z=\pm l/2} = 0$, is relaxed, and allowing slipping in the radial direction, equation (12.123b,c) may be solved in terms of the stream function $\Psi(r, z)$ defined such that

$$U = -\frac{1}{r}\frac{\partial \Psi}{\partial z}, \quad W = \frac{1}{r}\frac{\partial \Psi}{\partial r} \tag{12.125}$$

Introducing (12.125) into equations (12.123b, c), eliminating the pressure terms and using the continuity equation gives

$$\bar{\Delta}\left[\bar{\Delta} - \frac{i\Omega}{\nu}\right]\Psi = 0 \tag{12.126}$$

where the operator $\bar{\Delta} = \frac{\partial^2}{\partial r^2} - \frac{1}{r}\frac{\partial}{\partial r} + \frac{\partial^2}{\partial z^2}$. The solution of the partial differential equation (12.126) may be given in the form

$$\Psi(r,z) = \frac{i\nu}{\Omega}(A + Br^2) + \sum_{n=1}^{\infty}\left\{A_{2n-1}\frac{r}{R}I_1\left(\frac{r}{R}\sqrt{\frac{(2n-1)^2\pi^2R^2}{l^2} + \frac{i\Omega R^2}{\nu}}\right)\right.$$
$$\left. + \frac{i\nu}{\Omega}B_{2n-1}\frac{r}{R}I_1\left(\frac{r}{R}\frac{(2n-1)\pi R}{l}\right)\right\}\cos\left(\frac{(2n-1)\pi z}{l}\right) \tag{12.127}$$

where A, B, and A_{2n-1}, are integration constants, which depend on the excitation frequency Ω. In order to satisfy the upper and lower normal velocity boundary condition (12.124b), the constant B must take the value

$$B = -\frac{1}{2}\frac{\Omega^2}{\nu}z_0 \tag{12.128}$$

Using the definition of the velocity components (12.125), we write

$$U(r,z) = \sum_{n=1}^{\infty}\left\{A_{2n-1}\frac{(2n-1)\pi}{Rl}I_1\left(\frac{r}{R}\sqrt{\frac{(2n-1)^2\pi^2R^2}{l^2} + \frac{i\Omega R^2}{\nu}}\right)\right.$$
$$\left. + \frac{i\nu(2n-1)\pi}{\Omega Rl}B_{2n-1}I_1\left(\frac{r}{R}\frac{(2n-1)\pi R}{l}\right)\right\}\sin\left(\frac{(2n-1)\pi z}{l}\right) \tag{12.129a}$$

$$W(r,z) = -i\Omega z_0 + \sum_{n=1}^{\infty}\left\{\frac{A_{2n-1}}{R^2}\sqrt{\frac{(2n-1)^2\pi^2R^2}{l^2} + \frac{i\Omega R^2}{\nu}}I_0\right.$$
$$\times\left(\frac{r}{R}\sqrt{\frac{(2n-1)^2\pi^2R^2}{l^2} + \frac{i\Omega R^2}{\nu}}\right) + \frac{i\nu B_{2n-1}}{\Omega R^2}\frac{(2n-1)\pi R}{l}$$
$$\left.\times I_0\left(\frac{r}{R}\frac{2n-1\pi R}{l}\right)\right\}\cos\left(\frac{(2n-1)\pi z}{l}\right) \tag{12.129b}$$

The pressure equation (12.121b) is satisfied by the following solution

$$p = -\rho\Omega^2 z_0 z + \sum_{n=1}^{\infty}C_{2n-1}I_0\left(\frac{r}{R}\frac{(2n-1)\pi R}{l}\right)\sin\left(\frac{(2n-1)\pi z}{l}\right) \tag{12.130}$$

where C_{2n-1} is related to B_{2n-1} by introducing equations (12.128) and (12.130) into equations (12.123) from which one obtains

$$C_{2n-1} = \frac{\mu}{R} B_{2n-1} \tag{12.131}$$

In order to determine the natural frequencies of the liquid column, equations (12.129) and (12.130) are substituted into the free-surface boundary conditions (12.124c,d). The determinant of the coefficients A_{2n-1} and B_{2n-1} of the resulting equations gives the following frequency equation

$$\begin{aligned}
\Delta = & \left[S^2 + 4S\frac{(2n-1)^2\pi^2R^2}{l^2} + 4\frac{(2n-1)^4\pi^4R^4}{l^4}\right] I_0\left(\frac{(2n-1)\pi R}{l}\right) \\
& \times I_1\left(\sqrt{\frac{(2n-1)^2\pi^2R^2}{l^2} + S}\right)\left[\left(\frac{\sigma R}{\rho\nu^2}\right)\frac{(2n-1)\pi R}{l}\left\{\frac{(2n-1)^2\pi^2R^2}{l^2} - 1\right\}\right. \\
& \left. - 2S\frac{(2n-1)\pi R}{l}\right] I_1\left(\frac{(2n-1)\pi R}{l}\right) \\
& \times I_1\left(\sqrt{\frac{(2n-1)^2\pi^2R^2}{l^2} + S}\right) - 4\frac{(2n-1)^3\pi^3R^3}{l^3}\sqrt{\frac{(2n-1)^2\pi^2R^2}{l^2} + S} \\
& \times I_1\left(\frac{(2n-1)\pi R}{l}\right) \times I_0\left(\sqrt{\frac{(2n-1)^2\pi^2R^2}{l^2} + S}\right) = 0
\end{aligned} \tag{12.132}$$

where $S = \varpi R^2/\nu$ and ϖ is the complex natural frequency. For this axisymmetric case, the roots of equation (12.132) depend on the value of the surface tension parameter $\sigma R/(\rho\nu^2)$, this parameter is equivalent to the inverse of Ohnesorge number (Oh) or the inverse of the capillary number (Ca). Bauer (1990i) showed that equation (12.132) has damped oscillatory roots only if $2n\pi R/l > 1$, and diverging (unstable) roots if $2n\pi R/l \leq 1$. Figure 12.24 shows the dependence of real and imaginary parts of the damped natural frequencies on the axial wavelength parameter $\lambda_z/(2\pi R)$, for $\sigma R/(\rho\nu^2) = 10$. It is seen that for wavelength $\lambda_z/(2\pi R) > 1$, the liquid bridge becomes unstable. Depending on the value of the surface tension parameter, there exists a wavelength range over which no oscillatory damped motion exists. Decaying motion is the only possible liquid behavior in the range $1.6\pi R \leq \lambda_z \leq 2\pi R$. Below the value $\lambda_z \leq 1.6\pi R$, the liquid column experiences damped oscillatory motion. Figure 12.25 shows the boundary that separates between oscillatory and aperiodic decaying motion regimes on the plane of surface tension parameter versus axial wave parameter. If the fluid is inviscid the natural frequency may be obtained in the closed form (Bauer, 1990i),

$$\omega_{2n-1}^2 = \frac{\sigma}{\rho R^3}\frac{(2n-1)\pi R}{l}\left[\frac{(2n-1)^2\pi^2R^2}{l^2} - 1\right]\frac{I_1((2n-1)\pi R/l)}{I_0((2n-1)\pi R/l)}, n = 1, 2, \ldots \tag{12.133}$$

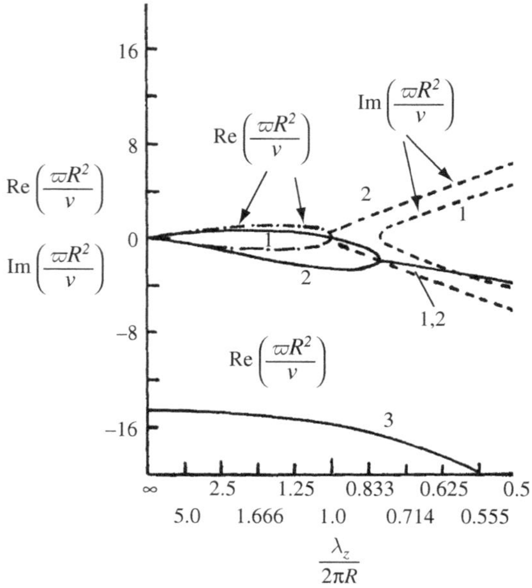

Figure 12.24 Dependence of damped natural frequency components on the axial wave length parameter of a liquid bridge for $\sigma R/(\rho\nu^2) = 10$. (Bauer, 1990f)

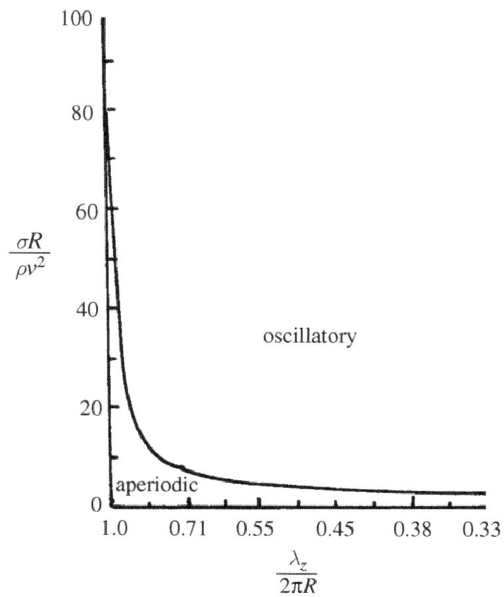

Figure 12.25 Boundary of oscillatory and aperiodic regimes of liquid bridge under axial excitation. (Bauer, 1990f)

The values of the coefficients A_{2n-1} and B_{2n-1} are

$$A_{2n-1} = \frac{8(-1)^{n-1}(\Omega R^2/\nu)^2(z_0\nu R/l)I_1((2n-1)\pi R/l)}{\bar{\Delta}} \tag{12.134a}$$

$$B_{2n-1} = \Big[4i(-1)^{n-1}((\Omega R^2)/\nu)^3((z_0\nu R)/l)\Big[(2(2n-1)^2\pi^2R^2)/l^2 + (i\Omega R^2)/\nu\Big] \times I_1\left(\sqrt{((2n-1)^2\pi^2R^2)/l^2 + (i\Omega R^2)/\nu}\right)\Big] \Big/ (\bar{\Delta}(2n-1)^2\pi^2R^4/l^2) \tag{12.134b}$$

where $\bar{\Delta} = \Delta$ by setting $S = i\Omega R^2/\nu$. The free-surface displacement is estimated from the kinematic condition $u = \partial\eta/\partial t$, which gives, upon integration, the following result

$$\eta(z,\tau) = -\frac{ie((i\Omega R^2)/\nu)\tau}{\nu((\Omega R^2)/\nu)}\sum_{n=1}^{\infty}\left\{A_{2n-1}\frac{(2n-1)\pi R}{l}I_1\left(\sqrt{\frac{(2n-1)^2\pi^2R^2}{l^2}+\frac{i\Omega R^2}{\nu}}\right) + \frac{iB_{2n-1}R^2}{(\Omega R^2/\nu)}\frac{(2n-1)\pi R}{l}I_1\left(\frac{(2n-1)\pi R}{l}\right)\right\}\sin\left(\frac{(2n-1)\pi z}{l}\right) \tag{12.135}$$

where $\tau = \nu t/R^2$ is a nondimensional time parameter.

Figures 12.26(a)–(d) show the dependence of the free-surface displacement, $|\eta(z,\tau)\nu/\bar{z}\, e^{i\bar{\Omega}\tau}|$, on the nondimensional excitation frequency, $\bar{\Omega} = \Omega R^2/\nu$, where $\bar{z} = z_0\nu R/l$. These figures are given for two different values of surface tension parameter, $\sigma R/(\rho\nu^2) = 400$ and 800, and at different locations, $z/l = 0.125$, 0.25, and 0.5 for two values of liquid bridge length ratio, $\Re = 0.5$, and 1.0. The first peak of the response corresponds to the first natural frequency of the liquid bridge. The problem of liquid bridges under an axial microgravity field was further studied by Bauer and Eidel (1991a, 1996a,b). The transient response of a viscous liquid bridge with anchored edges was determined by Komarenko and Lukorskii (1974). Eidel and Bauer (1988), Bauer (1994c), Bauer and Eidel (1988d) Lee, *et al.* (1996), and Petrov, *et al.* (1996) determined the nonlinear oscillations of an inviscid liquid column under zero gravity.

12.8.6 Axial excitation of spinning liquid bridges

Imposing a spinning motion to liquid bridges would suppress any oscillatory motion due to g-jitter, and Marangoni convection at a certain Marangoni number. Gillis (1961), Gillis and Kaufman (1961), and Gillis and Shuh (1962) considered the stability of a spinning viscous liquid column. The analysis presented in this section is based on the work of Bauer (1988a, 1989a, 1990a,b,c,d,e,g,h,i,j,k, 1991e,f,i, 1992a,b,c,g,h,j,k, l, 1993b,c,e, 1994a,b,d,e,f), Bauer and Straser (1993), and Bauer and Eidel (1995, 1996c,d,e, 1997d, 1998a), who studied the axisymmetric response of rotating and nonrotating finite liquid columns to axial excitation. Consider a frictionless incompressible liquid column of density ρ, length l and radius R, as shown in Figure 12.27. The column is rotating about the z-axis with constant angular velocity Ω_0 in a zero gravity environment and is axially excited with sinusoidal base excitation $z(t) = z_0\, e^{i\Omega t}$.

In terms of liquid velocity vector, $\mathbf{q}$, relative to the moving system, the Euler equation of motion is

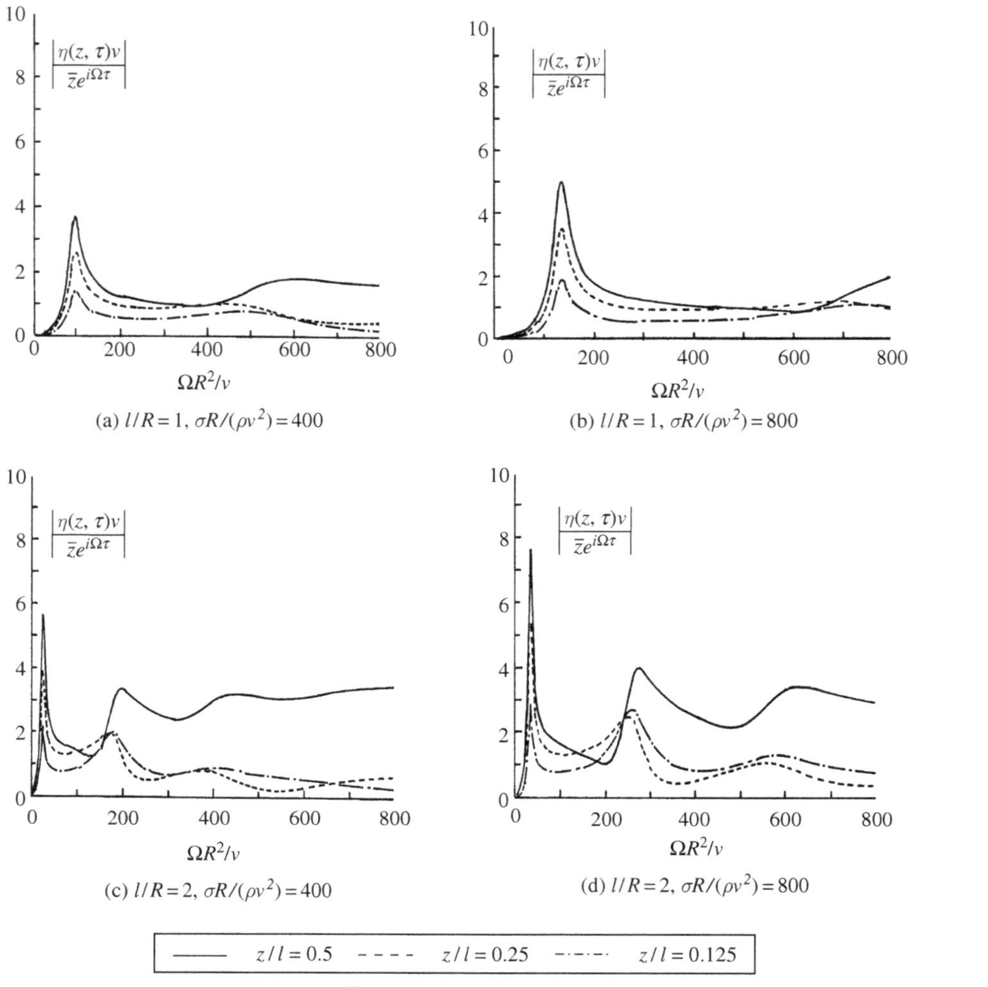

Figure 12.26 Amplitude–frequency response of damped free surface of liquid column for different values of length ratio and surface-tension parameter. (Bauer, 1990f)

$$\frac{\partial \mathbf{q}}{\partial t} + 2(\Omega_0 \times \mathbf{q}) + [\Omega_0 \times (\Omega_0 \times \mathbf{r})] = -\frac{1}{\rho}\nabla p + \ddot{z}(t)\mathbf{k} \tag{12.136}$$

where $\mathbf{r} = r\mathbf{e}_r + z\mathbf{e}_z$ is the position vector of a fluid particle in the moving coordinate frame, and $\mathbf{q} = u\mathbf{e}_r + v\mathbf{e}_\theta + w\mathbf{e}_z$ is the velocity of the fluid particle.
For axisymmetric flow, the continuity equation is the same as equation (12.117) of the previous section. The acceleration potential, Ψ, is introduced

$$\Psi = \frac{p - p_0}{\rho} - \frac{1}{2}\Omega_0^2(r^2 - R^2) - z\ddot{z}(t) \tag{12.137}$$

Equation (12.136) may be written in the scalar form in terms of the acceleration potential function

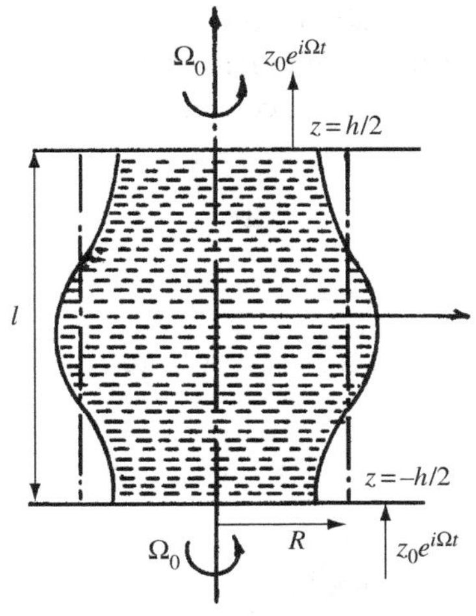

Figure 12.27 Geometry of a liquid bridge in spinning motion under sinusoidal g-jitter and coordinate system.

$$\frac{\partial u}{\partial t} - 2\Omega_0 v + \frac{\partial \Psi}{\partial r} = 0, \qquad \frac{\partial v}{\partial t} + 2\Omega_0 u = 0, \qquad \frac{\partial w}{\partial t} + \frac{\partial \Psi}{\partial z} = 0 \tag{12.138a,b,c}$$

The boundary conditions at the discs are

$$w|_{z=\pm l/2} = 0 \tag{12.139a}$$

At the free surface, $r = R + \eta(z, t)$, the kinematic boundary condition is

$$u|_{r=R+\eta} = \frac{\partial \eta}{\partial t} \tag{12.139b}$$

The dynamic condition, involving pressure discontinuity across the surface due to surface tension, is obtained from equation (12.137) at the free surface, $r = R + \eta(z, t)$, that is,

$$\Psi + R\Omega_0^2 \eta + z\ddot{z}(t) = \frac{\sigma}{\rho R} - \frac{\sigma}{\rho R^2}\left(\eta + R^2 \frac{\partial^2 \eta}{\partial z^2}\right) \tag{12.139c}$$

Differentiating equation (12.139c) with respect to time and introducing condition (12.139b) gives

$$\frac{\partial \Psi}{\partial t} + R\Omega_0^2 u + z\ddot{z}(t) + \frac{\sigma}{\rho R^2}\left(u + R^2 \frac{\partial^2 u}{\partial z^2}\right) \tag{12.139d}$$

where σ represents the surface tension between the liquid and gas interface. The solution of this boundary value problem gives the response of the rotating liquid column under axial excitation. Equations (12.138) may be written in the vector form

$$\frac{\partial \mathbf{q}}{\partial t} + 2(\mathbf{\Omega}_0 \times \mathbf{q}) + \nabla \Psi = 0 \tag{12.140}$$

Taking the curl of this equation gives

$$\frac{\partial}{\partial t}(\nabla \times \mathbf{q}) = 2\Omega \frac{\partial \mathbf{q}}{\partial z} \tag{12.141}$$

This equation reveals that an immediate consequence of Coriolis acceleration in equation (12.134) is that the velocity field cannot remain irrotational. Following the same steps as those of Section 11.4.3, upon taking the divergence of equation (12.134) and using the continuity equation, gives

$$\nabla^2 \Psi_{tt} + 4\Omega_0^2 \Psi_{zz} = 0 \tag{12.142}$$

where $\nabla^2 = \dfrac{\partial^2}{\partial r^2} + \dfrac{1}{r}\dfrac{\partial}{\partial r} + \dfrac{\partial^2}{\partial z^2}$

Let the velocity distribution and acceleration potential for the axisymmetric motion be written in the form

$$u(r, z, t) = e^{i\Omega t} \sum_{n=1}^{\infty} U_{2n-1}(r) \sin[(2n-1)\pi z/l] \tag{12.143a}$$

$$v(r, z, t) = e^{i\Omega t} \sum_{n=1}^{\infty} V_{2n-1}(r) \sin[(2n-1)\pi z/l] \tag{12.143b}$$

$$w(r, z, t) = e^{i\Omega t} \sum_{n=1}^{\infty} W_{2n-1}(r) \cos[(2n-1)\pi z/l] \tag{12.143c}$$

$$\Psi(r, z, t) = e^{i\Omega t} \sum_{n=1}^{\infty} P_{2n-1}(r) \sin[(2n-1)\pi z/l] \tag{12.143d}$$

The axial velocity (12.143c) satisfies the boundary condition (12.139a). Substituting equations (12.143) into equations (12.138) gives

$$i\Omega U_{2n-1} - 2\Omega_0 V_{2n-1} = -\frac{dP_{2n-1}}{dr} \tag{12.144a}$$

$$i\Omega V_{2n-1} + 2\Omega_0 U_{2n-1} = 0 \tag{12.144b}$$

$$i\Omega W_{2n-1} = -\frac{dP_{2n-1}}{dz} = \frac{(2n-1)\pi}{l} P_{2n-1} \tag{12.144c}$$

The velocity components in equations (12.144) can be written, in terms of the acceleration potential by simple elimination, in the form

$$\begin{aligned} U_{2n-1}(r) &= \frac{i\Omega(dP_{2n-1}/dr)}{\Omega^2 - 4\Omega_0^2} \\ V_{2n-1}(r) &= \frac{2\Omega_0(dP_{2n-1}/dr)}{\Omega^2 - 4\Omega_0^2} \\ W_{2n-1}(r) &= -\frac{i\pi(2n-1)}{l\Omega} P_{2n-1} \end{aligned} \tag{12.145}$$

Expressing z in the free-surface condition (12.139d) as a Fourier-sine series, that is,

$$z = \frac{4l}{\pi^2} \sum_{n=1}^{\infty} \frac{(-1)^{n-1} \sin[(2n-1)\pi z/l]}{(2n-1)^2} \tag{12.146}$$

and substituting (12.143) in the free-surface condition (12.139d) at $r = R$ gives

$$i\Omega P_{2n-1} + R\Omega_0^2 U_{2n-1} + \frac{\sigma}{\rho R^2}\left[1 - \frac{(2n-1)^2\pi^2 R^2}{l^2}\right] U_{2n-1} = \frac{i4z_0\Omega^3 l(-1)^n}{\pi^2(2n-1)^2} \tag{12.147}$$

Substituting (12.143d) into equation (12.142) gives

$$\frac{d^2 P_{2n-1}}{dr^2} + \frac{1}{r}\frac{dP_{2n-1}}{dr} + \left[\frac{(2n-1)^2\pi^2}{l^2}\left(1 - \frac{4\Omega_0^2}{\Omega^2}\right)\right] P_{2n-1} = 0 \tag{12.148}$$

The solution of this differential equation depends on whether the excitation frequency Ω is greater or less than twice the spin angular velocity Ω_0. In this case, we have to consider two cases, the hyperbolic case, $\Omega < 2\Omega_0$, and the elliptic case, $\Omega > 2\Omega_0$. For the hyperbolic case, let $\beta^2 = \frac{4\Omega_0^2}{\Omega^2} - 1$, and equation (12.146) has the following solution

$$P_{2n-1}(r) = B_{2n-1} J_0((2n-1)\pi\beta r/l) \tag{12.149a}$$

For the elliptic case, let $\alpha^2 = 1 - \frac{4\Omega_0^2}{\Omega^2}$, equation (12.146) has the following solution

$$P_{2n-1}(r) = A_{2n-1} I_0((2n-1)\pi\alpha r/l) \tag{12.149b}$$

where A and B are constants to be determined by substituting each solution in the free-surface condition (12.147) to give

$$A_{2n-1} = \frac{4z_0\Omega^2 l(-1)^n(\Omega^2 - 4\Omega_0^2)}{a_{2n-1}}, \quad B_{2n-1} = \frac{4z_0\Omega^2 l(-1)^n(\Omega^2 - 4\Omega_0^2)}{b_{2n-1}} \tag{12.150a,b}$$

where

$$a_{2n-1} = \pi^2(2n-1)^2\left\{(\Omega^2 - 4\Omega_0^2) I_0((2n-1)\pi\alpha R/l) + \lim_{x\to\infty}\frac{(2n-1)\pi\alpha \mathrm{R}}{l} I_1((2n-1)\pi\alpha R/l) \times\left[\Omega_0^2 - \frac{\sigma}{\rho R^3}\left(\frac{\pi^2(2n-1)^2 R^2}{l^2} - 1\right)\right]\right\}$$

$$b_{2n-1} = \pi^2(2n-1)^2\left\{(\Omega^2 - 4\Omega_0^2) J_0((2n-1)\pi\beta R/l) - \frac{(2n-1)\pi\beta R}{l} \times J_1((2n-1)\pi\beta R/l)\left[\Omega_0^2 - \frac{\sigma}{\rho R^3}\left(\frac{\pi^2(2n-1)^2 R^2}{l^2} - 1\right)\right]\right\}$$

It may be observed that in the absence of axial excitation, $z_0 = 0$, the denominators of equations (12.150) establish the natural frequency equation by setting, $\Omega = \omega_{2n-1}$, the natural frequency of the axisymmetric mode. In this case, the natural frequencies can be determined from the frequency equation.

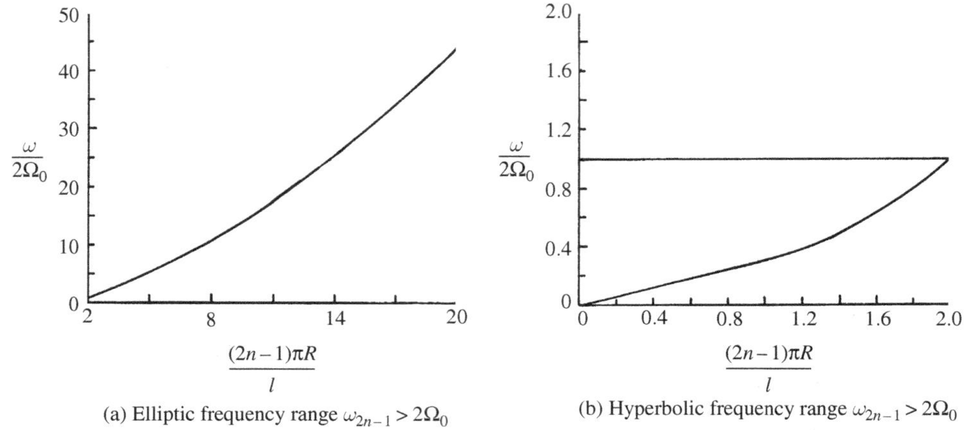

Figure 12.28 Natural frequencies of liquid column of axisymmetric mode for elliptic and hyperbolic frequency ranges, and $\sigma/(\rho R^3\Omega_0^2) = 1$. (Bauer, 1989a)

- For the elliptic case, $\Omega > 2\Omega_0$, the frequency equation is

$$\left(1 - \frac{4\Omega_0^2}{\omega^2}\right) I_0\left(\frac{(2n-1)\pi\alpha R}{l}\right) - \frac{(2n-1)\pi\alpha R}{4l} I_1\left(\frac{(2n-1)\pi\alpha R}{l}\right)$$
$$\times \left[\frac{\sigma}{\rho R^3\Omega_0^2}\left(\frac{\pi^2(2n-1)^2 R^2}{l^2} - 1\right) - 1\right]\frac{4\Omega_0^2}{\omega^2} = 0 \tag{12.151a}$$

- For the hyperbolic case, $\Omega > 2\Omega_0$, the frequency equation is

$$\left(1 - \frac{4\Omega_0^2}{\omega^2}\right) J_0\left(\frac{(2n-1)\pi\beta R}{l}\right) - \frac{(2n-1)\pi\beta R}{4l} I_1\left(\frac{(2n-1)\pi\beta R}{l}\right)$$
$$\times \left[\frac{\sigma}{\rho R^3\Omega_0^2}\left(\frac{\pi^2(2n-1)^2 R^2}{l^2} - 1\right) - 1\right]\frac{4\Omega_0^2}{\omega^2} = 0 \tag{12.151b}$$

Figures 12.28(a) and (b) show the dependence of the natural frequency ratio, $\omega/2\Omega_0$, on the column radius to length ratio, $(2n-1)\pi R/l$, for elliptic and hyperbolic cases, respectively, and bond number, $\mathsf{Bo} = \rho R^3\Omega_0^2/\sigma = 1$. As expected, as the aspect ratio, l/R, increases the natural frequency decreases.

The free-liquid-surface displacement is determined from the kinematic condition (12.139b) by integrating with respect to time

$$\eta(z,t) = \frac{e^{i\Omega t}}{(\Omega^2 - 4\Omega_0^2)} \sum_{n=1}^{\infty} \left\{ \begin{matrix} A_{2n-1}(((2n-1)\pi\alpha)/l) I_1(((2n-1)\pi\alpha R)/l) \\ -B_{2n-1}(((2n-1)\pi\beta)/l) J_1(((2n-1)\pi\beta R)/l) \end{matrix} \right\}$$
$$\times \sin\left(\frac{(2n-1)\pi z}{l}\right) \quad \text{for} \quad \left\{ \begin{matrix} \Omega > 2\Omega_0 \\ \Omega < 2\Omega_0 \end{matrix} \right\} \tag{12.152}$$

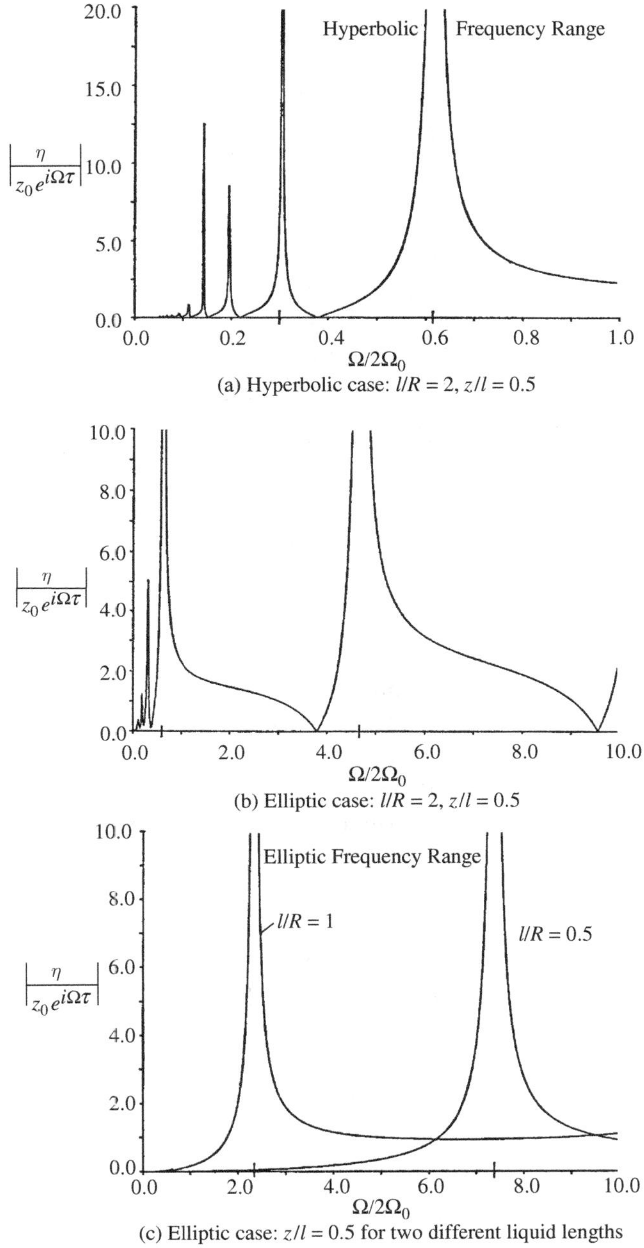

Figure 12.29 Amplitude–frequency response curves of free liquid surface under axial harmonic excitation for (a) hyperbolic and (b, c) elliptic cases. (Bauer, 1989a)

Figures 12.29(a)–(c) show the amplitude–frequency response of the free surface for the surface tension parameter, $\mathrm{Bo} = \rho R^3\Omega_0^2/\sigma = 1$. Figure 12.29(a) belongs to the hyperbolic case for $l/R = 2$ measured at $z/l = 0.5$, over an excitation frequency range $0 < \Omega/2\Omega_0 \leq 1$. Figures 12.29(b) and (c) are for the elliptic case. Figure 12.29(b) is for length to radius ratio $l/R = 2$,

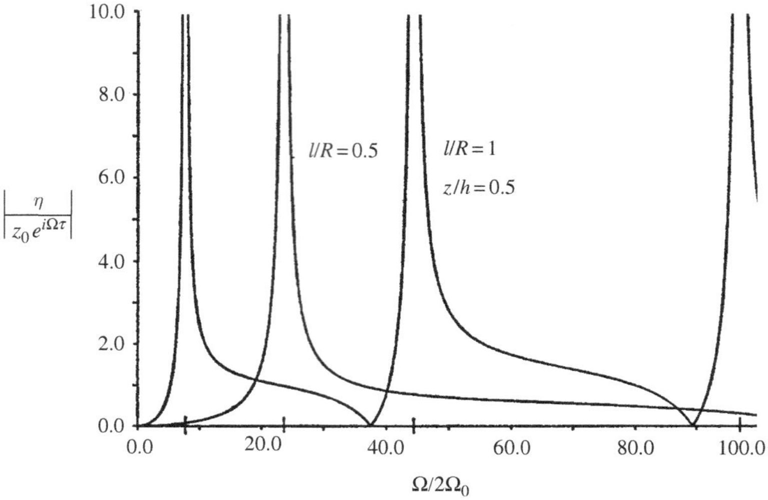

Figure 12.30 Amplitude–frequency response of the free liquid surface under axial harmonic excitation. (Bauer, 1989a)

while Figure 12.29(c) shows the amplitude–frequency responses for $l/R = 1$, and 0.5. Figure 12.30 shows the amplitude–frequency response for the elliptic case for $\mathsf{Bo} = 10$, and two different values of aspect ratio $l/R = 0.5$ and 1.

The static deflection of a cylindrical liquid column in support rotation Ω_0 and subjected to constant acceleration $\ddot{z} = -\Omega^2 z_0$ can be obtained from equation (12.152) after setting $\Omega = 0$. The resulting expression of the static deflection is

$$\frac{\eta_{\text{static}}(z)}{\ddot{z}(l/R)} = \frac{4}{\pi^2}\sum_{n=1}^{\infty}\frac{(-1)^2 \sin(((2n-1)\pi z)/l)}{(2n-1)^2\left[(\sigma/\rho R^3)((((2n-1)^2\pi^2 R^2)/l^2)-1)-\Omega_0^2\right]} \tag{12.153}$$

It is seen that the rotating bridge becomes unstable if

$$\Omega_0^2 \geq \frac{\sigma}{\rho R^3}\left(\frac{(2n-1)^2\pi^2 R^2}{l^2} - 1\right) \tag{12.154}$$

The distribution of the free-surface deflection, $\frac{\eta_{\text{static}}/(\ddot{z}l/R)}{(\sigma/\rho \mathrm{R}^3)}$, over the column length, z/l, is shown in Figure 12.31(a) and (b) for two values of length/radius ratio 1 and 0.2, respectively, each for two different values of Bond number $\mathsf{Bo} = \rho R^3\Omega_0^2/\sigma$. The response of liquid columns to pitching and roll excitations was later determined by Bauer (1991d,g) and Bauer and Komatsu (1994b). The Hamiltonian structure and three-dimensional instability of spinning liquid bridges were treated by Kruse, *et al.* (1999).

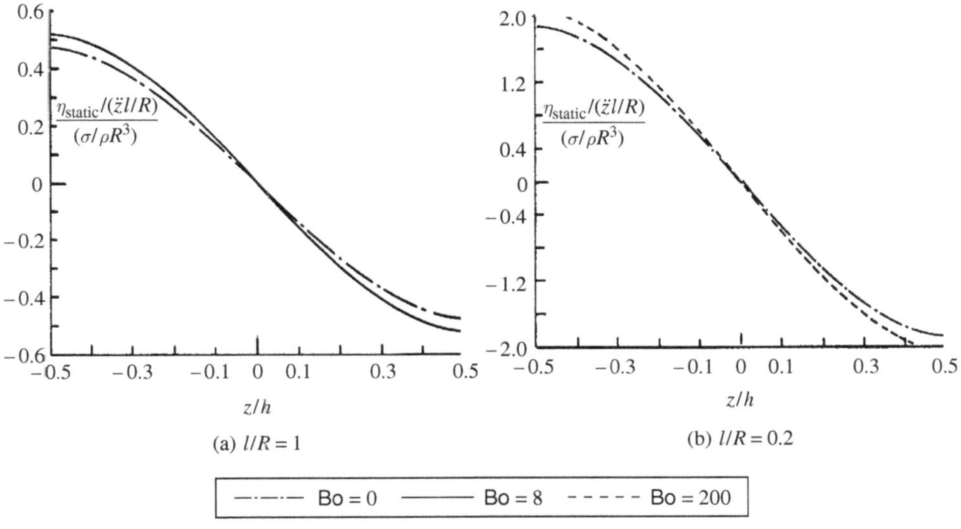

Figure 12.31 Static deflection of liquid column due to constant axial acceleration $\ddot{z}_0$ for two different length ratios, and for various Bond numbers. (Bauer, 1989a)

12.9 Thermocapillary convection

12.9.1 Thermocapillary instability of fluid flow

Thermocapillary convection plays an important role in microgravity fluid dynamics. Thermocapillary flows are driven by temperature-induced surface tension gradients at the interface between two immiscible fluids. An important feature of thermocapillary flow is the transition from steady to oscillatory flow under certain conditions. Much of the work done on oscillatory thermocapillary flow has been in the half-zone simulation of floating zone melt, a containerless crystal growth melt (Chang and Wilcox, 1975, 1976). Schwabe and Scharmann (1979) showed that oscillatory motion and instabilities of the system occurred by increasing the temperature gradient (that is, Marangoni number). Schwabe and Metzeger (1989) and Schwabe, *et al.* (1990) presented a nice review on oscillatory thermocapillary flow. However, the physical mechanism of oscillations is not fully understood, especially the importance of the free-surface deformability. Martinez (1976), Da-Riva and Martinez (1979), Davis (1987), and Chen and Chin (1995) presented linear stability analyses of thermocapillary effects in the floating zone. Hu and Tang (1989) presented some aspects of the excitation mechanism of thermocapillary oscillatory convection.

For most liquids, the surface tension decreases with increasing temperature. Thus, when the interface experiences a positive temperature gradient the bulk fluids on each side of the interface must balance an effective negative shear stress. Kamotani, *et al.* (1984), and Vrane and Smith (1994) suggested that a coupling among the velocity and temperature fields and the free-surface deformations was responsible for the oscillations. The free-surface deformation caused by the flow, although very small, changes the flow response time in an unsteady flow by a significant amount so that it triggers a three-way coupling among the velocity and temperature fields and surface deformation. According to Kamotani, *et al.* (1984), one oscillation cycle

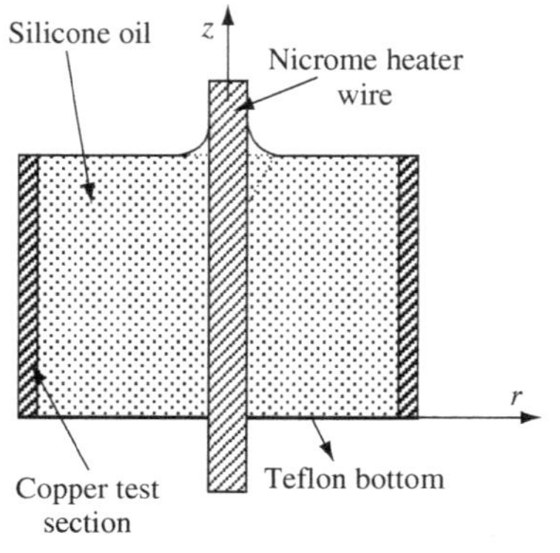

Figure 12.32 Schematic diagram of test section. (Lin, *et al.*, 1995)

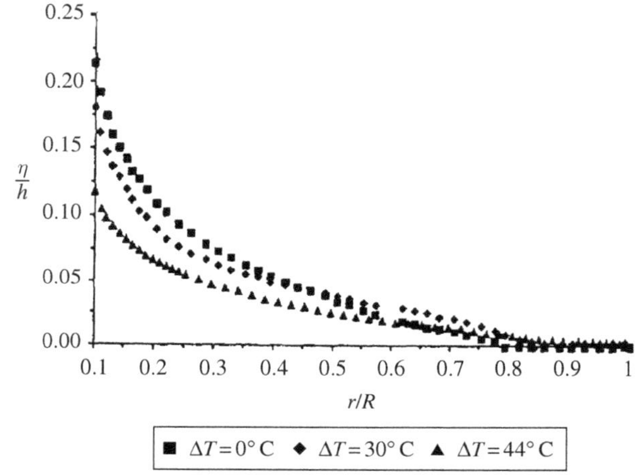

Figure 12.33 Comparison of measured free-surface shapes for a ratio fluid volume above the container edge (overfill volume) to the container volume of 0.0223 and three different temperature differentials. (Lin, *et al.*, 1995)

has an active period and a slow period of the surface flow. In the active period the surface flow is strong, the surface temperature increases, and the fluid is removed from the hot region and transported to the cold region. The opposite happens in the slow period. Lai (1990) and Chen and Hwu (1993) showed the existence of a time lag between the surface and return flows in unsteady thermocapillary flows caused by a deformable free surface. Kazarinoff and Wilkowski (1989, 1990) and Hu, *et al.* (1992a,b) analyzed axisymmetric flow in the half zone of silicon with a deformable free surface and showed the existence of oscillatory flow. Hu, *et al.* (1992b) found a sort of surface wave having a wave mode of large amplitude in one corner region of the liquid bridge. Anilkumar, *et al.* (1993) considered the control of thermocapillary convection in a liquid bridge by imposing vibration.

Gershuni and Zhukhovitskii (1976), Lin (1994) and Lin, *et al.* (1995) studied the free-surface motion in oscillatory thermocapillary flow in a cylindrical container of diameter 4.8 mm. The fluid is heated by a wire of diameter 0.48 mm placed along the centerline of the cylinder as shown in Figure 12.32. They measured the relative displacement and oscillation frequency of

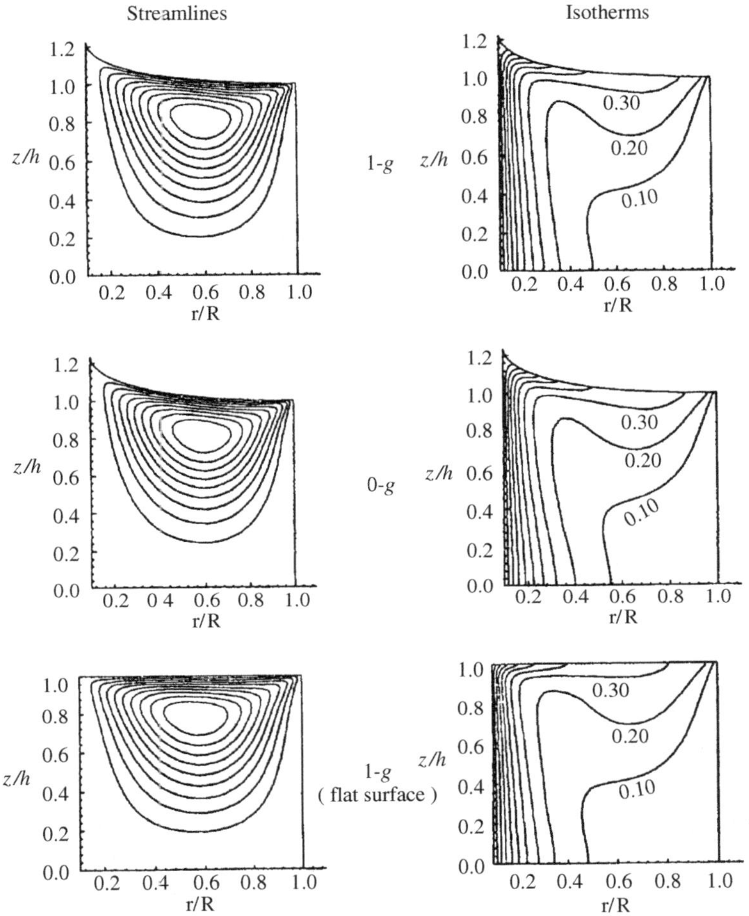

Figure 12.34 Computed streamlines and isotherms under 1-*g* and 0-*g* conditions and with curved and flat surface. (Lin, *et al.*, 1995)

the free surface. The static free surface shape was determined by the static Bond number, $\mathsf{Bo} = \rho g R^2/\sigma$, and the contact angle, ϑ_c, between the fluid and the container wall, and the fluid and heating wire. The fluid was filled to the edge of the container such that the contact angle is equal to or slightly greater than 90°. The heat transferred from the heater spread out quickly with increasing radius, which resulted in a large temperature gradient near the heater and thus the thermocapillary driving force acted mainly in a relatively small region near the heater. Figure 12.33 shows three different free-surface profiles each corresponding to three different values of the temperature difference, $\Delta T = T_h - T_c$, between heater and cold wall. Since the Marangoni number in this experiment was large, a thermal boundary layer existed along the heater as shown by the isotherms in Figure 12.34. As ΔT was increased, the free-surface height decreased as the thermocapillary pulled the fluid away from the heater region. The fluid was transported to the cold wall region and the surface height increased there although the increase was much less than the height reduction near the heater because the surface area was increased with radius.

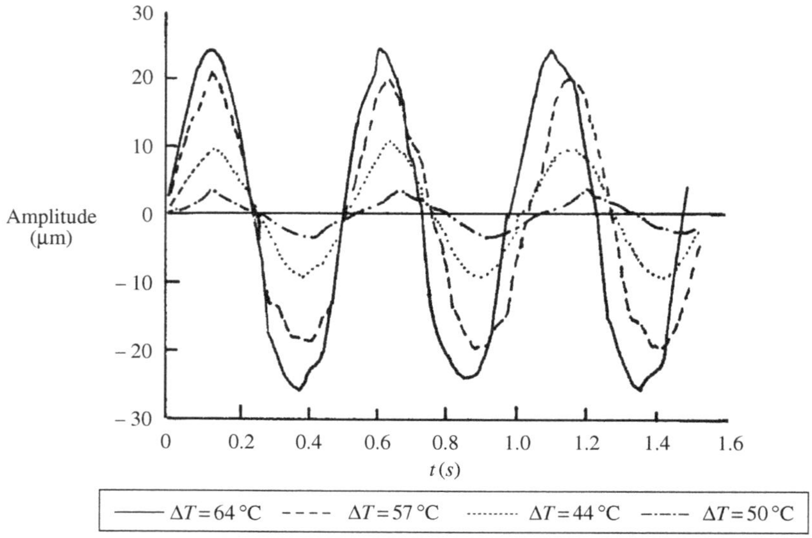

Figure 12.35 Time evolution of free-surface height during oscillations at $r/R \approx 0.84$ for different values of temperature difference. (Lin, *et al.*, 1995)

There is a critical temperature difference, ΔT_{cr}, above which the flow becomes oscillatory. Kamotani, *et al.* (1992) found that $\Delta T_{cr} = 41.2\,^\circ\text{C}$ for $\mathsf{Ma}_{cr} = 7.3 \times 10^4$. As ΔT was increased above the critical value the amplitude of the free-surface oscillations increased as well. According to the experimental measurements of Lin, *et al.* (1995), Figure 12.35 shows time history records of the free-surface oscillations at position $r/R = 0.84$ for four different values of temperature difference. The frequency of oscillations of each case is not the same, and it was found that $f_{\Delta T=44} = 1.75$ Hz, $f_{\Delta T=50} = 1.85$ Hz, $f_{\Delta T=57} = 1.92$ Hz, and $f_{\Delta T=64} = 2.03$ Hz. Schneider, *et al.* (1996) conducted another experimental investigation on an annular cylinder filled with a viscous fluid. Other experiments were conducted by Szymczyk and Siekmann (1989) on thermocapillary flow along a cylindrical interface under rotation.

Carruthers, *et al.* (1975) and Bauer (1989g) found that by rotating the liquid column about its longitudinal axis, a new vortex ring is established, which moves towards the free surface of the column as the spin increases. This movement is opposite to the vortex ring of the Marangoni convection. By proper choice of the spin speed, the velocity distribution of shear stress inside the column may be reduced. Chun and Wuest (1979) and Schwabe and Scharmann (1979) found that Marangoni convection is only stable below a critical Marangoni number. Above that critical number there is a transition into an oscillatory and the turbulent motion. Later, Chun (1980a) introduced rotation to the liquid column about its longitudinal axis to eliminate the unsteady motion. Bauer (1993f) considered the Marangoni convection in a rotating liquid container under the action of an axial temperature gradient. The rotating liquid formed an annular volume and the temperatures at the top and bottom of the container are different. Bauer's results revealed that the rotation of the liquid gives a reduction of the velocity distribution inside the liquid and in a shift of the Marangoni vortices toward the free surface of the core. Numerical simulations of three-dimensional thermocapillary flows in liquid bridges were presented by Castagnolo and Carotnuto (1999) and Lavalley, *et al.* (2001).

Bauer (1983e,g, 1985b, 1986d, 1987d, 1988d) considered the Marangoni convection due to time-wise oscillation of the temperature gradient applied to the free surface of a floating liquid column or liquid sphere. He found that the liquid free-surface displacement from the equilibrium position might, in the vicinity of the liquid resonance, reach a magnitude capable of disintegrating the system. In another study, Bauer (1983f) studied the transient convection due to a sudden change of an imposed axial temperature field at the surface of an infinite cylindrical liquid bridge. The combined thermocapillary and natural convection in the presence of g-jitter were considered by Bauer (1984h). The presence of g-jitter results in an unwanted velocity field in the liquid, which decreases its magnitude with increasing forcing frequency of the oscillatory temperature. Bauer and Buchholz (1998) analytically studied the Marangoni convection in a rectangular container subjected to a temperature gradient. A steady Marangoni convection takes place when one side of the tank walls is warmer than the opposite side. Time-oscillatory temperatures of the walls result in a time-dependent thermocapillary convection, which creates wave patterns on the free surface of the liquid (Bauer and Eidel, 1987c).

The transient thermo-fluid-dynamic fields in a differentially heated fluid cell in the presence of g-pulses of finite duration were analyzed by Monti and Savino (1994a, b, c), and Monti, Savino, and Alterio (1997a). They showed that for g-pulses orthogonal to the fluid density gradient, the g-tolerability could be formulated in terms of a maximum allowable g-level multiplied by the duration of the pulse. High frequency g-jitters in the presence of temperature and/or concentration gradients were found to give rise to time-average convective flows and in the range of frequencies of interest for the International Space Station (ISS).

Instabilities due to surface tension driven convection have been of great interest in recent years. For example, Marangoni flow (or thermo-capillary convection of the Benard type) occurs in a fluid layer with at least one free surface. It starts when a temperature gradient is established along the free surface such that there is no first transition. Along the free surface, surface tension can act to drive convection if the surface tension varies in magnitude along the surface due to its dependence upon a spatially varying temperature associated with a thermal disturbance.

For higher flow transition, a Hopf bifurcation similar to the Benard–Marangoni flow instability could exist in this configuration. Transition from steady state to oscillatory flow was experimentally observed in various container geometries (Ostrach, *et al.*, 1985, Monti and Fortezza, 1989, and Velten, *et al.*, 1991). Hsieh, *et al.* (1994) studied the onset of oscillations in terms of a modified dynamic Bond number and the container aspect ratio.

If a horizontal, unidirectional shear flow exists in the fluid layer that is heated, convection begins at the same value of a critical Rayleigh number (Ra_c) as without shear, but a well-defined pattern of rolls with axes in the flow direction, known as "longitudinal rolls," is predicted to be the preferred pattern of convection as experimentally observed by Kelly (1994). This means that the shear has a stabilizing effect upon all disturbances that have a nonzero wave number in the direction of the shear. Kelly and Hu (1993) and Hu and Kelly (1994) studied the case of oscillatory shear for small values of Reynolds number and concluded that stabilization does indeed occur. The degree of stabilization was found to increase with the Prandtl number. Kelly and Or (1994) indicated that the general problem of a deformable surface involves several nondimensional parameters such as the Marangoni number (Ma), the Prandtl number (Pr), the Bond number (Bo), and the crispation number (Cr), which tends to characterize the deformability of the surface. They showed that the nonplanar flow oscillations could have a significant effect upon the onset of Marangoni convection. In contrast to Rayleigh–Benard convection, the effect can be stabilizing or destabilizing, depending upon the operating conditions. Thevenard and Ben Hadid (1991) and Tao and Kou (1996) studied thermocapillary convection in a low Pr number in the presence and absence of g-jitter.

In an effort to control the oscillatory thermocapillary convection in microgravity, Neitzel (1994) conducted laboratory and numerical experiments in a thin layer of silicone oil. Neitzel was able to characterize the flow state in a limited range of Bond number–Marangoni number spaces of interest. Numerical studies revealed transitions in the pure thermocapillary problem where Grashof number (Gr) vanishes (Ohnishi and Azuma, 1992, Chen and Hwu, 1993, and Peltier and Biringen, 1993). For high Ma convection in a square cavity, only steady states were numerically predicted by Zebib, *et al.* (1985), Carpenter and Homsy (1990), and Buckle and Peric (1992). Mundrane and Zebib (1994) studied thermocapillary and buoyant thermo-capillary in rectangular cavities. They showed that while pure buoyant convection exhibits oscillatory behavior for Grashof numbers greater than the critical value (for the pure buoyant problem), pure thermocapillary convection is steady within the range of parameters tested. They considered the behavior of Hopf bifurcation, which exists in the pure buoyant case (when Reynolds number vanishes) but subsequently disappears under pure thermocapillary conditions.

12.9.2 Thermocapillary instability of liquid bridges

Thermocapillary convection in liquid bridges takes place in a fluid–fluid interface with a nonuniform interface tension, due to a temperature gradient along the interface itself. The studies of Marangoni flows in liquid bridges showed an oscillatory Marangoni convection. Schwabe, *et al.* (1978) showed the existence of a critical Marangoni number for the onset of the time-dependent asymmetric flow in high Prandtl number fluids. Chun and Wuest (1979), Schwabe, *et al.* (1979), Chun (1980a), Preisser, *et al.* (1983), Velten, *et al.* (1991), Monti *et al.* (1992, 1998), Frank and Schwabe (1998), and Carotenuto, *et al.* (1998) conducted a series of experiments to study the onset of oscillatory Marangoni flow. Their experimental observations revealed that for a liquid bridge of a given geometry, in the presence of sufficiently small temperature differences, the Marangoni convection is characterized by a steady, axisymmetric, toroidal flow pattern driven by the surface tension imbalance. Depending on the geometry, liquid properties, and boundary conditions, when the temperature difference exceeds a critical value, the flow experiences a transition to an oscillatory and three-dimensional pattern. Bauer (1982e, 1984c) and Bauer and Eidel (1991d, 1992) analytically determined the thermal convection inside a liquid bridge subjected to a purely angular temperature field at its free surface. The transition process from laminar to oscillatory Marangoni convection in a liquid bridge was considered by Hirata, *et al.* (1997). The stability of higher modes of the oscillatory Marangoni convection in liquid columns was determined by Chun and Wuest (1982) and Chun and Siekman (1994). Numerical studies of this problem (Schwabe, *et al.* 1978, 1991, Napolitano, 1982, Xu and Davis, 1983, Chun, 1984 and Shen, *et al.*, 1990) suggested that the onset conditions should be characterized by a critical value of the conditional Peclet number, known also as the critical Marangoni number (Ma_c)

$$\mathsf{Ma}_c = \frac{\Delta\sigma}{\mu}\frac{l}{\alpha} \tag{12.155}$$

where $\Delta\sigma$ is the variation of surface tension along a nonisothermal surface, μ is the kinematic viscosity, α is the thermal diffusivity, and l is the bridge length.

The liquid column free oscillations induced by temperature fluctuations or sudden change of temperature were determined by Bauer (1983d,h). Bauer and Eidel (1998d) studied the oscillatory thermocapillary convection in a spinning liquid bridge under a microgravity field. Rybicki and Florian (1987) studied the steady axisymmetric thermocapillary convection in cylindrical liquid bridges. They numerically analyzed the effects of the bridge aspect ratio and showed that a decrease in the aspect ratio (short bridges) would lead to the creation of several layers of vortices, with the strength of each layer decreasing approximately exponentially with the distance from the surface. Rupp, *et al.* (1989) developed a 3-D numerical algorithm to simulate the transition of Marangoni flow in cylindrical half zone configuration. They examined the behavior of the instability for several low Prandtl liquids and for a fixed aspect ratio. For liquid metals the first bifurcation is stationary (i.e., the supercritical 3-D state is steady) and that the regime becomes oscillatory only when the Marangoni number is further increased (second oscillatory bifurcation). For a fixed small Prandtl number ($\mathsf{Pr} = 0.01$) and for a fixed aspect ratio, Levenstam and Amberg (1995) analyzed the nature of the first steady and of the second oscillatory bifurcations of the Marangoni flow. Lappa and Savino (1999) showed that, for $\mathsf{Pr} = 0.04$, there is a strong relationship between the value of the aspect ratio and the critical azimuth wave number for instability to occur.

Kuhlmann and Rath (1993a,b), and Wanschura, *et al.* (1995) investigated the linear stability of steady axisymmetric thermocapillary flow in cylindrical liquid bridges. Kuhlman and Rath (1993a) found that for low Prandtl numbers, the instability breaks the spatial axisymmetry prior to the onset of time dependence (stationary bifurcation). On the other hand, for high Prandtl numbers, the first bifurcation is oscillatory (Hopf bifurcation). The three-dimensional supercritical state after the Hopf bifurcation point is interpreted as a superposition of two counter-propagating waves characterized by an axial and an azimuth component. The resulting disturbance exhibits an azimuth component running under a certain angle against the axis of the bridge. The ground experiment conducted by Velten, *et al.*, (1991) revealed an oscillatory instability with zero azimuth phase shift of the temperatures measured by thermocouples located in different azimuth positions. This instability identifies a third spatio-temporal model of oscillatory convection. This result seems to be in disagreement with the linear stability analysis by Kuhlman and Rath (1993a), who predicted instability waves with a nonzero azimuth component. Muehlner, *et al.* (1997) observed helical traveling-wave convection in a liquid bridge.

Shevtsova, *et al.* (1995, 1996, 1997) and Shevtsova and Legros (1997, 1998a, b) studied the effect of the liquid bridge volume on the steady thermocapillary flow by solving the axisymmetric Navier–Stokes equations in the domain occupied by the liquid. They transformed the physical space to a rectangular numerical domain using curvilinear body fitted coordinates. The transformed equations were numerically solved using a two-dimensional time-dependent finite difference technique. Later, Shevtsova, *et al.* (1999) examined the loss of stability of the steady convection in deformable liquid bridges of silicone oil with Prandtl number $\mathsf{Pr} = 105$. The deformation was found to be essential in the transition process from the steady axisymmetric two-dimensional basic state to the three-dimensional periodical one. The transition was characterized by two different oscillatory instability branches. Between these branches there is a small range of volume over which the steady flow is stable for very high values of Prandtl number. Chen and Hu (1998) used the linear stability analysis to study the influence of the liquid bridge volume on the instability of Marangoni flow for high Prandtl liquids. Monti, *et al.* (2000a) and Lappa, *et al.* (2001) developed a three-dimensional numerical simulation to study the influence of the shape factor of liquid bridges on the critical Marangoni number and critical azimuth wave number for semiconductor melts under zero gravity conditions. Their results showed that for semiconductor melts the first bifurcation was characterized by the loss of spatial symmetry rather than by the onset of time-dependence. When the basic axisymmetric flow field became unstable, after a short transient, a three-dimensional supercritical steady state was achieved. It was found that the Marangoni instability is sensitive to the shape factor, which measures the ratio of the volume of the liquid bridge to the volume of the cylindrical bridge.

Experiments conducted under microgravity conditions on large liquid bridges revealed that the oscillations occur at high values of the critical Marangoni number, Ma_c. Albanese, *et al.* (1995) reported some experimental measurements for the purpose of evaluating the onset (transition) condition from temperature data. They measured the dependence of the critical Marangoni numbers on the geometrical parameters of the liquid bridge and observed that Ma_c is an increasing function of the aspect ratio of the bridge. The comparisons of measured results in microgravity with those predicted by linear stability analyses, based on idealized analytical models are complicated due to the fact that the experiments typically involve time-depended boundary conditions, while the usual stability analysis assumes steady basic states. Monti, *et al.* (1997b) developed a numerical code to give three-dimensional and time-dependent simulation of the standard (zero surface deformation) half-zone model with time-dependent end-wall temperatures.

The results of the experiment PULSAR (Pulsating and Rotating Instabilities in Oscillatory Marangoni Flows) performed on the MAXUS 3 Sounding Rocket were reported by Monti, *et al.* (2000b). The oscillatory Marangoni convection in a cylindrical liquid bridge of silicon oil with a kinematic viscosity of 5 centistokes was reported. On-ground numerical simulations and micro-scale experimental studies showed that the oscillatory Marangoni instability occurs at the beginning in the form of a pulsating regime, caused by a hydrothermal standing wave, and then it turns to a rotating regime caused by traveling waves. Figures 12.36 and 12.37 show a comparison between numerical and experimental results of temperature–time history records near the hot disc and cold disc, respectively. The temperatures of two thermocouples azimuthally shifted at 90°. The numerical simulations of Figures 12.36(a) and 12.37(a) show that a standing wave is established at the onset of the periodic oscillations. These two records do not reveal any phase shift between the temperature signals by the two thermocouples at 90°. The experimental signals shown in Figures 12.36(b) and 12.37(b) reveal two different phases. During the time interval 100–450 s, a standing wave is established, and over the period 220–300 s a mixed regime is observed. The mixed regime is caused by the relative rotation of the pulsating plane around the bridge axis, probably induced by the strong angular acceleration at $t = 221$ s. The temperatures of the thermocouples near the hot disc, Figure 12.36(b), reveal that the amplitude of one temperature signal decreases while the other increases. This behavior was numerically simulated with a rotation of the pulsating plane around the bridge axis with angular speed of 0.15 deg/s as shown in Figures 12.36(c) and 12.37(c).

In crystal growth, it is well known that perturbations in the process of solidification are detrimental and that due to the heating of the crystal a temperature gradient exists. This temperature gradient affects the surface tension and thermal Marangoni convection takes place inside the liquid column (Wuest, 1976 and Chun, 1980a,b).

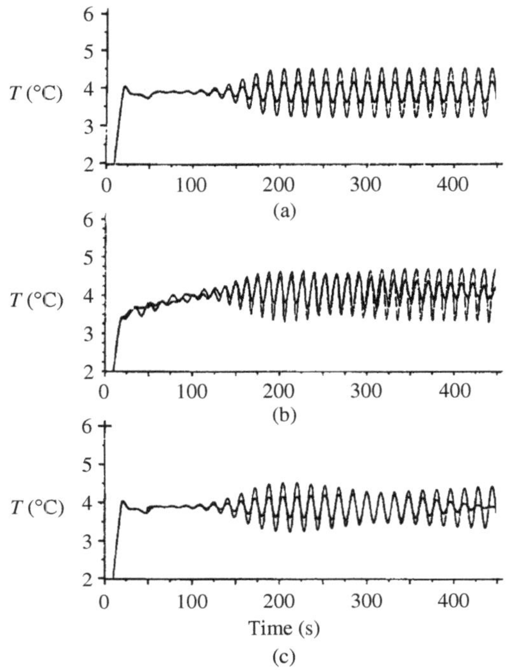

Figure 12.36 Comparison between numerical (a, c) and experimental (b) signals of temperatures measured by two thermocouples near the hot disc shifted at 90°. Figure (c) is obtained with a rotation of the pulsating plane around the bridge with an angular speed of 0.15 deg/s. Time axis shows the time after 200 seconds from the rocket launch. (Monti, *et al.*, 2000b)

Consider a cylindrical liquid bridge whose lower end, $z = 0$, is cold, and the upper end, $z = l$, is hot. It is assumed that the gravity is zero and the governing field equations: namely continuity, Navier–Stokes, and energy equations, are

$$\nabla \cdot \mathbf{v} = 0, \ \ \mathbf{v} = u\mathbf{e}_r + v\mathbf{e}_\theta + w\mathbf{e}_z \tag{12.156}$$

$$\frac{D\mathbf{v}}{Dt} = -\frac{1}{\rho}\nabla p + \nu \nabla^2 \mathbf{v} \tag{12.157}$$

$$\frac{DT}{Dt} = \alpha \nabla^2 T \tag{12.158}$$

where α is the thermal diffusivity. The initial conditions are

$$\mathbf{v}(r, z, \theta, t = 0) = T(r, z, \theta, t = 0) = 0 \tag{12.159}$$

The boundary conditions are
Cold disc

$$\mathbf{v}(r, z = 0, \theta, t) = 0, \ \ T(r, z = 0, \theta, t) = -\mathcal{R}t \tag{12.160a,b}$$

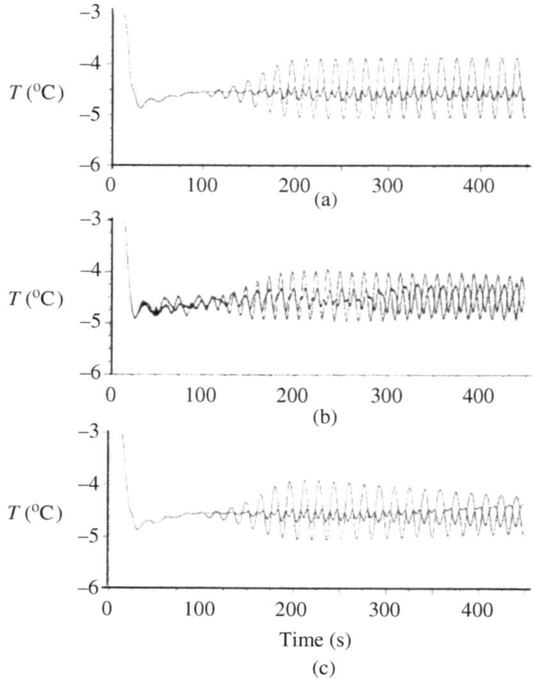

Figure 12.37 Comparison between numerical (a, c) and experimental (b) signals of temperatures measured by two thermocouples near the cold disc shifted at 90°. Figure (c) is obtained with a rotation of the pulsating plane around the bridge with an angular speed of 0.15 deg/s. Time axis shows the time after 200 seconds from the rocket launch. (Monti, *et al.* 2000b)

Hot disc

$$\mathbf{v}(r, z = l, \theta, t) = 0, \quad T(r, z = l, \theta, t) = \mathcal{R}t \tag{12.160c,d}$$

where t is the time, and $\mathcal{R}$ is the rate of temperature ramping. At the free surface, $r = R$, we have

$$\mathbf{u}(r = R, \theta, z, t) = 0 \tag{12.161a}$$

$$\mu \frac{\partial w(r = R, \theta, z, t)}{\partial r} = -\frac{\partial \sigma}{\partial T} \frac{\partial T(r = R, \theta, z, t)}{\partial z} \tag{12.161b}$$

$$r \frac{\partial v(r = R, \theta, z, t)}{\partial r} - v\theta = -\frac{1}{\mu} \frac{\partial \sigma}{\partial T} \frac{\partial T(r = R, \theta, z, t)}{\partial \theta} \tag{12.161c}$$

$$\frac{\partial T(r = R, \theta, z, t)}{\partial r} = 0 \tag{12.161d}$$

Monti, *et al.* (1997a) conducted numerical simulations of the above boundary value problem for a liquid bridge of silicon oil 2 cSt, length $l = 20$ mm and radius $R = 10$ mm. The temperature difference between the supporting discs was increased from zero to $\Delta T = 30$ K. As the temperature exceeds the critical temperature difference between the supporting discs, a

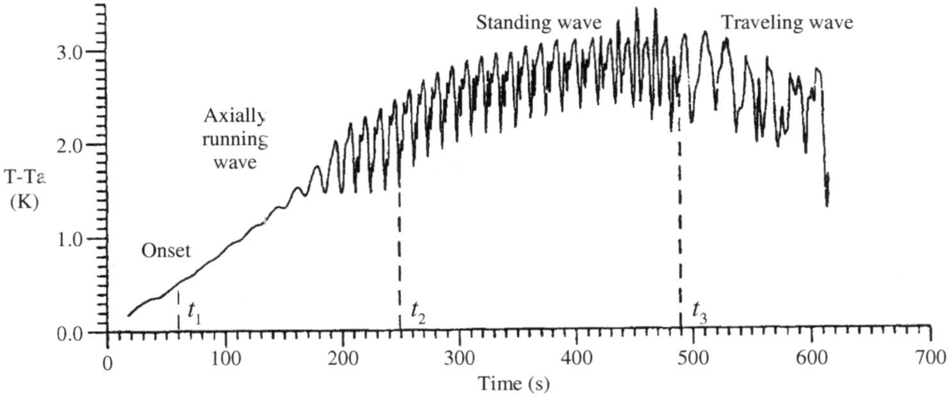

Figure 12.38 Time evolution of the temperature at $z = 0.75l$, $r = 0.9\ R$, $\theta = \pi$, with rate of temperature of $\Re = 0.6$ K/s. (Monti, *et al.*, 1997a)

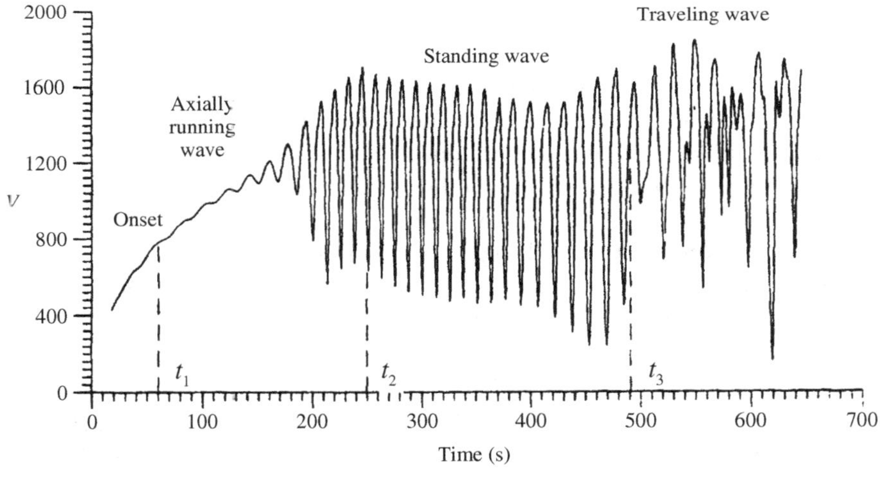

Figure 12.39 Time evolution of the nondimensional velocity at $z = 0.75l$, $r = 0.9\ R$, $\theta = \pi$, with rate of temperature of $R = 0.6$ K/s. (Monti, *et al.*, 1997a)

transient oscillatory phase develops until a stable supercritical oscillatory three-dimensional flow is reached. For example, a fluid particle located at $r = 0.9\ R$, $\theta = \pi$, and $z = 0.75l$, experiences an oscillatory behavior of temperature and velocity as shown in Figures 12.38 and 12.39. Note that the vertical axis of the temperature measures the deviation of the particle temperature from its temperature while the velocity axis is nondimensional. Thermocapillary convection became oscillatory when $\Delta T = T_{\text{hot}} - T_{\text{cold}} > 7$ K . When the imposed temperature was further increased, both the temperature and velocity of the liquid particle exhibited periodic oscillations.

Earlier, Preisser, *et al.* (1983) found that the amplitude of the temperature oscillation, δT, increases as a function of ΔT (or the instantaneous Marangoni number)

$$\mathsf{Ma} = \frac{\partial \sigma}{\partial T} \Delta T L / \mu \alpha \tag{12.162}$$

Velten, *et al.* (1991) found that the magnitude of the fluctuation could be generalized as

$$\delta T \cong \Delta T \left[\frac{\Delta T - \Delta T_c}{\Delta T_c} \right]^s \tag{12.163}$$

where the power s depends on the geometry of the bridge and the fluid properties. Monti, *et al.* (1997a) found that $s = 0.6$ fits very well their results. Monti and Savino (1995, 1996) identified a characteristic frequency that correlates experimental results obtained on ground and in microgravity, for $8 \leq \mathsf{Pr} \leq 74$, and aspect ratio $0.1 \leq \Re \leq 1.6$, by the following expression

$$f = \frac{\alpha}{2\pi} \frac{\mathsf{Ma}_c^{2/3}}{P^{3/2} \sqrt{2R}} \tag{12.164}$$

where Ma_c is the critical Marangoni number at which the onset of oscillatory instability occurs.

12.10 Sloshing problems of cryogenics

12.10.1 *Physical characteristics of superfluids*

Cryogenics is the science of producing and studying phenomena that occur at very low temperatures, that is, below 80 K. In 1898, James Dewar first liquefied hydrogen (Scurlock, 1991 and Weisend, 1998). Dewar also developed the vacuum-insulated flask with reflective walls for containing cryogenic liquids. His design is still at the heart of cryogenics containers. A comprehensive state-of-the-art of various cryogenic sub-fields is well documented in a handbook edited by Weisend (1998). In space technology, cryogenics is found in the liquid hydrogen and oxygen fuels used in rocket engines because of their performance advantages and in applications such as the Cosmic Background Explorer (COBE) satellite. Some experimental spacecraft use cryogenic cooling for telescope instrumentation, and occasionally to maintain temperatures near absolute zero for mechanical stability. Abdalla, *et al.* (1964) studied the performance of ullage surface with liquid hydrogen subjected to unsymmetrical heating. Aydelott (1995) addressed the technology requirement of cryogenic fluid management by NASA.

Superfluids experience an interactive mechanism if they possess velocities above a critical velocity (Liepmann, 1984, Van Sciver, 1986 and Donnelly, 1991). Superfluidity is a term used to refer to the frictionless flow of liquid helium at low temperature. Properties of the superfluid state (He-II) include, Parker (1983),

(1) Frictionless flow through holes as small as 2 nm in diameter for particle velocities up to approximately 50 cm/s.
(2) Formation of a highly mobile film over all surfaces touching the bulk liquid.
(3) Transmission of temperature waves in addition to pressure waves.
(4) Joint occurrence of temperature and pressure differences in connected regions of fluid (thermo-mechanical effects).
(5) Inability to carry heat when flowing through small capillaries.
(6) Inability to fully rotate in the manner of a solid body (irrotational flow).
(7) Failure of evaporation and condensation of helium vapor to produce any change in superfluid velocity.

At a temperature close to absolute zero (0 K = −273 °Celsius), quantum effects begin to be of importance in the properties of fluids. At a temperature of 2.17 K, liquid helium has a λ-point, which defines a second-order phase transition. At temperatures below this point liquid helium (He-II) has a number of remarkable properties, the most important of which is superfluidity. This is the property of being able to flow in low velocities without viscosity in very narrow capillaries or gaps. At temperatures other than zero, helium II behaves as if it were a mixture of two different liquids. One of these is a superfluid and moves with zero viscosity along a solid surface. The other is a normal viscous fluid. It is of great importance that friction occurs between these two parts of the liquid in their relative motion, that is, no momentum is transferred from one liquid to the other. Addriaans, *et al.* (1996) reported experimental measurements of the thermal gradient across the superfluid (He-II) – normal fluid (He-I) interface in helium under microgravity conditions. The experiments conducted by Mason, *et al.* (1978) showed that the classical fluid mechanics theory is applicable to cryogenic helium in large containers with sufficiently large velocities (see, Donnelly, 1988, 1991).

One of the major problems is the handling of these fluids in their containers during low-gravity maneuvering. These liquids may be reoriented and will not be able to move to the engine for re-starts. The screened channels that are commonly used for nonvolatile propellants, such as kerosene, are not convenient for cryogenics since they are susceptible to the generation of vapor within the channel by heat transfer.

12.10.2 Sloshing of cryogenics

The motion of helium-II in the low-gravity environment of an orbiting satellite is of critical importance for the design of the satellite attitude and translation control systems. Space vehicle designers are particularly interested in the frequency of oscillation and damping rate of the fundamental slosh mode as well as the cross-axis coupling induced by rotation of the satellite. The results of cryogenic liquid dynamic behavior in orbits were reported by Bowman (1966b). Hung, Lee, and Leslie (1992b) numerically studied the dynamical behavior of cryogenic liquid propellant under g-jitter. They included the effect of surface tension on partially rotating fluids. Their results revealed a train of large-amplitude oscillations in terms of wave mode propagation along the liquid–vapor interface. In addition to the Earth's gravitational force, the interaction between the fluid mass and the spacecraft mass due to gravity gradient accelerations is capable of exciting slosh waves in fluid systems. Under g-jitters, a group of waves with different frequencies and wavelengths of slosh waves are generated by the restoring force field of g-jitters.

Hung, Pan, and Long (1994) found that the equilibrium shape of the liquid–vapor interface is spherical for a Dewar tank with 70% liquid filling under a residual gravity environment below $10^{-7}g_0$, where g_0 is the regular gravitational acceleration. The time taken to reach this spherical shape was found to be 80 s after starting jitter acceleration associated with a slew motion. Hung (1990, 1993a) studied the sloshing of helium II in a rotating Dewar container.

Some concerns and uncertainties about the potential for persistent fluid sloshing due to the absence of viscosity in the super-fluid component of helium II are discussed by Ross (1994). Ross (1994) performed low-gravity verification experiments on the slosh behavior of He-II for use in the development of a computational fluid dynamics model that incorporates the two-fluid physics of He-II. He found that the two-fluid code predicts a different fluid motion response in a low-gravity environment from that predicted by a single-fluid model, while a 1-g

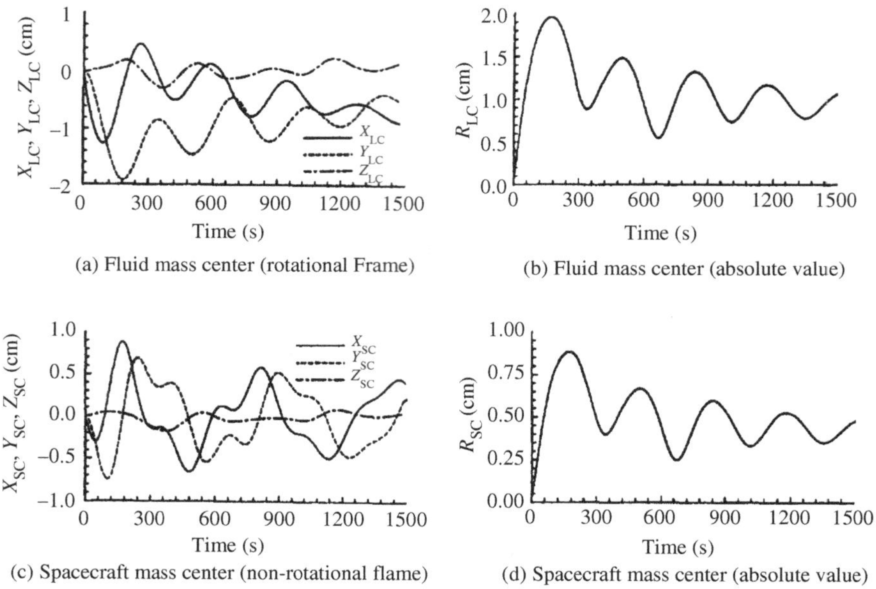

Figure 12.40 Time evolution of fluid and spacecraft mass centers due to coupling of lateral impulse and orbital dynamics. (Hung, Long and Zu, 1996)

response is identical for both types of model. Hung (1993b) conducted a simulation study to determine the forces and torques induced by cryogenic sloshing in a Dewar container. Hung (1994a,b, 1995a, 1996, 1997) presented a series of studies dealing with the sloshing of cryogenic helium driven by different forms of excitations in a microgravity field.

Liquid helium at a temperature of 1.8 K has been used on the Gravity Probe-B (GP-B). The equilibrium shape of the interface is governed by a balance of capillary, centrifugal, gravitational, and dynamic forces. Sloshing waves can be excited by longitudinal and lateral residual accelerations such as the Earth's gravity gradient and g-jitter. Hung, Long, and Zu (1996) studied the transient phenomena of coupling between slosh reaction torques driven by three types of excitations and spacecraft orbital dynamics. The three excitations are (a) lateral impulse, (b) gravity gradient and/or (c) g-jitter. They numerically solved the governing equations of slosh dynamics and orbital translational and rotational motions of spacecraft dynamics. The slosh dynamics were treated based on a rotational frame, while the orbital dynamics were associated with a nonrotational frame.

Under lateral impulse along the x-axis, the liquid–vapor interface (bubble) time evolution was determined. The bubble is first shifted to be in the positive x-direction and the liquid to be in the negative direction. The bubble experiences motion to the positive y-direction due to the Coriolis force with rotation along the z-axis in the rotational frame. The bubble also exhibits deformations with back-and-forth oscillations. Figure 12.40 shows the time evolutions of the growth and decay of fluid mass center fluctuations in response to a lateral impulse. The figure also shows the time evolution of the spacecraft mass center.

Under combined gravity gradient and g-jitter shown in Figure 12.41 over the entire orbit period of 6000 seconds. The gravity gradient acceleration, $\mathbf{a}_{gg}$, acting on the fluid mass of the spacecraft was given by Hung and Long (1995a) and Hung and Pan (1993, 1995a) in the form

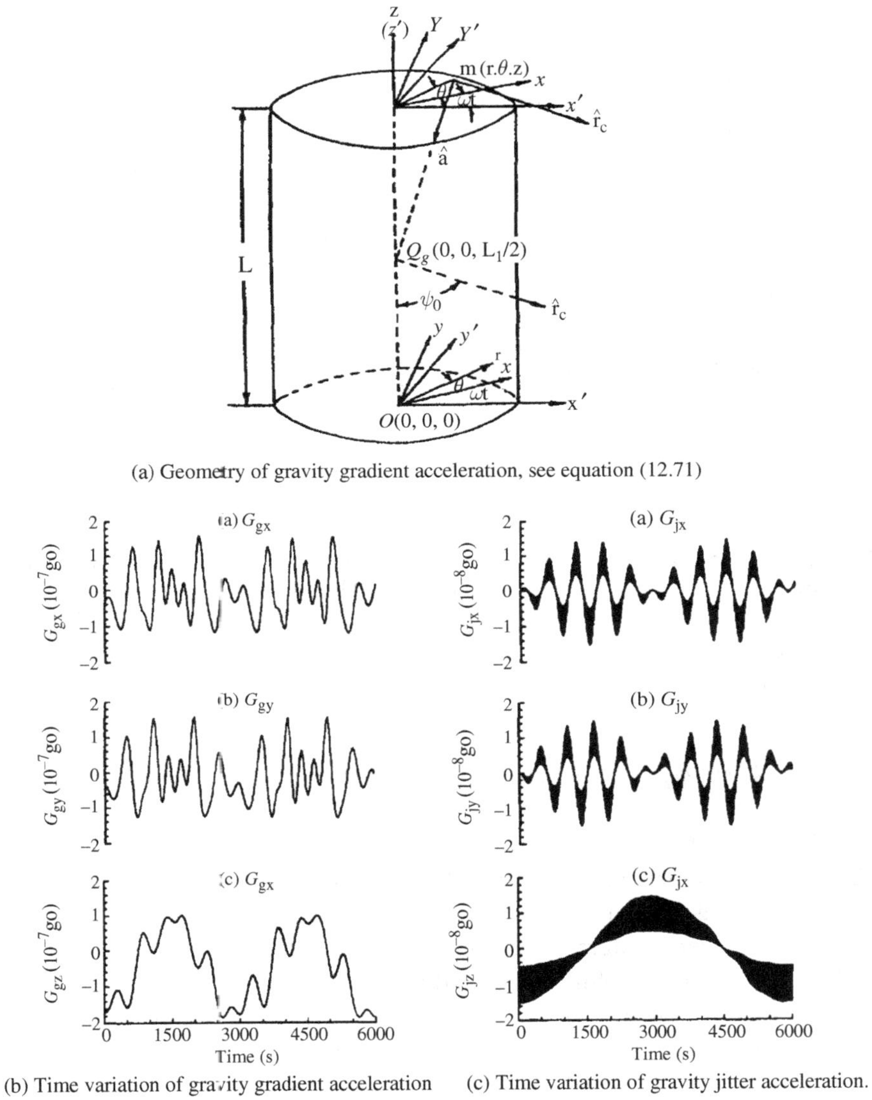

(a) Geometry of gravity gradient acceleration, see equation (12.71)

(b) Time variation of gravity gradient acceleration (c) Time variation of gravity jitter acceleration.

Figure 12.41 Time evolution of gravity gradient and g-jitter accelerations on the spacecraft during a full orbit period acting on a fluid element located at $(r, \theta, z) = (40\,\text{cm}, \pi/4, 10\,\text{cm})$. (Hung, Long and Zu, 1996)

$$\mathbf{a}_{gg} = -n^2[3(\mathbf{r}_c \cdot \mathbf{d})\mathbf{r}_c - \mathbf{d}] \tag{12.165}$$

where $n = 2\pi/\tau_0$ is the orbit rate, τ_0 is the orbit period (5856 s), $\mathbf{r}_c$ is a unit vector in the direction from the spacecraft mass center to the center of the Earth, $\mathbf{d}$ is a vector from the fluid element to the spacecraft mass center. Fluctuations in the residual gravity due to g-jitter acceleration were proposed in the form

$$g = g_B\left[1 + \frac{1}{2}\sin(2\pi f t)\right] \tag{12.166}$$

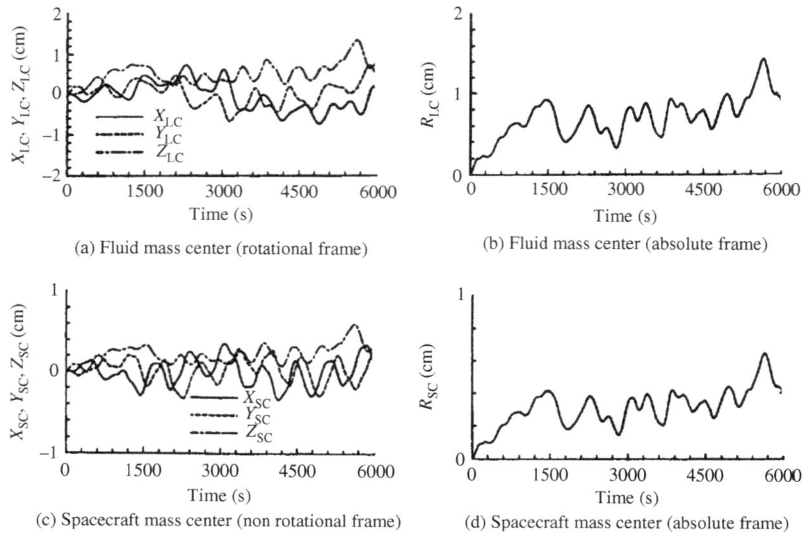

Figure 12.42 Time evolution of fluid and spacecraft mass centers driven by gravity gradient-dominated acceleration and orbital dynamics. (Hung, Long and Zu, 1996)

where g_B is the background gravity environment and f is the frequency of the g-jitter. Components of the g-jitter acceleration in the noninertial coordinate system are

$$\begin{aligned} \boldsymbol{a}_{gj} &= \{a_{gj,r},\ a_{gj,\theta},\ a_{gj,z}\} \\ &= \{g \sin \psi_E \cos(\theta + \omega t), -g \sin \psi_E \sin(\theta + \omega t), -g \cos \psi_E\} \end{aligned} \tag{12.167}$$

where $\psi_E = 2\pi t / \tau_0$ is the azimuth angle of the Earth to the spacecraft mass center as shown in Figure 12.41(a). The numerical simulation of Hung, Long, and Zu (1996) for the combined gravity gradient and g-jitter included two cases, see Figure 12.41(b) and (c). The first is when the gravity gradient is dominated while the second is when the g-jitter is dominated. Figure 12.42 shows the time evolutions of the fluid mass center fluctuations and the spacecraft mass center fluctuations due to gravity gradient-dominated acceleration. Figure 12.43 shows the time evolutions of the mass centers driven by g-jitter-dominated accelerations. It is clear that the coupling between He-II slosh dynamics and orbital dynamics can cause large-amplitude fluctuations to act on a spacecraft, resulting in deviations from normal operation.

Hung and Pan (1995c, 1996) developed an analytical modeling of orbital spacecraft sloshing dynamics for a partially filled cryogenic superfluid liquid He-II in a Dewar container. The container was actuated by the gravity gradient acceleration associated with slew motion. They carried out a numerical computation of sloshing dynamics described by time-dependent three-dimensional formulation of partial differential equations subject to initial and boundary conditions. Their results showed large-amplitude fluctuations of angular momentum along the y-axis, which reflected the characteristics and the trend of angular displacement along the y-axis due to slew motion. The bubble mass center was found to fluctuate as a result of the gravity gradient associated with the slew motion.

In view of the extreme low values of surface tension and viscosity of superfluids, the sloshing dynamics, in response to lateral impulsive excitation in Dewars rotating containers, lasts for a

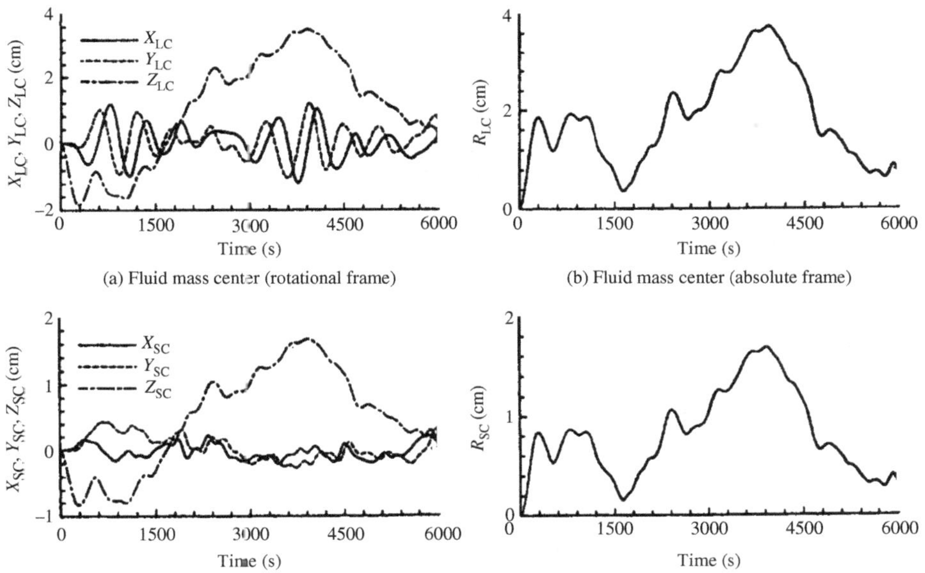

(a) Fluid mass center (rotational frame)
(b) Fluid mass center (absolute frame)
(c) Spacecraft mass center (non rotational frame)
(d) Spacecraft mass center (absolute frame)

Figure 12.43 Time evolution of fluid and spacecraft mass centers driven by g-jitter-dominated acceleration and orbital dynamics. (Hung, Long and Zu, 1996)

long period of time and gradually decays exponentially. Hung, Lee, and Leslie (1993a, b), Hung and Lee (1994), Hung (1995b), Hung and Long (1995b, 1996) studied the effect of baffles in providing an effective mechanism to suppress liquid oscillations under different rotating speeds.

12.11 Hydroelastic oscillations

The influence of container elasticity on the free-liquid oscillations under a microgravitational field was considered by some investigators. For example, the influence of tank wall elasticity and surface tension on the forced excitation of cylindrical liquid containers was studied by Tong and Fong (1965). Chiba and Bauer (1998) and Bauer and Chiba (1998) analyzed the free vibration of a liquid surface in a cylindrical sector shell under a zero gravity field. The case of an elastic rectangular container was considered by Bauer and Eidel (1993b). Other cases such as a container's membrane bottom were considered by Bauer, Chiba, and Sasaki (1998) and Bauer, Chiba, and Watanabe (2002). The influence of a flexible tank bottom on the hydroelastic sloshing was analytically examined by Chang (1966), Shih (1966), and Bauer and Eidel (1993a). The coupled frequencies of a rotating hydroelastic shell–liquid system under zero gravity were determined by Bauer (1988b).

Bauer and Komatsu (1994a, 1998) examined the coupled modes of a sector elastic shell with viscous liquid near zero gravity. The nonlinear coupling between liquid dynamics and spacecraft structural dynamics in a low gravitational field was studied by Peterson (1987). The nonlinear oscillations of an inviscid liquid–elastic cylinder system under zero gravity was studied analytically by Bauer and Eidel (1988a, 1989a).

Van Schoor, *et al.* (1990) and Van Schoor and Crawley (1992) experimentally studied the lateral sloshing dynamics and its nonlinear interaction with the spacecraft structural system on the shuttle Middeck. The nonlinear, nonplanar, and multiple response slosh behavior observed in the silicone-oil ground experiments, was found to be essentially linear in space. This was attributed to the very high damping ratio of the first sloshing mode in space. However, both the space uncoupled and coupled water experiments exhibited nonlinear and nonplanar characteristics similar to those observed in the ground experiments. This is due to the relatively high restoring force for water, which originates from the contact-angle hysteresis.

Peterson, *et al.* (1989) and Chiba, *et al.* (1998) studied the dynamics of linear spacecraft mode coupling with the nonlinear low-gravity liquid sloshing in a cylindrical container. The coupling includes two fundamental sloshing modes and three secondary sloshing modes. Under harmonic resonance the response exhibited a softening nonlinear behavior. Three additional stable nonlinear coupled responses were observed: (i) a secondary planar resonance, (ii) a skewed nonplanar motion, and (iii) a completely asymmetrical periodically modulated liquid-spacecraft motion. Weislogel (1991) designed an apparatus consisting of a programmable shaker table and a high-speed motion picture and video camera for viewing the fluid surface as drop tower tests were conducted. The purpose was to observe the stability of a variety of surface/vessel configurations.

Bauer (1985a, 1987a,b, 1992e) analyzed the coupled frequencies of hydroelastic systems in zero gravity. It was found that the interaction between the structure and viscous fluid may well be unstable, a fact that does not appear for an inviscid liquid. This surprising case indicates that the viscosity provides instability whereas for an ideal liquid there is always stability. Small perturbations of the structure lead to disturbances of the liquid surface and vice versa. In the case of a viscous liquid, for which different phase relations appear, the flow is quite different from that of an ideal liquid. An inviscid liquid exhibits slippage in the angular direction while a viscous liquid adheres to the wall and moves with it in the angular direction. Perturbations always result in liquid surface displacements that change the surface tension restoring force. For the case of a viscous liquid, an additional disturbance in the angular direction due to the motion of the elastic shell is not adequate to damp out the perturbations. These effects, that is, phase shifts followed by additional excitation of the liquid through angular motion of the elastic shell and vice versa, are not present in the system with an inviscid liquid and are responsible for instabilities.

12.12 Closing remarks

The area of fluid physics and sloshing dynamics in a microgravitational field is versatile and multidisciplinary. Some research monographs have been devoted to addressing some aspects of this area. These monographs include *Introduction to the Dynamics of Bodies Containing Liquids in the Weightless State* by Moiseev (1968); *The Hydrodynamics of Weightlessness* by Babskii, *et al.* (1976a); *Fluid Sciences and Materials Sciences in Space* by Walter (1987); *Low-Gravity Fluid Mechanics* and *Methods of Solving Fluid Mechanics Problems under Conditions of Weightlessness* both by Myshkis, *et al.* (1987, 1992); and *Physics of Fluids in Microgravity* by Monti (2001). The results of research activities on fluid physics in microgravity have been presented periodically every two years during the NASA biennial conference proceedings since 1992.

It was not possible to present all aspects of microgravity fluid physics and dynamics in one chapter. Instead, selected topics accompanied by extensive reviews have been presented. These topics are classified into four groups. The first group includes the kinematics and kinetics of free surface, modal analysis of liquid free surface and forcing sloshing with slipping and anchored contact lines, effect of g-jitter, liquid handling. The second group covers capillary systems such as

Marangoni flow, liquid bridges, axial excitation of liquid bridges with and without rotation of their supports, and thermocapillary convection. The third group deals with the sloshing of superfluids. The fourth group deals with hydroelastic oscillations in microgravity.

This chapter did not address the capillary problem in conical containers. This problem was treated by Ambrazyavichus (1981a,b). The stability of the configuration of liquid layers in containers describing rotational motion was studied by Capodano (2001). The modeling of low gravity liquid motion in spinning containers was discussed by Cutshall, *et al.* (1996) and Dodge (1996). Other issues not addressed in this chapter include numerical simulation of the surface tension effect on interface instability (Daly, 1969a,b), the influence of liquid sloshing on the dynamics of satellites (Guibert, *et al.*, 1978, Hung, Long, and Chi, 1996, and Monti, *et al.*, 2001), and magnetic effects on nonlinear Marangoni flow (Viviani, *et al.*, 1997).

Appendix

A1 Common dimensionless numbers

Biot Number, Bi, is the ratio of internal thermal resistance of a solid to the convective boundary layer thermal resistance, and is defined by the expression

$$\mathrm{Bi} = \frac{hL}{k_s}$$

Bond Number, Bo, is the ratio of gravitational to surface tension forces, and is defined by the expression

$$\mathrm{Bo} = \frac{\rho g L^2}{\sigma}$$

Froude Number, Fr, is the ratio of inertia to body (gravitational) forces, and is defined by the expression

$$\mathrm{Fr} = \frac{V^2}{gL} = \frac{V}{\sqrt{gL}}$$

Grashof Number, Gr, is the ratio of buoyancy to viscous forces

$$\mathrm{Gr} = \frac{\beta \mathrm{g} \mathrm{L}^3 \Delta \mathrm{T}}{\nu^2}$$

Marangoni Number, Ma, (also Thermal Marangoni Number) is the ratio of the imbalance in surface tension forces to liquid tangential forces, and is defined by the expression

$$\mathrm{Ma} = \frac{\partial \sigma}{\partial T}\left(\frac{L \Delta T}{\alpha \mu}\right)$$

Marangoni Concentration Number (also Solutal Marangoni Number), Ma_C, measures the ratio of the imposed concentration difference and the damping (the viscosity and solute diffusion),

$$\mathrm{Ma}_C = \left(\frac{\partial \sigma}{\partial C_0}\right)\frac{\Delta C_0 L}{D\mu}$$

Nusselt Number, Nu, is the ratio of the surface heat transfer into the fluid to the conduction heat transfer in the fluid, and is given by the expression

$$\mathsf{Nu} = \frac{hL}{k_f}$$

Ohnesorge Number, Oh, *or* ***Capillary Number***, Ca provides a measure of the surface deflection in response to thermocapillary induced stresses, and is defined by the expression

$$\mathsf{Oh} = \mathsf{Ca} = \frac{\mu\nu}{\sigma L}$$

Prandtl Number, Pr, is the ratio of the momentum transport to the thermal transport in the fluid (or the energy transferred by convection to the energy transferred by conduction), and is defined by the expression

$$\mathsf{Pr} = \frac{\mu c_p}{k} = \frac{\nu}{\alpha}$$

It is also defined as the ratio of the momentum to thermal diffusivities.

Rayleigh Number, $\Re$, for a fluid-filled space between two parallel horizontal planes a distance ℓ apart, measures the balance between the buoyant force associated with thermal expanson and the dissipation energy due to viscosity and thermal diffusion

$$\Re = \frac{\beta(\mathrm{T}_1 - \mathrm{T}_2)\mathrm{g}\ell^2}{\nu k}$$

Reynolds Number, Re, is the ratio of inertia to viscous forces, and is defined by the expression

$$\mathsf{Re} = \frac{\rho LV}{\mu} = \frac{LV}{\nu}$$

Weber Number, We, is the ratio of inertia to surface tension forces

$$\mathsf{We} = \frac{\rho LV^2}{\sigma}$$

where c_p is the specific heat at constant pressure (in joules/kilogram kelvin), D is the mass (Solutal) diffusivity, g is the gravitational acceleration (in m/s^2) h is the heat transfer coefficient from a surface to a fluid (in watts/meter^2 kelvin) and is defined in terms of the heat flow rate $Q = hA(T_0 - T_\infty)$, T_0 is the imposed temperature of the surface (in K), T_∞ is the ambient temperature (in K), k is the thermal conductivity (in watts/meter kelvin), L is a reference length, T is the absolute temperature (in K), $\Delta\tilde{T}_0 = \tilde{T}(y=1) - \tilde{T}(y=0)$, a tilde denotes a dimensionless quantity, V is the velocity (in m/s), $\alpha = k/(\rho c_p)$ (in m^2/s) is the thermal diffusivity, β is the volumetric thermal expansion coefficient $= -(1/\rho)(\partial\rho/\partial T)_p$, μ is the dynamic viscosity (in Pa s $=$ N s/m^2 $=$ kg/(m s), ν is the kinematic viscosity (in m^2/s), σ is the surface tension (in N/m $=$ kg/s^2), ρ is the density (in kg/m^3), C_0 is a dimensionless concentration of the denser component (S/ρ, S is the concentration of the denser component in kg/m^3, ΔC_0 is the concentration difference).

References

Aa E.V. and Sanders J.A. (1979), The 1:2:1-resonance, its periodic orbits and integrals, in *Asymptotic Analysis from Theory to Application*, Lecture Notes in Mathematics **711**: Editor F. Verhulst, 187–208, Springer-Verlag, Berlin.

Abdalla K.L., Flage R.A., and Jackson R.G. (1964), Zero gravity performance of ullage control surface with liquid hydrogen while subjected to unsymmetrical radiant heating, NASA TM X-1001.

Abdel Gawad A.F., Ragab S.A., Nayfeh A.H., and Mook D.T. (2001), Roll stabilization by anti-roll passive tanks, *Ocean Engrg.* **28**, 457–469.

Abramowitz G.L. and Segun I.A. (1968), *Handbook of Mathematical Functions*, New York, Dover Publications.

Abramson H.N. (1961a), Amazing motions of liquid propellant, *Astronaut* **6**, 35–37.

Abramson H.N. (1961b), Liquid dynamic behavior in rocket propellant tanks, Proc. ONR/AIAA Symp *Struct. Dynamics of High Speed Flight*, Los Angeles, April 1961, 287–318.

Abramson H.N. (1961c), Total force response resulting from liquid propellant sloshing in a rigid cylindrical tank with vertical center wall baffle, Tech. Rept. 9, SwRI, May.

Abramson H.N. (1961d), Theoretical and experimental studies of liquid sloshing in rigid cylindrical tanks, Tech. Rept. (Final Report), SwRI, May.

Abramson H.N. (1963a), Studies of liquid dynamics in rocket propellant tanks, SwRI, May.

Abramson H.N. (1963b), Dynamic behavior of liquid in moving containers, *ASME Appl. Mech. Rev.* **16**(7), 501–506.

Abramson H.N. (1964), Some measurements of the effects of ring baffles in cylindrical tanks, *J. Spacecraft Rock.* **1**(9/10), 560–562.

Abramson H.N. (1965), Further studies of liquid sloshing in rocket propellant tanks, Final Report, Contract NAS8–1555, SwRI, Dec 1965.

Abramson H.N. (Ed), (1966a), *The Dynamic Behavior of Liquids in Moving Containers*, NASA SP 106.

Abramson H.N. (1966b), Some current aspects of the dynamic behavior of liquids in rocket propellant tanks, in *Applied Mechanical Surveys*, H.N. Abramson, H Liebowitz, J.M. Crowley, and S. Juhasz, eds. Washington, DC, Sparton Books, 941–949.

Abramson H.N. (1968), Liquid propellant dynamics, *AGARD Manual of Aeroelasticity* (1), Chapter 8, Revised Edition, NATO, Paris.

Abramson H.N. (1969), *Slosh Suppression*, NASA SP-8031.

Abramson N.H. (2003), Dynamics of contained liquids: A personal odyssey, *ASME Appl. Mech. Rev.* **56**(1), R1–R7.

Abramson H.N., Bass R.L., Faltinsen O., and Olsen H.A. (1974), Liquid slosh in LNG carriers, *10th Symp. Naval Hydrodynamics*, MIT, Cambridge, MA.

Abramson H.N., Chu W.H., and Garza L.R. (1962a), Liquid sloshing in 45 degree sector compartmented cylindrical tanks, Tech. Rept. 3, SwRI, Nov 1962.

Abramson H. N., Chu W. H., Garza L. R., and Ransleben G. E. Jr (1962b), Some studies of liquid rotation and vortexing in rocket propellant tanks, NASA TN D-1212, June 1962.

Abramson H. N., Chu W. H., and Garza L. R. (1963), Liquid sloshing in spherical tanks, *AIAA J.* **1**(2), 384–389.

Abramson H. N., Chu W. H., and Kana D. D. (1966), Some studies of nonlinear lateral sloshing in rigid containers, *J. Appl. Mech.* **33**(4), also NASA CR-375, January.

Abramson H. N., Chu W. H., Kana D. D., and Lindholm U. S. (1962c), Bending vibrations of a circular cylindrical shell containing an internal liquid with a free surface, Tech. Rept. 4, SwRI, March.

Abramson H. N. and Garza L. R. (1964), Some measurements of the effects of ring baffles in cylindrical tanks, *J. Spacecraft Rock.* **1**(5), 560–562.

Abramson H. N. and Garza L. R. (1965), Some measurements of liquid frequencies and damping in compartmented cylindrical tanks, *J. Spacecraft Rock.* **2**(5/6), 453–455.

Abramson H. N., Garza L. R., and Kana D. D. (1962d), Some notes on liquid sloshing in compartmented cylindrical tanks, *Amer. Rocket Soc. J.* **32**(6), 978–980.

Abramson H. N., Garza L. R., and Squire W. (1961), An exploratory study of the effect of sloshing on heat transfer in similitude cryogenic liquid, Tech. Rept. 2, SwRI, February.

Abramson H. N. and Kana D. D. (1967), Some recent research on the vibrations of elastic shells containing liquids, Proc. Symp. *Shell Theory,* University of Houston, Texas.

Abramson H. N. and Kana D. D. (1970), Some experimental studies of the dynamic stability of thin shells containing liquid, 60th Anniversary volume in honor of V. Novozhilov, the USSR National Committee on Theor. and Appl. Mech., Academy of Science USSR, Moscow, May 18.

Abramson H. N., Martin R. J., and Ransleben G. E. Jr (1958), Application of similitude theory to the problem of fuel sloshing in rigid tanks, Tech. Rept. 1, SwRI, May.

Abramson H. N. and Nevill G. E. Jr (1963), Some modern developments in the application of scale models in dynamic testing, ASME Colloquium on Use of Models and Scaling in Shock and Vibration, November.

Abramson H. N. and Ransleben G. E. Jr (1959a), Simulation of fuel characteristics in missile tanks by use of small models, Tech. Rept. 3 SwRI, March.

Abramson H. N. and Ransleben G. E. Jr (1959b), A note on the effectiveness of two type slosh suppression devices, Tech. Rept. 6, SwRI, June.

Abramson H. N. and Ransleben G. E. Jr (1960), Simulation of fuel sloshing characteristics in missile tanks by use of small models, *Amer. Rocket Soc. J.* **30** (7), 603–612.

Abramson H. N. and Ransleben G. E. Jr (1961a), Some comparisons of sloshing behavior in cylindrical tanks having flat and conical bottoms, *Amer. Rocket Soc. J.* **31**, 542–544.

Abramson H. N. and Ransleben G. E. Jr (1961b), A note on wall pressure distribution during sloshing in rigid tanks, *Amer. Rocket Soc. J.* **31**, 545–547.

Abramson H. N. and Ransleben G. E. Jr (1961c), Representation of fuel sloshing in cylindrical tanks by an equivalent mechanical model, *Amer. Rocket Soc. J.* **31**(12), 1697–1705.

Abramson H. N. and Ransleben G. E. Jr (1961d), Some studies of a floating lid type device for suppression of liquid sloshing in rigid cylindrical tanks, Tech. Rept. 10, SwRI, May.

Abramson H. N. and Ransleben G. E. (1961e), Liquid sloshing in rigid cylindrical tanks undergoing pitching motion, Tech. Rept. 11, SwRI, May.

Abzug M. J. (1996), Fuel slosh in skewed tanks, *J. Guid. Contr. Dyn.* **19**(5), 1172–1177.

Ackerberg R. C. (1965), The viscous incompressible flow inside a cone, *J. of Fluid Mech.* **21**(1), 47–81.

Addington J. W. (1960), Dynamics of fuel in tanks, Note 99, College of Aeronautics, Cranfield, England.

Addriaans M. J., Moeu W. A., Boyd S. T. P., Strayer D. M., and Duncan R. V. (1996), Cryogenic design of the liquid helium experiment: Critical dynamics in microgravity, *Cryogenics* **36**, 787–794.

Adler W. F. (1979), The mechanics of liquid impact, in *Treatise on Material Science and Technology*, CM Proc., ed., **16**, Erosion, Academic Press 127–183.

Advani S. H. and Lee Y. C. (1970), Free vibrations of fluid-filled spherical shells, *J. Sound and Vib.* **12**(4), 453–462.

Aganovic I. (1981), On a spectral problem of hydroelasticity, *J. de Mec.* **20**(3) 409–414.

Agnon Y. and Golzman M. (1996), Periodic solutions for a complex Hamiltonian system: New standing water waves, *Wave Motion* **24**, 139–150.

Agrawal B. N. (1987), Interaction between liquid propellant slosh modes and attitude control in a dual-spin spacecraft, Proc. AIAA/ASME/ASCE/AFS 28th *Struct. Dyn. Mat. Conf.*, **Part 2B**, 774–780.

Agrawal B. N. (1993), Dynamic characteristics of liquid motion in partially filled tanks of a spinning spacecraft, *J. Guidance, Control, and Dynamics* **16**(4), 636–640.

Aita S. and Girbert R. J. (1986), Fluid-elastic instability of a flexible weir: a theoretical model, ASME Proc. *Flow-Induced Vib.*, **104**, 51–58.

Aita S., Tigeot Y., Bertaut C., and Serpantie J. P. (1986), Fluid-elastic instability of a flexible weir: experimental observations, ASME Proc. *Flow-Induced Vib.*, **104**, 41–50.

Aitta A. (1991), Nonlinear phenomena at an air–fluid interface in a horizontal rotating cylinder, *Eur. J. Mech. B/ Fluids* **10** (suppl 2), 175–180.

Akita Y. (1967), Dynamic pressure of cargo oil due to pitching and effectiveness of swash bulkhead in long tanks, *Japan Shipbuilding & Marine Eng.* **2**(5), 42–55.

Albanese C., Carotenuto L., Castagnolo D., Ceglia E., and Monti R. (1995), An investigation on the "onset" of oscillatory Marangoni flow, *Adv. Space Res.* **16**(7), 87–94.

Albright N. (1977), Mathematical and computational studies of the stability of axisymmetric annular capillary free surfaces, NASA, NTIS, Washington, DC.

Alexander J. I. D. (1990), Low-gravity experiment sensitivity to residual acceleration: a review, *Microgravity Sci. Techno.* **3**, 52–68.

Alexander J. I. D. (1997), Drops, jets and bubbles, in *Free Surface Flows*, Kuhlamann H. C., and Rath H. J., eds., New York, Springer-Verlag.

Alexander J. I. D., Garandet J. P., Favier J. J., and Lizee A. (1997), G-jitter effects on segregation during directional solidification of tin-bismuth in the Mephisto furnace facility, *J. Crystal Growth* **178**, 657–661.

Alexander J. I. D., Ouzzani J., and Rosenberger F. (1991), Analysis of the low gravity tolerance of Bridgman–Stockbarger crystal growth, *J. Crystal Growth* **112**, 21–38.

Alexander J. I. D, Slobozhanin L. A., Resnick A. H., Ramus J. F., and Delafontaine S. (2000), Stability limits and dynamics of nonaxisymmetric liquid bridges, Proc. Center for Microgravity Res., Cleveland, OH, 564–569.

Alfriend K. (1974), Partially filled viscous ring nutation damper, *J. Spacecraft* **11**, 456–462.

Alfriend K. T. and Spencer T. (1981), Comparison of filled and partly filled nutation dampers, AAS/AIAA Astrodyn Spec. Conf., Lake Tahoe, NV, Paper 81–141.

Aliabadi S., Johnson A., and Abedi J. (2003), Comparison of finite element and pendulum models for simulation of sloshing, *Comput. & Fluids* **32**, 535–545.

Allingham W. D. (1968), Zero gravity expulsion of cryogens with metal bellows, AIAA/Aerospace Corp. Symp. *Low Gravity Orientation and Expulsion*, Los Angeles, 199–208.

Amabili M. (1996a), Free vibration of partially filled, horizontal cylindrical shells, *J. Sound Vib.* **191**, 757–780.

Amabili M. (1996b), Effect of finite fluid depth on the hydroelastic vibrations of circular and annular plates, *J. Sound Vib.* **193**, 909–925.

Amabili M. (1997a), Bulging modes of circular bottom plates in rigid cylindrical containers filled with liquid, *Shock and Vibration* **4**, 51–68.

Amabili M. (1997b), Shell-plate interaction in the free vibrations of circular cylindrical tanks partially filled with a liquid: the artificial spring method, *J. Sound Vib.* **199**, 431–452.

Amabili M. (1997c), Ritz method and sub-structuring in the study of vibration with strong fluid–structure interaction, *J. Fluids Struct.* **11**, 507–523.

Amabili M. (2000a), Eigenvalue problems for vibrating structures coupled with quiescent fluids with free surface, *J. Sound Vib.* **231**, 79–97.

Amabili M. (2000b), Vibrations of fluid-filled hermetic cans, *J. Fluid Struct.* **14**, 235–255.

Amabili M. (2003), Theory and experiments for large-amplitude vibrations of empty and fluid-filled circular cylindrical shells with imperfections, *J. Sound Vib.* **262**, 921–975.

Amabili M. and Dalpiaz G. (1995), Breathing vibrations of a horizontal circular cylindrical tank shell, partially filled with liquid, *ASME J. Vib. Acoust.* **117**, 187–191.

Amabili M., and Dalpiaz G. (1998), Vibrations of base plates in annular cylindrical containers: theory and experiments, *J. Sound Vib.* **210**, 329–350.

Amabili M. and Garziera R. (2000), Vibrations of circular cylindrical shells with non-uniform constraints, elastic bed and added mass, part I: Empty and fluid-filled shells, *J. Fluid Struct.* **14**(1), 669–690.

Amabili M., and Garziera R. (2002), Vibrations of circular cylindrical shells with non-uniform constraints, elastic bed and added mass, part II: Shells containing or immersed in axial flow, *J. Fluid Struct.* **16**(1), 31–51.

Amabili M., Garziera R., and Negri A. (2002), Experimental study on large-amplitude vibrations of water-filled circular cylindrical shells, *J. Fluid Struct.* **16**(2), 213–227.

Amabili M. and Kwak M. K. (1996), Free vibration of circular plates coupled with liquid: revising the Lamb problem, *J. Fluids Struct.* **10**, 743–761.

Amabili M. and Kwak M. K. (1999), Vibration of circular plates on a free fluid surface: effect of surface waves, *J. Sound Vib.* **226**, 407–424.

Amabili M. and Paidoussis M. P. (2003), Review of studies on geometrically nonlinear vibrations and dynamics of circular cylindrical shells and panels, with and without fluid–structure interaction, *ASME Appl. Mech. Rev.* **56**(4), 349–381.

Amabili M., Paidoussis M. P., and Lakis A. A. (1998a), Vibrations of partially filled cylindrical tanks with ring-stiffeners and flexible bottom, *J. Sound Vib.* **213**, 259–299.

Amabili M., Pellicano F., and Paidoussis M. P. (1998b), Nonlinear vibrations of simply supported, circular cylindrical shells, coupled with quiescent fluid, *J. Fluids Struct.* **12**, 883–918.

Amabili M. Pellicano F., and Paidoussis M. P. (1999a), Nonlinear dynamics and stability of circular cylindrical shells containing flowing fluid, part I: Stability, *J Sound Vib.* **225**(4), 655–699.

Amabili M. Pellicano F., and Paidoussis M. P. (1999b), Nonlinear dynamics and stability of circular cylindrical shells containing flowing fluid, part II: large-amplitude vibrations without flow, *J. Sound Vib.* **228**(5), 1103–1124.

Amabili M. Pellicano F., and Paidoussis M. P. (1999c), Further comments on nonlinear vibrations of shells, *J. Fluids Struct.* **13**, 159–160.

Amabili M. Pellicano F., and Paidoussis M. P. (2000a), Nonlinear dynamics and stability of circular cylindrical shells containing flowing fluid, part III: truncation effect without flow and experiments, *J. Sound Vib.* **237**(4), 617–640.

Amabili M. Pellicano F., and Paidoussis M. P. (2000b), Nonlinear dynamics and stability of circular cylindrical shells containing flowing fluid, part IV: large-amplitude vibrations with flow, *J. Sound Vib.* **237**(4), 641–666.

Amabili M., Pellicano F., and Vakakis A. F. (2000c), Nonlinear vibrations and multiple resonances of fluid-filled, circular shells, part I: equations of motion and numerical results, *ASME J. Vib. Acoust.* **122**, 346–354.

Amabili M. Pellicano F., and Vakakis A. F. (2000d), Nonlinear vibrations and multiple resonances of fluid-filled, circular shells, part II: perturbation analysis, *ASME J. Vib. Acoust.* **122**, 355–364.

Amakhin V. M., Kalayzin E. I., and Voronokov A. V. (1975), Method of numerical calculation of emptying and filling a tank in a weak force field, *Tr. Moscow Aviats. Inst.* **323**, 110 (in Russian).

Amano K., and Iwano R. (1991), Experimental and analysis of jet-induced sloshing in a tank, *Tran. JSME* (in Japanese) **B57**, 1947–1954.

Amano K., Koizumi M., and Yamakawa M. (1989), Three dimensional analysis method for potential flow with a moving liquid surface using a boundary element method, *Sloshing and Fluid Structure Vibration*, ASME Pressure Vess. Piping Conf., PVP **157**, 127–132.

Ambrazyavichus A. P. (1981a), Solvability of problem on conical capillary, *Probl. Mat. Analiza. (Leningrad)* **8**, 3–9 (in Russian).

Ambrazyavichus A. P. (1981b), Construction of barrier functions for solving capillary problems in conical regions, *Lit. Mat. Sb.* **24**, 3–15 (in Russian).

Amick C. J. and Tolland J. F. (1987), The semi-analytic theory of standing waves, *Proc. Royal Soc. London A* **411**, 123–137.

Amieux J. C. and Dureigne M. (1972), Analytical design of optimal nutation damper, *J. Spacecraft Rock.* **9**(12), 934–935.

Amiro I. Y. and Prokopenko N. Y. (1997), Nonlinear oscillations of cylindrical shells, *Int Appl. Mech.* **33**(11), 903–908.

Amsden A. A. and Harlow F. H. (1971), The SMAC method: a numerical technique for calculating incompressible fluid flows, Los Alamos Scientific Laboratory Rept. LA-4370.

Ananthakrishnan P. and Yeung R. W. (1994), Nonlinear interaction of a vortex pair with clean and surfactant-covered free surface, *Wave Motion* **19**, 343–360.

Anderson J. G., Semercigil S. E., and Turan O. F. (2000a), A standing-wave-type sloshing absorber to control transient oscillations, *J. Sound Vib.* **232**(5), 839–856.

Anderson J. G., Semercigil S. E., and Turan O. F. (2000b), An improved standing-wave-type sloshing absorber, *J. Sound Vib.* **235**(4), 702–710.

Anderson J. G., Turan O. F., and Semercigil S. E. (2001), Experiments to control sloshing in cylindrical containers, *J. Sound Vib.* **240**(2), 394–404.

Andracchio C. R. and Abdalla K. L. (1968), An experimental study of liquid flow into a baffled spherical tank during weightlessness, NASA TM X-1526.

Ang T. Balendra Paramasivam P. K. K., and Lee S. L. (1982), Free vibration analysis of cylindrical liquid storage tanks, *Int. J. Mech. Sci.* **24**, 47–59.

Anilkumar A. V., Grugel R. N., Shen X. F., and Wang T. G. (1993), Control of thermocapillary convection in a liquid bridge by vibrations, *J. Appl. Phys.* **73**(9), 4165–4170.

Anisimov A. M. (1963), Axi-symmetrical vibrations of a spherical shell filled by a fluid, *IVUZ Aviatsionnaya Teknika* (*Proc. of Institutions of Higher Education, Aviation Technology*) **2**, 51–58.

Anisimov A. M. (1968), Application of finite-difference methods to the calculations of axially symmetric vibrations of shells of revolution with liquid, *IVUZ Aviatsionnaya Teknika* (*Proc. of Institutions of Higher Education, Aviation Technology*) **3**, 23–30.

Anosov Yu N. (1966), The nonlinear vibrations of a liquid in a cylindrical cavity, *Soviet Appl. Mech. (Prikl. Mekh.)* **2**(10), 22–28.

Anzai T., Ikeuchi M., Igarashi K., and Okanuma T. (1981), Active nutation damping system of engine test satellite-IV, 22nd Congress Int. Astronaut. Fed., IAF 81–350.

Arai M. (1986), Experimental and numerical studies of sloshing in liquid cargo tanks with internal structures, *Ishikawajima-harima Heavy Indust. Eng. Rev.* **19**(2), 51–56.

Arai M., Cheng L. Y., and Inoue Y. (1993), Numerical simulation of sloshing and swirling in cubic and cylindrical tank, *J. Kansai Soc. Naval Arch,, Japan N* **219**, 97–101.

Arai M. Cheng L. Y., and Inoue Y. (1994), 3-D numerical simulation of impact load due to liquid cargo sloshing, *J. Soc. Naval Arch. Japan* **171**, 177–184.

Arbell H. and Fineberg J. (1998), Spatial and temporal dynamics of two interacting modes in parametrically driven surface waves, *Phys. Rev. Lett.* **81**, 4384–4387.

Arbell H., and Fineberg J (2000a), Two-mode rhomboidal states in driven surface waves, *Phys. Rev. Lett.* **84**, 654–657.

Arbell H., and Fineberg J. (2000b), Temporally harmonic oscillations in Newtonian fluids, *Europhys. Lett.* Reprint.

Armenio V. and Rocca M. L. (1996), On the analysis of sloshing of water in rectangular containers: numerical study and experimental validation, *Ocean Eng.* **23**(8), 705–739.

Armstrong G. L. and Kachigan K. (1961), Propellant sloshing, in *Handbook of Astronautical Eng*, H. H. Koelle, ed., Chapter 14, sections 14–27.

Arnold R. N., and Warburton G. B. (1949), Flexural vibrations of the walls of thin cylindrical shells having freely supported ends, *Proc. Royal Soc. (London) A* **197**, 238.

Asaki T. J. and Marston P. L. (1995), Free decay of shape oscillations of bubbles acoustically trapped in water and seawater, *J. Fluid Mech.* **300**, 149–167.

Ashmore J., Hosoi A. E., and Stone H. A. (2003), The effect of surface tension on rimming flows in a partially filled rotating cylinder, *J. Fluid Mech.* **479**, 65–98.

Aslam M., Godden W. G., and Scanline D. T. (1979), Earthquake sloshing in annular and cylindrical tanks, *ASCE J. Eng. Mech. Div.* **105**, 371–389.

Atluri S. (1972), A perturbation analysis of nonlinear free flexural vibrations of a circular cylindrical shell, *Int. J. Solids & Struct.* **8**, 549–569.

Auli W., Fischer F. D., and Rammerstofer F. G. (1985), Uplifting of earthquake-loaded liquid-filled tanks, ASME PVP-987.

Au-Yang M. K. (1976), Free vibration of fluid-coupled coaxial cylindrical shells of different lengths, *J. Appl Mech.* **34**, 480–484.

Aydelott J. C. (1995), Technology requirements to be addressed by the NASA Lewis Research Center Cryogenic Fluid Management Facility Program, AIAA Paper AIAA-95–1229.

Azuma H. and Yoshinara S. (1999), Three-dimensional large-amplitude drop oscillations: experiments and theoretical analysis, *J. Fluid Mech.* **393**, 309–332.

Babenko K. I. and Yurev S. P. (1980), On the shape of the surface of a capillary liquid in a vertical cylinder of arbitrary cross section, *Dokl. Akad. Nauk. SSSR* **251**, 1326–1334.

Babitsky V. I. (1966), Vibro-impact motions of pendulum with inertial suspension in vibrating container, *Analysis and Synthesis of Automatic Machines*, Moscow, Nauka, 20–30.

Babitsky V. I. (1998), *Theory of Vibro-Impact Systems and Applications,* Berlin, Springer-Verlag.

Babskii V. G., Kopachevskii N. D., Myshkis A. D., Slobozhanin L. A., and Tyuptsov A. D. (1976a), *The Hydrodynamics of Weightlessness*, Moscow, Nauka.

Babskii V. G., Kopachevskii N. D., Myshkis A. D., Slobozhanin L. A., and Tyuptsov A. D. (1976b), Approximate methods in zero-gravity fluid mechanics, in *Problems of Mathematical Physics and Functional Analysis*, Kiev, Naukova Dumka, 83–94 (in Russian).

Babskii V. G. Kopachevskii N. D. Myshkis A. D. Slobozhanin L. A., and Tyuptsov A. D. (1980), On some unresolved problems of zero gravity hydromechanics, *Nonlin. Anal. Theo. Meth. Appl.* **4**(3), 607–621.

Babu S. S. and Bhattacharayya S. K. (1996), Finite element analysis of fluid-structure interaction effect on liquid retaining structure due to sloshing, *Comp. Struct.* **59**(6), 1165–1171.

Bagdasaryan G. E., and Gnuni V. T. (1966), The parametric vibrations of a cylindrical shell filled with a liquid to different depths, *Soviet Appl. Mech. (Prikl. Mekh.)* **2**(3), 21–26.

Bagno A. M. and Guz A. N. (1997), Elastic waves in prestressed bodies interacting with a liquid (review), *Soviet Appl. Mech. (Prikl. Mekh.)* **33**(6), 3–39.

Baines P. G. (1967), Forced oscillations of an enclosed rotating fluid, *J. Fluid Mech.* **30**(3), 533–546.

Baird M. H. I. (1963), Resonance bubbles in a vertical liquid column, *Can. J. Chem. Eng.* **41**, 52–56.

Balabukh L.I. (1966), Interaction of shells with a liquid and gas, Proc. All-Union Conf. on *Theory of Shells and Plates*, Baku, Moscow, Nauka, 935–944.

Balabukh L. I. and Molchanov A. G. (1967), Axi-symmetric oscillations of a spherical shell partially filled with fluid, *Inzh. Zh. Mekh. Tverd. Tela* **5**, 56–61.

Balakirev Y. G. (1967), Axi-symmetric oscillations of a shallow spherical shell containing a fluid, *Inzh. Zh. Mekh. Tverd. Tela* **5**, 116–123.

Balendra T., Ang K. K., Paramasivam P., and Lee S. (1982a), Free vibration analysis of cylindrical liquid storage tanks, *Int. J. Mech. Sci.* **24**(1), 47–59.

Balendra T., Ang K. K., Paramasivan P., and Lee S. L. (1982b), Seismic design of flexible cylindrical liquid storage tanks, *Earthquake Eng. Struct. Dyn.* **10**, 477–496.

Balendra T. and Nash W. A. (1978), Earthquake analysis of a cylindrical liquid storage tank with a dome by finite element method, University of Massachusetts, Amherst, MA.

Balendra T. and Nash W. A. (1980), Seismic analysis of a cylindrical liquid storage tank with a dome by the finite element method, Century 2, ASME Pressure Vess. Piping Conf., San Francisco, 1.

Balendra T., Wang C. M., and Cheong H. F. (1994), Effectiveness of tuned liquid column dampers for vibration control towers, *J. Eng. Struct.* **17**(9), 668–675.

Balendra T., Wang C. M., and Rakesh G. (1998), Vibration control of tapered building using TLCD, *J. Wind Eng. Indust. Aerodyn.* **77/78**, 245–257.

Balendra T., Wang C. M., and Rakesh G. (1999), Vibration control of various types of buildings using TLCD, *J. Wind. Eng. Indust. Aerodyn.* **83**, 197–208.

Balendra T., Wang C. M., and Yan N. (2001), Control of wind-excited towers by active tuned liquid column damper, *Eng. Struct.* **23**, 1054–1067.

Bandyopadhyay K. K. (1991), Overview of seismic panel activities, Proc. 3rd *DOE Natural Phenomena Hazards Mitigation Conf.*, St. Louis, MO, 423–429.

Banner M. L. and Peregrine D. H. (1993), Wave breaking in deep water, *Annu. Rev. Fluid. Mech.* **25**, 373–397.

Banner M. L., and Tian X. (1996), Energy and momentum growth rates in breaking water waves, *Phys. Rev. Lett.* **77**, 2953–2956.

Banning D. A., Hengeveld L. D., and Modi V. J. (1966), Apparatus for demonstrating dynamics of sloshing liquids, *Bull. Mech. Eng. Edu.* **5**, 65–70.

Bao G. W., and Pascal M. (1997), Stability of a spinning liquid-filled spacecraft, *Archive Appl. Mech.* **67**(6), 407–421.

Barber N. F. and Ghey G. (1969), *Water Waves,* The Wykeham Sci. Series, London.

Barnyak O. M. (1997), Normal oscillations of a viscous liquid partially filling a circular horizontal channel, *Int. Appl. Mech.* **33**(4), 335–343.

Barnyak M. Y. (1971a), Determination of natural frequencies and small oscillation forms of an ideal liquid in a vessel in weak gravitational field, *Mat. Fiz.* **9**, 3–11 (in Russian).

Barnyak M. Y. (1971b), Approximate methods of solving problems on the statics and dynamics of a liquid in a vessel under near zero-gravity conditions, Ph.D. thesis, Inst. Mat. Akad. Nauk. Ukraine, Kiev, SSSR (in Russian).

Barnyak M. Y. and Barnyak O. M. (1996), Normal oscillations of viscous liquid in a horizontal channel, *Int. Appl. Mech.* **32**(7), 560–566.

Baron M. L. and Bleich H. H. (1959), The dynamic analysis of empty and partially full cylindrical tanks and transient response by mode analysis, Final Report, DASA No 1123A, 22 May 1959, Paul Weidlinger, ASTIA No 220236.

Baron M. L. and Skalak R. (1962), Free vibrations of fluid filled cylindrical shells, *ASCE J. Eng. Mech.* **88**(1), 17–43.

Barr R. A. and Ankudinov V. (1977), Ship rolling: Its prediction and reduction using roll stabilization, *Marine Tech.* **14**(1), 19–41.

Barron R. and Chang S. W. R. (1989), Dynamic analysis and measurement of sloshing of fluid in containers, *Trans. ASME D.S.* **111**, 83–90.

Barton D. C. and Parker J. V. (1987), Finite element analysis of the seismic response of anchored and unanchored liquid storage tanks, *Earthquake Eng. Struct. Dyn.* **15**, 299–322.

Bass D. W. (1998), Roll stabilization for small fishing vessels using paravenes and anti-roll tanks, *Marine Technol.* **35**(2), 74–84.

Bass R. L. (1975), Dynamic slosh induced loads on liquid cargo tank bulkheads, Soc. Naval Archit. and Marine Eng., Rept. No R-19.

Basset A B. (1914), On the steady motion and stability of liquid in an ellipsoid vessel, *Quart. J. Math.* **45**.

Bateman H. (1944), *Partial Differential Equations of Mathematical Physics*, New York, Dover Publications.

Bauer H. F. (1957), Approximate effect of ring stiffener on the pressure distribution in an oscillating cylindrical tank partially filled with a liquid, ABMA, DA, Memo. No 264, DA-M-114, 12 September.

Bauer H. F. (1958a), Determination of approximate first natural frequencies of fluid in a spherical tank, ABMA, DA-TN-75–58.

Bauer H. F. (1958b), The influence of fluid in the tanks on the moment of inertia of Jupiter AM8, ABMA, DA-Memo No 333, DA-M-1–58, 31 March.

Bauer H. F. (1958c), Fluid oscillations in a circular cylindrical tank, Rept. No DA-TR-1–58, April.

Bauer H. F. (1958d), Fluid oscillations of a circular cylindrical tank performing lissajous-oscillations, ABMA, DA-TR-2–58, April.

Bauer H. F. (1958e), Fluid oscillations in a circular cylindrical tank due to bending of tank walls, ABMA, DA-TR-3–58, April.

Bauer H. F. (1958f), Fluid oscillations in a cylindrical tank with damping, ABMA, DA-TR-4–58, April.

Bauer H. F. (1958g), The moment of inertia of a liquid in a circular cylindrical tank, ABMA, Rept. No DA-TR-5–58, April.

Bauer H. F. (1958h), Damped fluid oscillations in circular cylindrical tank due to bending tank wall, ABMA, DA-TR-9–58, 16 May.

Bauer H. F. (1958i), Propellant sloshing, ABMA, DA-TR-18–58, 5 November.

Bauer H. F. (1959a), Force and moment of a liquid on a rigid fixed lid on the free fluid surface due to translational and rotational oscillation of a tank, ABMA, DA-TN-25–59, 20 March.

Bauer H. F. (1959b), Damped oscillations in a connected fluid system, ABMA, DA-TN-57–59, 1 May.

Bauer H. F. (1959c), The effective moment of inertia in roll of propellant and roll damping, ABMA, DA-TR-67–59, May.

Bauer H. F. (1960a), Mechanical model for the description of the liquid motion in a rectangular container, Lockheed Company, RN ER-8559, June.

Bauer H. F. (1960b), Theory of fluid oscillations in a circular cylindrical ring tank partially filled with liquid, NASA TN-D-557.

Bauer H. F. (1961a), The effects of interaction of structure, control, and propellant sloshing upon the stability of large space vehicles, MSFC, NASA, MTP-AERO-61–83.

Bauer H. F. (1961b), Parametric study of the influence of propellant sloshing on the stability of spacecraft, *Aero. /Space Sci. J.* **28**(10), 819–820.

Bauer H. F. (1961c), Mechanical analogy of fluid oscillations in cylindrical tanks with circular and annular cross-section, MSFC, NASA, MTP-AERO61–4.

Bauer H. F. (1961d), Dynamics of liquid propellant vehicles, Proc. ONR/AIAA Symp. on *Struct. Dynamics of High Speed Flight*, 319–355 (Office of Naval Res., Los Angeles, CA).

Bauer H. F. (1962a), Theory of fluid oscillations in partially filled cylindrical containers, MSFC, NASA, MTP-AERO-62–1, January.

Bauer H. F. (1962b), Mechanical model of fluid oscillations in cylindrical containers and introduction of damping, MTP-AERO-62–16.

Bauer H. F. (1962c), The damping factor provided by flat annular ring baffles for free surface oscillations, MTP-AERO-62–81, November.

Bauer H. F. (1963a), Tables and graphs of zero of cross product Bessel functions, MTP-AERO-63–50. Also *J. Math. Computation* **18**, 128–135, 1964.

Bauer H. F. (1963b), Stability boundaries of liquid propellant space-vehicles with sloshing, *AIAA J.* **1**(7), 1583–1589.

Bauer H. F. (1963c), Theory of liquid sloshing in compartmented cylindrical tanks due to bending excitation, *AIAA J.* **1**(7), 1590–1596.

Bauer H. F. (1963d), The effect of propellant sloshing on the stability of an accelerometer controlled rigid vehicle, NASA TN-D-1831.

Bauer H. F. (1963e), Liquid sloshing in a cylindrical quarter tank, *AIAA J.* **1**(11), 2601–2606.

Bauer H. F. (1964a), Fluid oscillations in the containers of a space vehicle and their influence on stability, NASA TR-R-187.

Bauer H. F. (1964b), Fuel vibration in rocket containers and their influence on the overall stability, *Zeit fur Flugweissenschaften* **12**(3/6), 85–101 and 222–229 (in German).

Bauer H. F. (1964c), Discussion of 'breathing vibrations of a partially filled cylindrical tank-linear theory', *J. Appl. Mech.* **31**(3), 569–570.

Bauer H. F. (1964d), Liquid sloshing in a 45 degree sector compartmented cylindrical tank, *AIAA J.* **2**(4), 768–770.

Bauer H. F. (1964e), Propellant oscillations in the containers of a roll oscillating space vehicle and moment of inertia of liquid, Proc. 5th Annual Struct. and Materials Conf., 184–190.

Bauer H. F. (1965a), The response of propellant in an arbitrary cylindrical tank due to single pulse excitation, in *Developments in Theoretical and Appl. Mech.* **2**, W. A. Shaw, ed., 351–383.

Bauer, H.F. (1965b), Nonlinear propellant sloshing in a rectangular container of infinite length, North Amer. Avia. Inc., S&ID Report, SID 64–1593.

Bauer H. F. (1966a), Liquid behavior in the reservoir of the sound-suppressor system, NASA TN-D-3165.

Bauer H. F. (1966b), Stability boundaries of liquid propelled elastic space-vehicles with sloshing, *J. Spacecraft Rock.* **3**(2), 240–246.

Bauer H. F. (1966c), Theory of liquid sloshing in a rectangular container, Rept. No ER-8390, Lockheed-Georgia Company, June.

Bauer H. F. (1966d), Comment on moment of inertia and damping of liquids in baffled cylindrical tanks, *J. Spacecraft Rock.* **3**(6), 957–959.

Bauer H. F. (1966e), Nonlinear mechanical model for the description of propellant sloshing, *AIAA J.* **4**(9), 1662–1668.

Bauer H. F. (1966f), Response of liquid in a rectangular container, *ASCE J. Eng. Mech. Div.* **92**, 1–23.

Bauer H. F. (1967), Nonlinear propellant sloshing in a rectangular container of infinite length, in *Developments in Theoretical and Appl Mech,* **3**, W. A. Shaw, ed., New York, Pergamon Press, 725–759.

Bauer H. F. (1968a), Response of the fuel in a rectangular container to a roll maneuver with numerical examples for C-5A-wing, Rept. No SMN-217, Lockheed-Georgia Company.

Bauer H. F. (1968b), Dynamics of the airplane with fuel sloshing, Rept. No SMN-246, Lockheed-Georgia Company.

Bauer H. F. (1969a), Fuel sloshing in accelerating rectangular container, Rept. No SMN-282, Lockheed-Georgia Company.

Bauer H. F. (1969b), Note on linear hydroelastic sloshing, *Zeit. Ang. Math. Meck. (ZAMM)* **49**(10), 577–589.

Bauer H. F. (1970), Hydroelastic oscillations in an upright circular cylindrical container, (in German), *Zeitsch. fur Flug.* **18**(4), 117–134.

Bauer H. F. (1971a), Hydroelastic vibrations of a uniformly rotating infinitely long circular cylindrical container, *Acta Mech.* **12**(3/4), 307–326.

Bauer H. F. (1971b), Migration of a large gas-bubble under the lack of gravity in a rotating liquid, *AIAA J.* **9**, 1426–1427.

Bauer H. F. (1972), On the destabilizing effect of liquids in various vehicles, part I, *Vehicle Syst. Dyn., Int. J. Vehicle Mech. and Mobility* **1**, 227–260.

Bauer H. F. (1973), On the destabilizing effect of liquids in various vehicles, part II, *Vehicle Syst. Dyn., Int. J. Vehicle Mech. and Mobility* **2**, 33–48.

Bauer H. F. (1981a), Dynamic behavior of an elastic separating wall in vehicle containers: part I, *Int. J. Vehicle Des.* **2**(1), 44–77.

Bauer H. F. (1981b), Hydroelastic vibrations in a rectangular container, *Int. J. Solids Struct.* **17**, 639–552.

Bauer H. F. (1981c), Flüssigkeitsschwingungen mit freir oberfläche in keilförmigen behälten, *Acta Mech.* **38**, 31–34.

Bauer H. F. (1982a), Coupled oscillations of a solidly rotating liquid bridge, *Acta Astron.* **9**, 547–563.

Bauer H. F. (1982b), Dynamic behavior of an elastic separating wall in vehicle container, Part I. I., *Int. J. Vehicle Des.* **3**, 307–332.

Bauer H. F. (1982c), Oscillations of immiscible liquids in free space or in spherical containers in zero gravity environment, *Ing. Arch.* **51**, 363–381.

Bauer H. F. (1982d), Rotating finite liquid systems under zero gravity, *Forschung im Ingenieurwesen* **48**, 159–179.

Bauer H. F. (1982e), Velocity distribution due to Marangoni-effect for angular temperature field along infinite liquid bridge, *Forschung im Ingenieurwesen* **48**, 50–55.

Bauer H. F. (1982f), Marangoni-convection in a freely floating liquid sphere due to axial temperature field, *Ing. Arch.* **52**, 263–273.

Bauer H. F. (1982g), Velocity distribution due to thermal Marangoni-effect in a liquid column under zero gravity environment, *Zeit. Angew. Math. Mech. (ZAMM)* **62**, 471–482.

Bauer H. F. (1982h), Velocity distribution in a liquid bridge due to thermal Marangoni-effect, *Zeit. fur Flugwissenschaften und Weltraumforschung* **6**, 252–260.

Bauer H. F. (1982i), Sloshing in conical tanks, *Acta Mech.* **43**(3/4), 185–200 (in German).

Bauer H. F. (1983a), Natural damped frequencies of an infinitely long column of immiscible viscous liquids, *Forsch. Ing. Wes.* **49**, 117–126.

Bauer H. F. (1983b), Surface and interface oscillations of freely floating spheres of immiscible viscous liquids, *Ing. Arch.* **53**, 371–383.

Bauer H. F. (1983c), Marangoni-effect velocity distribution due to time-oscillatory temperature gradients in zero gravity environment, *Acta Mech.* **46**, 167–187.

Bauer H. F. (1983d), Liquid surface oscillations in a viscous liquid column induced by temperature fluctuations, *Forschung im Ingenieurwesen* **49**, 58–65.

Bauer H. F. (1983e), Surface oscillations due to the Marangoni-effect in the freely floating sphere, *Ing. Arch.* **53**, 275–287.

Bauer H. F. (1983f), Transient thermal Marangoni-convection in a liquid bridge, *Zeit. fur Flugwissenschaften und Weltraumforschung* **7**, 120–133.

Bauer H. F. (1983g), Liquid surface oscillations induced by temperature fluctuations, *Zeit. fur Flugwissenschaften und Weltraumforschung* **7**, 274–278.

Bauer H. F. (1983h), Transient convection due to the sudden change of the temperature gradient, *Forschung im Ingenieurwesen* **49**, 181–188.

Bauer H. F. (1984a), Oscillations of immiscible liquids in a rectangular container: a new damper for excited structures, *J. Sound Vib.* **93**(1), 117–133.

Bauer H. F. (1984b), Natural damped frequencies of an infinitely long column of immiscible viscous liquids, *Zeit. Angew. Math. Mech. (ZAMM)* **64**, 475–490.

Bauer H. F. (1984c), Free liquid surface response induced by fluctuations of thermal Marangoni convection, *AIAA J.* **22**(3), 421–428.

Bauer H. F. (1984d), Forced liquid oscillations in paraboloid-containers, *Zeit. fur Flugwissenschaften und Weltraumforschung* **8**, 49–55.

Bauer H. F. (1984e), Surface and interface oscillations of a rotating viscous liquid column of immiscible liquids, *Forsch. Ing. Wes.* **50**, 21–31.

Bauer H. F. (1984f), Combined Marangoni and natural convection in a variable micro-gravity field, *Mech. Res. Comm.* **11**, 11–20.

Bauer H. F. (1984g), A theoretical study of Marangoni convection in a liquid column in zero gravity, *Acta Astron.* **11**, 301–311.

Bauer H. F. (1984h), Combined thermo-capillary and natural convection and g-jitter in a constant micro-gravity field, *Forsch. Ing. Wes.* **50**, 169–200.

Bauer H. F. (1985a), Surface and interface oscillations in an immiscible spherical visco-elastic system, *Acta Mech.* **55**, 127–149.

Bauer H. F. (1985b), Induced free liquid surface oscillations in a visco-elastic liquid column due to angular temperature fluctuations, *Forsch. Ing. Wes.* **51**, 133–140.

Bauer H. F. (1985c), Combined residual natural and Marangoni convection in a liquid sphere subjected to a constant and variable micro-gravity field, *Zeito Angew. Math. Mech. (ZAMM)* **65**, 461–470.

Bauer H. F. (1986a), Free surface- and interface oscillations of an infinitely long visco-elastic liquid column, *Acta Astron.* **13**(1), 9–22.

Bauer H. F. (1986b), Coupled frequencies of a hydroelastic viscous liquid system, *Int. J. Solids and Structures* **22**, 1471–1484.

Bauer H. F. (1986c), Induced free surface oscillations in a freely floating visco-elastic liquid sphere imposed to an oscillatory temperature gradient, *Forsch. Ing. Wes.* **52**, 81–88.

Bauer H. F. (1986d), Thermo-capillary-induced axisymmetric free liquid surface oscillations in a visco-elastic column, *Zeit. Angew. Math. Mech. (ZAMM)* **66**, 283–295.

Bauer H. F. (1987a), Coupled frequencies of a hydroelastic system consisting of an elastic shell and frictionless liquid, *J. Sound Vib.* **113**, 217–232.

Bauer H. F. (1987b), Hydroelastic oscillations of a viscous infinitely long liquid column, *J. Sound Vib.* **119**(2), 249–265.

Bauer H. F. (1987c), Natural frequencies and stability of immiscible cylindrical z-independent liquid systems, *Applied Micro Gravity Techn.* **1**, 11–26.

Bauer H. F. (1987d), Thermo-capillary and residual natural convection in an orbiting spherical liquid system, *Forsch. Ing. Wes.* **53**, 83–93.

Bauer H. F. (1988a), Nonlinear oscillations of axially independent liquid column under zero gravity environment, *Forsch. Ing. Wes.* **54**, 82–93.

Bauer H. F. (1988b), Coupled frequencies of a rotating hydroelastic shell–liquid system under zero gravity, *J. Fluids Struct.* **2**, 407–423.

Bauer H. F. (1988c), Natural frequencies and stability of immiscible spherical liquid systems, *Applied Micro Gravity Techn.* **1**, 90–102.

Bauer H. F. (1988d), Marangoni convection in finite cylindrical liquid bridges, *Zeit. fur Flugwissenschaften und Weltraumforschung* **12**, 332–340.

Bauer H. F. (1989a), Response of a spinning liquid column to axial excitation, *Acta Mech.* **77**(1–4), 153–170.

Bauer H. F. (1989b), Natural frequencies and stability of circular cylindrical immiscible liquid systems, *Appl. Microgravity Tech. II.*, 27–44.

Bauer H. F. (1989c), Response of an annular cylindrical liquid column in zero gravity, *Forschung im Ingenieurwesen* **55**, 79–88.

Bauer H. F. (1989d), Damped response of an axially excited rotating liquid bridge under zero gravity, *Acta Mech.* **79**, 295–301.

Bauer H. F. (1989e), Response of a finite rotating annular liquid layer to axial excitation, *Forschung im Ingenieurwesen* **55**, 120–127.

Bauer H. F. (1989f) Vibrational behavior of a viscous column with a free liquid surface, *Zeit. fur Flugwissenschaften und Weltraumforschung* **13**, 248–253.

Bauer H. F. (1989g), Marangoni convection in rotating liquid systems, *Applied Micro Gravity Techn.* **2**, 142–157.

Bauer H. F. (1990a) Response of a liquid column to one-sided axial excitation in zero gravity, *Forschung im Ingenieurwesen* **56**, 14–21.

Bauer H. F. (1990b), Axial response and transient behavior of a cylindrical liquid column in zero gravity, *Zeit. fur Flugwissenschaften und Weltraumforschung* **14**, 174–182.

Bauer H. F. (1990c), Response of a liquid column to unequal axial excitations under zero gravity, *Applied Micro Gravity Techn.* **3**, 34–40.

Bauer H. F. (1990d), Oscillatory response of a liquid column to counter rotational excitation, *J. Sound Vib.* **142**, 125–133.

Bauer H. F. (1990e), Response of a liquid column with respect to oscillatory rotational top and bottom excitation, *J. Sound Vib.* **142**, 379–390.

Bauer H. F. (1990f), Response of a viscous liquid to various pitch excitations, *Acta Astron.* **21**, 553–569.

Bauer H. F. (1990g), Response of a liquid column to counter-directional excitation under zero gravity, *J. Spacecraft Rock.* **27**, 675–680.

Bauer H. F. (1990h), Response of an annular cylindrical liquid column to various axial excitations in zero gravity, *Forschung im Ingenieurwesen* **56**, 183–188.

Bauer H. F. (1990i), Response of a viscous liquid column to axial excitation in zero gravity, *Zeit. Angew. Math. Mech. (ZAMM)* **70**, 359–369.

Bauer H. F. (1990j), Response of differently axial-excited spinning liquid columns, *Acta Mech.* **84**, 155–173.

Bauer H. F. (1990k), Response of a spinning liquid column to pitch excitation, *Acta Mech.* **84**, 1–12.

Bauer H. F. (1991a), Axi-symmetric natural frequencies and response of a spinning liquid column under strong surface tension, *Acta Mech.* **90**, 21–35.

Bauer H. F. (1991b), Liquid oscillations under strong surface tension in a circular cylindrical container, Tech. Rept. LRT-WE-9-FB-26–1991, Universitat der Bundeswehr Muenchen, Neubiberg.

Bauer H. F. (1991c), Liquid sloshing response in a spinning container due to pitching excitation, *Zeit. fur Flugwissenschaften und Weltraumforschung* **15**, 386–392.

Bauer H. F. (1991d), Response of a viscous liquid column to pitching and roll excitations, *Zeit. Angew. Math. Mech. (ZAMM)* **71**, 479–491.

Bauer H. F. (1991e), Response of a viscous liquid layer around a center-core to axial excitation in zero gravity, *Forschung im Ingenieurwesen* **57**, 14–21.

Bauer H. F. (1991f), Response of a rotating finite annular liquid layer to various axial excitations in zero gravity, *J. Sound Vib.* **149**, 219–234.

Bauer H. F. (1991g), Response of a viscous liquid layer to pitching- and roll excitations in zero gravity environment, *Forschung im Ingenieurwesen* **57**, 18–131.

Bauer H. F. (1991h), Natural axisymmetric frequencies of cylindrical liquid column with anchored free surface, *JSME Int. J. Ser. III* **34**, 475–480.

Bauer H. F. (1991i), Response of a non-viscous liquid column and layer with anchored liquid surface to axial excitation, *JSME Int. J. Ser. II* **34**, 474–481.

Bauer H. F. (1991j), Liquid sloshing response in a spin-stabilized missile or satellite due to axial excitation, *Zeit. fur Flugwissenschaften und Weltraumforschung* **15**, 252–256.

Bauer H. F. (1992a), Response of a viscous liquid column with anchored edges to axial excitation, *Zeit. fur Flugwissenschaften und Weltraumforschung* **16**, 42–48.

Bauer H. F. (1992b), Asymmetric natural damped frequencies of a rotating infinitely long viscous liquid column, *Zeit. for Flugwissenshaften and Weltraumforschung* **16**, 317–324.

Bauer H. F. (1992c), Natural damped frequencies and axial response of a rotating finite viscous liquid column, *Acta Mech.* **93**, 29–52.

Bauer H. F. (1992d), Liquid oscillations in a circular container with "sliding" contact line, *Forsch. Ingenieurwes. – Engineering Research* **58**, 240–251.

Bauer H. F. (1992e), Hydroelstic oscillations of rotating liquid–structure systems in zero gravity, *J. Fluid Struct.* **6**, 603–632.

Bauer H. F. (1992f), Liquid oscillations of spherical and conical systems with anchored edges under zero gravity, *Archive Appl. Mech.* **62**, 517–529.

Bauer H. F. (1992g), Response of a liquid bridge with rotationally excited bottom, *ASME J. Appl Mech.* **59**, 191–195.

Bauer H. F. (1992h), Response of a viscous annular liquid layer in zero gravity to different excitations of the rigid boundary plates, *Appl Sci. Res.* **49**, 283–309.

Bauer H. F. (1992i), Response of a viscous spherical and conical liquid system to axial excitation, *Micro Gravity Sci. Techn.* **5**, 35–42.

Bauer H. F. (1992j), The effect of rotation on the vibrational and thermocapillary behavior of liquid columns, *Micro Gravity Sci. Techn.* **3**, 124–133.

Bauer H. F. (1992k), Asymmetric natural damped frequencies of a rotating infinitely long viscous liquid column, *Zeit. fur Flugwissenschaften und Weltraumforschung* **6**, 317–324.

Bauer H. F. (1992l), Response of axially excited spherical and conical liquid systems with anchored edges, *Forsch. Ingenieurwes.* **58**, 96–103.

Bauer H. F. (1993a), Frequencies of a hydroelastic rectangular system, *Forsch. Ingenieurwes. – Engineering Research* **59**, 18–28.

Bauer H. F. (1993b), Natural frequencies and response of spinning liquid column with apparently sliding contact line, *Acta Mech.* **97**, 115–122.

Bauer H. F. (1993c), Natural damped frequencies of a rotating viscous cylindrical liquid layer, *Forsch. Ingenieurwes. – Engineering Research* **59**, 217–227.

Bauer H. F. (1993d), Theoretical micro-gravity fluid dynamics, Proc. Symp. *Aerospace and Fluid Science*, Inst. Fluid Sci., Tokyo University, 30–49.

Bauer H. F. (1993e), Axial response of differently excited anchored viscous liquid bridges in zero gravity, *Ing. Archiv. – Arch. Appl. Mech.* **63**, 322–336.

Bauer H. F. (1993f), Marangoni-convection in a rotating liquid container, *Wärme-und Stoffübertragung* **28**, 131–138.

Bauer H. F. (1994a), Axisymmetric natural frequencies of a rotating infinitely long viscous liquid column, *Zeit. Angew. Math. Mech. (ZAMM)* **74**, 201–210.

Bauer H. F. (1994b), Response of an anchored viscous liquid layer to axial excitation in zero gravity, *Zeit. Angew. Math. Mech. (ZAMM)* **74**, 369–383.

Bauer H. F. (1994c), Transient response of a viscous liquid bridge of finite length and anchored edges, *Micro Gravity Sci. Techn.* **VII**(3), 258–265.

Bauer H. F. (1994d), Response of a rotating anchored finite liquid layer to axial excitation, *Acta Astron.* **32**, 191–197.

Bauer H. F. (1994e), Natural damped frequencies and axial response of a sloshing rotating finite viscous liquid layer in a cylindrical container, *Forsch. Ingenieurwes. – Engineering Research* **60**, 193–205.

Bauer H. F. (1994f), Transient response of a viscous and anchored liquid bridge to an axial pulse, *Zeit. fur Flugwissenschaften und Weltraumforschung* **18**, 113–119.

Bauer H. F. (1995a), Coupled frequencies of a liquid in a circular cylindrical container with elastic liquid surface cover, *J. Sound Vib.* **180**, 689–704.

Bauer H. F. (1995b), Liquid oscillations in a spherical system with sliding edges bounded by two conical solid surfaces in zero gravity, *Zeit. Angew. Math. Mech. (ZAMM)* **75**, 141–151.

Bauer H. F. (1995c), Free fall liquid column dynamics in a fall tower, Festschrift zum 70. Geburtstag von Herrn Professor J. Siekmann, Universitat Gesamthochschule Essen, 229–242.

Bauer H. F. (1995d), Oscillations of a frictionless liquid with anchored edges in a circular cylindrical ring in zero gravity, *Micrograv. Quart.* **5**, 75–90.

Bauer H. F. (1999), Oscillations of non-viscous liquid in various container geometries, Forschungbericht LRT-WE-9-FB-1.

Bauer H. F. and Buchholz A. (1998), Marangoni convection in a rectangular container, *Forsch. Ingenieurwes. – Engineering Research* **63**, 339–348.

Bauer H. F., Chang S. S., and Wang J. T. S. (1971), Nonlinear-liquid motion in a longitudinally excited container with elastic bottom, *AIAA J.* **9**(12), 2333–2339.

Bauer H. F. and Chiba M. (1998), Free hydroelastic vibrations of a liquid attached to a cylindrical sector shell in a zero gravity condition, *ASME J. Vib. Acoust.* **120**, 54–62.

Bauer H. F. and Chiba M. (2000a), Hydroelastic viscous oscillations in a circular cylindrical container with an elastic cover, *J. Fluids Struct.* **14**, 917–936.

Bauer H. F. and Chiba M. (2000b), Interaction of elastic container bottom with free surface of a viscous liquid in a cylindrical container, *ISTS–Japan Soc Aeron. Space Sciences* **C-5**, 1–6.

Bauer H. F. and Chiba M. (2001), Viscous hydroelastic vibrations in a cylindrical container with an elastic bottom, *J. Sound Vib.* **247**, 33–57.

Bauer H. F., Chiba M., and Sasaki H. (1998), Influence of attached liquid layers on free hydroelastic vibrations of a cylindrical sector in a zero gravity condition, *J. Sound Vib.* **214**(2), 245–260.

Bauer H. F., Chiba M., and Watanabe H. (2002), Hydroelastic coupled vibrations in a cylindrical container with a membrane bottom containing liquid with surface tension, *J. Sound Vib.* **251**, 717–740.

Bauer H. F., Clark C. D., and Woodward J. H. (1965), Analytical mechanical model for the description of the rotary propellant sloshing motion, Final Report, Contract NAS8–11159, Eng. Experiment Station, Georgia Tech., Atlanta.

Bauer H. F. and Eidel W. (1986), Nonlinear liquid oscillations in spherical systems under zero gravity, *Acta Mech.* **65**, 107–126.

Bauer H. F. and Eidel W. (1987a), Vibration of a visco-elastic spherical immiscible liquid system, *Zeit. Ang. Math. Mech. (ZAMM)* **67**, 525–535.

Bauer H. F. and Eidel W. (1987b), Marangoni-convection in a spherical liquid system, *Acta Astron.* **15**, 275–290.

Bauer H. F. and Eidel W. (1987c), Induced surface oscillations due to time oscillatory temperature distribution in a viscoelastic spherical system, *Ing. Arch.* **57**, 209–222.

Bauer H. F. and Eidel W. (1988a), Non-linear hydroelastic vibrations in rectangular containers, *J. Sound Vib.* **125**(1), 93–114.

Bauer H. F. and Eidel W. (1988b), Non-linear liquid motion in conical containers, *Acta Mech.* **73**, 11–31.

Bauer H. F. and Eidel W. (1988c), Non-linear liquid oscillations in paraboloidal containers, *Zeit. fur Flugwissenschaften und Weltaumforschung* **12**, 246–252.

Bauer H. F. and Eidel W. (1988d), Nonlinear oscillations of an inviscid liquid column under zero gravity, *Ing. Archiv.* **58**, 276–284.

Bauer H. F. and Eidel W. (1989a), Nonlinear oscillations of a non-viscous cylindrical liquid system in weightlessness condition, *Zeit. Ang. Math. Phys. (ZAMP)* **40**(4), 510–525.

Bauer H. F. and Eidel W. (1989b), Liquid oscillations in a prolate spheroidal container, *Ing. Archiv.* **59**(5–6), 371–381.

Bauer H. F. and Eidel W. (1989c), Small amplitude liquid oscillations in a rectangular container under zero gravity, *Aeron. J.*, 379–386, also *L'Aeronautique et L'Astronautique* **139**, 35–42.

Bauer H. F. and Eidel W. (1990), Linear liquid oscillations in a cylindrical container under zero gravity, *Appl. Microgravity Tech. II* **4**, 212–220.

Bauer H. F. and Eidel W. (1991a), Vibrations of a cylindrical liquid column under the influence of a steady axial micro gravity field, *Micro Gravity Sci. Techn.* **4**, 238–245.

Bauer H. F. and Eidel W. (1991b), Axial response of an amphora-type liquid column, *Acta Astron.* **25**, 699–707.

Bauer H. F. and Eidel W. (1991c), Vibrational behavior of amphora liquid column in micro-gravity fields, *Micro Gravity Sci. Techn.* **4**, 72–74.

Bauer H. F. and Eidel W. (1991d), Stationary geometry of liquid column and liquid drop under the influence of a temperature gradient in zero gravity, *Zeit. fur Flugwissenschaften und Weltaumforschung* **15**, 297–299.

Bauer H. F. and Eidel W. (1992), Nonlinear axial vibrations of a cylindrical liquid column in zero gravity, *Zeit. Ang.Math. Phys. (ZAMP)* **43**, 726–741.

Bauer H. F. and Eidel W. (1993a), Hydroelastic vibrations in a circular cylindrical container with flexible bottom in zero gravity, *J. Fluid. Struct.* **7**, 783–802.

Bauer H. F. and Eidel W. (1993b), Hydroelastic vibrations in a rectangular container under zero gravity, *Micrograv. Quart.* **3**, 7–16.

Bauer H. F. and Eidel W. (1993c), Natural frequencies of a spinning amphora-type liquid column, *J. Sound Vib.* **167**, 331–346.

Bauer H. F. and Eidel W. (1994), Response of a solidly rotating amphora-type liquid column to axial excitation, *Acta Astron.* **32**, 113–119.

Bauer H. F. and Eidel W. (1995), Anchored-edge pitching response of a spinning frictionless liquid column due to bottom and/or top excitation, *Forschung im Ingenieurwesen – Engineering Research* **61**, 258–270.

Bauer H. F. and Eidel W. (1996a), Response of an anchored frictionless rotating liquid column to various excitations, *Zeit. fur Flugwissenschaften und Weltaumforschung* **20**, 39–48.

Bauer H. F. and Eidel W. (1996b), Pitching response of an anchored liquid bridge to bottom and/or top disc excitation, *Zeit. Angew. Math. Mech. (ZAMM)* **76**, 127–143.

Bauer H. F. and Eidel W. (1996c), Axisymmetric damped frequencies of a rotating finite visco-elastic liquid column, *Acta Astron.* **38**, 689–698.

Bauer H. F. and Eidel W. (1996d), Viscous response of an anchored spinning liquid column to various axial excitations, *Acta Mech.* **116**, 153–170.

Bauer H. F. and Eidel W. (1996e), Natural damped frequencies of a rotating visco-elastic infinite column, *Forschung im Ingenieurwesen – Engineering Research* **62**(7/8), 214–227.

Bauer H. F. and Eidel W. (1997a), Axisymmetric viscous liquid oscillations in a cylindrical container, *Forschung im Ingenieurwesen – Engineering Research* **63**, 189–201.

Bauer H. F. and Eidel W. (1997b), Axisymmetric oscillations in a slowly rotating cylindrical container filled with viscous liquid, *Forschung im Ingenieurwesen – Engineering Research* **63**, 215–223.

Bauer H. F. and Eidel W. (1997c), Oscillations of a viscous liquid in a cylindrical container, *Aerospace Sci. Tech.* **8**, 519–532.

Bauer H. F. and Eidel W. (1997d), Oscillations and response of a viscous liquid with anchored edges in a cylindrical ring in zero gravity, *Micrograv. Quart.* **7**, 59–88.

Bauer H. F. and Eidel W. (1998a), Axisymmetric thermocapillary convection in a slowly rotating annular cylindrical container, *Heat Mass Trans.* **34**(1), 79–90.

Bauer H. F. and Eidel W. (1998b), Transient motion of a viscous liquid with free surface in a cylindrical container, *Zeit. Angew. Math. Mech. (ZAMM)* **79**, 97–106.

Bauer H. F. and Eidel W. (1998c), Marangoni convection in an annular cylindrical tank, *Heat Mass Transf.* **33**, 457–464.

Bauer H. F. and Eidel W. (1998d), Oscillatory thermo-capillary convection in a spinning cylindrical liquid bridge under micro-gravity condition, *Forsch. Ingenieurwes. – Engineering Research* **64**, 197–214.

Bauer H. F. and Eidel W. (1999a), Axisymmetric natural damped frequencies of a viscous liquid in a circular cylindrical container: an alternative semi-analytical solution, *Forschung im Ingenieurwesen – Engineering Research* **65**, 191–199.

Bauer H. F. and Eidel W. (1999b), Free oscillations and response of a viscous liquid in a cylindrical container, *Aerospace Sci. Tech.* **3**, 495–512.

Bauer H. F. and Eidel W. (1999c), Frictionless liquid sloshing in circular cylindrical container configurations, *Aerospace Sci. Tech.* **3**(5), 301–311.

Bauer H. F. and Eidel W. (2000), Free and forced oscillations of a frictionless liquid in a long rectangular tank with structural obstructions at the free liquid surface, *Arch. Appl. Mech.* **70**(8/9) 550–560.

Bauer H. F. and Eidel W. (2002a), Linear response of a viscous liquid to translational oscillations, *Forsch. Ing. Wes.* **67**, 72–83.

Bauer H. F. and Eidel W. (2002b), Axisymmetric damped natural frequencies in a slowly spinning cylindrical container filled with viscous liquid, *Forsch. Ing. Wes.* **67**, 93–99.

Bauer H. F., Hsu T. M., and Wang J.T.S. (1967), Liquid sloshing in elastic containers, NASA CR-882, September.

Bauer H. F., Hsu T. M., and Wang J. T. S. (1968), Interaction of a sloshing liquid with elastic containers, *J. Basic Eng., Series D* **68**, 373–377.

Bauer H. F. and Kammerer L. (1992), Natural frequencies and axial response of a liquid layer under the influence of an axial steady micro-gravity field, *Forschung im Ingenieurwesen* **58**, 26–34.

Bauer H. F. and Kammerer L. (1993), Natural frequencies and axial response of a spinning liquid layer in micro-gravity field, *Zeit. Ang. Math. Phys. (ZAMP)* **44**, 348–369.

Bauer H. F. and Knölker E. (1981), Theoretische und experimentelle bestimmung der eigenfrequenzen in einem teilweise gefüllten elliptischen zylinder. Diploma thesis, Hochschule der Bundeswehr München, Intitute für Raumfahrttechnik.

Bauer H. F. and Komatsu K. (1994a), Coupled frequencies of a hydroelastic system of an elastic two-dimensional sector-shell and frictionless liquid in zero gravity, *J. Fluids Struct.* **8**, 817–831.

Bauer H. F. and Komatsu K. (1994b), Response of an anchored frictionless liquid bridge to pitching bottom and/or top excitation, *Forschung im Ingenieurwesen – Engineering Research* **60**, 237–243.

Bauer H. F. and Komatsu K. (1998), Vibration of a hydroelastic system consisting of a sector shell and viscous liquid in zero gravity, *J. Fluids Struct.* **12**, 367–385.

Bauer H. F. and Komatsu K. (2000), Coupled frequencies of a frictionless liquid in a circular cylindrical tank with an elastic partial surface cover, *J. Sound Vib.* **230**(5), 1147–1163.

Bauer H. F. Langbein D., Falk F., Grossbach R., Heide W., Meseguer J., Perales J. M., and Sanz A. (1994), LICOR-Liquid columns' resonance, Proc. Norderney Symp. *Scientific Results of German Spacelab Mission D-2*, P. R. Sahm, M. H. Keller, and B. Schiene, eds., 209–219.

Bauer H. F. and Offenbach D. (1986), Thermo-capillary convection in a semi-spherical drop, *Zeit. fur Flugwissenschaften und Weltaumforschung* **10**, 92–96.

Bauer H. F. and Reinfurth, MH (1959), Oscillations in a connected fluid system, ABMA DA-TN-52–59, 9 April.

Bauer H. F. and Siekmann J. (1969), Note on linear hydroelastic sloshing, *Zeit. Angew. Math. Mech. (ZAMM)* **49**, 577–589.

Bauer H. F. and Siekmann J. (1971a), Dynamic interaction of liquid with elastic structure of a circular cylindrical container, *Ing. Archiv.* **40**, 266–280.

Bauer H. F. and Siekmann J. (1971b), On the shape of a rotating fluid system consisting of a gas bubble enclosed in a liquid globe, *Zeit. Ang. Math. Phys. (ZAMP)* **22**, 532–542.

Bauer H. F. and Siekmann J., and Wang J. T. S. (1968), Axi-symmetric hydroelastic sloshing in an annular cylindrical container, *J. Spacecraft Rock.* **5**(8), 981–983.

Bauer H. F. and Strasser U. (1993), Marangoni-convection in a rotating liquid column, *Micro Gravity Sci. Techn.* **VI**(3), 164–175.

Bauer H. F. and Villenueva J. (1966), Theory of liquid sloshing in a rectangular container with numerical examples for C-45A wing, Lockheed-Georgia Co., Rept. No ER 8390.

Bauer H. F. and Villenueva J. (1967a), Theory of liquid sloshing in a rectangular container with numerical examples for C-45A wing, Lockheed-Georgia Co., Marietta, G Rept. No ER-8790.

Bauer H. F. and Villenueva J. (1967b), Mechanical model for the description of the liquid motion in a rectangular container with numerical examples for C-45A wing, Lockheed-Georgia Co., Marietta, G Rept No. ER-8559.

Bauer H. F., Wang J. T. S., and Chen P. Y. (1972), Axisymmetric hydroelastic sloshing in a circular cylindrical container, *The Aeronautical J. of the Aeron. Soc.* **76**, 704–712.

Beal T. R., Coale C. W., and Nagano M. (1965), Influence of shell inertia and bending stiffness on the axi-symmetric modes of a partially filled cylindrical tank, AIAA Paper No 65–412, AIAA Annual Meeting.

Beam R. M. and Guist L. R. (1967), The axially symmetric response of an elastic shell partially filled with liquid, NASA TN-D-3877, February.

Bechhoefer J. (1996), A simple model of Faraday waves, *Am. J. Phys.* **64**(12), 1482–1487.

Bechhoefer J., Ego V., Manneville S., and Johnson B. (1995), An experimental study of the onset of parametrically pumped surface waves in viscous fluid, *J. Fluid Mech.* **288**, 325–350.

Beig H. G. (1984), Study of slosh dynamics of fluid filled containers on 3-axis spin stabilized spacecraft, **1**, *Fluid Slosh Studies* 5328/83/NL/Bi(Sc), ESA.

Belik O., Bishop R. E. D., and Price W. G. (1988), Influence of bottom and flare slamming on structural responses, *Trans. RINA* **130**, 325–337.

Belik O. and Price W. G. (1982), Comparison of slamming theories in the time simulation of ship responses, *Trans. RINA* **130**, 325–337.

Belinfante J. G., Kolman B., and Smith H. (1966), An introduction to Lie groups algebras with applications, *SIAM Review* **8**(1), 11–46.

Bell J. and Walker W. P. (1966), Activated and passive controlled fluid tank system for ship stabilization, *Soc. Naval Archit. Marine Engrg. (SNAME) Transactions* **74**, 150–193.

Bellman R. and Pennington R. H. (1954), Effects of surface tension and viscosity on Taylor instability, *Quart. J. Mech. Appl. Math.* **12**(2), 151–160.

Benard H. (1901), *Les tourbillons cellulars dans une nappe liquide propageant de la chaleu par convection*, Paris, Gouthier-Villers, 88.

Benedict E. T. (1959), Scale of separation phenomena in liquids under conditions of nearly free fall, *J. Amer. Rocket. Soc.* **29**, 150–155.

Benedict E. T. (1960), General behavior of a liquid in a zero or near-zero gravity environment, in E. T. Benedict, ed., *Weightlessness – Physical Phenomena and Biological Effect*, New York, Plenum Press, Proc. Symp. Physical and Biological Phenomena under Zero G Conditions.

Benjamin T. B. and Feir J. E. (1967), The disintegration of wave trains on deep water, *J. Fluid Mech.* **27**, 417–430.

Benjamin T. B. and Scott J. C. (1979), Gravity-capillary waves with edge constraints, *J. Fluid Mech.* **92**, 241–267.

Benjamin T. B. and Ursell F. (1954), The stability of a plane free surface of a liquid in vertical periodic motion, *Proc. Royal Soc. (London) A* **225**, 505–515.

Berger H., Boujot J., and Ohayon R. (1975), On a spectral problem in vibration mechanics computation of elastic tanks partially filled with liquids, *J. Math. Anal. Appl.* **51**, 272–298.

Berlamont J. and Vanderstappen N. (1979), The effect of baffles in a water tower tank, Proc. 5th Int. Conf. *Wind Engineering*, Fort Collins, CO.

Berlot R. R. (1959), Production of rotation in a confined liquid through translational motion of the boundaries, *ASME J. Appl. Mech.*, 513–516.

Berlot R. R., Birkhoff G., and Miles J. (1957), Slosh damping in a rigid cylindrical tank, RWC, Electronics Lab., Rept. No GM-TR-263, October.

Berlot, R. R. and Freed L. F. (1956), Comparison of experimental sloshing results with spherical pendulum theorem, Space Technol. Labs, Memo. GM42–6–54, January.

Bermann A. S., Lundgren T. S., and Cheng A. (1985), Asynchronous whirl in a rotating cylinder partially filled with liquid, *J. Fluid Mech.* **15**, 311–427.

Bermudez A. and Rodriguez R. (1994), Finite element computation of the vibration modes of a fluid-solid system, *Comp. Methods in Mech. Eng.* **119**, 355–370.

Bernard B. J. S., Mahony J. J., and Pritchard W. G. (1977), The excitation of surface waves near cut-off frequency, *Phil. Trans. R. Soc. Lon. A* **286**, 87–123.

Bernard B. J. S. and Pritchard W. G. (1972), Cross-waves, part 2: experiment, *J. Fluid Mech.* **55**, 245–255.

Berry J. G. and Reissner E. (1958), The effect of an internal compressible fluid column on the breathing vibrations of a thin pressurized cylindrical shell, *J. of Aero./Space Sci.* **25**(5), 288–294.

Berry R. I., Demchak L. J., and Tegart J. R. (1981), Analytical tool for simulating large amplitude propellant slosh, AIAA Paper 81–0500-CP.

Bessem T., Edwards W. S., and Tuckerman L. (1996), Two-frequency parametric excitation of surface waves, *Phys. Rev.* ***E*** **54**, 507–513.

Betchov R. (1958), Nonlinear oscillations of a column of gas, *Phys. Fluids* **1**, 3.

Beyer J. and Friedrich R. (1995), Faraday instability: linear analysis for viscous fluids, *Phys. Rev. E* **51**, 1162–1168.

Bezdenezhnikh N. K., Briskman B. A., Lapin A. Yu, Lyubimov D. V., Lyubimov T. P., Cherepanov A. A., and Zakharov I. V. (1991), The influence of high frequency tangential vibrations on stability of the fluid in microgravity, Proc. IUTAM Symp. *Microgravity Fluid Mechanics*, Bremen, Berlin, Springer, 1992, 137–144.

Bezdenezhnikh N. K., Briskman B. A., Lyubimov D. V., Cherepanov A. A., and Sharov M. T. (1984), Control of stability of two-fluid interface by means of vibrations, electric and magnetic fields, All-Union Seminar on *Hydromechanics and Heat and Mass Transfer: Abstracts*, Chernogolovka, 18–20.

Bhattacharyya R., (1978), *Dynamics of Marine Vehicles*, New York, John Wiley & Sons.

Bhuta P. G., Hutton R. E., Abramson H. N., and Stephens D. G. (1968), Propellant slosh loads, NASA-SP-8009, August.

Bhuta P. G. and Koval I. R. (1964a), Coupled oscillations of a liquid in a tank having a flexible bottom, *Zeit. Ang. Math. Phys. (ZAMP)* **15**(4), 466–488.

Bhuta P. G. and Koval L. R. (1964b), Hydroelastic solution of the sloshing of a cylindrical tank, *J. Acoust. Soc. Amer.* **36**(11), 2071–2079.

Bhuta P. G. and Koval L. R. (1965), Sloshing of a liquid in a draining or filling tank under variable "g" conditions, Symp. *Fluid Mech. and Heat Transfer under Low Gravitational Conditions*, Lockheed Missiles & Space Co, June 24–25.

Bhuta P. G., Pravin G., and Koval L. R. (1964), Coupled oscillations of a liquid with a free surface in a tank, *Zeit. Ang. Math. Phys. (ZAMP)* **15**, 466–480.

Bhuta P. G. and Yeh G. C. K. (1965), Liquid sloshing due to a time dependent discontinuous boundary, *Int. J. Mech. Sci.*, 475–488.

Biesel F. (1949), Calculation of wave damping in a viscous liquid of unknown depth, *La Houille Blanche* **4**, 630–634.

Binks D. and Van de Water W. (1997), Nonlinear pattern formation of Faraday waves, *Phys. Rev. Lett.* **78**, 4043–4046.

Binks D., Westra M. T., and Van de Water W. (1997), Effect of depth on the pattern formation of Faraday waves, *Phys. Rev. Lett.* **79**, 5010–5013.

Binnie A. M. (1941), Waves in an open oscillating tank, *Eng.* **151**, 224–226.

Binnie A. M. (1955), Self-excited oscillations in an open circular water tank, *Phil. Mag.* **46**, 327–337.

Birkhoff G. (1956), Liquid oscillations in static containers, Space Technol. Labs, Memo. GM-TN 12, 18 April, Los Angeles, CA.

Biswal K. C., Bhattacharayya S. K. and Sinha P. K. (2002), Coupled dynamic response of liquid filled composite cylindrical tanks with baffles, Proc. ASME WAM IMECE2002–32953.

Biswal K. C., Bhattacharayya S. K., and Sinha P. K. (2003), Free vibration analysis of liquid filled tank with baffles, *J. Sound Vib.* **259**(2), 177–192.

Blackhear W. T. and Eide D. G. (1964), The equilibrium free surface of a contained liquid under low gravity and centrifugal forces, NASA TN-D-2471, October.

Blacklock J. H. (1965), Effects of propellant sloshing, in *Automatic Control of Aircraft and Missiles*, NY, John Wiley, 254–259.

Blak A. M. (1996), Lagrangian for water waves, *Phys. Fluids* **8**, 416–420.

Bleich H. H. (1956a), Longitudinal forced vibrations of cylindrical fuel tanks, *Jet. Prop. ARS*, 109–111.

Bleich H. H. (1956b), Effect of vibrations on the motion of small gas bubble in a liquid, *Amer. Rocket Soc. J.* **26**(10), 958–964.

Block M. J. (1956), Surface tension as the cause of Benard cells and surface deformation in a liquid film, *Nature* (London) **178**, 650–651.

Bogoryad I. B. (1962), On the solution by a variational method of a problem on oscillations of a fluid partially filled cavity, *Prikl. Math. Mekh. (PMM)* **26**, 1122–1127.

Bogoryad I. B. (1980), *The Dynamics of a Viscous Liquid with a Free Surface*, Izd. Tomskogo University, Tomskogo (in Russian).

Bogoryad I. B. (1990), Damping factors for damping due to the presence of fluid with a free surface in a moving vessel, *Prikl. Mekhaniks* **26**(4), 397–402.

Bogoryad I. B., Druzhinin I. A., Druzhinina G. Z., and Libin E. E. (1977), *Introduction to the Dynamics of Vessels Filled with a Liquid*, Izd. Tomskogo University, Tomskogo (in Russian).

Bolotin V. V. (1956), On the motion of a fluid on an oscillating vessel, *Prikl. Math. Mekh. (PMM)* **20**(2), 293–294.

Bolotin V. V. (1964), *The Dynamic Stability of Elastic Systems*, San Francisco, Holden-Day, Inc.

Bonneau E. (1964), Sloshing of liquids in missiles, *Bull. Assoc. Tech. Marit. Aeronaut.* **64**, 163–177 (in French).

Bontozoglou V. and Hanratty T. J. (1990), Capillary-gravity Kelvin–Helmholtz waves close to resonance, *J. Fluid Mech.* **217**, 71–78.

Borisova E. P. (1962), Free oscillations of a fluid in an inclined cylinder, in *Variational Methods in Problems of Oscillations of a Fluid and of a Body with a Fluid*, Vychislit. Tsentr. Akad. Nauk. SSSR, 203–210.

Bosch E., Lambermont H., and van de Water W. (1994), Average pattern in Faraday waves, *Phys. Rev. E* **4**, R3580–R3583.

Bosch E. and van de Water W. (1993), Spatio-temporal intermittency in the Faraday experiment, *Phys. Rev. Lett.* **2**, 3420–3423.

Boudet M. (1968), The frequency spectra of sloshing liquids, Laboratoire de Recherches Balistiques et Aerodynamiques, Vernon, (France), NT-4/68/EN, 2 May (in French).

Boujot J. (1973), Sur le probleme spectral associe aux vibrations d'un reservoir deformable, *C. R. Acad. Sci. Paris,* ***T*** **277**, Serie A.

Boussinesq J. (1878), *J. Math. Pures. Appl.* **4**, 335.

Bowman T. E. (1966a), Response of the free surface of a cylindrically contained liquid to off-axis accelerations, Proc. 1966 *Heat Transfer and Fluid Mech. Inst.*, University of Santa Clara, CA, M. A. Saad and J. A. Miller, eds., Stanford University Press, 295–314.

Bowman T. E. (1966b), Cryogenic liquid experiment in orbit, I – Liquid Settling and Interface Dynamics, NASA CR-651.

Box G. E. P. and Muller M. E. (1958) A note on the generation of random normal deviatas, *Ann. Math. Stat.* **28**, 610–611.

Boyarshina L. G. (1984), Resonance effects in the nonlinear vibrations of cylindrical shells containing a liquid, *Soviet Appl. Mech.* **20**(8), 765–770.

Boyarshina L. G. (1988), Nonlinear wave modes of an elastic cylindrical shell partially filled with a liquid under conditions of resonance, *Soviet Appl. Mech.* **24**(5), 528–534.

Boyarshina L. G. (1994), Nonlinear-wave motions of a liquid in a cylindrical vessel with elastic bottom, *Int. Appl. Mech.* **30**(5), 385–391.

Boyarshina L. G. and Ganiev R. F. (1976), On nonlinear oscillations of deformable bodies with fluid performing motion in space, *Soviet Appl. Mech.* **12**(7), 62–69.

Boyarshina L. G. and Kholopova V. V. (1994), On resonance in the nonlinear oscillations of a liquid bilayer, *Int. Appl. Mech.* **30**(10), 810–814.

Boyarshina L. G. and Koval'chuk P. S. (1986), The wave modes of the free surface of a liquid contained in a cylindrical vessel, *Soviet Appl. Mech. (Prikl Mekh)* **22**(3), 113–116.

Boyarshina L. G. and Koval'chuk P. S. (1990), Analysis of nonlinear wave motions of a fluid in a cylindrical vessel performing given angular oscillations, *Soviet Appl. Mech.* **26**(6), 601–607.

Boyarshina L. G., Koval'chuk P. S., Puchka G. N., and Kholopova, V. V. (1987), Dynamic instability of the free surface of a fluid in a cylindrical container subjected to a longitudinal excitation, *Prikl. Mekh.* **23**(7), 82–87.

Boyarshina L. G., Koval'chuk P. S., Puchka G. N., and Kholopova V. V. (1998), Constructing periodic solutions of one class of nonautonomous systems of nonlinear second-order equations in a resonance case, Proc. M Kravchuk 7th Int. Sci. Conf. Kiev, May 14–16, 66–70 (in Russian).

Boyarshina L. G. and Puchka G. N. (1985), On nonlinear oscillations of a stratified liquid in a cylindrical container, *Appl. Mech. (Prikl. Mekh.)* **21**(11), 118–123.

Brand R. P. and Nyborg W. L. (1965), Parametrically excited surface waves, *J. Acoust. Soc. Amer.* **37**(3), 509–515.

Bratu C. (1971), Oscillations of liquid masses in reservoirs, (in French), *Bulletin de l' Assoc. Tech. Maritime et Aeronaut.* **71**, 221–241.

Brathu M. C., Huther M., and Planeix J. M. (1972), Computer calculations of liquid motions in tanks, *Shipping World and Shipbuilder*, December.

Bredmose H., Brocchini M., Peregrine D. H., and Thais L. (2003), Experimental investigation and numerical modeling of steep forced water waves, *J. Fluid Mech.* **490**, 217–249.

Bridges T. J. (1982), A numerical simulation of large amplitude sloshing, Proc. 3rd Int. *Numerical Ship Hydrodynamics*, 269–281.

Bridges T. J. (1987), Secondary bifurcation and change of type of three-dimensional standing waves in a finite depth, *J. Fluid Mech.* **179**, 137–153.

Briskman V. A., Ivanova A. A., and Shaidurov G. F. (1976), Parametric oscillations of fluid in communicating vessels, *Izvestiya Akademii Nauk. SSSR, Mekhanik Zhidkosti i Gaza*, **2**, 36–42.

Briskman V. A. and Shaidurov G. F. (1968), Parametric instability of a fluid surface in an alternating electric field, *Sov. Phys.-Doklady*. **13**(6), 540–542.

Brocchini M., Beregrine D. H., and Thais L. (1997), Violent free-surface motion: Some wave-tank experiments, Tech. Rept. AM-97–01, School of Mathematics, University of Bristol.

Brodkey R. S. (1967), *The Phenomena of Fluid Motions*, Massachusetts, Addison-Wesley.

Brooks J. E. (1959), Dynamics of fluids in moving containers, Space Tech. Lab. Rept. No GM-TR-4–26, 9 February.

Brosset L. (1995), Numerical simulation of liquid sloshing in moving tanks, *J. Marine Merchande, Nouveautes Techniques Maritimes* **9**.

Brown K. M. (1954), Laboratory test of fuel sloshing, Douglas Aircraft, CA, Rep. Dev 782, 18 March.

Brown S. J. (1980), Theoretical-experimental seismic tests of fluid coupled co-axial cylinders, ASME, Century 2, Pressure Vess. Piping Conf., San Francisco, PVPD, 80-C/PVP-45, 12–24.

Brown S. J. (1982a), Hydrodynamic response of fluid coupled cylinders: simplified damping and inertia coefficients, Welding Research Council Bulletin, No 281.

Brown S. J. (1982b), A survey of studies into the hydrodynamic response of fluid coupled cylinders, *ASME J. Pressure Vessel Tech.* **104**(1) 2–20.

Brown S. J. (1982c), The effect of fluid axial forces on flexible finite length co-axial cylinders, in *Dynamic and Seismic Analysis of Systems and Components*, M. J. Yan, ed., PVP69.

Brown S. J. and Chu M. (1983), A case study of the seismic response of fluid-coupled-flexible cylinders, *Exp. Mech.*, 270–281.

Brown S. J. and Hsu K. H. (1978), On the use of the finite element displacement method to solve solid–fluid interaction vibration problems, ASME Symp. on *Fluid Transients and Acoust. in the Power Industry*.

Bryant P. J. (1989), Nonlinear progressive waves in a circular basin, *J. Fluid Mech.* **205**, 453–467.

Bryant P. J. (1993), Breakdown to chaotic motion of a forced, damped, spherical pendulum, *Physica D* **64**, 324–339.

Bryant P. J. and Stiassnie M. (1994), Different forms for nonlinear standing waves in deep water, *J. Fluid Mech.* **272**, 135–156.

Bublik B. N. and Merkulov V. N. (1960), On the dynamic stability of thin elastic shells filled with fluid, *Prikl. Math. Mekh. (PMM)* **24**, 941–946.

Buchanan H. J. (1966), Effect of Reynolds number on slosh damping by flat ring baffles, NASA TM X-53559.

Buchanan H. J. (1968), Drag on flat plates oscillating in incompressible fluid at low Reynolds numbers, M.S. thesis, University of Alabama, Also NASA TM-X-53759, 22 July.

Buchanan H. J. and Bugg F. M. (1966), Orbital investigation of propellant dynamics in a large rocket booster, NASA TM X-53442.

Buchanan R. H., Jameson G., and Oedjoe D. (1962), Cyclic migration of bubbles in vertically vibrating liquid columns, *Indust. Eng. Chem. Fund.* **1**(2), 82–86.

Buchner B. (1995), On the impact of green water loading on ship and offshore unit design, The 6th Int. Sym. *Practical Design of Ships and Mobile Units* (PRADS), September.

Buckle U. and Peric M. (1992), Numerical simulation of buoyant and thermocapillary convection in a square cavity, *Numer. Heat Trans., A* **21**, 121–141.

Budiansky B. (1960), Sloshing of liquids in circular canals and spherical tanks, *J. Aero./Space Sciences* **27**(3), 161–172.

Bustamante J. I. and Flores A. (1966), Water pressure on dams subjected to earthquakes, *ASCE J. Eng. Mech. Div.* **92**, 115–127.

Burris L., *et al.* (1987), The application of electro-refining for recovery and purification of fuel discharged from the integral fast reactor, AICHE Symp. Series **83**(254), 1986, 36 pages, available from NTIS, PC A03/MF A01;1: Conference AICHE Winter Ann. Meeting, Miami, 2 Nov 1986; 87(6):37472 Energy.

Buzhinskii V. A. (1990), The energy of vortex formation for oscillations in a fluid of a body with sharp edges, *Dkl. Akad. Nauk. SSSR* **313**(2), 1072–1074.

Buzhinskii V. A. (1998a), Vortex drag of a plate in the case of oscillations in a low viscous fluid, *J. Appl. Mat. Mech. (PMM)* **54**(2), 233–238.

Buzhinskii V. A. (1998b), Vortex damping of sloshing in tanks with baffles, *J. Appl. Math. Mech. (PMM)* **62**, 217–224.

Buzhinskii V. A. and Stolbetsov V. I. (1987), Determination of hydrodynamic characteristics of a cavity, partially filled with a fluid, with a pendulum inside, *Izv. Akad. Nauk. SSSR, MZhG* **6**, 91–100.

Campbell I. J. (1953), Wave motion in an annular tank, *Phil. Mag. Series* **7**(44), 845–854.

Canard A. (1956), Hydrodynamic forces induced in fluid containers by earthquakes, Thesis, Caltech, Pasadena.

Candless C. E. and Walls J. C. (1969), Slosh test analysis for the 200 inch multicell tank, NASA-CR-61268.

Capodano P. (2001), The stability of configuration of a liquid layer in an axially symmetric container uniformly rotating under conditions of weightlessness, *J. Appl. Mat. Mech.* **65**(4), 605–615.

Cariou A. and Casella G. (1999), Liquid sloshing in ship tanks: a comparative study of numerical simulation, *Marine Struct.* **12**(3), 183–198.

Carotenuto L., Castagnolo D., Albanese C., and Monti R. (1998), Instability of thermocapillary convection in liquid bridges, *Phys. Fluids* **10**, 1–11.

Carpenter B. M. and Homsy G. M. (1990), High Marangoni number convection in a square cavity, part 2, *Phys. Fluids A* **2**(12), 137–148.

Carrier G. F. and Di Pima R. C. (1956), On the torsional oscillations of a solid sphere in a viscous fluid, *ASME J. Appl Mech.* **23**, 601–605.

Carruthers J. R., Gibson E. G., Klett M. G., and Facemire B. R. (1975), Studies of rotating liquid floating zones on Skylab IV, AIAA 10th Thermophysics Conf., Denver CO, AIAA-Paper 75–692.

Carruthers J. R. and Grasso M. (1972), Studies of floating liquid zones in simulated zero gravity, *J. Appl. Phys.* **43**(2), 436–445.

Casademunt J., Zhang W., Viñals J., and Sekerka R. F. (1993), Stability of a fluid in a microgravity environment, *AIAA Journal* **31**, 2027.

Case K. and Parkinson W. (1956a), On the motion of a vessel containing fluid with a free surface, Memo. Space Tech. Lab., Feb.

Case K. and Parkinson W. (1956b), The damping of liquid in a right circular cylindrical tank, Memo. GM45–75, Space Tech. Lab., 17 September.

Case K. and Parkinson W. (1957), Damping of surface waves in an incompressible liquid, *J. Fluid Mech.* **2**(2), 172–184.

Casella G. and Dogliani M. (1996), Evaluation of sloshing-induced fatigue damage on a FSO tanker, 6th ISOPE Conf., Los Angeles, May, 304–314.

Casella G., Sebastiani L., and Valdenazzi F. (1996), Fluid–structure interaction in numerical simulation of liquid sloshing, Catena International Seminar on *Hydroelasticity for Ship Structural Design*, Genoa, Italy, February Rept. No 5882.

Castagnolo D. and Carotenuto L. (1999), A numerical simulation of three-dimensional thermocapillary flows in liquid bridges, *Num. Heat Transfer: Part A Applications* **36**(8), 859–877.

Castagnolo D. and Monti R. (2001), Thermal Marangoni flows, in *Physics of Fluids in Microgravity*, R Monti, ed., London, Taylor & Francis, 78–125.

Celebi M. S. and Akyildiz H. (2002), Nonlinear modeling of liquid sloshing in a moving rectangular tank, *Ocean Eng.* **29**(12), 1527–1553.

Celebi M. E., Kim M. H., and Beck R. F. (1998), Fully nonlinear 3-D numerical wave tank simulation, *J. Ship Res.* **42**(1), 33–45.

Cerda E. A. and Tirapegul E. L. (1997), Faraday's instability for viscous fluids, *Phys. Rev. Lett.* **78**, 859–862.

Cerda E. A. and Tirapegul E. L. (1998), Faraday's instability in viscous fluid, *J. Fluid Mech.* **368**, 195–228.

Cesari L. (1963), *Asymptotic Behavior and Stability Problems in Ordinary Differential Equations*, NY, Academic Press.

Chadwick J. H. and Klotter K. (1954), On the dynamics of anti-roll tanks, *Schiffstechnik* **2**, 23–45.

Chaisen P., Fujino Y., Pacheco B. M., and Sun L. M. (1989), Interaction of tuned liquid damper (TLD) and structure – theory, experimental verification and application, *JSCE Struct. Eng./Earthquake Eng.* **6**(2), 103–112.

Chalhoub M. S. and Kelly J. M. (1990), Shake table test of cylindrical water tanks in base-isolated structure, *J. Engrg. Mech.* **116**, 1451–1472.

Chan E. S. (1994), Mechanics of deep-water plunging-wave impacts on vertical structures, *Coast. Engrg.* **22**, 115–133.

Chan K. C. and Street R. L. (1970), SUMMAC – A numerical method for water waves, Report 135, Stanford University.

Chandrasekhar S. (1961), *Hydrodynamic and Hydromagnetic Stability*, New York, Dover.

Chang C. C. and Gu M. (1999), Suppression of vortex-excited vibration of tall buildings using tuned liquid dampers, *J. Wind Eng. Indust. Aerodyn.* **83**, 225–237.

Chang C. C. and Hsu C. T. (1998), Control performance of liquid column vibration absorbers, *Eng. Struct.* **20**(7), 580–586.

Chang C. C. and Shen M. C. (2000), Chaotic motion and internal resonance of forced surface waves in a water-filled circular basin, *Wave Motion* **31**(4), 317–331.

Chang C. E. and Wilcox W. R. (1975), Inhomogeneities due to thermocapillary flow in floating zone melting, *J. Crystal Growth* **28**, 8–12.

Chang C. E. and Wilcox W. R. (1976), Analysis of surface tension driven flow in floating zone melting, *Int. J. Heat Mass Transfer* **19**, 355–366.

Chang P. M., Lou J. Y. K., and Lutes L. D. (1998), Model identification and control of tuned liquid damper, *Eng Struct.* **20**(3), 155–163.

Chang S. C. (1966), On sub-gravity hydroelastic sloshing of a liquid with a curved free surface in a cylindrical tank having flexible bottom, Thesis, University of Florida.

Chang S. S. (1969), Longitudinally excited nonlinear liquid motion in a circular cylindrical tank with elastic bottom, Ph.D. Dissertation., Georgia Institute of Tech., Atlanta, Georgia.

Chang Y. W., Ma D. C., and Gvildys J. (1987), EPRI/CRIEPI/joint program on seismic sloshing of LMR reactors. part II-Numerical solutions, Trans. Struct. Mech. in Reactor Techn.-9 Conf., Lausanne, Switzerland, E, 253–258.

Chao L., Kamotani Y., and Ostrach S. (1992), G-jitter effects on the capillary surface motion in an open container under weightless condition, ASME WAM *Symp. Fluid Mech. Phen. Microgravity*, **AMD-154/FED-142**, 133–143.

Chato D. J., Dalton P. J., Dodge F. T., and Green S. T. (1998) Liquid motion experiment flight-test results, AIAA Paper 98–3197, 34th AIAA/ASME/SAE/ASE Joint Propulsion Conf. and Exhibit., Cleveland, OH.

Chautard J. P. (1963), Effect on stability of the axial rotation of a missile containing a liquid mass, *Rech. Aeronaut.* **93**, 3–9 (in French).

Chen B. F. (1994), Nonlinear hydrodynamic pressures by earthquakes on dam faces with arbitrary reservoir shapes, *J. Hyd. Res.* **32**(3), 401–413.

Chen B. F. (1997), 3-D nonlinear hydrodynamic analysis of vertical cylinder during earthquakes, I: Rigid motion, *J. Eng. Mech.* **123**(5), 458–465.

Chen B. F. and Chiang H. W. (2000), Complete two-dimensional analysis of sea-wave-induced fully nonlinear sloshing fluid in a rigid floating tank, *Ocean Eng.* **27**(9), 953–977.

Chen B. and Saffman P. G. (1979), Steady gravity-capillary waves on deep water I: Weakly nonlinear waves, *Stud. Appl. Maths.* **60**, 183–210.

Chen C. C. (1961), On the hydrodynamic pressure on a dam caused by its aperiodic or impulsive vibrations and vertical vibrations on the earth surface, *Prikl. Math. Mekh. (PMM)* **25**, 1060–1076.

Chen H. H., Kelecy F. J. and Pletcher R. J. (1994), Numerical and experimental study of three-dimensional liquid sloshing-flows, *J. Thermophys. Heat Trans.* **8**(3), 507–513.

Chen H. H. and Pletcher R. H. (1993), Simulation of three-dimensional liquid sloshing flows using a strongly implicit calculation procedure, *AIAA J.* **31**(5), 901–910.

Chen J. C. and Babcock C. D. (1975), Nonlinear vibration of cylindrical shells, *AIAA J.* **13**, 868–876.

Chen J. C. and Chin G. H. (1995), Linear stability analysis of thermocapillary convection in the floating zone, *J. Cryst. Growth* **154**, 98–107.

Chen J. C. and Hwu F. S. (1993), Oscillatory thermocapillary flow in a rectangular cavity, *Int. J. of Heat and Mass Transfer* **36**(15), 3743–3749.

Chen P. and Viñals J. (1999), Amplitude equations and pattern selection in Faraday waves, *Phys. Rev.E* **60**, 559–570.

Chen P. Y. (1977), Nonlinear characteristics of liquid sloshing in rigid tanks, Proc. Int. Conf. *Pressure Vessel Technology* 3, P 1: Anal. Des. Insp., Tokyo, 357–387.

Chen Q. S. and Hu W. R. (1998), Influence of liquid bridge volume on instability of floating half zone convection, *Int. J. Heat Mass Transfer* **41**(6–7), 825.

Chen S. S. (1981), Fluid damping for circular cylindrical structures, *Nuclear Eng. and Design* **63**, 81–100.

Chen S. S. (1987), *Flow-Induced Vibration of Circular Cylindrical Tubes*, New York, Hemisphere Publications.

Chen S. S. and Rosenberg G. S. (1975), Dynamics of coupled shell-filled system, *Nuclear Eng. Des.* **32**, 302–310.

Chen S. S., Johnson D. B., and Raad P. B. (1994), Velocity boundary conditions for the simulation of free surface fluid flow. Individual Papers in Fluid Eng., ASME FED-**202**, 55–66.

Chen W. Q. and Ding H. J. (1999), Natural frequencies of fluid-filled transversely isotropic cylindrical shells, *Int. J. Mech. Sci.* **41**, 677–684.

Chen W. Q., Ding H. J., Guo Y. H., and Yang Q. D. (1997), Free vibrations of fluid-filled transversely orthotropic cylindrical shells, *ASCE J. Engrg. Mech.* **123**, 1130–1133.

Chen W. W. (1991), Seismic analysis of a sodium-filled tank, *Fluid–Structure Vibration and Sloshing, ASME Pressure Vess. Piping Conf.*, PVP-**223**, 35–43.

Chen W., Haroun M. A., and Liu F. (1996), Large amplitude liquid sloshing in seismically excited tanks, *Earthquake Eng. Struct. Dyn.* **25**, 653–669.

Chen Y. H., Hwang W. S., Chiu L. T., and Sheu S. M. (1995), Flexibility of LTD to high-rise building by simple experiment and comparison, *Comput. Struct.* **57**(5), 855–861.

Chermykh S. V. (1966), On the stability of rigid body filled with liquid, *Inzh. Zh. Mekh. Tv. Tela.* **3**, 33–39 (in Russian).

Chernous'ko F. L. (1965), Motion of a rigid body with cavities filled with a viscous fluid for low Reynolds numbers, *Zh. Vychisl. Matem. i Matem. Fiz.* **5**(6), 1049–1070.

Chernous'ko F. L. (1966), The motion of a body with a cavity partly filled with a viscous liquid, *Prikl. Math. Mekh. (PMM)* **30**(6), 1167–1184.

Chernousko F. L. (1967a), Motion of a body with a cavity filled with a viscous fluid at large Reynolds numbers, *Prikl. Math. Mekh. (PMM)* **30**(3), 568–589.

Chernousko F. L. (1967b), Oscillations of a rigid body with a cavity filled with a viscous fluid, *Inzh. Zh.* MTT, **1**.

Chernousko F. L. (1967c), On free oscillations of a viscous fluid in a vessel, *Prikl. Math. Mekh. (PMM)* **30**(5), 990–1003.

Chernousko F. L. (1969), *The Motion of a Rigid Body with Cavities Containing a Viscous Liquid*, Vychisl. Tsentr. Akad. Nauk. SSSR, Moscow.

Chester W. (1964), Resonant oscillations in closed tubes, *J. Fluid Mech.* **18**, 44–64.

Chester W. (1968), Resonance oscillations of water waves, I: Theory, *Proc. of the Royal Soc.* **A306**, 5–22.

Chester W. and Bones, J. A (1968), Resonance oscillations of water waves, I: Experiment, *Proc. Royal Soc.* **A306**, 23–39.

Chetayev N. G. (1957), On the stability of rotational motions of a rigid body with a cavity with an ideal fluid, *Prikl. Math. Mekh. (PMM)* **21**(2) 419–421.

Cheung Y. K. and Zhou D. (2000), Coupled vibratory characteristics of a rectangular container bottom plate, *J. Fluid Struct.* **14**, 339–357.

Cheung Y. K., and Zhou D. (2002), Hydroelastic vibration of a circular container bottom plate using the Galerkin method, *J. Fluids Struct.* **16**(4), 561–580.

Chiba M. (1986), Dynamic stability of liquid-filled cylindrical shells under horizontal excitation, part I: Experiment, *J. Sound Vib.* **104**(2), 301–319.

Chiba M. (1989), Nonlinear axisymmetric free vibration of a cylindrical tank with an elastic bottom, Forschungsberichte, Universitat der Bundeswehr, Munich, LRT-WE-9-FB-17.

Chiba M. (1992), Nonlinear hydroelastic vibration of a cylindrical tank with an elastic bottom, containing liquid, part I Experiment, *J. Fluids Struct.* **6**(2), 181–206.

Chiba M. (1993a), Nonlinear hydroelastic vibration of a cylindrical tank with an elastic bottom, containing liquid, part II: Linear axi-symmetric vibration analysis, *J. Fluid Struct.* **7**(1), 57–73.

Chiba M. (1993b), Non-linear hydroelastic vibration of a cantilever cylindrical tank – I: Experiment (empty case), *Int. J. Non-Linear Mech.* **28**(5), 591–599.

Chiba M. (1993c), Non-linear hydroelastic vibration of a cantilever cylindrical tank – II: Experiment (liquid-filled case), *Int. J. Non-Linear Mech.* **28**(5), 601–612.

Chiba M. (1993d), Nonlinear hydroelastic vibration of a cylindrical tank with an elastic bottom, containing liquid, part I: Experiment, *J. Fluid Struct.* **6**, 181–206.

Chiba M. (1993e), Experimental studies on a nonlinear hydroelastic vibration of a clamped cylindrical tank partially filled with liquid, *J. Pressure Vessel Tech.* **115**(4), 381–388.

Chiba M. (1994), Axi-symmetric free hydroelastic vibration of a flexural bottom plate in a cylindrical tank supported on an elastic foundation, *J. Sound Vib.* **169**(3), 387–394.

Chiba M. (1995), Free vibration of a clamped-free circular cylindrical shell partially submerged in a liquid, *J. Acoust. Soc. Amer.* **97**(4), 2238–2248.

Chiba M. (1996a), Free vibration of a partially liquid-filled and partially submerged, clamped-free circular cylindrical shell, *J. Acoust. Soc. Amer.* **100**(4), 2170–2180.

Chiba M. (1996b), Non-linear hydroelastic vibration of a cylindrical tank with an elastic bottom containing liquid – III: Nonlinear analysis with Ritz averaging method, *Int. J. Non-Linear Mech.* **31**(2), 155–165.

Chiba M. (1997), The influence of elastic bottom plate motion on the resonant response of a liquid free surface in a cylindrical container: A linear analysis, *J. Sound Vib.* **202**(3), 417–426.

Chiba M. and Abe K. (1999), Nonlinear hydroelastic vibration of a cylindrical tank with an elastic bottom containing liquid – analysis using harmonic balance method, *Thin-Walled Struct.* **34**(3), 233–260.

Chiba M. and Bauer H. F. (1998), Free hydroelastic vibrations of a liquid attached to a cylindrical sector shell in a zero gravity condition, *J. Vib. Acoust.* **120**(1), 54–62.

Chiba M., Bauer H. F., and Sasaki H. (1998), Influence of attached liquid layers on free hydroelastic vibrations of a cylindrical sector shell in a zero gravity condition, *J. Sound Vib.* **214** (2), 245–260.

Chiba M., Kiuchi H., and Tani J. (1997), Unstable sloshing vibration in a thin cylindrical weir, *J. Press. Vessel Tech.* **119**(1), 68–73.

Chiba M. and Tani J. (1987), Dynamic Stability of liquid-filled cylindrical shells under vertical excitation, part I: Experimental results, *J. Earthquake Eng. Struct. Dyn.* **15**(1), 23–26.

Chiba M., Tani J., Hashimoto H., and Sudo S. (1986), Dynamic stability of liquid-filled cylindrical shells under horizontal excitation, part I: Experiment, *J. Sound Vib.* **104**, 301–319.

Chiba M., Tani J., and Yamaki N. (1987), Dynamic stability of liquid-filled cylindrical shells under vertical excitation, part II: Theoretical results, *J. of Earthquake Eng. Struct. Dyn.* **15**(1), 37–51.

Chiba M., Watanabe H., and Bauer H. F. (2002), Hydroelastic coupled vibrations in a cylindrical container with a membrane bottom containing liquid with surface tension, *J. Sound Vib.* **251**(4), 717–740.

Chiba M., Yamaki N., and Tani J. (1984a), Free vibration of a clamped-free circular cylindrical shell partially filled with liquid, part I: Theoretical analysis, *Thin Walled Struct.* **2**, 265–284.

Chiba M., Yamaki N., and Tani J. (1984b), Free vibration of a clamped-free circular cylindrical shell partially filled with liquid, part II: Numerical results, *Thin Walled Struct.* **2**, 307–324.

Chiba M., Yamaki N., and Tani J. (1985), Free vibration of a clamped-free circular cylindrical shell partially filled with liquid, part III: Experimental results, *Thin Walled Struct.* **3**, 1–14.

Chiba M., Yamashita T., Sugiyama H., and Tani J. (1989), Dynamic stability of liquid-filled cylindrical shells under periodic shearing forces, *J. Pressure Vessel Tech.* **111**(4), 420–427.

Chiba T., Nakajima S., Mieda T., Ogawa N., and Shibata H. (1995), The sloshing behavior of high viscous liquid in cylindrical tanks, in *Fluid-Sloshing and Fluid-Structure Interaction*, ASME Press Vess Piping Conf, PVP-**314**, 57–82.

Chin J. H. and Gallagher L. W. (1964), Effect of fluid motion on free surface shape under reduced gravity, *AIAA J.* **2**(12), 2215–2217.

Cho J. R., Lee H. W., and Kim K. W. (2002), Free vibration analysis of baffled liquid-storage tanks by the structural-acoustic finite element formulation, *J. Sound Vib.* **258**(5), 847–866.

Cho J. R., Song J. M., and Lee J. K. (2001), Finite element techniques for the free-vibration and seismic analysis of liquid-storage tanks, *Finite Elements in Analysis and Design* **37**, 467–483.

Chobotov V. and Fowler J. (1957), Experimental investigation of the moment of inertia in a cylindrical tank, Space Technol. Labs, Memo. GM61–2–3, 18 February.

Chopra A. K. (1967), Hydrodynamic pressures on dams during earthquakes, *J. of Eng. Mech. Div.* **93**(EM6), 205–223.

Chorin A. J. (1967), A numerical method for solving incompressible viscous flow problems, *J. Compt. Phys.* **2**, 12–26.

Chorin A. J. (1976), Random choice solution of hyperbolic systems, *J. Comput. Phys.* **22**(4), 517–533.

Chorin A. J. (1977), Random choice methods with application to reacting gas flow, *J. Comput Phys.* **25**(3), 253–272.

Crandall S. H. and Mroszczyk J. (1986), Whirling of flexible cylinder filled with fluid, Proc. Int. Conf. *Rotordynamics*, Tokyo, JSME.

Christiansen B., Alstrom P., and Levinson M. T. (1992), Ordered capillary-wave states: quasi-crystals, hexagons, and radial waves, *Phys. Rev. Lett.* **68**, 2157–2160.

Chu B. T. and Ying S. J. (1963), Thermally driven nonlinear oscillations in a pipe with traveling shock waves, *Phys. Fluids* **6**, 11.

Chu H. N. (1961), Influence of large amplitudes on flexural vibrations of a thin circular cylindrical shell, *J. Aerospace Sci.* **28**, 602–609.

Chu M. L. and Brown S. J. (1981), Experiments on the dynamic behavior of fluid-coupled concentric cylinders, *Exper. Mech.* **21**, 129–137.

Chu M. L., Brown S. J., Lieb E., and Lestingi J. P. (1979), An experimental study of gap and thickness influence on the vibration response and damping of flexible fluid-coupled coaxial cylinders, ASME Pressure Vess. Piping Conf., PVP-39, 1–19.

Chu W. H. (1960a), Sloshing of liquids in cylindrical tanks of elliptic cross-section, *Amer. Rocket Soc J.* **30**, 360–363.

Chu W. H. (1960b), Free surface conditions for sloshing resulting from pitching and some corrections, *Amer. Rocket Soc. J.* **30**(11), 1093–1094.

Chu W. H. (1963), Breathing vibrations of a partially filled cylindrical tank – linear theory, *ASME J. Appl. Mech.* **30**, 532–536.

Chu W. H. (1964a), Fuel sloshing in a spherical tank filled to an arbitrary depth, *AIAA J.* **2**, 1972–1979.

Chu W. H. (1964b), Liquid sloshing in a spherical tank filled to an arbitrary depth, *AIAA J.* **2**(11), 1972–1979.

Chu W. H. (1964c), An analysis of the quasi-steady liquid-gas interface in an axi-symmetric tank during low gravity transfer, SwRI, Final Rept. (Contract NAS8–5468).

Chu W. H. (1968), Subharmonic oscillations in arbitrary axi-symmetric tank resulting from axial excitation, *J. Appl Mech.* **35**(1), 148–154.

Chu W. H. (1970), Low-gravity fuel sloshing in an arbitrary axi-symmetric rigid tank, *J. Appl Mech.* **37**(3), 828–837.

Chu W. H. (1971), Sloshing of an arbitrary two-dimensional tank with flat mean free surface, *CASI Transactions* **4**(7), (Technical Forum), 48–60.

Chu W. H. and Gonzales R. (1964), Supplement to breathing vibrations of a partially filled cylindrical tank – linear theory, *ASME J. Appl Mech* **31**(4), 722–723.

Chu W. H. and Kana D. D. (1967), A theory for nonlinear transverse vibrations of a partially filled elastic tank, *AIAA J.* **5**(10), 1828–1835.

Chun C H. (1980a), Experiments on steady and oscillatory temperature distribution in a floating zone due to the Marangoni convection, *Acta Astron.* **7**, 479–488.

Chun C. H. (1980b), Marangoni convection in a floating zone under reduced gravity, *J. Crystal Growth* **48**, 600–610.

Chun C H. (1984), Numerical study on the thermal Marangoni convection and comparison with experimental results from the TEXUS-rocket program, *Acta Astron.* **11**(3–4).

Chun C. H., Ehmann M., Siekmann J., Wozniak G. (1987), Vibrations of rotating menisci, Proc. 6th Europ. Symp. *Material Sciences under Microgravity Conditions*, Borde France ESA-SP-256, 226–234.

Chun C. H. and Siekmann J. (1994), Higher modes and their instabilities of the oscillatory Marangoni convection in a large cylindrical liquid column, Proc. Norderney Symp. *Scientific Results of the German Spacelab Mission* D-2, 235–241.

Chun C. H. and Wuest W. (1979), Experiments on the transition from steady to the oscillatory Marangoni-convection of a floating zone under reduced gravity effects, *Acta Astron.* **6**, 1073–1082.

Chun C. H. and Wuest W. (1982), Suppression of temperature oscillations of thermal Marangoni convection in a floating zone by superimposing of rotating flow, *Acta Astron.* **9**, 225–230.

Ciliberto S. and Gollub J. P. (1984), Pattern competition leads to chaos, *Phys. Rev. Lett.* **52**(11), 922–925.

Ciliberto S. and Gollub J. P. (1985a), Phenomenological model of chaotic mode competition in surface waves, *IL Nouvo Cimento* **6**(4), 309–316.

Ciliberto S. and Gollub J. P. (1985b), Chaotic mode competition in parametrically forced surface waves, *J. Fluid Mech.* **158**, 381–398.

Clark C. D. (1959), Effect of propellant oscillations on SATURN roll moment of inertia, Army Ballistic Missile Agency, Rept. No DA-TN-87–59.

Clark L. V. and Stephens D. G. (1967), Simulation and scaling of low-gravity slosh frequencies and damping, Second ASTM, IES and AIAA Space Simulation Conf., ASTM, 43–49.

Clary R. R. and Turner L. J. Jr (1970), Measured and predicted longitudinal vibrations of a liquid propellant two stage launch vehicle, NASA TN-D-5943, August.

Clough D. P. (1977), Experimental evaluation of seismic design methods for broad cylindrical tanks, Rept. No UCB/EERC-77/10, University of California, Berkeley.

Clough D. P. (1979), Experimental evaluations of seismic design method for broad cylindrical tanks, UCA, Berkeley/EERC-79/06.

Clough D. P. and Niwa A. (1979), Static tilt tests of a tall cylindrical liquid storage tank, Rept. No UCB/EERC-79/06, UCA, Berkeley.

Clough R. W., Niwa A., and Clough D. P. (1979), Experimental seismic study of cylindrical tanks, *ASCE J. Struct. Division* **105**(ST12).

Coale C. W. and Nagano M. (1969), Axi-symmetric vibrations of a cylindrical hemispherical tank partially filled with a liquid, *AIAA J.* **7**(2), 235–243.

Cocciaro B., Faetti S., and Nobili M. (1991), Capillary effects on surface gravity waves in a cylindrical container: wetting boundary conditions, *J. Fluid Mech.* **231**, 325–343.

Cocciaro B., Faetti S., and Festa C. (1993), Experimental investigation of capillary effects on surface gravity waves: non-wetting boundary conditions, *J. Fluid Mech.* **246**, 43–66.

Cochran W. G (1934), The flow due to a rotating disc, *Proc. Camb. Phil. Soc.* **30**, 365–375.

Cointe R. (1989), Two-dimensional fluid–solid impact, *ASME J. Offshore Mech. & Arctic Engrg.* **111**(2), 109–114.

Cointe R. and Armand J. L. (1987), Hydrodynamic impact analysis of a cylinder, *ASME J. Offshore Mech. & Arctic Engrg.* **109**(3), 237–243.

Cokonis T. J., Tomassoni J. E., and Seiferth R. W. (1963), Dome impact analysis – an approximate solution, Martin Co. Rept., TN LV 211, May.

Cole H. A. Jr (1966a), On a fundamental damping law for fuel sloshing, NASA TN-D-3240, March.

Cole H. A. Jr (1966b), Baffle thickness effects in fuel sloshing experiments, NASA TN-D-3716.

Cole H. A. Jr and Gambucci B. J. (1961a), Measured two dimensional damping effectiveness of fuel sloshing baffles applied to ring baffles in cylindrical tanks, NASA TN-D-694.

Cole H. A. Jr and Gambucci B. J. (1961b), Tests of an asymmetrical baffle for fuel sloshing suppression, NASA TN-D-1036.

Concus P. (1963), Capillary stability in an inverted rectangular tank, Proc. Symp. *Physical and Biological Phenomena in a Weightless State,* E. T. Bendikt and R. W. Halliburton, eds., AAS, **14**, 21–37.

Concus P. (1964), Capillary stability in an inverted rectangular channel for free surfaces with curvature of changing sign, *AIAA J.* **2**(12), 2228.

Concus P. (1968), Static menisci in a vertical right circular cylinder, *J. Fluid Mech.* **34**, 481.

Concus P., Crane G. E., and Perko L. M. (1965), Inviscid fluid flow in an accelerating axi-symmetric container, Symp. *Fluid Mechanics and Heat Transfer under Low Gravitational Conditions,* Lockheed Missiles & Space Co.

Concus P., Crane G. E., and Satterlee H. M. (1967), Small-amplitude lateral sloshing in a cylindrical tank with a hemispherical bottom under low gravitational conditions, NASA – CR 54700.

Concus P. Crane G. E., and Satterlee H. M. (1968), Low gravity lateral sloshing in a hemispherically bottomed cylindrical tank, Proc. 1968 Heat Transfer and Fluid Mechanics Institute held at Seattle, A. F. Emery and C. A. Depew, eds., Stanford University Press, (June17–18), 80–97.

Concus P. Crane G. E., and Satterlee H. M. (1969), Small-amplitude lateral sloshing in spheroidal containers under low gravitational conditions, Final Report, NASA – CR 72500.

Concus P. and Finn R. (1974a), On capillary free surfaces in the absence of gravity, *Acta Math.* **132**, 177–198.

Concus P., and Finn R. (1974b), On capillary free surfaces in gravitational field, *Acta Math.* **132**, 207–218.

Concus P., and Finn R. (1976), On the height of a capillary surface, *Math. Z.* **147**, 93–108.

Concus P., and Finn R. (1987), Continuous and discontinuous disappearance of capillary surfaces, in *Variational Methods for Free Surface Interface,* P Concus and R. Finn, eds., New York, Springer-Verlag, 197–204.

Concus P. and Finn R., (1990a), Capillary surfaces in microgravity, in *Low-Gravity Fluid Dynamics and Transport Phenomena*, JN Koster and R. L. Sani, eds., 130, Progress in Astronaut. and Aeronaut., AIAA, Washington, DC, 183–205.

Concus P., and Finn R. (1990b), Dichotomous behavior of capillary surfaces in zero gravity, *Microgravity Science and Technology* **III**, 87–92, Errata, **III**, (1991), 230.

Concus P., and Finn R. (1991a), Exotic containers for capillary surfaces, *J. Fluid Mech.* **224**, 383–394.

Concus P. and Finn R. (1991b), On accurate determination of contact angle, Microgravity Science and Technology, **IV**, 69–70.

Concus P., and Finn R. (1992), On canonical cylinder sections for accurate determination of contact angle in microgravity, ASME WAM Symp. *Fluid Mech. Phenomena in Microgravity*, D. A. Siginer and M. M. Weislogel, AMD-**154**/FED-**142**, 125–131.

Concus P., and Finn R. (1998), Discontinuous behavior of liquids between parallel tilted plates, *Phys. Fluids* **10**, 39–43.

Concus P., Finn R., and Weislogel M. (1999), Capillary surfaces in an exotic container: results from space experiments, *J. Fluid Mech.* **394**, 119–135.

Concus P., Crane G. E., and Satterlee H. M. (1968), Low gravity lateral sloshing in a hemispherically bottomed cylindrical tank, Proc. Heat Transf. Fluid Mech. Inst., Seattle, WA, (University Press, Stanford, CA) 80–90.

Coney T. A. and Masica W. J. (1969), Effect of flow rate on the dynamic contact angle for wetting liquids, NASA TN-D-5115.

Coney T. A. and Salzman J. A. (1971), Lateral sloshing in oblate spheroidal tanks under reduced- and normal-gravity conditions, NASA TN-D-6250.

Conly J. F. (1963), Surface waves on a rotating fluid, Ph.D. dissertation, Columbia University, New York.

Cooker M. J. (1994), Water waves in a suspended container, *Wave Motion* **20**, 385–395.

Cooker M. J. and Peregrine D. H. (1992), Wave impact pressure and its effect upon bodies lying on the bed, *Coastal Engrg.* **18**, 205–229.

Cooker M. J. and Peregrine D. H. (1995), Pressure-impulse theory for liquid impact problems, *J. Fluid Mech.* **297**, 193–214.

Cooper R. M. (1960), Dynamics of liquids in moving containers, *Amer. Rocket Soc. J.* **30**(8), 725–729.

Courant R. and Friedrichs K. O. (1948), *Supersonic Flow and Shock Waves*, New York, Wiley Interscience.

Cox P. A., Bowles E. B., and Bass R. L. (1980), Evaluation of liquid dynamic loads in slack LNG cargo tanks, Tech. Rept. SSC-297, SwRI.

Craik, A.D.D. (1986), *Wave Interactions and Fluid Flows*, Cambridge, Cambridge University Press.

Craik, A.D.D. (1995), The stability of some three-dimensional and time-dependent flows, IUTAM Symp. *Nonlinear Instability of Nonparallel Flows*, Lin S.P., Pillips W. R. C., and Valentine D.T., eds., Springer, 382–396.

Craik A. D. D. and Armitage J (1995), Faraday excitation, hysteresis and wave instability in a narrow rectangular wave tank, *Fluid Dyn. Research* **15**, 129–143.

Crandall S. H. and Mroszczyk J. (1986), Whirling of a flexible cylinder filled with liquid, Proc. Int. Conf. *Rotordynamics*, Tokyo, 449–452.

Crawford J. D. (1991a), Normal forms for driven surface waves: Boundary conditions, symmetry, and genericity, *Physica D* **52**, 429–457.

Crawford J. D. (1991b), Surface waves in nonsquare containers with square symmetry, *Phys. Rev. Lett.* **67**, 441–444.

Crawford J. D., Gollub J. P., and Lane D. (1993), Hidden symmetries of parametrically forced waves, *Nonlinearity* **6**, 119–164.

Crawford J. D. and Knobloch E. (1991), Symmetry and symmetry-breaking in fluid dynamics, *Ann. Rev. Fluid Mech.* **23**, 341–387.

Crawford J. D., Knobloch E., and Riecke H. (1990), Period-doubling mode interactions with circular symmetry, *Physica D* **44**, 340–396.

Crews A. A. and Earth R. L. (1963), Instrument to measure water wave height, *J. Acoust. Soc. Amer.* **35**(5), 737–738.

Curvelier C. (1985), A capillary free boundary problem governed by the Navier–Stokes equations, *Comp. Methods Appl. Mech. Eng.* **48**, 45–80.

Curvelier C. and Schulkes R. M. S. M. (1990), Some numerical methods for the computation of capillary free boundaries governed by the Navier–Stokes equations, *SIAM Rev.* **32**, 355–423.

Cutshall W. K., Dodge F. T., Gree S. T., and Unruh J. F. (1996), Modeling low-g liquid motions in spinning spacecraft tanks, Final Report, SwRI Project 04–9769.

Daich I. M. (1971), Nonconservative problems on oscillations of a rigid partially filled with ideal liquid, *Appl. Mech. (Prikl. Mech.)* **11**(7), 44–48.

Daich I. M. and Bar I. L. (1973), Oscillations of rotating rigid body having cavity filled with viscous fluid, *Appl. Mech. (Prikl. Mech.)* **13**(5), 64–69.

Daich I. M., Bar I. L., and Korzovskaya I. D. (1976), Nonlinear oscillations of rotating rigid body partially filled with ideal liquid, *Appl. Mech. (Prikl. Mech.)* **12**(6), 111–115.

Daly B. J. (1969a), Numerical study of the effect of surface tension on interface instability, *Phys. Fluids* **12**, 1340.

Daly B. J. (1969b), A technique for including surface tension effects in hydrodynamic calculations, *J. Comput. Phys.* **4**, 97.

Dalzell J. F. (1967a), Exploratory studies of liquid behavior in randomly excited tanks: longitudinal excitation, SwRI, Tech. Rept. 1.

Dalzell J. F. (1967b), Exploratory studies of liquid behavior in randomly excited tanks: lateral excitation, SwRI, Tech. Rept. 2, Contract NAS8–20319.

Dalzell J. F., Chu W. H., and Modisette J. E. (1964), Studies of ship roll stabilization tanks, SwRI, Tech. Rept. 1.

Dalzell J. F. and Garza I. R. (1964), An exploratory study of simulation of liquid impact in space vehicle and booster tanks, SwRI, Tech. Rept. 9, Contract NAS8–1555.

Danabasoglu G. and Biringen S. (1988), Computational of convective flow with g-jitter in rectangular cavities, AIAA Paper 88–3727.

Da-Riva I. and Martinez I. (1979), Floating zone stability, Proc. 3rd Europ. Symp. *Material Science in Space*, Grenoble, ESA-SP-142, 67–73.

Davis S. H. (1987), Thermocapillary instabilities, *Ann. Rev. Fluid Mech.* **19**, 403–435.

Dean R. G. and Dalrymple R. A. (1992), *Water Wave Mechanics for Engineering and Scientists*, Singapore, World Scientific.

Debnath L. (1994), *Nonlinear Water Waves*, New York, Academic Press.

Debonginie J. F. (1986), On a purely Lagrangian formulation of sloshing and fluid-induced vibrations of tanks eigenvalue problems, *Com. Meth. Appl. Mech. Eng.* **58**, 1–18.

Decent S. P. (1995), Hysteresis and mode competition in Faraday waves, Ph.D. thesis, University of St Andrews, Scotland.

Decent S. P. (1997), The nonlinear damping of parametrically excited two-dimensional gravity waves, *Fluid Dyn. Res.* **19**, 201–217.

Decent S. P. and Craik A. D. D. (1995), Hysteresis in Faraday resonance, *J. Fluid Mech.* **293**, 237–268.

Decent S. P. and Craik A. D. D. (1997), On limit cycles arising from the parametric excitation of standing waves, *Wave Motion* **25**(3), 275–294.

Decent S. P., and Craik A. D. D. (1999), Sideband instability and modulations of Faraday waves, *Wave Motion* **30**(1), 43–55.

Defelice D. M. (1987), Cryogenic fluid management flight experiment (CFMFE), in R. Vernon, ed., Microgravity Fluid Management Symposium, NASA Lewis Research Center, Cleveland O. H., 119–124, NASA Conference Publication NASA-CP-2465.

Deffenbaugh D. M., Dodge F. T., and Green S. T. (1998), Liquid motion in a rotating tank experiment (LME), Final Report, SwRI Project 04–6322, Document No. 6322-FNL-01, NASA Contract NAS3–27252.

De Gennes P. G. and Prost J. (1994) *The Physics of Liquid Crystals*, Oxford, Clarendon.

DeLombard R., McPherson K., Moskowitz M., and Hrovat K. (1997), Comparison tools for assessing the microgravity environment of missions, carriers and conditions, NASA Tech. Rept. TM 107446.

Den Hartog J. P. (1985), *Mechanical Vibrations*, New York, Dover Publications.

Derdul J. D., Grubb L. S., and Petrash D. A. (1966), Experimental investigation of liquid outflow from cylindrical tanks during weightlessness, NASA TN D-3746.

Derendyayev N. V. (1988), Andronov–Hopf bifurcation in dynamics of rotor system contained liquid, Rept. Academy Sciences USSR **301**, 789–801.

Derendyayev N. V. and Sanalov V. M. (1982), On stationary rotation stability of the cylinder partially filled with viscous incompressible liquid, *Appl. Math. Mech.* (PMM) **4**.

Desyatov V. T. (1986), Experimental investigation of stability of rotary motion of bodies filled with liquid, in *Space Apparatus Dynamics and Outer Space Investigation*, Moscow, Mashinostroyeniye.

Devitt E. B. and Melcher J. R. (1965), Surface electrodynamics with high-frequency fluids, *Phys. Fluids* **8**(6), 1193–1195.

Dias F. and Kharif C. (1999), Nonlinear gravity and capillary-gravity waves, *Annu. Rev. Fluid Mech.* **31**, 301–346.

Dillingham J. (1981), Motion studies of a vessel with water on deck, *Marine Technology* **18**(1), 38–50.

DiMaggio F. L. (1975), Dynamic response of fluid-filled shells, *Shock Vib. Digest* **7**(5), 5–12.

DiMaggio F. L. (1978), Recent research on the dynamic response of fluid-filled shells, *Shock. Vib. Dig* **10**(7), 15–19.

DiMaggio F. L. (1981), Dynamic response of fluid-filled shells – an update, *Shock Vib. Digest* **13**(6), 3–6.

DiMaggio F. L. (1984), Dynamic response of shells containing fluid, *Shock Vib. Digest* **16**(6), 3–9.

DiMaggio O. D. and Rehm A. S. (1965), Nonlinear free oscillations of a perfect fluid in a cylindrical container, AIAA Symp. *Structural Dyn. Aeroelasticity* **30**, 156–161.

Dimentberg M. F. (1988), *Statistical Dynamics of Non-linear and Time-varying Systems*, New York, John Wiley & Sons.

Dodge F. T. (1963), A review of research studies on bubble motion in liquids contained in vertically vibrating tanks, SwRI Tech. Rept. 1, (Contract NAS8–11045).

Dodge F. T. (1968), Discussion of bubble trajectory and equilibrium levels in vibrated columns, *J. Basic Eng.* **90**(4), 629.

Dodge F. T. (1971), Study of flexible baffles for slosh suppression, NASA CR-1880.

Dodge F. T. (1990), Fluid management in low gravity, *Prog. Astronautics and Aeronautics, Low-Gravity Fluid Dynamics and Transport Phenomena* **130**, 3–14.

Dodge F. T. (1991), Fluid management technology liquid slosh dynamics and control: Interim report, SwRI, NASA-NTIS Washington, DC.

Dodge F. T. (1996), Liquid motion analytical models and pre-flight simulations for the liquid motion in a rotating tank experiment (LME), SwRI Project 04–6322.

Dodge F. T. (1999), Modeling liquid motions in spinning spacecraft tanks, Final Report SwRI Project 18-R9946.

Dodge F. T. and Abramson N. H. (2000), Liquid propellant dynamics in the SATURN/APOLLO vehicles – a look back, AIAA Paper 2000–1676.

Dodge F. T. and Bowles E. B. (1984), Vapor flow into a capillary propellant-acquisition device, *J. Spacecraft and Rock.* **21**(3), 267–273.

Dodge F. T. and Garza L. R. (1967a), Experimental and theoretical studies of liquid sloshing at simulated low gravity, *ASME J. Appl. Mech.* **34**(3), 555–562.

Dodge F. T. and Garza L. R. (1967b), Simulated low-gravity sloshing in cylindrical tanks including effects of damping and small liquid depth, NASA Tech. Rept. No 5, Contract NAS8–20290, 1–32.

Dodge F. T. and Garza L. R. (1968), Simulated low-gravity sloshing in cylindrical tanks including effects of damping and small liquid depth, Proc. 1968 Heat Transfer and Fluid Mechanics Institute held at Seattle, A. F. Emery and C. A. Depew, eds., Stanford University, 67–79.

Dodge F. T. and Garza L. R. (1970a), Simulated low-gravity sloshing in spherical, ellipsoidal, and cylindrical tanks, *J. Spacecraft Rock.* **7**(1), 204–206.

Dodge F. T. and Garza L. R. (1970b), Magnetic fluid simulation of liquid sloshing in low gravity, NASA Tech. Rep. 9, 1–29.

Dodge F. T. and Garza L. R. (1971), Propellant dynamics in an aircraft type launch vehicle, NASA CR-119891.

Dodge F. T. and Garza L. R. (1972), Free surface vibrations of a magnetic liquid, *J. Eng. for Indust.*, 103–108.

Dodge F. T. and Green S. T. (2000), Spinning slosh test rig: Data analysis and determination of nutation time constant model parameters for the genesis spacecraft, Final Report, **1**, SwRI Project 18.03513, Boeing/NASA-KSC Contract K59000001.

Dodge F. T., Green S. T., and Cruise M. W. (1991), Analysis of small-amplitude low gravity sloshing in axisymmetric tanks, *Micrograv. Sci. and Technol.* **IV**(4), 228–234.

Dodge F. T., Green S. T., Petullo S. P., and Kuhl C. A., (1999), A forced motion spin table for launch vehicle slosh, Final Report SwRI Project 18–03303, Boeing Contract L5999006.

Dodge F. T. and Kana D. D. (1966), Moment of inertia and damping of liquid in baffled cylindrical tanks, *J. Spacecraft Rock.* **3**(1), 153–155. (Also NASA CR-383, February).

Dodge F. T., and Kana D. D. (1987), Dynamics of liquid sloshing in upright and inverted bladdered tanks, *ASME J. Fluids Engineering* **109**(1), 58–63.

Dodge F. T., Kana D. D., and Abramson H. N. (1965), Liquid surface oscillations in longitudinally excited rigid cylindrical containers, *AIAA J.* **3**(4), 685–695. (Also SwRI Tech Rept. No 2, 30 April, 1964.)

Dodge F. T., Unruh S. T., and Green S. T. (2000), A boundary layer estimate of viscous energy dissipation in spinning spacecraft tanks, AIAA Paper 2000–5215, AIAA Space 2000 Conference.

Dodge F. T., Unruh S. T., Green S. T., and Cruse M. W. (1994), A mechanical model of liquid inertial waves for use with spinning spacecraft, ASME FED **198**, 13021.

Doğangün A. Durmuş A., and Ayvaz Y. (1996), Static and dynamic analysis of rectangular tanks by using the Lagrangian fluid finite element, *Computers Struct.* **59**(1), 547–552.

Dokuchaev L. V. (1964), On the solution of a boundary value problem on the sloshing of a liquid in conical cavities, *J. Appl. Math. Mech. (PMM)* **28**(1), 151–154.

Dommermuth D. G. (1993), The laminar interaction of a pair of vortex tubes with a free surface, *J. Fluid Mech.* **246**, 91–112.

Dong R. G. (1978), Sloshing in pools under earthquake-like ground motions, *J. Power Div.* **104**(2), 157–167.

Donnell L. H. (1976), *Beams, Plates, and Shells*, New York, McGraw-Hill.

Donnelly R. J. (1988), Superfluid turbulence, *Scientific American* **259**(5), 100–108.

Donnelly R. J. (1991), *Quantized Vortices in Helium I. I.*, Cambridge, Cambridge University Press.

Douady S. (1990), Experimental study of the Faraday instability, *J. Fluid Mech.* **221**, 383–409.

Douady S., Fauve S., and Thual O. (1989), Oscillatory phase modulation of parametrically forced surface waves, *Europhys. Lett.* **62**, 309.

Dowell E. H. (1967), On the nonlinear flexural vibrations of rings, *AIAA J.* **5**, 1508–1509.

Dowell E. H. (1998), Comments on the nonlinear vibrations of cylindrical shells, *J. Fluids and Struct.* **12**, 1087–1089.

Dowell E. H. and Ventres C. S. (1968), Modal equations for nonlinear flexural vibrations of a cylindrical shell, *Int. J. Solids Struct.* **4**, 975–991.

Drake K. R. (1999), The effect of internal pipes on the fundamental frequency of liquid sloshing in a circular tank, *Appl. Ocean Res.* **21**(3), 133–143.

Drolet F. and Viñals J. (1998), Fluid physics in a fluctuating acceleration environment, Proc. 4th Microgravity Fluid Phys. and Transport Phenomena Conf., Cleveland, Ohio, August 12–14.

Druz B. I. and Magula V. Z. (1967), Small longitudinal vibrations of a fluid in a perfectly flexible, elastic, cylindrical shell, (in Russian) *Inzh. Zh. Mekh. Tverdogo Tela.* **6**, 63–67.

Dupont P. E. (1992), The effect of Coulomb friction on the existence and uniqueness of the forward dynamics problems, Proc. 1992 *IEEE Int. Conf. on Robotics and Automation* 1442–1447.

Dussan V. E. B. (1976), The moving contact line: The slip boundary condition, *J. Fluid Mech.* **77**, 665–668.

Dussan V. (1979), On the spreading of liquids on solid surfaces: Static and dynamic contact-lines, *Annals Rev. Fluid Mech.* **11**, 371–400.

Dussan V. and Davis S. H. (1974), On the motion of fluid–fluid interface along a solid surface, *J. Fluid Mech.* **77**, 71–95.

Dzyuba V. V. and Kubenko V. D. (2002), Axisymmetric interaction problem for a sphere pulsating inside an elastic cylindrical shell filled with and immersed into a liquid, *Int. Appl. Mech.* **38**(10), 1210–1219.

Easton C. R. and Catton I. (1970), Nonlinear free surface effects in tank draining at low gravity, *AIAA J.* **8**(12), 2195–2199.

Ebert E. (1984), Study of slosh dynamics of fluid filled containers on a 3-axis spin stabilized spacecraft, **2**, Fluid Slosh Studies 5328/83/NL/Bi(Sc), ESA.

Ebert E. (1989), Modeling of liquid sloshing effects in multi-body systems, Proc. Int. Conf. *Spacecraft Structures and Mechanical Testing*, Noordwijk, NL, 19–21 Oct 1988, ESA SP-289, 269–275.

Eckhardt K. and Netter G. (1982), Experiment for investigation of the dynamic behavior of fluid in a surface tension tank under microgravity condition, *Acta Astronautica* **9**(9), 565–571.

Edwards N. W. (1969), A procedure for dynamic analysis of thin walled cylindrical liquid storage tanks subjected to lateral ground motions, Ph.D. dissentation., University of Michigan, Ann Arbor, MI.

Edwards W. S. and Fauve S. (1993), Parametrically excited quasicrystalline surface waves, *Phys. Rev. E* **47**, R788–791.

Edwards W. S. and Fauve S. (1994), Patterns and quasi-patterns in the Faraday experiment, *J. Fluid Mech.* **278**, 123–148.

Eggleston D. M. (1968), Dynamic stability of space vehicles, **XIV**: Testing for booster propellant sloshing parameters, NASA CR-948.

Ehrlich L. W. (1959), Exact solutions for sloshing problems, Space Tech. Lab., Memo-PA 2450/79, 22 September.

Ehrlich L. W., Riley J. D., Strange W. G., and Troesch B. A. (1961), Finite difference techniques for a boundary problem with an eigenvalue in a boundary condition, *J. Soc. Indust. Appl. Math.* **9**, 149–164.

Eide D. G. (1964), Preliminary analysis of variation of pitch motion of a vehicle in a space environment due to fuel sloshing in a rectangular tank, NASA TN-D-2336.

Eidel W. (1989), Nonlinear liquid oscillations in prolate spheroidal containers, *Z. Flugwiss. Weltraumorsch* **13**, 159–165.

Eidel W. and Bauer H. F. (1988), Nonlinear oscillations of an inviscid liquid column under zero gravity, *Ing. Arch.* **58**, 276–284.

Einfeldt J., Kutschke K. H., and Weber S. V. (1969), Computation of the shape of the free liquid surface in containers with axial symmetry, *Monatsberichte der Deutschen Akademic der Wissenschaften Zu Berlin* **11**(8/9), 610–615 (in German).

Elliott G. E. P. and Riddiford A. C. (1967), Dynamic contact angles, I: The effect of increased motion, *J. Colloid. Interface Sci.* **23**, 389–398.

El-Rahib M. and Wagner P. (1981), Vibration of a liquid with a free surface in a spinning spherical tank, *J. Sound Vib.* **76**, 83–93.

Engin A. E. (1969a), Vibrations of fluid-filled spherical shells, *J. Acoust. Soc. Amer.* **46**(1), part 2, 184–190.

Engin A. E. (1969b), The axi-symmetric response of a fluid-filled spherical shell to a local radial impulse – a model for head injury, *J. Biomech.* **2**, 325–341.

Enright P. J. and Wong E. C. (1994), Propellant slosh models for the Cassini spacecraft, AIAA Paper 94–3730.

Epperson T. B. and Brown R. (1957), Dynamic loads due to fuel motion in missile tanks, SwRI, Final Report, Contract No DA-23–072-ORD-1062.

Epperson T. B., Brown R., and Abramson H. N. (1961), Dynamic loads resulting from fuel motion in missile tanks, in *Advances in Ballistic Missile & Space Tech*, **II**, NY, Pergamon Press, 313–327.

Ergin A. (1997), An approximate method for the free vibration analysis of partially filled and submerged, horizontal cylindrical shells, *J. Sound Vib.* **207**, 761–767.

Ergin A. and Temarel P. (2002), Free vibrations of a partially liquid-filled and submerged, horizontal cylindrical shell, *J. Sound Vib.* **254**(5), 951–965.

Erneux T. and Mandel P. (1986), Imperfect bifurcation with a slowly-varying control parameter, *SIAM J. Appl. Math.* **46**(1), 1–15.

Espinosa F. M. and Gallego-Juarez J. A. (1984), On the resonance frequencies of water-loaded circular plates, *J. Sound Vib.* **94**, 217–222.

Eulitz W. R. (1957), The sloshing phenomenon and the mechanics of a liquid in motion in an oscillating missile container, Army Ballistic Missile Agency, Rept. No. DS-R-31-Div. Operation.

Eulitz W. R. (1958), A Can Type anti-slosh device derived from basis slosh studies, Army Ballistic Missile Agency, Rept. No. DSD-TR-4–58.

Eulitz W. R. (1961), Analysis and control of liquid propellant sloshing during missile flight, NASA-MSFC, MTP-P and VE-P-61, December.

Eulitz W. R. (1963), Practical consequences of liquid propellant slosh characteristics derived by nomographic methods, NASA-MSFC, MTP-P and VE-P-63–7.

Eulitz W. R. and Glaser R. F. (1961), Comparative experimental and theoretical considerations on the mechanism of fluid oscillations in cylindrical containers, Army Ballistic Missile Agency, Rept. No. MTP-M-S and M-61, 29 May.

Evan-Iwanowski R. M. (1976), *Resonance Oscillations in Mechanical Systems*, New York, Elsevier Science.

Evans D. V. (1990), The wide-spacing approximation applied to multiple scattering and sloshing problems, *J. Fluid Mech.* **210**, 647–658.

Evans D. V. and Linton C. M. (1993), Sloshing frequencies, *Quart. J. Mech. and Appl. Math.* **46**, 71–87.

Evans D. V. and McIver P. (1987), Resonant frequencies in a container with vertical baffles, *J. Fluid Mech.* **175**, 295–307.

Evensen D. A. (1963), Some observations on the nonlinear vibration of thin cylindrical shells, *AIAA J.* **1**, 2857–2858.

Evensen D. A. (1964), Nonlinear flexural vibrations of thin circular rings, Ph.D. thesis, California Inst. Tech.

Evensen D. A. (1966), A theoretical and experimental study of the nonlinear flexural vibrations of thin circular rings, *ASME J. Appl. Mech.* **33**, 553–560.

Evensen D. A. (1967), A nonlinear flexural vibrations of thin-walled circular cylinders, NASA TND 4090.

Evensen D. A. (1974), Nonlinear vibrations of circular cylindrical shells, in *Thin Walled Structures: Theory, Experiment and Design*, Y. C. Fung, and E. E. Sechler, eds., 133–155. Englewood Cliffs, Prentice-Hall.

Evensen D. A. (1999), Nonlinear vibrations of cylindrical shells – logical rationale, *J. Fluids & Struct.* **13**, 161–164.

Ewart D. G. (1956), Fuel oscillations in cylindrical tanks and the forces produced thereby, De Havilland Propellers Ltd, G. W. Dynamics Dept, Tech. Note No 2050, 8 November.

Faltinsen O. M. (1974), A nonlinear theory of sloshing in rectangular tanks, *J. Ship Res.* **18**(4), 224–241.

Faltinsen O. M. (1978), A numerical nonlinear method of sloshing in tanks with two-dimensional flow, *J. Ship Res.* **18**(4), 224–241.

Faltinsen O. M. and Rognebakke O. F. (1999), Sloshing and slamming in tanks, in *Hydona'99-Manoeuvring'99*, Gdansk-Ostroda, Poland Technical University of Gdansk.

Faltinsen O. M., Olsen H. A., Abramson H. N., and Bass R. L. (1974), Liquid slosh in LNG carriers, Det. Norske Veritas, Norway, Publication No 85.

Faltinsen O. M., Rognebakke O. F., Lukovskii I. A., and Timokha A. N. (2000), Multidimensional modal analysis of nonlinear sloshing in a rectangular tank with finite water depth, *J. Fluid Mech.* **407**, 201–234.

Faltinsen O. M., Rognebakke O. F., and Timokha A. N. (2003), Resonant three-dimensional nonlinear sloshing in a square base basin, *J. Fluid Mech.* **487**, 1–42.

Faltinesn O. M. and Timokha A. N. (2001), An adaptive multidimensional approach to nonlinear sloshing in a rectangular tank, *J. Fluid Mech.* **432**, 167–200.

Faltinesn O. M. and Timokha A. N. (2002a), Asymptotic modal approximation of nonlinear resonant sloshing in a rectangular tank with small fluid depth, *J. Fluid Mech.* **470**, 319–357.

Faltinesn O. M. and Timokha A. N. (2002b), Analytically-oriented approaches to two-dimensional fluid sloshing in a rectangular tank (survey), in *Problems of Analytical Mechanics and its Applications* 44, 321–345, Proc. Inst. Math. Ukrakian National Akad. Sci.

Faraday M. (1831), On the forms and states assumed by fluids in contact with vibrating elastic surfaces, *Phil. Trans. Royal Soc. (London)* **121**, 319–340.

Farooq A. and Homsy G. M. (1994), Streaming flows due to g-jitter-induced natural convection, *J. Fluid Mech.* **271**, 351–378.

Fauve S., Kumar K., Laroche C., Beysens D., and Garabos Y. (1992), Parametric instability of a liquid-vapor interface close to the critical point, *Phys. Rev. Lett.* **68**, 3160–3163.

Fediw A. A., Isyumov N., and Vickery B. J. (1995), Performance of a tuned sloshing water damper, *J. Wind Eng. Indust. Aerodyn.* **57**(2/3), 237–247.

Felippa C. (1991), *Mixed Variational Formulations of Finite Element Analysis of Elastoacoustic/Slosh Fluid Structure Interaction*, NASA, NTIS, Washington, DC.

Felippa C. A. and Ohayon R. (1990), Mixed variational formulation of finite element analysis of acoustoelastic/slosh fluid-structure interaction, *Int. J. Fluid Struct.* **4**, 35–57.

Feng G. C. (1973), Dynamic loads due to moving liquid, AIAA Paper No 73–409.

Feng G. C. and Robertson S. J. (1971), Study on propellant dynamics during docking, Interim Rept., NASA-CR-119904, June.

Feng G. C., and Robertson S. J. (1972), Study on propellant dynamics during docking, NASA-CR-124055, 15 March.

Feng Z. C. (1990), Bifurcation analysis of surface waves, Ph.D. dissertation, University of Minnesota.

Feng Z. C. and Sethna P. R. (1989), Symmetry-breaking bifurcations in resonance surface waves, *J. Fluid Mech.* **199**, 495–518.

Ferrant P. and Touze D. (2001), Simulation of sloshing waves in a 3D tank based on a pseudo-spectral method, Proc. 16th Int. Workshop on *Water Waves and Floating Bodies*, Hiroshima, Japan.

Field S. B. and Martin J. P. (1976), Comparative effects of U-tube and free surface type passive roll stabilization systems, *Royal Inst. Naval Archit.* **2**, 73–92.

Figarov N. G. (1971), Vibrations of fluids in elliptical reservoirs, in *Structural Dynamics*, B. G. Koreneva, ed., Moscow, Izdatelsvo Literatury po Stroitelstvu.

Filippov A. (1964), Differential Equations with Discontinuous Right-Hand Sides, Rhode Island: *Amer. Math. Soc. Translations* **42**, Series 2.

Filstead C. G. Jr (1972), The design and operation of LNG ships with regard to safety, *Shipping World and Shipbuilder* **165**(3866), 259–262.

Finn R. (1974), A note on capillary problem, *Acta Math.* **132**, 199–207.

Finn R. (1978), Some comparison properties and bounds for capillary surfaces, in *Complex Analysis and its Applications* (Moscow, Nauka), 555–562.

Finn R. (1979), Existence and nonexistence of capillary surfaces, *Manuscr. Math.* **28**, 1–9.

Finn R. (1983), Existence criteria for capillary free surfaces without gravity, *Indiana University Math. J.* **32**, 439–460.

Finn R. (1984), A subsidiary variational problem and existence criteria for capillary surfaces, *J. Reine. Angew. Math.* **353**, 196–214.

Finn R. (1986), *Equilibrium Capillary Surfaces*, New York, Springer-Verlag.

Finn R. (1999), Capillary surface interface, *Notices AMS* **46**, 770–781.

Fischer D. F. (1979), Dynamic fluid effects in liquid filled flexible cylindrical tanks, *Earthquake Eng. Struct. Dyn.* **7**, 587–601.

Fischer D. F. and Rammerstorfer F. G. (1982), The stability of liquid-filled cylindrical shells under dynamic loading, in *Buckling of Shells*, Proc. State-of-the-Art Colloq., Berlin, 569–597.

Fischer D. F. and Rammerstorfer F. G. (1984), Stability of liquid storage tanks under earthquake excitation, Proc. 8th World Conf. Earthquake Eng., San Francisco, CA 5, 215–222.

Fischer D. F., and Rammerstorfer F. G. (1999), A refined analysis of sloshing effects in seismically excited tanks, *Int. J. Pressure Vess. Piping* **76**(10), 693–709.

Fischer D. F., Rammerstorfer F. G., and Scharf K. (1991), Earthquake resistant design of anchored and unanchored liquid storage tanks under three-dimensional earthquake excitation, in *Structural Dynamics: Recent Advances*, G. I. Schueller, ed., Berlin, Springer, 317–371.

Floryan J. M. and Rasmussen H. (1989), Numerical methods for viscous flows with moving boundaries, *ASME Appl. Mech. Rev.* **42**, 323–349.

Flugrad D. R. and Obermaier L. A. (1992), Computer simulation of a test rig to model sloshing in spin-stabilized satellites, *J. Dyn. Syst., Measurement and Control* **114**, 689–698.

Fogiel M. (1994), *Handbook of Mathematical, Scientific, and Engineering Formulas, Tables, Functions, Graphs, Transforms*, Piscataway, NJ, Research & Education Association.

Fontenot L. L. (1968), The dynamics of liquids in fixed and moving containers, **VII** in *Dynamic Stability of Space Vehicles*, NASA CR-941.

Fontenot L. L. and Lianis G. (1963), The free vibrations of thin elastic pressurized cylindrical shells filled with a perfect and incompressible liquid having a free surface, Int. Symp. on *Space Technology and Science*, Tokyo.

Fontenot L. L. and Lomen D. O. (1964), Dynamic behavior of partially filled containers of arbitrary geometry executing prescribed vertical excitation, Convair Rep. EER-AN-533.

Fontenot L. L., McDonough G. F., and Lomen D. O. (1965), Liquid free surface instability resulting from random vertical acceleration, Proc. 6th Int. Symp. on *Space Technology and Science*, Tokyo.

Fossen T. L. (1994), *Guidance and Control of Ocean Vehicles*, New York, John Wiley & Sons.

Foster G. K. and Craik ADD (1997), Second-harmonic resonance with Faraday excitation, *Wave Motion* **26**, 361–377.

Fox D. W. and Kuttler J. R. (1981), Upper and lower bounds for sloshing frequencies by intermediate problems, *J. Appl. Math. Phys.* (ZAMP) **32**, 667–682.

Fox D. W., and Kuttler J. R. (1983), Sloshing frequencies, *J. Appl. Math. Phys.* (ZAMP) **34**, 668–696.

Frahm H. (1911), Results of trials of the anti-rolling tanks at sea, *Transactions of the Inst. Naval Architecture* **53**, 183–197.

Francescutto A. and Contento G. (1994), An experimental study of the coupling between roll motion and sloshing in a compartment, ISOPE'94, Osaka, Japan, **3**, 283–288.

Frandsen J. B. (2002), Sloshing effects in periodically and seismically excited tanks, 5th World Cong. Comput. Mech., Vienna, Austria.

Frandsen J. B. and Borthwick AGL (2002), Nonlinear sloshing in fixed and vertically excited containers, Proc. ASME WAM IMECE 2002–32948.

Frank S. and Schwabe D. (1998), Temporal and spatial elements of thermocapillary convection in floating zones, *Experiments in Fluids* **23**, 234–251.

Frasier J. T. and Scott W. E. (1971), Stability of a liquid-filled gyroscope: Inviscid analysis, viscous corrections and experiments, *J. Spacecraft and Rock.* **8**(5), 523–526.

Freed L. E. (1957), Stability of motion of conical pendulums, Space Technol. Labs. Memo. GM 45.3–434, October.

Froude W. (1874), Considerations respecting the rolling of ships at sea, *Transactions of the Inst Naval. Architecture* **14**, 96–116

Fu R. D. (1993), Finite element analysis of lateral sloshing response in axi-symmetric tanks with triangular elements, *Comp. Mech.* **12**, 51–58.

Fujii K., Tamura Y., Sato T., and Wakahara T. (1988), Wind-induced vibration of tower and practical applications of tuned sloshing damper, *J. Wind Eng.* **37**, 537–646.

Fujii K., Tamura Y., Sato T., and Wakahara T. (1990), Wind-induced vibration of tower and practical applications of tuned sloshing damper, *J. Wind Eng. and Indus. Aerodyn.* **33**, 263–272.

Fujino Y. and Abe M. (1993), Design formulas for tuned dampers based on a perturbation technique, *Earthquake Eng. Struct. Dyn.* **22**, 833–854.

Fujino Y., Pacheco B. M., Chaiseri P., and Fujii K. (1988a), An experimental study on tuned liquid damper using circular containers, *JSCE J. Struct. Eng.* **34A**, 603–616.

Fujino Y., Pacheco B. M., Chaiseri P., and Sun L. M. (1988b), Parametric study on tuned liquid damper (TLD) using circular containers by free-oscillation experiments, *JSCE Struct. Eng./Earthquake Eng.* **5**(2), 177–187.

Fujino Y., Pacheco B. M., Chaiseri P., Sun L. M., and Koga K. (1990), Understanding of LTD properties based on TMD analogy, *JSCE J. Struct. Eng.* **36A**, 577–590.

Fujino Y., Pacheco B. M., Sun L. M., Chaiseri P., and Isobe M. (1989), Simulation of nonlinear waves in rectangular tuned liquid damper (TLD) and its verification, *JSCE J. Struct. Eng.* **35A**, 561–574.

Fujino Y. and Sun L. M. (1993), Vibration control by multiple tuned liquid dampers (MTLDs), *ASCE J. Struct. Eng.* **119**(12), 3482–3502.

Fujino Y., Sun L. M., and Koga K. (1991), Simulation and experiment on tuned liquid damper subjected to pitching motion, *JSCE J. Struct. Eng.* **37A**, 805–814.

Fujino Y., Sun L. M., Pacheco B. M., and Chaiseri P. (1992), Tuned liquid damper (LTD) for suppressing horizontal motion of structures, *ASCE J. Eng. Mech.* **118**, 2017–2030.

Fujino Y. *et al.* (1992), Tuned liquid dampers for suppressing horizontal motion of structures, *ASCE J. Eng. Mech.* **118**, 2017–2030.

Fujita K. (1981), A seismic response analysis of a cylindrical liquid storage tank including the effect of sloshing, *Bull. JSME* **24**(195), 1634–1641.

Fujita K. (1985), Vibration analysis of fluid-coupled two coaxial axisymmetric shells containing fluid, *Trans. JSME* (in Japanese) **51**, 1170–1179.

Fujita K., Ito T., Kodama T., Adachi S., Sekine K., Ozaki H, Eguchi Y., and Yamamoto K. (1993), A study on flow-induced vibration of a flexible weir due to fluid discharge: effect of weir stiffness, ASME Proc. *Flow-Induced Vib. and Fluid–Structure Interaction* **258**, 143–150.

Fujita K., Ito T., Kodama T., Eguchi Y., and Yamamoto K. (1996), Flow-induced vibration of a flexible weir due to fluid discharge: effect of weir stiffness, *J. of Fluid and Structures* **10**, 79–98.

Fujita K., Ito T., Shiraishi T., Kodama T., Akutsu M., and Okabe Y. (1991), Study on sloshing characteristics in the reactor vessel of top entry system of FBR, *Fluid–Structure Vibration and Sloshing*, ASME Pressure Vess. Piping Conf., PVP- **223**, 91–96.

Fujita K., Ito T., and Wada H. (1990), Experimental study on the dynamic buckling of cylindrical shell due to seismic excitation, in *Flow-Structure Vibration and Sloshing*, ASME Pressure Vess. Piping Conf., PVP- **191**, 31–36.

Fukaya M., Baba M., Okamoto K., and Madarame H. (1995), Self-induced sloshing caused by a jet from a rectangular tank wall, Proc. HYDRA-2000, 26th Congress IAHR 1, 468–473.

Fukaya M., Madarame H., and Okamoto K. (1996), Growth mechanism of self-induced sloshing caused by vertical plane jet, Proc. ASME– JSME 4th Int. Conf. *Nuclear Engineering*, A. S. Rao, R. B. Duffey, and D. Elias, eds., New Orleans, **1**, 781–787.

Fukaya M., Okamot K., Madarame H., and Iida M. (1993), Effect of tank geometries on self-induced sloshing caused by upward plane jet, Proc Asia-Pacific Vib. Conf.,93, 1, 271–276.

Fuller C. R. and Fahy F. J. (1982), Characteristics of wave propagation and energy distribution in cylindrical elastic shells filled with fluid, *J. Sound Vib.* **81**(4), 501–518.

Fultz D. (1962), An experimental note on finite-amplitude standing gravity waves, *J. Fluid Mech.* **13**, 193–212.

Funakoshi M. and Inoue S. (1987), Chaotic behavior of resonantly forced surface waves, *Phys. Lett* A **121**, 229–232.

Funakoshi M. and Inoue S. (1988), Surface waves due to resonant horizontal oscillation, *J. Fluid Mech.* **192**, 219–247.

Funakoshi M. and Inoue S. (1991), Bifurcations in resonantly forced water waves, *European J. Mech. B/Fluids* **10**, 31–36.

Fung F. W. (1965), Dynamic response of liquids in partially filled containers suddenly experiencing weightlessness, Symp. on *Fluid Mechanics and Heat Transfer under Low Gravitational Conditions*, Lockheed Missiles & Space Co, 24–25 June.

Fung Y. C. Sechler E. E., and Kaplan A. (1957), On the vibration of thin cylindrical shells under internal pressure, *J. Aeron. Sci.* **24**(9), 650–660.

Galiev Sh. U. (1977), *The Dynamics of the Interaction of Structural Elements with a Pressure Wave in a Liquid*, Kiev, Naukova Dumka (in Russian).

Galiev Sh. U. (1981), *The Dynamics of Hydroelastoplastic Systems*, Kiev, Naukova Dumka (in Russian).

Ganapathi M. and Vardan T. K. (1996), Large amplitude vibrations of circular cylindrical shells, *J. Sound Vib.* **192**, 1–14.

Ganiev R. F. (1977), Nonlinear resonance oscillations of bodies with a liquid, *Soviet Appl. Mech.* **13**(10), 978–984.

Ganiev R. F. and Kholopova, V. V. (1975), Nonlinear oscillations of a body with a fluid performing motion in space, *Soviet Appl. Mech.* **11**(11), 65–77.

Ganiev R. F. and Koval'chuk P. S. (1980), *The Dynamics of Systems of Rigid and Elastic Bodies*, Moscow, Mashinostroyeniye (in Russian).

Ganiev R. F., Lakiza V. D., and Tsapenko A. S. (1976), On resonance effects for vibrating bodies containing a liquid under zero gravity, in *Space Studies in the Ukrane* **9**, 26–37, Kiev, Naukova Dumka.

Ganiev R. F. Lakiza V. D., and Tsapenko A. S. (1977), On dynamic behavior of a free liquid surface under low gravity and vibrational effects, *Soviet Appl. Mech.* **13**(5), 102–107.

Gans R. F. and Yalisover S. M. (1981), Observations and measurements of flow in a partially-filled horizontally rotating cylinder, ASME Paper 81-WA/FE-21.

Gao H. and Kwok K. C. S. (1997), Optimization of tuned liquid column dampers, *Eng. Struct.* **19**(6), 476–486.

Gao H., Kwok K. C. S., and Samali B. (1999), Characteristics of multiple tuned liquid column dampers in suppressing structural vibration, *Eng. Struct.* **21**, 316–331.

Gardarsson S. (1997), Shallow-water sloshing, Ph.D. thesis, University of Washington, Seattle.

Gardarsson S., Yeh H., and Reed D. (2001), Behavior of sloped-bottom tuned liquid dampers, *ASCE J. Eng. Mech.* **127**(3), 266–271.

Garrett C. (1970), On cross-waves, *J. Fluid Mech.* **41**(4), 837–849.

Garza L. R. (1963), A brief comparison of ring and asymmetrical baffle characteristics, SwRI Tech. Rept. 6.

Garza L. R. (1964), Measurements of liquid natural frequencies and damping in compartmented cylindrical tanks, SwRI Tech. Rept. 8.

Garza L. R. (1966), Theoretical and experimental pressures and forces on a ring baffle under sloshing conditions, *J. Spacecraft Rock.* **3**(2), 276–278.

Garza L. R. and Abramson H. N. (1963), Measurements of liquid damping provided by ring baffles cylindrical tanks, SwRI, Tech. Rept. No 5, April.

Garza L. R. and Dodge F. T. (1967), Comparison of flexible and rigid ring baffles for slosh suppression, *J. Spacecraft Rock.* **4**(6), 805–806.

Gau J. S. (1985), Deterministic behavior of elevated water tanks under vertical motion, Master thesis, Texas Tech University, Lubbock, TX.

Gauss K. F. (1829), Principia generalia theoriae figurae fluidorum in statu aequilibrii, Gött. Gelehrte Anz. 1829: 1641–1648, Werke 5, 287–292. Göttingen: K. Ges. Wiss. Gött. 1867, Hildesheim: Olms, 1973.

Gavrilyuk I., Lukovsky I. A., and Timokha A. N. (2000), A multimodal approach to nonlinear sloshing in a circular cylindrical tank, *Hybrid Methods in Engineering* **2**(4), 469–484.

Gavrilyuk, I. Lukovsky I. A., and Timokha A. N. (2001), Nonlinear sloshing in a circular conical tank, *Hybrid Methods in Engineering* **3**(4), 1–39.

Gedikli A., and Ergüven M. E. (1999), Seismic analysis of a liquid storage tank with a baffle, *J. Sound Vib.* **223**(1), 141–155.

Genevaux J. M., and Lu D. (2000), On sloshing in a container with moving wall, *J. Fluid Struct.* **14**(2), 275–278.

Gerlach C. R. (1968), Surface disintegration of liquid in longitudinally excited containers, *J. Spacecraft Rock.* **5**(5), 553–560.

Gerrits J. (2001), Dynamics of liquid-filled spacecraft: Numerical simulation of coupled solid–liquid dynamics, Ph.D. thesis, Rijksaniversiteit Groningen.

Gerrits J., Loots G. E., Fekken G., and Veldman AEP (1999), Liquid sloshing in space and on earth, in *Moving Boundaries* V, B. Sarler, C. A. Brebbia, and H. Power, eds., Southampton, WIT Press, 111–120.

Gershuni G. Z. and Lyubimov D. V. (1998), *Thermal Vibrational Convection*, New York, Wiley.

Gershuni G. Z. and Zhukhovitskii E. M. (1963), On parametric excitation of convective instability, *Prikl. Mat. Mekh. (PMM)* **27**(5), 1197–1204.

Gershuni G. Z. and Zhukhovitskii E. M. (1964), On parametric excitation of a rigid rotating fluid, *Prikl. Mat. Mekh. (PMM)* **28**(5), 1010–1016.

Gershuni G. Z. and Zhukhovitskii E. M. (1976), *Convective Stability of Incompressible Fluids*, Jerusalem, Keter .

Gershuni G. Z. and Zhukhovitskii E. M. (1986), Vibrational thermal convection in zero gravity, *Fluid Mech. – Sov. Res.* **15**, 63–84.

Ghali S. A. (1965), The dynamics of liquids in rectangular moving containers, Ph.D. thesis, University of Leeds, Dept. Civil Eng.

Giancaglia G. E. O. (1972), *Perturbation Methods in Nonlinear Systems*, Berlin, Springer.

Gilbert C. R. (1985), Gas payload No G-025: Study of liquid sloshing behavior in microgravity, NASA Goddard Space Flight Center's 1985 *Get Away Special Experimenter's Symposium* 165–176, NASA CP-2401.

Gilbert C. R., Netter G., and Vits P. (1984), Study of liquid sloshing behavior in microgravity, Goddard Space Flight Center's 1984 *Get Away Special Experimenter's Symposium* 38–54.

Gillard P. (1963), Theoretical and experimental research of nonlinear oscillations of liquids in containers and channels of constant depth, Publ. Sci. Tech. Minist. l'Air, France, No 412 (in French).

Gillis J. (1961), Stability of a column of rotating viscous liquid, *Proc. Cambridge Phil. Soc.* **57**, 152–159.

Gillis J. and Kaufman B. (1961), The stability of a rotating viscous jet, *Quart. Appl. Math.* **19**(4), 301–308.

Gillis J. and Shuh K. S. (1962), Stability of a rotating liquid column, *Phys. Fluids* **5**, 1149–1155.

Ginsberg J. H. (1973), Large amplitude forced vibrations of simply supported thin cylindrical shells, *ASME J. Appl. Mech.* **40**, 471–477.

Ginsberg J. H. (1974), Nonlinear axisymmetric free vibration in simply supported cylindrical shells, *ASME J. Appl. Mech.* **41**, 310–311.

Ginsberg J. H. (1975), Multidimensional nonlinear acoustic wave propagation, part II: The nonlinear interaction of an acoustic fluid and plate under harmonic excitation, *J. Sound Vib.* **40**, 359–379.

Glaser R. F. (1967), Axi-symmetric vibrations of partially liquid filled cylindrical containers, NASA TN-D-4026, July.

Glimm J. (1965), Solutions in the large for nonlinear-hyperbolic systems of equations, *Comm. on Pure and Appl. Math.* **18**.

Gluck D. F. and Gille J. P. (1965), Fluid mechanics of zero-g propellant transfer in spacecraft propulsion systems, *J. Eng. for Industry* **87**(1), 1–8.

Gluck D. F., Gille J. P., Simkin D. J., and Zukoski E. E. (1966), Distortion of the liquid surface during tank discharge under low-g conditions, *J. Spacecraft Rock.* **3**(11), 1691–1692.

Gold H., McArdle J. G., and Petrash D. A. (1967), Slosh dynamics study in near zero gravity, description of vehicle and spacecraft, NASA TN-D-3985.

Goldberg Z. A. (1972), Parametric amplification of standing waves in fluids, *Sov. Phys. Dokl.* **16**(11), 949–950.

Goldsborough G. H. (1930), The tidal oscillations in an elliptic basin of variable depth, *Proc. Royal Soc. (London)* A **130**, 157–167.

Goldstein S. (1938), *Modern Developments in Fluid Dynamics*, **2**, London, Oxford University Press.

Gollub J. P. and Meyer C. W. (1983), Symmetry-breaking instabilities on a fluid surface, *Physica* **6**D, 337–346.

Gollub J. P. and Simonelli F. (1989), Bifurcation and modal interactions in fluid mechanics: surface waves, Proc. XVII Int. Congress of *Theor. Appl. Mech*, Grenoble, France, 21–27 August 1988, P Germain, M Piau, and D. Caillerie, eds., North Holland, Elsevier Sciences, 73–82.

Goltser I. M. and Kuntisyn A. L. (1975), On the stability of autonomous systems with intrinsic resonance, *J. Appl. Math. Mech. (PMM)* **39**(6), 974–984.

Golubitsky M., Stewart I., and Schaeffer D. G. (1988), *Singularity and Groups in Bifurcation Theory*, **II** (69) in Appl. Math. Sci. Series, New York, Springer-Verlag.

Gonçalves P. B. and Batista R. C. (1988), Nonlinear vibration analysis of fluid-filled cylindrical shells, *J. Sound Vib.* **127**, 133–143.

Gonçalves P. B. and Ramos N. R. S. S. (1996), Free vibration analysis of cylindrical tanks partially filled with liquid, *J Sound Vib.* **195**, 429–444.

Goodrich C. L., Shi W. T., Hentschel H. C. E., and Lathrop D. P. E (1997), Viscous effects in droplet-ejecting capillary waves, *Phys. Rev.* **56**, 472–475.

Goodrich C. L., Shi W. T., and Lathrop D. P. E. (1996), Threshold dynamics of singular gravity-capillary waves, *Phys. Rev. Lett.* **76**, 1824–1827.

Gorbunov Yu A. (1965), Experimental investigation of the vibrations of spherical and cylindrical shells with a liquid in the presence of pressure, Abstracts of Papers read at the 5th All-Union Conf. on *the Theory of Plates and Shells*, Moscow, Nauka, 57–61.

Görtler H. (1957), On forced oscillations in rotating fluids, Proc. 5th Midwestern Conf. Fluid Mechanics, University of Michigan, 1–10.

Gossard M. L. (1965), Axi-symmetric dynamic response of liquid-filled hemispherical, thin walled, elastic tanks, *AIAA Symp. Struct. Dyn. Aeroelasticity*, 30 August – 1September.

Gradshteyn I. S. and Ryzhik I. M. (1980), *Tables of Integrals, Series, and Products*, New York, Academic Press.

Graham E. W. (1951), The forces produced by fuel oscillations in a rectangular tank, Douglas Aircraft Co, SM-13748.

Graham E. W. (1960), Discussion on 'production of rotation in a confined liquid through translation motion of the boundaries' by Berlot,*ASME J. Appl. Mech.* **27**(2), 365.

Graham E. W. and Rodriguez A. M. (1952), The characteristics of fuel motion which affect airplane dynamics,*ASME J.Appl.Mech.* **19**(3), 381–388.

Graham-Eagle J. (1983), A new method for calculating eigenvalues with application to gravity-capillary waves with edge constraints, *Math. Proc. Cambridge Phil. Soc.* **94**, 553–564.

Graham-Eagle J. (1984), Gravity-capillary waves with edge constraints, Ph.D. thesis, University of Oxford.

Grassia P. and Homsy G. M. (1998a), Thermocapillary and buoyant flows with low frequency jitter, I: Jitter confined to the plane, *Phys. Fluids* **10**, 1273–1290.

Grassia P., and Homsy G. M. (1998b), Thermocapillary and buoyant flows with low frequency jitter, II: Span wise jitter, *Phys. Fluids* **10**, 1291.

Greaves D. M., Borthwick A. G. L., Wu G. X., and Eatock Taylor R. (1997), A moving boundary finite element method for fully nonlinear wave simulations, *J. Ship Res.* **41**(3), 181–194.

Grebogi C., Ott E., and Yorke J. A. (1982), Chaotic attractors in crisis, *Phys. Rev. Lett.* **48**(22), 1507–1510.

Grebogi C., Ott E., Romeiras F., and Yorke J. A. (1987), Critical exponents for crisis-induced intermittency, *Phys. Rev.* **36**, 5365–5380.

Green J. W. (1957), Further remarks on conical sloshing, Memo. PA/M-553/2, STL.

Green J. W. (1959), On the approximation to the eigenvalues in a sloshing problem by the eigenvalues of the corresponding discrete problem, Space Tech. Lab., Memo. PA/M-245011, 21 August.

Green J. W. and Landau H. J. (1957), A third note on conical sloshing, Memo. PA/M-553/3, STL.

Greenhill A. G. (1880), On the general motion of a liquid ellipsoid, *Proc. Cambridge. Phil Soc.* **4**.

Greenhill A. G. (1887), Wave motion in hydrodynamics, *Amer. J. Math.* **9**, 62–112.

Greenspan H. P. (1968), *The Theory of Rotating Fluids*, Cambridge, Cambridge University Press.

Gribkov V. A. (1982), *The Basic Results of Investigations of the Dynamic Characteristics of Shells Filled with a Liquid: Analysis and Systemization of Studies*, Deposit in VINITO No 4577–82, Moscow.

Grigoliuk E. I. (1970), Problems of interaction of shells with a liquid, Proc. 7th All Union Conf. on the *Theory of Shells and Plates*, Dnepropetrovsk, Moscow, Nauka, 755–778.

Grigoliuk E. I. (Edi) (1980), *Thin-Walled Cylindrical Shells*, Moscow, Mashinostroenie (in Russian).

Grigoliuk E. I., and Gorshkov A. G. (1976), *The Interaction of Elastic Structures with a Liquid*, Moscow, Sudostryeniye.

Grigoliuk E. I., Gorshkov A. G., and Shkliarchuk F. N. (1968), On a method of calculating the vibrations of a liquid partially filling an elastic shell of revolution, *Izv. Akad. Nauk. SSSR*, Mekh. zhid Gaza, No **3**, 74–80.

Grigoliuk E. I. and Shkliarchuk F. N. (1970), Equations of perturbed motion of a solid with a thin-walled elastic shell partly filled with fluid, *J. Appl. Mat. Mech.(PMM)* **34**(3).

Grigorev E. T. (1966), The axisymmetric vibrations of a shell with a liquid, *Sov. Applied. Mech. (Prikl. Mekh.)* **2**(4), 39–49.

Grigorev E. T. (1967), The stability of longitudinal vibrations of a shell containing a fluid, *Sov. Appl. Mech. (Prik. Mekh.)* **3**(6), 23–30.

Grigoryeva N. B. (1999), The stability of the Jacobi ellipsoids of a rotating liquid, *J. Appl. Mat. Mech.* **63**(6), 1052–1054.

Grigoryeva N. B. (2000), The stability of the equilibrium ellipsoids of a rotating liquid, *J. Appl. Mat. Mech.* **64**(6), 925–936.

Grinchenko V. T. and Komissarova G. L. (1984), Wave propagation in a hollow elastic cylinder with a liquid, *Soviet Appl. Mech. (Prikl. Mekh.)* **20**(1), 21–26.

Grinchenko V. T., and Komissarova G. L. (1988), Properties of normal non-axisymmetric waves in a thick-walled cylinder filled with a liquid, *Sov Appl. Mech. (Prikl. Mekh.)* **24**(10), 15–20.

Grinchenko V. T. and Meleshko V. V. (1981), Harmonic vibrations and waves in elastic bodies, Kiev, Naukova Dumka (in Russian).

Grodzka P. G. and Bannister T. C. (1972), Heat flow and convection demonstration experiments aboard Apollo 14, *Science* **176**, 506.

Grubb L. S. and Petrash D. A. (1967), Experimental investigation of interfacial behavior following termination of outflow in weightlessness, NASA TN D-3896.

Gu X. M. (1986), Nonlinear surface waves of a fluid in rectangular containers subjected to vertical periodic excitations, Ph.D. dissertation, University of Minnesota.

Gu X. M. and Sethna P. R. (1987), Resonance surface waves and chaotic phenomena, *J. Fluid Mech.* **183**, 543–565.

Gu X. M., Sethna P. R., and Narain A. (1988), On three-dimensional nonlinear subharmonic resonance surface waves in a fluid, part I: theory, *J. Appl. Mech.* **55**, 213–219.

Guadagnoli D. (1991), The description of a typical spacecraft, monopropellant reaction control system, using spherical propellant tanks with surface tension propellant management devices, GE Astro Space Division Memo.

Guarino J. C. and Elger D. F. (1992), Modal analysis of a fluid-filled elastic shell containing an elastic sphere, *J. Sound Vib.* **156**, 461–479.

Guibert J. P., Huynh H. T., and Marce J. L. (1978), Etude general de l'influence des liquids contenus dans les satellites, ONERA RT 1/3307 SY, September.

Gunter N. M. (1965), *Potential Theory and Its Application to Basic Problems of Mathematical Physics*, New York, Frederick Unger Publishing Co.

Gupta M. R. K. (1995a), Sloshing in shallow cylindrical tanks, *J. Sound Vib.* **180**(3), 397–415.

Gupta M. R. K. (1995b), Free vibrations of partially filled cylindrical tanks, *Engrg. Struct.* **17**(3), 221–230.

Gupta M. R. K. and Hutchinson G. L. (1988), Free vibration analysis of liquid storage tanks, *J. Sound Vib.* **122**, 491–506.

Gupta M. R. K., and Hutchinson G. L. (1989), Solid-water interaction in liquid storage tanks, *J. Sound Vib.* **135**(3), 357–374.

Gupta M. R. K., and Hutchinson G. L. (1991), Effects of a wall flexibility on the dynamic response of liquid storage tanks, *Engineering Structures* **13**, 253–267.

Gustafson D. A. and Gustafson G. R. (1969), Heavy vehicle overturning problems, Chalmers' Institute Tech., Gothenburg, Sweden (in Swedish).

Guthar G. S. and Wu T. Y. (1991), Observation of a standing-kink cross wave parametrically excited, *Proc. Royal Soc. London* A **434**, 430–440.

Guyan R. J., Ujihara B. H., and Welch P. W. (1968), Hydroelastic analysis of axi-symmetric systems by a finite element method, Proc. of the 2nd Conf. on *Matrix Methods in Structural Mechanics*, Ohio, Wright-Patterson AFB, October.

Guyett P. R. (1967), Measurements of the unsteady forces acting on a circular cylindrical tank containing liquid during harmonic motion, ARE TR 67098, April.

Guz A. N., Kubenko V. D., and Babaev A. E. (2002), Dynamics of shell systems interacting with a liquid, *Int. Appl. Mech.* **38**(3), 260–301.

Haberman R. (1979), Slowly-varying jump and transition phenomena associated with algebraic bifurcation problems, *SIAM J. Appl. Math.* **37**, 69–105.

Habip L. M. (1965), On the mechanics of liquids in sub-gravity, *Acta Astronaut.* **11**(6), 401–409.

Hagiuda H. (1989), Oscillation control system exploiting fluid force generated by water sloshing, (in Japanese), *Mitsui Zosen Technical Review* **137**, 13–20.

Hale J. K. (1969), *Ordinary Differential Equations*, New York, Wiley-Interscience.

Hamann F. H. and Datton C. (1971), The forces on a cylinder oscillating sinusoidally in water, *J. Eng. for Industry* **9**(4), 1197–1202.

Hamdi M. and Ousset V. (1978), A displacement method for the analysis of vibrations of coupled fluid–structure systems, *Int. J. Num. Methods in Eng.* **13**, 139–150.

Hamlin N. A., Lou Y. K., Maclean W. M., Seibold F., and Chandras L. M. (1986), Liquid sloshing in slack ship tanks: theory, observations and experiments, *SNAME Trans.* **94**, 159–195.

Hammack J. L. and Henderson D. M. (1993), Resonant interactions among surface water waves, *Annu. Rev. Fluid Mech.* **25**, 55–97.

Hanson R. D. (1973), Behavior of liquid storage tanks: the great Alaska earthquake of 1964, National Academy Sci. Washington D. C., 331–339.

Hara F. (1990), Experimental study on sloshing characteristics of a flowing liquid in a tank, *Trans. JSME Int. J.,* Series III **33**(3), 330–338.

Hara F. (1992), Identification of sloshing suppression force generated by intermittent gas-bubble injection, Proc. 1st MOVIC, 1110–1115.

Hara F. and Saito O. (1988), Active control for earthquake-induced sloshing of a liquid contained in a circular tank, Proc. 9th World Conf. *Earthquake Engineering* **V**, 889–894, Tokyo–Kyoto, Japan.

Hara F. and Shibata H. (1986), Experimental study on active control of sloshing of a liquid contained in a tank by intermittent gas-bubble injection, *Trans. JSME* **52**(C-481), 2392–2396 (in Japanese).

Hara F. and Shibata H. (1987), Experimental study on active suppression by gas bubble injection for earthquake induced sloshing in tanks, *JSME Int. J.* **30**(260), 318–323.

Hara F. and Suzuki T. (1992), Dynamic instability analysis of self-excited vibration of a flexible weir due to fluid discharge, ASME Symp. *Flow-Induced Vibration and Noise*, **5**: *Axial and Annular Flow-Induced Vibration and Instabilities*, PVP- **244**, 45–58.

Harlow F. H., Amsden A. A., and Nix J. R. (1976), *J. Comput. Phys.* 20, 119

Harlow F. H. and Welch J. E. (1965), Numerical calculation of time-dependent viscous incompressible flow of fluid with free surface, *Phys. of Fluids* **8**, 2152–2196.

Harlow F. H., and Welch J. E. (1966), Numerical study of large-amplitude free surface motions, *Phys. of Fluids* **9**, 842.

Haroun M. A. (1980), Dynamic analysis of liquid storage tanks, Ph.D. diss., Caltech, Pasadena, CA.

Haroun M. A. (1983), Vibration studies and tests of liquid storage tanks, *Earthquake Eng. Struct. Dyn.* **11**, 179–206.

Haroun M. A. (1991), Implications of observed earthquake-induced damage on seismic codes and standards, *Fluid–Structure Vibration and Sloshing*, ASME Pressure Vess. Piping Conf., PVP-**223**, 1–7.

Haroun M. A. and Elliathy H. M. (1985a), Seismically induced fluid forces on elevated tanks, *ASCE J. Tech. Topics in Civil Eng.* **111**, 1–15.

Haroun M. A., and Elliathy H. M. (1985b), Model for flexible tanks undergoing rocking, *ASCE J. Eng. Mech.* **111**(2), 143–157.

Haroun M. A. and Housner G. W. (1981a), Earthquake response of deformable liquid storage tanks, *ASME J. Appl. Mech.* **48**, 411–418.

Haroun M. A. and Housner G. W. (1981b), Dynamic interaction of liquid storage tanks and foundation soil, Proc. 2nd ASCE/EMD Specialty Conf. *Dynamic Response of Structures*, Atlanta, Georgia.

Haroun M. A. and Housner G. W. (1982), Complications in free vibration analysis of tanks, *ASCE J. Eng. Mech. Div.* **108**(EM5), 801–818.

Haroun M. A., Lee L. R., and Elliathy H. M. (1989), Dynamic behavior of shell towers supporting liquid filled tanks, *Sloshing and Fluid Structure Vibration*, ASME Pressure Vess. Piping Conf., PVP- **157**, 1–8.

Haroun M. A. and Mourad S. A. (1990), Buckling behavior of liquid filled shells under lateral seismic shear, in *Flow–Structure Vibration and Sloshing*, ASME Pressure Vess. Piping Conf., PVP- **191**, 11–17.

Haroun M. A., Mourad S. A., and Pence P. W. (1991), Vibration suppression through liquid oscillations, *Mechanics Computing* in 1990s and Beyond, EMD Special Conf., Columbus, Ohio, **2**, 656–660.

Haroun M. A., Pires J. A., and Won A. Y. J. (1994a), Hybrid liquid column dampers for suppression of environmentally-induced vibrations in tall buildings, Proc. 3rd Conf. *Tall Buildings in Seismic Regions*, Los Angeles, CA.

Haroun M. A. Pires J. A., and Won A. Y. J. (1994b), Active orifice control in hybrid liquid column dampers, 1st World Conf. *Structural Control* 3, FA1–69–78, Los Angeles, CA.

Haroun M. A. Pires J. A., and Won A. Y. J. (1996), Suppression of environmentally-induced vibration in tall buildings by hybrid liquid column dampers, *Struct. Des. Tall Buildings* **5**, 45–54.

Haroun M. A. and Tayel M. A. (1984), Dynamic behavior of cylindrical liquid storage tanks under vertical earthquake excitation, Proc. 8th World Conf. Earthquake Eng., San Francisco, CA **7**, 421–428.

Haroun M. A. and Tayel M. A. (1985a), Axi-symmetric vibrations of tanks – numerical, *ASCE. J. Eng. Mech. Div.* **111**, 329–345.

Haroun M. A. and Tayel M. A. (1985b), Axi-symmetric vibrations of tanks – analytical, *ASCE J. Eng. Mech. Div.* **111**, 346–358.

Haroun M. A. and Tayel M. A. (1985c), Response of tanks to vertical seismic excitations, *Earthquake Eng. and Struct. Dyn.* **13**, 583–595.

Harper J. (1958), Propellant sloshing in conical tank undergoing arbitrary forced translational motion, Convair, Astronautics, Rep. ZU-7–089-TN, 2 January.

Harrison J. V. and Murphy J. (1987), Sub-scale testing of liquid propellant tanks for the RCA USSB spacecraft, Final Rept. 529460–80-F50, RCA.

Hasegawa E., Umehara S., and Atsumi M. (1983), The critical condition for the onset of waves on the free surface of a horizontal liquid layer under a vertical oscillation, *Trans. JSME B* **49**(448), 2901–2907 (in Japanese).

Hashimoto H. and Sudo S. (1980), Surface disintegration and bubble formation in vertically vibrated liquid column, *AIAA J.* **18**(4), 442–449.

Hashimoto H. and Sudo S. (1982), Frequency characteristics of pressure induced by vertical vibration of a cylindrical container containing liquid, Proc. 1982 Joint Conf. on Experimental Mech. (SESA), 353–358.

Hashimoto H. and Sudo S. (1984), Basic behavior of liquid free surface in a cylindrical container subject to vertical vibration, *Bulletin JSME* **27**(227) 923–930.

Hashimoto H. and Sudo S. (1985a), Frequency characteristic of a bubble cluster in a vibrated liquid column, *J. Spacecraft Rock.* **22**(6), 649–655.

Hashimoto H. and Sudo S. (1985b), Dynamic behavior of stratified fluids in a rectangular container subjected to vertical excitation, *Bull. JSME* **28**, 1910–1917.

Hashimoto H. and Sudo S. (1987), Drop formation mechanism in a vertically vibrated liquid column, *AIAA J.* **25**(5), 727–732.

Hashimoto H. and Sudo S. (1988), Violent liquid sloshing in vertically excited cylindrical containers, *Exp. Therm. Fluid. Sci.* **1**, 159–169.

Hastings L. J. and Rutherford R. III (1968), Low gravity liquid–vapor interface shapes in axi-symmetric containers and a computer simulation, NASA TM X-53790.

Hastings L. J. and Toole L. E. (1968), An experimental study of the behavior of a sloshing liquid subjected to a sudden reduction in acceleration, NASA TM-X-5375.

Hatano T. and Konno H. (1966), Numerical solution of hydrodynamic pressures during earthquakes on arch dam, *Trans. JSCE* **131**, 19–23 (in Japanese).

Hattori M., Arami A., and Yui T. (1994), Wave impact pressure on vertical walls under breaking waves of various types, *Coastal Engrg.* **22**, 79–114.

Haxton R. S. and Barr A. D. S. (1972), Autoparametric vibration absorber, *ASME J. Eng. Indust.* **94**, 119–125.

Hayama S., Aruga K., and Watanabe T. (1983), Nonlinear response of sloshing in rectangular tanks (1st report, nonlinear response of surface elevation), *Bulletin JSME* **26**(219), 1641–1648.

Hayama S., Inoue Y., and Watanabe T. (1989), The suppression of sloshing in a liquid tank by means of a reversed u-tube, *Sloshing and Fluid Structure Vibration*, ASME Pressure Vess. Piping Conf., PVP- **157**, 95–102. (Also *JSME Int. J. Series III* **33**(3) 339–345, 1990.)

Hayama S. and Iwabuchi M. (1985), Sloshing suppression using inverse u-tube, *Trans. JSME* **51**(470), 2505–2511 (in Japanese).

Hayama S. and Iwabuchi M. (1986), A study on the suppression of sloshing in a liquid tank: 1st report: suppression of sloshing by means of a reversed u-tube, *Bull. JSME* **29**, 1834–1841.

Heinrich K. and Kaufman F. M. (1956), Sloshing stability for vehicles with one free surface, Space Technol. Labs. Memo. GM45–3–45, 12 July.

Heinrich R. T. (1986), Experimental investigation of random parametric liquid sloshing, Master thesis, Texas Tech University, Lubbock, TX.

Helmholtz H. V. (1860), Über reibung tropfbarer flüssigkeiten, *Sitzungsberichte der K Akademie der Wissenschaften zu Wien* **40**.

Henderson D. M. (1998), Effects of surfactants on Faraday-wave dynamics, *J. Fluid Mech.* **365**, 89–107.

Henderson D. M. and Miles J. W. (1990), Single mode Faraday waves in small cylinders, *J. Fluid Mech.* **213**, 95–109.

Henderson D. M. and Miles J. W. (1994), Surface-wave damping in a circular cylinder with a fixed contact-line, *J. Fluid Mech.* **275**, 285–299.

Hendricks P. C. (1986), Stability of a clamped-free rotor partially filled with liquid, *ASME J. Appl. Mech.* **53**, 166–172.

Hendricks P. C. and Klauber R. D. (1984), Optimal control of a rotor partially filled with an inviscid incompressible fluid, *ASME J. Appl. Mech.* **51**, 863–868.

Hendricks S. L. and Morton J. B. (1979), Stability of a rotor partially filled with a viscous incompressible fluid, *ASME J. Appl. Mech.* **46**, 913–918.

Henrici P., Troesch B. A., and Wuytack L. (1970), Sloshing frequencies for a half-space with circular or stripe-like aperture, *Zeit. Ang. Math. Phys. (ZAMP)* **21**(3), 285–318.

Henstock W. and Sani R. L. (1974), On the stability of the free surface of a cylindrical layer of fluid in vertical periodic motion, *Lett. Heat Transfer* **1**, 95–102.

Hieatt J. L. and Riley J. D. (1959), Digital program for fluid sloshing in tanks with axial symmetry, Rept. TM-59-0000-00389, Space Tech. Labs. (Now TRW), available from DDC as A. D. 607548.

Higuera H. and Nicolás J. A. (1997), Linear non-axisymmetric oscillations of nearly-inviscid liquid bridges, *Phys. Fluids* **9**, 276–285.

Higuera H., Nicolás J. A., and Vega J. M. (1994), Linear oscillations of weakly dissipative axisymmetric liquid bridges, *Phys. Fluids A* **6**, 438–450.

Hill D. E. and Baumgarten J. R. (1992), Control of spin-stabilized spacecraft with sloshing fluid stores, *J. Dyn. Syst., Measurement and Control* **114**, 728–731.

Hill D. E., Baumgarten J. R., and Miller J. T. (1988), Dynamic simulation of spin-stabilized spacecraft with sloshing fluid stores, *J. Guidance* **11**(6), 597–599.

Hiramatsu T., Komura Y., and Shintani M. (1989a), Theoretical analysis for the dynamic behavior of nuclear power reactor internals, *Sloshing and Fluid Structure Vibration*, ASME Pressure Vess. Piping Conf., PVP- **157**, 149–156.

Hiramatsu T., Komura Y., Shintani M., and Shintaku S. (1989b), Dynamic behavior of nuclear power reactor internals-model by experiments, *Sloshing and Fluid Structure Vibration*, ASME Pressure Vess. Piping Conf., PVP- **157**, 157–163.

Hirata A., Sakurai M., Ohishi N., Koyama M., Morita T., and Kawasaki H. (1997), Transition process from laminar to oscillatory Marangoni convection in a liquid bridge under normal and microgravity, *J. Japan Soc. Microgravity Appl.* **14**(2), 137–143.

Hirt C. W., Amsed A. A., and Cook J. L. (1974), An arbitrary Lagrangian–Eulerian computing method of all speeds, *J. Comput. Phys.* **14**, 227–253.

Hirt C. W. and Nichols B. D. (1981), Volume of fluid (VOF) method for the dynamics of free boundaries, *J. Comp. Phys.* **39**, 201–225.

Hirt C. W., Nichols B. D., and Romero N. C. (1975), SOLA – A numerical solution algorithm for transient fluid flows, Report LA-5852, Los Alamos Scientific Laboratory.

Hitchcock P. A., Kwok K. C. S., Watkins R. D., and Samali B. (1997a), Characteristics of liquid column vibration absorbers (LCVA) – I, *Eng. Struct.* **19**(2), 126–134.

Hitchcock P. A., Kwok K. C. S., Watkins R. D., and Samali B. (1997b), Characteristics of liquid column vibration absorbers (LCVA) – II, *Eng. Struct.* **19**(2), 135–144.

Hoard C. Q., Robertson C. R., and Acrivos A. (1970), Experiments on the cellular structure in Benard convection, *Int. J. Heat Mass Transf.* **13**, 849.

Hocking L. M. (1965), On the unsteady motion of a rotating fluid in a cavity, *Mathematica* **12**, 97–106.

Hocking L. M. (1976), A moving fluid interface on a rough surface, *J. Fluid Mech.* **76**, 801–817.

Hocking L. M. (1977), A moving fluid interface, Part 2: The removal of the force singularity by a slip flow, *J. Fluid Mech.* **79**, 209–229.

Hocking L. M. (1987a), The damping of capillary gravity waves at a rigid boundary, *J. Fluid Mech.* **179**, 253–266.

Hocking L. M. (1987b), Waves produced by a vertically oscillating plate, *J. Fluid Mech.* **179**, 267–281.

Hocking L. M. and Mahdmina D. (1991), Capillary-gravity waves produced by a wave maker, *J. Fluid Mech.* **224**, 217–226.

Hocking L. M. and Michael D. H. (1959), The stability of a column of rotating liquid, *Mathematika* **6**, 25–32.

Hodges B. R. and Street R. L. (1999), On simulation of turbulent nonlinear free surface flows, *J. Comput. Phys.* **151**, 425–457.

Hollister M. P. and Satterlee H. M. (1965), Low gravity liquid reorientation, Proc. of a Symp. on *Fluid mechanics and heat transfer under low gravitational conditions*, Lockheed Missiles & Space Co.

Holm-Christenson G. and Träger K. (1991), A note on rotor instability caused by liquid motions, *ASME J. Appl. Mech.* **58**, 804–811.

Holmes P. J. (1986), Chaotic motion in a weakly nonlinear model for surface waves, *J. Fluid Mech.* **162**, 365–388.

Homicz G. F. and Gerber N. (1987), Numerical model for fluid spin-up from rest in a partially filled cylinder, *ASME J. Fluids Eng.* **109**, 194–197.

Honda K. and Matsushita T. (1913), An investigation of the oscillations of tank water, *Scientific Reports*, Tohoku, Imperial University, First Series **21**, 131–148.

Honda K. and Tajima K. (1979), Sloshing of two superposed liquid layer in a rectangular tank, *Trans. JSME* 45B, 1450–1457.

Hori G. I. (1966), Theory of the general perturbation with unspecified canonical variables, *Astron. Soc. Japan* **18**(4).

Hori Y., Kanai M., and Fujisawa F. (1994), Two-dimensional coupling vibration analysis of fluid and structure using FEM displacement method (in Japanese), *Trans. JSME C* **60**(572), 1138–1143.

Hoskins L. M. and Jacobsen L. S. (1957), Water pressure in a tank caused by a simulated earthquake, *Bull. Seismol. Soc. Amer.* **47**(1), 1–32.

Hough (1895), The oscillations of a rotating ellipsoidal shell containing fluid, *Phil. Trans. A* **186**(1).

Houghton G. (1968), Particle retardation in vertically oscillating fluids, *Canadian J. Chem. Eng.* **46**, 79–91.

Housner G. W. (1957), Dynamic pressures on accelerated fluid containers, *Bull. Seis. Soc. Amer.* **47**(1), 15–35.

Housner G. W. (1963a), The dynamic behavior of water tanks, *Bull. Seis. Soc. Amer.* **53**(2), 381–387.

Housner G. W. (1963b), Dynamic pressure on fluid containers, TID-7024, *Nuclear Reactors and Earthquakes*, Chapter 6, (US Atomic Energy Commission), 183–209.

Housner G. W. (1963c), Dynamic analysis of fluids in containers subjected to acceleration, TID-7024, *Nuclear Reactors and Earthquakes*, Appendix F, (US Atomic Energy Commission), 368–390.

Housner J. (1980), Hydroelastic vibration analysis of partially liquid-filled shells using a series representation of the liquid, NASA, Sci. and Tech. Information Center, Washington, DC.

Howard L. N. (1963), Fundamentals of the theory of rotating fluids, *ASME J. Appl. Mech.* **30**, 481–485.

Howell E. and Ebler F. G. (1956), Experimental investigation of the influence of mechanical baffles on fundamental sloshing mode of water in a cylindrical tank, Space Tech. Lab. Rep. GM-TR-69, 6 July.

Howell J. V. (1957), Motion of a conical pendulum, Space Tech. Lab. Memo. GM 42.4–4.

Hsieh D. Y., Sun S. M., and Shen M. C. (1995), Nonlinear theory of forced surface waves in a circular basin, *Wave Motion* **21**(4), 331–341.

Hsieh K. C., Thompson R. L., Zandt D. V., DeWitt K., and Nash J. (1994), Oscillatory/chaotic thermocapillary flow induced by radiant heating, NASA 2nd *Microgravity Fluid Physics* Conf., Cleveland, Ohio, 57–63.

Hsiung H. C. H. and Weingarten V. I. (1973), Dynamic analysis of hydroelastic systems using the finite element method, University of Southern C. A., Dept. Civil Eng., Report USCCE 013, November.

Hu H., Kobayashi T., Saga T., Segawa T., Taniguch N., Nagoshi M., and Okamoto K. (1999), A PIV study on the self-induced sloshing in a tank with circulating flow, 2nd Pacific Symp. on *Flow Visualization and Image Processing*, PF-152.

Hu H. C. and Kelly R. E. (1994), The effect of finite-amplitude nonplanar flow oscillations upon the onset of Rayleigh–Bernard convection, Proc. 10th *Int. Heat Transfer* Conf. **7**, 79.

Hu W. C. L. (1964), A survey of the literature of the vibrations of thin shells, SwRI Tech. Rept. 1.

Hu W. and Tang Z. M. (1989), Excitation mechanism of thermocapillary oscillatory convection, *Sciences in China* **33**(8), 934–938.

Hu W. R., You H. T., and Cao Z. H. (1992a), Free surface oscillation of thermocapillary convection in liquid bridge of half-floating zone, *Sciences in China* **35**(9), 1101–1109.

Hu W. R., Shu J. Z., and Chen Q. S. (1992b), Space experiment on free surface oscillation in liquid bridge of half floating zone, *Microgravity Quart.* **2**, 153–158.

Huang D. T. and Soedel W. (1993), On the free vibrations of multiple plates welded to a cylindrical shell with special attention to mode pairs, *J. Sound Vib.* **166**, 315–339.

Huang Y. Y. (1991), Orthogonality of wet modes in coupled vibrations of cylindrical shells containing liquids, *J. Sound Vib.* **145**, 51–60.

Huang Z. J. and Hsiung C. C. (1996), Nonlinear shallow-water flow on deck, *J. Ship Res.* **40**(4), 303–315.

Huerta A., Liu W. K., Gvildys J. (1989), Large-amplitude sloshing with submerged blocks, *Sloshing and Fluid Structure Vibration*, ASME Pressure Vess. Piping Conf., PVP- **157**, 143–148.

Hughes T. J. R, Liu W. K., and Zimmermann T. K. (1981), Lagrangian–Eulerian finite element formulation for incompressible viscous flows, Proc. Interdisciplinary Finite Element Anal, 179–216, J. F. Abel, T. Kawai, and S. F. Shen, eds., Cornell University.

Huleux A. (1964), Water oscillations in a container in uniformly accelerated translation, *Bull. Acad. Royal Sci.* **50**(11), 1315–1330 (in French).

Hung R. J. (1990), Superfluid and normal fluid helium I. I. in a rotating tank under low and microgravity environments, *Proc. National Sci. Council A* **14**, 289–297.

Hung R. J. (1993a), Final report on sloshing dynamics on rotating helium Dewar tank, NASA NTIS, Washington, DC.

Hung R. J. (1993b), Simulation of sloshing dynamics induced forces and torques actuated on Dewar container driven by gravity gradient and g-jitter accelerations in microgravity, NASA NTIS, Washington, DC.

Hung R. J. (1994a), Superfluid helium sloshing dynamics induced oscillations and fluctuations of angular momentum, force and moment actuated on spacecraft driven by gravity gradient or jitter acceleration associated with slew motion, University of Alabama in Huntsville, NASA, NTIS, Washington, DC.

Hung R. J. (1994b), Annual report on numerical studies of the surface tension effect of cryogenic liquid helium, University of Alabama in Huntsville, NASA, NTIS, Washington, DC.

Hung R. J. (1995a), Mathematical model of bubble sloshing dynamics for cryogenic liquid helium in orbital spacecraft Dewar container, University of Alabama in Huntsville, NASA, NTIS, Washington, DC.

Hung R. J. (1995b), Effect of baffle on slosh reaction forces in rotating liquid helium subjected to a lateral impulse in microgravity, University of Alabama in Huntsville, NASA, NTIS, Washington, DC.

Hung R. J. (1996), Sloshing of cryogenic helium driven by lateral impulsive/gravity gradient-dominated/or g-jitter-dominated accelerations and orbital dynamics, University of Alabama in Huntsville, NASA, NTIS, Washington, DC.

Hung R. J. (1997), DynAmerl models for sloshing dynamics of helium II under low-g conditions, Final Research Report, University of Alabama in Huntsville, NASA, NTIS, Washington, DC.

Hung R. J. and Lee C. C. (1994), Effect of baffle on gravity gradient acceleration excited slosh waves in microgravity, *J. Spacecraft Rock.* **31**(6), 1107–1114.

Hung R. J., Lee C. C., and Leslie F. W. (1990a), Slosh wave excitation in gravity probe-b spacecraft experiment, *Develop. Theor. Appl. Mech.*, 197–205, Georgia Tech., Atlanta.

Hung R. J. Lee C. C., and Leslie F. W. (1990b), Effects of g-jitters on the stability of rotating bubble under microgravity environment, *Acta Astron.* **21**, 309–321.

Hung R. J. Lee C. C., and Leslie F. W. (1991a), Gravity-jitter response slosh waves excitation on the fluid in a rotating Dewar, *Advances in Space Res.* **11**, 201–208.

Hung R. J. Lee C. C., and Leslie F. W. (1991b), Gravity-jitter effected slosh waves on the stability of rotating bubble under microgravity environment, *Advances in Space Res.* **11**, 209–216.

Hung R. J. Lee C. C., and Leslie F. W. (1991c), Slosh wave excitation in a partially filled rotating tank due to gravity jitters in a microgravity environment, *Acta Astron.* **25**, 523–551.

Hung R. J. Lee C. C., and Leslie F. W. (1992a), Spacecraft dynamical distributions of fluid stresses activated by gravity jitters-induced slosh waves, *J. Guidance, Control and Dynamics* **15**(4), 817–824.

Hung R. J., andLee C. C., and Leslie F. W. (1992b), Similarity rules in gravity jitter-related spacecraft liquid propellant slosh waves excitation, *J. Fluid Struct.* **6**, 493–522.

Hung R. J. Lee C. C., and Leslie F. W. (1993a), Effect of the baffle on the asymptotic gravity-jitter excited slosh waves and spacecraft moment and angular momentum fluctuations, *J. Aerospace Eng.* **207**, 105–120.

Hung R. J. Lee C. C., and Leslie F. W. (1993b), Effect of the baffle on the spacecraft fluid propellant viscous stress and moment fluctuations, *Trans. Japanese Soc. Aeronaut. Space Sci.* **35**, 187–207.

Hung R. J., Lee C. C., and Shyu K. L. (1990), Reorientation of rotating fluid in microgravity environment, *Develop. Theor. Appl. Mech.*, 206–212, Georgia Tech., Atlanta.

Hung R. J. Lee C. C., and Shyu K. L. (1991), Reorientation of rotating fluid in microgravity environment with and without gravity jitters, *J. Spacecraft Rock.* **28**(1), 71–78.

Hung R. J., Lee C. C., and Tsao Y. D. (1990), Fluid behavior in microgravity environment, *Proc. National Sci. Council A* **14**, 21–34.

Hung R. J., Lee C. C., and Wu J. L. (1990), Gravity-jitters and excitation of slosh waves, *Develop. Theor. Appl. Mech.*, 213–220, Georgia Tech, Atlanta.

Hung R. J. and Leslie F. W. (1988), Bubble shape in a liquid filled rotating container under low gravity, *J. Spacecraft Rock.* **26**, 70–74.

Hung R. J. and Long T. Y. (1995a), Response and decay of rotating cryogenic liquid reacted to impulsive accelerations in microgravity, *Trans. Japanese Soc. Aeron. Space Sci.* **37**, 291–310.

Hung R. J. and Long Y. T. (1995b), Effect of baffle on slosh reaction forces in rotating liquid helium subjected to the lateral impulse in microgravity, *Cryogenics* **35**, 589–597.

Hung R. J. and Long Y. T. (1996), Response of lateral impulse on liquid helium sloshing with baffle effect in microgravity, *Int. J. Mech. Sci.* **38**(8/9), 951–965.

Hung R. J., Long Y. T., and Chi Y. M. (1996), Slosh dynamics coupled with spacecraft attitude dynamics, Part 1: Formulation and theory, Part 2: Orbital environment application, *J. Spacecraft Rock.* **33**(4), 575–581, 582–593.

Hung R. J., Long Y. T., and Pan H. L. (1994), Sloshing dynamics induced angular momentum fluctuations driven by jitter accelerations associated with slew motion in microgravity, *Trans. Japn Soc. Aeronaut. Space Sci.* **37**, 217–233.

Hung R. J., Long Y. T., and Zu G. J. (1996), Sloshing of cryogenic helium driven by lateral impulse/gravity gradient-dominated/or g-jitter dominated accelerations or orbital dynamics, *Cryogenics* **36**(10), 829–841.

Hung R. J. and Pan H. L. (1993), Differences in gravity gradient and gravity jitter excited slosh waves in microgravity, *Trans. Japan Soc. Aeronaut. and Space Sci.* **36**(1), 153–169.

Hung R. J. and Pan H. L. (1994a), Asymmetric slosh wave excitation in liquid–vapor interface under microgravity, *Acta Mech. Sinica* **9**(2), 298–311.

Hung R. J. and Pan H. L. (1994b), Gravity gradient or gravity jitter induced viscous stress and moment fluctuations in microgravity, *Fluid Dyn. Res.* **14**, 29–51.

Hung R. J. and Pan H. L. (1995a), Fluid force activated spacecraft dynamics driven by gravity gradient and jitter accelerations, *J. Guidance, Control and Dyn.* **18**(5), 1190–1196.

Hung R. J. and Pan H. L. (1995b), Sloshing-modulated liquid–vapor interface fluctuations activated by orbital accelerations associated with spinning and /or slew motions, *J. Colloid. and Interface Sci.* **170**, 538–549.

Hung R. J. and Pan H. L. (1995c), Mathematical model of bubble sloshing dynamics for cryogenic liquid helium in orbital spacecraft Dewar container, *Appl. Math. Modeling* **19**(8), 483–498.

Hung R. J. and Pan H. L. (1996), Modeling of sloshing modulated angular momentum fluctuations actuated by gravity gradient associated with spacecraft slow motion, *Appl. Math. Modeling* **20**(5), 399–409.

Hung R. J., Pan H. L., and Leslie F. W. (1994a), Gravity gradient and gravity jitter induced viscous stress and moment fluctuations in microgravity, *Fluid Dyn. Res.* **34**(1), 29–44.

Hung R. J. Pan H. L., and Leslie F. W. (1994b), Fluid system angular momentum and moment fluctuations driven by gravity gradient or gravity g-jitter in microgravity, *J. Flight Sci. Space Res.* **18**, 195–202.

Hung R. J., Pan H. L., and Long Y. T. (1994), Peculiar behavior of helium II disturbance due to sloshing dynamics driven by jitter acceleration associated with slew motion in microgravity, *Cryogenics* **34**(8), 641–648.

Hung R. J. and Shyu K. L. (1991a), Cryogenic hydrogen reorientation and geyser initiation at various liquid-filled levels in microgravity, *Advances in Space Research* **11**, 217–226.

Hung R. J. and Shyu K. L. (1991b), Cryogenic liquid hydrogen reorientation activated by high frequency impulsive acceleration of geyser initiation, *Microgravity Quart.* **1**, 81–92.

Hung R. J. and Shyu K. L. (1991c), Space-based cryogenic liquid hydrogen reorientation activated by low frequency impulsive reverse thruster of geyser initiation, *Acta Astron.* **25**, 709–719.

Hung R. J. and Shyu K. L. (1992a), Constant reverse thrust activated reorientation of liquid hydrogen with geyser initiation, *J. Spacecraft Rock.* **29**, 279–285.

Hung R. J. and Shyu K. L. (1992b), Excitation of slosh waves associated with low frequency impulsive reverse gravity acceleration of geyser initiation, *Acta Astron.* **26**, 425–433.

Hung R. J. and Shyu K. L. (1992c), Medium frequency impulsive thrust activated liquid hydrogen reorientation with geyser, *J. Propulsion Power* **8**, 987–994.

Hung R. J. and Shyu K. L. (1995), Slosh wave and geyser excitations due to liquid hydrogen shut-off during draining in microgravity, *Acta Astron.* **35**(8), 509–523.

Hung R. J. and Shyu K. L. (1996), Liquid resettlement and slosh wave excitation during fluid reorientation in microgravity, *Acta Astron.* **36**.

Hung R. J., Shyu K. L., and Lee C. C. (1991), Slosh wave excitation associated with high frequency impulsive reverse gravity acceleration of geyser initiation, *Microgravity Quart.* **1**, 125–133.

Hung R. J., Shyu K. L., and Lee C. C. (1992a), Liquid hydrogen slosh waves excited by constant reverse gravity acceleration of geyser initiation, *J. Spacecraft Rock.* **29**(4) 523–528.

Hung R. J., Shyu K. L., and Lee C. C. (1992b), Medium frequency impulsive thrust excited slosh waves during propellant reorientation with geyser, *J. Propulsion and Power* **8**, 778–785.

Hung R. J., Tsao Y. D., and Hong B. B. (1988a), Effect of surface tension on the dynamical behavior of bubble in rotating fluids under low gravity environment, Proc. 1st National Fluid Dyn. Cong., 1399–1409.

Hung R. J., Tsao Y. D., Hong B. B., and Leslie F. W. (1988b), Time dependent dynamical behavior of surface tension on rotating fluids under microgravity environment, *Advances in Space Research* **8**(12), 205–213.

Hung R. J., Tsao Y. D., Hong B. B., and Leslie F. W. (1988c), Surface tension and bubble shapes in a partially filled rotating cylinder under low gravity, AIAA Paper No 88–0455.

Hung R. J., Tsao Y. D., Hong B. B., and Leslie F. W. (1989a), Dynamical behavior of surface tension on rotating fluids in low and microgravity environments, *Int. J. for Microgravity Res. Appl.* **11**, 81–95.

Hung R. J., Tsao Y. D., Hong B. B., and Leslie F. W. (1989b), Axisymmetric bubble profiles in slowly rotating helium Dewar under low and microgravity environment, *Acta Astron.* **19**, 411–426.

Hung R. J., Tsao Y. D., Hong B. B., and Leslie F. W. (1989c), Bubble behaviors in a slowly rotating helium Dewar in Gravity Probe-B spacecraft experiment, *J. Spacecraft Rock.* **26**, 167–172.

Hung R. J., Tsao Y. D., Hong B. B., and Leslie F. W. (1990), Dynamics of surface tension in microgravity environment, *Progress in Aeronaut. and Astronaut.* **127**, 124–150.

Hunt J. B. (1979), *Dynamic Vibration Absorbers*, Letchworth, The Garden City Limited.

Hunt J. N. (1952), Viscous damping waves over an inclined bed in a channel of finite width, *La Houille Blanche* **6**, 836–841.

Hunt J. N. (1964), The viscous damping of gravity waves in shallow water, *La Houille Blanche* **18**, 685–692.

Hunt K. H. and Crossley F. R. E. (1975), Coefficient of restitution interpreted as damping in vibro-impact, *ASME J. Appl. Mech.* **97**, 440–445.

Huntly I. D. (1972), Observations on a spatial-resonance phenomenon, *J. Fluid Mech.* **53**, 209–216.

Huntly I. D. and Smith R. W. (1973), Hysteresis and nonlinear detuning in a spatial-resonance phenomenon, Report No 42, University of Essex, Fluid Mech. Res. Inst.

Hurwitz M. and Melcher J. R. (1968), Dielectrophoretic control of propellant slosh in low gravity, Final Report, NASA-CR-98168, 31 March.

Huther M., Dubois M., and Planeix J. M. (1973), Model studies on the movement of liquid in tanks, *Marine Eng. Rev.*, January 5–16.

Hutton R. E. (1962), An investigation of resonance, nonlinear and nonplanar free surface oscillations of fluid, Ph.D. diss., UCLA, also as NASA TN-D-1870, (1963).

Hutton R. E. (1964), Fluid particle motion during rotary sloshing, *ASME J. Appl. Mech.*, 123–130.

Hutton R. E. and Bhuta P. G. (1965), Propellant slosh loads, TRW Syst., Redondo Beach, CA, EM-15–10 (4388–6001–50000).

Hutton R. E. and Yeh G. C. K. (1964), Dynamic pressure force on a tank of arbitrary shape during liquid sloshing, TRW Syst., Redondo Beach, CA, Memo. 64–9713–6–11.

Hwang C. (1965), Longitudinal sloshing of a liquid in a flexible hemispherical tank, *ASME J. Appl. Mech.*, 665–670.

Hwang J. H., Kim I. S., Soel Y. S., Lee S. C., and Chon Y. K. (1992), Numerical simulation of liquid sloshing in three-dimensional tanks, *Computers and Structures* **44**(1/2) 339–342.

Ibrahim R. A. (1969), Unsteady motion of liquid propellants in moving containers, Master thesis, Department of Aeron. Engrg., Cairo University.

Ibrahim R. A. (1974), Autoparametric interaction in a structure containing a liquid, Ph.D. thesis, Dept. of Mech. Engrg., University of Edinburgh, Scotland.

Ibrahim R. A. (1976), Multiple internal resonance in a structure–liquid system, *J. of Eng. Indust.* **98**(3), 1092–1098.

Ibrahim R. A. (1985), *Parametric Random Vibration*, New York, Wiley.

Ibrahim R. A. (1992), Nonlinear random vibration: Experimental results, *ASME Appl. Mech. Rev.* **44**(10), 423–446, 1991.

Ibrahim R. I. and Barr A. D. S. (1975a), Autoparametric resonance in a structure containing a liquid, part I: Two mode interaction, *J. Sound Vib.* **42**(2), 159–179.

Ibrahim R. I. and Barr A. D. S. (1975b), Autoparametric resonance in a structure containing a liquid, part II: Three mode interaction, *J. Sound Vib.* **42**(2), 181–200.

Ibrahim R. I. and El-Sayad M. A. (1999), Simultaneous parametric and internal resonances in systems involving strong nonlinearities, *J. Sound Vib.* **225**(5), 857–885.

Ibrahim R. I., Gau J. S., and Soundararajan A. (1988), Parametric and autoparametric vibrations of an elevated water tower, part I: Parametric response, *J. Sound Vib.* **121**(3), 413–428.

Ibrahim R. I. and Heinrich R. T. (1988), Experimental investigation of liquid sloshing under parametric random excitation, *ASME J. Appl. Mech.* **55**(2), 467–473.

Ibrahim R. I. and Latorre G. (1988), Experimental investigation of dynamic parameters of viscous fluids in unsteady flow, *ASME J. Pressure Vessel Tech.* **110**(1), 29–35.

Ibrahim R. I. and Li W. (1988), Parametric and autoparametric vibrations of an elevated water tower, part II: Autoparametric response, *J. Sound Vib.* **121**(3), 429–444.

Ibrahim R. A., Pilipchuk V. N., and Ikeda T. (2001), Recent advances in liquid sloshing dynamics, *ASME Appl. Mech. Rev.* **54**(2), 133–199.

Ibrahim R. I. and Soundararajan A. (1983), Nonlinear liquid sloshing under random vertical excitation, *J. Sound Vib.* **93**(1), 119–134.

Ibrahim R. I. and Soundararajan A. (1985), An improved approach for random parametric response of dynamic systems with nonlinear inertia, *Int. J. Nonlinear Mech.* **20**(4), 309–323.

Ibrahim R. I., Soundararajan A., and Heo H. (1985), Stochastic response of nonlinear dynamic systems based on a non-Gaussian closure, *ASME J. Appl. Mech.* **52**, 965–970.

Ifrim M. and Bratu C. (1969), The effect of seismic action on the dynamic behavior of elevated water tanks, Proc. 4th World Conf. *Earthquake Eng.*, Santiago, B-4, 127–142.

Iglesias A. S., Rojas L. P., and Rodriguez R. Z. (2004), Simulation of anti-roll tanks and sloshing types problems with smoothed particle hydrodynamics, *Ocean Engrg.* **31**, 1169–1192.

Iida M. (2000), Numerical analysis of self-induced free surface flow oscillation by fluid dynamics computer code splash-ale, *Nuclear Engrg. Design* **200**, 127–138.

Iida M., Madarame H., Okamoto K., and Fukaya M. (1993), Self-induced oscillation of cylindrical upward jet impinging on the free surface, Proc. Asia-Pacific Vib. Conf. '93, **1**, 260–264.

Ikeda T. (1997), Nonlinear vibrations in an elastic structure subjected to vertical excitation and coupled with liquid sloshing in a rectangular tank, ASME Proc. DET '97, Sept 14–17, Sacramento, CA, 1–12.

Ikeda T. (2003), Nonlinear parametric vibrations of an elastic structure with a rectangular liquid tank, *Nonlin. Dyn.* **33**(1), 43–70.

Ikeda T. and Ibrahim R. A. (2002), Random excitation of a structure interaction with liquid sloshing dynamics, Proc. ASME FSA, AE & FIV + N Symposium: *Sloshing*, Paper Number IMECE 2002–32934.

Ikeda T. and Murakami S. (1999), Nonlinear vibrations of an elastic structure subjected to vertical excitation and coupled with liquid sloshing in a cylindrical tank (resonance with an axi-symmetric mode), ASME 1999 Design Eng. Tech. Conf., Sept 12–16, Las Vegas, NV.

Ikeda T. and Murakami S. (2001), Nonlinear parametric vibrations of an elastic structure subjected to vertical excitation and coupled with liquid sloshing in a cylindrical tank (resonance with an anti-symmetric mode), ASME 2001 Design Eng. Tech. Conf., DETC2001/VIB-21650, Sept 9–12, Pittsburgh, PA.

Ikeda T. and Nakagawa N. (1995), Nonlinear vibrations of a structure caused by water sloshing in a cylindrical tank, ASME PVP **310**, *Fluid–Structure Interaction and Struct. Mechanics*, 63–76.

Ikeda T. and Nakagawa N. (1997), Nonlinear vibrations of a structure caused by water sloshing in a rectangular tank, *J. Sound Vib.* **201**(1), 23–41.

Ikegawa M. (1974), Finite element analysis of fluid motion in a container, in *Finite Element Methods in Flow Problems*, J. T. Oden *et al., eds.*, Huntsville, AL, UAH Press, 737–738.

Ikegawa M. and Washizu K. (1973), Finite element method applied to analysis of flow over a spillway crest, *Int. J. Num. Meth. Engrg.* **6**, 179–189.

Inagaki T., Umeoka T., Fujita K., Nakamura T., Shiraishi T., Kiyokawa T., and Sigiyama Y. (1987), Flow induced vibration of inverted U-shaped piping containing flowing fluid of top entry system for LMFBR, Trans. Intl. Conf. Struct. Mech. in Reactor Technology, Lausanne, F. H. Wittman, ed., **9-E**, 295–302, Balkema.

Ilgamov M. A. (1969), *The Vibrations of Elastic Shells Containing a Liquid and Gas*, Moscow, Nauka (in Russian).

Ilgamov M. A. (1975), Boundary conditions on the shell–liquid interface in the Euler–Lagrange form, Proc. 10th All-Union Conf. on *Theory of Shells and Plates* (in Russian), **2**, Tbilisi, Metsniereba, 170–180.

Irons F. E. (1990), Concerning the nonlinear behavior of the forced spherical pendulum including the dowsing pendulum, *Eur. J. Phys.* **11**, 107–115.

Isaacson M. and Subbiach K. (1991), Earthquake-induced sloshing in a rigid circular tank, *Canadian J. Civil Eng.* **18**, 904–915.

Isermann H. (1970), Die kippgrenze von sattelkraftfahrzeugen mit fester und flussiger ladung, Deutsche Kraftfahrtforschung und Strassenverkehrstechnik, Heft 200.

Ishibashi H. and Hayama S. (1989), Nonlinear response of sloshing based on the shallow water wave theory (2nd report, nonlinear response in rectangular tanks and cylindrical tanks), (in Japanese), *Trans. JSME C* **55**(511), 663–670.

Ishida K. and Kobayashi N. (1985), An effective method of analyzing rocking motion for unanchored cylindrical tanks including uplift, ASME Pressure Vess. Piping Conf., PVP- 98, 7.

Ishida K. and Mieda T. (1989), Calculation method of shear stress in bottom plate of anchored cylindrical tank under seismic loading, *Sloshing and Fluid Structure Vibration*, ASME Pressure Vess. Piping Conf., PVP- **157**, 23–27.

Ishlinskiy A.Yu., Gorbachuk M. L., and Temochenko M.Ye. (1986), On the small rotation stability of axially symmetric bodies supported by strings and having cavities filled with liquid, in *Space Apparatus Dynamics and Outer Space Investigation*, Moscow, Mashinostroyeniye (in Russian).

Ishlinskiy A. Yu. and Temochenko M. Ye. (1960), On small vibrations of the vertical axis of a top with a cavity completely filled with an ideal incompressible fluid, *Prikl. Mekh. i Tekh. Fiz.* **3**.

Ishlinskiy A.Yu., and Temochenko M. Ye. (1966), On the stability of rigid body supported by a string and having ellipsoidal cavity completely filled with ideal incompressible liquid, *PMM* **30**(1), 30–41.

Ivanova A. A., Kozlov V. G., and Evoux P. (2001a), Interfacial dynamics of two-liquid system under horizontal vibrations, *Izvestiya RAN Mekhanika Zhidkosti I Gasa (Fluid Dynamics)* **3**, 28–35.

Ivanova A. A., Kozlov V. G., and Legros J. C. (1997), Mean dynamics of two liquid system in a cavity subjected to rotational vibrations, Proc. Joint 10th Europ. and 6th Russian Symp. Phys. Sci. in Microgravity, St Petersburg, Moscow Inst. Probl. Mech. RAS **1**, 260–273.

Ivanova A. A., Kozlov V. G., Lyubomov D. V., Lyubimova T. P., Merady S., and Roux B. (1998), Structure of averaged flow driven by a vibrating body with a large-curvature edge, *Fluid Dyn.* **33**(5), 656–665.

Ivanova A. A., Kozlov V. G., and Tasjkinov S. I. (2001b), Interfacial dynamics of two-liquid system due to rotational vibrations (experiments), *Izvestiya RAN Mekhanika Zhidkosti I Gasa (Fluid Dynamics)* **6**, 114–121.

Jacobsen L. S. (1949), Impulsive hydrodynamics of fluid inside a cylindrical container, *Bull. Seismol. Soc. Amer.* **39**, 189–202.

Jacobsen L. S. and Ayre R. S. (1951), Hydrodynamic experiments with rigid cylindrical tanks subjected to transient motions, *Bull. Seismol. Soc. Amer.* **41**(4), 313–346.

Jacqmin D. (1990), Stability of an oscillated fluid with a uniform density gradient, *J. Fluid Mech.* **219**, 449.

Jacqmin D. and Duval W. M. B. (1988), Instabilities caused by oscillating acceleration normal to a viscous fluid–fluid interface, *J. Fluid Mech.* **196**, 495.

Jahsman W. E. (1961), The equilibrium shape of the surface of a fluid in a cylindrical tank, in *Developments in Mechanics*, **1**, J. E. Lay and J. E. Malven, eds., New York, Plenum Press, 603–612.

Jain R. K. (1974), Vibration of fluid-filled, orthotropic cylindrical shells, *J. Sound Vib.* **37**, 379–388.

Jeffries H. (1924), Free oscillations of water in an elliptical lake, *Proc. London Math. Soc.* **23**.

Jeffries H. (1951), The surface elevation in cellular convection, *Quart. J. Mech.* **4**, 283.

Jen Y. (1968), Laboratory study of inertia forces on a pile, *ASCE J. Waterways and Harbors Div.* **94**(ww1), 59–76.

Jiang L., Ting C. L., Perlin M., and Schultz W. W. (1996), Moderate and steep Faraday waves: instabilities, modulation and temporal asymmetries, *J. Fluid. Mech.* **329**, 275–307.

Jiang L., Ting C. L., Perlin M., and Schultz W. W. (1998), Period tripling and energy dissipation of breaking standing waves, *J. Fluid Mech.* **369**, 273–299.

Jitu K., Chiba T., and Mieda T. (1994), An experimental study of the effect of liquid viscosity on dynamic response of the fluid-filled co-axial cylinder, *Sloshing, Fluid-Structure Interaction and Struct. Response Due to Shock and Impact Loads*, ASME Pressure Vess. Piping Conf., PVP-**272**, 101–110.

Johnson D. B., Raad P. E., and Chen S. (1994), Simulation of impacts of fluid free surfaces with solid boundaries, *Int. J. Num. Meth. Fluids* **19**, 153–176.

Jones A. F. and Hulme A. (1987), The hydrodynamics of water on deck, *J. Ship Res.* **31**(2), 125–135.

Joo S. W., Schultz W. W., and Messiter A. F. (1990), An analysis of the initial wave maker problem, *J. Fluid Mech.* **214**, 161–183.

Kachigan K. (1955), Forced oscillations of a fluid in a cylindrical tank, Convair, San Diego, CA, Rept. ZU-7–046.

Kagawa K., Fujita N., Nonaka M., Matsuo M., and Watanabe E. (1990), Development of tuned liquid damper for ship vibration (2nd report), *Trans. West.-Japan Soc. Naval Arch.* **81**, 181–188.

Kagawa K., Koukawa H., Fujita K., Zensho Y., and Matsuo M. (1989), Development of tuned liquid damper for ship vibration, *Trans. West.-Japan Soc. Naval Arch.* **78**, 251–258.

Kaliadasis S., Kiyashko A., and Demerkhin E. A. (2003), Marangoni instability of a thin film heated from below by a local heat source, *J. Fluid Mech.* **475**, 377–408.

Kalinichenko V. A., Kravtsov A. V., Rodriguez-Mijangos R., Sekerz-Zen'kovich S. Ya., and Flores-Espinoza R. (2000), Harmonic instability of a low-viscosity liquid in a vertically oscillating vessel, *J. Appl. Mat. Mech. (PMM)* **64**(2), 275–282.

Kalinichenko V. A., Nesterov S. V., Sekerz-Zen'kovich S. Ya., and Chaykovskii A. V. (1995), Experimental study of surface waves with Faraday resonance excitation, *Fluid Dyn.* **30**(1), 101–106.

Kalinowski A. J. (1975), Fluid/struct interactions, in *Shock and Vibration Computer Programs: Reviews and Summaries*, W Pilky and B. Pilky, eds., Washington, DC, The Shock and Vib. Information Center.

Kamke T. and Umeki M. (1990), Nonlinear dynamics of two-mode interaction in parametric excitation of surface waves, *J. Fluid Mech.* **139**, 461–471.

Kamotani Y., Lee J. H., Ostrach S., and Pline A. (1992), An experimental study of oscillatory thermocapillary convection in cylindrical containers, *Phys. Fluid* **4**, 955–962.

Kamotani Y. and Ostrach S. (1987), Design of a thermocapillary flow experiment in reduced gravity, *J. Thermophys. Heat Transfer* **1**(1), 83–89.

Kamotani Y., Ostrach S., and Vargas M. (1984), Oscillatory thermocapillary convection in a simulated floating zone configuration, *J. Crystal Growth* **66**, 83–90.

Kamotani Y. and Platt J. (1992), Effect of free surface shape on combined thermocapillary and natural convection, *J. Thermophys. Heat Transfer* **6**, 721–726.

Kamotani Y., Prasad A., and Ostrach S. (1981), Thermal convection in an enclosure due to vibration aboard spacecraft, *AIAA J.* **19**(4), 511–516.

Kana D. D. (1963a), Longitudinal forced vibration of a partially filled tank, SwRI, Tech. Rept. No 6.

Kana D. D. (1963b), Vertical oscillation of partially full spherical tanks, SwRI NASw-146, April.

Kana D. D. (1964a), Experiments on liquid dynamics in Titan I. I. propellant tanks, SwRI, Final Rept.

Kana D. D. (1964b), A resistive Wheatstone bridge liquid wave height transducer, SwRI, Tech. Rept. No 3.

Kana D. D. (1966), An experimental study of liquid surface oscillations in longitudinally excited compartmented cylindrical and spherical tanks, NASA CR-545.

Kana D. D. (1967), Parametric oscillations of a longitudinally excited cylindrical shell containing liquid, Ph.D. dissentation, University of Texas, Austin. (Also *J. Spacecraft Rock.* **5**(1), 13–21.)

Kana D. D. (1987), A model for nonlinear rotary slosh in propellant tanks, *J. Spacecraft Rock.* **24**(3,4), 169–177.

Kana D. D. (1989), Validated spherical pendulum model for rotary liquid slosh, *J. Spacecraft Rock.* **26**(3), 188–195.

Kana D. D. and Abramson H. N. (1966), Longitudinal vibration of ring stiffened cylindrical shells containing liquids, SwRI, Tech. Rept. No 7.

Kana D. D. and Chu W. H. (1969), Influence of a rigid top mass on the response of a pressurized cylinder containing liquid, *J. Spacecraft Rock.* **6**(2), 103–110.

Kana D. D. and Chu W. H. (1970), Dynamic stability of cylindrical propellant tanks, *J. Spacecraft Rock.* **7**(5), 587–597.

Kana D. D., and Craig R. R. Jr (1968), Parametric oscillations of a longitudinally excited cylindrical shell containing liquid, *J. Spacecraft Rock.* **5**(1), 13–21.

Kana D. D. and Dodge F. T. (1966), Bubble behavior in liquids contained in vertically vibrated tanks, *J. Spacecraft Rock.* **3**(5), 760–763.

Kana D. D. and Fox D. J. (1995), Distinguishing the transition to chaos in a spherical pendulum, *Chaos* **5**(1), 298–310.

Kana D. D. and Gormley, J. E. (1967), Longitudinal vibration of a model space vehicle propellant, *J. Spacecraft Rock.* **4**(12), 1585–1591.

Kana D. D., Glaser R. F., Eulitz W. R., and Abramson H. N. (1968), Longitudinal vibration of spring supported cylindrical membrane shells containing liquid, *J. Spacecraft Rock.* **5**(2), 189–196.

Kana D. D., Ko W. L., Francis P. H., and Nagy A. (1971), Coupling between structure and liquid propellant in a parallel stage shuttle design, *J. Spacecraft Rock.* **9**(11), 789–790.

Kana D. D., Lindholm U. S., and Abramson H. N. (1966), An experimental study of liquid instability in a vibrating elastic tank, *J. Spacecraft Rock.* **3**(8), 1183–1188.

Kana D. D., Unruh J. F., Fey T. A., and Dodge F. T. (1985), Development of simple models for Centaur g-prime/g model tank propellant slosh, Report 06–8190–001, SwRI, 14 June.

Kaneko S. (1992), Dampers for structures utilizing the resonance of fluid in a tank, (in Japanese), *J. Marine Eng. Soc. Japan* **27**(4), 323–329.

Kaneko S. and Ishikawa M. (1999), Modeling of tuned liquid damper with submerged nets, *J. Pressure Vess. Tech.* **121**(3), 334–343.

Kaneko S. and Mizota Y. (1996), Dynamic modeling of deep-water type cylindrical tuned liquid damper with a submerged net, ASME Pressure Vess. Pipes Conf. *Fluid–Struct. Interaction*, PVP- **337**, 19–30.

Kaneko S., Nagakura H., and Nakano R. (1999), Analytical model for self-excited vibration of an overflow flexible plate weir, *J. Pressure Vess. Tech.* **121**(3), 296–303.

Kaneko S. and Yoshida O. (1999), Modeling of deep water-type rectangular tuned liquid damper with submerged nets, *J. Pressure Vess. Tech.* **121**(4), 413–422.

Karapetyan A. V. (1972), The stability of regular precession of a symmetrical rigid body with an ellipsoidal cavity, *Vestnik Mosk Gos Univ, Ser 1: Matem Mekh* **6**, 122–125.

Karapetyan A. V. (2001), The steady motions of a fluid-filled spheroid on a plane with friction, *J. Appl. Mat. Mech.* **65**(4), 631–637.

Karapetyan A. V. and Prokomina O. V. (2000), The stability of permanent rotations of a top with a cavity filled with liquid on a plane with friction, *ASME J. Appl. Math. Mech.* **64**(1), 81–86.

Kareem A. (1993), Liquid tuned mass dampers: past, present and future, Proc. 7th U. S. National Conf. Wind Eng., UCLA.

Kareem A. (1994), The next generation of tuned liquid dampers, 1st World Conf. *Structural Control* **3**, FP5–19–26.

Kareem A. and Sun W. J. (1987), Stochastic response of structures with fluid-containing appendages, *J. Sound Vib.* **119**(3), 398–408.

Kawakami M., Michimoto J., and Kobayashi K. (1977), Prediction of long term whipping vibration stress due to slamming of large full ships in rough seas, *Int. Shipbuild. Prog.* **24**, 83–110.

Kayuk Ya F (1995), The dynamics of a shell of revolution with a viscous incompressible liquid, *Sov. Appl. Mech. (Prikl. Mekh.)* **31**(11), 58–64.

Kayuk Ya F, Ivanov Yu A, and Shekera M. K. (1996), Nonstationary dynamic processes in a cylindrical reservoir containing a viscous liquid, *Sov. Appl. Mech. (Prikl. Mekh.)* **32**(8), 72–79.

Kazarinoff N. D. and Wilkowski J. S. (1989), A numerical study of Marangoni flows in zone-refined silicone crystals, *Phys. Fluids A* **1**, 625–627.

Kazarinoff N. D. and Wilkowski J. S. (1990), Bifurcations of numerically simulated thermocapillary flows in axially symmetric float zones, *Phys. Fluids A* **2**, 1797–1807.

Kelland H. (1839), On waves, *Trans. Royal Soc. Edinburgh* **X**.

Kelly R. E. (1994), The onset and development of thermal convection in fully developed shear flows, *Advanced Appl. Mech.* **30**, 35–112.

Kelly R. E. and Hu H. C. (1993), The onset of Rayleigh–Benard convection in non-planar oscillatory flows, *J. Fluid Mech.* **249**, 373–390.

Kelly R. E. and Or A. (1994), Stabilization of thermocapillary convection by means of nonplanar flow oscillations, NASA 2nd *Microgravity Fluid Physics Conf.*, Cleveland, Ohio, 15–20.

Kelvin L. (1877), On the precessional motion of a liquid, *Nature* **15**, 297–329.

Kelvin L. (1880), On an experimental illustration of minimal energy in vortex motion, *Nature* **23**, 69–70.

Kelvin L. (1882), *Mathematical and Physical Papers*, **4**, Cambridge, Cambridge University Press.

Keolian R. and Rudnick J. A. (1984), The sole of phase locking in quasi-pesiodic surface waves on liquid helium and water, in Frontiers in Physical Acoustics, D Sette, ed.

Keolian R., Turkevich L. A., Putterman S. J., Rudnick I., and Rudnick J. A. (1981), Subharmonic sequences in the Faraday experiment: departures from period doubling, *Phys. Rev. Lett.* **47**, 1133–1136.

Kestin J. and Persen L. N. (1954), Slow oscillations of an infinite plate and an infinite disk in a viscous fluid, Brown University, Report No AF-891/1, Contract AF18 (600)-891.

Keulegan G. H. (1959,) Energy dissipation in standing waves in rectangular basins, *J. Fluid Mech.* **6**(1), 33–50.

Keulegan G. H. and Carpenter L. H. (1958), Forces on cylinders and plates in an oscillating fluid, *J. Res. Nat. Bur. Stand.* **80**, 423–440.

Khabbaz G. R. (1971), Dynamic behavior of liquids in elastic tanks, *AIAA J.* **9**(10), 1985–1990.

Khandelwal R. S. (1980), Some problems in the sloshing of liquid in moving containers, Ph.D. thesis, Indian Inst. Tech., Madras, India, Dept. Aeron. Eng.

Khandelwal R. S. and Nigam N. C. (1981), Parametric instabilities of a liquid free surface in a flexible container under vertical periodic motion, *J. Sound Vib.* **74**(2), 243–249.

Khazina G. G. (1974), Certain stability in the presence of resonances, *J. Appl. Math. Mech. (PMM)* **38**(1), 43–51.

Khosropour R., Cole S. L., and Strayer T. D. (1995), Resonant free surface waves in a rectangular basin, *Wave Motion* **22**(2), 187–199.

Kikura H., Sawada T., and Tanahashi T. (1991), Surface behavior of magnetic fluid sloshing and its frequency response, *Trans. JSME* **57B**, 1629–1634.

Kim Y. (2001), Numerical simulation of sloshing flows with impact load, *Appl. Ocean Res.* **23**(1), 53–62.

Kim Y., Park Y. J., and Lee H. R. (1994), A numerical study on the prediction of sloshing impact pressure, *Trans. Soc. Naval Arch. Korea* **30**(4), 61–73.

Kimura K., Ogura K., Mieda T., Yamamoto K., Eguchi Y., Moriya S., Hagiwara Y., Takakuwa M., Kodama T., and Koike K. (1995), Experimental and analytical studies on the multi-surface sloshing characteristics of a top entry loop type FBR, *Nuc. Eng. Des.* **157**(1/2), 49–63.

Kimura N. and Ohashi H. (1978), Nonlinear sloshing in containers with arbitrary axi-symmetric geometries, 1st report, derivation of governing equations and the behavior of their solutions, *Trans. JSME* **44**, 3024–3033 (in Japanese).

Kimura K. and Takahara H. (1997), Parametric vibration of liquid surface in a rectangular tank subjected to pitching excitation (1st report, asymmetric mode), *Trans. JSME* C **63**(608), 1052–1060 (in Japanese).

Kimura K., Takahara H., Ito T., and Sakata M. (1992), Nonlinear liquid oscillation in a circular cylindrical tank subjected to pitching excitation, *Trans. JSME C* **58**(556), 3564–3571 (in Japanese).

Kimura K., Takahara H., and Ogura H. (1996b), Three-dimensional sloshing analysis in a rectangular tank subjected to pitching excitation, *Trans. JSME C* **62**(596), 1285–1294 (in Japanese).

Kimura, K., Takahara H., and Oomori N., and Sakata M. (1994a), Free vibration analysis of liquid motion in a rotating circular cylindrical tank (elliptic mode), *Trans. JSME C* **60**(570), 374–379 (in Japanese)

Kimura K., Takahara H., and Sakata M. (1993), Sloshing in a rigid tank subjected to pitching excitation (condition of excitation for liquid surface remain planar), *Trans. JSME C* **59**(565), 2606–2612 (in Japanese).

Kimura K., Takahara H., and Sakata M. (1994b), Effects of higher order radial modes upon nonlinear sloshing in a circular cylindrical tank subjected to vertical excitation, *Trans. JSME C* **60**(578), 3259–3267 (in Japanese).

Kimura K., Takahara H., and Sakata M. (1996a), Sloshing in a circular cylindrical tank subjected to pitching excitation (condition for liquid surface remaining plane), *Proc. 73rd JSME Spring Ann. Meeting*, **96**–1(VI), 87–90.

Kimura K., Utsumi M., and Sakata, M (1995), Nonstationary responses of nonlinear liquid motion in a circular cylindrical tank to stochastic earthquake excitation with two-directional components, *Proc. Int. Conf. Nonlinear Stoch. Dyn.*, Hanoi, Vietnam, 159–168.

Kirkham D. (1958), Graphs and formulas of zeros of cross product Bessel functions, *J. Math. Phys.* **36**, 371–377.

Kit E., Shemer L., and Miloh T. (1987), Experimental and theoretical investigation of nonlinear sloshing waves in a rectangular channel, *J. Fluid Mech.* **181**, 265–291.

Kitchens C. W. (1980), Navier–Stokes equations for spin-up in a filled cylinder, *AIAA J.* **18**, 929–934.

Kleefsman T. (2000), Numerical simulation of ship motion stabilization by an activated U-tube anti-roll tank, Master thesis, Department of Mathematics, University of Groningen, the Netherlands.

Kobayashi N. (1980), Impulsive acting on the tank roofs caused by sloshing liquid, Proc. 7th World Conf. *Earthquake Eng.* **5**, 315–322.

Kobayashi N. (1982), Anti sloshing device for oil storage tank using wire rope, *Proc. JSME Kansai* **55**, 78–80 (in Japanese).

Kobayashi N. (1986), *A Study on the Behavior of Cylindrical Liquid Storage Tanks During Earthquakes*, Ph.D. thesis, University of Tokyo, Dept. Mechanical Eng. (in Japanese).

Kobayashi N. and Chiba T. (1983), A comparison of the vibrational characteristics of fluid-coupled cylindrical shells and coaxial cylinder, *Tech. Rept. Ishikawajima-Harima Heavy Industries* **23**, 233–237 (in Japanese).

Kobayashi N., Mieda N., Shibata H., and Shinozaki Y. (1989), A study of the liquid slosh response in horizontal cylindrical tanks, *Trans. ASME, PVT*, **111**(1), 32–38.

Kobayashi N., Mieda T., Jitsu K., and Shibata H. (1995), Sloshing suppression by bulkhead in cylindrical and co-axial cylindrical liquid vessels, *Fluid-Sloshing and Fluid–Structure Interaction*, ASME Pressure Vess. Piping Conf., PVP- **314**, 1–6.

Kobrinsky A. E. and Kobrinsky A. A. (1973), *Vibro-impact Systems*, Moscow, Nauka (in Russian).

Kochin N. E., Kibel I. A., and Rose N. V. (1964), *Theoretical Hydrodynamics*, New York, Interscience Publishers.

Kochupillai J., Ganesan N., and Padmanabhan C. (2002), A semi-analytical coupled finite element formulation for shells conveying fluids, *Comput. Struct.* **80**, 271–286.

Koga T. and Tsushima M. (1990), Breathing vibrations of a liquid-filled circular cylindrical shell, *Int. J. Solids Struct.* **26**(9/10), 1005–1015.

Koh C. G., Mahatma S., and Wang C. M. (1994), Theoretical and experimental studies on rectangular liquid dampers with arbitrary excitations, *Earthquake Eng. Struct. Dyn.* **23**, 17–31.

Koh C. G., Mahatma S., and Wang C. M. (1995), Reduction of structural vibrations by multiple-mode liquid dampers, *Engrg. Struct.* **17**(2), 122–128.

Koide T. and Goto Y. (1988), Fundamental study on feedback control system regarding sloshing in storage tank, Proc. 9th World Conf. *Earthquake Engineering* **V**, 883–888, Tokyo Kyoto, Japan.

Koleshov V. B. and Shveiko Yu Yu (1971), Asymmetric oscillations of cylindrical shells partially filled with a liquid, Izv. Academii Nauk. SSSR, *Mekh. Tverdogo Tela.* **3**, 126–136.

Kolesinkov K. S. (1971), *Longitudinal Vibrations of a Rocket with a Liquid Propellant Engine*, Moscow, Mashinostroyeniye (in Russian).

Kolesnikov N. N. (1962), On the stability of an unconstrained rigid body with a cavity filled with an incompressible viscous fluid, *J. Appl. Math. Mech. (PMM)* **26**(4), 914–923.

Kollmann F. G. (1962), Experimentelle und theoretische untersuchungen über die kritischen drehzahlen flussigkeitsgefüllter Hohlkörper, Forsch Ing Wes H.4–5, Bd 28, 115–153.

Komarenko A. N., Lukovskii I A (1974), Nonlinear oscillations of a liquid in a cylindrical vessel under low gravity, in *Dynamics and Stability of Multidimensional Systems*, Inst. Metem. Akad. Nauk. Ukraine, Kiev, SSSR, 86–97.

Komatsu K. (1987), Nonlinear slosh analysis of liquid in tanks with arbitrary geometries, *Int. J. Nonlin. Mech.* **22**(3), 193–207.

Komissarova G. L. (1990), Solution of the problem on wave propagation in a cylinder with a liquid, *Soviet Appl. Mech. (Prikl. Mekh.)* **26**(8), 25–29.

Komissarova G. L. (2002), Propagation of normal waves through a fluid contained in a thin-walled cylinder, *Int. Appl. Mech.* **38**(1), 103–112.

Kondo H. (1981a), Axi-symmetric free vibration analysis of a circular cylindrical tank, *Bull. JSME* **24**(187), 215–221.

Kondo H. (1981b), Response of a circular cylindrical tank subjected to vertical excitation, *Theor. and Appl. Mech.* **31**, 311–319.

Kondo H. (1989), Seismic response of a reactor vessel subjected to horizontal excitation, *Sloshing and Fluid Structure Vibration*, ASME Pressure Vess. Piping Conf., PVP- **157**, 55–60.

Kondo H., Yamamoto S., and Sasaki Y. (1989), Fluid–structure interaction analysis program for axisymmetric structures, *Sloshing and Fluid Structure Vibration*, ASME Pressure Vess. Piping Conf., PVP- **157**, 179–185.

Konstantinov Akh., and Lapshin A. L. (2001), Stability of the oscillating free liquid surface in a cylindrical cavity under random perturbations, *Int. Appl. Mech.* **37**(10), 1332–1340.

Konstantinov Akh., Mikityuk Yu I, and Pal'ko L. S. (1978), Study of the dynamic stability of a liquid in a cylindrical cavity, in *Dynamics of Elastic Systems with Continuous-Discrete Parameters*, Kiev, Naukova Dumka, 69–72 (in Russian).

Korobkin A. (1997), *Liquid–Solid Impact*, Siberian Branch of the Russian Academy, Novosibrisk.

Korobkin A. A. and Pukhachov V. V. (1988), Initial stage of water impact, *Annu. Rev. Fluid Mech.* **20**, 159–185.

Kornecki A. (1983), Dynamics of a mobile tank partially filled with liquid: equations of motion and their linearization, *SM Archives* **8**, 217–241.

Koschmieder E. L. (1967), On convection under air surface, *J. Fluid Mech.* **30**, 9–15.

Koval L. R. (1968), A zero slosh problem, *J. Spacecraft Rock.* **5**(7), 865–868.

Koval R. and Bhuta P. G. (1968), A direct solution for capillary-gravity waves in a cylindrical tank, *Devel. in Mechanics*, 3(2): *Dynamics and Fluid Mechanics*, T. C. Huang and M. N. Johnson Jr, eds., 277–290.

Koval'chuk P. S. (1990), The dynamic instability of thin-walled cylindrical shells carrying a liquid, Abstracts of papers read at the All-Union Conf. on *the Nonlinear Vibrations of Mechanical Systems*, Gor'kii, 42 (in Russian).

Koval'chuk P. S. and Boyarshina L. G. (1998), Spatial self-oscillations of a solid with a cylindrical cavity partially filled with liquid, *Int. Appl. Mech.* **34**(4), 377–384.

Koval'chuk P. S. and Filin V. G. (2003a), Circumferential traveling waves in filled cylindrical shells, *Int. Appl. Mech.* **39**(2), 192–196.

Koval'chuk P. S., and Filin V. G. (2003b), On modes of flexural vibrations of initially bent cylindrical shells partially filled with a liquid, *Int. Appl. Mech.* **39**(4), 464–471.

Koval'chuk P. S. and Kholopova V. V. (1997), Analysis of energy interchange between the natural modes of a liquid in a cylindrical tank, *Dokl. Akad. Nauk.* Ukrain., Ser. A, No 1, 36–41.

Koval'chuk P. S. and Kholopova V. V. (2001), Analysis of the nonstationary slow passage of a cylindrical vessel partially filled with a liquid through resonant zones, *Int. Appl. Mech.* **37**(6), 790–797.

Koval'chuk P. S., Krishtalev V. V., and Podchasov N. P. (1997), Analysis of the energy interchange between the natural modes of a liquid in a cylindrical vessel attached to elastic supports, *Sov. Appl. Mech. (Prikl. Mekh.)* **33**(4), 54–61.

Koval'chuk P. S. and Kruk L. A. (2000), Nonlinear energy interchange between the natural modes of freely vibrating circular cylindrical shells filled with liquid, *Int. Appl. Mech.* **36**(1), 103–110.

Koval'chuk P. S. and Kruk L. A. (2002), Forced nonlinear vibrations of cylindrical shells filled with a liquid, *Int. Appl. Mech.* **38**(11), 1388–1393.

Koval'chuk P. S. and Kubenko V. D. (1991), Interaction of vibrating cylindrical shells with its liquid content, in *Dynamics of Bodies Interacting with a Medium*, A. N. Guz, ed., Kiev, Naukova Dumka, 168–214.

Koval'chuk P. S. and Pavlovskii V. S. (1992), The wave modes of the nonlinear deformation of liquid-containing shells under radial dynamic pressure, *Dokl. Akad. Nauk. Ukrain., Ser. A*, No 11, 24–28.

Koval'chuk P. S. and Pavlovskii V. S. (1993), The nonlinear wave deformation of cylindrical modes of liquid-containing shells under impulse radial pressure, *Sov. Appl. Mech. (Prikl. Mekh.)* **29**(12), 43–50.

Koval'chuk P. S. and Pavlovskii V. S. (1994), Nonlinear wave processes in cylindrical liquid-containing shells under short-term dynamic loading, in *Technical–Physical Problems of New Engineering*: Abstract Book, Moscow, Izd MVTU (in Russian).

Koval'chuk P. S., Pavlovskii V. S., and Filin V. G. (1993), The nonlinear dynamics of cylindrical bodies carrying liquid masses and subject to periodic and impulsive radial perturbation, in *Fundamental and Applied Problems of Space Exploration*, Moscow, Zhitomir, 23 (in Russian).

Koval'chuk P. S. Pavlovskii V. S., and V. G. Filin (1994), Analysis of resonance regimes of the wave deformation of cylindrical shells carrying a liquid, *Sov. Appl. Mech. (Prikl. Mekh.)* **30**(2), 49–54.

Koval'chuk P. S. Pavlovskii V. S., and V. G. Filin (2000), Stability analysis of resonance regimes of the nonlinear vibrations of liquid-containing cylindrical shells under longitudinally transversal periodic excitation, *Dokl. NAN Ukrain.*, No **3**, 31–34.

Koval'chuk P. S. and Podchasov N. P. (1988), Nonlinear flexural waves in cylindrical shells with periodic excitation, *Soviet Appl. Mech.* **24**(11), 1086–1090.

Koval'chuk P. S. and Podchasov N. P. (1996), Nonstationary wave motions of a liquid in a vibrating cylindrical vessel when passing through a zone of resonance, *Int. Appl. Mech. (Prikladnaya Mekhanika)* **32**(1), 30–34.

Koval'chuk P. S. and Podchasov N. P. (1999), Simulation of the nonlinear multimodal vibrations of thin cylindrical shells with close fundamental frequencies, Int. Conf. *Modeling and Investigation of System Stability*, Kiev, May, 40.

Koval'chuk P. S., Podchasov N. P., and Kholopova V. V. (2002), Periodic modes in the forced nonlinear vibrations of filled cylindrical shells with an imperfection, *Int. Appl. Mech.* **38**(6), 716–722.

Koval'chuk P. S., Puchka G. N., and Kholopova V. V. (1987), The nonlinear resonance vibrations of a liquid in a cylindrical vessel under longitudinal and transverse periodic excitation, *Soviet Appl. Mech. (Prikl. Mekh.)* **23**(8), 95–100.

Koval'chuk P. S., Puchka, G.N., and Kholopova V. V. (1989), Nonlinear wave motions in axisymmetric cavities subjected to vibrations, Izv. Akad. Nauk. SSSR Mekh. Zhidk., Gaza **3**, 120–125.

Koval'chuk P. S. Puchka, G. N., and Kholopova V. V. (1999), Nonstationary wave motion of a liquid in longitudinally oscillating cylindrical vessel, *Dokl. NAN Ukrain.*, No **11**, 26–31.

Krasnopolskaya T. S. and Lavrov K. A. (1988), The nonlinear vibrations of a liquid-containing cylindrical shell under limited excitation, *Soviet Appl. Mech. (Prikl. Mekh.)* **24**(11), 50–58.

Krasnopolskaya T. S. and Podchasov N. P. (1992a), Waves in a liquid between two coaxial cylindrical shells due to the vibrations of the inner cylinder, *Soviet Appl. Mech. (Prikl. Mekh.)* **28**(3), 61–70.

Krasnopolskaya T. S. and Podchasov N. P. (1992b), The forced vibrations of a liquid between two cylinders excited by the vibrations of the inner shell, *Soviet Appl. Mech. (Prikl. Mekh.)* **28**(4), 42–48.

Krasnopolskaya T. S. and Podchasov N. P. (1993), Resonances and chaos in nonaxisymmetric dynamic processes in hydroelastic systems, *Soviet Appl. Mech. (Prikl. Mekh.)* **29**(12), 72–77.

Krasnopolskaya T. S. and Shvets A. Yu. (1990), Regular and chaotic surface waves in a liquid in a cylindrical tank, *Int. Appl. Mech., Priklad. Mekh.* **26**(8), 787–794.

Krasnopolskaya T. S. and Shvets A. Yu. (1992), Properties of chaotic oscillations of a liquid in cylindrical tanks, *Int. Appl. Mech., Priklad. Mekh.* **28**(6), 52–58.

Krasnopolskaya T. S. and Shvets A. Yu. (1993), Parametric resonance in the system: liquid in tank + electric motor, *Int. Appl. Mech., Priklad. Mekh.* **29**(9), 722–730.

Kravtsov A. V. and Sekerzhi-Zenkovich S. Y. (1993), Parametric excitation of the oscillations of a viscous two-layer fluid in a closed vessel, *Zh. Vychisl. Mat. Fiz.* **33**(4), 611–619.

Kravtsov A. V. and Sekerzhi-Zenkovich S. Y. (1996), Parametric excitation of the oscillations of a viscous continuously stratified fluid in a closed vessel, *J. Appl. Math. Mech. (PMM)* **60**(3), 449–454.

Krein S. G. (1964), On the oscillations of a viscous fluid in a vessel, *Soviet Math. Doklady.* **5**(6), 1467–1471.

Krein S. G. and Kan N. Z. (1969a), The problem of small motions of a body with a cavity partially filled with a viscous fluid, *J. Appl. Math. Mech. (PMM)* **30**(1), 110–117.

Krein S. G., and Kan N. Z. (1969b), Asymptotic method in the problem of oscillations of a strongly viscous fluid, *J. Appl. Math. Mech. (PMM)* **33**(3), 442–450.

Krein S. G. and Laptev G. I. (1968), On the problem of motion of a viscous fluid in an open vessel, *Funkts. Analiz.* **2**(1), 38–47.

Krein S. G. and Moiseev N. N. (1957), On vibrations of a rigid body containing liquid having free surface, *J. Appl. Math. Mech. (PMM)* **21**, 169–174.

Kreis A. and Klein M. (1991), On the analysis of liquid filled free–free tanks, Proc. Int. Conf. *Spacecraft Structures and Mechanical Testing*, Noordwijk, the Netherlands, 437–442.

Krishnamurti R. (1968), Finite amplitude convection with changing mean temperature, *J. Fluid Mech.* **33**, 445–457.

Kruse H. P., Mahalov A., and Marsden J. E. (1999), On the Hamiltonian structure and three-dimensional instabilities of rotating liquid bridges, *Fluid Dyn. Res.* **24**, 37–59.

Krushinskaya S. I. (1965), Vibrations of a heavy viscous fluid in a moving vessel, *Zh. Vychisl. Matem. i Matem. Fiz.* **3**.

Kubenko V. D. (1979), *The Nonstationary Interaction of Structural Elements with a Medium*, (in Russian) Kiev, Naukova Dumka.

Kubenko V. D. and Koval'chuk P. S. (1987), The nonlinear vibrations and wave modes of deformation of cylindrical shells filled with a liquid, 11th Int. Conf on *Nonlinear Oscillations*, Budapest, 17–23 August, 44.

Kubenko V. D. and Koval'chuk P. S. (2000), Nonlinear problems of the dynamics of elastic shells partially filled with a liquid, *Int. Appl. Mech.* **36**(4), 421–448.

Kubenko V. D., Koval'chuk P. S., and Boyarshina L. G. (1992), *Nonlinear Dynamics of Axisymmetric Bodies Carrying a Liquid*, Kiev, Naukova Dumka (in Russian).

Kubenko V. D., Koval'chuk P. S., and Krasnopo'skaya T. S. (1984), *Nonlinear Modal Interaction of Flexible Vibrations of Cylindrical Shells*, Kiev, Naukova Dumka.

Kubenko V. D., Koval'chuk P. S., and Kruk L. A. (2003), On multimode nonlinear vibrations of filled cylindrical shells, *Int. Appl. Mech.* **39**(1), 85–92.

Kubenko V. D., Koval'chuk P. S., and Podchsov N. P. (1989), *Nonlinear Vibrations of Cylindrical Shells*, (in Russian) Kiev, Vyshcha Shkola.

Kubenko V. D. and Kruk L. A. (1994), On the oscillations of an incompressible liquid in an infinite cylindrical shell containing a spherical body oscillating along the shell axis, *Soviet Appl. Mech. (Prikl. Mekh.)* **30**(4), 31–37.

Kubenko V. D. and Lakiza V. D. (1996), Dynamic behavior of a gas–liquid media in ellipsoidal shells subjected to vibration, *Int. Appl. Mech.* **32**(2), 89–94.

Kubenko V. D. and Lakiza V. D. (1998), Vibrational resonance modes of motion of gas–liquid media in shell structures under polyharmonic excitation, *Int. Appl. Mech.* **34**(1), 54–59.

Kubenko V. D. and Lakiza V. D. (1999), Dynamic behavior of liquids and gases in conical shells in a vibrational force field, *Int. Appl. Mech.* **35**(2), 128–133.

Kubenko V. D., Lakiza V. D., Pavolvskii V. S., and Pelykh N. A. (1988), *Dynamics of Elastic Gas–Liquid Systems under Vibratory Actions*, Kiev, Naukova (in Russian).

Kubenko V. D. and Savin V. A. (2000), Construction of the velocity potential of a semi-infinite liquid column excited by a spherical calotte oscillating inside a cylindrical cavity, *Int. Appl. Mech.* **36**(3), 358–366.

Kubenk V. D. and Savin V. A. (2002), Dynamics of a semi-infinite cylindrical shell with a liquid containing a vibrating spherical segment, *Int. Appl. Mech.* **38**(3), 974–982.

Kuhlmann H. C. and Rath H. J. (1993a), Hydrodynamic instabilities in cylindrical thermocapillary liquid bridges, *J. Fluid Mech.* **247**, 247–274.

Kuhlmann H. C. and Rath H. J. (1993b), On the interpretation of phase measurements of oscillatory convection in liquid bridges, *Phys. Fluids A* **5**, 2117–2120.

Kuipers M. (1964), On the stability of a flexible mounted rotating cylinder partially filled with liquid, *Appl. Sci. Res. A* **13**(2/3), 121–137.

Kuipers M. (1966), Die instabilität eines federnd gelagerten und teilweise mit flussigkeit gefüllten umlaufenden hohlzylinders, *Forsch. Ing. Wes.* **32**(6), 194–195.

Kukner A. and Baykal M. A. (1999), Wave sloshing in corrugated bottom tanks with two-dimensional flow, *Ocean Eng.* **26**(8), 703–712.

Kumar K. (1996), Linear theory of Faraday instability in viscous liquids, *Proc. Royal Soc. London A* **452**, 1113–1126.

Kumar K. and Bajaj K. M. S. (1995), Competing patterns in the Faraday experiment, *Phys. Revs. E* **52**, R4606–4609.

Kumar K. and Tuckerman L. S. (1994), Parametric instability of the interface between two fluids, *J. Fluid Mech.* **279**, 49–68.

Kumar R. (1971), Flexural vibrations of fluid-filled circular cylindrical shells, *Acoustica* **24**(3), 137–146.

Kumar R. (1972), Dispersion of axially symmetric waves in empty and fluid-filled circular cylindrical shells, *Acoustica* **25**(6), 317–329.

Kumok Yu.Z. and Novgorodtseva L. Z. (1965), Small free oscillations of a rotating cylinder partially filled with a liquid, *Prikl. Mekh.* **1**(12), 87–94.

Kunitsyn A. L. (1971), On the stability of pure imaginary roots in the critical case in the presence of intrinsic resonance, *J. Appl. Math. Mech. (PMM)* **35**, 164–167.

Kunitsyn A. L. and Markeev A. P. (1979), Stability in resonance cases, in *Achievements of Science and Technology, General Mechanics* **4**, Moscow, Viniti.

Kunitsyn A. L. and Matveyev M. V. (1991), The stability of a class of reversible systems, *J. Appl. Math. Mech. (PMM)* **55**(6), 780–788.

Kunitsyn A. L. and Muratov A. S. (1993), The stability of a class of quasi-autonomous periodic systems with internal resonance, *J. Appl. Math. Mech. (PMM)* **57**(2), 247–255.

Kunitsyn A. L. and Perezhogin A. A. (1985), The stability of neutral systems in the case of a multiple fourth-order resonance, *J. Appl. Math. Mech. (PMM)* **49**(1), 53–57.

Kunitsyn A. L. and Tuyakbayev A. A. (1992), The stability of Hamiltonian systems in the case of a multiple fourth-order resonance, *J. Appl. Math. Mech. (PMM)* **56**(4), 572–576.

Kurdolli A. and Gollub J. P. (1996a), Patterns and spatio-temporal chaos in parametrically forced surface waves: a systematic survey at large aspect ratio, *Physica D* **97**, 133–154.

Kurdolli A. and Gollub J. P. (1996b), Localized spatio-temporal chaos in surface waves, *Phys. Revs. E* **54**, R1052–1055.

Kurdolli A., Pier B., and Gollub J. P. (1998), Superlattice patterns in surface waves, *Physica D* **123**, 99–111.

Kurihara C., Masuko Y., and Sakurai A. (1992), Sloshing impact pressure in roofed liquid tanks, *Fluid–Structure Vibration and Sloshing, ASME Pressure Vess. Piping Conf.*, PVP- **232**, 19–24.

Kuttler J. R. and Sigillito V. G. (1969), Lower bounds for sloshing frequencies, *Quart. Appl. Math.* **27**(3), 405–408.

Kuttler J. R., and Sigillito V. G. (1978), Bounding eigenvalues of elliptic operators, *SIAM J. Math. Anal.* **9**, 768–773.

Kuttler J. R. and Sigillito V. G. (1984), Sloshing of liquids in cylindrical tanks, *AIAA J.* **22**(2), 309–311.

Kuz'ma V. M. and Kholopova V. V. (1983), Oscillations of the free surface of a liquid in a cylindrical container with longitudinal vibrations, *Prikl. Mekh.* **19**(3), 249–253.

Kwak M. K. (1991), Vibration of circular plates in contact with water, *ASME J. Appl. Mech.* **58**, 480–483.

Kwak M. K. (1997), Hydroelastic vibration of circular plates, *J. Sound Vib.* **201**, 293–303.

Kwok K. C. S. (1984), Damping increase in building with tuned mass damper, *ASCE J. Eng. Mech.* **110**, 1645–1649.

Kwok K. C. S., Samali B., and Xu Y. L. (1991), Control of wind induced vibration of tall structures by optimized tuned liquid column dampers, Proc. Asia-Pacific Conf. *Computational Mechanics*, Hong Kong, 569–574.

Labus T. L. (1969), Natural frequency of liquids in annular cylinders under low gravitational conditions, NASA TN D-5412.

Lacovic R. F., Yeh F. C., Szabo S. V., Brun R. J., Stofan A. J., and Berns J. A. (1968), Management of cryogenic propellants in a full-scale orbiting space vehicle, NASA TN D-4571.

Lai C. L. (1990), Unsteady thermocapillary flows and free surface oscillations in reduced gravity environments, *Acta Astron.* **21**, 171–181.

Lakis A. A. and Laveau A. (1991), Nonlinear dynamic analysis of anisotropic cylindrical shells containing a flowing fluid, *Int. J. Solids & Struct.* **28**, 1079–1094.

Lakis A. A. and Sinno M. (1992), Free vibrations of axisymmetric and beam-like cylindrical shells, partially filled with liquid, *Int. J. for Num. Methods in Engrg.* **33**, 235–268.

Lakis A. A. and Païdoussis M. P. (1971), Free vibrations of cylindrical shells partially filled with liquid, *J. Sound Vib.* **19**(1), 1–15.

Lakiza V. D. (1995), Experimental studies of the dynamic behavior of a gas–liquid media in spherical shells under the action of vibration, *Int. Appl. Mech.* **31**(1), 61–65.

Lamb H. (1945), *Hydrodynamics*, Cambridge, Cambridge University Press.

Landau L. D. and Lifshitz E. M. (1959), *Fluid Mechanics*, Reading, MA, Addison.

Langbein D. (1992), Stability of liquid bridges between parallel plates, *Microgravity Sci. Technol.* V/**1**, 2.

Langbein D., Falk F., and Grossbach R. (1995), Oscillations of liquid columns under microgravity, *Adv. Space Res.* **16**, 23–26.

Langner C. G. (1963), A preliminary analysis for optimum design of ring and vertical wall antislosh baffles, SwRI, TR 7, April.

Lappa M. and Savino R. (1999), Parallel solution of the three-dimensional Marangoni flow instabilities in liquid bridges, *Int. J. Num. Meth. Fluids* **31**(5).

Lappa M., Savino R., and Monti R. (2001), Three-dimensional numerical simulation of Marangoni instabilities in liquid bridges: Influence of geometrical aspect ratio, *Int. J. Heat Mass Transfer* **44**(10), 1983–2003.

Larraza A. P. and Putterman S. (1984), Theory of non-propagating surface-wave solitons, *J. Fluid Mech.* **148**, 443–449.

Lau D. T. and Zeng X. (1992), Hydrodynamic forces in uplifting cylindrical tanks, *Fluid–Structure Vibration and Sloshing, ASME Pressure Vess. Piping Conf.*, PVP- **232**, 39–44.

Lavalley R., Amberg G., and Alfredsson H. (2001), Experimental and numerical investigation of nonlinear thermocapillary oscillations in an annular geometry, *Eur. J. Mech. B-Fluids* **20**, 771–797.

Laverón-Simavilla A. and Perales J. M. (1995), Equilibrium shapes of non-axisymmetric liquid bridges of arbitrary volume in gravitational fields and their potential energy, *Phys. Fluids* **7**, 1204–1213.

Lawrence H. R., Wang C. J., and Reddy R. B. (1958), Variational solution of fuel sloshing modes, *Jet Propul., Publication of Amer. Rocket Soc.* **28**(11), 728–736.

Lee B. S. and Vassalos D. (1996), Investigation into the stabilization effects of anti-roll tanks with flow obstructions, *Int. Shipbuilding Progress* **43**, 70–88.

Lee C. P., Anilkumar A. V., and Wang T. G. (1996), Streaming generated in a liquid bridge due to nonlinear oscillations driven by the vibrations of an endwall, *Phys. Fluids* **8**(12), 3234–3246.

Lee D. Y. and Choi H. S. (1999), Study of sloshing in cargo tanks including hydroelastic effects, *J. Marine Sci. Technol.* **4**(1), 27–34.

Lee S. C. and Reddy D. V. (1982), Frequency tuning of offshore platforms by liquid sloshing, *Appl. Ocean Res.* **4**, 226–231.

Lee S. H., *et al.* (1995), Simulation of 3-D sloshing and structural response in ship's tanks taking account of fluid–structure interaction, *SNAME Trans.* **103**, 321–342.

Lee T. H., Zhou Z., and Cao Y. (2002), Numerical simulations of hydraulic jumps in water sloshing and water impacting, *ASME J. Fluids Engrg.* **124**(1), 215–226.

Leissa A. W. (1973), Vibration of shells, NASA SP-288.

Leonard H. W. and Walton W. C. Jr (1961), An investigation of the natural frequencies and mode shapes of liquids in oblate spheroidal tanks, NASA TN-D-904.

Leonov G. A. and Morozov A. V. (1988), On the global stability of the forced motions of a liquid within an ellipsoid, *J. Appl. Math. Mech. (PMM)* **52**(1), 133–135.

Lepelletier T. G. and Raichlen F. (1988), Nonlinear oscillation in rectangular tanks, *ASCE J. Eng Mech* **114**(1), 1–23.

Leroy J. (1963), On breathing vibrations of thin cylinders partially full of liquid, *Comptes Rendus* **257**(18), 2607–2609 (in French).

Leslie F. W. (1985), Measurements of rotating bubble shapes in a low gravity environment, *J. Fluid Mech.* **161**, 269–279.

Levenstam M. and Amberg G. (1995), Hydrodynamic instability of thermocapillary flow in a half-zone, *J. Fluid Mech.* **297**, 357–372.

Levich V. G. and Krylov V. S. (1969), Surface tension driven phenomena, *Ann. Rev. Fluid Mech.* **1**, 293–316.

Levin E. (1957), Conical sloshing, Memo. PA/M-553/1, STL.

Levin E. (1963), Oscillations of a fluid in rectilinear conical containers, *AIAA J.* **1**(6), 1447.

Levine N. B., Krainman H., and Green J. (1966), Positive expulsion bladders for storable propellants, *AFML-TR-65-379*.

Levleva O. B. (1964), Small oscillations of a pendulum having a spherical cavity filled with a viscous fluid, *J. Appl. Math. Mech. (PMM)* **28**(6), 1359–1361.

Levleva O. B. (1966), Oscillations of a body filled with a viscous fluid, *J. Appl. Mech. Tech. Phys.* (English translation) **7**(6), 18–22.

Levy A. M., Chin J. H., Donaldson J. O., Gallagher L. W., Harper E. Y., Hurd S. E., and Satterlee H. M. (1964), Analytical and experimental study of liquid orientation and stratification in standard and reduced gravity fields, Lockheed Missiles & Space Co., Rept. No 2-05-64-1.

Lewin L., Burger M., and Souli M. (1997), Simulation of dynamic fuel sloshing using an explicit finite element approach, *ASME Symp. Adv. Analytical, Experimental and Computational Technologies in Fluids, Structures, Transients and Natural Hazards*, PVP- **355**, 103–111.

Lewis D. J. (1950), The instability of liquid surfaces when accelerated in a direction perpendicular to their planes, *Proc. Royal Soc. (London) A* **202**, 81–96.

Lewison B. A. (1975), Optimum design of passive roll stabilizer tanks, *Royal Inst. Naval Archit.* **78**, 31–45.

Li L. (1983a), On the stability of the rotational motion of a rigid body having a liquid-filled cavity under finite initial disturbance, *Appl. Math. Mech.* **4**, 667–680.

Li L. (1983b), On the global stability of rotational motion and the qualitative analysis of the behavior of distributed motion of a rigid body having a liquid-filled cavity, *Appl. Math. Mech.* **4**, 865–874.

Li W. (1985), Response of a nonlinear two-degree-of-freedom system to a horizontal harmonic excitation, Master thesis, Texas Tech University, Lubbock, TX.

Lichter S. and Shemer L. (1986), Experiments on nonlinear cross-waves, *Phys. Fluids* **29**, 3971–3975.

Lichtenberg G. (1982), Vibration of an elastically mounted spinning rotor partially filled with liquid, *J. Mech. Des.* **104**(2) 389–396.

Lidström M. (1976), Road tankers overturning limit: the steady-state case, (in Swedish) National Road and Traffic Research Institute, Linköping, Sweden, Working Paper 1976–12–17.

Lidström M. (1977), Road tanker overturning: with and without longitudinal baffles, (in Swedish) National Road and Traffic Research Institute, Linköping, Sweden, Report No 115.

Liepmann H. W. and Lagina G. A. (1984), Nonlinear interaction in the fluid mechanics of helium II, *Annu. Rev. Fluid Mech.* **16**, 139–177.

Lifshitz R. and Petrich D. M. (1997), Theoretical model for Faraday waves with multiple-frequency forcing, *Phys. Rev. Lett.* **79**, 1261–1264.

Lighthill M. J. (1966), Dynamics of rotating fluids: A Survey, *J. Fluid Mech.* **26**, 411–431.

Limarchenko O. S. (1978a), Variational formulation of the problem on the motion of a tank with fluid, *Dopovidi Akademii Nauk. Ukrains'koi RSR*, A **10**, 903–907 (in Ukranian).

Limarchenko O. S. (1978b), Direct method for solution of nonlinear dynamic problems for a tank with fluid, *Dopovidi Akademii Nauk. Ukrains'koi RSR*, A **11**, 999–1002 (in Ukranian).

Limarchenko O. S. (1980), Variational method investigation of problems of nonlinear dynamics of a reservoir with a liquid, *Soviet Appl. Mech.* **16**(1), 74–79.

Limarchenko O. S. (1981), Effect of capillarity on the dynamics of a container liquid system, *Soviet Appl. Mech.* **17**(6), 601–604.

Limarchenko O. S. (1983a), Direct method of solving problems on the combined spatial motions of a body fluid system, *Soviet Appl. Mech.* **19**(8), 715–721.

Limarchenko O. S. (1983b), Application of a variational method to the solution of nonlinear problems of the dynamics of combined motions of a tank with fluid, *Soviet Appl. Mech.* **19**(11), 1021–1025.

Limarchenko O. S. (1987), Application of direct methods to the study of applied problems of nonlinear dynamics of bodies partially filled by a liquid, Inst. Mathematics, Ukraine, Academy of Sciences, No 87.8, Kiev.

Limarchenko O. S. (1991), Modeling of dynamics of structures hosting a liquid with a free surface in cavities of revolution, Inst. Math., Ukraine, Academy of Sciences, No 91.16, Kiev.

Limarchenko O. S. (1995), Non-steady rotation of a cylindrical storage tank partially filled with a viscose fluid, *Int. Appl. Mech. (Prikl. Mekh.)* **31**(5), 406–411.

Limarchenko O. S., Lukovskii I. A., and Timokha A. N. (1992), Nonlinear models in applied dynamics problems of bodies with a free surface liquid, *Prikl. Mekh.* **28**(11), 757–763.

Limarchenko O. S. and Yasinskii V. V. (1996), *Nonlinear Dynamics of Constructions with a Fluid*, Kiev Polytechnic University (in Russian).

Lin C. C. (1954), On a perturbation theory based on the method of characteristics, *J. Math. Phys.* **33**, 117.

Lin J. (1994), Measurement of free surface deformation in oscillatory flow, M.Sc., thesis, Case Western Reserve University, Cleveland, OH.

Lin J. D. and Howard L. N. (1960), Nonlinear standing waves in a rectangular tank due to forced oscillations, MIT Hydrodynamics Lab., TR 44.

Lin J., Kamotani Y., and Ostrach S. (1995), An experimental study of free surface deformation in oscillatory flow, *Acta Astronaut.* **35**(8), 525–536.

Lin T. C. and Morgan G. W. (1956), Wave propagation through fluid contained in a cylindrical elastic shell, *J. Acoust. Soc. Amer.* **28**(6), 1165–1176.

Lin Y. K. (1967), *Probabilistic Theory of Struct Dynamics*, New York, McGraw-Hill.

Lindholm U. S., Chu W. H., Kana D. D., and Abramson H. N. (1963), Bending vibrations of a circular cylindrical shell containing an internal liquid with a free surface, *AIAA J.* **1**(9), 2092–2099.

Lindholm U. S., Kana D. D., and Abramson H. N. (1962a), Breathing vibrations of a circular cylindrical shell with an internal liquid, *J. of Aero./Space Sciences* **29**(9), 1052–1059.

Lindholm U. S., Kana D. D., Chu W. H., and Abramson H. N. (1962b), Research in liquid dynamics in missile fuel tanks, SwRI, TR 7.

Lindstrom K., Kjellander H., and Jonsson C. (1969), A new instrument for the measurement of liquid level, *Rev. Sci. Instruments* **4**, 1083–1087.

Liu A. P. C. and Lou J. Y. K. (1990), Dynamic coupling of a liquid–tank system under transient excitation, *Ocean Engrg.* **17**(3), 263–277.

Liu F. C. (1964), Pressure on baffle rings due to sloshing in a cylindrical tank, NASA MSFC, R-AERO-4–64.

Liu W. K. and Belytschko T. B. (1983), Fluid structure interaction with sloshing, *Trans. 7th Int. Conf. Struct. Mech. in Reactor Tech.* B, 11–18.

Liu W. K., Chen Y. J., Tsukimori K., and Uras R. A. (1991), Recent advances in dynamic buckling analysis of liquid-filled shells, *ASME J. Pressure Vessel Tech.* **113**, 314–320.

Liu W. K. and Lam D. (1983), Nonlinear analysis of liquid-filled tank, *J. Eng. Mech.* **109**, 1344–1357.

Liu W. K. and Ma D. C. (1982), Coupling effect between liquid sloshing and flexible fluid-filled systems, *Nuclear Engrg. Des.* **72**, 345–357.

Liu W. K. and Uras R. A. (1988), Variational approach to fluid–structure interaction with sloshing, *Nuclear Eng. and Design* **106**, 69–85.

Liu W. K., and Uras R. A. (1989a), Transient buckling analysis of liquid-storage tanks – part I: Theory, *Sloshing and Fluid Structure Vibration*, ASME Pressure Vess. Piping Conf., PVP- **157**, 35–40.

Liu W. K., and Uras R. A. (1989b), Transient buckling analysis of liquid-storage tanks – part II: Applications, *Sloshing and Fluid Structure Vibration*, ASME Pressure Vess. Piping Conf., PVP-**157**, 41–46.

Liu Y. Z. (1992), The stability of a fluid-filled top rotating on a horizontal plane, *Arch. Appl. Mech.* **62**, 487–494.

Liu Z. P. and Huang Y. (1994), A new method for large amplitude sloshing problems, *J. Sound Vib.* **175**(2), 185–195.

Lloyed A. R. J. M. (1989), *Seakeeping – Ship Behavior in Rough Weather*, Chichester, Ellis Harwood.

Lomen D. O. (1965a), Liquid propellant sloshing in mobile tanks of arbitrary shape, NASA CR-222, April.

Lomen D. O. (1965b), Digital analysis of liquid propellant sloshing in mobile tanks with rotational symmetry, *NASA* CR-230.

Lomen D. O. and Fontenot L. L. (1967), Fluid behavior in parabolic containers undergoing vertical excitation, *J. Math. Phys.* **46**(1), 43–53.

Longuet-Higgins M. S. (1985), Bifurcation in gravity waves, *J. Fluid Mech.* **151**, 457–475.

Longuet-Higgins M. S. (2000), Theory of water waves derived from a Lagrangian, Part 1: standing waves, *J. Fluid Mech.* **423**, 275–292.

Lorell J. (1951), Forces produced by fuel oscillations, Jet Prop. Lab. Progress Report 20–149.

Lorell J. (1952), Effect of fuel sloshing on the position of the center of rotation of a missile, Jet Propulsion Labs., Caltech, Rept. 20–65.

Lorenz E. N. (1963), Deterministic nonperiodic flow, *J. Atmos. Sci.* **20**, 130–141.

Lou J. Y. K, Lutes L. D., and Li J. (1994), Active tuned liquid damper for structural control, Proc 1st World Conf. *Structural Control*, Los Angeles 2, TP1–70–79.

Lou Y. K., Wu M. C., and Lee C. K. (1985), Further studies on liquid sloshing, Maritime Admins. Rept. No MA-RD-760–85009, March.

Lowry B. J. and Steen P. H. (1995), Flow-induced stabilization of liquid columns, *J. Colloid. Interface Sci.* **170**, 38–43.

Lu D., Kondo S., and Takizawa A. (1995), Analysis of overflow-induced sloshing coupled with oscillation of elastic wall, *Fluid-Sloshing and Fluid–Structure Interaction*, ASME Pressure Vess. Piping Conf., PVP- **314**, 147–152.

Lu D., Takizawa A., and Kondo S. (1997), Overflow-induced vibration of a weir coupled with sloshing in a downstream tank, *J. Fluids Struct.* **11**, 367–393.

Lu D. Takizawa A. and Kondo S. (1998), Extension of physical component BFC method for the analysis of free-surface flows coupled with moving boundaries, *Comput. Mech.* **21**(1), 71–80.

Lübeck G. (1873), Über den einfluss, welchen auf die bewegung eimes pendels mit einem kugelförmigen hohlraume ein in ihm enthaltens reibende flüssigkeit ausübt, *Journal f, reine u ang Math* 77.

Lubin B. T. and Springer G. S. (1967), The formulation of a dip on the surface of a liquid draining from a tank, *J. Fluid Mech.* **29**, 385–390.

Ludwieg H. (1951), Die ausgebildete kanalströmung in einem rotierenden system, *Ing. Archiv.* **19**, 296–308.

Lught H. J. and Ohring S. (1992), The oblique ascent of a viscous vortex pair toward a free surface, *J. Fluid Mech.* **236**, 461–480.

Lugovski V. V. (1961), On an investigation of the hydrodynamic forces in oscillations of finite amplitude, *Tsentral Nauchnoissled Inst Morskogo Flota, Trudy, Vyp* **351**, 40–48.

Lui A. P. and Lou J. Y. K. (1990), Dynamic coupling of a liquid tank system under transient excitation, *Ocean Eng.* **17**(3), 263–277.

Luk C. H. (1969), Finite element analysis for liquid sloshing problems, MIT, Aeroelastic and Struct. Res. Lab., Tech. Rept. TR-144–3, AFOSR 69–1504, May.

Luke J. C. (1967), A variational principle for a fluid with a free surface, *J. Fluid Mech.* **27**, 395–397.

Lugovski V. V. (1961), On an investigation of the hydrodynamic forces in oscillations of finite amplitude, *Tsentral Nauchnoissled Inst Morskogo Flota, Trudy, Vyp* **351**, 40–48.

Lukovskii I. A. (1961a), Small wave motions of homogeneous incompressible fluid in containers which are solids of revolution, *Dopovidi Akad Nauk Ukrain RSR*, 1013–1017.

Lukovskii I. A. (1961b), Perturbed motion of a solid body with spherical cavity partly filled with liquid, *Dopovidi Akad Nauk Ukrain RSR*, 1418–1423.

Lukovskii I. A. (1967), Variational method in the nonlinear problems of the dynamics of a limited liquid volume with free surface, in *Oscillations of Elastic Constructions with Liquid*, 260–264, Moscow, Volna (in Russian).

Lukovskii I. A. (1975), *Nonlinear Fluid Oscillations in Vessels of Complex Geometrical Shapes*, Kiev, Naukova Dumka (in Russian).

Lukovskii I. A. (1990), *Introduction to the Nonlinear Dynamics of Aircraft with Liquid*, Kiev, Nauk Dumka (in Russian).

Lukovskii I. A., Barnyak M. Y., and Komarenko A. N. (1984), Approximate methods of solving the problems of the dynamics of a limited liquid volume, Kiev, Naukova Dumka (in Russian).

Lukovskii I. A. and Bilyk A. N. (1985), Forced nonlinear oscillations of a liquid in moving axial-symmetric conical tanks, in *Numerical-Analytical Methods of Studying the Dynamics and Stability of Multidimensional Systems*, Institute of Math 12–26, Kiev. (in Russian)

Lukovskii I. A. and Timokha A. N. (1991a), Acoustic action on a free surface bounded by a liquid volume, *Akust Zhurnal* **37**(1), 144–149.

Lukovskii I. A. and Timokha A. N. (1991b), Stabilization of a separation surface of a liquid and a gas during interaction with acoustic fields in a gas, *Izv. Akad. Nauk. SSSR, Mekh. Zhidk. Gaz* **3**, 80–86.

Lukovskii I. A. and Timokha A. N. (1992), Variational formulations of the problem of the interaction of a "liquid–gas" interface with acoustic fields in the gas, in *Math. Methods of Investigating Applied Problems of the Dynamics of Bodies Carrying a Liquid*, Kiev, Inst. Math. 3–10.

Lukovskii I. A. and Timokha A. N. (1995), *Variational Methods in Nonlinear Dynamics of a Limited Liquid Volume*, Kiev, Inst. Math. (in Russian).

Lukovskii I. A. and Timokha A. N. (1999), Nonlinear theory of sloshing in mobile tanks: classical and non-classical problems (survey), in *Problems of Analytical Mechanics and its Applications*, Kiev, Institute of Mathematics of NASU, 169–200.

Lukovskii I. A. and Timokha A. N. (2000a), Steady state nonlinear sloshing in a rectangular tank: passage to shallow water, *Dopovidi of National Academy of Sciences of Ukraine* 11, 48–51.

Lukovskii I. A. and Timokha A. N. (2000b), Modal modeling of nonlinear fluid sloshing in tanks with non-vertical walls: Non-conformal mapping technique, *Applied Hydromechanics (Prykl. Hydromechanika)* **2**(74), No 4.32–47 (in Russian).

Lukovskii I. A. and Timokha A. N. (2001), Asymptotic and variational methods in non-linear problems of the interaction of surface waves with acoustic fields, *J. Appl. Mat. Mech.* **65**(3), 463–470.

Lukovskii I. A., Trotsenko V. A., and Usyukin V. I. (1989), *Interaction of Thin-Walled Elastic Elements with a Liquid in Mobile Cavities*, Kiev, Naukova Dumka (in Russian).

Luskin H. and Lapin E. (1952), An analytical approach to the fuel sloshing and buffeting problems of aircraft, *J. Aeronaut. Sci.* **19**(4), 217–228.

Lyubimov D. B. and Cherepanov A. A. (1986), Development of a steady relief at the interface of fluids in a vibrational field, *Fluid Dyn.* **21**, 849–854.

Lyubimov D. B. and Cherepanov A. A. (1989), The occurrence of a stationary relief on the interface between fluids in a vibration field, *Izv. Akad. Nauk. SSSR, MZhG* **6**, 8–13.

Lyubimov D. B., Cherepanov A. A., and Briskan V. A. (1981), Control of the fluid free surface by fluctuating fields: II, All Union Seminar on Hydromechanics and Heat and Mass Transfer, Abstracts, Perm 112–114.

Lyubimov D. B., Lyubimova T. P., Skuridin R. V., Chen G., and Roux B. (2002), Numerical investigation of meniscus deformation and flow in an isothermal liquid bridge subject to high-frequency vibrations under zero gravity conditions, *Comput. Fluids* **31**, 663–682.

Ma D. C., Gvildys J., Chang Y. W., and Liu W. K. (1982), Seismic behavior of liquid-filled shells, *Nuclear Eng. Design* **70**, 437–455.

Macdonald H. M. (1894), Waves in Canals, *Proc. Lond. Math. Soc.* **XXV**, 101–111.

Machida T., Kaneko S., and Watanabe T. (1998), Nonlinear characteristics of sloshing with inlet and outlet flow in a rectangular tank (in Japanese), *Trans. JSME* C **64**(620), 1184–1192.

Mack L. R. (1960), Finite amplitude gravity waves of an ideal liquid in a right circular cylindrical tank with horizontal bottom, Ph.D. dissertation, University of Michigan, Ann Arbor.

Mack L. R. (1962), Periodic finite-amplitude, axi-symmetric gravity waves, *J. Geophys. Res.* **67**(2), 829–843.

Mack L. R., Jay B. E., and Sattler D. E. (1967), Standing gravity waves of finite amplitude, *Dev. Theor. Appl. Mech.* **3**, 3, Pergamon Press.

Madarame H., Okamoto K., and Hagiwara T. (1992), Self-induced sloshing in a tank with circulating flow, *Fluid–Structure Vibration and Sloshing*, *ASME Pressure Vess. Piping Conf.*, PVP- **232**, 5–11.

Madarame H., Fukaya M., Iida M., and Okamoto K. (1993a), Self-induced oscillations of upward plane jet impinging on free surface, *ASME Pressure Vess. Piping*, PVP- **258**, 113–119.

Madarame H., Iida M., Okamoto K., and Fukaya M. (1993b), Jet-flutter: self-induced oscillations of upward plane jet impinging on free surface, Proc. Asia-Pacific Vib. Conf. '93, **1**, 265–270.

Madarame H. and Ida M. (1998), Mechanism of jet-flutter: self-induced oscillation of an upward plane jet impinging on a free surface, *JSME Int. J. B* **41**, 610–617.

Madarame H., Okamoto K., and Ida M. (2002), Self-induced sloshing caused by an upward round jet impinging on the free surface, *J. Fluid Struct.* **16**(3), 417–433.

Madsen R. A. and Easton C. R. (1970), Numerical analysis of low-g propellant flow problems, *J. Spacecraft Rock.* **7**(1), 89–91.

Magnus K. (1971), *Kreisel* (*Gyroscope*), Berlin, Springer-Verlag.

Mahajan M. P., Tsige M., Taylor P. L., and Rosenblatt C. (1998), Paramagnetic liquid bridge in a gravity-compensating magnetic field, *Phys. Fluids* **10**(9), 2208–2211.

Mahajan M. P. Tsige M. Taylor P. L., and Rosenblatt C. (1999a), Stability of magnetically levitated liquid bridges of arbitrary volume subjected to axial and lateral gravity, *J. Colloid Interface Sci.* **213**, 592–595.

Mahajan M. P. Tsige M. Taylor P. L., and Rosenblatt C. (1999b), Stability of liquid crystalline bridges, *Phys. Fluids* **11**(2), 491–493.

Mahajan M. P., Tsige M., Zhang S., Alexander J. I. D., Taylor P. L., and Rosenblatt C. (2000), Collapse dynamics of liquid bridges investigated by time-varying levitation, *Phys. Rev. Lett.* **84**(2), 338–341.

Mahajan M. P. Tsige M. Zhang S. Alexander J. I. D. Taylor P. L., and Rosenblatt C. (2002), Resonance behavior of liquid bridges under axial and lateral oscillating total body forces, *Experiments in Fluids* **33**, 503–507.

Mahony J. J. and Smith R. (1972), On a model representation for certain spatial-resonance phenomena, *J. Fluid Mech.* **53**, 193–208.

Malashenko S. V. (1960), Some experimental investigations relating to rotation of bodies, *PMTPH* **3**, 205–211 (in Russian).

Malhotra P. K. (1995), Base uplifting analysis of flexibly supported liquid storage tanks, *Earthquake Engg. and Struct. Dyn.* **24**(12), 1591–1607.

Manasseh, R. (1993), Visualization of the flows in precessing tanks with internal baffles, *AIAA J.* **31**(2), 312–318.

Manevich L. I., Prokopalo E. F., and Shukurov A. Kh. (1976), Investigation of parametric oscillations of fluid-filled shells, in *Vibrations of Elastic Structures with Fluids*, Volna Scientific Information Bureau, Moscow (in Russian).

Manos G. C. (1989), Study of the behavior of cylindrical liquid storage tanks when subjected to lateral loads, *Sloshing and Fluid Structure Vibration*, ASME Pressure Vess. Piping Conf., PVP-**157**, 75–82.

Manos G. C. (1990), Testing of cylindrical liquid-storage tank scaled models subjected to lateral loads, *Flow–Structure Vibration and Sloshing*, ASME Pressure Vess. Piping Conf., PVP- **191**, 19–24.

Manos G. C. and Clough R. W. (1982), Further study of the earthquake response of a broad cylindrical liquid-storage tank model, Report UCB/EERC-82/7, University of California, Berkeley.

Manos G. C., and Clough R. W. (1985), Tank damage during the May 1983 Coalinga earthquake, *Earthquake Eng. Struct. Dyn.* **13**, 449–466.

Manos G. C. and Talaslidis D. (1986), Experimental and numerical study of unanchored cylindrical tanks subjected to lateral load, Proc. 3rd Int. Conf. *Comp. Methods and Exp. Meas.*, Greece.

Marangoni C. G. (1871), Ueber die ausbreitung der tropfen einer flüssigkeit auf der oberfläche einer anderen, *Ann. Phys. Chem. (Poggendorf)* **143**(7), 337–354.

Markeyev A. P. (1985a), The stability of the rotation of a top filled with fluid, *Izv. Akad. Nauk. SSSR, MTT* **3**, 19–26.

Markeyev A. P. (1985b), The oscillations of rigid body with a cavity containing liquid on an absolutely rough plane, in *Some Problems and Methods for Investigating the Dynamics of Mechanical Systems*, Mosk. Aviats. Inst., Moscow, 19–25.

Markus S. (1988), *Mechanics of Vibrations of Cylindrical Shells,* Amsterdam, Elsevier.

Marr-Lyon M. J., Thiessen D. B., and Marston P. L. (1997), Stabilization of a cylindrical capillary bridge far beyond the Rayleigh–Plateau limit using acoustic radiation pressure and active feedback, *J. Fluid Mech.* **351**, 345–357.

Marston P. L., Marr-Lyon M. J., Morse S. F., and Thiessen D. B. (1996), Stabilization and low frequency oscillation of capillary bridges with modulated acoustic radiation pressure, 3rd Microfluid Physics Conf., Cleveland, Ohio, NASA Conference Publication 3338, 475.

Martel C., Nicolas J. A., and Vega J. M. (1998), Surface-wave damping in a brimful circular cylinder, *J. Fluid Mech.* **360**, 213–228.

Martin E. R. (1971), Fuel slosh and dynamic stability of Intelsat IV, AIAA Paper 71–954.

Martinez I. (1976), Floating zone under reduced gravity-axisymmetric equilibrium shapes, 2nd Europ. Symp. *Material Sciences in Space*, Frascati, Italy, ESA-SP-114, 277–282.

Martinez I. (1987), Stability of liquid bridges, Results of SL-D1 experiment, *Acta Astron.* **15**, 449–453.

Martinez I., Haynes J. M., and Langbein D. (1987), Fluid statics and capillary, in *Fluid Sciences and Materials Sciences in Space*, H. U. Walter, ed., New York, Springer-Verlag.

Martinez I. and Perales J. M. (2001), Mechanical behavior of liquid bridges in microgravity, in *Physics of Fluid in Microgravity* R. Monti, ed., Taylor & Francis, 21–45.

Maschek A., *et al.* (1992), Investigation of sloshing fluid motions in pools related to recriticalities in liquid-metal fast breeder reactors, *Nuclear Techn.* **98**(1), 27–35.

Maschek A., Flad M., Arnecke G., and Lo Pinto P. (1998), Mitigation of core disruptive accident energetics in burner cores, *Prog. Nuclear Energy* **32**(3/4), 639–646.

Masica W. J. (1967), Experimental investigation of liquid surface motion in response to lateral acceleration during weightlessness, NASA TN-D-4066, July.

Masica W. J., Derdul J. D., and Petrash D. A. (1964a), Hydrostatic stability of the liquid-vapor interface in a low-acceleration field, NASA TN-D-2444.

Masica W. J., Petrash D. S., and Otto E. W. (1964b), Hydrostatic instability of the liquid-vapor interface in a gravitational field, NASA TN-D-2267.

Masica W. J. and Salzman J. A. (1965), An experimental investigation of the dynamic behavior of the liquid-vapor interface under adverse low-gravitational conditions, AFOSR/LMSC Symp. *Fluid Mechanics and Heat Transfer under Low Gravitational Conditions,* Lockheed Missile & Space Co., (June 24–25), 25–41.

Masica W. J., and Salzman J. A. (1969), Lateral sloshing in cylinders under low gravity conditions, NASA TN D-5058m.

Mason P., Collins D., Petrac D., Yang L., Edeskuty F., Schuch A., and Williamson K. (1978), The behavior of superfluid helium in zero gravity, Proc. 7th Int. *Cryogenic Engineering* Conf., Surrey, England, Science and Technology Press.

Mathiessen J. (1976), Sloshing loads due to random pitch motion, *Norwegian Maritime Research* **4**(3), Det. Norske Veritas.

Mathiessen L. (1868), Akustische versuche, die klieinsten transversalivellen der flussigkeiten betreffend, *Annalen der Physik* **134**, 107–117.

Mathiessen L. (1870), Uber die transversal-schwingungen tonender tropharer und elastisher flussigkeiten, *Annalen der Physik* **141**, 375–393.

Mathieu E. (1868), Memoire sur le movement vibratoire d'une membrane de forme elliptique, *J. Math. Pures Appli.* (J Liouville) **13**, 137–203.

Matsuura F., Matsubara Y., Sawada T., and Tanahashi T. (1995), Surface behaviors of two-layers liquid sloshing under non-uniform magnetic field, *Advanced Computational and Design Techniques in Applied Electronic Systems*, S. Y. Han, ed., Amsterdam, Elsevier Sciences, 517–520.

Matsuura Y., Matsumoto K., Mizuuchi M., Arima K., Jouuchi H., and Hayashi S. (1986), On a mean to reduce excited-vibration with the sloshing in a tank, *J. Soc. Naval. Archit. Japan* **160**, 424–432.

Matsuzaki Y. and Fung Y. C. (1977), Unsteady fluid dynamic forces on a simply-supported circular cylinder of finite length conveying a flow, with application to stability analysis, *J. Sound Vib.* **54**, 317–330.

Matsuzaki Y. and Kobayashi S. (1969a), An analytical study of the nonlinear flexural vibration of thin circular cylindrical shells, *J. Japan Soc. Aeron. Space Sci.* **17**, 308–315 (in Japanese).

Matsuzaki Y. and Kobayashi S. (1969b), A theoretical and experimental study of the nonlinear flexural vibration of thin circular cylindrical shells with clamped ends, *Trans. J. Japan Soc. Aeron. Space Sci.* **12**, 55–62.

Maulard J. and Jourdin A. (1966), Experimenting on liquid behavior at the weightlessness laboratory, *Rech. Aero-Spatiale* **110**, 29–37 (in French).

McCarty J. L., Leonard H. W., and Walton W. C. Jr (1960), Experimental investigation of the natural frequencies of liquids in toroidal tanks, NASA TN D-531.

McCarty J. L. and Stephens D. G. (1960), Investigation of the natural frequencies of fluid in spherical and cylindrical tanks, NASA TN D-252.

McCaughan F. E. and Bedir H. (1992), Marangoni convection with a deformable free surface, ASME WAM Symp. on *Fluid Mechanics Phenomena in Microgravity*, ed. D. A. Siginer and M. M. Weislogel, AMD- 154/FED- **142**, 77–84.

McEwan A. D. (1970), Inertial oscillations in a rotating fluid cylinder, *J. Fluid Mech.* **40**, 603–640.

McGoldrick L. F. (1970), On Wilton's ripples: a special case of resonance interactions, *J. Fluid Mech.* **42**, 193–200.

McIver P. (1989), Sloshing frequencies for cylindrical and spherical containers filled to an arbitrary depth, *J. Fluid Mech.* **201**, 243–257.

McIver P. and McIver M. (1993), Sloshing frequencies of longitudinal modes for a liquid contained in a trough, *J. Fluid Mech.* **252**, 525–541.

McLachlan N. W. (1947), *Theory and Application of Mathieu Functions*, New York, Oxford University Press.

McMahon J. (1894), On the roots of the Bessel and certain related function, *Ann. Mathematics* **9**, 23–30.

McNeil W. A. and Lamb J. P. (1970), Fundamental sloshing frequency for an inclined fluid-filled right circular cylinder, *J. Spacecraft Rock.* **7**(8), 1001–1002.

McRobie F. A., Popov A. A., and Thomson J. M. T. (1999), Autoparametric resonance in cylindrical shells using geometric averaging, *J. Sound Vib.* **227**, 65–84.

Mei C. C. and Liu L. F. (1973), The damping of surface gravity waves in a bounded liquid, *J. Fluid Mech.* **59**(2), 239–256.

Meirovitch L. (1970), *Methods of Analytical Dynamics*, New York, McGraw-Hill.

Meron E. and Procaccia I. (1986a), Theory of chaos in surface waves: the reduction from hydrodynamics to few-dimensional dynamics, *Phys. Rev. Lett.* **56**, 1323–1326.

Meron E. and Procaccia I. (1986b), Low-dimensional chaos in surface waves: theoretical analysis of an experiment, *Phys. Revs. A* **34**, 3221–3237.

Merten K. F. and Stephenson B. H. (1952), Some dynamic effects of fuel motion in simplified model tip tanks on suddenly excited bending oscillations, NACA TN-2789.

Meseguer J. (1983), The breaking of axisymmetric slender liquid bridges, *J. Fluid Mech.* **130**, 123–152.

Meseguer J., Slobozhanin L. A., and Perales J. M. (1995), A review on the stability of liquid bridges, *Adv. Space Res.* **16**(7), (7)5–(7)14.

Meserole J. S. and Fortini A. (1987), Slosh dynamics in a toroidal tank, *J. Spacecraft* **24**(6), 523–531.

Mesquita G. N., Kane S., and Gollub J. P. (1992), Transport by capillary waves: fluctuating Stokes drift, *Phys. Revs. A* **45**, 3700–3705.

Mettler E. (1968), Combination resonance in mechanical systems under harmonic excitations, Proc. 4th Int. Conf. *Nonlinear Oscillations*, Prague, Academia, Publishing House of Czechoslovak Academy of Science, 51–70.

Mieda T., Ishida K., Jitu K., and Chiba T. (1993), An experimental study of viscous damping in sloshing mode of cylindrical tanks, *Struct. Mech. in Reactor Techn.*, 122–131.

Mikelis N. E., Miller J. K., and Taylor K. V. (1984), Sloshing in partially filled tanks and its effect on ship motions: numerical simulations and experimental verification, *Proc. Royal Inst. Naval Archit.*, Spring Meeting.

Mikhlin S. G. (1964), Variational methods in mathematical physics, 2nd edition, New York, Pergamon Press.

Mikishev G. N. and Dorozhkin N. (1961), Experimental investigation of free oscillations of a liquid in tanks, *Izv. Akad. Nauk. SSSR Otd. Tekhn. Nauk. Mekh. i Mashinostr.* **4**, 48–53. (Translated into English by D. D. Kana, SwRI, June 30, 1963).

Mikishev G. N. and Rabinovich B. I. (1968), *Dynamics of a rigid body with cavities partly filled with fluid*, Moscow, Mashinostroeie (in Russian).

Mikishev G. N., and Rabinovich B. I. (1971), *Dynamics of Thin-Walled Constructions with Compartments Containing a Liquid*, Moscow, Mashinostroyeniye.
Miles J. W. (1956), On the sloshing of liquid in a cylindrical tank, Ramo-Wooldridge Corp. Guided Missile Research Div., GM-TR-18.
Miles J. W. (1958a), On the sloshing of liquid in a flexible tank, *J. Appl. Mech.* **25**(2), 277–283.
Miles J. W. (1958b), Ring damping of free surface oscillations in a circular tank, *J. Appl Mech.* **25**(2), 274–276.
Miles J. W. (1959a), On the free surface oscillations in a rotating liquid, Space Technol. Lab., GM-TR-0165–00458.
Miles J. W. (1959b), Centrifugal slosh, Space Technol. Lab., Memo. GM-59, 8021, 6–1&602, February.
Miles J. W. (1959c), Free surface oscillations in a rotating liquid, *Phys. Fluids* **2**(3), 297–305.
Miles J. W. (1962a), Stability of forced oscillations of a spherical pendulum, *Quart. Appl. Math.* **20**, 21–32.
Miles J. W. (1962b), Note on the damping of free surface oscillations due to draining, *J. Fluid Mech.* **12**(3), 438–440.
Miles J. W. (1963), The Cauchy–Poisson problem for a rotating liquid, *J. Fluid Mech.* **17**, 75–88.
Miles J. W. (1964), Free-surface oscillations in a slowly rotating liquid, *J. Fluid Mech.* **18**(2), 187–194.
Miles J. W. (1967), Surface-wave damping in closed basins, *Proc. Royal Soc. (London), A, Math. Phys.* **297**, 459–475.
Miles J. W. (1972), On the eigenvalue problem for fluid sloshing in a half-space, *Zeit. Ang. Math. Phys. (ZAMP)* **23**, 861–869.
Miles J. W. (1976a), Nonlinear surface waves in closed basins, *J. Fluid Mech.* **75**, 419–448.
Miles J. W. (1976b), On internal resonance of two damped oscillators, *Stud. Appl. Math.* **55**(4), 351–359.
Miles J. W. (1984a), Nonlinear Faraday resonance, *J. Fluid Mech.* **146**, 285–302.
Miles J. W. (1984b), Internally resonance surface waves in a circular cylinder, *J. Fluid Mech.* **149**, 1–14.
Miles J. W. (1984c), Resonant motion of a spherical pendulum, *Physica* **11**(D), 309–323.
Miles J. W. (1984d), Resonantly forced surface waves in a circular cylinder, *J. Fluid Mech.* **149**, 15–31.
Miles J. W. (1984e), Parametrically excited solitary waves, *J. Fluid Mech.* **148**, 451–460.
Miles, J. W. (1984f), Resonant non-planar motion of a stretched string, *J. Acoust. Soc. Amer.* **75**, 1505–1510.
Miles J. W. (1984g), Strange attractors in fluid dynamics, *Adv. Appl Mech.* **24**, J. W. Hutchinson and T. Y. Wu, eds., 189–214, New York, Academic Press.
Miles J. W. (1985a), Note on a parametrically excited trapped cross-wave, *J. Fluid Mech.* **151**, 391–394.
Miles J. W. (1985b), Resonantly forced, nonlinear gravity waves in a shallow rectangular tank, *Wave Motion* **7**(1985), 291–297.
Miles J. W. (1988a), Resonance and symmetry breaking for the pendulum, *Physica D* **31**, 252–268.
Miles J. W. (1988b), Parametrically excited, standing cross-waves, *J. Fluid Mech.* **186**, 119–127.
Miles J. W. (1989), The pendulum from Huygens' horologium to symmetry breaking and chaos, Proc. XVII Int. Congress of *Theo. Appl. Mech.*, Grenoble, France, P Germain, M Piau, and D. Caillerie, eds., North Holland, Elsevier Science, 193–215.
Miles J. W. (1990a), Parametrically excited standing edge waves, *J. Fluid Mech.* **214**, 43–57.
Miles J. W. (1990b), Capillary-viscous forcing surface waves, *J. Fluid Mech.* **219**, 635–646.
Miles J. W. (1991a), The capillary boundary layer for standing waves, *J. Fluid Mech.* **222**, 197–205.
Miles J. W. (1991b), Wave motion in a viscous fluid of variable depth: moving contact-line, *J. Fluid Mech.* **223**, 47–55.
Miles J. W. (1992) On surface waves with zero contact angle, *J. Fluid Mech.* **245**, 485–492.
Miles J. W. (1993), On Faraday waves, *J. Fluid Mech.* **248**, 571–683.

Miles J. W. (1994) Faraday waves: rolls versus squares, *J. Fluid Mech.* **269**, 353–371.

Miles J. W. (1996), On forced capillary-gravity waves in a circular cylinder, *Wave Motion* **23**(4), 387–391.

Miles J. W. and Ball F. K. (1963), On free-surface oscillations in a rotating paraboloid, *J. Fluid Mech.* **17**(2), 257–266.

Miles J. W. and Becker J. (1988), Parametrically excited, progressive cross-waves, *J. Fluid Mech.* **186**, 129–146.

Miles J. W. and Henderson D. M. (1990), Parametrically forced surface waves, *Ann. Rev. Fluid Mech.* **22,** 143–165.

Miles J. W., and Henderson D. M. (1998), A note on interior vs boundary layer damping of surface waves in a circular cylinder, *J. Fluid Mech.* **364**, 319–323.

Miles J. W. and Troesch B. A. (1961), Surface oscillations of a rotating liquid, *ASME J. Appl Mech.* **28**, 491–496.

Miles J. W. and Young D. (1958), Generalized missile dynamics analysis in sloshing, Space Tech. Lab., GM-TR-165–00361, 7 April.

Miles J. W. and Zhou Q. P. (1993), Parametric excitation of a tuned spherical pendulum, *J. Sound Vib.* **164**, 237–250.

Milgram J. H. (1969), The motion of a fluid in a cylindrical container with a free surface following vertical impact, *J. Fluid Mech.* **37**(3), 435–448.

Milner S. T. (1991), Square patterns and secondary instabilities in driven capillary waves, *J. Fluid Mech.* **225**, 81–100.

Minowa C. (1980), The dynamic analysis for rectangular water tanks, recent advances in lifeline earthquake eng. in Japan, ASME Pressure Vess. Piping Conf., PVP- **43**, 135–141.

Minowa C. (1989), Surface-sloshing behaviors of liquid storage tanks, *Sloshing and Fluid Structure Vibration*, ASME Pressure Vess. Piping Conf., PVP- **157**, 165–171.

Minowa C. (1994), Sloshing impact of rectangular water tank (water tank damage caused by Kobe earthquake), *Trans. JSME C* **63**(612), 2643–2649.

Minowa C. and Kiyosumi K. (1997), Sloshing impact analysis of roof damaged water tank in Kobe earthquake, ASME Symp. Adv. Anal. Exper. Comp. Tech. Fluids Struct. Trans. Nat. Hazards, PVP- **355**, 271–278.

Minowa C., Ogawa N., Harada I., and Ma D. C. (1994), Sloshing roof impact tests of a rectangular tank, *Sloshing, Fluid–Structure Interaction and Struct. Response Due to Shock and Impact Loads*, ASME Pressure Vess. Piping Conf., PVP- **272**, 13–21.

Minrosky N. (1935), Problems of anti-rolling stabilization of ships by the activated tank method, *American Society of Naval Engineers (ASNE)* **47**, 87–119.

Mitchell R. R. (1968), Stochastic stability of the liquid free surface in vertically excited cylinders, NASA-CR-98009, May.

Mixon J. S. and Catherine J. J. (1964a), Experimental lateral vibration characteristics of a 1/5-scale model of Saturn SA-I with an eight-cable suspension system, NASA TN D-2214.

Mixon J. S., and Catherine J. J. (1964b), Comparison of experimental vibration characteristics obtained from a 1/5-scale model and from a full-scale Saturn SA-I, NASA TN D-2215.

Mixon J. S., Catherine J. J., and Arman A. (1963), Investigation of the lateral vibration characteristics of a 1/5-scale model of Saturn SA-I, NASA TN D-1593.

Mixon J. S. and Herr R. W. (1962), An investigation of the vibration characteristics of pressurized thin walled circular cylinders partially filled with liquid, NASA TR R-145.

Miyata H. (1986), Finite difference simulation of breaking waves, *J. Comput. Phys.* **65**, 179–214.

Miyata H., Inui T., and Kajitani H. (1980), Free surface shock waves around ships and their effects on ship resistance, *J. Soc. Naval Archit. Japan* **147**, 1–9.

Miyata T., Yamada H., and Saito Y. (1988), Suppression of tower-like structures vibration by damping effect of sloshing water contained (in Japanese), *Trans. JSCE* **34A**, 617–626.

Miyata T., Yamada H., and Saito Y. (1989), Feasibility study of the sloshing damper system using rectangular containers (in Japanese), *Trans. JSCE* **35A**, 553–560.

Modi V. J. and Munchi R. S. (1998), An efficient liquid sloshing damper for vibration control, *J. Fluids Struct.* **12**, 1055–1071.

Modi V. J. and Seto M. L. (1997), Suppression of flow-induced oscillations using sloshing liquid dampers, *J. Wind Eng. Indust. Aerodyn.* **67/68**, 611–625.

Modi V. J. and Seto M. L. (1998), Passive control of flow-induced oscillations using rectangular nutation dampers, *J. Vib. Control* **4**(4), 381–404.

Modi V. J. and Welt F. (1984), Nutation dampers and suppression of wind induced instabilities, ASME Joint Multidivisional Symp. on *Flow-Induced Oscillations* **1**, 173–187.

Modi V. J. and Welt F. (1987), Vibration control using nutation dampers, Proc. Int. Conf. Flow-Induced Vib., England, 369–376.

Modi V. J. and Welt F. (1988), Damping of wind induced oscillation through liquid sloshing, *J. Wind Eng. Indust. Aerodyn.* **30**, 85–94.

Modi V. J., Welt F. and Irani M. B. (1988), On the suppression of vibrations using nutation dampers, *J. Wind Eng.* **37**, 547–556.

Modi V. J., Welt F., and Seto M. L. (1995), Control of wind-induced instabilities through application of nutation dampers: a brief review, *Eng. Structures* **17**(9), 626–638.

Moiseev N. N. (1952a),The motion of a rigid body with cavities partially filled with an ideal liquid, *Dokl. Akad. Nauk. SSSR* **85**(4), 719–722.

Moiseev N. N. (1952b), On oscillations of a heavy ideal and incompressible liquid in a container, *Dokl. Akad. Nauk. SSSR* **85**(5), 963–966.

Moiseev N. N. (1952c), Dynamics of a ship having a liquid load, *Izv. Akad. Nauk. SSSR Otd. Tekhn. Naud.* **7**, 27–45.

Moiseev N. N. (1952d), The problem of small oscillations of an open vessel with a fluid under the action of an elastic force, *Ukrainian Mat. Zh.* **4**, 168–173.

Moiseev N. N. (1953), The problem of the motion of a rigid body filled with a liquid having a free surface, *Matemticheskii Sbornik* **47**(7), issue 1, Acad. Sci. SSSR, 61–96.

Moiseev N. N. (1954), Some questions on the theory of oscillations of vessel with a fluid, *Inzhenernyi Sbornik* **19**, 167–170.

Moiseev N. N. (1956), Studies of the motion of a solid body containing fluid masses with free surfaces, *Vestnik Acad. Sci. SSSR* **5**.

Moiseev N. N. (1958), On the theory of nonlinear vibration of a liquid of finite volume, *Prikl. Math. Mekh. (PMM)* **22**, 612–621.

Moiseev N. N. (1959), On the theory of vibration of elastic bodies with liquid cavities, *Prkl. Math. Mekh.* **23**(5), 862–878.

Moiseev N. N. (1960), On the theory of elastic oscillations of a fluid filled body, *Dokl. Akad. Nauk. SSSR* **27**(1), 53–56. Also *Soviet Physics* **4**(4), February.

Moiseev N. N. (1961), On boundary-value problem for linearized Navier–Stokes equations for the case of low viscosity, *Zh. Vychisl. Matem. i Matem. Fiz.* **1**(3).

Moiseev N. N. (1964), Introduction to the theory of oscillations of liquid-containing bodies, *Adv. Appl. Mech.* **8**, 233–289, H. L. Dryden and Th von Karman, eds., New York, Academic Press.

Moiseev N. N. (1968), *Introduction to the Dynamics of Bodies Containing Liquids in the Weightless State*, Vychislitelnyi Tsentr. AN SSSR, Moscow (in Russian).

Moiseev N. N. (1970), Some problems in the hydrodynamics of surface waves, in *Advances in Mechanics in USSR over 50 years*, **2**, Moscow, Nauka, 55–72.

Moiseev N. N. and Chernousko F. L. (1965), Problems of oscillations of a liquid subject to surface tension forces, *Zh. Vychisl. Mat. Mat. Fiz.* **5**, 1071–1082.

Moiseev N. N. and Petrov A. A. (1965), *Numerical Methods of Calculating the Natural Frequencies of Vibrations of a Bounded Volume of Fluid*, Moscow, Acad. Sci. USSR Press.

Moiseev N. N. and Petrov A. A. (1966), The calculation of free oscillations of a liquid in a motionless container, *Adv. Appl. Mech.* **9**, 91–154, NY, Academic Press.

Moiseev N. N. and Rumnyantsev V. V. (1968), *Dynamic Stability of Bodies Containing Fluid*, Applied Physics and Eng./An. Int. Series, **6**, H. N. Abramson, ed., Springer-Verlag.

Molin B. (2001), On the piston and sloshing modes in moonpools, *J. Fluid Mech.* **430**, 27–50.

Molin B., Cointe R., and Fontaine E. (1996), On energy arguments applied to slamming force, Proc. 11th Int. Workshop on *Water Waves and Floating Bodies*, Hamburg, Germany.

Mololkov L. A. and Krauklis P. V. (1967), Oscillations of a cylindrical shell filled with a fluid and surrounded by an elastic medium, *Prikl. Math. Mekh. (PMM)* **31**(5), 922–925.

Momoi T. A. (1965), A motion of water excited by earthquake, part 1: Rectangular basin (one dimensional), *Bull. Earthquake Res. Inst., Tokyo University* **43**(1), 111–127.

Monti R. (1990a), On the onset of the oscillatory regimes in Marangoni flows, *Acta Astronaut.* **15**, 557–560.

Monti R. (1990b), Gravity jitters: Effects on typical fluid science experiments, in *Low Gravity Fluid Dynamics and Transport Phenomena*, J. N. Koster and R. L. Sani, eds., *Progress in Astronautics and Aeronautics* **130**, 275–307.

Monti R. (2001), *Physics of Fluids in Microgravity*, London, Taylor & Francis.

Monti R. and Fortezza R. (1989), Oscillatory Marangoni flow in a floating zone: design of a Telescience experiment for Texus 23, Proc. 7th *European Symp. Materials and Fluid Sci. in Microgravity*, 285–289.

Monti R., Fortezza R., and Desiderio G. (1992), Onset of oscillatory Marangoni flow: Scientific results of the experiment performed in Telescience on Texus 23, Final Report of Sounding Rocket Experiments, in Fluid Science and Material Science, ESA SP-1132, **2**.

Monti R. and Savino R. (1994a), A new approach to the g-level tolerability for fluid and material science experiments, *Acta Astron.* **37**, 313–331.

Monti R., and Savino R. (1994b), Transient developments of thermocapillary, natural and vibrational convection in floating zone, *Microgravity Quart.* **4**(4), 247–258.

Monti R., and Savino R. (1994c), Effect of unsteady thermal boundary condition on Marangoni flow in liquid bridges, *Microgravity Quart.* **4**, 163–170.

Monti R. and Savino R. (1995), G-jitter effect in fluid-physics microgravity experimentation, Proc. Int. Congress *In Space* '95, 16–17 October, Tokyo, 112–175.

Monti R., and Savino R. (1996), Influence of g-jitter on fluid physics experimentation on-board the International Space Station, ESA-SP-385, 215–224.

Monti R., Savino R., and Alterio G. (1997a), Modeling and simulation of g-jitter effects on fluid science experimentation: Impact on the utilization of the ISS, *Acta Astron.* **40**(2–8), 369–381.

Monti R., Savino R., and Lappa M. (1997b), Oscillatory thermocapillary flows in simulated floating-zones with time-dependent temperature boundary conditions, *Acta Astron.* **41**(12), 863–875.

Monti R., Savino R., and Lappa M. (1998), Influence of geometrical aspect ratio on the oscillatory Marangoni convection in liquid bridges, 49th IAF Congress, Melbourne, Australia.

Monti R., Savino R., and Lappa M. (2000a), Influence of geometrical aspect ratio on the oscillatory Marangoni convection in liquid bridges, *Acta Astron.* **47**, 10, 753–761.

Monti R., Savino R., and Lappa M. (2001), On the convective disturbances induced by g-jitter on the space station, *Acta Astron.* **48**(5–12), 603–615.

Monti R., Savino R., Lappa M., Carotenuto L., Castagnolo D., and Fortezza R. (2000b), Flight results on Marangoni flow instability in liquid bridges, *Acta Astron.* **47**(2–9), 325–334.

Moody F. J. and Reynolds W. C. (1972), Liquid surface motion induced by accelerated and external pressure, *ASME J. Basic Eng.* **94**, 606–612.

Mooney J., Ryan J., and Walls J. (1964b), Slosh handbook, NSL Tech. Memo. 1, 21 February.

Mooney J., Walls J., and Ryan J. (1964a), Liquid natural frequencies of a tank with partially submerged sector walls, NSL Tech. Memo. 14, 7 April.

Moore R. E. and Perko L. M. (1965), Inviscid fluid flow in an accelerating cylindrical container, *J. Fluid Mech.* **22**(2), 305–320.

Morand J. J. P. (1977), Deux theorems de Congruence relatifs aux Vibrations de Liquides Couples a des structures, Proc. Int. Conf. ESA-CNES, Publ. SP-129, Toulouse.

Morand H. J. P. and Ohayon R. (1975), Internal pressure effects on the vibration of partially filled elastic tanks, Proc. *World Cong. Finite Element Methods in Struct. Mech.*, Bournemouth, TP ONERA 66.

Morand H. J. P. and Ohayon R. (1989), Finite element method applied to the prediction of the vibrations of liquid-propelled launch vehicles, Proc. ASME-PVP Conf. **176**, Hawaii.

Morand H. J. P. and Ohayon R. (1995), *Fluid Structure Interaction: Applied Numerical Methods*, New York, John Wiley & Sons.

Morgan G. W. (1951), A Study of motion in a rotating liquid, *Proc. Royal Soc. (London) A* **206**, 108–130.

Morison J. P., O'Brien M. P., Johnson J. W., and Schaaf S. A. (1950), The force exerted by surface waves on a pile, *Petroleum Trans.* **189**, 149–154.

Morris B. T. (1938), A laboratory model study of the behavior of liquid filled cylindrical tanks in earthquakes, Thesis, Stanford University, June.

Morse P. M. and Feshbach H. (1953), *Methods of Theoretical Physics*, New York, McGraw-Hill.

Morse S. F., Thiessen D. B., and Marston P. L. (1996), Capillary bridge modes driven with modulated ultrasonic radiation pressure, *Phys. Fluids* **8**, 3–5.

Muehlner K. A., Schatz M., Petrov V., McCormic W. D., Swift J. B., and Swinney H. L. (1997), Observation of helical traveling-wave convection in a liquid bridge, *Phys. Fluids* **9**, 1850–1852.

Muenz K. and Marcello J. M. (1964), Technique for measuring amplitudes of small surface waves, *Rev. Sci. Instrum.* **35**(8), 953–957.

Müller H. W. (1993), Periodic triangular patterns in the Faraday experiment, *Phys. Rev. Lett.* **71**, 3287–3290.

Müller H. W., Friedrich R., and Papathanassiou D. (1998), Theoretical and experimental investigations of the Faraday instability, *Evolution Spontaneous Structures in Dissipative Continuous Systems*, F Buses and S. C. Müller, eds., Lecture Notes in Physics, 231–265, Berlin, Springer-Verlag.

Müller H. W., Wittmer C., Wagner C., Albers J., and Knorr K. (1997), Analytic stability theory for Faraday waves and the observation of the harmonic surface response, *Phys. Rev. Lett.* **78**, 2357–2360.

Mundrane M. R. and Zebib A. (1994), Oscillatory thermocapillary convection, NASA 2nd *Microgravity Fluid Physics Conf.*, Cleveland, Ohio, 27–32.

Mushtari Kh. M. and Galimov K. Z. (1957), *The Nonlinear Theory of Elastic Shells*, Tatknigoizdat, Kazan (in Russian).

Muto K., Kasai Y., and Nakahara M. (1988), Experimental tests for suppression effects of water restraint plates on sloshing of a water pool, *ASME J. Pressure Vess. Tech.* **110**, 240–246.

Myshkis A. D., Babskii V. G., Kopachevskii N. D., Slobozhanin L. A., and Tyuptsov A. D. (1987), *Low-Gravity Fluid Mechanics*, New York, Springer-Verlag.

Myshkis A. D., Babskii V. G., Zhukov M. Y., Kopachevskii N. D., Slobozhanin L. D., and Typsov A. D. (1992), *Methods of Solving Fluid Mechanics Problems under Conditions of Weightlessness*, A. D. Myshkis, ed., Kiev, Naukova Dumka.

Nachtigall I., Gebbeken N., and Urrutia-Galicia J. L. (2003), On the analysis of vertical circular cylindrical tanks under earthquake excitation at its base, *Engrg. Struct.* **25**, 201–213.

Nagakura H. and Kaneko S. (1995), A study on self-excited sloshing due to the fluid discharge over a flexible weir, *ASME Fluid–Structure Interaction and Struct. Mechanics* **310**, 15–22.

Nagakura H. and Kaneko S. (1998a), Self-excited sloshing due to the fluid discharge over a flexible weir (1st report, excitation mechanism of instability in the rectangular model), *Trans. JSME C* **64**(623), 2437–2445 (in Japanese).

Nagakura H. and Kaneko S. (1998b), Self-excited sloshing due to the fluid discharge over a flexible weir (2nd report, excitation mechanism of instability in the cylindrical model), *Trans. JSME C* **64**(623), 2446–2454 (in Japanese).

Nagakura H. and Kaneko S. (1998c), Self-excited sloshing due to the fluid discharge over a flexible weir (3rd report, comparison between experimental and analytical results and discussion on the parameters which effect on the stability of the system), *Trans. JSME C* **64**(624), 2882–2885 (in Japanese).

Nagakura H. and Kaneko S. (2000a), Self-excited sloshing due to the fluid discharge over a flexible cylindrical weir, *ASME. J. Pressure Vessel Tech.* **122**, 33–39.

Nagakura H. and Kaneko S. (2000b), Self-excited sloshing due to the fluid discharge over a flexible plate weir, *ASME J. Pressure Vessel Tech.* **122**, 192–197.

Nagata M. (1989), Nonlinear Faraday resonance in a box with a square base, *J. Fluid Mech.* **209**, 265–284.

Nagata M. (1991), Behavior of parametrically excited surface waves in square geometry, *Eur. J. Mech. B/Fluids* **10**(2). 61–66.

Nagay K. and Takeuchi J. (1984), Vibration of a plate with arbitrary shape in contact with a fluid, *J. Acoust Soc. Amer.* **74**(5), 1511–1518.

Nakagaki K., Arima K., Ueda T., and Kadou H. (1990), On natural vibration and damping effect of tuned sloshing damper, *JSCE J. Struct Eng.* **36A**, 591–602.

Nakagawa K. (1955), On the vibration of an elevated water-tank – II, Tech. Rept. Osaka University **5**(170), 317–336.

Nakagawa K. (1956), On the vibration of an elevated water-tank – III, Tech. Rept. Osaka University **6**(193), 53–62.

Nakayama T. and Washizu K. (1980), Nonlinear analysis of liquid motion in a container subjected to forced pitching oscillations, *Int. J. Numer. Methods Eng.* **15**, 1207–1220.

Nakayama T. and Washizu K. (1981), The boundary element method applied to the analysis of two-dimensional nonlinear sloshing problems, *Int. J. Numer. Methods Eng.* **17**(11), 1631–1646.

Napolitano L. G. (1982), Surface tension and buoyancy driven free convection, *Acta Astron.* **9**, 199–215.

Narimanov G. S. (1956), Concerning the motion of a rigid body with a cavity partially filled with a liquid, *Prikl. Math. Mekh. (PMM)* **20**(1), 21–38.

Narimanov G. S. (1957a), Concerning the motion of a symmetrical gyroscope with cavity partially filled with liquid, *Prikl. Math. Mekh. (PMM)* **21**, 696–700.

Narimanov G. S. (1957b), Concerning the motion of a container partially filled with a liquid taking into account large motion of the latter, *Prikl. Math. Mekh. (PMM)*, **21**(4), 696–700 (Space Tech. Lab. Trans. T-RU-18.).

Narimanov G. S. (1957c), Concerning the vibration of fluids in moving cavities, Izv. Acad. Sci. SSSR, OTN 10.

Narimanov G. S., Dokuchaev L. V., and Lukovskii I. A. (1977), *Nonlinear Dynamics of Aircraft with Liquid*, Moscow, Mashinostroenie (in Russian).

Nash W. A., Yu B. Q., and Kirchhoff R. H. (1989), Struct. behavior of liquid-elastic solid systems, *Sloshing and Fluid Structure Vibration*, ASME Pressure Vess. Piping Conf., PVP- **157**, 89–94.

Natsiavas S. (1987), *Response and failure of fluid-filled tanks under base excitation,* Ph. D. dissertation, Caltech, Pasadena, CA.

Natsiavas S. (1988), An analytical model for unanchored fluid-filled tanks under base excitation, *ASME J. Appl. Mech.* **110**, 648–653.

Natsiavas S. (1989), Simplified models for the dynamic response of tall unanchored liquid storage tanks, *Sloshing and Fluid Structure Vibration*, D. C. Ma, J Tani, Chenss, and Liu WK, eds., ASME Pressure Vess. Piping Conf., PVP- **157**,15–22.

Natsiavas S. and Babcock C. D. (1988), Behavior of unanchored fluid-filled tanks subjected to ground excitation, *ASME J. Appl. Mech.* **55**, 654–659.

Natushkin V. F. and Rakhimov I. S. (1964), Oscillation of a cylindrical shell partially filled with a fluid, *Aviat. Tekh.* **17**(3), IAAA 64–28276.

Nayfeh A. H. (1985), *Problems in Perturbation*, New York, Wiley Interscience.

Nayfeh A. H. (1987), Surface waves in closed basins under parametric and internal resonances, *Phys. Fluids* **30**, 2976–2983.

Nayfeh A. H. (1993), *Methods of Normal Forms*, New York, John Wiley and Sons.

Nayfeh A. H. and Mook D. (1979), *Nonlinear Oscillations*, New York, Wiley.

Nayfeh A. H. and Nayfeh J. F. (1990), Surface waves in closed basins under principal and autoparametric resonances, *Phys. Fluids A* **2**(9), 1635–1648.

Nayfeh A. H. and Raouf R. A. (1987), Nonlinear oscillations of circular cylindrical shells, *Int. J. Solids Struct.* **23**, 1625–1638.

Neitzel G. P. (1994), Control of oscillatory thermocapillary convection in microgravity, NASA 2nd *Microgravity Fluid Physics Conf.*, Cleveland, OH, 21–26.

Nelson E. S. (1991), An experimentation of anticipated g-jitter on space station and its effects on materials process, NASA TM 103775.

Nelson R. H. Jr (1960), The sloshing of a fluid draining from a flexible tank, Ph.D. thesis, MIT, Cambridge, MA.

Nesterov A. V. (1982), Parametric excitation of internal waves in a continuously stratified fluid, *Izv. Akad. Nauk. SSSR, MZhG* **5**, 167–169.

Netter G. and Eckhardt K. (1981), Fluid dynamic experiment in a surface tension tank: Phase 1/Phase 2A, Final Report, MBB/ERNO, Rept. No BMFT-FB-W-83–002 (in German).

Neu J. T. and Good R. J. (1963), Equilibrium behavior of fluids in containers at zero gravity, *AIAA J.* **1**(4), 814–819.

Nevolin V. G. (1983), Parametric excitation of oscillations of a fluid flowing out of a container, *Izv. Akad. Nauk. SSSR, Mekh. Zhidkosti i Gaza* **2**, 293–295.

Nevolin V. G. (1985), Parametric excitation of surface waves, *J. Eng. Phys.* **49**, 1482–1494. Translation 1984 from *Inzh. Fiz. Zh.* **47**, 1028–1042.

Newmark N. M. and Rosenbluetch E. (1971), *Fundamentals of Earthquake Engineering*, Englewood Cliffs, Prentice Hall.

Ngan C. G. and Dussan E. B. (1982), On the nature of the dynamic contact-angle: an experimental study, *J. Fluid Mech.* **118**, 27–40.

Nichols B. D. and Hirt C. W. (1971), Improved free surface conditions for numerical incompressible flow computations, *J. Comput Phys.* **8**, 434–448.

Nichols B. D. and Hirt C. W. (1975), Method for calculating multi-dimensional transient free surface flow past bodies, Proc.1st Int. Conf. *Numerical Ship Hydrodynamics*, Gaithersburg, MD.

Nichols B. D., Hirt C. W., and Hotchkiss R. S. (1980), SOLA-VOF: A solution algorithm for transient fluid flow with multiple free boundaries, LA 8355, Los Alamos Scientific Laboratory, Los Alamos, NM.

Nicolás J. A. (1992), Magnetohydrodynamic stability of cylindrical liquid bridges under a uniform axial magnetic field, *Phys. Fluids A* **4**, 2573–2577.

Nicolás J. A. and Vega J. M. (1996), Weakly nonlinear oscillations of axisymmetric liquid bridges, *J. Fluid Mech.* **328**, 95–128.

Nicolás J. A., Rivas D., and Vega J. M. (1997), The interaction of thermocapillary convection and low-frequency vibration near-inviscid liquid bridges, *Z. Angew. Math. Phys.* **48**, 389–423.

Nicolás J. A., Rivas D., and Vega J. M. (1998), On the steady streaming flow due to high-frequency vibration in nearly inviscid liquid bridges, *J. Fluid Mech.* **354**, 147–174.

Nikitin S. K. (1987), The nonstationary interaction of elastic solids of revolution partially filled with a liquid, Ph. D. dissertation, S. P. Timoshenko Institute of Machanics, National Academy of Sciences of Ukraine, Kiev (in Russian)

Nishino H. and Mochio T. (1995), Sloshing analysis of cylindrical shell with rubber-covered surface, *Fluid-Sloshing and Fluid–Structure Interaction*, ASME Pressure Vess. Piping Conf., PVP-**314**, 43–55.

Niwa A. (1978), Seismic behavior of tall liquid storage tanks, Rept. EERC 78–04, Earthquake Eng. Res. Center, UCA, Berkeley.

Niwa A. and Clough R. W. (1982), Buckling of cylindrical liquid-storage tanks under earthquake excitation, *Earthquake Eng. and Struct. Dyn.* **10**, 107–122.

Noji T., Yoshida H., Tatsumi E., Kosaka H., and Haguida H. (1988), Study on vibration control damper utilizing sloshing of water, *J. Wind Eng.* **37**, 557–566.

Noji T., Yoshida H., Tatsumi E., Kosaka H., and Haguida H. (1990), Study of water-sloshing vibration control damper, *J. Struct. Const. Eng. AIJ* **411**, 97–105.

Novozhilov V. V. (1962), *The Theory of Thin Shells*, Leningrad, Sudpromgiz (in Russian).

Nowinski J. (1963), Nonlinear transverse vibrations of orthotropic cylindrical shells, *AIAA J.* **1**, 617–620.

Nussle R. G., Derdul J. D., and Petrash D. A. (1965), Photographic study of propellant weightlessness, NASA TN D-2572.

Obraztsova E. I. (1976a), Nonlinear parametric vibrations of a cylindrical shell with a liquid under longitudinal excitation, *Izv. Vyssh. Uchebn. Zaved. Aviats. Tekh.* **3**, 87–93.

Obraztsova E. I. (1976b), The forced nonlinear vibrations of a cylindrical shell with a liquid under longitudinal excitation, Proc. 3rd Symp. on *Vibration of Elastic Structures with a Liquid*, Moscow, 314–319 (in Russian).

Obraztsova E. I. and Shklyarchuk F. N. (1979), Nonlinear parametric vibrations of a cylindrical tank containing a liquid, *Izv. Akad. Nauk. SSSR Mekh. Tverd. Tela.* **4**, 115–126.

Obraztsova E. I. and Shklyarchuk F. N. (1980), The nonlinear parametric vibrations of an orthotropic cylindrical tank partially filled with liquid, in *The Vibrations of Elastic Structures with a Liquid*, Tsentr. Byuro Nauch.-Tekhn. Inform., Voln, Moscow, 216–224 (in Russian).

Ochi M. K. and Motter L. E. (1971), A method to estimate the slamming characteristics for ship design, *Mar. Technol.* **8**, 219–232.

Ockendon J. R. and Ockendon H. (1973), Resonance surface waves, *J. Fluid Mech.* **59**(2), 397–413.

Ockendon J. R. and Ockendon H. (2001), Nonlinearity in fluid resonances, *Meccanica* **36**(3).

Ockendon H., Ockendon J. R., and Johnson A. D. (1986), Resonant sloshing in shallow water, *J. Fluid Mech.* **167**, 465–479.

Ockendon H., Ockendon J. R., and Waterhouse D. D. (1996), Multimode resonances in fluids, *J. Fluid Mech.* **315**, 317–344.

Ogawa N., Chiba T., Nakajima S., and Ma D. C. (1996), An experimental study of the sloshing behavior of the worm tank, ASME Pressure Vess. Piping Conf. *Fluid–Structure Interaction*, PVP-**337**, 141–146.

Ogawa N., Chiba T., Nakajima S., Shibata H., and Ma D. C. (1997), An experimental study of the fluid–structure interaction behavior of the worm tank, ASME Symp. *Advances in Analytical, Exp. and Comp. Tech. Fluids, Struct., Transients and Natural Hazards*, PVP-**355**, 237–243.

Ohaba M., Sawada T., Saito S., Sudo S., and Tanahashi T. (1998), Axisymmetric interfacial oscillations of a magnetic fluid column, *JSME Int. J. B* **41**(2), 511–523.

Ohaba M. and Sudo S. (1995), Liquid surface behavior of a magnetic liquid in a container subjected to magnetic field and vertical vibration, *J. Magnetism Magnetic Materials* **149**, 38–41.

Ohaba M., Sugimoto T., Sawada T., Sudo S., and Tanahashi T. (1998), Magnetic field effects on axi-symmetrically oscillating magnetic liquid columns, *J. Japan Soc. Microgravity Appl.* **15**, 243–248.

Ohaba M., Sugimoto T., Sawada T., Sudo S., and Tanahashi T. (1999), Oscillations of magnetic liquid column subjected to random vibration, *J. Magnetism and Magnetic Materials* **201**, 293–296.

Ohaba M., Tomonori M., Sawada T., Sudo S., and Tanahashi T. (1998), Axisymmetric oscillation of magnetic liquid columns in a magnetic field, *Studies in Appl. Electromagn. Mech.* **13**, Nonlinear Electromagnetic Systems, 863.

Ohayon R. (1989), Alternative formulations for static and modal analysis of structures containing fluids, Proc. ASME- PVP Conf. **176**, Hawaii.

Ohayon R. and Felippa C. A. (1990), The effect of wall motion on the governing equations of contained fluids, *J. Appl. Mech.* **57**, 783–785.

Ohayon R. and Morand H. (1995), Mechanical and numerical modeling of a fluid–structure vibration instabilities of liquid propelled launch vehicles, *Chaos, Solitons, Fractals* **5**(9), 1705–1724.

Ohnishi M. and Azuma H. (1992), Computer simulation of oscillatory Marangoni flow, *Acta Astron.* **26**(8–10), 685–696.

Ohyama T. and Fuji K. (1989), A boundary element analysis for two-dimensional nonlinear sloshing problem, *JSCE J. Struct. Eng.* **35A**, 575–584 (in Japanese).

Okamoto K., Fukaya M., and Madarame H. (1993), Self-induced sloshing caused by flow in a tank, ASME PV&P **258**, 105–111.

Okamoto K. and Kawahara M. (1990), Two-dimensional sloshing analysis by Lagrangian finite element method, *Int. J. Num. Methods Fluids* **11**, 453–477.

Okamoto K. and Madarame H. (1998), Fluid dynamics of free surface in liquid metal fast breeder reactors, *Prog. Nuclear Energy* **32**, 195–207.

Okamoto K., Madarame H., and Hagiwara T. (1991), Self-induced oscillation of free surface in a tank with circulating flow, Proc. IMech E *Flow-Induced Vibration* **1991**–6, 539–545.

Okamoto K., Madarame H., and Hagiwara T. (1992), Self-induced water level oscillation in a tank with flow pattern transformation, ASME Pressure Vess. Piping Conf., PVP-**232**, 13–17.

Okamoto K., Saeki S., Madarame H., Saga T., and Kobayashi T. (2000), Analysis on the self-induced sloshing using particle image velocimetry, http://www.egr.msu.edu/~huhui/paper/2000/2000/okamoto.pdf.

Okazaki K., Nozaki Y., Tani J., and Shimizu N. (1990), Seismic response analysis of a multi-walled coaxial cylindrical tank under horizontal excitation, *Flow–Structure Vibration and Sloshing*, ASME Pressure Vess. Piping Conf., PVP- **191**, 55–61.

Okazaki K. and Tani J. (1991), Seismic response analysis of a multi-walled coaxial cylindrical tank under vertical excitation, *Fluid–Structure Vibration and Sloshing*, ASME Pressure Vess. Piping Conf., PVP-**223,223**, 9–16.

Okazaki K., Watanabe K., and Tani J. (1995), Experimental study on chaos of a liquid-filled tank under vertical excitation, *Fluid-Sloshing and Fluid–Structure Interaction*, ASME Pressure Vess. Piping Conf., PVP- **314**, 99–108.

Okazaki K., Zhu Q., Tani J., and Watanabe K. (1993), Experimental study on chaos of a liquid-filled tank under horizontal excitation, ASME Pressure Vess. Piping Conf., Denver, CO.

Okhotsimski D. E. (1956), On the motion of a body with cavities partly filled with a liquid, NASA TTF-33, (May, 1960), (NASA translation from *Prikl. Math. Mekh. (PMM)* **20**(1)).

Okubo M., Ishibashi Y., Oshima S., Katakura H., and Yamane R. (1990), Interfacial waves of the magnetic fluid in vertical alternating magnetic fields, *J. Magn. Mat.* **85**, 163–166.

Olson L. G. and Bathe K. J. (1983), A study of displacement-based fluid finite elements for frequencies of fluid and fluid–structure system, *Nuc. Engrg. Des.* **76**, 137–151.

Olson M. D. (1965), Some experimental observations on the nonlinear vibration of cylindrical shells, *AIAA J.* **3**, 1775–1777.

Omura N., Yamashita K., and Sawada T. (1999), Dynamic characteristic of a magnetic fluid column formed by magnetic field, *J. Magn. Magnet. Mater.* **201**, 297.

O'Neill J. P. (1960), Final report on an experimental investigation of sloshing, Space Tech. Lab. STL/TR-59–000–09960 March.

Or A. C. (1992), Rotor-pendulum model for the Perigee assist module nutation anomaly, *J. Guidance Cont. Dyn.* **15**(2), 297–303.

Or A. C., Challoner A. D., and Yip P. P. (1994), Stability of a liquid-filled spinning top: A numerical approach, *J. Sound Vib.* **175**(1), 17–37.

Ortiz J. L. and Barhorst A. A. (1997), Closed-form modeling of fluid–structure interaction with nonlinear sloshing potential flow, *AIAA J.* **35**(9), 1510–1517.

Ostrach S. (1982), Low gravity fluid flows, *Annual Review Fluid Mechanics* **14**, 313–345.

Oser H. (1958), Experimentelle Untersuchung uber harmonische schwingungen in rotierenden flussigkeiten, *Z. Angew. Math. Mech. (ZAMM)* **38**, 386–391.

Ostrach S., Kamotani Y., and Lai C. L. (1985), Oscillatory thermocapillary flows, *Phys. Chem. Hydro.* **6**(5/6) 585–599.

Padday J. F. (1969), Theory of surface tension, in *Surface and Colloid Science*, E. Matijevic, ed., 1, 39–251, New York, Wiley.

Padday J. F. (1971), The profiles of axially symmetric menisci, *Phil. Trans. Royal Soc. London A* **269**, 265–472.

Padday J. F. (1976), Capillary forces and stability in zero gravity environments, Proc. 2nd European Symp. *Material Sciences in Space*, Frascati, Italy, ESA-SP-114, 443–454.

Paidoussis M. P. (1998), *Fluid–Structure Interactions: Slender Structures and Axial Flow*, **1**, London, Academic Press.

Paidoussis M. P., Chan S. P., and Misra A. K. (1984), Dynamics and stability of coaxial cylindrical shells containing flowing fluid, *J. Sound Vib.* **97**, 201–235.

Paidoussis M. P. and Denise J. P. (1972), Flutter of thin cylindrical shells conveying fluid, *J. Sound Vib.* **20**, 9–26.

Paidoussis M. P., Misra A. K., and Chan S. P. (1985), Dynamics and stability of coaxial cylindrical shells conveying viscous fluid, *ASME J. Appl. Mech.* **52**, 389–396.

Pal N. C., Bhattacharyya S. K., and Sinha P. K. (2003), Nonlinear coupled slosh dynamics of liquid-filled laminated composite containers: a two-dimensional finite element approach, *J. Sound Vib.* **261**, 729–749.

Palmer J. H. and Asher G. W. (1965), Calculation of axi-symmetric longitudinal modes for fluid-elastic-ullage gas system and comparison with model test results, Proc. AIAA *Syst. Struct. Dyn. Aeroelasticity*, 189–193.

Palmer J. H. and Berge J. C. (1971), Convective instability in liquid pools heated from below, *J. Fluid Mech.* **47**, 779–787.

Pantazonpoulos M. S. (1988), Three-dimensional sloshing of water on decks, *Marine Technology* **25**(4), 253–261.

Pantazonpoulos M. S. and Adee B. H. (1987), Three-dimensional shallow water waves in an oscillating tank, Proc. ASCE Conf. *Coastal Hydrodynamics*, University of Delaware, Newark, DE, 399–412.

Papell S. S. (1965), Low viscosity magnetic fluid obtained by the colloidal suspension of magnetic particles. US Patent 3215572.

Park J. E. and Rezvani M. A. (1995), In-tank fluid sloshing impact effects during earthquakes: a preliminary computational simulation, in *Fluid-Sloshing and Fluid–Structure Interaction*, ASME Pressure Vess. Piping Conf., PVP- **314**, 73–78.

Park J. H., Koh H. M., and Kim J. K. (2000), Seismic isolation of pool-type tanks for the storage of nuclear spent fuel assemblies, *Nuclear Engrg. Des.* **199**, 143–154.

Park Y. J., Choi J. K., Kim Y. S., and Bae Y. S. (1995), A practical prediction method of sloshing loads in cargo ship tanks on board ships, PRADS'95, Souel, September, 1418–1425.

Parker S. P. (ed.) (1983), *McGraw-Hill Encyclopedia of Physics*, New York, McGraw-Hill.

Parks P. C. (1979), Stability of liquid-filled spinning spheroids via Liapunov's second method, *J. Appl Mech.* **46**, 259–262.

Parkus H. (1982), Modes and frequencies of vibrating liquid-filled cylindrical tanks, *Int. J. Eng. Sci.* **20**(2), 319–326.

Passerone A., Liggieri L., and Ravera F. (2001), Interfacial phenomena, in *Physics of Fluids in Microgravity*, R Monti, ed., London, Taylor & Francis, 46–77.

Patel N. M., Dodge M. R., Alexander J. I. D., Slobozhanin L. A., Taylor P. L., and Rosenblatt C. (2002), Stability of connected cylindrical liquid bridges, *Phys. Rev. E* **65**(2), Paper No 026306, 4 pages.

Patrom L. S. (1985), Numerical calculation of equivalent moment of inertia for a fluid in a cylindrical container with partitions, *J. Numer. Meth. Fluids* **5**, 25–42.

Patrom L. S. (1987), Application of the VOF method to the sloshing of a fluid in a partially filled cylindrical container, *J. Numer. Meth. Fluids* **7**, 535–550.

Pavolvskii V. S. (2000), The nonlinear vibrations and stability of orthotropic cylindrical liquid-carrying shells under periodic actions, *Int. Appl. Mech.* **36**(10), 1369–1379.

Pavlovskii V. S. and Filin V. G. (1979), Stability of vibrations of a cylindrical shell filled with a liquid under conditions of nonlinear resonances, *Prikl. Mekh.* **15**(8), 709–716.

Pavlovskii V. S. and Filin V. G. (1981), The nonlinear resonant vibrations of cylindrical shells with a liquid under transversal loads, 5th seminar on the *Dynamics of Elastic and Rigid Bodies Interacting with a Liquid*, Izd. Tomsk. University, Tomskogo, 158–163 (in Russian).

Pavlovskii V. S. and Filin, V. G. (1985), Nonlinear parametric vibrations of liquid-filled cylindrical shells with an initial deflection, *Prikl. Mekh.* **21**(3), 249–258.

Pawell A. (1997), Free surface waves in a wave tank, *Int. Series Num. Math.* **124**, 311–320, Birkhauser.

Paynter H. L. (1964a), Special zero gravity fluid problems, *Rocket Propellant and Pressurization Systems,* E Ring, ed., Englewood Cliffs, NJ, Prentice-Hall.

Paynter H. L. (1964b), Time for a totally wetting liquid to deform from a gravity-dominated to a nulled-gravity equilibrium state, *AIAA J.* **2**(9), 1627–1634.

Pearson J. R. (1958), On convection cells induced by surface tension, *J. Fluid Mech.* **4**, 489–500.

Peek R. (1986), Analysis of unanchored liquid storage tanks under seismic loads, Ph. D. dissentation, Caltech, Pasadena, CA.

Peek R. (1988), Analysis of unanchored liquid storage tanks under lateral loads, *Earthquake Eng. Struct. Dyn.* **16**, 1087–1100.

Peek R. and Jennings P. C. (1988), Simplified analysis of unanchored tanks, *Earthquake Eng. Struct. Dyn.* **16**, 1073–1085.

Peltier L. J. and Biringen S. (1993), Time-dependent thermocapillary convection in a rectangular cavity: numerical results for a moderate Prandtl number, *J. Fluid Mech.* **257**, 339–357.

Pengelley C. D. (1968), Natural frequency of longitudinal modes of liquid propellant space launch vehicles, *J. Spacecraft Rock.* **5**(12), 1425–1431.

Penney W. G. and Price AT(1952), Finite periodic stationary waves in a perfect liquid, part II, *Phil. Trans. Royal Soc. (London)* **244**, 254–284.

Perales J. M. and Meseguer J. (1992), Theoretical and experimental-study of the vibration of axisymmetrical viscous-liquid bridges, *Phys. Fluids. A* **4**, 1110–1130.

Perko L. M. (1969), Large-amplitude motions of a liquid-vapor interface in an accelerating container, *J. Fluid Mech.* **35**(1), 77–96.

Perlin M. and Schultz W. W. (1996), On the boundary conditions at an oscillating contact-line: a physical/ numerical experimental program, Proc. NASA 3rd *Microgravity Fluid Physics* Conf., Cleveland, OH, 615–620.

Perlin M. and Schultz W. W. (2000), Capillary effects on surface waves, *Annu. Rev. Fluid Mech.* **32**, 241–274.

Peterson L. D. (1987), The nonlinear coupled dynamics of fluids and spacecraft in low gravity, Ph. D. dissentation, Dept. Aeronaut. and Astronaut., MIT, Cambridge, MA, SSL Rep 22–87.

Peterson L. D., Crawley E. F., and Hansman R. J. (1989), Nonlinear fluid slosh coupled to the dynamics of a spacecraft, *AIAA J.* **27**(9), 1230–1240.

Petrash D. A. and Nelson T. M. (1963), Effect of surface energy on the liqui–vapor interface during weightlessness, NASA TN D-1582.

Petrash D. A., Nussel R. C., and Otto E. W. (1963), Effect of the acceleration disturbances encountered in the MA-7 spacecraft on the liquid–vapor interface in a baffled tank during weighlessness, NASA TN D-2075.

Petrash D. A. and Otto E. W. (1962), Studies of the liquid–vapor interface configuration in weighlessness, Amer. Rocket Soc. Space Power Systems Conf., 25–28 September.

Petrash D. A. and Otto E. W. (1964), Controlling the liquid–vapor interface under weightlessness, *AIAA J.* **2**(3), 56.

Petrash D. A., Zappa R. F., and Otto E. W. (1962), Experimental study of the effects of weightlessness on the configurations of mercury and alcohol in spherical tanks, NASA TN-D-1197.

Petrenko M. O. (1969), The natural vibrations of a liquid with a free surface and the elastic bottom of a cylindrical cavity, *Soviet Appl. Mech. (Prikl. Mekh.)* **5**(6), 44–50.

Petrenko M. O. (1970), The concurrent vibrations of a liquid and the elastic bottom of a cylindrical tank, Proc. 7th All-Union Conf. on *The Theory of Shells and Plates*, Moscow, Nauka, 474–478 (in Russian).

Petrenko M. O. (1978), Resonance vibrations of the free liquid surface in a tank with an elastic bottom, Proc. 3rd Workshop *on the Dynamics of Elastic and Rigid Bodies Interacting with a Liquid*, Tomsk, 91–97 (in Russian).

Petrov A. A. (1961), Oscillations of a fluid in an annular cylindrical vessel with horizontal generators, *Zh. Vychislit. Matem. Fiz.* **1**, 741–746.

Petrov A. A. (1962a), Oscillations of fluids in cylindrical basins with horizontal generators – variational methods in problems of oscillations of a fluid and of a body with a fluid, Vych. Tsenter. Akad. Nauk. SSSR, Moscow, 179–202.

Petrov A. A. (1962b), Approximate solution of the problem of oscillations of a fluid in a cylindrical vessel with horizontal generators – variational methods in problems of oscillations of a fluid and of a body with a fluid, Vych. Tsentr. Akad. Nauk. SSSR, Moscow, 213–230.

Petrov A. A. (1962c), Equations of motion of an airplane carrying tanks with liquids – variational methods in problems of oscillations of a fluid and of a body with a fluid, Vych. Tsentr. Akad. Nauk. SSSR, Moscow, 231–236.

Petrov A. A. (1963), Approximate methods for calculating the natural vibrations of a fluid in vessels of arbitrary shape and of Zhukovskiy's potentials for these vessels, *Zh. Vychisl. Matem. Fiz.* **5**.

Petrov V., Schatz M., Muehlner K. A., Van Hook S. J., McCormic W. D., Swift J. B., and Swinney H. L. (1996), Nonlinear control of remote unstable states in a liquid bridge convection experiment, *Phys. Rev. Lett.* **77**, 3779.

Pfeiffer F. (1967a), On the fuel sloshing in rotational symmetric rocket tanks with arbitrary internal boundary, *Jahrbuch der Wissenschaftlichen Gesellschaft fur Luft Raumfahrt* E V, 408–413 (in German).

Pfeiffer F. (1967b), Linearized free oscillations of fuel in arbitrary containers, (in German), *Zeitschrift fur Flugwissenschaften* **16**(6), 188–194.

Pfeiffer F. (1977), Problems of contained rotating fluids with respect to aerospace applications, CNESS-ESA Conf. on Attitude Control of Space Vehicles, Toulouse, ESA SP-129, 51–62.

Pih H. and Wu C. G. (1969), Experimental studies of vibrations of axially excited circular shells containing cylindrical shells containing fluid, *ASME J. Eng. Industry* **91**(4), 1119–1127.

Pilipchuk V. N. and Ibrahim R. A. (1997), The dynamics of a nonlinear system simulating liquid sloshing impact in moving structures, *J. Sound Vib.* **205**(5), 593–615.

Pilipchuk V. N. and Ibrahim R. A. (1999), Application of the Lie group transformations to nonlinear dynamical systems, *J. Appl Mech.* **66**(2), 439–447.

Pilipchuk V. N. and Ibrahim R. I. (2000), Dynamics of a two-pendulum model with impact interaction and an elastic support, *Nonlinear Dyn.* **21**(3), 221–247.

Pinson L. D. (1963), Propellant-dome impact analysis, NASA Internal Memo., November.

Pinson L. D. (1964), Longitudinal spring constants for liquid propellant tanks with ellipsoidal tanks, NASA TN D-2220, November.

Pinson L. D. and Leonard H. W. (1969), Longitudinal vibration characteristics of 1/10-scale Apollo/Saturn V replica model, NASA TN D-5159, April.

Plank W. S. Beardsley G. F., and Burt W. V. (1972), An exprimental evaluation of a passive anti-roll tank system, Ocean Engrg. **2**(3), 131–139.

Plateau J. A. F (1863), Experimental and theoretical researches on the figures of equilibrium of a liquid mass withdrawn from the action of gravity, Smithsonian Inst. Ann. Rep., 207–285; also (1864), 286–369; (1865), 411–435, (1866), 255–289, Washington, Government Printing Office.

Platt G. K. (1967), Space vehicle low gravity fluid mechanics problems and the feasibility of their experimental investigation, NASA TM X-53589.

Pocha J. J., (ed) (1986), *An Experimental Investigation of Spacecraft Sloshing*, ESA Proc. Intl. Symp. *Spacecraft Flight Dynamics*, ESA SP-255, 307–314.

Podesta G. (1956), Acceleration and differential feed-back methods for controlling the instability due to propellant sloshing in missiles, Convair, Report No ZU–7–070–TN, 15 November.

Poincaré H. (1885), Sur l'équilibra d'une masse fluide animée d'un mouvement de rotation, *Acta Math.* **7**, 259.

Poincaré H. (1910), Sur la precession des corps deformable, *Bull. Astronomique* **27**, 321–356.

Poisson S. D. (1828–9), Sur les petites oscillations de l'eau contenue dans un cylindre, *Ann. De Gergonne* **XIX**, 2225.

Pomeau Y. and Manneville P. (1980), Intermittent transition to turbulence in dissipative dynamical systems, *Com. Math. and Phys.* **74**, 189–197.

Ponder C. A., Blount D. H., and Fritz C. G. (1964), Bubble coalescence in a longitudinal vibrated liquid column, part 1, NASA TM X-53180, December.

Popov A. A. (2003), Parametric resonance in cylindrical shells: A case study in the nonlinear vibration of structural shells, *Engrg. Struct.* **25**, 789–799.

Popov A. A., Thomson J. M. T., and McRobie F. A. (2001), Chaotic energy exchange through autoparametric resonance in cylindrical shells, *J. Sound Vib.* **248**(3), 395–411.

Popov G. V., Sankar S., and Sankar T. S. (1992), Liquid sloshing in rectangular road containers, *Computers and Fluids* **21**(4), 551–569.

Popov G., Sankar S., and Sankar T. S. (1993a), Dynamics of liquid sloshing in baffled and compartmented road containers, *J. Fluid Struct.* **7**, 803–821.

Popov G., Vatistas G. H., Sankar S., and Sankar T. S. (1993b), Numerical simulation of viscous liquid sloshing in arbitrary shaped reservoirs, *AIAA J.* **31**(1), 10–11.

Popplewell N., Lu M. L., and Shah A. H. (2002), Behavior of a nutation damper undergoing a coupled horizontal/rotational motion, Proc. ASME WAM IMECE2002–32861.

Porter A. W. (1933), The calculation of surface tension from experiment, *Phil. Magazine* **15**, 163–169.

Pozhalostin A. A. (1967), Free oscillations of a fluid in a rigid circular cylindrical container with an elastic shallow, spherical shell bottom, IZV Vys Uchebnykh Zavedenii, *Aviatsion Tekh.* **2**, 21–27 (in Russian).

Pozharitskiy G. K. (1962), The problem of a minimum in the problem of stability of a rigid body partially filled with a fluid, *Prikl. Math. Mekh. (PMM)* **26**(4), 895–913.

Pozharitskiy G. K. (1964), On the effect of viscosity on the stability of equilibrium and of steady-state rotations of a rigid body with a cavity partially filled with a viscous fluid, *Prikl. Math. Mekh. (PMM)* **28**(1), 67–76.

Pozharitskiy G. K. and Rumyantsev V. V. (1963), The problem of the minimum in the problem of stability of motion of a rigid body with a fluid-filled cavity, *Prikl. Math. Mekh. (PMM)* **27**(1), 18–32.

Poznyak E. (1994), Paradoxes and analogies in the problems of dynamics for rotating containers partially filled with liquid, unpublished review article.

Preisser F., Schwabe D., and Scharmann A. (1983), Steady and oscillatory thermocapillary convection in liquid columns with free cylindrical surface, *J. Fluid Mech.* **126**, 545–567.

Prokopalo E. F., Gruzhin V. I., and Shchukurov Akh. (1980), Study of the parametric vibrations of glass-fiber-reinforced plastic shells filled with a liquid, in *The Vibrations of Elastic Structures with a Liquid*, Tsentr Byuro Nauch-Tekhn Inform "Volna", Moscow, 243–247 (in Russian).

Pshenichnov G. I. (1969), Application of asymptotic method of integration to problems of free oscillations of a thin elastic shell of revolution partly filled with fluid, Tr VII A 11 Union Conf. on the *Theory of Shells and Plates* Dnepropetrovsk.

Pshenichnov G. I. (1971), Exact solutions of some problems concerned with oscillations of fluid contained in an elastic momentless shell, *Prikl. Math. Mekh. (PMM)* **35**(4), 689–694.

Pshenichnov G. I. (1972), Free oscillations of liquid in rigid vessels, *Prikl. Math. Mekh. (PMM)* **36**(2), 229–235.

Puchka G. N. and Kholopova V. V. (1996), Nonlinear surface waves in liquid within a cylindrical vessel, *Int. Appl. Mech.* **32**(4), 307–309.

Pursi A. and Malik S. K. (1995), Parametrically excited nonlinear surface waves and chaos in magnetic fluids, *J. Magn. Magnet. Mater.* **149**, 132–136.

Qinque W. and Lidu H. (1992), Eigen-problem of liquid-container coupling, *Comput. Struct.* **44**(1/2), 353–355.

Qiu S., and Scherer R. J. (1994), Active control of liquid sloshing in tank under seismic excitation, 1st World Conf. *Structural Control* **1**, WP1–108–113, Los Angeles, CA.

Raake D. and Siekmann J. (1991), Axially excited natural frequencies in a rapidly rotating cylindrical container, *Z. Flugwiss. Weltraumforsch.* **15**, 289–296.

Rabinovich B. I. (1956), Concerning the equations of perturbed motion of a solid body with a cylindrical cavity partially filled with a fluid, *PMM* **20**(1), 39–50.

Rabinovich B. I. (1959), Concerning the equations of elastic vibrations of thin-walled, fluid-filled rods with free fluid surfaces, Izvestiya Acad. Sci. SSSR, OTN, Mekhanika I Mashinostroyniye, No. 3; English transl by Space Technol. Lab., STL-T-RV-19.

Rabinovich B. I. (1964), Concerning the equations of transverse vibrations of fluid-filled shells, Izvestiya Acad. Sci. SSSR, series Mekhanika I Mashinostroyeniye, No. 1, 166–169, English translation NASA TT-F-216.

Rabinovich B. I. (1980), Oscillation of structures with fluid-filled cavities, *Handbook Vibration in Engineering* **3**, 61–89.

Raco R. J. (1968), Stability of a liquid in a longitudinal time varying electric field, *AIAA J.* **6**, 979–980.

Radwan H. R. and Genin J. (1975), Nonlinear modal equations for thin elastic shells, *Int. J. Nonlin. Mech.* **10**, 15–29.

Radwan H. R., and Genin J. (1976), Nonlinear vibrations of thin cylinders, *ASME J. Appl. Mech.* **43**, 370–372.

Rajappa N. R. (1970a), A new approach to the study of standing surface waves of finite amplitude, *Acta Mech.* (91/2), 130–136.

Rajappa N. R. (1970b), On the instability of fluid surfaces when accelerated perpendicular to their planes, *Acta Mech.* **10**, 193–205.

Raju K. K. and Rao V. G (1976), Large amplitude asymmetric vibrations of some thin shells of revolution, *J. Sound Vib.* **44**, 327–333.

Rakesh G. (1997), Tuned liquid column dampers for vibration control of tall structures, M. Eng. thesis, National University of Singapore.

Rakheja S., Sankar S., and Ranganathan R. (1988), Roll plane analysis of articulated tank vehicles during steady turning, *Vehicle Sys. Dyn.* **17**(1–2), 81–104.

Rammerstofer F. G., Scharf K., and Fischer F. D. (1989), On problems in the use of earthquake response spectrum methods for fluid–structure–soil interaction, *Sloshing and Fluid Structure Vibration*, D. C. Ma, *et al.*, editors., ASME Pressure Vess. Piping Conf., PVP-**157**, 61–68.

Rammerstofer F. G., Scharf K., and Fischer F. D. (1990), Storage tanks under earthquake loading, *ASME Appl. Mech. Rev.* **43**(11), 261–283.

Ramos J., Incecik A., and Guedes Soares C. (2000), Experimental study of slam-induced stresses in a containership, *Marine Struct.* **13**, 25–51.

Rand R. and DiMaggio F. (1967), Vibrations of fluid-filled spherical and spheroidal shells, *J. Acoustical Soc. Amer.* **42**(6), 1278–1286.

Ranganathan R., Rakheja S., and Sankar S. (1989a), Dynamic analysis of a b-train carrying liquid cargo – part I: Response to steady steer inputs, ASME Proc. WAM *Advan. Automot. Technol.*, DSC-**13**, 1–18.

Ranganathan R., Rakheja S., and Sankar S. (1989b), Dynamic analysis of a b-train carrying liquid cargo – part II: Response to transient steer inputs, ASME Proc. WAM. *Advan. Automot. Technol.*, DSC-**13**, 19–30.

Rankine W. J. M. (1865), Supplement to a paper on streamlines, *Phil. Mag.* **29**(4), 25–28.

Ransleben G. E. Jr and Abramson H. N. (1960), Discussion of the Berlot paper: production of rotation in a confined liquid through translation motion of the boundaries, *ASME J. Appl. Mech.* **27**(2), 265–270.

Rapoport I. M. (1968), *Dynamics of Elastic Containers Partially Filled with Liquids*, NY, Springer-Verlag, Applied Physics and Eng./an Int. Series **5**, H. N. Abramson, ed., translation by Technical S. Inc.

Rashed M. I. I. (1965), Inertia and damping effects due to the flow in a cylindrical vessel filled with fluid and subjected to a simple harmonic oscillatory motion, Tech. Rept., Cairo University, Dept. Aeronaut. Eng., Giza.

Rattayya J. V. (1965), Sloshing of liquids in axi-symmetric ellipsoidal tanks, AIAA Paper 65–114, AIAA 2nd Aerospace Science Meeting, New York.

Rayleigh J. W. S. (Lord) (1879), On the capillarity phenomena of jets, *Proc. Royal Soc.* **XXIX**, 71–97; On instability of jets, *Proc. London Math. Soc.* **X**, 4–13.

Rayleigh J. W. S. (Lord) (1883a), On the crispations of fluid resting upon a vibrating support, *Phil. Mag.* **15**, 229–235.

Rayleigh J. W. S. (Lord) (1883b), On maintained vibrations, *Phil. Mag.* **15**, 229–235 (Scientific Papers, **2**, 188–193).

Rayleigh J. W. S. (Lord) (1887), On the maintenance of vibrations by forces of double frequency and on the propagation of waves through a medium endowed with a periodic structure, *Phil. Mag.* **24**, 145–159.

Rayleigh J. W. S. (Lord) (1892), On the instability of cylindrical surfaces, Collected papers, 594, **3**, Cambridge, England, Cambridge University Press.

Rayleigh J. W. S. (Lord) (1916), On the convection currents in a horizontal layer of fluid when the higher temperature is on the under side, *Phil. Mag.* **32**, 529–546.

Rayleigh J. W. S. (Lord) (1945), *The Theory of Sound*, New York, Dover.

Reed D., Yu J., Yeh H., and Gardarsson S. (1998a), Investigation of tuned liquid dampers under large amplitude of excitation, *ASCE J. Eng. Mech.* **124**, 405–413.

Reed D., Yeh H., Yu J., and Gardarsson S. (1998b), Tuned liquid dampers under large amplitude excitation, *J. Wind Eng. Indust. Aerodyn.* **74–76**, 923–930.

Reed F. E. (1961), Dynamic vibration absorbers and auxiliary mass dampers, Chapter 6, *Shock and Vibration Handbook*, CM Harris, and C. E. Crede, eds., **1**, New York, McGraw-Hill.

Reese J. R. (1955), Some effects of fluid on pylon-mounted tanks on flutter, NACA RML 55F10.

Reese J. R. and Sewal J. L. (1955), Effective moments of inertia of a fluid in offset and swept wing tanks undergoing pitching oscillations, NACA TN-3353.

Rehbinder G. and Apazidis N. (1995), Motion of a torsion pendulum immersed in a linear viscous liquid, *J. Sound Vib.* **184**(5), 745–758.

Reifel M. D. (1964), Vibrations of pressurized thin elastic circular cylinders filled with liquid subjected to translational excitations, Report ERR AN-653, General Dynamics/Astronautics.

Reismann H. (1958), Liquid propellant inertia and damping due to airframe roll, *Amer. Rocket Soc. J.* **28**(11), 746–747.

Reissner E. (1955), Nonlinear effects in vibrations of cylindrical shells, Ramo-Wooldridge Co. Rept. AM5–6.

Reissner E. (1956), Notes on forced and free vibrations of pressurized cylindrical shells which contains a heavy liquid with a free surface, Space Technol. Lab., AM 6–15, GM TR87, Los Angeles.

Reissner E. (1957), Complementary energy procedure for vibrations of liquid filled circular cylindrical tanks, Space Technol. Lab., GM TR, EM7–9.

Reynolds J. M. (1965), Stability of an electrostatically supported fluid column, *Phys. Fluids* **8**(1), 161–170.

Riecke H. (1990), Stable wave number kinks in parametrically excited standing waves, *Europhys. Lett.* **11**(3), 213–218.

Riley J. D. and Trembath N. W. (1961), Sloshing of liquid in spherical tanks, *J. Aer./Space Sci.* **28**(3), 245–246.

Ring E. (1964), Propellant sloshing, in *Rocket Propellant and Pressurization System*, Englewood Cliffs, NJ, Prentice Hall.

Roberts J. B. and Spanos P. D. (1990), *Random Vibration and Statistical Linearization*, New York, Wiley.

Roberts J. R., Basurto E. R., and Chen P. Y. (1966), Slosh design handbook, NASA CR-406.

Rocca M. L., Mele P., and Armenio V. (1997), Variational approach to the problem of sloshing in a moving container, *J. Theoret. Appl. Fluid Mech.* **1**(4), 280–310.

Rocca M. L., Sciortino G., and Boniforti M. A. (2000), A fully nonlinear model for sloshing in a rotating container, *Fluid Dyn. Res.* **27**(1), 23–52.

Rocca M. L., Sciortino G., and Boniforti M. A. (2002), Interfacial gravity waves in a two-fluid system, *Fluid Dyn. Res.* **30**(1), 31–66.

Rogge T. R. and Weiss H. J. (1965), An approximate nonlinear analysis of the stability of sloshing modes under translational and rotational excitation, NASA CR-220.

Rosenblat S. (1959), Torsional oscillations of a plane in a viscous fluid, *J. Fluid Mech.* **6**(2), 206–220.

Rosenhead L. (1963), *Laminar Boundary Layer*, Oxford, Clarendon Press.

Rosensweig R. E. (1987), Magnetic fluids, *Annu. Rev. Fluid Mech.* **19**, 417–463.

Ross G. O. (1994), Helium II. slosh in low gravity, 2nd Microgravity Fluid Physics Conf., NASA Lewis Research Center, Cleveland, 213–218.

Rudenko T. V. (2002), The stability of the steady motion of a gyrostat with a liquid in a cavity, *J. Appl. Mat. Mekh.* **66**(2), 171–178.

Rumold W. (1998), Modeling and simulation of sloshing liquids, *Z. Angew. Math. Mech. (ZAMM)* **78**, S691–S692.

Rumold W. (2001), Modeling and simulation of vehicles carrying liquid cargo, *Multibody System Dynamics* **5**, 375–386.

Rumyantsev V. V. (1954), Equations of motion of a rigid body having cavities partially filled with a liquid, *Prikl. Math. Mekh. (PMM)* **18**(6), 719–728.

Rumyantsev V. V. (1955), On equations of motion of a rigid body with a cavity filled with liquid, *Prikl. Math. Mekh. (PMM)* **19**(1), 1–12.

Rumyantsev V. V. (1959a), The stability of Maclaurin ellipsoids of a rotating liquid, *Prikl. Math. Mekh. (PMM)* **23**, 494–504.

Rumyantsev V. V. (1959b), On the stability of rotational motion of a rigid body filled with a fluid, *Prikl. Math. Mekh. (PMM)* **23**(6), 1512–1524.

Rumyantsev V. V. (1960), On the stability of motion of a top with a cavity filled with a viscous fluid, *Prikl. Math. Mekh. (PMM)* **24**(4), 903–912.

Rumyantsev V. V. (1962), On the stability of steady motions of rigid bodies with fluid-filled cavities, *Prikl Math. Mekh. (PMM)* **26**(6).

Rumyantsev V. V. (1964a), On the motion of a rigid body containing cavities with a viscous fluid, *Prikl. Math. Mekh. (PMM)* **28**(6), 1353–1358.

Rumyantsev V. V. (1964b), Lyapunov method in the study of stability of motion of rigid bodies with fluid-filled cavities, *Adv. Appl. Math.* **8**, 183–232, HL Dryden and Th von Karman, eds., New York, Academic Press.

Rumyantsev V. V. (1964c), On the stability of motion of a solid with a liquid having surface tension, *Prikl. Math. Mekh. (PMM)* **28**, 746–751.

Rumyantsev V. V. (1966), On the motion of rigid bodies with fluid-filled cavities, *Prikl. Math. Mekh. (PMM)* **30**(1), 57–77.

Rumyantsev V. V. (1968), On the motion and stability of a solid with a rotor and liquids having surface tension, in *Introduction to the Dynamics of a Body Containing a Liquid under Zero-Gravity Conditions*, Vychisl. Tsentr. Akad. Nauk. SSSR, Moscow, 259–268 (in Russian).

Rumyantsev V. V. (1969a), Collision with obstacle of a body containing a viscous fluid, *Prikl. Math. Mekh. (PMM)* **33**(5), 876–881.

Rumyantsev V. V. (1969b), On the motion and stability of an elastic body with a cavity containing a fluid, *Prikl. Math. Mekh. (PMM)* **33**(6), 927–937.

Rumyantsev V. V., Rubanovsky V. N., and Stepanov S. Y. (1979) Oscillation and stability of rigid bodies with cavities containing a fluid, *Handbook Vibration in Engineering*, **2**, 280–305.

Rundgren L. (1958), Water wave forces, A theoretical and laboratory study, Ph.D. thesis, Royal Inst. Tech., Stockholm.

Rupp R., Muller G., and Neumann G. (1989), Three-dimensional time-dependent modeling of the Marangoni convection in zone melting configuration for GaAs, *J. Crystal Growth* **97**, 34.

Russell S. (1844), Report on waves, Brit. Ass. Rept.

Ryan R. S. and Buchanan H. (1966), An evaluation of the low-g propellant behavior of a space vehicle during waiting orbit, NASA TM X-53476.

Rybicki A. and Florian J. M. (1987), Thermocapillary effects in liquid bridges, I: Thermocapillary convection, *Phys. Fluids* **30**, 1956–1965.

Saad M. A. and Oliver D. A. (1964), Linearized time-dependent free-surface flow in rectangular and circular tanks, Heat Transfer and Fluid Mechanics Institute, California, Stanford University Press.

Saeki S., Madarame H., and Okamoto K. (2001), Self-induced sloshing excited by a horizontally injected plane jet, *J. Fluid Mech.* **448**, 81–114.

Saga T., Hu H., Kobayashi T., Murata S., Okamoto K., and Nishio S. (2000b), A comparative study of the PIV and LDV measurement on a self-induced sloshing flow, *J. Visualization* **3**(2), 145–156.

Saga T., Hu H., Kobayashi T., Segawa S., and Taniguchi N. (2000a), Research on the self-induced sloshing phenomena in a rectangular tank, 9th Int. Symp. *Flow Visualization*, Paper number **259**, 1–10.

Saito S. and Someya T. (1980), Self-excited vibration of a rotating hollow shaft partially filled with liquid, *ASME J. Mech. Des.* **102**, 185–192.

Sakai F. (1985), The state-of-the-art and future subjects on seismic design and study, *Trans. JSCE* **362**(I-4), 1–6 (in Japanese).

Sakai F. and Isoe A. (1989), Computation and experiment on base-uplift behavior of cylindrical oil storage tanks during earthquakes, *Sloshing and Fluid Structure Vibration*, ASME Pressure Vess. Piping Conf., PVP-**157**, 9–14.

Sakai F., Isoe A., Hirakawa H., and Mentani Y. (1989), Large scale static tilt tests of seismic behavior of flat-bottomed cylindrical liquid storage tanks, *Sloshing and Fluid Structure Vibration*, ASME Pressure Vess. Piping Conf., PVP- **157**, 69–74.

Sakai F.,Takaeda S., and Tamaki T. (1989), Tuned liquid column damper, new type device for suppression of building vibrations, Proc. Int. Conf. *Highrise Buildings*, Nanjing, China, 926–931.

Sakai F., Takaeda S., and Tamaki T. (1991a), Tuned liquid column dampers (TLCD) for cable-stayed bridges, Proc. Specialty Conf. *Innovation in Cable-Stayed Bridges*, Fukuoka, Japan, 197–205.

Sakai F., Takaeda S., and Tamaki T. (1991b), Experimental study on tuned liquid column dampers (TLCD), Proc. Colloq. *Structural Control*, JSCE, Tokyo, 189–196 (in Japanese).

Sakata M., Kimura K., and Utsumi M. (1983), Non-stationary random responses of nonlinear liquid motion in a circular cylindrical tank, *Trans. JSME* C **49**(442), 963–970 (in Japanese).

Sakata M., Kimura K., and Utsumi M. (1984a), Non-stationary response of nonlinear liquid motion in a cylindrical tank subjected to random base excitation, *J. Sound Vib.* **94**(3), 351–363.

Sakata M., Kimura K., and Utsumi M. (1984b), Non-stationary random vibration of an elastic circular cylindrical liquid storage tank to a simulated earthquake excitation, *Trans. JSME C* **50**(458), 2004–2012 (in Japanese).

Sakata M., Kimura K., and Utsumi M. (1991), Non-stationary random responses of an elastic circular cylindrical liquid storage container to a simulated earthquake excitation, *Trans. 11*th *Int. Conf. Struct. Mech. Reactor Technol.* K, K35/6, 571–576.

Sakata M., Kimura K., Utsumi M., and Okada T. (1986), The coupled oscillations of a liquid and the elastic side walls of a rectangular tank, *Trans. JSME C* **52**(479), 1915–1921 (in Japanese).

Saleme E. and Liber T. (1965), Breathing vibrations of pressurized partially filled tanks, *AIAA J.* **3**(1), 132–136.

Salzman J. A., Coney T. A., and Masica W. J. (1968), Effects of liquid depth on lateral sloshing under weightless conditions, NASA TN D-4458.

Salzman J. A., Labus T. L., and Masica W. J. (1967), An experimental investigation of the frequency and viscous damping of liquids during weightlessness, NASA TN D-4132.

Salzman J. A. and Masica W. J. (1967), Experimental investigation of liquid propellant reorientation, NASA TN D-3789.

Salzman J. A. and Masica W. J. (1969), An experimental investigation of the frequency and damping of liquids during weightlessness, NASA TN D-5058.

Salzman J. A., Masica W. J., and Lacovie R. E. (1973), Low gravity reorientation in a scale-model Centaur liquid hydrogen tank, NASA TN-7168.

Samali B., Kwok K. C. S., and Tapner D. (1992a), Vibration control of structures by tuned liquid column dampers, Proc. IABSE Congress, New Delhi, India, 461–466.

Samali B., Kwok K. C. S, Parsanejad S., and Xu Y. L. (1992c), Vibration control of buildings by tuned liquid column dampers, Proc. 2nd Int. Conf. *Highrise Buildings*, Nanjing, China, 425–430.

Samali B., Kwok K. C. S., Young G., and Xu Y. L. (1992b), Effectiveness of optimized tuned liquid column dampers in controlling vibration of tall buildings subject to strong ground motions, Proc. 2nd Int. Conf. *Highrise Buildings*, Nanjing, China, 402–407.

Samali B., Lee P., and Kwok K. C. S. (1992d), Wind-induced vibration control of tall buildings with tuned liquid column dampers, 1st Int. Conf. *Motion and Vibration Control*, Yokohama, Japan, 182–187.

Sames P. C., Marcouly D., and Schellin T. E. (2002), Sloshing in rectanagular and cylindrical tanks, *J. Ship Res.* **46**(3), 186–200.

Samoylov Y. A. and Pavlov B. S. (1964), Vibrations of a fluid-filled semi-spherical shell, IVUZ, Aviatson-naya Tekh. **7**(3), 79–86, IAAA64–28277.

Sandorff P. E. (1960), Principles of design of dynamerlly similar models for large propellant tanks, NASA TN D-99.

Sankaran S. and Saville D. (1993), Experiments on the stability of a liquid bridge in an axial electric field, *Phys. Fluids* A **5**, 1081–1087.

Sanz A. (1985), The influence of outer bath in the dynamics of axisymmetric liquid bridges, *J. Fluid Mech.* **156**, 101–140.

Sanz A. and Diez J. L. (1989), Non-axisymmetric oscillations of liquid bridges, *J. Fluid Mech.* **205**, 503–521.

Sanz A. and Martinez I. (1983), Minimum volume for a liquid bridge between equal disks, *J. Colloid Interface Sci.* **93**, 1–9.

Saoka Y., Tamaki T., Sakai F., and Takaeda S. (1989), A proposal for suppression of structural vibrations by tuned liquid dampers, Proc. 43rd Ann. Meeting JSCE (in Japanese).

Satterlee H. M. (1962), Propellant control at zero-g, *Space/Aeronaut.* 72–75.

Satterlee H. M. and Chin J. L. (1965), Meniscus shape under reduced gravity conditions, symp. on fluid mechanics and heat transfer under low gravitational conditions, Lockheed Missiles & Space Co.

Satterlee H. M. and Reynolds W. C. (1964), The dynamics of the free liquid surface in cylindrical containers under strong capillary and weak gravity conditions, Stanford University, Dept. Mech. Eng., Tech. Rept. LG2.

Savino R. (1997), Order of magnitude analysis of g-jitter induced disturbances in fluid-dynamic microgravity experimentation, *Microgravity Quart.* **7**(4), 141–154.

Savino R. (2000), Influence of residual-g and g-jitter on the measurement of thermophysical properties in microgravity, 33rd COSPAR Scientific Assembly, 16–23 July, Warsaw, Poland.

Savino R. and Monti R. (1996), Oscillatory convection in a cylindrical liquid bridge, *Phys. Fluids* **8**, 2906–2920.

Savino R. and Monti R. (1999), Convection induced by residual-g and g-jitters in diffusion experiments, *Int. J. Heat Mass Trans.* **42**, 111–126.

Savino R., and Monti R. (2001), Fluid-dynamic experiment sensitivity to accelerations prevailing on microgravity platforms, in *Physics of Fluids in Microgravity*, R Monti, ed., London, Taylor & Francis, 513–559.

Savino R., Paterna D., and Lappa M. (2003), Marangoni flotation of liquid droplets, *J. Fluid Mech.* **479**, 307–326.

Sawada T., Fujiwara Y., Omura N., Ohaba M., and Tanahashi T. (1998), Lateral vibration of magnetic fluid column, *Studies in Appl. Electromagn. Mech.* **13**, Nonlinear Electromagnetic Systems, 871.

Sawada T., Hashimoto H., and Katagiri K. (1987), Interfacial instability of magnetic fluid in a rectangular container, *JSME Int J.* **30**, 1086–1092.

Sawada T., Hikura H., Shibata S., and Tanahashi T. (1993), Lateral sloshing of a magnetic fluid in a container, *J. Magn. Magnetic Mater.* **122**, 424–427.

Sawada T., Hikura H., Shibata S., and Tanahashi T. (1999), Kinematic characteristics of magnetic fluid sloshing in a rectangular container subjected to nonuniform magnetic fields, *Exp. Fluids* **26**(3), 215–221.

Sawada T., Ohira Y., and Houda H. (2002a), Sloshing motion of a magnetic fluid in a cylindrical container due to horizontal oscillation, *Energy Conv. Manag.* **43**(3), 299–308.

Sawada T., Ohira Y., and Houda H. (2002b), Sloshing behavior of a magnetic fluid in a cylindrical container, *Exp. Fluids* **32**(2), 197–203.

Sayad M. A. El- (1999), Parametric excitation of nonlinear coupled oscillators simulating liquid sloshing impact loading, Ph.D. thesis, Alexandria University, Egypt.

Sayad M. A., El-Hanna S. N., and Ibrahim R. A. (1999), Parametric excitation of nonlinear elastic systems involving hydrodynamic sloshing impact, *Nonlinear Dyn.* **18**(1), 25–50.

Sayar B. A. (1971), *Fluid slosh dampers*, Ph.D. dissertation, Georgia Tech., Atlanta.

Sayar B. A. and Baumgarten J. R. (1981), Pendulum analogy for nonlinear fluid oscillations in spherical containers, *J. Appl Mech.* **48**, 769–772.

Sayar B. A. and Baumgarten J. R. (1982), Linear and nonlinear analysis of fluid slosh dampers, *AIAA J.* **20**(11), 1534–1538.

Scarsi G. (1971), Natural frequencies of viscous liquids in rectangular tanks, *Meccanica* **6**(4), 223–234.

Scarsi G. and Brizzolara E. (1970), On the behavior of liquids in a rectangular tank motion, *Int. Shipbuilding Progress* **17**(194), 316–329.

Schilling U. and Siekmann J. (1980), Calculation of the free oscillations of a heavy incompressible fluid in an axially symmetrical vessel by means of panel methods, *Israel J. Tech.* **18**, 13–20.

Schilling U. and Siekmann J. (1981), Numerical calculation of the natural frequencies of a sloshing liquid in axial symmetrical tanks under strong capillary and weak gravity conditions, *Israel J. Techn.* **19**, 44–50.

Schilling U. and Siekmann J. (1982), Numerical calculation of the translational forced oscillations of a sloshing liquid in axially symmetric tanks, *Israel J. Techn.* **20**, 201–205.

Scharf K. (1990), *Beitraege zur erfassung des Verhaltens von erdbebebenerregten, oberirdischen tankbauwerken*, Ph.D. thesis, Fortschritt-Berichte VDI, Reihe 4 (97), Duesseldorf, FRG.

Schlichting H. (1960), *Boundary Layer Theory*, New York, McGraw-Hill.

Schluter A., Lortz D., and Busse F. H. (1965), On the stability of steady finite amplitude convection, *J. Fluid Mech.* **23**, 129–144.

Schmidt G. (1975), *Parametric Oscillation* VEB Deutscher Verlarg der Wissenschaften, Berlin (in German).

Schmidt G. and Weidenhammer F. (1961), Instability of damped rheolinear vibrations, *Mathematische Nachtichten* **23**, 301–318 (in German).

Schmitt A. F. (1956), Forced oscillations of a fluid in a cylindrical tank undergoing both translation and rotation, Rept. No ZU-7–079, Convair San Diego, CA.

Schmitt A. F. (1957), Forced oscillations of a fluid in cylindrical tank oscillating in a carried acceleration field: a correction, Rept. No ZU-7–074 Convair San Diego, CA, 4 February.

Schneider C. C. and Likins P. W. (1973), Nutation damper versus precession dampers for axisymmetric spinning spacecraft, *J. Spacecraft Rock.* **10**, 218–222.

Schneider J., Schwabe D., and Scharmann A. (1996), Experiments on the surface waves in dynamic thermocapillary liquid layers, *Microgravity Sci. Technol.* **IX**(2), 86–97.

Schoenhals R. J. and Overcamp T. J. (1965), Pressure distribution and bubble formation induced by longitudinal vibration of flexible liquid-filled cylinder, NASA TM-X-53353, MSFC.

Schoenhals R. J., Winter E. R. F, and Griggs E. I. (1967), Effects of longitudinal vibration on discharge of liquids from propellant tanks, Proc. 1967 Heat Transfer and Fluid Mech. Inst., San Diego, 277–297, Stanford University Press.

Scholl H. F., Pinson L. D., and Stephens D. G. (1971), Investigation of slosh anomaly in Apollo lunar module propellant gage, NASA-TM-X-2362, L-7757.

Scholl H. F., Stephens D. G., and Davis P. K. (1972), Ring-baffle pressure distribution and slosh damping in large cylindrical tanks, NASA Tech. Note TND-6870.

Schotte J. S. (2001), Effect of gravity on the linear vibrations of an elastic structure containing an incompressible liquid, Ph.D. thesis, CNAM Paris (in French).

Schotte J. S. and Ohayon R. (1999), Vibration analysis of elastic tanks partially filled with incompressible liquids in presence of a gravity field, Proc. Launch Vehicle Vibrations, 1st European Conf. Launcher Technology, CNES Toulouse, France.

Schotte J. S. and Ohayon R. (2001), Interactions between sloshing modes and incompressible hydroelastic modes for internal fluid–structure systems, Proc. 1st MIT Conf. Comput. Fluid and Solid Mech., Elsevier Science.

Schotte J. S. and Ohayon R. (2002), Effect of gravity on a free–free fluid–structure system, Proc. ASME WAM IMECE2002–32868.

Schulkes R. M. and Cuvelier C. (1991), On the computation of normal modes of a rotating, viscous incompressible fluid with a capillary free boundary, *Comp. Methods Appl. Mech. Eng.* **92**(1), 97–120.

Schultz W. W., Vanden-Broeck J., Jiang L., and Perlin M. (1998), Highly nonlinear standing water waves with small capillary effects, *J. Fluid Mech.* **369**, 253–272.

Schwabe D. and Metzeger J. (1989), Coupling and separation of the buoyant thermocapillary convection, *J. Crystal Growth* **97**, 23–33.

Schwabe D., Möller U., Schneider J., and Scharmann A. (1992), Instabilities of shallow dynamic thermocapillary liquid layers, *Phys. Fluids A* **4**, 2368–2381.

Schwabe D. and Scharmann A. (1979), Some evidence for the existence and magnitude of a critical Marangoni number for the onset of oscillatory flow in crystal growth melts, *J. Crystal Growth* **46**, 125–131.

Schwabe D., Scharmann A., and Preisser F. (1979), Steady and oscillatory Marangoni convection in floating zones under microgravity, Proc. 3rd European Symp. *Material Sciences in Space*, Grenoble.

Schwabe D., Scharmann A., Preisser F., and Oeder R. (1978), Experiments on surface tension driven flow in floating zone melting, *J. Crystal Growth* **43**, 305–312.

Schwabe D., Scharmann A., and Xiadong D. (1991), Prandtl number dependence on the onset of thermocapillary flow in floating zones, *Adv. Space Res.* **XI**, 233–240.

Schwabe D., Velten R., and Scharmann A. (1990), The instability of surface tension driven floating zones under normal and reduced gravity, *J. Crystal Growth* **99**, 1258–1266.

Schwartz E. (1961), Study of liquid behavior under zero and reduced gravity, in *Weightlessness – Physical Phenomena and Biological Effects*, New York, Plenum, 138–149.

Schwind R. G., Scotti R. S., and Skogh J. (1964), The effect of baffles on tank sloshing, Lockheed Missiles Space Co., LMSCA642961.

Schwind R. G., Scotti R. S., and Skogh J. (1967), Analysis of flexible baffles for damping tank sloshing, *J, Spacecraft Rock.* **4**(1), 47–53.

Schy A. A. (1952), A theoretical analysis of the effects of fuel motion on airplane dynamics, NACA Rept. 1080.

Scolan Y. M. and Korobkin A. A. (2001), Three-dimensional theory of water impact, Part q: Inverse Wagner problem, *J. Fluid Mech.* **440**, 293–326.

Scolan Y. M. and Korobkin A. A. (2003), Energy distribution from vertical impact of a three-dimensional solid body onto the flat free surface of an ideal fluid, *J Fluids Struct* **17**, 275–286.

Scott W. E. (1973), The dynamical effect of inertial waves on the gyroscopic motion of a body containing several eccentrically located liquid-filled cylinders, *J. Fluid Mech.* **57**(1), 161–165.

Scott W E. (1975), The large amplitude motion of liquid-filled gyroscope and the non-interaction of inertial and Rossby waves, *J. Fluid Mech.* **72**(4), 649–660.

Scriven L. E. and Sternling L. V. (1960), Marangoni effects, *Nature* **187**, 186–188.

Scriven L. E. and Sterling L. V. (1964), On cellular convection driven by surfactant gradients: effect of mean surface tension and viscosity, *J. Fluid Mech.* **19**, 321–340.

Scurlock R. G. (1991), *History and Origins of Cryogenics*, Oxford, Oxford University Press.

Seebold G., Hollister M. P., and Satterlee H. M. (1967), Capillary hydrostatics in annular tanks, *J. Spacecraft* **4**(1), 101–105.

Sekerzhi-Zenkovich S. Y. (1983), Parametric resonance in a stratified fluid when there are vertical oscillations of the vessel, *Dokl. Akad. Nauk. SSSR* **270**(5), 1089–1091.

Sen B. M. (1927), Waves in canals and basins, *Proc. London Math. Soc. Series* **2**(26), 363–376.

Selmane A. and Lakis A. A. (1997a), Nonlinear dynamic analysis of orthotropic open cylindrical shells subjected to a flowing fluid, *J. Sound Vib.* **202**, 67–93.

Selmane A. and Lakis A. A. (1997b), Vibration analysis of anisotropic open cylindrical shells subjected to a flowing fluid, *J. Fluids Struct.* **11**, 111–134.

Semenov V. O. (1975), On the flow of a liquid from an oscillating cylindrical tank with a curvilinear bottom, in *Differential Equations and Their Applications* **3** Dneprropetrvsk, 106–1114.

Senda K. and Nakagawa K. (1954), On the vibration of an elevated water-tank – I, Tech. Rept. Osaka University **4**(117), 247–264.

Seto M. L. (1996), An investigation on the suppression of flow-induced vibrations of baffled bodies, Ph.D. thesis, University of British Columbia, Vancouver, Canada.

Seto M. L. and Modi V. J. (1997), A numerical approach to liquid sloshing dynamics and control of fluid–structure interaction instabilities, ASME Fluids Eng. Div. Summer Meeting, Paper No FEDSM 97–3302.

Sewal J. L. (1957), An experimental and theoretical study of the effects of fuel on pitching-translation flutter, NACA TN-4166.

Shaaban S. H. and Nash W. A. (1975), Finite element analysis of seismically excited cylindrical storage tank, ground supported and partially filled with liquid, University of Massachusetts, Amherst, MA.

Shang S. P. and Lei G. P. (1988), The free vibration of orthotropic moderately thick shells conveying fluid, *Applications of Mechanics in Engineering, Hangshou*, 231–241.

Shankar K. and Balendra T. (2002), Application of the energy flow method to vibration control of buildings with multiple liquid dampers, *J. Wind Engrg & Indust. Aerodyn.* **90**, 1893–1906.

Shankar P. N. and Kidambi R. (2002), A modal method for finite amplitude nonlinear sloshing, *Pramana-J. Physics* **59**(4), 631–651.

Sharma C. B., Darvizeh M., and Darvizeh A. (1996), Free vibration response of multilayered orthotropic fluid-filled circular cylindrical shells, *Compos. Struct.* **34**, 349–355.

Sharma C. B., Darvizeh M., and Darvizeh A. (1998), Natural frequency response of vertical cantilever composite shells containing fluid, *Engrg Struct.* **20**, 732–737.

Sharma R., Semercill S. E., and Turan O. F. (1992), Floating and immersed plates to control sloshing in a cylindrical container at the fundamental mode, *J. Sound and Vib.* **115**(2), 365–370.

Shemer L. (1990), On the directly generated resonant standing waves in a rectangular tank, *J. Fluid Mech.* **217**, 143–165.

Shemer L. and Chamesse M. (1990), On the hysteresis phenomenon in the directly excited nonlinear resonance sloshing waves in a tank, *Acta Mech.* **81**(1/2), 47–58.

Shemer L., Chamesse M., and Kit E. (1989), Measurements of the dissipation coefficient at the wave maker in the process of generation of the resonant standing waves in a tank, *Exp. Fluids* **7**, 506–512.

Shemer L. and Kit E. (1988), Study of the role of dissipation in evolution of nonlinear sloshing waves in a rectangular channel, *Fluid Dyn. Res.* **4**(2), 89–105.

Shemer L., Kit E., and Miloh T. (1987), Measurements of two- and three-dimensional waves in a channel including the vicinity of cut-off frequencies, *Exper. Fluids* **5**, 66–72.

Shen M. C., Sun S. M., and Hsieh D. Y. (1993), Forced capillary–gravity waves in a circular basin, *Wave Motion* **18**, 401–402.

Shen Y., Neitzel G. P., Jankowki D. F., and Mittelmann H. D. (1990), Energy stability of thermocapillary convection in a model of float-zone crystal-growth process, *J. Fluid Mech.* **217**, 639–669.

Shepherd R. (1969), Earthquake resistant design of petroleum storage tanks, Proc. 2nd Austral. Conf. *Mech. Struct. and Mat.*, Adelaide, 8.1–8.11.

Shepherd R. (1972), The two mass representation of a water tower structure, *J. Sound Vib.* **23**(3), 391–396.

Shevtsova V. M., Ermakov M. K., Ryabitskii E., and Legros J. C. (1997), Oscillations of a liquid bridge free surface due to the thermal convection, *Acta Astron.* **41**(4–10), 471–479.

Shevtsova V. M., Kuhlmann H. C., and Rath H. J. (1995), Thermocapillary convection in liquid bridges with a deformed surface, 46th Cong. Int. Astronaut. Federation, Oslo, Norway, 2–6 Oct.

Shevtsova V. M. and Legros J. C. (1997), Influence of the free surface shape on stability of liquid bridges, Proc. 10th European and 6th Russian Symp. *Physical Sciences in Microgravity*, St. Petersburg, 15–21 June.

Shevtsova V. M. and Legros J. C. (1998a), Convection motion and stability of strongly deformed liquid bridges, *Phys. Fluids* **7**.

Shevtsova V. M. and Legros J. C. (1998b), Oscillatory convective in deformed liquid bridges, *Phys. Fluids* **10**, 1621–1634.

Shevtsova V. M., Mojahed M., Legros J. C. (1999), The loss of stability in ground based experiments in liquid bridges, *Acta Astron.* **44**(7–12), 625–634.

Shevtsova V. M., Wanschura M., Kuhlmann H. C., Shu J. Z., and Legros J. C. (1996), Thermocapillary motion and stability of liquid bridges, 2nd European Symp. *Fluids in Space*, Naples.

Shi J. J. (1987), Fluid sloshing analysis for toroidal tank, Proc. AIAA/ASME/ASCE/AFS 28th *Struct. Struct. Dyn. Mat. Conf.*, Part 2A, 40–44.

Shibata H., Sato H., and Shigeta T. (1965), A seismic design of machine structure, *Proc. 3 WCEE* II, 552–562.

Shibata H., Shinozaki Y., Kobayashi N., and Mieda T. (1986), A study of the liquid slosh response in horizontal cylindrical tanks, ASME *Proc. Pressure Vess. Piping Conf.*, PVP-**108**, 137–142.

Shih C. C. (1966), On subgravity hydroelastic sloshing of a liquid with a curved free surface in a cylindrical tank having a flexible bottom, Thesis, University of Florida.

Shih C. C. and Buchanan H. J. (1971), The drag oscillating flat plate in liquids at low Reynolds numbers, *J. Fluid Mech.* **48**(2, 229–239.

Shih C. F. (1981), Failure of liquid storage tanks due to earthquake excitation, Ph.D. dissertation, Caltech, Pasadena, CA.

Shimizu N. (1983), Advancements and trends of seismic design of liquid storage tanks (in Japanese), *Trans. JSME* **49**(438, C), 145–159.

Shimizu N. (1990), Advances and trends in design of cylindrical liquid storage tanks, *Flow–Structure Vibration and Sloshing*, ASME Pressure Vess. Piping Conf., PVP- **191**, 1–9. (Also *JSME Int. J.* Series III, **33**(2), 111–124.).

Shimizu T. and Hayama S. (1987), Nonlinear response of sloshing based on the shallow water wave theory, *JSME Int. J.* **30**(263), 806–813.

Shinkai A., Mano M., and Tamai S. (1994), Numerical analysis of sloshing problems for the middle sized double hull tanker, *J. Soc. Naval Arch. Japan* **176**, 387–396.

Shinkai A., Tamia S., and Mano M. (1995), Sloshing impact pressure induced on cargo oil tank walls on the middle-sized double hull tanker, *Trans. Soc. Naval Arch. West. Japan* **90**, 91–98.

Shiojiri H. and Hagiwara Y. (1990), Development of a computational method for nonlinear sloshing by BEM, in *Flow–Structure Vibration and Sloshing*, ASME Pressure Vess. Piping Conf., PVP-**191**, 149–154.

Shkenev Yu. S. (1964) The dynamics of an elastic and elastoplastic shell filled with an ideal liquid, Proc. 4th All-Union Conf. on *The Theory of Shells and Plates*, Izd. Akad. Nauk. SSSR, Moscow, 997–1007 (in Russian).

Shkenev Yu. S. (1970), On the stability of an elastic shell filled with an ideal liquid, *Prikl. Mekh.* **6**(8), 128–132.

Shklyarchuk F. N. (1965a), The axisymmetric vibrations of a liquid inside an elastic cylindrical shell with an elastic bottom, *Izv. Vuzov., Aviats.* **4**, 75–83.

Shklyarchuk F. N. (1965b), Approximate method of calculating symmetric oscillations of fluid-filled shells of revolution, *Izv. AN SSSR Mekhanika* **6**, 123–129.

Shklyarchuk F. N. (1966), Variational methods of calculating axi-symmetric oscillations of shells of revolution, partially filed with fluid, Tr V. I. Vses Konferentsii Po Teorii Obolocheki Plastemok, Moscow, Nauka, 835–840 (in Russian).

Shklyarchuk F. N. (1970a), The dynamic stability of a cylindrical shell partially filled with a liquid and subject to longitudinal vibrations, Proc. 7th All-Union Conf. on *The Theory of Shells and Plates*, Moscow, Nauka, 619–624 (in Russian).

Shklyarchuk F. N. (1970b), Parametric vibrations of a cylindrical shell partially filled with a liquid and subject to longitudinal excitation, Proc. 7th All-Union Conf. on *the Theory of Shells and Plates*, Moscow, Nauka, 314–329 (in Russian).

Shklyarchuk F. N. (1973), The effect of the compressibility of a liquid on the longitudinal vibrations of a cylindrical tank, in *Vibrations of Elastic Structures with a Liquid*, Novosib, Elektrotekh Inst, Novosibirk, 291–313.

Shklyarchuk F. N. (1983), *The Dynamics of Structures of Aircrafts*, Izd. Moskovskogo Aviats. Inst., Moscow (in Russian).

Shmakov V. P. (1964), Concerning the equations of axi-symmetrical vibrations of a fluid cylindrical shell, *Izvestiya Acad. Sci. SSSR, Series Mekh. i Mash.* **1**, 170–173, Translated as NASA TT-F-219.

Shrimali M. K. and Jangid R. S. (2002), Nonlinear seismic response of base-isolated liquid storage tanks to bi-directional excitation, *Nuclear Engrg Des.* **217**, 1–20.

Shved G. L. (1959), Small oscillations of the free surface of an ideal fluid in a plane moving vessel with smooth vertical walls, Trudy Odessk, *Teknol. Inst. Pislch., ikholodil'n Promysh* **8**, 150–160.

Shuleikin V. V. (1962), Some remarks on the behavior of a liquid approaching zero-g conditions, *Dokl. Akad. Nauk. SSSR* **147**, 1075–1083.

Shul'ga N. A., and Shul'ga O. M. (2003), Wave propagation in an orthotropic cylinder with a liquid, *Int. Appl. Mech.* **39**(6), 736–740.

Shveiko Yu. Yu. and Kuleshov V. B. (1973), Investigation of natural oscillations of cylindrical shells with fluid, in *Oscillations of Elastic Structures with Fluid*, Izd-vo Novosib Elektrotekhn in-ta, Novosibirska (in Russian).

Shyu K. L. and Kuo H. C. (1996) Dynamic behavior of a U-type tuned liquid damper, *Int. Shipbuilding Prog.* **43**(436), 331–345.

Siegel R. (1961), Transient capillary rise in reduced and zero gravity fields, *ASME J. Appl. Mech.* **28**(2), 165–186.

Siegert C. E., Petrash D. A., and Otto E. W. (1964), Time response of liquid–vapor interface after entering weightlessness, NASA TN D-2458.

Siegert C. E. Petrash D. A., and Otto E. W. (1965), Behavior of liquid–vapor interface of cryogenic liquids during weightlessness, NASA TN D-2658.

Siekmann J. and Chang S. C. (1966), On liquid sloshing in a cylindrical tank with a flexible bottom, Proc. Southeastern Symp. Missile and Aerospace Vehicles Sciences, *Am. Astronaut. Soc.* **2**(1–8), 98–105.

Siekmann J. and Chang S. C. (1967), On liquid sloshing in a cylindrical tank with a flexible bottom under strong capillary and weak gravity conditions, *J. Astronaut. Sci.* **XIV**(4), 167–172.

Siekmann J. and Chang S. C. (1968), On the dynamics of liquids in a cylindrical tank with a flexible bottom, *Ing. Arch.* **37**, 99–109.

Siekmann J. and Chang S. C. (1971a), Dynamic interaction of liquid with elastic structure of a circular cylindrical container, *Ing. Arch.* **4**, 266–280.

Siekmann J. and Chang S. C. (1971b), On the change of natural frequencies of a sloshing liquid by movable devices, *Acta Mech.* **11**(1/2), 73–86.

Siekmann J., Scheideler W., and Tietze R. (1981), Static meniscus configurations in propellant tanks under reduced gravity, *Comput. Methods Appl. Mech. Engrg.* **28**, 103–108.

Siekmann J. and Schilling U. (1974a), Calculation of the free oscillations of a liquid in axi-symmetric motionless containers of arbitrary shape, *Zeit. Flugwiss.* **22,** 168–173.

Siekmann J. and Schilling U. (1974b), Calculation of the free oscillations of a liquid in motionless axisymmetric containers of arbitrary shape, *Zeit. Flugwiss.* **22**, 168–173.

Silber M. and Knobloch E. (1989), Parametrically excited surface waves in square geometry, *Phys. Lett.* **137**, 349–354.

Silber M. and Proctor M. R. E. (1998), Nonlinear competition between small and large hexagonal patterns, *Phys. Rev. Lett.* **81**, 2450–2453.

Silber M. and Skeldon A. C. (1999), Parametrically excited surface waves: two-frequency forcing, normal form symmetries, and pattern selection, *Phys. Revs. E* **59**, 5446–5456.

Silber M., Topaz C. M., and Skeldon A. C. (2000), Two-frequency forced Faraday waves: weakly damped modes and pattern selection, *Physica D* **143**(1–4), 205–225.

Silveira M. A., Stephens D. C., and Leonard H. W. (1961), An experimental investigation of damping of liquid oscillations in cylindrical tanks with various baffles, NASA TN D-715.

Simonelli F. and Gollub, J. P. (1989), Surface wave mode interactions: effects of symmetry and degeneracy, *J. Fluid Mech.* **199**, 471–494.

Sivak V. F. (1998), Experimental study of the resonance and dissipative properties of a liquid-filled glass fiber composite reinforced plastic cylindrical shell, *Int. Appl. Mech.* **34**(2), 136–139.

Sivak V. F. (2001), Experimental investigation of the natural vibrations of shells of revolution with a liquid and added masses, *Int. Appl. Mech.* **37**(1), 122–125.

Sivak V. F. and Telalov A. I. (1991), Experimental investigation of the vibrations of a cylindrical shell contacting with a liquid, *Soviet Appl. Mech. (Prikl. Mekh.)* **27**(10), 121–123.

Sivak V. F. and Telalov A. I. (1992), Experimental investigation of the vibration damping of a longitudinally reinforced cylindrical shell contacting with a liquid, *Soviet Appl. Mech. (Prikl. Mekh.)* **28**(4), 77–80.

Sivak V. F. and Telalov A. I. (1998), Experimental investigation of the resonance and dissipative properties of a glass-fiber-reinforced plastic cylindrical shell filled with a liquid, *Sov. Appl. Mech. (Prikl. Mekh.)* **34**(2), 39–42.

Skalak M. A. and Conly J. F. (1964), Surface waves on a rotating fluid, *AIAA J.* **2**(2), 306–312.

Skalak R. and Yarymovich M. I. (1962), Forced large amplitude surface waves, Proc. 4th US Nat. Congress Appl. Mech., 1411–1418.

Slabinski V. J. (1977), Intelsat I. V. in-orbit liquid slosh tests and problems in the theoretical analysis of the data, ESA SP-129, 87–102.

Slafer L. I. and Challoner A. D. (1988), Propellant interaction with the payload control system of dual-spin spacecraft, *J. Guidance* **11**(4), 343–351.

Slibar A. and Troger H. (1975), Dynamic steady state behavior of a tractor-semitrailer-system carrying liquid load, Proc. IUTAM Symp. 1975-08-18-22, *The Dynamics of Vehicles on Roads & Railway Tracks*, Hans B. Pacejka, ed., the Netherlands, Swts & Zeitlinger BV.

Slibar A. and Troger H. (1976), Die kritischen fahrzustande des tank-aufliegerzuges bei verschiedenen baladungsgraden und beruhrbedingungen im stationaren fahrbetrieb. Bundes-miniserium fur Bauten und Technik Wien, Strassenforschung, Heft 55.

Slibar A. and Troger H. (1977), The steady state behavior of tank trailer system carrying rigid or liquid cargo, VSD-IUTAM Symp. *Dynamics of Vehicles on Roads & Trucks*, Vienna, 256–264.

Sloane M. N. (1960), The dynamics of the sloshing phenomenon: a literature survey, Rept. GM-60–5111–5, Space Tech. Lab., Inc.

Slobozhanin L. A. (1966), Hydrostatics in weak force fields: equilibrium shapes of rotating liquid surface at zero-g, Fluid Dynamics **1**(5), 113–116. (Translation of *Izvestiya Acad. Nauk. SSSR Mekh. Zhid. Gaza* **1**(5), 157–160, 1966, by New York, Faraday Press.

Slobozhanin L. A. and Alexander J. I. D. (2000), Stability of disconnected free surfaces in a cylindrical container under zero gravity: Simple cases, *Phys. Fluids* **12**(11), 2800–2808.

Slobozhanin L. A., Alexander J. I. D, and Fedoseyev A. I. (1999a), Doubly connected axisymmetric free surface in a cylindrical container: Shape, stability and application, AIAA Paper AIAA-99–1029, 37th AIAA Aerospace Sciences Meeting, Reno, NV.

Slobozhanin L. A., Alexander J. I. D, and Fedoseyev A. I. (1999b), Shape and stability of doubly connected axisymmetric free surface in a cylindrical container, *Phys. Fluids* **11**, 3668–3677.

Slutskii F. (1895), De la rotation de la terre supposeé fluide a son intérieur, *Bull. de la Soc. des Naturalistes de Moscow* **9**.

Smith C. (1948), The effect of fuel sloshing on the lateral stability of a free-flying airplane model, NACA RM L8C.

Smith K. A. (1966), On convective instability induced by surface tension gradients, *J. Fluid Mech.* **24**, 401–414.

Smith K. W. (1947), Fuel sloshing relation between surface movement and moving masses displacement for a circular tank, RAE, Great Britain, G W File Ref GW/55 027/KWS.

Smith K. W. (1956), Presented work on the effects of fuel sloshing in the control of ballistic missile, RAE, Great Britain, Note GW/5/507/LKWS.

Sneddon I. H. (1972), *The Use of Integral Transforms*, New York, McGraw-Hill.

Snowdon J. C. (1968), *Vibration and Shock in Damped Mechanical Systems*, New York, John Wiley.

Snyder H. A. (1999), Sloshing in microgravity, *Cryogenics* **39**(12), 1047–1055.

Snyder H. A. (2001), Effect of sloshing on the mechanics of Dewar systems in low-gravity, *Cryogenics* **41**(11/12), 825–832.

Sobolev S. L. (1960), On the motion of a symmetrical top with a fluid-filled cavity, *Prikl. Mekh. i Tekh. Fiz.* **3**, 20–55.

Soedel S. M. and Soedel W. (1994), On the free and forced vibration of a plate supporting a freely sloshing surface liquid, *J. Sound Vib.* **171**(2), 159–171.

Sogabe K. and Shibata H. (1974a), Response analysis on sloshing of liquid in a cylindrical storage i: basic equation and response to sinusoidal input, *J. Inst. Indust. Scie., University of Tokyo* 26(3), 119–122 (in Japanese).

Sogabe K. and Shibata H. (1974b), Response analysis on sloshing of liquid in a cylindrical storage ii: transient response to sinusoidal input, *J. Inst. Indust. Scie., University of Tokyo* **26**(4), 152–155 (in Japanese).

Sokolov, VI (1954), The influence of rotor contained fluid mass on the critical velocity of centrifugal machine shafts, Proc. Moscow Inst. Meat and Milk Indust. 11, 71–72.

Solaas F. (1995), Analytical and numerical studies of sloshing in tanks, Ph.D. thesis, The Norwegian Institute of Technology, Trondheim.

Solaas F. and Faltinsen O. M. (1997), Combined numerical and analytical solution for sloshing in two-dimensional tanks of general shape, *J. Ship Research* **41**, 118–129.

Someya S., Okamoto K., and Madarame H. (1995), Self-induced sloshing caused by upward jet impinging on upper-inner-structure, in *Fluid-Sloshing and Fluid–Structure Interaction*, ASME Pressure Vess. Piping Conf., PVP-**314**, 115–120.

Someya S., Okamoto K. and Madarame H. (2000), The self-induced free-surface "swell flapping" caused by the interaction among a jet, a free surface and a structure, *J. Fluids Struct.* **14**(4), 511–528.

Sonobe Y. and Nishikawa T. (1969), Study of the earthquake proof design of elevated water tanks, Proc. 4th World Conf. *Earthquake Eng.*, Santiago **B-4**, 11–24.

Soong T. T. and Constantinou M. C. (1994), *Passive and Active Structural Vibration Control in Civil Engineering*, New York, Springer.

Sorokin V. I. (1957), The effect of fountain formation at the surface of a vertically oscillating liquid, *Soviet Physics-Dokl.* **3**(3), 281–291.

Souli M. and Zolesio J. P. (2001), Arbitrary Lagrangian–Eulerian and free surface methods in fluid mechanics, *Comput. Meth. Appl. Mech. Engrg.* **191**, 451–466.

Soundararajan A. (1983), Stationary response of liquid free surface under wide-band random parametric excitation, Master thesis, Texas Tech. University, Lubbock, TX.

Soundararajan A. and Ibrahim R. A. (1988), Parametric and autoparametric vibrations of an elevated water tower, part iii: random response, *J. Sound Vib.* **121**(3), 445–462.

Souto A. and Gonzalez V. (2001), Passive stabilizer simulation using SPH models, Proc. *Fluid–Structure Interaction* 2001, Halkidiki, Greece **1**, 35–44.

Souto A., Laioz I., Abril S., and Abad R. (2001), Passive stabilizer tanks dimensioning using SPH simulations, Proc. 4th Int. Conf. *Marine Technology*, Sczcecin, Poland **1**, 171–181.

Spradley L. W., Bourgeois S. L., and Lin F. N. (1975), Space processing convection evaluation, g-jitter convection of confined fluids in low gravity, AIAA Paper 75–695.

Squire H. B. (1958), *Aero. Res. Counc., London* **16**, 021.

Sretanskii L. N. (1956), Propagation of waves of finite amplitudes in a circular channel, *Trudy. Mor. Gidrofiz. in-ta Akad. Nauk. SSSR* **6**, 3–6.

Sretanskii L. N. (1957), The oscillations of liquid in a moving container, *Izv. Akad. Nauk. SSSR Otd. Tekh. Nauk.* **10**, 1494.

Standberg L. (1978), *Lateral Stability of Road Tankers*, National Road & Traffic Res. Inst. Rept. 138A, Sweden.

Steel Structure & Civil Eng Division, 1991, Oscillation control of pylon and cables in cable stayed Sakitama, Bridge (up-route side), *Mitui Zosen Tech. Rev.* **143**, 17–25.

Steklov M.W. (1902), Sur les problemes fondamentaux de la physique mathematique, *Ann. Sci. Ecole Norm.* Sup. 19, 455–490.

Stephens D.G. (1965), Experimental investigation of liquid impact in a model propellant tank, NASA TN D-2913.

Stephens D.G. (1966), Flexible baffles for slosh damping, *J. Spacecraft Rock.* **3**(5), 765–766.

Stephens D.G. and Leonard H.W. (1963), The coupled dynamic response of a tank partially filled with a liquid undergoing free and forced planner oscillations, NASA TN D-1945.

Stephens D.G., Leonard H W., and Perry T.W. (1962), Investigation of the damping of liquids in right circular tanks including the effects of a time-variant liquid depth, NASA TN D-1367.

Stephens D.G., Leonard H.W., and Silveira M.A. (1963), An experimental investigation of the damping of liquid oscillations in an oblate spheroidal tank with and without baffles, NASA TN D-808.

Stephens D.G. and Scholl H.F. (1967), Effectiveness of flexible and rigid ring baffles for damping liquid oscillations in large-scale cylindrical tanks, NASA TN D-3878.

Sternling C.V. and Scriven L.E. (1959), Interfacial tension hydrodynamic instability and the Marangoni effect, *AICHE J.* **5**, 514.

Steven G.P. (1976), Consistent mass matrix in fluid sloshing problems, *AIAA J.* **14**(2), 245–247.

Stewartson K. (1959), On the stability of a spinning top containing liquid, *J. Fluid Mech.* **5**(4), 577–592.

Stigter C. (1966), The performance of U-tanks as a passive anti-rolling device, *Royal Inst. Naval Archit.* **13**, 249–275.

Stillman W.E. (1973), Free vibration of cylinders containing liquid, *J. Sound Vib.* **30**, 509–524.

Stofan A.J. and Armstead A.L. (1962), Analytical and experimental investigation of forces and frequencies resulting from liquid sloshing in a spherical tank, NASA TN D-1281.

Stofan A.J. and Irvine E.S. (1963), Experimental investigation of the slosh-damping effectiveness of positive-expulsion bags and diaphragms in spherical tanks, NASA TN D-1712.

Stofan A.J. and Pauli A.J. (1962), Experimental damping of liquid oscillations in a spherical tank by positive expulsion bags and diaphragms, NASA TN D-1311.

Stofan A.J. and Sumner I.E. (1963), Experimental investigation of the slosh-damping effectiveness of positive-expulsion bags and diaphragms in spherical tanks, NASA TN D-1712.

Stoker J.J. (1957), *Water Waves*, London, Interscience Publishers Inc.

Stokes Sir G.G. (1851), On the effect of the internal friction of fluids on the motion of pendulums, *Trans. Cambridge Phil. Soc.* **9**(11), 8–106.

Stokes Sir G.G. (1880), Considerations relative to the greatest height of oscillatory waves which can propagated without change of form, *Math. Phys. Pap.* **1**, 225–228.

Stokes Sir G.G. (1905), *Mathematical and Physical Paper* (1813–1903), London, Cambridge University Press, **1**, 1880–1905.

Stokes Sir G.G. (1947), On the theory of oscillatory waves, *Trans. Camb. Phil. Soc.* **8**, 441–455, Reprinted in *Math. Phys.*, Papers **1**, 197–219.

Stokes Sir G.G. (1949a), On some cases of fluid motion, *Trans. Camb. Phil. Soc.* **8**.

Stokes Sir G.G. (1949b), On the critical values of sums of periodic series, *Trans. Camb. Phil. Soc.* **8**.

Struble R.A. (1962), *Nonlinear Differential Equations*, New York, McGraw-Hill.

Sturtevant B. (1966), Optical depth gauge for laboratory studies of water waves, *Review of Scientific Instruments* **37**(11), 1460–1463.

Su T.C. (1981), The effect of viscosity on free oscillations of fluid-filled spherical shells, *J. Sound. Vib.* **74**(2), 205–220.

Su T.C. (1983), The effect of viscosity on the forced vibrations of a fluid-filled elastic shell, *ASME J. Appl. Mech.* **50**, 517–524.

Su T. C. (1987), Sloshing due to acceleration in the rectangular tanks, Proc. ASME Symp. *Fluid–Structure Vibration and Liquid Sloshing*, PVP-**128**, 69–79.

Su, T. C. (1992), Nonlinear sloshing and the coupled dynamics of liquid propellants and spacecraft, Tech. Rept. Florida Atlantic University, NTIS ADA250023.

Su T. C. and Kang S. Y. (1984a), Analysis of liquid impact of moving containers, *Dev. Appl. Mech.*, **12**.

Su T. C. and Kang S. Y. (1984b), Numerical simulation of liquid sloshing, *Eng. Mech. Civil Eng.*, A. P. Boresi and K. P. Chong, eds., ASCE **2**, 1069–1072.

Su T. C. and Kang S. Y. (1986), Analysis and testing of the large amplitude liquid sloshing in rectangular containers, Seismic Eng. Piping Systems, Tanks and Power Plant Equip., PVP- **108**, ASME, 149–154.

Su T. C. and Lian Q. X. (1992), Some preliminary observations of vortex formation during draining, *Fluid–Structure Vibration and Sloshing*, ASME Pressure Vess. Piping Conf., PVP- **232**, 1–4.

Su T. C., Lou Y. K., Flipse J. E., and Bridges T. (1982a), A numerical analysis of large amplitude liquid sloshing in baffled containers, Rept. MA-RD-940–82046, US DoT, Maritime Admin, 290 pages.

Su T. C., Lou Y. K. Flipse J. E., and Bridges T. (1982b), A nonlinear analysis of liquid sloshing in rigid containers, Rept. DOT/RSPA/DMA-50/82/1, Office of University Research, US DoT.

Su T. C. and Wang Y. (1989), Numerical simulation of three-dimensional large amplitude liquid sloshing in rectangular containers subjected in vertical excitation, *Sloshing and Fluid Structure Vibration*, ASME Pressure Vess. Piping Conf., PVP- **157**, 115–126.

Su T. C. and Wang Y. (1990), Numerical simulation of three-dimensional large amplitude liquid sloshing in cylindrical tanks subjected to arbitrary excitations, *Flow-Structure Vibration and Sloshing*, ASME Pressure Vess. Piping Conf., PVP- **191**, 127–148.

Su T. C. and Wang H. (1991), The control of large amplitude liquid sloshing in rectangular containers by active baffles, *Fluid–Structure Vibration and Sloshing*, ASME Pressure Vess. Piping Conf., PVP- **223**, 81–89.

Sudo S. and Hashimoto H. (1986), Dynamic behavior of a liquid in a cylindrical container subject to horizontal vibration (on nonlinear response of liquid surface), *Trans. JSME B* **52**(483), 3655–3659 (in Japanese).

Sudo S. and Hashimoto H. (1987), Unsteady pressure response of a liquid in a cylindrical container subject to vertical vibration, *Trans. JSME B* **53**(491), 2047–2053 (in Japanese).

Sudo S., Hashimoto H., Ikeda A., and Katagiri K. (1987c), Some studies of magnetic liquid sloshing, *J. Magn. Magn. Mat.* **65**(2/3), 219–222.

Sudo S., Hashimoto H., and Katagiri K. (1987a), Interfacial instability of magnetic fluids in a rectangular container, *JSME Int. J.* **30**, 1086–1092.

Sudo S., Hashimoto H., Katagiri K., and Ikeda A. (1987b), Experimental study of surface behavior in magnetic liquid sloshing, *Trans. JSME B* **53**(495), 3242–3246 (in Japanese).

Sudo S., Hashimoto H., Katagiri K., and Shibuya T. (1988), Nonlinear response of a liquid surface in lateral sloshing, *Trans. JSME B* **54**(507), 3242–3246 (in Japanese).

Sudo S., Hashimoto H., Tani J., and Chiba M. (1985), Dynamic behavior of a liquid in a cylindrical container subject to horizontal vibration (liquid surface response in an elastic container subject to vibration with higher frequencies), *Trans. JSME B* **51**(472), 4155–4161 (in Japanese).

Sudo S. and Ohaba M. (1998), Interfacial phenomena of liquids in moving container, Res. Bull. Iwaki Meisei University **11**, 45–59 (in Japanese).

Sudo S., Ohaba M., Katagiri K., and Hashimoto H. (1993), Dynamic behavior of magnetic liquid drop on a solid base subject to magnetic field and vertical vibration, *J. Magn. Magn. Mat.* **122**, 248–258.

Sudo S., Ohaba M., Katagiri K., and Hashimoto H. (1997), Magnetic fluid sloshing in moving container, *Magnetohydrodynamics* **33**(4), 480–485.

Sumner I. E. (1963), Preliminary experimental investigation of frequencies and forces resulting from liquid sloshing in toroidal tanks, NASA TN D-1709.

Sumner I. E. (1964), Experimental investigation of slosh suppression effectiveness of annular ring baffles in spherical tanks, NASA TN D-2519.

Sumner I. E. (1965), Experimentally determined pendulum analogy of liquid sloshing in spherical and oblate-spheroidal tanks, NASA TN D-2737.

Sumner I. E. (1966), Experimental investigation of stability boundaries for planar and nonplanar sloshing in spherical tanks, NASA TN D-3210.

Sumner I. E., Lacovic R. F., and Stofan A. J. (1966), Experimental investigation of liquid sloshing in a scale-model Centaur liquid hydrogen tank, NASA TN X-1313.

Sumner I. E. and Stofan A. J. (1963), An experimental investigation of the viscous damping of liquid sloshing in spherical tanks, NASA TN D-1991.

Sumner I. E., Stofan A. J., and Shramo D. J. (1964), Experimental sloshing characteristics and a mechanical analogy of liquid sloshing in a scale-model centaur liquid oxygen tank, NASA TM X-999.

Sun K. (1994), Earthquake responses of buildings with liquid column dampers, Proc. 5th U. S. National Conf. *Earthquake Eng.* **II**, 411–420, Chicago.

Sun L. M. and Fujino Y. (1994), A semi-analytical model for tuned liquid damper (TLD) with wave breaking, *J. Fluid Struct.* **8**(5), 471–488.

Sun L. M., Fujino Y., Chaiseri P., and Pacheco B. M. (1995), The properties of tuned liquid dampers using a TMD analogy, *Earthquake Eng. Struct. Dyn.* **24**, 967–976.

Sun L. M., Fujino Y., Pacheco B. M., and Isobe M. (1989), Nonlinear waves and dynamic pressures in rectangular tuned liquid damper (LTD) – simulation and experimental verification, *JSCE Struct. Eng./Earthquake Eng.* **6**(2) 251s–262s.

Sun S. M., Shen M. C., and Hsieh D. Y. (1995), Nonlinear theory of forced waves in a circular basin, *Wave Motion* **21**(4), 331–341.

Sun Tsao (1960), On fluid surface waves influenced by centrifugal force pressure, PMTPH **3**, 90–96.

Swalley F. E., Platt G. K., and Hastings L. J. (1965), Sturn V low-gravity fluid mechanics problems and their investigation by full-scale experiment, AFOSR/LMSC Symp. *Fluid Mechanics and Heat Transfer under Low-g*, Palo Alto, 1–24.

Symans M. D. and Constantinou M. C. (1999), Semi-active control systems for seismic protection of structures: a state-of-the art review, *Eng. Struct.* **21**, 469–487.

Szymczyk J. A. and Siekmann J. (1989), Experimental investigation of the thermocapillary flow along cylindrical interfaces under rotation, Proc. 7th Europ. Symp. *Materials and Fluid Sciences in Microgravity*, Oxford, ESA-SP-295, 315–320.

Tadjbakhsh I. and Keller J. B. (1960), Standing waves of finite amplitude, *J. Fluid Mech.* **8**(3), 442–451.

Takahara H. and Kimura K. (1997), Parametric vibration of liquid surface in rectangular tank subjected to pitching excitation (2nd report, symmetric mode), *Trans. JSME C* **63**(610), 1861–1868 (in Japanese).

Takahara H., Kimura K., Ito T., and Sakata M. (1993a), Nonlinear liquid oscillation in an elastic circular tank subjected to pitching excitation, *Trans. JSME C* **59**(561), 1378–1385 (in Japanese).

Takahara H., Kimura K., and Sakata M. (1993b), Nonlinear liquid oscillation in a circular cylindrical tank subjected to pitching excitation, *Proc. of Asia-Pacific Vibration Conf. '93* **1**, 294–299.

Takahara H., Kimura K., and Sakata M. (1994b), Effects of higher order radial modes upon nonlinear sloshing in a circular cylindrical tank subjected to pitching excitation, *Trans. JSME C* **60**(574), 1924–1932 (in Japanese).

Takahara H., Kimura K., and Sakata M. (1994c), Parametric vibration of liquid surface in a circular cylindrical tank subjected to pitching excitation, *Trans. JSME C* **60**(579), 3727–3732 (in Japanese).

Takahara H., Kimura K., and Sakata M. (1995), Frequency response of sloshing in a circular cylindrical tank subjected to pitching excitation, *Proc. Asia-Pacific Vibration Conf. '95* **II**, 703–708.

Takahara H., Kimura K., Tsutsui M., and Sakata M. (1994a), Frequency response of sloshing in a circular cylindrical tank subjected to pitching excitation, *Trans. JSME C* **60**(572), 1210–1216 (in Japanese).

Takahashi S. and Hirano Y. (1970), Vibration of a combination of circular plates and cylindrical shells: first report, a cylindrical shell with circular plates at ends, *Bull. JSME* **13**, 240–247.

Takayangi M. (1990), Parametric resonance of liquid storage axi-symmetric shell under horizontal excitation, *Flow–Structure Vibration Sloshing*, ASME Pressure Vess. Piping Conf., PVP-**191**, 63–68.

Takemoto H., Oka S., *et al.* (1994), Experimental study on sloshing impact loads of middle sized tankers with double hull, *J. Soc. Naval Arch. Japan* **176**, 399–410.

Takizawa A. and Kondo S. (1993), Flow-induced sloshing of in-vessel free-surface flow: Computer discovery using PCBFC method, Proc. 5th Int. Symp. *Computational Fluid Dynamics* **3**, 193–198.

Takizawa A. and Kondo S. (1995), Computer discovery of the mechanism of flow-induced sloshing, *Fluid-Sloshing and Fluid–Structure Interaction*, ASME Pressure Vess. Piping Conf., PVP-**314**, 153–158.

Takizawa A., Koshisuka S., and Kondo S. (1992), Generalization of physical component boundary filled coordinate (PCBFC) method for the analysis of free surface flow, *Int. J. Numer. Meth. Fluids* **15**, 1213–1237.

Tam B. T. (1998), A displacement-based formulation of nearly-incompressible fluid element for analysis of large amplitude liquid sloshing for tuned liquid damper applications, Ph.D. dissertation, Asian Inst. Tech., Bangkok, Thailand.

Tamura Y. (1990), Suppression of wind-induced vibrations of buildings, *J. Wind Engrg.* **44**, 71–84.

Tamura Y., Fujii K., Sato T., Wakahara T., and Kosugi M. (1988), Wind-induced vibration of tall towers and practical applications of tuned sloshing damper, Proc. Workshop *Serviceability of Buildings*, Ottawa, 228–241.

Tamura Y., Kohsaka R., and Modi V. J. (1992), Practical application of nutation damper for suppressing wind induced vibrations of airport towers, *J. Wind Eng. Indust. Aerodyn.* **41–44**, 1919–1930.

Tamura Y., Kohsaka R., Nakamura O., Miyashita K. I., and Modi V. J. (1996), Wind-induced response of an airport tower – efficiency of tuned liquid damper, *J. Wind. Eng. Indust. Aerodyn.* **65**, 121–131.

Tan DGH (1994), Dynamics of a spinning spacecraft containing liquid propellant, in *Aerospace Vehicle Dynamics and Control*, M. V. Cook and M. J. Rycroft, eds., Clarendon, 417–438.

Tanaka S., Kaneko S., and Watanabe T. (1999), Self-excited oscillations of dual cylindrical flexible weir shells due to over flow fluid, *Trans. JSME C* **65**(634), 2226–2234 (in Japanese).

Tanaka S., Kaneko S., and Watanabe T. (2000), Self-excited oscillations of dual cylindrical flexible weir shells due to over flow fluid, *J. Vib. Control* **6**(3), 351–373.

Tanaka H. and Nakayama T. (1991), A boundary element method for the analysis of nonlinear sloshing in three-dimensional containers, *Trans. JSME B* **57**(538), 1934–1940 (in Japanese).

Tanaka N. and Fukuhara K. (1996), Numerical prediction of self-excited free surface sloshing, in *Flow Modeling Turbulence Measurements V. I.*, Proc. 6th Int. Symp., Tallahassee, FL (Chen, Shih, and King, eds.,) 671–678, Balkema.

Tang Y. (1992), Hydrodynamic pressure in a tank containing two liquids, *Fluid–Structure Vibration and Sloshing*, ASME Pressure Vess. Piping Conf., PVP- **232**, 25–30.

Tang Y. (1993), Sloshing displacements in a tank containing two liquids, Proc. ASME Pressure Vessel Piping Conf., PVP- **258**, 143–149.

Tang Y. (1994), SSI effects for a tank containing two liquids, *Sloshing, Fluid–Structure Interaction and Struct. Response Due to Shock and Impact Loads*, ASME Pressure Vess. Piping Conf., PVP-**272**, 6771.

Tang Y. and Chang Y. W. (1993a), The exact solutions to the dynamic response of tanks containing two liquids, Tech. Rept. ANL/RE-93/2, ANL, Argonne, IL.

Tang Y. and Chang Y. W. (1993b), Rocking response of tanks containing two liquids, Technical Report ANL/RE-93/5, ANL, Argonne, IL.

Tang Y., Ma D. C., and Chang Y. W. (1991), Sloshing response in a tank containing two liquids, *Fluid–Structure Vibration and Sloshing*, ASME Pressure Vess. Piping Conf., PVP-**223**, 97–104.

Tang Y., Uras R. A., and Chang Y. W. (1993), Effect of viscosity on dynamic response of a liquid storage tank, Proc. ASME. Pressure Vess. Piping Conf., PVP- **258**, 135–142.

Tani J., Chonan S., Zhang H., and Kondo K. (1989a), Free vibrations of partially liquid-filled rotating cylindrical shells, *Sloshing and Fluid Structure Vibration*, ASME Pressure Vess. Piping Conf., PVP- **157**, 173–178.

Tani J. and Haiji H. (1986), The free vibration of free-clamped fluid-coupled coaxial cylindrical shells (in Japanese), *Trans. JSME* C **52**, 3137–3144.

Tani J., Kodama S., Sakai T., and Chiba M. (1989b), Dynamic stability of fluid-coupled coaxial cylindrical shells under vertical excitation, *Sloshing and Fluid Structure Vibration*, ASME Pressure Vess. Piping Conf., PVP- **157**, 41–46.

Tani J., Ohtomo K., Sakai T, and Chiba M. (1989c), Hydroelastic vibration of partially liquid-filled coaxial shells, *Sloshing and Fluid Structure Vibration*, ASME Pressure Vess. Piping Conf., PVP-**157**, 29–34.

Tao M. and Zhang W. (2002a), Dynamic stability of a flexible spinning cylinder partially filled with a liquid, *ASME J. Appl. Mech.* **69**, 708–710.

Tao M. and Zhang W. (2002b), Dynamic stability of a rotor partially filled with a viscous liquid, *ASME J. Appl. Mech.* **69**, 705–707.

Tao Y. and Kou S. (1996), Thermocapillary convection in a low Pr material under simulated reduced gravity condition, 3rd Microgravity Fluid Phys. Conf., NASA cp 3338, 301.

Taylor G. I. (1950), The instability of liquid surfaces when accelerated in a direction perpendicular to their planes, *Proc. Royal Soc. (London) A* **201**, 192–196.

Taylor G. I. (1954), An experimental study of standing waves, *Proc. Royal Soc. (London) A* **218**, 44–59.

Temochenko Mye (1969), On the investigation of stability criteria of rigid body supported by a string and the top both having ellipsoidal cavity filled with liquid, Proc. Acad. Sci. USSR, MTT 1, 26–31 (in Russian).

Terashima K. and Yano K. (2001), Sloshing analysis and suppression control of tilting-type automatic pouring machine, *Control Eng. Practice* **9**(6), 607–620.

Thevenard D. and Ben Hadid H. (1991), Low Prandtl number convection in a rectangular cavity with longitudinal thermal gradient and transverse g-jitters, *Int. J. Heat Mass Transfer* **34**(8), 2167–2173.

Thomson J. R., Casademunt J., and Viñals J. (1995), Cavity flow induced by a fluctuating acceleration field, *Phys. Rev. Lett.* **74**, 690.

Thomson J. R., Drolet F., and Viñals J. (1996), Fluid physics in a fluctuating acceleration environment, Proc. 3rd Microgravity Fluid Physics Conf., NASA Conf. Publications 3338, 429–434.

Thomson M. M. (1965), *Theoretical Hydrodynamics*, New York, MacMillan.

Thoroddsen S. T. and Mahadevan L. (1997), Experimental study of coating flows in a partially-filled horizontally rotating cylinder, *Experiments in Fluids* **23**(1), 1–13.

Timokha A. N. (1992), Behavior of the free surface of a fluid in a vibrating bounded vessel, Akad. Nauk. Ukraine Inst. Mat. 22, 46 pages.

Timokha A. N. (1997), The effect of transverse vibrations of the container on the free surface of a liquid, *Tekh. Mekh.* **5**, 33–41.

Ting C. L. and Perlin M. (1995), Boundary conditions of the contact-line at a vertically oscillating plate: an experimental investigation, *J. Fluid. Mech.* **295**, 263–300.

Tokuda N. and Sakurai N. (1994), Sloshing analysis method using exciting FEM structure analysis codes, *Trans. JSME C* **60**(572), 1217–1222 (in Japanese).

Tome M. F., Castelo A., Murakami J., Cuminato J. A. Minghim R., Oliveira M. C. F., Mangiavacchi N., and Mckee S. (2000), Numerical simulation of axisymmetric free surface flows, *J. Comput. Phys.* **157**, 441–472.

Tondl A. (1965), *Some Problems of Rotor Dynamics*, London, Chapman and Hall.

Tong P. (1966), Liquid motion in an elastic container, Ph.D. dissertation, Caltech.

Tong P. (1967), Liquid motion in a circular cylindrical container with a flexible bottom, *AIAA J.* **5**(10), 1842–1848.

Tong P. and Fung Y. C. (1965), The effect of wall elasticity and surface tension on the forced oscillations of a liquid in a cylindrical container, Paper 11 in *Fluid Mechanics and Heat Transfer Under Low Gravity*, H Cohan and H. Rogers, eds., Lockheed Missiles and Space Company A. D. 633580.

Toole L. E. and Hastings L. J. (1968), Behavior of a sloshing liquid subjected to a sudden reduction in axial acceleration, AIAA/Aerospace Corp. Symp. *Low Gravity Propellant Orientation and Expulsion*, Los Angeles, 1–17.

Toorani M. H. and Lakis A. A. (2001), Dynamic analysis of anisotropic cylindrical shells containing flowing fluid, *ASME J. Presure Vess. Tech.* **123**, 454–460.

Toorani M. H., and Lakis A. A. (2003), Dynamic behavior of axisymmetric and beam-like anisotropic cylindrical shells conveying fluid, *J. Sound Vib.* **259**(2), 265–298.

Topliss M. E. (1994), Water wave impact on structures, Ph.D. thesis, School of Mathematics, University of Bristol.

Townsond A. A. (1976), *The Structure of Turbulent Shear Flow*, Cambridge, Cambridge University Press.

Trembath N. W. (1957), Fluid sloshing in tanks of arbitrary shape, R-GM-45, 3–378, STL 28 August.

Triol J. (1965), Study of the breathing modes of a cylindrical container for cryogenic fluid (in French), *J. Mec.* **4**(2), 133–150.

Tritton D. (1986a), Chaos in the swing of a pendulum, *New Scientist* **24**, 37–40.

Tritton D. J. (1986b), Ordered and chaotic motion of a forced spherical pendulum, *Eur. J. Phys.* **7**, 162–169.

Troesch B. A. (1957), Pendulum analogy, Space Tech. Lab. Memo. GM 42.4–4.

Troesch B. A. (1960), Free oscillations of a fluid in a container, in *Boundary Problems in Differential Equations*, (RE Langner, ed.), Madison, University of Wisconsin Press, 279–299.

Troesch B. A. (1967), Fluid motion in a shallow trapezoidal container, *SIAM J. Appl. Math.* **15**(3), 627–637.

Troesch B. A. (1970), An isoperimetric problem for sloshing in a shallow convex, *Zeit. Ang. Math. Phys. (ZAMP)* **21**(4), 501–513.

Troesch B. A. and Troesch H. R. (1972), A remark on the sloshing frequencies for a half-space, *Z. Angew. Math. Phys.* **23**, 703–711.

Trotsenko V. A. (1967), Free oscillations of a liquid in a cylinder with annular ribs, *Prikl. Mekh.* **3**(12), 73–77.

Troung T. D., Semercigil S. E., and Turan Ö F. (2003), Experiments for control of structural oscillations using free and restricted sloshing waves, *J. Sound Vib.* **264**, 235–244.

Tsai W. T. and Yue D. K. P. (1996), Computation of nonlinear free surface flows, *Annu. Rev. Fluid Mech.* **28**, 249–265.

Tsai W. T., Yue D. K. P., and Yip K. M. K. (1990), Resonantly excited regular and chaotic motions in a rectangular wave tank, *J. Fluid Mech.* **216**, 343–380.

Tsui T. Y. and Small N. C. (1968), Hydroelastic oscillations of a liquid surface in an annular circular cylindrical tank with flexible bottom, *J. Spacecraft Rock.* **5**(2), 202–206.

Tsukimori K., Liu W. K., and Uras R. A. (1991), Dynamic buckling analysis of liquid-filled shells with imperfections, *Fluid–Structure Vibration and Sloshing*, ASME Pressure Vess. Piping Conf., PVP- **223**, 71–80.

Tufillaro N. B., Ramshankar R., and Gollub J. P. (1989), Order-disorder transition in capillary ripples, *Phys. Rev. Lett.* **62**, 422.

Ueda T., Nakagaki R., and Koshida K. (1992), Suppression of wind-induced vibration by dynamic dampers in tower-like structure, *J. Wind Eng. Ind. Aerodyn.* **41/44**, 1907–1918.

Ulitin G. M. (1986), The dynamic stability of a cylindrical shell filled with a liquid to different levels, *Teor. Prikl. Mekh.* **17**, 93–98.

Umeki M. (1991), Faraday resonance in rectangular geometry, *J. Fluid Mech.* **227**, 161–192.

Umeki M. and Kambe T. (1989), Nonlinear dynamics and chaos in parametrically excited surface waves, *J. Phys. Soc. Japan* **58**, 140–154.

Unruh J. F., Kana D. D., Dodge F. T., and Fey T. A. (1986), Digital data analysis techniques for extraction of slosh model parameters, *J. Spacecraft Rock.* **23**(3,4), 171–177.

Uras R. A. (1989), Dynamic stability of fluid-filled shells, Proc. AIAA/ASME/ASCE/ AHS/ASC 30th Struct. Struct. Dyn. Materials Conf., Mobile, Alabama, Paper 89–1207–CP, 447–453.

Uras R. A. (1995), Sloshing analysis of viscous liquid storage tanks, *Fluid-Sloshing and Fluid–Structure Interaction*, ASME Pressure Vess. Piping Conf., PVP- **314**, 63–72.

Uras R. A. and Liu W. K. (1990), Dynamic buckling of liquid-filled shells under horizontal excitation, *J. Sound Vib.* **141**(3), 389–408.

Uras R. A., Liu W. K., and Chen Y. J. (1990), Study of the influence of imperfections on the dynamic stability of tanks, *Flow–Structure Vibration and Sloshing*, ASME Pressure Vess. Piping Conf., PVP- **191**, 47–54.

Ursell F. (1952), Edge waves on a sloping beach, *Proc. Royal Soc. London A* **214**, 79–97.

Ushima S. (1998), Three-dimensional arbitrary Lagrangian–Eulerian numerical prediction method for nonlinear free surface oscillations, *Int. J. Numer. Methods Fluids* **26**, 605–623.

Utsumi M. (1988), Liquid sloshing in an axi-symmetric container in low-gravity environments, *Trans. JSME C* **54** (505), 2041–2050 (in Japanese).

Utsumi M. (1989), The meniscus and sloshing of a liquid in an axi-symmetric container at low-gravity, in *Sloshing and Fluid Structure Vibration* (Ma D. C., Tani J., Chen S. S., and Liu W. K., eds.), PVP- **157**, 103–113, New York, ASME. (Also *JSME Intl. J.* III **33**(3), 346–356, 1990).

Utsumi M. (1995), Liquid sloshing in an axisymmetric container in low-gravity environments (2nd Report, Forced vibration analysis), *Trans. JSME C* **61** (585),1784–1791.

Utsumi M. (1997), Position of propellant in teardrop tank systems, *J. Spacecraft Rock.* **34**(6), 799–804.

Utsumi M. (1998), Low-gravity propellant slosh analysis using spherical coordinates, *J. Fluid Struct.* **12**(1), 57–83.

Utsumi M., Kimura K., and Sakata M. (1983), Nonstationary responses of an elastic rectangular container with an internal liquid to a random base excitation, *Trans. JSME* C **49**(446), 1670–1680 (in Japanese).

Utsumi M., Kimura K., and Sakata M. (1984), Stochastic response analysis of an elastic rectangular container with an internal liquid to simulated seismic excitation, *J. Sound Vib.* **96**(1), 83–99.

Utsumi M., Kimura K., and Sakata M. (1986), Non-stationary vibration of an elastic circular cylindrical liquid storage tank to a simulated earthquake excitation with two-directional horizontal components, *Trans. JSME C* **52**(473), 279–287 (in Japanese).

Utsumi M., Kimura K., and Sakata M. (1987), The nonstationary random vibration of an elastic circular cylindrical liquid storage tank in simulated earthquake excitation (straightforward analysis of tank wall deformation), *JSME Int. J.* III **30**(261), 467–475.

Utsumi M. and Kondo H. (1987), Liquid sloshing in a spherical container at low-gravity environments (1st Report, Static configuration of liquid surface), *Trans. JSME C* **53** (492), 1683–1690.

Van Daalen E. F. G, Kleefsman K. M. T, Gerrits J., Luth H. R., and Veldman AEP (2000), Anit-roll tank simulations with a volume of fluid (VoF) based Navier–Stokes solver, Symp. Naval Hydrodynamics, Val de Rueil, September.

Van Daalen E. F. G, Van Doeveren A. G., Driessen P. C. M, and Visser C. (1999), Two-dimensional free surface anti-roll simulations with a volume of fluid based NavierStokes solver, Report No. 15306–1–OE, Maritime Research Inst., the Netherlands, October.

Van Den Bosch J. J. and Vugts J. H. (1964), Some notes on the performance of free surface tanks as passive anti-rolling devices, *The Report of the Shipbuilding Laboratory* No 119, Technological University of Delft, Delft, 352.

Van Den Bosch J. J. and Vugts J. H. (1966), On roll damping by free-surface tanks, *Transactions of the Inst. Naval Architecture* **69**, 345–361.

Vanden-Broeck J. M. (1984), Nonlinear gravity-capillary standing waves in water of arbitrary uniform depth, *J. Fluid Mech.* **139**, 97–104.

Van Dorn W. G. (1966), Boundary dissipation of oscillating waves, *J. Fluid Mech.* **24**, 769–779.

Van Erp F. C. (1969), Oscillation problems with a water tower, *Ingenieur* **81**(42), 148–152 (in Dutch).

Van Schoor M. C. and Crawley E. F. (1992), The nonlinear forced response characteristics of contained fluids in microgravity, ASME WAM Symp. *Fluid Mechanics Phenomena in Microgravity*, AMD- **154**/FED- **142**, 145–160. Also in *J. Spacecraft Rock.* **32**(3), 521–532, 1995.

Van Schoor M. C., Peterson L. D., and Crawley E. F. (1990), The coupled nonlinear dynamic characteristics of contained fluids in zero gravity, Proc. 31st AIAA/ASME/ASCE/AHS *Struct. Struct. Dyn. Materials* Conf., Longbeach, CA.

Van Sciver S. W. (1986), *Helium Cryogenics*, New York, Plenum Press.

Varadan T. K., Prathap G., and Ramani H. V. (1989), Nonlinear free flexural vibration of thin circular cylindrical shells, *AIAA J.* **27**, 1303–1304.

Vasta J., Giddings A. J., Taplin A., and Stillwell J. J. (1961), Roll stabilization by means of passive tanks, *Trans. SNAME* **69**, 411–460.

Velarde M. and Nepomnyashchy A. A. (2001), Interfacial patterns and waves, in *Physics of Fluids in Microgravity*, R Monti, ed., London, Taylor & Francis, 126–148.

Veldman A. E. P. and Vogels M. E. S. (1984), Axi-symmetric liquid sloshing under low gravity conditions, *Acta Astron.* **11**, 641–649.

Veletsos A. S. (1974), Seismic effects in flexible liquid storage tanks, Proc. 5th World Conf. *Earthquake Eng.* **1**, Rome, Italy, 630–639.

Veletsos A. S. and Kumar A. (1984), Dynamic response of vertically excited liquid storage tanks, Proc. 8th World Conf. *Earthquake Eng.*, San Francisco, VII, 453–459.

Veletsos A. S. and Shivakumar P. (1993), Sloshing response of layered liquids in rigid tanks, *Earthquake Eng. Struct. Dyn.* **22**, 801–821.

Veletsos A. S. and Tang Y. (1986), Dynamics of vertically excited liquid storage tanks, *ASCE J. Struct. Eng.* **112**(6), 1228–1246.

Veletsos A. S., and Tang Y. (1990), Soil–structure interaction effects for laterally excited liquid storage tanks, *Earthquake Engg. and Struct. Dyn.* **19**, 473–496.

Veletsos A. S. and Turner J. W. (1979), Dynamics of out-of-round liquid storage tanks, Proc. 3rd EMD Special Conf., Austin, 471–474.

Veletsos A. S. and Yang J. Y. (1977), Earthquake response of liquid storage tanks, advances in civil eng. through eng. mech., Proc. Ann. ASCE EMD Specialty Conf., Raleigh, NC, 1–24.

Velten R., Schwabe D., and Scharmann A. (1991), The periodic instability of thermocapillary convection in cylindrical liquid, *Physics of Fluids A* **3**(2), 267–279.

Venediktov B. L. and Shibanov R. A. (1986), Self-excitation of low-frequency vibrations in a liquid due to the high-frequency vibrations of a vessel, in *The Dynamics of Spacecraft and Extraterrestrial Investigations*, Moscow, Mashinostroyniya, 215–227.

Verhagen J. H. G. and WijnGaarden L. V. (1965), Nonlinear oscillations of fluid in a container, *J. Fluid Mech.* **22**(4), 737–751.

Verma G. R. and Keller J. B. (1962), Three-dimensional standing surface waves of finite amplitude, *Phys. Fluids* **5**(1), 52–56.

Victorov E. D. (1965), Computation of the damping factor for free oscillations of a viscous liquid in a cylindrical vessel, *J. Appl. Mech. Tech. Phys. (PMTF)* **2**, 138–146.

Viecelli J. A. (1969), A method for including arbitrary external boundaries in the MAC incompressible fluid computing technique, *J. Comput. Phys.* **4**, 543–551.

Viring J. C., Berman A. S., and Sethna P. R. (1988), On three-dimensional nonlinear subharmonic resonance surface waves in a fluid, part ii: experiment, *J. Appl. Mech.* **55**, 220–224.

Viviani A., Golia C., and Cioffi M. (1997), Magnetic effects on nonlinear Marangoni flow in plane liquid bridge, *Acta Astron.* **41**(2), 113–120.

Vladimirov V. A. (1981), Parametric resonance in a stratified liquid, *Zh. Prikl. Mekh. Tekh. Fiz.* **6**, 168–174.

Vladimirov V. A. and Ilin K. I. (1999), On the stability of the dynamerl system: rigid body + inviscid fluid, *J. Fluid Mech.* **386**, 43–75.

Vlasov V. Z. (1949), *The General Theory of Shells and Its Application in Engineering*, Moscow, Gostekhizdat (in Russian).

Volfson D. and Viñals J. (2001), Flow induced by a randomly vibrating boundary, *J. Fluid Mech.* **432**, 387–408.

Vogel T. I. (1982), Asymmetric unbounded liquid bridges, *Ann. Scuola Norm. Super. Pisa. Cl. Sci.* **9**, 433–442.

Vollman J., Breu R., and Dual J. (1997), High-resolution analysis of the complex spectrum in a cylindrical shell containing a viscoelastic medium, part II: Experimental results versus theory, *J. Acoust. Soc. Amer.* **102**(2), 909–920.

Vollman J. and Dual J. (1997), High-resolution analysis of the complex spectrum in a cylindrical shell containing a viscoelastic medium, part I: Theory and numerical results, *J. Acoust. Soc. Amer.* **102**(2), 896–908.

Vol'mir A. S. (1972), *Nonlinear Dynamics of Plates and Shells*, Moscow, Nauka (in Russian).

Vol'mir AS(1979), *Shells in Fluid Flow: Hydroelastic Problems*, Moscow, Nauka (in Russian).

Von Karman T. (1921), *Z. Angew. Math. Mech. (ZAMM)* **1**, 233.

Vrane D. R. and Smith M. K. (1994), The behavior of unsteady thermocapillary flows, NASA 2nd *Microgravity Fluid Physics Conf.*, Cleveland, Ohio, 51–56.

Vreeburg J. P. B. (1986), Observations on behavior of liquid in weightlessness, Proc. Symp. *Fluid Dynamics and Space*, Rhode-St-Genese, Belgium, ESA SP-265, 129–136.

Vreeburg J. P. B. (1992), Free motion of an unsupported tank that is partially filled with liquid, in *Microgravity Fluid Dynamics*, H. R. Rath, ed., Springer, 519–528.

Vreeburg J. P. B. (1999), Simulation of liquid dynamics onboard slosh sat FLEVO, Proc. STAIF-99, Albuquerque N. M., Am. Inst. Phys. CP 458, 836–841.

Vreeburg J. P. B. and Veldman AEP (2001), Transient and sloshing in an unsupported container, in *Physics of Fluids in Microgravity*, R. Monti, ed., London, Taylor & Frances, 293–321.

Vreeburg J. P. B. and Vogels M. E. S. (1986), Liquid motion in partially filled containers: preliminary results of the D-1 mission, *Adv. Space Res.* **6**(5), 85–92.

Vugts J. H. (1969), A comparative study on four different passive roll damping tanks – Part I. I., *Int. Shipbldg Prog.* **16**, 212–223.

Wagner C., Müller H. W., and Knorr K. (1999), Faraday waves on a viscoelastic liquid, *Phys. Rev. Lett.* **83**, 308–311.

Wagner H. (1932), Über stoss – und gleitvorgänge an der oberfläche von flüssigkeiten, *Z. Angew. Math. Mech. (ZAMM)* **12**, 193–215.

Wakahara T. (1993), Wind-induced response of TLD-structure coupled system considering nonlinearity of liquid motion, *Shimizu Tech. Res. Bull.* **12**, 41–52.

Wakahara T., Ohyama T., and Fujii K. (1991), Wind response analysis of LTD structure system considering nonlinearity of liquid motion, *J. Struct. Construct. Eng. AIJ* **426**, 79–88.

Wakahara T., Ohyama T., and Fujii K. (1992), Suppression of wind-induced vibration of a tall building using tuned liquid damper, *J. Wind Eng. Ind. Aerodyn.* **41/44**, 1895–1906.

Walter H. U., (Ed) (1987), *Fluid Sciences and Materials Sciences in Space*, New York, Springer-Verlag.

Wang M. and Hung T. (1986), A nonlinear hydrodynamic analysis on Pine Flat dam, ASCE *Advan. Aerodynamics, Fluid Mech. and Hydraulics*, 629–636.

Wang W., Wang X., Wang J., and Wei R. (1996), Dynamical behavior of parametrically excited solitary waves in Faraday's water trough experiment, *Phys. Lett. A* **219**, 74–78.

Wanschura M., Shevtsova V. M., Kuhlmann H. C., and Rath H. J. (1995), Convective instability mechanism in thermocapillary liquid bridges, *Phys. Fluids* **5**, 912–925.

Ward C. A., Yee D., Sasges M. R., Pataki L., and Stangam D. (1992), Configurationally stability of fluid systems in near-weightlessness, ASME WAM Symp. *Fluid Mechanics Phenomena in Microgravity*, D. A. Siginer and M. M. Weislogel, AMD- **154**/FED- **142**, 111–123.

Warnitchai P. and Pinkaew T. (1998), Modeling of liquid sloshing in rectangular tanks with flow-damping devices, *Engrg Structures* **20**, 593–600.

Washizu K., Nakayama T., Ikegawa M., and Tanaka Y. (1977), Application of the finite element method to some free surface fluid problems, in *Finite Elements in Water Resources*, W. G. Gray, *et al.*, eds., London, Pentech Press, 4.247–4.266.

Washizu K., Nakayama T., Ikegawa M., Tanaka Y., and Adachi T. (1984), Some finite element techniques for the analysis of nonlinear sloshing problems, Chapter 17, 357–376, in *Finite Elements in Fluids*, **5**, R. H. Gallagher, J. T. Oden, O. C. Zienkiewicz, and M. Kawahara, eds., New York, John Wiley.

Watanabe S. (1969), Methods of vibration reductions, *Proc. Japan Naval Arch. Soc. Symp.*, 156–179.

Waterhouse D. D. (1994), Resonant sloshing near a critical depth, *J. Fluid Mech.* **281**, 313–318.

Watkins R. D. (1991), Tests on various arrangements of liquid column vibration absorbers, Research Rept. R639, School of Civil and Mining Eng., University of Sidney.

Watkins R. D. and Hitchcock P. A. (1992), Tests on various liquid column vibration absorbers, Proc. Int. Conf. *Motion and Vibration Control*, Yokohama, Japan, 1130–1134.

Watts P. (1883), On a method of reducing the rolling of ship at sea, *Transactions of the Inst. Naval Architecture* **1**, 165.

Watts P. (1885), The use of water chambers for reducing the rolling of ships at sea, *Transactions of the Inst. Naval Architecture* **2**, 30.

Webster W. C. (1967), Analysis of the control of activated anti-roll tanks, *Soc. Naval Archit. Marine Engrg (SNAME) Transactions* **75**, 296–331.

Webster W. C., Dalzell J. F., and Barr R. A. (1988), Prediction and measurement of the performance of free flooding ship anti-rolling tanks, *Soc. Naval Archit. Marine Engrg (SNAME) Transactions* **96**, 333–364.

Wedemeyer E. H. (1964), The unsteady flow within a spinning cylinder, *J. Fluid Mech.* **20**(3), 383–399.

Wedemeyer E. H. and Reese J. R. (1953), Moment of inertia and damping of fluid in tanks undergoing pitching oscillations, NACA RM L 53E, June.

Wehausen J. V. and Laitone E. V. (1960), Surface waves, Article in *Encyclopedia of Physics*, S Flügge, ed., **IX**, p 446, Berlin, Springer-Verlag.

Weihs D. and Dodge F. T. (1991), Liquid motions in nonaxisymmetric, partially filled containers rotating at zero gravity, *J. Spacecraft Rock.* **28**(4), 425–432.

Weiqiu C., Ding H. J., Guo Y. M., and Yang Q. D. (1997), Free vibrations of fluid-filled orthotropic cylindrical shells, *ASCE. J. Eng. Mech.* **123**, 1130–1133.

Weisend J. G. II (1998), *Handbook of Cryogenic Engineering*, Philadelphia, PA, Taylor & Frances.

Weislogel M. (1991), Stability of capillary surfaces, *ASME Forum on Microgravity Flows*, 1st ASME/JSME Fluids Eng. Conf. FED- **111**, 11–13.

Weiss H. J. and Rogge T. R. (1965), A nonlinear analysis for sloshing forces and moments on a cylindrical tank, NASA CR-221.

Welch J. E., Harlow F. H., Shannon J. P., and Daly B. J. (1965), The MAC method: a computing technique for solving viscous, incompressible, transient fluid-flow problems involving free surfaces, Los Alamos Sci. Lab. Rept. LA-3425.

Welt F. and Modi V. J. (1989a), Vibration damping through liquid sloshing, part I: A non-linear analysis, Proc. ASME Design Technical Conf. *Fluid Structure Interaction* **H0508E**.

Welt F., and Modi V. J. (1989b), Vibration damping through liquid sloshing, part II: experimental results, Proc. ASME Design Technical Conf. *Fluid Structure Interaction* **H0508E**.

Welt F., and Modi V. J. (1992a), Vibration damping through sloshing, part I: A nonlinear analysis, *ASME J. Vib. and Acoust.* **114**(1), 10–16.

Welt F., and Modi V. J. (1992b), Vibration damping through sloshing, part II: Experimental results, *ASME J. Vib. and Acoust.* **114**(1), 17–23.

Weng C. (1992), Roll motion stabilization for small fishing vessels, Ph.D. thesis, Memorial University of Newfoundland, Canada.

Werner P. W. and Coldwell J. T. (1961), Experimental evaluation of analytical models for the inertias and natural frequencies of fuel sloshing in a circular cylindrical tanks, NASA TN D-865.

Werner P. W. and Sundquist K. J. (1949), On hydrodynamic earthquake effects, *Trans. Amer. Geophys. Union* **30**(5), 636–657.

Westera M. T., Binks D. J., and DeWater W. V. (2003), Patterns of Faraday waves, *J. Fluid Mech* **496**, 1–32.

Westergaard H. M. (1933), Water pressures on dams during earthquakes, *Trans. ASCE* **98**, 418–472.

White W. F. (1964), LOX tank four-ring slosh baffle test, DSV-IVB, Rept. SM-46762, Douglas Aircraft Co., Inc.

Whitham G. B. (1967), Variational methods and applications to water waves, *Proc. Royal Soc. A* **299**(1456), 6–25.

Whittakar E. T. and Watson G. N. (1962), *A Course of Modern Analysis*, 4th edn. Cambridge, Cambridge University Press.

Whittenburry C. G., Huber E. A., and Newell G. S. (1959), Instrument for measuring water waves, *Rev. Sci. Instrum.* **30**(8), 674–676.

Wichterle K. and Wichterle O. (1970), Surface shapes of fluids in rotating vessels, *Appl. Sci. Res.* **22**(2), 150–158.

Wilde P. and Vanyo J. (1993), Spin-up and spin-over transients of liquid fuels: Flow visualization and energy dissipation, AIAA Paper 93–1184.

Williams A. N. and Wang X. (1992), Nonlinear transient wave motions in base-excited rectangular tanks, *J. Fluid Struct.* **6**(4), 471–491.

Williams R. E. and Hussey R. G. (1972), Oscillating cylinders and Stokes' paradox, *Phys. of Fluids* **15**(2), 2083–2088.

Wilner L. B., Morrison W. L., and Brown A. E. (1960), An instrument for measuring liquid and slosh in the tanks of a liquid propellant rocket, *J. Proc. IRE* **48**(4), 786–788.

Wilson E. L. and Khalvati M. (1983), Finite elements for the dynamic analysis of fluid–solid systems, *Int. J. Num. Methods Eng.* **19**, 1657–1668.

Winch D. M. (1962), An investigation of the liquid level at the wall of a spinning tank, NASA TN D-1536.

Wohlen R. L., Park A. C., and Warner D. M. (1975), Finite element solution of low bond number sloshing, Martin-Marietta, Denver, Co., Contractor Rep. MCR–75–139, Contract NAS8–29946.

Wohlsbruck R. (1985), Stability of a rotor whose cavity has an arbitrary meridian and is partially filled with fluid, *ASME J. Vib. Acoust. Stress and Reliab. in Des.* **4**, 440–445.

Wolf I. A. (1968), Whirl dynamics of a rotor partially filled with liquid, *ASME J. Appl. Mech.* **35**, 676–682.

Won A. Y. J., Pires J. A., and Haroun M. A. (1996), Stochastic seismic performance evaluation of tuned liquid column dampers, *ASCE J. Eng. Mech.* **118**(1), 20–39.

Won A. Y. J., Pires J. A., and Haroun M. A. (1997), Performance assessment of tuned liquid column dampers under random seismic loading, *Int. J. Nonlin. Mech.* **32**(4), 745–758.

Wood D. J., Beregrine D. H., and Bruce T. (2000), Study of wave impact against a wall with pressure-impulse theory: Part 1, Trapped air, *J. Waterway, Port, Coastal Ocean Engrg* **126**, 182–190.

Woodward J. H. (1966), *Fluid Motion in a Circular Tank of Sector-Annular Cross-Section when Subjected to Longitudinal Excitation*, Ph.D. dissertation, Georgia Tech.

Woodward J. H. and Bauer H. F.1970, Fluid behavior in a longitudinally excited cylindrical tank of arbitrary sector-annular cross section, *AIAA J.* **8**(4), 713–719.

Wright J. H. (1999), Numerical simulations of two-dimensional Faraday oscillations, Doctoral dissertation, Department of Mathematics, UCA, Berkeley.

Wright J. H., Yon S., and Pozrikidis C. (2000), Numerical studies of two-dimensional Faraday oscillations of inviscid fluids, *J. Fluid Mech.* **402**, 1–32.

Wu G. X., Ma Q. W., and Taylor R. E. (1998), Numerical simulation of sloshing waves in a 3-D tank based on a finite element method, *Appl Ocean Res.* **20**(6), 337–355.

Wu J., Keolian R., and Rudnick I. (1984), Observation of a non-propagating hydrodynamic soliton, *Phys. Rev. Lett.* **52**, 1421–1424.

Wu J. S. and Hsieh M. (2002), Study on the dynamic characteristic of a U-type tuned liquid damper, *Ocean Eng.* **29**, 689–709.

Wu W. F. and Lin Y. K. (1984), Cumulant neglect closure for nonlinear oscillators under random parametric excitation, *Int. J. Nonlin. Mech.* **19**, 349–362.

Wuest W. (1976), Fluid dynamics of a floating zone, Proc. 2nd European Symp. *Material Sciences in Space*, Frascati, Italy ESA-SP-114, 455–465.

Wunderlich W. and Seiler C. (2000), Nonlinear treatment of liquid-filled storage tanks under earthquake excitation by a quasistatic approach, *Comput. Struct.* **78**, 385–395.

Xu J. J. and Davis S. H. (1983), Liquid bridges with thermocapillary, *Phys. Fluids* **26**, 2880–2886.

Xu Y. L., Kwok K. C. S., and Samali B. (1992b), The effect of tuned mass dampers and liquid dampers on cross-wind response of tall/slender structures, *J. Wind Eng. Indust. Aerodyn.* **10**, 33–54.

Xu Y. L., Samali B., and Kwok K. C. S. (1992a), Control of along wind response of structures by mass and liquid dampers, *ASCE J. Eng. Mech.* **118**, 20–39.

Xue S. D. (1999), Torsional vibration control of suspension bridge decks using tuned liquid column dampers, Ph.D. thesis, The Hong Kong Polytechnic University, Hong Kong.

Xue S. D., Ko J. M., and Xu Y. L. (2000), Tuned liquid column damper for suppressing pitching motion of structures, *Engrg. Struct.* **23**,1538–1551.

Yakubovich V. A. and Starzhinskii V. M. (1975), *Linear Differential Equations with Periodic Coefficients*, (two volumes), New York, John Wiley & Sons, Israel Program for Scientific Translations.

Yalla S. and Kareem A. (1999), Modeling tuned liquid dampers as sloshing-slamming dampers, in *Wind Engineering into the 21st Century*, Rotterdam, Netherlands, Balkema Press.

Yalla S. and Kareem A. (2000), Optimal absorber parameters for tuned liquid column dampers, *ASCE J. Struct. Eng.* **126**, 906–915.

Yalla S. and Kareem A. (2001a), Semi-active tuned liquid column dampers for vibration control of structures, *Eng. Struct.* **23**, 1469–1479.

Yalla S. and Kareem A. (2001b), Beat phenomenon in combined structure–liquid damper systems, *Eng. Struct.* **23**, 622–630.

Yalla S., Kareem A., and Kantor J. C. (1998), Semi-active control strategies for tuned liquid column dampers to reduce wind and seismic response of structures, Proc. 2nd World Conf. *Structural Control*, Kyoto, Japan, John Wiley, 559–568.

Yam Y., Mingori D. L., and Halsmer D. M. (1997), Stability of a spinning axisymmetric rocket with dissipative internal mass motion, *J. Guid. Contr. Dyn.* **20**(2), 306–313.

Yamada Y., Iemura H., Noda S. and Shimada S. (1987), Long-period response spectra from nonlinear sloshing analysis under horizontal and vertical excitations, *Natural Disaster Sci.* **9**(2), 39–54.

Yamaguchi A. (1996), SPLASH program for three-dimensional fluid dynamics with free surface boundaries, *Comput. Mech.* **18**(1), 12–23.

Yamaguchi S. and Shinkai A. (1995), An advanced adaptive control system for activated anti-rolling tank, *Int. J. Offshore Polar Engrg* **5**(1), 17–22.

Yamaki N., Tani J., and Yamaji T. (1984), Free vibration of a clamped–clamped circular cylindrical shell partially filled with liquid, *J. Sound Vib.* **94**, 531–550.

Yamamoto K. and Kawahara M. (1999), Structural oscillation control using tuned liquid damper, *Comput. Struct.* **71**, 435–446.

Yamamoto S., Kataoka F., Shioda S., Ashitani Y. (1995), Study on impact pressure due to sloshing in midsized LNG, *Int. J. Offshore Polar Engrg.*, **5**(1), 10–23.

Yamamoto Y. (1981), A variational principle for a solid–water interaction system, *Int. J. Eng. Sci.* **19**(12), 1757–1763.

Yamamoto T. and Basoglu B. (1995), Optimal geometry for fuel solution sloshing based on the boundary perturbation theory, *Annals Nuc. Energy* **22**(10), 649–658.

Yang J. Y. (1976), *Dynamic Behavior of Fluid–Tank Systems*, Ph.D. disentation, Rice University, Texas.

Yangyi O. and Jingliang M. (1985), Dynamic interaction of elastic container with fluid, Proc. ASME 10th Biennial Conf. Mech. Vib. Noise, Cincinnati, OH, Symp. *Fluid–Structure Interaction and Aerodynamic Damping*, 225–230.

Yarymovich M. I. (1959), Forced large amplitude surface waves, D.Sc. thesis, Columbia University.

Yasirc T., Kawashima H., and Sakurai A. (1987), A study on nonlinear sloshing of pool type LMFBR coolant, Proc. 9th Int. Conf. *Struct. Mech. in Reactor Tech.*, North-Holland, **K**. 459–464.

Ye Z. and Birk A. M. (1994), Fluid pressure in partially liquid-filled horizontal cylindrical vessels undergoing impact acceleration, *J. Pressure Vessel Tech.* **116**, 449–459.

Yeh G. C. K. (1965), Liquid sloshing in a moving tank with time dependent discontinuous boundary, *Developments in Mechanics*, **3** (2), (Dynamics and Fluids), T. C. Huang and M. N. Johnson, Jr, eds., 265–276.

Yeh G. C. K. (1966), Sloshing of a liquid in connected cylindrical tanks owing to u-tube free oscillations, *J. Acoust. Soc. Amer.* **40**(4), 807–812.

Yeh G. C. K. (1967), Free and forced oscillations of a liquid in an axi-symmetric tank at low gravity environments, *J. Appl. Mech.* **34**(1), 23–28.

Yeh G. C. K. and Graham D. J. (1969), Draining of a liquid from a transversely moving cylindrical tank, AIAA Paper 679.

Yeh H. H. (1986), Experimental study of standing edge waves, *J. Fluid Mech.* **168**, 291–304.

Yepishev L. V. (1959), On dynamic instability of spinning rotor incompletely filled with liquid, Scientific Reports of High School, Service Mechanical Engineering and Instrument-Making Industry 2, 66–79.

Yi W. and Natsiavas S. (1990), Seismic response of anchored fluid-filled tanks using finite elements, in *Flow-Structure Vibration and Sloshing*, ASME Pressure Vess. Piping Conf., PVP- **191**, 25–30.

Yih C. S. (1968a), Instability of unsteady flows or configurations, part I: Instability of a horizontal liquid layer on an oscillatory plane, *J. Fluid Mech.* **31**(4), 737–751.

Yih C. S. (1968b), Stability of a horizontal fluid interface in a periodic vertical electric field, *Phys. Fluid* **11**(7), 1447–1449.

Yilmaz O., Incecik A., and Han J. C. (2003), Simulation of green water flow on deck using nonlinear dam breaking theory, *Ocean Engrg.* **30**, 601–610.

Yin L., Wang B., Ma X., and Zou J. (1999), The nonlinear sloshing of liquid in tank with pitching, *ASME J. Appl. Mech.* **66**, 1032–1034.

Yaussef K. S., Mook D. T., Nayfeh A. H., and Ragab S. A. (2003), Roll stabilization by passive anti-roll tanks using an improved model of tank–liquid motion, *J. Vibration and Control* **9**, 839–862.

Youssef K. S., Ragab S. A., Nayfeh A. H., and Mook D. T. (2002), Design of passive anti-roll tanks for roll stabilization in the nonlinear range, *Ocean Engrg.* **29**, 177–192.

Yoneda M. (1989), A practical study of tuned liquid damper with application to Sakitama bridge, *JAWE J. Wind Eng.* **41**, 105–106.

Yoneda M., Chaiseri P., Maeda K., and Fujino Y. (1991), A practical study on tuned liquid damper with application to cable-stayed bridge tower, *JSCE J. Struct. Eng.* **37A**, 1019–1028.

Yoshida S. and Miyoshi T. (1987), Seismic response analysis of a multi-walled coaxial cylindrical tank, *Trans. JSME* **53**, 1670–1675 (in Japanese).

Yoshida S. and Miyoshi T. (1989), Seismic response analysis of a multi-walled coaxial cylindrical tank under vertical excitation, *Trans. JSME* **55**, 1638–1643 (in Japanese).

Yoshimatsu K. and Funakoshi M. (1998), Primary patterns in Faraday surface waves at high aspect ratio, *J. Phys. Soc. Japan* **67**, 451–461.

Young G. W. and Davis S. H. (1987), A plate oscillating across a liquid interface: effect of contact-angle hysteresis, *J. Fluid Mech.* **174**, 327–356.

Young P. G. (2002), A parametric study on the axisymmetric modes of vibration of multi-layered spherical shells with liquid cores of relevance to head impact modeling, *J. Sound Vib.* **256**(4), 665–680.

Young-Sun C. and Chung-Bang Y. (1996), Sloshing characteristics in rectangular tanks with a submerged block, *Comp. Struct.* **61**(3), 401–413.

Young T. (1805), An essay on the cohesion of fluids, *Trans. Royal Soc. London* **95**, 65–87.

Yu B., Nash W. A., and Kirchhoff R. H. (1987), A nonlinear analysis of sloshing in circular cylindrical tanks by perturbation method, Proc. ASME Symp. *Fluid–Structure Vibration and Liquid Sloshing*, PVP- **128**, 63–68.

Yu J. (1997), Nonlinear characteristics of tuned liquid dampers, Ph.D. dissertation, University of Washington.

Yu Y. Y. (1955), Free vibration of thin cylindrical shells, *ASME J. Appl. Mech.* **22**, 547–552.

Yuan J. and Dickinson S. M. (1992), On the use of artificial springs in the study of the free vibrations of systems comprised of straight and curved beams, *J. Sound Vib.* **153**, 203–216.

Yuan J. and Dickinson S. M. (1994), The free vibration of circularly cylindrical shell and plate systems, *J. Sound Vib.* **175**, 241–263.

Zakharov V. E. (1968), Stability of periodic waves on the surface of a deep fluid, *J. Appl. Mech. Tech. Phys.* **2**, 190–194.

Zang Y., Street R. L., and Koseff J. R. (1994), A non-staggered grid, fractional step method for time-dependent incompressible Navier–Stokes equations in curvilinear coordinates, *J. Comput. Phys.* **114**, 18–39.

Zebib A., Homsy G. M., and Meiburg E. (1985), High Marangoni number convection in a square cavity, *Phys. Fluids* **28**(12), 3467–3476.

Zelazo R. E. and Melcher J. R. (1969), Dynamics and stability of ferrofluids surface interaction, *J. Fluid Mech.* **39**, 1–24.

Zenkevich V. B. (1965), Behavior of a fluid in conditions of weightlessness (in Russian), *Teplofiz Vysokikh Temperatur* **2**(2), 230–237, Ref Zh Mekh 2, Rev. 2B364.

Zhang S., Yue D. K. P., and Tanizawa K. (1996), Simulation of plunging wave impact on a vertical wall, *J. Fluid Mech.* **327**, 221–254.

Zhang W., Casademunt J., and Vincaronals J. (1993), Study of the parametric oscillator driven by narrow-band noise to model the response of a fluid surface to time-dependent accelerations, *Phys. Fluids* A **5**(12) 3147–3161.

Zhang W., Tang J., and Tao M. D. (1996), Dynamic stability of a rotor filled or partially filled with liquid, *ASME J. Appl. Mech.* **63**, 101–105.

Zhang W. and Viňals J. (1997a), Pattern formation in weakly damped parametric surface waves, *J. Fluid Mech.* **336**, 301–330.

Zhang W. and Viňals J. (1997b), Pattern formation in weakly damped parametric surface waves driven by two frequency components, *J. Fluid Mech.* **341**, 225–244.

Zhang Y. L., Reese J. M., and Gorman D. G. (2001), A comparative study of axisymmetric finite elements for the vibration of cylindrical shells conveying fluids, *Int. J. Num. Methods Engrg.* **54**, 89–110.

Zhang Y. L., Reese J. M., and Gorman D. G. (2002), Initially tensioned orthotropic cylindrical shells conveying fluid: a vibration analysis, *J. Fluid Struct.***16**(1), 53–70.

Zhao Z. and Fujino Y. (1993), Numerical simulation and experimental study of deeper-water TLD in the presence of screens, *J. Struct Eng., Tokyo* **39**, 699–711.

Zhong Z., Falzarano J., and Fithen R. (1998), Numberical study of U-tube passive anti-rolling tanks, *Proc. 8th Int. Offshore and Polar Engrg. Conf., Montreal*, **3**, 504–512.

Zhou D. and Cheung Y. K. (2000), Vibration of vertical rectangular plate in contact with water on one side, *Earthquake Eng. Struct. Dyn.* **29**, 693–710.

Zhou Q. and Graebel W. P. (1990), Axisymmetric draining of a cylindrical tank with a free surface, *J. FluidMech.* **221**, 511–532.

Zhou Z., deKat J. O., and Buchner B. (1999), A nonlinear 3-D approach to simulate green water dynamics on deck, Proc. 7th Int. Symp. *Numerical Ship Hydrodynamics*, Rept. No 82000-NSH 7, May, France.

Zhu F. (1991), Orthogonality of wet modes in coupled vibration, *J. Sound Vib.* **146**, 439–448.

Zhu Q., Liu Y., and Yue D. K. P. (2003), Three-dimensional instability of standing waves, *J. Fluid Mech.* **496**, 213–242.

Zhu R. Z. (1984), Stability theory of steady rotation of a gyroscope with liquid-filled cavity, *Scientia Sinica* A **27**, 624–633.

Zhu R. Z. (1988), Spectrum analysis and potential theory of operators in state space for stability of spinning solid body with liquid-filled cavity on plane, *Scientia Sinica* A **31**, 687–693.

Zhu R. Z. (1992), Distribution, stability, bifurcations and catastrophe of steady rotation of a symmetric heavy gyroscope with viscous-liquid-filled cavity, *Int. J. Nonlinear Mech.* **27**, 477–488.

Zhu W. Q. and Huang T. C. (1984), Dynamic instability of liquid free surface in a container with elastic bottom under combined harmonic and stochastic longitudinal excitation, in *Random Vibrations*, ASME WAM Proc AMD- **65**, 195–220.

Zhukovskii N. Y. (1936), On the motion of a rigid body having cavities filled with a homogeneous liquid, *Collected Works* **3**, Moscow, Nauka.

Zhukovskii N. Y. (1948), Concerning the motion of a solid body containing cavities filled with a homogeneous liquid, *Collected Works*, **1**, Moscow, Gostekhizdat.

Zhukovskii N. Y. (1949), The motion of a rigid body having cavities filled with a homogeneous liquid, in *Collected Papers*, **2**, Moscow, Gostekhizdat, 152–309.

Zhuravlev V. F. (1977), Investigation of certain vibro-impact systems by the method of non-smooth transformation, *Prikl. Mat. Mekh. (PMM)* **49**, 572–578.

Zhuravlev V. F. (1986), The application of monomial lie groups to the problem of asymmetrically integrating equations of mechanics, *Prikl. Mat. Meks. (PMM)* **50**, 346–352.

Zhuravlev V. F. (1992), Oscillations shape control in resonant systems, *J. Appl. Math. Mech. (PMM)* **56**(5), 725–735.

Zhuravlev V. F. and Klimov D. M. (1988), *Applied Methods in Vibration Theory*, Moscow, Nauka.

Zienkiewicz O. C. and Bettes P. (1978), Fluid–structure dynamic interaction and wave forces: an introduction to numerical treatment, *Int. J. Numer. Methods Engrg.* **13**, 1–16.

Zinchenko A. Z., Rother M. A., and Davis R. H. (1999), Cusping, capture, and breakup of interacting drops by a curvatureless boundary-integral algorithm, *J. Fluid Mech.* **391**, 249–292.

Zui H. and Shinke T. (1985), Seismic deformation analysis of unanchored cylindrical tanks accompanied with the uplift of the bottom plate, *Proc. Struct. Eng. Symp.* **31A.**

Index

For EU product safety concerns, contact us at Calle de José Abascal, 56–1°,
28003 Madrid, Spain or eugpsr@cambridge.org.

www.ingramcontent.com/pod-product-compliance
Lightning Source LLC
Chambersburg PA
CBHW061153120925
32404CB00029B/64
9780521838856